# Lecture Notes in Computer Science 12333

More information about this series at http://www.springer.com/series/7408

Helmut Simonis (Ed.)

# Principles and Practice of Constraint Programming

## 26th International Conference, CP 2020
## Louvain-la-Neuve, Belgium, September 7–11, 2020
## Proceedings

Springer

*Editor*
Helmut Simonis
Insight Centre for Data Analytics,
School for Computer Science
and Information Technology
University College Cork
Cork, Ireland

ISSN 0302-9743 ISSN 1611-3349 (electronic)
Lecture Notes in Computer Science
ISBN 978-3-030-58474-0 ISBN 978-3-030-58475-7 (eBook)
https://doi.org/10.1007/978-3-030-58475-7

LNCS Sublibrary: SL2 – Programming and Software Engineering



This Springer imprint is published by the registered company Springer Nature Switzerland AG
The registered company address is: Gewerbestrasse 11, 6330 Cham, Switzerland

# Preface

This volume contains the papers presented at the 26th International Conference on Principles and Practice of Constraint Programming (CP 2020), held during September 7–11, 2020. The conference was originally planned to be held at UC Louvain, Belgium. Due to the COVID-19 pandemic, we were forced to run the conference as a virtual event instead.

There were 122 submissions for the conference. The Program Committee decided to accept 55 papers. As is customary for the CP conference series, we offered multiple tracks for submitted papers to widen the scope of the conference and attract papers at the interface of Constraint Programming and other disciplines. This year there were tracks for:

- Technical Track
- Application Track
- Constraint Programming, Data Science and Machine Learning
- Testing and Verification
- Constraint Programming and Operations Research
- Computational Sustainability

I want to thank the application track chair, Andreas Schutt (Data61, Australia), the CP/ML track chair, Michele Lombardi (DISI, University of Bologna, Italy), and the chair for the Testing and Verification track, Nadjib Lazaar (LIRMM, France), for their dedicated work. They selected and invited track specific Program Committees, helped with the paper assignment across the different tracks, and managed the discussion and final paper selection after the review phase.

The paper selection in the other tracks was supported by a Senior Program Committee, who not only encouraged and guided the discussion of papers during reviewing, but also provided meta-reviews as summaries and basis for the paper selection.

The Program Committee for this year's conference faced a special challenge as the time given for reviewing was very compressed. The original submission deadline was shifted by more than five weeks to counteract the impact of the COVID-19 lock-down in many countries around the world. I was very impressed by, and thankful for, the quality and detail of the reviews provided.

The reviewer assignment was decided for all tracks at the same time, to balance the workload of all reviewers. For this, we used an extended version of a Constraint Programming model developed and used last year by the CP 2019 program chairs, Thomas Schiex and Simon de Givry.

Besides the paper tracks, the conference also had many other elements, handled by special chairs. Maria-Andreina Francisco Rodriguez (Uppsalla University, Sweden) organized the workshops on the first day of the conference, Lars Kotthoff (University of Wyoming, USA) selected three tutorials for the main conference, and Edward Lam

(Monash University, Australia) and Kuldeep Meel (National University of Singapore, Singapore) organized the Doctoral Program. All of these events were heavily affected by the virtualization of the conference, requiring new ideas and methods.

The biggest change, caused by the virtual conference, was for the conference chairs, Tias Guns (VUB, Belgium), Siegfried Nijssen, and Pierre Schaus (both UC Louvain). Having planned a beautiful conference on the campus of UC Louvain, they had to completely rethink the organization, planning, and finances for a very different type of conference.

Special thanks go to the publicity chair, Hélène Verhaeghe (UC Louvain), and local publicity chair, Bernard Fortz (ULB, Belgium), for dealing with information updates and handling communications with the overall community. The help of the president of the Association for Constraint Programming, Maria Garcia de la Banda (Monash University), and the conference coordinator of the ACP Executive Committee, Claude-Guy Quimper (Laval University, Canada), especially during the initial lock-down phase due to the COVID-19 pandemic, is gratefully acknowledged.

My final thanks go to the authors of paper submissions for CP 2020, writing their papers in a very challenging situation: without your work, for both successful and unsuccessful papers, we would not had this interesting conference program.

July 2020 Helmut Simonis

# Organization

## Senior Program Committee

| | |
|---|---|
| Maria Garcia De La Banda | Monash University, Australia |
| Emmanuel Hebrard | LAAS, CNRS, France |
| Ines Lynce | INESC-ID/IST, Universidade de Lisboa, Portugal |
| Laurent Michel | University of Connecticut, USA |
| Michela Milano | DISI, University of Bologna, Italy |
| Claude-Guy Quimper | Laval University, Canada |
| Pierre Schaus | UC Louvain, Belgium |
| Christine Solnon | CITI Inria, INSA Lyon, France |
| Kostas Stergiou | University of Western Macedonia, Greece |
| Peter J. Stuckey | Monash University, Australia |
| Roland Yap | National University of Singapore, Singapore |

## Program Committee

| | |
|---|---|
| Ignasi Abío | Barcelogic, Spain |
| Deepak Ajwani | University College Dublin, Ireland |
| Özgür Akgün | University of St Andrews, UK |
| Carlos Ansótegui | Universitat de Lleida, Spain |
| Ekaterina Arafailova | Tesco, UK |
| Behrouz Babaki | Polytechnique Montreal, Canada |
| Sébastien Bardin | CEA LIST, France |
| Roman Barták | Charles University, Czech Republic |
| Chris Beck | University of Toronto, Canada |
| Nicolas Beldiceanu | IMT Atlantique (LS2N), France |
| Gleb Belov | Monash University, Australia |
| Stefano Bistarelli | Università di Perugia, Italy |
| Miquel Bofill | Universitat de Girona, Spain |
| Andrea Borghesi | University of Bologna, Italy |
| Ken Brown | University College Cork, Ireland |
| David Browne | Insight Centre for Data Analytics, Ireland |
| Hadrien Cambazard | G-SCOP, Grenoble INP, CNRS, Joseph Fourier University, France |
| Roberto Castañeda Lozano | The University of Edinburgh, UK |
| Berthe Y. Choueiry | University of Nebraska-Lincoln, USA |
| Andre Augusto Cire | University of Toronto, Canada |
| Laura Climent | University College Cork, Ireland |
| Martin Cooper | IRIT, Université Paul Sabatier, France |
| Patrick De Causmaecker | UC Leuven, Belgium |
| Allegra De Filippo | DISI, University of Bologna, Italy |

## Additional Reviewers

Hoenicke, Jochen
Hu, Hao
Hurley, Barry
Isoart, Nicolas
Koshimura, Miyuki
Kotthoff, Lars
Liu, Fanghui
Lo Bianco, Giovanni
Mattmüller, Robert
Mercanti, Ivan
Nordström, Jakob
Ojeda, Jesus
Ornek, Arslan
Ouali, Abdelkader
Palak, Gokce
Peitl, Tomáš
Pettersson, William
Philipp, Tobias
Prestwich, Steve
Quimper, Claude-Guy
Refalo, Philippe
Ruffini, Manon
Santini, Francesco
Savvaris, Al
Serret, Michel Fabrice
Sioutis, Michael
Tardivo, Fabio
Trimble, James
Tsoupidi, Rodothea Myrsini
van Driel, Ronald
Vilim, Petr
Villaret, Mateu
Vinyals, Marc
Wilson, Nic
Woodward, Robert

# Contents

**Application Track**

# Technical Track

# Dashed Strings and the Replace(-all) Constraint

Roberto Amadini[1(✉)], Graeme Gange[2], and Peter J. Stuckey[2]

[1] University of Bologna, Bologna, Italy
roberto.amadini@unibo.it
[2] Monash University, Melbourne, VIC, Australia
{graeme.gange,peter.stuckey}@monash.edu

**Abstract.** Dashed strings are a formalism for modelling the domain of string variables when solving combinatorial problems with string constraints. In this work we focus on (variants of) the REPLACE constraint, which aims to find the *first* occurrence of a query string in a target string, and (possibly) replaces it with a new string. We define a REPLACE propagator which can also handle REPLACE-LAST (for replacing the *last* occurrence) and REPLACE-ALL (for replacing *all* the occurrences). Empirical results clearly show that string constraint solving can draw great benefit from this approach.

## 1 Introduction

In the past decade the interest in solving string constraints has considerably grown in disparate application domains such as test-case generation, software analysis and verification, model checking, web security, database query processing, bionformatics (see, e.g., [3,8,9,11–15,18,24,29]).

Among the most effective paradigms for string constraint solving we mention Satisfiability Modulo Theory (SMT), which relies on (extensions of) the theory of *word equations*, and Constraint Programming (CP), which basically uses an *unfolding*-based approach to eventually decompose a string variable into sequences of integer variables representing the string characters.

A promising CP approach, on which this paper focuses, is based on *dashed strings*. Given a fixed finite alphabet $\Sigma$ and a maximum string length $\lambda$, dashed strings are a particular class of regular expressions denoting sets of concrete strings $W \subseteq \{w \in \Sigma^* \mid |w| \leq \lambda\}$ through the concatenation of distinct sets of characters called *blocks*. Dashed strings are used to represent in a compact way the domain of string variables without eagerly unfolding them into $\lambda$ integer variables. This can make a huge difference, especially when $\lambda$ is big.

Several algorithms for dashed strings have been defined to propagate string constraints like (in-)equality, concatenation, length, or regular expression membership. Most of them are based on the concept of dashed strings *equation* [7].

In this paper, we focus on a well-known constraint frequently occurring in problems derived from software verification and testing: the REPLACE constraint.


H. Simonis (Ed.): CP 2020, LNCS 12333, pp. 3–20, 2020.
https://doi.org/10.1007/978-3-030-58475-7_1

Despite being a construct heavily used in modern programming, very few solvers have implemented REPLACE, and its variants, due its non trivial semantics. For this reason, few benchmarks exist.

Given an input string $x$, a query string $q$ and a new string $q'$ we have that REPLACE$(x, q, q')$ is the string obtained by replacing the *first* occurrence of $q$ in $x$ with $q'$ (REPLACE$(x, q, q') = x$ if $q$ does not occur in $x$). Common variants are REPLACE-LAST (that replaces the *last* occurrence of the query string) and REPLACE-ALL (that replaces *all* the occurrences of the query string).

At present, no dashed string propagator has been defined for REPLACE (in [5] it is simply *decomposed* into other string constraints). This approach is sound but certainly improvable. In this work we define a unified propagation algorithm that *(i)* improves the precision and the efficiency of REPLACE propagation, and *(ii)* is general enough to propagate REPLACE-LAST and REPLACE-ALL. The latter in particular cannot be decomposed into basic constraints because we do not know *a priori* how many times the query string occurs.

Empirical results show that the approach we propose significantly outperforms the performance of the decomposition approach and state-of-the-art SMT solvers.

*Paper structure.* In Sect. 2 we give preliminary notions. In Sect. 3 we explain how we propagate REPLACE and related constraints. In Sect. 4 we report the empirical results, before discussing the related works in Sect. 5 and concluding in Sect. 6.

## 2 Dashed Strings

Let $\Sigma$ be a finite alphabet and $\Sigma^*$ the set of all the strings over $\Sigma$. We denote with $\epsilon$ the empty string and with $|w|$ the length of $w \in \Sigma^*$. We use 1-based notation for (dashed) strings: $w[i]$ is the $i$-th symbol of $w$ for $i = 1, \dots, |w|$.

The concatenation of $v, w \in \Sigma^*$ is denoted by $v \cdot w$ (or simply $vw$) while $w^n = ww^{n-1}$ is the concatenation of $w$ for $n > 0$ times ($w^0 = \epsilon$). The concatenation of $V, W \subseteq \Sigma^*$ is denoted with $V \cdot W = \{vw \mid v \in V, w \in W\}$ (or simply $VW$) while $W^n = WW^{n-1}$ is the concatenation of $W$ for $n$ times ($W^0 = \{\epsilon\}$). In this work we focus on *bounded-length* strings, i.e., we fix an upper bound $\lambda \in \mathbb{N}$ and only consider strings in $\mathbb{S}_\Sigma^\lambda = \{w \in \Sigma^* \mid |w| \leq \lambda\}$.

A *dashed string* $X = S_1^{l_1,u_1} \cdots S_k^{l_k,u_k}$ is a concatenation of $k = |X| > 0$ blocks such that, for $i = 1, \dots, k$: *(i)* $S_i \subseteq \Sigma$; *(ii)* $0 \leq l_i \leq u_i \leq \lambda$; *(iii)* $u_i + \sum_{j \neq i} l_j \leq \lambda$. For each block $X[i] = S_i^{l_i,u_i}$ we call $S_i$ the *base* and $(l_i, u_i)$ the *cardinality*, where $l_i = lb(S_i^{l_i,u_i})$ is the *lower bound* while $u_i = ub(S_i^{l_i,u_i})$ is the *upper bound* of $X[i]$. We extend this notation to dashed strings: $lb(X) = \sum_{i=1}^n lb(X[i])$ and $ub(X) = \sum_{i=1}^n ub(X[i])$. Intuitively, $lb(X)$ and $ub(X)$ are respectively the lower and the upper bound on the length of each concrete string denoted by $X$.

The primary goal of dashed strings is to compactly represent sets of strings, so we define a function $\gamma$ such that $\gamma(S^{l,u}) = \{w \in S^* \mid l \leq |w| \leq u\} \subseteq \mathbb{S}_\Sigma^\lambda$ is the *language denoted* by block $S^{l,u}$ (in particular $\gamma(\emptyset^{0,0}) = \{\epsilon\}$ for the *null block* $\emptyset^{0,0}$). We extend $\gamma$ to dashed strings: $\gamma(S_1^{l_1,u_1} \cdots S_k^{l_k,u_k}) =$

$(\gamma(S_1^{l_1,u_1})\dots\gamma(S_k^{l_k,u_k}))\cap\mathbb{S}_\Sigma^\lambda$. Blocks of the form $S^{0,u}$ are called *nullable* because $\epsilon\in\gamma(S^{0,u})$. A dashed string $X$ is *known* if $\gamma(X)=\{w\}$, i.e., it represents a single, "concrete" string $w\in\mathbb{S}_\Sigma^\lambda$.

We call $\mathbb{DS}_\Sigma^\lambda$ the set of all the dashed strings. Note that, while $\mathbb{S}_\Sigma^\lambda$ is finite, $\mathbb{DS}_\Sigma^\lambda$ is *countable*: $\lambda$ bounds the cardinalities of the blocks, but not the number of blocks that may appear in a dashed string. For example, given distinct characters $a,b\in\Sigma$ we can generate an infinite sequence of dashed strings in $\mathbb{DS}_\Sigma^\lambda$: $\{a\}^{0,1}\{b\}^{0,1}$, $\{a\}^{0,1}\{b\}^{0,1}\{a\}^{0,1}$, $\{a\}^{0,1}\{b\}^{0,1}\{a\}^{0,1}\{b\}^{0,1},\dots$.

A dashed string $X=S_1^{l_1,u_1}S_2^{l_2,u_2}\cdots S_k^{l_k,u_k}$ is *normalised* if: *(i)* $S_i\neq S_{i+1}$, *(ii)* $S_i=\emptyset\Leftrightarrow l_i=u_i=0$, *(iii)* $X=\emptyset^{0,0}\vee S_i\neq\emptyset$ for $i=1,\dots,k$. The operator NORM normalises a dashed string (note $\gamma(X)=\gamma(\text{NORM}(X))$). We denote the set of normalised dashed strings as $\overline{\mathbb{DS}_\Sigma^\lambda}=\{\text{NORM}(X)\mid X\in\mathbb{DS}_\Sigma^\lambda\}$. Normalisation is fundamental to remove a large number of "spurious" dashed strings from $\mathbb{DS}_\Sigma^\lambda$, i.e., distinct dashed strings denoting the same language. If not otherwise specified, we will always refer to normalised dashed strings.

We define the partial order $\sqsubseteq$ such that $X\sqsubseteq Y\Leftrightarrow(X=Y\vee\gamma(X)\subset\gamma(Y))$ to model the relation *"is more precise than or equal to"* between dashed strings, i.e., we consider $X$ more precise than $Y$ if $\gamma(X)\subset\gamma(Y)$.[1] Note that the poset $(\overline{\mathbb{DS}_\Sigma^\lambda},\sqsubseteq)$ is well-founded but *not a lattice*: in general, we may not be able to determine the dashed string which best represents two or more concrete strings. For example, the set $\{ab,ba\}$ has two minimal representations $\{a\}^{0,1}\{b\}^{1,1}\{a\}^{0,1}$ and $\{b\}^{0,1}\{a\}^{1,1}\{b\}^{0,1}$ which are not comparable according to $\sqsubseteq$.

From a graphical perspective, we can see a block $S^{l,u}$ as a continuous segment of length $l$ followed by a dashed segment of length $u-l$. The continuous one indicates that exactly $l$ characters of $S$ *must* occur in each concrete string of $\gamma(X)$; the dashed one indicates that $k\le u-l$ characters of $S$ *may* occur. Consider dashed string $X=\{\texttt{B},\texttt{b}\}^{1,1}\{\texttt{o}\}^{2,4}\{\texttt{m}\}^{1,1}\{\texttt{!}\}^{0,3}$ of Fig. 1. Each string of $\gamma(X)$ must start with 'B' or 'b', followed by 2 to 4 'o's, one 'm', and 0 to 3 '!'s.

## 2.1 Positions and Equation

A convenient way to refer a dashed string is through its *positions.* Given $X=S_1^{l_1,u_1}\cdots S_k^{l_k,u_k}$, a positions is a pair $(i,j)$ where *index* $i\in\{1,\dots,k\}$ refers to block $X[i]$ and *offset* $j$ indicates how many characters from the beginning of $X[i]$ we are considering. As shown in Fig. 1, indexes are 1-based while offsets are 0-based: position $(i,0)$ refers to the beginning of $X[i]$ and can be equivalently identified with the end of $X[i-1]$, i.e., with position $(i-1,u_{i-1})$. For convenience, we consider $(k+1,0)$ equivalent to $(k,u_k)$ and $(0,0)$ to $(1,0)$. Given $X=S_1^{l_1,u_1}\cdots S_k^{l_k,u_k}$ and positions $(i,j),(i',j')$ we denote with $X[(i,j),(i',j')]$

[1] $\sqsubseteq$ is not defined as $X\sqsubseteq Y\iff\gamma(X)\subseteq\gamma(Y)$ because otherwise $\sqsubseteq$ would be a *pre-order* but not a partial order in general, e.g., if $\lambda=2$ then $X=\{a,b,c\}^{0,2}$ and $Y=\{a,b\}^{0,2}\{b,c\}^{0,2}\{a,c\}^{0,2}$ are such that $\gamma(X)\subseteq\gamma(Y)$ and $\gamma(Y)\subseteq\gamma(X)$ so $X\sqsubseteq Y$ and $Y\sqsubseteq X$ but $X\neq Y$.

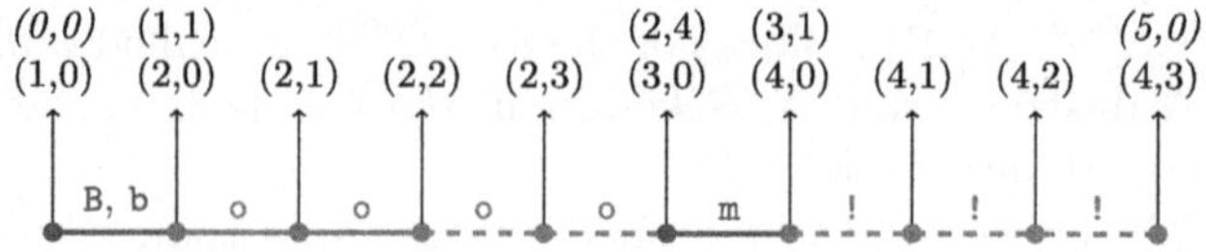


**Fig. 1.** Representation of $\{\mathtt{B,b}\}^{1,1}\{\mathtt{o}\}^{2,4}\{\mathtt{m}\}^{1,1}\{\mathtt{!}\}^{0,3}$.

the *region* of $X$ between $(i,j)$ and $(i',j')$, defined by:

$$\begin{cases} \emptyset^{0,0} & \text{if } (i,j) \succeq (i',j') \\ S_i^{\max(0,l_i-j),j'-j} & \text{else if } i=i' \\ S_i^{\max(0,l_i-j),u_i-j} S_{i+1}^{l_{i+1},u_{i+1}} \cdots S_{i'-1}^{l_{i'-1},u_{i'-1}} S_{i'}^{\min(l_{i'},j'),j'} & \text{otherwise} \end{cases}$$

For brevity, we also define $X[...,P] = X[(0,0),P]$ and $X[P,...] = X[P,(k,u_k)]$.

For example, in Fig. 1 we have that region $X[(2,1),(2,4)]$ is $\{\mathtt{o}\}^{1,3}$ while $X[(2,3),(4,2)] = \{\mathtt{o}\}^{0,1}\{\mathtt{m}\}^{1,1}\{\mathtt{!}\}^{0,2}$. Region $X[(2,2),...]$ is $\{\mathtt{o}\}^{0,2}\{\mathtt{m}\}^{1,1}\{\mathtt{!}\}^{0,3}$ while $X[...,(4,1)]$ is $\{\mathtt{B,b}\}^{2,4}\{\mathtt{o}\}^{2,4}\{\mathtt{m}\}^{1,1}\{\mathtt{!}\}^{0,1}$.

Positions are used to implement a fundamental operation between dashed strings, on which the propagator REPLACE propagator relies: the dashed string *equation.* Equating $X$ and $Y$ means looking for a *refinement* of $X$ and $Y$ including *all* the strings of $\gamma(X)\cap\gamma(Y)$ and removing *the most* strings not belonging to $\gamma(X)\cap\gamma(Y)$. In other words, we look for a minimal—or at least small enough—dashed string denoting all the concrete strings of $\gamma(X)$ and $\gamma(Y)$ (the smallest dashed string does not always exist because $(\overline{\mathbb{DS}_\Sigma^\lambda}, \sqsubseteq)$ is not a lattice).

In [7] the authors introduced a sweep-based algorithm, that here we will call EQUATE[2], such that: *(i)* if $\text{EQUATE}(X,Y) = \bot$, then $\gamma(X)\cap\gamma(Y) = \emptyset$; *(ii)* if $\text{EQUATE}(X,Y) = (X',Y') \neq \bot$, then $X' \sqsubseteq X$, $Y' \sqsubseteq Y$ and $\gamma(X)\cap\gamma(Y) = \gamma(X')\cap\gamma(Y')$. In a nutshell, $\text{EQUATE}(X,Y)$ matches each block $X[i]$ against each block of $Y$ to determine its earliest/latest start/end positions in $Y$, and uses this information to possibly refine $X[i]$ into more precise block(s) $X'[i] \sqsubseteq X[i]$.

For example, consider matching block $B = \{\mathtt{o,m,g}\}^{2,6}$ against dashed string $X$ of Fig. 1 starting from position $(0,0)$. The earliest start of $B$ is $(2,0)$ (it cannot match $X[1] = \{\mathtt{B,b}\}^{1,1}$ because its base does not contain any B or b) and the earliest end is $(2,2)$ ($B$ contains at least 2 characters); the latest start is $(2,3)$ because $B$ cannot finish after latest end position $(4,0)$ (it cannot match $\{!\}^{0,3}$). This means that if $B$ is the $i$-th block of a dashed string $Y$ that we are equating with $X$, then the prefix $Y[...,i-1]$ has to match $\{\mathtt{B,b}\}^{1,1}$ and the suffix $Y[i+1,...]$ has to match $\{!\}^{0,3}$ (otherwise, $\text{EQUATE}(X,Y) = \bot$).

The region between earliest start and latest end is called *feasible region,* while the region between latest start and earliest end is the *mandatory region.* In the above example, the feasible region is $X[(2,0),(4,0)] = \{\mathtt{o}\}^{2,4}\{\mathtt{m}\}^{1,1}$ while mandatory region is null (latest start comes after earliest end). This information enables the refinement of $B$ into the more precise blocks $\{\mathtt{o}\}^{2,4}\{\mathtt{m}\}^{1,1}$.

2 In [7] it was called SWEEP to distinguish it the COVER equation algorithm.

```
function Push+(S^{l,u}, Y, (i, j))
  (i0, j0) ← (i, j)
  k ← l                          ▷ k is the number of "consumed" characters for S^{l,u}
  while k > 0 do
    if i > |Y| then
      return (i, j), (i, j)
    S'^{l',u'} ← Y[i]
    if S ∩ S' = ∅ then                  ▷ Incompatible block, move to next one
      (i, j) ← (i + 1, 0)
      if l' > 0 then
        (i0, j0) ← (i, j)                    ▷ Updating earliest start position
        k ← l
    else if k ≤ u' − j then       ▷ We can consume all the characters of S^{l,u}
      return (i0, j0), (i, j + k)
    else
      k ← k − (u' − j)                 ▷ Consuming u' − j characters of S^{l,u}
      (i, j) ← (i + 1, 0)
  return (i0, j0), (i, j)
```

**Fig. 2.** Push$^+$ algorithm.

```
function Stretch−(S^{l,u}, Y, (i, j))
  k ← u
  if j > lb(Y[i]) then
    j ← lb(Y[i])                       ▷ Skip the "optional part" of Y[i]
  while i > 0 do
    S'^{l',u'} ← Y[i]
    if l' = 0 then                                       ▷ Skip block Y[i]
      if i > 1 then
        (i, j) ← (i − 1, lb(Y[i − 1]))
      else
        return (0, 0)
    else if S ∩ S' = ∅ then                           ▷ We cannot go further
      return (i, j)
    else if k < l' then
      return (i, j − k)
    else
      k ← k − l'                 ▷ Consuming the least characters of S^{l,u}
      if i > 1 then
        (i, j) ← (i − 1, lb(Y[i − 1]))
      else
        return (0, 0)
  return (i, j)
```

**Fig. 3.** Stretch$^-$ algorithm

Equate$(X, Y)$ uses auxiliary functions Push$^-$ and Push$^+$ to detect the "earliest" positions. Given block $X[i]$, dashed string $Y$ and initial position $P_0$ of $Y$, Push$^+(X[i], Y, P_0)$ returns a pair $(P_s, P_e)$ where $P_s$ (resp., $P_e$) is the earliest start (resp. end) of $X[i]$ in $Y[P_0, ...]$ (see pseudo-code in Fig. 2). If $X[i]$ cannot match any block of $Y[P_0, ...]$, then $P_s = P_e = (|Y| + 1, 0)$. Dually, $\text{PUSH}^-$ works "backwards" to find $(P_s, P_e)$ in $Y[..., P_0]$ from right to left.

In Fig. 1, if $P_0 \preceq (3,0)$ then $\text{PUSH}^+(\{\mathtt{m}\}^{1,2}, X, P_0) = ((3,0),(4,0))$ and $\text{PUSH}^-(\{\mathtt{m}\}^{1,2}, X, P_0) = ((0,0),(0,0))$; otherwise, if $P_0 \succ (3,0)$, then we have $\text{PUSH}^+(\{\mathtt{m}\}^{1,2}, X, P_0) = ((5,0),(5,0))$ and $\text{PUSH}^-(\{\mathtt{m}\}^{1,2}, X, P_0) = ((4,0),(3,0))$.

Analogously, a pair of STRETCH functions are used for "latest" positions: $\text{STRETCH}^+(X[i], Y, P_0)$ is the latest position (left-to-right) where $X[i]$ can match $Y$ starting from $P_0$ (while $\text{STRETCH}^-$ works right-to-left, see Fig. 3).[3]

In Fig. 1, for example, if $P_0 = (2,1)$ then $\text{STRETCH}^+(\{\mathtt{o,m,g}\}^{1,8}, X, P_0) = (3,1)$ and $\text{STRETCH}^-(\{\mathtt{o,m,g}\}^{1,8}, X, P_0) = (1,1)$. If instead $P_0 = (0,0)$ then both functions return $P_0$ because $\text{STRETCH}^-$ cannot go further left and $\text{STRETCH}^+$ cannot go further right since $\{\mathtt{o,m,g}\}^{1,8}$ does not match $\{\mathtt{B,b}\}^{1,1}$.

## 3 Propagating REPLACE

If $y = \text{REPLACE}(x, q, q')$, then $y$ is the string obtained by replacing the first occurrence of $q$ in $x$ with $q'$ (if $q$ does not occur in $x$, then $x = y$). The only implementation of REPLACE with dashed strings, defined in [5], simply rewrites $y = \text{REPLACE}(x, q, q')$ into:[4]

$$\exists\, n, p, s. \begin{pmatrix} n = \text{FIND}(q, x) \wedge \text{FIND}(q, p) = [\![|q| = 0]\!] \\ x = p \cdot q^{[\![n>0]\!]} \cdot s \wedge y = p \cdot q'^{[\![n>0]\!]} \cdot s \wedge \\ |p| = \max(0, n-1) \end{pmatrix} \quad (1)$$

where $\text{FIND}(q, x)$ returns the index of the first occurrence of $q$ in $x$ ($\text{FIND}(\epsilon, x) = 1$ and $\text{FIND}(q, x) = 0$ if $q$ does not occur in $x$) and $[\![b]\!] \in \{0,1\}$ is the integer value of Boolean variable $b$, i.e., $[\![b]\!] = 1 \iff b$.

This rewriting is sound but, as we shall see, has three main disadvantages: *(i)* for each REPLACE constraint, all the constraints of decomposition (1) are added; *(ii)* it may lose precision; *(iii)* it cannot be generalised to REPLACE-ALL.

To overcome the above drawbacks we defined a unified REPLACE propagator which can also handle REPLACE-LAST and REPLACE-ALL. The pseudo-code of the propagator is shown in Fig. 4, where $x, q, q', y$ are the variables involved in the constraint and the Boolean flag *LST* (resp. *ALL*) is *true* if we are propagating REPLACE-LAST (resp. REPLACE-ALL). REPLACE* denotes all three of REPLACE. The PROP-REPLACE function returns: *(i)* UNS, if REPLACE* is *infeasible*; *(ii)* OK, if REPLACE* *may* be feasible (in this case the domains are possibly refined); *(iii)* REWRITE($C$) if REPLACE* is *rewritten* into constraint $C$.

[3] PUSH and STRETCH are not dual. For example, when incompatible blocks are found, PUSH moves on (Fig. 2, line 9) while STRETCH returns (Fig. 3, line 13).

[4] In [5] the order of REPLACE arguments is different. In this paper we change the notation to be consistent with SMT-LIB specifications.

```
1: function PROP-REPLACE(x, q, q′, y, LST, ALL)
2:   (X, Q, Q′, Y) ← dom(x, q, q′, y),    m ← 0
3:   if γ(Q) = {q} then                                          ▷ q is fixed
4:     if q = ϵ then
5:       if ALL then
6:         if γ(X) = {x} then                                    ▷ x is fixed
7:           return REWRITE( REPALLFIXED(x, ϵ, q′, y) )
8:       else
9:         if LST then
10:          return REWRITE( y = x · q′ )
11:        return REWRITE( y = q′ · x )
12:    if γ(Q′) = {q′} ∧ q = q′ then                             ▷ q′ is fixed
13:      return REWRITE( y = x )
14:    if γ(X) = {x} then                                        ▷ x is fixed
15:      if ALL then
16:        return REWRITE( REPALLFIXED(x, q, q′, y) )
17:      n ← LST ? FINDLAST(q, x) : FIND(q, x)
18:      if n = 0 then
19:        return REWRITE( y = x )
20:      else
21:        return REWRITE( y = x[..., n − 1] · q′ · x[n + |q|, ...] )
22:    m ← COUNTMIN(q, X, ALL)
23:  if m = 0 ∧ ¬CHECK-EQUATE(X, Y) then                         ▷ x ≠ y
24:    m ← 1
25:  if m > 0 ∧ ¬ALL then
26:    C ← x = p · q · s ∧ y = p · q′ · s ∧ LST ? FIND(q, s) = 0 : FIND(q, p) = 0
27:    return REWRITE( C )
28: ──────────────────────────────────────────────────────────────────────
29:  (es, le) ← CHECK-FIND(Q, X)
30:  if es = ⊥ then                                              ▷ q not occurs in x
31:    if m > 0 then
32:      return UNS
33:    return REWRITE( x = y )
34:  else                                                        ▷ q may occur in x
35:    B ← (⋃X ∪ ⋃Q′)^{0,ub(Y)}
36:    Y′ ← NORM(X[..., es] · (B · Q′)^m · B · X[le, ...])
37:    if EQUATE(Y, Y′) = ⊥ then
38:      return UNS
39:  (es, le) ← CHECK-FIND(Q′, Y)
40:  if es = ⊥ then                                              ▷ q′ not occurs in y
41:    if m > 0 then
42:      return UNS
43:    return REWRITE( FIND(q, x) = 0 ∧ x = y )
44:  else                                                        ▷ q′ may occur in y
45:    B ← (⋃Q ∪ ⋃Y)^{0,ub(X)}
46:    X′ ← NORM(Y[..., es] · (B · Q)^m · B · Y[le, ...])
47:    if EQUATE(X, X′) = ⊥ then
48:      return UNS
49:  dom(x) ← X,   dom(y) ← Y
50:  return OK
```

**Fig. 4.** Generalised propagator for REPLACE*. Line 28 separates the particular cases from the general case.

PROP-REPLACE first retrieves the dashed strings $X, Q, Q', Y$ corresponding to the domains of variables $x, q, q', y$ respectively (line 2). From now on we indicate in bold font the *fixed* variables, i.e., those representing a single, concrete string. We can divide the pseudo-code of PROP-REPLACE in two main blocks. The first one (until line 28) refers to the *particular* cases, where either $q$ is fixed or $x \neq y$. Here we do not refine domains, but we possibly rewrite REPLACE*. The second one (after line 28) models instead the *general* case. Here we possibly refine the variables' domain with a sweep-based approach relying on the EQUATE algorithm.

The "if" statement between lines 3 and 22 handles the cases where $q$ is fixed by rewriting REPLACE* into basic constraints without any loss of precision. In particular, as shown in Fig. 5, REPALLFIXED is used to break down REPLACE-ALL$(x, q, q')$ into concatenations when both $x$ and $q$ are fixed. Note that REPLACE-ALL assumes *no overlaps*, e.g., REPLACE-ALL$(aaaaa, aa, b) = bba$. We assume that $\epsilon$ occurs in each position of a string, e.g., REPLACE$(ab, \epsilon, c) = cab$, while REPLACE-LAST$(ab, \epsilon, c) = abc$, and REPLACE-ALL$(ab, \epsilon, c) = cacbc$ (see lines 4–11).

1: **function** REPALLFIXED(**x**, **q**, $q'$, $y$)
2: **if** $\mathbf{q} = \epsilon$ **then**
3: **return** $y = q' \cdot \mathbf{x}[1] \cdot q' \cdot \mathbf{x}[2] \cdot q' \cdots q' \cdot \mathbf{x}[\,|\mathbf{x}|\,] \cdot q'$
4: $i \leftarrow$ FIND$(\mathbf{q}, \mathbf{x})$
5: **if** $i = 0$ **then**
6: **return** $y = x$
7: $j \leftarrow 1$
8: $Y \leftarrow [\,]$
9: **while** $i \neq 0$ **do**
10: PUSH-BACK$(Y, \mathbf{x}[j..i-1])$
11: $j \leftarrow i + |\mathbf{q}|$
12: $i \leftarrow$ FIND$(\mathbf{q}, \mathbf{x}[j, ...])$
13: **return** $y = Y[1] \cdot q' \cdot Y[2] \cdot q' \cdots q' \cdot Y[\,|Y|\,] \cdot q' \cdot \mathbf{x}[j, ...]$

**Fig. 5.** REPALLFIXED algorithm.

If $q = q'$ then $x = y$ (line 13), and if $x$ is fixed then we can safely rewrite REPLACE* into simpler constraints (lines 14–21, note that FINDLAST$(q, x)$ returns the *last* occurrence of $q$ in $x$).

In line 22, COUNTMIN (see Fig. 6) is used to count the occurrences of $q$ in $x$. Precisely, COUNTMIN$(q, X, ALL) = m$ if $q$ surely occurs in each string of $\gamma(X)$ at least $m$ times. If $ALL = true$, we want to maximize the returned value $m$ (for REPLACE-ALL); otherwise, COUNTMIN returns 1 if $q$ must occur in $\gamma(X)$ and 0 otherwise (for REPLACE and REPLACE-LAST). For example, with $X$ as in Fig. 1, we have that: COUNTMIN$(\mathtt{b}, X, true)$ = COUNTMIN$(\mathtt{b}, X, false) = 0$, COUNTMIN$(\mathtt{o}, X, true) = 2$, and COUNTMIN$(\mathtt{o}, X, false) = 1$.

```
function CountMin(q, X, ALL)
  curr ← ϵ
  k ← 0
  for i ← 1, ..., |X| do
    S^{l,u} ← X[i]
    if S = {a} then
      curr ← curr · a^l
      j ← Find(q, curr)
      while j ≠ 0 do
        k ← k + 1
        if ¬ALL then
          return 1
        curr ← curr[j + |q| + 1, ...]
        j ← Find(q, curr)
      if l < u then
        curr ← a^l
    else
      curr ← ϵ
  return k
```

**Fig. 6.** CountMin algorithm.

In line 24, if $m = 0$, we call Check-Equate$(X, Y)$ to check if $X$ and $Y$ are equatable. Check-Equate is a lightweight version of Equate that does not refine $X$ and $Y$; it returns *false* if $X$ and $Y$ are surely not equatable (i.e., $\gamma(X) \cap \gamma(Y) = \emptyset$), otherwise *true* is returned. Hence, $\neg$Check-Equate$(X, Y)$ means that $x \neq y$ and thus $q$ must occur at least once in $x$ (otherwise we would have $x = y$) so $m$ is set to 1.

If $q$ surely occurs in $x$ and we are not propagating Replace-All, then we can safely rewrite as done in lines 25–27. This also implies a better precision. Consider for example $y = \text{Replace}(x, bb, \epsilon)$ where the domains of $x$ and $y$ are respectively $X = \{a\}^{2,\lambda}\{b\}^{0,\lambda}$ and $Y = \{a\}^{1,1}\{b\}^{0,\lambda}$ with $a, b \in \Sigma$. We have that $\neg$Check-Equate$(X, Y)$ because $X$ has at least two $a$'s and $Y$ has at most one, so Replace is rewritten in $x = p \cdot bb \cdot s \wedge y = p \cdot s \wedge \text{Find}(bb, p) = 0$ with $p, s$ fresh string variables. From $x = p \cdot bb \cdot s$ the concatenation propagator infers that $p$ must start with $aa$, and then $y = p \cdot s$ fails because $y$ has at most one $a$.

Note that the decomposition approach of [5] cannot do this because from $n = \text{Find}(bb, x)$ one can only infer that $n \notin \{1, 2\}$ so $[\![n > 0]\!]$ stays unknown. This dramatically affects the resolution which, in this case, has to prove unsatisfiability via systematic search. In general, the decomposition approach tends to lose precision when we do not know if $n = \text{Find}(q, x) > 0$: if the Boolean variable $[\![n > 0]\!]$ is unknown, we can say little about $x = p \cdot q^{[\![n>0]\!]} \cdot s$ and $y = p \cdot q'^{[\![n>0]\!]} \cdot s$.

Lines 29–50 handle the general case. Check-Find$(X, Y)$ (see Fig. 7) returns, if feasible, the earliest and the latest position where $X$ can match $Y$ (otherwise $(\bot, \bot)$ is returned). To do so, it uses a pair of vectors $ESP, LEP$ such that, for each $i = 1, \dots, |X|$, $ESP[i]$ and $LEP[i]$ are respectively the earliest start and

```
function Check-Find(X, Y)
  n ← |X|
  LEP[0] ← (0,0)
  ESP[n+1] ← (|Y|+1,0)
  for i ← 1,...,n do                                  ▷ Initialising positions.
    ESP[i] ← (0,0)
    LEP[i] ← (|Y|+1,0)
  ESP ← Push-ESP(X,Y,ESP)
  if ESP = ⊥ then
    return (⊥,⊥)
  LEP ← Push-LEP(X,Y,LEP)
  if LEP = ⊥ then
    return (⊥,⊥)
  for i ← 1,...,n do
    if LEP[i−1] ≺ ESP[i] ∨ ESP[i+1] ≺ LEP[i] then
      return (⊥,⊥)
  return (ESP[1], LEP[n])
```

**Fig. 7.** Check-Find algorithm.

```
function Push-ESP(X, Y, ESP)
  gap ← false
  end ← (0,0)
  repeat
    for i ← 1,...,|Y| do
      if gap then
        end ← ESP[i]
      (ESP[i], end) ← Push+(X[i], Y, end)
      if end = (|Y|+1,0) then
        return ⊥
    gap ← false
    for i ← n,...,1 do
      end ← Stretch−(X[i], Y, end)
      if ESP[i] ≺ end then
        ESP[i] ← end
        gap ← true
  until gap
```

**Fig. 8.** Push-ESP algorithm.

the latest end position for block $X[i]$ in $Y$. After $ESP$ and $LEP$ are initialised Check-Find calls Push-ESP to possibly improve the $ESP$ positions.

Push-ESP (see Fig. 8) uses $\text{Push}^+$ to possibly push forward the earliest start positions. Variable *end* keeps track of the earliest ends. If *end* for a block $X[i]$ cannot be stretched backward until its earliest start $ESP[i]$ then we have a "gap" between $ESP[i]$ and *end*; in this case we update $ESP[i]$ with *end* and we repeat the procedure until there are no more gaps. Push-LEP is symmetrical to Push-ESP (it uses $\text{Push}^-$ for latest end positions).

For example, if $X = \{\mathtt{o},\mathtt{p}\}^{1,3}\{\mathtt{q}\}^{0,1}\{\mathtt{m}\}^{1,2}$ and $Y$ as in Fig. 4, we have CHECK-FIND $(X, Y) = ((2,0),(4,0))$ because $X$ cannot match $\{\mathtt{B},\mathtt{b}\}^{1,1}$ and $\{!\}^{0,3}$.

Assuming that set operations are performed in constant time, the worst-case complexity of PUSH-ESP is quadratic in the number of blocks because function $\text{PUSH}^+$, costing $O(|X| + |Y|)$ in the worst case, may be called $\max(|X|, |Y|)$ times. Note that, as done e.g. in [7], assuming that set operations cost $O(1)$ is reasonable for string solving because: (1) $\Sigma$ is constant, and (2) the base of each block it is typically either a small set or a large range. So, with a proper interval representation, the time complexity of set operations can be safely considered $O(1)$ in our context—we experimentally verified in all the empirical evaluations conducted so far that it is highly unlikely to have blocks with a large number of disjoint intervals. Furthermore, even if in principle the worst-case complexity of PUSH-ESP is $O(|X| \max(|X|, |Y|))$, in practice we experimentally verified that it is almost always $O(\max(|X|, |Y|))$. In other words, gaps are very rare.

If CHECK-FIND$(Q, X)$ returns $(\perp, \perp)$ then $q$ cannot be a substring of $x$, so it must be that $x = y$ (lines 30–33). Otherwise, in lines 34–38 we try to refine the domain of $y$. Precisely, we replace the portion of $X$ that can be matched by $Q$ with a sequence of blocks approximating the replacement of $Q$ with $Q'$. Note that at this stage we cannot know in general if and where $Q$ occurs in $X$ so we have to be conservative by adding a "buffer" block $B$ containing all the characters of $X$ and $Q'$. The resulting dashed string $Y'$ is the normalised alternate repetition of $m + 1$ blocks $B$ and $m$ dashed strings $Q'$. Once built $Y'$, we call EQUATE$(Y, Y')$ to possibly refine $Y$ (or we return UNS if $Y$ and $Y'$ are not equatable).[5]

The above reasoning is applied symmetrically to possibly refine $X$, then OK is returned after refining the domain of $x$ and $y$ with the corresponding dashed strings (lines 49). Unfortunately, there is little room for incremental propagation because—unless fixed—dashed strings can both shrink and expand due to normalization. Also, it is hard to define filtering properties or consistency notions for the dashed string propagators because, as seen in Sect. 2, dashed strings do not form a lattice w.r.t. the $\sqsubseteq$ relation.

*Example 1.* Consider propagating REPLACE-ALL with $X$ as in Fig. 1, $Y = \Sigma^{0,3}$, $Q = \{\mathtt{o},\mathtt{p}\}^{1,3}\{\mathtt{q}\}^{0,1}\{\mathtt{m}\}^{1,2}$, and $Q' = \{\mathtt{r}\}^{2,5}$. Because $X$ cannot match $Y$, we have $\neg$CHECK-EQUATE$(X, Y)$, so $m = 1$, and CHECK-FIND$(Q, Y) = ((2,0),(4,0))$. Once defined the "buffer" block $B = \{\mathtt{B},\mathtt{b},\mathtt{o},\mathtt{m},!,\mathtt{r}\}^{0,3}$ we equate $Y$ and $\{\mathtt{B},\mathtt{b}\}^{1,1}\, B\, \{\mathtt{r}\}^{2,5}\, B\, \{!\}^{0,3}$. The EQUATE function will thus refine $Y$ into $\{\mathtt{B},\mathtt{b}\}^{1,1}\{\mathtt{r}\}^{2,2}$ (because $Y$ can have at most 3 characters). By reapplying the same logic, $X$ will be refined into $\{\mathtt{B},\mathtt{b}\}^{1,1}\{\mathtt{o}\}^{2,4}\{\mathtt{m}\}^{1,1}$.

Under the reasonable assumption that set operations cost $O(1)$, the overall worst-case complexity of PROP-REPLACE is dominated by PUSH-ESP, i.e., $O(|X||Y|)$. However, our experience is that that for almost all the problems we encounter the cost of propagation is linear in the number of blocks.

[5] For simplicity, in Fig. 4 we assume that, when EQUATE$(X, Y) \neq \perp$, EQUATE$(X, Y)$ modifies its arguments instead of returning a pair of refined dashed strings $(X', Y')$.

**Table 1.** Results on STRANGER benchmarks.

| | MIN | MAX | AVG | PSI |
|---|---|---|---|---|
| G-STRINGS | **0.00** | 0.11 | **0.01** | **100** |
| G-DECOMP | **0.00** | 17.16 | 0.23 | **100** |
| CVC4 | **0.00** | **0.06** | **0.01** | **100** |
| Z3 | 0.01 | $T$ | 66.74 | 78.48 |
| Z3STR3 | 0.00 | $T$ | 197.60 | 17.72 |

# 4 Evaluation

We extended G-STRINGS [26], a state-of-the-art string solver extending CP solver GECODE [17] with (dashed) strings solving capabilities, with the REPLACE propagator defined earlier. We browsed the literature to look for known benchmarks containing REPLACE, but unfortunately we found that only some of the STRANGER benchmarks used in [6] contain it—precisely, only 79 problems (77 satisfiable, 2 unsatisfiable).

We compared G-STRINGS against state-of-the-art SMT solvers supporting the theory of strings, i.e., CVC4 [23] and Z3 [25] (its default version using the theory of sequences, and the one using the Z3STR3 [10] string solver). Note that SMT solvers do not support REPLACE-LAST while REPLACE-ALL is only supported by CVC4. We are not aware of how REPLACE* is handled by these SMT solvers.

We also include in the evaluation G-DECOMP, the version of G-STRINGS decomposing REPLACE into basic constraints as in Sect. 3. For both G-STRINGS and G-DECOMP we used the default maximum string length $\lambda = 2^{16} - 1$.

Results are shown in Table 1, where we report the minimum (MIN), maximum (MAX) and average (AVG) runtime in seconds to solve an instance (timeout $T = 300$ s is assigned for unsolved instances) and the percentage of solved instances (PSI).[6] As can be seen, these problems are not challenging for CVC4 and the 2 versions of G-STRINGS: they solve almost all of them instantaneously. G-STRINGS outperforms G-DECOMP on average and for one problem it does considerably better (about 17 s faster) thanks to its tighter propagation.

Conversely, Z3 and Z3STR3 are slower. However, the performance of solvers almost certainly depends on the overall problem structure rather than on REPLACE itself, that always come in the form REPLACE$(x, \mathbf{q}, \mathbf{q}', y)$ with $\mathbf{q}, \mathbf{q}'$ fixed.

We also evaluated $2\times 79 = 158$ more instances by replacing all the occurrences of REPLACE with REPLACE-ALL and REPLACE-LAST respectively, but we did not notice any performance difference.

---

[6] We used a Ubuntu 15.10 machine with 16 GB of RAM and 2.60 GHz Intel® i7 CPU. The benchmarks and the source code of the experiments is available at https://bitbucket.org/robama/exp_cp_2020.

**Table 2.** Results on encoding benchmarks.

| | SAT. | | | | UNS. | | | |
|---|---|---|---|---|---|---|---|---|
| | MIN | MAX | AVG | PSI | MIN | MAX | AVG | PSI |
| G-STRINGS | **0.00** | **0.50** | **0.14** | **100** | **0.00** | **0.25** | **0.09** | **100** |
| G-DECOMP | **0.00** | 0.87 | 0.27 | **100** | **0.00** | 0.53 | 0.20 | **100** |
| CVC4 | 0.24 | $T$ | 56.85 | 84.00 | $T$ | $T$ | $T$ | 0.00 |

To have more insights on REPLACE we defined the following problem: given a word $\omega \in \Sigma^*$ and a sequence of bits $\beta \in \{0,1\}^*$, find if there exists a non-ambiguous encoding $f : \Sigma \to \{0,1\}^*$ such that $f(\omega[1] \cdots \omega[n]) = \beta$ where $n = |\omega|$. We first modelled this problem as:

$$
\begin{aligned}
&X[0] = \omega \ \wedge \ X[n] = \beta \ \wedge \\
&\forall\, i = 1, \ldots, n : X[i] = \text{REPLACE}(X[i-1], \omega[i], F[i]) \\
&\forall\, 1 \le i < j \le n : \omega[i] = \omega[j] \iff F[i] = F[j]
\end{aligned}
$$

where $X$ is an array of string variables in $(\Sigma \cup \{0,1\})^*$ and $F$ an array of string variables in $\{0,1\}^*$ s.t. $F[i] = f(\omega[i])$ and $X[i] = f(\omega[1] \cdots \omega[i])\omega[i+1] \cdots \omega[n]$. We then generated 50 problems (25 satisfiable and 25 unsatisfiable) with random bits $\beta$ and random words $\omega$ having alphabet $\Sigma = \{a, \ldots, z, A, \ldots, Z\}$.

For simplicity, we had to impose a fixed-length 8-bits encoding, i.e., $|F[i]| = 8$ and $|X[i]| = |X[i-1]| + 7$ for $i = 1, \ldots, 25$ and a limited word length $|\omega| = 5i$ to be able to include the SMT solvers in the comparison (otherwise the problem would have been too difficult for them). Unsatisfiable instances were generated with reverse engineering starting from satisfiable instances.

As Table 2 shows, G-STRINGS clearly achieves the best performance (we omit the results of Z3, solving only 1 instance, and Z3STR3, which cannot solve any instance). The performance of G-DECOMP is rather close. CVC4 can solve almost half of the satisfiable instances, but it fails to detect unsatisfiability.

To have a more significant comparison between G-STRINGS and G-DECOMP, we generated 150 more instances by varying $|\omega|$ in $\{130, 135, \ldots, 500\}$. Figure 9 shows the cumulative runtime distributions of the two approaches for $|\omega| = 5, 10, \ldots, 500$ (we use cumulative runtimes to better display the distributions). As we can see, as $|\omega|$ grows the performance gap between G-DECOMP and G-STRINGS increases accordingly.

To evaluate REPLACE-ALL, we modelled the "most frequent pattern" problem: given a string $x \in \Sigma^*$ and an integer $k > 0$, find the substring $q$ of $x$ with $|q| \ge k$ occurring the most times in $x$. This generalises the problem of finding the most frequent character (where $k = 1$). We modelled this problem by replacing all the occurrences of $q$ in $x$ with a string of length $|q| + 1$, and then by

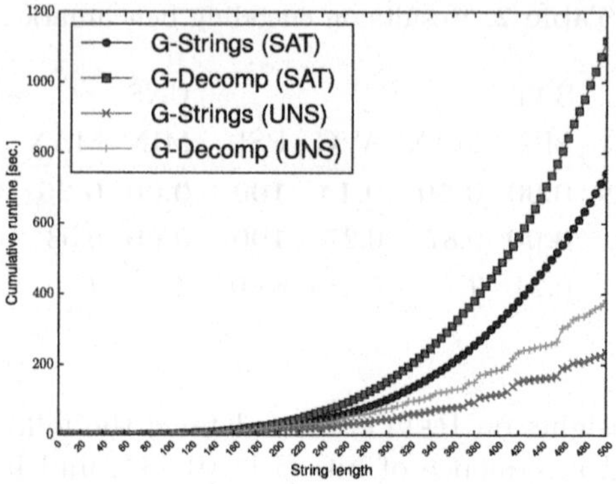


**Fig. 9.** Cumulative runtimes for G-STRINGS variants.

**Table 3.** Results on most frequent pattern benchmarks.

| Solver | OPT | SAT | TTF | TIME | SCORE |
|---|---|---|---|---|---|
| CVC4 | 0 | **100** | 0.09 | 300.00 | 49.00 |
| G-STRINGS | 22 | **100** | 0.06 | 248.44 | 80.50 |
| G-STRINGS$^+$ | **72** | **100** | **0.02** | **141.31** | **93.00** |

maximising the number of occurrences $|y| - |x|$ of $q$ in $x$.

$$\max(|y| - |x|) \ s.t.$$
$$k \leq |q| \ \wedge \ |q'| = |q| + 1 \ \wedge$$
$$y = \text{REPLACE-ALL}(x, q, q') \ \wedge \ \text{FIND}(q, x) > 0.$$

We generated 100 problems by varying $|x| \in \{50i \mid i \in [1, 25]\}$ and $k \in \{1, 2, 4, 8\}$ and we added to the evaluation G-STRINGS$^+$, a variant of G-STRINGS which search strategy assigns the minimum value to FIND$(q, x)$ integer variable, i.e., it looks for the first occurrence of $q$ in $x$. We recall that G-DECOMP cannot be evaluated because REPLACE-ALL is not decomposable into basic constraints.

Table 3 shows for each solver the number of problems where an optimal solution is proven (OPT) or at least one is found (SAT), the average time to find the first solution (TTF) and to prove optimality (TIME), where timeout $T = 300$ s is set if optimality is not proven.

The SCORE metric evaluates the solution quality. For each problem, a solver scores 0 for finding no solution, 0.25 for the worst known solution, 0.75 for the best known solution, a linear scale value in (0.25, 0.75) for other solutions, and 1 for proving the optimal solution (in Table 3 we report the sum of all the scores).

As we can see, the dashed string approach of G-STRINGS clearly outperforms CVC4, especially when it comes to proving optimality or, in other words, in detecting unsatisfiability.[7] The performance of G-STRINGS$^+$ confirms the importance of defining good search heuristics for CP solving. In this particular case, we argue that G-STRINGS$^+$ achieves a better performance w.r.t. G-STRINGS because its search heuristics somehow mimics a left-to-right search of $q$ in $x$.

## 5 Related Work

Although string solvers are still in their infancy, a large number of string solving approaches have been proposed. We can classify them into three rough families: automata-based, word-based, and unfolding-based.

*Automata-based* approaches use automata to handle string variables and related operations. Examples of automata-based solvers are STRANGER [32], PASS [22], STRSOLVE [20]. The advantage of automata is that they can handle unbounded-length strings and precisely represent infinite sets of strings. However, automata face performance issues due to state explosion and the integration with other domains (e.g., integers).

*Word-based* solvers are basically SMT solvers treating strings without abstractions or representation conversions. Among the most known word-based solvers, mainly based on the DPLL(T) paradigm [16], we mention: CVC4 [23], the family of solvers Z3STR [34], Z3STR2 [33], and Z3STR3 [10] based on the Z3 solver [25], S3 [30] and its evolutions S3P [30] and S3# [31], NORN [2]. More recent proposals are SLOTH [19] and TRAU [1]. These solvers can reason about unbounded strings and take advantage of already defined theories. However, most of them are incomplete and face scalability issues due to disjunctive case-splitting.

*Unfolding-based* approaches basically select a length bound $k$, and consider each string variable as a vector of $k$ elements. String constraints can be compiled down to bit-vector constraints (e.g., [21,27] solvers) or integer constraints (e.g., [4]). GECODE+$\mathbb{S}$ [28] instead defines dedicated propagators for string constraints. The unfolding approach adds flexibility but may sacrifice high-level relationships between strings, cannot deal with unbounded-length strings and may be very inefficient when the length bound is large—even if the generated solutions have short length. The dashed string approach implemented in G-STRINGS [7] can be seen as a *lazy unfolding* performing an higher level reasoning over the blocks of the corresponding dashed string domain. The well-known *regular* global constraint [6] is impractical for string solving because it assumes *fixed-length* strings.

## 6 Conclusions

String solving is of growing interest, since its a basis of many important program analyses. It has already been demonstrated that the G-STRINGS solver is

[7] Since optimization is not a native feature for CVC4 we implemented it by iteratively solving decision problems with refined objective bounds.

highly competitive in solving problems involving strings arising from program analysis [3].

In this paper[8], we devise and implement a unified propagator for handling (variants of) the Replace constraint with dashed strings. Empirical results confirm the effectiveness of such propagator when compared to the decomposition approach and state-of-the-art SMT approaches. While on simple examples the propagator is often not that much better than the decomposition, it scales much better, and allows us to handle Replace-All constraints which cannot be defined by decomposition.

String solving is by no means a closed question, and their is much scope for further work in this direction, in particular: handling more exotic string constraints such as extended regular expression matching, or transducers. An important direction for investigation is to extend a dashed string solver to use nogood learning, which can exponentially reduce the amount of work required. This raises significant questions for nogood learning both in practice and theory.

## References

1. Abdulla, P.A., et al.: Flatten and conquer: a framework for efficient analysis of string constraints. In: Proceedings of the 38th ACM SIGPLAN Conference on Programming Language Design and Implementation, PLDI 2017, Barcelona, Spain, 18–23 June 2017, pp. 602–617 (2017)
2. Abdulla, P.A., et al.: Norn: an SMT solver for string constraints. In: Kroening, D., Păsăreanu, C.S. (eds.) CAV 2015. LNCS, vol. 9206, pp. 462–469. Springer, Cham (2015). https://doi.org/10.1007/978-3-319-21690-4_29
3. Amadini, R., Andrlon, M., Gange, G., Schachte, P., Søndergaard, H., Stuckey, P.J.: Constraint programming for dynamic symbolic execution of JavaScript. In: Rousseau, L.-M., Stergiou, K. (eds.) CPAIOR 2019. LNCS, vol. 11494, pp. 1–19. Springer, Cham (2019). https://doi.org/10.1007/978-3-030-19212-9_1
4. Amadini, R., Flener, P., Pearson, J., Scott, J.D., Stuckey, P.J., Tack, G.: MiniZinc with strings. In: Hermenegildo, M.V., Lopez-Garcia, P. (eds.) LOPSTR 2016. LNCS, vol. 10184, pp. 59–75. Springer, Cham (2017). https://doi.org/10.1007/978-3-319-63139-4_4
5. Amadini, R., Gange, G., Stuckey, P.J.: Propagating lex, find and replace with dashed strings. In: van Hoeve, W.-J. (ed.) CPAIOR 2018. LNCS, vol. 10848, pp. 18–34. Springer, Cham (2018). https://doi.org/10.1007/978-3-319-93031-2_2
6. Amadini, R., Gange, G., Stuckey, P.J.: Propagating regular membership with dashed strings. In: Hooker, J. (ed.) CP 2018. LNCS, vol. 11008, pp. 13–29. Springer, Cham (2018). https://doi.org/10.1007/978-3-319-98334-9_2
7. Amadini, R., Gange, G., Stuckey, P.J.: Sweep-based propagation for string constraint solving. In: Proceedings 32nd AAAI Conference Artificial Intelligence, pp. 6557–6564. AAAI Press (2018)
8. Amadini, R., et al.: Combining string abstract domains for JavaScript analysis: an evaluation. In: Legay, A., Margaria, T. (eds.) TACAS 2017. LNCS, vol. 10205, pp. 41–57. Springer, Heidelberg (2017). https://doi.org/10.1007/978-3-662-54577-5_3

[8] This work is supported by Australian Research Council through Linkage Project Grant LP140100437 and Discovery Early Career Researcher Award DE160100568.

9. Barahona, P., Krippahl, L.: Constraint programming in structural bioinformatics. Constraints **13**(1–2), 3–20 (2008)
10. Berzish, M., Ganesh, V., Zheng, Y.: Z3str3: a string solver with theory-aware heuristics. In: Stewart, D., Weissenbacher, G. (eds.) Proceedings of 17th Conference Formal Methods in Computer-Aided Design, pp. 55–59. FMCAD Inc. (2017)
11. Bisht, P., Hinrichs, T.L., Skrupsky, N., Venkatakrishnan, V.N.: WAPTEC: white-box analysis of web applications for parameter tampering exploit construction. In: Proceedings of ACM Conference on Computer and Communications Security, pp. 575–586. ACM (2011)
12. Bjørner, N., Tillmann, N., Voronkov, A.: Path feasibility analysis for string-manipulating programs. In: Kowalewski, S., Philippou, A. (eds.) TACAS 2009. LNCS, vol. 5505, pp. 307–321. Springer, Heidelberg (2009). https://doi.org/10.1007/978-3-642-00768-2_27
13. Costantini, G., Ferrara, P., Cortesi, A.: A suite of abstract domains for static analysis of string values. Softw.: Pract. Exp. **45**(2), 245–287 (2015)
14. Emmi, M., Majumdar, R., Sen, K.: Dynamic test input generation for database applications. In: Proceedings of the ACM SIGSOFT International Symposium on Software Testing and Analysis (ISSTA), pp. 151–162. ACM (2007)
15. Gange, G., Navas, J.A., Stuckey, P.J., Søndergaard, H., Schachte, P.: Unbounded model-checking with interpolation for regular language constraints. In: Piterman, N., Smolka, S.A. (eds.) TACAS 2013. LNCS, vol. 7795, pp. 277–291. Springer, Heidelberg (2013). https://doi.org/10.1007/978-3-642-36742-7_20
16. Ganzinger, H., Hagen, G., Nieuwenhuis, R., Oliveras, A., Tinelli, C.: DPLL($T$): fast decision procedures. In: Alur, R., Peled, D.A. (eds.) CAV 2004. LNCS, vol. 3114, pp. 175–188. Springer, Heidelberg (2004). https://doi.org/10.1007/978-3-540-27813-9_14
17. Gecode Team: Gecode: generic constraint development environment (2016). http://www.gecode.org
18. Hojjat, H., Rümmer, P., Shamakhi, A.: On strings in software model checking. In: Lin, A.W. (ed.) APLAS 2019. LNCS, vol. 11893, pp. 19–30. Springer, Cham (2019). https://doi.org/10.1007/978-3-030-34175-6_2
19. Holík, L., Janku, P., Lin, A.W., Rümmer, P., Vojnar, T.: String constraints with concatenation and transducers solved efficiently. PACMPL **2**(POPL), 4:1–4:32 (2018)
20. Hooimeijer, P., Weimer, W.: StrSolve: solving string constraints lazily. Autom. Softw. Eng. **19**(4), 531–559 (2012). https://doi.org/10.1007/s10515-012-0111-x
21. Kieżun, A., Ganesh, V., Artzi, S., Guo, P.J., Hooimeijer, P., Ernst, M.D.: HAMPI: a solver for word equations over strings, regular expressions, and context-free grammars. ACM Trans. Softw. Eng. Methodol. **21**(4), article 25 (2012)
22. Li, G., Ghosh, I.: PASS: string solving with parameterized array and interval automaton. In: Bertacco, V., Legay, A. (eds.) HVC 2013. LNCS, vol. 8244, pp. 15–31. Springer, Cham (2013). https://doi.org/10.1007/978-3-319-03077-7_2
23. Liang, T., Reynolds, A., Tinelli, C., Barrett, C., Deters, M.: A DPLL($T$) theory solver for a theory of strings and regular expressions. In: Biere, A., Bloem, R. (eds.) CAV 2014. LNCS, vol. 8559, pp. 646–662. Springer, Cham (2014). https://doi.org/10.1007/978-3-319-08867-9_43
24. Loring, B., Mitchell, D., Kinder, J.: ExpoSE: practical symbolic execution of standalone JavaScript. In: Proceedings of 24th ACM SIGSOFT International SPIN Symposium Model Checking of Software (SPIN'17), pp. 196–199. ACM Press (2017)

25. de Moura, L., Bjørner, N.: Z3: an efficient SMT solver. In: Ramakrishnan, C.R., Rehof, J. (eds.) TACAS 2008. LNCS, vol. 4963, pp. 337–340. Springer, Heidelberg (2008). https://doi.org/10.1007/978-3-540-78800-3_24
26. Amadini, R.: G-strings: gecode with string variables (2020). https://bitbucket.org/robama/g-strings
27. Saxena, P., Akhawe, D., Hanna, S., Mao, F., McCamant, S., Song, D.: A symbolic execution framework for JavaScript. In: Proceedings of 2010 IEEE Symposium Security and Privacy, pp. 513–528. IEEE Computer Society (2010)
28. Scott, J.D., Flener, P., Pearson, J., Schulte, C.: Design and implementation of bounded-length sequence variables. In: Salvagnin, D., Lombardi, M. (eds.) CPAIOR 2017. LNCS, vol. 10335, pp. 51–67. Springer, Cham (2017). https://doi.org/10.1007/978-3-319-59776-8_5
29. Thomé, J., Shar, L.K., Bianculli, D., Briand, L.C.: Search-driven string constraint solving for vulnerability detection. In: ICSE 2017, Buenos Aires, Argentina, 20–28 May 2017, pp. 198–208 (2017)
30. Trinh, M., Chu, D., Jaffar, J.: S3: a symbolic string solver for vulnerability detection in web applications. In: SIGSAC, pp. 1232–1243. ACM (2014)
31. Trinh, M.-T., Chu, D.-H., Jaffar, J.: Model counting for recursively-defined strings. In: Majumdar, R., Kunčak, V. (eds.) CAV 2017. LNCS, vol. 10427, pp. 399–418. Springer, Cham (2017). https://doi.org/10.1007/978-3-319-63390-9_21
32. Yu, F., Alkhalaf, M., Bultan, T.: STRANGER: an automata-based string analysis tool for PHP. In: Esparza, J., Majumdar, R. (eds.) TACAS 2010. LNCS, vol. 6015, pp. 154–157. Springer, Heidelberg (2010). https://doi.org/10.1007/978-3-642-12002-2_13
33. Zheng, Y., Ganesh, V., Subramanian, S., Tripp, O., Dolby, J., Zhang, X.: Effective search-space pruning for solvers of string equations, regular expressions and length constraints. In: Kroening, D., Păsăreanu, C.S. (eds.) CAV 2015. LNCS, vol. 9206, pp. 235–254. Springer, Cham (2015). https://doi.org/10.1007/978-3-319-21690-4_14
34. Zheng, Y., Zhang, X., Ganesh, V.: Z3-str: a Z3-based string solver for web application analysis. In: Proceedings of 9th Joint Meeting on Foundations of Software Engineering, pp. 114–124. ACM (2013)

# Combinatorial Search in CP-Based Iterated Belief Propagation

Behrouz Babaki[1], Bilel Omrani[2], and Gilles Pesant[2(✉)]

[1] HEC Montréal, Montreal, Canada
behrouz.babaki@hec.ca
[2] Polytechnique Montréal, Montreal, Canada
{bilel.omrani,gilles.pesant}@polymtl.ca

**Abstract.** Compared to most other computational approaches to solving combinatorial problems, Constraint Programming's distinctive feature has been its very high-level modeling primitives which expose much of the combinatorial substructures of a problem. Weighted counting on these substructures (i.e. constraints) can be used to compute beliefs about certain variable-value assignments occurring in a solution to the given constraint. A recent proposal generalizes the propagation mechanism of constraint programming to one sharing such beliefs between constraints. These beliefs, even if not computed exactly, can be very revealing for search. In this paper we investigate how best to guide combinatorial search in this CP-based belief propagation framework. We empirically evaluate branching heuristics on a wide set of benchmark constraint satisfaction problems.

## 1 Introduction

Compared to most other computational approaches to solving combinatorial problems, Constraint Programming's (CP) distinctive feature has been its very high-level modeling primitives which bring out much of the combinatorial structure of a problem explicitly in the statement of the model. These primitives take the form of so-called global constraints [3] which encapsulate powerful algorithms for both inference and search. On the side of inference, filtering algorithms remove a value from the domain of possible values for a variable if it does not appear in any tuple satisfying a given constraint (i.e. is unsupported) and then share that information between constraints by propagating it through common variables. On the side of search, explaining and recording the sequence of decisions that led to a failure avoids wasting search effort by repeatedly rediscovering that same dead-end [4].[1] As well, work on model counting at the level of individual constraints led to counting-based search [10], a family of effective branching heuristics. Generalizing this to weighted counting allowed one to handle combinatorial *optimization* problems in which individual costs are associated with each variable assignment [8]. The close relationship between weighted counting

[1] One could argue that this is also a manifestation of inferring redundant constraints.


H. Simonis (Ed.): CP 2020, LNCS 12333, pp. 21–36, 2020.
https://doi.org/10.1007/978-3-030-58475-7_2

and probabilistic inference had already been pointed out in Chavira and Darwiche [2]. From a CP perspective it can be interpreted for a given constraint as the belief in a certain variable-value assignment computed by summing, over satisfying tuples of the constraint that feature that assignment, the likelihood of each tuple expressed as the combined beliefs of its component assignments. Pesant [9] investigates a richer propagation medium for CP which does not simply share unsupported variable-value assignments but shares instead for each variable-value assignment the corresponding quantitative belief that it occurs in some solution. The resulting process is akin to iterated belief propagation [7].

These computed beliefs can be viewed as approximate marginals onto individual variables of the joint probability distribution defined by the set of solutions. Such information, even if not exact, can be very revealing for search. In this paper we investigate how best to guide combinatorial search given the approximate marginals computed by CP-based belief propagation. We empirically evaluate generic branching heuristics on a wide set of benchmark constraint satisfaction problems.

The rest of the paper is organized as follows. Section 2 recalls the recently proposed CP-BP framework and presents the search heuristics we investigate. Section 3 reports the results of our initial experiment and analyzes them for further improvement. Section 4 investigates two benchmark problems on which we did not perform well. Section 5 investigates an oscillation phenomenon observed on the computed marginals. Section 6 concludes with final recommendations about combinatorial search in the CP-BP framework and points to future directions for research.

## 2 The CP-BP Framework

In this section we briefly introduce the CP-BP framework and refer the reader to Pesant [9] for details. A *factor graph* is a bipartite graph with variable nodes, factor nodes, and edges between some variable-factor pairs. A *message-passing* algorithm on a factor graph iteratively exchanges messages between variable nodes and factor nodes. The message from a factor node is a real-valued function of the variables it is connected to. The message from a variable node is a real-valued function of that variable. The *sum-product algorithm*, also known as belief propagation, is one instantiation of message passing where the message from a variable to a factor is the product of the messages received from the other factors, and the message from a factor to a variable is essentially a sum of products of messages received from the other variables. It has been used to compute the *marginals* of individual variables for the factored function. That computation is exact for trees but not for general graphs with cycles.

A constraint graph can be viewed as a special case of a factor graph with factors (constraints) issuing 0/1 messages. In that sense standard constraint propagation in CP is a form of message passing, exchanging information about which variable-value assignments are supported by each constraint. The CP-BP framework proposes to go back to real-valued messages providing information about the likelihood of a given variable-value assignment in a solution to the problem. It does so by performing weighted counting of solutions within each constraint. The approximate marginals resulting from message passing can inform branching heuristics.

To illustrate this CP-BP framework consider the following example taken from [9]: constraints ALLDIFFERENT$(a, b, c)$, $a + b + c + d = 7$, and $c \leq d$, with variables $a, b, c, d \in \{1, 2, 3, 4\}$. This CSP is domain consistent but only admits two solutions: $\langle 2, 3, 1, 1\rangle$ and $\langle 3, 2, 1, 1\rangle$.

In CP-BP message passing is synchronized in two phases: in the first phase constraints receive messages from the variables; in the second phase constraints send messages to the variables, which are multiplied and then normalized. This process is repeated for some number of iterations. Initially variables send identical (i.e. uniform) messages in the first phase. For the second phase, consider variable $a$ that appears in the ALLDIFFERENT and the sum constraints: if we count the proportion of solutions to the former constraint in which $a$ takes value 1, it is $\frac{1}{4}$ and the same is true for the other values in its domain; the proportion from the latter constraint is $\frac{10}{20}$ for value 1 and $\frac{6}{20}, \frac{3}{20}, \frac{1}{20}$ respectively for the other values 2, 3, 4. The combined information, obtained by multiplying these "messages" from each of the two constraints to variable $a$, indicates a strong positive bias (or marginal) toward assigning value 1 to $a$. The situation is identical for variable $b$. Variable $c$ appears in the same two constraints and the messages it receives from them are the same as those for $a$ and $b$, but it also appears in the binary inequality: from that one the proportions are $\langle \frac{4}{10}, \frac{3}{10}, \frac{2}{10}, \frac{1}{10}\rangle$. So there is an even stronger bias toward assigning value 1 to $c$. Variable $d$ gets conflicting information: $\langle \frac{10}{20}, \frac{6}{20}, \frac{3}{20}, \frac{1}{20}\rangle$ from the sum constraint and $\langle \frac{1}{10}, \frac{2}{10}, \frac{3}{10}, \frac{4}{10}\rangle$ from the binary inequality.

**Table 1.** Biases (marginals) after ten iterations of iterated belief propagation on our running example (left) and true biases (right).

| | 1 | 2 | 3 | 4 |
|---|---|---|---|---|
| $a$ | .01 | .52 | .46 | .01 |
| $b$ | .01 | .52 | .46 | .01 |
| $c$ | .98 | .02 | .00 | .00 |
| $d$ | .90 | .10 | .00 | .00 |

| | 1 | 2 | 3 | 4 |
|---|---|---|---|---|
| $a$ | 0 | 1/2 | 1/2 | 0 |
| $b$ | 0 | 1/2 | 1/2 | 0 |
| $c$ | 2/2 | 0 | 0 | 0 |
| $d$ | 2/2 | 0 | 0 | 0 |

For the next iteration of message passing and all others that follow, the messages sent by each variable (first phase) reflect its current biases. And from now on, each solution to a constraint that we count (second phase) is weighted by the product of the biases of the individual assignments that make up that solution. Note that typically we do not explicitly enumerate the solutions to a constraint and compute their weight: efficient algorithms specialized for each type of constraint have been designed, much like what has been done for domain filtering.

After ten iterations we have converged to the biases on the left in Table 1, which are quite close to the true ones (Table 1, on the right) derived from the two solutions we previously identified. Because from a factor graph perspective, in a typical CP model each constraint represents a bigger chunk of the problem involving more variables, there tends to be fewer cycles in the graph and it was observed in [9] that iterated belief propagation converges better. Initial experiments suggested that applying constraint propagation followed by a fixed number (five) of BP iterations works well.

### 2.1 Candidate Search Heuristics

As is commonplace in CP, we consider two-way branching heuristics that build a search tree. We can choose to use the computed marginals in two complementary ways: either we first explore subtrees defined by making highly likely assignments or subtrees defined by excluding highly unlikely assignments. Heuristic max-marginal first tries (left branch) to assign the variable-value pair exhibiting the largest marginal and disallows that assignment upon backtracking (right branch). Heuristic max-strength first tries (left branch) to assign the variable-value pair exhibiting the largest (positive) difference between its marginal and the reciprocal of the domain size (i.e. its *marginal strength*) and disallows that assignment upon backtracking (right branch). To illustrate how these two heuristics may differ, consider variable $x \in \{a, b\}$ with respective marginals $\langle .8, .2\rangle$ and variable $y \in \{a, b, c, d\}$ with marginals $\langle .7, .1, .1, .1\rangle$. Heuristic max-marginal will first branch on $x = a$ whereas max-strength will branch on $y = a$ since $.7 - 1/4 = .45 > .8 - 1/2 = .3$. Heuristic min-marginal identifies the variable-value pair exhibiting the smallest marginal, first removes that value from the domain of that variable and then assigns it upon backtracking. Heuristic min-strength identifies the variable-value pair exhibiting the smallest marginal strength, first removes that value from the domain of that variable and then assigns it upon backtracking.

## 3 Initial Experiments

The recommendations about search in [9], namely branching in a depth-first search tree according to maximum marginal strength (max-strength), were based on a small set of problems and instances. We first wish to revisit this on a larger and more diverse set of benchmark problems. The current implementation of MiniCPBP[2] dictates that these should be constraint satisfaction problems[3] on integer variables and with constraints among ALLDIFFERENT, SUM, (positive) TABLE, REGULAR, AMONG, and CARDINALITY. The xcsp.org website features a selection engine that enables one to select instances from a large set according to such requirements. We selected the following problems (excluding some

[2] Available at https://github.com/PesantGilles/MiniCPBP.

[3] The search guidance provided by its branching heuristics currently ignores solution cost and so it is unlikely to perform well on constraint optimization problems.

series of instances that were trivially solved by all heuristics or that caused an out-of-memory error): CarSequencing, Dubois, Kakuro (sumdiff/medium), LatinSquare (m1, xcsp2/bqwh⋆), MagicSquare (sum/s1), Partially-Filled MagicSquare (m1/gp), MultiKnapsack, Nonogram (regular), Ortholatin, PigeonPlus, Primes, and PseudoBoolean (dec). In all, 1336 instances. The experiments were executed on Intel E5-2683 2.1 Ghz machines running CentOS Linux 7. The memory budget was set to 10 GB and the timeout to 10 h.

### 3.1 Comparison Between Our Candidate Branching Heuristics

Figure 1 compares the search guidance of the four heuristics presented in Sect. 2.1 using depth-first search (DFS) and limited-discrepancy search (LDS). Each plot presents for a benchmark problem the percentage of instances solved within a number of failed search tree nodes by each heuristic. One interesting observation is that max-marginal is performing better overall than what was previously thought best, max-strength. That recommendation was based on three problems (LatinSquare, Partially-Filled MagicSquare, Primes) for which we see that their performance is comparable. This broader experiment points us in a slightly different direction. One possible advantage of max-marginal is that it may be biased toward smaller domains, as illustrated in Sect. 2.1.

Generally LDS performs better than DFS, which is usually the case with a good search heuristic since it prioritizes search tree nodes reached by branching against the heuristic as little as possible. Figure 2 shows the proportion of instances solved using at most a given number of such discrepancies. Interestingly for some problems a large proportion are solved in very few or even no discrepancies, directly reaching a solution.

Based on these results we continue with combination max-marginal / LDS and now turn to comparing with other heuristics and solvers.

### 3.2 Comparison with Default Search of State-of-the-Art Solvers

We compare to the best CP solvers in the latest XCSP competition, `Choco-4.10.2`[4] and `Abscon-19.06`.[5] We executed these solvers using the same command line options as those used in the competition. As a result, solvers use their default search strategies. The default heuristics for variable and value selection are dom-wdeg and lexicographic for both solvers. Such a comparison will not allow us to isolate perfectly the effects of search heuristics on guidance and runtime because the solvers are different and the filtering algorithms implemented for each constraint may be different as well. For example an inspection of Choco's source code indicates that for the ALLDIFFERENT constraint, the domain consistency algorithm is called probabilistically according to how much/often it performs more filtering than the bounds consistency algorithm on

[4] Available at https://github.com/chocoteam/choco-solver/releases/tag/4.10.2.

[5] Available at https://www.cril.univ-artois.fr/~lecoutre/#/softwares.

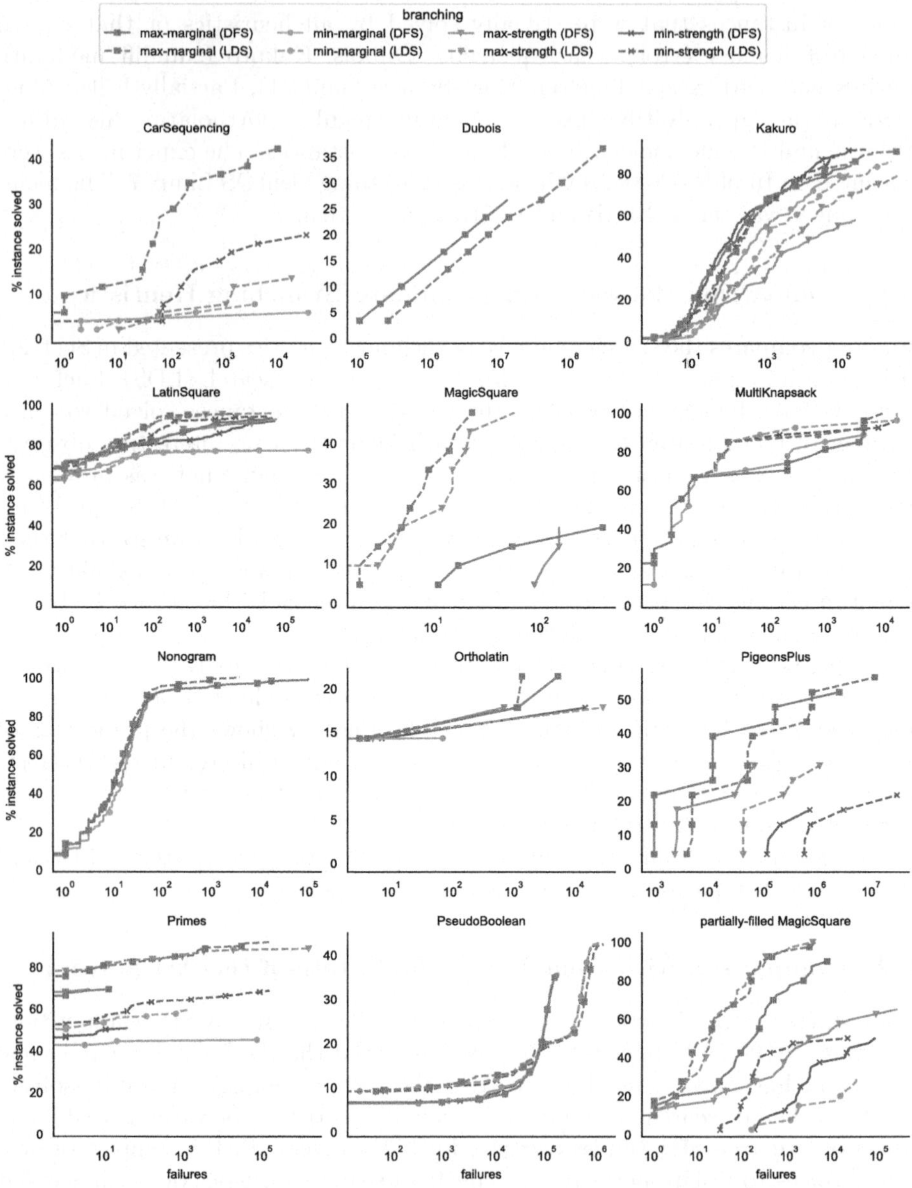


**Fig. 1.** %instances solved vs #fails for CP-BP branching heuristics on 12 benchmark problems

a given instance—in contrast MiniCPBP exclusively uses the domain consistency algorithm, which may provide a bit more filtering but also may put us at a disadvantage for computation time. Nevertheless it provides some indication of where we stand compared to the implementation choices of state-of-the-art solvers.

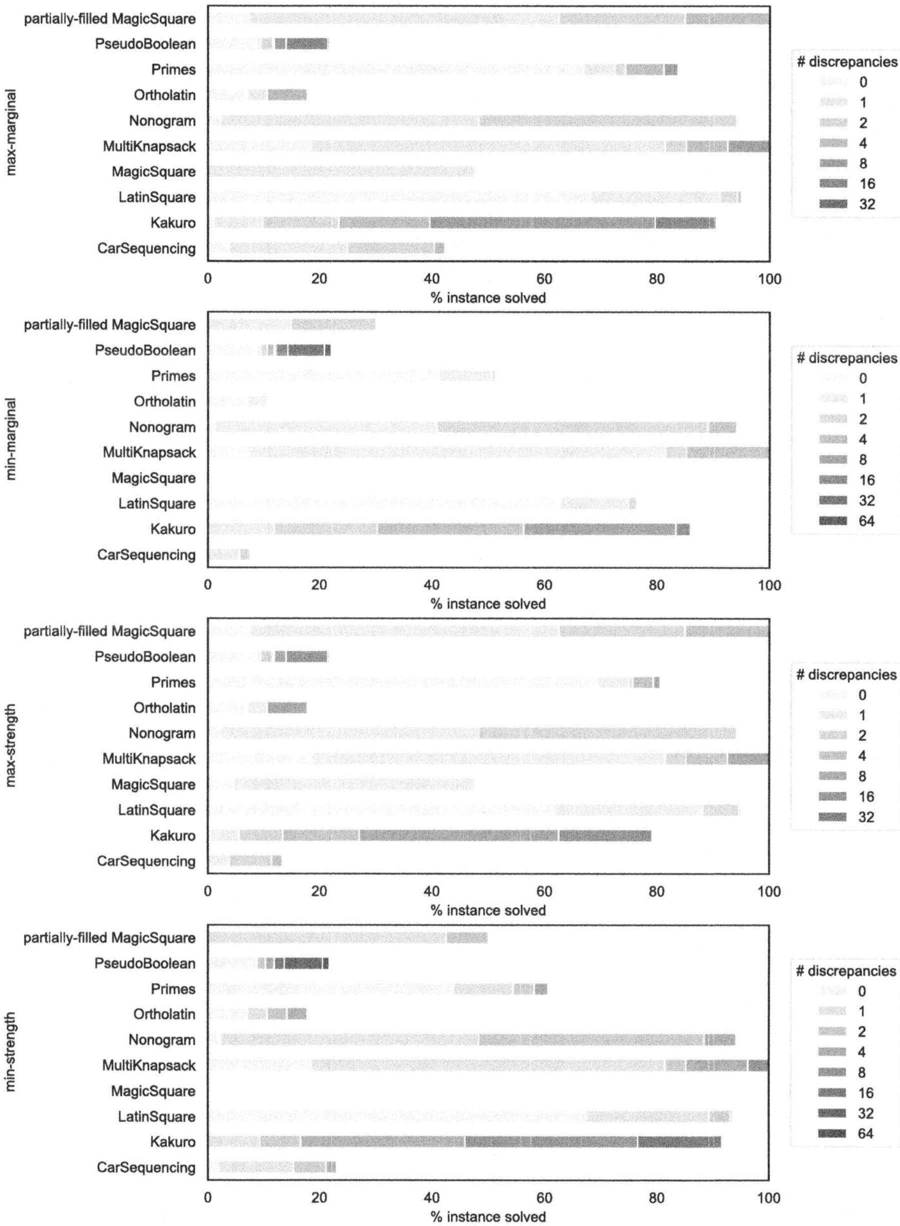


**Fig. 2.** %instances solved given #discrepancies allowed in the search tree

As a more easily comparable baseline we also use MiniCP[6] with min-domain variable ordering and random value ordering and report the median result over ten runs. Here the underlying solver is essentially the same.

[6] Available at http://www.minicp.org/.

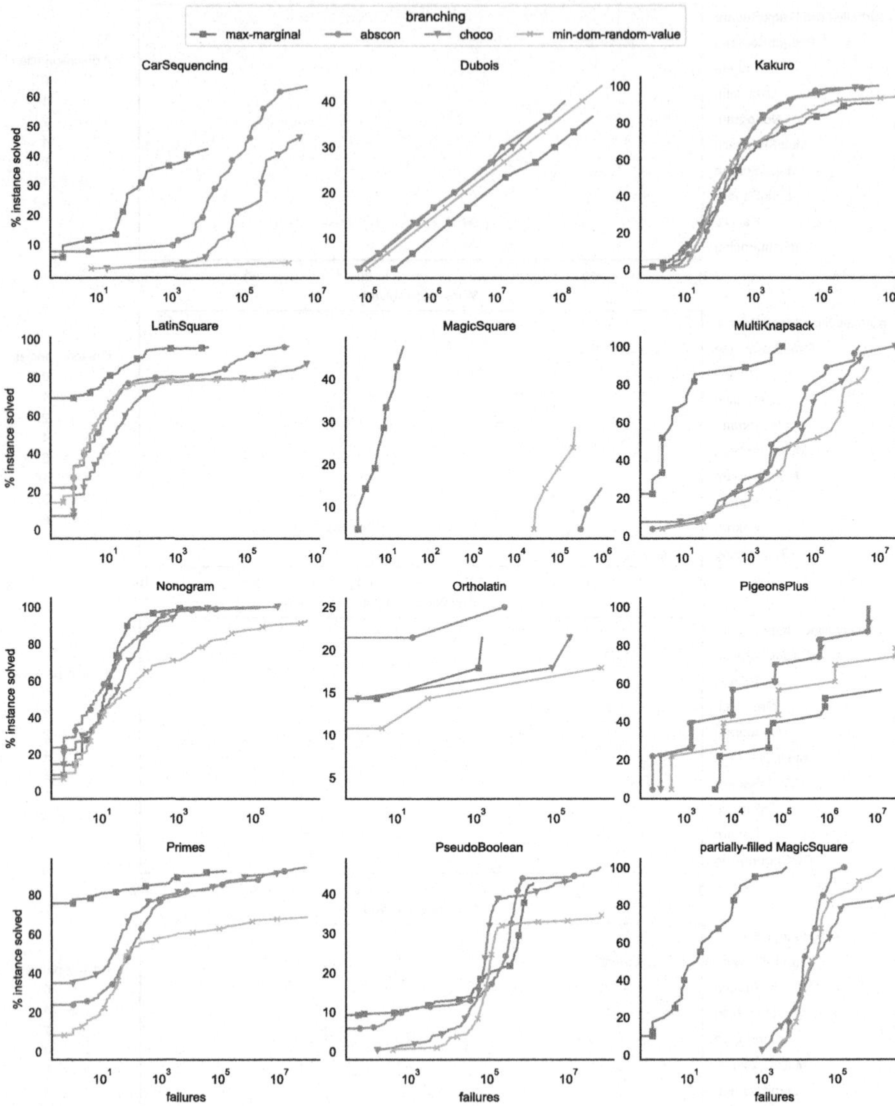


**Fig. 3.** %instances solved vs #fails for our best-performing search against Abscon and Choco

Figure 3 compares search guidance: we observe that we perform better than these state-of-the-art solvers on six problems (LatinSquare, MagicSquare, MultiKnapsack, Nonogram, Primes, Partially-Filled MagicSquare) and about as well on three of them (CarSequencing, OrthoLatin, PseudoBoolean). But there are three problems on which we do not perform as well: Dubois, Kakuro, and PigeonPlus.

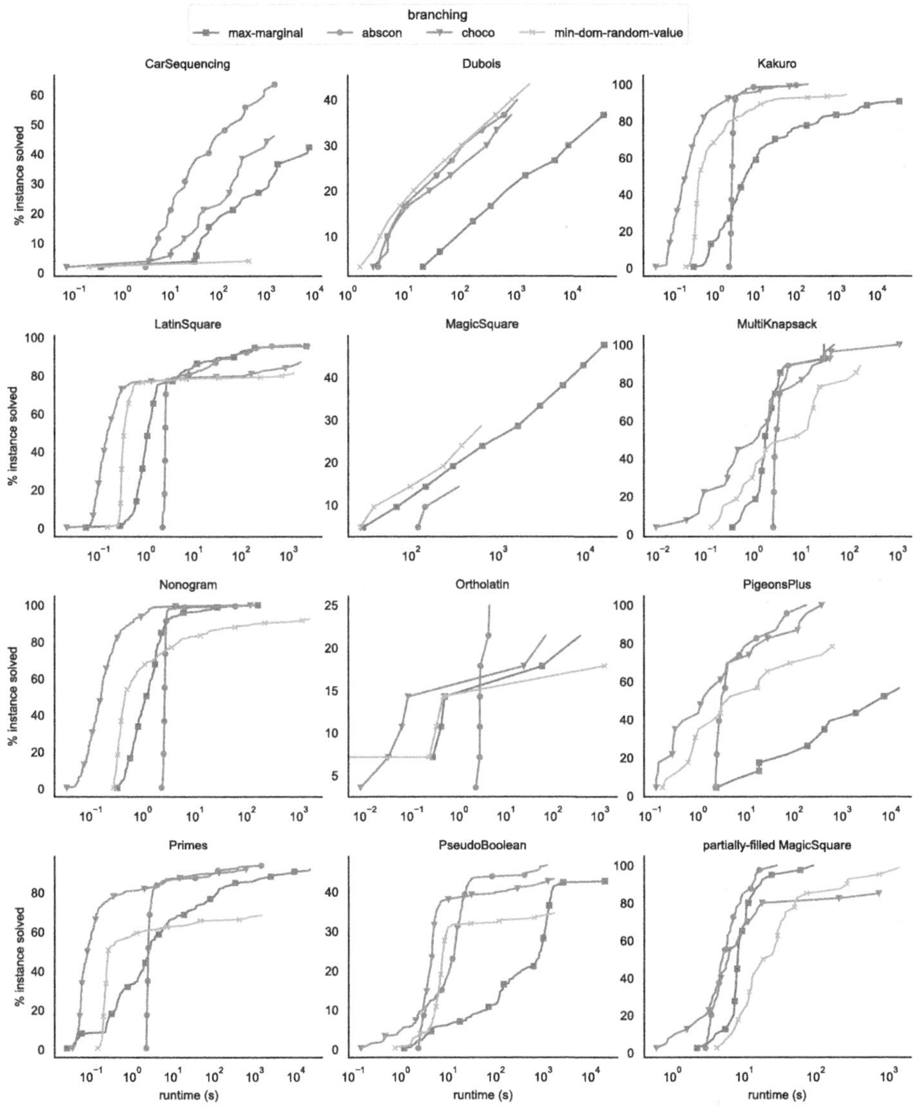


**Fig. 4.** %instances solved vs runtime for our best-performing search against Abscon and Choco

Comparing runtimes is difficult here even when using the same computers for the reasons cited above. In addition MiniCPBP is implemented on top of MiniCP, an academic solver not as optimized as the ones we are comparing to. Nevertheless Fig. 4 presents such a comparison. We find that our search heuristic remains competitive for the problems on which guidance was superior. It is difficult to give a general estimation of the additional computational effort of belief propagation compared to standard support propagation since it depends on the number, arity, and type of constraints in the model (the weighted counting algorithms are different for each type). If we take a problem for which search effort is similar for max-marginal and min-domain (both basically running MiniCP), such as Kakuro (see Fig. 3), there is in this case approximately a one order of magnitude difference (see Fig. 4).

In the following sections we look in detail at some of the problems on which we do not perform well in an effort to improve our combinatorial search heuristics further.

## 4 Dubois and PigeonPlus—Uniform Marginals

Dubois are crafted instances with binary variables and ternary constraints. They have a special structure, essentially hiding in an increasingly long chain of ternary constraints a tiny unsatisfiable instance: one constraint says $(x, y, z)$ should exhibit an even number of 1's whereas another says the opposite. Even on such a tiny instance, we essentially enumerate potential solutions. That behaviour is confirmed by the straight lines we observe for all solvers/heuristics when we plot the number of instances solved against the number of failed search tree nodes in log scale (the number of variables in these instances increases smoothly). The marginals are uniform (all at 0.5) and remain so even once we start branching. Therefore there is no heuristic information to allow us to discriminate between values. We likely perform worse than the rest because we are using LDS so some search tree nodes are visited multiple times (see also Fig. 1).

The same is true for PigeonPlus: instances are unsatisfiable, so we need to exhaust the search tree, and marginals are either uniform ($x$ variables) or almost uniform ($y$ variables), remaining the same throughout BP iterations. The branching for max-marginal tends to alternate between $x$ and $y$ variables, which may not be optimal for the size of the refutation tree because $x$ variables have a slightly smaller domain—hence min-domain would be expected to perform better here. The staircase appearance of search effort is linked to the size of the instances partly increasing in discrete steps.

The lesson here is that when marginals are (almost) uniform and remain so even after we branch, there is no discriminative information to exploit in the marginals so we should use some other branching heuristic and spare the effort of computing marginals. Fortunately this is easy to check at the beginning of search.

### 4.1 An Opportunity for Higher Forms of Consistency

Note that Dubois is a problem for which the generalization to belief propagation of marginals on *subsets* of variables would be worthwhile. Consider marginals on pairs of variables: as soon as we set one of the three variables, say $x$ to 1, unsatisfiability is uncovered: the marginals on $yz \in \{00, 01, 10, 11\}$ from the "even" constraint are $\langle 0, .5, .5, 0\rangle$ and from the "odd" one are $\langle .5, 0, 0, .5\rangle$, whose combination makes every pair of values vanish. The current CP-BP framework could be extended to offer such a consistency level reminiscent of path consistency, though at some computational expense.

# 5 CarSequencing—Handling Dramatic Marginal Oscillation

We noticed that for CarSequencing we were initially branching on null marginals, a disconcerting behaviour. We studied the progression of marginals during BP iterations on a small instance (CarSequencing-m1-jcr/CarSequencing-dingbas.xml) and observed that marginals of the variables start an extreme oscillation between two values in their domain after a few iterations that eventually drives all marginals to zero. Even if we stop after fewer iterations, that oscillation makes the information we use for branching very unstable. Marginal oscillation (and, more generally, failure to converge) is a documented phenomenon [6] that is linked to the presence of cycles in the constraint graph but as we wrote in Sect. 2 CP models with large-arity constraints tend to have marginals converge more. Still, some oscillation was observed in [9] but attributed to an alternation between values taken by a given variable in the multiple solutions to an instance. This cannot be the case here: there are six solutions to that small instance but some variables are backbone (i.e. are assigned the same value in all solutions) and yet their marginals oscillate as well.

## 5.1 Studying a Tiny Instance with a CARDINALITY Constraint

One reason for this oscillation seems to be the way MINICPBP decomposes CARDINALITY constraints into a set of AMONG constraints, one per counted value, and how the indicator variables are linked to the original variables. That decomposition is used in MINICPBP because of the availability of an efficient exact weighted counting algorithm for SUM constraints whereas (unweighted) counting on CARDINALITY as a whole is already quite challenging and time consuming [1].

Consider this tiny instance:

$$x_1, x_2, x_3 \in \{0, 1, 2\}$$

$$\text{CARDINALITY}(\{x_1, x_2, x_3\}, \langle 0, 1, 2\rangle, \langle c_0, c_1, c_2\rangle)$$

for some fixed integer parameters $c_i$ prescribing the number of occurrences of value $i$ in $\{x_1, x_2, x_3\}$. This CARDINALITY constraint is decomposed into

$$\text{AMONG}(\{x_1, x_2, x_3\}, \langle i\rangle, c_i) \qquad 0 \leq i \leq 2$$

each AMONG itself decomposed into

$$y_j^i = 1 \iff x_j = i \qquad 1 \leq j \leq 3$$
$$\sum_{j=1}^{3} y_j^i = c_i$$

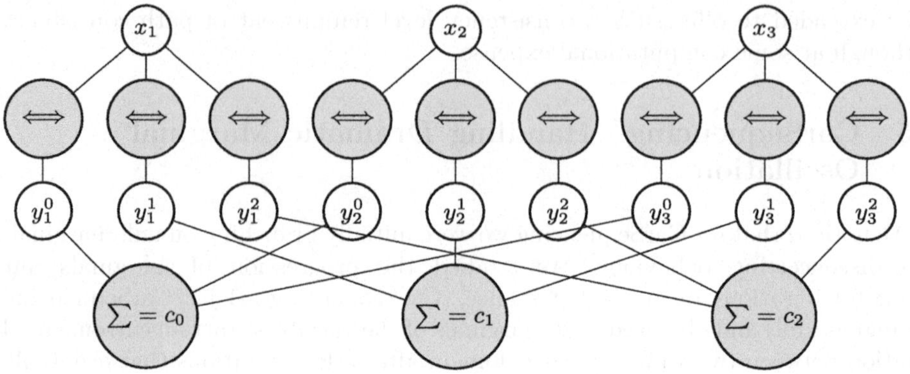


**Fig. 5.** Constraint graph for decomposed CARDINALITY. White vertices represent variables and shaded ones, constraints.

introducing an indicator variable $y_j^i$ for each original variable $x_j$. The decomposition of a single AMONG does not introduce cycles in the constraint graph but the combination of all of them for CARDINALITY does: Fig. 5 shows the corresponding constraint graph for the whole CARDINALITY constraint using that decomposition. It contains many long cycles (length 12). One drawback of the decomposition is the many binary bijective constraints it contains. We can aggregate them into a single TABLE constraint per $x_j$ variable linking it to its indicator variables as shown below and in Fig. 6: shorter cycles (length 8), fewer constraints, larger-arity constraints. In general if there are $d$ values in a domain, we replace $d$ binary constraints by a single arity-$(d+1)$ constraint.

$$\texttt{table}(\langle x_j, \langle y_j^i \rangle_{0 \le i \le 2} \rangle, \mathcal{T}) \qquad 1 \le j \le 3$$

$$\sum_{j=1}^{3} y_j^i = c_i \qquad 0 \le i \le 2$$

At Fig. 7 the dotted curves in the top and middle plots show the behaviour of the old and new decompositions, respectively, for $\langle c_0, c_1, c_2 \rangle = \langle 1, 1, 1 \rangle$ and $\langle 1, 2, 0 \rangle$. For the former, the marginals computed by each decomposition for the $x_j$'s are identical and immediately stabilize to 1/3, which is the true value. For the latter there is oscillation with both decompositions and their amplitude is the same but the new decomposition features a shorter period, which can translate into faster convergence.

Observe that there are some redundant constraints in the CARDINALITY decomposition: the number of occurrences of a given value, say $c_0$, is equal to the number of $x_j$ variables minus the sum of the other $c_i$'s. So we could leave out one of the SUM constraints from the decomposition, thereby introducing fewer cycles. We see at Fig. 7 that doing this in the uniform case, $\langle 1, 1, x \rangle$ meaning that the sum for value 2 is left out, does not degrade much the accuracy of the computed marginal which soon moves very close to the true value, while significantly improving accuracy and showing convergence in the non-uniform case

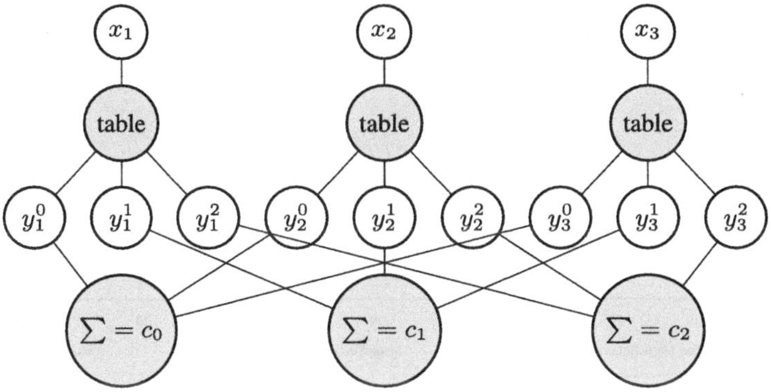


**Fig. 6.** Constraint graph for improved decomposition of CARDINALITY.

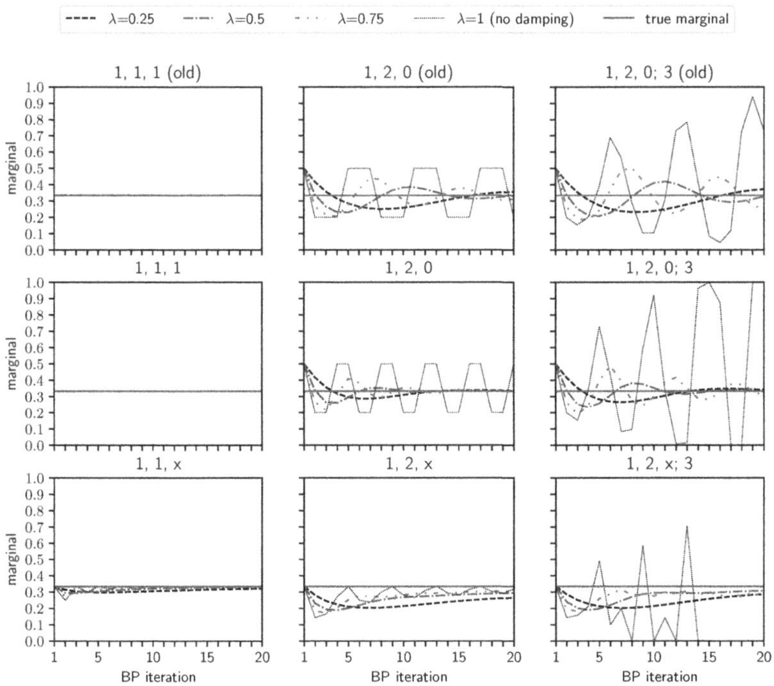


**Fig. 7.** The effect of message damping with different values of parameter $\lambda$

(see $\langle 1, 2, x \rangle$). Surprisingly, adding the redundant constraint that the sum over all $y_j^i$'s is equal to 3 (thus introducing more cycles) makes things much worse, as exemplified by plots $\langle 1, 2, 0; 3 \rangle$ and $\langle 1, 2, x; 3 \rangle$ that increasingly oscillate until eventually stabilizing to 0 or 1. So it may be the case that adding redundant constraints to a model hurts belief propagation and should be avoided. Note that the CarSequencing instances include redundant constraints.

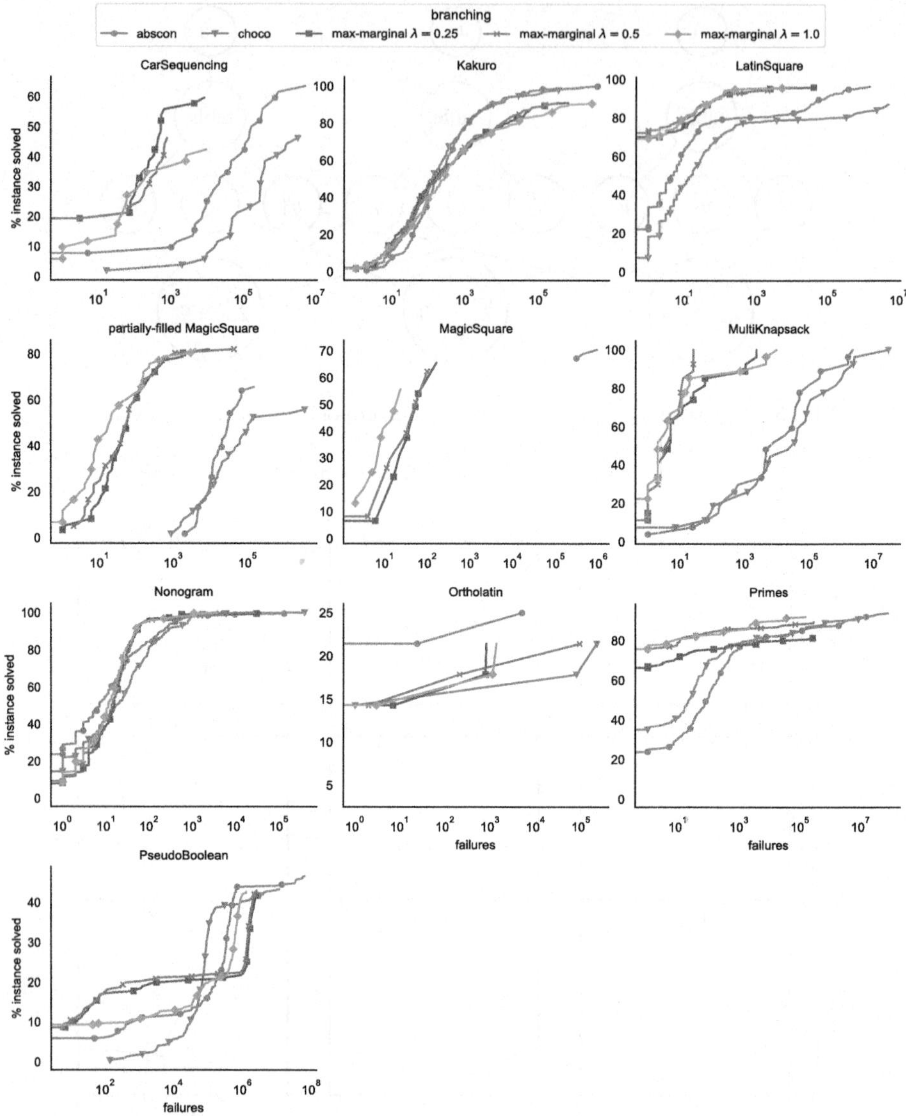


**Fig. 8.** Final comparison between Abscon, Choco, and max-marginal with LDS and damping

The fact that redundant constraints, which typically improve inference, may be deteriorating our search is unsettling. Fortunately there are known remedies to oscillation.

### 5.2 Incorporating Message Damping

Convergence of belief propagation can only be theoretically guaranteed for simple graphs. In practice, it is possible that the beliefs keep getting updated without converging to a value. This results in oscillating messages and beliefs. *Damping* is a common method for reducing the chance of such oscillations [5]. It involves taking a weighted average of the new and old beliefs and using that to update the belief. We apply damping to the messages sent from variables to constraints. At iteration $t$, those messages are calculated as:

$$\mu^{(t)}_{x \rightarrow c}(v) = \lambda \mu_{x \rightarrow c}(v) + (1 - \lambda)\mu^{(t-1)}_{x \rightarrow c}(v)$$

where the *damping factor* $(0 \leq \lambda \leq 1)$ balances the old and new beliefs. The solid curves at Fig. 7 show the effect of damping for three values of $\lambda$: all eventually cancel oscillation. Note that another form of message damping was investigated in [9] without any significant benefit being observed.

Figure 8 presents a final comparison of search guidance over the benchmark problems[7] in which message damping has been added to max-marginal / LDS. For several problems, notably those on which we already performed better, damping does not change the outcome of the comparison and even helps further in the case of MultiKnapsack. For those where the comparison was close it shows a greater impact, in particular for CarSequencing which provided the original motivation to look into message damping and whose solving is improved.

## 6 Conclusion

In this paper we undertook an empirical investigation of combinatorial search heuristics for the CP-BP framework using a wide set of benchmark problems and deepened our understanding of their behaviour, which led to an improved search: LDS with max-marginal as branching heuristic and message damping. It appears to guide search better than state-of-the-art search heuristic dom-wdeg, enough to remain often competitive in terms of computing time. In Sect. 4 we identified a situation under which no benefit can come from computing marginals and a simple way to identify it.

There are other opportunities for improved search that we would like to pursue. There are sometimes a large number of (quasi-)ties for branching: should we proceed differently in that case? Other times several domain values are tied for maximum marginal but one stands out for minimum marginal: instead of choosing a priori to branch according to max-marginal or min-marginal, should we decide between them at each search tree node depending on the context? The number of BP iterations to compute the marginals is currently fixed to five: should that number be adapted according to how marginals progress while iterating?

[7] We no longer report on Dubois and PigeonPlus now that they have been settled.

**Acknowledgements.** The authors wish to thank the anonymous referees for their constructive criticism that helped improve this work. Financial support for this research was provided by IVADO through the Canada First Research Excellence Fund (CFREF) grant, the *Fonds de recherche du Québec–Nature et technologies* (FRQNT), and NSERC Discovery Grant 218028/2017. This research was enabled in part by support provided by Calcul Québec and Compute Canada.

## References

1. Bianco, G.L., Lorca, X., Truchet, C., Pesant, G.: Revisiting counting solutions for the global cardinality constraint. J. Artif. Intell. Res. **66**, 411–441 (2019). https://doi.org/10.1613/jair.1.11325
2. Chavira, M., Darwiche, A.: On probabilistic inference by weighted model counting. Artif. Intell. **172**(6–7), 772–799 (2008). https://doi.org/10.1016/j.artint.2007.11.002
3. van Hoeve, W., Katriel, I.: Global constraints. In: Rossi, F., van Beek, P., Walsh, T. (eds.) Handbook of Constraint Programming, Foundations of Artificial Intelligence, vol. 2, pp. 169–208. Elsevier (2006). https://doi.org/10.1016/S1574-6526(06)80010-6
4. Katsirelos, G., Bacchus, F.: Generalized NoGoods in CSPs. In: Veloso, M.M., Kambhampati, S. (eds.) Proceedings, The Twentieth National Conference on Artificial Intelligence and the Seventeenth Innovative Applications of Artificial Intelligence Conference, 9–13 July 2005, Pittsburgh, Pennsylvania, USA, pp. 390–396. AAAI Press/The MIT Press (2005). http://www.aaai.org/Library/AAAI/2005/aaai05-062.php
5. Murphy, K.P.: Machine Learning - A Probabilistic Perspective. Adaptive Computation and Machine Learning Series. MIT Press, Cambridge (2012)
6. Murphy, K.P., Weiss, Y., Jordan, M.I.: Loopy belief propagation for approximate inference: an empirical study. In: Laskey, K.B., Prade, H. (eds.) Proceedings of the Fifteenth Conference on Uncertainty in Artificial Intelligence, UAI 1999, Stockholm, Sweden, July 30–August 1, pp. 467–475. Morgan Kaufmann (1999)
7. Pearl, J.: Probabilistic Reasoning in Intelligent Systems - Networks of Plausible Inference. Morgan Kaufmann Series in Representation and Reasoning. Morgan Kaufmann (1989)
8. Pesant, G.: Counting-based search for constraint optimization problems. In: Schuurmans, D., Wellman, M.P. (eds.) Proceedings of the Thirtieth AAAI Conference on Artificial Intelligence, 12–17 February 2016, Phoenix, Arizona, USA, pp. 3441–3448. AAAI Press (2016). http://www.aaai.org/ocs/index.php/AAAI/AAAI16/paper/view/12065
9. Pesant, G.: From support propagation to belief propagation in constraint programming. J. Artif. Intell. Res. **66**, 123–150 (2019). https://doi.org/10.1613/jair.1.11487
10. Pesant, G., Quimper, C.G., Zanarini, A.: Counting-based search: branching heuristics for constraint satisfaction problems. J. Artif. Intell. Res. **43**, 173–210 (2012). https://doi.org/10.1613/jair.3463

# Replication-Guided Enumeration of Minimal Unsatisfiable Subsets

Jaroslav Bendík(✉) and Ivana Černá

Faculty of Informatics, Masaryk University, Brno, Czech Republic
{xbendik,cerna}@fi.muni.cz

**Abstract.** In many areas of computer science, we are given an unsatisfiable Boolean formula $F$ in CNF, i.e. a set of clauses, with the goal to identify minimal unsatisfiable subsets (MUSes) of $F$. The more MUSes are identified, the better insight into $F$'s unsatisfiability is obtained. Unfortunately, finding even a single MUS can be very time consuming since it naturally subsumes repeatedly solving the satisfiability problem, and thus a complete MUS enumeration is often practically intractable. Therefore, contemporary MUS enumeration algorithms tend to identify as many MUSes as possible within a given time limit. In this work, we present a novel MUS enumeration algorithm. Compared to existing algorithms, our algorithm is much more frugal in the number of performed satisfiability checks. Consequently, our algorithm is often able to find substantially more MUSes than contemporary algorithms.

## 1 Introduction

Given an unsatisfiable set $F = \{c_1, \dots, c_n\}$ of Boolean clauses, a minimal unsatisfiable subset (MUS) of $F$ is a set $M \subseteq F$ such that $M$ is unsatisfiable and for all $c \in M$ the set $M \setminus \{c\}$ is satisfiable. MUSes represent the minimal reasons for $F$'s unsatisfiability, and as such, they find applications in a wide variety of domains including, e.g., formal equivalence checking [18], Boolean function bi-decomposition [17], counter-example guided abstraction refinement [1], circuit error diagnosis [21], type debugging in Haskell [37], and many others [2,23–25,32].

The more MUSes are identified, the better insight into the unsatisfiability of $F$ is obtained. However, there can be, in general, up to exponentially many MUSes w.r.t. $|F|$, and thus, the complete MUS enumeration is often practically intractable. Consequently, there have been proposed several algorithms, e.g., [4, 5,9,12,27,30,33,36], that enumerate MUSes *online*, i.e., one by one, and attempt to identify as many MUSes as possibly within a given time limit.

Many of the algorithms can be classified as *seed-shrink* algorithms [11]. A seed-shrink algorithm gradually explores subsets of $F$; *explored* subsets are those,

This research was supported by ERDF "CyberSecurity, CyberCrime and Critical Information Infrastructures Center of Excellence" (No. CZ.02.1.01/0.0/0.0/16_019/0000822).


H. Simonis (Ed.): CP 2020, LNCS 12333, pp. 37–54, 2020.
https://doi.org/10.1007/978-3-030-58475-7_3

whose satisfiability is already known by the algorithm, and *unexplored* are the others. To find each single MUS, a seed-shrink algorithm first identifies an unsatisfiable unexplored subset, called a *u-seed*, and then *shrinks* the u-seed into a MUS via a single MUS extraction subroutine. The exact way of finding and shrinking u-seeds differ for individual algorithms. In general, the algorithms find a u-seed by repeatedly checking unexplored subsets for satisfiability, via a SAT solver, until they find an unsatisfiable one. Naturally, the performance of the algorithms highly depends on the number these satisfiability checks.

In this paper, we propose a novel seed-shrink algorithm called UNIMUS. The algorithm employs two novel techniques for finding u-seeds. One of the techniques works on the same principle as the existing seed-shrinks algorithms do: it checks unexplored subsets for satisfiability until it finds a u-seed. The novelty is in the selection of subsets to be checked; we use the union of already explored MUSes to identify a search-space where we can quickly find new u-seeds. The other technique for finding u-seeds works on a fundamentally different principle; we cheaply (in polynomial time) *deduce* that some unexplored subsets are unsatisfiable. We experimentally compare UNIMUS with 4 contemporary MUS enumeration algorithms on a standard collection of benchmarks. UNIMUS outperforms all of its competitors on majority of the benchmarks. Remarkably, UNIMUS often finds 10–100 times more MUSes than its competitors.

## 2 Preliminaries

A Boolean formula $F = \{c_1, \dots, c_n\}$ in a conjunctive normal form is a set of clauses over a set of variables $Vars(F)$. A clause $c = \{l_1, \dots, l_k\}$ is a set of literals. A literal is either a variable $x \in Vars(F)$ or its negation $\neg x$. A truth assignment $I$ is a mapping $Vars(F) \to \{\top, \bot\}$. A clause $c \in F$ is satisfied by an assignment $I$ iff $I(x) = \top$ for some $x \in c$ or $I(y) = \bot$ for some $\neg y \in c$. The formula $F$ is satisfied by $I$ iff $I$ satisfies every clause $c \in F$; in such a case $I$ is a *model* of $F$. Finally, $F$ is *satisfiable* if it has a model; otherwise $F$ is *unsatisfiable*. Hereafter, we use $F$ to denote the input formula of interest, capital letters, e.g. $N, T, K$ to denote subsets of $F$, small letters, e.g., $c, d, c_i$ to denote clauses of $F$, and small letters, e.g., $x, y, x_i$ to denote variables of $F$. Given a set $X$, we use $|X|$ to denote the cardinality of $X$, and $\mathcal{P}(X)$ to denote the power-set of $X$.

### 2.1 Minimum Unsatisfiability

**Definition 1 (MUS).** *A set $N$, $N \subseteq F$, is a* minimal unsatisfiable subset (MUS) *of $F$ iff $N$ is unsatisfiable and for all $c \in N$ the set $N \setminus \{c\}$ is satisfiable.*

Note that the minimality refers to a set minimality, not to minimum cardinality. Therefore, there can be MUSes with different cardinalities and in general, there can be up to exponentially many MUSes of $F$ w.r.t. $|F|$ (see [35]). We write $\texttt{UMUS}_F$ to denote the union of all MUSes of $F$.

*Example 1.* Assume that we are given a formula $F = \{c_1 = \{x_1\}, c_2 = \{\neg x_1\}, c_3 = \{x_2\}, c_4 = \{\neg x_1, \neg x_2\}\}$. There are two MUSes: $\{c_1, c_2\}$ and $\{c_1, c_3, c_4\}$, and $\texttt{UMUS}_F = F$. The power-set of $F$ is illustrated in Fig. 1a.

**Definition 2 (critical clause).** *Let $U$ be an unsatisfiable subset of $F$. A clause $c \in U$ is* critical *for $U$ iff $U \setminus \{c\}$ is satisfiable.*

Note that if $c$ is critical for $U$ then $c$ has to be contained in every unsatisfiable subset of $U$, and especially in every MUS of $U$. Furthermore, note that $U$ is a MUS if and only if every clause $c \in U$ is critical for $U$.

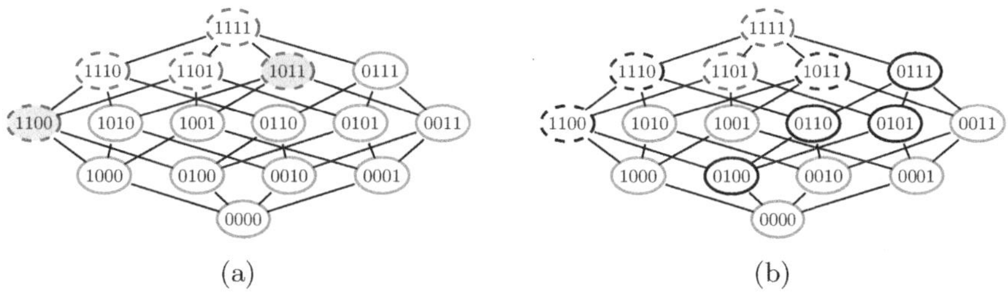


**Fig. 1.** a) illustrates the power-set of the set $F$ of clauses from Example 1. We encode individual subsets of $F$ as bit-vectors; for example, the subset $\{c_1, c_3, c_4\}$ is written as 1011. The subsets with a dashed border are the unsatisfiable subsets, and the others are satisfiable subsets. The MUSes are filled with a background color. b) illustrates the explored and unexplored subsets from Example 2. (Color figure online)

Finally, we exploit capabilities of contemporary SAT solvers. Given a set $N \subseteq F$, many of SAT solvers are able to provide an *unsat core* of $N$ when $N$ is unsatisfiable, and a *model* $I$ of $N$ when $N$ is satisfiable. The unsat core is often small, yet not necessarily minimal, unsatisfiable subset of $N$. The model $I$, on the other hand, induces the *model extension* $E$ of $N$ defined as $E = \{c \in F \mid I$ satisfies $c\}$. Note that $N \subseteq E$ and $E$ is satisfiable ($I$ is its model).

## 2.2 Unexplored Subsets

Every online MUS enumeration algorithm during its computation gradually *explores* satisfiability of individual subsets of $F$. *Explored* subsets are those whose satisfiability is already known by the algorithm and *unexplored* are the other ones. We write `Unexplored` to denote the set of all unexplored subsets. Furthermore, we classify the unexplored subsets either as *s-seeds* or *u-seeds*; *s-seeds* are satisfiable unexplored subsets and *u-seeds* are unsatisfiable unexplored subsets.

Note that if a subset $U$ of $F$ is unsatisfiable, then also every superset of $U$ is unsatisfiable. Therefore, when $U$ becomes explored, then also all supersets of $U$ become explored. Dually, when a satisfiable $S$, $S \subseteq F$, becomes explored, then also all subsets of $S$ become explored since they are necessarily also satisfiable.

**Algorithm 1:** Seed-Shrink Scheme

```
Unexplored ← P(F)
while there is a u-seed do
    S ← find a u-seed
    S_mus ← shrink(S)
    output S_mus
    Unexplored ← Unexplored \ {X | X ⊆ S_mus ∨ X ⊇ S_mus}
```

**Observation 1.** *If N is a u-seed, then every unsatisfiable subset of N is also a u-seed. Dually, if N is an s-seed then every satisfiable superset of N is an s-seed.*

*Example 2* Figure 1b shows a possible state of exploration of the power-set of $F$ from Example 1. There are seven explored satisfiable subsets (green with solid border), two explored unsatisfiable subsets (red with dashed border), four s-seeds (black with solid border), and three u-seeds (black with dashed border).

Note that based on `Unexplored`, we can *mine* some critical clauses for some u-seeds. For instance, in Example 2, we can see that $c_2$ is critical for the u-seed $U = \{c_1, c_2, c_3\}$ since $U \setminus \{c_2\}$ is explored and thus satisfiable (Observation 1).

**Definition 3 (minable critical).** *Let $U$ be a u-seed and $c$ a critical clause for $U$. The clause $c$ is a* minable critical clause *for $U$ if $U \setminus \{c\} \notin$* `Unexplored`.

Details on how exactly we represent and perform operations over `Unexplored` are postponed to Sect. 3.5.

### 2.3 Seed-Shrink Scheme

Many of existing MUS enumeration algorithms, e.g., [6,9,12,28,36], and including the algorithm we present in this paper, can be classified as *seed-shrink* algorithms [11]. The base scheme (Algorithm 1) works iteratively. Each iteration starts by identifying a u-seed $S$. Then, the u-seed $S$ is *shrunk* into a MUS $S_{mus}$ via a single MUS extraction subroutine. The iteration is concluded by removing all subsets and all supersets of the MUS from `Unexplored` since none of them can be another MUS. The computation terminates once there is no more u-seed.

The exact way $S$ is found differs for individual seed-shrink algorithms. In general, existing algorithms identify $S$ by repeatedly picking and checking an unexplored subset for satisfiability until they find a u-seed. The algorithms vary in *which* and *how* many unexplored subsets they check. In general, it is worth to minimize the number of these checks as they are very expensive. Also, it generally holds that the closer (w.r.t. set containment) the u-seed is to a MUS, the easier it is to shrink the u-seed into a MUS. As for the shrinking, all the algorithms can collect and exploit all the minable critical clauses for $S$; these clauses have to be contained in every MUS of $S$ and thus their prior knowledge can significantly speed up the MUS extraction. However, the exact way the algorithms find the MUS differ for individual algorithms. See Sects. 3.4 and 4 for more details.

**Algorithm 2:** UNIMUS

`Unexplored` $\leftarrow \mathcal{P}(F)$; $B \leftarrow \emptyset$
**while** `Unexplored` $\neq \emptyset$ **do**
    $B \leftarrow$ refine$(B)$
    UNIMUSCore$(B)$

# 3 Algorithm

Our MUS enumeration algorithm, called UNIMUS, is based on the seed-shrink scheme. It employs a novel shrinking procedure. Moreover, it employs two novel approaches for finding u-seeds. One of the approaches is based on the same idea as the contemporary seed-shrink algorithms: to find a u-seed, we are repeatedly picking and checking for satisfiability (via a SAT solver) an unexplored subset until we identify a u-seed. The novelty is in the choice of unexplored subsets to be checked. Briefly, we maintain a *base* $B$ and a *search-space* $\texttt{Unex}_B = \{X \mid X \in \texttt{Unexplored} \wedge X \subseteq B\}$ that is induced by the base. UNIMUS searches for u-seeds only within $\texttt{Unex}_B$. The base $B$ (and thus $\texttt{Unex}_B$) is maintained in a way that allows identifying u-seeds with performing only few satisfiability checks. Moreover, the u-seeds are very close to MUSes and thus relatively easy to shrink.

Our other approach for finding u-seeds is based on a fundamentally different principle. Instead of checking unexplored subsets for satisfiablity via a SAT solver, we *deduce* that some unexplored subsets are unsatisfiable. The deduction is based on already identified MUSes and it is very cheap (polynomial).

## 3.1 Main Procedure

UNIMUS (Algorithm 2) first initializes `Unexplored` to $\mathcal{P}(F)$ and the base $B$ to $\emptyset$. Then, it in a while-loop repeats two procedures: refine that updates the base $B$, and UNIMUSCore that identifies MUSes in the search-space $\texttt{Unex}_B = \{X \mid X \in \texttt{Unexplored} \wedge X \subseteq B\}$. The algorithm terminates once $\texttt{Unexplored} = \emptyset$.

UNIMUSCore (Algorithm 3) works iteratively. Each iteration starts by picking a *maximal element* $N$ of $\texttt{Unex}_B$, i.e., $N \in \texttt{Unex}_B$ such that $N \cup \{c\} \notin \texttt{Unex}_B$ for every $c \in B \setminus N$. Subsequently, a procedure isSAT is used to determine, via a SAT solver, the satisfiability of $N$. Moreover, in dependence of $N$'s satisfiability, isSAT returns either an unsat core $K$ or a model extension $E$ of $N$. If $N$ is unsatisfiable, the algorithm shrinks the core $K$ into a MUS $K_{mus}$ and removes from `Unexplored` all subsets and all supersets of $K_{mus}$. Subsequently, a procedure replicate is invoked which attempts to identify additional MUSes in $\texttt{Unex}_B$.

In the other case, when $N$ is satisfiable, we remove all subsets of $E$ from `Unexplored`. Then, $N$ is used to guide the algorithm into a search-space with more minable critical clauses. In particular, we know that for every $c \in B \setminus N$ the set $N \cup \{c\}$ is explored and thus unsatisfiable (Observation 1). Consequently, every clause $c \in B \setminus N$ is critical for $N \cup \{c\}$ and especially for every u-seed contained in $N \cup \{c\}$. Moreover, all these critical clauses are minable critical as

**Algorithm 3:** UNIMUSCore($B$)

**while** $\{N \in \texttt{Unexplored} \mid N \subseteq B\} \neq \emptyset$ **do**
    $N \leftarrow$ a maximal element of $\{X \mid X \in \texttt{Unexplored} \land X \subseteq B\}$ // $\texttt{Unex}_B$
    $(sat?, E, K) \leftarrow$ isSAT($N$)
    **if** ***not*** $sat?$ **then**
        $K_{mus} \leftarrow$ shrink($K$)
        **output** $K_{mus}$
        $\texttt{Unexplored} \leftarrow \texttt{Unexplored} \setminus \{X \mid X \subseteq K_{mus} \lor X \supseteq K_{mus}\}$
        replicate($K_{mus}, N$)
    **else**
        $\texttt{Unexplored} \leftarrow \texttt{Unexplored} \setminus \{X \mid X \subseteq E\}$
        **if** $|B \setminus N| > 1$ **then**
            **for** $c \in B \setminus N$ **do** UNIMUSCore($N \cup \{c\}$)

all subsets of $N$ were removed from `Unexplored`. If $N \cup \{c\} = B$ then $c$ is minable critical for every u-seed in the current search-space. Otherwise, if $|B \setminus N| > 1$, we recursively call UNIMUSCore with a base $B' = N \cup \{c\}$ for every $c \in B \setminus N$.

The procedures refine, shrink, and replicate are described in Sects. 3.2, 3.4, and 3.3, respectively. All procedures of UNIMUS follow two rules about `Unexplored`. First, `Unexplored` is *global*, i.e., shared by the procedures. Second, we remove an unsatisfiable set from `Unexplored` only if the set is a superset of an explicitly identified MUS. Consequently, no MUS can be removed from `Unexplored` without being explicitly identified. Thus, when Algorithm 2 terminates (i.e., $\texttt{Unexplored} = \emptyset$), it is guaranteed that all MUSes were identified.

**Heuristics.** According to the above description, UNIMUSCore terminates once all subsets, and especially all MUSes, of $B$ become explored. However, based on our empirical experience, UNIMUSCore can get into a situation such that there is a lot of s-seeds in $\texttt{Unex}_B$ but only few or even no u-seed. Consequently, the MUS enumeration can get stuck for a while. To prevent such a situation, we track the number of subsequent iterations of UNIMUSCore in which the set $N$ was satisfiable. If there are 5 such subsequent iterations, we backtrack from the current recursive call of UNIMUSCore.

### 3.2 The Base and the Search-Space

The base $B$ is modified in two situations. The first situation is the case of recursive calls of UNIMUSCore which was described in the previous section. Here, we describe the second situation which is an execution of the procedure refine before each top-level call of UNIMUSCore. The goal is to identify a base $B$ such that u-seeds in $\texttt{Unex}_B$ can be easily found and are relatively easy to shrink.

We exploit the union $\texttt{UMUS}_F$ of all MUSes of $F$. Assume that we set $B$ to $\texttt{UMUS}_F$. Since every MUS of $F$ is contained in $\texttt{UMUS}_F$, the induced search-space

**Algorithm 4:** refine($B$)

**while** `Unexplored` $\neq \emptyset$ **do**
  $T \leftarrow$ a maximal element of `Unexplored`
  $(sat?, E, K) \leftarrow$ isSAT($T$)
  **if** *not* $sat?$ **then**
    $K_{mus} \leftarrow$ shrink($K$)
    **output** $K_{mus}$
    `Unexplored` $\leftarrow$ `Unexplored` $\setminus \{X \mid X \subseteq K_{mus} \vee X \supseteq K_{mus}\}$
    **return** $B \cup K_{mus}$
  **else** `Unexplored` $\leftarrow$ `Unexplored` $\setminus \{X \mid X \subseteq T\}$
**return** $B$

$\texttt{Unex}_B$ would contain all MUSes. Moreover, compared to the whole $F$, the cardinality of $\texttt{UMUS}_F$ can be relatively small and thus the u-seeds in $\texttt{Unex}_B$ would be easy to shrink. Unfortunately, based on recent studies [14, 31], computing $\texttt{UMUS}_F$ is often practically intractable even for small input formulas. Thus, instead of initially computing $\texttt{UMUS}_F$, we use as the base $B$ just an under-approximation of $\texttt{UMUS}_F$. Initially, we set $B$ to $\emptyset$ (Algorithm 2, line 1) and in each call of refine we attempt to refine (enlarge) the under-approximation. Eventually, $B$ becomes $\texttt{UMUS}_F$ and, thus, eventually the search-space will contain all MUSes of $F$.

The procedure refine (Algorithm 4) attempts to enlarge $B$ with an unexplored MUS. In each iteration, it picks a maximal element $T$ of `Unexplored`, i.e., $T \in$ `Unexplored` such that $T \cup \{c\} \notin$ `Unexplored` for every $c \in F \setminus T$. Then, $T$ is checked for satisfiability via the procedure isSAT. If $T$ is unsatisfiable, then isSAT also returns an unsat core $K$ of $T$, refine shrinks the core $K$ into a MUS $K_{mus}$ and based on $K_{mus}$ updates the set `Unexplored`. Subsequently, refine terminates and returns an updated base $B' = B \cup \{K_{mus}\}$. Otherwise, if $T$ is satisfiable, refine removes all subsets of $T$ from `Unexplored` and continues with a next iteration.

There is no guarantee that each call of refine indeed enlarges $B$. One possibility is the corner case when all MUSes are already explored, but there are some s-seeds left. Another possibility is that the search-space $\texttt{Unex}_B$ was not completely explored in the last call of UNIMUSCore due to the preemptive termination heuristic. Thus, refine might identify a MUS that is a subset of $B$. Also, note that the procedure refine is very similar to a MUS enumeration algorithm MARCO [28]. The difference is that refine finds only a single unexplored MUS whereas MARCO finds them all (see Sect. 4 for details).

### 3.3 MUS Replication

We now describe the procedure replicate($K_{mus}, N$) that based on an identified MUS $K_{mus}$ of $N$ attempts to identify additional unexplored MUSes. The procedure follows the seed-shrink scheme, i.e., it searches for u-seeds and shrinks them to MUSes. However, contrary to existing seed-shrink algorithms which identify

**Algorithm 5:** replicate($K_{mus}, N$)

$\mathcal{M} \leftarrow \{K_{mus}\}$; $rStack \leftarrow \langle K_{mus} \rangle$
**while** *rStack is not empty* **do**
  $M \leftarrow rStack.pop()$
  **for** $c \in M$ **do**
    **if** *c is minable critical for N* **then continue**
    $S \leftarrow$ propagate($M, c, N, \mathcal{M}$)
    **if** *S is* null **then continue**
    $S_{mus} \leftarrow$ shrink($S$)
    **output** $S_{mus}$
    `Unexplored` $\leftarrow$ `Unexplored` $\setminus \{X \mid X \subseteq S_{mus} \vee X \supseteq S_{mus}\}$
    $\mathcal{M} \leftarrow \mathcal{M} \cup \{S_{mus}\}$
    $rStack.push(S_{mus})$

u-seeds via a SAT solver, replicate identifies u-seeds with a cheap (polynomial) deduction technique; we call the technique *MUS replication.*

Each call of replicate possibly identifies several unexplored MUSes and all these MUSes are subsets of $N$. Note that since $N$ is a subset of the base $B$ in Algorithm 3, all MUSes identified by replicate are contained in the search-space $\texttt{Unex}_B$. Also, note that when replicate is called, $K_{mus}$ is the only explored MUS of $N$ (since $N$ was a u-seed that we shrunk to $K_{mus}$). In the following, we will use $\mathcal{M}$ to denote the set of all explored MUSes of $N$, i.e., initially, $\mathcal{M} = \{K_{mus}\}$.

**Main Procedure.** The main procedure of replicate (Algorithm 5) maintains two data-structures: the set $\mathcal{M}$ and a stack *rStack*. The computation starts by initializing both $\mathcal{M}$ and *rStack* to contain the MUS $K_{mus}$. The rest of replicate is formed by two nested loops. In each iteration of the outer loop, replicate pops a MUS $M$ from the stack. In the nested loop, $M$ is used to identify possibly several unexplored MUSes. In particular, for each clause $c \in M$ the algorithm attempts to identify a u-seed $S$ such that $M \setminus \{c\} \subset S \subseteq N \setminus \{c\}$. Observe that if $c$ is minable critical for $N$ then such a u-seed cannot exist; thus, we skip such clauses. The attempt to find such $S$ is carried out by a procedure propagate. If propagate fails to find the u-seed, the inner loop proceeds with a next iteration. Otherwise, the u-seed $S$ is shrunk into a MUS $S_{mus}$ and the set `Unexplored` is appropriately updated. The iteration is concluded by adding $S_{mus}$ to $\mathcal{M}$ and also pushing $S_{mus}$ to *rStack*, i.e., each identified MUS is used to possibly identify additional MUSes. The computation terminates once *rStack* becomes empty.

**Propagate.** The procedure propagate is based on the well-known concepts of *backbone literals* and *unit propagation* [16,26]. Given a formula $P$, a literal $l$ is a *backbone literal* of $P$ iff every model of $P$ satisfies $\{l\}$. A *backbone* of $P$ is a set of backbone literals of $P$. If $A$ is a backbone of $P$ then $A$ is also a backbone of every superset of $P$. A clause $d$ is a *unit clause* iff $|d| = 1$. Note that if $d$ is a unit clause of $P$ then the literal $l \in d$ is a backbone literal of $P$.

Given a backbone $A$ of $P$, the *backbone propagation* can simplify $P$ and possibly show that $P$ is unsatisfiable. In particular, for every $l \in A$ and every clause $d \in P$ such that $\neg l \in d$, we remove the literal $\neg l$ from $d$ (since no model of $P$ can satisfy $\neg l$). If a new unit clause emerges during the propagation, the backbone literal that forms the unit clause will be also propagated. If the propagation reduces a clause to an empty clause, then the original $P$ is unsatisfiable.

The procedure propagate employs backbone propagation to identify a u-seed $S$ such that $M \setminus \{c\} \subset S \subseteq N \setminus \{c\}$. Observe that since $M$ is unsatisfiable and $M \setminus \{c\}$ is satisfiable, then the set $A = \{\neg l \,|\, l \in c\}$ is a backbone of every such $S$. Thus, one can pick such $S$ and attempt to show, via propagating $A$, that $S$ is unsatisfiable. However, there are too many such $S$ to choose from. Moreover, we need to guarantee that we find $S$ that is both unsatisfiable and unexplored. Thus, instead of fixing a particular $S$ and then trying to show its unsatisfiability, we attempt to gradually build such $S$. Initially, we set $S$ to $M \setminus \{c\}$ and we step-by-step add clauses from $N \setminus \{c\}$ to $S$. The addition of the clauses is driven by a currently known backbone $A$ of $S$ and also by the set $\mathcal{M}$ of explored MUSes to ensure that the resulting $S$ is a u-seed.

**Observation 2.** *For every* ***unsatisfiable*** *$S$, $S \subseteq N \setminus \{c\}$, it holds that $S \in$* `Unexplored` *(i.e., $S$ is a u-seed) if and only if $\forall_{X \in \mathcal{M}} S \not\supseteq X$.*

*Proof.* In UNIMUS, we remove from `Unexplored` only unsatisfiable sets that are supersets of explored MUSes and all explored MUSes of $N$ are stored in $\mathcal{M}$.

Observation 2 shows which clauses *can be added* to the initial $S$ while ensuring that if we finally obtain an unsatisfiable $S$, then the final $S$ will be unexplored. Note that the initial $S = M \setminus \{c\}$ trivially satisfies $\forall_{X \in \mathcal{M}} M \setminus \{c\} \not\supseteq X$ since it is satisfiable ($M$ is a MUS). In the following, we show which clauses *should be added* to $S$ to eventually make it unsatisfiable.

**Definition 4 (operation $\setminus\!\setminus$).** *Let $d$ be a clause and $A$ be a set of literals. The the binary operation $d \setminus\!\setminus A$ creates the clause $d \setminus\!\setminus A = \{l \mid l \in d \text{ and } \neg l \notin A\}$.*

**Definition 5 (units, violated).** *Let $S$ be a set such that $M \setminus \{c\} \subset S \subseteq N \setminus \{c\}$, and let $A$ be a backbone of $S$. We define the following sets:*

$$units(S, A) = \{d \in N \setminus \{c\} \mid \forall_{X \in \mathcal{M}} S \cup \{d\} \not\supseteq X \wedge |d \setminus\!\setminus A| = 1\}$$
$$violated(S, A) = \{d \in N \setminus \{c\} \mid \forall_{X \in \mathcal{M}} S \cup \{d\} \not\supseteq X \wedge |d \setminus\!\setminus A| = 0\}$$

Informally, a clause $d \in N \setminus \{c\}$ belongs to *units*$(S, A)$ (*violated*$(S, A)$) if the propagation of $A$ would simplify $d$ to a unit clause (empty clause) and, simultaneously, $d \in S$ or $d$ can be added to $S$ in a harmony with Observation 2.

**Observation 3.** *For every $S$ such that $M \setminus \{c\} \subset S \subseteq N \setminus \{c\}$, a backbone $A$ of $S$, and a clause $d \in units(S, A)$, it holds that $A \cup d \setminus\!\setminus A$ is a backbone of $S \cup \{d\}$.*

*Proof.* Assume that $d = \{l, l_0, \ldots, l_k\}$ where $l = d \setminus\!\setminus A$ and $\{\neg l_0, \ldots, \neg l_k\} \subseteq A$. Since $A$ is a backbone of $S$ then every model of $S$ satisfies $\{\{\neg l_0\}, \ldots, \{\neg l_k\}\}$. Consequently, every model of $S \cup \{d\}$ satisfies $\{l\}$.

**Algorithm 6:** propagate($M$,$c$,$N$,$\mathcal{M}$)

$S \leftarrow M \setminus \{c\}$; $A \leftarrow \{\neg l \mid l \in c\}$; $H \leftarrow \{c\}$<br>
**while** $units(S, A) \setminus H \neq \emptyset \wedge violated(S, A) = \emptyset$ **do**<br>
    $d \leftarrow$ choose $d \in units(S, A) \setminus H$<br>
    $S \leftarrow S \cup \{d\}$; $H \leftarrow H \cup \{d\}$; $A \leftarrow A \cup d \setminus\setminus A$<br>
**if** $violated(S, A) = \emptyset$ **then return** *null*<br>
**else**<br>
    $d \leftarrow$ choose $d \in violated(S, A)$<br>
    **return** $S \cup \{d\}$

**Observation 4.** *For every $S$ such that $M \setminus \{c\} \subset S \subseteq N \setminus \{c\}$, a backbone $A$ of $S$, and a clause $d \in violated(S, A)$, it holds that $S \cup \{d\}$ is unsatisfiable.*

*Proof.* Assume that $d = \{l_0, \ldots, l_q\}$. As $d \in violated(S, A)$ then $\{\neg l_0, \ldots, \neg l_q\} \subseteq A$. Since $A$ is a backbone of $S$ then every model of $S$ satisfies $\{\{\neg l_0\}, \ldots, \{\neg l_q\}\}$, i.e., no model of $S$ satisfies $d$.

The procedure propagate (Algorithm 6) maintains three data structures: the sets $S$ and $A$, and an auxiliary set $H$ for storing clauses that were used to enlarge $A$. Initially, $S = M \setminus \{c\}$, $A = \{\neg l \mid l \in c\}$ and $H = \{c\}$. In each iteration, propagate picks a clause $d \in units(S, A) \setminus H$ and, based on Observation 3, adds $d$ to $S$ and to $H$, and the literal of $d \setminus\setminus A$ to $A$. The loop terminates once there is no more backbone literal to propagate ($units(S, A) \setminus H = \emptyset$), or once $violated(S, A) \neq \emptyset$. If $violated(S, A) = \emptyset$, propagate failed to find a u-seed. Otherwise, propagate picks a clause $d \in violated(S, A)$ and returns the u-seed $S \cup \{d\}$.

Finally, note that backbone propagation is cheap (polynomial) but it is not a *complete* technique for deciding satisfiability. Consequently, it can happen that there is a u-seed $S$, $M \setminus \{c\} \subset S \subseteq N \setminus \{c\}$, but MUS replication fails to find it.

### 3.4 Shrink

Existing seed-shrink algorithms can be divided into two groups. Algorithms from one group, e.g. [4,5,13], implement the shrinking via a custom single MUS extractor that fully shares information with the overall MUS enumeration process. Consequently, all information obtained during shrinking can be exploited by the overall MUS enumeration algorithm and vice versa. Algorithms from the other group, e.g. [9,12,28], implement the shrinking via an external, black-box, single MUS extraction tool. The advantage is that one can always use the currently best available single MUS extractor to implement the shrinking. On the other hand, the black-box extractor cannot fully share information with the overall MUS enumeration algorithm. The only output of the extractor is a MUS of a given u-seed $N$. As for the input, besides the u-seed $N$, contemporary single MUS extractors, e.g., [4,8], allow the user to provide also a set $C$ of clauses that are critical for $N$ since a prior knowledge of $C$ can significantly speed up the

**Algorithm 7:** criticalExtension($N$)

$C \leftarrow$ collect all minable critical clauses for $N$
$Q \leftarrow C$
**while** $Q \neq \emptyset$ **do**
  $c \leftarrow$ pick $c \in Q$
  $Q \leftarrow Q \setminus \{c\}$
  **for** $l \in c$ **do**
    $M \leftarrow \{d \in N \mid \neg l \in d\}$
    **if** $|M| = 1$ *and* $M \cap C = \emptyset$ **then**
      $C \leftarrow C \cup M$; $Q \leftarrow Q \cup M$
**return** $C$

extraction. Thus, contemporary algorithms [9,12,28] collect all minable critical clauses for $N$ and pass them to the single MUS extractor together with $N$.

In our work, we follow the black-box approach, i.e., to find a MUS of a u-seed $N$, we first identify a set $C$ of clauses that are critical for $N$ and then pass $N$ and $C$ to an external single MUS extractor (e.g., [4,8]). However, contrary to existing algorithms, we identify more than just *minable* critical clauses for $N$. We introduce a technique that, based on the minable critical clauses for $N$, can cheaply *deduce* that some other clauses are critical for $N$. We call the deduction technique *critical extension* and it is based on the following observation.

**Observation 5.** *Let $N$ be a u-seed, $c \in N$ a critical clause for $N$, and $l \in c$. Moreover, let $M \subseteq N$ be the set of all clauses of $N$ that contain the literal $\neg l$. If $|M| = 1$ then the clause $d \in M$ is critical for $N$, i.e. $N \setminus \{d\}$ is satisfiable.*

*Proof.* Assume that $N \setminus \{d\}$ is unsatisfiable. Since $c$ is critical for $N$, then $c$ is critical also for $N \setminus \{d\}$. Thus, $N \setminus \{c, d\}$ has a model and every its model satisfies $\{\neg l\}$ (as $l \in c$) which contradicts that $\neg l$ is contained only in $d$.

The critical extension technique (Algorithm 7) takes as an input a u-seed $N$ and outputs a set $C$ of clauses that are critical for $N$. The algorithm starts by collecting (see Sect. 3.5) all minable critical clauses for $N$ and stores them to $C$ and also to an auxiliary set $Q$. The rest of the computation works iteratively. In each iteration, the algorithm picks and removes a clause $c$ from $Q$ and employs Observation 5 on $c$. In particular, for each literal $l \in c$, the algorithm builds the set $M = \{d \in N \mid \neg l \in d\}$. If $M$ contains only a single clause, say $d$, and $d \notin C$, then $d$ is a new critical clause for $N$ and thus it is added to $C$ and to $Q$. The computation terminates once $Q$ becomes empty.

Our technique is similar to *model rotation* [4,7] which identifies additional critical clauses based on a critical clause $c$ of $N$ and a model of $N \setminus \{c\}$. The difference is that we do not need the model. Another approach [38] that also does not need the model is based on *rotation edges* in a *flip graph* of $F$.

### 3.5 Representation of Unexplored Subsets

To maintain the set Unexplored, we adopt a representation that was originally proposed by Liffiton et al. [28] and nowadays is used by many MUS enumeration algorithms (e.g., [5,9,12,33]). Given a formula $F = \{c_1, \ldots, c_n\}$, we introduce a set $X = \{x_1, \ldots, x_n\}$ of Boolean variables. Note that every valuation of $X$ corresponds to a subset of $F$ and vice versa. To represent Unexplored, we maintain two formulas, $map^+$ and $map^-$, over $X$ such that every model of $map^+ \wedge map^-$ corresponds to an element of Unexplored and vice versa. In particular:

- Initially, Unexplored $= \mathcal{P}(F)$, thus we set $map^+ = map^- = \top$.
- To remove a set $U$, $U \subseteq F$, and all supersets of $U$ from Unexplored, we add to $map^-$ the clause $\bigvee_{c_i \in U} \neg x_i$.
- Dually, to remove a set $S$, $S \subseteq F$, and all subsets of $S$ from Unexplored, we add to $map^+$ the clause $\bigvee_{c_i \notin S} x_i$.

To get an arbitrary element of Unexplored, one can ask a SAT solver for a model of $map^+ \wedge map^-$. However, in UNIMUS, we need to obtain more specific unexplored subsets. Given a set $B$, we require a *maximal* element $N$ of $\texttt{Unex}_B = \{X \mid X \in \texttt{Unexplored} \wedge X \subseteq B\}$. One of SAT solvers that allows us to obtain such $N$ is miniSAT [20]. To obtain $N$, we instruct miniSAT to fix the values of the variables $\{x_i \mid c_i \notin B\}$ to $\bot$ and ask it for a maximal model of $map^+ \wedge map^-$.

Finally, given a u-seed $U$, to collect all minable critical clauses of $U$ we check for each $c \in U$ whether $U \setminus \{c\}$ corresponds to a model of $map^+ \wedge map^-$. To do it efficiently, observe that the information represented by $map^-$ is irrelevant. Intuitively, $map^-$ *requires an absence* of clauses, and since $U$ satisfies $map^-$ (it is a u-seed), the set $U \setminus \{c\}$ also satisfies $map^-$. Thus, $c$ is minable critical for $U$ iff $U \setminus \{c\}$ does not correspond to a model of $map^+$.

## 4 Related Work

MUS enumeration was extensively studied in the past decades and many various algorithms were proposed (see e.g., [4–6,9,10,12,19,21,22,27,29,33,34,36]). In the following, we briefly describe contemporary online MUS enumeration algorithms.

FLINT [33] computes MUSes in *rounds* and each round consists of two phases: *relaxing* and *strengthening*. In the relaxing phase, the algorithm starts with an unsatisfiable formula $U$ and weakens it by iteratively relaxing its unsat core until it gets a satisfiable formula $S$. The intermediate unsat cores are used to extract MUSes. The resulting satisfiable formula $S$ is passed to the second phase, where the formula is again strengthened to an unsatisfiable formula that is used in the next round as an input for the relaxing phase.

MARCO [28] and ReMUS [12] are algorithms based on the seed-shrink scheme. That is, similarly as UNIMUS, to find each single MUS, the algorithms first identify a u-seed and then shrink the u-seed into a MUS. Before the shrinking, the algorithms first reduce the u-seed to its unsat core (provided by a SAT solver) and they also collect the minable critical clauses for the u-seed. The shrinking is performed via an external, black-box subroutine. The main difference between the two algorithms is how they find the u-seed. MARCO is iteratively picking and checking for satisfiability a maximal unexplored subset of $F$, until it finds a u-seed $S$. Since unsatisfiable subsets of $F$ are naturally more concentrated among the larger subsets, MARCO usually performs only few checks to find the u-seed. However, large u-seeds are generally hard to shrink. Thus, the efficiency of MARCO crucially depends on the capability of the SAT solver to provide a reasonably small unsat core of $S$. ReMUS, in contrast to MARCO, tends to identify u-seeds that are relatively small and thus easy to shrink. In particular, the initial u-seed $S$ is found among the maximal unexplored subsets of $F$ and then shrunk into a MUS $S_{mus}$. To find another MUS, ReMUS picks some $R$ such that $S_{mus} \subset R \subset S$, and recursively searches for u-seeds among the maximal unexplored subsets of $R$. The (expected) size of the u-seeds thus decreases with each recursive call. The disadvantage of ReMUS is that it was designed as a domain-agnostic MUS enumeration algorithm, i.e., $F$ can be a set of constraints in an arbitrary logic (e.g., SMT or LTL). Consequently, ReMUS does not directly employ any techniques that are specific for the SAT (Boolean) domain (such as the MUS replication and the critical extension that we use in UNIMUS).

MCSMUS [5] can be seen as another instantiation of the seed-shrink scheme. Contrary to MARCO, ReMUS, and UNIMUS, MCSMUS implements the shrinking via a custom single MUS extraction procedure that fully shares information and works in a synergy with the overall MUS enumeration algorithm. For example, satisfiable subsets of $F$ that are identified during shrinking are remembered by MCSMUS and exploited in the further computation.

## 5 Experimental Evaluation

We have experimentally compared UNIMUS with four contemporary MUS enumeration algorithms: MARCO [28], MCSMUS [5], FLINT [33], and ReMUS [12]. A precompiled binary of FLINT was kindly provided to us by its author, Nina Narodytska. The other three tools are available at https://sun.iwu.edu/~mliffito/marco/, https://bitbucket.org/gkatsi/mcsmus/src, and https://github.com/jar-ben/mustool. The implementation of UNIMUS is available at: https://github.com/jar-ben/unimus.

We used the best (default) settings for all evaluated tools. Note that individual tools use different SAT solvers and different shrinking subroutines. ReMUS, MARCO and FLINT use the tool muser2 [8] for shrinking, MCSMUS uses its custom shrinking subroutine, and in UNIMUS we employ the shrinking procedure from the MCSMUS tool. As for SAT solvers, MARCO and ReMUS use miniSAT [20], MCSMUS uses glucose [3] and UNIMUS uses CaDiCaL [15].

As benchmarks, we used a collection of 291 CNF formulas from the MUS track of the SAT 2011 Competition.[1] This collection is standardly used in MUS related papers, including the papers that presented our four competitors. All

[1] http://www.cril.univ-artois.fr/SAT11/.

experiments were run using a time limit of 3600 s and computed on an AMD 16-Core Processor and 1 TB memory machine running Debian Linux. Complete results are available in an online appendix: https://www.fi.muni.cz/~xbendik/research/unimus.

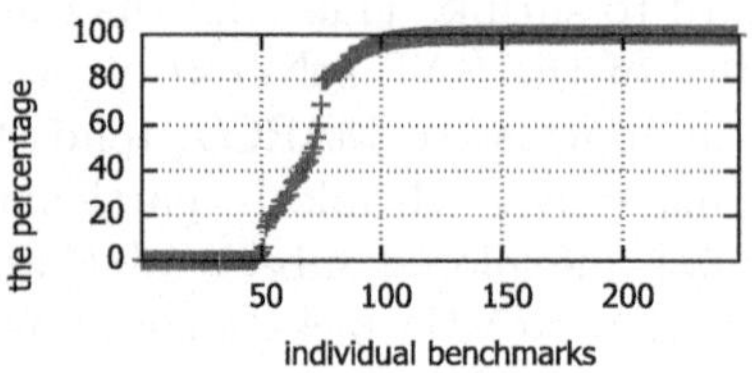


**Fig. 2.** Percentage of MUSes found by MUS replication.

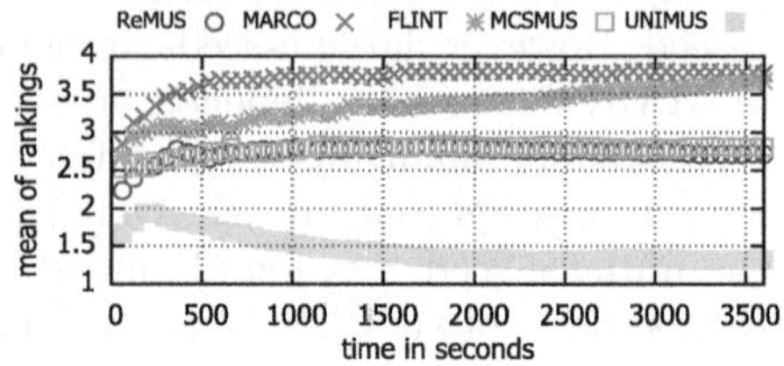


**Fig. 3.** 5% truncated mean of rankings after each 60 s.

**Manifestation of MUS Replication.** MUS replication is a crucial part of UNIMUS as, to the best of our knowledge, it is the first existing technique that identifies u-seeds in polynomial time. Therefore, we are interested in what is the percentage of u-seeds, and thus MUSes, that UNIMUS identifies via MUS replication. Figure 2 shows this percentage (y-axis) for individual benchmarks (x-axis); the benchmarks are sorted by the percentage. We computed the percentage only for the 248 benchmarks where UNIMUS found at least 5 MUSes. Remarkably, in case of 161 benchmarks, the percentage is higher than 90%, and in case of 130 benchmarks, it is higher than 99%.[2] Unfortunately, there are 49 benchmarks where MUS replication found no u-seed at all. Let us note that 40 of the 49 benchmarks are from the *same family* of benchmarks, called "fdmus". The MUS benchmarks from the SAT competition consist of several families and benchmarks in a family often have very similar structure. Most of the families contain only few benchmarks, however, there are several larger families and the "fdmus" family is by far the largest one.

**Number of Indentified MUSes.** We now examine the number of identified MUSes by the evaluated algorithms on individual benchmarks within the time limit of 3600 s. In case of 28 benchmarks, all the algorithms completed the enumeration, and thus found the same number of MUSes. Therefore, we focus here only on the remaining 263 benchmarks.

Scatter plots in Fig. 4 pair-wise compare UNIMUS with its competitors. Each point in the plot shows a result from a single benchmark. The x-coordinate of a point is given be the algorithm that labels the x-axis and the y-coordinate by the algorithm that labels the y-axis. The plots are in a log-scale and hence cannot show points with a zero coordinate, i.e., benchmarks where at least one

[2] Thus, in those benchmarks, SAT solver calls are performed almost only by the shrinking procedure (which uses glucose in our implementation).

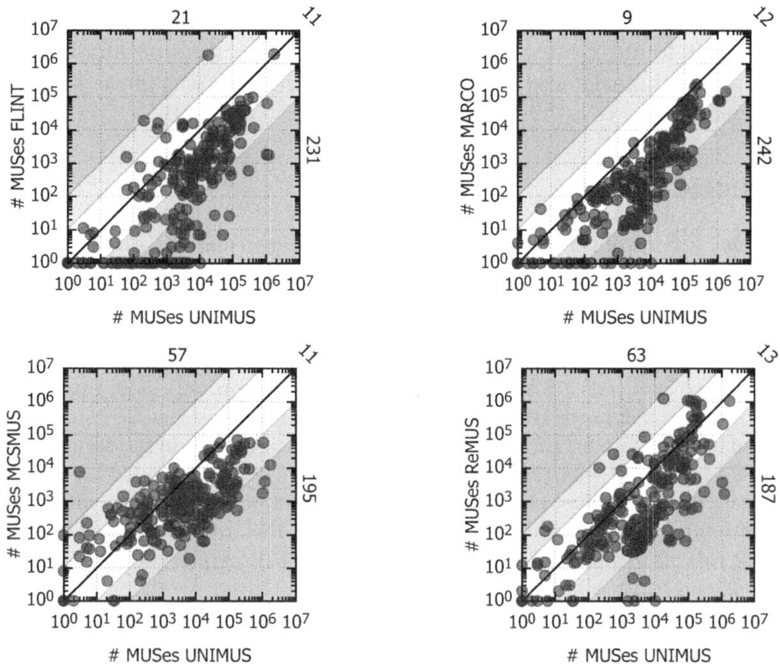


**Fig. 4.** Scatter plots comparing the number of produced MUSes.

algorithm found no MUS. Therefore, we lifted the points with a zero coordinate to the first coordinate. Moreover, we provide three numbers right/above/in the right corner of the plot, that show the number of points below/above/on the diagonal. For example, UNIMUS found more/less/equal number of MUSes than MARCO in case of 242/9/12 benchmarks. We also use green and red colors to highlight individual orders of magnitude (of 10).

In Fig. 3, we examine the overall *ranking* of the algorithms. In particular, assume that for a benchmark B both UNIMUS and ReMUS found 100 MUSes, MCSMUS found 80 MUSes, and MARCO and FLINT found 50 MUSes. In such a case, UNIMUS and ReMUS share the 1st (best) rank for B, MCSMUS is 3rd, and MARCO and FLINT share the 4th position. For each algorithm, we computed an arithmetic mean of the ranking on all benchmarks. To eliminate the effect of outliers (benchmarks with an extreme ranking), we computed the 5% truncated arithmetic mean, i.e., for each algorithm we discarded the 5% of benchmarks where the algorithm achieved the best and the worst ranking. Moreover, to capture the performance stability of the algorithms in time, we computed the mean for each subsequent 60 s of the computation.

UNIMUS conclusively dominates all its competitors. It maintained the best ranking during the whole time period and gradually improved the ranking towards the final value 1.3. The closest, yet still very distant, competitors are ReMUS and MCSMUS who maintained ranking around 2.75. FLINT and MARCO achieved the final raking around 3.7. UNIMUS also dominated in the pair-wise comparison. It found more MUSes than all its competitors on an overwhelming majority of benchmarks and, remarkably, the difference was often several orders of magnitude.

## References


1. Andraus, Z.S., Liffiton, M.H., Sakallah, K.A.: Cegar-based formal hardware verification: a case study. Technical report, University of Michigan, CSE-TR-531-07 (2007)
2. Arif, M.F., Mencía, C., Ignatiev, A., Manthey, N., Peñaloza, R., Marques-Silva, J.: BEACON: an efficient sat-based tool for debugging $\mathcal{EL}^+$ ontologies. In: Creignou, N., Le Berre, D. (eds.) SAT 2016. LNCS, vol. 9710, pp. 521–530. Springer, Cham (2016). https://doi.org/10.1007/978-3-319-40970-2_32
3. Audemard, G., Simon, L.: Predicting learnt clauses quality in modern SAT solvers. In: IJCAI, pp. 399–404 (2009)
4. Bacchus, F., Katsirelos, G.: Using minimal correction sets to more efficiently compute minimal unsatisfiable sets. In: Kroening, D., Păsăreanu, C.S. (eds.) CAV 2015. LNCS, vol. 9207, pp. 70–86. Springer, Cham (2015). https://doi.org/10.1007/978-3-319-21668-3_5
5. Bacchus, F., Katsirelos, G.: Finding a collection of MUSes incrementally. In: Quimper, C.-G. (ed.) CPAIOR 2016. LNCS, vol. 9676, pp. 35–44. Springer, Cham (2016). https://doi.org/10.1007/978-3-319-33954-2_3
6. Bailey, J., Stuckey, P.J.: Discovery of minimal unsatisfiable subsets of constraints using hitting set dualization. In: Hermenegildo, M.V., Cabeza, D. (eds.) PADL 2005. LNCS, vol. 3350, pp. 174–186. Springer, Heidelberg (2005). https://doi.org/10.1007/978-3-540-30557-6_14
7. Belov, A., Marques-Silva, J.: Accelerating MUS extraction with recursive model rotation. In: FMCAD, pp. 37–40. FMCAD Inc. (2011)
8. Belov, A., Marques-Silva, J.: MUSer2: an efficient MUS extractor. JSAT **8**, 123–128 (2012)
9. Bendík, J., Beneš, N., Černá, I., Barnat, J.: Tunable online MUS/MSS enumeration. In: FSTTCS, LIPIcs, vol. 65, pages 50:1–50:13. Schloss Dagstuhl - Leibniz-Zentrum fuer Informatik (2016)
10. Jaroslav Bendík and Ivana Černá. Evaluation of domain agnostic approaches for enumeration of minimal unsatisfiable subsets. In: LPAR, EPiC Series in Computing, vol. 57, pp. 131–142. EasyChair (2018)
11. Bendík, J., Černá, I.: MUST: minimal unsatisfiable subsets enumeration tool. TACAS 2020. LNCS, vol. 12078, pp. 135–152. Springer, Cham (2020). https://doi.org/10.1007/978-3-030-45190-5_8
12. Bendík, J., Černá, I., Beneš, N.: Recursive online enumeration of all minimal unsatisfiable subsets. In: Lahiri, S.K., Wang, C. (eds.) ATVA 2018. LNCS, vol. 11138, pp. 143–159. Springer, Cham (2018). https://doi.org/10.1007/978-3-030-01090-4_9
13. Bendík, J., Ghassabani, E., Whalen, M., Černá, I.: Online enumeration of all minimal inductive validity cores. In: Johnsen, E.B., Schaefer, I. (eds.) SEFM 2018. LNCS, vol. 10886, pp. 189–204. Springer, Cham (2018). https://doi.org/10.1007/978-3-319-92970-5_12
14. Bendík, J., Meel, K.S.: Approximate counting of minimal unsatisfiable subsets. In: Lahiri, S.K., Wang, C. (eds.) CAV 2020. LNCS, vol. 12224, pp. 439–462. Springer, Cham (2020). https://doi.org/10.1007/978-3-030-53288-8_21

15. Biere, A.: Cadical, lingeling, plingeling, treengeling and yalsat entering the sat competition 2018. In: Proceedings of SAT Competition, pp. 13–14 (2018)
16. Bollobás, B., Borgs, C., Chayes, J.T., Kim, J.H., Wilson, D.B.: The scaling window of the 2-sat transition. Random Struct. Algorithms **18**(3), 201–256 (2001)
17. Chen, H., Marques-Silva, J.: Improvements to satisfiability-based Boolean function bi-decomposition. In: VLSI-SoC, pp. 142–147. IEEE (2011)
18. Cohen, O., Gordon, M., Lifshits, M., Nadel, A., Ryvchin, V.: Designers work less with quality formal equivalence checking. In: Design and Verification Conference (DVCon). Citeseer (2010)
19. de la Banda, M.J.G., Stuckey, P.J., Wazny, J.: Finding all minimal unsatisfiable subsets. In: PPDP, pp. 32–43. ACM (2003)
20. Eén, N., Sörensson, N.: An extensible SAT-solver. In: Giunchiglia, E., Tacchella, A. (eds.) SAT 2003. LNCS, vol. 2919, pp. 502–518. Springer, Heidelberg (2004). https://doi.org/10.1007/978-3-540-24605-3_37
21. Han, B., Lee, S.-J.: Deriving minimal conflict sets by CS-trees with mark set in diagnosis from first principles. IEEE Trans. Syst. Man Cybern. Part B **29**(2), 281–286 (1999)
22. Hou, A.: A theory of measurement in diagnosis from first principles. AI **65**(2), 281–328 (1994)
23. Hunter, A., Konieczny, S.: Measuring inconsistency through minimal inconsistent sets. In: KR, pp. 358–366. AAAI Press (2008)
24. Ivrii, A., Malik, S., Meel, K.S., Vardi, M.Y.: On computing minimal independent support and its applications to sampling and counting. Constraints **21**(1), 41–58 (2015). https://doi.org/10.1007/s10601-015-9204-z
25. Jannach, D., Schmitz, T.: Model-based diagnosis of spreadsheet programs: a constraint-based debugging approach. Autom. Softw. Eng. **23**(1), 105–144 (2014). https://doi.org/10.1007/s10515-014-0141-7
26. Kilby, P., Slaney, J.K., Thiébaux, S., Walsh, T.: Backbones and backdoors in satisfiability. In: AAAI, pp. 1368–1373. AAAI Press/The MIT Press (2005)
27. Liffiton, M.H., Malik, A.: Enumerating infeasibility: finding multiple MUSes Quickly. In: Gomes, C., Sellmann, M. (eds.) CPAIOR 2013. LNCS, vol. 7874, pp. 160–175. Springer, Heidelberg (2013). https://doi.org/10.1007/978-3-642-38171-3_11
28. Liffiton, M.H., Previti, A., Malik, A., Marques-Silva, J.: Fast, flexible MUS enumeration. Constraints **21**(2), 223–250 (2015). https://doi.org/10.1007/s10601-015-9183-0
29. Liffiton, M.H., Sakallah, K.A.: Algorithms for computing minimal unsatisfiable subsets of constraints. JAR **40**(1), 1–33 (2008). https://doi.org/10.1007/s10817-007-9084-z
30. Luo, J., Liu, S.: Accelerating MUS enumeration by inconsistency graph partitioning. Sci. China Inf. Sci. **62**(11), 212104 (2019). https://doi.org/10.1007/s11432-019-9881-0
31. Mencía, C., Kullmann, O., Ignatiev, A., Marques-Silva, J.: On computing the union of MUSes. In: Janota, M., Lynce, I. (eds.) SAT 2019. LNCS, vol. 11628, pp. 211–221. Springer, Cham (2019). https://doi.org/10.1007/978-3-030-24258-9_15
32. Kedian, M.: Formulas free from inconsistency: an atom-centric characterization in priest's minimally inconsistent LP. J. Artif. Intell. Res. **66**, 279–296 (2019)
33. Narodytska, N., Bjørner, N., Marinescu, M.-C., Sagiv, M.: Core-guided minimal correction set and core enumeration. In: IJCAI, pp. 1353–1361 (2018). https://www.ijcai.org/

34. Previti, A., Marques-Silva, J.: Partial MUS enumeration. In: AAAI. AAAI Press (2013)
35. Sperner, E.: Ein satz über untermengen einer endlichen menge. Math. Z. **27**(1), 544–548 (1928). https://doi.org/10.1007/BF01171114
36. Stern, R.T., Kalech, M., Feldman, A., Provan, G.M.: Exploring the duality in conflict-directed model-based diagnosis. In: AAAI. AAAI Press (2012)
37. Stuckey, P.J., Sulzmann, M., Wazny, J.: Interactive type debugging in haskell. In: Haskell, pp. 72–83. ACM (2003)
38. Wieringa, S.: Understanding, improving and parallelizing MUS finding using model rotation. In: Milano, M. (ed.) CP 2012. LNCS, pp. 672–687. Springer, Heidelberg (2012). https://doi.org/10.1007/978-3-642-33558-7_49

# Solving Satisfaction Problems Using Large-Neighbourhood Search

Gustav Björdal[1(✉)], Pierre Flener[1], Justin Pearson[1], Peter J. Stuckey[2], and Guido Tack[2]

[1] Department of Information Technology, Uppsala University, Uppsala, Sweden
{Gustav.Bjordal,Pierre.Flener,Justin.Pearson}@it.uu.se

[2] Faculty of Information Technology, Monash University, Melbourne, Australia
{Peter.Stuckey,Guido.Tack}@monash.edu

**Abstract.** Large-neighbourhood search (LNS) improves an initial solution, hence it is not directly applicable to satisfaction problems. In order to use LNS in a constraint programming (CP) framework to solve satisfaction problems, we usually soften some hard-to-satisfy constraints by replacing them with penalty-function constraints. LNS is then used to reduce their penalty to zero, thus satisfying the original problem. However, this can give poor performance as the penalties rarely cause propagation and therefore do not drive each CP search, and by extension the LNS search, towards satisfying the replaced constraints until very late. Our key observation is that entirely replacing a constraint is often overkill, as the propagator for the replaced constraint could have performed *some* propagation without causing backtracking. We propose the notion of a *non-failing propagator*, which is subsumed just before causing a backtrack. We show that, by only making a few changes to an existing CP solver, any propagator can be made non-failing without modifying its code. Experimental evaluation shows that non-failing propagators, when used in conjunction with penalties, can greatly improve LNS performance compared to just having penalties. This allows us to leverage the power of the many sophisticated propagators that already exist in CP solvers, in order to use LNS for solving hard satisfaction problems and for finding initial solutions to hard-to-satisfy optimisation problems.

## 1 Introduction

Large-neighbourhood search (LNS) [19] is a popular method for local search. It often uses constraint programming (CP) for neighbourhood exploration and has been successfully applied to a vast variety of optimisation problems. LNS starts from a feasible assignment and explores a large neighbourhood of similar assignments by forcing most of the variables to take their current values while performing CP search in order to find better values for the other variables. This process is repeated in the hope of finding a good enough feasible assignment. *However, as LNS requires a feasible assignment to start from, LNS cannot be directly applied to satisfaction problems*, because the initial feasible assignment


H. Simonis (Ed.): CP 2020, LNCS 12333, pp. 55–71, 2020.
https://doi.org/10.1007/978-3-030-58475-7_4

would be an acceptable solution. Furthermore, an optimisation problem that is hard to satisfy, that is where finding a feasible assignment is difficult, cannot be solved by LNS until an initial feasible assignment is obtained. We emphasise that finding an initial feasible assignment is in fact a satisfaction problem. Therefore, *if we can efficiently solve satisfaction problems using LNS, then we can also solve hard-to-satisfy optimisation problems using* (two rounds of) *LNS.*

One approach to using LNS for satisfaction problems is to (manually) identify and *soften* the constraints that make the problem hard to satisfy. Traditionally, soft constraints for CP have been investigated mostly for *over-constrained problems* [10], that is for problems where not all the constraints can be satisfied. There is little previous work on softening constraints in order to enable LNS (see Sect. 6). Still, there are generic methods for softening a constraint, such as replacing it by using a penalty function and minimising the penalty via the objective function (see Sect. 2.1 for examples). However, these methods tend to give poor performance in practice, as they significantly increase the size of the CP search space and provide little propagation to drive the CP search towards a zero-penalty solution (as we show in Sect. 3).

In this paper, we argue that entirely replacing constraints by using penalty functions in order to enable LNS is overkill, because it means that we lose *all* their propagation, including the propagation that would *not* have caused failure but would have avoided unnecessarily high penalties.

Based on this observation, we propose the notion of a *non-failing propagator*: the inconsistent domain values of a variable are only pruned as long as doing so does *not* cause a failure. As soon as propagation would cause failure, the propagator is disabled. This prevents the propagator from directly causing backtracking and, when used in conjunction with a penalty function, helps the CP search to quickly reach low-penalty solutions.

After giving some definitions on soft constraints and LNS (Sect. 2) and a motivating example (Sect. 3), our contributions and impact are as follows:

- the concept and theory of non-failing propagators (Sects. 4.1 and 4.3);
- a recipe for modifying a CP solver so that any propagator can automatically become non-failing, without modifying its code (Sect. 4.2);
- an empirical evaluation of the often drastic effect of non-failing propagators on solving satisfaction problems by LNS, as well as of their use when solving hard-to-satisfy optimisation problems by LNS (Sect. 5).

We discuss related work (Sect. 6) and future work and conclude (Sect. 7).

## 2 Definitions

A *constraint satisfaction problem* is a triple $\langle \mathcal{X}, D, \mathcal{C} \rangle$ where $\mathcal{X}$ is a set of variables, the function $D$ maps each variable $x$ of $\mathcal{X}$ to a finite set $D(x)$, called the *current domain* of $x$, and $\mathcal{C}$ is a set of constraints. An *assignment* is a mapping $\sigma$ where $\sigma(x) \in D(x)$ for each $x$ of $\mathcal{X}$. A *feasible assignment* is an assignment that satisfies all the constraints in $\mathcal{C}$. A *constrained optimisation problem* $\langle \mathcal{X}, D, \mathcal{C}, o \rangle$ has a designated variable $o$ of $\mathcal{X}$ constrained in $\mathcal{C}$ to take the value of an objective function that is to be minimised (without loss of generality). An *optimal assignment* is a feasible assignment where $o$ is minimal.

### 2.1 Soft Constraints

Given a satisfaction problem $\langle \mathcal{X}, D, \mathcal{C} \rangle$ and an assignment $\sigma$, the *penalty under* $\sigma$ of a constraint $C(\mathcal{V})$ in $\mathcal{C}$, where $\mathcal{V}$ is an ordered (multi)subset of $\mathcal{X}$, is given by a function $\pi$, called the *penalty function*, such that $\pi(\sigma)$ is 0 if $C(\mathcal{V})$ is satisfied under $\sigma$, and otherwise a positive number proportional to the degree that $C(\mathcal{V})$ is violated. For example, for the constraint $x + y = z$, the penalty could be $|\sigma(x) + \sigma(y) - \sigma(z)|$, which is the distance between $x+y$ and $z$ under $\sigma$. See [20] for a variety of penalty functions in the context of constraint-based local search.

For a constraint $C(\mathcal{V})$ and a penalty function $\pi$, the *soft constraint* $C_\pi(\mathcal{V}, p)$ constrains a new variable $p$, called the *penalty variable*, to take the value $\pi$ takes.

*Example 1.* In our experiments of Sect. 5, the soft constraint for a linear equality constraint $\sum_i \mathcal{A}_i \mathcal{V}_i = c$ is $p = |\sum_i \mathcal{A}_i \mathcal{V}_i - c|$, that is $p = |x + y - z|$ for the unweighted equality constraint $x + y = z$ we considered above. For a linear inequality constraint $\sum_i \mathcal{A}_i \mathcal{V}_i \leq c$, we use $p = \max(0,\ \sum_i \mathcal{A}_i \mathcal{V}_i - c)$. For a global cardinality constraint $\text{GCC}(\mathcal{V}, \mathcal{A}, \mathcal{L}, \mathcal{U})$, constraining every value $\mathcal{A}_i$ to be taken between $\mathcal{L}_i$ and $\mathcal{U}_i$ times by the variables $\mathcal{V}_j$, which cannot take any other values, we use $p = \sum_i \max(0,\ \mathcal{L}_i - \sum_j [\mathcal{V}_j = \mathcal{A}_i],\ \sum_j [\mathcal{V}_j = \mathcal{A}_i] - \mathcal{U}_i) + \sum_{d \notin \mathcal{A}} \sum_j [\mathcal{V}_j = d]$, where $[\alpha]$ denotes value 1 if constraint $\alpha$ holds and value 0 otherwise. □

We say that we *soften* a constraint $C(\mathcal{V})$ when we replace it in $\mathcal{C}$ by $C_\pi(\mathcal{V}, p)$ for some $\pi$, with variable $p$ added to $\mathcal{X}$ and used in an objective function. We call $C(\mathcal{V})$ the *replaced constraint*, not to be mixed up with $C_\pi(\mathcal{V}, p)$.

We define $\text{SOFT}(\langle \mathcal{X}, D, \mathcal{C} \rangle, \mathcal{S}, \pi, \lambda)$ as the softening of a subset $\mathcal{S} \subseteq \mathcal{C}$ of $n$ constraints in the satisfaction problem $\langle \mathcal{X}, D, \mathcal{C} \rangle$ into the optimisation problem $\langle \mathcal{X} \cup \{p_i \mid i \in 1..n\} \cup \{o\}, D', \mathcal{C} \setminus \mathcal{S} \cup \{C^i_{\pi_i}(\mathcal{V}, p_i) \mid C^i(\mathcal{V}) \in \mathcal{S}\} \cup \{o = \sum_{i=1}^n \lambda_i p_i\}, o \rangle$ by using the penalty functions $\pi_i$ and weights $\lambda_i \geq 0$, where $D'$ is $D$ extended to give the initial domain $0 \mathrel{..} \infty$ to the introduced objective variable $o$ and each introduced penalty variable $p_i$.

**Definition 1.** For any $\text{SOFT}(\mathcal{P}, \mathcal{S}, \pi, \lambda)$ of a satisfaction problem $\mathcal{P}$, we define:

- a *pseudo-solution* is a feasible assignment where at least one introduced penalty variable takes a positive value, and therefore the non-penalty variables do not form a feasible assignment for $\mathcal{P}$; and
- a *solution* is a feasible assignment where all penalty variables take the value 0, and therefore the non-penalty variables do form a feasible assignment for $\mathcal{P}$.

Note that a solution to $\text{SOFT}(\mathcal{P}, \mathcal{S}, \pi, \lambda)$ is in fact an optimal solution to it, as the introduced objective variable takes its lower bound 0, no matter what the weights $\lambda_i$ are, and thereby that solution establishes the satisfiability of $\mathcal{P}$.

Consider a soft constraint $C_\pi(\mathcal{V}, p)$ and a variable $v$ in $\mathcal{V}$: we say that a value $d$ in $D(v)$ *imposes a penalty* when $\min(D(p))$ would increase if $D(v)$ became $\{d\}$.

### 2.2 Large-Neighbourhood Search

Large-neighbourhood search (LNS) [19] is a local-search method for solving an optimisation problem $\langle \mathcal{X}, D, \mathcal{C}, o \rangle$. It starts from a feasible assignment $\sigma$, which evolves as what we call the *current assignment.* At each iteration, an LNS heuristic selects a non-empty strict subset $\mathcal{M}$ of the variables $\mathcal{X}$, called the *fragment,* where $o \in \mathcal{M}$. The optimisation problem $\langle \mathcal{X}, D, \mathcal{C} \cup \{x = \sigma(x) \mid x \in \mathcal{X} \setminus \mathcal{M}\}, o \rangle$, where all but the variables of the fragment take their current values, is solved, not necessarily to optimality. If an improving feasible assignment is found, then it replaces the current assignment, otherwise the current assignment is unchanged. The search continues from the current assignment by selecting a new fragment.

LNS can in principle be used to solve a satisfaction problem $\mathcal{P}$ that has been softened by some $\textsc{Soft}(\mathcal{P}, \mathcal{S}, \pi, \lambda)$ into an optimisation problem, but we show in Sects. 3 and 5 that the performance can be poor in practice.

The optimisation problem at each LNS iteration, as well as the satisfaction problem of finding an initial feasible assignment, can in principle be solved by using any technology, but we here only consider CP.

There exists extensive literature on the challenge of *heuristically* selecting a fragment at each LNS iteration: either by exploiting the structure of the underlying problem [19] or by using more generic methods [9,13,14]. Both approaches can have a significant impact on the performance of LNS.

*In this paper, we focus on an orthogonal challenge of LNS, namely efficiently solving* (hard) *satisfaction problems* (*and even optimisation problems that are hard to satisfy, that is where finding a feasible assignment is difficult*), *so that there is a need to improve the propagation in each LNS iteration in a new way.*

## 3 Motivation

To make some motivating observations, we consider as a running example the satisfaction problem of subset sum, as solved by CP. We then soften the problem in order to show how it can in principle be solved by LNS.

*Example 2.* Given an integer set $\mathcal{S}$ and an integer $t$, the subset-sum problem is to find a subset $\mathcal{S}' \subseteq \mathcal{S}$ such that $\sum_{s \in \mathcal{S}'} s = t$. We can express this as a satisfaction problem using a 0/1 variable $x_i$ for each element of $\mathcal{S}$, and a single constraint, say for $\mathcal{S} = \{11, -3, 2, 5, 9, -6\}$ and $t = 1$:

$$11x_1 - 3x_2 + 2x_3 + 5x_4 + 9x_5 - 6x_6 = 1 \tag{1}$$

Using the classical idempotent bounds-consistency propagator in Algorithm 1 for the linear equality (1)—because achieving domain consistency is NP-hard [2]—and the CP search strategy that branches on the variables in order from $x_1$ to $x_6$, always with $x_i = 1$ as the left-branch decision, we obtain the following CP search tree. At the root node, the value 1 is pruned from $D(x_1)$. Upon the decision $x_2 = 1$, no value is pruned. Upon the decision $x_3 = 1$, propagation first prunes the value 1 from both $D(x_5)$ and $D(x_6)$, but then fails as all values of $D(x_4)$ must

**Algorithm 1.** Bounds-consistency propagator for $\sum_i \mathcal{A}_i \mathcal{V}_i = c$, where $\mathcal{A}$ is an array of integers, $\mathcal{V}$ an equally long array of integer variables, and $c$ an integer; it updates the domain function $D$. (Idempotent due to lines 1–3 and 12–13.)

1: *done* $\leftarrow$ *false*
2: **while not** *done* **do**
3:     *done* $\leftarrow$ *true*   *// this might be the last iteration*
4:     $\ell \leftarrow \sum_{i=1}^{|\mathcal{V}|} \mathcal{A}_i \cdot \min(D(\mathcal{V}_i))$;   $u \leftarrow \sum_{i=1}^{|\mathcal{V}|} \mathcal{A}_i \cdot \max(D(\mathcal{V}_i))$
5:     **if** $\ell = c = u$ **then**
6:         **return** SUBSUMED
7:     **else if** $c < \ell$ **or** $u < c$ **then**
8:         **return** FAILED
9:     **for** $i = 1$ **to** $|\mathcal{V}|$ **do**
10:         $D(\mathcal{V}_i) \leftarrow D(\mathcal{V}_i) \cap \left\lceil \frac{c-u+\max(D(\mathcal{V}_i))}{\mathcal{A}_i} \right\rceil .. \max(D(\mathcal{V}_i))$   *// tighten lower bound*
11:         $D(\mathcal{V}_i) \leftarrow D(\mathcal{V}_i) \cap \min(D(\mathcal{V}_i)) .. \left\lfloor \frac{c-\ell+\min(D(\mathcal{V}_i))}{\mathcal{A}_i} \right\rfloor$   *// tighten upper bound*
12:     **if** some $D(\mathcal{V}_i)$ has changed **then**
13:         *done* $\leftarrow$ *false*   *// continue iterating*
14: **return** ATFIXPOINT

be pruned. The search backtracks, failing at two more nodes, until it finds the only feasible assignment (namely $x_1 = x_2 = x_5 = 0$ and $x_3 = x_4 = x_6 = 1$, corresponding to $\mathcal{S}' = \{2, 5, -6\}$) upon the decisions $x_2 \neq 1$ and $x_3 = 1$. □

*To solve a satisfaction problem with LNS, one must soften some constraints in order to turn it into an optimisation problem where a zero-penalty solution corresponds to a feasible assignment to the satisfaction problem.* If no such softening is performed, then the initial feasible assignment will be an acceptable solution, which means that LNS adds no benefit.

Although constraints can be softened in a generic way by using penalty functions (as explained in Sect. 2.1), *softening will in practice significantly increase the size of the CP search space for each LNS iteration, as soft constraints usually only cause propagation towards the bottom of the search tree, where most variables are fixed, and provide little to no propagation that drives the CP search towards an* (optimal) *solution*, as shown in the following example.

*Example 3.* In order to solve the subset-sum satisfaction problem of Example 2 by LNS, its constraint (1) must be softened, say as:

$$p = |11x_1 - 3x_2 + 2x_3 + 5x_4 + 9x_5 - 6x_6 - 1| \tag{2}$$

where $p \in 0..26$, and the objective is to minimise $p$. By Definition 1, if $p = 0$ then the solution corresponds to a feasible assignment to the satisfaction problem. Consider the CP search tree while finding an initial feasible assignment for LNS, using the search strategy of Example 2. Since $p$ is essentially unconstrained, there is no propagation (no matter what consistency is targeted) and search descends the left branch, arriving at a pseudo-solution where all $x_i = 1$ and $p = 17$. Improving it into a solution requires $x_1 = 0$: this was achieved by root-node propagation in Example 2, but is here only achievable by $x_1$ being in a fragment. However, even that may not be enough: if the first fragment is $\{x_1, x_3\}$, then the next pseudo-solution will have $x_3 = 0$ and $p = 15$. Clearly, even in a small instance, early bad decisions severely degrade performance. □

# 4 Avoiding Bad CP Search Decisions

We are here concerned with constraints that must be softened to enable the use of LNS for satisfaction problems. *We want to improve the CP search for an initial (pseudo-) solutions for the first LNS iteration, as well as the CP search within each LNS iteration for better pseudo-solutions and eventually a solution.*

We saw in Example 3 that bad CP search decisions can be made early if the propagation from the *soft* constraints does not prune values that impose high penalties. It could therefore be beneficial to prune *some* of those values, using *some* propagation from the *replaced* constraints, so that the CP search avoids those bad decisions. However, it would be counterproductive to prune all values that impose a penalty, as that would again make the constraints (and by extension the problem) hard to satisfy, namely by causing significant backtracking.

Still, given the crucial role that propagation plays in the effectiveness of CP search, we argue that only replacing some constraints by using penalty functions (and thereby effectively removing all the propagation for those constraints) does not fully utilise the decades of research on efficient and powerful propagators.

Based on this observation, we propose non-failing propagators (Sect. 4.1), show how to modify a CP solver such that any propagator can be made non-failing without modifying its code (Sect. 4.2), and discuss how the scheduling of non-failing propagators can impact backtracking (Sect. 4.3).

## 4.1 Non-failing Propagators

We want to extend an optimisation problem $\textsc{Soft}(\langle \mathcal{X}, D, \mathcal{C} \rangle, \mathcal{S}, \pi, \lambda)$ by prescribing *additional* propagators for $\mathcal{S}$ to prune values that would impose a penalty, but without causing backtracking. For this, we propose non-failing propagators:

**Definition 2.** For a constraint $C(\mathcal{V})$, a *non-failing propagator* prunes domain values that are inconsistent with $C(\mathcal{V})$, but only until the pruning would empty the domain of some variable $v$ in $\mathcal{V}$; at that point, the propagator is subsumed instead of pruning the last domain value (or values) of $v$ and being failed, which would cause backtracking.

For example, the propagator of Algorithm 1 can be turned into a non-failing propagator by rewriting line 8 to return the status SUBSUMED instead of FAILED. In Sect. 4.2, we achieve this at the *solver* level instead of the propagator level.

We use the notation $C(\mathcal{V}) :: \textsc{NonFailing}$ to indicate that the propagator used for $C(\mathcal{V})$ should be non-failing.

A non-failing propagator never causes backtracking itself, but it can do so *indirectly* by pruning values that make the normal propagators for the constraints $\mathcal{C} \backslash \mathcal{S}$ be failed. Just like any propagator that is disabled during CP search, a non-failing propagator is restored upon backtracking and restarts. Non-failing propagators are safe to use on problems that are satisfiable:

**Theorem 1.** *If a problem is satisfiable, then non-failing propagators for any of its constraints will never remove any feasible assignments from the search space.*

*Proof.* Consider a feasible assignment $\sigma$ for a problem. A non-failing propagator prunes no more than a normal propagator, by Definition 2. Therefore, no propagator for any constraint of the problem can prune a value occurring in $\sigma$ if all values in $\sigma$ are in the domains of the corresponding variables. □

We define $\textsc{Soft}_{\text{nonfail}}(\mathcal{P}, \mathcal{S}, \pi, \lambda)$ as the softening of the constraints $\mathcal{S}$ in the satisfaction problem $\mathcal{P}$ via $\textsc{Soft}(\mathcal{P}, \mathcal{S}, \pi, \lambda)$ but with the addition of the constraints $\{C(\mathcal{V})::\textsc{NonFailing} \mid C(\mathcal{V}) \in \mathcal{S}\}$: that is, the constraints $\mathcal{S}$ are implemented *both* by non-failing propagators for themselves *and* by normal propagators (or decompositions) for their soft versions.

*Example 4.* The application of $\textsc{Soft}_{\text{nonfail}}$ to the subset-sum problem of Example 2 uses *both* a non-failing propagator for its constraint (1) *and* normal propagation for its soft constraint from Example 3:

$$\begin{aligned}
&\text{minimise } p \\
&\text{such that } (11x_1 - 3x_2 + 2x_3 + 5x_4 + 9x_5 - 6x_6 = 1)::\textsc{NonFailing} \\
&\qquad p = |11x_1 - 3x_2 + 2x_3 + 5x_4 + 9x_5 - 6x_6 - 1| \\
&\qquad x_i \in \{0, 1\}, \qquad \forall i \in 1\,..\,6
\end{aligned}$$

Root-node propagation of the non-failing propagator prunes value 1 from $D(x_1)$, as in Example 2 and unlike in Example 3. Given the same CP search strategy as in Example 2, the first decision, $x_2 = 1$, does not trigger any propagation. The second decision, $x_3 = 1$, causes the non-failing propagator to first prune value 1 from $D(x_5)$, then prune value 1 from $D(x_6)$, and then infer failure. However, rather than failing, the propagator is subsumed. The next decision is then $x_4 = 1$, at which point the CP search is at a pseudo-solution (namely $x_1 = x_5 = x_6 = 0$ and $x_2 = x_3 = x_4 = 1$). The propagator of the soft constraint from Example 3 so far had no impact on the decisions, but gives $p = 3$ at this pseudo-solution. Given this initial LNS assignment, if a fragment consisting of $x_2$ and $x_6$ is selected in the first LNS iteration, then root-node propagation of the non-failing propagator will immediate solve the problem by inferring $x_2 = 0$ and $x_6 = 1$, giving $p = 0$. Unlike in Example 3, where only the soft constraint is used, we here see how a non-failing propagator can help the CP searches avoid bad decisions and also allow the LNS search to quickly arrive at a zero-penalty assignment. While this example was specifically constructed to showcase this behaviour, our experimental evaluation in Sect. 5 shows that this seems also to be beneficial in practice, across a variety of benchmarks. □

**Algorithm 2.** FIXPOINT($P, Q, D$), where $P$ is the set of all non-disabled propagators (initially those for the constraints $\mathcal{C}$ of the problem), $Q$ is the priority queue of propagators not known to be at fixpoint (initially those for $\mathcal{C}$, later those of a CP search decision), and $D$ is the function giving the current domains.

1: **while** $Q$ is not empty **do**
2: $\quad p \leftarrow Q$.dequeue()
3: $\quad$ *status* $\leftarrow p$.propagate($D$) // *note that this can enqueue propagators of P*
4: $\quad$ **if** *status* = SUBSUMED **then**
5: $\quad\quad p$.disable() // *this achieves* $P \leftarrow P \setminus \{p\}$
6: $\quad$ **else if** *status* = FAILED **then**
7: $\quad\quad$ **return** FAILURE // *fail and cause backtracking*
8: $\quad$ **else**
9: $\quad\quad$ ... // *other status messages are not relevant here*
10: **return** ATCOMMONFIXPOINT

### 4.2 Implementation

In principle, any propagator can be made non-failing by modifying its code (such as in the example after Definition 2). However, this can be both tedious and error-prone. Fortunately, we can instead modify the propagation engine of a CP solver to treat as non-failing any propagator tagged as ::NONFAILING.

Algorithm 2 shows a typical fixpoint algorithm, based on a queue of propagators that need to be executed. The only change required to support non-failing propagators is to replace line 4 by

**if** *status* = SUBSUMED **or** (non-failing($p$) **and** *status* = FAILED) **then**

so that when a propagator tagged as non-failing returns the status FAILED, then the status is instead treated as SUBSUMED. However, in order for this modification to Algorithm 2 to be correct, the CP solver and its propagators must guarantee that the domains of the variables are always in a *consistent state*: when a propagator returns FAILED, the domain of each variable must be a *non-empty* subset of the domain before running the propagator.

For our experiments in Sect. 5, we modified Gecode [6] in this way. Domain updates in Gecode are not guaranteed to leave the domains in a consistent state after failure: we therefore modified the domain update functions to check whether an update would result in a domain wipe-out *before* they modify that domain.

### 4.3 Scheduling of Non-failing Propagators

Non-failing propagators are non-monotonic: the amount of propagation they achieve (and whether they propagate or are subsumed) depends on the order in which all propagators are executed [18]. Many CP solvers order propagators using a *priority queue*, for example based on their algorithmic complexity [17].

The priority assigned to non-failing propagators can therefore determine if a node fails or succeeds. However:

**Theorem 2.** *There is no static priority order (independent of the internal state of the fixpoint algorithm) of propagators that guarantees that non-failing propagators only cause failure when failure occurs for all possible priority orders.*

*Proof.* Consider the following problem:

$$D(x) = D(y) = D(z) = 0\,..\,5,\ D(b_0) = D(b_1) = D(b_2) = 0\,..\,1$$
$$x + y + z = 5,\ b_1 \rightarrow x \geq 2,\ b_1 \rightarrow y \geq 2,\ b_2 \rightarrow x \leq 1,\ b_2 \rightarrow y \leq 1$$
$$b_0 \rightarrow b_1{::}\text{NONFAILING},\ b_0 \rightarrow b_2{::}\text{NONFAILING}$$

where the non-failing propagators are $p_1$ for $b_0 \rightarrow b_1 :: \text{NONFAILING}$ and $p_2$ for $b_0 \rightarrow b_2 :: \text{NONFAILING}$. We assume that both non-failing propagators are always propagated last, as that decreases the probability of failure.

Upon the initial CP search decision $z \geq 2$, we reach the node where $D(z) = 2\,..\,5$ and $D(x) = D(y) = 0\,..\,3$. If we make the decision $b_0 = 1$, then both $p_1$ and $p_2$ are enqueued. If $p_1$ is dequeued first, then we propagate $b_1 = 1$, $x \geq 2$, $y \geq 2$, $b_2 = 0$, and then the propagator for $x + y + z = 5$ is dequeued and fails. If $p_2$ is dequeued first, then we propagate $b_2 = 1$, $x \leq 1$, $y \leq 1$, $b_1 = 0$, $z \geq 1$, and then $p_1$ is dequeued and subsumed (because it fails). That is, the node only succeeds when $p_2$ runs before $p_1$.

Upon the opposite CP search decision $z < 2$, we reach the node where $D(z) = 0\,..\,1$. If we make the decision $b_0 = 1$, then both $p_1$ and $p_2$ are enqueued. If $p_1$ is dequeued first, then we propagate $b_1 = 1$, $x \geq 2$, $y \geq 2$, $b_2 = 0$, and then $p_2$ is dequeued and subsumed (because it fails). If $p_2$ is dequeued first, then we propagate $b_2 = 1$, $x \leq 1$, $y \leq 1$, $b_1 = 0$, and then the propagator for $x+y+z = 5$ is dequeued and fails. That is, the node only succeeds when $p_1$ runs before $p_2$.

So, no static priority order can always avoid indirect failures caused by non-failing propagators, even when the indirect failure could have been avoided. □

Empirically, we found it beneficial to always run non-failing propagators at the lowest priority. Intuitively, this makes sense: consider two propagators $p_1$ and $p_2$, where running $p_1$ causes $p_2$ to fail, and vice versa: if only $p_1$ is non-failing, then backtracking is only avoided when running $p_1$ after $p_2$. We therefore modified Gecode to schedule all non-failing propagators to run as late as possible, with a first-in-first-out tie-breaking between non-failing propagators.

## 5 Experimental Evaluation

This section presents an empirical evaluation of the benefit of non-failing propagators for both satisfaction and optimisation problems.

### 5.1 Setup

We compare three approaches to finding a solution to a satisfaction problem $\mathcal{P}$:

**hard** treat all constraints as hard, that is: solve $\mathcal{P}$ by CP;

**soft** soften some constraints $\mathcal{S}$ of $\mathcal{P}$, that is: solve $\textsc{Soft}(\mathcal{P}, \mathcal{S}, \pi, \lambda)$ by LNS;
**non-failing** soften the same constraints $\mathcal{S}$ of $\mathcal{P}$, but also use non-failing propagators for those constraints, that is: solve $\textsc{Soft}_{\text{nonfail}}(\mathcal{P}, \mathcal{S}, \pi, \lambda)$ by LNS.

For all problems, we used published MiniZinc (version 2.4.3) [12] models.[1]

We modified Gecode (version 6.2.0) [6] as described in Sects. 4.2 and 4.3. We modified its MiniZinc interface so that some constraints, when tagged with a new `soften` annotation for MiniZinc, are automatically softened under the **soft** and **non-failing** approaches with $\lambda = \bar{1}$ and using the penalty functions $\pi$ in Example 1. For the **non-failing** approach, the `soften` annotation also tags the constraint with the ::NonFailing annotation, that is, we have *both* a soft version *and* a non-failing version of the tagged constraint when using the **non-failing** approach. To propagate a soft constraint, we use a decomposition of its penalty function rather than a specialised propagator. We used Gecode's built-in LNS via its MiniZinc interface[2]: it selects a variable *not* to be in the fragment under a given probability; we prescribed a probability of 70% or 80%, depending on the size of the problem instances, (See footnote 1) and used the `constant` restart strategy.

For each problem instance, we report the average time for finding a first solution by LNS, over 10 independent runs, each allocated 10 minutes. The average was only computed over the runs where a solution was actually found.

We used Linux Ubuntu 18.04 (64 bit) on an Intel Xeon E5520 of 2.27 GHz, with 4 processors of 4 cores each, with 24 GB RAM and an 8 MB L2 cache. Note that we only run Gecode on a single core for our experiments.

### 5.2 Satisfaction Problems

We want to see whether our new generic **non-failing** approach allows LNS to outperform the classical generic **soft** approach to LNS for satisfaction problems, and whether both beat the **hard** approach via only CP. We look at instances of three satisfaction problems that are difficult to solve with Gecode via MiniZinc.

*Nurse Rostering.* We use the model for a simple nurse rostering problem from the MiniZinc Handbook[3] but modify it by using global-cardinality constraints on the daily numbers of nurses on each shift. We handcrafted $10 + 10 = 20$ satisfiable instances to be either *easy* (by having many nurses available) or *difficult* (by being at the border of unsatisfiability in terms of available nurses), both for Gecode under the **hard** approach. We prescribe softening for all the global-cardinality constraints. In Fig. 1a, we see that **soft** solves fewer instances than **hard** and often needs over an order of magnitude more time (both only solve the easy instances), while **non-failing** solves all but one (difficult) instance and does so with seemingly no overhead compared to **hard** (on the easy instances).

[1] We modified the models in order to deploy more global constraints and better CP search strategies. Our versions of the MiniZinc models, the instances, and the Gecode library are available at https://github.com/astra-uu-se/CP2020.

[2] See https://www.minizinc.org/doc-latest/en/lib-gecode.html.

[3] Section 2.3.1.4 of https://www.minizinc.org/doc-latest/en/predicates.html.

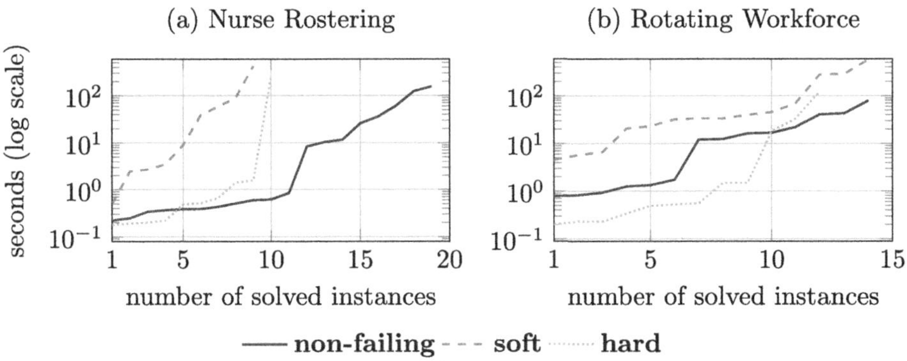


**Fig. 1.** Number of instances that are each solved within a given time.

*Rotating Workforce.* In the rotating workforce problem [11], a roster is to be built for employees, satisfying complex rules on sequences of shifts. We use the model and 50 instances in [11]: these are difficult for Gecode under the **hard** approach, although not too difficult for mixed-integer-programming and lazy-clause-generation solvers [11]. We prescribe softening for the global-cardinality constraints on the daily numbers of assigned shifts. In Fig. 1b, we see that each approach only solves at most 14 of the 50 instances: **soft** is slowest but solves as many instances (though not the same) as **non-failing**, while **hard** is arguably fastest but solves fewer instances than **non-failing**.

*Car Sequencing.* In this problem [4], a sequence of cars of various classes, each class having a set of options, is to be produced, satisfying capacity constraints on the options over a sliding window on the sequence and occurrence constraints on the classes. We use the MiniZinc Benchmark model[4] and the classic 78 instances for sequences of 100 to 200 cars.[5] We prescribe softening for the capacity constraints, which are expressed by linear inequalities. A problem-specific softening, which does *not* rely on penalties in the spirit of those in Example 1, was successfully used with LNS in [13]: we model it in MiniZinc and solve it by LNS with Gecode calling this the **reformulation** approach. In Fig. 2, we see that **hard** only solves 4 instances, while **soft** solves 65 instances but takes an order of magnitude more time than **non-failing**, which solves 69 of the 78 instances; **reformulation** solves 67 instances and takes time between **soft** and **non-failing**.

### 5.3 Hard-to-Satisfy Optimisation Problems

For a hard-to-satisfy optimisation problem, we solve the *satisfaction* problem of finding a *first* solution, which enables another round of LNS to find better ones.

---

[4] Available at https://github.com/MiniZinc/minizinc-benchmarks.

[5] Available also at http://www.csplib.org/Problems/prob001/data/data.txt.html.

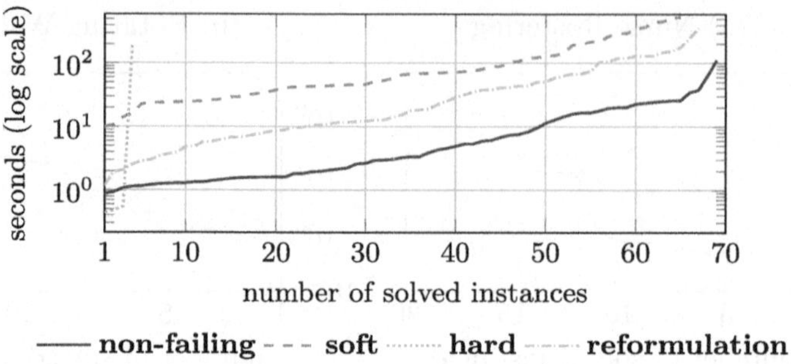


**Fig. 2.** Number of car-sequencing instances that are each solved within a given time.

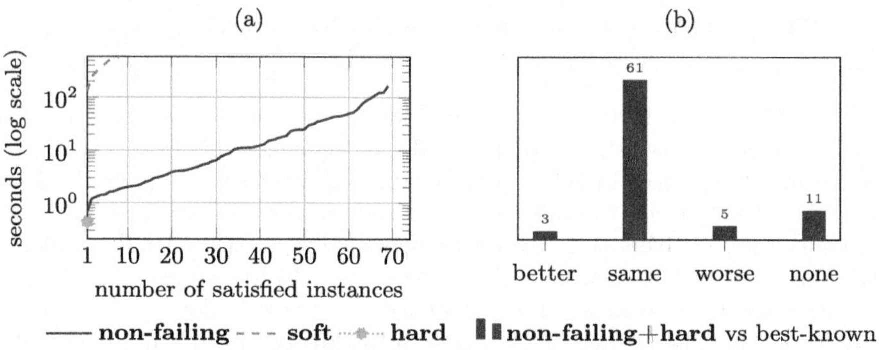


**Fig. 3.** (a) Number of TSPTW instances that are each satisfied within a given time. (b) Numbers of TSPTW instances where our best-found minimum is better, the same, or worse than the best-known one; 'none' means we found no feasible assignment.

*TSPTW.* In the travelling salesperson problem (TSP) with time windows, a shortest tour visiting all nodes in a graph during their time windows is to be found. We use the 80 satisfiable n40, n60, and n80 instances of the Gendreau-DumasExtended benchmark,[6] as they are very difficult to satisfy under the **hard** approach. We prescribe softening for the linear inequalities that require the arrival time at a node to be at least the arrival time at its predecessor plus the travel time in-between. In Fig. 3a, we see that **hard** only satisfies 1 instance and **soft** only 8 instances, while **non-failing** satisfies 69 of the 80 instances. When **non-failing** finds a solution within the allocated 10 minutes, we switch to **hard**, but with LNS, in order to try and improve it for the remaining time. We call this the **non-failing+hard** approach. In Fig. 3b, we compare the best solutions found by **non-failing+hard** to the best known solutions from the literature. (See footnote 6) Our new approach can improve these bounds for three instances.

[6] Available at http://lopez-ibanez.eu/tsptw-instances.

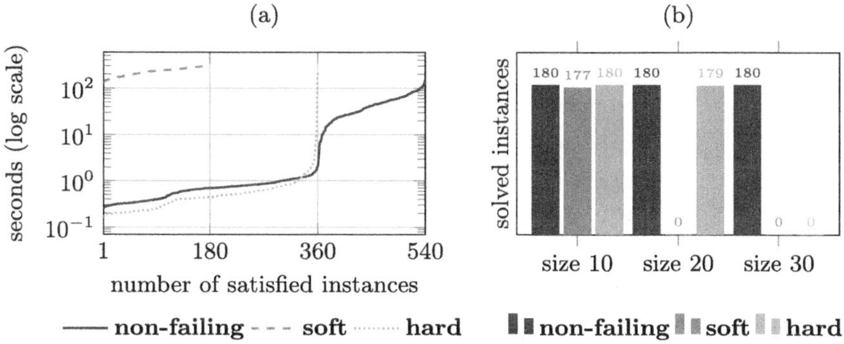


**Fig. 4.** (a) Number of TDTSP instances that are each satisfied within a given time. (b) Numbers of satisfied TDTSP instances for the three batches of instances.

*TDTSP.* In the time-dependent TSP [1], a shortest tour visiting all nodes in a graph during their time windows is to be found, similarly to TSPTW, but the travel time between two nodes depends on the arrival time at the first node. We use the model of the MiniZinc Challenge 2017,[7] but modify it to constrain the tour using successor variables and a `circuit` global constraint [8]. We prescribe softening as for TSPTW. By private communication with the author of the original MiniZinc model, we received 540 generated instances of 10, 20, and 30 nodes, with 180 instances for each batch; some of the size-10 and size-20 instances were used in the Challenge, but none of the size-30 ones as, for most of them, no MiniZinc backend ever found a feasible solution so that they were deemed too hard. In Fig. 4a, we see that **soft** only satisfies 177 instances and is over two orders of magnitude slower than the other approaches. Note that **hard** satisfies 359 instances about as fast as **non-failing**, but as the instances become more difficult to satisfy its runtime quickly increases. In Fig. 4b, we see that **soft** only satisfies instances of size 10, whereas **hard** only satisfies instances of size at most 20, and **non-failing** satisfies all the instances.

*HRC.* For the hospitals/residents matching problem with couples (HRC), we use the model and 5 instances of the MiniZinc Challenge 2019. (See footnote 7) We prescribe softening for the linear inequalities on the hospital capacities. In Table 1, we see that **soft** satisfies all the instances, whereas **non-failing** completely backfires and satisfies no instance (and actually does not even find a pseudo-solution for any instance). The non-failing propagators prevent the CP search from reaching a pseudo-solution due to too many *indirect* failures. Furthermore, since **hard** can solve 3 instances, it seems that the CP search space for **non-failing** is larger than for **hard**. We describe how to address indirect failures in Sect. 7.

[7] Available at https://www.minizinc.org/challenge.html.

**Table 1.** Runtimes (in seconds, or '–' if more than 600) to satisfy each HRC instance: the **non-failing** approach backfires.

| Instance | **non-failing** | **soft** | **hard** |
|---|---|---|---|
| exp1-1-5110 | – | 83.47 | 26.96 |
| exp1-1-5425 | – | 21.98 | 240.74 |
| exp1-1-5460 | – | 341.01 | 40.89 |
| exp2-1-5145 | – | 65.10 | – |
| exp2-1-5600 | – | 208.22 | – |

## 6 Related Work

We now discuss three areas of related work: soft constraints, variable-objective LNS, and streamliners.

Traditionally, soft constraints for CP have almost exclusively been researched in the context of over-constrained problems: see [10] and [16, Chapter 3] for extensive overviews. In this paper, we *assume* the problem is *not* over-constrained but hard to satisfy and therefore requires softening to enable the use of LNS. To the best of our knowledge, there exists very little research on softening problems that are *not* over-constrained, and replacing constraints by using penalties seems to originate from the over-constrained setting. The only other work we found is [5], which generalises Lagrangian relaxations to CP models.

Variable-objective LNS (VO-LNS) [15] is based on the observation that the penalty variables introduced by softening are usually connected to the objective variable by a linear equality and any new bounds on the objective variable result in little to no propagation on the penalty variables. Therefore, VO-LNS eagerly bounds penalty variables during branch-and-bound. This achieves more propagation from the soft constraints. This is conceptually related to our approach: we improve the poor propagation from the soft constraints and reduce their negative impact on LNS, by pruning more. VO-LNS satisfies Theorem 1: it never removes a solution from the CP search space if the problem is satisfiable. But, like our approach, VO-LNS can remove pseudo-solutions from the CP search space, and might therefore remove all pseudo-solutions with the lowest positive penalty. VO-LNS and non-failing propagators can be complementary: first experiments, where both approaches are used together, indicate that there can be a synergy.

Streamliners [7] are constraints added in order to remove a large portion of the CP search space while ideally not removing all solutions. Streamliners are identified by empirically observing structures in solutions to easy instances and hoping that those structures, and thereby constraints, extend to difficult instances. However, while streamliners are ideally safe, by not removing non-dominated solutions, they are not always guaranteed to be safe; their addition can even make a satisfiable instance unsatisfiable. Non-failing propagators, when added to a problem $\textsc{Soft}(\mathcal{P}, \mathcal{S}, \pi, \lambda)$, can be thought of as streamliners since they remove a large portion of the CP search space. But, unlike streamliners, non-failing propagators are always safe to use as they never make a satisfiable instance unsatisfiable, due to Theorem 1, and they are not based on empirical observation, but rather on the actual constraints of the problem.

## 7 Conclusion and Future Work

LNS is a powerful approach to solving difficult problems, but is typically only applied to (easy-to-satisfy) optimisation problems. We show that by using our non-failing propagators we can apply LNS to effectively tackle hard satisfaction problems (including those arising from hard-to-satisfy optimisation problems). Implementing non-failing propagators is not difficult in a CP solver, and can be done at the engine level with some care. Experimental results show that non-failing propagators can drastically improve the solving of hard-to-satisfy problems, although they are not universally beneficial.

Future work includes the design of constraint-specific non-failing propagators. For example, consider $\textsc{AllDifferent}([x_1, x_2, x_3, \ldots, x_n])$: rather than disabling its propagator upon detecting $x_1 = x_2$, one can replace it by a propagator for $\textsc{AllDifferent}([x_3, \ldots, x_n])$, thereby still avoiding the failure but now without losing the propagation on the remaining variables.

A weakness of non-failing propagators is that they can cause too many *indirect* failures, via normal propagators, as seen in Table 1. Indirect failures can sometimes be avoided by giving non-failing propagators the right priority (see Sect. 4.3). As future work, we can address this weakness with the following two orthogonal ideas. First, when using a learning CP solver such as Chuffed [3], which explains failures, we can detect that a failure is indirect and we can identify a node where a responsible non-failing propagator ran: the CP search can then backtrack and disable that propagator at that node, thus avoiding the failure. Second, non-failing propagators can be made more cautious about the values they prune: for example, rather than eagerly pruning values that impose a penalty, a non-failing propagator can prune only the values that impose a penalty above some threshold, which can be adjusted during CP search. Initial experiments show that these ideas work in principle, but do not yet outperform our generic implementation (of Sect. 4.2), as they bring their own sets of challenges, which require further investigation.

**Acknowledgements.** We thank the reviewers for their constructive feedback, and Graeme Gange for discussions on how to implement these ideas in LCG solvers. This work is supported by the Swedish Research Council (VR) through Project Grant 2015-04910, by the Vice-Chancellor's travel grant from the Wallenberg Foundation for a visit by the first author to Monash University, and by the Australian Research Council grant DP180100151.

## References

1. Aguiar Melgarejo, P., Laborie, P., Solnon, C.: A time-dependent no-overlap constraint: application to urban delivery problems. In: Michel, L. (ed.) CPAIOR 2015. LNCS, vol. 9075, pp. 1–17. Springer, Cham (2015). https://doi.org/10.1007/978-3-319-18008-3_1
2. Choi, C.W., Harvey, W., Lee, J.H.M., Stuckey, P.J.: Finite domain bounds consistency revisited. In: Sattar, A., Kang, B. (eds.) AI 2006. LNCS (LNAI), vol. 4304, pp. 49–58. Springer, Heidelberg (2006). https://doi.org/10.1007/11941439_9
3. Chu, G.: Improving Combinatorial Optimization. Ph.D. thesis, Department of Computing and Information Systems, University of Melbourne, Australia (2011). http://hdl.handle.net/11343/36679; the Chuffed solver and MiniZinc backend are available at https://github.com/chuffed/chuffed
4. Dincbas, M., Simonis, H., Van Hentenryck, P.: Solving the car-sequencing problem in constraint logic programming. In: Kodratoff, Y. (ed.) ECAI 1988, pp. 290–295. Pitman (1988)
5. Fontaine, D., Michel, L., Van Hentenryck, P.: Constraint-based Lagrangian relaxation. In: O'Sullivan, B. (ed.) Principles and Practice of Constraint Programming, CP 2014. LNCS, vol. 8656, pp. 324–339. Springer, Heidelberg (2014). https://doi.org/10.1007/978-3-319-10428-7_25
6. Gecode Team: Gecode: A generic constraint development environment (2020). the Gecode solver and its MiniZinc backend are available at https://www.gecode.org
7. Gomes, C., Sellmann, M.: Streamlined constraint reasoning. In: Wallace, M. (ed.) CP 2004. LNCS, vol. 3258, pp. 274–289. Springer, Heidelberg (2004). https://doi.org/10.1007/978-3-540-30201-8_22
8. Laurière, J.L.: A language and a program for stating and solving combinatorial problems. Artif. Intell. **10**(1), 29–127 (1978)
9. Lombardi, M., Schaus, P.: Cost impact guided LNS. In: Simonis, H. (ed.) CPAIOR 2014. LNCS, vol. 8451, pp. 293–300. Springer, Cham (2014). https://doi.org/10.1007/978-3-319-07046-9_21
10. Meseguer, P., Rossi, F., Schiex, T.: Soft constraints. In: Rossi, F., van Beek, P., Walsh, T. (eds.) Handbook of Constraint Programming, chap. 9, pp. 281–328. Elsevier (2006)
11. Musliu, N., Schutt, A., Stuckey, P.J.: Solver independent rotating workforce scheduling. In: van Hoeve, W.-J. (ed.) CPAIOR 2018. LNCS, vol. 10848, pp. 429–445. Springer, Cham (2018). https://doi.org/10.1007/978-3-319-93031-2_31
12. Nethercote, N., Stuckey, P.J., Becket, R., Brand, S., Duck, G.J., Tack, G.: MiniZinc: towards a standard CP modelling language. In: Bessière, C. (ed.) CP 2007. LNCS, vol. 4741, pp. 529–543. Springer, Heidelberg (2007). https://doi.org/10.1007/978-3-540-74970-7_38
13. Perron, L., Shaw, P., Furnon, V.: Propagation guided large neighborhood search. In: Wallace, M. (ed.) CP 2004. LNCS, vol. 3258, pp. 468–481. Springer, Heidelberg (2004). https://doi.org/10.1007/978-3-540-30201-8_35
14. Prud'homme, C., Lorca, X., Jussien, N.: Explanation-based large neighborhood search. Constraints **19**(4), 339–379 (2014). https://doi.org/10.1007/s10601-014-9166-6
15. Schaus, P.: Variable objective large neighborhood search: a practical approach to solve over-constrained problems. In: Brodsky, A. (ed.) ICTAI 2013, pp. 971–978. IEEE Computer Society (2013)

16. Schiendorfer, A.: Soft constraints in MiniBrass: foundations and applications. Ph.D. thesis, Universität Augsburg, Germany (2018). https://doi.org/10.13140/RG.2.2.10745.72802
17. Schulte, C., Stuckey, P.J.: Efficient constraint propagation engines. ACM Trans. Program. Lang. Syst. **31**(1), 1–43 (2008)
18. Schulte, C., Tack, G.: Weakly monotonic propagators. In: Gent, I.P. (ed.) CP 2009. LNCS, vol. 5732, pp. 723–730. Springer, Heidelberg (2009). https://doi.org/10.1007/978-3-642-04244-7_56
19. Shaw, P.: Using constraint programming and local search methods to solve vehicle routing problems. In: Maher, M., Puget, J.-F. (eds.) CP 1998. LNCS, vol. 1520, pp. 417–431. Springer, Heidelberg (1998). https://doi.org/10.1007/3-540-49481-2_30
20. Van Hentenryck, P., Michel, L.: Constraint-Based Local Search. The MIT Press, Cambridge (2005)

# Quantum-Accelerated Global Constraint Filtering

Kyle E. C. Booth[1,2(✉)], Bryan O'Gorman[1,3], Jeffrey Marshall[1,2], Stuart Hadfield[1,2], and Eleanor Rieffel[1]

[1] Quantum AI Laboratory (QuAIL), NASA Ames Research Center, Moffett Field, CA 94035, USA
{kyle.boothjeffrey.s.marshall,stuart.hadfield,eleanor.rieffel}@nasa.gov
[2] USRA Research Institute for Advanced Computer Science (RIACS), Mountain View, CA 94043, USA
[3] University of California, Berkeley, CA 94720, USA
bogorman@berkeley.edu

**Abstract.** Motivated by recent advances in quantum algorithms and gate-model quantum computation, we introduce quantum-accelerated filtering algorithms for global constraints in constraint programming. We adapt recent work in quantum algorithms for graph problems and identify quantum subroutines that accelerate the main domain consistency algorithms for the `alldifferent` constraint and the global cardinality constraint (`gcc`). The subroutines are based on quantum algorithms for finding maximum matchings and strongly connected components in graphs, and provide speedups over the best classical algorithms. We detail both complete and bounded-probability frameworks for quantum-accelerated global constraint filtering algorithms within backtracking search.

**Keywords:** Quantum computing · Constraint programming · Backtracking search · Logical inference · Global constraints

## 1 Introduction

Quantum computers are designed to leverage quantum-mechanical phenomena to outperform classical computers for certain tasks. While early quantum devices, such as quantum annealers, were limited to the implementation of specialized algorithms, the past decade has seen the advent of general-purpose gate-model quantum computers, capable of implementing any algorithm that can be expressed as a series of quantum logic gates. In this model, quantum gates are applied to qubits, the basic memory unit of quantum processors, reminiscent of classical computation where logic gates are applied to bits. While the current gate-model processors remain small, in the noisy intermediate-scale quantum (NISQ) regime, they have already enabled exciting developments, such as

---


H. Simonis (Ed.): CP 2020, LNCS 12333, pp. 72–89, 2020.
https://doi.org/10.1007/978-3-030-58475-7_5

the availability of quantum computers accessible in the cloud [10,13], and the achievement of quantum supremacy in the context of sampling random quantum circuits [3]. Additionally, a well-developed theory of quantum error correction and quantum fault tolerance provides the underpinnings of extensive engineering efforts to realize fault-tolerant, scalable quantum computers [33].

Concurrent work in advancing quantum algorithms is critical to extend the known applications of quantum computing independent of processor design. Recent efforts indicate speedups for a number of problems in graph theory [14], mathematical programming [35], constraint satisfaction [9], and search [25,26].

Building on these results, we investigate quantum subroutines to accelerate filtering algorithms for global constraints. We argue that the search paradigm in constraint programming (CP) represents an attractive framework for deployment of quantum subroutines that accelerate inference algorithms at each node of a search tree. Encapsulation of combinatorial substructure in global constraints provides an elegant mechanism for carving off portions of complex problems into subproblems that can be solved by a quantum co-processor. These smaller subproblems require fewer qubits, making them promising candidates for early fault-tolerant quantum chips. While CP has been used recently to efficiently compile quantum circuits [6], the use of quantum algorithms to accelerate global constraint filtering in CP, has, to the authors' knowledge, not been investigated.

The primary contributions of this paper are as follows:

i. A quantum-accelerated $O(|X|\sqrt{(|X|+|V|)|V|}\log^2|V|)$-time bounded-error algorithm for domain consistency of the `alldifferent` constraint, where $|X|$ is the number of variables and $|V|$ is the number of unique domain values. Our approach follows the main classical algorithm, accelerating the basic subroutines performed at each iteration with quantum analogs. The complexity is dominated by that for finding maximum matchings in bipartite graphs. The best deterministic and randomized classical algorithms known take $O(|X|\sqrt{|X|}|V|)$ and $O(|X|^{\omega-1}|V|)$ time, respectively, where $\omega$ corresponds to the asymptotic cost of classical matrix multiplication; the best upper bound known on $\omega$ is 2.373.[1] Our approach improves over these time-complexity upper bounds by factors on the order of $\sqrt{|X||V|/(|X|+|V|)}$ and $\sqrt{|X|^{2\omega-4}|V|/(|X|+|V|)}$, respectively, and up to polylogarithmic terms.

ii. A quantum-accelerated $O(|X|\sqrt{(|X|+|V|)|V|}\log^2|V|)$-time bounded-error algorithm for domain consistency of the global cardinality constraint (`gcc`), providing speedups over the best classical approach known.

iii. We discuss complete and bounded-probability frameworks for using quantum-accelerated global constraint filtering in backtracking tree search.

## 2 Background

Quantum computers work in a fundamentally different way than classical computers: they process quantum information, a generalization of classical informa-

[1] We note that the instance size at which the asymptotic scaling becomes relevant is so large that, in practice, matrix multiplication takes cubic time.

tion, and can use uniquely quantum operations to carry out computations. At an abstract level there are similarities between the two paradigms: quantum computers have quantum registers that hold quantum states, and a quantum computation acts on these states by quantum gates. Some of these gates are directly analogous to classical gates such as NOT and CNOT, but others are uniquely quantum. The fundamental unit of information on which quantum computers act is a qubit, a generalization of the classical bit. At the end of a quantum computation, during which quantum gates are applied to qubits in the quantum registers, quantum measurements are made to the qubits to extract classical information, such that a string of bits is returned. The interested reader is referred to a number of sources for a more thorough review of the subject [33,38].

We adopt Dirac's "ket" notation [11], universally used in quantum mechanics and quantum computing, in which a column vector is represented by a "ket" such as $|x\rangle$, where $x$ is a label, equivalent to $\overrightarrow{x}$. Generally, there will be some preferred basis of the vector space referred to as the "computational basis". For example, a 3-dimensional vector space can be spanned by $|0\rangle$, $|1\rangle$, and $|2\rangle$, corresponding to the unit vectors $(1,0,0)^T$, $(0,1,0)^T$, and $(0,0,1)^T$, respectively. Finally, the notation $|x\rangle\,|y\rangle$ is shorthand for the tensor product $|x\rangle \otimes |y\rangle$.

**Definition 1 (Qubit).** *A qubit is a quantum system whose state is represented by a two-dimensional complex vector. The computational basis consists of two orthonormal vectors denoted* $|0\rangle$ *and* $|1\rangle$. *Unlike a classical bit, which must be either 0 or 1, a qubit can in general be in a superposition of these states* $a\,|0\rangle + b\,|1\rangle$, *subject to the normalization condition* $|a|^2 + |b|^2 = 1$, *where* $a$ *and* $b$ *are complex numbers referred to as "amplitudes".*

**Definition 2 (Quantum register).** *An* $n$*-qubit quantum register holds the state of* $n$ *qubits, represented as a vector in a* $2^n$*-dimensional complex vector space. The computational basis* $\{|\mathbf{x}\rangle : \mathbf{x} \in \{0,1\}^n\}$ *consists of* $2^n$ *orthonormal vectors, labeled by the* $2^n$ *classical* $n$*-bit strings* $\{0,1\}^n$ *or the corresponding integers* $\{0, 1, \ldots, 2^n - 1\}$. *Any* $n$*-qubit state can be written as a superposition (linear combination) of the form* $|\phi\rangle = \sum_{\mathbf{x}\in\{0,1\}^n} a_{\mathbf{x}}\,|\mathbf{x}\rangle$, *subject to* $\sum_{\mathbf{x}} |a_{\mathbf{x}}|^2 = 1$.

**Definition 3 (Quantum measurement).** *A measurement of a quantum register in the computational basis returns classical information. Specifically, it will return each bit string with probability proportional to the amplitude squared. For example, measuring a qubit in state* $a|0\rangle + b|1\rangle$ *returns state* $|0\rangle$ *with probability* $|a|^2$ *and state* $|1\rangle$ *with probability* $|b|^2$. *The qubits of a multi-qubit register may be measured independently, and in general the outcomes will be correlated.*

Quantum information processing is an interesting mix of quantum states, which can take on a continuum of values, and quantum measurement, which enforces discrete outcomes. The design of quantum algorithms, the topic of this paper, involves transforming quantum systems. These transformations can be represented as unitary matrices, and each of these can be decomposed into a sequence of one- and two-qubit transformations called quantum gates.

**Definition 4 (Quantum gate and quantum circuit).** *A quantum state transformation (i.e., a unitary operator) acting on a quantum register is called a quantum gate. A quantum circuit is a sequence of quantum gates.*

In gate-model quantum computation, the resources by which algorithms are compared include the number of qubits used (the *space complexity*) and the number of primitive gates used (the *gate complexity* or *time complexity*).

### 2.1 Grover's Algorithm and Quantum Search

In this section, we introduce Grover's algorithm for unstructured search [18], a well-known quantum algorithm, and essential for the speedups in this work.

**Definition 5 (Unstructured search problem).** *Given an $N$ element unstructured list and blackbox access to predicate $P : \{0, \dots, N-1\} \to \{0, 1\}$, find a* solution*, $x \in \{0, 1, \dots, N-1\}$, such that $P(x) = 1$, with the fewest queries to $P$.*

For a predicate $P$ with $m$ solutions, Grover's algorithm finds a solution to the unstructured search problem with constant success probability using $O(\sqrt{N/m})$ queries to $P$, even when $m$ is unknown. Classical (including randomized) algorithms require $\Omega(N/m)$ queries, and so Grover's algorithm provides a quadratic speedup in the oracle model. In the quantum case, the predicate is instantiated as an operator $U_P : |x\rangle |0\rangle \mapsto |x\rangle |P(x)\rangle$ that computes $P(x)$ in the $|0\rangle$ register. This speedup is due to quantum computing's ability to evaluate $P$ on a superposition of states according to $U_P \sum_x a_x |x\rangle |0\rangle = \sum_x a_x |x\rangle |P(x)\rangle$. By linearity, the operator $U_P$ computes $P(x)$ over all $x$ in superposition. Grover's algorithm is optimal in the sense that any quantum algorithm for the unstructured search problem must have query complexity $\Omega(\sqrt{N/m})$. With a modification, all $m$ solutions can be retrieved using $O(\sqrt{mN})$ queries.

For the unstructured search problem, it is evident that Grover's algorithm quadratically improves over the best possible classical algorithm. For more complicated problems, however, this is not always the case, as unconditional complexity lower bounds are difficult to obtain. As such, throughout the paper, we claim a speedup for a quantum algorithm when an improvement in time complexity, for a given problem, is shown over the best classical algorithm known.

**A Sketch of Grover's Algorithm.** Leveraging the ability of quantum computing to compute on quantum superposition states, Grover's algorithm exploits quantum interference to concentrate amplitude on the target states. We use Fig. 1 to provide a pictorial representation of the algorithm [33]. In this example it suffices to consider real-valued amplitudes only.

*Initialization.* The algorithm starts with a uniform superposition state, $|\psi\rangle = \frac{1}{\sqrt{N}} \sum_x |x\rangle$, of all $N$ values of the search space. In Fig. 1a, this uniform superposition corresponds to a uniform amplitude histogram (top), where each bar in the histogram represents the amplitude associated with an element in the search,

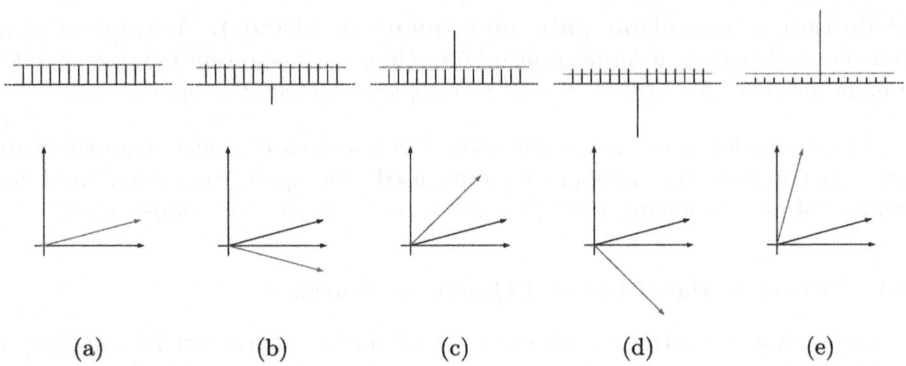


**Fig. 1.** Grover's algorithm for unstructured search [18], illustrating initialization and two rotations. **Top:** concentration of amplitude on search states (vertical lines) with each iteration, where the dotted horizontal line represents the average amplitude. **Bottom:** the changing superposition state (red vector) as reflection operations are applied, moving towards the target $y$-axis [33]. (Color figure online)

and is represented by a red vector (bottom) at some angle away from the target state ($y$-axis). In these diagrams, the red vector represents some superposition state of the $N$ values.

*Amplitude Amplification.* This process seeks to amplify the amplitude associated with the target state, while diminishing the amplitudes associated with all the other states. It is accomplished with $O(\sqrt{N/m})$ applications of two operations which, using a geometrical interpretation, can be seen as reflections of the superposition state vector about axes. The first operation reflects the superposition state about the x-axis, representing a state orthogonal to the target states, through the application of the oracle $U_P$. As in Fig. 1b, this reflection yields a new superposition state (red vector), a negative amplitude for the target states, and a lowering of the average amplitude. The second operation reflects the superposition state about the uniform superposition state, $|\psi\rangle$. As shown in Fig. 1c, this results in a positive, more concentrated amplitude associated with the target states and a superposition state (red vector) closer to the $y$-axis.

Each iteration of Grover's algorithm rotates the initial state, $|\psi\rangle$, towards the target states ($y$-axis). A straightforward calculation shows that quantum interference effects facilitate transfer of probability amplitude from non-target states to targets states (Fig. 1d and 1e provide one more iteration.). At the end of the algorithm, a measurement yields one of the solutions with probability proportional to the amplitude squared. We note that Grover's algorithm can be expressed as a quantum circuit consisting of logical gates and queries to $U_P$.

A few comments are in order. The rotation perspective correctly suggests that choosing the number of iterations is critical; otherwise over-rotation may occur, resulting in less amplitude in the target states. The number of iterations depends only on $m$, the number of target states, not on which states these are. Specifically, the optimal number of rotations is $\frac{\pi}{4}\sqrt{\frac{N}{m}}$ [7]. If $m$ is not known, it can be estimated using a quantum counting algorithm, or by running the algorithms a number of times, with an increasing number of iterations until discovery of a marked state. Both preserve the overall query complexity of $O(\sqrt{N/m})$ [7]. The algorithm is a bounded-probability algorithm, meaning that it can fail to find a solution element even if one exists, an important aspect discussed in Sect. 5. To ensure the failure probability of each Grover search is polynomially small in $N$, we repeat the algorithm $O(\log N)$ times, contributing only to the logarithmic terms noted in many of the complexity results.

### 2.2 Related Work

The use of quantum search to realize speedups for various problems has seen a surge of activity in recent years. Previous work provides lower and upper bounds for the bounded-error query complexity of various graph problems, including connectivity, minimum spanning tree, and single-source shortest path [5,14]. Related work, leveraged heavily in these papers, investigated the query complexity for various matching problems [2,12]. More recently, quantum search has been applied to problems within mathematical programming, such as semidefinite programming [8,35] and the acceleration of the simplex method [28]. The latter, in a similar fashion to this work, uses quantum search to accelerate the subroutines of the simplex method, such as variable pricing. There also exist recent efforts to use algorithms based on quantum search to speed up tree search methods, including backtracking search [25], and branch-and-bound [26].

## 3 Quantum Subroutines for `alldifferent`

A constraint satisfaction problem (CSP) consists of a set of decision variables $X = \{x_1, \ldots, x_n\}$, with domains $\mathcal{D} = \{D_1, \ldots, D_n\}$, and constraints $\mathcal{C} = \{C_1, \ldots, C_\ell\}$. The domain of a variable is the set of values the variable can be assigned. Each constraint $C \in \mathcal{C}$ acts on a subset of $X$. A solution is an assignment to the variables of values that satisfies the constraints.

**Definition 6 (`alldifferent` constraint).** *alldifferent*$(x_1, \ldots, x_k)$ *is a constraint that requires that all of the variables in its scope take on different values (i.e., in a solution to the constraint,* $x_i \neq x_j, \forall i \neq j \in \{1, \ldots, k\}$*).*

The `alldifferent` global constraint is widely used in CP, and arises naturally in many problems. The main domain consistency filtering algorithm for `alldifferent` was proposed by Régin (see Algorithm 1) and consists of two primary subroutines, FindMaximumMatching and RemoveEdges, leveraging existing graph algorithms [32]. Our approach uses a classical processor that follows Régin's high-level algorithm, accelerating each subroutine using quantum graph algorithms. While recent work has investigated practical optimizations for Régin's algorithm, these do not improve upon its worst-case complexity [16,39].

**Algorithm 1.** The `alldifferent` filtering algorithm of Régin [32]

**Result:** *False* if no solution, otherwise filtered domains $D^*$
Build $G = (X, V, D)$;
$M \leftarrow$ FINDMAXIMUMMATCHING$(G)$;
**if** $|M| < |X|$ **then**
  **return** *False*;
**end**
$D^* \leftarrow D \setminus$ REMOVEEDGES$(G, M)$;
**return** $D^*$;

The algorithm, as in Algorithm 1, begins by constructing a bipartite variable/value graph $G = (X, V, D)$, with vertices $X \cup V$ and edges $D$. Such a graph has $n = |X| + |V|$ vertices and $m = |D| = \sum_{i=1}^{|X|} |D_i|$ edges, where $m \leq |X||V|$. FINDMAXIMUMMATCHING finds a matching of maximum size in $G$ and REMOVEEDGES finds edges in $G$ that can never participate in a maximum matching. If FINDMAXIMUMMATCHING returns a matching $M$ whose number of edges $|M| < |X|$, then the constraint cannot be satisfied and the algorithm terminates. If a matching exists with $|M| = |X|$, the algorithm prunes domains based on the output of REMOVEEDGES.

The FINDMAXIMUMMATCHING subroutine bears the brunt of the computational complexity [36]. The best deterministic classical algorithms known for finding maximum matchings run in $O(m\sqrt{n})$ time; the algorithm of Hopcroft and Karp (HK) is for bipartite graphs [19], while the algorithm of Micali and Vazirani (MV) applies to general graphs [24,37]. Alt et al. proposed an $O(n^{3/2}\sqrt{m/\log n})$ algorithm [1], however, it only improves upon the aforementioned algorithms for dense graphs. There is also a randomized $O(n^{\omega})$-time algorithm [20,27], where $\omega$ corresponds to the classical asymptotic cost of matrix multiplication; the best upper bound known on $\omega$ is approximately 2.373 [22].

In order to remove edges which participate in no maximum matching, and thus cannot satisfy the constraint, REMOVEEDGES finds strongly connected components (SCCs) in a directed transformation of $G$ using Tarjan's $O(n + m)$ algorithm [34]. While this subroutine evidently does not bear the computational brunt of `alldifferent` filtering, its acceleration can still be valuable in practice.

In this section, we introduce the quantum query model and definitions needed to describe our approaches. We then detail quantum algorithms for the FINDMAXIMUMMATCHING and REMOVEEDGES subroutines to accelerate the filtering of the `alldifferent` constraint. For the former, we detail a quantum algorithm proposed by Dörn for finding maximum matchings in general graphs [12]. For the latter, we combine a number of quantum graph algorithms, including an extension of work that identified strong connectivity in graphs [14].

### 3.1 Input Models, Accounting, and Definitions

Many quantum algorithms are posed in the "oracle model", in which *black box* access is given to the quantum operation $U_f : |w\rangle |0\rangle \mapsto |w\rangle |f(w)\rangle$ in unit time, where $f : W \to Y$ is a classical function encoding the input. ($U_f$ generalizes the operator $U_P$ from Sect. 2.1 for a non-Boolean function $f$.) The *query complexity* of such algorithms is the number of calls to the oracle $U_f$. The time complexity is always at least the query complexity. Because the calls to the oracle are often the most significant part of the computation, the two are often the same (up to polylogarithmic factors), but this isn't always the case.

In this work, we aim to provide a practical speedup, and so must account for the cost of implementing any quantum queries used by our algorithms. We address this by using quantum random access memory (QRAM) [17], a data structure with which oracle queries and updates (including initialization of the QRAM) to the quantum data structure can be made in time polylogarithmic in the size of the database. There are proposals for special-purpose hardware QRAM, with small-scale experiments demonstrating a proof of principle [21], as well as several ways of implementing QRAM directly in the standard circuit model [23]. These circuit implementations assume the same availability of fault-tolerant, gate-model quantum computers as the algorithms that call the QRAM.

The main template employed here takes an algorithm posed in the oracle model, and uses QRAM to implement the queries with logarithmic overhead, taking care to account for the cost of QRAM initialization. Henceforth, by "time complexity", we mean number of logical gates and queries or updates to QRAM. With a circuit implementation of QRAM, this time complexity upper bounds the overall circuit depth, assuming that the QRAM circuits are parallelized. Parallelization of the logical (non-QRAM) parts of the circuit, which we do not attempt here, can only improve this overall depth.

For our quantum algorithms, we store and access the graph in the "list" (or "array") model. For each vertex $v \in \{X \cup V\}$, we query the oracle:

$$U_{N_v} : |i\rangle |j\rangle \mapsto |i\rangle |j + N_v(i) \bmod d_v\rangle \tag{1}$$

where $d_v$ is the degree of $v$, $i \in \{1, \dots, d_v\}$, $j \in \{1, \dots, n\}$, and $N_v(i)$ is the $i$th neighbor of vertex $v$. Using QRAM, the quantum data structure in Eq. 1 can be initialized in time $O(d_v \log d_v)$ and quantum queries made in time $O(\log d_v)$. When the graph is directed, $U_{N_v}$ queries only the outgoing neighbors. There are other ways of formulating quantum access to a graph, but in this paper we use only the list model.

**Definitions.** Given a graph $G$ and a matching $M$, a vertex is *exposed* if no edge in the matching $M$ is incident to it. A path (resp., cycle), consisting of a sequence of vertices, is *alternating* if its edges are alternately in the matching $M$ and not in $M$. The length of the path (resp., cycle) is the number of edges in the path (resp., cycle). A path is *augmenting* if it is alternating and the first and last vertices in the path are exposed.

### 3.2 Subroutine: FindMaximumMatching

The essence for a quantum filtering algorithm is simple: use a quantum algorithm to solve the maximum matching problem. Recent work proposed a series of algorithms for finding maximum matchings in terms of calls to a quantum oracle [2,12]; however, to the authors' knowledge, this work has never been linked to accelerating global constraint filtering in CP. In the list model, an initially proposed algorithm is capable of finding maximum matchings in $O(n\sqrt{m}+n\log^2 n)$ time [2], while the second, improved algorithm runs in $O(n\sqrt{m}\log^2 n)$ time [12]. The latter improves over both existing deterministic and randomized algorithms for the majority of parameter values, and follows the classical MV algorithm for finding maximum matchings in general graphs [24], but accelerates its primary subroutines with quantum search. We give an overview of this algorithm by tracing the classical algorithm in the context of a simple example, and comment on the processes that are speed-up with quantum algorithms.

*Example 1.* Consider a CSP with variables $X = \{x_1, x_2\}$; domains $D_1 = \{v_1, v_2\}, D_2 = \{v_1\}$; and the constraint `alldifferent`$(x_1, x_2)$. A trace of the classical MV algorithm for finding a maximum matching is shown in Fig. 2.

We follow the exposition of the MV algorithm of Peterson and Loui [29]. The input to the algorithm is a graph, such as the bipartite variable/value graph $G$ shown in Fig. 2a. Initialization starts with an empty matching, $M = \emptyset$. In phases, the algorithm looks for a set of minimum-length, vertex-disjoint augmenting paths to iteratively extend the current matching until a maximum matching is found. Each phase is performed in $O(m)$ time, and the number of phases is bounded by $O(\sqrt{n})$ [19], resulting in an $O(m\sqrt{n})$-time classical algorithm [24,37]. Each phase of the classical algorithm begins with a matching and conducts three subroutines: Search, BlossAug, and FindPath [24]. As a property of their structure, bipartite graphs do not contain blossoms, precluding the need to cover the algorithmic details associated with dealing with them and, as a consequence, the need for the FindPath subroutine [24]. As such, our FindMaximumMatching subroutine needs only the Search and BlossAug lower-level subroutines and their quantum implementations.

**Search.** This subroutine performs a simultaneous breadth-first search (BFS) from each exposed vertex (E.g., in Fig. 2b all vertices are exposed since $M = \emptyset$.). The subroutine labels each vertex with a value pair, (*EvenLevel*, *OddLevel*), where *EvenLevel* (resp., *OddLevel*) is the length of the minimum even- (resp., odd-) length alternating path from an exposed vertex to the current vertex, if any, and $\infty$ otherwise. Exposed vertices have *EvenLevel* $= 0$ (and infinite *OddLevel*), resulting in the labeling at the top of Fig. 2b. The subroutine then searches for *bridges*, edges whose vertices both have finite *EvenLevel* or both have finite *OddLevel* values. In Phase 1 of the example (Fig. 2b), all edges are bridges. In Phase 2 of the example (Fig. 2c), only edge $(x_1, v_1)$ is a bridge. The Search subroutine passes each discovered bridge to the BlossAug subroutine.

*Quantum Algorithm.* The quantum acceleration of the Search subroutine, instead of a classical BFS, uses a quantum BFS search [12] from each exposed

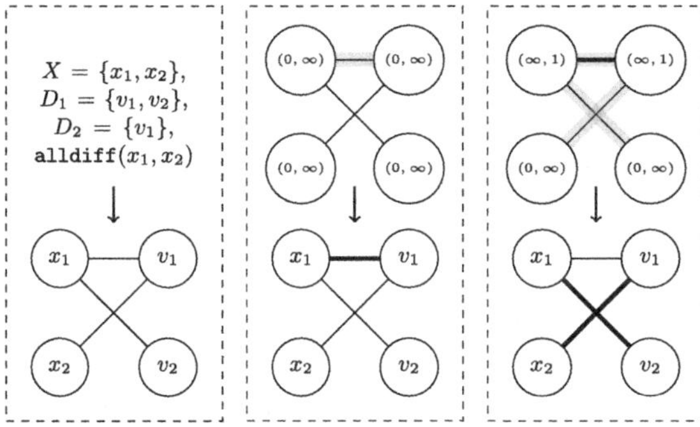


(a) Initialization. (b) Phase 1. (c) Phase 2.

**Fig. 2.** Finding maximum matchings in bipartite graphs [24]. Bold, black arcs: edges in matching. Shaded, gray arcs: minimum-length augmenting paths.

vertex, using query access to a list-model representation of the graph, and a stack bookkeeping the discovered vertices, both implemented with QRAM. From a given vertex $v$, one needs to obtain all $n_v$ neighbors not yet discovered. If the degree of $v$ is $d_v$, the quantum time complexity to find all neighbors is $O(\sqrt{n_v d_v} \log d_v)$, as discussed in Sect. 2.1. Because $O(n)$ searches are run, each is repeated $O(\log n)$ times to get the aggregate error down to constant. Noting that each vertex is discovered once only (i.e., $\sum_v n_v = O(n)$) and that $\sum_v d_v = O(m)$, the full quantum time complexity is $O(\sum_v \sqrt{d_v n_v} \log d_v \log n) = O(\sqrt{nm} \log^2 n)$, by the Cauchy-Schwarz inequality.

**BlossAug.** This subroutine takes a bridge and performs a simultaneous double depth-first search (DDFS) from each vertex in the bridge until it finds two different exposed vertices. The subroutine ensures that a given vertex can only take part in at most one of the two DFSs. In Phase 1 of the example (Fig. 2b), BlossAug$(x_1, v_1)$ would find exposed vertices $x_1$ and $v_1$. In Phase 2 of the example (Fig. 2c), BlossAug$(x_1, v_1)$ would find vertex $v_2$ in the DFS from $x_1$, and $x_2$ in the DFS from $v_1$. Since we are concerned only with bipartite graphs, which do not contain blossoms [24], the results of the DFSs are then concatenated to generate an augmenting path (e.g., augmenting path $(x_1, v_1)$ for Phase 1 of the example, and augmenting path $(v_2, x_1, v_1, x_2)$ for Phase 2).

*Quantum Algorithm.* To accelerate BlossAug, Dörn use quantum search to speedup the DDFS [12]. Each DFS has a time complexity of $O(\sqrt{nm} \log^2 n)$. The derivation is similar to that for the BFS complexity of Search above.

For each augmenting path found by BlossAug, the algorithm extends the matching along the augmenting path (i.e., an edge in $M$ is removed from $M$ and an edge not in $M$ is added to it) and marks the vertices in the path as 'visited' in the current phase. The marking of vertices ensures the augmenting paths found during a phase are disjoint (i.e., do not share a vertex). Once the set of bridges is empty, the phase is over and the next phase begins. Figure 2 provides a trace of the algorithm, starting with matching $M = \emptyset$, extending this

**Algorithm 2.** RemoveEdges($G$, $M$)

**Data:** Bipartite graph $G = (X, V, D)$ and matching $M$
**Result:** Set of edges to prune
$G_M \leftarrow$ DirectGraph$(G, M)$;
$D_{\text{used}} \leftarrow$ FindSimplePaths$(G_M)$ ; /* Set of 'used' edges */
$\mathcal{S} \leftarrow$ FindSCC$(G_M)$; /* SCC for each vertex */
**return** IdentifyEdges$(G, M, D_{\text{used}}, \mathcal{S})$;

to $M = \{(x_1, v_1)\}$ in the first phase via bridge $(x_1, v_1)$, and then to the maximum matching $M = \{(x_1, v_2), (v_1, x_2)\}$ in the second phase via bridge $(x_1, v_1)$.

In the interest of space, the details pertaining to the vertex deletion subroutine in the algorithm of Dörn are not included here; however, the time complexity of the subroutine is the same as for Search and BlossAug. The time complexity of each phase is $O(\sqrt{nm}\log^2 n)$, because that is the aggregate time complexity of each subroutine (e.g., Search and BlossAug). Since the number of phases required is at most $O(\sqrt{|X|}) = O(\sqrt{n})$ [19], the complexity of the overall algorithm is $O(n\sqrt{m}\log^2 n)$ or, in terms of our graph properties, $O(|X|\sqrt{(|X|+|V|)|V|}\log^2 |V|)$, for a constant $\Omega(1)$ success probability [12]. The classical $O(m\sqrt{n})$-time algorithm of MV, in terms of our graph properties, has time complexity of $O(|X|\sqrt{|X||V|})$, indicating an improvement by a factor of $\sqrt{|X||V|/(|X|+|V|)}$, up to polylogarithmic terms.

### 3.3 Subroutine: RemoveEdges

If a maximum matching is found such that $|M| = |X|$, Algorithm 1 proceeds to initiate the RemoveEdges subroutine with graph $G$ and matching $M$ as input (Fig. 3a). The steps of the subroutine are detailed in Algorithm 2.

From Berge [4], an edge belongs to some maximum matching if and only if, for an arbitrary given maximum matching, it belongs to either an even-length alternating path which begins at an exposed vertex, or to an even-length alternating cycle. If an edge does not satisfy Berge's property, it should be pruned. Instead of applying Berge's conditions directly, the problem has been previously translated into a search for edges in directed simple paths and strongly connected components (SCC) in a directed transformation of the graph [32,36].

The input to the RemoveEdges subroutine is the variable/value graph $G$ and a matching $M$ (found with FindMaximumMatching). In DirectGraph the edges in $G$ are directed depending upon whether or not they are in matching $M$, producing directed graph $G_M$. Edges in the matching are directed from variables to values ('right-facing') and the remaining edges from values to variables ('left-facing'). For the running example, this is illustrated in Fig. 3b. The output of the subroutine is the set of edges to prune. The RemoveEdges subroutine has classical time complexity $O(m)$ stemming from three lower-level subroutines: FindSimplePaths, FindSCC, and IdentifyEdges. We provide an overview of these subroutines and comment on their quantum analogs.

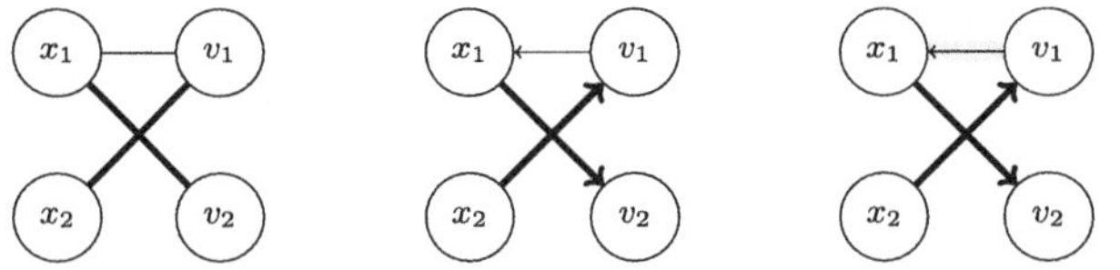


(a) Input. (b) Directed graph. (c) Edges to remove.

**Fig. 3.** Removing edges. Bold, black arcs: edges in matching. Shaded, red arcs: edges removed. (Color figure online)

**FINDSIMPLEPATHS.** To satisfy the first condition of Berge's property, it suffices to find all edges present in at least one directed simple path starting at an exposed vertex. This is achieved by a BFS starting collectively at the exposed vertices, marking each edge considered as 'used'. This will output a set of edges $D_{\text{used}}$ with the label 'used'. The number of edges processed during this subroutine is $O(|D_{\text{used}}|)$, therefore giving overall complexity $O(|D_{\text{used}}|)$, which is on the order of $m$ in the worst case. When there are only a few edges to mark as used, the run time of this step can of course be significantly less than $O(m)$. In Fig. 3b, there are in fact no exposed vertices in $G_M$, and no BFS is even initiated.

*Quantum Algorithm.* Since the classical algorithm strictly takes time linear in the output size, $O(|D_{\text{used}}|)$, there is no possible asymptotic speedup (quantum or otherwise). This step is therefore implemented with a classical BFS.

**FINDSCC.** To satisfy the second condition of Berge's property, we compute the SCC in $G_M$. A directed graph is *strongly connected* if there is a path between all pairs of vertices in the graph. A *strongly connected component* (SCC) of a directed graph is a maximal strongly connected subgraph. Classically, the SCCs can be computed in time $O(n + m)$ with Tarjan's algorithm [34].

Tarjan's algorithm is a modified DFS. When a node is discovered in the DFS, it is put on a stack, and obtains index and low-link values, which are initially equal to infinity and updated during the search. On the backtrack, low-link values can be updated: of all forward neighbors on the stack, update the current low-link value to the minimum index value. After this completes, the low-link value of a vertex is the SCC to which it belongs.

This can be achieved by looping over all neighbors at a vertex $v$ where there are two possible outcomes: i) if the neighbor $w$ has not been discovered, advance the DFS to $w$, ii) if $w$ is on the stack (i.e., already discovered), update the current low-link value of $v$ to the minimum of $v$'s low-link value, and $w$'s index value. Upon looping over all neighbors, if $v$'s low-link is equal to its index value, then this value is the SCC to which it belongs. Moreover, all nodes with this low-link value make up the SCC, and these can be removed from the stack.

FINDSCC labels the SCC to which each vertex belongs. The directed graph $G_M$ illustrated in Fig. 3b contains four trivial SCCs.

*Quantum Algorithm.* Existing work has produced quantum algorithms for determining if a graph is strongly connected [14], noting that an adaptation of the approach can yield the identification of SCCs. Here, we describe such an adaptation based directly on Tarjan's algorithm, observing that it conducts essentially two searches at each step. In particular, from a vertex one needs to: i) find an undiscovered neighbor, and ii) find the minimum index value over the neighbors. While backtracking through the DFS, one can perform a search over the index values of all forward neighbours. The complexity of implementing quantum searches is the same as quantum DFS, plus the cost to perform a quantum minimum finding at each node which is $O(\sqrt{d_v} \log d_v)$ [15]. Overall the quantum time complexity is therefore $O(\sqrt{nm} \log^2 n)$, with each search repeated $O(\log n)$ times. In addition to the graph and stack QRAM data structures to perform the DFS, one also needs to maintain a QRAM data structure for the index values of each vertex, which is used in the quantum minimum finding.

**IDENTIFYEDGES.** The output of FINDSIMPLEPATHS and FINDSCC is, respectively, a set of edges $D_{\text{used}} \subseteq D$ marked as 'used', and the SCC to which each vertex belongs. IDENTIFYEDGES identifies the set of edges to be removed, satisfying three conditions: i) the edge is not in $D_{\text{used}}$, ii) the edge is not between vertices in the same SCC, and iii) the edge is not in the current matching $M$. The set $R$ of edges to remove is easy to construct by iterating over all edges and checking condition i-iii), giving a complexity of $O(m)$. In the context of our example, since $D_{\text{used}} = \emptyset$, and all SCCs are trivial, the subroutine returns $G \backslash M$, namely the edge $(x_1, v_1)$ as illustrated in Fig. 3c.

*Quantum Algorithm.* A quantum search can be used to find the edges $R$ that need to be removed. From each variable vertex $x \in X$, a quantum search over the $d_x$ incident edges (of the original, undirected graph) can determine which need to be removed, subject to the three criteria as in the classical version of IDENTIFYEDGES. If there are $r_x$ edges to be removed, where $\sum_x r_x = |R|$, the quantum time complexity is $O(\sqrt{r_x d_x} \log d_x)$ for each $x$, resulting in a total of $O(\log n \sum_x \sqrt{r_x d_x} \log d_x) = O(\sqrt{m|R|} \log^2 n)$, with each search repeated $O(\log n)$ times. To perform such a quantum search, QRAM access to the graph is required, as is QRAM access to the classical data: $M, D_{\text{used}}$, and the SCCs. Each of these can be set up prior to and maintained during the execution of the above subroutines without changing the overall complexity of REMOVEEDGES.

Up to logarithmic factors and for constant error probability, the time complexity of quantum REMOVEEDGES is $O(|D_{\text{used}}| + \sqrt{mn} + \sqrt{m|R|})$, reflecting the three lower-level subroutines. In the worst case this is $O(m)$, occurring when $|D_{\text{used}}| = O(m)$ and/or $|R| = O(m)$, and equivalent to the classical runtime. In cases where $|R| = O(n)$ (or lower), and $|D_{\text{used}}| = O(\sqrt{mn})$ (or lower), the full complexity is $O(\sqrt{mn})$. This upper bound on the quantum runtime improves over the $O(m)$ classical runtime by a factor of up to $\sqrt{m/n}$, or $\sqrt{|X||V|/(|X| + |V|)}$.

## 4 Extensions to the Global Cardinality Constraint

The global cardinality constraint (`gcc`) is an extension of the `alldifferent` constraint, commonly used in scheduling, rostering, and timetabling problems [31].

**Definition 7 (gcc constraint).** *Given a set of variables* $X = \{x_1, \ldots, x_n\}$*, a set of values* $V = \{v_1, \ldots, v_m\}$*, and a set of cardinality bounds* $\Delta = \{\delta_1, \ldots, \delta_m\}$*, where each* $\delta_i \in \Delta$ *is defined by* $[\ell_i, u_i]$*, the constraint* $\texttt{gcc}(X, V, \Delta)$ *requires that value* $v_i$ *take place in the solution between* $\ell_i$ *and* $u_i$ *times, inclusively.*

State-of-the-art classical `gcc` filtering employs a $O(|X|\sqrt{|X|}|V|)$-time algorithm for achieving domain consistency [31], leveraging previous work for `alldifferent` [32]. The first stage of the algorithm enforces the domain consistency of `gcc` when all cardinality intervals are fixed to $u_i$, while the second stage enforces domain consistency when the cardinality is fixed to $\ell_i$, following a previous result that this is sufficient for the domain consistency of `gcc` [30].

For each stage, the classical filtering algorithm constructs a bipartite variable/value graph, $G = (X, V, D)$. Then, a capacity, $cap(x_i) = 1$, is associated with each variable node, and $cap(v_i) \geq 0$, with each value node. A matching in this graph is a subset of edges such that no more than the capacity of a given node is adjacent to that node. The algorithm then finds matchings of maximum cardinality for $cap(v_i) = u_i$ (first stage) and for $cap(v_i) = \ell_i$ (second stage). To do this, the algorithm of HK [19] can be used on an augmented graph, $G'$, where value nodes are duplicated $cap(v_i)$ times, and the capacity of each node in $G'$ is set to one; however, the complexity with this naive implementation will scale with the number of edges in the augmented graph. Instead, Quimper et al. describe an alteration of HK that runs on $G$ by ensuring that in the DFS each free vertex $v$ is visited at most $c(v) - d_M(v)$ times, where in each phase, $d_M(v)$ is the number of edges in the current matching $M$ adjacent to node $v$ [31]. This ensures the complexity is bounded by the number of edges in $G$, yielding an $O(|X|\sqrt{|X|}|V|)$ algorithm. Pruning the domains using Tarjan's algorithm [34], with the matchings at each stage, is sufficient to prune the domains for the domain consistency of `gcc` [30]. Given that Tarjan's algorithm is less computationally expensive than the maximum matching algorithm, the best-known overall classical complexity for achieving domain consistency for `gcc` is $O(|X|\sqrt{|X|}|V|)$.

*Quantum Algorithm.* The filtering algorithm for `gcc` utilizes the same subroutines used in `alldifferent`, indicating that the detailed quantum subroutines can also be used to accelerate `gcc` filtering. For each of the two stages in the filtering algorithm, we can use the quantum-accelerated version of FindMaximumMatching. For this, the algorithm of Dörn [12] is modified, following the modification Quimper et al. made to HK, to ensure that, in each stage, free vertices can be visited $c(v) - d_M(v)$ times. This is done by associating an integer counter with each vertex, in QRAM. The quantum-accelerated RemoveEdges subroutine can then be applied to the maximum matchings found at both stages. The time complexity of the overall `gcc` filtering algorithm follows that of finding maximum matchings, which is $O(|X|\sqrt{(|X| + |V|)|V|}\log^2 |V|)$.

## 5 Integration in Backtracking Search

As noted, the quantum search algorithms we employ have some probability of failure. For the purposes of the discussion here, the quantum FindMaximum-

MATCHING subroutine is extended such that the output is a valid matching with $|M| = |X|$ (as verified with a classical check) or *False*; the quantum subroutine is said to fail if such a matching exists but is not found. If we let $\epsilon$ be the failure probability of a given quantum subroutine, the acceptable value of $\epsilon$ depends on how the quantum subroutine is used in the search.

*Exact Method.* When the quantum subroutine has perfect soundness (e.g., quantum FINDMAXIMUMMATCHING), then one approach is to require perfect completeness.[2] This can be achieved by running the classical subroutine whenever the quantum one does not return a satisfying item (e.g., a valid matching of sufficient size). Let $t(n) = \mathsf{poly(n)}$ be the runtime of the classical subroutine. By repeating the quantum subroutine $O(\log(n))$ times, $\epsilon$ can be brought to $o(1/t(n))$, so that when a satisfying item exists, the expected cost of running the classical subroutine because the quantum one fails to find it is $o(1)$; therefore, at all nodes with a satisfying item, we get a quantum speedup on average. If no such item exists, the quantum subroutine will not return one. At nodes without a satisfying item, we run the classical algorithm, yielding no speedup for those nodes.

*Bounded-Error and Heuristic Methods.* Alternatively, suppose we want the overall tree search to fail (i.e., not find a solution if one exists) with at most a constant probability $O(1)$ (which can be made arbitrarily small without changing the asymptotic runtimes). Let $T_{\mathsf{Q}}$ be the number of tree search nodes at which the quantum subroutine is run. It suffices then to repeat the quantum subroutine $O(\log T_{\mathsf{Q}})$ times to get $\epsilon = O(1/T_{\mathsf{Q}})$. However, if $T_{\mathsf{Q}}$ is exponential in $n$, this can overwhelm the quantum speedup. To preserve the speedup, we could restrict the tree search to calling the quantum subroutine only $T_{\mathsf{Q}} = \mathsf{poly(n)}$ times. In practice, tree search algorithms often only explore a polynomial number of nodes, either due to limited resources, or because that is sufficient for the problem instance at hand. In cases that the tree search explores more than a polynomial number of nodes, the quantum filtering can be disabled; in this case, quantum search benefits a large number of nodes early in the tree. The search algorithm can also be run in "heuristic mode", using the quantum subroutine at every node. In this case, the effect of subroutine failures on the overall tree search is strongly dependent on the tree search algorithm used and in general cannot be bounded.

## 6 Conclusions

We introduce quantum-accelerated filtering algorithms for global constraints, with subroutines for the `alldifferent` constraint and the global cardinality constraint (`gcc`). This work is intended to be a first step towards a larger effort

[2] For an algorithm intended to find an item with a certain property, we say that the algorithm has perfect completeness if it always finds such an item, if one exists, and the algorithm has perfect soundness if it never returns an item without the property.

of using quantum algorithms to accelerate constraint programming. In the long-term, quantum computing is a promising technology for approaching hard computational problems, and we demonstrate here that the constraint programming community is well-positioned to benefit from this progress.

**Acknowledgements.** K.B., J.M., and S.H. were supported by NASA Academic Mission Services (NAMS), contract number NNA16BD14C. K.B. was also supported by the NASA Advanced Exploration Systems (AES) program. B.O. was supported by a NASA Space Technology Research Fellowship. We thank the anonymous reviewers and Prof. J. Christopher Beck whose valuable feedback helped improve the final version of the manuscript.

## References

1. Alt, H., Blum, N., Mehlhorn, K., Paul, M.: Computing a maximum cardinality matching in a bipartite graph in time $O(n^{1.5}m\log n)$. Inf. Process. Lett. **37**(4), 237–240 (1991)
2. Ambainis, A., Špalek, R.: Quantum algorithms for matching and network flows. In: Durand, B., Thomas, W. (eds.) STACS 2006. LNCS, vol. 3884, pp. 172–183. Springer, Heidelberg (2006). https://doi.org/10.1007/11672142_13
3. Arute, F., et al.: Quantum supremacy using a programmable superconducting processor. Nature **574**(7779), 505–510 (2019)
4. Berge, C.: Graphs and Hypergraphs. North-Holland, Amsterdam (1973)
5. Berzina, A., Dubrovsky, A., Freivalds, R., Lace, L., Scegulnaja, O.: Quantum query complexity for some graph problems. In: Van Emde Boas, P., Pokorný, J., Bieliková, M., Štuller, J. (eds.) SOFSEM 2004. LNCS, vol. 2932, pp. 140–150. Springer, Heidelberg (2004). https://doi.org/10.1007/978-3-540-24618-3_11
6. Booth, K.E.C., Do, M., Beck, J.C., Rieffel, E., Venturelli, D., Frank, J.: Comparing and integrating constraint programming and temporal planning for quantum circuit compilation. In: Twenty-Eighth International Conference on Automated Planning and Scheduling (ICAPS) (2018)
7. Boyer, M., Brassard, G., Høyer, P., Tapp, A.: Tight bounds on quantum searching. Fortschritte der Physik Prog. Phys. **46**(4–5), 493–505 (1998)
8. Brandao, F.G., Svore, K.M.: Quantum speed-ups for solving semidefinite programs. In: 2017 IEEE 58th Annual Symposium on Foundations of Computer Science (FOCS), pp. 415–426. IEEE (2017)
9. Campbell, E., Khurana, A., Montanaro, A.: Applying quantum algorithms to constraint satisfaction problems. Quantum **3**, 167 (2019)
10. Devitt, S.J.: Performing quantum computing experiments in the cloud. Phys. Rev. A **94**(3), 032329 (2016)
11. Dirac, P.A.M.: A new notation for quantum mechanics. In: Mathematical Proceedings of the Cambridge Philosophical Society, vol. 35, pp. 416–418. Cambridge University Press (1939)
12. Dörn, S.: Quantum algorithms for matching problems. Theory Comput. Syst. **45**(3), 613–628 (2009)
13. Dumitrescu, E.F., et al.: Cloud quantum computing of an atomic nucleus. Phys. Rev. Lett. **120**(21), 210501 (2018)
14. Dürr, C., Heiligman, M., Høyer, P., Mhalla, M.: Quantum query complexity of some graph problems. SIAM J. Comput. **35**(6), 1310–1328 (2006)

15. Dürr, C., Høyer, P.: A quantum algorithm for finding the minimum. arXiv preprint arXiv:quant-ph/9607014 (1996)
16. Gent, I.P., Miguel, I., Nightingale, P.: Generalised arc consistency for the alldifferent constraint: an empirical survey. Artif. Intell. **172**(18), 1973–2000 (2008)
17. Giovannetti, V., Lloyd, S., Maccone, L.: Quantum random access memory. Phys. Rev. Lett. **100**(16), 160501 (2008)
18. Grover, L.K.: A fast quantum mechanical algorithm for database search. In: Proceedings of the Twenty-Eighth Annual ACM Symposium on Theory of Computing, pp. 212–219 (1996)
19. Hopcroft, J.E., Karp, R.M.: An $n^{5/2}$ algorithm for maximum matchings in bipartite graphs. SIAM J. Comput. **2**(4), 225–231 (1973)
20. Ibarra, O.H., Moran, S.: Deterministic and probabilistic algorithms for maximum bipartite matching via fast matrix multiplication. Inf. Process. Lett. **13**(1), 12–15 (1981)
21. Jiang, N., Pu, Y.F., Chang, W., Li, C., Zhang, S., Duan, L.M.: Experimental realization of 105-qubit random access quantum memory. npj Quantum Inf. **5**(1), 1–6 (2019)
22. Le Gall, F.: Powers of tensors and fast matrix multiplication. In: Proceedings of the 39th International Symposium on Symbolic and Algebraic Computation, pp. 296–303 (2014)
23. Matteo, O.D., Gheorghiu, V., Mosca, M.: Fault-tolerant resource estimation of quantum random-access memories. IEEE Trans. Quantum Eng. **1**, 1–13 (2020)
24. Micali, S., Vazirani, V.V.: An $O(\sqrt{|V|}|E|)$ algorithm for finding maximum matching in general graphs. In: 21st Annual Symposium on Foundations of Computer Science (SFCS 1980), pp. 17–27. IEEE (1980)
25. Montanaro, A.: Quantum walk speedup of backtracking algorithms. arXiv preprint arXiv:1509.02374 (2015)
26. Montanaro, A.: Quantum speedup of branch-and-bound algorithms. Phys. Rev. Res. **2**(1), 013056 (2020)
27. Mucha, M., Sankowski, P.: Maximum matchings via Gaussian elimination. In: 45th Annual IEEE Symposium on Foundations of Computer Science, pp. 248–255. IEEE (2004)
28. Nannicini, G.: Fast quantum subroutines for the simplex method. arXiv preprint arXiv:1910.10649 (2019)
29. Peterson, P.A., Loui, M.C.: The general maximum matching algorithm of Micali and Vazirani. Algorithmica **3**(1–4), 511–533 (1988). https://doi.org/10.1007/BF01762129
30. Quimper, C.G., Golynski, A., López-Ortiz, A., Van Beek, P.: An efficient bounds consistency algorithm for the global cardinality constraint. Constraints **10**(2), 115–135 (2005). https://doi.org/10.1007/s10601-005-0552-y
31. Quimper, C.-G., López-Ortiz, A., van Beek, P., Golynski, A.: Improved algorithms for the global cardinality constraint. In: Wallace, M. (ed.) CP 2004. LNCS, vol. 3258, pp. 542–556. Springer, Heidelberg (2004). https://doi.org/10.1007/978-3-540-30201-8_40
32. Régin, J.C.: A filtering algorithm for constraints of difference in CSPs. In: AAAI 1994, pp. 362–367 (1994)
33. Rieffel, E.G., Polak, W.H.: Quantum Computing: A Gentle Introduction. MIT Press, Cambridge (2011)
34. Tarjan, R.: Depth-first search and linear graph algorithms. SIAM J. Comput. **1**, 146–160 (1972)

35. Van Apeldoorn, J., Gilyén, A., Gribling, S., de Wolf, R.: Quantum SDP-solvers: better upper and lower bounds. Quantum **4**, 230 (2020)
36. Van Hoeve, W.J.: The alldifferent constraint: a survey. arXiv preprint arXiv:cs/0105015 (2001)
37. Vazirani, V.V.: A simplification of the MV matching algorithm and its proof. arXiv preprint arXiv:1210.4594 (2012)
38. Yanofsky, N.S., Mannucci, M.A.: Quantum Computing for Computer Scientists. Cambridge University Press, Cambridge (2008)
39. Zhang, X., Li, Q., Zhang, W.: A fast algorithm for generalized arc consistency of the alldifferent constraint. In: IJCAI, pp. 1398–1403 (2018)

# Pure MaxSAT and Its Applications to Combinatorial Optimization via Linear Local Search

Shaowei Cai[1,2](✉) and Xindi Zhang[1,2]

[1] State Key Laboratory of Computer Science, Institute of Software, Chinese Academy of Sciences, Beijing, China
{caisw,zhangxd}@ios.ac.cn
[2] School of Computer Science and Technology, University of Chinese Academy of Sciences, Beijing, China

**Abstract.** Maximum Satisfiability (MaxSAT) is a general model for formulating combinatorial optimization problems. MaxSAT formulas encoded from different domains have different features, yet most MaxSAT solvers are designed for general formulas. This work considers an important subclass of MaxSAT, named as Pure MaxSAT, which characterizes a wide range of combinatorial optimization problems particularly subset problems. We design a novel local search method for Pure MaxSAT, which combines the idea of linear search and local search and is dubbed as linear local search. Our algorithm LinearLS significantly outperforms state of the art MaxSAT solvers on Pure MaxSAT instances, including instances from MaxSAT Evaluations and those encoded from three famous NP hard combinatorial optimization problems. Moreover, LinearLS outperforms state of the art algorithms for each tested combinatorial optimization problem on the popular benchmarks.

**Keywords:** Pure MaxSAT · Combinatorial optimization · Linear local search

## 1 Introduction

Maximum Satisfiability (MaxSAT) is an optimisation version of Boolean Satisfiability (SAT), and its general form contains both hard clauses and soft clauses, where the soft clauses can be unweighted or weighted. Solving such a MaxSAT instance involves finding a truth assignment that satisfies the hard clauses along with a maximum number (resp. weight) of soft clauses. MaxSAT is a natural model for formulating combinatorial optimization problems, and has been used to efficiently solve many combinatorial optimization problems that appear in industrial domains.

Most MaxSAT algorithms are developed for general purpose and focus on achieving better performance on a wide range of benchmarks which come from


H. Simonis (Ed.): CP 2020, LNCS 12333, pp. 90–106, 2020.
https://doi.org/10.1007/978-3-030-58475-7_6

diverse domains, and they are usually tested on benchmarks from MaxSAT Evaluations (MSEs). The most popular and effective approach for MaxSAT is the SAT-based approach [1,17,35], which reformulates the MaxSAT optimization problem into a sequence of SAT decision problems and solves them by iteratively calling a SAT solver. SAT-based MaxSAT algorithms can be divided into two types: *linear* [6,7,26,32] and *core-guided* [2,21,36,39]. SAT-based algorithms are essentially complete: they can prove the optimality of the solutions they find when they terminate. Some SAT-based solver such as the linear search ones and the hybrid ones [3,4], refine the upper bound during the search, and can be used for incomplete solving. SAT-based solvers have shown strong performance in the MSEs.

There has been growing interest in incomplete MaxSAT algorithms in recent years, with a surge of new methods at the two recent MSEs. A main incomplete approach for MaxSAT is local search, which aims to find high quality solutions quickly. Local search algorithms typically maintain a complete assignment and modify it iteratively by flipping the value of variables to quickly visit the search space and look for solutions of increasing quality. Local search for MaxSAT has witnessed significant progress during recent years [10,20,27,30]. Particularly, a dynamic local search algorithm named SATLike [27] is competitive with SAT-based solvers on solving unweighted industrial instances.

When solving combinatorial optimization problems by MaxSAT, most works utilize the general solvers off the shelf [5,13,14,22]. However, MaxSAT instances from different domains have their own characteristics, which we believe should be taken into account. Very limited works have been done on developing MaxSAT solvers for specific problems such as Maximum Weight Clique [16,23]. But such algorithms are limited to just one specific problem. An important fact is that many combinatorial optimization problems share the same feature when they are formulated in MaxSAT. Therefore, a significant direction is to develop effective algorithms for important subclasses of MaxSAT, which can have better performance than general MaxSAT algorithms while at the same time can be applied to a wide range of problems.

In this work, we introduce an important subclass of MaxSAT called Pure MaxSAT (PureMS for short), which characterizes a wide range of combinatorial optimization problems, particularly including subset problems. In fact, a considerable portion of the benchmarks in recent MSEs belong to this subclass. For solving PureMS, we propose a new search paradigm named linear local search, which is inspired by the great success of the linear SAT-based solvers. The core idea is that, whenever finding a better solution, the algorithm only visits assignments with strictly lower soft cost (i.e., with smaller weight of falsified soft clauses). Thus, every feasible solution visited during the search has a lower cost than previously found solutions. This is the first time that the idea of linear search is integrated to local search for MaxSAT, and our experiments show that the linear local search is powerful for solving PureMS formulas.

Our linear local search is a two-phase local search algorithm, which consists of two phases in each iteration. The first one is dedicated to decrease the soft cost, while the second focuses on satisfying hard clauses, subject to keeping the soft cost lower than the cost of the previously found best solution. To improve the local search, we propose a variant of the Variable Neighbourhood Descent (VND) method [34]. VND is a variant of Variable Neighbourhood Search (VNS), which benefits from the advantages of large neighbourhoods without incurring a high time complexity of the search steps. VND employs small neighbourhoods until a local optimum is encountered, at which point the search process switches to a larger neighbourhood (corresponding to flipping more variables in one iteration in the context of MaxSAT), which might allow further improvement. Different from previous VND or VNS methods which only consider the number of elements to change values, we also take into account a structure parameter, i.e., the total degree of the chosen variables.

We carry out experiments to evaluate our algorithm dubbed LinearLS on a wide range of benchmarks, including all PureMS instances in recent MSEs, as well as the benchmarks from three famous combinatorial optimization problems, namely maximum clique (MaxClq), minimum vertex cover (MinVC) and set cover problem (SCP). Our results show that LinearLS is significantly better than state of the art MaxSAT solvers, including SAT-based and local search ones on all the benchmarks. Moreover, LinearLS outperforms state of the art algorithms for MaxClq, MinVC and SCP. Note that, our algorithm is general for combinatorial optimization problems that can be formulated as PureMS, while the competitors are tailored for each specific problem respectively.

The remainder of this paper is structured as follows. The next section introduces background knowledge. Section 3 introduces the Pure MaxSAT problem, and Sect. 4 presents the linear local search method. Experiment results are presented in Sect. 5. Finally, Sect. 6 concludes the paper.

## 2 Preliminary

Given a set of $n$ Boolean variables $X = \{x_1, x_2, ..., x_n\}$, a *literal* is either a variable $x_i$ (positive literal) or its negation $\neg x_i$ (negative literal). The *polarity* of a positive literal is 1, while the polarity of a negative literal is 0. A *clause* is a disjunction of literals (i.e. $C_i = l_{i1} \vee l_{i2} \vee ... \vee l_{ij}$), and can be viewed as a set of literals. A *unit clause* is a clause with only one literal. A Conjunctive Normal Form (CNF) formula $F = C_1 \wedge C_2 \wedge ... \wedge C_m$ is a conjunction of clauses.

A mapping $\alpha : X \rightarrow \{0, 1\}$ is called an *assignment*, and a *complete assignment* is a mapping that assigns to each variable either 0 or 1. Given an assignment $\alpha$, a clause is satisfied if it has at least one true literal, and is falsified if all its literals are false under $\alpha$.

Given a CNF formula, the Maximum Satisfiability (MaxSAT) problem concerns about finding an assignment to satisfy the most clauses. In its general form, the clauses are divided into hard clauses and soft clauses, where the soft clauses can be unweighted or weighted, and the goal is to find an assignment that satisfies all hard clauses and maximizes the number (resp. weight) of satisfied soft clauses. Such formulas are referred to as (Weighted) Partial MaxSAT. Hereafter, when we say MaxSAT, we refer to this kind of formulas.

Given a MaxSAT formula $F$ and an assignment $\alpha$ to its variables, two important sets are defined here.

- $H_f(\alpha) = \{c|c$ is a hard clause falsified under $\alpha\}$.
- $S_f(\alpha) = \{c|c$ is a soft clause falsified under $\alpha\}$.

The cost functions are defined below.

- the *hard cost* of $\alpha$, denoted as $cost_h(\alpha)$, is the number of falsified hard clauses under $\alpha$.
- the *soft cost* of $\alpha$, denoted as $cost_s(\alpha)$, is the number (or total weight) of falsified soft clauses under $\alpha$.
- the *cost* of $\alpha$ is $cost(\alpha) = +\infty \cdot cost_h(\alpha) + cost_s(\alpha)$.

An assignment $\alpha$ is feasible iff it satisfies all hard clauses in $F$. It is easy to see $cost(\alpha) = cost_s(\alpha)$ for feasible assignments, while $cost(\alpha) = +\infty$ for infeasible assignments.

Two variables are *neighbors* if they occur in at least one clause, and $N(x)$ denotes all the neighboring variables of $x$. The degree of $x$ is denoted as $degree(x) = |N(x)|$. We use $\overline{\Delta}(F)$ to denote the averaged degree of formula $F$.

Below we give the definitions of the three combinatorial optimization problems studied in our experiments.

**MaxClq and MinVC:** Given an undirected graph $G = (V, E)$, a clique is a subset $K \subseteq V$ whose elements are pairwise adjacent, while a vertex cover is a subset $C \subseteq V$ such that every edge has at least one endpoint in $C$. Given a graph, the Maximum Clique (MaxClq) problem is to find a maximum sized clique, while the Minimum Vertex Cover (MinVC) problem is to find the minimum sized vertex cover.

**SCP:** Given an ground set $U$ and a set $S$ of subsets of $U$ with $\cup_{\forall s \in S} = U$, where each element in $S$ is associated with a weight $w(s)$, the goal of Set Cover Problem (SCP) is to find a set $F \subseteq S$ of the smallest total weight but still contains all elements in $U$, that is, $\cup_{\forall s \in F} = U$. In the unweighted version of SCP, also known as uniform cost SCP, each element in $S$ has the same weight 1, and thus the goal is to find $F \subseteq S$ such that the cardinality of $F$ is the smallest.

## 3 The Pure MaxSAT Problem

We propose a new variant of MaxSAT named Pure MaxSAT, which is an important subclass of the MaxSAT problem.

**Definition 1 (pure clause).** *A clause is a pure clause if all its literals are of the same polarity (either positive or negative). The polarity of a pure clause is defined as the polarity of its literals.*

**Definition 2 (Pure MaxSAT).** *The Pure MaxSAT problem is a special type of Partial MaxSAT where all hard clauses are pure clauses with the same polarity, and all soft clauses are pure clauses with the opposite polarity. In the weighted Pure MaxSAT, each soft clause has a positive number as its weight.*

When formulated as the language of MaxSAT, many combinatorial optimization problems naturally fall into the class of Pure MaxSAT. We give three examples, which are famous NP hard problems with wide applications of their own.

- MaxClq: For each vertex $i \in V$, the PureMS instance has a boolean variable $x_i$ that indicates whether vertex $i$ is chosen in the clique. For each vertex pair $(i, j) \notin E$ ($E$ is the edge set), generate a hard clause $\neg x_i \vee \neg x_j$; for each vertex $i \in V$, generate a unit soft clause $\{x_i\}$.
- MinVC: For each vertex $i \in V$, the PureMS instance has a boolean variable $x_i$ that indicates whether vertex $i$ is chosen in the vertex cover. For each edge $(i, j) \in E$, generate a hard clause $x_i \vee x_j$; for each vertex $i \in V$, generate a unit soft clause $\{\neg x_i\}$.
- SCP: For each element (a subset) $s \in S$, the PureMS instance has a boolean variable $x_s$ that indicates whether $s$ is chosen in the solution. For each element $e \in U$, generate a hard clause $\{x_s | s \in S, e \in s\}$, to ensure that each element in $U$ is covered by at least one subset in $S$. For each element $s \in S$, generate a soft clause $\{\neg x_s\}$ and its weight is equal to $w(s)$.

Observing the feature of PureMS, we can gain some insights on local search algorithms for this problem. Since the polarity of hard clauses is opposite to that of the soft clauses, the goal of satisfying more hard clauses and the goal of satisfying more soft clauses are obviously conflicting. Whenever we flip a variable to reduce $cost_s$, it causes an increase on $cost_h$, although sometimes the increment can be 0 (rarely happens). Similarly, a flip of a variable which reduces $cost_h$ potentially goes along with an increase on $cost_s$. This observation leads us to design linear local search for PureMS.

## 4 Linear Local Search for Pure MaxSAT

In this section, we propose a linear local search algorithm for PureMS. We firstly introduce the linear local search framework and the scoring function, and then present the algorithm.

### 4.1 Linear Local Search and Its Scoring Function

We propose a two-phase local search framework (Algorithm 1), which allows to implement a linear search that visits solutions with monotonically decreasing *cost*. First, we note that, for PureMS, since the polarity of literals in hard clauses are the same, we can produce a feasible initial assignment with guarantee. At the trivial case, we can simply assign 0 to all variables if hard clauses consist of negative literals, and 1 on the contrary. Nevertheless, there are better initialization algorithms. After the initialization, the algorithm executes a loop until reaching a time limit. Each iteration of the loop consists of two phases.

**Algorithm 1:** A Linear Local Search Framework for PureMS

**Input**: MaxSAT instance $F$, the *cutoff* time
**Output**: A feasible assignment $\alpha^*$
**begin**
    $\alpha \leftarrow InitAssignment()$;
    **while** *elapsed time* $<$ *cutoff* **do**
        **if** $H_f(\alpha) = \emptyset$ **then**
            $\alpha^* \leftarrow \alpha, cost^* \leftarrow cost(\alpha)$;
        flip some variables to decrease $cost_s(\alpha)$;
        **while** $H_f(\alpha) \neq \emptyset$ **do**
            choose a variable $y$ from falsified hard clauses;
            **if** *flipping* $y$ *would cause* $cost_s(\alpha) \geq cost^*$ **then break**;
            $flip(\alpha, y)$;
    **return** $(\alpha^*, cost^*)$

- In the first phase, the algorithm chooses some variables to flip, with the purpose of decreasing the soft cost. This phase produces some newly falsified hard clauses.
- In the second phase, the algorithm tries to satisfy as many hard clauses as possible, with a constraint that keeps the soft cost strictly lower than $cost^*$ (the cost of the best found solution). Thus, if all hard clauses are satisfied (i.e., $H_f(\alpha) = \emptyset$), that means a better solution is found.

Local search algorithms are typically guided by scoring functions, which are used to evaluate variables and critical in selecting the variable to flip. We design a scoring function which cooperates well with the linear local search framework.

Our scoring function is related to a clause weighting scheme. Most local search algorithms for MaxSAT employ constraint weighting techniques, which serve as a form of diversification. Our algorithm utilizes a clause weighting scheme that works on hard clauses (the details will be described in the LinearLS algorithm in Sect. 4.2). We associate each hard clause $c$ with a positive integer weight $hw(c)$[1], which are initialized to 1 and updated during the search. Note that the weights of soft clauses are not changed in our algorithm. Our scoring function relies on two basic functions which are defined below.

[1] To distinguish with the weight of soft clauses $w(c)$ in the original formula, we use $hw(c)$ to denote the hard clause weight introduced by our method.

**Definition 3 (hard score, soft score).** *Given a MaxSAT formula and let $\alpha$ be a complete assignment, the hard score of a variable $x$ w.r.t. $\alpha$ is defined as*

$$hscore(\alpha, x) = \sum_{c \in H_f(\alpha)} hw(c) - \sum_{c \in H_f(\alpha')} hw(c),$$

*and the soft score of $x$ w.r.t. $\alpha$ is defined as*

$$sscore(\alpha, x) = cost_s(\alpha) - cost_s(\alpha'),$$

*where $\alpha'$ differs from $\alpha$ only in the value of $x$.*

In this work, the $\alpha$ in scoring functions always refers to the current assignment and can be omitted. Hence, $hscore(\alpha, x)$ and $sscore(\alpha, x)$ can be written as $hscore(x)$ and $sscore(x)$. Intuitively, $hscore(x)$ and $sscore(x)$ are the incremental changes in the objective for flipping $x$ w.r.t. the current assignment.

**Lemma 1.** *Given any PureMS formula $F$, and $\alpha$ is a complete assignment to $F$, for any variable $x$, we have*

$$hscore(\alpha, x) \cdot sscore(\alpha, x) \leq 0.$$

**Proof:** According to the definition of PureMS, all clauses in $F$ are pure clauses and the literals in hard clauses have the opposite polarity to those in soft clauses. Without loss of generality, let us assume the hard clauses contain only positive literals, and the soft clauses contain only negative literals. If $hscore(\alpha, x) > 0$, which indicates that the flip of $x$ make at least one falsified hard clause become satisfied, then it must flip the value of $x$ from 0 to 1. Such a $0 \rightarrow 1$ flip never makes any falsified soft clause become satisfied, as all soft clauses have negative literals. Therefore, $hscore(\alpha, x)$ and $sscore(\alpha, x)$ cannot be positive at the same time. □

Based on these two basic functions and Lemma 1, we derive a novel scoring function as follows.

**Definition 4 (ratio score).** *The ratio score of a variable $x$ is defined as*

$$rscore(x) = \frac{hscore(x)}{|sscore(x)| + 1}.$$

This $rscore$ measures the ratio of $hscore$ and $sscore$. We add one to the denominator to avoid the "divide by 0" error. Also, we adopt the absolute value of $sscore$ for convenient usage—by doing this, we can prefer larger $rscore$ no matter in the first or second phase. In the first phase, we focus on satisfying soft clauses, and the chosen variables have $sscore(x) > 0$ and $hscore(x) \leq 0$, and thus $rscore(x) \leq 0$. For equal $hscore$, a larger $rscore$ means a larger $sscore$, which means satisfying more soft clauses; for equal $sscore$, a larger $rscore$ means breaking fewer hard clauses. In the second phase, we focus on satisfying hard clauses, and the chosen variables have $hscore(x) > 0$ and $sscore(x) \leq 0$, and thus

$rscore(x) \geq 0$. For equal $sscore$, a larger $rscore$ means a larger $hscore$, which leads to more satisfied hard clauses; for equal $hscore$, a larger $rscore$ means a smaller $|sscore|$, which means breaking fewer soft clauses. Our algorithm employs $rscore$ in both phases of local search, and prefers to pick the variables with larger $rscore$.

## 4.2 The LinearLS Algorithm

Based on the linear local search framework and the $rscore$ function, we develop an algorithm named LinearLS (Algorithm 2). The algorithm is described in details below.

*Initialization*: Unlike previous local search algorithms for SAT/MaxSAT which generate the initial solution randomly, our algorithm employs a greedy strategy. Firstly, all variables are assigned with the value equal to the polarity of the soft clauses. This makes all hard clauses falsified and all soft clauses satisfied. Then, the algorithm iteratively picks a random falsified hard clause and flips a variable with highest $hscore$ in the clause, until there is no falsified hard clause. Thus the initial assignment is feasible. At the worst case, all variables are flipped in order to make all hard clauses satisfied, then the cost would be the largest among feasible solutions as all soft clauses are falsified in this situation. In practice, however, this initialization procedure usually finds a much better solution than the worst case.

After initialization, the local search loop (lines 5–20) is executed until a given cutoff time is reached. During the search, the best found solution $\alpha^*$ and its cost are updated. An important feature of our linear local search algorithm is that, whenever we find a feasible solution, we are sure that it is better than $\alpha^*$, as the algorithm always keeps $cost_s(\alpha)$ strictly lower than $cost^*$. Thus, whenever $\alpha$ becomes feasible, $\alpha^*$ is updated to $\alpha$ and $cost^*$ is updated accordingly (lines 6–7). When the algorithm reaches the time limit, it returns the best found solution $\alpha^*$ and its cost.

The local search is based on the two-phase framework, and we propose a variant of Variable Neighbourhood Descent (VND) method for striking a good balance between exploitation and exploration.

In the first phase (lines 8–13), we flip $K$ variables, where $K$ is adjusted according to the algorithm's behavior. If the algorithm has not found a better solution for a long period (which is set to $2{\cdot}10^4$ steps for SCP benchmarks and $10^4$ for the others in LinearLS), then $K$ increases by 1 for improving exploration, while once the algorithm finds a solution, $K$ is reset to 1 for fast converge. This is implemented in the $Adjust\_flip\_num\_phase1(K)$ function. Each flipping variable in the first phase is chosen from all falsified soft clauses by picking the variable with the highest $rscore$ (line 10), breaking ties by preferring the one that is least recently flipped. Additionally, we set a dynamic maximum limit to $K$ by considering the total degree of the flipping variables in the first phase (line 12). Once this value achieves a threshold ($t \times \overline{\Delta}(F)$, where $t$ is set to 1 for

**Algorithm 2:** LinearLS

1 **Input**: Pure MaxSAT instance $F$, the *cutoff* time
2 **Output**: A feasible assignment $\alpha^*$ and its cost
3 **begin**
4     $\alpha \leftarrow InitSolution()$;
5     **while** *elapsed time* $<$ *cutoff* **do**
6         **if** $H_f(\alpha) = \emptyset$ **then**
7             $\alpha^* \leftarrow \alpha, cost^* \leftarrow cost(\alpha)$;
8         $SumDegree \leftarrow 0$;
9         **for** $i \leftarrow 1$ *to* $K$ **do**
10             $x \leftarrow$ a variable in $S_f(\alpha)$ with highest *rscore*;
11             $flip(\alpha, x)$;
12             $SumDegree{+}= degree(x)$;
13             **if** $SumDegree \geq t \times \overline{\Delta}(F)$ **then** **break** ;
14         **while** $H_f(\alpha) \neq \emptyset$ **do**
15             $c \leftarrow$ a random falsified hard clause;
16             $y \leftarrow$ a variable in $c$ with highest *rscore*;
17             **if** $cost_s(\alpha) - sscore(\alpha, y) \geq cost^*$ **then** **break** ;
18             $flip(\alpha, y)$;
19         $Adjust\_flip\_num\_phase1(K)$;
20         $Update\_hard\_clause\_weights()$;
21     **return** $(\alpha^*, cost^*)$

MaxClq and SCP benchmarks, 2 for the rest), the first phase is stopped and the algorithm goes to the second phase (line 13).

Here we provide the intuition of limiting VND with a degree based upper bound. Generally, the more variables flipped in the first phase, the more candidate variables are generated for the second phase. However, the other factor to the number of candidate flipping variables (thus the size of search area) in the second phase is the degree of the variables flipped in the first phase. We take into account both factors in our VND method.

The second phase (lines 14–18) is dedicated to satisfy hard clauses, and thus each flipping variable is chosen from a random falsified hard clause. The variable with highest *rscore* is picked, breaking ties by preferring the one that is least recently flipped. For each selected variable, LinearLS checks whether its flip would cause $cost_s(\alpha)$ greater than or equal to $cost^*$, and if this is the case, it leaves the second phase immediately without flipping the variable. By doing this, we guarantee that $cost_s(\alpha) < cost^*$ always holds during the search.

In the end of each iteration, the $K$ value is updated when necessary according to our VND method. Also, the hard clause weights are updated (line 20): the weight of each falsified hard clause is increased by 1, and when the average weight achieves a threshold (which is set to $\frac{n}{2}$), early weighting decisions are forgotten as $hw(c) \leftarrow \rho \cdot hw(c)$, where $\rho \in (0, 1)$ is a constant factor and set to 0.3.

### 4.3 More Optimizations

An effective strategy to avoid the cycle phenomenon (i.e., revisiting some search areas) in local search is the Configuration Checking (CC) strategy [12], which forbids flipping a variable $x$, if after the last time $x$ was flipped, none of its neighboring variables has changed its value. The CC strategy has proved effective in local search for SAT [11] and MaxSAT [30,31]. LinearLS also employs CC. Variables flipped in last iteration are also forbidden to be flipped again. These are common techniques in local search to reduce the cycle phenomenon.

## 5 Experiments

We carry out experiments to compare LinearLS with state of the art algorithms on a wide range of benchmarks. LinearLS is implemented in C++ and compiled by g++ with -O3 option. Our all experiments were conducted on a server using Intel Xeon Platinum 8153 @2.00 GHz, 512G RAM, running Centos 7.7.1908 Linux operation system. The time limit for all algorithms is 300 s, except that we additionally test an exact MaxSAT solver for one hour.

### 5.1 Results on PureMS Benchmarks from MSEs

We collect all PureMS instances from both unweighted and weighted benchmarks in MaxSAT Evaluations (MSEs) 2017, 2018 and 2019. There are several duplicate instances in the unweighted benchmarks of the three MSEs. We compare LinearLS with 4 state of the art MaxSAT solvers, from which are 1 local search solver and 3 SAT-based solvers.

**Table 1.** Results on Pure MaxSAT benchmarks from MaxSAT Evaluations 2017–2019, including unweighted benchmarks and weighted benchmarks.

| Benchmark | #inst. | *LinearLS* | | *SATLike(_w)* | | *Loandra* | | *TT-OpenWBO* | | *RC2* | | *RC2(1h)* | |
|---|---|---|---|---|---|---|---|---|---|---|---|---|---|
| | | #win | Time | #win | Time | #win | Time | #win | Time | #win | Time | #win | Time |
| Unweighted | | | | | | | | | | | | | |
| MSE17 | 113 | **111** | 6.1 | 48 | 31.6 | 57 | 27.5 | 57 | 40.9 | 57 | 15.1 | 64 | 153.9 |
| MSE18 | 110 | **100** | 16.2 | 46 | 29.5 | 46 | 23.7 | 36 | 44.1 | 61 | 56.0 | 70 | 235.2 |
| MSE19 | 101 | **88** | 22.8 | 28 | 17.0 | 33 | 51.6 | 25 | 59.4 | 42 | 84.2 | 49 | 264.4 |
| MSEall | 284 | **261** | 14.2 | 112 | 29.6 | 121 | 28.0 | 105 | 41.9 | 148 | 50.4 | 168 | 206.7 |
| Weighted | | | | | | | | | | | | | |
| MSE17 | 40 | **40** | <0.1 | 31 | 0.1 | 33 | 16.7 | 20 | 0.1 | 25 | 25.5 | 37 | 372.8 |
| MSE18 | 15 | **15** | <0.1 | 15 | 0.1 | 14 | 29.5 | 0 | N/A | 4 | 151.7 | 14 | 859.2 |
| MSE19 | 51 | **51** | 30.7 | 30 | 6.1 | 21 | 10.7 | 8 | 27.4 | 19 | 45.7 | 25 | 238.7 |
| MSEall | 106 | **106** | 14.8 | 76 | 2.5 | 68 | 17.5 | 28 | 7.9 | 48 | 44.0 | 76 | 418.3 |

- SATLike [27] is the best local search MaxSAT solver, which won the two unweighted categories of incomplete track in MSE 2018 and placed 2nd in the 300 s unweighted category of incomplete track in MSE 2019. SATLike has another version optimized for weighted categories, which is denoted as SATLike_w.
- Loandra [6] won the two unweighted categories, and was ranked 2nd in two weighted categories of incomplete track in MSE 2019.
- TT-Open-WBO-inc [38] won the two weighted categories of incomplete track in MSE 2019.
- RC2 (implementing the relaxable cardinality constraints method) [37] won both weighted and unweighted categories of complete track in MSE 2019. Since RC2 is an exact solver, in our experiments, we test RC2 with 2 time limits, 300 s (as with other solvers) and one hour. We note that local search and exact solvers have different advantages and it is better to see them as complementary alternatives. The comparisons with exact solvers are just for reference, which may give us some insights.

**Table 2.** Averaged SCORE results of MaxSAT solvers on each family of MSE benchmarks. We also report the results of the complete solver RC2 with 1 h time limit for reference.

| Domain(#inst.) | *LinearLS* | *Linear_init* | *SATLike(_w)* | *Loandra* | *TTOpenWBO* | *RC2* | *RC2*(1h) |
|---|---|---|---|---|---|---|---|
| Unweighted | | | | | | | |
| maxclique(68) | **1.0** | 0.978 | 0.995 | 0.995 | 0.997 | 0.441 | 0.485 |
| aes(14) | **1.0** | 0.882 | 0.895 | 0.769 | 0.303 | 0.143 | 0.143 |
| frb(40) | **1.0** | 0.978 | 0.994 | 0.999 | **1.0** | 0.975 | **1.0** |
| bcp-syn(53) | **1.0** | 0.469 | 0.996 | 0.955 | 0.92 | 0.396 | 0.509 |
| optic(69) | **1.0** | 0.904 | 0.989 | 0.961 | 0.835 | 0.333 | 0.377 |
| drmx-cryptogen(40) | 0.991 | 0.827 | 0.984 | 0.955 | 0.881 | 0.825 | **1.0** |
| Weighted | | | | | | | |
| auc-paths(35) | **1.0** | 0.98 | **1.0** | **1.0** | 0.993 | 0.257 | 0.886 |
| auc-scheduling(20) | **1.0** | 0.996 | 0.996 | 0.999 | **1.0** | **1.0** | **1.0** |
| MinimumWeight DominatingSetProblem(7) | **1.0** | 0.966 | 0.402 | 0.708 | 0.705 | 0.0 | 0.0 |
| auctions(16) | **1.0** | 0.985 | 0.996 | 0.996 | 0.995 | 0.312 | 0.625 |
| set-covering(28) | **1.0** | 0.888 | 0.996 | 0.981 | 0.949 | 0.5 | 0.536 |

For each algorithm on each instance family, we report the number of instances where the solver finds the best solution among all solvers ("#win") and the mean time of doing so over such winning instances. These results (Table 1) clearly show the superiority of LinearLS over other MaxSAT solvers. Particularly, the "#win" number of LinearLS is always significantly larger than other solvers.

To show how far the solution provided by a solver are from the best solution found by all the solvers, for each algorithm $\mathcal{A}$ on each fomula $F$, we calculate a metric measured as $SCORE(F, \mathcal{A}) = \frac{BEST\_COST(F)+1}{COST(F,\alpha^{\mathcal{A}})+1}$, where $\alpha^{\mathcal{A}}$ is the solution found by algorithm $\mathcal{A}$ while $BEST\_COST(F)$ denotes the lowest cost found in the time limit by any of the solvers. These benchmarks consists formulas encoded from different domains, and we report the averaged *SCORE* for each algorithm on each domain in Table 2. The *SCORE* of LinearLS is 1.0 (full score) for all domains except 0.991 for the *cryptogen* domain. Nevertheless, the best SCORE is obtained by RC2 for one hour time limit. If we compare all the solvers with the time limit of 300 s, then LinearLS is still the best, achieving a full score, which indicates its strong performance on a wide range of benchmarks from diverse domains.

We also calculate the *SCORE* of the initial solutions of LinearLS. As can be seen from the table, the initial solutions of LinearLS are better than the solutions returned by TTOpenWBO and RC2 on most of the Pure MaxSAT instances. Besides, although the initial solutions are not as good as those returned by the incomplete solvers SATLike and Loandra, the gaps are not large. This indicates that the design of the problem is an important factor to the good performance on Pure MaxSAT. By executing the local search procedure of LinearLS, the solutions are further improved.

**Table 3.** Results on MaxClq benchmarks. This table reports results for three MaxClq benchmarks, including Kidney Exchange (Kidney), Research Excellence Network (REN) and DIMACS. The error-correcting codes (ECC) benchmark instances are too easy that all algorithms find the optimal solution quickly and not reported.

| Solvers | Kidney(120) | | REN(129) | | DIMACS(37) | | MaxClq_all(286) | |
|---|---|---|---|---|---|---|---|---|
| | Win | Time | Win | Time | Win | Time | Win | Time |
| *LinearLS* | **120** | 0.1 | **129** | **0.1** | **33** | 11.1 | **282** | 1.4 |
| *LSCC* | 118 | 0.8 | 127 | <0.1 | **33** | **<0.1** | 278 | 0.3 |
| *BBMS* | 98 | 27.4 | **129** | 4.8 | 26 | 9.3 | 253 | 14.0 |
| *IncMaxCLQ* | 106 | 36.9 | 110 | 5.2 | 26 | 32.1 | 242 | 22.0 |
| *IncMC2* | 110 | 3.4 | 107 | <0.1 | 26 | 10.9 | 243 | 2.7 |
| *MaxClqDyn* | 74 | 3.7 | 114 | 0.6 | 21 | 15.6 | 209 | 3.2 |
| *MCS* | 108 | 2.8 | 115 | 0.6 | 20 | 13.6 | 243 | 2.7 |
| *SATLike* | 84 | 13.3 | 115 | 10.6 | 8 | 19.0 | 207 | 12.0 |
| *Loandra* | 88 | 5.3 | 101 | 2.4 | 16 | 19.7 | 205 | 5.0 |
| *TT-OpenWBO* | 86 | 16.5 | 102 | 8.3 | 17 | 56.6 | 205 | 15.8 |

### 5.2 Results on Maximum Clique Benchmarks

We evaluate LinearLS on 4 popular MaxClq benchmarks which are mostly from applications [33]:

- Kidney Exchange, where the clique stands for a maximally desirable set of donor/patient exchanges. The instances were generated using data from [15].

- Error-correcting Codes (ECC), where the clique stands for a set of words maximally pair-wise distant [40].
- Research Excellence Network (REN) [33], where the clique stands for the optimal set of publications that a university department can provide to the authority assessing it.
- DIMACS, the MaxClq benchmark from Second DIMACS Implementation Challenge (1992–1993)[2]. Thirty seven graphs were selected by the organizers to be the Second DIMACS Challenge Test Problems.

Besides the MaxSAT algorithms, we compare with the following MaxClq algorithms. According to [28], state of the art MaxClq algorithms include IncMC2 [28], BBMC [42,43], IncMaxCLQ [29], MCS[44], MaxCliqueDyn [25]. We also compare with LSCC [46], which is a recent local search algorithm that performs well on both unweighted and weighted MaxClq benchmarks.

The ECC instances are so easy that all algorithms find the optimal solution quickly, and the local search algorithms do so within one second, and thus are not reported. The results (Table 3) show that Our LinearLS gives the best performance in terms of the solution quality, and is the best algorithm for the two application benchmarks namely Kidney and REN. Although the other local search LSCC is fast, it fails to find the best solution for some instances in these two benchmarks. The MaxSAT solvers, including SATLike, Loandra and TT-OpenWBO, perform much worse than LinearLS.

### 5.3 Results on Minimum Vertex Cover Benchmark

Recently, MinVC algorithms focus on solving massive graphs. Particularly, the Network Repository [41], which collects massive graphs from various areas, has become the most popular benchmark for testing MinVC algorithms in recent years [8,9,24,45]. FastVC [8] is an efficient local search for massive graphs, and afterwards it is improved by a preprocessor, resulting in the FastVC2+p algorithm [9]. Seen from the literature, FastVC2+p is currently the best algorithm for solving MinVC on the Network Repository instances.

On these massive MinVC instances, all the MaxSAT solvers perform significantly worse than LinearLS on almost all instances, and thus we do not report their results here. We focus on the comparison with MinVC algorithms FastVC and FastVC2+p. For fair comparison, when compared with FastVC2+p, LinearLS also utilizes the preprocessor in FastVC2+p to preprocess the graphs. We choose the graphs with at least $10^5$ vertices, resulting in 65 graphs. Each algorithm is performed 10 runs on each graph with random seed from 1 to 10. We report the number of instances where the algorithm gives better results in terms of the minimum cost ('min') and the averaged cost ('avg') among the 10 runs. Seen from Table 4, the performance of LinearLS is surprisingly good on these massive MinVC instances, pushing the state of the art in MinVC solving on massive graphs.

[2] ftp://dimacs.rutgers.edu/pub/challenges.

**Table 4.** Results on large MinVC instances in Network Repository

| | *LinearLS* | | *FastVC* | |
|---|---|---|---|---|
| | Min | Avg | Min | Avg |
| #better-solution | **42** | **42** | 12 | 13 |
| #equal-solution | 11 | 10 | 11 | 10 |
| | *LinearLS+p* | | *FastVC2+p* | |
| | Min | Avg | Min | Avg |
| #better-solution | **22** | **27** | 17 | 14 |
| #equal-solution | 26 | 24 | 26 | 24 |

### 5.4 Results on Set Cover Benchmarks

We evaluate LinearLS on 2 important SCP benchmarks, including (1) the STS benchmark [18], which contains unweighted SCP instances from Steiner triple systems; (2) the Rail benchmark[3], which contains weighted SCP instances that arise from an application in Italian railways and have been contributed by Paolo Nobili.

LinearLS is compared with SATLike(_w) and the best SAT-based solver for each SCP benchmark. Also, it is compared with the SCP algorithm from [19], which is the best SCP algorithm on both unweighted and weighted SCP (the algorithm was not given a name, and is denoted by the paper [19]). Since the number of instances is limited, each algorithm is performed 10 runs on each instance, and we report the minimum cost and averaged cost, and the aver-

**Table 5.** Results on unweighted and weighted SCP instances

| Instance | *LinearLS* | | [19] | | *SATLike* | | *Loandra* | |
|---|---|---|---|---|---|---|---|---|
| | Min(avg) | Time | Min(avg) | Time | Min(avg) | Time | Min(avg) | Time |
| STS135 | **103(103.0)** | 3.3 | **103(103.0)** | 3.7 | 104(104.0) | 292.6 | 104(104.0) | 62.3 |
| STS243 | **198(198.0)** | <0.1 | **198(198.0)** | 0.1 | 198(201.8) | 285.4 | 202(202.0) | 258 |
| STS405 | **335(335.0)** | 3.8 | **335**(335.8) | 31.0 | 342(344.0) | 288.4 | 347(347.0) | 4.9 |
| STS729 | **617(617.0)** | 7.0 | **617**(619.0) | 26.5 | 646(647.4) | 270.3 | 643(643.0) | 5.7 |
| | *LinearLS* | | [19] | | *SATLike_w* | | *TT-OpenWBO* | |
| rail507 | **176(176.0)** | 3.6 | **176**(176.3) | 101.5 | 194(196.0) | 114.4 | 248(248.0) | 64.8 |
| rail516 | **182(182.0)** | 33.8 | **182**(182.1) | 128.8 | 188(189.4) | 170.4 | 226(226.0) | 21.1 |
| rail582 | **213(213.0)** | 1.9 | **213**(213.9) | 136.4 | 223(225.2) | 176.5 | 286(286.0) | 25.4 |
| rail2536 | **716(716.0)** | 93.3 | 764(772.1) | 288.1 | N/A(N/A) | N/A | 1125(1125.0) | 3.9 |
| rail2586 | **990(990.0)** | 30.2 | 1036(1055.0) | 292.8 | N/A(N/A) | N/A | 1463(1463.0) | 3.2 |
| rail4284 | **1117(1117.0)** | 173.8 | 1203(1221.8) | 287.5 | N/A(N/A) | N/A | 1734(1734.0) | 4.1 |
| rail4872 | **1589(1589.0)** | 55.8 | 1688(1733.2) | 295.0 | N/A(N/A) | N/A | 2355(2355.0) | 3.5 |

[3] http://people.brunel.ac.uk/~mastjjb/jeb/orlib/files/.

aged run time to find the final solution in each run. The results (Table 5) show that LinearLS outperforms the MaxSAT competitors significantly. Moreover, LinearLS finds better solutions than the SCP algorithm [19] and is much faster.

## 6 Conclusions

We introduced the Pure MaxSAT problem, which is an important subclass of MaxSAT and characterizes many combinatorial optimization problems particularly subset problems. We proposed the linear local search method for PureMS, which is the first work exploiting linear search in local search for MaxSAT problems. Experiments on benchmarks from MaxSAT Evaluations and benchmarks of three famous NP hard problems showed that our algorithm significantly outperforms previous MaxSAT algorithms, and achieves better results than state of the art specific algorithms for the three problems.

It is interesting to develop exact algorithms for Pure MaxSAT that can achieve better results than general MaxSAT solvers. Also, we would like to study the inference rules and reduction rules for Pure MaxSAT, which can be used to further improve the performance of Pure MaxSAT solvers.

**Acknowledgments.** This work is partially supported by Youth Innovation Promotion Association of Chinese Academy of Sciences [No. 2017150] and Beijing Academy of Artificial Intelligence (BAAI).

## References

1. Ansótegui, C., Bonet, M.L., Levy, J.: SAT-based MaxSAT algorithms. Artif. Intell. **196**, 77–105 (2013)
2. Ansótegui, C., Didier, F., Gabàs, J.: Exploiting the structure of unsatisfiable cores in MaxSAT. In: Proceedings of IJCAI 2015, pp. 283–289 (2015)
3. Ansótegui, C., Gabàs, J.: WPM3: an (in)complete algorithm for weighted partial MaxSAT. Artif. Intell. **250**, 37–57 (2017)
4. Ansótegui, C., Gabàs, J., Levy, J.: Exploiting subproblem optimization in SAT-based MaxSAT algorithms. J. Heuristics **22**(1), 1–53 (2016). https://doi.org/10.1007/s10732-015-9300-7
5. Benedetti, M., Mori, M.: On the use of Max-SAT and PDDL in RBAC maintenance. Cybersecurity **2**(1) (2019). Article number: 19. https://doi.org/10.1186/s42400-019-0036-9
6. Berg, J., Demirović, E., Stuckey, P.J.: Core-boosted linear search for incomplete MaxSAT. In: Rousseau, L.-M., Stergiou, K. (eds.) CPAIOR 2019. LNCS, vol. 11494, pp. 39–56. Springer, Cham (2019). https://doi.org/10.1007/978-3-030-19212-9_3
7. Berre, D.L., Parrain, A.: The SAT4J library, release 2.2. JSAT **7**(2–3), 59–64 (2010)
8. Cai, S.: Balance between complexity and quality: local search for minimum vertex cover in massive graphs. In: Proceedings of IJCAI 2015, pp. 747–753 (2015)
9. Cai, S., Lin, J., Luo, C.: Finding a small vertex cover in massive sparse graphs: construct, local search, and preprocess. J. Artif. Intell. Res. **59**, 463–494 (2017)

10. Cai, S., Luo, C., Thornton, J., Su, K.: Tailoring local search for partial MaxSAT. In: Proceedings of AAAI 2014, pp. 2623–2629 (2014)
11. Cai, S., Su, K.: Local search for Boolean Satisfiability with configuration checking and subscore. Artif. Intell. **204**, 75–98 (2013)
12. Cai, S., Su, K., Sattar, A.: Local search with edge weighting and configuration checking heuristics for minimum vertex cover. Artif. Intell. **175**(9–10), 1672–1696 (2011)
13. Demirovic, E., Musliu, N.: MaxSAT-based large neighborhood search for high school timetabling. Comput. OR **78**, 172–180 (2017)
14. Demirovic, E., Musliu, N., Winter, F.: Modeling and solving staff scheduling with partial weighted MaxSAT. Ann. OR **275**(1), 79–99 (2019). https://doi.org/10.1007/s10479-017-2693-y
15. Dickerson, J.P., Procaccia, A.D., Sandholm, T.: Optimizing kidney exchange with transplant chains: theory and reality. In: AAMAS 2012, pp. 711–718 (2012)
16. Fang, Z., Li, C., Xu, K.: An exact algorithm based on MaxSAT reasoning for the maximum weight clique problem. J. Artif. Intell. Res. **55**, 799–833 (2016)
17. Fu, Z., Malik, S.: On solving the partial MAX-SAT problem. In: Biere, A., Gomes, C.P. (eds.) SAT 2006. LNCS, vol. 4121, pp. 252–265. Springer, Heidelberg (2006). https://doi.org/10.1007/11814948_25
18. Fulkerson, D.R., Nemhauser, G.L., Trotter, L.: Two computationally difficult set covering problems that arise in computing the 1-width of incidence matrices of Steiner triple systems. In: Balinski, M.L. (ed.) Approaches to Integer Programming. MATHPROGRAMM, vol. 2, pp. 72–81. Springer, Heidelberg (1974). https://doi.org/10.1007/BFb0120689
19. Gao, C., Weise, T., Li, J.: A weighting-based local search heuristic algorithm for the set covering problem. In: Proceedings of the IEEE Congress on Evolutionary Computation, CEC 2014, pp. 826–831 (2014)
20. Guerreiro, A.P., Terra-Neves, M., Lynce, I., Figueira, J.R., Manquinho, V.: Constraint-based techniques in stochastic local search MaxSAT solving. In: Schiex, T., de Givry, S. (eds.) CP 2019. LNCS, vol. 11802, pp. 232–250. Springer, Cham (2019). https://doi.org/10.1007/978-3-030-30048-7_14
21. Heras, F., Morgado, A., Marques-Silva, J.: Core-guided binary search algorithms for maximum satisfiability. In: Proceedings of AAAI 2011 (2011)
22. Huang, W., et al.: Finding and proving the exact ground state of a generalized Ising model by convex optimization and MAX-SAT. Phys. Rev. B **94**, 134424 (2016)
23. Jiang, H., Li, C., Liu, Y., Manyà, F.: A two-stage MaxSAT reasoning approach for the maximum weight clique problem. In: Proceedings of AAAI 2018, pp. 1338–1346 (2018)
24. Katzmann, M., Komusiewicz, C.: Systematic exploration of larger local search neighborhoods for the minimum vertex cover problem. In: Proceedings of AAAI 2017, pp. 846–852 (2017)
25. Konc, J., Janezic, D.: An improved branch and bound algorithm for the maximum clique problem. Commun. Math. Comput. Chem. **58**, 569–590 (2007)
26. Koshimura, M., Zhang, T., Fujita, H., Hasegawa, R.: QMaxSAT: a partial Max-SAT solver. JSAT **8**(1/2), 95–100 (2012)
27. Lei, Z., Cai, S.: Solving (weighted) partial MaxSAT by dynamic local search for SAT. In: Proceedings of IJCAI 2018, pp. 1346–1352 (2018)
28. Li, C., Fang, Z., Jiang, H., Xu, K.: Incremental upper bound for the maximum clique problem. INFORMS J. Comput. **30**(1), 137–153 (2018)
29. Li, C., Fang, Z., Xu, K.: Combining MaxSAT reasoning and incremental upper bound for the maximum clique problem. In: ICTAI 2013, pp. 939–946 (2013)

30. Luo, C., Cai, S., Su, K., Huang, W.: CCEHC: an efficient local search algorithm for weighted partial maximum satisfiability. Artif. Intell. **243**, 26–44 (2017)
31. Luo, C., Cai, S., Wu, W., Jie, Z., Su, K.: CCLS: an efficient local search algorithm for weighted maximum satisfiability. IEEE Trans. Comput. **64**(7), 1830–1843 (2015)
32. Martins, R., Manquinho, V., Lynce, I.: Open-WBO: a modular MaxSAT solver'. In: Sinz, C., Egly, U. (eds.) SAT 2014. LNCS, vol. 8561, pp. 438–445. Springer, Cham (2014). https://doi.org/10.1007/978-3-319-09284-3_33
33. McCreesh, C., Prosser, P., Simpson, K., Trimble, J.: On maximum weight clique algorithms, and how they are evaluated. In: Beck, J.C. (ed.) CP 2017. LNCS, vol. 10416, pp. 206–225. Springer, Cham (2017). https://doi.org/10.1007/978-3-319-66158-2_14
34. Mladenovic, N., Hansen, P.: Variable neighborhood search. Comput. OR **24**(11), 1097–1100 (1997)
35. Morgado, A., Heras, F., Liffiton, M.H., Planes, J., Marques-Silva, J.: Iterative and core-guided MaxSAT solving: a survey and assessment. Constraints Int. J. **18**(4), 478–534 (2013). https://doi.org/10.1007/s10601-013-9146-2
36. Morgado, A., Heras, F., Marques-Silva, J.: Improvements to core-guided binary search for MaxSAT. In: Cimatti, A., Sebastiani, R. (eds.) SAT 2012. LNCS, vol. 7317, pp. 284–297. Springer, Heidelberg (2012). https://doi.org/10.1007/978-3-642-31612-8_22
37. Morgado, A., Ignatiev, A., Marques-Silva, J.: MSCG: robust core-guided MaxSAT solving. JSAT **9**, 129–134 (2014)
38. Nadel, A.: Tt-Open-WBO-Inc.: Tuning polarity and variable selection for anytime SAT-based optimizatio. In: Proceedings of MaxSAT Evaluation 2019: Solver and Benchmark Description, p. 29 (2019)
39. Narodytska, N., Bacchus, F.: Maximum satisfiability using core-guided MaxSAT resolution. In: Proceedings of AAAI 2014, pp. 2717–2723 (2014)
40. Östergård, P.R.J.: A new algorithm for the maximum-weight clique problem. Electron. Notes Discrete Math. **3**, 153–156 (1999)
41. Rossi, R.A., Ahmed, N.K.: The network data repository with interactive graph analytics and visualization. In: Proceedings of AAAI 2015, pp. 4292–4293 (2015)
42. Segundo, P.S., Lopez, A., Batsyn, M., Nikolaev, A., Pardalos, P.M.: Improved initial vertex ordering for exact maximum clique search. Appl. Intell. **45**(3), 868–880 (2016). https://doi.org/10.1007/s10489-016-0796-9
43. Segundo, P.S., Rodríguez-Losada, D., Jiménez, A.: An exact bit-parallel algorithm for the maximum clique problem. Comput. OR **38**(2), 571–581 (2011)
44. Tomita, E., Sutani, Y., Higashi, T., Wakatsuki, M.: A simple and faster branch-and-bound algorithm for finding a maximum clique with computational experiments. IEICE Trans. **96-D**(6), 1286–1298 (2013)
45. Wagner, M., Friedrich, T., Lindauer, M.: Improving local search in a minimum vertex cover solver for classes of networks. In: IEEE Congress on Evolutionary Computation, CEC 2017, pp. 1704–1711 (2017)
46. Wang, Y., Cai, S., Yin, M.: Two efficient local search algorithms for maximum weight clique problem. In: Proceedings of AAAI 2016, pp. 805–811 (2016)

# Tractable Fragments of Temporal Sequences of Topological Information

Quentin Cohen-Solal(✉)

LAMSADE, Université Paris-Dauphine, PSL, CNRS, Paris, France
cohen-solal@cril.fr

**Abstract.** In this paper, we focus on qualitative temporal sequences of topological information. We firstly consider the context of topological temporal sequences of length greater than 3 describing the evolution of regions at consecutive time points. We show that there is no Cartesian subclass containing all the basic relations and the universal relation for which the algebraic closure decides satisfiability. However, we identify some tractable subclasses, by giving up the relations containing the non-tangential proper part relation and not containing the tangential proper part relation.

We then formalize an alternative semantics for temporal sequences. We place ourselves in the context of the topological temporal sequences describing the evolution of regions on a partition of time (i.e. an alternation of instants and intervals). In this context, we identify large tractable fragments.

**Keywords:** Qualitative spatio-temporal reasoning · Satisfiability decision

## 1 Introduction

The reasoning on temporal and spatial qualitative information is necessary to solve many problems that are found in the context of planning, simulation, robotics, intelligent environments and human-computer interaction [8,12,27,31, 41,47]. For this reason, many *spatio-temporal formalisms* have been proposed [5, 7,11,22,30,33,42–44,48]. Spatio-temporal formalisms generally decompose into a spatial formalism and a temporal formalism. The point algebra is a formalism describing the relative positions of points on a line (the timeline or a line of space). $RCC_8$ is another formalism, more expressive than the point algebra, expressing the *topological relations* between regions. It expresses the notions of contact and inclusion.

The *qualitative temporal sequences* [7,48] are the simplest spatio-temporal descriptions, in the sense that there is no uncertainty about temporal information. However, strong negative results have been identified for one of the simplest spatial formalisms: deciding the satisfiability of a temporal sequence over the point algebra is NP-complete (even while restricting the language to *basic*


H. Simonis (Ed.): CP 2020, LNCS 12333, pp. 107–125, 2020.
https://doi.org/10.1007/978-3-030-58475-7_7

*relations* and *the universal relation*) [48]. One can then wonder whether deciding satisfiability is necessarily NP-hard within the framework of spatio-temporal formalisms. However, the complexity of fragments of $RCC_8$ has not been studied within the context of temporal sequences. There could be fragments, not expressing the point algebra, which are tractable.

We therefore study in this paper the complexity of deciding the satisfiability of the topological temporal sequences. We identify in particular a negative result: the classical procedure to decide the satisfiability of polynomial fragments, the *algebraic closure*, does not decide the satisfiability in this context (even while being limited to the basic relations and to the universal relation, if the length of the sequence is greater than 3). We also identify a positive result, by considering semantics different from the classical semantics of temporal sequences. More precisely, we no longer consider that temporal sequences describe the evolutions of entities at neighboring instants. We consider instead that they describe the evolutions of entities on a partition of time (i.e. on an alternation of instants and intervals). In the context of this semantics, we identify large tractable fragments.

In the next section, we present related work, the $RCC_8$ formalism and temporal sequences over $RCC_8$. In Sect. 3, we introduce our negative result and we identify tractable fragments that do not contain all the basic relations. Finally, in Sect. 4, we formalize the alternative semantics of temporal sequences, present the new reasoning operators, and then identify the large tractable fragments.

## 2 Background

### 2.1 Related Work

Many works deal with spatio-temporal reasoning and its complexity. Tractable fragments have been identified in the context of topological temporal sequences describing the evolution of *constant-size* regions at non-*neighboring* instants (regions can satisfy any relations between the instants) [7]. The temporal sequences that we consider, like those of the NP-completeness result of the point algebra, describe regions at neighboring instants. *Temporal sequence ordering* (at neighboring instants) is an NP-complete problem for several fundamental formalisms, such as $RCC_8$ [44]. Formalisms with a higher temporal expressivity have also been proposed. For example, $RCC_8$ has been combined with *Allen's interval algebra* [22]. The cardinal direction calculus has also been combined with the Allen's interval algebra [33].

In general, a qualitative spatio-temporal formalism is based on a transition graph, i.e. a graph representing the possible evolutions of basic relations. It can be a *neighbourhood graph* [16] or a *dominance graph* [20,21]. In a neighbourhood graph, two relations $b$, $b'$ are *neighbour* (i.e. adjacent in the graph), if there exists a pair of evolving entities $(e, e')$ satisfying $b$ at an instant $t$ and $b'$ at an instant $t'$, and satisfying $b$ or $b'$ between $t$ and $t'$. In a dominance graph, a relation $b$ dominates another relation $b'$ (i.e. there is an arc from $b'$ to $b$ in the graph), denoted by $b \vdash b'$, if there exists $t, t' \in \mathbb{R}$ and a pair of evolving entities $(e, e')$

satisfying $b$ at $t$ and satisfying $b'$ at each instant of $]t, t']$. Many transition graphs have been determined [9,13–17,26,29,33,34,36,36,40,46,52,53].

Spatio-temporal qualitative reasoning is also studied in the context of logics [1–4,18,19,23,24,32,43,50,51]. Deciding the satisfiability of these logics is generally PSPACE-hard. Ontologies of time based on points or/and intervals have been studied [45]. There is, in particular, the Event Calculus, a logic of action and change, which can express properties at instant and interval [25].

## 2.2 Region Connection Calculus $\mathrm{RCC}_8$

$\mathrm{RCC}_8$ [28,30,35] is a classical qualitative formalism [5,11,30]. Thus, it is a triplet $(\mathcal{A}, \mathcal{U}, \varphi)$ where $\mathcal{A}$ is a set of relations forming a finite *non-associative binary relation algebra*, $\mathcal{U}$ is the *universe*, i.e. the set of considered entities, and $\varphi$ is a particular interpretation function associating with each relation of $\mathcal{A}$ a relation over $\mathcal{U}$. We denote by $\mathrm{RCA}_8$ the algebra of $\mathrm{RCC}_8$. The universe $\mathcal{U}$ of $\mathrm{RCC}_8$ is the set of regions of a certain topological space $\mathfrak{T}$ (i.e. the non-empty, *closed* and *regular* subsets of $\mathfrak{T}$). $\mathfrak{T}$ is generally $\mathbb{R}^n$. Any algebra $\mathcal{A}$ has special relations, called *basic relations*. Every relation of $\mathcal{A}$ is a union of basic relations. The 8 basic relations of $\mathrm{RCC}_8$: DC (disconnected), EC (externally connected), PO (partially overlapping), EQ (equal), TPP (tangential proper part), NTPP (non-tangential proper part), and the converse of the two previous relations are described in Fig. 1 and defined in Table 1. We denote by $\boldsymbol{\mathcal{B}}_{\mathrm{RCC}_8}$ the set of the basic relations of $\mathrm{RCC}_8$. We denote by $\mathcal{B}_{\mathrm{RCC}_8}$ the *universal relation* (i.e. the union of all relations) and by $\varnothing$ the empty relation. Any algebra $\mathcal{A}$ has several operators: the *union* $\cup$, the *intersection* $\cap$, the *converse* $\bar{\cdot}$, and the *(abstract) composition* $\circ$. These operators are used to infer new relations: $x\ r\ y \implies y\ \bar{r}\ x$, $x\ r\ y \wedge x\ r'\ y \implies x\ (r \cap r')\ y$, and $x\ r\ y \wedge y\ r'\ z \implies x\ (r \circ r')\ z$ (with $r, r' \in \mathcal{A}$ and $x, y, z$ being entity variables). The abstract composition $\circ$ of $\mathrm{RCC}_8$ is the *weak composition*: $r \circ r' = \bigcup \{b \in \boldsymbol{\mathcal{B}}_{\mathrm{RCC}_8} \mid \varphi(b) \cap (\varphi(r) \circ \varphi(r')) \neq \varnothing\}$ with $r, r' \in \mathcal{A}$. For example, the composition of relations TPP $\cup$ EQ and TPP is the relation TPP. The composition of basic relations is described in a so-called *composition table* [28].

Generally, a description based on $\mathrm{RCC}_8$ is a qualitative constraint network, i.e. a conjunction of relations between different entities. For instance, $x$ DC $\cup$ EC $y \wedge z$ TPP $\cup$ NTPP $\cup$ EQ $y$ is such a description, which means that the interiors of regions $x$ and $y$ are disjoint and that the region $z$ is included in $y$. Deciding the satisfiability of qualitative constraint networks whose relations belong to $\mathrm{RCA}_8$ is an NP-complete problem [37]. Tractable fragments have been identified. They consist in restricting the relations of constraint networks to a particular subset of $\mathrm{RCA}_8$. Three large tractable subsets containing all the basic relations and the universal relation have been identified: $\mathcal{H}_8$, $\mathcal{Q}_8$, and $\mathcal{C}_8$ [37]. They are moreover maximal for tractability. They are defined in Table 2. On these subclasses, applying the *algebraic closure*, which is a reasoning operator on networks using the algebra operators, decides satisfiability.

**Table 1.** Definitions of RCC relations ($C(x, y)$ is the *contact relation*, it means that the closedregions $x$ and $y$ intersect; $P$ is the *part relation*; O the *overlap relation*; variables $x, y, z$ are closed regions).

| Relation | Definition |
|---|---|
| $x$ DC $y$ | $\neg(x \text{ C } y)$ |
| $x$ P $y$ | $\forall z \quad z \text{ C } x \implies z \text{ C } y$ |
| $x$ PP $y$ | $x \text{ P } y \wedge \neg(y \text{ P } x)$ |
| $x$ EQ $y$ | $x \text{ P } y \wedge y \text{ P } x$ |
| $x$ O $y$ | $\exists z \quad z \text{ P } x \wedge z \text{ P } y$ |
| $x$ PO $y$ | $x \text{ O } y \wedge \neg(x \text{ P } y) \wedge \neg(y \text{ P } x)$ |
| $x$ EC $y$ | $x \text{ C } y \wedge \neg(x \text{ O } y)$ |
| $x$ TPP $y$ | $x \text{ PP } y \wedge (\exists z\ z \text{ EC } x \wedge z \text{ EC } y)$ |
| $x$ NTPP $y$ | $x \text{ PP } y \wedge \neg(\exists z\ z \text{ EC } x \wedge z \text{ EC } y)$ |
| $x$ $\overline{\text{TPP}}$ $y$ | $y$ TPP $x$ |
| $x$ $\overline{\text{NTPP}}$ $y$ | $y$ NTPP $x$ |

**Table 2.** Definitions of the relations sets $\mathcal{H}_8$, $\mathcal{Q}_8$, and $\mathcal{C}_8$.

| | Definition |
|---|---|
| $\mathcal{N}$ | $\left\{r \in \text{RCA}_8 \mid \text{PO} \not\subseteq r \wedge r \cap (\text{TPP} \cup \text{NTPP}) \neq \varnothing \wedge r \cap \left(\overline{\text{TPP}} \cup \overline{\text{NTPP}}\right) \neq \varnothing\right\}$ |
| $\mathcal{NP}_8$ | $\mathcal{N} \cup \left\{r_1 \cup \text{EC} \cup r_2 \cup \text{EQ} \mid r_1 \in \{\varnothing, \text{DC}\} \wedge r_2 \in \left\{\text{NTPP}, \overline{\text{NTPP}}\right\}\right\}$ |
| $\mathcal{P}_8$ | $\text{RCA}_8 \setminus \mathcal{NP}_8$ |
| $\mathcal{H}_8$ | $\mathcal{P}_8 \cap \left\{r \in \text{RCA}_8 \mid \left(\text{NTPP} \cup \text{EQ} \subseteq r \implies \text{TPP} \subseteq r \wedge \overline{\text{NTPP}} \cup \text{EQ} \subseteq r\right) \implies \overline{\text{TPP}} \subseteq r\right\}$ |
| $\mathcal{Q}_8$ | $\mathcal{P}_8 \cap \left\{r \in \text{RCA}_8 \mid \left(\text{EQ} \subseteq r \wedge r \cap \left(\text{TPP} \cup \text{NTPP} \cup \overline{\text{TPP}} \cup \overline{\text{NTPP}}\right) \neq \varnothing\right) \implies \text{PO} \subseteq r\right\}$ |
| $\mathcal{C}_8$ | $\mathcal{P}_8 \cap \left\{r \in \text{RCA}_8 \mid \left(\text{EC} \subseteq r \wedge r \cap \left(\text{TPP} \cup \text{NTPP} \cup \overline{\text{TPP}} \cup \overline{\text{NTPP}} \cup \text{EQ}\right) \neq \varnothing\right) \implies \text{PO} \subseteq r\right\}$ |

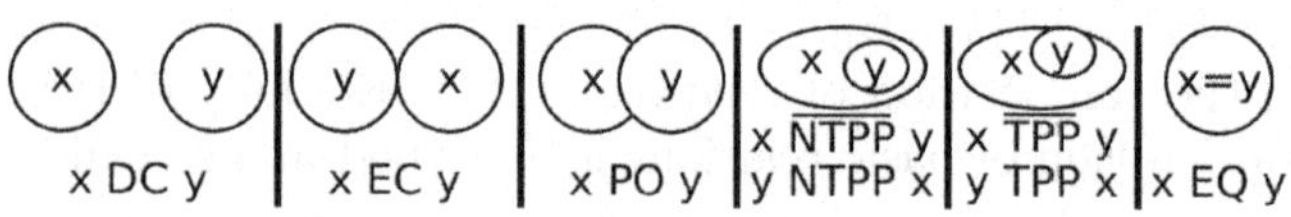


**Fig. 1.** The 8 basic relations of $\text{RCC}_8$ in the plane.

## 2.3 Link with Finite CSP

We briefly discuss the links between qualitative formalisms and finite CSP. On the one hand, for some qualitative formalisms, the algebraic closure enforces path-consistency [38]. On the other hand, a qualitative constraints network can be translated into a network of finite quantitative constraints [49]. The CSP variables are the relations between the qualitative variables. More precisely, there is a CSP variable $v_{xy}$ for each pair of qualitative variables $(x, y)$. The set of possible values for the CSP variable $v_{xy}$ is the set of basic relations contained in the relation between $x$ and $y$. The CSP constraints between the CSP variables encode the composition operator. These constraints are ternary and of the form $\{(b'', b, b') \in \mathcal{B}^3 \mid b'' \subseteq b \circ b'\}$.

### 2.4 Semantics of Continuously Evolving Regions

Before presenting temporal sequences over $\mathrm{RCC}_8$, we must formally define what we call a region evolving continuously over time. A region evolving continuously during a time interval $I$ (i.e. a real closed interval) is naturally defined as a continuous function $f$ from $I$ to the set of considered regions $\mathfrak{R}$ of a topological space (for instance, $\mathfrak{R}$ can be the regions of $\mathbb{R}^n$ with $n \geq 1$ and can possibly be restricted to convex or connected regions). However, this standard mathematical definition requires that $\mathfrak{R}$ be associated with a topology. Thus, we require the following concept:

**Definition 1.** *A* topological region space $(\mathfrak{R}, T)$ *is a set of regions* $\mathfrak{R}$ *of a topological space associated with a topology* $T$ *(i.e.* $(\mathfrak{R}, T)$ *is also a topological space).*

There are several possible topologies for the regions of $\mathbb{R}^n$ [10,20]. In particular, choosing a metric between regions amounts to choosing a topology. Depending on the choice of the topological region space, the evolution of regions satisfies or violates certain properties, such as continuity of particular functions (area, distance, union, projection, convex hull, ...) [10]. In fact, solids, gases, shadows, ... do not evolve continuously in the same way [20]. The usual metric of the regions of $\mathbb{R}^n$ is the *Hausdorff distance*. Unfortunately, the corresponding evolution of the relations of regions is not *compatible* with the classical neighbourhood graph of $\mathrm{RCC}_8$ (Fig. 2a) [10]. The *dual-Hausdorff distance* [10] corrects this problem: the evolution of regions according to this metric is compatible with the classical neighbourhood graph of $\mathrm{RCC}_8$. Note that other metrics also correct it.

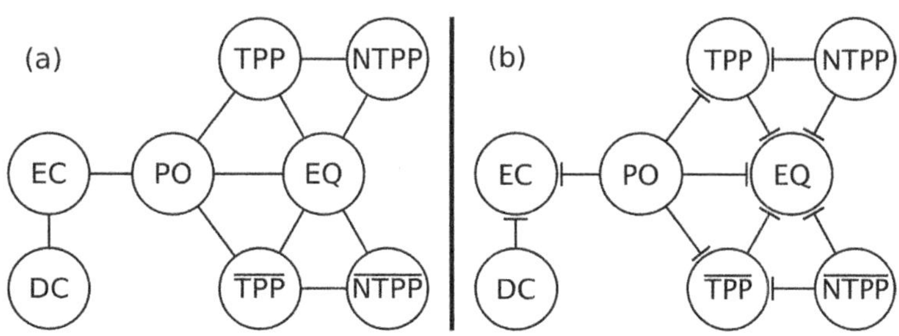


**Fig. 2.** Neighbourhood graph of $\mathrm{RCC}_8$ (a) and dominance graph of $\mathrm{RCC}_8$ (b).

### 2.5 Topological Sequences at Neighboring Instants

We present in this section the topological temporal sequences describing the continuous evolution of regions at neighboring instants [7], that we denote $\mathrm{TRCC}_8^{\mathrm{n}}$.

For this, we recall the basics of the framework of *multi-algebras* [6,7] from which it is defined. It is an abstract framework that includes several extensions of classical qualitative formalisms, such as temporal sequences.

**Projections and Relations.** Multi-algebras generalize non-associative binary relation algebras. *A multi-algebra* is a Cartesian product $\mathcal{A} = \mathcal{A}_1 \times \cdots \times \mathcal{A}_m$ of relation algebras satisfying certain properties. We denote by $\mathtt{I}$ the index set of the multi-algebra, i.e. $\{1, \ldots, m\}$. In the context of temporal sequences, each algebra $\mathcal{A}_i$ corresponds to the same relation algebra but to a different time period. The *set of basic relations* of $\mathcal{A}$, denoted $\boldsymbol{\mathcal{B}}$, is $\boldsymbol{\mathcal{B}}_1 \times \cdots \times \boldsymbol{\mathcal{B}}_m$ where $\boldsymbol{\mathcal{B}}_i$ is the set of basic relations of $\mathcal{A}_i$. Multi-algebras are equipped with a set of additional operators $\Rsh_i^j$ from $\mathcal{A}_i$ to $\mathcal{A}_j$ for all distinct $i, j \in \mathtt{I}$, called *projections.* Any projection $\Rsh$ satisfies by definition $\Rsh (r \cup r') = (\Rsh r) \cup (\Rsh r')$ and $\Rsh (\overline{r}) = \overline{\Rsh (r)}$. In the context of temporal sequences, projections describe the possible evolution of relations over time.

**Table 3.** Neighboring relations of the basic relations of $\mathrm{RCC}_8$.

| $b$ | $\updownarrow b$ |
|---|---|
| DC | DC ∪ EC |
| EC | DC ∪ EC ∪ PO |
| PO | EC ∪ PO ∪ TPP ∪ $\overline{\mathrm{TPP}}$ ∪ EQ |
| TPP | PO ∪ TPP ∪ NTPP ∪ EQ |
| NTPP | TPP ∪ NTPP ∪ EQ |
| EQ | PO ∪ TPP ∪ NTPP ∪ $\overline{\mathrm{TPP}}$ ∪ $\overline{\mathrm{NTPP}}$ ∪ EQ |

**Definition 2.** *The operator* $\updownarrow$ *from* $\mathrm{RCA}_8$ *to* $\mathrm{RCA}_8$ *is the projection satisfying the Table 3.*

The projection $\updownarrow$ encodes the neighbourhood graph of $\mathrm{RCC}_8$ described in Fig. 2a (i.e. $b \subseteq \updownarrow b'$ if and only if $b$ and $b'$ are neighbours). For instance, $\mathrm{PO} \not\subseteq \updownarrow \mathrm{DC}$ because it is not possible to have a continuous transition from DC to PO. A *relation* of a multi-algebra is an $m$-tuplet of classical relations. By adding semantics, that is to say a universe $\mathcal{U}$ and a specific interpretation function $\varphi$, we get a qualitative formalism said *loosely combined*, also called *sequential formalism*.

*Example 1.* To illustrate the preceding concepts and to give intuition concerning $\mathrm{TRCC}_8^{\mathrm{n}}$, we give some examples ($\mathrm{TRCC}_8^{\mathrm{n}}$ will be formalized in the next subsection). The Cartesian product of the multi-algebra $\mathcal{A}$ of $\mathrm{TRCC}_8^{\mathrm{n}}$ is $\mathrm{RCA}_8^m$ where $m$ is the length of considered temporal sequences (the sequences describe regions at instants $t_1, \ldots, t_m$). The component $i \in \mathtt{I}$ of a relation $R$ of $\mathrm{TRCC}_8^{\mathrm{n}}$, $R_i$, is the relation of $\mathrm{RCC}_8$ which must be satisfied at the instant $t_i$. An example of relations of $\mathrm{TRCC}_8^{\mathrm{n}}$, with $m = 3$, is $(\mathrm{TPP} \cup \mathrm{NTPP} \cup \overline{\mathrm{TPP}} \cup \overline{\mathrm{NTPP}}, \mathrm{PO} \cup \mathrm{EQ}, \mathrm{EC} \cup \mathrm{DC})$. This relation means on the one hand that one of the two regions is first included in the other ($R_1$ is satisfied at $t_1$), then they overlap or are equal ($R_2$ is satisfied at $t_2$) and finally they are disjoint ($R_3$ is satisfied at $t_3$). Since the instants of the sequence are *neighbors*, this relation means, on the other hand, that between $t_1$ and $t_2$ the regions satisfy either the basic relation being satisfied at $t_1$ (which is TPP or NTPP or $\overline{\mathrm{TPP}}$ or $\overline{\mathrm{NTPP}}$) or the one being satisfied at $t_2$ (which is PO or EQ) and between $t_2$ and $t_3$ they satisfy either the basic relation being satisfied at $t_2$ (which is PO or EQ) or the one being satisfied at $t_3$ (which is EC or DC). This additional constraint is called *continuity without intermediary relation* and also *continuous qualitative change* [7,48]. This constraint enforces that each sequence of relations $R$ describes all changes of relations between regions. In other words, between two instants $t_i$ and $t_{i+1}$, there must be no change of relation, other than the transition from the basic relation satisfied at $t_i$ towards the basic relation satisfied at $t_{i+1}$.

**Multi-algebra and Relation Operators.** We recall the definition of $\mathrm{TRCC}_8^{\mathrm{n}}$ (originally denoted $\mathrm{TT}_{\mathrm{wir}}$ [7]).

**Definition 3.** *Let $t_1, \ldots, t_m \in \mathbb{R}$ be consecutive instants and $(\mathfrak{R}, T)$ be a topological region space.* $\mathrm{TRCC}_8^{\mathrm{n}}$ *is the triplet $(\mathcal{A}, \mathcal{U}, \varphi)$ where:*

- *$\mathcal{A}$ is $\mathrm{RCA}_8^m$ equipped with the projections $\restriction_i^j$ fully defined by $\restriction_i^j b = \updownarrow b$ if $|j-i| = 1$ and $\restriction_i^j b = \mathcal{B}_{\mathrm{RCC}_8}$ otherwise, for all $b \in \boldsymbol{\mathcal{B}}_{\mathrm{RCC}_8}$ and $i, j \in \mathtt{I}$,*
- *$\mathcal{U}$ is the set of continuous functions from $[t_1, t_m]$ to $\mathfrak{R}$, and*
- *$\varphi$ is the function from $\mathcal{A}$ to $2^{\mathcal{U} \times \mathcal{U}}$ such that for all $R \in \mathcal{A}$, $\varphi(R)$ is the set of pairs of functions $(f, f') \in \mathcal{U} \times \mathcal{U}$ satisfying at each instant $t_i$ the relation $R_i$ (i.e. $\forall i \in \mathtt{I}\ (f(t_i), f'(t_i)) \in \varphi_{\mathrm{RCC}_8}(R_i)$) and satisfying no intermediary relations between each instants $t_i$ and $t_{i+1}$ (i.e. during each $[t_i, t_{i+1}]$ a basic relation is satisfied, then another, formally: for all $i \in \mathtt{I} \setminus \{m\}$ there exist $\tau \in [t_i, t_{i+1}]$ and $b, b' \in \boldsymbol{\mathcal{B}}_{\mathrm{RCC}_8}$ such that at each instant $t \in [t_i, \tau[$, $(f(t), f'(t)) \in \varphi_{\mathrm{RCC}_8}(b)$, at each instant $t \in ]\tau, t_{i+1}]$, $(f(t), f'(t)) \in \varphi_{\mathrm{RCC}_8}(b')$, and that $(f(\tau), f'(\tau)) \in \varphi_{\mathrm{RCC}_8}(b \cup b')$), with $\varphi_{\mathrm{RCC}_8}$ the interpretation function of* $\mathrm{RCC}_8$.

*Remark 1.* $\mathrm{TRCC}_8^{\mathrm{n}}$ depends on a set of regions but also on a topology for the regions (i.e. a notion of continuity). Note that $\mathrm{TRCC}_8^{\mathrm{n}}$ is not necessarily a sequential formalism (i.e. its reasoning operators are not necessarily correct: reasoning operators could remove some solutions). To be a sequential formalism, the evolution of regions corresponding to the chosen topological region space must be compatible with the classical neighbourhood graph of $\mathrm{RCC}_8$ (see Sect. 2.4). Thus, if $\mathfrak{R}$ is $\mathbb{R}^n$ equipped with the dual-Hausdorff distance, $\mathrm{TRCC}_8^{\mathrm{n}}$ is a sequential formalism.

Every multi-algebra has operators on its relations, namely *composition*, *union*, *intersection*, and *converse*. They are defined componentwise. For example, the composition of $R$ and $R'$, $R \circ R'$, is defined by $(R \circ R')_i = R_i \circ R'_i$ for all $i \in \mathtt{I}$.

There is another operator on relations: the *projection closure* of a relation $R$, denoted $\Rsh(R)$. It consists in sequentially applying the following operation until reaching a fixed point: for all $j \in \mathtt{I}$, $R_j \leftarrow R_j \cap \bigcap_{i \neq j} \Rsh_i^j R_i$. Projection closure refines relations by removing classical basic relations that are impossible to satisfy. In the context of $\mathrm{TRCC}_8^n$, projection closure enforces continuity without intermediary relation. For example, the projection closure of the following relation, with $m = 3$, $\left(\mathrm{TPP} \cup \mathrm{NTPP} \cup \overline{\mathrm{TPP}} \cup \overline{\mathrm{NTPP}}, \mathrm{PO} \cup \mathrm{EQ}, \mathrm{EC} \cup \mathrm{DC}\right)$ is $\left(\mathrm{TPP} \cup \overline{\mathrm{TPP}}, \mathrm{PO}, \mathrm{EC}\right)$. Indeed, in particular, there is no transition from the relation PO or from the relation EQ to the relation DC without intermediary relation. In addition, there is no transition from NTPP to DC or EC in just two qualitative changes. Projection closure removes such impossibilities. Relations closed under projection, i.e. satisfying $\Rsh(R) = R$, are said $\Rsh$*-closed*. Note that projection closure can be seen as a kind of arc-consistency.

**Constraint Networks and Algebraic Closure.** A description in the context of multi-algebras is a *(qualitative constraint) network*. A network over a multi-algebra $\mathcal{A}$ is a set of variables $\mathtt{E}$ and a function $N$ associating with each pair of variables $(x, y) \in \mathtt{E}^2$ such that $x \neq y$ a relation of $\mathcal{A}$ and satisfying $N(x, y) = \overline{N(y, x)}$ for all distinct $x, y \in \mathtt{E}$. A sequence of classical constraint networks is thus represented by a single constraint network whose relations are sequences of relations, i.e. relations of a multi-algebra. We denote $N(x, y)$ more succinctly by $N^{xy}$. It is sometimes useful to refer to the "subnetwork" corresponding to the index $i \in \mathtt{I}$ of a network $N$, denoted $N_i$, called *slice*. $N_i$ is defined by $(N_i)^{xy} = (N^{xy})_i$ for all distinct $x, y \in \mathtt{E}$. In the context of temporal sequences, the slice $i$ of a network $N$, $N_i$, describes the relations of the sequence at the instant $t_i$. Similarly, the *slice* $i \in \mathtt{I}$ of a subset $\mathcal{S} \subseteq \mathcal{A}$, denoted $\mathcal{S}_i$, is $\{R_i \mid R \in \mathcal{S}\}$. A network is said to be *satisfiable* (or *consistent*) if there is a *solution* to this network, that is, an assignment for the variables $\{u_x\}_{x \in \mathtt{E}} \subseteq \mathcal{U}$ satisfying the relations of the constraint network, i.e. $(u_x, u_y) \in \varphi(N^{xy})$. A network $N$ is said *over* a subset of relations $\mathcal{S} \subseteq \mathcal{A}$ if for all distinct $x, y \in \mathtt{E}$, $N^{xy} \in \mathcal{S}$. A *scenario* is a network over $\boldsymbol{\mathcal{B}}$.

The reasoning operator on networks is the *algebraic closure*, which applies the operators of the multi-algebra. It propagates information within the network, makes inferences, by refining relations. In the context of topological temporal sequences, the algebraic closure propagates information over regions at each instant and between the different instants. A relation $R$ *refines* a relation $R'$ if $R_i \subseteq R'_i$ for all $i \in \mathtt{I}$. More generally, $N$ *refines* $N'$, denoted $N \subseteq N'$, if for all distinct $x, y \in \mathtt{E}$, $N^{xy} \subseteq (N')^{xy}$. Algebraic closure closes networks under composition and under projection. Algebraic closure thus applies the two following operations until reaching a fixed point: $N^{xz} \leftarrow N^{xz} \cap (N^{xy} \circ N^{yz})$ and $N^{xz} \leftarrow \Rsh(N^{xz})$ for all distinct $x, y, z \in \mathtt{E}$. We denotes by $\mathfrak{C}(N)$ the algebraic closure of $N$. In the context of topological temporal sequences, the composition operator makes inferences at every instant and the projection operator makes inferences between the instants. A network $N$ is called *algebraically closed* if it is closed under composition, i.e. for all distinct $x, y, z \in \mathtt{E}$, $N^{xz} \subseteq N^{xy} \circ N^{yz}$, and if each of its relations $N^{xy}$ is closed under projection, i.e. for all distinct $i, j \in \mathtt{I}$, $N_j^{xy} \subseteq \Rsh_i^j N_i^{xy}$.

***Consistency and Satisfiability.*** A sequential formalism is said *complete* if all its algebraically closed scenarios are satisfiable. It is a fondamental property for deciding satisfiability in an algebraic way.

*Remark 2.* To know if $\mathrm{TRCC}_8^{\mathrm{n}}$ is complete for the regions of $\mathbb{R}^n$ equipped with the dual-Hausdorff distance or for another topological region space is a complex problem. In fact, perhaps there is no topological region space such that $\mathrm{TRCC}_8^{\mathrm{n}}$ is a complete sequential formalism. For this reason, several studies have dealt with a "weak satisfiability", i.e. satisfiability with a weaker notion of continuity [2,22, 44,48]. Formally, a network $N$ over $\mathrm{TRCC}_8^{\mathrm{n}}$ is *weakly satisfiable* if it contains an algebraically closed scenario (i.e. if there exists a sequence of satisfiable classical scenarios satisfying the constraints of the networks $N_i$ *and* the neighbourhood graph). The tractable subclasses that we identify in this article are tractable for $\mathrm{TRCC}_8^{\mathrm{n}}$ associated with a topological region space such that $\mathrm{TRCC}_8^{\mathrm{n}}$ is a complete sequential formalism. They are also tractable for this notion of weak satisfiability.

A relation $R$ is said *trivially unsatisfiable* if there exists $i \in \mathtt{I}$ such that $R_i = \varnothing$. Note that a relation which is not trivially unsatisfiable can be *unsatisfiable*, i.e. $\varphi(R) = \varnothing$. This is the case of (PO, PO, DC). A relation is said $\Rsh$*-consistent* if it is $\Rsh$-closed and it is not trivially unsatisfiable. A network is said *trivially unsatisfiable* if there exists distinct $x, y \in \mathtt{E}$ such that $N^{xy}$ is a trivially unsatisfiable relation. An algebraically closed network that is not trivially unsatisfiable is said to be *algebraically consistent.*

**Tractable Subclasses.** By restricting networks to certain subsets of relations $\mathcal{S}$, we get the following property: if the algebraic closure of a network over $\mathcal{S}$ is algebraically consistent, then this network is satisfiable. Such subsets are said to be *algebraically tractable.* In other words, with an algebraically tractable subset $S$, to decide the satisfiability of a network over $S$, it suffices to verify that its algebraic closure is not trivially inconsistent. The search for algebraically tractable subsets has focused on particular subsets [30]. A subset $\mathcal{S} \subseteq \mathcal{A}$ is said a *subclass* if it is closed under intersection, composition, and converse (i.e. for all $R, R' \in \mathcal{S}$, we have $R \cap R' \in \mathcal{S}$, $R \circ R' \in \mathcal{S}$, and $\overline{R} \in \mathcal{S}$). Subclasses containing all basic relations (i.e. $\boldsymbol{\mathcal{B}} \subseteq \mathcal{S}$) are called *subalgebras.* A subset $\mathcal{S} \subseteq \mathcal{A}$ is said $\Rsh$*-closed* if for all $R \in \mathcal{S}$, $\Rsh(R) \in \mathcal{S}$. Finally, we say that a subset $\mathcal{S}$ is *Cartesian* if $\mathcal{S} = \mathcal{S}_1 \times \cdots \times \mathcal{S}_m$.

Note that a list of conditions guaranteeing algebraic tractability has been identified [6] (see the slicing and refinement theorems). One of these conditions is algebraic stability by a refinement $H$. A *refinement* $H$ is a function from $\mathcal{A}$ to $\mathcal{A}$ satisfying $H(R) \subseteq R$. A subset $\mathcal{S} \subseteq \mathcal{A}$ is *algebraically stable by* $H$ if for any algebraically consistent network $N$ over $\mathcal{S}$, the network $H(N)$ is algebraically consistent, where $H(N)$ is the network obtained from $N$ by substituting each relation $N^{xy}$ by $H(N^{xy})$. $\text{RCC}_8$ has two fundamental refinements: $h_{\mathcal{H}_8}(r) = \text{a}_{\overline{\text{TPP}}}(\text{a}_{\text{TPP}}(\text{a}_{\text{PO}}(\text{a}_{\text{EC}}(\text{a}_{\text{DC}}(r)))))$ and $h_{\mathcal{C}_8}(r) = \text{a}_{\overline{\text{TPP}}}(\text{a}_{\text{TPP}}(\text{a}_{\overline{\text{NTPP}}}(\text{a}_{\text{NTPP}}(\text{a}_{\text{PO}}(\text{a}_{\text{DC}}(r))))))$ with $\text{a}_b$, $b \in \boldsymbol{\mathcal{B}}$, the function from $\text{RCA}_8$ to $\text{RCA}_8$ defined by $\text{a}_b(r) = b$ if $b \subseteq r$ and $\text{a}_b(r) = r$ otherwise. $\mathcal{H}_8$ and $\mathcal{Q}_8$ are algebraically stable by $h_{\mathcal{H}_8}$ and $\mathcal{C}_8$ is algebraically stable by $h_{\mathcal{C}_8}$ [37]. Moreover, for every $r \in \mathcal{H}_8 \cup \mathcal{Q}_8$ such that $r \neq \varnothing$, $h_{\mathcal{H}_8}(r) \in \boldsymbol{\mathcal{B}}_{\text{RCC}_8}$ and for every $r \in \mathcal{C}_8 \backslash \{\varnothing\}$, $h_{\mathcal{C}_8}(r) \in \boldsymbol{\mathcal{B}}_{\text{RCC}_8}$ [37]. In the following, we are interested in the refinement $H_{\mathcal{S}}$ defined by $H_{\mathcal{S}}(R)_i = h_{\mathcal{H}_8}(R_i)$ if $\mathcal{S}_i \subseteq \mathcal{H}_8$ or $\mathcal{S}_i \subseteq \mathcal{Q}_8$ and $H_{\mathcal{S}}(R)_i = h_{\mathcal{C}_8}(R_i)$ otherwise, for all $i \in \text{I}$ and $R \in \text{RCA}_8^m$, with $\mathcal{S} \subseteq \text{RCA}_8^m$.

## 3 Study of $\text{TRCC}_8^{\text{n}}$ Subclasses

In this section, we are interested in temporalized $\text{RCC}_8$ at neighboring instants, i.e. $\text{TRCC}_8^{\text{n}}$ (see Sect. 2.5). More precisely, we search for subclasses that are algebraically tractable. Unfortunately, as the following proposition shows, there are no algebraically tractable Cartesian subalgebras (at least for $m \geq 4$).

**Proposition 1.** *Let* $\mathfrak{R}$ *be a topological region space such that* $\text{TRCC}_8^{\text{n}}$ *is a sequential formalism.*

> *No Cartesian subalgebra of* $\text{TRCC}_8^{\text{n}}$ *is algebraically tractable (when* $m \geq 4$*).*
> *No* $\Rsh$*-closed Cartesian subalgebra of* $\text{TRCC}_8^{\text{n}}$ *is algebraically tractable (if* $m \geq 2$*).*

*Proof.* We show the case $m = 4$. The idea is that the algebraic closure can produce relations $R$ verifying $\text{TPP} \cup \text{EQ} \subseteq R_i \subseteq \text{TPP} \cup \text{NTPP} \cup \text{EQ}$ and $\overline{\text{TPP}} \cup \text{EQ} \subseteq R_j \subseteq \overline{\text{TPP}} \cup \overline{\text{NTPP}} \cup \text{EQ}$ with $|i-j| = 1$ which causes that some unsatisfiable networks are algebraically consistent. Let $\mathcal{S}$ be any Cartesian subalgebra (thus $\mathcal{S}$ contains the closure of the basic relations and the universal relation of $\text{RCC}_8$ under intersection, composition, and converse). We show that there exists an unsatisfiable network over $\mathcal{S}$ whose algebraic closure is algebraically consistent. Consider the network $N$ satisfying: $\text{E} = \{u, v, w, x, y, z\}$, $N_1^{xy} = \text{NTPP}$, $N_4^{xy} = \overline{\text{NTPP}}$, $N_1^{xz} = \text{NTPP}$, $N_3^{wz} = \text{NTPP}$, $N_1^{yz} = \overline{\text{TPP}}$, $N_2^{yz} = \text{PO} \cup \text{TPP} \cup \overline{\text{TPP}} \cup \text{EQ}$, $N_1^{wx} = \overline{\text{TPP}}$, $N_2^{wx} = \text{PO} \cup \text{TPP} \cup \overline{\text{TPP}} \cup \text{EQ}$, $N_2^{wy} = \text{PO} \cup \text{TPP}$, $N_4^{xu} = \overline{\text{NTPP}}$, $N_2^{vu} = \overline{\text{NTPP}}$, $N_4^{yu} = \text{TPP}$, $N_3^{yu} = \text{PO} \cup \text{TPP} \cup \overline{\text{TPP}} \cup \text{EQ}$, $N_4^{vx} = \text{TPP}$, $N_3^{vx} = \text{PO} \cup \text{TPP} \cup \overline{\text{TPP}} \cup \text{EQ}$, $N_3^{vy} = \text{PO} \cup \overline{\text{TPP}}$, and $N_i^{ab} = \mathcal{B}_{\text{RCC}_8}$ in the other cases. The network is over $\mathcal{S}$ (indeed, we have $\text{DC} \circ \text{DC} = \mathcal{B}_{\text{RCC}_8}$, $\overline{\text{TPP}} \circ \text{TPP} = \text{PO} \cup \text{TPP} \cup \overline{\text{TPP}} \cup \text{EQ}$, and $(\text{EC} \circ \text{EC}) \cap (\text{EC} \circ \text{NTPP}) = \text{PO} \cup \text{TPP}$). Its algebraic closure $\mathfrak{C}(N)$ is the algebraically consistent network satisfying:

- $\mathfrak{C}(N)^{xy} = (\text{NTPP}, \text{TPP} \cup \text{NTPP} \cup \text{EQ}, \overline{\text{TPP}} \cup \overline{\text{NTPP}} \cup \text{EQ}, \overline{\text{NTPP}})$,
- $\mathfrak{C}(N)^{yz} = (\overline{\text{TPP}}, \text{PO} \cup \overline{\text{TPP}} \cup \text{EQ}, \mathcal{B} \backslash (\text{DC} \cup \text{EC}), \mathcal{B} \backslash \text{DC})$,
- $\mathfrak{C}(N)^{xz} = (\text{NTPP}, \text{TPP} \cup \text{NTPP} \cup \text{EQ}, \mathcal{B} \backslash (\text{DC} \cup \text{EC}), \mathcal{B} \backslash (\text{DC} \cup \text{EC}))$,

- $\mathfrak{C}(N)^{wx} = (\overline{\text{TPP}}, \text{PO} \cup \overline{\text{TPP}} \cup \text{EQ}, \mathcal{B} \backslash (\text{DC} \cup \overline{\text{NTPP}}), \mathcal{B})$,
- $\mathfrak{C}(N)^{wy} = (\text{PO} \cup \text{TPP} \cup \text{NTPP}, \text{PO} \cup \text{TPP}, \mathcal{B} \backslash (\text{DC} \cup \overline{\text{NTPP}}), \mathcal{B})$,
- $\mathfrak{C}(N)^{wz} = (\text{PO}\cup\text{TPP}\cup\text{NTPP}, \text{TPP}\cup\text{NTPP}\cup\text{EQ}, \text{NTPP}, \text{TPP}\cup\text{NTPP}\cup\text{EQ})$,
- $\mathfrak{C}(N)^{ux} = (\mathcal{B}\backslash\text{DC}, \mathcal{B}\backslash(\text{DC} \cup \text{EC}), \text{TPP} \cup \text{NTPP} \cup \text{EQ}, \text{NTPP})$,
- $\mathfrak{C}(N)^{uy} = (\mathcal{B}\backslash(\text{DC} \cup \text{EC}), \mathcal{B}\backslash(\text{DC} \cup \text{EC}), \text{PO} \cup \overline{\text{TPP}} \cup \text{EQ}, \overline{\text{TPP}})$,
- $\mathfrak{C}(N)^{uz} = (\mathcal{B}\backslash(\text{DC} \cup \text{EC}), \mathcal{B}\backslash(\text{DC} \cup \text{EC}), \mathcal{B}\backslash\text{DC}, \mathcal{B}\backslash\text{DC})$,
- $\mathfrak{C}(N)^{uw} = (\mathcal{B}\backslash\text{DC}, \mathcal{B}, \mathcal{B}, \mathcal{B})$, $\mathfrak{C}(N)^{vx} = (\mathcal{B}\backslash\text{DC}, \mathcal{B}\backslash(\text{DC} \cup \text{EC} \cup \text{NTPP}), \text{PO} \cup \text{TPP} \cup \text{EQ}, \text{TPP})$,
- $\mathfrak{C}(N)^{vy} = (\mathcal{B}\backslash(\text{DC} \cup \text{EC}), \mathcal{B}\backslash(\text{DC} \cup \text{EC} \cup \text{NTPP}), \text{PO} \cup \overline{\text{TPP}}, \text{EC} \cup \text{PO} \cup \overline{\text{TPP}} \cup \overline{\text{NTPP}})$,
- $\mathfrak{C}(N)^{vz} = (\mathcal{B}\backslash(\text{DC} \cup \text{EC}), \mathcal{B}\backslash(\text{DC} \cup \text{EC}), \mathcal{B}\backslash\text{DC}, \mathcal{B})$,
- $\mathfrak{C}(N)^{vw} = (\mathcal{B}\backslash\text{DC}, \mathcal{B}, \mathcal{B}, \mathcal{B})$,
- $\mathfrak{C}(N)^{vu} = (\overline{\text{TPP}}\cup\overline{\text{NTPP}}\cup\text{EQ}, \overline{\text{NTPP}}, \overline{\text{TPP}}\cup\overline{\text{NTPP}}\cup\text{EQ}, \text{PO}\cup\overline{\text{TPP}}\cup\overline{\text{NTPP}})$.

However, $N$ is not satisfiable since $\mathfrak{C}(N)$ is not satisfiable. Indeed, to refine $\mathfrak{C}(N)_2^{xy}$ by EQ, TPP, or NTPP and then to apply the algebraic closure gives a trivially unsatisfiable network. This can be seen by the fact that the only satisfiable basic relations $B$ of $\mathfrak{C}(N)^{xy}$ satisfy $B_2 = \text{EQ}$ or $B_3 = \text{EQ}$ and that there exists neither algebraically closed scenario $S \subseteq \mathfrak{C}(N)_2$ satisfying $S^{xy} = \text{EQ}$ nor algebraically closed scenario $S \subseteq \mathfrak{C}(N)_3$ satisfying $S^{xy} = \text{EQ}$. For example, by setting $\mathfrak{C}(N)_2^{xy} = \text{EQ}$, we get $\mathfrak{C}(N)_2^{xy} = \mathfrak{C}(N)_2^{yz} = \mathfrak{C}(N)_2^{xz} = \text{EQ}$ and therefore $\mathfrak{C}(N)_2^{wx} = \mathfrak{C}(N)_2^{wy} = \mathfrak{C}(N)_2^{wz} = \mathfrak{C}(N)_2^{wx} \cap \mathfrak{C}(N)_2^{wy} \cap \mathfrak{C}(N)_2^{wz}$ i.e. $(\text{PO} \cup \overline{\text{TPP}} \cup \text{EQ}) \cap (\text{PO} \cup \text{TPP}) \cap (\text{TPP} \cup \text{NTPP} \cup \text{EQ}) = \varnothing$.

Let $\mathcal{S}$ be a $\Gamma$-closed Cartesian subalgebra of $\text{TRCC}_8^{\text{n}}$ with $m = 2$. The network $N'$ satisfying $\mathtt{E} = \{u, v, w, x, y, z\}$, $N'_1 = \mathfrak{C}(N)_2$, and $N'_2 = \mathfrak{C}(N)_3$ is algebraically consistent, unsatisfiable, and over $\mathcal{S}$.

For this reason, we are looking for Cartesian subclasses that do not contain all basic relations. In particular, we are interested in the following subset of $\text{RCA}_8$: $\mathcal{H}_8^{\text{NTPP}\Rightarrow\text{TPP}}$ defined by $\{r \in \mathcal{H}_8 | \text{NTPP} \subseteq r \implies \text{TPP} \subseteq r \wedge \overline{\text{NTPP}} \subseteq r \implies \overline{\text{TPP}} \subseteq r\}$. It is easy to prove that this subset is a subclass.

**Lemma 1.** *The subset* $\mathcal{H}_8^{\text{NTPP}\Rightarrow\text{TPP}}$ *is a subclass.*

We show that we can obtain algebraically tractable Cartesian subclasses of $\text{TRCC}_8^{\text{n}}$ satisfying $\mathcal{S}_i \in \{\mathcal{Q}_8, \mathcal{H}_8^{\text{NTPP}\Rightarrow\text{TPP}}\}$. For this, we apply the refinement theorem (by using the refinement $H_{\mathcal{S}}$; see Sect. 2.5). We begin by showing the conditions of the theorem.

**Lemma 2.** *We have the following properties:*

- $\forall r \in \mathcal{Q}_8$, $\updownarrow r \in \mathcal{H}_8^{\text{NTPP}\Rightarrow\text{TPP}}$ *and*
- $\forall r \in \mathcal{H}_8^{\text{NTPP}\Rightarrow\text{TPP}}$, $\updownarrow r \in \mathcal{Q}_8$.

*Proof.* See the extended version of this paper at https://arxiv.org/abs/2007.07711.

**Lemma 3.** *Let* $\mathcal{S} \in \{\mathcal{Q}_8 \times \mathcal{Q}_8, \mathcal{Q}_8 \times \mathcal{H}_8, \mathcal{H}_8 \times \mathcal{Q}_8\}$ *be a subclass of* $\mathrm{TRCC}_8^{\mathrm{n}}$ $(m = 2)$ *and let* $R \in \mathcal{S}$.

*If* $R$ *is* $\Rsh$*-consistent then* $H_{\mathcal{S}}(R)$ *is* $\Rsh$*-consistent.*

*Proof.* See the extended version of this paper at https://arxiv.org/abs/2007.07711.

Note the following proposition, before identifying the tractable subclasses. It shows that although the tractable subalgebras of $\mathrm{RCC}_8$ cannot be combined to obtain algebraically tractable Cartesian subalgebras, the algebraically consistent networks over the majority of these combinations are satisfiable.

**Proposition 2.** *Let* $\mathfrak{R}$ *be a topological region space such that* $\mathrm{TRCC}_8^{\mathrm{n}}$ *is a complete sequential formalism. Let* $\mathcal{S}$ *be a subset of* $\mathrm{TRCC}_8^{\mathrm{n}}$ *satisfying* $\mathcal{S}_i \in \{\mathcal{H}_8, \mathcal{Q}_8\}$ *and* $\mathcal{S}_i = \mathcal{H}_8 \implies (i = m \vee \mathcal{S}_{i+1} = \mathcal{Q}_8) \wedge (i = 1 \vee \mathcal{S}_{i-1} = \mathcal{Q}_8)$ *for all* $i \in \mathtt{I}$.

*Algebraically consistent networks over* $\mathcal{S}$ *are satisfiable.*

*Proof.* We apply the refinement theorem [6]. $H_{\mathcal{S}}$ is a refinement from $\mathcal{S}$ to the set $\boldsymbol{\mathcal{B}} \cup \{(\varnothing, \ldots, \varnothing)\}$ (since $h_{\mathcal{H}_8}$ is a refinement from $\mathcal{H}_8 \cup \mathcal{Q}_8$ to $\boldsymbol{\mathcal{B}}_{\mathrm{RCC}_8} \cup \{\varnothing\}$ [37]). $\mathcal{S}$ is algebraically stable by $H_{\mathcal{S}}$. Indeed, since on the one hand, $\mathcal{H}_8$ and $\mathcal{Q}_8$ are algebraically stable by $h_{\mathcal{H}_8}$ [37]. Since, on the other hand, for any $\Rsh$-consistent relation $R \in \mathcal{S}$, $H_{\mathcal{S}}(R)$ is $\Rsh$-consistent (by Lemma 3). Algebraically consistent networks over $\boldsymbol{\mathcal{B}} \cup \{(\varnothing, \ldots, \varnothing)\}$ are satisfiable ($\mathrm{TRCC}_8^{\mathrm{n}}$ is complete). By the refinement theorem, algebraically consistent networks over $\mathcal{S}$ are satisfiable.

Satisfiability of algebraically consistent networks is, in general, a weaker property than algebraic tractability. It is not equivalent for subclasses that are not $\Rsh$-closed. Indeed, applying the algebraic closure on a network over a subclass which is not $\Rsh$-closed can move the network out of the subclass. In that case, we cannot therefore conclude that the network is satisfiable (if it is not trivially inconsistent). The previous subclasses are not $\Rsh$-closed: the projection of NTPP, $\updownarrow \mathrm{NTPP} = \mathrm{TPP} \cup \mathrm{NTPP} \cup \mathrm{EQ}$ does not belong to $\mathcal{Q}_8$.

We end this section by showing that the subclasses of the following particular forms $(\mathcal{Q}_8 \times \mathcal{H}_8^{\mathrm{NTPP}\Rightarrow\mathrm{TPP}})^\star$, $(\mathcal{H}_8^{\mathrm{NTPP}\Rightarrow\mathrm{TPP}} \times \mathcal{Q}_8)^\star$, $\mathcal{H}_8^{\mathrm{NTPP}\Rightarrow\mathrm{TPP}} \times (\mathcal{Q}_8 \times \mathcal{H}_8^{\mathrm{NTPP}\Rightarrow\mathrm{TPP}})^\star$, and $\mathcal{Q}_8 \times (\mathcal{H}_8^{\mathrm{NTPP}\Rightarrow\mathrm{TPP}} \times \mathcal{Q}_8)^\star$ are algebraically tractable.

**Proposition 3.** *Let* $\mathfrak{R}$ *be a topological region space such that* $\mathrm{TRCC}_8^{\mathrm{n}}$ *is a complete sequential formalism. Let* $\mathcal{S}$ *be a subset of* $\mathrm{TRCC}_8^{\mathrm{n}}$ *satisfying one of the two following properties:*

- $\mathcal{S}_i = \mathcal{H}_8^{\mathrm{NTPP}\Rightarrow\mathrm{TPP}}$ *if* $i$ *is even and* $\mathcal{S}_i = \mathcal{Q}_8$ *otherwise, for all* $i \in \mathtt{I}$,
- $\mathcal{S}_i = \mathcal{H}_8^{\mathrm{NTPP}\Rightarrow\mathrm{TPP}}$ *if* $i$ *is odd and* $\mathcal{S}_i = \mathcal{Q}_8$ *otherwise, for all* $i \in \mathtt{I}$.

*The subclass* $\mathcal{S}$ *is algebraically tractable.*

*Proof.* Let $\mathcal{S}$ be a subset of $\mathrm{TRCC}_8^{\mathrm{n}}$ as described in the statement. $\mathcal{S}$ satisfies the conditions of the first part of the refinement theorem (see the proof of Proposition 2). $\mathcal{S}$ also satisfies the conditions of the second part. Indeed, on the one hand,

$\mathcal{S}$ is a subclass (since $\mathcal{S}$ is Cartesian and each $\mathcal{S}_i$ is a subclass [37]; Lemma 1). On the other hand, $\mathcal{S}$ is $\Rsh$-closed (by Lemma 2 and since $\mathcal{S}$ is a Cartesian subclass). $\mathcal{S}$ is thus algebraically tractable (refinement theorem [6]).

Note that the tractable subclasses identified by Proposition 3 do not contain all the basic relations (thus, they are not subalgebras).

## 4 Topological Sequences on a Partition of Time

We have shown that there is no algebraically tractable Cartesian subalgebra in the context of $\mathrm{TRCC}_8^{\mathrm{n}}$, the context of regions described at different time points between which there are no intermediary relations (i.e. at time points which characterize all the qualitative changes). Does this mean that there are no large tractable subclasses in the context of (topological) temporal sequences? We show that this is not the case, by considering topological temporal sequences describing the evolution of regions on a time partition (i.e. on a contiguous alternation of instants and open intervals).

### 4.1 Formalization

We begin by defining the formalism of topological temporal sequences on a partition of time, which we denote $\mathrm{TRCC}_8^{\mathrm{d}}$. Without loss of generality, we consider only the partitions of the interval $[t_0, t_l[$ of the form $(t_0, ]t_0, t_1[, \ldots, t_{l-1}, ]t_{l-1}, t_l[)$ with $m = 2l$, $t_{i-1} < t_i$ for all $i \in \{1, \ldots, l\}$, $t_i \in \mathbb{R}$ for all $i \in \{0, \ldots, l\}$, and $l \in \mathbb{N}^*$. Thus, the sequences of $\mathrm{TRCC}_8^{\mathrm{d}}$ describe the topological relations at each time periods $t_i$ and during each interval $]t_i, t_{i+1}[$.

**Definition 4.** *Let $(\mathfrak{R}, T)$ be a topological region space. Let $t_0, \ldots, t_l \in \mathbb{R}$ be consecutive instants.*

*The formalism $\mathrm{TRCC}_8^{\mathrm{d}}$ (associated with $(\mathfrak{R}, T)$) is the triplet $(\mathcal{A}, \mathcal{U}, \varphi)$ where:*

- *$\mathcal{U}$ is the set of continuous functions from $[t_1, t_m[$ to the set of regions $\mathfrak{R}$,*
- *$\mathcal{A}$ is the multi-algebra whose Cartesian product is $\mathrm{RCA}_8^m$ and whose projections satisfy $\Rsh_i^j b = \uparrow b$ if $|j - i| = 1$ and $i$ is even, $\Rsh_i^j b = \downarrow b$ if $|j - i| = 1$ and $i$ is odd, and $\Rsh_i^j b = \mathcal{B}_{\mathrm{RCC}_8}$ otherwise, for all $b \in \boldsymbol{\mathcal{B}}_{\mathrm{RCC}_8}$ and $i, j \in \mathrm{I}$ with $\uparrow$ and $\downarrow$ defined by Table 4, and*
- *$\varphi$ is the function from $\mathcal{A}$ to $2^{\mathcal{U} \times \mathcal{U}}$ such that for all $R \in \mathcal{A}$, $\varphi(R)$ is the set of pairs of functions $(f, f') \in \mathcal{U} \times \mathcal{U}$ satisfying at each instant $t_i$ the relation $R_{2i+1}$ (i.e. $\forall i \in \{0, \ldots, l-1\}$ $(f(t_i), f'(t_i)) \in \varphi_{\mathrm{RCC}_8}(R_{2i+1})$) and satisfying, for each $i \in \{0, \ldots, l-1\}$, one (and only one) basic relation $b \subseteq R_{2i+2}$ at each instant between $t_i$ and $t_{i+1}$ (i.e. $\exists b \in \boldsymbol{\mathcal{B}}_{\mathrm{RCC}_8}$ $b \subseteq R_{2i+2}$ $\forall t \in ]t_i, t_{i+1}[$ $(f(t), f'(t)) \in \varphi_{\mathrm{RCC}_8}(b)$), with $\varphi_{\mathrm{RCC}_8}$ the interpretation function of $\mathrm{RCC}_8$.*

*Remark 3.* The operator $\uparrow$ encodes the dominance graph of $\mathrm{RCC}_8$ described in Fig. 2b, i.e. the possible evolutions of relations being satisfied during an open interval ($\uparrow$ returns the corresponding relations possibly satisfied at the limits of the interval). The operators $\uparrow$ and $\downarrow$ enforces continuity (when $\mathrm{TRCC}_8^{\mathrm{d}}$ is a complete sequential formalism). Remark 1 on $\mathrm{TRCC}_8^{\mathrm{n}}$ also applies to $\mathrm{TRCC}_8^{\mathrm{d}}$.

*Example 2.* An example of relations of $\mathrm{TRCC}_8^{\mathrm{d}}$, with $m = 4$, is the relation $R = (\mathrm{TPP} \cup \mathrm{NTPP}, \mathrm{PO} \cup \mathrm{EQ}, \mathrm{EC} \cup \mathrm{DC}, \mathrm{DC})$. This relation means that the first region is included in the second at the instant $t_0$ ($R_1$ is satisfied), then they overlap during the interval $]t_0, t_1[$ or are equal during the interval $]t_0, t_1[$ ($R_2$ is satisfied), they are disjoined at the instant $t_1$ ($R_3$ is satisfied), and finally they are disconnected at every instant of $]t_1, t_2[$ ($R_4$ is satisfied). The only satisfiable basic relation included in $R$ is $(\mathrm{TPP}, \mathrm{PO}, \mathrm{EC}, \mathrm{DC}) = \Rsh(R)$.

**Table 4.** Dominant and dominated relations of the basic relations of $\mathrm{RCC}_8$.

| $b$ | $\uparrow b$ | $\downarrow b$ |
|---|---|---|
| DC | $\mathrm{DC} \cup \mathrm{EC}$ | DC |
| EC | EC | $\mathrm{DC} \cup \mathrm{EC} \cup \mathrm{PO}$ |
| PO | $\mathrm{EC} \cup \mathrm{PO} \cup \mathrm{TPP} \cup \overline{\mathrm{TPP}} \cup \mathrm{EQ}$ | PO |
| TPP | $\mathrm{TPP} \cup \mathrm{EQ}$ | $\mathrm{PO} \cup \mathrm{TPP} \cup \mathrm{NTPP}$ |
| NTPP | $\mathrm{NTPP} \cup \mathrm{TPP} \cup \mathrm{EQ}$ | NTPP |
| EQ | EQ | $\mathcal{B}_{\mathrm{RCC}_8} \setminus (\mathrm{DC} \cup \mathrm{EC})$ |

### 4.2 Tractability Results

We now identify large algebraically tractable Cartesian subalgebras, by applying again the refinement theorem. We begin by showing its conditions.

**Lemma 4.** *Let* $\mathcal{S} \in \{\mathcal{H}_8 \times \mathcal{H}_8, \mathcal{H}_8 \times \mathcal{Q}_8, \mathcal{Q}_8 \times \mathcal{H}_8, \mathcal{Q}_8 \times \mathcal{Q}_8, \mathcal{H}_8 \times \mathcal{C}_8, \mathcal{Q}_8 \times \mathcal{C}_8\}$ *be a subclass of* $\mathrm{TRCC}_8^{\mathrm{d}}$ $(m = 2)$ *and* $R \in \mathcal{S}$.

*If* $R$ *is* $\Rsh$*-consistent then* $H_{\mathcal{S}}(R)$ *is* $\Rsh$*-consistent.*

*Proof.* See the extended version of this paper at https://arxiv.org/abs/2007.07711.

Note that, as in the context of $\mathrm{TRCC}_8^{\mathrm{n}}$, algebraically consistent networks over most combinations of the subalgebras $\mathcal{Q}_8$ and $\mathcal{H}_8$, but *also* of $\mathcal{C}_8$, are satisfiable.

**Proposition 4.** *Let* $\mathfrak{R}$ *be a topological region space such that* $\mathrm{TRCC}_8^{\mathrm{d}}$ *is a complete sequential formalism. Let* $\mathcal{S}$ *be a subset of* $\mathrm{TRCC}_8^{\mathrm{d}}$ *satisfying* $\mathcal{S}_{2i-1} \in \{\mathcal{H}_8, \mathcal{Q}_8\}$ *and* $\mathcal{S}_{2i} \in \{\mathcal{H}_8, \mathcal{Q}_8, \mathcal{C}_8\}$ *for all* $i \in \{1, \ldots, l\}$.

*Algebraically consistent networks over* $\mathcal{S}$ *are satisfiable.*

*Proof.* We apply the refinement theorem [6]. $H_{\mathcal{S}}$ is a refinement from $\mathcal{S}$ to the set $\boldsymbol{\mathcal{B}} \cup \{(\varnothing, \ldots, \varnothing)\}$ (since $h_{\mathcal{H}_8}$ (resp. $h_{\mathcal{C}_8}$) is a refinement from $\mathcal{H}_8 \cup \mathcal{Q}_8$ (resp. $h_{\mathcal{C}_8}$) to $\boldsymbol{\mathcal{B}}_{\mathrm{RCC}_8} \cup \{\varnothing\}$ [37]). $\mathcal{S}$ is algebraically stable by $H_{\mathcal{S}}$. Indeed, since on the one hand, $\mathcal{H}_8$ (resp. $\mathcal{Q}_8$ ; resp. $\mathcal{C}_8$) is algebraically stable by $h_{\mathcal{H}_8}$ (resp. $h_{\mathcal{H}_8}$ ; resp. $h_{\mathcal{C}_8}$) [37]. Since, on the other hand, for any $\Rsh$-consistent relation $R \in \mathcal{S}$, $H_{\mathcal{S}}(R)$ is $\Rsh$-consistent (by Lemma 4). Algebraically consistent networks over $\boldsymbol{\mathcal{B}} \cup \{(\varnothing, \ldots, \varnothing)\}$ are satisfiable ($\mathrm{TRCC}_8^{\mathrm{d}}$ is complete). By the refinement theorem, algebraically consistent networks over $\mathcal{S}$ are satisfiable.

**Lemma 5.** *Let $r \in \mathrm{RCA}_8 \backslash \mathcal{N}$. We have:*

- $\uparrow r \in \mathcal{H}_8$,
- $\downarrow r \in \mathcal{H}_8 \cap \mathcal{Q}_8 \cap \mathcal{C}_8$.

*Proof.* From the definitions of $\mathcal{H}_8$, $\mathcal{Q}_8$, $\mathcal{C}_8$ and the projections of $\mathrm{TRCC}_8^{\mathrm{d}}$, we derive the lemma. Let $r \in \mathrm{RCA}_8 \backslash \mathcal{N}$ (see Table 2). On the one hand, we show $\uparrow r \in \mathcal{H}_8$. For this, we show the three following properties: $\uparrow r \in \mathrm{RCA}_8 \backslash \mathcal{N}$, $\mathrm{NTPP} \subseteq \uparrow r \implies \mathrm{TPP} \subseteq \uparrow r$, and $\uparrow r \notin V = \{\mathrm{EC} \cup \mathrm{NTPP} \cup \mathrm{EQ}, \mathrm{DC} \cup \mathrm{EC} \cup \mathrm{NTPP} \cup \mathrm{EQ}, \mathrm{EC} \cup \overline{\mathrm{NTPP}} \cup \mathrm{EQ}, \mathrm{DC} \cup \mathrm{EC} \cup \overline{\mathrm{NTPP}} \cup \mathrm{EQ}\}$. We show $\uparrow r \in \mathrm{RCA}_8 \backslash \mathcal{N}$. If $\uparrow r \cap (\mathrm{TPP} \cup \mathrm{NTPP}) = \varnothing$ or $\uparrow r \cap (\overline{\mathrm{TPP}} \cup \overline{\mathrm{NTPP}}) = \varnothing$ then $\uparrow r \in \mathrm{RCA}_8 \backslash \mathcal{N}$. Suppose $\uparrow r \cap (\mathrm{TPP} \cup \mathrm{NTPP}) \neq \varnothing$ and $\uparrow r \cap (\overline{\mathrm{TPP}} \cup \overline{\mathrm{NTPP}}) \neq \varnothing$. Therefore, $r \cap (\mathrm{PO} \cup \mathrm{TPP} \cup \mathrm{NTPP}) \neq \varnothing$ and $r \cap (\mathrm{PO} \cup \overline{\mathrm{TPP}} \cup \overline{\mathrm{NTPP}}) \neq \varnothing$. If $r \cap (\mathrm{TPP} \cup \mathrm{NTPP}) \neq \varnothing$ and $r \cap (\overline{\mathrm{TPP}} \cup \overline{\mathrm{NTPP}}) \neq \varnothing$ then $\mathrm{PO} \subseteq r$ (since $r \in \mathrm{RCA}_8 \backslash \mathcal{N}$). Thus, $\mathrm{PO} \subseteq r$ and therefore $\mathrm{PO} \subseteq \uparrow r$. We have $\uparrow r \in \mathrm{RCA}_8 \backslash \mathcal{N}$. We show $\mathrm{NTPP} \subseteq \uparrow r \implies \mathrm{TPP} \subseteq \uparrow r$. If $\mathrm{NTPP} \subseteq \uparrow r$, then $\mathrm{NTPP} \subseteq r$ and therefore $\uparrow \mathrm{NTPP} \subseteq \uparrow r$. Thus, $\mathrm{TPP} \subseteq \uparrow r$. We show $\uparrow r \notin V$. If $\uparrow r \cap (\mathrm{NTPP} \cup \overline{\mathrm{NTPP}}) = \varnothing$ then $\uparrow r \notin V$. If $\mathrm{NTPP} \subseteq \uparrow r$, then $\mathrm{TPP} \subseteq \uparrow r$. If $\overline{\mathrm{NTPP}} \subseteq \uparrow r$ then $\overline{\mathrm{TPP}} \subseteq \uparrow r$. Thus, in all cases, $\uparrow r \notin V$.

On the other hand, we show $\downarrow r \in \mathcal{H}_8 \cap \mathcal{Q}_8 \cap \mathcal{C}_8$. If $\downarrow r \cap (\mathrm{TPP} \cup \mathrm{NTPP}) = \varnothing$ or $\downarrow r \cap (\overline{\mathrm{TPP}} \cup \overline{\mathrm{NTPP}}) = \varnothing$ then $\downarrow r \in \mathrm{RCA}_8 \backslash \mathcal{N}$. Suppose $\downarrow r \cap (\mathrm{TPP} \cup \mathrm{NTPP}) \neq \varnothing$ and $\downarrow r \cap (\overline{\mathrm{TPP}} \cup \overline{\mathrm{NTPP}}) \neq \varnothing$. We have $r \cap (\mathrm{TPP} \cup \mathrm{NTPP} \cup \mathrm{EQ}) \neq \varnothing$ and $r \cap (\overline{\mathrm{TPP}} \cup \overline{\mathrm{NTPP}} \cup \mathrm{EQ}) \neq \varnothing$. If $\mathrm{EQ} \subseteq r$, then $\mathrm{PO} \subseteq \downarrow r$ and therefore $\downarrow r \in \mathrm{RCA}_8 \backslash \mathcal{N}$. If $\mathrm{EQ} \nsubseteq r$, then $r \cap (\mathrm{TPP} \cup \mathrm{NTPP}) \neq \varnothing$ and $r \cap (\overline{\mathrm{TPP}} \cup \overline{\mathrm{NTPP}}) \neq \varnothing$. Since $r \in \mathrm{RCA}_8 \backslash \mathcal{N}$, $\mathrm{PO} \subseteq r$. Thus, $\mathrm{PO} \subseteq \downarrow r$ and therefore $\downarrow r \in \mathrm{RCA}_8 \backslash \mathcal{N}$. Thus, in all cases, $\downarrow r \in \mathrm{RCA}_8 \backslash \mathcal{N}$. Moreover, we have $\downarrow r \in \mathcal{P}_8$ since $\downarrow r \notin V$. Indeed, if $\mathrm{EC} \subseteq \downarrow r$, then $\mathrm{EC} \subseteq r$ and therefore $\mathrm{PO} \subseteq \downarrow r$. By the same argument, we have $\downarrow r \in \mathcal{C}_8$. In addition, we have $\downarrow r \in \mathcal{H}_8 \cap \mathcal{Q}_8$ and therefore $\downarrow r \in \mathcal{H}_8 \cap \mathcal{Q}_8 \cap \mathcal{C}_8$, since if $\mathrm{EQ} \subseteq \downarrow r$ then $\mathrm{EQ} \subseteq r$ and therefore $\mathrm{PO} \cup \mathrm{TPP} \cup \overline{\mathrm{TPP}} \subseteq \downarrow r$.

We end by showing that the subalgebras of the form $(\mathcal{H}_8 \times \{\mathcal{H}_8, \mathcal{Q}_8, \mathcal{C}_8\})^{\star}$ are algebraically tractable.

**Theorem 1.** *Let $\mathfrak{R}$ be a topological region space such that $\mathrm{TRCC}_8^{\mathrm{d}}$ is a complete sequential formalism.*

*Subalgebras $\mathcal{S}$ of $\mathrm{TRCC}_8^{\mathrm{d}}$ satisfying $\mathcal{S}_{2i-1} = \mathcal{H}_8$ and $\mathcal{S}_{2i} \in \{\mathcal{H}_8, \mathcal{Q}_8, \mathcal{C}_8\}$ for all $i \in \{1, \ldots, l\}$ are algebraically tractable.*

*Proof.* Let $\mathcal{S}$ be a subset of $\mathrm{TRCC}_8^{\mathrm{d}}$ satisfying $\mathcal{S}_{2i-1} = \mathcal{H}_8$ and $\mathcal{S}_{2i} \in \{\mathcal{H}_8, \mathcal{Q}_8, \mathcal{C}_8\}$ for all $i \in \{1, \ldots, l\}$. $\mathcal{S}$ satisfies the conditions of the first part of the refinement theorem (i.e. $\mathcal{S}$ satisfies the conditions of the first implication ; see the proof of Proposition 4). $\mathcal{S}$ also satisfies the conditions of the second part (i.e. the conditions of the second implication). Indeed, on the one hand, $\mathcal{S}$ is a subclass (since $\mathcal{S}$ is Cartesian and each $\mathcal{S}_i$ is a subclass [37]). On the other hand, $\mathcal{S}$ is $\Rsh$-closed (by Lemma 5 and since $\mathcal{S}$ is a Cartesian subclass). $\mathcal{S}$ is thus algebraically tractable (refinement theorem [6]).

## 5 Conclusion

First, we have focused on $\mathrm{TRCC}_8^{\mathrm{n}}$, the qualitative formalism of topological temporal sequences describing the evolution of regions at instants between which there are no intermediary relations (i.e. at time points which characterize all the qualitative changes). We have shown that there is no algebraically tractable Cartesian subalgebra (subclass containing all basic relations) for $\mathrm{TRCC}_8^{\mathrm{n}}$ when the length of sequences is longer than 3. However, we have identified some tractable subclasses. The price of tractability has been to give up the relations containing NTPP not containing TPP and thus to give up the basic relation NTPP.

Then, we have formalized $\mathrm{TRCC}_8^{\mathrm{d}}$, the qualitative formalism of topological temporal sequences describing the evolution of regions on a partition of time (i.e. on a contiguous alternation of instants and open intervals). In this context, we have identified large algebraically tractable Cartesian subalgebras.

It is possible to identify other algebraically tractable subclasses for $\mathrm{TRCC}_8^{\mathrm{n}}$ and $\mathrm{TRCC}_8^{\mathrm{d}}$. The tractability limit of the subclasses of these two formalisms should be precisely determined. In particular, a definitive answer to the question of the existence of polynomial Cartesian subalgebra for $\mathrm{TRCC}_8^{\mathrm{n}}$ should be given. Note that the identification of universes ensuring the completeness of $\mathrm{TRCC}_8^{\mathrm{n}}$ and of $\mathrm{TRCC}_8^{\mathrm{d}}$ remains an open problem, on which we are working.

Concerning the applications, $\mathrm{TRCC}_8^{\mathrm{n}}$ and $\mathrm{TRCC}_8^{\mathrm{d}}$ can be used to decide if it is possible to go from a topological scenario $S$ to another $S'$, with at most $m$ qualitative changes, while satisfying at each instant the constraints of a network $N$ and to determine one of the corresponding intermediate temporal sequences. This problem should be useful for spatial planning. When $S$, $S'$, and $N$ correspond to one of the previous algebraically tractable subclasses (for instance when $S$, $S'$, and $N$ are over $\mathcal{H}_8$), the problem is polynomial. Otherwise, it is possible that using tractable subclasses still speeds up the resolution of the problem, as in the classic case [39]. Note that $\mathrm{TRCC}_8^{\mathrm{d}}$ is more interesting than $\mathrm{TRCC}_8^{\mathrm{n}}$ for this problem since it allows to find a more expressive intermediate sequence while having larger tractable subclasses.

## References

1. Bennett, B., Cohn, A.G., Torrini, P., Hazarika, S.M.: Describing rigid body motions in a qualitative theory of spatial regions (2000)

2. Bennett, B., Cohn, A.G., Wolter, F., Zakharyaschev, M.: Multi-dimensional modal logic as a framework for spatio-temporal reasoning. Appl. Intell. **17**(3), 239–251 (2002). https://doi.org/10.1023/A:1020083231504
3. Burrieza, A., Muñoz-Velasco, E., Ojeda-Aciego, M.: A PDL approach for qualitative velocity. Int. J. Uncertainty Fuzziness Knowl.-Based Syst. **19**(01), 11–26 (2011)
4. Burrieza, A., Ojeda-Aciego, M.: A multimodal logic approach to order of magnitude qualitative reasoning with comparability and negligibility relations. Fundam. Informaticae **68**(1–2), 21–46 (2005)
5. Chen, J., Cohn, A.G., Liu, D., Wang, S., Ouyang, J., Yu, Q.: A survey of qualitative spatial representations. Knowl. Eng. Rev. **30**(01), 106–136 (2015)
6. Cohen-Solal, Q., Bouzid, M., Niveau, A.: Checking the consistency of combined qualitative constraint networks. In: Proceedings of AAAI (2017)
7. Cohen-Solal, Q., Bouzid, M., Niveau, A.: Temporal sequences of qualitative information: reasoning about the topology of constant-size moving regions. In: Twenty-Sixth International Joint Conference on Artificial Intelligence, pp. 986–992 (2017)
8. Cohn, A.G., Gotts, N.M., Cui, Z., Randell, D.A., Bennett, B., Gooday, J.: Exploiting temporal continuity in qualitative spatial calculi. In: Spatial and Temporal Reasoning in Geographic Information Systems, pp. 5–24 (1998)
9. Cohn, A.G., Hazarika, S.M.: Qualitative spatial representation and reasoning: an overview. Fundam. informaticae **46**(1–2), 1–29 (2001)
10. Davis, E.: Continuous shape transformation and metrics on regions. Fundam. Informaticae **46**(1–2), 31–54 (2001)
11. Dylla, F., et al.: A survey of qualitative spatial and temporal calculi: algebraic and computational properties. ACM Comput. Surv. (CSUR) **50**(1) (2017). Article no. 7
12. Dylla, F., Moratz, R.: Exploiting qualitative spatial neighborhoods in the situation calculus. In: Freksa, C., Knauff, M., Krieg-Brückner, B., Nebel, B., Barkowsky, T. (eds.) Spatial Cognition 2004. LNCS (LNAI), vol. 3343, pp. 304–322. Springer, Heidelberg (2005). https://doi.org/10.1007/978-3-540-32255-9_18
13. Dylla, F., Wallgrün, J.O.: Qualitative spatial reasoning with conceptual neighborhoods for agent control. J. Intell. Robot. Syst. **48**(1), 55–78 (2007). https://doi.org/10.1007/s10846-006-9099-4
14. Egenhofer, M.J.: The family of conceptual neighborhood graphs for region-region relations. In: Fabrikant, S.I., Reichenbacher, T., van Kreveld, M., Schlieder, C. (eds.) GIScience 2010. LNCS, vol. 6292, pp. 42–55. Springer, Heidelberg (2010). https://doi.org/10.1007/978-3-642-15300-6_4
15. Egenhofer, M.J.: Qualitative spatial-relation reasoning for design. In: Gero, J.S. (ed.) Studying Visual and Spatial Reasoning for Design Creativity, pp. 153–175. Springer, Dordrecht (2015). https://doi.org/10.1007/978-94-017-9297-4_9
16. Freksa, C.: Conceptual neighborhood and its role in temporal and spatial reasoning. In: Proceedings of the IMACS Workshop on Decision Support Systems and Qualitative Reasoning, pp. 181–187 (1991)
17. Freksa, C.: Temporal reasoning based on semi-intervals. Artif. Intell. **54**(1–2), 199–227 (1992)
18. Gabelaia, D., Kontchakov, R., Kurucz, A., Wolter, F., Zakharyaschev, M.: On the computational complexity of spatio-temporal logics. In: FLAIRS Conference, pp. 460–464 (2003)
19. Galton, A.: Towards an integrated logic of space, time, and motion. In: Proceedings of IJCAI, pp. 1550–1555 (1993)

20. Galton, A.: Qualitative Spatial Change. Oxford University Press, Oxford (2000)
21. Galton, A.: Dominance diagrams: a tool for qualitative reasoning about continuous systems. Fundam. Informaticae **46**(1–2), 55–70 (2001)
22. Gerevini, A., Nebel, B.: Qualitative spatio-temporal reasoning with RCC-8 and Allen's interval calculus: computational complexity. In: Proceedings of ECAI, pp. 312–316 (2002)
23. Hazarika, S.M., Cohn, A.G.: Qualitative spatio-temporal continuity. In: Montello, D.R. (ed.) COSIT 2001. LNCS, vol. 2205, pp. 92–107. Springer, Heidelberg (2001). https://doi.org/10.1007/3-540-45424-1_7
24. Hazarika, S.M.: Qualitative spatial change: space-time histories and continuity. Ph.D. thesis, University of Leeds (2005)
25. Kowalski, R., Sergot, M.: A logic-based calculus of events. In: Schmidt, J.W., Thanos, C. (eds.) Foundations of Knowledge Base Management. TINF, pp. 23–55. Springer, Heidelberg (1989). https://doi.org/10.1007/978-3-642-83397-7_2
26. Kurata, Y., Egenhofer, M.J.: The 9+-intersection for topological relations between a directed line segment and a region
27. Landsiedel, C., Rieser, V., Walter, M., Wollherr, D.: A review of spatial reasoning and interaction for real-world robotics. Adv. Robot. **31**(5), 222–242 (2017)
28. Li, S., Ying, M.: Region connection calculus: its models and composition table. Artif. Intell. **145**(1–2), 121–146 (2003)
29. Ligozat, G.: Towards a general characterization of conceptual neighborhoods in temporal and spatial reasoning. In: AAAI 1994 The Twelfth National Conference on Artificial Intelligence (1994)
30. Ligozat, G.: Qualitative Spatial and Temporal Reasoning. Wiley, Hoboken (2013)
31. Mansouri, M., Pecora, F.: A robot sets a table: a case for hybrid reasoning with different types of knowledge. J. Exp. Theor. Artif. Intell. **28**(5), 801–821 (2016)
32. Muller, P.: Topological spatio-temporal reasoning and representation. Comput. Intell. **18**(3), 420–450 (2002)
33. Ragni, M., Wölfl, S.: Temporalizing cardinal directions: from constraint satisfaction to planning. In: Proceedings of KR, pp. 472–480 (2006)
34. Ragni, M., Wölfl, S.: Reasoning about topological and positional information in dynamic settings. In: Proceedings of the FLAIRS Conference, pp. 606–611 (2008)
35. Randell, D.A., Cui, Z., Cohn, A.G.: A spatial logic based on regions and connection. In: Proceedings of KR, pp. 165–176 (1992)
36. Reis, R.M., Egenhofer, M.J., Matos, J.L.: Conceptual neighborhoods of topological relations between lines. In: Ruas, A., Gold, C. (eds.) Headway in Spatial Data Handling. LNGC, pp. 557–574. Springer, Heidelberg (2008). https://doi.org/10.1007/978-3-540-68566-1_32
37. Renz, J.: Maximal tractable fragments of the region connection calculus: a complete analysis. In: Proceedings of IJCAI, pp. 448–455 (1999)
38. Renz, J., Ligozat, G.: Weak composition for qualitative spatial and temporal reasoning. In: van Beek, P. (ed.) CP 2005. LNCS, vol. 3709, pp. 534–548. Springer, Heidelberg (2005). https://doi.org/10.1007/11564751_40
39. Renz, J., Nebel, B.: Efficient methods for qualitative spatial reasoning. J. Artif. Intell. Res. **15**, 289–318 (2001)
40. Santos, M.Y., Moreira, A.: Conceptual neighborhood graphs for topological spatial relations. In: Proceedings of the World Congress on Engineering, vol. 1, pp. 12–18 (2009)
41. Sioutis, M., Alirezaie, M., Renoux, J., Loutfi, A.: Towards a synergy of qualitative spatio-temporal reasoning and smart environments for assisting the elderly at home. In: IJCAI Workshop on Qualitative Reasoning, pp. 901–907 (2017)

42. Sioutis, M., Condotta, J.-F., Salhi, Y., Mazure, B.: A qualitative spatio-temporal framework based on point algebra. In: Agre, G., Hitzler, P., Krisnadhi, A.A., Kuznetsov, S.O. (eds.) AIMSA 2014. LNCS (LNAI), vol. 8722, pp. 117–128. Springer, Cham (2014). https://doi.org/10.1007/978-3-319-10554-3_11
43. Sioutis, M., Condotta, J.-F., Salhi, Y., Mazure, B.: Generalized qualitative spatio-temporal reasoning: complexity and tableau method. In: De Nivelle, H. (ed.) TABLEAUX 2015. LNCS (LNAI), vol. 9323, pp. 54–69. Springer, Cham (2015). https://doi.org/10.1007/978-3-319-24312-2_5
44. Sioutis, M., Condotta, J.-F., Salhi, Y., Mazure, B., Randell, D.A.: Ordering spatio-temporal sequences to meet transition constraints: complexity and framework. In: Chbeir, R., Manolopoulos, Y., Maglogiannis, I., Alhajj, R. (eds.) AIAI 2015. IAICT, vol. 458, pp. 130–150. Springer, Cham (2015). https://doi.org/10.1007/978-3-319-23868-5_10
45. Van Benthem, J.: The Logic of Time: A Model-Theoretic Investigation into the Varieties of Temporal Ontology and Temporal Discourse, vol. 156. Springer, Heidelberg (2013)
46. Van de Weghe, N., De Maeyer, P.: Conceptual neighbourhood diagrams for representing moving objects. In: Akoka, J., et al. (eds.) ER 2005. LNCS, vol. 3770, pp. 228–238. Springer, Heidelberg (2005). https://doi.org/10.1007/11568346_25
47. Westphal, M., Dornhege, C., Wölfl, S., Gissler, M., Nebel, B.: Guiding the generation of manipulation plans by qualitative spatial reasoning. Spat. Cogn. Comput. **11**(1), 75–102 (2011)
48. Westphal, M., Hué, J., Wölfl, S., Nebel, B.: Transition constraints: a study on the computational complexity of qualitative change. In: Proceedings of IJCAI, pp. 1169–1175 (2013)
49. Westphal, M., Wölfl, S.: Qualitative CSP, finite CSP, and SAT: comparing methods for qualitative constraint-based reasoning. In: Twenty-First International Joint Conference on Artificial Intelligence (2009)
50. Wolter, F., Zakharyaschev, M.: Spatio-temporal representation and reasoning based on RCC-8. In: KR, pp. 3–14 (2000)
51. Wolter, F., Zakharyaschev, M.: Qualitative spatio-temporal representation and reasoning: a computational perspective. In: Exploring Artificial Intelligence in the New Millenium, pp. 175–216 (2002)
52. Wu, J., Claramunt, C., Deng, M.: Towards a qualitative representation of movement. In: Indulska, M., Purao, S. (eds.) ER 2014. LNCS, vol. 8823, pp. 191–200. Springer, Cham (2014). https://doi.org/10.1007/978-3-319-12256-4_20
53. Zimmermann, K., Freksa, C.: Enhancing spatial reasoning by the concept of motion. In: Prospects for Artificial Intelligence, pp. 140–147 (1993)

# Strengthening Neighbourhood Substitution

Martin C. Cooper(✉)

ANITI, IRIT, University of Toulouse III, Toulouse, France
cooper@irit.fr

**Abstract.** Domain reduction is an essential tool for solving the constraint satisfaction problem (CSP). In the binary CSP, neighbourhood substitution consists in eliminating a value if there exists another value which can be substituted for it in each constraint. We show that the notion of neighbourhood substitution can be strengthened in two distinct ways without increasing time complexity. We also show the theoretical result that, unlike neighbourhood substitution, finding an optimal sequence of these new operations is NP-hard.

## 1 Introduction

Domain reduction is classical in constraint satisfaction. Indeed, eliminating inconsistent values by what is now known as arc consistency [27] predates the first formulation of the constraint satisfaction problem [23]. Maintaining arc consistency, which consists in eliminating values that can be proved inconsistent by examining a single constraint together with the current domains of the other variables, is ubiquitous in constraint solvers [1]. In binary CSPs, various algorithms have been proposed for enforcing arc consistency in $O(ed^2)$ time, where $d$ denotes maximum domain size and $e$ the number of constraints [3,24]. Generic constraints on a number of variables which is unbounded are known as global constraints. Arc consistency can be efficiently enforced for many types of global constraints [18]. This has led to the development of efficient solvers providing a rich modelling language. Stronger notions of consistency have been proposed for domain reduction which lead to more eliminations but at greater computational cost [1,2,28].

In parallel, other research has explored methods that preserve satisfiability of the CSP instance but do not preserve the set of solutions. When searching for a single solution, all but one branch of the explored search tree leads to a dead-end, and so any method for faster detection of unsatisfiability is clearly useful. An important example of such methods is the addition of symmetry-breaking constraints [4,17]. In this paper we concentrate on domain-reduction methods. One family of satisfiability-preserving domain-reduction operations is value merging. For example, two values can be merged if the so-called broken

This work was partially funded by ANITI, funded by the French program "Investing for the Future – PIA3" under Grant agreement n° ANR-19-PI3A-0004.


H. Simonis (Ed.): CP 2020, LNCS 12333, pp. 126–142, 2020.
https://doi.org/10.1007/978-3-030-58475-7_8

triangle (BT) pattern does not occur on these two values [11]. Other value-merging rules have been proposed which allow less merging than BT-merging but at a lower cost [22] or more merging at a greater cost [12,25]. Another family of satisfiability-preserving domain-reduction operations are based on the elimination of values that are not essential to obtain a solution [15]. The basic operation in this family which corresponds most closely to arc consistency is neighbourhood substitution: a value $b$ can be eliminated from a domain if there is another value $a$ in the same domain such that $b$ can be replaced by $a$ in each tuple in each constraint relation (reduced to the current domains of the other variables) [14]. In binary CSPs, neighbourhood substitution can be applied until convergence in $O(ed^3)$ time [7]. In this paper, we study notions of substitutability which are strictly stronger than neighbourhood substitutability but which can be applied in the same $O(ed^3)$ time complexity. We say that one elimination rule $R_1$ is stronger than (subsumes) another rule $R_2$ if any value in a non-trivial instance (an instance with more than one variable) that can be eliminated by $R_2$ can also be eliminated by $R_1$, and is strictly stronger (strictly subsumes) if there is also at least one non-trivial instance in which $R_1$ can eliminate a value that $R_2$ cannot. Two rules are incomparable if neither is stronger than the other.

To illustrate the strength of the new notions of substitutability that we introduce in this paper, consider the instances shown in Fig. 1. These instances are all globally consistent (each variable-value assignment occurs in a solution) and neighbourhood substitution is not powerful enough to eliminate any values. In this paper, we introduce three novel value-elimination rules, defined in Sect. 2: SS, CNS and SCSS. We will show that snake substitution (SS) allows us to reduce all domains to singletons in the instance in Fig. 1(a). Using the notation $\mathcal{D}(x_i)$ for the domain of the variable $x_i$, conditioned neighbourhood-substitution (CNS), allows us to eliminate value 0 from $\mathcal{D}(x_2)$ and value 2 from $\mathcal{D}(x_3)$ in the instance shown in Fig. 1(b), reducing the constraint between $x_2$ and $x_3$ to a null constraint (the complete relation $\mathcal{D}(x_2) \times \mathcal{D}(x_3)$). Snake-conditioned snake-substitution (SCSS) subsumes both SS and CNS and allows us to reduce all domains to singletons in the instance in Fig. 1(c) (as well as in the instances in Fig. 1(a), (b)).

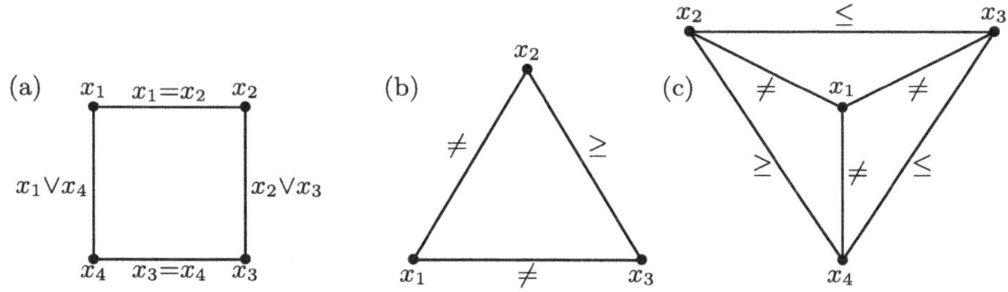


**Fig. 1.** (a) A 4-variable CSP instance over boolean domains; (b) a 3-variable CSP instance over domains $\{0, 1, 2\}$ with constraints $x_1 \neq x_2$, $x_1 \neq x_3$ and $x_2 \geq x_3$; (c) A 4-variable CSP instance over domain $\{0, 1, 2, 3\}$ with constraints $x_1 \neq x_2$, $x_1 \neq x_3$, $x_1 \neq x_4$, $x_2 \leq x_3$, $x_2 \geq x_4$ and $x_4 \leq x_3$.

In Sect. 2 we define the substitution operations SS, CNS and SCSS. In Sect. 3 we prove the validity of these three substitution operations, in the sense that they define satisfiability-preserving value-elimination rules. In Sect. 4 we explain in detail the examples in Fig. 1 and we give other examples from the semantic labelling of line drawings. Section 5 discusses the complexity of applying these value-elimination rules until convergence: the time complexity of SS and CNS is no greater than neighbourhood substitution (NS) even though these rules are strictly stronger. However, unlike NS, finding an optimal sequence of value eliminations by SS or CNS is NP-hard: this is shown in Sect. 6.

## 2 Definitions

We study binary constraint satisfaction problems.

A *binary CSP instance* $I = (X, \mathcal{D}, R)$ comprises

- a set $X$ of $n$ variables $x_1, \ldots, x_n$,
- a domain $\mathcal{D}(x_i)$ for each variable $x_i$ $(i = 1, \ldots, n)$, and
- a binary constraint relation $R_{ij}$ for each pair of distinct variables $x_i, x_j$ $(i, j \in \{1, \ldots, n\})$

For notational convenience, we assume that there is exactly one binary relation $R_{ij}$ for each pair of variables. Thus, if $x_i$ and $x_j$ do not constrain each other, then we consider that there is a *trivial constraint* between them with $R_{ij} = \mathcal{D}(x_i) \times \mathcal{D}(x_j)$. Furthermore, $R_{ji}$ (viewed as a boolean matrix) is always the transpose of $R_{ij}$. A *solution* to $I$ is an $n$-tuple $s = \langle s_1, \ldots, s_n \rangle$ such that $\forall i \in \{1, \ldots, n\}$, $s_i \in \mathcal{D}(x_i)$ and for each distinct $i, j \in \{1, \ldots, n\}$, $(s_i, s_j) \in R_{ij}$.

We say that $v_i \in D(x_i)$ has a *support* at variable $x_j$ if $\exists v_j \in D(x_j)$ such that $(v_i, v_j) \in R_{ij}$. A binary CSP instance $I$ is *arc consistent (AC)* if for all pairs of distinct variables $x_i, x_j$, each $v_i \in D(x_i)$ has a support at $x_j$ [21].

In the following we assume that we have a binary CSP instance $I = (X, \mathcal{D}, R)$ over $n$ variables and, for clarity of presentation, we write $j \neq i$ as a shorthand for $j \in \{1, \ldots, n\} \setminus \{i\}$. We use the notation $b \xrightarrow{ij} a$ for

$$\forall c \in \mathcal{D}(x_j),\ (b, c) \in R_{ij} \Rightarrow (a, c) \in R_{ij}$$

(i.e. $a$ can be substituted for $b$ in any tuple $(b, c) \in R_{ij}$).

**Definition 1** [14]. *Given two values $a, b \in \mathcal{D}(x_i)$, $b$ is* neighbourhood substitutable (NS) *by $a$ if $\forall j \neq i$, $b \xrightarrow{ij} a$.*

It is well known and indeed fairly obvious that eliminating a neighbourhood substitutable value does not change the satisfiability of a binary CSP instance. We will now define stronger notions of substitutability. The proofs that these are indeed valid value-elimination rules are not directly obvious and hence are delayed until Sect. 3. We use the notation $b \overset{ik}{\rightsquigarrow} a$ for

$$\forall d \in \mathcal{D}(x_k),\ (b, d) \in R_{ik}\ \Rightarrow\ \exists e \in \mathcal{D}(x_k)((a, e) \in R_{ik}\ \wedge\ \forall \ell \notin \{i, k\}, d \xrightarrow{k\ell} e).$$

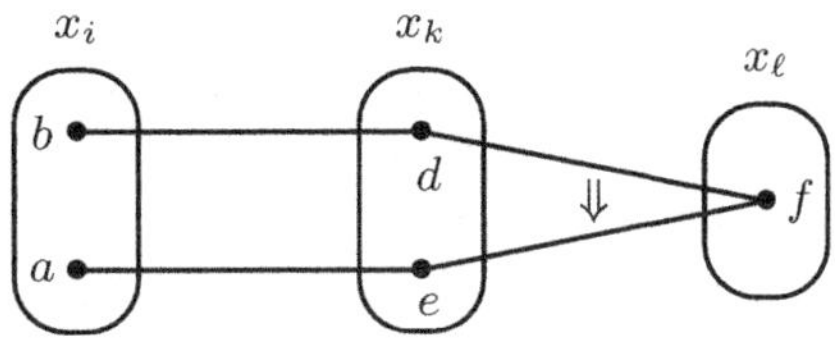


**Fig. 2.** An illustration of the definition of $b \overset{ik}{\rightsquigarrow} a$.

This is illustrated in Fig. 2, in which ovals represent domains, bullets represent values, a line joining two values means that these two values are compatible (so, for example, $(a, e) \in R_{ik}$), and the $\Downarrow$ means that $(d, f) \in R_{k\ell} \Rightarrow (e, f) \in R_{k\ell}$. Since $e$ in this definition is a function of $i, k, a$ and $d$, if necessary, we will write $e(i, k, a, d)$ instead of $e$. In other words, the notation $b \overset{ik}{\rightsquigarrow} a$ means that $a$ can be substituted for $b$ in any tuple $(b, d) \in R_{ik}$ provided we also replace $d$ by $e(i, k, a, d)$. It is clear that $b \xrightarrow{ik} a$ implies $b \overset{ik}{\rightsquigarrow} a$ since it suffices to set $e(i, k, a, d) = d$ since, trivially, $d \xrightarrow{k\ell} d$ for all $\ell \notin \{i, k\}$. In Fig. 1(a), the value $0 \in \mathcal{D}(x_1)$ is snake substitutable by 1: we have $0 \overset{12}{\rightsquigarrow} 1$ by taking $e(1, 2, 1, 0) = 1$ (where the arguments of $e(i, k, a, d)$ are as shown in Fig. 2), since $(1, 1) \in R_{12}$ and $0 \xrightarrow{23} 1$; and $0 \overset{14}{\rightsquigarrow} 1$ since $0 \xrightarrow{14} 1$. Indeed, by a similar argument, the value 0 is snake substitutable by 1 in each domain.

**Definition 2.** *Given two values $a, b \in \mathcal{D}(x_i)$, $b$ is* snake substitutable (SS) *by $a$ if $\forall k \neq i$, $b \overset{ik}{\rightsquigarrow} a$.*

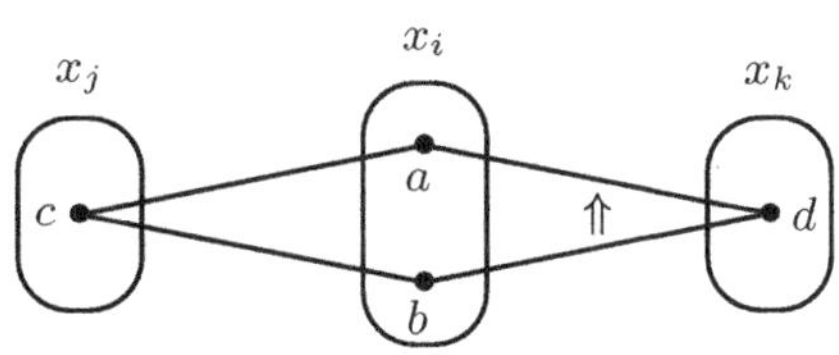


**Fig. 3.** An illustration of the definition of conditioned neighbourhood-substitutability of $b$ by $a$ (conditioned by $x_j$).

In the following two definitions, $b$ can be eliminated from $\mathcal{D}(x_i)$ because it can be substituted by some other value in $\mathcal{D}(x_i)$, but this value is a function of the value assigned to another variable $x_j$. Definition 3 is illustrated in Fig. 3.

**Definition 3.** *Given $b \in \mathcal{D}(x_i)$, $b$ is* conditioned neighbourhood-substitutable (CNS) *if for some $j \neq i$, $\forall c \in \mathcal{D}(x_j)$ with $(b, c) \in R_{ij}$, $\exists a \in \mathcal{D}(x_i) \setminus \{b\}$ such that $((a, c) \in R_{ij} \;\wedge\; \forall k \notin \{i, j\}, b \xrightarrow{ik} a)$.*

A CNS value $b \in \mathcal{D}(x_i)$ is substitutable by a value $a \in \mathcal{D}(x_i)$ where $a$ is a function of the value $c$ assigned to some other variable $x_j$. In Fig. 1(b), the value $0 \in \mathcal{D}(x_2)$ is conditioned neighbourhood-substitutable (CNS) with $x_1$ as the conditioning variable (i.e. $j = 1$ in Definition 3): for the assignments of 0 or 1 to $x_1$, we can take $a = 2$ since $0 \xrightarrow{23} 2$, and for the assignment 2 to $x_1$, we can take $a = 1$ since $0 \xrightarrow{23} 1$. By a symmetrical argument, the value $2 \in \mathcal{D}(x_3)$ is CNS, again with $x_1$ as the conditioning variable. We can note that in the resulting CSP instance, after eliminating 0 from $\mathcal{D}(x_2)$ and 2 from $\mathcal{D}(x_3)$, all domains can be reduced to singletons by applying snake substitutability.

Observe that CNS subsumes arc consistency; if a value $b \in \mathcal{D}(x_i)$ has no support $c$ in $\mathcal{D}(x_j)$, then $b$ is trivially CNS (conditioned by the variable $x_j$). It is easy to see from their definitions that SS and CNS both subsume NS (in instances with more than one variable), but that neither NS nor SS subsume arc consistency.

We now integrate the notion of snake substitutability in two ways in the definition of CNS: the value $d$ (see Fig. 3) assigned to a variable $k \notin \{i, j\}$ may be replaced by a value $e$ (as in the definition of $b \overset{ik}{\rightsquigarrow} a$, above), but the value $c$ (see Fig. 3) assigned to the conditioning variable $x_j$ may also be replaced by a value $g$. This is illustrated in Fig. 4.

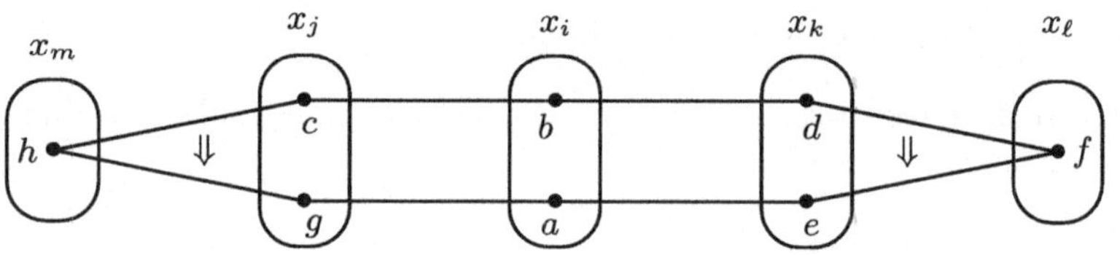


**Fig. 4.** An illustration of snake-conditioned snake-substitutability of $b$ by $a$.

**Definition 4.** *A value* $b \in \mathcal{D}(x_i)$ *is* snake-conditioned snake-substitutable (SCSS) *if for some* $j \neq i$, $\forall c \in \mathcal{D}(x_j)$ *with* $(b, c) \in R_{ij}$, $\exists a \in \mathcal{D}(x_i) \setminus \{b\}$ *such that* $(\forall k \notin \{i, j\}, b \overset{ik}{\rightsquigarrow} a \ \wedge\ (\exists g \in \mathcal{D}(x_j)((a, g) \in R_{ij} \ \wedge\ \forall m \notin \{i, j\}, c \xrightarrow{jm} g)))$.

In Fig. 1(c), the value $3 \in \mathcal{D}(x_1)$ is snake-conditioned snake-substitutable (SCSS) with $x_2$ as the conditioning variable: for the assignment of 0 or 2 to $x_2$, we can take $a = 1$ since $3 \overset{13}{\rightsquigarrow} 1$ (taking $e(1, 3, 1, d) = 3$ for $d = 0, 1, 2$) and $3 \overset{14}{\rightsquigarrow} 1$ (taking $e(1, 4, 1, d) = 0$ for $d = 0, 1, 2$), and for the assignment of 1 to $x_2$, we can take $a = 2$ since $3 \overset{13}{\rightsquigarrow} 2$ (again taking $e(1, 3, 2, d) = 3$ for $d = 0, 1, 2$) and $3 \overset{14}{\rightsquigarrow} 2$ (again taking $e(1, 4, 2, d) = 0$ for $d = 0, 1, 2$). By similar arguments, all domains can be reduced to singletons following the SCSS elimination of values in the following order: 0 from $\mathcal{D}(x_1)$, 0, 1 and 2 from $\mathcal{D}(x_3)$, 0, 1 and 2 from $\mathcal{D}(x_2)$, 1, 2 and 3 from $\mathcal{D}(x_4)$ and 2 from $\mathcal{D}(x_1)$.

We can see that SCSS subsumes CNS by setting $g = c$ in Definition 4 and by recalling that $b \xrightarrow{ik} a$ implies that $b \overset{ik}{\rightsquigarrow} a$. It is a bit more subtle to see that SCSS subsumes SS: if $b$ is snake substitutable by some value $a$, it suffices to choose $a$ in Definition 4 to be this value (which is thus constant, i.e. not dependent on the value of $c$), then the snake substitutability of $b$ by $a$ implies that $b \overset{ik}{\rightsquigarrow} a$ for all $k \neq i, j$ and $b \overset{ij}{\rightsquigarrow} a$, which in turn implies that $(a, g) \in R_{ij} \wedge \forall m \notin \{i, j\}, c \xrightarrow{jm} g$ for $g = e(i, j, a, c)$; thus $b$ is snake-conditioned snake-substitutable.

## 3 Value Elimination

It is well-known that NS is a valid value-elimination property, in the sense that if $b \in \mathcal{D}(x_i)$ is neighbourhood substitutable by $a$ then $b$ can be eliminated from $\mathcal{D}(x_i)$ without changing the satisfiability of the CSP instance [14]. In this section we show that SCSS is a valid value-elimination property. Since SS and CNS are subsumed by SCSS, it follows immediately that SS and CNS are also valid value-elimination properties.

**Theorem 1.** *In a binary CSP instance $I$, if $b \in \mathcal{D}(x_i)$ is snake-conditioned snake-substitutable then $b$ can be eliminated from $\mathcal{D}(x_i)$ without changing the satisfiability of the instance.*

*Proof.* By Definition 4, for some $j \neq i$, $\forall c \in \mathcal{D}(x_j)$ with $(b, c) \in R_{ij}$, $\exists a \in \mathcal{D}(x_i) \setminus \{b\}$ such that

$$\forall k \notin \{i, j\}, b \overset{ik}{\rightsquigarrow} a \tag{1}$$

$$\wedge \quad \exists g \in \mathcal{D}(x_j)((a, g) \in R_{ij} \ \wedge \ \forall m \notin \{i, j\}, c \xrightarrow{jm} g). \tag{2}$$

We will only apply this definition for fixed $i, j$, and for fixed values $a$ and $c$, so we can consider $g$ as a constant (even though it is actually a function of $i, j, a, c$). Let $s = \langle s_1, \ldots, s_n \rangle$ be a solution to $I$ with $s_i = b$. It suffices to show that there is another solution $t = \langle t_1, \ldots, t_n \rangle$ with $t_i \neq b$. Consider $c = s_j$. Since $s$ is a solution, we know that $(b, c) = (s_i, s_j) \in R_{ij}$. Thus, according to the above definition of SCSS, there is a value $a \in \mathcal{D}(x_i)$ that can replace $b$ (conditioned by the assignment $x_j = c = s_j$) in the sense that (1) and (2) are satisfied. Now, for each $k \notin \{i, j\}$, $b \overset{ik}{\rightsquigarrow} a$, i.e.

$$\forall d \in \mathcal{D}(x_k), \ (b, d) \in R_{ik} \ \Rightarrow \ \exists e \in \mathcal{D}(x_k)((a, e) \in R_{ik} \ \wedge \ \forall \ell \notin \{i, k\}, d \xrightarrow{k\ell} e).$$

Recall that $e$ is a function of $i, k, a$ and $d$. But we will only consider fixed $i, a$ and a unique value of $d$ dependant on $k$, so we will write $e(k)$ for brevity. Indeed, setting $d = s_k$ we can deduce from $(b, d) = (s_i, s_k) \in R_{ik}$ (since $s$ is a solution) that

$$\forall k \neq i, j, \ \exists e(k) \in \mathcal{D}(x_k)((a, e(k)) \in R_{ik} \ \wedge \ \forall \ell \notin \{i, k\}, s_k \xrightarrow{k\ell} e(k)). \tag{3}$$

Define the $n$-tuple $t$ as follows:

$$t_r = \begin{cases} a & \text{if } r = i \\ s_r & \text{if } r \neq i \ \wedge \ (a, s_r) \in R_{ir} \\ g & \text{if } r = j \ \wedge \ (a, s_r) \notin R_{ir} \\ e(r) & \text{if } r \neq i, j \ \wedge \ (a, s_r) \notin R_{ir} \end{cases}$$

Clearly $t_i \neq b$ and $t_r \in \mathcal{D}(x_r)$ for all $r \in \{1, \ldots, n\}$. To prove that $t$ is a solution, it remains to show that all binary constraints are satisfied, i.e. that $(t_k, t_r) \in R_{kr}$ for all distinct $k, r \in \{1, \ldots, n\}$. There are three cases: (1) $k = i$, $r \neq i$, (2) $k = j$, $r \neq i, j$, (3) $k, r \neq i, j$.

(1) There are three subcases: (a) $r = j$ and $(a, s_j) \notin R_{ij}$, (b) $r \neq i$ and $(a, s_r) \in R_{ir}$, (c) $r \neq i, j$ and $(a, s_r) \notin R_{ir}$. In case (a), $t_i = a$ and $t_j = g$, so from Eq. 2, we have $(t_i, t_r) = (a, g) \in R_{ij}$. In case (b), $t_i = a$ and $t_r = s_r$ and so, trivially, $(t_i, t_r) = (a, s_r) \in R_{ir}$. In case (c), $t_i = a$ and $t_r = e(r)$, so from Eq. 3, we have $(t_i, t_r) = (a, e(r)) \in R_{ir}$.

(2) There are four subcases: (a) $(a, s_r) \in R_{ir}$ and $(a, s_j) \in R_{ij}$, (b) $(a, s_r) \notin R_{ir}$ and $(a, s_j) \in R_{ij}$, (c) $(a, s_r) \in R_{ir}$ and $(a, s_j) \notin R_{ij}$, (d) $(a, s_r) \notin R_{ir}$ and $(a, s_j) \notin R_{ij}$. In case (a), $t_j = s_j$ and $t_r = s_r$, so $(t_j, t_r) \in R_{jr}$ since $s$ is a solution. In case (b), $t_j = s_j$ and $t_r = e(r)$; setting $k = r$, $\ell = j$ in Eq. 3, we have $(t_j, t_r) = (s_j, e(r)) \in R_{jr}$ since $(s_j, s_r) \in R_{jr}$. In case (c), $t_j = g$ and $t_r = s_r$; setting $c = s_j$ and $m = r$ in Eq. 2 we can deduce that $(t_j, t_r) = (g, s_r) \in R_{jr}$ since $(s_j, s_r) \in R_{jr}$. In case (d), $t_j = g$ and $t_r = e(r)$. By the same argument as in case 2(b), we know that $(s_j, e(r)) \in R_{jr}$, and then setting $c = s_j$ and $m = r$ in Eq. 2, we can deduce that $(t_j, t_r) = (g, e(r)) \in R_{jr}$.

(3) There are three essentially distinct subcases: (a) $(a, s_r) \in R_{ir}$ and $(a, s_k) \in R_{ik}$, (b) $(a, s_r) \notin R_{ir}$ and $(a, s_k) \in R_{ik}$, (c) $(a, s_r) \notin R_{ir}$ and $(a, s_k) \notin R_{ik}$. In cases (a) and (b) we can deduce $(t_k, t_r) \in R_{kr}$ by the same arguments as in cases 2(a) and 2(b), above. In case (c), $t_k = e(k)$ and $t_r = e(k)$. Setting $\ell = r$ in Eq. 3, we have $s_k \xrightarrow{kr} e(k)$ from which we can deduce that $(e(k), s_r) \in R_{kr}$ since $(s_k, s_r) \in R_{kr}$. Reversing the roles of $k$ and $r$ in Eq. 3 (which is possible since they are distinct and both different to $i$ and $j$), we also have that $s_r \xrightarrow{rk} e(r)$. We can then deduce that $(t_k, t_r) = (e(k), e(r)) \in R_{kr}$ since we have just shown that $(e(k), s_r) \in R_{kr}$.

We have thus shown that any solution $s$ with $s_i = b$ can be transformed into another solution $t$ that does not assign the value $b$ to $x_i$ and hence that the elimination of $b$ from $\mathcal{D}(x_i)$ preserves satisfiability.

**Corollary 1.** *In a binary CSP instance $I$, if $b \in \mathcal{D}(x_i)$ is snake-substitutable or conditioned neighbourhood substitutable, then $b$ can be eliminated from $\mathcal{D}(x_i)$ without changing the satisfiability of the instance.*

## 4 Examples

We have already illustrated the power of SS, CNS and SCSS using the examples given in Fig. 1.

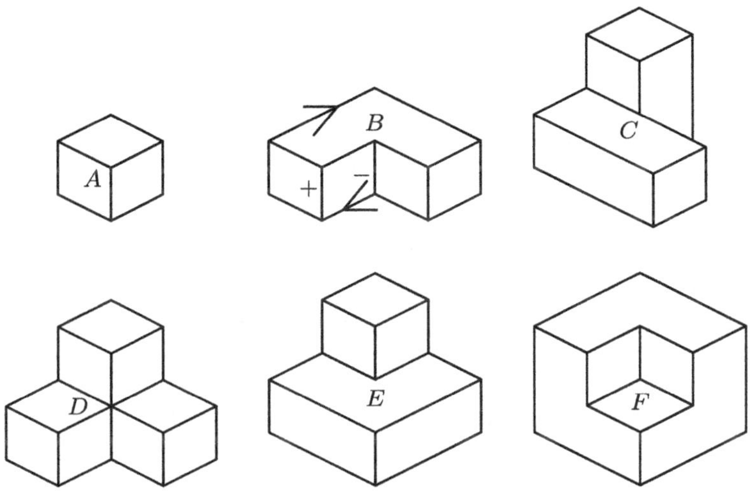


**Fig. 5.** The six different types of trihedral vertices: $A$, $B$, $C$, $D$, $E$, $F$.

To give a non-numerical example, we considered the impact of SS and CNS in the classic problem of labelling line-drawings of polyhedral scenes composed of objects with trihedral vertices [5,19,27]. There are six types of trihedral vertices: $A$, $B$, $C$, $D$, $E$ and $F$, shown in Fig. 5. The aim is to assign each line in the drawing a semantic label among four possibilities: convex (+), concave (−) or occluding (← or → depending whether the occluding surface is above or below the line). Some lines in the top middle drawing in Fig. 5 have been labelled to illustrate the meaning of these labels. This problem can be expressed as a binary CSP by treating the junctions as variables. The domains of variables are given by the catalogue of physically realisable labellings of the corresponding junction according to its type. This catalogue of junction labellings is obtained by considering the six vertex types viewed from all possible viewpoints [5,19]. For example, there are 6 possible labellings of an L-junction, 8 for a T-junction, 5 for a Y-junction and 3 for a W-junction [9]. There is a constraint between any two junctions joined by a line: this line must have the same semantic label at both ends. We can also apply binary constraints between distant junctions: the 2Reg constraint limits the possible labellings of junctions such as $A$ and $D$ in Fig. 6, since two non-colinear lines, such as $AB$ and $CD$, which separate the same two regions cannot both be concave [8,9].

The drawing shown in Fig. 6 is ambiguous. Any of lines $AB$, $BC$ or $CD$ could be projections of concave edges (meaning that the two blocks on the left side

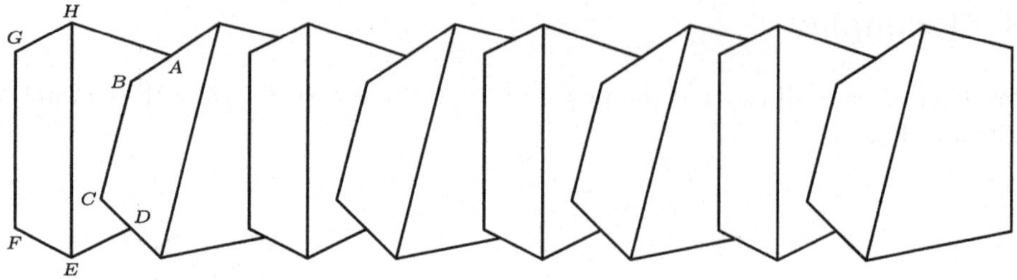


**Fig. 6.** An example from a family of line drawings whose exponential number of labellings is reduced to one by snake substitution.

of the figure are part of the same object) or all three could be projections of occluding edges (meaning that these two blocks are, in fact, separate objects). The drawing shown in Fig. 6 is an example of a family of line drawings. In this figure there are four copies of the basic structure, but there is a clear generalisation to drawings containing $n$ copies of the basic structure. The ambiguity that we have pointed out above gives rise to an exponential number of valid labellings for this family of drawings. However, after applying arc consistency and snake substitution until convergence, each domain is a singleton in this family of line drawings. We illustrate this by giving one example of a snake substitution. After arc consistency has been established, the labelling $(-,+,-)$ for junction $E$ in Fig. 6 is snake substitutable by $(\leftarrow,+,\leftarrow)$: snake substitutability follows from the fact that the labelling $(-,+,-)$ for $E$ can be replaced by $(\leftarrow,+,\leftarrow)$ in any global labelling, provided the labelling $(\uparrow,-)$ for $F$ is also replaced by $(\uparrow,\leftarrow)$ and the labelling $(\leftarrow,-,\leftarrow)$ for $D$ is also replaced by $(\leftarrow,\leftarrow,\leftarrow)$.

Of course, there are line drawings where snake substitution is much less effective than in Fig. 6. Nevertheless, in the six drawings in Fig. 5, which are a representative sample of simple line drawings, 22 of the 73 junctions have their domains reduced to singletons by arc consistency alone and a further 20 junctions have their domains reduced to singletons when both arc consistency and snake substitution are applied. This can be compared with neighbourhood substitution which eliminates no domain values in this sample of six drawings. It should be mentioned that we found no examples where conditioned neighbourhood substitution could lead to the elimination of labellings in the line-drawing labelling problem.

## 5 Complexity

In a binary CSP instance $(X,\mathcal{D},R)$, we say that two variables $x_i,x_j \in X$ constrain each other if there is a non-trivial constraint between them (i.e. $R_{ij} \neq \mathcal{D}(x_i) \times \mathcal{D}(x_j)$). Let $E \subseteq \{1,\ldots,n\} \times \{1,\ldots,n\}$ denote the set of pairs $\{i,j\}$ such that $x_i,x_j$ constrain each other. We use $d$ to denote the maximum size of the domains $\mathcal{D}(x_i)$ and $e = |E|$ to denote the number of non-trivial binary constraints. We have designed algorithms for applying CNS, SS and SCSS until convergence using classical propagation techniques and data structures. The proofs of the following results can be found in the long version of this paper [10].

**Theorem 2.** *Value eliminations by snake substitution or conditioned neighbourhood substitution can be applied until convergence in $O(ed^3)$ time and $O(ed^2)$ space.*

**Theorem 3.** *It is possible to verify in $O(ed^3)$ time and $O(ed^2)$ space whether or not any value eliminations by SCSS can be performed on a binary CSP instance. Value eliminations by SCSS can then be applied until convergence in $O(end^5)$ time and $O(ed^2)$ space.*

We now investigate the interaction between arc consistency and the substitution operations we have introduced in this paper. It is well known that arc consistency eliminations can provoke new eliminations by neighbourhood substitution (NS) but that NS eliminations cannot destroy arc consistency [7]. It follows that arc consistency eliminations can provoke new eliminations by SS, CNS and SCSS (since these notions subsume NS). It is easily seen from Definition 3 that eliminations by CNS cannot destroy arc consistency, since each value $c$ for $x_j$ which had $b$ as a support at $x_i$ has another support $a$ at $x_i$ and this value $a$ is also a support for all values $d$ for other variable $x_k$ which had $b$ as a support at $x_i$. We therefore establish arc consistency before looking for eliminations by any form of substitution. Nonetheless, unlike CNS, eliminations by SS (or SCSS) can provoke new eliminations by arc consistency; however, these eliminations cannot themselves propagate. To see this, suppose that $b \in \mathcal{D}(x_i)$ is eliminated since it is snake-substitutable by $a$. If $b$ is the only support of $d \in \mathcal{D}(x_k)$ at $x_i$, then $d$ can then be eliminated by arc consistency. However, the elimination of $d$ cannot provoke any new eliminations by arc consistency. To see this, recall that, by Definition 2 of SS, there is a value $e \in \mathcal{D}(x_k)$ such that for all $\ell \neq i, k$, for all $f \in \mathcal{D}(x_\ell)$, if $d$ was a support for $f$ at $x_k$ then so was $e$ (as illustrated in Fig. 2). Furthermore, since $b$ was the only support for $d$ at $x_i$, no other value in $\mathcal{D}(x_i)$ can lose its support when $d$ is eliminated from $\mathcal{D}(x_k)$. In conclusion, the algorithm for applying SS has to apply this limited form of arc-consistency (without propagation) whereas the algorithm to apply CNS does not need to test for arc consistency since we assume that it has already been established. Furthermore, since AC is, in fact, subsumed by SCSS we do not explicitly need to test for it in the algorithm to apply SCSS.

We now consider the interaction between neighbourhood substitution and CNS. Recall that CNS subsumes neighbourhood substitution. It is also clear from Definition 3 of CNS that eliminating values by neighbourhood substitution cannot prevent elimination of other values by CNS. However, the converse is not true: eliminations by CNS can prevent eliminations of other values by NS. To see this, consider a 2-variable instance with constraint $(x_1 = x_2) \vee (x_2 = 0)$ and domains $\mathcal{D}(x_1) = \{1, \ldots, d-1\}$, $\mathcal{D}(x_2) = \{0, \ldots, d-1\}$. The value $0 \in \mathcal{D}(x_2)$ can be eliminated by CNS (conditioned by the variable $x_1$) since $\forall c \in \mathcal{D}(x_1)$, $\exists a = c \in \mathcal{D}(x_2) \setminus \{0\}$ such that $(a, c) \in R_{12}$. After eliminating 0 from $\mathcal{D}(x_2)$, no further eliminations are possible by CNS or neighbourhood substitution. However, in the original instance we could have eliminated all elements of $\mathcal{D}(x_2)$ except 0 by neighbourhood substitution. Thus, in our algorithm to apply CNS, we give priority to eliminations by NS.

In this section we have seen that it is possible to apply CNS and SS until convergence in $O(ed^3)$ time and that it is possible to check SCSS in $O(ed^3)$ time. Thus, the complexity of applying the value-elimination rules CNS, SS and SCSS is comparable to the $O(ed^3)$ time complexity of applying neighbourhood substitution (NS) [7]. This is interesting because (in instances with more than one variable) CNS, SS and SCSS all strictly subsume NS.

## 6 Optimal Sequences of Eliminations

It is known that applying different sequences of neighbourhood operations until convergence produces isomorphic instances [7]. This is not the case for CNS, SS or SCSS. Indeed, as we show in this section, the problems of maximising the number of value-eliminations by CNS, SS or SCSS are all NP-hard. These intractability results do not detract from the utility of these operations, since any number of value eliminations reduces search-space size regardless of whether or not this number is optimal.

**Theorem 4.** *Finding the longest sequence of CNS value-eliminations or SCSS value-eliminations is NP-hard.*

*Proof.* We prove this by giving a polynomial reduction from the set cover problem [20], the well-known NP-complete problem which, given sets $S_1, \ldots, S_m \subseteq U$ and an integer $k$, consists in determining whether there are $k$ sets $S_{i_1}, \ldots, S_{i_k}$ which cover $U$ (i.e. such that $S_{i_1} \cup \ldots \cup S_{i_k} = U$). We can assume that $S_1 \cup \ldots \cup S_m = U$ and $k < m$, otherwise the problem is trivially solvable. Given sets $S_1, \ldots, S_m \subseteq U$, we create a 2-variable CSP instance with $\mathcal{D}(x_1) = \{1, \ldots, m\}$, $\mathcal{D}(x_2) = U$ and $R_{12} = \{(i, u) \mid u \in S_i\}$. We can eliminate value $i$ from $\mathcal{D}(x_1)$ by CNS (with, of course, $x_2$ as the conditioning variable) if and only if $S_1, \ldots, S_{i-1}, S_{i+1}, \ldots, S_m$ cover $U$. Indeed, we can continue eliminating elements from $\mathcal{D}(x_1)$ by CNS provided the sets $S_j$ ($j \in \mathcal{D}(x_1)$) still cover $U$. Clearly, maximising the number of eliminations from $\mathcal{D}(x_1)$ by CNS is equivalent to minimising the size of the cover. To prevent any eliminations from the domain of $x_2$ by CNS, we add variables $x_3$ and $x_4$ with domains $\{1, \ldots, m\}$, together with the three equality constraints $x_2 = x_3$, $x_3 = x_4$ and $x_4 = x_2$. To complete the proof for CNS, it is sufficient to observe that this reduction is polynomial.

It is easily verified that in this instance, CNS and SCSS are equivalent. Hence, this proof also shows that finding the longest sequence of SCSS value-eliminations is NP-hard.

In the proof of the following theorem, we need the following notion: we say that a sequence of value-eliminations by snake-substitution (SS) is *convergent* if no more SS value-eliminations are possible after this sequence of eliminations is applied.

**Theorem 5.** *Finding a longest sequence of snake-substitution value-eliminations is NP-hard.*

*Proof.* It suffices to demonstrate a polynomial reduction from the problem MAX 2-SAT which is known to be NP-hard [16]. Consider an instance $I_{2SAT}$ of MAX 2-SAT with variables $X_1, \ldots, X_N$ and $M$ binary clauses: the goal is to find a truth assignment to these variables which maximises the number of satisfied clauses. We will construct a binary CSP instance $I_{CSP}$ on $O(N + M)$ variables, each with domain of size at most four, such that the convergent sequences $S$ of SS value-eliminations in $I_{CSP}$ correspond to truth assignments to $X_1, \ldots, X_N$ and the length of $S$ is $\alpha N + \beta m$ where $\alpha, \beta$ are constants and $m$ is the number of clauses of $I_{2SAT}$ satisfied by the corresponding truth assignment.

We require four constructions (which we explain in detail below):

1. the construction in Fig. 7 simulates a MAX 2-SAT literal $X$ by a path of CSP variables joined by greater-than-or-equal-to constraints.
2. the construction in Fig. 8 simulates the relationship between a MAX 2-SAT variable $X$ and its negation $\overline{X}$.
3. the construction in Fig. 9 allows us to create multiple copies of a MAX 2-SAT literal $X$.
4. the construction in Fig. 10 simulates a binary clause $X \vee Y$ where $X, Y$ are MAX 2-SAT literals.

In each of these figures, each oval represents a CSP variable with the bullets inside the oval representing the possible values for this variable. If there is a non-trivial constraint between two variables $x_i, x_j$ this is represented by joining up with a line those pairs of values $a, b$ such that $(a, b) \in R_{ij}$. Where the constraint has a compact form, such as $x_1 \geq x_2$ this is written next to the constraint. In the following, we write $b \overset{x_i}{\rightsquigarrow} a$ if $b \in \mathcal{D}(x_i)$ is snake substitutable by $a \in \mathcal{D}(x_i)$. Our constructions are such that the only value that can be eliminated from any domain by SS is the value 2.

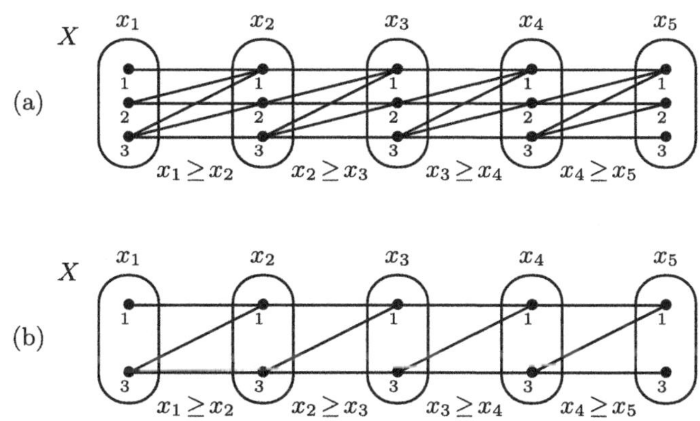


**Fig. 7.** A construction to simulate a MAX 2-SAT variable $X$: (a) $X = 0$, (b) $X = 1$.

Figure 7(a) shows a path of CSP variables constrained by greater-than-or-equal-to constraints. The end variables $x_1$ and $x_5$ are constrained by other variables that, for clarity of presentation, are not shown in this figure. If value 2 is eliminated from $\mathcal{D}(x_1)$, then we have $2 \overset{x_2}{\rightsquigarrow} 3$. In fact, 2 is neighbourhood substitutable by 3. Once the value 2 is eliminated from $\mathcal{D}(x_2)$, we have $2 \overset{x_3}{\rightsquigarrow} 3$. Indeed, eliminations of the value 2 propagate so that in the end we have the situation shown in Fig. 7(b). By a symmetrical argument, the elimination of the value 2 from $\mathcal{D}(x_5)$ propagates from right to left (this time by neighbourhood substitution by 1) to again produce the situation shown in Fig. 7(b). It is easily verified that, without any eliminations from the domains $\mathcal{D}(x_1)$ or $\mathcal{D}(x_5)$, no values for the variables $x_2, x_3, x_4$ are snake-substitutable. Furthermore, the values 1 and 3 for the variables $x_2, x_3, x_4$ are not snake-substitutable even after the elimination of the value 2 from all domains. So we either have no eliminations, which we associate with the truth assignment $X = 0$ (where $X$ is the MAX 2-SAT literal corresponding to this path of variables in $I_{CSP}$) or the value 2 is eliminated from all domains, which we associate with the truth assignment $X = 1$.

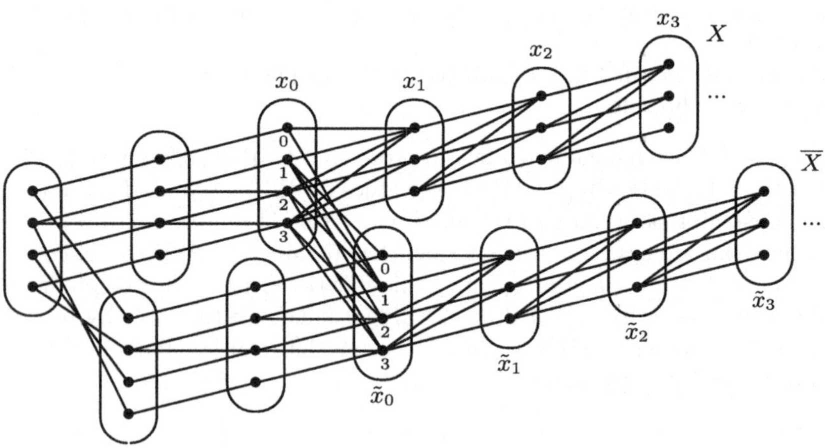


**Fig. 8.** A construction to simulate a MAX 2-SAT variable $X$ and its negation $\overline{X}$.

The construction in Fig. 8 joins the two path-of-CSP-variables constructions corresponding to the literals $X$ and $\overline{X}$. This construction ensures that exactly one of $X$ and $\overline{X}$ are assigned the value 1. It is easy (if tedious) to verify that the only snake substitutions that are possible in this construction are $2 \overset{x_0}{\rightsquigarrow} 3$ and $2 \overset{\tilde{x}_0}{\rightsquigarrow} 3$, but that after elimination of the value 2 from either of $\mathcal{D}(x_0)$ or $\mathcal{D}(\tilde{x}_0)$, the other snake substitution is no longer valid. Once, for example, 2 has been eliminated from $\mathcal{D}(x_0)$, then this elimination propagates along the path of CSP variables $(x_1, x_2, x_3, \ldots)$ corresponding to $X$, as shown in Fig. 7(b). By a symmetrical argument, if 2 is eliminated from $\mathcal{D}(\tilde{x}_0)$, then this elimination propagates along the path of CSP variables $(\tilde{x}_1, \tilde{x}_2, \tilde{x}_3, \ldots)$ corresponding to $\overline{X}$.

Thus, this construction simulates the assignment of a truth value to $X$ and its complement to $\overline{X}$.

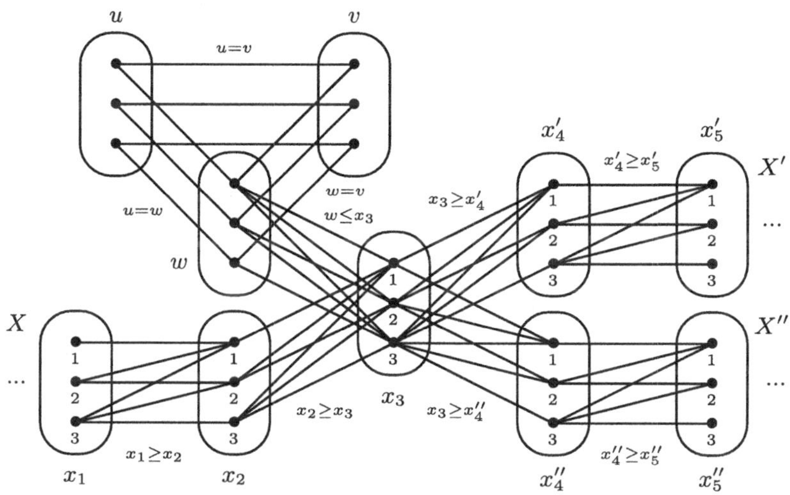


**Fig. 9.** A construction to create two copies $X'$ and $X''$ of the MAX 2-SAT variable $X$.

Since any literal of $I_{2SAT}$ may occur in several clauses, we need to be able to make copies of any literal. Figure 9 shows a construction that creates two copies $X', X''$ of a literal $X$. This construction can easily be generalised to make $k$ copies of a literal, if required, by having $k$ identical paths of greater-than-equal-to constraints on the right of the figure all starting at the pivot variable $x_3$. Before any eliminations are performed, no snake substitutions are possible in this construction. However, once the value 2 has been eliminated from $\mathcal{D}(x_1)$, eliminations propagate, as in Fig. 7: the value 2 can successively be eliminated from the domains of variables $x_2$, $x_3$, $x'_4$, $x'_5$, and $x''_4$, $x''_5$. Each elimination is in fact by neighbourhood substitution, as in Fig. 7. These eliminations mean that we effectively have two copies $X'$, $X''$ of the literal $X$. The triangle of equality constraints at the top left of this construction is is there simply to prevent propagation in the reverse direction: even if the value 2 is eliminated from the domains of $x'_5, x'_4$ and $x''_5, x''_4$ by the propagation of eliminations from the right, this cannot provoke the elimination of the value 2 from the domain of the pivot variable $x_3$.

Finally, the construction of Fig. 10 simulates the clause $X \vee Y$. In fact, this construction simply joins together the paths of CSP-variables corresponding to the two literals $X, Y$, via a variable $z$. It is easily verified that the elimination of the value 2 from the domain of $x_1$ allows the propagation of eliminations of the value 2 from the domains of $x_2$, $z$, $y_2$, $y_1$ in exactly the same way as the propagation of eliminations in Fig. 7. Similarly, the elimination of the value 2 from the domain of $y_1$ propagates to all other variables in the opposite order $y_2$,

$z$, $x_2$, $x_1$. Thus, if one or other of the literals $X$ or $Y$ in the clause is assigned 1, then the value 2 is eliminated from all domains of this construction. Eliminations can propagate back up to the pivot variable ($x_3$ in Fig. 9) but no further, as explained in the previous paragraph.

Putting all this together, we can see that there is a one-to-one correspondence between convergent sequences of SS value-eliminations and truth assignments to the variables of the MAX 2-SAT instance. Furthermore, the number of SS value-eliminations is maximised when this truth assignment maximises the number of satisfied clauses, since it is $\alpha N + \beta m$ where $\alpha$ is the number of CSP-variables in each path of greater-than-or-equal-to constraints corresponding to a literal, $\beta$ is the number of CSP-variables in each clause construction and $m$ is the number of satisfied clauses. This reduction is clearly polynomial.

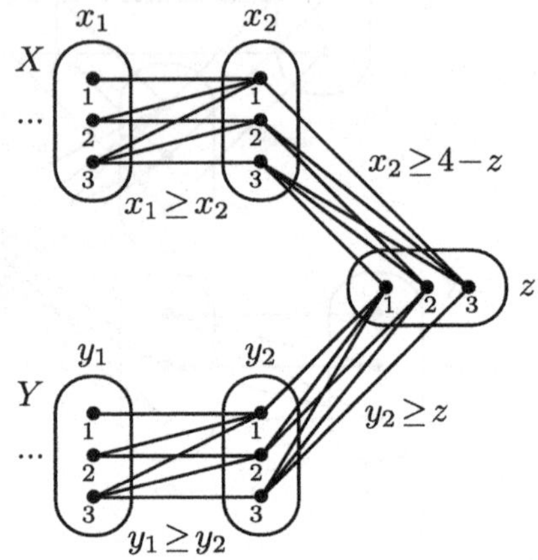


**Fig. 10.** A construction to simulate a MAX 2-SAT clause $X \vee Y$.

## 7 Discussion and Conclusion

Both snake substitutability (SS) and conditioned neighbourhood substitutability (CNS) strictly subsume neighbourhood substitution but nevertheless can be applied in the same $O(ed^3)$ time complexity. We have also given a more general notion of substitution (SCSS) subsuming both these rules that can be detected in $O(ed^3)$ time. The examples in Fig. 1 show that these three rules are strictly stronger than neighborhood substitution and that SS and CNS are incomparable.

An avenue of future research is the generalisation to valued CSPs. The notion of snake substitutability has already been generalised to binary valued CSPs and it has been shown that it is possible to test this notion in $O(ed^4)$ time if the aggregation operator is addition over the non-negative rationals [13]. However, further research is required to determine the complexity of applying this operation until convergence.

It is possible to efficiently find all (or a given number of) solutions to a CSP after applying NS: given the set of all solutions to the reduced instance, it is possible to find $K \geq 1$ solutions to the original instance $I$ (or to determine that $I$ does not have $K$ solutions) in $O(K(de + n^2))$ time [7]. This also holds for CNS, since, as for NS, for each solution $s$ found and for each value $b$ eliminated from some domain $\mathcal{D}(x_i)$, it suffices to test each putative solution obtained by replacing $s_i$ by $b$. Unfortunately, the extra strength of snake substitution (SS) is here a drawback, since, by exactly the same argument as for the $\exists 2$*snake* value-elimination rule (which is a weaker version of SS) [6], we can deduce that determining whether a binary CSP instance has two or more solutions is NP-hard, even given the set of solutions to the reduced instance after applying SS.

## References


1. Bessiere, C.: Constraint propagation. In: Rossi et al. [26], pp. 29–83. https://doi.org/10.1016/S1574-6526(06)80007-6
2. Bessière, C., Debruyne, R.: Optimal and suboptimal singleton arc consistency algorithms. In: Kaelbling, L.P., Saffiotti, A. (eds.) IJCAI-2005, Proceedings of the Nineteenth International Joint Conference on Artificial Intelligence, pp. 54–59. Professional Book Center (2005). http://ijcai.org/Proceedings/05/Papers/0495.pdf
3. Bessière, C., Régin, J., Yap, R.H.C., Zhang, Y.: An optimal coarse-grained arc consistency algorithm. Artif. Intell. **165**(2), 165–185 (2005). https://doi.org/10.1016/j.artint.2005.02.004
4. Chu, G., Stuckey, P.J.: Dominance breaking constraints. Constraints Int. J. **20**(2), 155–182 (2014). https://doi.org/10.1007/s10601-014-9173-7
5. Clowes, M.B.: On seeing things. Artif. Intell. **2**(1), 79–116 (1971). https://doi.org/10.1016/0004-3702(71)90005-1
6. Cohen, D.A., Cooper, M.C., Escamocher, G., Zivny, S.: Variable and value elimination in binary constraint satisfaction via forbidden patterns. J. Comput. Syst. Sci. **81**(7), 1127–1143 (2015). https://doi.org/10.1016/j.jcss.2015.02.001
7. Cooper, M.C.: Fundamental properties of neighbourhood substitution in constraint satisfaction problems. Artif. Intell. **90**(1–2), 1–24 (1997). https://doi.org/10.1016/S0004-3702(96)00018-5
8. Cooper, M.C.: Constraints between distant lines in the labelling of line drawings of polyhedral scenes. Int. J. Comput. Vis. **73**(2), 195–212 (2007). https://doi.org/10.1007/s11263-006-9783-7
9. Cooper, M.C.: Line Drawing Interpretation. Springer, Heidelberg (2008). https://doi.org/10.1007/978-1-84800-229-6
10. Cooper, M.C.: Strengthening neighbourhood substitution. CoRR abs/2007.06282 (2020). https://arxiv.org/abs/2007.06282
11. Cooper, M.C., Duchein, A., Mouelhi, A.E., Escamocher, G., Terrioux, C., Zanuttini, B.: Broken triangles: from value merging to a tractable class of general-arity constraint satisfaction problems. Artif. Intell. **234**, 196–218 (2016). https://doi.org/10.1016/j.artint.2016.02.001
12. Cooper, M.C., El Mouelhi, A., Terrioux, C.: Extending broken triangles and enhanced value-merging. In: Rueher, M. (ed.) CP 2016. LNCS, vol. 9892, pp. 173–188. Springer, Cham (2016). https://doi.org/10.1007/978-3-319-44953-1_12
13. Cooper, M.C., Jguirim, W., Cohen, D.A.: Domain reduction for valued constraints by generalising methods from CSP. In: Hooker, J. (ed.) CP 2018. LNCS, vol. 11008, pp. 64–80. Springer, Cham (2018). https://doi.org/10.1007/978-3-319-98334-9_5
14. Freuder, E.C.: Eliminating interchangeable values in constraint satisfaction problems. In: Dean, T.L., McKeown, K.R. (eds.) Proceedings of the 9th National Conference on Artificial Intelligence, vol. 1, pp. 227–233. AAAI Press/The MIT Press (1991). http://www.aaai.org/Library/AAAI/1991/aaai91-036.php
15. Freuder, E.C., Wallace, R.J.: Replaceability and the substitutability hierarchy for constraint satisfaction problems. In: Benzmüller, C., Lisetti, C.L., Theobald, M. (eds.) GCAI 2017, 3rd Global Conference on Artificial Intelligence. EPiC Series in Computing, vol. 50, pp. 51–63. EasyChair (2017). http://www.easychair.org/publications/paper/mKkF

16. Garey, M.R., Johnson, D.S., Stockmeyer, L.J.: Some simplified NP-complete graph problems. Theor. Comput. Sci. **1**(3), 237–267 (1976). https://doi.org/10.1016/0304-3975(76)90059-1
17. Gent, I.P., Petrie, K.E., Puget, J.: Symmetry in constraint programming. In: Rossi et al. [26], pp. 329–376. https://doi.org/10.1016/S1574-6526(06)80014-3
18. van Hoeve, W., Katriel, I.: Global constraints. In: Rossi et al. [26], pp. 169–208. https://doi.org/10.1016/S1574-6526(06)80010-6
19. Huffman, D.A.: Impossible objects as nonsense sentences. In: Meltzer, B., Michie, D. (eds.) Machine Intelligence, vol. 6, pp. 295–323. Edinburgh University Press (1971)
20. Karp, R.M.: Reducibility among combinatorial problems. In: Miller, R.E., Thatcher, J.W. (eds.) Proceedings of Symposium on the Complexity of Computer Computations. The IBM Research Symposia Series, pp. 85–103. Plenum Press, New York (1972). http://www.cs.berkeley.edu/%7Eluca/cs172/karp.pdf
21. Lecoutre, C.: Constraint Networks Techniques and Algorithms. ISTE/Wiley, Hoboken (2009)
22. Likitvivatanavong, C., Yap, R.H.C.: Many-to-many interchangeable sets of values in CSPs. In: Shin, S.Y., Maldonado, J.C. (eds.) Proceedings of the 28th Annual ACM Symposium on Applied Computing, SAC 2013, pp. 86–91. ACM (2013). https://doi.org/10.1145/2480362.2480382
23. Mackworth, A.K.: Consistency in networks of relations. Artif. Intell. **8**(1), 99–118 (1977). https://doi.org/10.1016/0004-3702(77)90007-8
24. Mohr, R., Henderson, T.C.: Arc and path consistency revisited. Artif. Intell. **28**(2), 225–233 (1986). https://doi.org/10.1016/0004-3702(86)90083-4
25. Naanaa, W.: New schemes for simplifying binary constraint satisfaction problems. Discrete Math. Theor. Comput. Sci. (2019)
26. Rossi, F., van Beek, P., Walsh, T. (eds.): Handbook of Constraint Programming, Foundations of Artificial Intelligence, vol. 2. Elsevier, Amsterdam (2006). http://www.sciencedirect.com/science/bookseries/15746526/2
27. Waltz, D.: Understanding line drawings of scenes with shadows. In: Winston, P.H. (ed.) The Psychology of Computer Vision. Computer Science Series, pp. 19–91. McGraw-Hill, New York (1975)
28. Woodward, R.J., Karakashian, S., Choueiry, B.Y., Bessiere, C.: Revisiting neighborhood inverse consistency on binary CSPs. In: Milano, M. (ed.) CP 2012. LNCS, pp. 688–703. Springer, Heidelberg (2012). https://doi.org/10.1007/978-3-642-33558-7_50

# Effective Encodings of Constraint Programming Models to SMT

Ewan Davidson[1], Özgür Akgün[1], Joan Espasa[1](✉), and Peter Nightingale[2]

[1] School of Computer Science, University of St Andrews, St Andrews, UK
{eld5,ozgur.akgun,jea20}@st-andrews.ac.uk

[2] Department of Computer Science, University of York, York, UK
peter.nightingale@york.ac.uk

**Abstract.** Satisfiability Modulo Theories (SMT) is a well-established methodology that generalises propositional satisfiability (SAT) by adding support for a variety of theories such as integer arithmetic and bit-vector operations. SMT solvers have made rapid progress in recent years. In part, the efficiency of modern SMT solvers derives from the use of specialised decision procedures for each theory. In this paper we explore how the Essence Prime constraint modelling language can be translated to the standard SMT-LIB language. We target four theories: bit-vectors (QF_BV), linear integer arithmetic (QF_LIA), non-linear integer arithmetic (QF_NIA), and integer difference logic (QF_IDL). The encodings are implemented in the constraint modelling tool Savile Row. In an extensive set of experiments, we compare our encodings for the four theories, showing some notable differences and complementary strengths. We also compare our new encodings to the existing work targeting SMT and SAT, and to a well-established learning CP solver. Our two proposed encodings targeting the theory of bit-vectors (QF_BV) both substantially outperform earlier work on encoding to QF_BV on a large and diverse set of problem classes.

**Keywords:** Constraint modelling · SMT · Automated reformulation

## 1 Introduction

Constraint programming (CP) is a powerful paradigm for solving constraint satisfaction and optimisation problems, with many diverse applications. Essence Prime [22] is a solver-independent CP modelling language that offers decision variables of Boolean and integer domains as well as arrays of these decision variables, arithmetic and logical operators for expressing constraints, and global constraints. Essence Prime is comparable in its modelling capabilities to MiniZinc [17], OPL [25], Simply [6] and other similar languages. To solve a problem instance described in Essence Prime, a problem class model and a parameter file are translated by Savile Row [21] into input suitable for a backend solver. Savile Row applies automated reformulation steps such as common sub-expression elimination to improve the model [21].


H. Simonis (Ed.): CP 2020, LNCS 12333, pp. 143–159, 2020.
https://doi.org/10.1007/978-3-030-58475-7_9

Satisfiability Modulo Theories (SMT) is a problem solving methodology that decides the satisfiability of a propositional formula with respect to a selection of theories in first order logic with equality [20]. SMT has its roots in the field of hardware and software verification, but lately it is being used to solve a wider range of problems. The Satisfiability Modulo Theories Library (SMT-LIB) [4] provides a standard for the specification of benchmarks and theories for SMT. Most SMT solvers are restricted to decidable quantifier free fragments of their logics, which is sufficient for many applications. SMT solvers have made rapid progress in recent years. In part, the efficiency of modern SMT solvers derives from the use of specialised decision procedures for each theory. In this work we will focus on four theories: *QF_LIA*, *QF_NIA*, *QF_IDL*, and *QF_BV*.

*QF_LIA* and *QF_NIA* are the theories of quantifier free linear and non-linear integer arithmetic respectively. The formulas are Boolean combinations of constraints comparing expressions to a constant with respect to $\leq$ and $\geq$. Linear expressions are typically enough to naturally express common problems in formal verification and scheduling problems. SMT solvers typically use variants of the Simplex algorithm to implement integer arithmetic theories [12].

*QF_IDL* is the theory of quantifier free integer difference logic. It supports Boolean combinations of inequalities of the form $x - y < b$ where $x$ and $y$ are integer variables and $b$ is an integer constant. The renewed interest in this fragment came from timed automata, where the verification conditions arising take the form of difference logic formulas [18].

*QF_BV* is the theory of quantifier-free formulas over fixed-size bit-vectors. Bit-vector arithmetic is very commonly used for verification and equivalence checking in the hardware industry. Current solvers [13] typically apply heavy preprocessing techniques that ultimately flatten the formula to SAT, also known as *bit-blasting*.

The main contribution of this paper is a new SMT backend for Savile Row that is able to produce any one of the four theories listed above and at two different levels of flattening, thus providing eight distinct SMT encodings. We exhaustively evaluate the performance of these encodings, showing that our encodings perform significantly better than existing SMT approaches [9] and that they are complementary to the existing SAT and CP backends of Savile Row.

## 2 Related Work

There has been a great deal of work on translating declarative constraint modelling languages to lower-level languages such as SAT, SMT, and MIP. We cannot cover it all here so we cover all the most relevant work (on translation to SMT) and give examples for the rest. One example of SAT encoding is FznTini [14], which translates FlatZinc into SAT. The MiniZinc compiler is able to translate the MiniZinc language into MIP [5]. There are several encodings of a constraint language to SMT: Simply [6,8], fzn2smt [7], and FZN2OMT [9].

Simply [6] is a compiler from a custom constraint modelling language (comparable to Essence Prime) to SMT. It supports translations to QF_LIA and

QF_IDL logics and was later extended to support meta-constraints and weighted CSPs [8]. The same authors presented fzn2smt [7], an approach that translates FlatZinc models to SMT, supporting the standard data types and constraints of FlatZinc. The logic used for solving each instance is determined automatically during the translation, and it handles optimisation problems by means of a dichotomic (binary) search on the domain of the optimisation variable. As in our approach, search annotations are ignored, as they do not make sense in the context of SMT. Only the allDifferent and Cumulative global constraints are supported by decomposing them into SMT.

FZN2OMT [9] can translate standard FlatZinc into suitable input for SMT solvers, and back to FlatZinc. It supports the Z3 [16] and OptiMathSAT [24] solvers, using either the QF_BV or QF_LIRA logics. When using OptiMathSAT, it takes advantage of the fact that OptiMathSAT natively supports a subset of the FlatZinc 1.6 language. Unlike fzn2smt, FZN2OMT is compatible with the current MiniZinc toolchain. We compare our proposed SMT encodings to all four configurations of FZN2OMT in Sect. 4 below.

## 3 Translating Essence Prime to SMT

The modelling tool Savile Row works by reading a problem class model written in Essence Prime and optionally a parameter file that contains values defining a problem instance. It instantiates the problem class model, unrolls comprehensions and quantified expressions, and performs expression *flattening* where the target solver language does not support nested expressions. Flattening is the task of introducing *auxiliary* variables to represent nested subexpressions. Section 3.3 explains flattening in the SMT backend.

The input to Savile Row is a solver-independent model with nested expressions. Savile Row already has existing backends to several CP/SAT/MIP solvers, some of these via FlatZinc. Some of these backends require a lower level output than others. For example, Minion's input language allows a limited form of nesting (for example, constraints may be contained in conjunctions and disjunctions) but still requires a mostly flat structure, whereas FlatZinc-based solvers do not allow this kind of nesting at all. SMT-LIB allows more kinds of nested expressions than any other Savile Row backend. We implement a fully-flat variant as well as a *nested* variant that only flattens when strictly necessary. It is difficult to know which variant will perform better for a given theory, SMT solver and problem class. Flattening introduces potentially useful new variables, and preserving nesting gives solvers access to the high-level structure.

A model in Essence Prime consists of three main components: decision variable declarations, constraint expressions, and an optional objective function. The following sections explain how these components are translated to SMT.

### 3.1 Decision Variables and Their Domains

Essence Prime is a finite-domain language, where every decision variable is associated with a finite domain of discrete values. It supports Booleans and integers as atomic types, and matrices of arbitrary dimensions of these atomic types. SAVILE ROW expands matrices to lists of atomic types so its output languages do not need to support matrices or arrays. For example, given a matrix $M$ with three rows and columns, it will create 9 declarations $M_{1,1}, M_{1,2}, \ldots M_{3,3}$ of its declared type.

When targeting SMT, decision variable declarations are translated to their equivalent variable declarations in SMT-LIB together with unary constraints to define the bounds of the domain. For example, when using the LIA encoding, a `find x : int(1..10)` declaration would be translated as the SMT integer variable `x` and the constraint $1 \leq x \leq 10$. When using the QF_BV theory, numerical variables are represented using fixed-size signed bit-vectors (i.e. binary numbers in two's-complement). We use as few bits as we can, governed by the largest domain value in the union of all decision variables. SAVILE ROW performs domain shaving by enforcing a strong level of consistency (Singleton Arc Consistency on the bounds of the variables) via Minion. This step helps in removing any unnecessarily large values in the domains and reducing the number of variables required after bit-blasting [15].

### 3.2 Constraints and the Objective

Constraint expressions are at the heart of a constraint model. We translate them to SMT-LIB using specialised implementations per expression type and by falling back to the standard SAT encodings available in SAVILE ROW where necessary. Boolean expressions ($\wedge$, $\vee$, $\rightarrow$, ...) and relational operators over integers ($<$, $\leq$, $=$, ...) are translated to use the corresponding operators in SMT-LIB. Arithmetic operators ($+$, $-$, $/$, mod, abs, ...) are similarly translated depending on the theory we use. Quantified expressions, such as the universally and existentially quantified expression or quantified sums are unrolled and turned into $\wedge$, $\vee$, and weighted sums. AllDifferent and Global Cardinality (GCC) are decomposed: for each value (or each constrained value in GCC) a linear constraint is used to restrict the number of occurrences of the value. Table constraints use the Bacchus encoding [3] and short tables are encoded similarly [1].

SMT solvers are typically implemented by augmenting an existing SAT solver with implementations of theories. Hence, they allow SAT clauses in the input. We implement our SMT encodings when they are specifically supported by a backend theory and fall back to the existing SAT encodings in SAVILE ROW for the rest of the cases.

When targeting the QF_BV logic, we use *bvadd*, *bvor* and *bvneg* as appropriate for arithmetic and logical expressions. These operators are not restricted to linear expressions, they support nested expressions with sums, multiplications, divisions, and modulo operations. Modern QF_BV solvers typically use a technique called bit-blasting (amongst others). Bit-blasting converts the problem to SAT by introducing a SAT variable for each bit in a bit vector. This approach can generate a smaller encoding when compared to the direct or order encodings used by the SAT backend of SAVILE ROW [23], since the direct and order encodings scale linearly in domain size and bit-blasting scales sub-linearly.

The objective function (if present) is represented by a single decision variable and dichotomic search is applied to find the optimal solution. However, when Z3 is the target solver we use its built-in optimisation (regardless of the theory). In our experiments we use Z3 for QF_NIA and other solvers for the other three theories, so solver built-in optimisation is only used with QF_NIA.

### 3.3 Flat vs Nested Encodings

ESSENCE PRIME (like MiniZinc, OPL and other comparable constraint modelling languages) allows stating *nested* constraints such as allDiff($[x+1, y+2, z+3]$) directly. This constraint is considered nested because each element of the list inside the allDiff constraint is an expression and not a decision variable.

Some of the backend solvers for SAVILE ROW do not support nested constraints, and therefore constraints like these have to be *flattened*. In this case, the flattening process would create auxiliary variables for each sub-expression ($aux_1$, $aux_2$, $aux_3$), post additional constraints $aux_1 = x + 1$, $aux_2 = y + 2$, $aux_3 = z + 3$, and replace the constraint with allDiff($[aux_1, aux_2, aux_3]$). This transformation will introduce additional variables and constraints.

Since SMT-LIB allows nested expressions we have an opportunity to evaluate the effect of using SAVILE ROW's standard flattening process vs maintaining the nested expressions and letting the SMT solver flatten them if needed. A nested encoding is likely to be smaller in size and may allow the SMT solver to do a better job if and when it chooses to flatten. We evaluate flat and nested encodings for each of the four theories in Sect. 4.

## 4 Empirical Evaluation

In this section we compare our set of encodings (SAVILE ROW-SMT, or *SR-SMT*) with the state of the art on a wide range of problems, showing the advantages of our approach. We compare SR-SMT to the SAT encoding implemented in SAVILE ROW [21] using the SAT solver CaDiCaL 1.3.0, and also to the well-established learning constraint solver Chuffed 0.10.3. In each experiment we use the same large set of problem instances and the same basic settings of SAVILE ROW. We have not been able to compare our approach to fzn2smt [7] or FznTini [14], as they have become obsolete due to changes in the FlatZinc language. We compare our approach to FZN2OMT [9], described in Sect. 2.

SR-SMT has eight configurations and FZN2OMT has four, which presents a challenge when attempting to compare the two systems. We resolve this by using the *virtual best solver* (VBS) approach for each system; i.e. for each problem instance the best configuration is selected. We also look into which configurations are contributing most to the VBS for each system.

### 4.1 Experimental Setup

As SR-SMT outputs standard SMT-LIB2 files, we are able to target many solvers easily. For the QF_LIA and QF_IDL encodings we use Yices 2.6.2 [11], for the

QF_BV encoding we target Boolector 3.2.1 [19] and finally for QF_NIA we use Z3 4.8.8 [16]. Experience of using SMT solvers for planning and scheduling problems (expressed in QF_LIA or QF_IDL) led us to use Yices for those theories; Boolector is known to perform well on QF_BV; and Z3 was chosen for QF_NIA because of its advanced preprocessing and heuristics (for example, as noted below, Z3 converts the QF_NIA Discrete Tomography problem into QF_BV). Regarding optimisation, we handle it by means of a binary search on the domain of the variable to optimise. The only exception is when the Z3 solver is used, as it has native support for optimisation.

The set of problem classes used is an expanded version of the one used in [21], with 63 problem classes and 757 instances in total[1]. Both satisfaction and optimization problems are included, and optimization problems are reported as solved when optimality of the last solution has been proven.

We used a time limit of 1 h total time (i.e. Savile Row time plus solver time). We use PAR2 to summarise the performance of each configuration. PAR2 is the mean of total time, where the instances that timed out are assumed to have taken two times the time limit (i.e. 2 h). Configurations with a lower PAR2 score are considered to be better.

The default set of optimisations in Savile Row are used for all solvers and encodings: domain filtering, variable unification, aggregation and active CSE [21]. A standard FlatZinc backend was added to Savile Row to be used for experiments with FZN2OMT. Decompositions of global constraints closely follow those in the MiniZinc `std` library. For example, allDifferent is decomposed into a clique of pairwise not-equal constraints, and global cardinality into one sum for each constrained value. The existing Chuffed FlatZinc backend is very similar but does not decompose allDifferent or lexicographic ordering constraints.

### 4.2 Comparison to FZN2OMT

Figure 1 gives an overview of the results comparing the SR-SMT virtual best solver (SR-SMT-VBS) to the FZN2OMT virtual best solver (FZN2OMT-VBS). For SR-SMT-VBS, 685 instances were solved and for FZN2OMT-VBS 643 were solved. In Table 1 we report the number of instances solved by the two virtual best solvers as well as each configuration of both systems. The results show a clear advantage for SR-SMT-VBS on the bulk of the instances. However, there are confounding factors. First, SR-SMT-VBS is constructed from 8 configurations rather than 4, potentially giving it an advantage. Second, different SMT solvers are used. For example, Boolector was used to solve QF_BV encodings in SR-SMT-VBS, whereas Z3 and OptiMathSAT were used in FZN2OMT.

Many problem classes contain the allDifferent constraint. For these, SR-SMT uses a decomposition where each value is constrained to have at most one occurrence using a linear constraint (in common with Savile Row's SAT backend),

[1] Experiment scripts, model and parameter files and raw results can be found at: https://github.com/stacs-cp/CP2020-SRSMT [10].

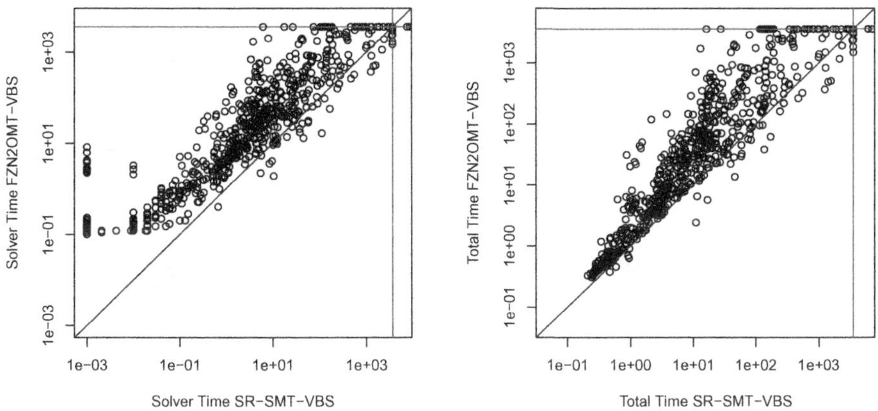


**Fig. 1.** SR-SMT-VBS vs FZN2OMT-VBS, solver time (left) and total time (right).

**Table 1.** Number of instances solved and PAR2 for each configuration. Bold indicates the overall best configuration (by instances solved), red indicates the best configuration of one system (SR-SMT or FZN2OMT).

| Encoding | Instances solved | Mean PAR2 |
|---|---|---|
| **SR-SMT-VBS** | **685** | **803** |
| SR-SMT-Nested-VBS | 682 | 840 |
| SR-SMT-Flat-VBS | 682 | 839 |
| FZN2OMT-VBS | 643 | 1283 |
| SR-SMT-BV-Nested | 663 | 1092 |
| SR-SMT-BV-Flat | 661 | 1106 |
| SR-SMT-NIA-Nested | 499 | 2620 |
| SR-SMT-NIA-Flat | 511 | 2522 |
| SR-SMT-LIA-Nested | 590 | 1695 |
| SR-SMT-LIA-Flat | 586 | 1729 |
| SR-SMT-IDL-Nested | 592 | 1714 |
| SR-SMT-IDL-Flat | 591 | 1716 |
| SR-SMT-BV-Nested-Z3 | 657 | 1138 |
| SR-SMT-BV-Nested-Z3-PA | 643 | 1228 |
| FZN2OMT-LIA-Z3 | 508 | 2638 |
| FZN2OMT-LIA-OptiMathSAT | 517 | 2609 |
| FZN2OMT-BV-Z3 | 587 | 1921 |
| FZN2OMT-BV-OptiMathSAT | 533 | 2454 |
| SAT | 671 | 931 |
| Chuffed | 637 | 1248 |

whereas the standard FlatZinc backend used with FZN2OMT decomposes to pairwise not-equal constraints. We compare the two decompositions below.

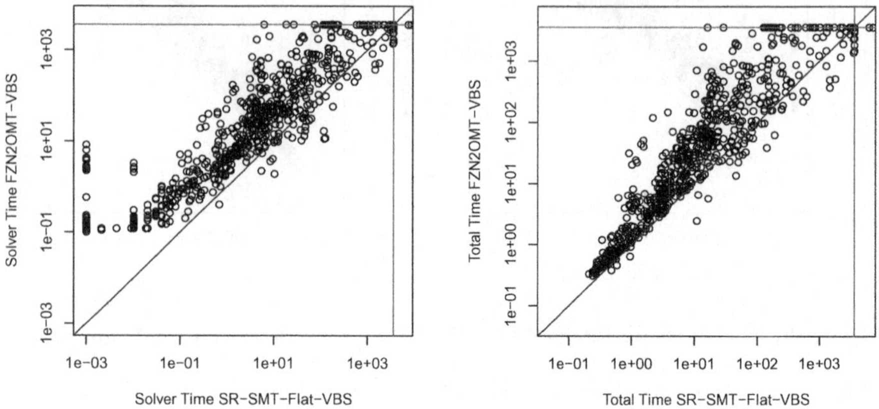


**Fig. 2.** SR-SMT-Flat-VBS (i.e. virtual best solver of the 4 Flat configurations) vs FZN2OMT-VBS, solver time (left) and total time (right).

To provide a fairer comparison between two portfolios of the same size, we constructed two further virtual best solvers for SR-SMT, one using the four Flat configurations (SR-SMT-Flat-VBS) and the other using the four Nested configurations (SR-SMT-Nested-VBS). As Table 1 shows, the Flat configuration solves more instances for one theory (QF_NIA) and is quite close to Nested in performance for the other theories. On the other hand, Nested solves more instances with the most promising theory (QF_BV) and also QF_LIA and QF_IDL. It turns out that SR-SMT-Flat-VBS is very slightly stronger (Table 1). Figure 2 compares SR-SMT-Flat-VBS to FZN2OMT-VBS, and Table 1 contains the number of instances solved and PAR2 score for both. Even with a smaller portfolio it is clear that SR-SMT is performing better than FZN2OMT on these benchmarks.

Also, there is the issue that the solvers do not match for any theory. SR-SMT uses Yices with QF_LIA and Boolector with QF_BV, whereas FZN2OMT uses Z3 and OptiMathSAT. To be able to compare just the encodings, we ran SR-SMT-BV-Nested (the strongest configuration of SR-SMT measured by instances solved and PAR2) with Z3 instead of Boolector, creating a configuration called SR-SMT-BV-Nested-Z3. Table 1 shows that SR-SMT-BV-Nested-Z3 is somewhat weaker than SR-SMT-BV-Nested on these benchmarks, solving 6 fewer instances within 1 h and having a higher PAR2 score. Figure 3 compares SR-SMT-BV-Nested-Z3 to FZN2OMT-BV-Z3. The results are mixed, with some instances solving much faster with FZN2OMT-BV-Z3, and 15 solved only by FZN2OMT-BV-Z3. However, the overall trend is that SR-SMT-BV-Nested-Z3 is stronger, it solves 70 more instances and has a lower PAR2 score.

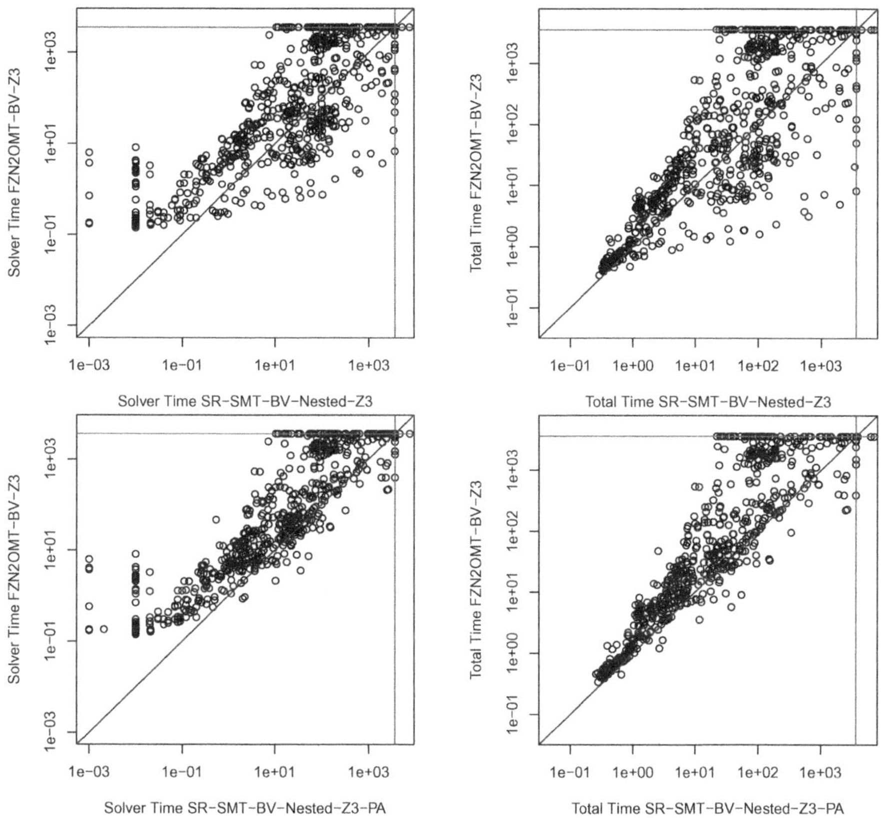


**Fig. 3.** SR-SMT-BV-Nested-Z3 (i.e. SR-SMT-BV-Nested with Z3 instead of Boolector) vs FZN2OMT-BV-Z3, solver time (upper left) and total time (upper right). The lower plots are the same but with the pairwise AllDifferent decomposition (PA).

Finally, there is the issue that the decompositions of AllDifferent do not match. We created another configuration SR-SMT-BV-Nested-Z3-PA, which is SR-SMT-BV-Nested-Z3 with the pairwise decomposition of AllDifferent (matching FZN2OMT). As Table 1 shows, the new configuration is weaker than SR-SMT-BV-Nested-Z3, solving 14 fewer instances in total. Figure 3 compares SR-SMT-BV-Nested-Z3-PA to FZN2OMT-BV-Z3. The two are quite strongly correlated but SR-SMT is stronger, solving 56 more instances in total.

### 4.3 Analysis of SR-SMT

In this section we look at which configurations of SR-SMT are most effective, and how each configuration contributes to the virtual best solver. First we compare Nested to Flat configurations, then we compare the four theories. Also, Table 2 shows how each configuration contributed to the virtual best solver SR-SMT-VBS, and how many additional instances were solved within multiples of 2 and 5 of the VBS time. This gives an overview of how strong the configurations are relative to the VBS.

**Table 2.** Number of instances solved by each configuration within multiples of SR-SMT-VBS or FZN2OMT-VBS total time. An instance is counted for configuration X and multiple $f$ if total time for X is within $f\times$ total time for the VBS. The highest value in each column is highlighted in bold.

| Configuration | Multiple of SR-SMT-VBS Total Time | | | |
|---|---|---|---|---|
| | 1 | 2 | 5 | Any |
| SR-SMT-BV-Nested | 76 | 350 | 465 | **663** |
| SR-SMT-BV-Flat | 62 | 319 | 473 | 661 |
| SR-SMT-NIA-Nested | 25 | 244 | 351 | 499 |
| SR-SMT-NIA-Flat | 60 | 294 | 365 | 511 |
| SR-SMT-LIA-Nested | 146 | **434** | 530 | 590 |
| SR-SMT-LIA-Flat | 86 | 433 | **534** | 586 |
| SR-SMT-IDL-Nested | 120 | 401 | 488 | 592 |
| SR-SMT-IDL-Flat | **149** | 403 | 486 | 591 |
| Configuration | Multiple of FZN2OMT-VBS Total Time | | | |
| | 1 | 2 | 5 | Any |
| FZN2OMT-LIA-Z3 | 148 | 318 | 362 | 508 |
| FZN2OMT-LIA-OptiMathSAT | 114 | 222 | 308 | 517 |
| FZN2OMT-BV-Z3 | **239** | **439** | **498** | **587** |
| FZN2OMT-BV-OptiMathSAT | 150 | 282 | 378 | 533 |

**Comparing Nested to Flat Translation.** Figure 4 compares Nested to Flat configurations for each theory. With the theory of bit-vectors the two configurations are remarkably similar. Total number of instances solved and PAR2 score from Table 1 suggest that Nested is slightly better, and Nested is also selected more often in SR-SMT-VBS (Table 2).

With the QF_NIA theory, Flat translation performs better overall, solving 12 more instances and with a lower PAR2 score. Some problem classes were solved much better with Flat, such as BIBD (in green). For some instances of EFPA (highlighted in blue), the two encodings are similar, and for others Flat is significantly more efficient. BIBD exhibits increasing gains for Flat as the instances become more difficult. However, Nested is more efficient for Langford's Problem, solving several instances that Flat cannot.

For the QF_LIA theory, we found that almost all problem classes were very similar. There were three exceptions: Killer Sudoku (where the Flat encoding is somewhat better), Car Sequencing, and Peg Solitaire (where the instances are scattered and neither Nested nor Flat seem to have an advantage).

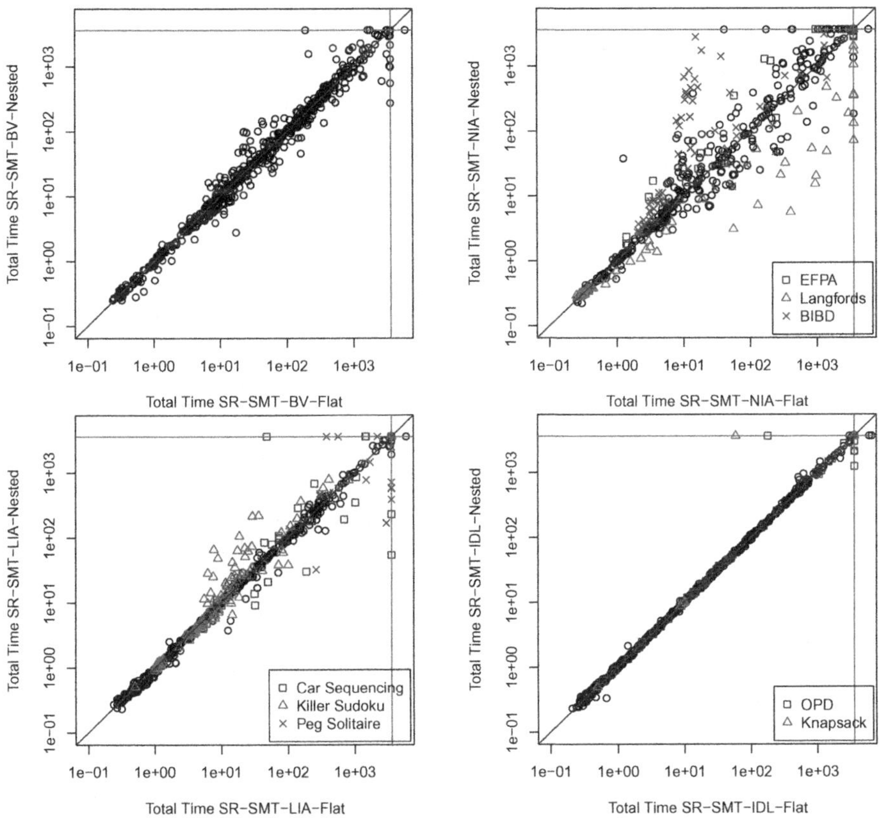


**Fig. 4.** Nested vs Flat translation for each theory, total time.

Finally, with QF_IDL for almost all problem classes the Nested and Flat encodings were very similar in performance. There are no problem classes where one encoding consistently outperforms the other by a substantial margin.

**Comparing Theories.** The four theories have quite different characteristics and the ideal choice of theory might vary by problem class. To compare the four theories, for each one we take the better configuration (of Nested or Flat), so we compare SR SMT-BV-Nested, SR-SMT-NIA-Flat, SR-SMT-LIA-Nested, and SR-SMT-IDL-Nested. BV is the strongest theory in terms of instances solved within 1 h and PAR2 score, therefore we use BV as the gold standard and compare the other theories to it.

Figure 5 (top left) plots NIA-Flat against BV-Nested. Many instances time out with NIA-Flat, for example the majority of Aces-Up (in red), and all instances of the block party metacube problem (BPMP, in blue). Langford's Problem is an example where the two theories are correlated, but BV is more efficient on the bulk of the instances. In contrast, Discrete Tomography is solved more efficiently by NIA-Flat. In this case, Z3 internal heuristics decide to convert the QF_NIA formula into a QF_BV formula before solving, and this is more efficient than Boolector directly applied to the BV-Nested formula. Also, both NIA encodings contribute a relatively small number of instances to the VBS (Table 2).

LIA-Nested is plotted against BV-Nested in Fig. 5 (top right). Langford's Problem and Discrete Tomography are examples where LIA-Nested performs substantially better than BV-Nested. The constraints in these problems are well suited to the QF_LIA theory (in particular Discrete Tomography, where all constraints are linear). Car Sequencing is split, some instances are solved more quickly by LIA-Nested while others time out. All instances of JPEncoding are solved more efficiently by LIA-Nested. Overall BV-Nested has a substantial edge: it is able to solve 73 more instances and its PAR2 score is much lower.

Finally, we compare IDL-Nested to BV-Nested in Fig. 5 (lower). MRCPSP (the multi-mode resource-constrained project scheduling problem) has precedence constraints that are naturally expressed in IDL. The most difficult instances of MRCPSP are solved more efficiently by IDL-Nested. Langford's Problem is also solved more efficiently by IDL-Nested. OPD is mixed but shows large speed-ups for BV for some of the most difficult instances. Many instances time out for IDL-Nested and are solved with BV-Nested, e.g. all instances of the JPEncoding problem. In total, 71 more instances are solved by BV-Nested.

In summary, we have seen several cases where NIA, LIA, or IDL performs well on a problem class and in these cases the problem class has constraints that are naturally expressed in the theory. For example, Discrete Tomography (with linear constraints) is solved very well by LIA-Nested (and also by NIA-Flat, where the solver converts it into a QF_BV formula). In contrast, BV seems the most robust. It performs well on a wide range of problem classes and in each comparison solves many instances that the other configuration did not.

### 4.4 Analysis of FZN2OMT

Table 2 gives an overview of how each configuration of FZN2OMT contributes to the virtual best solver, FZN2OMT-VBS. The BV encoding with Z3 is clearly the strongest combination on these benchmarks, contributing the most instances to the VBS and also solving the most within 1 h. Z3 was more effective than OptiMathSAT on the BV formulas, however the picture is not so simple with LIA. OptiMathSAT seems to be stronger overall on LIA formulas, solving more instances within 1 h, however Z3 contributes more instances to the VBS.

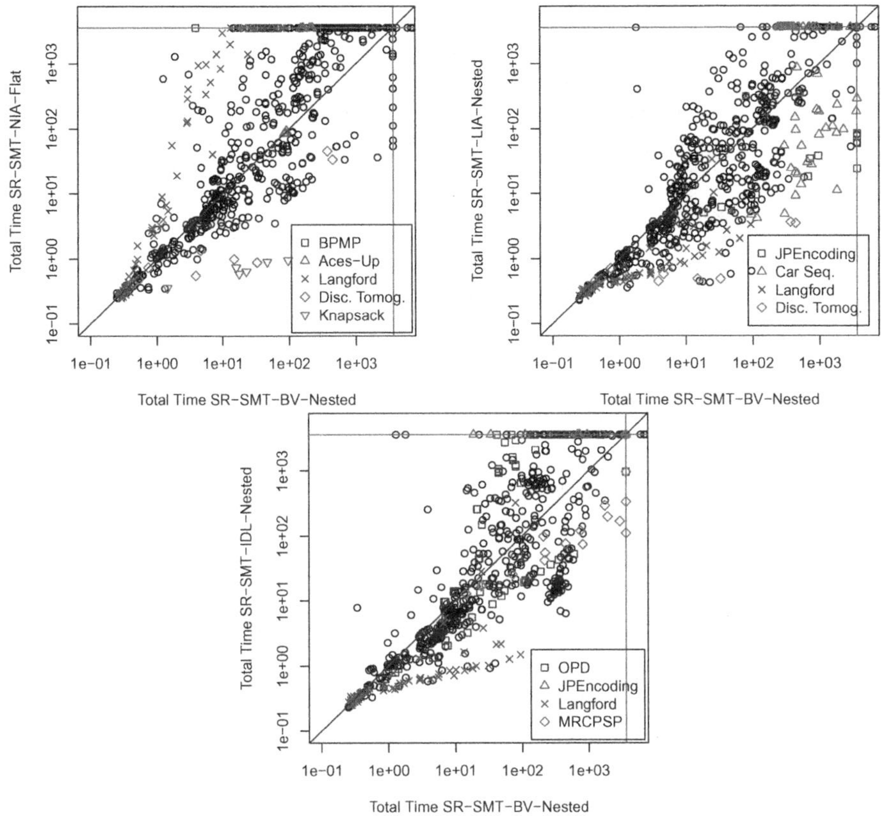


**Fig. 5.** Comparison of the four theories using the better configuration for each one (of Nested or Flat). (Color figure online)

## 4.5 Comparison to SAT

As described in Sect. 3, all SR-SMT encodings are built on the default SAT encoding of Savile Row, so comparison between SR-SMT and SAT is particularly relevant. We use CaDiCaL as the SAT solver and use the default options for the SAT backend. Figure 6 (top left) plots SAT against the strongest SMT configuration, SR-SMT-BV-Nested. The two are quite similar in overall performance, with similar PAR2 scores. SAT solves 8 more of the benchmark instances than SR-SMT-BV-Nested (Table 1). Some problem classes are solved more efficiently by SAT, such as Discrete Tomography, Langford, and MRCPSP. In contrast, SR-SMT-BV-Nested is able to solve 6 out of 10 instances of JPEncoding whereas SAT solves none. Car Sequencing is mixed but the majority of instances are solved more efficiently by SAT. 644 instances are solved by both SAT and BV-Nested, and of those the majority (510) are solved faster by SAT.

Figure 6 (top right) plots SAT against SR-SMT-VBS. The VBS solves a further 22 instances compared to BV-Nested (14 more than SAT), and improves

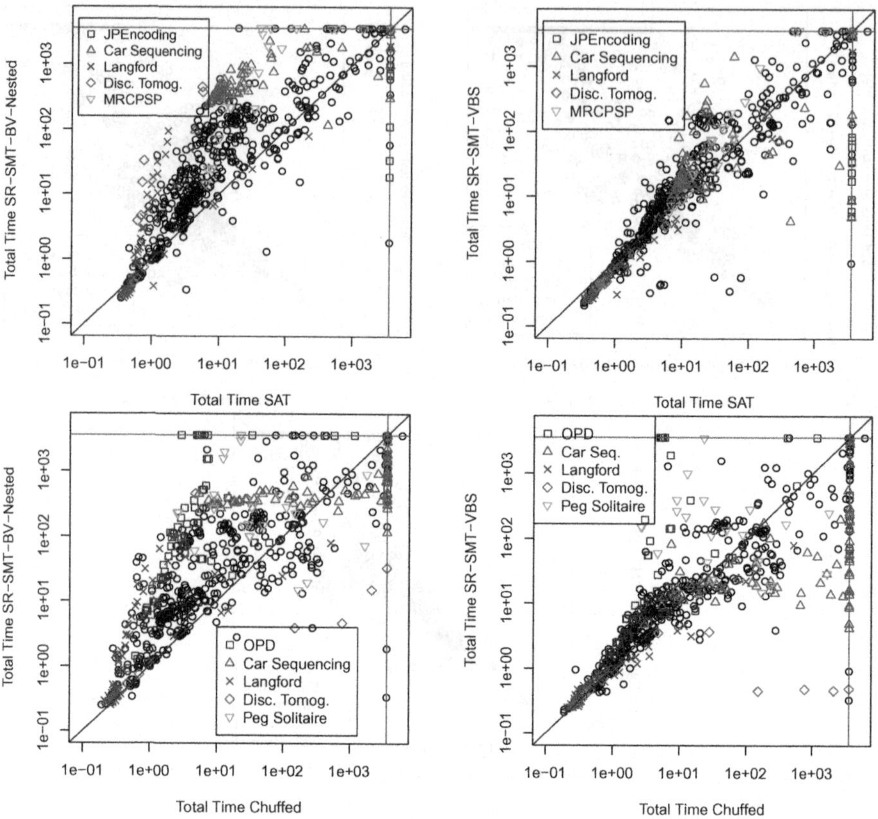


**Fig. 6.** Comparison of SAT to SR-SMT-BV-Nested (top left) and SR-SMT-VBS (top right). Comparison of Chuffed to SR-SMT-BV-Nested (lower left) and SR-SMT-VBS (lower right).

on various problem classes including Langford, Discrete Tomography, MRCPSP, and Car Sequencing. For the 657 instances that are solved by both the VBS and SAT, the majority (337) are solved more efficiently by SAT, however the VBS has a better PAR2 score (by over 100 s). The VBS is of course a theoretical solver, but these results show the value of selecting an appropriate theory for a given instance. As part of our future work we intend to investigate algorithm selection methods to construct a portfolio of the 8 SR-SMT configurations.

We found that the QF_IDL encodings were the largest, with a median clause ratio of 104.5% for both IDL-Nested and IDL-Flat compared to SAT. The QF_LIA encodings were more compact, with median clause ratios of 15.7% for LIA-Flat and 5.2% for LIA-Nested. The more expressive theories of QF_BV and QF_NIA allowed for the smallest encodings, with median clause ratios of 7.5% for BV-Flat, 2.8% for BV-Nested, 7.3% for NIA-Flat, and 2.8% for NIA-Nested. It is notable that one of the two smallest encodings (BV-Nested) has the highest performance. Also, Nested encodings are smaller than Flat with the exception of IDL where they are very similar in size.

### 4.6 Comparison to Chuffed

Chuffed is a well-established learning CP solver that uses a similar learning scheme as CDCL SAT and SMT solvers. The Chuffed backend of SAVILE ROW produces FlatZinc that is specific to Chuffed, i.e. it uses the global constraints implemented in Chuffed. We use Chuffed's free search option and also provide a search annotation given in the model or a default search annotation (variable declaration order). Figure 6 (lower left) plots Chuffed against SR-SMT-BV-Nested.

Results are mixed, with several problem classes solved much more efficiently by Chuffed (such as OPD and Langford for a large majority of their instances). Others are split, several Car Sequencing instances are solved substantially faster by Chuffed but others time out for Chuffed and are only solved by BV-Nested. The BV-Nested encoding performs well on Discrete Tomography (with linear constraints) and some instances of Peg Solitaire. In terms of instances solved, SR-SMT-BV-Nested performs better than Chuffed, solving 26 more instances within 1 h. However Chuffed is more efficient for 510 of the 615 instances that they both solve. The timeouts cause Chuffed to have a relatively high PAR2 score of 1248, compared to 1092 for BV-Nested.

Figure 6 (lower right) plots Chuffed against SR-SMT-VBS. The VBS solves 48 more instances than Chuffed, but Chuffed remains more efficient for the majority of instances (380 out of 627) that they both solve. Comparing the two plots, the VBS is more efficient than BV-Nested on many instances (including large numbers of OPD, Langford and Car Sequencing instances) and has narrowed the gap between SMT and Chuffed on instances where Chuffed is faster.

## 5 Conclusions and Future Work

We have presented SR-SMT, an SMT backend for SAVILE ROW that is able to target four SMT theories, each with two levels of flattening. We have performed an extensive set of experiments comparing our encodings to each other and also to FZN2OMT, SAT, and Chuffed. We found that SR-SMT with the QF_BV theory is a very robust approach: it solves more instances than other theories, Chuffed, and FZN2OMT within the time limit. However, SR-SMT with QF_BV is not always the fastest approach, suggesting that it would be a useful component of a portfolio of solvers.

While we found QF_BV to be particularly robust, other theories (QF_NIA, QF_LIA, and QF_IDL) performed strongly when problem constraints are naturally expressed in the theory, for example the LIA theory applied to the Discrete Tomography problem (which is linear). Consequently, the virtual best solver composed of all 8 SR-SMT configurations is significantly stronger than any one configuration. As part of future work, we will look at algorithm selection methods (such as the SUNNY algorithm used in the SUNNY-CP portfolio solver [2]) to construct a portfolio of SMT encodings, and investigate whether such a portfolio has similar performance to the VBS.

In summary, encoding constraint problems to SMT via SR-SMT is a successful approach, solving more instances of our benchmark set than the mature learning CP solver Chuffed and the existing FZN2OMT system.

**Acknowledgements.** We thank Marc Roig Vilamala who worked on an early version of the SMT backend of Savile Row. This work is supported by EPSRC grant EP/P015638/1.

## References

1. Akgün, Ö., Gent, I.P., Jefferson, C., Miguel, I., Nightingale, P.: Exploiting short supports for improved encoding of arbitrary constraints into SAT. In: Principles and Practice of Constraint Programming - 22nd International Conference, CP, Toulouse, France, pp. 3–12 (2016)
2. Amadini, R., Gabbrielli, M., Mauro, J.: SUNNY-CP and the MiniZinc challenge. Theory Pract. Log. Programm. **18**(1), 81–96 (2018). https://doi.org/10.1017/S1471068417000205
3. Bacchus, F.: GAC via unit propagation. In: Principles and Practice of Constraint Programming, 13th International Conference, CP Providence, RI, USA, pp. 133–147 (2007)
4. Barrett, C., Fontaine, P., Tinelli, C.: The SMT-LIB standard: version 2.6. Technical report, Department of Computer Science, The University of Iowa (2017). www.SMT-LIB.org
5. Belov, G., Stuckey, P.J., Tack, G., Wallace, M.: Improved linearization of constraint programming models. In: Rueher, M. (ed.) CP 2016. LNCS, vol. 9892, pp. 49–65. Springer, Cham (2016). https://doi.org/10.1007/978-3-319-44953-1_4
6. Bofill, M., Palahı, M., Suy, J., Villaret, M.: SIMPLY: a compiler from a CSP modeling language to the SMT-LIB format. In: Proceedings of the 8th International Workshop on Constraint Modelling and Reformulation, pp. 30–44 (2009)
7. Bofill, M., Palahí, M., Suy, J., Villaret, M.: Solving constraint satisfaction problems with SAT modulo theories. Constraints **17**(3), 273–303 (2012). https://doi.org/10.1007/s10601-012-9123-1
8. Bofill, M., Palahí, M., Suy, J., Villaret, M.: Solving intensional weighted CSPs by incremental optimization with BDDs. In: O'Sullivan, B. (ed.) CP 2014. LNCS, vol. 8656, pp. 207–223. Springer, Cham (2014). https://doi.org/10.1007/978-3-319-10428-7_17
9. Contaldo, F., Trentin, P., Sebastiani, R.: From minizinc to optimization modulo theories, and back (extended version). CoRR abs/1912.01476 (2019). http://arxiv.org/abs/1912.01476
10. Davidson, E., Akgün, Ö., Espasa, J., Nightingale, P.: Experiments for CP2020 SR-SMT paper (2020). https://doi.org/10.5281/zenodo.3953600
11. Dutertre, B.: Yices 2.2. In: Computer Aided Verification - 26th International Conference, CAV Vienna, Austria, pp. 737–744 (2014)
12. Dutertre, B., de Moura, L.M.: A fast linear-arithmetic solver for DPLL(T). In: Computer Aided Verification, 18th International Conference, CAV Seattle, WA, USA, pp. 81–94 (2006)

13. Hadarean, L., Bansal, K., Jovanovic, D., Barrett, C.W., Tinelli, C.: A tale of two solvers: eager and lazy approaches to bit-vectors. In: Computer Aided Verification - 26th International Conference, CAV Vienna, Austria, pp. 680–695 (2014)
14. Huang, J.: Universal booleanization of constraint models. In: Stuckey, P.J. (ed.) CP 2008. LNCS, vol. 5202, pp. 144–158. Springer, Heidelberg (2008). https://doi.org/10.1007/978-3-540-85958-1_10
15. Kroening, D., Strichman, O.: Decision Procedures - An Algorithmic Point of View. Texts in Theoretical Computer Science. An EATCS Series, 2nd edn. Springer, Heidelberg (2016). https://doi.org/10.1007/978-3-662-50497-0
16. de Moura, L.M., Bjørner, N.: Z3: an efficient SMT solver. In: Tools and Algorithms for the Construction and Analysis of Systems, 14th International Conference, TACAS, Budapest, Hungary, pp. 337–340 (2008)
17. Nethercote, N., Stuckey, P.J., Becket, R., Brand, S., Duck, G.J., Tack, G.: MiniZinc: towards a standard CP modelling language. In: Bessiere, C. (ed.) Principles and Practice of Constraint Programming - CP 2007, 13th International Conference, Providence, RI, USA, vol. 4741, pp. 529–543 (2007). https://doi.org/10.1007/978-3-540-74970-7_38
18. Niebert, P., Mahfoudh, M., Asarin, E., Bozga, M., Maler, O., Jain, N.: Verification of timed automata via satisfiability checking. In: Formal Techniques in Real-Time and Fault-Tolerant Systems, 7th International Symposium, FTRTFT Oldenburg, Germany, pp. 225–244 (2002)
19. Niemetz, A., Preiner, M., Wolf, C., Biere, A.: Btor2, BtorMC and Boolector 3.0. In: Computer Aided Verification - 30th International Conference, CAV Oxford, UK, pp. 587–595 (2018)
20. Nieuwenhuis, R., Oliveras, A., Tinelli, C.: Solving SAT and SAT modulo theories: from an abstract Davis-Putnam-Logemann-Loveland procedure to DPLL(T). J. ACM **53**(6), 937–977 (2006). https://doi.org/10.1145/1217856.1217859
21. Nightingale, P., Akgün, Ö., Gent, I.P., Jefferson, C., Miguel, I., Spracklen, P.: Automatically improving constraint models in Savile Row. Artif. Intell. **251**, 35–61 (2017). https://doi.org/10.1016/j.artint.2017.07.001
22. Nightingale, P., Rendl, A.: Essence' description. Computing Research Repository (CoRR) abs/1601.02865 (2016), http://arxiv.org/abs/1601.02865
23. Nightingale, P., Spracklen, P., Miguel, I.: Automatically improving SAT encoding of constraint problems through common subexpression elimination in Savile row. In: Pesant, G. (ed.) CP 2015. LNCS, vol. 9255, pp. 330–340. Springer, Cham (2015). https://doi.org/10.1007/978-3-319-23219-5_23
24. Sebastiani, R., Trentin, P.: OptiMathSAT: a tool for optimization modulo theories. In: Computer Aided Verification - 27th International Conference, CAV, San Francisco, CA, USA, pp. 447–454 (2015)
25. Van Hentenryck, P.: The OPL Optimization Programming Language. MIT Press (1999)

# Watched Propagation of 0-1 Integer Linear Constraints

Jo Devriendt[1,2](✉)

[1] Lund University, Lund, Sweden
jo.devriendt@cs.lth.se
[2] University of Copenhagen, Copenhagen, Denmark

**Abstract.** Efficient unit propagation for clausal constraints is a core building block of conflict-driven clause learning (CDCL) Boolean satisfiability (SAT) and lazy clause generation constraint programming (CP) solvers. Conflict-driven pseudo-Boolean (PB) solvers extend the CDCL paradigm from clausal constraints to 0-1 integer linear constraints, also known as (linear) PB constraints. For PB solvers, many different propagation techniques have been proposed, including a counter technique which watches all literals of a PB constraint. While CDCL solvers have moved away from counter propagation and have converged on a two watched literals scheme, PB solvers often simultaneously implement different propagation algorithms, including the counter one.

The question whether watched propagation for PB constraints is more efficient than counter propagation, is still open. Watched propagation is inherently more complex for PB constraints than for clauses, and several sensible variations on the idea exist. We propose a new variant of watched propagation for PB constraints and provide extensive experimental results to verify its effectiveness. These results indicate that our watched propagation algorithm is superior to counter propagation, but when paired with specialized propagation algorithms for clauses and cardinality constraints, the difference is fairly small.

## 1 Introduction

Although the Boolean satisfiability (SAT) problem is NP-complete [7,19] these days so-called *conflict-driven clause learning (CDCL)* solvers [20,23] routinely solve problems with up to millions of variables. Independently, a similar technique was developed for constraint programming (CP) solvers [2]. These solvers *learn* a propositional disjunction (a *clause*) from each failing search branch, over time accumulating huge databases of clauses that further constrain the search. For example, if, during search, all but one literals of a clause are set to false, the last remaining literal should be *propagated* to true.

To efficiently detect which clauses in the database propagate a literal, modern SAT solvers settled on the *watched literal propagation* technique [23]. Its core idea is to only *watch* two literals of a clause at a time, replacing these *watches* when one or both are set to false. If no two non-falsified watches can be found, the clause either propagates a literal, or it is falsified, indicating a search *conflict*.


H. Simonis (Ed.): CP 2020, LNCS 12333, pp. 160–176, 2020.
https://doi.org/10.1007/978-3-030-58475-7_10

The conflict-driven paradigm has been transferred to *linear pseudo-Boolean (PB) solving*, where solvers deal with linear inequalities over 0-1 integer variables, or *PB constraints* for short.[1] Formulas of PB constraints are a straightforward generalization of the conjunctive normal form (CNF) used for SAT solvers. Crucially, such *conflict-driven PB solvers* learn PB constraints instead of clauses [6,14,18,28], which allows them to construct *cutting planes proofs* [8] instead of the exponentialy weaker *resolution proofs* [4,9,10,25] underlying CDCL.

As the database of learned PB constraints grows during search, and as conflict-driven PB search endeavors to make the learned constraints as strong as possible, the propagation routine forms the main computational bottleneck of PB solvers. Similar to CDCL SAT solvers, a hypothetical doubling the efficiency of PB propagation could translate to almost halving the total run time for conflict-driven PB solvers. Unlike CDCL SAT solvers however, conflict-driven PB solvers have not settled on a dominant propagation scheme.

The *Galena* solver investigated a highly involved watched literal scheme for PB constraints, but finally settled on a three-tiered approach where clauses and *cardinality* constraints were handled with specialized watched propagation techniques, but propagation of general PB constraints was done by *counter* propagation, watching all literals at once [6]. The *Pueblo* solver initially employed the same three-tiered approach [27], but later opted for a custom watched literal scheme [28]. The *Sat4J* system also uses the three-tiered approach by default, but has the option to use watched propagation for general PB constraints similar to the *Galena* watched literal scheme [18]. Finally, the *RoundingSat* solver employs watched propagation, sharing similarities with both the *Pueblo* and original *Galena* approach, but adding its own twists [14].

Unsurprisingly, efficient watched PB propagation is still an open question:

> PB solvers get slower when dealing with pseudo-Boolean constraints because we have not yet found an efficient lazy data structure similar to [...] watched literals for those constraints. This is especially the case for the cutting-planes-based solver because the number of pseudo-Boolean constraints grows during the search [18].

In this paper, we propose a novel efficient watched PB propagation algorithm, and contribute extensive experimental data to shed light on key issues. The general conclusion is that watched PB propagation is more efficient than counter propagation on its own, but that the difference between a counter-based and a watched-based three-tiered approach is fairly small.

This paper continues with preliminaries in Sect. 2 followed by a description of our proposed watched PB propagation algorithm in Sect. 3. Section 4 highlights differences and similarities with the approaches used by the above PB solvers. Experimental results are presented in Sect. 5 and the paper concludes with Sect. 6.

[1] In general, PB constraints can be non-linear, but we restrict our attention to linear PB constraints.

## 2 Preliminaries

Throughout this paper, we use the term *pseudo-Boolean (PB) constraint* to refer to a 0–1 linear inequality. We identify 1 with *true* and 0 with *false*. A *literal* $\ell$ denotes either a variable $x$ or its negation $\overline{x}$, where $\overline{x} = 1-x$. We assume without loss of generality that all constraints $\sum_i c_i\ell_i \geq w$ are written in *normalized form*, where literals $\ell_i$ are over pairwise distinct variables, coefficients $c_i$ are non-negative integers, and $w$ is a positive integer called the *degree (of falsity)*. For a constraint $C$, $\mathit{lits}(C)$ denotes its set of literals, $\mathit{size}(C)$ its number of literals, and $\mathit{maxcf}(C)$ its largest coefficient. A PB constraint $C$ where $\mathit{maxcf}(C) = 1$ is a *cardinality* constraint and a constraint with degree 1 is a *clause*.

A *(partial) assignment* $\rho$ is a set of literals over pairwise distinct variables. A literal $\ell$ is *assigned to true* by an assignment $\rho$ if $\ell \in \rho$, *assigned to false* or *falsified* if $\overline{\ell} \in \rho$, and is *unassigned* otherwise. The *slack* of a constraint $C \doteq \sum_i c_i\ell_i \geq w$ under a partial assignment $\rho$ is

$$\mathit{slack}(C,\rho) = -w + \sum_{\overline{\ell}_i \notin \rho} c_i \ , \tag{1}$$

i.e., the maximal value the left-hand side can attain under any partial assignment $\rho' \supseteq \rho$ minus the degree. We say that $\rho$ *falsifies* $C$ if $\mathit{slack}(C,\rho) < 0$ and *satisfies* $C$ if for any $\rho' \supseteq \rho$ it holds that $\mathit{slack}(C,\rho') \geq 0$. A *pseudo-Boolean formula* $\varphi$ is a set of PB constraints. An assignment $\rho$ is a *solution* to $\varphi$ if $\rho$ satisfies all constraints in $\varphi$. A formula is *satisfiable* if it has a solution.

A *sequence* $(e_1, \ldots, e_n)$ is a finite ordered collection of elements allowing repetitions.[2] In programming fashion, $\mathit{seq}[i]$ denotes the $i$th element of $\mathit{seq}$. The *size* of a sequence is denoted as $\mathit{size}(\mathit{seq})$. A *tuple* is a fixed size sequence with named elements and $\mathit{tup}.e$ refers to the element with name $e$ of tuple $\mathit{tup}$.[3]

### 2.1 Conflict-Driven Pseudo-Boolean Solving

We present the bare essentials of conflict-driven PB solving necessary for the discussions in this paper (referring the reader to, e.g., [5] for more details). Conflict-driven PB solving is very similar to the CDCL algorithm for Boolean satisfiability, but uses PB constraints instead of clauses.

The state of a PB solver can be abstractly represented by a tuple $(\psi, \rho)$, where $\psi$ is a set of constraints called the *constraint database* and $\rho$ is a sequence of pairwise distinct literals representing the *current assignment*.[4] Initially, $\psi$ is the input formula $\varphi$ and $\rho$ is the empty sequence ().

---

[2] Common data structures for sequences are arrays, lists, and vectors.

[3] Tuples abstract the record data type.

[4] Slightly abusing notation, we defined an assignment as a set, but we often operate on the current assignment $\rho$ as a sequence, *pushing* and *popping* literals from the back.

Given a solver state, the search loop starts with a *propagation* phase, which checks for any constraint $C \in \psi$ whether it is falsified:

$$slack(C, \rho) < 0, \tag{2}$$

or whether a literal $\ell_i$, not yet assigned by $\rho$, in $C$ with coefficient $c_i$, is implied by $C$ under $\rho$:

$$slack(C, \rho) < c_i \text{ with } \ell_i \notin \rho, \overline{\ell_i} \notin \rho. \tag{3}$$

If condition (3) holds, $C$ is falsified by $\rho \cup \overline{\ell}_i$, so $\ell_i$ is implied by $C$ under $\rho$. Hence, $\rho$ can be safely extended with the implied $\ell_i$, which is called a *propagation*, while also saying that $C$ *propagates* $\ell_i$. Each propagation can enable new propagations, continuing the propagation phase until condition (3) does not hold for any constraint in the database $\psi$, or until condition (2) holds for at least one.

If condition (2) holds for some constraint, it is considered a *conflict*, and the solver enters a *conflict analysis* phase. During this phase, the solver derives a *learned constraint* that is a logical consequence of the input formula and would have propagated a literal at some earlier state, preventing the same conflict from happening. This new constraint is added to $\psi$, after which the solver *backjumps* to the earlier state. Alternatively, if no conflict is detected, the solver extends $\rho$ by making a heuristic *decision* to assign some currently unassigned variable. In either case, the solver continues with a new iteration of the search loop.

The PB solver reports unsatisfiability whenever it learns a constraint equivalent to the trivial inconsistency $0 \geq 1$. If propagation does not lead to a conflict and all variables have been assigned, the solver reports that the input formula is satisfiable. Conflict-driven PB solvers, like their CDCL counterparts, frequently backjump to the root search node, clearing the current assignment from any decision literals and consequent propagations, which is called a *restart*.

In this paper, we focus on the propagation phase, ensuring that after each decision and each backjump, the current assignment is extended with implied literals until fixpoint, or until a conflict arises.

### 2.2 Counter Pseudo-Boolean Propagation

A straightforward propagation algorithm is the *counter* approach. It takes its inspiration from early SAT propagation algorithms and eagerly computes the slack of each constraint under changes to the current assignment $\rho$. I.e., each time a literal $\ell$ is pushed to resp. popped from $\rho$, due to decisions or propagations resp. backjumps or restarts, each constraint $C$ containing $\overline{\ell}$ has its slack decreased resp. increased with the coefficient of $\overline{\ell}$ in $C$. When the slack of $C$ is decreased, condition (2) and (3) are checked as well to detect propagations and conflicts.

*Example 1.* Consider a freshly initialized solver where the input formula consists only of the constraint $C \doteq 3x+2y+z+w \geq 3$. Initially, $\rho = ()$, so $slack(C, \rho) = 4$ and neither condition (2) or (3) hold.

If the solver decides $x = 0$, then $\rho = (\overline{x})$, and counter propagation decreases the slack with 3: $slack(C, \rho) = 1$. Now, condition (3) holds: $slack(C, \rho) < 2$ and $y \notin \rho, \overline{y} \notin \rho$. Hence, $y$ is propagated and $\rho = (\overline{x}, y)$. No further slack decreases are triggered, so counter propagation does not need to check whether condition (2) or (3) hold.

The solver can now decide a new variable, say $z = 0$, so $\rho = (\overline{x}, y, \overline{z})$, at which point counter propagation again decreases the slack of $C$ with 1 to $slack(C, \rho) = 0$, and propagates $w$, leaving $C$ satisfied by $\rho$. If the solver instead executes a restart, the current assignment is reset to $\rho = ()$, and counter propagation increases the slack of $C$ with $1 + 3$ to its original value $slack(C, \rho) = 4$.

Unfortunately, counter propagation has the potential for a lot of overhead:

*Example 2.* Consider a freshly initialized solver where the input formula consists only of the constraint $C \doteq 3x + 2y + z + \sum_{i=1}^{1000} w_i \geq 3$. Let the literals $\overline{w_i}$ be prioritized by the solver's decision heuristic. Initially, $slack(C, \rho) = 1003$, which decreases by 1 after each decision of some $w_i$. For each of these thousand slack decrements, condition (2) and (3) are never met, since as long as none of $x$, $y$ and $z$ are falsified by the current assignment, $slack(C, \rho) \geq 3$.

This phenomenon of large amounts of slack decrements (and increments during backjumps) can occur in thousands of constraints simultaneously, considerably slowing down the solver. The watched literal technique attempts to significantly reduce the number of times the slack of a constraint is calculated.

## 3 Watched Pseudo-Boolean Propagation

Similar observations to those in Example 2 led to the development of *watched (literal)* propagation in SAT solvers [23,30]. This watched approach has been generalized to pseudo-Boolean solving [6,14,18,28]. The central idea of watched PB propagation is to track (*watch*) for each constraint only a subset of its literals – the *watched* literals. The subset is chosen sufficiently large to ensure that as long as none of the watched literals are assigned to false, the constraint is not propagating or conflicting. If one of the watches is assigned false, a search for new non-falsified watches is triggered. If insufficient new watches are found, the constraint may be propagating or conflicting, which is calculated only then.

More formally, we associate each constraint $C$ with a set of watched literals $watches(C)$. For a constraint with watched literals, the *watch slack* of a constraint $C \doteq \sum_i c_i \ell_i \geq w$ under a partial assignment $\rho$ is

$$watchslack(C, \rho) = -w + \sum_{\overline{\ell_i} \notin \rho, \ell_i \in watches(C)} c_i. \tag{4}$$

Clearly, for any $C$, $\rho$ and watches for $C$, $watchslack(C, \rho) \leq slack(C, \rho)$, and $watchslack(C, \rho) = slack(C, \rho)$ if all non-watched literals are falsified by $\rho$. Hence, condition (2) and (3) will never hold (so $C$ will not propagate or be conflicting) if for some set of watches

$$watchslack(C, \rho) \geq maxcf(C). \tag{5}$$

As a result, no exact slack needs to be calculated for constraints for which condition (5) holds, and only by falsifying a watched literal can condition (5) become violated. However, efficiently maintaining appropriate watched literal sets during backjumps, decisions and propagations is a highly non-trivial matter.

To describe our proposed watched PB propagation algorithm in detail, we abstract the state of a constraint $C$ to a tuple $(terms, w, wslk)$, where $terms$ is a sequence of terms, $w$ a positive integer representing the degree, and $wslk$ an integer storing the watch slack of the constraint. The state of a term in $terms$ is abstractly represented by a tuple $(coef, lit, wflag)$ where $coef$ is the coefficient of the term, $lit$ the literal, and $wflag$ a flag denoting whether the literal is watched for the constraint, i.e., whether $lit \in watches(C)$. We fix $terms$ to be sorted in decreasing coefficient order, so $maxcf(C) = C.terms[1]$ – the first term of $C$ contains its largest coefficient.

We also extend the abstraction of the solver state to a tuple $(\psi, \rho, q, wlist)$, where the *propagation index* $q$ is an integer s.t. $0 \leq q \leq size(\rho)$[5], and the *watch list* $wlist$ is a function mapping literals to the set of constraints that currently watch the literal combined with the index of the literal in the constraint's term list: $(C, i) \in wlist(\ell)$ iff $\ell \in watches(C)$ with $\ell = C.terms[i].lit$. We define $\rho^i \doteq (\rho[1], \ldots, \rho[i])$ as the *subassignment* up to index $i$, with $0 \leq i \leq size(\rho)$. The propagation index indicates which part of the current assignment has already been processed for propagation: constraints watching literals in $\rho \setminus \rho^q$ will need to be checked for propagation. Initially, $q = 0$.

### 3.1 Detailed Algorithm

We now have the necessary abstractions in place to describe our proposed watched PB propagation algorithm in detail. For simplicity, we assume that initially, none of the constraints $C$ are propagating or conflicting, and that their initial watched literals can be chosen to satisfy $watchslack(C, ()) = C.wslk \geq maxcf(C)$.

Procedures `processWatches`, `propagate` and `backjump` present the proposed watched PB propagation algorithm.

Procedure `processWatches` iterates over all literals $\ell$ in the current assignment, adjusting the watch slack for each constraint $C$ watching $\overline{\ell}$, maintaining the invariant that $watchslack(C, \rho^q) = C.wslk$. It subsequently checks whether $C$ can propagate (or is conflicting) by calling `propagate` for $C$. If $C$ is conflicting, it is returned. However, breaking out of the loop at line 5 leaves behind a semi-processed set of constraints watching $\ell$. To repair this, `processWatches` decreases the propagation index by one, and increases the watch slack for those constraints still watching $\overline{\ell}$ that had their watch slack decreased.

To check whether a constraint is conflicting or propagating, `propagate` first attempts to find non-falsified non-watched literals to use as watches, in the

[5] In *MiniSAT* [13] parliance, $q$ is the *qhead*.

**Procedure.** propagate(constraint $C$, integer $idx$)

**External data:** watch list $wlist$, current assignment $\rho$
**Result:** OK if $C$ is not falsified, otherwise CONFLICT

```
1  i ← 1
2  while i ≤ size(C) and C.wslk < maxcf(C) do
3  |   ℓ ← C[i].lit
4  |   if ℓ̄ ∉ ρ and C[i].wflag = 0 then
5  |   |   C[i].wflag = 1
6  |   |   wlist(ℓ) ← wlist(ℓ) ∪ {(C, i)}
7  |   |   C.wslk ← C.wslk + C[i].coef
8  |   i ← i + 1
9  if C.wslk ≥ maxcf(C) then
10 |   C[idx].wflag = 0
11 |   wlist(C[idx].lit) ← wlist(C[idx].lit) \ {(C, idx)}
12 |   return OK
13 if C.wslk < 0 then return CONFLICT
14 j ← 1
15 while j ≤ size(C) and C.wslk < C[j].coef do
16 |   ℓ ← C[j].lit
17 |   if ℓ ∉ ρ and ℓ̄ ∉ ρ then ρ.push(ℓ)
18 |   j ← j + 1
19 return OK
```

loop at line 2. If a sufficient amount of watches is found such that $C.wslk \geq maxcf(C)$, no propagation or conflict occurs, the old watch can be discarded at lines 10 and 11, and the routine returns. If all non-falsified literals are employed as watches, yet the watch is still less than zero, the constraint is conflicting, which is returned at line 13. Finally, if the watch slack is non-negative but less than the largest coefficient, the constraint may propagate unassigned literals, which is checked in the loop at line 14. Recall that the terms of a constraint are sorted in decreasing coefficient order, allowing the loop at line 14 to conclude when $C.wslk < C[j].coef$, avoiding a full linear scan. In case $C.wslk < maxcf(C)$, the constraint keeps watching the falsified literal. This allows procedure `backjump` to increase the watch slack of a constraint during backjumps, to a point where $C.wslk \geq maxcf(C)$ without searching for new watches.

### 3.2 An Extensive Example

*Example 3.* As in Example 2, consider a freshly initialized solver where the input formula consists only of the constraint $C \doteq 3x + 2y + z + \sum_{i=1}^{1000} w_i \geq 3$. Let the initial watches for $C$ be $\{x, y, z\}$, and hence, $watchslack(C, \rho^q) = C.wslk = 3$. Let the literals $\overline{w_1}$ to $\overline{w_{997}}$ be prioritized by the solver's decision heuristic, so the current assignment $\rho$ is incrementally extended by deciding the literals $\overline{w_1}$ to $\overline{w_{997}}$, and after each decision, procedure `processWatches` is called, incrementing $q$ to 997. As no constraint watches any $w_i$, `propagate` is never called.

**Procedure.** processWatches

**External data:** database $\psi$, current assignment $\rho$, propagation index $q$, watch list *wlist*

**Result:** OK if no constraint is falsified, otherwise a falsified constraint

**while** $q < \mathit{size}(\rho)$ **do**
  $q \leftarrow q + 1$
  $\ell \leftarrow \rho[q]$
  $\mathit{visited} \leftarrow \emptyset$
  **foreach** $(C, \mathit{idx}) \in \mathit{wlist}(\overline{\ell})$ **do**
    $\mathit{visited} \leftarrow \mathit{visited} \cup \{(C, \mathit{idx})\}$
    $C.\mathit{wslk} \leftarrow C.\mathit{wslk} - C[\mathit{idx}].\mathit{coef}$
    **if** `propagate`$(C, \mathit{idx})= \mathit{CONFL}$ **then**
      $q \leftarrow q - 1$
      **foreach** $(C', \mathit{idx}') \in \mathit{visited} \cap \mathit{wlist}(\overline{\ell})$ **do**
        $C'.\mathit{wslk} \leftarrow C'.\mathit{wslk} + C'[\mathit{idx}'].\mathit{coef}$
      **return** $C$
**return** $\mathit{OK}$

**Procedure.** backjump(integer $s$)

**External data:** database $\psi$, assignment $\rho$, propagation index $q$

**while** $\mathit{size}(\rho) > s$ **do**
  $\ell \leftarrow \rho[\mathit{size}(\rho)]$
  **if** $q = \mathit{size}(\rho)$ **then**
    $q \leftarrow q - 1$
    **foreach** $(C, \mathit{idx}) \in \mathit{wlist}(\overline{\ell})$ **do**
      $C.\mathit{wslk} \leftarrow C.\mathit{wslk} + C[\mathit{idx}].\mathit{coef}$
  $\rho$.pop()

Let $\overline{x}$ be the 998st decision, leading to $\rho = (\overline{w_1}, \dots, \overline{w_{997}}, \overline{x})$. `processWatches` increases $q$ to 998, and as $x$ is watched by $C$, decreases $C.\mathit{wslk}$ to 0 and calls `propagate`$(C, 1)$. `propagate` iterates over $C$, picking $w_{998}$, $w_{999}$ and $w_{1000}$ as new watches in the loop at line 2. After exiting the loop, the watch slack has increased to $C.\mathit{wslk} = 3$, so $C.\mathit{wslk} \geq \mathit{maxcf}(C)$ and $x$ is dropped as watch at lines 10 and 11. The watched literals for $C$ now are $\mathit{watches}(C) = \{y, z, w_{998}, w_{999}, w_{1000}\}$.

Let $\overline{z}$ be the 999th literal decision, leading to $\rho = (\overline{w_1}, \dots, \overline{w_{997}}, \overline{x}, \overline{z})$. Running `processWatches` increases $q$ to 999, and as $z$ is watched by $C$, decreases $C.\mathit{wslk}$ to 2 and calls `propagate`$(C, 3)$. `propagate` cannot find further watches, so $0 \leq C.\mathit{wslk} < \mathit{maxcf}(C)$ and the while loop at line 14 looks for literals to propagate. The only literal for which $C.\mathit{wslk} < C[j].\mathit{coef}$, $x$, is already assigned to false, so no literals can be propagated, and both $\rho$ and $\mathit{watches}(C)$ remain unchanged.

Let $\overline{w}_{998}$ be the next literal decision, leading to $\rho = (\overline{w_1}, \dots, \overline{w_{997}}, \overline{x}, \overline{z}, \overline{w_{998}})$. `processWatches` increases $q$ to 1000, and as $w_{998}$ is watched, decreases $C.\mathit{wslk}$ to 1 and calls `propagate`$(C, 1001)$. `propagate` cannot find further watches, so

$0 \leq C.wslk < maxcf(C)$ and the while loop at line 14 checks which literals can be propagated. The only literals for which $C.wslk < C[j].coef$ are $x$ and $y$. The latter is still unassigned, so it is propagated at line 17. Returning to `processWatches`, $\rho = (\overline{w_1}, \ldots, \overline{w_{997}}, \overline{x}, \overline{z}, \overline{w_{998}}, y)$, so $q = 1000 < size(\rho) = 1001$ – the loop at line 5 continues. $q$ is incremented to 1001, but as $y$ is not watched by $C$, `propagate` is not called again.

Finally, let the solver backjump to the root by calling `backjump`(0). First, literal $y$ is unassigned, which is not watched by $C$ (even though its negation $\overline{y}$ is) so its watch slack is not updated. Then, $\overline{w_{998}}$ is unassigned, which is watched by $C$, so $C.wslk$ is incremented to 2. Next, $\overline{z}$ is unassigned, which is watched by $C$, so $C.wslk$ is incremented to 3. At this point, $q$ decreased from 1001 to 998. For the remaining 998 iterations, no further adjustments to $C.wslk$ are needed, as none of its watches $\{y, z, w_{998}, w_{999}, w_{1000}\}$ are falsified.

### 3.3 Algorithm Analysis

The following two invariants underpin the soundness and completeness of our approach. Short proof sketches are available online [12].

**Lemma 4 (Watch slack invariant).** *The procedures* `processWatches` *(calling* `propagate`*) and* `backjump` *preserve the property*

$$C.wslk = watchslack(C, \rho^q) \tag{6}$$

**Lemma 5 (Watch set invariant).** *The procedures* `processWatches` *(calling* `propagate`*) and* `backjump` *preserve the property*

$$C.wslk < maxcf(C) \Rightarrow \forall \ell \in lits(C) \setminus watches(C) \colon \overline{\ell} \in \rho \tag{7}$$

*for a constraint $C$ if the argument of* `backjump` *is chosen in such a way that for all constraints $C$ where $C.wslk < maxcf(C)$, either all of its falsified watches become unassigned, or none of its non-watched literals become unassigned.*

To maintain the watch set invariant, the solver has to take care where to backjump. Withholding detail, the well-known technique of partitioning the current assignment in contiguous *decision levels* and backjumping over each level as a whole maintains the watch set invariant.

**Lemma 6.** *If the watch set and watch slack invariants hold, calling the procedure* `processWatches` *(calling* `propagate`*) propagates literal $\ell_i$ with coefficient $c_i$ in constraint $C$ only if it is unassigned and $slack(C, \rho) < c_i$, and reports that $C$ is conflicting only if $slack(C, \rho) < 0$. I.e.,* `processWatches` *is sound.*

**Lemma 7.** *Assuming the watch set and watch slack invariant hold, if the procedure* `processWatches` *(calling* `propagate`*) returns OK, no conflicting constraint under $\rho$ exists, and no further propagations under $\rho$ are possible. I.e.,* `processWatches` *is complete.*

### 3.4 Two Optimizations

The datastructures needed for our proposed algorithm are fairly simple and should have a linear memory footprint and take (amortized) constant time for each operation. However, a performance bottleneck resides in the loops at lines 2 and 15 in `propagate`. E.g., Example 3 frequently iterates over the full size of the constraint to find new watches, even though it was clear no new watches were available. By reducing the time spent in those loops or even avoiding to enter them at all, we can improve efficiency.

First observe that when calling `propagate` because $\overline{\rho[q]}$ is watched by $C$, if $C.\mathit{wslk} < \mathit{maxcf}(C)$ holds at the end of the loop at line 2, all potential watches have been exhausted per the watch set invariant. Hence, when calling `propagate` for $\overline{\rho[q']}$ with $q' > q$ without backjumping over $q$, the loop at line 2 can be skipped. To detect this situation, we check whether $C.\mathit{wslk} + C[\mathit{idx}].\mathit{coef} < \mathit{maxcf}(C)$. If it holds, there was an earlier call to `propagate` that exited the loop at line 2 with $C.\mathit{wslk} < \mathit{maxcf}(C)$, so the loop can now be safely skipped.[6]

Next observe that, for a given constraint, any literal that is checked to become a watch by the loop at line 2, but that was not available as watch because it was falsified or already a watch (line 4 fails), can only become available as watch after a backjump occurs, since without a backjump the current assignment is only extended. Similarly, literals checked to be propagated by the loop at line 15 cannot be propagated later without a backjump occurring. To exploit this, we permanently store the indices $i$ and $j$ of the loops at lines 2 and 15 for each constraint (e.g., $C.i$ and $C.j$) and only reset them to 0 if a backjump happened. The latter condition is simple to check: keep a global variable (e.g., *bkjmps*) that increments by 1 at each backjump. For each constraint, check whether the global *bkjmps* matches a local copy *C.lastbkjmp* that is set at each `propagate` call.

Procedure `propagateOpt` extends `propagate` with these two optimizations. Remark that as a result of these optimizations, in between backjumps, all calls to `propagateOpt` with a given constraint $C$, perform only $O(\mathit{size}(C))$ operations in aggregate.

# 4 Related Work

A PB propagation algorithm closely related to our work is that of the *Pueblo* solver [28]. It also sorts the terms of a constraint in decreasing coefficient order and checks for propagation if the slack over the watched literals of the constraint is less than the maximum coefficient of the constraint. However, except in the case of a conflicting constraint, it does not keep falsified literals as watch, as per `propagate`. It also does not update the watch slack during backjumps, as per `backjump`. Hence, it is not clear how *Pueblo* would restore the watches in the restart scenario described at the end of Example 3.[7] Also, *Pueblo* does not

[6] Note that this first optimization depends on the watch set invariant, and thus on an appropriate backjump scheme.

[7] After inquiring with the authors, the source code of *Pueblo* no longer seems available.

**Procedure.** propagateOpt(constraint $C$, integer $idx$)

```
External data: watch list wlist, current assignment ρ, backjump count bkjmps
Result: OK if C is not falsified, otherwise CONFLICT
if C.lastbkjmp < bkjmps then
    C.i ← 1
    C.j ← 1
    C.lastbkjmp ← bkjmps
if C.wslk + C[idx].coef ≥ maxcf(C) then
    while C.i ≤ size(C) and C.wslk < maxcf(C) do
        ℓ ← C[C.i].lit
        if ℓ̄ ∉ ρ and C[C.i].wflag = 0 then
            C[C.i].wflag = 1
            wlist(ℓ) ← wlist(ℓ) ∪ {(C, C.i)}
            C.wslk ← C.wslk + C[C.i].coef
        C.i ← C.i + 1
if C.wslk ≥ maxcf(C) then
    C[idx].wflag = 0
    wlist(C[idx].lit) ← wlist(C[idx].lit) \ {(C, idx)}
    return OK
if C.wslk < 0 then return CONFLICT
while C.j ≤ size(C) and C.wslk < C[C.j].coef do
    ℓ ← C[C.j].lit
    if ℓ ∉ ρ and ℓ̄ ∉ ρ then ρ.push(ℓ)
    C.j ← C.j + 1
return OK
```

implement the optimizations described in Sect. 3.4, and does not store the index of a watched literal of a constraint in the watch lists, which might lead to a linear lookup overhead or require a cache-inefficient associative array.

Before *Pueblo*, work on the *Galena* solver [6] also prompted PB propagation investigation. It uses a watched propagation scheme where the number of watches of a constraint depends on a dynamic maximum coefficient $a_{max}$ of the literals currently not assigned to true. This minimizes the number of watched literals, but according to [27], two thirds of the run time of the *Galena* propagation procedure is spent updating $a_{max}$ for each constraint. Because of this, it was proposed to keep $a_{max}$ fixed to the highest coefficient (i.e., $maxcf(C)$), but *Galena* eventually settled on a three-tiered approach with watched propagation only for clauses and cardinality constraints, and counter propagation for general PB constraints [6].

The more recent *Sat4J* uses this three-tiered approach by default, but provides the option to enable a less efficient watched propagation [18].

Finally, the *RoundingSat* solver [14] implements a watched propagation algorithm which, as in our approach, uses a static maximum coefficient to calculate the number of needed watches and keeps watching falsified literals, but swaps watched literals to the front of the constraint [26]. This makes calculating the watch slack after every call relatively efficient, as only the watched literals in the front of the constraint need to be iterated over, rendering the update of the watch slack during backjumps obsolete. As the watch swaps alter the order of the literals of the constraint, the index of a watched literal cannot be stored in the watchlist, as in our approach, and is recalculated during watch slack calculation. To check for propagating literals, *RoundingSat* again always iterates over all watched literals. However, the number of watches of a PB constraint can grow linearly in the size of the constraint, which leads to a potentially large overhead for constraints that require lots of watches.

On the CP side, to the best of our knowledge, the constraint in the global constraint catalog most closely related to PB constraints is *sum_set* [29], which constrains an integer variable $V$ to take the sum of a variable subset of a set of values. In the special case where $V$ is constrained only by one fixed bound, sum_set is equivalent to a PB constraint. The propagator for sum_set in the CP solver *Gecode* [15] relies on counter propagation, though the comparison is not fully fair as not only literals have to be propagated, but bounds on $V$ as well.

# 5 Experimental Evaluation

To experimentally evaluate our proposed propagation algorithm, we implemented it in the *RoundingSat* PB solver [26]. Source code, a binary, and raw experimental data are available online [12]. As hardware we used AMD Opteron 6238 nodes having 6 cores and 16 GiB of memory each. Each run was executed as a single thread on a node with a 5000 s timeout limit.

To make a sufficiently broad comparison, we present experiments on instances from the linear small coefficient decision and optimization tracks from the most recent PB competition [24], referred to as PB16dec and PB16opt. Additionally, we investigate 0-1 integer linear programming instances from the MIPLIB libraries [1,3,16,17,21,22]. Since these sets contain few decision instances, we also created decision versions of the optimization problems. For this, we constructed a first instance by replacing the objective function $f$ with a constraint stating that $f$ should be at least the best known value, and a second where $f$ should be strictly better. As *RoundingSat* can currently only deal with integer coefficients of magnitude at most $10^9$, some of the instances were rescaled and rounded. We refer to the corresponding MIPLIB decision and optimization problems as MIPLIBdec and MIPLIBopt. These instances are available online [11].

## 5.1 Two Optimizations to Watched PB Propagation

Let's start with a simple question: how effective are the two optimizations described in Sect. 3.4? For this, we implemented in *RoundingSat* watched propagation per Procedure `propagate` (*watch*) and per Procedure `propagateOpt` (*watch-opt*), and compare the propagation speed defined as the total propagations performed divided by the solve time. As *watch* and *watch-opt* do not differ in the order in which propagations happen, the runs for both *watch* and *watch-opt* have the same conflict and decision counts and any difference in propagation speed is solely due to algorithmic efficiency. Figure 1 plots the result for the instances that were solved by both *watch* and *watch-opt* within resource limits and took at least 1 s to solve. The result is clear: the optimizations can increase the propagation speed by an order of magnitude and never incur significant overhead.

### 5.2 Expensive Backjumps?

One advantage of watched propagation in SAT solvers is that no work needs to be done during backjumps, a feature preserved by the original propagation implementation of *RoundingSat*. Our approach updates the watch slack during backjumps, though only for those constraints $C$ that have falsified watches, which only happens if $C.wslk < maxcf(C)$. Figure 2 plots the number of times *watch-opt* looked up a constraint when backjumping over a falsified watched literal (line 6 in `backjump` and 11 in `processWatches`) versus the number of times it looked up a constraint during propagation of a watch (lines 7 and 8 in `processWatches`), for instances solved within resource limits.

Backjump lookups happen frequently, but never more than propagation lookups. Often, backjump lookups happen significantly less than propagation lookups, up to two orders of magnitude. The median number of backjump watch lookups is also less than half the median of propagation watch lookups. As backjump lookups perform few operations compared to propagation lookups, the resulting overhead does not seem to induce a performance bottleneck.

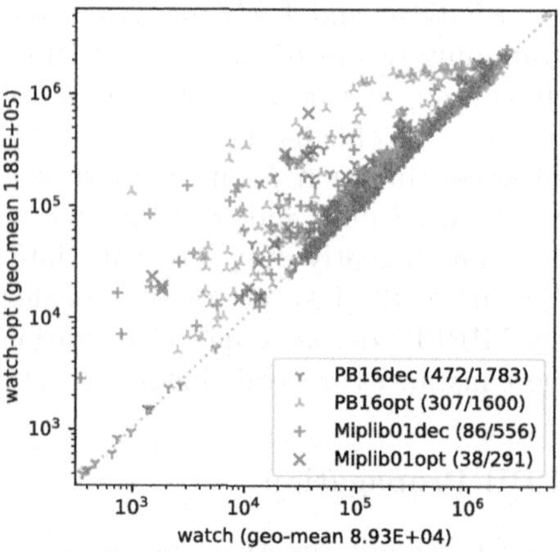


**Fig. 1.** Propagations per second for *watch* and *watch-opt*.

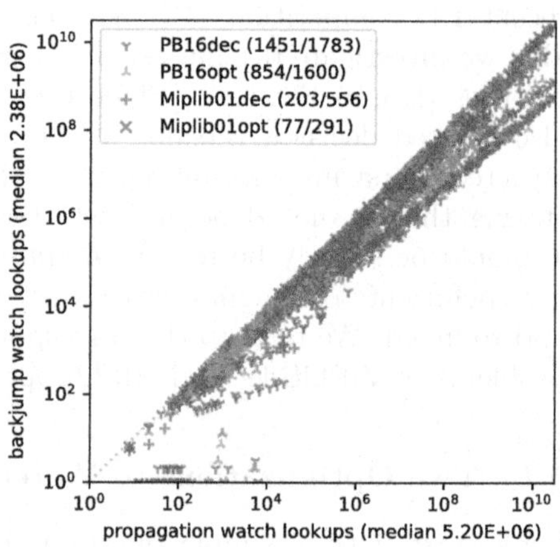


**Fig. 2.** Watch lookups for *watch-opt*.

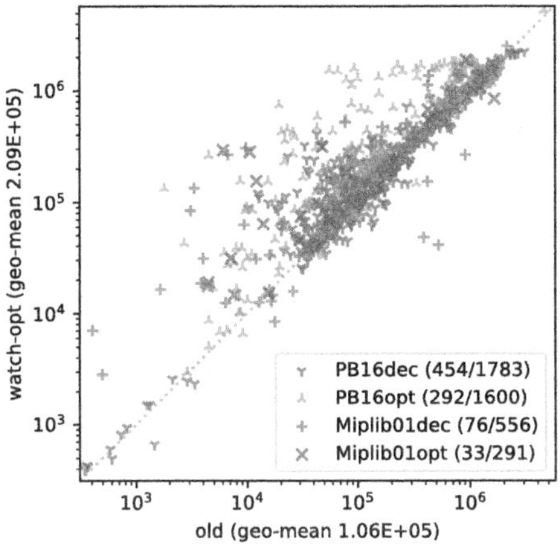


**Fig. 3.** Propagations per second for *old* and *watch-opt*.

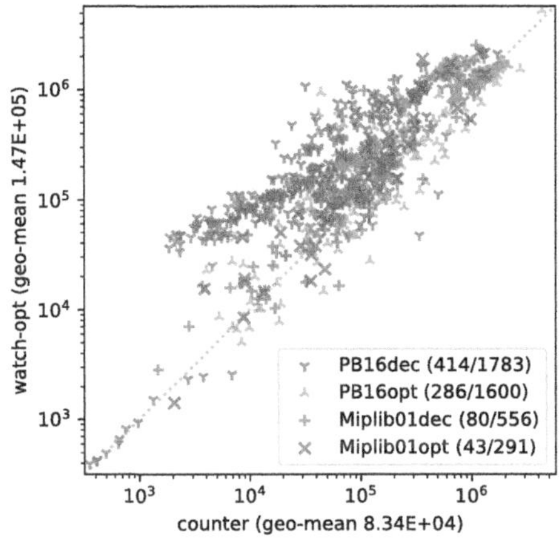


**Fig. 4.** Propagations per second for *counter* and *watch-opt*.

### 5.3 Performance Evaluation

To evaluate the performance of our approach, we compare *watch-opt* to:

- *counter*: an implementation of PB counter propagation (see Sect. 2.2)
- *old*: the original propagation algorithm of *RoundingSat* (see Sect. 4)
- *counter-cc*: *counter*, but **c**lauses and **c**ardinality constraints are handled with specialized watched propagation routines – the three-tiered approach default in *Sat4J* (see Sect. 4)
- *old-cc*: three-tiered *old* with the same specialized routines
- *watch-opt-cc*: three-tiered *watch-opt* with the same specialized routines

Figures 3, 4, 5 and 6, compare the propagation speed of *watch-opt* to the above alternatives, based on the instances succesfully solved by the compared approaches within resource limits and taking at least 1 s to solve. Table 1 presents the total number of succesfully solved instances by each approach.

Often, the propagation speed of *watch-opt* is orders of magnitude faster than of *old* and *counter*, with the reverse being true only infrequently. This translates to significantly more solved instances compared to *old* and *counter*. The specialized propagation for clauses and cardinality constraints improves performance in general, with most *-cc* configurations solving more instances than their counterparts. *watch-opt-cc* solves the most instances overall, while *counter-cc* seems to profit most from the specialized routines, almost fully closing the gap with *watch-opt-cc*. The propagation speed plots in Figs. 5 and 6 tell a similar tale: *old-cc* propagates significantly slower than *watch-opt-cc*, but it becomes harder to judge that *watch-opt-cc* propagates faster. The geometric means of their propagation speed in Fig. 6 still give the edge to *watch-opt-cc*.

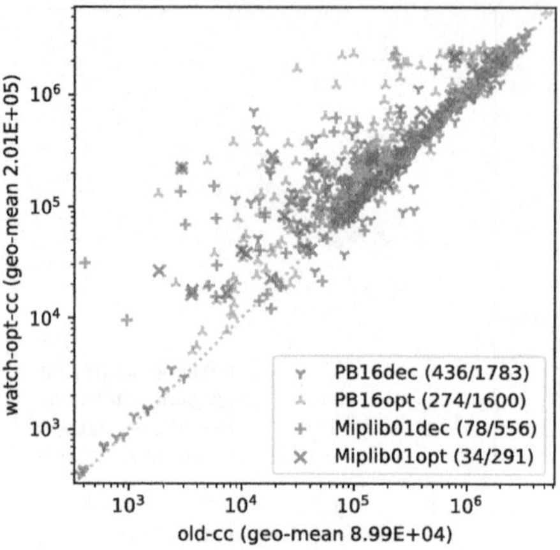


**Fig. 5.** Propagations per second for *old-cc* and *watch-opt-cc*.

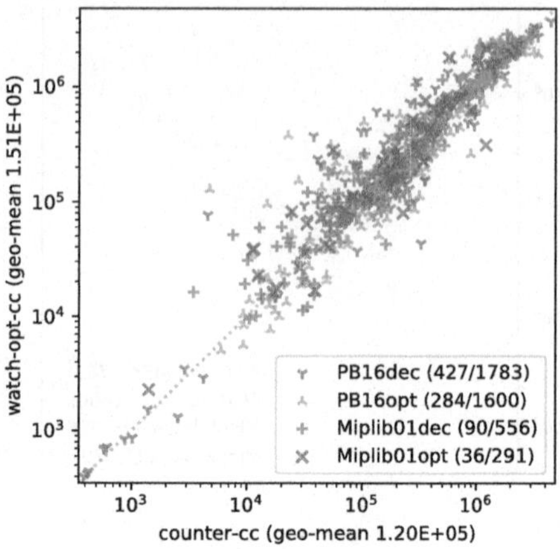


**Fig. 6.** Propagations per second for *counter-cc* and *watch-opt-cc*.

**Table 1.** Solved instance counts for different propagation implementations

| | *old* | *counter* | *watch-opt* | *old-cc* | *counter-cc* | *watch-opt-cc* |
|---|---|---|---|---|---|---|
| PB16dec (1783) | 1429 | 1385 | 1451 | 1444 | 1456 | **1472** |
| MIPLIBdec (556) | 182 | 196 | 203 | 187 | 204 | **205** |
| PB16opt (1600) | 820 | 846 | 854 | 825 | **862** | 854 |
| MIPLIBopt (291) | 69 | 76 | 77 | 71 | 75 | **79** |

To explain the relative difference between *old*/*old-cc* and *counter*/*counter-cc*, it is useful to characterize when *counter* and *old* accrue the most overhead. A counter algorithm induces most overhead for constraints with low watch count as continually updating the high slacks for these constraints is often unnecessary. Inversely, *old* incurs more overhead for constraints that have a relatively high number of watches, as its eager recalculation of watch indices, watch slacks, and propagating watches, are linear operations in the number of watches. Since clauses and low-degree cardinality constraints are frequently generated constraints with low watch counts, this can explain why *counter* profits a lot more from the specialized propagation routines than *old*.

We conclude that *watch-opt* is indeed more efficient than its *counter* counterpart. However, adding specialized clause and cardinality constraint propagation into the mix strongly diminishes its advantage – *counter-cc*, the *Sat4J* default approach, is definitely a close second.

## 6 Conclusion

We present an optimized watched propagation algorithm for PB or 0-1 integer linear constraints. Our experiments indicate it is more efficient than counter propagation used by *Sat4J* and the watched propagation used by *RoundingSat*. Hence, our approach seems a good candidate to replace PB counter propagation with PB watched propagation, though the performance gains are moderate in the three-tiered setting. Nonetheless, the results are sufficiently convincing to consider *watch-opt-cc* as a new default propagation algorithm for *RoundingSat*.

An interesting avenue to speed up PB propagation would be to pinpoint which PB constraints propagate more efficiently with a counter approach and which favor the watched approach. Maybe those constraints which most of the time have a relatively large number of watched literals are better off with the counting approach? Other future work may reconsider the idea of *Galena*: track the largest coefficient of non-true literals to reduce the number of watches for a constraint. Our work can also prove useful to improve CP propagators for constraints closely related to PB constraints, such as the sum_set constraint. Finally, the order in which constraints propagate strongly influences what a conflict-driven solver will learn. Prioritizing certain types of constraints during propagation may yield better learned constraints.

**Acknowledgments.** The Swedish National Infrastructure for Computing (SNIC) at the High Performance Computing Center North (HPC2N) at Umeå University provided computational resources. The author is supported by the Swedish Research Council grant 2016-00782.

We are grateful to Emir Demirovic, Jan Elffers, Stephan Gocht, Daniel Le Berre and Jakob Nordström for discussions on PB propagation.

## References

1. Achterberg, T., Koch, T., Martin, A.: MIPLIB 2003. Oper. Res. Lett. **34**(4), 361–372 (2006). https://doi.org/10.1016/j.orl.2005.07.009, http://www.zib.de/Publications/abstracts/ZR-05-28/
2. Bayardo Jr., R.J., Schrag, R.: Using CSP look-back techniques to solve real-world SAT instances. In: Proceedings of the 14th National Conference on Artificial Intelligence (AAAI 1997), pp. 203–208 (1997)
3. Bixby, R., Ceria, S., McZeal, C., Savelsbergh, M.: An updated mixed integer programming library: MIPLIB 3.0 (1998)
4. Blake, A.: Canonical expressions in Boolean algebra. Ph.D. thesis, University of Chicago (1937)
5. Buss, S., Nordström, J.: Proof complexity and SAT solving. In: Handbook of Satisfiability, 2nd edn. (2020, to appear). Draft version. http://www.csc.kth.se/~jakobn/research/
6. Chai, D., Kuehlmann, A.: A fast pseudo-Boolean constraint solver. IEEE Trans. Comput.-Aided Des. Integr. Circuits Syst. **24**(3), 305–317 (2005). Preliminary version in DAC 2003
7. Cook, S.A.: The complexity of theorem-proving procedures. In: Proceedings of the 3rd Annual ACM Symposium on Theory of Computing (STOC 1971), pp. 151–158 (1971)
8. Cook, W., Coullard, C.R., Turán, G.: On the complexity of cutting-plane proofs. Discrete Appl. Math. **18**(1), 25–38 (1987)

9. Davis, M., Logemann, G., Loveland, D.: A machine program for theorem proving. Commun. ACM **5**(7), 394–397 (1962)
10. Davis, M., Putnam, H.: A computing procedure for quantification theory. J. ACM **7**(3), 201–215 (1960)
11. Devriendt, J.: Miplib 0–1 instances in OPB format (2020). https://doi.org/10.5281/zenodo.3870965
12. Devriendt, J.: Online Repository for "Watched Propagation of 0–1 Integer Linear Constraints" (2020). https://doi.org/10.5281/zenodo.3952444
13. Eén, N., Sörensson, N.: An extensible SAT-solver. In: Giunchiglia, E., Tacchella, A. (eds.) SAT 2003. LNCS, vol. 2919, pp. 502–518. Springer, Heidelberg (2004). https://doi.org/10.1007/978-3-540-24605-3_37
14. Elffers, J., Nordström, J.: Divide and conquer: towards faster pseudo-Boolean solving. In: Proceedings of the 27th International Joint Conference on Artificial Intelligence (IJCAI 2018), pp. 1291–1299 (2018)
15. Gecode: Generic constraint development environment. https://www.gecode.org/
16. Gleixner, A., et al.: MIPLIB 2017: data-driven compilation of the 6th mixed-integer programming library. Technical report, Optimization Online (2019). http://www.optimization-online.org/DB_HTML/2019/07/7285.html
17. Koch, T., et al.: MIPLIB 2010. Math. Programm. Comput. **3**(2), 103–163 (2011). https://doi.org/10.1007/s12532-011-0025-9, http://mpc.zib.de/index.php/MPC/article/view/56/28
18. Le Berre, D., Parrain, A.: The Sat4j library, release 2.2. J. Satisfiability Boolean Model. Comput. **7**, 59–64 (2010)
19. Levin, L.A.: Universal sequential search problems. Problemy peredachi informatsii **9**(3), 115–116 (1973). (in Russian). http://mi.mathnet.ru/ppi914
20. Marques-Silva, J.P., Sakallah, K.A.: GRASP: a search algorithm for propositional satisfiability. IEEE Trans. Comput. **48**(5), 506–521 (1999). Preliminary version in ICCAD 1996
21. MIPLIB 2.0 (1996). http://miplib2010.zib.de/miplib2/miplib2.html
22. MIPLIB 2017 (2018). http://miplib.zib.de
23. Moskewicz, M.W., Madigan, C.F., Zhao, Y., Zhang, L., Malik, S.: Chaff: engineering an efficient SAT solver. In: Proceedings of the 38th Design Automation Conference (DAC 2001), pp. 530–535 (2001)
24. Pseudo-Boolean competition 2016 (2016). http://www.cril.univ-artois.fr/PB16/
25. Robinson, J.A.: A machine-oriented logic based on the resolution principle. J. ACM **12**(1), 23–41 (1965)
26. RoundingSat. https://gitlab.com/miao_research/roundingsat
27. Sheini, H.M., Sakallah, K.A.: Pueblo: a modern pseudo-Boolean SAT solver. In: Proceedings of the Design, Automation and Test in Europe Conference (DATE 2005), pp. 684–685 (2005)
28. Sheini, H.M., Sakallah, K.A.: Pueblo: a hybrid pseudo-Boolean SAT solver. J. Satisfiability Boolean Model. Comput. **2**(1–4), 165–189 (2006). Preliminary version in DATE 2005
29. Global constraint catalog: sum_set. https://sofdem.github.io/gccat/gccat/Csum_set.html
30. Zhang, H., Stickel, M.: Implementing the Davis-Putnam method. J. Autom. Reasoning **24**(1), 277–296 (2000). https://doi.org/10.1023/A:1006351428454, https://doi.org/10.1023/A:1006351428454

# Bounding Linear Programs by Constraint Propagation: Application to Max-SAT

Tomáš Dlask(✉) and Tomáš Werner

Faculty of Electrical Engineering, Czech Technical University in Prague, Prague, Czech Republic
{dlaskto2,werner}@fel.cvut.cz

**Abstract.** The Virtual Arc Consistency (VAC) algorithm by Cooper et al. is a soft local consistency technique that computes, in linear space, a bound on the basic LP relaxation of the Weighted CSP (WCSP). We generalize this technique by replacing arc consistency with a (problem-dependent) constraint propagation in a system of linear inequalities over the reals. When propagation detects infeasibility, the infeasibility certificate (a solution to the alternative system in Farkas' lemma) provides a dual improving direction. We illustrate this approach on the LP relaxation of Weighted Max-SAT. We show in experiments that the obtained bounds are often not far from global LP optima and we prove that they are exact for known tractable subclasses of Weighted Max-SAT.

**Keywords:** Linear programming relaxation · Constraint propagation · Weighted CSP · Virtual Arc Consistency · Weighted Max-SAT

## 1 Introduction

Although the linear programming (LP) problem is solvable in polynomial time, solving very large sparse linear programs can be challenging in practice. Such linear programs occur in many areas, a prominent example being the computation of bounds in branch-and-bound search by LP relaxation. To solve such LPs, the classical simplex and interior point methods may not always be suitable, if only for their worst-case space complexity which is super-linear in the number of non-zeros of the problem matrix (to the best of our knowledge, not much is known about worst-case complexity of solving sparse linear programs [18]). First-order methods such as subgradient, smoothing or augmented Lagrangian methods have linear space complexity but tend to be slow (see experimental comparison [8] of methods for large-scale WCSP) and need a long time to re-converge when warm-started after a small change of the problem. This is a motivation to search for problem-specific (possibly approximate) solvers that would be more efficient than classical methods.

This work has been supported by the Czech Science Foundation (grant 19-09967S), the OP VVV project CZ.02.1.01/0.0/0.0/16_019/0000765, and the Grant Agency of the Czech Technical University in Prague (grant SGS19/170/OHK3/3T/13).


H. Simonis (Ed.): CP 2020, LNCS 12333, pp. 177–193, 2020.
https://doi.org/10.1007/978-3-030-58475-7_11

One such approach is known as the *primal-dual method*[1] [17], which is efficient for LP formulations of some tractable combinatorial optimization problems. Given a feasible dual solution, we consider the *restricted problem*, which minimizes infeasibility of the complementary slackness conditions. Optimal solutions of the *dual restricted problem* turn out to be dual-improving directions. The restricted problem is a linear program simpler than the original one, thus often amenable to combinatorial algorithms. Many classical algorithms for, e.g., flow and assignment problems can be seen as examples of the primal-dual method.

A similar idea has been employed in the VAC algorithm [3] (and the closely related Augmenting DAG algorithm [12,25]), which computes an upper-bound on the basic LP relaxation of the WCSP [19,20,23,25]. Strictly speaking, this is not a primal-dual method since the restricted problem is the LP relaxation of a CSP, which is a feasibility rather than optimization problem. Another difference is that the restricted problem is solved only approximately by arc consistency (AC), which not always detects infeasibility. Consequently, the method only obtains an upper bound on the LP relaxation of WCSP.

We propose a generalization of this technique. To detect infeasibility of the restricted problem, we propose to use a suitable (problem dependent) form of *constraint propagation in a system of linear inequalities*. If infeasibility is detected, a *certificate of infeasibility* (a solution to the alternative system in Farkas' lemma) is constructed, which provides a dual-improving direction. Since propagation may not always detect infeasibility, the approach yields only an upper bound on the global optimum of the LP. Note, while constraint propagation in CSP with infinite domains is well-known [2], the novelty of our approach is in using infeasibility certificates to iteratively improve the dual solution.

To illustrate the approach on a problem different than WCSP, we chose the LP relaxation of the Weighted Max-SAT problem [22]. We experimentally show that the obtained bounds are often not far from global LP optima and we prove that they are exact for known tractable subclasses of the Weighted Max-SAT.

## 2 Linear Optimization by Constraint Propagation

### 2.1 Constraint Propagation for Linear Inequalities

In the CSP, we are given a set of relations (constraints) $\phi_1, \dots, \phi_m \subseteq D^n$ and seek to find $x = (x_1, \dots, x_n) \in \phi_1 \cap \dots \cap \phi_m$ or prove that no such solution exists. A heuristic that can help achieve this is *constraint propagation*, where we iteratively generate new constraints that are implied by (i.e., inferred from) the constraint set and add them to the constraint set. By this, we make explicit some knowledge about the solution set, which before was only implicit in the constraints. As exhaustive enumeration of all implied constraints is usually impossible, only a small predefined set of simple inference (or propagation) rules is used. Since we are not doing complete inference, the procedure is refutation-incomplete: it need not infer a contradiction even if the CSP is infeasible.

---

[1] As remarked in [17], this name is a misnomer as it is in fact a purely dual method.

Deciding feasibility, finding a solution and, more generally, deciding if the constraints imply a given relation, is usually intractable. The situation is much simpler if $D = \mathbb{R}$ (the reals) and $\phi_i$'s are linear inequalities. We write a linear inequality $\phi_i$ as $a_i^T x \leq b_i$ and the system $\phi_1, \dots, \phi_m$ as $Ax \leq b$ where $x \in \mathbb{R}^n$, $A = \begin{bmatrix} a_1 \cdots a_m \end{bmatrix}^T \in \mathbb{R}^{m \times n}$ and $b = (b_1, \dots, b_m) \in \mathbb{R}^m$. In this case, the above tasks can be solved by linear programming. In particular, the logic of linear inequalities over $\mathbb{R}$ is described by the affine form of Farkas' lemma [7,21]:

**Theorem 1.** *A system $Ax \leq b$ implies an inequality $c^T x \leq d$ iff some non-negative combination of the inequalities $Ax \leq b$ implies $c^T x \leq d$, i.e., there is $y \geq 0$ such that $A^T y = c$ and $b^T y \leq d$.*

In particular (Farkas' lemma), the system $Ax \leq b$ is infeasible iff some non-negative combination of the inequalities equals $0^T x \leq d$ where $d < 0$, i.e., there is a vector $y \geq 0$ such that $A^T y = 0$ and $b^T y < 0$. The vector $y$ can be seen as a proof (certificate, cause) for the inequality $c^T x \leq d$ resp. infeasibility.

Thus, constraint propagation for linear inequalities works as follows. Using a fixed set of inference rules (which depends on the problem solved), we generate new linear inequalities until either no new inequality can be generated or a contradiction is found. Each time a new inequality is generated, its 'cause' vector is stored, encoding how the inequality was created from the existing inequalities. When a contradiction is found, a certificate of infeasibility can be computed by tracking the newly generated inequalities back to the original system and composing the cause vectors.

## 2.2 Computing Certificate of Infeasibility

Let us focus on obtaining the certificate of infeasibility. As an example, consider the system of $m = 5$ initial inequalities on the left in Fig. 1. From inequalities $\phi_2$ and $\phi_3$, we infer inequality $\phi_6 = 2\phi_2 + \phi_3$. Next, we gradually infer inequalities $\phi_7 = \phi_1 + 3\phi_6$, $\phi_8 = \phi_4 + \phi_6$, $\phi_9 = \phi_6 + \phi_7$, and finally $\phi_{10} = \phi_5 + \phi_7 + \phi_8$. Since $\phi_{10}$ reads $0 \leq -2$, the initial system $\phi_1, \dots, \phi_5$ is infeasible.

The history of propagation is represented by a directed acyclic graph (DAG) $E \subseteq V \times V$ with edge weights $\alpha\colon E \to \mathbb{R}_+$, where $V$ is the set of all (initial and inferred) inequalities and each inferred inequality is given by $\phi_i = \sum_{j \in N_i} \alpha_{ij} \phi_j$ where $N_i = \{\, j \in V \mid (i,j) \in E \,\}$. By composing the inferences, each inequality $\phi_i$ can be expressed in terms of the initial inequalities as $\phi_i = \sum_{j=1}^m y_j^i \phi_j$, where we call $y^i = (y_1^i, \dots, y_m^i) \in \mathbb{R}^m$ the *cause vector* of $\phi_i$. For $i \leq m$, we have $y^i = e^i$ where $e^i$ is the $i$th standard-basis vector of $\mathbb{R}^m$. For $i > m$, we have

$$y^i = \sum_{j \in N_i\colon N_j = \emptyset} \alpha_{ij} e^j + \sum_{j \in N_i\colon N_j \neq \emptyset} \alpha_{ij} y^j. \tag{1}$$

In the example, $V = \{1, \dots, 10\}$, $y^6 = 2e^2 + e^3 = (0,2,1,0,0)$, $y^7 = e^1 + 3y^6$, $y^8 = e^4 + y^6$, $y^9 = y^6 + y^7$, and $y^{10} = e^5 + y^7 + y^8 = (1,8,4,1,1)$. Since $b^T y^{10} = -2$ and $A^T y^{10} = 0$, vector $y^{10}$ is a certificate of infeasibility by Theorem 1.

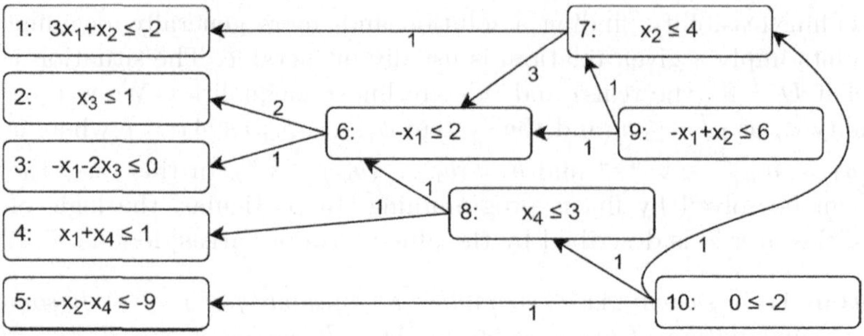


**Fig. 1.** Propagation in a simple system of linear inequalities. The inequalities are indexed by 1–10. Inequalities 1–5 are initial, inequalities 6–10 are inferred. Edge weights indicate the coefficients of non-negative combinations.

As we need the cause vector only for the final (contradictory) inequality ($\phi_{10}$ in the example), storing all cause vectors explicitly in the memory is wasteful. In addition, some inferred inequalities may not be needed for the proof of infeasibility ($\phi_9$ in the example). We show that any single cause vector can be computed more efficiently by dynamic programming.

For any initial inequality $\phi_i$ and any derived inequality $\phi_k$, $y_i^k$ is the sum of weight-products[2] of all directed paths from node $k$ to node $i$ in the DAG. Suppose we want to compute $y^k$ for some single $k$. We can consider only the subgraph of the DAG reachable from node $k$ along directed paths. We introduce auxiliary variables $z_j$, which are to equal the sum of weight-products of all directed paths from node $k$ to node $j$. Initially, we set $y^k = 0$, $z_k = 1$, and $z_j = 0$ for all $j \neq k$. Then we process the nodes $i$ of the subgraph in a topological order as follows: if $N_i = \emptyset$ then set $y_i^k := z_i$, otherwise update $z_j := z_j + \alpha_{ij} z_i$ for all $j \in N_i$. Eventually, we have $y_i^k = z_i$ for all $i = 1, \dots, m$. The time and space complexity of this algorithm is linear in the size of the graph.

### 2.3 Application to Linear Programming

Now we show how constraint propagation can be used to possibly improve a feasible dual solution of a linear program. Consider a pair of mutually dual linear programs (the primal on the left, the dual on the right)

$$c^T x \to \max \qquad b^T y \to \min \tag{2a}$$

$$Ax \leq b \qquad y \geq 0 \tag{2b}$$

$$x \lesseqgtr 0 \qquad A^T y = c \tag{2c}$$

where[3] $x \lesseqgtr 0$ denotes that the components of $x$ can have arbitrary signs (as in [17]). By the complementary slackness theorem, a primal feasible solution $x$ and a dual

[2] The weight-product of a path is the product of all edge weights along the path.

[3] $A, b$ in (2) denote different matrices than $A, b$ in the previous sections.

feasible solution $y$ are simultaneously optimal iff for every $i$ we have $a_i^T x = b_i$ or $y_i = 0$ (or both). Denoting by $I = \{\, i \mid y_i = 0 \,\}$ the set of dual constraints (2b) active at $y$, this condition can be written as the left-hand system of the pair

$$\begin{aligned} & & b^T\bar{y} < 0 & & \quad (3a)\\ a_i^T x \le b_i & & \bar{y}_i \ge 0 & \quad \forall i \in I & \quad (3b)\\ a_i^T x = b_i & & \bar{y}_i \lesseqgtr 0 & \quad \forall i \notin I & \quad (3c)\\ x \lesseqgtr 0 & & A^T\bar{y} = 0. & & \quad (3d) \end{aligned}$$

On the right of (3), we wrote the Farkas alternative system to these conditions[4]. Thus, a point $y$ feasible for the dual in (2) is optimal iff the left-hand system in (3) is feasible, which holds iff the right-hand system in (3) is infeasible. Moreover, any solution $\bar{y}$ to the right-hand system is an improving direction for the dual in (2), i.e., there is $\epsilon > 0$ such that $b^T(y + \epsilon\bar{y}) < b^T y$, $y + \epsilon\bar{y} \ge 0$ and $A^T(y + \epsilon\bar{y}) = c$.

The method thus proceeds as follows. Having a feasible solution $y$ for the dual in (2), try to prove infeasibility of the left-hand system in (3) by constraint propagation and find a certificate of infeasibility $\bar{y}$, i.e., a solution to the right-hand system in (3). Then choose (by exact or approximate line search) a step size $\epsilon$ and update $y := y + \epsilon\bar{y}$. By repeating this iteration, a better and better upper bound on linear program (2) is obtained. Terminate when the propagation fails to detect infeasibility of the left-hand system in (3).

In the rest of the paper, we apply this approach to LP relaxations of WCSP and Max-SAT. These LPs will involve equality constraints and non-negative variables. Though they could be transformed to the general form (2) by well-known tricks (such as replacing an equality with two inequalities or adding slack variables), it will be more convenient to adapt the basic approach described in Sect. 2 to these cases, resulting in somewhat different and more complex algorithms.

## 3 LP Relaxation of Weighted CSP

In the (binary) WCSP, we are given a graph $E \subseteq \binom{V}{2}$, a finite domain $D$, and weights $c_\emptyset \in \mathbb{R}$, $c_{uk} \le 0$ ($u \in V$, $k \in D$) and $c_{uk,vl} \le 0$ ($uv \in E$, $k, l \in D$). We maximize

$$f_c(\lambda) = c_\emptyset + \sum_{u\in V} c_{u\lambda(u)} + \sum_{uv\in E} c_{u\lambda(u),v\lambda(v)} \quad (4)$$

over all assignments $\lambda\colon V \to D$. We abbreviated $\{u, v\}$ by $uv$ and adopted that $c_{uk,vl} = c_{vl,uk}$. The basic LP relaxation of WCSP can be written as[5]

[4] The two systems (3) correspond to a more general form of Farkas' lemma than Theorem 1, allowing for equality constraints and non-negative variables [14, §6.4].

[5] The basic LP relaxation of WCSP can be written in several different ways, see e.g. [19,23,25]. We chose the one that is closest to the VAC paper [3].

$$
\begin{aligned}
c^T x &\to \max & h &\to \min && \text{(5a)}\\
Ax &= 0 & y &\lesseqgtr 0 && \text{(5b)}\\
x_\emptyset &= 1 & h &\lesseqgtr 0 && \text{(5c)}\\
x &\geq 0 & A^T y + e^\emptyset h &\geq c && \text{(5d)}
\end{aligned}
$$

where the system $Ax = 0$ reads

$$
\begin{aligned}
\textstyle\sum_{l \in D} x_{uk,vl} - x_{uk} &= 0 \qquad \forall u \in V,\ v \in N_u,\ k \in D && \text{(6a)}\\
\textstyle\sum_{k \in D} x_{uk} - x_\emptyset &= 0 \qquad \forall u \in V. && \text{(6b)}
\end{aligned}
$$

The system $A^T y + e^\emptyset h \geq c$ reads

$$
\begin{aligned}
y_u - \textstyle\sum_{v \in N_u} y_{uk,v} &\geq c_{uk} \qquad \forall u \in V,\ k \in D && \text{(7a)}\\
y_{uk,v} + y_{vj,u} &\geq c_{uk,vl} \qquad \forall uv \in E,\ k,l \in D && \text{(7b)}\\
\textstyle\sum_{u \in V} y_u + h &\geq c_\emptyset && \text{(7c)}
\end{aligned}
$$

where $e^\emptyset$ is the standard-basis vector such that $x^T e^\emptyset = x_\emptyset$.

For any $y$, replacing the weight vector $c$ with the vector $c' = c - A^T y$ is an *equivalent transformation* [3] (a.k.a. a *reparameterization* [19,24]) of the WCSP objective[6]. Indeed, $c'^T x = (c^T - y^T A)x = c^T x$ for all feasible $x$, hence also $f_{c'}(\lambda) = f_c(\lambda)$ for all assignments $\lambda$. A reparameterization is feasible (satisfying (7a) and (7b) if $c'_{uk} \leq 0$ and $c'_{uk,vl} \leq 0$. After eliminating variable $h$, the dual thus minimizes $c'_\emptyset$ over feasible reparameterizations. Note that for feasible reparameterizations, $c'_\emptyset$ is an upper bound on the WCSP optimal value.

Given a feasible dual solution $(y, h)$, let $J$ denote the set of indices of dual inequalities (5d) that are active at $(y, h)$. Then, the complementary slackness conditions read as the left-hand system of

$$
\begin{aligned}
 & & \bar{h} &< 0 & & \text{(8a)}\\
Ax &= 0 & \bar{y} &\lesseqgtr 0 & & \text{(8b)}\\
x_\emptyset &= 1 & \bar{h} &\lesseqgtr 0 & & \text{(8c)}\\
x_j &\geq 0 & A_j^T \bar{y} + \bar{h} e_j^\emptyset &\geq 0 & \forall j \in J \quad & \text{(8d)}\\
x_j &= 0 & & & \forall j \notin J \quad & \text{(8e)}
\end{aligned}
$$

where $A_j$ denotes the $j$th column of $A$ and $e_j^\emptyset = [\![ j = \emptyset ]\!]$ is the $j$th component of $e^\emptyset$ (where $[\![\cdot]\!]$ denotes the Iverson bracket). On the right, we wrote the Farkas alternative system.

The left-hand system in (8) is the LP relaxation of the CSP instance formed by the active tuples of the reparameterized WCSP instance. The WCSP is *virtually arc-consistent (VAC)* if this CSP has a non-empty AC closure [3]. Indeed, propagating zero values of the primal variables $x_j$ in (8) using the marginalization constraint (6a) is equivalent to the AC algorithm in this CSP.

[6] Note, $c'$ is 'almost' (up to variable $h$) the reduced cost vector of the primal of (5).

When propagation detects a contradiction, we construct a certificate $(\bar{y}, \bar{h})$ satisfying the right-hand system of (8). If a variable $x_j$ is inferred to be zero, we store a cause vector $y^j$ of the coefficients of linear combination $x_j + t_j = 0$ of the primal constraints (6), where $t_j$ is a non-negative combination of $x_i$, $i \in J$, and may contain $x_i$, $i \notin J$, with arbitrary sign. Under constraints (8e)+(8d), $t_j$ is non-negative, therefore $x_j + t_j = 0$ implies $x_j = 0$. For $j \notin J$, we initialize $y^j = 0$. All vectors $y^j$ have the same dimension, equal to the number of primal constraints (6). We denote the standard-basis vector of this space corresponding to (6a) (resp. (6b)) by $e^{uk,v}$ (resp. $e^u$).

We now describe the propagation rules in detail, including the cause vectors $y^j$. If $j \in J$, each $y^j$ will represent an equality $x_j + t_j = 0$. If $j \notin J$, $y^j$ is initialized to zero and thus represent equality $0 = 0$; in this case the excess variable $-x_j$ on the left-hand side will be included into $t_i$ on the right-hand side, which is allowed by the definition of $t_i$. The propagation rules are as follows:

- If $x_{uk} = 0$ for some $u \in V$ and $k \in D$, constraints (6a) imply $x_{uk,vl} = 0$ for all $v \in N_u$ and $l \in D$. Inference in terms of equalities:
$$\big(x_{uk} + t_{uk} = 0\big) + \big(\sum_{l \in D} x_{uk,vl} - x_{uk} = 0\big) = \big(\sum_{l \in D} x_{uk,vl} + t_{uk} = 0\big)$$
The cause vectors are given by $y^{uk,vl} = y^{uk} + e^{uk,v}$, for $v \in N_u$ and $l \in D$.
- If for some $uv \in E$ and $k \in D$ we have $x_{uk,vl} = 0$ for all $l \in D$, constraint (6a) implies $x_{uk} = 0$. Inference in terms of equalities:
$$\sum_{l \in D} \big(x_{uk,vl} + t_{uk,vl} = 0\big) - \big(\sum_{l \in D} x_{uk,vl} - x_{uk} = 0\big) = \big(x_{uk} + \sum_{l \in D} t_{uk,vl} = 0\big)$$
The cause vector is given by $y^{uk} = \sum_{l \in D} y^{uk,vl} - e^{uk,v}$.
- If for some $u \in V$ we have $x_{uk} = 0$ for all $k \in D$, constraint (6b) implies a contradiction (domain wipe-out). Inference in terms of equalities:
$$\sum_{k \in D} \big(x_{uk} + t_{uk} = 0\big) - \big(\sum_{l \in D} x_{ul} - x_{\emptyset} = 0\big) - \big(x_{\emptyset} = 1\big) = \big(\sum_{k \in D} t_{uk} = -1\big)$$
The certificate of infeasibility is given by $\bar{y} = \sum_{k \in D} y^{uk} - e^u$, $\bar{h} = -1$.

By properties of $t_j$ and the fact that coefficients $(\bar{y}, \bar{h})$ encode an equality in the form $\sum_j t_j = -1$, it is not hard to show that $(\bar{y}, \bar{h})$ are feasible for the right-hand side (8) and therefore constitute an improving direction for the dual in (5) from the current point $(y, h)$.

The described algorithm is 'almost' equivalent to the VAC / Augmenting DAG algorithm [3,12,25]. The fixed points of both algorithms are characterized by the same property, namely VAC. However, our improving directions $\bar{y}$ are in general different from the ones in [3,12,25], sometimes having larger absolute values of their components (and thus allowing smaller step size $\epsilon$). It is subject to further research to clarify the relation between them.

## 4 LP Relaxation of Weighted Max-SAT

In the Weighted Max-SAT problem, we are given a set $V$ of variables and a set $C$ of clauses with positive weights $w\colon C \to \mathbb{R}_{++}$ and we seek to maximize the weighted sum of satisfied clauses. Let $V_c^+$ (resp. $V_c^-$) denote the set of variables that occur in clause $c \in C$ non-negated (resp. negated). Let $C_i^\pm = \{\, c \in C \mid i \in V_i^\pm \,\}$ denote the set of clauses where variable $i \in V$ occurs non-negated/negated. We denote $n_c = |V_c^-|$. For any $S \subseteq V$, we will use the shortcut $x(S) = \sum_{i \in S} x_i$, similarly for $y$. The classical LP relaxation of Weighted Max-SAT [22] reads

$$
\begin{array}{llrl}
w^T z \to \max & n^T y + q(C) + p(V) \to \min & & \text{(9a)}\\
z_c \le x(V_c^+) + n_c - x(V_c^-) & y_c \ge 0 & \forall c \in C & \text{(9b)}\\
x_i \ge 0 & p_i - y(C_i^+) + y(C_i^-) \ge 0 & \forall i \in V & \text{(9c)}\\
x_i \le 1 & p_i \ge 0 & \forall i \in V & \text{(9d)}\\
z_c \ge 0 & q_c + y_c \ge w_c & \forall c \in C & \text{(9e)}\\
z_c \le 1 & q_c \ge 0 & \forall c \in C & \text{(9f)}
\end{array}
$$

where we wrote also the dual LP on the right. The primal variables $x_i$ represent the (relaxed) original Boolean variables. Clearly, at dual optimum we have

$$
\begin{aligned}
p_i &= \max\{y(C_i^+) - y(C_i^-), 0\} && \forall i \in V && \text{(10a)}\\
q_c &= \max\{w_c - y_c, 0\} && \forall c \in C. && \text{(10b)}
\end{aligned}
$$

Substituting (10) into the dual objective together with $n^T y = \sum_{i \in V} y(C_i^-)$ results in a simpler form of the dual,

$$
\min_{y \ge 0} \sum_{c \in C} \max\{w_c - y_c, 0\} + \sum_{i \in V} \max\{y(C_i^+), y(C_i^-)\} \tag{11}
$$

which minimizes a convex piecewise-affine function of non-negative variables.

**Theorem 2.** *Point $y \in \mathbb{R}_+^C$ is optimal for* (11) *iff there exists $x \in \mathbb{R}^V$ satisfying the left-hand system of*

$$
\begin{array}{rrll}
x(V_c^+) + n_c - x(V_c^-) \ge 1 & \bar{y}_c \ge 0 & \forall c \in C^{\ge 1} & \text{(12a)}\\
x(V_c^+) + n_c - x(V_c^-) = 1 & \bar{y}_c \lesseqgtr 0 & \forall c \in C^{=1} & \text{(12b)}\\
x(V_c^+) + n_c - x(V_c^-) \le 1 & \bar{y}_c \le 0 & \forall c \in C^{\le 1} & \text{(12c)}\\
x(V_c^+) + n_c - x(V_c^-) = 0 & \bar{y}_c \lesseqgtr 0 & \forall c \in C^{=0} & \text{(12d)}\\
x_i = 1 & \bar{p}_i \lesseqgtr 0 & \forall i \in X^1 & \text{(12e)}\\
x_i = 0 & \bar{p}_i \lesseqgtr 0 & \forall i \in X^0 & \text{(12f)}\\
x_i \lesseqgtr 0 & \bar{p}_i + \bar{y}(C_i^+) - \bar{y}(C_i^-) = 0 & \forall i \in X^0 \cup X^1 & \text{(12g)}\\
x_i \ge 0 & \bar{p}_i + \bar{y}(C_i^+) - \bar{y}(C_i^-) \le 0 & \forall i \in X^U & \text{(12h)}\\
x_i \le 1 & \bar{p}_i \le 0 & \forall i \in X^U & \text{(12i)}
\end{array}
$$

*where*

$$X^0 = \{\, i \in V \mid y(C_i^+) < y(C_i^-)\,\} \qquad C^{\geq 1} = \{\, c \in C \mid y_c = 0\,\} \tag{13a}$$
$$X^1 = \{\, i \in V \mid y(C_i^+) > y(C_i^-)\,\} \qquad C^{=1} = \{\, c \in C \mid 0 < y_c < w_c\,\} \tag{13b}$$
$$X^U = \{\, i \in V \mid y(C_i^+) = y(C_i^-)\,\} \qquad C^{\leq 1} = \{\, c \in C \mid y_c = w_c\,\} \tag{13c}$$
$$C^{=0} = \{\, c \in C \mid y_c > w_c\,\} \tag{13d}$$

*are partitions of* $V$ *and* $C$*. If the left-hand system is infeasible, then there exist values* $(\bar{p}, \bar{y})$ *for the right-hand system*[7] (12) *such that*

$$\sum_{c \in C} ([\![c \notin C^{=0}]\!] - n_c)\bar{y}_c + \bar{p}(X^1 \cup X^U) > 0, \tag{14}$$

*where* $\bar{y}$ *is an improving direction for* (11) *from* $y$.

*Proof.* The left-hand system in (12) is the complementary slackness condition for the primal-dual pair (9), where we used (10) and eliminated variables $z_c$.

The right-hand system of (12) and (14) form the alternative system to the left-hand system of (12) in Farkas' lemma. To show that $\bar{y}$ is improving for (11), realize that $\bar{p}$ can w.l.o.g. satisfy $\bar{p}_i = \bar{y}(C_i^-) - \bar{y}(C_i^+)$ for all $i \in X^0 \cup X^1$ and $\bar{p}_i = \min\{\bar{y}(C_i^-) - \bar{y}(C_i^+), 0\}$ for all $i \in X^U$. Then, (14) can be reformulated (after multiplying by $-1$ and substituting $-\bar{p}_i$ terms) as

$$-\bar{y}(C - C^{=0}) + \sum_{i \in X^U} \max\{\bar{y}(C_i^+), \bar{y}(C_i^-)\} + \sum_{i \in X^1} \bar{y}(C_i^+) + \sum_{i \in X^0} \bar{y}(C_i^-) < 0,$$

which states that (11) decreases in terms of the affine functions that are active[8] in the current point $y$, as defined by the sets (13). □

We now define propagation rules for the left-hand system (12). These rules set the values of some of the undecided variables $x_i$, $i \in X^U$, to 0 or 1. Precisely, we iteratively visit each constraint (12a)–(12d) and look whether with the already decided variables it permits only a single value of some so-far undecided variable. If so, we fix the value of this variable (i.e., make it decided). If some constraint (12a)–(12d) cannot be satisfied by any assignment subject to the already decided variables, the left-hand system in (12) is infeasible. During propagation, we update the dual variables of (12), so that if infeasibility is detected, we are able to construct an improving direction $\bar{y}$ for (11).

We now need a technical definition. For $j \in V$, we call a cause vector $(p^j, y^j)$

- *1-j-defining* if it satisfies the right-hand system in (12), except for $i = j$ when it satisfies $p_j^j + y^j(C_j^+) - y^j(C_j^-) = 1$ and left-hand side of (14) equals 1. This cause vector defines an inequality $x_j + t_j \geq 1$ derived from the left-hand system in (12), where $t_j$ is a non-positive weighted sum of $x_i$, $i \in X^U$. Clearly, this inequality implies $x_j = 1$.

---

[7] Note, the right-hand system (12) has opposite direction of inequalities. This is due to writing left-hand inequalities (12a)–(12d) with opposite directions than in (9b).

[8] For $f(x) = \max\{a^T x, b^T x\}$ and a fixed $x$, we say that the function $a^T x$ is active and $b^T x$ is inactive at $x$ if $a^T x > b^T x$. If $a^T x = b^T x$, both functions are active at $x$.

**Table 1.** Propagation rules for system (12). The first column determines the type of constraints to which the rule applies.

| Constraint | Rule | |
|---|---|---|
| $C^{\geq 1} \cup C^{=1}$ | A1 | If there is only one undecided variable $x_k$, $k \in V_c$, and $x_i^c = 0$ for all other $i \in V_c - \{k\}$, we set $x_k = [\![k \in V_c^+]\!]$ and $y^k = e^c + \sum_{i \in V_c - \{k\}} y^i$ |
| | A2 | If all the variables $x_i$ for $i \in V_c$ are decided and satisfy $x_i^c = 0$, then we obtain a contradiction and set $\bar{y} = e^c + \sum_{i \in V_c} y^i$ |
| $C^{=1} \cup C^{\leq 1}$ | B1 | If there is exactly one decided variable $x_i, i \in V_c$, with $x_i^c = 1$, then we set $x_k = [\![k \in V_c^-]\!]$ and $y^k = -e^c + y^i$ for each undecided $x_k, k \in V_c$ |
| | B2 | If there are two (or more) decided variables $x_i, x_j$ for $i, j \in V_c$ with $x_i^c = x_j^c = 1$, then we obtain a contradiction and set $\bar{y} = -e^c + y^i + y^j$ |
| $C^{=0}$ | C1 | If there is no decided variable $x_i, i \in V_c$, with $x_i^c = 1$, then set all undecided variables $x_k, k \in V_c$, as $x_k = [\![k \in V_c^-]\!]$ and $y^k = -e^c$ |
| | C2 | If there is a decided variable $x_i, i \in V_c$, with $x_i^c = 1$, then we obtain a contradiction and set $\bar{y} = -e^c + y^i$ |

- *0-j-defining* if it satisfies the right-hand system in (12) and $p_j^j + y^j(C_j^+) - y^j(C_j^-) = -1$, and the left-hand side of (14) equals 0. This cause vector defines an inequality $-x_j + t_j \geq 0$, which implies $x_j = 0$.

Note that the sign constraints on the right-hand side (12) ensure that the inequalities on left-hand side (12) are combined in correct directions.

As mentioned in Theorem 2, it is sufficient to store vectors $y^j \in \mathbb{R}^C$ because these will be used to construct the improving direction $\bar{y}$ if a contradiction is detected. Thus, for each decided variable $x_j$ that is set to a value $v \in \{0, 1\}$, we can store only the $y^j$ component of a $v$-$j$-defining vector $(p^j, y^j)$.

The propagation rules are listed in Table 1, divided into groups according to which set (13) clause $c$ belongs. For each rule, we also specify how to construct the cause vector $y^i$ for each inferred variable $x_i$. For $i \in X^1 \cup X^0$, we define $y^i = 0$ so that it can be referred to in the equations for creation of other $y^j$ or $\bar{y}$. To simplify the explanation of the rules, for any $i \in V$ and $c \in C$ we denote

$$x_i^c = \begin{cases} x_i, & \text{if } i \in V_c^+, \\ 1 - x_i, & \text{if } i \in V_c^-. \end{cases} \tag{15}$$

We denote $e^c \in \mathbb{R}^C$ to be the standard-basis vector with 1 in the place corresponding to clause $c$. We also define $V_c = V_c^+ \cup V_c^-$.

The derivation of the updates for cause vectors $y^i$ is technical and must be done for each rule separately. The proof relies on the fact that for each initially decided variable $x_j$, $j \in X^1$, (resp. $j \in X^0$), we can initialize a 1-$j$-defining (resp. 0-$j$-defining) cause vector $(p^j, y^j)$ as $(e^j, 0)$ (resp. $(-e^j, 0)$) where $e^j \in \mathbb{R}^V$ is a standard-basis vector. This corresponds to setting $y^j = 0$ for the initially decided variables. Then it is possible to show how to derive a $v$-$k$-defining vector for a newly decided variable $x_k$ from the previous ones. We are going to show this in detail for rule B1, which we believe is most complicated.

**Theorem 3.** *Let $c \in C^{=1} \cup C^{\leq 1}$ such that there is exactly one decided variable $x_i$, $i \in V_c$, with $x_i^c = 1$ and $x_k, k \in V_c$, is an undecided variable. Let $x_i$ be decided to a value $v \in \{0, 1\}$, and let $(p^i, y^i)$ be $v$-$i$-defining. Then there exists a vector $p^k$ such that $(p^k, -e^c + y^i)$ is $[\![k \in V_c^-]\!]$-$k$-defining.*

*Proof.* First of all, see that $x(V_c^+) + n_c - x(V_c^-) = \sum_{j \in V_c} x_j^c$ by definition (15). The inequality encoded by the $v$-$i$-defining cause vector $(p^i, y^i)$ can be compactly rewritten as $x_j^c + t_j \geq 1$ and analogously, each defining equality for $j \in V_c \cap (X^0 \cup X^1)$ with $x_j^c = 0$ can be expressed as $-x_j^c = 0$, hence $x_j^c \geq 0$. Then, the derivation of the defining inequality for undecided variable $x_k$ is given as follows:

$$
\begin{aligned}
-\sum_{j \in V_c} x_j^c &\geq -1 && && (0, -e^c) && \text{(16a)}\\
x_i^c + t_i &\geq 1 && && (p^i, y^i) && \text{(16b)}\\
x_j^c &\geq 0 && \forall j \in (V_c \cap X^0) - \{i\} && (e^j, 0) && \text{(16c)}\\
x_j^c &\geq 0 && \forall j \in (V_c \cap X^1) - \{i\} && (-e^j, 0) && \text{(16d)}\\
x_j^c &\geq 0 && \forall j \in (V_c^- \cap X^U) - \{i, k\} && (-e^j, 0) && \text{(16e)}\\
-x_k^c + t_k &\geq 0 && && (p^k, y^k) && \text{(16f)}
\end{aligned}
$$

where

$$t_k = -\sum_{j \in (X^U \cap V_c^+) - \{i,k\}} x_j + t_i. \tag{17}$$

Inequality (16a) is (12b) or (12c) multiplied by $-1$, (16b), (16c), and (16d) are inequalities determining the values of already decided variables, and (16e) is (12i). Inequality (16f) is given as the sum of the inequalities above it. Each row in (16) is marked on the right by the coefficients $(p, y)$ with which it was derived from the original system (12). It is easy to check that the sign constraints in the right-hand system (12) are satisfied for each pair $(p, y)$ in (16).

The coefficients for the last row are determined by summing the above coefficients, i.e., $y^k = -e^c + y^i$ (which is the same equation as in Table 1) and $p^k = p^i + \sum_{j \in J'} e^j - \sum_{j \in J''} e^j$ where $J'$ (resp. $J''$) is set used on line (16c) (resp. union of the sets on lines (16d) and (16e). To show that the vector $(p^k, y^k)$ is $[\![k \in V_c^-]\!]$-$k$-defining, substitute the definition (15) into (16f) and see that $t_k$ is again a non-positive combination of other variables from $X^U$ as this held for $t_i$ by the assumption of the theorem. □

Each propagation rule can be formulated in a general form similar to (16) that defines vectors $(y^k, p^k)$ as described in Table 1. Using these vectors, a contradiction defined by $(\bar{p}, \bar{y})$ encodes an inequality $\bar{t} \geq 1$ where $\bar{t}$ is a non-positive sum of $x_i, i \in X^U$, and it is possible to show that the pair $(\bar{p}, \bar{y})$ satisfies conditions of Theorem 2.

*Remark 1.* One can ask whether it is possible to infer other values of undecided variables than 0 or 1 (such as $\frac{1}{2}$). Assuming that inference is done only from a single constraint from (12a)–(12d), this is impossible because the polyhedron defined by a single (in)equality from (12a)–(12d) subject to $0 \leq x_i \leq 1$, where some of the variables may be already set to 0 or 1, has integral vertices.

*Remark 2.* For general Max-SAT, the propagation rules in Table 1 do not always prove infeasibility of the left-hand system in (12). However, for Weighted Max-2SAT they do: if no more propagation is possible and no contradiction is detected, setting all undecided variables $x_i$ to $\frac{1}{2}$ satisfies all constraints of (12).

*Remark 3.* The restricted system (12) is the LP relaxation of a CSP with Boolean variables. The propagation corresponds to enforcing arc consistency of this CSP. The whole algorithm seeks to find a feasible dual solution of the LP relaxation of Max-SAT that enforces this CSP to have a non-empty AC closure. Compare this with the WCSP case, where the restricted system (8) is the LP relaxation of the CSP formed by the active tuples and the VAC algorithm seeks to find an equivalent transformation (a linear transformation of the weight vector that preserves the objective function) that makes this CSP arc-consistent. Note that, in contrast to WCSP, there is no obvious analogy of equivalent transformations for Weighted Max-SAT.

### 4.1 Finding Step Size by Approximate Line Search

If a contradiction is detected in (12) and improving direction $\bar{y}$ at the point $y$ is constructed, we need to find a feasible step size $\epsilon > 0$, as mentioned in Sect. 2.3. The optimal way (exact line search) would be to minimize $f(y + \epsilon\bar{y})$ over $\epsilon > 0$ subject to $y + \epsilon\bar{y} \geq 0$, where $f$ is the objective of (11). Since this is too costly for large instances, we do only approximate line search: we find the first breakpoint of the univariate convex piece-wise affine function $\epsilon \mapsto f(y + \epsilon\bar{y})$ , i.e., the smallest $\epsilon$ at which at least one previously inactive affine function becomes active. This ensures a non-zero improvement of $f$. Such $\epsilon$ is the maximum number satisfying the following constraints:

- To stay within the feasible set, we need $y_c + \epsilon\bar{y} \geq 0$, therefore $\epsilon \leq -y_c/\bar{y}_c$ for all $c \in C - C^{\geq 1}$ with $\bar{y}_c < 0$.
- For terms $\max\{w_c - y_c, 0\}$, if $w_c - y_c > 0$ (resp. $w_c - y_c < 0$) and $w_c - y_c - \epsilon\bar{y}_c$ decreases (resp. increases), then we need $\epsilon \leq (w_c - y_c)/\bar{y}_c$ where the bound is the point where the terms equalize. This is for all $c \in C$ such that $(w_c - y_c)\ \bar{y}_c > 0$.

– For terms $\max\{y(C_i^+), y(C_i^-)\}$, if $y(C_i^+) > y(C_i^-)$ and $\bar{y}(C_i^+) < \bar{y}(C_i^-)$ (resp. with inverted inequalities), we need $\epsilon \leq (y(C_i^+) - y(C_i^-))/(\bar{y}(C_i^-) - \bar{y}(C_i^+))$ where the bound is the point where the terms equal. This is for all $i \in V$ with $(y(C_i^+) - y(C_i^-))(\bar{y}(C_i^-) - \bar{y}(C_i^+)) > 0$.

Using the conditions on $\bar{y}$ determined by Theorem 2, it can be shown that there always exists $\epsilon > 0$ satisfying these bounds.

### 4.2 Algorithm Overview

Let us summarize the algorithm. We start with $y = 0$ (which is dual-feasible) and repeat the following iteration: From the current $y$, construct system (12). Apply rules in Table 1 to fix values of undecided variables. During that, construct the DAG defining each $y^i$ until no rule is applicable or contradiction is detected. If no contradiction is detected, stop. If contradiction is detected, compute $\bar{y}$ from the DAG, similarly as in Sect. 2.2. Calculate step size $\epsilon$ as in Sect. 4.1 and update $y := y + \epsilon\bar{y}$.

To speed up the algorithm and facilitate convergence[9], we redefine sets (13) up to a tolerance $\delta > 0$, by replacing $y_c > 0$ with $y_c > \delta$, $y(C_i^+) < y(C_i^-)$ with $y(C_i^+) + \delta < y(C_i^-)$, etc. We start with some large value of $\delta$. When the algorithm achieves a fixed point, we keep the current $y$ and set $\delta := \delta/10$. We continue until $\delta$ is not very small ($10^{-6}$).

All data structures used by the algorithm need space that is linear in the input size, i.e., in the number $\sum_{c \in C} |V_c|$ of non-zeros in linear program (9). In particular, it can be shown that the DAG (used to calculate $\bar{y}$) can be conveniently stored as a subgraph of the bipartite clause-variable incidence graph.

### 4.3 Results

We compared the upper bound on the optimal value of (9) obtained by our algorithm with the exact optimal value of (9) obtained by an off-the-shelf LP solver (we used Gurobi with default parameters) on the Max-SAT Evaluations 2018 benchmark [1]. This benchmark contains 2591 instances of Weighted Max-SAT. Gurobi was able to optimize (without memory overflow) the smallest 2100 instances, the largest of which had up to 600 thousand clauses, 300 thousand variables and 1.6 million non-zeros. The largest instances in the benchmark have up to 27 million clauses, 19 million variables and 77 million non-zeros and was still manageable by our algorithm.

From the smallest 2100 instances, 154 instances were Max-2SAT and 91 instances did not contain any unit clause. As discussed in Remark 2, the algorithm attained the exact optimum of the LP on instances of Max-2SAT. Similarly, if an instance does not contain any unit clause, then setting $x_i = \frac{1}{2}$ for all

---

[9] Though we do not present any convergence analysis of our method, it is known that the VAC / Augmenting DAG algorithm with $\delta = 0$ can converge to a point that does not satisfy virtual arc consistency [3,12,26].

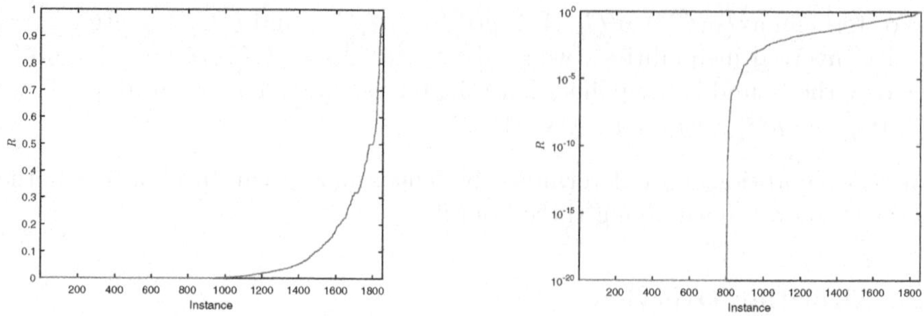


**Fig. 2.** Sorted values of $R$ with linear (left) and logarithmic (right) scale.

$i \in V$ yields an optimal solution of (9) with objective value $w(C)$. The algorithm also attains optimality on these instances because $y = 0$ is already optimal for the dual. These instances are excluded from the evaluation.

Each of the remaining 1855 instances contains a clause of length at least 3 and also contain a unit clause, thus the bound is not guaranteed to be optimal. We measure the quality of the bound by the criterion $R = (U - U^*)/(w(C) - U^*)$ where $U^*$ is the globally optimal value of (9) and $U$ is the upper bound computed by our algorithm. This criterion is invariant to scaling the weights and shows how tight the bound is relative to the trivial bound $w(C)$.

The sorted numbers $R$ for the selected 1855 instances are plotted in Fig. 2. For 802 instances the bound was tight ($U = U^*$). Due to this, the vertical (logarithmic) axis in the right-hand plot is trimmed, starting from $10^{-20}$. The left-hand plot shows that the obtained upper bound is informative in at least 1000–1100 cases. In fact, $R$ was higher than 0.6 only in 35 instances.

We computed also another criterion $Q = (U - U^*)/U^*$, which was lower than $10^{-6}$ (resp. $10^{-8}$) on 1644 (resp. 1308) from the 1855 instances. Overall, $Q$ was always lower than 0.029.

For 152 out of the 2100 considered instances, the integrality gap of the LP relaxation is known to be tight. In 133 of them, our algorithm attained this optimum. Only 2 of these were Max-2SAT and each contained a unit clause, so optimality was not guaranteed trivially.

An unoptimized implementation of our algorithm was on average 3.3 times faster than Gurobi. We believe a significant speed-up could be achieved by warm-starting. The part of the DAG needed to explain the found contradiction (see Sect. 2.2) is usually very small. If the DAG is built in every iteration from scratch, most of it is therefore thrown away. Since the system (12) changes only slightly between consecutive updates, it makes sense to re-use a part of the DAG in the next iteration. Such warm-starting was presented for the VAC algorithm in [16] and for the Augmenting DAG algorithm in [26] with significant speed-ups.

### 4.4 Tightness of the Bound on Tractable Classes

We show that the constraint propagation in system (12) is refutation-complete for tractable subclasses of Weighted Max-SAT that either use tractable clause types (language) or have acyclic structure (clause-variable incidence graph). For these instances, integrality gap of the LP relaxation (9) is zero and all fixed points of our algorithm are the optima of the unrelaxed problem.

It was shown in [9, Theorem 1] that a subclass of generalized Max-SAT (i.e., Max-CSP with Boolean variables) defined by restricting constraint types (language) is tractable if and only if one of the following holds:

- All constraints are 0-valid (resp. all are 1-valid). In this case, if the constraints are given as clauses (i.e., we restrict ourselves to the ordinary Weighted Max-SAT), the optimal value is $w(C)$, which coincides with the optimum of the LP and our algorithm attains this optimum already at $y = 0$.
- All constraints are 2-monotone. Again, restricting these constraints to clauses results in clauses with at most two literals where at most one of them is positive (resp. negative). In this case, Max-SAT can be reduced to minimum $st$-cut problem [9, Lemma 3] and the optimum of its LP formulation equals (up to a trivial recalculation) the optimum of the LP relaxation of Max-SAT which is thus tight. Since this is an instance of Weighted Max-2SAT, by Remark 2 all fixed points of our algorithm are the optima of the LP.

If we view (12) as the LP relaxation of a CSP with Boolean variables, then the propagation rules in Table 1 enforce arc consistency of this CSP. If the factor graph of this CSP is acyclic, arc consistency solves this CSP exactly [5, Theorem 1]. Hence, if the clause-variable incidence graph is acyclic, our constraint propagation rules are refutation-complete and the fixed points of our algorithm are optimal. Additionally, if no contradiction is detected, an integral solution to the left-hand system (12) can be constructed, so the integrality gap is zero.

## 5 Conclusion

We have proposed a technique to compute, with small space complexity, bounds on certain large sparse linear programs with suitable structure. Having a feasible dual solution, infeasibility of the complementary slackness conditions (a system of linear inequalities) is detected by constraint propagation and the infeasibility certificate is recovered, providing a dual improving direction. This technique can be seen as a generalization of the VAC algorithm [3] for WCSP. We have newly applied it to the LP relaxation of the Weighted Max-SAT.

The main purpose of soft local consistencies in WCSP, such as FDAC, EDAC, VAC and OSAC [3], is to bound the optimal value of WCSP during search. Each local consistency has a different trade-off point between bound tightness and computational complexity. In this view, our approach can be seen as a soft local consistency technique for other problems than WCSP. It is open whether the trade-off point of our method for Max-SAT will allow designing better algorithms to compute exact or approximate solutions of the unrelaxed Max-SAT problem.

Though in principle our approach can also be applied to other LPs (if an initial dual-feasible solution is available), the existence of good propagation rules and the quality of obtained bounds critically depends on the problem structure in a so-far unknown way. It is yet to be seen if there are other such 'propagation friendly' classes of LPs beyond the LP relaxation of WCSP and Max-SAT.

In comparison with the first-order optimization methods (such as subgradient methods or ADMM, see [8]), our approach may have the advantage that it reconverges faster after a small change of the problem instance. Though this claim would need more experimental support, evidence can be found in [15] for the VAC algorithm and in [11] for the Augmenting DAG algorithm.

VAC in WCSP is closely related to convergent message-passing methods developed within computer vision, such as [6,10,13,25]. Their fixed points are also characterized by a local consistency (usually AC in disguise) and they can be seen as versions of block-coordinate descent applied to the dual LP relaxation. This suggests there is a close connection between our approach and block-coordinate descent methods. We clarify this connection in [4].

## References

1. Bacchus, F., Järvisalo, M., Martins, R.: MaxSAT Evaluation 2018: new developments and detailed results. J. Satisfiability Boolean Model. Comput. **11**(1), 99–131 (2019). Instances https://maxsat-evaluations.github.io/
2. Benhamou, F., Granvilliers, L.: Continuous and interval constraints. In: Handbook of Constraint Programming, chap. 16. Elsevier (2006)
3. Cooper, M.C., de Givry, S., Sanchez, M., Schiex, T., Zytnicki, M., Werner, T.: Soft arc consistency revisited. Artif. Intell. **174**(7–8), 449–478 (2010)
4. Dlask, T., Werner, T.: On relation between constraint propagation and block-coordinate descent in linear programs. In: Simonis, H. (ed.) International Conference on Principles and Practice of Constraint Programming. LNCS, vol. 12333, pp. 194–210. Springer, Heidelberg (2020). https://doi.org/10.1007/978-3-030-58475-7_12
5. Freuder, E.C.: A sufficient condition for backtrack-free search. J. ACM (JACM) **29**(1), 24–32 (1982)
6. Globerson, A., Jaakkola, T.S.: Fixing max-product: convergent message passing algorithms for MAP LP-relaxations. In: Advances in Neural Information Processing Systems, pp. 553–560 (2008)
7. Hooker, J.: Logic-Based Methods for Optimization: Combining Optimization and Constraint Satisfaction. Wiley Series in Discrete Mathematics and Optimization, Wiley (2000)
8. Kappes, J.H., et al.: A comparative study of modern inference techniques for structured discrete energy minimization problems. Intl. J. Comput. Vis. **115**(2), 155–184 (2015)
9. Khanna, S., Sudan, M.: The optimization complexity of constraint satisfaction problems. In: Electronic Colloquium on Computational Complexity. Citeseer (1996)
10. Kolmogorov, V.: Convergent tree-reweighted message passing for energy minimization. IEEE Trans. Pattern Anal. Mach. Intell. **28**(10), 1568–1583 (2006)

11. Komodakis, N., Paragios, N.: Beyond loose LP-relaxations: optimizing MRFs by repairing cycles. In: Forsyth, D., Torr, P., Zisserman, A. (eds.) ECCV 2008. LNCS, vol. 5304, pp. 806–820. Springer, Heidelberg (2008). https://doi.org/10.1007/978-3-540-88690-7_60
12. Koval, V.K., Schlesinger, M.I.: Dvumernoe programmirovanie v zadachakh analiza izobrazheniy (Two-dimensional programming in image analysis problems). Autom. Telemech. **8**, 149–168 (1976). in Russian
13. Kovalevsky, V., Koval, V.: A diffusion algorithm for decreasing energy of max-sum labeling problem. Glushkov Institute of Cybernetics, Kiev, USSR (1975, unpublished)
14. Matoušek, J., Gärtner, B.: Understanding and Using Linear Programming. Universitext. Springer, Heidelberg (2006). https://doi.org/10.1007/978-3-540-30717-4
15. Nguyen, H., de Givry, S., Schiex, T., Bessiere, C.: Maintaining virtual arc consistency dynamically during search. In: International Conference on Tools with Artificial Intelligence (ICTAI), pp. 8–15. IEEE Computer Society (2014)
16. Nguyen, H., Schiex, T., Bessiere, C.: Dynamic virtual arc consistency. In: The 28th Annual ACM Symposium on Applied Computing, pp. 98–103 (2013)
17. Papadimitriou, C.H., Steiglitz, K.: Combinatorial Optimization: Algorithms and Complexity. Courier Corporation (1998)
18. Pardalos, P.M., Vavasis, S.A.: Open questions in complexity theory for numerical optimization. Math. Program. **57**, 337–339 (1992)
19. Savchynskyy, B.: Discrete graphical models - an optimization perspective. Found. Trends Comput. Graph. Vis. **11**(3–4), 160–429 (2019)
20. Schlesinger, M.: Sintaksicheskiy analiz dvumernykh zritelnikh signalov v usloviyakh pomekh (Syntactic analysis of two-dimensional visual signals in noisy conditions). Kibernetika **4**(113–130), 2 (1976)
21. Schrijver, A.: Theory of Linear and Integer Programming. Wiley, Hoboken (1986)
22. Vazirani, V.V.: Approximation Algorithms. Springer, Heidelberg (2003). https://doi.org/10.1007/978-3-662-04565-7
23. Živný, S.: The Complexity of Valued Constraint Satisfaction Problems. Cognitive Technologies. Springer, Heidelberg (2012). https://doi.org/10.1007/978-3-642-33974-5
24. Wainwright, M.J., Jordan, M.I.: Graphical models, exponential families, and variational inference. Found. Trends Mach. Learn. **1**(1–2), 1–305 (2008)
25. Werner, T.: A linear programming approach to max-sum problem: a review. IEEE Trans. Pattern Anal. Mach. Intell. **29**(7), 1165–1179 (2007)
26. Werner, T.: On coordinate minimization of piecewise-affine functions. Technical report. CTU-CMP-2017-05, Department of Cybernetics, Faculty of Electrical Engineering, Czech Technical University in Prague (2017)

# On Relation Between Constraint Propagation and Block-Coordinate Descent in Linear Programs

Tomáš Dlask(✉) and Tomáš Werner

Faculty of Electrical Engineering, Czech Technical University in Prague, Prague, Czech Republic
{dlaskto2,werner}@fel.cvut.cz

**Abstract.** Block-coordinate descent (BCD) is a popular method in large-scale optimization. Unfortunately, its fixed points are not global optima even for convex problems. A succinct characterization of convex problems optimally solvable by BCD is unknown. Focusing on linear programs, we show that BCD fixed points are identical to fixed points of another method, which uses constraint propagation to detect infeasibility of a system of linear inequalities in a primal-dual loop (a special case of this method is the Virtual Arc Consistency algorithm by Cooper et al.). This implies that BCD fixed points are global optima iff a certain propagation rule decides feasibility of a certain class of systems of linear inequalities.

**Keywords:** Block-coordinate descent · Constraint propagation · Primal-dual method · Linear programming · Virtual Arc Consistency

## 1 Introduction

Block-coordinate descent (BCD) is a popular method in large-scale optimization which in every iteration optimizes the problem over a subset (block) of variables, keeping the remaining variables constant. Unfortunately, BCD fixed points can be arbitrarily far from global optima even for convex problems. The class of convex optimization problems for which BCD provably converges to global optima is currently quite narrow, revolving around unconstrained minimization of convex function whose non-differentiable part is separable [16].

For general (non-differentiable and/or constrained) convex problems, the set of block-optimizers in a BCD iteration can contain more than one element. It has been recently argued [21,22] that in that case, one should choose an optimizer from the relative interior of this set. BCD updates satisfying this *relative interior*

This work has been supported by the Czech Science Foundation (grant 19-09967S), the OP VVV project CZ.02.1.01/0.0/0.0/16_019/0000765, and the Grant Agency of the Czech Technical University in Prague (grant SGS19/170/OHK3/3T/13).


H. Simonis (Ed.): CP 2020, LNCS 12333, pp. 194–210, 2020.
https://doi.org/10.1007/978-3-030-58475-7_12

*rule* are not worse than any other rule to choose block-wise minimizers. Of course, this rule does not guarantee convergence to global optima.

BCD methods known as convex message passing are state-of-the art for approximately solving the dual linear programming (LP) relaxation of the MAP inference problem in graphical models [8,15,18] in computer vision and machine learning, which is equivalent to the Weighted (or Valued) CSP [17]. Examples are max-sum diffusion [11,19], TRW-S [9] or MPLP [7]. These methods comply to the relative interior rule [21] (except for MPLP) and their fixed points are characterized by local consistencies (equivalent to arc consistency) of the active tuples.

Another approach to tackle the dual LP relaxation of Weighted CSP is the Virtual Arc Consistency (VAC) algorithm [2] and the similar Augmenting DAG algorithm [10,19]. Though these are not BCD, their fixed points are also characterized by arc consistency of the active tuples. In [5] we show that this approach is related to the primal-dual method [14, §5] in linear programming and propose its generalization to any[1] linear program by replacing the arc-consistency algorithm with general constraint propagation in a system of linear inequalities.

It has been observed [4] that when BCD with the relative interior rule is applied to the dual LP relaxation of SAT, it corresponds to unit propagation. Moreover, there also exists a connection between a form of the dominating unit-clause rule and BCD with the relative interior rule applied to the dual LP relaxation of Weighted Max-SAT [4].

The above results suggest there is a close relation between BCD applied to a linear program and constraint propagation in a system of linear inequalities (and possibly equalities). In this paper we describe this relation precisely. While constraint propagation in a linear inequality system can be done in many ways, we consider the particular propagation rule that infers from a subset of inequalities that some of them are active (i.e., hold with equalities). For this rule, we show that the primal-dual approach [5] and BCD with the relative interior rule have the same fixed points. Thus, the question if a given linear program is exactly solvable by BCD can be translated to the question if feasibility of a certain system of linear inequalities is decidable by this propagation rule.

To fix notation, we consider the primal-dual pair of linear programs (LPs)

$$\max c^T x \qquad\qquad \min b^T y \tag{1a}$$

$$Ax = b \qquad\qquad y \in \mathbb{R}^m \tag{1b}$$

$$x \geq 0 \qquad\qquad A^T y \geq c \tag{1c}$$

where $A \in \mathbb{R}^{m\times n}$, $b \in \mathbb{R}^m$, $c \in \mathbb{R}^n$ are constants and $x \in \mathbb{R}^n, y \in \mathbb{R}^m$ are variables. We denote by $x_i$ the $i$-th component of vector $x$ (similarly for $y, b, c$), by $A^j$ (resp. $A_i$) the $j$-th row (resp. $i$-th column) of $A$, where $i \in [n] = \{1, \dots, n\}$ and $j \in [m] = \{1, \dots, m\}$. We assume both linear programs are feasible and bounded. We assume a feasible dual solution $y$ is given, so that $b^T y$ is an upper

[1] Under the assumption that an initial dual feasible solution is provided.

bound on the joint optimal value of the pair. The goal is to improve this feasible dual solution, ideally to make it dual-optimal. We further assume a finite collection $\mathcal{B} \subseteq 2^{[m]}$ of subsets (blocks) of dual variables is given.

## 2 Block-Coordinate Descent with Relative Interior Rule

We start by describing BCD applied to the dual LP (1), taking into account the result [21,22]. For convenience, we include the dual constraints into the function $f\colon \mathbb{R}^m \to \mathbb{R} \cup \{\infty\}$ by defining

$$f(y) = \begin{cases} b^T y & \text{if } A^T y \geq c, \\ \infty & \text{otherwise.} \end{cases} \tag{2}$$

One BCD iteration improves a feasible dual solution $y$ by choosing a block $B \in \mathcal{B}$ and optimizing over variables $y_B = (y_i)_{i \in B}$, keeping the remaining variables $y_{-B} = (y_i)_{i \in [m]-B}$ constant. That is, it changes $y_B$ to satisfy

$$y_B \in \operatorname*{argmin}_{y'_B \in \mathbb{R}^B} f(y'_B, y_{-B}). \tag{3}$$

The set $\operatorname{argmin}_{y'_B \in \mathbb{R}^B} f(y'_B, y_{-B}) \subseteq \mathbb{R}^B$ of block-wise minimizers is a non-empty convex polyhedron. If this polyhedron contains more than one point, we need to choose a single element from this polyhedron. To satisfy the relative interior rule, the update must be modified to

$$y_B \in \operatorname{ri} \operatorname*{argmin}_{y'_B \in \mathbb{R}^B} f(y'_B, y_{-B}) \tag{4}$$

where $\operatorname{ri} X$ denotes the relative interior of a convex set $X$ [12, §2.1]. The following are the main results of [21,22]:

**Definition 1.** *A point $y$ feasible to the dual in* (1) *is*

- *a local minimum (LM) of $f$ w.r.t. $\mathcal{B}$ if* (3) *holds for all $B \in \mathcal{B}$,*
- *an interior local minimum (ILM) of $f$ w.r.t. $\mathcal{B}$ if* (4) *holds for all $B \in \mathcal{B}$,*
- *a pre-interior local minimum (pre-ILM) of $f$ w.r.t. $\mathcal{B}$ if there is an ILM $y'$ such that $y$ is in a face of the polyhedron $\{ y \mid A^T y \geq c \}$ containing $y'$ in its relative interior.*

**Theorem 1.** *Let $(B_i)_{i=1}^\infty$ be a sequence of blocks $B_i \in \mathcal{B}$ that contains each element of $\mathcal{B}$ an infinite number of times. Let $(y^i)_{i=1}^\infty$ be a sequence produced by the BCD method, where the blocks are visited in the order given by $(B_i)_{i=1}^\infty$.*

*A. If $(y^i)_{i=1}^\infty$ satisfies* (4) *and $y^1$ is an ILM, then $y^i$ is an ILM for all $i$.*
*B. If $(y^i)_{i=1}^\infty$ satisfies* (4) *and $y^1$ is a pre-ILM, then $y^i$ is an ILM for some $i$.*
*C. If $(y^i)_{i=1}^\infty$ satisfies* (3) *and $y^1$ is a pre-ILM, then $b^T y^i = b^T y^1$ for all $i$.*
*D. If $(y^i)_{i=1}^\infty$ satisfies* (4) *and $y^1$ is not a pre-ILM, then $b^T y^i < b^T y^1$ for some $i$.*

Thus, when we are at a pre-ILM, the objective cannot be improved by any further BCD iterations. When we are not at a pre-ILM, BCD with the relative interior rule inevitably improves the objective in a finite number of iterations.

# 3 Primal-Dual Approach

Let us now focus on the second of the two approaches we consider in this paper.

By the *complementary slackness* theorem [13,14], a primal feasible solution $x$ and a dual feasible solution $y$ to (1) are optimal if and only if $x_i(A_i^T y - c) = 0$ for all $i \in [n]$. In addition, $x^*$ is in the relative interior of the primal optimizers and $y^*$ is in the relative interior of the dual optimizers if and only if they satisfy *strict complementary slackness* condition [23] $(x_i^* = 0) \oplus (A_i^T y^* = c)$ for all $i \in [n]$ where $\oplus$ denotes exclusive disjunction. If both primal and dual are feasible and bounded, there always exist such $x^*, y^*$ [12, Theorem 2.1.3].

The iteration of the primal-dual approach proceeds as follows. Denoting

$$K(y) = \{i \in [n] \mid A_i^T y = c_i\}, \tag{5}$$

the index set of dual constraints active at $y$, the complementary slackness condition reads

$$Ax = b \tag{6a}$$
$$x_i \geq 0 \qquad \forall i \in K(y) \tag{6b}$$
$$x_i = 0 \qquad \forall i \in [n] - K(y) \tag{6c}$$

Thus, $y$ is dual-optimal for (1) if and only if system (6) is feasible. By Farkas' lemma [13, §6] (or by LP duality), (6) is infeasible if and only if the system

$$b^T \bar{y} < 0 \tag{7a}$$
$$A_i^T \bar{y} \geq 0 \qquad \forall i \in K(y) \tag{7b}$$

is feasible. In that case, any solution $\bar{y}$ to (7) is an improving direction for the dual (1) from point $y$, i.e., there is $\epsilon > 0$ such that $b^T(y + \epsilon\bar{y}) < b^T y$ and $A^T(y + \epsilon\bar{y}) \geq c$. Updating $y \leftarrow y + \epsilon\bar{y}$ yields a better dual feasible solution.

The described approach is similar to the well-known primal-dual method [14, §5], where complementary slackness (6) is not required strictly but only its violation is minimized. The motivation for the method is that problem (6) may be easier to solve than (1), possibly by combinatorial algorithms [14, §6].

## 3.1 Constraint Propagation

Deciding feasibility of a system of linear inequalities (such as (6)) can be too costly for large instances. Therefore, we proposed in [5, §2] to do it by constraint propagation: using a small fixed set of inference (or propagation) rules, we iteratively infer new linear inequalities from the system and add them to the system. If a contradictory inequality is inferred, the initial system was infeasible; then an infeasibility certificate (such as $\bar{y}$ in (7)) is constructed from the propagation history. The drawback of this method is that it is in general refutation-incomplete: it may not infer a contradiction even if the system is infeasible.

While in [5, §2] we did not restrict the form of the used inference rules, here we consider one particular form: *choose a subset of the inequalities and infer which of them are active (i.e., hold with equality).* For the particular case of system (6), this means we choose a subset $B \in \mathcal{B}$ of equalities (6a) and decide if they, together with (6b) and (6c), imply that some of the inequalities (6b) is active, i.e., $x_i = 0$. Indeed, this can be seen as inferring the inequality $x_i \leq 0$ from the system. It is our key observation in this paper that with this propagation rule the primal-dual approach has the same fixed points as BCD (see Sect. 4).

Precisely, the algorithm first initializes $K = K(y)$ and then repeats the following iteration: choose $B \in \mathcal{B}$, find all indices $i \in K$ for which the system

$$A^j x = b_j \qquad \forall j \in B \tag{8a}$$

$$x_i \geq 0 \qquad \forall i \in K \tag{8b}$$

$$x_i = 0 \qquad \forall i \in [n] - K \tag{8c}$$

implies[2] $x_i = 0$, and remove these indices from $K$. If the set $K$ shrinks so much that system (8) becomes infeasible for some $B \in \mathcal{B}$, then clearly the original system (6) is infeasible. Next, we analyze this algorithm, showing that its properties are analogous to the well-known arc-consistency algorithm.

**Definition 2.** *For $B \subseteq [m]$, a set $K \subseteq [n]$ is $B$-consistent if for every $i \in K$ system* (8) *does not imply $x_i = 0$, i.e., if the system*

$$A^j x = b_j \qquad \forall j \in B \tag{9a}$$

$$x_i > 0 \qquad \forall i \in K \tag{9b}$$

$$x_i = 0 \qquad \forall i \in [n] - K \tag{9c}$$

*is feasible. For $\mathcal{B} \subseteq 2^{[m]}$, $K$ is $\mathcal{B}$-consistent if it is $B$-consistent for all $B \in \mathcal{B}$.*

**Proposition 1.** *If $K$ and $K'$ are $\mathcal{B}$-consistent, then $K \cup K'$ is $\mathcal{B}$-consistent.*

*Proof.* If (9) for some $B \in \mathcal{B}$ is satisfied by $x$ (resp. $x'$) for $K$ (resp. $K'$), then it is satisfied by $(x + x')/2$ for $K \cup K'$. □

By Proposition 1, the $\mathcal{B}$-consistent sets form a join-semilattice w.r.t. the inclusion. Therefore, for any $K \subseteq [n]$, either there is no $\mathcal{B}$-consistent subset of $K$ or there exists the unique maximal $\mathcal{B}$-consistent subset of $K$.

**Definition 3.** *The* propagator *over block $B \subseteq [m]$ is the map $P_B \colon 2^{[n]} \cup \{\bot\} \to 2^{[n]} \cup \{\bot\}$ defined by*[3]*:*

[2] By saying that (8) implies $x_i = 0$ we mean that $x_i = 0$ holds for all $x$ satisfying (8). This can be decided by, e.g., projecting polyhedron (8) onto the $i$-th coordinate. The projection is a singleton set $\{0\}$ if and only if (8) implies $x_i = 0$. The projection can be computed by the Fourier-Motzkin elimination or by maximizing $x_i$ subject to (8) (which equals 0 if and only if (8) implies $x_i = 0$).

[3] Note that $P_B(K) = \bot$ is different from $P_B(K) = \emptyset$, since system (8) can be feasible even for $K = \emptyset$ (if $b = 0$).

- *If* $K = \bot$, *then* $P_B(K) = \bot$.
- *If* $K \subseteq [n]$ *and* (8) *is infeasible, then* $P_B(K) = \bot$.
- *If* $K \subseteq [n]$ *and* (8) *is feasible, then* $P_B(K) \subseteq [n]$ *and* $i \in P_B(K)$ *if and only if* (8) *does not imply* $x_i = 0$.

Clearly, a set $K \subseteq [n]$ is $B$-consistent if and only if $P_B(K) = K$, and $K$ is $\mathcal{B}$-consistent if and only if $P_B(K) = K$ for all $B \in \mathcal{B}$.

**Proposition 2.** *Map* $P_B(\cdot)$ *satisfies the axioms of a closure operator unless*[4] $P_B(\cdot) = \bot$, *i.e., for all* $K, K' \subseteq [n]$ *such that* $P_B(K), P_B(K') \neq \bot$ *we have*

- $P_B(P_B(K)) = P_B(K)$ *(idempotence)*
- $P_B(K) \subseteq K$ *(intensivity)*
- $K' \subseteq K \implies P_B(K') \subseteq P_B(K)$ *(monotonicity).*

*Proof.* Idempotence and intensivity are straightforward. To prove monotonicity, let $K' \subseteq K$ and let $H'$ (resp. $H$) be the polyhedron defined by (8) for $K'$ (resp. $K$). Clearly, $\emptyset \neq H' \subseteq H$. If $i \in [n] - P_B(K)$, the projection of $H$ onto $x_i$ contains only 0. Therefore, the projection of $H'$ onto $x_i$ also contains only 0, i.e., (8) for $K'$ implies $x_i = 0$, hence $i \in [n] - P_B(K')$. Thus $P_B(K') \subseteq P_B(K)$. □

**Definition 4.** *Given* $K \in 2^{[n]} \cup \{\bot\}$, *the* propagation algorithm *repeats the following iteration: find* $B \in \mathcal{B}$ *such that* $P_B(K) \neq K$ *and set* $K \leftarrow P_B(K)$. *If no such* $B \in \mathcal{B}$ *exists, return the final* $K$.

The propagation algorithm terminates in a finite number of steps. If at any iteration we get $K = \bot$, the algorithm terminates due to $P_B(\bot) = \bot$ for all $B \in \mathcal{B}$. Otherwise, by intensivity of $P_B(\cdot)$, $K$ can decrease only a finite number of times.

**Proposition 3.** *If* $K \in [n]$ *has a* $\mathcal{B}$*-consistent subset, the propagation algorithm returns the maximal* $\mathcal{B}$*-consistent subset of* $K$.

*Proof.* The propagation algorithm creates a finite decreasing chain $K_1 \supset K_2 \supset K_3 \supset \cdots$ where $K_1 = K$ and $K_{l+1} = P_{B_l}(K_l)$ where $B_l \in \mathcal{B}$ is the block chosen in the $l$-th step. Let $L$ be arbitrary $\mathcal{B}$-consistent subset of $K$. We will prove by induction that $L \subseteq K_l$ for all $l$. Clearly, $L \subseteq K = K_1$. If $L \subseteq K_l$, then

$$L = P_{B_l}(L) \subseteq P_{B_l}(K_l) = K_{l+1} \tag{10}$$

where the first equality follows from $\mathcal{B}$-consistency of $L$ and the inclusion follows from monotonicity of $P_{B_l}(\cdot)$ by Proposition 2. See that it cannot happen that $P_{B_l}(K_l) = \bot$ for any $l$ because (8) is feasible for all $B \in \mathcal{B}$ for $L$ and $L \subseteq K_l$. □

By Proposition 3, the result of the propagation algorithm does not depend on the order in which the elements of $\mathcal{B}$ are visited. Thus, we can introduce the operator $P_{\mathcal{B}}$: $2^{[n]} \cup \{\bot\} \to 2^{[n]} \cup \{\bot\}$ where $P_{\mathcal{B}}(K)$ is the unique result of the algorithm with input $K$.

[4] The exception of $\bot$ could be removed by augmenting the set $2^{[n]}$, partially ordered by set inclusion, with $\bot$ as its least element.

**Proposition 4.** *The operator $P_{\mathcal{B}}(\cdot)$ satisfies the axioms of a closure operator unless $P_{\mathcal{B}}(\cdot) = \bot$.*

*Proof.* Idempotence and intensivity follow directly from the proof of Proposition 3. We will prove monotonicity by contradiction: let $K' \subseteq K$ and $P_{\mathcal{B}}(K) \subset P_{\mathcal{B}}(K')$. By intensivity, we obtain $P_{\mathcal{B}}(K') \subseteq K' \subseteq K$. However, $P_{\mathcal{B}}(K)$ is $\mathcal{B}$-consistent and $P_{\mathcal{B}}(K') \subseteq K$. Since $P_{\mathcal{B}}(K) \subset P_{\mathcal{B}}(K')$, $P_{\mathcal{B}}(K)$ is not the maximal $\mathcal{B}$-consistent subset of $K$, which is contradictory with Proposition 3. □

Due to Proposition 4, $P_{\mathcal{B}}(K)$ can be called the *$\mathcal{B}$-consistency closure* of $K$. Observe that the properties of $P_B(\cdot)$ and $P_{\mathcal{B}}(\cdot)$ are analogous to the properties of the arc-consistency propagator and arc-consistency closure, respectively. In more general view, the propagator resembles domain-based constraint propagation [1], where stability under union corresponds to the property given by Proposition 1 and $\Phi$-closure corresponds to $\mathcal{B}$-consistent closure.

If $P_{\mathcal{B}}(K(y)) = \bot$, then system (6) is infeasible. Then there exists an improving direction $\bar{y}$ satisfying (7). Such an improving direction can be constructed from the history of the propagation, as we describe in Appendix B. But note that improving directions are not necessary for our analysis in this paper as we only consider the fixed points of the primal-dual approach.

As propagation is not refutation-complete, $P_{\mathcal{B}}(K(y)) \neq \bot$ does not in general imply that (6) is feasible. Consequently, $b^T y$ is not the optimal value of the pair (1) but only its upper bound.

The primal-dual approach with the described propagation is used, under various names, in several existing methods. One example is the VAC algorithm [2] and the Augmenting DAG algorithm [10,19], where the primal problem (1) is the basic LP relaxation of the Weighted CSP and our propagation is equivalent to the arc-consistency algorithm [5]. An approach proposed in [5] to upper-bound the LP relaxation of Weighted Max-SAT is (up to technical details) another example. If the minimization of a convex piecewise-affine function is expressed as an LP, then our method subsumes the sign relaxation technique introduced in [20] and further developed in [3].

## 4 Relation Between the Approaches

We now state the relation between BCD with the relative interior rule (Sect. 2) and the primal-dual approach in which system (6) is solved by constraint propagation as described in Sect. 3.1. The proof of the theorem is in Appendix A.

**Theorem 2.** *Let $y$ be a feasible point for dual* (1). *Then:*

- *$y$ is an LM of dual* (1) *w.r.t. $\mathcal{B}$ if and only if $P_B(K(y)) \neq \bot$ for all $B \in \mathcal{B}$,*
- *$y$ is an ILM of dual* (1) *w.r.t. $\mathcal{B}$ if and only if $P_{\mathcal{B}}(K(y)) = K(y)$,*
- *$y$ is a pre-ILM of dual* (1) *w.r.t. $\mathcal{B}$ if and only if $P_{\mathcal{B}}(K(y)) \neq \bot$.*

Theorem 2 characterizes the previously introduced types of local minima in BCD by local consistency conditions. It also shows that BCD with relative interior rule cannot improve the fixed points of the primal-dual approach based on propagation and vice versa. This yields the following corollary.

**Corollary 1.** *The following statements are equivalent:*

- *For all feasible $y$ for the dual* (1), *if* (6) *is infeasible then* $P_{\mathcal{B}}(K(y)) = \bot$ *(i.e., propagation is refutation-complete).*
- *Any ILM $y$ of the dual* (1) *w.r.t.* $\mathcal{B}$ *is a global optimum.*

This result shows that the question whether BCD fixed points are global minima for a given class of LPs can be reformulated as the question whether constraint propagation decides feasibility of a certain class of linear inequalities.

## 5 Other Forms of Linear Programs

It is well-known that linear programs come in different forms [14, §2.1] which can be easily transformed to each other, preserving global optima. One can ask if the propagation algorithm can be formulated and the equivalence with BCD holds also for different forms than (1). This question is non-trivial because transformations that preserve global optima do not necessarily preserve (pre-)interior local optima [6]. We show that independently of the formulation, if we use the propagation rule that infers activity of inequality constraints (as we mentioned in the beginning of Sect. 3.1), the two approaches remain equivalent.

### 5.1 Primal LP with Inequalities and Non-negative Variables

Consider for example the primal-dual pair

$$\max c^T x \qquad\qquad \min b^T y \tag{11a}$$
$$Ax \le b \qquad\qquad y \ge 0 \tag{11b}$$
$$x \ge 0 \qquad\qquad A^T y \ge c \tag{11c}$$

that can be equivalently reformulated [13] by introducing slack variables $s_j \ge 0$, $j \in [m]$ as

$$\max c^T x \qquad\qquad \min b^T y \tag{12a}$$
$$Ax + s = b \qquad\qquad y \in \mathbb{R}^m \tag{12b}$$
$$x \ge 0 \qquad\qquad A^T y \ge c \tag{12c}$$
$$s \ge 0 \qquad\qquad y \ge 0 \tag{12d}$$

which is in the form (1). See that BCD in the duals (11) and (12) is identical.

The propagation rules presented previously in Sect. 3.1 for the LP (12) correspond to deciding which $s_j$ and $x_i$ are forced to be zero. Clearly, setting $s_j = 0$

corresponds to setting $A^j x = b_j$ and enforcing $s_j > 0$ implies $A^j x < b_j$. Thus, instead of rewriting (11) into (12), we can apply propagation directly on the primal (11) except that when considering the system (9) for some $B \in \mathcal{B}$, we will instead of a single set $K$ use two sets $K^X \subseteq [n]$ and $K^S \subseteq [m]$ that indicate which inequalities need to be satisfied strictly and which with equality, i.e., we will use

$$A^j x < b_j \quad \forall j \in K^S \cap B \qquad\qquad x_i > 0 \quad \forall i \in K^X \tag{13a}$$

$$A^j x = b_j \quad \forall j \in ([m] - K^S) \cap B \qquad\qquad x_i = 0 \quad \forall i \in [n] - K^X \tag{13b}$$

instead of (9). Clearly, deciding which inequalities among $Ax \le b$ in primal (11) need to be satisfied with strict inequality (resp. with equality) by considering a set $K^S \subseteq [m]$ is in one-to-one correspondence with deciding which slack variables $s_j$ in (12) can be non-zero (resp. are forced to be zero).

### 5.2 Primal LP with Inequalities and Unconstrained Variables

Another general primal-dual pair that we are going to consider is

$$\max c^T x \qquad\qquad \min b^T y \tag{14a}$$

$$Ax \le b \qquad\qquad y \ge 0 \tag{14b}$$

$$x \in \mathbb{R}^n \qquad\qquad A^T y = c \tag{14c}$$

where $y$ is optimal for the dual if and only if there exists $x \in \mathbb{R}^n$ such that

$$A^j x \le b_j \qquad\qquad \forall j \in K'(y) \tag{15a}$$

$$A^j x = b_j \qquad\qquad \forall j \in [m] - K'(y) \tag{15b}$$

where $K'(y) = \{j \in [m] \mid y_j = 0\}$ and (15) again follows from complementary slackness. From this point, we could completely repeat the reasoning in Sect. 3.1 and prove the same theorem as in Sect. 4 except that we would replace $K(y)$ by $K'(y)$, replace condition $A_i^T y > c$ (resp. $A_i^T y = c$) by $y_j > 0$ (resp. $y_j = 0$) and infer whether the inequality $A^j x \le b_j$ should hold strictly or with equality instead of inferring it for $x_i \ge 0$. This is based on similarity between (15) and (6).

### 5.3 Redundant Constraints

It was observed in [6] that adding redundant constraints into an LP has significant influence on its solvability by (block-)coordinate descent. Using our results from this paper, we are able to explain this quite naturally.

As an example, consider the following LP relaxation of weighted vertex cover on a graph $(V, E)$ with vertex weights $w : V \to \mathbb{R}^+$ together with its dual

$$\min w^T x \qquad\qquad \max \textstyle\sum_{\{i,j\} \in E} y_{ij} \tag{16a}$$

$$x_i + x_j \ge 1 \qquad\qquad y_{ij} \ge 0 \qquad\qquad \forall \{i,j\} \in E \tag{16b}$$

$$x_i \ge 0 \qquad\qquad \textstyle\sum_{j \in N_i} y_{ij} \le w_i \qquad\qquad \forall i \in V \tag{16c}$$

where $N_i$ is the set of neighbors of vertex $i$ in the graph. If we optimized the primal or the dual (16) by coordinate descent along individual variables (i.e., blocks of size 1), there are interior local optima[5] that are not global optima [6]. However, if we add redundant constraints $x \le 1$ to the primal, we obtain

$$\begin{aligned}
\min w^T x \qquad & \max \textstyle\sum_{\{i,j\}\in E} y_{ij} + \sum_{i\in V} z_i && && (17a)\\
x_i + x_j \ge 1 \qquad & y_{ij} \ge 0 && \forall \{i,j\} \in E && (17b)\\
x_i \ge 0 \qquad & z_i + \textstyle\sum_{j\in N_i} y_{ij} \le w_i && \forall i \in V && (17c)\\
x_i \le 1 \qquad & z_i \le 0 && \forall i \in V. && (17d)
\end{aligned}$$

By the result in [6], any interior local optimum of dual (17) w.r.t. blocks[6] consisting of variables $y_{ij}$, $z_i$, and $z_j$ for each $\{i,j\} \in E$ is a global optimum.

The explanation for the difference between (non-)optimality for the different formulations lies in the fact that in case of (16), we can only propagate equality in constraints (16b) and $x_i = 0$. However, in (17), we are also able to propagate $x_i = 1$ due to the added constraint $x_i \le 1$. This results in a stronger propagation algorithm which is even refutation-complete for this case.

This also holds for the LP formulation of min-$st$-cut and its dual, maximum flow, which was also considered in [6, §4.3]. Adding redundant bounds $0 \le x_i \le 1$ for variables in min-$st$-cut results in optimality of BCD on its dual. However, the dual of the usual formulation of min-$st$-cut (i.e., without these bounds) is not amenable to BCD [6, §4.3]. This difference is now explained by the possibility of the underlying propagation algorithm to set these variables to their bounds, i.e., set $x_i = 0$ or $x_i = 1$ which is not possible if variables $x$ are unbounded.

The result in this paper therefore also sheds light on which constraints are useful in terms of propagation or BCD even though they are redundant from the point of global optimality of the original linear program.

## 6 Conclusion

Even though propagation in a system of linear inequalities can be performed in many ways, we have defined a propagation algorithm which not only has natural and useful properties, but it also allows full characterization of types of local minima in BCD. Additionally, there is a tight connection between the fixed points of BCD with relative interior rule and the fixed points of primal-dual approach based on this propagation algorithm. Despite the fact that both

[5] In case of the dual, we maximize, so we should talk about interior local maxima, but this relation is straightforward by inverting the sign in the criterion and changing maximization to minimization.

[6] In analogy with [6, §3 equation (7)], $z_i = \min\{w_i - \sum_{j\in N_i} y_{ij}, 0\}$ $\forall i \in V$ holds in any optimal solution of dual (17) and so the dual can be equivalently reformulated as maximization of a concave piecewise-affine function with non-negative variables, which makes optimization along these blocks simpler. In detail, variables $z$ were eliminated and thus we update only each $y_{ij}$ separately.

algorithms may not reach a global optimum, none of the algorithms can improve the fixed points of the other.

We argued that the propagation algorithm can be generalized to linear programs in any form. In detail, BCD in the dual for a given set of blocks $\mathcal{B}$ corresponds to propagating which primal constraints given by complementary slackness should be active and which inactive while inferring only from subsets of the constraints given by sets in $\mathcal{B}$.

We believe that our findings are interesting for the theory of BCD as they explain what kind of local consistency is reached by any BCD algorithm (both with or without relative interior rule) on any LP. E.g., As shown in [21], since both TRW-S [9] and max-sum diffusion [11,19] satisfy the relative interior rule, their fixed point conditions are equivalent to the proposed local consistency condition if applied to the specific LP formulations which these algorithms optimize.

This tight connection between the decidability of feasibility of a system of linear inequalities by refutation-incomplete propagation and BCD may provide theoretical ground for analysis of BCD in terms of constraint propagation. Moreover, it may result in newly discovered classes of problems optimally solvable by BCD or better design for choices of blocks of variables so that the propagation is more effective and BCD may reach better local optima. This connection also precisely explains the differences in applicability of BCD caused by minor changes in the formulation of the optimized LP, as discussed in Sect. 5.3.

The practical impact of these results is mainly focused on approximately optimizing challenging large-scale LPs which are not solvable by off-the-shelf LP solvers due to their super-linear space complexity. Propagation algorithms subsumed (up to technical details) by the proposed one were previously derived ad-hoc for specific LPs [2,3,5,10,20] where they provided useful solutions which were often close to global optima. Presenting all of these algorithms in a single framework may simplify design of similar algorithms in the future.

## A Proofs

**Proposition 5.** *Let $y$ be feasible for the dual* (1) *and let $B \subseteq [m]$. Block of variables $y_B$ satisfies* (4) *if and only if $P_B(K(y)) = K(y)$.*

*Proof.* For the 'only-if' direction, construct the dual (1) restricted only to the variables $y_B$ as follows:

$$\max k^T x \qquad \min \textstyle\sum_{j\in B} b_j y_j \tag{18a}$$

$$A^j x = b_j \qquad y_j \in \mathbb{R} \qquad \forall j \in B \tag{18b}$$

$$x_i \geq 0 \qquad \textstyle\sum_{j\in B} A_{ji} y_j \geq k_i \qquad \forall i \in [n] \tag{18c}$$

where $k_i = c_i - \sum_{j\in[m]-B} A_{ji} y_j$ are viewed as constants determined by the remaining variables that are not in the block and $A_{ji}$ is the entry of matrix $A$ on $j$-th row and $i$-th column. The problem on left is the corresponding primal.

Since $y_B$ is in the relative interior of optimizers of the dual (18) by our assumption, there must exist a solution $x \in \mathbb{R}^n_+$ for the primal (18) such that strict complementary slackness holds. The condition for this case reads

$$\textstyle\sum_{j \in B} A_{ji} y_j = k_i \iff x_i > 0 \qquad \forall i \in [n], \tag{19}$$

therefore $x$ satisfies $x_i = 0\ \forall i \in [n] - K(y)$ and $x_i > 0\ \forall i \in K(y)$ by definition of $K(y)$ and $k_i$. By feasibility of $x$ for primal (18), we have that $i \in P_B(K(y))$ for all $i \in K(y)$. By intensivity of $P_B(\cdot)$, we obtain $P_B(K(y)) = K(y)$.

For the 'if' direction, assume $P_B(K(y)) = K(y)$, then there must exist a solution $x \in \mathbb{R}^n$ for (9) where $K = K(y)$. This vector $x$ is a feasible solution for the primal (18). By definition of $K(y)$ in (5) and definition of $k_i$, it follows that strict complementary slackness (19) is satisfied in (18), therefore both $x$ and $y_B$ lie in the relative interior of optimizers of the primal-dual pair (18). □

**Corollary 2.** *Let $y$ be feasible for dual* (1). *Then $y$ is an ILM of dual* (1) *w.r.t. $\mathcal{B}$ if and only if $P_{\mathcal{B}}(K(y)) = K(y)$.*

*Proof.* By definition, $y$ is an ILM of dual (1) w.r.t. $\mathcal{B}$ if (4) holds $\forall B \in \mathcal{B}$. Applying Proposition 5, this is equivalent to $P_B(K(y)) = K(y)\ \forall B \in \mathcal{B}$, i.e., $P_{\mathcal{B}}(K(y)) = K(y)$. □

**Proposition 6.** *Let $y$ be feasible for the dual* (1) *and let $B \subseteq [m]$. Block of variables $y_B$ satisfies* (3) *if and only if $P_B(K(y)) \neq \bot$.*

*Proof.* Block $y_B$ satisfies (3) if and only if it is optimal for the dual (18), which happens if and only if there exists $x \in \mathbb{R}^n_+$ satisfying complementary slackness. By definition of $K(y)$, complementary slackness conditions are equivalent to (8) for $K = K(y)$ which is feasible if and only if $P_B(K(y)) \neq \bot$. □

**Proposition 7.** *If point $x$ is in the relative interior of optimizers of the primal* (1), *then the set $\{i \in [n] \mid x_i = 0\}$ is minimal w.r.t. inclusion among all optimal solutions and is unique.*

*Proof.* By contradiction: let $x$ (resp. $y$) be from the relative interior of optimizers of primal (resp. dual) (1). Let $x'$ be also optimal for the primal and let $\{i \in [n] \mid x'_i = 0\}$ be smaller and/or different. Then, there is $k \in [n]$ such that $x'_k > 0$ and $x_k = 0$. Since $x$ and $y$ are in the relative interior of optimizers, they satisfy strict complementary slackness, thus $A_k^T y > c_k$. Complementary slackness is satisfied by all pairs of primal and dual optimal solutions, but $x'$ and $y$ do not satisfy it because $x'_k > 0$ and $A_k^T y > c_k$, hence $x'$ is not optimal. □

**Proposition 8.** *Let $y$ be a feasible point for dual* (1) *and let $B \subseteq [m]$ so that $P_B(K(y)) = K \neq \bot$. Then, there exists a feasible point $y'$ such that $b^T y = b^T y'$ and $P_B(K(y')) = K(y') = K$.*

*Proof.* Consider the primal-dual pair

$$
\begin{array}{rrll}
\max 0 & \min b^T \bar{y} & & \text{(20a)}\\
A^j x = b_j & \bar{y}_j \in \mathbb{R} & \forall j \in B & \text{(20b)}\\
x_i = 0 & - & \forall i \in [n] - K(y) & \text{(20c)}\\
x_i \geq 0 & A_i^T \bar{y} \geq 0 & \forall i \in K(y) & \text{(20d)}\\
- & \bar{y}_j = 0 & \forall j \in [m] - B. & \text{(20e)}
\end{array}
$$

which was simplified in the sense that if some primal (resp. dual) variable equals zero, then we can omit the corresponding dual (resp. primal) constraint without changing the problem since the whole column (resp. row) of $A$ can be set to zero.

Let $x$ (resp. $\bar{y}$) be in the relative interior of optimizers for the primal (resp. dual) (20). By Proposition 7 applied on matrix $A$ with only rows in $B$ and only columns in $K(y)$, if some $x_i = 0$, then this is the only value $x_i$ can take, therefore $i \notin K$ because primal (20) is (8) for $K(y)$. If some variable $x_i$ can take a non-zero value in (20), then it is non-zero again by Proposition 7 and $i \in K$.

Since the pair of optimal solutions $x, \bar{y}$ lies in the relative interior of optimizers, they satisfy strict complementary slackness in this form:

$$x_i = 0 \wedge A_i^T \bar{y} > 0 \quad \forall i \in K(y) - K \tag{21a}$$

$$x_i > 0 \wedge A_i^T \bar{y} = 0 \quad \forall i \in K. \tag{21b}$$

We will now choose any $\epsilon$ such that

$$0 < \epsilon < \frac{c_i - A_i^T y}{A_i^T \bar{y}} \qquad \forall i \in [n] - K(y) \text{ such that } A_i^T \bar{y} < 0 \tag{22}$$

where the upper bound is positive because $A_i^T \bar{y} < 0$ by the condition in the upper bound and for all $i \in [n] - K(y)$, $c_i - A_i^T y < 0$ by feasibility of $y$ and definition of $K(y)$. Therefore, $(c_i - A_i^T y)/(A_i^T \bar{y})$ is positive for all $i$ considered in (22) and there exists some $\epsilon$ satisfying (22). We choose any $\epsilon$ satisfying (22) and claim that $y' = y + \epsilon \bar{y}$ satisfies the required conditions.

- If $i \in K(y) - K$, then $A_i^T \bar{y} > 0$ by (21) and $A_i^T y = c_i$ by definition of $K(y)$. Therefore, $A_i^T y' = c_i + \epsilon A_i^T \bar{y} > c_i$, so $i \notin K(y')$, i.e., $i \in [n] - K(y')$.
- If $i \in K$, then $A_i^T y = c_i$ because $i \in K = P_B(K(y)) \subseteq K(y)$ and $A_i^T \bar{y} = 0$ by (21). Therefore, $A_i^T y' = c_i + \epsilon \cdot 0 = c_i$ and $i \in K(y')$.
- If $i \in [n] - K(y)$, then $A_i^T y > c_i$ by definition of $K(y)$ and $A_i^T \bar{y}$ can have any sign. We distinguish the following cases:

• If $A_i^T \bar{y} \geq 0$, then $A_i^T y' > c_i + \epsilon A_i^T \bar{y} \geq c_i$, so $i \notin K(y')$, i.e., $i \in [n] - K(y')$.
• If $A_i^T \bar{y} < 0$, then by definition of $\epsilon$, $\epsilon < (c_i - A_i^T y)/(A_i^T \bar{y})$ which implies $A_i^T y' = A_i^T y + \epsilon A_i^T \bar{y} > c_i$, hence $i \notin K(y')$, i.e., $i \in [n] - K(y')$.

Therefore, $[n] - K(y) \subseteq [n] - K(y')$ and $K(y) - K \subseteq [n] - K(y')$, which results in $[n] - K \subseteq [n] - K(y')$. This together with $K \subseteq K(y')$ yields $K(y') = K$.

Point $y'$ is feasible since $A_i^T y' \geq c_i$ for all $i \in [n]$ as just shown above. Because the optimal value of the primal (20) is 0 and $\bar{y}$ is an optimal dual solution, it follows from strong duality that $b^T \bar{y} = 0$, therefore $b^T y' = b^T y + \epsilon b^T \bar{y} = b^T y$. By idempotency of $P_B(\cdot)$, it follows that $P_B(K) = K$, i.e., $P_B(K(y')) = K(y')$. □

*Remark 1.* The point $y'$ constructed in Proposition 8 can in fact be obtained by updating block $y_B$ to satisfy (4). By construction of $y'$ from $y$ derived in Proposition 8, $y_j = y'_j \; \forall j \in [m] - B$, therefore only the variables in block $B$ change. Combining Proposition 5 with $P_B(K(y')) = K(y')$ implies that block $y'_B$ is in the relative interior of optimizers of the dual (1) restricted to this block.

**Proposition 9.** *Let $y$ be a feasible point for dual* (1) *such that $P_{\mathcal{B}}(K(y)) \neq \bot$, then $y$ is a pre-ILM of dual* (1) *w.r.t. $\mathcal{B}$.*

*Proof.* By the definition of $P_{\mathcal{B}}(\cdot)$, there must exist a finite sequence $(B_l)_{l=1}^{L}$ for $B_l \in \mathcal{B}, l \in [L]$ such that

$$P_{B_L}(P_{B_{L-1}}(P_{B_{L-2}}(\cdots P_{B_2}(P_{B_1}(K(y)))\cdots))) = K \tag{23}$$

and $P_{\mathcal{B}}(K) = K$. In other words, the sequence corresponds to the order of the blocks $B$ applied in the propagation algorithm until a fixed point is reached.

We construct a sequence $(y^l)_{l=1}^{L+1}$, $y^l \in \mathbb{R}^m$ where $y^1 = y$ and $y^{l+1}$ is constructed from $y^l$ as in the proof of Proposition 8 for $B = B_l$. By induction and properties of the construction, since $y^1$ is feasible, the other points $y^2, y^3, ..., y^{L+1}$ are also feasible. Also, $b^T y^1 = b^T y^2 = ... = b^T y^{L+1}$ because the construction preserves objective. Finally, $P_{B_l}(K(y^l)) = K(y^{l+1})$ for all $l \in [L]$, therefore $P_{B_L}(K(y^L)) = K(y^{L+1}) = K$ and $P_{\mathcal{B}}(K) = K$, so $P_{\mathcal{B}}(K(y^{L+1}) = K(y^{L+1})$. By Corollary 2, $y^{L+1}$ is an ILM w.r.t. $\mathcal{B}$.

By Remark 1, the sequence $y^1, ..., y^{L+1}$ can be obtained by updating the corresponding blocks into the relative interior of optimizers. Because the objective did not improve during these updates and $y^{L+1}$ is ILM, it follows from Theorem 1 (statements A,C,D) that $y$ is a pre-ILM. □

**Proposition 10.** *If $y$ is a pre-ILM of dual* (1) *w.r.t. $\mathcal{B}$, then $P_{\mathcal{B}}(K(y)) \neq \bot$.*

*Proof.* Proof by contradiction. Suppose $y$ is pre-ILM and $P_{\mathcal{B}}(K(y)) = \bot$, then there must exist a finite sequence $(B_l)_{l=1}^{L}$ for $B_l \in \mathcal{B}, l \in [L]$ such that

$$P_{B_L}(P_{B_{L-1}}(P_{B_{L-2}}(\cdots P_{B_2}(P_{B_1}(K(y)))\cdots))) = \bot \tag{24}$$

which consists of the used blocks $B$ in the propagation algorithm.

As discussed in Remark 1, we can imitate this propagation by creating a sequence of dual feasible points $y^1, y^2, \cdots, y^L$ where $y^1 = y$ and $y^{l+1}$ is created from $y^l$ by changing block of variables $B_l$ to be in the relative interior of optimizers. This is given by construction in the proof of Proposition 8 and it holds that $P_{B_l}(K(y^l)) = K(y^{l+1})$ for all $l \in [L-1]$. Since $P_{B_L}(K(y^L)) = \bot$, $y^L$ is not a local minimum by Proposition 6. Therefore, updating the block of variables $B_L$ in $y^L$ by (4) (or even (3)) to obtain a point $y^{L+1}$ improves objective.

Thus, we applied BCD with relative interior rule to obtain the sequence $(y^l)_{l=1}^{L+1}$ and the objective improved. This is contradictory with Theorem 1 (statement C) which states that block updates that choose any optimizer (even without relative interior rule) cannot improve the objective from a pre-ILM. □

*Proof (**Theorem** 2).* For the first part, point $y$ is an LM of dual (1) w.r.t. $\mathcal{B}$ by its definition if $y_B$ satisfies (3) for all $B \in \mathcal{B}$. By Proposition 6, this is equivalent to $P_{\mathcal{B}}(K(y)) \neq \bot \; \forall B \in \mathcal{B}$. The second part is given in Corollary 2 and the third part follows from Proposition 9 and Proposition 10. □

## B Constructing an Improving Feasible Direction

As discussed in Sect. 3, if (6) is infeasible, there exists an improving feasible direction (7), we are going to describe how to obtain it based on the propagation algorithm defined in Sect. 3.1. We remark that conditions (7) define a whole convex cone of improving directions and our algorithm finds one of them based on the specific implementation of the construction.

Let us have a set of blocks $\mathcal{B} \subseteq 2^{[m]}$ and a dual feasible point $y$ such that $P_{\mathcal{B}}(K(y)) = \bot$, which implies infeasibility of (6). Consider sequences $(B_l)_{l=1}^{L}$ and $(K_l)_{l=1}^{L}$ where $K_1 \supset K_2 \supset \cdots \supset K_L$, $K_1 = K(y)$, $K_{l+1} = P_{B_l}(K_l)$ for every $l \in [L-1]$, and $P_{B_L}(K_L) = \bot$. To construct $\bar{y}$, we use the primal-dual pair

$$\begin{array}{lll l}
\max 0 & \min b^T \bar{y}^l & & \text{(25a)}\\
A^j x = b_j & \bar{y}^l_j \in \mathbb{R} & \forall j \in B_l & \text{(25b)}\\
x_i = 0 & - & \forall i \in [n] - K_l & \text{(25c)}\\
x_i \geq 0 & A_i^T \bar{y}^l \geq 0 & \forall i \in K_l & \text{(25d)}\\
- & \bar{y}^l_j = 0 & \forall j \in [m] - B_l. & \text{(25e)}
\end{array}$$

and proceed as follows:

1. Initialize $\bar{y} \leftarrow \bar{y}^L$ where $\bar{y}^L$ is any feasible dual solution of (25) for $l = L$ with[7] $b^T \bar{y}^L < 0$.
2. For all $l \in \{L-1, L-2, ..., 2, 1\}$ in descending order:
   (a) If $A_i^T \bar{y} \geq 0$ for all $i \in K_l - K_{l+1}$, continue with $l \leftarrow l - 1$.
   (c) Else, find $\bar{y}^l$ from the relative interior of optimizers of dual (25) for current $l$, update $\bar{y} \leftarrow \bar{y} + \delta_l \bar{y}^l$ where $\delta_l = \max_{\substack{i \in K_l - K_{l+1} \\ A_i^T \bar{y} < 0}} -\frac{A_i^T \bar{y}}{A_i^T \bar{y}^l}$, and set $l \leftarrow l - 1$.
3. Return $\bar{y}$ as improving feasible direction satisfying (7).

Due to lack of space, we omit the proof of this procedure. We will only state that it is based on induction, i.e., after some index $l \in [L]$ is processed, it holds

[7] Such $\bar{y}^L$ exists because primal (25) is infeasible for $l = L$ due to $P_{B_L}(K_L) = \bot$ and the dual (25) is therefore unbounded since the dual always has a feasible solution.

that $A_i^T \bar{y} \geq 0$ for all $i \in K_l$ and $b^T \bar{y} = b^T \bar{y}^L < 0$ is maintained during the whole algorithm. Thus, eventually $A_i^T \bar{y} \geq 0$ holds for all $i \in K_1 = K(y)$.

After $\bar{y}$ is calculated, we can find a step size $\epsilon > 0$ and perform update of $y$ as discussed in Sect. 3. Even though this approach may seem complicated, it is easy to see that in cases when the blocks $B$ are small, the problem (25) is also small and could even be solvable in closed-form for some special cases.

## References


1. Bessiere, C.: Constraint propagation. In: Handbook of Constraint Programming, chap. 3. Elsevier (2006)
2. Cooper, M.C., de Givry, S., Sanchez, M., Schiex, T., Zytnicki, M., Werner, T.: Soft arc consistency revisited. Artif. Intell. **174**(7–8), 449–478 (2010)
3. Dlask, T.: Minimizing convex piecewise-affine functions by local consistency techniques. Master's thesis (2018)
4. Dlask, T.: Unit propagation by means of coordinate-wise minimization. In: International Conference on Machine Learning, Optimization, and Data Science. Springer, Heidelberg (2020)
5. Dlask, T., Werner, T.: Bounding linear programs by constraint propagation: application to Max-SAT. In: Simonis, H. (ed.) International Conference on Principles and Practice of Constraint Programming. LNCS, vol. 12333, pp. 177–193. Springer, Heidelberg (2020). https://doi.org/10.1007/978-3-030-58475-7_11
6. Dlask, T., Werner, T.: A class of linear programs solvable by coordinate-wise minimization. In: Kotsireas, I.S., Pardalos, P.M. (eds.) LION 2020. LNCS, vol. 12096, pp. 52–67. Springer, Cham (2020). https://doi.org/10.1007/978-3-030-53552-0_8
7. Globerson, A., Jaakkola, T.S.: Fixing max-product: convergent message passing algorithms for MAP LP-relaxations. In: Advances in Neural Information Processing Systems, pp. 553–560 (2008)
8. Kappes, J.H., Kappes, J.H., et al.: A comparative study of modern inference techniques for structured discrete energy minimization problems. Int. J. Comput. Vis. **115**(2), 155–184 (2015)
9. Kolmogorov, V.: Convergent tree-reweighted message passing for energy minimization. IEEE Trans. Pattern Anal. Mach. Intell. **28**(10), 1568–1583 (2006)
10. Koval, V.K., Schlesinger, M.I.: Dvumernoe programmirovanie v zadachakh analiza izobrazheniy (Two-dimensional programming in image analysis problems). Autom. Telemec. **8**, 149–168 (1976). (in Russian)
11. Kovalevsky, V., Koval, V.: A diffusion algorithm for decreasing energy of max-sum labeling problem. Glushkov Institute of Cybernetics, Kiev, USSR (1975, unpublished)
12. Lemaréchal, C., Hiriart-Urruty, J.: Fundamentals of Convex Analysis. Springer, New York (2004). https://doi.org/10.1007/978-3-642-56468-0
13. Matoušek, J., Gärtner, B.: Understanding and Using Linear Programming. Universitext. Springer, Heidelberg (2006). https://doi.org/10.1007/978-3-540-30717-4
14. Papadimitriou, C.H., Steiglitz, K.: Combinatorial Optimization: Algorithms and Complexity. Courier Corporation (1998)
15. Savchynskyy, B.: Discrete graphical models - an optimization perspective. Found. Trends Comput. Graph. Vis. **11**(3–4), 160–429 (2019)

16. Tseng, P.: Convergence of a block coordinate descent method for nondifferentiable minimization. J. Optim. Theory Appl. **109**(3), 475–494 (2001)
17. Živný, S.: The Complexity of Valued Constraint Satisfaction Problems. Cognitive Technologies. Springer, Heidelberg (2012). https://doi.org/10.1007/978-3-642-33974-5
18. Wainwright, M.J., Jordan, M.I.: Graphical models, exponential families, and variational inference. Found. Trends Mach. Learn. **1**(1–2), 1–305 (2008)
19. Werner, T.: A linear programming approach to max-sum problem: a review. IEEE Trans. Pattern Anal. Mach. Intell. **29**(7), 1165–1179 (2007)
20. Werner, T.: On coordinate minimization of piecewise-affine functions. Technical report. CTU-CMP-2017-05, Department of Cybernetics, Faculty of Electrical Engineering, Czech Technical University in Prague (2017)
21. Werner, T., Průša, D., Dlask, T.: Relative interior rule in block-coordinate descent. In: Proceedings of the IEEE/CVF Conference on Computer Vision and Pattern Recognition, pp. 7559–7567 (2020)
22. Werner, T., Průša, D.: Relative interior rule in block-coordinate minimization. arXiv.org (2019)
23. Zhang, S.: On the strictly complementary slackness relation in linear programming. In: Du, D.Z., Sun, J. (eds.) Advances in Optimization and Approximation, vol. 1, pp. 347–361. Springer, Boston (1994). https://doi.org/10.1007/978-1-4613-3629-7_19

# DPMC: Weighted Model Counting by Dynamic Programming on Project-Join Trees

Jeffrey M. Dudek, Vu H. N. Phan(✉), and Moshe Y. Vardi

Rice University, Houston, TX 77005, USA
{jmd11,vhp1,vardi}@rice.edu

**Abstract.** We propose a unifying dynamic-programming framework to compute exact literal-weighted model counts of formulas in conjunctive normal form. At the center of our framework are project-join trees, which specify efficient project-join orders to apply additive projections (variable eliminations) and joins (clause multiplications). In this framework, model counting is performed in two phases. First, the planning phase constructs a project-join tree from a formula. Second, the execution phase computes the model count of the formula, employing dynamic programming as guided by the project-join tree. We empirically evaluate various methods for the planning phase and compare constraint-satisfaction heuristics with tree-decomposition tools. We also investigate the performance of different data structures for the execution phase and compare algebraic decision diagrams with tensors. We show that our dynamic-programming model-counting framework `DPMC` is competitive with the state-of-the-art exact weighted model counters `Cachet`, `c2d`, `d4`, and `miniC2D`.

**Keywords:** Treewidth · Factored representation · Early projection

## 1 Introduction

*Model counting* is a fundamental problem in artificial intelligence, with applications in machine learning, probabilistic reasoning, and verification [24,34,50]. Given an input set of constraints, with the focus in this paper on Boolean constraints, the model-counting problem is to count the number of satisfying assignments. Although this problem is #P-Complete [69], a variety of tools exist that can handle industrial sets of constraints, e.g., [19,46,52,58].

Dynamic programming is a powerful technique that has been applied across computer science [7], including to model counting [4,36,56]. The key idea is to solve a large problem by solving a sequence of smaller subproblems and then incrementally combining these solutions into the final result. Dynamic programming provides a natural framework to solve a variety of problems defined on sets

Work supported in part by NSF grants IIS-1527668, CCF-1704883, IIS-1830549, and DMS-1547433.


H. Simonis (Ed.): CP 2020, LNCS 12333, pp. 211–230, 2020.
https://doi.org/10.1007/978-3-030-58475-7_13

of constraints: subproblems can be formed by partitioning the constraints. This framework has been instantiated into algorithms for database-query optimization [48], satisfiability solving [2,54,68], and QBF evaluation [11].

Dynamic programming has also been the basis of several tools for model counting [25–27,31]. Although each tool uses a different data structure–algebraic decision diagrams (ADDs) [26], tensors [25,27], or database tables [31]–the overall algorithms have similar structure. The goal of this work is to unify these approaches into a single conceptual framework: *project-join trees*. Project-join trees are not an entirely new idea. Similar concepts have been used in constraint programming (as join trees [22]), probabilistic inference (as cluster trees [60]), and database-query optimization (as join-expression trees [48]). Our original contributions include the unification of these concepts into project-join trees and the application of this unifying framework to model counting.

We argue that project-join trees provide the natural formalism to describe execution plans for dynamic-programming algorithms for model counting. In particular, considering project-join trees as *execution plans* enables us to decompose dynamic-programming algorithms such as the one in [26] into two phases, following the breakdown in [27]: a *planning* phase and an *execution* phase. This enables us to study and compare different planning algorithms, different execution environments, and the interplay between planning and execution. Such a study is the main focus of this work. While the focus here is on model counting, our framework is of broader interest. For example, in [65], Tabajara and Vardi described a dynamic-programming, binary-decision-diagram-based framework for functional Boolean synthesis. Refactoring the algorithm into a planning phase followed by an execution phase is also of interest in that context.

The primary contribution of the work here is a dynamic-programming framework for weighted model counting based on project-join trees. In particular:

1. We show that several recent algorithms for weighted model counting [25,26, 31] can be unified into a single framework using project-join trees.
2. We compare the one-shot[1] constraint-satisfaction heuristics used in [26] with the anytime[2] tree-decomposition tools used in [25] and observe that tree-decomposition tools outperform constraint-satisfaction heuristics.
3. We compare (sparse) ADDs [5] with (dense) tensors [38] and find that ADDs outperform tensors on single CPU cores.
4. We find that project-join-tree-based algorithms contribute to a portfolio of model counters containing `Cachet` [58], `c2d` [19], `d4` [46], and `miniC2D` [52].

These conclusions have significance beyond model counting. The superiority of anytime tree-decomposition tools over classical one-shot constraint-satisfaction heuristics can have broad applicability. Similarly, the advantage of compact data structures for dynamic programming may apply to other optimization problems.

---

[1] A *one-shot* algorithm outputs exactly one solution and then terminates immediately.

[2] An *anytime* algorithm outputs better and better solutions the longer it runs.

## 2 Preliminaries

***Pseudo-Boolean Functions and Early Projection.*** A *pseudo-Boolean function* over a set $X$ of variables is a function $f : 2^X \to \mathbb{R}$. Operations on pseudo-Boolean functions include *product* and *projection*. First, we define product:

**Definition 1.** *Let $X$ and $Y$ be sets of Boolean variables. The* product *of functions $f : 2^X \to \mathbb{R}$ and $g : 2^Y \to \mathbb{R}$ is the function $f \cdot g : 2^{X \cup Y} \to \mathbb{R}$ defined for all $\tau \in 2^{X \cup Y}$ by $(f \cdot g)(\tau) \equiv f(\tau \cap X) \cdot g(\tau \cap Y)$.*

Next, we define (additive) projection, which marginalizes a single variable:

**Definition 2.** *Let $X$ be a set of Boolean variables and $x \in X$. The* projection *of a function $f : 2^X \to \mathbb{R}$ w.r.t. $x$ is the function $\sum_x f : 2^{X \setminus \{x\}} \to \mathbb{R}$ defined for all $\tau \in 2^{X \setminus \{x\}}$ by $(\sum_x f)(\tau) \equiv f(\tau) + f(\tau \cup \{x\})$.*

Note that projection is commutative, i.e., that $\sum_x \sum_y f = \sum_y \sum_x f$ for all variables $x, y \in X$ and functions $f : 2^X \to \mathbb{R}$. Given a set $X = \{x_1, x_2, \ldots, x_n\}$, define $\sum_X f \equiv \sum_{x_1} \sum_{x_2} \cdots \sum_{x_n} f$. Our convention is that $\sum_\varnothing f \equiv f$.

When performing a product followed by a projection, it is sometimes possible to perform the projection first. This is known as *early projection* [48].

**Theorem 1 (Early Projection).** *Let $X$ and $Y$ be sets of variables. For all functions $f : 2^X \to \mathbb{R}$ and $g : 2^Y \to \mathbb{R}$, if $x \in X \setminus Y$, then $\sum_x (f \cdot g) = (\sum_x f) \cdot g$.*

Early projection is a key technique in symbolic computation in a variety of settings, including database-query optimization [40], symbolic model checking [9], satisfiability solving [54], and model counting [26].

***Weighted Model Counting.*** We compute the total weight, subject to a given weight function, of all models of an input propositional formula. Formally:

**Definition 3.** *Let $X$ be a set of Boolean variables, $\varphi : 2^X \to \{0, 1\}$ be a Boolean function, and $W : 2^X \to \mathbb{R}$ be a pseudo-Boolean function. The* weighted model count *of $\varphi$ w.r.t. $W$ is $W(\varphi) \equiv \sum_{\tau \in 2^X} \varphi(\tau) \cdot W(\tau)$.*

The weighted model count of $\varphi$ w.r.t. $W$ can be naturally expressed in terms of pseudo-Boolean functions: $W(\varphi) = (\sum_X (\varphi \cdot W))(\varnothing)$. The function $W : 2^X \to \mathbb{R}$ is called a *weight function*. In this work, we focus on literal-weight functions, which can be expressed as products of weights associated with each variable. Formally, a *literal-weight function* $W$ can be factored as $W = \prod_{x \in X} W_x$ for pseudo-Boolean functions $W_x : 2^{\{x\}} \to \mathbb{R}$.

***Graphs.*** A *graph* $G$ has a set $\mathcal{V}(G)$ of vertices, a set $\mathcal{E}(G)$ of (undirected) edges, a function $\delta_G : \mathcal{V}(G) \to 2^{\mathcal{E}(G)}$ that gives the set of edges incident to each vertex, and a function $\epsilon_G : \mathcal{E}(G) \to 2^{\mathcal{V}(G)}$ that gives the set of vertices incident to each edge. Each edge must be incident to exactly two vertices, but multiple edges can exist between two vertices. A *tree* is a simple, connected, and acyclic graph. A *leaf* of a tree $T$ is a vertex of degree one, and we use $\mathcal{L}(T)$ to denote the set of leaves of $T$. We often refer to a vertex of a tree as a *node* and an edge as an *arc* to avoid confusion. A *rooted tree* is a tree $T$ together with a distinguished node $r \in \mathcal{V}(T)$ called the *root*. In a rooted tree $(T, r)$, each node $n \in \mathcal{V}(T)$ has a (possibly empty) set of *children*, denoted $\mathcal{C}(n)$, which contains all nodes $n'$ adjacent to $n$ s.t. all paths from $n'$ to $r$ contain $n$.

## 3 Using Project-Join Trees for Weighted Model Counting

In model counting, a Boolean formula is often given in conjunctive normal form (CNF), i.e., as a set $\varphi$ of clauses. For each clause $c \in \varphi$, define $\texttt{Vars}(c)$ to be the set of variables appearing in $c$. Then $c$ represents a Boolean function over $\texttt{Vars}(c)$. Similarly, $\varphi$ represents a Boolean function over $\texttt{Vars}(\varphi) \equiv \bigcup_{c \in \varphi} \texttt{Vars}(c)$.

It is well-known that weighted model counting can be performed through a sequence of projections and joins on pseudo-Boolean functions [25,26]. Given a CNF formula $\varphi$ and a literal-weight function $W$ over a set $X$ of variables, the corresponding weighted model count can be computed as follows:

$$W(\varphi) = \left( \sum_{X} \left( \prod_{c \in \varphi} c \cdot \prod_{x \in X} W_x \right) \right) (\varnothing) \tag{1}$$

By taking advantage of the associative and commutative properties of multiplication as well as the commutative property of projection, we can rearrange Eq. (1) to apply early projection. It was shown in [26] that early projection can significantly reduce computational cost. There are a variety of possible rearrangements of Eq. (1) of varying costs. Although [26] considered several heuristics for performing this rearrangement (using bucket elimination [20] and Bouquet's Method [8]), they did not attempt to analyze rearrangements.

In this work, we aim to analyze the quality of the rearrangement, in isolation from the underlying implementation and data structure used for Eq. (1). This approach has been highly successful for database-query optimization [48], where the central object of theoretical reasoning is the *query plan*. The approach has also seen similar success in Bayesian network inference [18].

We model a rearrangement of Eq. (1) as a *project-join tree*:

**Definition 4.** *Let $X$ be a set of Boolean variables and $\varphi$ be a CNF formula over $X$. A* project-join tree *of $\varphi$ is a tuple $(T, r, \gamma, \pi)$ where:*

- *$T$ is a tree with root $r \in \mathcal{V}(T)$,*
- *$\gamma : \mathcal{L}(T) \to \varphi$ is a bijection between the leaves of $T$ and the clauses of $\varphi$, and*
- *$\pi : \mathcal{V}(T) \setminus \mathcal{L}(T) \to 2^X$ is a labeling function on internal nodes.*

*Moreover, $(T, r, \gamma, \pi)$ must satisfy the following two properties:*

1. *$\{\pi(n) : n \in \mathcal{V}(T) \setminus \mathcal{L}(T)\}$ is a partition of $X$, and*
2. *for each internal node $n \in \mathcal{V}(T) \setminus \mathcal{L}(T)$, variable $x \in \pi(n)$, and clause $c \in \varphi$ s.t. $x$ appears in $c$, the leaf node $\gamma^{-1}(c)$ must be a descendant of $n$ in $T$.*

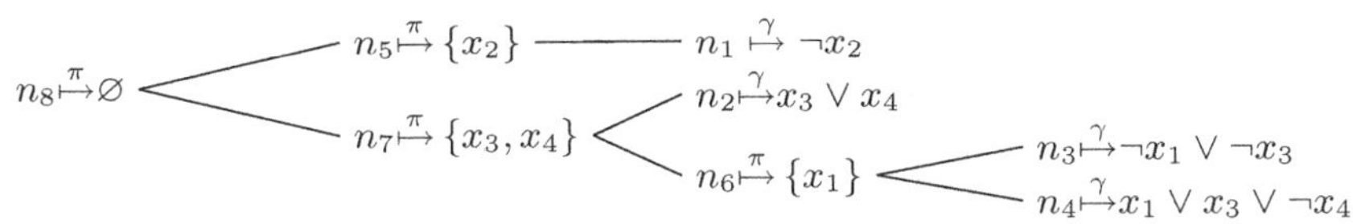


**Fig. 1.** A project-join tree $(T, n_8, \gamma, \pi)$ of a CNF formula $\varphi$. Each leaf node is labeled by $\gamma$ with a clause of $\varphi$. Each internal node is labeled by $\pi$ with a set of variables of $\varphi$.

If $n$ is a leaf node, then $n$ corresponds to a clause $c = \gamma(n)$ in Eq. (1). If $n$ is an internal node, then $n$'s children $\mathcal{C}(n)$ are to be multiplied before the projections of variables in $\pi(n)$ are performed. The two properties ensure that the resulting expression is equivalent to Eq. (1) using early projection. See Fig. 1 for a graphical example of a project-join tree.

Project-join trees have previously been studied in the context of database-query optimization [48]. Project-join trees are closely related to contraction trees in the context of tensor networks [25,28]. Once a rearrangement of Eq. (1) has been represented by a project-join tree, we can model the computation process according to the rearrangement. In particular, given a literal-weight function $W = \prod_{x \in X} W_x$, we define the *W-valuation* of each node $n \in \mathcal{V}(T)$ as a pseudo-Boolean function associated with $n$. The $W$-valuation of a node $n \in \mathcal{V}(T)$ is denoted $f_n^W$ and defined as follows:

$$f_n^W \equiv \begin{cases} \gamma(n) & \text{if } n \in \mathcal{L}(T) \\ \sum_{\pi(n)} \left( \prod_{o \in \mathcal{C}(n)} f_o^W \cdot \prod_{x \in \pi(n)} W_x \right) & \text{if } n \notin \mathcal{L}(T) \end{cases} \tag{2}$$

Note that the $W$-valuation of a leaf node $n \in \mathcal{L}(T)$ is a clause $c = \gamma(n) \in \varphi$, interpreted in this context as an associated function $\lambda_c : 2^{\texttt{Vars}(c)} \to \{0, 1\}$ where $\lambda_c(\tau) = 1$ if and only if the truth assignment $\tau$ satisfies $c$. The main idea is that the $W$-valuation at each node of $T$ is a pseudo-Boolean function computed as a subexpression of Eq. (1). The $W$-valuation of the root is exactly the result of Eq. (1), i.e., the weighted model count of $\varphi$ w.r.t. $W$:

**Theorem 2.** *Let $\varphi$ be a CNF formula over a set $X$ of variables, $(T, r, \gamma, \pi)$ be a project-join tree of $\varphi$, and $W$ be a literal-weight function over $X$. Then $f_r^W(\varnothing) = W(\varphi)$.*

This gives us a two-phase algorithm for computing the weighted model count of a formula $\varphi$. First, in the *planning* phase, we construct a project-join tree $(T, r, \gamma, \pi)$ of $\varphi$. We discuss algorithms for constructing project-join trees in Sect. 4. Second, in the *execution* phase, we compute $f_r^W$ by following Eq. (2). We discuss data structures for computing Eq. (2) in Sect. 5.

When computing a $W$-valuation, the number of variables that appear in each intermediate pseudo-Boolean function has a significant impact on the runtime.

The set of variables that appear in the $W$-valuation of a node is actually independent of $W$. In particular, for each node $n \in \mathcal{V}(T)$, define $\texttt{Vars}(n)$ as follows:

$$\texttt{Vars}(n) \equiv \begin{cases} \texttt{Vars}(\gamma(n)) & \text{if } n \in \mathcal{L}(T) \\ \left(\bigcup_{o \in \mathcal{C}(n)} \texttt{Vars}(o)\right) \setminus \pi(n) & \text{if } n \notin \mathcal{L}(T) \end{cases} \tag{3}$$

For every literal-weight function $W$, the domain of the function $f_n^W$ is $2^{\texttt{Vars}(n)}$. To characterize the difficulty of $W$-valuation, we define the *size* of a node $n$, $\texttt{size}(n)$, to be $|\texttt{Vars}(n)|$ for leaf nodes and $|\texttt{Vars}(n) \cup \pi(n)|$ for internal nodes. The *width* of a project-join tree $(T, r, \gamma, \pi)$ is $\texttt{width}(T) \equiv \max_{n \in \mathcal{V}(T)} \texttt{size}(n)$. We see in Sect. 6 how the width impacts the computation of $W$-valuations.

## 4 Planning Phase: Building a Project-Join Tree

In the planning phase, we are given a CNF formula $\varphi$ over Boolean variables $X$. The goal is to construct a project-join tree of $\varphi$. In this section, we present two classes of techniques that have been applied to model counting: using constraint-satisfaction heuristics (in [26]) and using tree decompositions (in [25,31]).

### 4.1 Planning with One-Shot Constraint-Satisfaction Heuristics

A variety of constraint-satisfaction heuristics for model counting were presented in a single algorithmic framework by [26]. These heuristics have a long history in constraint programming [21], database-query optimization [48], and propositional reasoning [54]. In this section, we adapt the framework of [26] to produce project-join trees. This algorithm is presented as Algorithm 1, which constructs a project-join tree of a CNF formula using constraint-satisfaction heuristics. The functions `ClusterVarOrder`, `ClauseRank`, and `ChosenCluster` represent heuristics for fine-tuning the specifics of the algorithm. Before discussing the various heuristics, we assert the correctness of Algorithm 1 in the following theorem.

**Theorem 3.** *Let $X$ be a set of variables and $\varphi$ be a CNF formula over $X$. Assume that* `ClusterVarOrder` *returns an injection $X \to \mathbb{N}$. Furthermore, assume that all* `ClauseRank` *and* `ChosenCluster` *calls satisfy the following conditions:*

1. $1 \leq \texttt{ClauseRank}(c, \rho) \leq m$,
2. $i < \texttt{ChosenCluster}(n_i) \leq m$, *and*
3. $X_s \cap \texttt{Vars}(n_i) = \varnothing$ *for all integers $s$ where* $i < s < \texttt{ChosenCluster}(n_i)$.

*Then Algorithm 1 returns a project-join tree of $\varphi$.*

By Condition 1, we know that $\{\Gamma_i\}_{i=1}^m$ is a partition of the clauses of $\varphi$. Condition 2 ensures that lines 11–12 place a new internal node $n_i$ in a cluster that has not yet been processed. Also on lines 11–12, Condition 3 prevents the node $n_i$ from skipping a cluster $\kappa_s$ if there exists some $x \in X_s \cap \texttt{Vars}(n_i)$, since $x$

**Algorithm 1.** Using combined constraint-satisfaction heuristics to build a project-join tree

**Input:** $X$: set of $m \geq 1$ Boolean variables
**Input:** $\varphi$: CNF formula over $X$
**Output:** $(T, r, \gamma, \pi)$: project-join tree of $\varphi$

$(T, \texttt{null}, \gamma, \pi) \leftarrow$ empty project-join tree
$\rho \leftarrow \texttt{ClusterVarOrder}(\varphi)$ /* injection $\rho : X \to \mathbb{N}$ */
**for** $i = m, m-1, \ldots, 1$
  $\Gamma_i \leftarrow \{c \in \varphi : \texttt{ClauseRank}(c, \rho) = i\}$ /* $1 \leq \texttt{ClauseRank}(c, \rho) \leq m$ */
  $\kappa_i \leftarrow \{\texttt{LeafNode}(T, c) : c \in \Gamma_i\}$
  /* for each $c$, a leaf $l$ with $\gamma(l) = c$ is constructed and put in cluster $\kappa_i$ */
  $X_i \leftarrow \texttt{Vars}(\Gamma_i) \setminus \bigcup_{j=i+1}^{m} \texttt{Vars}(\Gamma_j)$ /* $\{X_i\}_{i=1}^{m}$ is a partition of $X$ */
**for** $i = 1, 2, \ldots, m$
  **if** $\kappa_i \neq \varnothing$
    $n_i \leftarrow \texttt{InternalNode}(T, \kappa_i, X_i)$ /* $\mathcal{C}(n_i) = \kappa_i$ and $\pi(n_i) = X_i$ */
    **if** $i < m$
      $j \leftarrow \texttt{ChosenCluster}(n_i)$ /* $i < j \leq m$ */
      $\kappa_j \leftarrow \kappa_j \cup \{n_i\}$
**return** $(T, n_m, \gamma, \pi)$

is projected in iteration $s$, i.e., $x$ is added to $\pi(n_s)$. These invariants are sufficient to prove that Algorithm 1 indeed returns a project-join tree of $\varphi$. All heuristics we use in this work satisfy the conditions of Theorem 3.

There are a variety of heuristics to fine-tune Algorithm 1. For the function `ClusterVarOrder`, we consider the heuristics **Random**, **MCS** (*maximum-cardinality search* [67]), **LexP/LexM** (*lexicographic search for perfect/minimal orders* [42]), and **MinFill** (*minimal fill-in* [21]) as well as their inverses (**InvMCS**, **InvLexP**, **InvLexM**, and **InvMinFill**). Heuristics for `ClauseRank` include **BE** (*bucket elimination* [20]) and **BM** (*Bouquet's Method* [8]). For `ChosenCluster`, the heuristics we use are **List** and **Tree** [26]. We combine `ClauseRank` and `ChosenCluster` as *clustering heuristics*: **BE − List**, **BE − Tree**, **BM − List**, and **BM − Tree**. These heuristics are described in [26].

### 4.2 Planning with Anytime Tree-Decomposition Tools

A central technique in join-query optimization uses *tree decompositions* to compute join trees [17,48]. Tree decompositions [55] offer a way to decompose a graph into a tree structure. Formally:

**Definition 5.** *A* tree decomposition $(T, \chi)$ *of a graph* $G$ *is an unrooted binary tree* $T$ *together with a labeling function* $\chi : \mathcal{V}(T) \to 2^{\mathcal{V}(G)}$ *where:*

1. *for all* $v \in \mathcal{V}(G)$, *there exists* $n \in \mathcal{V}(T)$ *s.t.* $v \in \chi(n)$,
2. *for all* $e \in \mathcal{E}(G)$, *there exists* $n \in \mathcal{V}(T)$ *s.t.* $\epsilon_G(e) \subseteq \chi(n)$, *and*
3. *for all* $n, o, p \in \mathcal{V}(T)$, *if* $o$ *is on the path from* $n$ *to* $p$, *then* $\chi(n) \cap \chi(p) \subseteq \chi(o)$.

*The* treewidth, *or simply* width, *of* $(T, \chi)$ *is* $\texttt{tw}(T, \chi) \equiv \max_{n \in \mathcal{V}(T)} |\chi(n)| - 1$.

**Algorithm 2.** Using a tree decomposition to build a project-join tree

**Input:** $X$: set of Boolean variables
**Input:** $\varphi$: CNF formula over $X$
**Input:** $(S, \chi)$: tree decomposition of the Gaifman graph of $\varphi$
**Output:** $(T, r, \gamma, \pi)$: project-join tree of $\varphi$

$(T, \texttt{null}, \gamma, \pi) \leftarrow$ empty project-join tree
$\textit{found} \leftarrow \varnothing$ /* clauses of $\varphi$ that have been added to $T$ */
**function** $\texttt{Process}(n, \ell)$:
    **Input:** $n \in \mathcal{V}(S)$: node of $S$ to process
    **Input:** $\ell \subseteq X$: variables that must not be projected out here
    **Output:** $N \subseteq \mathcal{V}(T)$
    $\textit{clauses} \leftarrow \{c \in \varphi : c \notin \textit{found} \text{ and } \texttt{Vars}(c) \subseteq \chi(n)\}$
    $\textit{found} \leftarrow \textit{found} \cup \textit{clauses}$
    $\textit{children} \leftarrow \{\texttt{LeafNode}(T, c) : c \in \textit{clauses}\} \cup \bigcup_{o \in \mathcal{C}(n)} \texttt{Process}(o, \chi(n))$
        /* constructing new leaf nodes $p$ in the tree $T$ with $\gamma(p) = c$ */
    **if** $\textit{children} = \varnothing$ or $\chi(n) \subseteq \ell$
        **return** $\textit{children}$
    **else**
        **return** $\{\texttt{InternalNode}(T, \textit{children}, \chi(n) \setminus \ell)\}$
            /* new internal node $o$ with label $\pi(o) = \chi(n) \setminus \ell$ */
$s \leftarrow$ arbitrary node of $S$ /* fixing $s$ as root of tree $S$ */
$r \leftarrow$ only element of $\texttt{Process}(s, \varnothing)$
**return** $(T, r, \gamma, \pi)$

In particular, join-query optimization uses tree decompositions of the *join graph* to find optimal join trees [17,48]. The *join graph* of a project-join query consists of all attributes of a database as vertices and all tables as cliques. In this approach, tree decompositions of the join graph of a query are used to find optimal project-join trees; see Algorithm 3 of [48]. Similarly, tree decompositions of the *primal graph* of a factor graph, which consists of all variables as vertices and all factors as cliques, can be used to find variable elimination orders [37]. This technique has also been applied in the context of tensor networks [25,49].

Translated to model counting, this technique allows us to use tree decompositions of the *Gaifman graph* of a CNF formula to compute project-join trees. The Gaifman graph of a CNF formula $\varphi$, denoted $\texttt{Gaifman}(\varphi)$, has a vertex for each variable of $\varphi$, and two vertices are adjacent if the corresponding variables appear together in some clause of $\varphi$. We present this tree-decomposition-based technique as Algorithm 2. The key idea is that each clause $c$ of $\varphi$ forms a clique in $\texttt{Gaifman}(\varphi)$ between the variables of $c$. Thus all variables of $c$ must appear together in some label of the tree decomposition. We identify that node with $c$.

The width of the resulting project-join tree is closely connected to the width of the original tree decomposition. We formalize this in the following theorem:

**Theorem 4.** *Let $\varphi$ be a CNF formula over a set $X$ of variables and $(S, \chi)$ be a tree decomposition of $\texttt{Gaifman}(\varphi)$ of width $w$. Then Algorithm 2 returns a project-join tree of $\varphi$ of width at most $w + 1$.*

The key idea is that, for each node $n \in \mathcal{V}(S)$, the label $\chi(n)$ is a bound on the variables that appear in all nodes returned by $\texttt{Process}(n, \cdot)$. Theorem 4 allows us to leverage state-of-the-art anytime tools for finding tree decompositions [1, 64,66] to construct project-join trees, which we do in Sect. 6.1.

On the theoretical front, it is well-known that tree decompositions of the Gaifman graph are actually equivalent to project-join trees [48]. That is, one can go in the other direction as well: given a project-join tree of $\varphi$, one can construct a tree decomposition of $\texttt{Gaifman}(\varphi)$ of equivalent width. Formally:

**Theorem 5.** *Let $\varphi$ be a CNF formula and $(T, r, \gamma, \pi)$ be a project-join tree of $\varphi$ of width $w$. Then there is a tree decomposition of $\texttt{Gaifman}(\varphi)$ of width $w - 1$.*

Theorem 5 is Lemma 1 of [48] and can be seen as the inverse of Theorem 4.

## 5 Execution Phase: Performing the Valuation

The execution phase involves a CNF formula $\varphi$ over variables $X$, a project-join tree $(T, r, \gamma, \pi)$ of $\varphi$, and a literal-weight function $W$ over $X$. The goal is to compute the valuation $f_r^W$ using Eq. (2). Several data structures can be used for the pseudo-Boolean functions that occur while using Eq. (2). In this work, we consider two data structures that have been applied to weighted model counting: ADDs (as in [26]) and tensors (as in [25]).

### 5.1 Algebraic Decision Diagrams

An *algebraic decision diagram (ADD)* is a compact representation of a pseudo-Boolean function as a directed acyclic graph [5]. For functions with logical structure, an ADD representation can be exponentially smaller than the explicit representation. Originally designed for matrix multiplication and shortest path algorithms, ADDs have also been used for Bayesian inference [12,33], stochastic planning [35], model checking [44], and model counting [26,29].

Formally, an ADD is a tuple $(X, S, \sigma, G)$, where $X$ is a set of Boolean variables, $S$ is an arbitrary set (called the *carrier set*), $\sigma : X \to \mathbb{N}$ is an injection (called the *diagram variable order*), and $G$ is a rooted directed acyclic graph satisfying the following three properties. First, every leaf node of $G$ is labeled with an element of $S$. Second, every internal node of $G$ is labeled with an element of $X$ and has two outgoing edges, labelled 0 and 1. Finally, for every path in $G$, the labels of internal nodes must occur in increasing order under $\sigma$. In this work, we only need to consider ADDs with the carrier set $S = \mathbb{R}$.

An ADD $(X, S, \sigma, G)$ is a compact representation of a function $f : 2^X \to S$. Although there are many ADDs representing $f$, for each injection $\sigma : X \to \mathbb{N}$, there is a unique minimal ADD that represents $f$ with $\sigma$ as the diagram variable order, called the *canonical ADD*. ADDs can be minimized in polynomial time, so it is typical to only work with canonical ADDs.

Several packages exist for efficiently manipulating ADDs. For example, `CUDD` [63] implements both product and projection on ADDs in polynomial time (in the size of the ADD representation). CUDD was used as the primary data structure for weighted model counting in [26]. In this work, we also use ADDs with CUDD to compute $W$-valuations.

**MCS** was the best diagram variable order on a set of standard weighted model counting benchmarks in [26]. So we use **MCS** as the diagram variable order in this work. Note that all other heuristics discussed in Sect. 4.1 for cluster variable order could also be used as heuristics for diagram variable order.

### 5.2 Tensors

A *tensor* is a multi-dimensional generalization of a matrix. Tensor are widely used in data analysis [13], signal and image processing [14], quantum physics [3], quantum chemistry [62], and many other areas of science. Given the diverse applications of tensors and tensor networks, a variety of tools [6,38] exist to manipulate them efficiently on a variety of hardware architectures, including multi-core and GPU-enhanced architectures.

Tensors can be used to represent pseudo-Boolean functions in a dense way. Tensors are particularly efficient at computing the contraction of two pseudo-Boolean functions: given two functions $f : 2^X \to \mathbb{R}$ and $g : 2^Y \to \mathbb{R}$, their *contraction* $f \otimes g$ is the pseudo-Boolean function $\sum_{X \cap Y} f \cdot g$. The contraction of two tensors can be implemented as matrix multiplication and so leverage significant work in high-performance computing on matrix multiplication on CPUs [47] and GPUs [30]. To efficiently use tensors to compute $W$-valuations, we follow [25] in implementing projection and product using tensor contraction.

First, we must compute the weighted projection of a function $f : 2^X \to \mathbb{R}$, i.e., we must compute $\sum_x f \cdot W_x$ for some $x \in X$. This is exactly equivalent to $f \otimes W_x$. Second, we must compute the product of two functions $f : 2^X \to \mathbb{R}$ and $g : 2^Y \to \mathbb{R}$. The central challenge is that tensor contraction implicitly projects all variables in $X \cap Y$, but we often need to maintain some shared variables in the result of $f \cdot g$. In [25], this problem was solved using a reduction to tensor networks. After the reduction, all variables appear exactly twice, so one never needs to perform a product without also projecting all shared variables.

In order to incorporate tensors in our project-join-tree-based framework, we take a different strategy that uses copy tensors. The *copy tensor* for a set $X$ represents the pseudo-Boolean function $\blacksquare_X : 2^X \to \mathbb{R}$ s.t. $\blacksquare_X(\tau)$ is 1 if $\tau \in \{\varnothing, X\}$ and 0 otherwise. We can simulate product using contraction by including additional copy tensors. In detail, for each $z \in X \cap Y$ make two fresh variables $z'$ and $z''$. Replace each $z$ in $f$ with $z'$ to produce $f'$ and replace each $z$ in $g$ with $z''$ to produce $g'$. Then one can check that $f \cdot g = f' \otimes g' \otimes \bigotimes_{z \in X \cap Y} \blacksquare_{\{z, z', z''\}}$.

When a product is immediately followed by the projection of shared variables (i.e., we are computing $\sum_Z f \cdot g$ for some $Z \subseteq X \cap Y$), we can optimize this procedure. In particular, we skip creating copy tensors for the variables in $Z$ and instead eliminate them directly as we perform $f' \otimes g'$. In this case, we do not ever fully compute $f \cdot g$, so the maximum number of variables needed in each intermediate tensor may be lower than the width of the project-join tree. In the context of tensor networks and contraction trees, the maximum number of variables needed after accounting for this optimization is the *max-rank* of the contraction tree [25,43]. The max-rank is often lower than the width of the corresponding project-joint tree. On the other hand, the intermediate terms in the computation of $f \cdot g$ with contractions may have more variables than either $f$, $g$, or $f \cdot g$. Thus the number of variables in each intermediate tensor may be higher than the width of the project-join tree (by at most a factor of 1.5).

## 6 Empirical Evaluation

We are interested in the following experimental research questions, where we aim to answer each research question with an experiment.

(RQ1) In the planning phase, how do constraint-satisfaction heuristics compare to tree-decomposition solvers?
(RQ2) In the execution phase, how do ADDs compare to tensors as the underlying data structure?
(RQ3) Are project-join-tree-based weighted model counters competitive with state-of-the-art tools?

To answer RQ1, we build two implementations of the planning phase: `HTB` (for Heuristic Tree Builder, based on [26]) and `LG` (for Line Graph, based on [25]). `HTB` implements Algorithm 1 and so is representative of the constraint-satisfaction approach. `HTB` contains implementations of four clustering heuristics (**BE-List**, **BE-Tree**, **BM-List**, and **BM-Tree**) and nine cluster-variable-order heuristics (**Random**, **MCS**, **InvMCS**, **LexP**, **InvLexP**, **LexM**, **InvLexM**, **MinFill**, and **InvMinFill**). `LG` implements Algorithm 2 and so is representative of the tree-decomposition approach. In order to find tree decompositions, `LG` leverages three state-of-the-art heuristic tree-decomposition solvers: `FlowCutter` [64], `htd` [1], and `Tamaki` [66]. These solvers are all *anytime*, meaning that `LG` never halts but continues to produce better and better project-join trees when given additional time. On the other hand, `HTB` produces a single project-join tree. We compare these implementations on the planning phase in Sect. 6.1.

To answer RQ2, we build two implementations of the execution phase: `DMC` (for Diagram Model Counter, based on [26]) and `tensor` (based on [25]). `DMC` uses ADDs as the underlying data structure with `CUDD` [63]. `tensor` uses tensors as the underlying data structure with `NumPy` [51]. We compare these implementations on the execution phase in Sect. 6.2. Since `LG` is an anytime tool, each execution tool must additionally determine the best time to terminate `LG` and begin performing the valuation. We explore options for this in Sect. 6.2.

To answer RQ3, we combine each implementation of the planning phase and each implementation of the execution phase to produce model counters that use project-join trees. We then compare these model counters with the state-of-the-art tools `Cachet` [58], `c2d` [19], `d4` [46], and `miniC2D` [52] in Sect. 6.3.

We use a set of 1976 literal-weighted model counting benchmarks from [26]. These benchmarks were gathered from two sources. First, the **Bayes** class[3] con-

[3] https://www.cs.rochester.edu/u/kautz/Cachet/Model_Counting_Benchmarks.

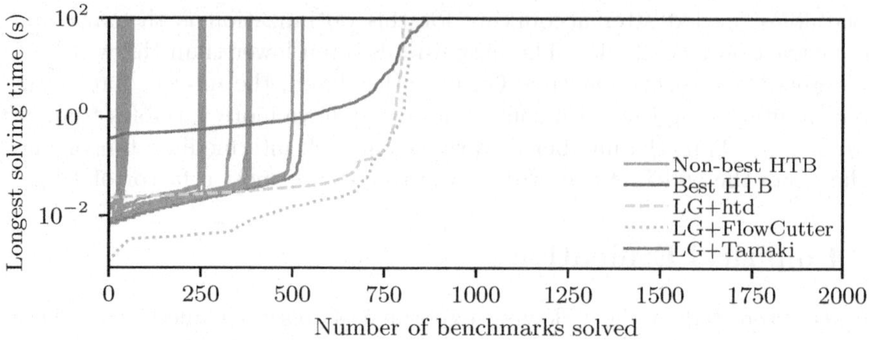


**Fig. 2.** A cactus plot of the performance of various planners. A planner "solves" a benchmark when it finds a project-join tree of width 30 or lower.

sists of 1080 CNF benchmarks[4] that encode Bayesian inference problems [59]. All literal weights in this class are between 0 and 1. Second, the **Non-Bayes** class[5] consists of 896 CNF benchmarks[6] that are divided into eight families: *Bounded Model Checking (BMC)*, *Circuit*, *Configuration*, *Handmade*, *Planning*, *Quantitative Information Flow (QIF)*, *Random*, and *Scheduling* [15,39,53,61]. All **Non-Bayes** benchmarks are originally unweighted. As we focus in this work on weighted model counting, we generate weights for these benchmarks. Each variable $x$ is randomly assigned literal weights: either $W_x(\{x\}) = 0.5$ and $W_x(\varnothing) = 1.5$, or $W_x(\{x\}) = 1.5$ and $W_x(\varnothing) = 0.5$. Generating weights in this particular fashion results in a reasonably low amount of floating-point underflow and overflow for all model counters.

We ran all experiments on single CPU cores of a Linux cluster with Xeon E5-2650v2 processors (2.60-GHz) and 30 GB of memory. All code, benchmarks, and experimental data are available in a public repository (https://github.com/vardigroup/DPMC).

### 6.1 Experiment 1: Comparing Project-Join Planners

We first compare constraint-satisfaction heuristics (HTB) and tree-decomposition tools (LG) at building project-join trees of CNF formulas. To do this, we ran all 36 configurations of HTB (combining four clustering heuristics with nine cluster-variable-order heuristics) and all three configurations of LG (choosing a tree-decomposition solver) once on each benchmark with a 100-second timeout. In Fig. 2, we compare how long it takes various methods to find a high-quality (meaning width at most 30) project-join tree of each benchmark. We chose 30 for Fig. 2 since [25] observed that tensor-based approaches were unable to handle trees whose widths are above 30, but Fig. 2 is qualitatively similar for

[4] Excluding 11 benchmarks double-counted by [26].

[5] http://www.cril.univ-artois.fr/KC/benchmarks.html.

[6] Including 73 benchmarks missed by [26].

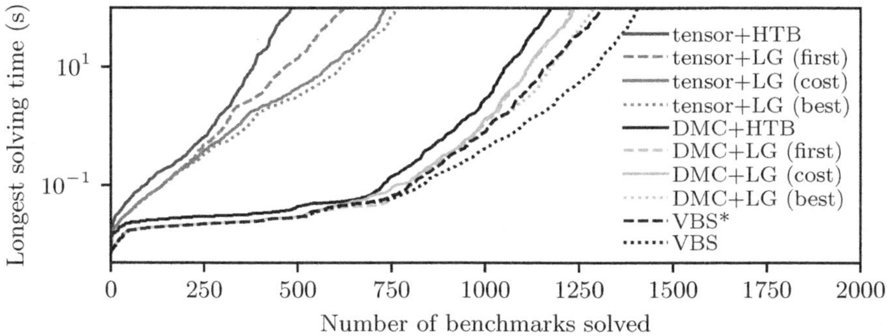


**Fig. 3.** A cactus plot of the performance of various planners and executors for weighted model counting. Different strategies for stopping LG are considered. "(first)" indicates that LG was stopped after it produced the first project-join tree. "(cost)" indicates that the executor attempted to predict the cost of computing each project-join tree. "(best)" indicates a simulated case where the executor has perfect information on all project-join trees generated by LG and valuates the tree with the shortest total time. VBS* is the virtual best solver of DMC+HTB and DMC+LG (cost). VBS is the virtual best solver of DMC+HTB, DMC+LG (cost), tensor+HTB, and tensor+LG (cost).

other choices of widths. We observe that LG is generally able to find project-join trees of lower widths than those HTB is able to find. We therefore conclude that tree-decomposition solvers outperform constraint-satisfaction heuristics in this case. We observe that **BE-Tree** as the clustering heuristic and **InvLexP** as the cluster-variable-order heuristic make up the best-performing heuristic configuration from HTB. This was previously observed to be the second-best heuristic configuration for weighted model counting in [26]. We therefore choose **BE-Tree** with **InvLexP** as the representative heuristic configuration for HTB in the remaining experiments. For LG, we choose FlowCutter as the representative tree-decomposition tool in the remaining experiments.

### 6.2 Experiment 2: Comparing Execution Environments

Next, we compare ADDs (DMC) and tensors (tensor) as a data structure for valuating project-join trees. To do this, we ran both DMC and tensor on all project-join trees generated by HTB and LG (with their representative configurations) in Experiment 1, each with a 100-second timeout. The total times recorded include both the planning stage and the execution stage.

Since LG is an anytime tool, it may have produced more than one project-join tree of each benchmark in Experiment 1. We follow [25] by allowing tensor and DMC to stop LG at a time proportional to the estimated cost to valuate the best-seen project-join tree. The constant of proportionality is chosen to minimize the PAR-2 score (i.e., the sum of the running times of all completed benchmarks plus twice the timeout for every uncompleted benchmark) of each executor. tensor and DMC use different methods for estimating cost. Tensors are a dense data

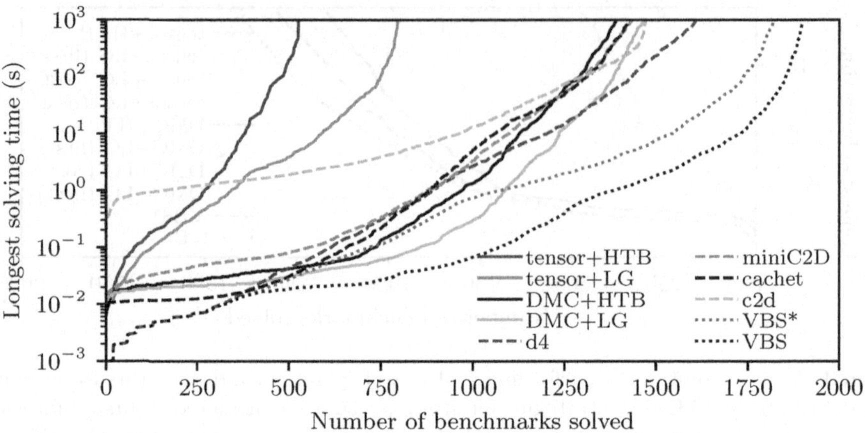


**Fig. 4.** A cactus plot of the performance of four project-join-tree-based model counters, two state-of-the-art model counters, and two virtual best solvers: `VBS*` (without project-join-tree-based counters) and `VBS` (with project-join-tree-based counters).

structure, so the number of floating-point operations to valuate a project-join tree can be computed exactly as in [25]. We use this as the cost estimator for `tensor`. ADDs are a sparse data structure, and estimating the amount of sparsity is difficult. It is thus hard to find a good cost estimator for `DMC`. As a first step, we use $2^w$ as an estimate of the cost for `DMC` to valuate a project-join tree of width $w$.

We present results from this experiment in Fig. 3. We observe that the benefit of `LG` over `HTB` seen in Experiment 1 is maintained once the full weighted model count is computed. We also observe that `DMC` is able to solve significantly more benchmarks than `tensor`, even when using identical project-join trees. We attribute this difference to the sparsity of ADDs over tensors. Nevertheless, we observe that `tensor` still outperforms `DMC` on some benchmarks; compare `VBS*` (which excludes `tensor`) with `VBS` (which includes `tensor`).

Moreover, we observe significant differences based on the strategy used to stop `LG`. The executor `tensor` performs significantly better when cost estimation is used than when only the first project-join tree of `LG` is used. In fact, the performance of `tensor` is almost as good as the hypothetical performance if `tensor` is able to predict the planning and valuation times of all trees produced by `LG`. On the other hand, `DMC` is not significantly improved by cost estimation. It would be interesting in the future to find better cost estimators for `DMC`.

### 6.3 Experiment 3: Comparing Exact Weighted Model Counters

Finally, we compare project-join-tree-based model counters with state-of-the-art tools for weighted model counting. We construct four project-join-tree-based model counters by combining `HTB` and `LG` (using the representative configurations from Experiment 1) with DMC and tensor (using the cost estimators for LG from Experiment 2). Note that HTB+DMC is equivalent to ADDMC [26] and LG+tensor is equivalent to TensorOrder [25]. We compare against the state-of-the-art model counters Cachet [58], c2d [19], d4 [46], and miniC2D [52]. We ran each benchmark once with each model counter with a 1000-second timeout and recorded the total time taken. For the project-join-tree-based model counters, time taken includes both the planning stage and the execution stage.

We present results from this experiment in Fig. 4. For each benchmark, the solving time of VBS* is the shortest solving time among all pre-existing model counters (Cachet, c2d, d4, and miniC2D). Similarly, the time of VBS is the shortest time among all model counters, including those based on project-join trees. We observe that VBS performs significantly better than VBS*. In fact, DMC+LG is the fastest model counter on 471 of 1976 benchmarks. Thus project-join-tree-based tools are valuable for portfolios of weighted model counters.

## 7 Discussion

In this work, we introduced the concept of project-join trees for weighted model counting. These trees are at the center of a dynamic-programming framework that unifies and generalizes several model counting algorithms, including those based on ADDs [26], tensors [25], and database management systems [31]. This framework performs model counting in two phases. First, the planning phase produces a project-join tree from a CNF formula. Second, the execution phase uses the project-join tree to guide the dynamic-programming computation of the model count of the formula w.r.t. a literal-weight function. The current implementation of our dynamic-programming model-counting framework DPMC includes two planners (HTB and LG) and two executors (DMC and tensor).

For the planning phase, we implemented HTB based on constraint-satisfaction heuristics [8,20,21,42,67] and LG based on tree-decomposition tools [1,64,66]. Our empirical work indicates that tree-decomposition tools tend to produce project-join trees of lower widths in shorter times. This is a significant finding with applications beyond model counting, e.g., in Boolean functional synthesis [65].

For the execution phase, we implemented DMC based on ADDs [26,63] and tensor based on tensors [25,51]. Empirically, we observed that (sparse) ADDs outperform (dense) tensors on single CPU cores. Whether this holds for richer architectures as well is a subject for future work. We will also consider adding to our framework an executor based on databases (e.g., [31]).

We showed that our dynamic-programming model-counting framework DPMC is competitive with state-of-the-art tools (Cachet [58], c2d [19], d4 [46], and miniC2D [52]). Although no single model counter dominates, DPMC considerably improves the virtual best solver and thus is valuable as part of the portfolio.

In this work, we did not consider preprocessing of benchmarks. For example, [25] found that preprocessing (called FT, based on a technique to reduce variable occurrences using tree decompositions of the incidence graph [57]) significantly improved tensor-network-based approaches for weighted model counting. Moreover, [32] and [27] observed that the `pmc` preprocessor [45] notably improved the running time of some dynamic-programming-based model counters. We expect these techniques to also improve `DPMC`.

A promising future research direction is multicore programming. Our planning tool `LG` can be improved to run back-end tree-decomposition solvers in parallel, as in [27]. We can also make the execution tool `DMC` support multicore ADD packages (e.g., `Sylvan` [23]). Our other executor, `tensor`, is built on top of `NumPy` [51] and should be readily parallelizable (e.g., using techniques from [27]). We can then compare `DPMC` to parallel solvers (e.g., [10,16]).

Finally, decision diagrams have been widely used in artificial intelligence in the context of *knowledge compilation*, where formulas are compiled into a tractable form in an early phase to support efficient query processing [19,41,46,52]. Our work opens up an investigation into the combination of knowledge compilation and dynamic programming. The focus here is on processing a single model-counting query. Exploring how dynamic programming can also be leveraged to handle several queries is another promising research direction.

## References

1. Abseher, M., Musliu, N., Woltran, S.: htd – a free, open-source framework for (customized) tree decompositions and beyond. In: Salvagnin, D., Lombardi, M. (eds.) CPAIOR 2017. LNCS, vol. 10335, pp. 376–386. Springer, Cham (2017). https://doi.org/10.1007/978-3-319-59776-8_30
2. Aguirre, A.S.M., Vardi, M.Y.: Random 3-SAT and BDDs: the plot thickens further. In: Walsh, T. (ed.) CP 2001. LNCS, vol. 2239, pp. 121–136. Springer, Heidelberg (2001). https://doi.org/10.1007/3-540-45578-7_9
3. Arad, I., Landau, Z.: Quantum computation and the evaluation of tensor networks. SICOMP **39**(7), 3089–3121 (2010). https://doi.org/10.1137/080739379
4. Bacchus, F., Dalmao, S., Pitassi, T.: Solving #SAT and Bayesian inference with backtracking search. JAIR **34**, 391–442 (2009). https://doi.org/10.1613/jair.2648
5. Bahar, R.I., et al.: Algebraic decision diagrams and their applications. Formal Method Syst. Des. **10**(2–3), 171–206 (1997). https://doi.org/10.1023/A:1008699807402
6. Baumgartner, G., et al.: Synthesis of high-performance parallel programs for a class of ab initio quantum chemistry models. IEEE **93**(2), 276–292 (2005). https://doi.org/10.1109/JPROC.2004.840311
7. Bellman, R.: Dynamic programming. Science **153**(3731), 34–37 (1966). https://doi.org/10.1126/science.153.3731.34
8. Bouquet, F.: Gestion de la dynamicité et énumération d'impliquants premiers: une approche fondée sur les Diagrammes de Décision Binaire. Ph.D. thesis, Aix-Marseille 1 (1999), https://www.theses.fr/1999AIX11011
9. Burch, J.R., Clarke, E.M., Long, D.E.: Symbolic model checking with partitioned transition relations. In: VLSI, pp. 49–58 (1991). http://citeseerx.ist.psu.edu/viewdoc/summary?doi=10.1.1.23.1932
10. Burchard, J., Schubert, T., Becker, B.: Laissez-Faire caching for parallel #SAT solving. In: Heule, M., Weaver, S. (eds.) SAT 2015. LNCS, vol. 9340, pp. 46–61. Springer, Cham (2015). https://doi.org/10.1007/978-3-319-24318-4_5

11. Charwat, G., Woltran, S.: BDD-based dynamic programming on tree decompositions. Technical report, Technische Universität Wien, Institut für Informationssysteme (2016)
12. Chavira, M., Darwiche, A.: Compiling Bayesian networks using variable elimination. In: IJCAI, pp. 2443–2449 (2007). https://dl.acm.org/doi/10.5555/1625275.1625669
13. Cichocki, A.: Era of big data processing: a new approach via tensor networks and tensor decompositions. arXiv preprint arXiv:1403.2048 (2014)
14. Cichocki, A., et al.: Tensor decompositions for signal processing applications: from two-way to multiway component analysis. IEEE SPM **32**(2), 145–163 (2015). https://doi.org/10.1109/MSP.2013.2297439
15. Clarke, E., Biere, A., Raimi, R., Zhu, Y.: Bounded model checking using satisfiability solving. Formal Method Syst. Des. **19**(1), 7–34 (2001). https://doi.org/10.1023/A:1011276507260
16. Dal, G.H., Laarman, A.W., Lucas, P.J.: Parallel probabilistic inference by weighted model counting. In: PGM, pp. 97–108 (2018). http://proceedings.mlr.press/v72/dal18a.html
17. Dalmau, V., Kolaitis, P.G., Vardi, M.Y.: Constraint satisfaction, bounded treewidth, and finite-variable logics. In: Van Hentenryck, P. (ed.) CP 2002. LNCS, vol. 2470, pp. 310–326. Springer, Heidelberg (2002). https://doi.org/10.1007/3-540-46135-3_21
18. Darwiche, A.: Dynamic jointrees. In: UAI, pp. 97–104 (1998). https://dl.acm.org/doi/10.5555/2074094.2074106
19. Darwiche, A.: New advances in compiling CNF to decomposable negation normal form. In: ECAI, pp. 318–322 (2004). https://dl.acm.org/doi/10.5555/3000001.3000069
20. Dechter, R.: Bucket elimination: a unifying framework for reasoning. AIJ **113**(1–2), 41–85 (1999). https://doi.org/10.1016/S0004-3702(99)00059-4
21. Dechter, R.: Constraint Processing. Morgan Kaufmann (2003). https://doi.org/10.1016/B978-1-55860-890-0.X5000-2
22. Dechter, R., Pearl, J.: Tree clustering for constraint networks. AIJ **38**(3), 353–366 (1989). https://doi.org/10.1016/0004-3702(89)90037-4
23. van Dijk, T., van de Pol, J.: Sylvan: multi-core decision diagrams. In: Baier, C., Tinelli, C. (eds.) TACAS 2015. LNCS, vol. 9035, pp. 677–691. Springer, Heidelberg (2015). https://doi.org/10.1007/978-3-662-46681-0_60
24. Domshlak, C., Hoffmann, J.: Probabilistic planning via heuristic forward search and weighted model counting. JAIR **30**, 565–620 (2007). https://dl.acm.org/doi/10.5555/1622637.1622652
25. Dudek, J.M., Dueñas-Osorio, L., Vardi, M.Y.: Efficient contraction of large tensor networks for weighted model counting through graph decompositions. arXiv preprint arXiv:1908.04381 (2019)
26. Dudek, J.M., Phan, V.H., Vardi, M.Y.: ADDMC: weighted model counting with algebraic decision diagrams. In: AAAI, vol. 34, pp. 1468–1476 (2020). https://doi.org/10.1609/aaai.v34i02.5505
27. Dudek, J.M., Vardi, M.Y.: Parallel weighted model counting with tensor networks. In: MCW (2020). https://mccompetition.org/assets/files/2020/MCW_2020_paper_1.pdf
28. Evenbly, G., Pfeifer, R.N.: Improving the efficiency of variational tensor network algorithms. Phys. Rev. B **89**(24), 245118 (2014). https://doi.org/10.1103/PhysRevB.89.245118

29. Fargier, H., Marquis, P., Niveau, A., Schmidt, N.: A knowledge compilation map for ordered real-valued decision diagrams. In: AAAI (2014). https://dl.acm.org/doi/10.5555/2893873.2894036
30. Fatahalian, K., Sugerman, J., Hanrahan, P.: Understanding the efficiency of GPU algorithms for matrix-matrix multiplication. In: SIGGRAPH/EUROGRAPHICS, pp. 133–137 (2004). https://doi.org/10.1145/1058129.1058148
31. Fichte, J.K., Hecher, M., Thier, P., Woltran, S.: Exploiting database management systems and treewidth for counting. In: Komendantskaya, E., Liu, Y. (eds.) PADL 2020. LNCS, vol. 12007, pp. 151–167. Springer, Cham (2020). https://doi.org/10.1007/978-3-030-39197-3_10
32. Fichte, J.K., Hecher, M., Zisser, M.: An improved GPU-based SAT model counter. In: Schiex, T., de Givry, S. (eds.) CP 2019. LNCS, vol. 11802, pp. 491–509. Springer, Cham (2019). https://doi.org/10.1007/978-3-030-30048-7_29
33. Gogate, V., Domingos, P.: Approximation by Quantization. In: UAI, pp. 247–255 (2011). https://dl.acm.org/doi/10.5555/3020548.3020578
34. Gomes, C.P., Sabharwal, A., Selman, B.: Model counting. In: Handbook of Satisfiability (2009). https://doi.org/10.3233/978-1-58603-929-5-633. (Chap. 20)
35. Hoey, J., St-Aubin, R., Hu, A., Boutilier, C.: SPUDD: stochastic planning using decision diagrams. In: UAI, pp. 279–288 (1999). https://arxiv.org/abs/1301.6704
36. Jégou, P., Kanso, H., Terrioux, C.: Improving exact solution counting for decomposition methods. In: ICTAI, pp. 327–334. IEEE (2016). https://doi.org/10.1109/ICTAI.2016.0057
37. Kask, K., Dechter, R., Larrosa, J., Dechter, A.: Unifying tree decompositions for reasoning in graphical models. AIJ **166**(1–2), 165–193 (2005). https://doi.org/10.1016/j.artint.2005.04.004
38. Kjolstad, F., Kamil, S., Chou, S., Lugato, D., Amarasinghe, S.: The tensor algebra compiler. PACMPL **1**(OOPSLA), 1–29 (2017). https://doi.org/10.1145/3133901
39. Klebanov, V., Manthey, N., Muise, C.: SAT-based analysis and quantification of information flow in programs. In: Joshi, K., Siegle, M., Stoelinga, M., D'Argenio, P.R. (eds.) QEST 2013. LNCS, vol. 8054, pp. 177–192. Springer, Heidelberg (2013). https://doi.org/10.1007/978-3-642-40196-1_16
40. Kolaitis, P.G., Vardi, M.Y.: Conjunctive-query containment and constraint satisfaction. JCSS **61**(2), 302–332 (2000). https://doi.org/10.1006/jcss.2000.1713
41. Koriche, F., Lagniez, J.M., Marquis, P., Thomas, S.: Knowledge compilation for model counting: affine decision trees. In: IJCAI (2013). https://dl.acm.org/doi/10.5555/2540128.2540265
42. Koster, A.M., Bodlaender, H.L., Van Hoesel, S.P.: Treewidth: computational experiments. Electron Notes Discrete Math. **8**, 54–57 (2001). https://doi.org/10.1016/S1571-0653(05)80078-2
43. Kourtis, S., Chamon, C., Mucciolo, E., Ruckenstein, A.: Fast counting with tensor networks. SciPost Phys. **7**(5) (2019). https://doi.org/10.21468/SciPostPhys.7.5.060
44. Kwiatkowska, M., Norman, G., Parker, D.: Stochastic model checking. In: Bernardo, M., Hillston, J. (eds.) SFM 2007. LNCS, vol. 4486, pp. 220–270. Springer, Heidelberg (2007). https://doi.org/10.1007/978-3-540-72522-0_6
45. Lagniez, J.M., Marquis, P.: Preprocessing for propositional model counting. In: AAAI (2014). https://dl.acm.org/doi/10.5555/2892753.2892924
46. Lagniez, J.M., Marquis, P.: An improved decision-DNNF compiler. In: IJCAI, pp. 667–673 (2017). https://doi.org/10.24963/ijcai.2017/93

47. Lawson, C.L., Hanson, R.J., Kincaid, D.R., Krogh, F.T.: Basic linear algebra subprograms for Fortran usage. TOMS **5**(3), 308–323 (1979). https://doi.org/10.1145/355841.355847
48. McMahan, B.J., Pan, G., Porter, P., Vardi, M.Y.: Projection pushing revisited. In: Bertino, E., et al. (eds.) EDBT 2004. LNCS, vol. 2992, pp. 441–458. Springer, Heidelberg (2004). https://doi.org/10.1007/978-3-540-24741-8_26
49. Morgenstern, A., Schneider, K.: From LTL to symbolically represented deterministic automata. In: Logozzo, F., Peled, D.A., Zuck, L.D. (eds.) VMCAI 2008. LNCS, vol. 4905, pp. 279–293. Springer, Heidelberg (2008). https://doi.org/10.1007/978-3-540-78163-9_24
50. Naveh, Y., et al.: Constraint-based random stimuli generation for hardware verification. AI Mag. **28**(3), 13–13 (2007). https://dl.acm.org/doi/10.5555/1597122.1597129
51. Oliphant, T.E.: A guide to NumPy, vol. 1. Trelgol Publishing, USA (2006). https://dl.acm.org/doi/book/10.5555/2886196
52. Oztok, U., Darwiche, A.: A top-down compiler for sentential decision diagrams. In: IJCAI (2015). https://dl.acm.org/doi/10.5555/2832581.2832687
53. Palacios, H., Geffner, H.: Compiling uncertainty away in conformant planning problems with bounded width. JAIR **35**, 623–675 (2009). https://dl.acm.org/doi/10.5555/1641503.1641518
54. Pan, G., Vardi, M.Y.: Symbolic techniques in satisfiability solving. J. Autom. Reasoning **35**(1–3), 25–50 (2005). https://doi.org/10.1007/s10817-005-9009-7
55. Robertson, N., Seymour, P.D.: Graph minors. X. Obstructions to tree-decomposition. J. Comb. Theory B **52**(2), 153–190 (1991). https://doi.org/10.1016/0095-8956(91)90061-N
56. Samer, M., Szeider, S.: Algorithms for propositional model counting. J. Discrete Algorithms **8**(1), 50–64 (2010). https://doi.org/10.1007/978-3-540-75560-9_35
57. Samer, M., Szeider, S.: Constraint satisfaction with bounded treewidth revisited. JCSS **76**(2), 103–114 (2010). https://doi.org/10.1016/j.jcss.2009.04.003
58. Sang, T., Bacchus, F., Beame, P., Kautz, H.A., Pitassi, T.: Combining component caching and clause learning for effective model counting. SAT **4**, 20–28 (2004). http://www.satisfiability.org/SAT04/accepted/65.html
59. Sang, T., Beame, P., Kautz, H.A.: Performing Bayesian inference by weighted model counting. In: AAAI, vol. 1, pp. 475–482. AAAI Press (2005). https://dl.acm.org/doi/10.5555/1619332.1619409
60. Shachter, R.D., Andersen, S.K., Szolovits, P.: Global conditioning for probabilistic inference in belief networks. In: UAI, pp. 514–522. Elsevier (1994). https://doi.org/10.1016/B978-1-55860-332-5.50070-5
61. Sinz, C., Kaiser, A., Küchlin, W.: Formal methods for the validation of automotive product configuration data. AI EDAM **17**(1), 75–97 (2003). https://doi.org/10.1017/S0890060403171065
62. Smilde, A., Bro, R., Geladi, P.: Multi-way Analysis: Applications in the Chemical Sciences. Wiley, Hoboken (2005). https://doi.org/10.1002/0470012110
63. Somenzi, F.: CUDD: CU decision diagram package-release 3.0.0. University of Colorado at Boulder (2015). https://github.com/ivmai/cudd
64. Strasser, B.: Computing tree decompositions with FlowCutter: PACE 2017 submission. arXiv preprint arXiv:1709.08949 (2017)
65. Tabajara, L.M., Vardi, M.Y.: Factored Boolean functional synthesis. In: FMCAD, pp. 124–131. IEEE (2017). https://dl.acm.org/doi/10.5555/3168451.3168480

66. Tamaki, H.: Positive-instance-driven dynamic programming for treewidth. J. Comb. Optim. **37**(4), 1283–1311 (2019). https://doi.org/10.1007/s10878-018-0353-z
67. Tarjan, R.E., Yannakakis, M.: Simple linear-time algorithms to test chordality of graphs, test acyclicity of hypergraphs, and selectively reduce acyclic hypergraphs. SICOMP **13**(3), 566–579 (1984). https://doi.org/10.1137/0213035
68. Uribe, T.E., Stickel, M.E.: Ordered binary decision diagrams and the Davis-Putnam procedure. In: Jouannaud, J.-P. (ed.) CCL 1994. LNCS, vol. 845, pp. 34–49. Springer, Heidelberg (1994). https://doi.org/10.1007/BFb0016843
69. Valiant, L.G.: The complexity of enumeration and reliability problems. SICOMP **8**(3), 410–421 (1979). https://doi.org/10.1137/0208032

# Aggregation and Garbage Collection for Online Optimization

Alexander Ek[1,2](✉), Maria Garcia de la Banda[1], Andreas Schutt[2], Peter J. Stuckey[1,2], and Guido Tack[1,2]

[1] Monash University, Melbourne, Australia
{Alexander.Ek,Maria.GarciadelaBanda,Peter.Stuckey,Guido.Tack}@monash.edu
[2] Data61, CSIRO, Melbourne, Australia
Andreas.Schutt@data61.csiro.au

**Abstract.** Online optimization approaches are popular for solving optimization problems where not all data is considered at once, because it is computationally prohibitive, or because new data arrives in an ongoing fashion. Online approaches solve the problem iteratively, with the amount of data growing in each iteration. Over time, many problem variables progressively become *realized*, i.e., their values were fixed in the past iterations and they can no longer affect the solution. If the solving approach does not remove these realized variables and associated data and simplify the corresponding constraints, solving performance will slow down significantly over time. Unfortunately, simply removing realized variables can be incorrect, as they might affect unrealized decisions. This is why this complex task is currently performed manually in a problem-specific and time-consuming way. We propose a problem-independent framework to identify realized data and decisions, and remove them by summarizing their effect on future iterations in a compact way. The result is a substantially improved model performance.

## 1 Introduction

*Online optimization* tackles the solving of problems that evolve over time. In online optimization problems, in some areas also called *dynamic* or *reactive*, the set of input data is only partially known a priori and new data, such as new customers and/or updated travel times in a dynamic vehicle routing problem, continuously or periodically arrive while the current solution is executed. This new data and the current execution state must be incorporated into the problem in an ongoing fashion to revise previous decisions and to take new decisions.

Online optimization problems are solved iteratively as a *sequence* of *offline* optimization problems, where each problem represents the available information state of the online problem at a particular point in time. In each iteration or *session*, the online optimization approach must create an *update instance* to update the current solution. An *update model* is a model that can be repeatedly

Partly funded by Australian Research Council grant DP180100151.


H. Simonis (Ed.): CP 2020, LNCS 12333, pp. 231–247, 2020.
https://doi.org/10.1007/978-3-030-58475-7_14

instantiated to create new update instances in the sequence; it incorporates the new data, the old data (possibly modified by external sources), and the results from solving the previous update instance (also possibly modified).

One inherent challenge of online problems is their continuous growth in size as time passes, which can yield an unnecessary and prohibitive performance decay. As a result, online optimization approaches often try to remove, or *garbage collect*, data that is irrelevant for current and future decisions to be made, such as completed delivery trips in dynamic vehicle routing problems. Unfortunately, garbage collection is a complex task, as it requires understanding the interaction between time, data and the variables and constraints used to model the problem. For this reason, existing removal methods (see related work in Sect. 6) are highly problem-specific and, thus, not easily transferable to other problems, or naïve, thus still causing significant performance issues as shown in Sect. 5.

We propose a problem-independent framework, in which the modeler provides minimal information about the relation between time, data, and variables. The framework performs three main steps. First, the modeler's information is used to analyze each constraint that would be part of the update instance, with the aim to automatically identify the data and variables that are now *realized*, i.e., can no longer affect future decisions, either because they can no longer change, or because any change is now irrelevant. Second, once the realized information for a constraint is inferred, the constraint is modified to aggregate as much of the realized information as it can. And third, once all constraints have been analyzed and aggregated, the garbage from all constraints is collected and removed to ensure it does not form part of the update instance. We note that this paper significantly extends our previous problem-independent online modeling approach [7], with a much more sophisticated and effective method for garbage collection. We also note that our online approach can also be used for solving large-scale offline optimization problem, when applying an iterative solving approach over a rolling horizon [19].

To sum up, our main **contributions** are as follows: (1) A systematic way of modeling and inferring when each part of a model is realized (Sect. 4.1), and how to utilize this. (2) For several kinds of constraints, methods of summarizing the effect that realized parts will have on the future without keeping them (Sect. 4.2). This is done via incrementally aggregating the realized values and slightly reformulating the constraints to use these aggregated values. (3) A garbage collection mechanism that analyzes a high-level constraint model, identifies which parts are never used again, and safely removes them from future sessions (Sect. 4.3). (4) An empirical evaluation of the proposed approach (Sect. 5).

## 2 Preliminaries

A *constraint satisfaction problem* (CSP) $P = (\mathbf{X}, \mathbf{D}, \mathbf{C})$ consists of a set of variables $\mathbf{X}$, a function $\mathbf{D}$ mapping each variable $x \in \mathbf{X}$ to its domain (usually a finite set of values) $D_x$, and a set of constraints $\mathbf{C}$ over $\mathbf{X}$. For optimization problems, we add a special variable $o$ to $\mathbf{X}$, called the *objective*, to be w.l.o.g.

minimized. A *dynamic constraint satisfaction problem* (DCSP) [6] is a (possibly infinite) list of CSPs $DP = (P_1, P_2, \dots)$, where each $P_i$ represents the state after the $i$th problem change is considered, where a change can remove and/or add constraints. It allows us to consider each session $s_i$ as solving the stand-alone CSP $P_i$. A solving method is *offline* if it is designed to solve one CSP, and *online* if it is designed to iteratively solve and generate the CSPs in a DCSP.

We distinguish between a problem *model*, where the input data is described in terms of *parameters* (i.e., data that will be known before the search starts), and a model *instance*, where the values of the parameters are added to the model. While an instance can be directly represented as a CSP, a model can be represented as a *parameterized CSP* [13] $P[Data]$, that is, a family of CSPs $P[\delta]$ for every $\delta \in Data$. The parameterization also applies to its components $(\mathbf{X}[Data], \mathbf{D}[Data], \mathbf{C}[Data])$. This allows us to extend the online approach of [7] by representing a DCSP as the list $DP = (P[\delta_1], P[\delta_2], \cdots)$, where data can change over time, while the underlying model remains unchanged.

We assume DCSPs have *complete recourse* [5], that is, all their CSPs are feasible. We also assume DCSPs have *uncertainty in execution*, that is, the values of the parameters between consecutive CSPs, $P[\delta_i]$ and $P[\delta_{i+1}]$, can be modified by external sources. As an example for external modifications, consider a dynamic vehicle routing problem. The distance between two locations may change (e.g., due to road works or traffic congestion), or a vehicle may spend more time at a customer due to unforeseen delays. External modifications are, however, no longer possible once the parameters in question become *realized*, that is, once their actual values are either observed (indicating they refer to the past) or guaranteed never to be observed (indicating they will never be used and thus can be safely ignored). For example, once a vehicle has actually left a customer site, the time it left is now realized and can be assumed as fixed forever.

Note that, for simplicity, we do not use stochastic information during online optimization [2,20,21], nor predict future changes based on historical data [3]. We simply react to any changes that occur in what [4] calls *pure reaction*, and resolve each CSP in its entirety for each session. Thus, no information about $P[\delta_{i+1}]$ is used by $P[\delta_i]$. Also, we assume *non-preemption* for our online solving methods, that is, once a $P[\delta_i]$ starts to be solved, the execution cannot abort (either to later resume it or to restart it). However, we believe that the concepts presented in this paper can be generalized in a straightforward way to the stochastic, preemptive, incomplete recourse case.

We denote by *aggregation* the process of simplifying a constraint by replacing any fixed decision variables by their value. Faster solving can be achieved if we precompute the *aggregated value* of the fixed variables in each constraint and remove these variables from it. Consider for example a constraint containing the sum $\sum_i c_i \times x_i$. We can easily partition the coefficients $c_i$ and variables $x_i$ into two sets: those containing fixed decision variables $F$ and the rest $V$. Then, we can substitute the sum $\sum_i c_i \times x_i$ by $fixcost + \sum_{i \in V} c_i \times x_i$, where the value $fixcost = \sum_{i \in F} c_i \times x_i$ is precomputed. This technique is common in many solvers, particularly copying solvers (e.g., Gecode [9]), which aggressively simplify all constraints before copying them. In our online setting, we aim to *remember* (and not redo) the simplifications done in past sessions, and thus incrementally *aggregate the results* over time (see Sect. 4.2).

## 3 Basics of Our Framework and Running Example

Given a DCSP $DP = (P[\delta_1], P[\delta_2], \dots)$, our framework consists of a series of (solving) sessions, where each session $s_i$ starts to execute instance $P[\delta_i]$ after session $s_{i-1}$ finishes. The result of session $s_i$ contains the best solution to $P[\delta_i]$ that could be found within a given time limit. As we will see later, the result also contains information about which parts of $\delta_i$ have become garbage, and this information will be used to generate $P[\delta_{i+1}]$.

Time, quite obviously, plays an important role in online optimization. We denote by $\tau_i$ the *deadline* for session $s_i$, i.e., the latest time by which the result of $s_i$ must have been produced (which can be calculated as the start time of $s_i$ plus the time limit). Importantly, $\tau_i$ is available as a parameter in $\delta_i$, essentially representing the value "now", as seen from the outside world. For example, if the current session $s_i$ makes a decision to start a task on a particular machine, the earliest possible time for that job to start is $\tau_i$, since any earlier time will be in the past by the time the solution can be implemented. Similarly, if an earlier session made the decision to start a task on a particular machine, then the current session may be able to revise that decision if the start time is still in the future, i.e., greater than $\tau_i$. The parameter $\tau_i$ thus synchronizes the session with the outside world.

Each $\delta \in \mathit{Data}$ is a tuple $(OS_\delta, DV_\delta)$, where $OS_\delta$ is the set of *object set* parameters and $DV_\delta$ is the set of *data value* parameters. The former contains a set per type of dynamic object in the problem, i.e., per type of object that can arrive as time progresses, such as the set of jobs for a job scheduling problem, the set of product orders for a product manufacturing problem, and the set of customer requests for a vehicle routing problem. Typically, these sets are the backbone of the DCSP models, as they index most loops and parameter arrays. For example, the set of jobs will index the array of precedences among jobs, and the set of customer requests will index the array containing the amount of time required by each customer request. This is why object sets are the key to our garbage collection method. It is also why the objects in each object set of $OS_\delta$ must be uniquely named, with their number determining the size of the instance.

The set of data value parameters denotes the particular givens in the current state of the instance. For example, the current location of all vehicles to route, the current expected travel times between points in a traffic network, or the durations of the jobs that must be run. Each element of the set can be defined as a singleton or as a multi-dimensional array possibly indexed by object sets.

Many of the data value parameters, decision variables and constraints in instance $P[\delta_{i+i}]$ will need to refer to the values obtained by session $s_i$ when solving $P[\delta_i]$. Thus, every $\delta_{i+1} \in \mathit{Data}$ contains *new* data, i.e., objects and parameter values for session $s_{i+1}$, and *old* data, i.e., the decisions, objects and parameter values either produced by session $s_i$ when solving $P[\delta_i]$ or modified by an external source after $s_i$ (e.g., if a task previously scheduled required more time to be performed than expected). We will distinguish between old and new data sets by prefixing them with the $\omega$ and $\nu$ symbols, respectively.

### 3.1 Running Example: Dynamic Vehicle Routing

The following problem is used throughout the rest of the paper as an illustrative example. Consider a dynamic vehicle routing problem (DVRP), where a fleet of vehicles has to be routed to attend to upcoming customer requests as soon as possible, while minimizing the total travel time. The problem model is as follows.

**Object Sets:** The arrival of a new customer request $c$ at time $\tau_c$, is modeled by adding object $c$ to the object set of *customer requests* $C$ at time $\tau_c$. We model vehicle availability by means of the availability of a new *tour*. This allows us to ignore whether the vehicle is new, or just finished its previous tour. Thus, a new tour $t$ becoming available at time $\tau_t$, is modeled by adding tour object $t$ to the object set of *tours* $T$ at time $\tau_t$. Each tour $t \in T$ starts at its *start depot* $S(t)$, services some customer requests and ends at its *end depot* $E(t)$. The set of customer requests $C$, start depots $S(T)$ and end depots $E(T)$, yield a set of locations modeled as *nodes* $N$. As a result, $C$ and $T$ are input object sets, while $N$ is an object set constructed from them as $N = S(T) \cup E(T) \cup NC(C)$, where $NC(C)$ maps the customer requests to their associated nodes.

**Data Value Parameters:** They are provided by two tables. The first, $\texttt{wait}\colon N \rightarrow \mathbb{Z}^{\geq}$, provides the amount of time $\texttt{wait}_n$ a vehicle has to wait after reaching node $n \in N$ before it can go on to the next node (this models, e.g., the time required to provide some service at node $n$). The second, $\texttt{dist}\colon N \times N \rightarrow \mathbb{Z}^{\geq}$, provides the traveling time $\texttt{dist}_{nm}$ from every node $n$ to any node $m$ in $N$. We assume $\texttt{wait}_n = 0$ if $n$ corresponds to a depot. In addition, the special parameter $\tau$ gives the end of solving time for the current session, and hence the earliest time after which decisions can be modified.

**Variables:** While the set of customer requests serviced by a tour must be decided before the tour begins (i.e., leaves its starting depot), the time of service (and thus the order in which customer requests are serviced) can change. Therefore, the exact routes are modeled using three arrays of decision variables indexed by the set of nodes $N$: $serv_n$ determines the tour that serves node $n$, $succ_n$ the node that is served after node $n$, and $arri_n$ the time at which the tour arrives at node $n$.

**Problem Constraints:** The problem is modeled by the following six constraints. PC1 ensures each tour visits its start and end depot. PC2 records the time at which each tour that is new for the current session becomes available, by setting its arrival time at its start depot to $\tau$. This ensures it is ready to start by the end of the current solving session. PC3 connects the nodes serviced by each tour. PC4 ensures the successor of a node $n$ is not reached until $n$ has been serviced for the required amount of time, and the time taken to travel to the successor is considered. The last two constraints ensure the whole successor array forms a single circuit. PC5 establishes a circuit for each tour using the global constraint on the successor array, thus ensuring every node belongs to one tour. SBC1 is a symmetry breaking constraint that closes the circuit by ensuring the successor of the end depot of tour $t$ is the start depot of tour $t+1$ (in a fixed order), and the successor of the last tour is the start depot of the first one.

$$\textbf{forall}_{t \text{ in } T}\ serv_{S(t)} = serv_{E(t)} = t \tag{PC1}$$

$$\textbf{forall}_{t \text{ in } \nu T}\ arri_{S(t)} = \tau \tag{PC2}$$

$$\underset{n \text{ in } S(T) \cup NC(C)}{\textbf{forall}}\ serv_n = serv_{succ_n} \tag{PC3}$$

$$\underset{n \text{ in } S(T) \cup NC(C)}{\textbf{forall}}\ arri_n + \texttt{wait}_n + \texttt{dist}_{n,succ_n} \leq arri_{succ_n} \tag{PC4}$$

$$\textsc{Circuit}(succ) \tag{PC5}$$

$$succ_{E(\max(T))} = S(\min(T)) \wedge \underset{t \text{ in } T \setminus \max(T)}{\textbf{forall}}\ succ_{E(t)} = S(t+1) \tag{SBC1}$$

**Overlap Constraints:** Online solving methods need to build an *update model* that solves the problem while taking into account earlier decisions. The *overlap constraints* add information about how earlier information affects the current session. In the approach of [7], which we extend, these are constructed automatically from annotations of the original model. The overlap constraints focus on old decisions that must be *committed*, that is, decisions taken by the previous session (or modified by external sources) that cannot change in the current one. OC1 commits the customer requests serviced by a tour if the tour has left the start depot by the time the current solving session ends. OC2 commits the successor of a node as soon as the tour starts heading towards it. Successors of end depots are not included in this constraint, as that would interfere with SBC1. Finally, OC3 commits the arrival time if the node has started to be serviced by the time the session ends.

$$\begin{aligned} &\textbf{forall}_{t \text{ in } \omega T}\ \textbf{if}\ \tau \geq \omega arri_{\omega succ_{S(t)}} - \texttt{dist}_{S(t),\omega succ_{S(t)}} \\ &\quad \textbf{then}\ \textbf{forall}_{n \text{ in } \omega N}\ serv_n = t \leftrightarrow \omega serv_n = t \end{aligned} \tag{OC1}$$

$$\underset{n \text{ in } \omega N \setminus E(\omega T)}{\textbf{forall}}\ \textbf{if}\ \tau \geq \omega arri_n + \texttt{wait}_n\ \textbf{then}\ succ_n = \omega succ_n \tag{OC2}$$

$$\underset{n \text{ in } \omega N}{\textbf{forall}}\ \textbf{if}\ \tau \geq \omega arri_n\ \textbf{then}\ arri_n = \omega arri_n\ \textbf{else}\ arri_n \geq \tau \tag{OC3}$$

**Objective:** We minimize the total travel cost, computed as the sum of arrival times at the end depot minus departure times from the start depot:

$$\begin{aligned} &cost = \textbf{sum}_{t \text{ in } T}\ arri_{E(t)} - depart_{S(t)}, \\ &\quad \textbf{where}\ depart_{S(t)} = arri_{succ_{S(t)}} - dist_{S(t),succ_{S(t)}}. \end{aligned} \tag{OBJ}$$

# 4 Automated Garbage Collection

Our *garbage collection* method aims to remove as many *garbage* objects as possible from our object sets, together with their associated data value parameters, decision variables and constraints. An object is considered garbage if it can be removed without affecting the solutions, that is, if all decisions associated with the object are realized, and their values can be safely aggregated by all constraints in the instance. Recall that a data value parameter or decision is realized if it has already been observed (and can therefore no longer change) or will never be observed (and can thus be safely ignored). As we will see, our garbage collection method cannot simply consider the instance being compiled by the current session. Instead, it must become an incremental and cumulative process that operates across multiple sessions, aggregating realized data and decisions.

The method consists of the following three main steps: (1) Identify realized data value parameters and decision variables, in order to determine the fully realized objects (Sect. 4.1); (2) Aggregate the constraints together with the realized data value parameters and decision variables used in these constraints (Sect. 4.2); (3) Collect the garbage resulting from the aggregation and remove it from future instances (Sect. 4.3).

Note that the above automated garbage collection method is performed at the beginning of each session, when constructing the constraint problem *instance* $P[\delta]$ that will be sent to the solver. This is achieved by reformulating the update model to reason about garbage, as we describe below. While it may be more efficient to deal with aggregation and garbage collection outside of the model, this has the downside of making the method problem-specific. In contrast, our proposed automated method deduces what is garbage from the realization information in the model alone and can be used for any problem, since both the aggregation of each constraint and the garbage collection method are independent of the model in which they occur.

## 4.1 Phase One: Identifying Realizations

As time moves forward, objects, data value parameters, and decision variables become realized. We define a realization function $\mathcal{R}$ for a session solving instance $P[\delta] = (\mathbf{X}[\delta], \mathbf{D}[\delta], \mathbf{C}[\delta])$ as $\mathcal{R}\colon \bigcup OS_\delta \cup DV_\delta \cup \mathbf{X}[\delta] \rightarrow \{true, false\}$ to indicate whether a given object, data value parameter, or decision variable is realized. For ease of notation we define the complement to indicate whether an object, data parameter, or decision variable is not realized $\mathcal{NR}(z) = \neg\mathcal{R}(z)$, and extend both functions in the obvious way to operate on sets of elements.

If online optimization is used to solve a static problem, then any data value parameter that becomes known and any decision variable that is committed is automatically realized. For dynamic problems, however, the modeler's expert knowledge is needed to annotate the data value parameters and decision variables in the model to indicate when they are realized. Object realization can be automatically inferred from this. Let us illustrate this inference by means of our

DVRP model. We will assume that newly added data and variables are never realized, and focus on the old data and variables (identified by $\omega$).

**Variable and Data Realizations:** As shown in Sect. 3.1, our DVRP model has three arrays of decision variables and two arrays of data value parameters. Their realization is provided by the modeler as the functions (for all $n, m \in N$):

$$\mathcal{R}(\omega serv_n) = \omega arri_{\omega succ_{S(\omega serv_n)}} - \mathtt{dist}_{S(\omega serv_n),\omega succ_{S(\omega serv_n)}} \leq \tau; \quad (\mathcal{R}\text{V1})$$

$$\mathcal{R}(\omega succ_n) = \omega arri_{\omega succ_n} \leq \tau; \quad (\mathcal{R}\text{V2})$$

$$\mathcal{R}(\omega arri_n) = \omega arri_n \leq \tau; \quad (\mathcal{R}\text{V3})$$

$$\mathcal{R}(\omega \mathtt{wait}_n) = \omega arri_n + \omega \mathtt{wait}_n \leq \tau; \text{ and} \quad (\mathcal{R}\text{D1})$$

$$\mathcal{R}(\omega \mathtt{dist}_{nm}) = \omega arri_{\omega succ_n} \leq \tau \vee \omega arri_m \leq \tau. \quad (\mathcal{R}\text{D2})$$

$\mathcal{R}$V1 marks the tour that serves node $n$ as realized if the vehicle of that tour has left the depot before the end of the current session (i.e., by $\tau$). $\mathcal{R}$V2 marks the successor of node $n$ as realized, if the tour arrives at $n$'s successor by $\tau$. $\mathcal{R}$V3 marks the arrival time at node $n$ as realized, if it is less or equal than $\tau$. $\mathcal{R}$D1 marks the waiting time at node $n$ as realized, if the tour has already arrived at $n$ and finished waiting by $\tau$. $\mathcal{R}$D2 marks the distance from node $n$ to node $m$ as realized, if the tour has already arrived either at the successor of $n$ or at $m$ by $\tau$. Note that if $m$ is not $n$'s successor, distance $\omega \mathtt{dist}_{nm}$ will never be used.

This example illustrates why realization needs to be defined by the modeler: The current model assumes that we cannot change the allocation of customers to vehicles after a vehicle has left the depot. This may be suitable for vehicles that need to pick up goods from the depot and deliver to the customers. But it may be unnecessarily restrictive for vehicles that provide a service, in which case we could change rule $\mathcal{R}$V1 to $\mathcal{R}(\omega serv_n) = \omega arri_n \leq \tau$.

**Object Realizations and Correspondence:** An object is realized when all the data value parameters and decisions about that object are realized. Since all five arrays of data value parameters and variables are indexed by object sets, we can automatically infer which objects are realized. Note that, since $N$ is constructed from $C$ and $T$, the realization relationships between them must be examined. We will come back to this later. We introduce the local realization function $\mathcal{R}^L\colon \delta \cup \mathbf{X}[\delta] \to \{\mathit{true}, \mathit{false}\}$, to reason about direct usage. The (for now manual) analysis determines that $C$ and $T$ are not used to index any of the five arrays. Therefore, all their objects are **locally** realized, i.e., $\mathcal{R}^L(c) = \mathcal{R}^L(t) = \mathit{true}$, for all $c \in \omega C$ and $t \in \omega T$. $N$ is however used as index set in all arrays of decision variables and data value parameters. Its local realization is defined for all $n \in N$ as:

$$\mathcal{R}^L(n) = \mathcal{R}(\omega serv_n) \wedge \mathcal{R}(\omega succ_n) \wedge \mathcal{R}(\omega arri_n) \wedge \mathcal{R}(\omega \mathtt{wait}_n) \\ \wedge (\forall m \in \omega N\colon \mathcal{R}(\omega \mathtt{dist}_{nm}) \wedge \mathcal{R}(\omega \mathtt{dist}_{mn})),$$

which marks object $n \in N$ as locally realized if all variables and data parameter arrays indexed by $n$ (in one or more dimensions) are also realized.

Direct usage is extended to indirect usage $\forall c \in \omega C$, $t \in \omega T$, and $n \in \omega N$ as:

$$\mathcal{R}(c) = \mathcal{R}^L(c) \wedge \mathcal{R}^L(NC(c));$$
$$\mathcal{R}(t) = \mathcal{R}^L(t) \wedge \mathcal{R}^L(S(t)) \wedge \mathcal{R}^L(E(t)); \text{ and}$$
$$\mathcal{R}(n) = \mathcal{R}^L(n) \wedge ((n = S(t) \vee n = E(t)) \rightarrow \mathcal{R}^L(t)) \wedge (n = NC(c) \rightarrow \mathcal{R}^L(c));$$

indicating a customer request $c$ is realized if both $c$ and its associated node $NC(c)$ are locally realized; a tour $t$ is realized if $t$, its start node, and its end node are locally realized; and a node $n$ is realized if $n$ is locally realized and its associated tour or customer are locally realized (which in this case is always true). For brevity, $\mathcal{R}(S)$ denotes the set of all realized objects in object set $S$.

### 4.2 Phase Two: Aggregation

Realized data, variables and objects can never change in future uses, or are never used in the future. Thus, a constraint can be removed if all its elements are realized. Consider, e.g., Eq. PC3: when $succ_n$ is realized the constraint must already hold and can thus be ignored. If a constraint cannot be removed, any subexpression containing only realized elements can be replaced with an aggregated value. In general, the aggregation consists of three steps: (1) identifying the set of realized objects that can be aggregated, (2) redefining the constraint to use the aggregations, and (3) introducing and defining the aggregated value(s).

In most constraints, the realized elements are not used to access any other values (e.g., as array index). For them aggregation is straightforward. Consider, for example, the objective OBJ of the DVRP model, which loops over object set $T$. Partitioning $T$ into realized and non-realized objects gives:

$$cost = \mathbf{sum}_{t \text{ in } \mathcal{R}(T)}(arri_{E(t)} - depart_{S(t)}) + \mathbf{sum}_{t \text{ in } \mathcal{NR}(T)}(arri_{E(t)} - depart_{S(t)}).$$

To aggregate the realized part, we simply compute and keep its value by introducing a new data parameter, $\mathtt{agg_{obj}} : \mathbb{Z}^{\geq}$ to represent this value, and then transform the objective constraint OBJ into:

$$cost = \mathtt{agg_{obj}} + \mathbf{sum}_{t \text{ in } \mathcal{NR}(T)}(arri_{E(t)} - depart_{S(t)}), \quad \text{(AC-OBJ)}$$

where $\mathtt{agg_{obj}}$ sums up everything aggregated so far, $\omega\mathtt{agg}$ (corresponding to objects no longer in $T$), plus everything aggregated in the current session:

$$\mathtt{agg_{obj}} = \omega\mathtt{agg_{obj}} + \mathbf{sum}_{t \text{ in } \mathcal{R}(\omega T)}(arri_{E(t)} - depart_{S(t)}). \quad \text{(AV-OBJ)}$$

For constraints where the realized values are indeed used to access other values, aggregation can be quite complex. Consider, for example, the circuit constraint for a graph with six nodes $a, \ldots, f$. Suppose the previous session obtained the solution illustrated in Fig. 1(a). Suppose that nodes $c$, $d$ and $e$ are

**Fig. 1.** Visualization of circuit aggregation via short-circuiting. The current solution (a) has realized nodes shown in double circles. Nodes $d$ and $e$ can be removed as garbage and node $c$ is short-circuited to point at $f$, but must remain since (b) some non-garbage node will point at it, in the next session.

completely realized. Aggregation for this circuit can then be achieved by short-circuiting it as visualized in Fig. 1(b). Note we cannot remove $c$ because some variable must point to $c$ in future sessions. We can however remove $d$ and $e$ from the model, and they become garbage.

In general aggregation is constraint specific. Thankfully it is well understood, if not well publicized, and modern solvers, in particular Gecode [9], aggressively aggregate constraints. The extra challenge that arises in automatic garbage collection is that the result of the aggregation (e.g., $\texttt{agg}_{\texttt{obj}}$) needs to be communicated *across* sessions. Our automated garbage collection modifies the update model to both make use of the modified form of constraints with aggregation (e.g., AC-OBJ) values (e.g., AV-OBJ) as well as output this new aggregate value, so that it is available to the next session.

We can create a library of aggregating versions for common global constraints, which may require adding new arguments to transmit the aggregate values. We can also create a library of functions to compute the aggregate values required.

Finally, some constraints can be ignored throughout the aggregation process. In particular, redundant (or implied) constraints aim to speed up solving, but are not necessary for defining the problem, and are only relevant to the current instance. Hence, redundant constraints can be safely ignored during aggregation, and kept as-is for solving. Similarly, the overlap constraints are only required to communicate the effects from the previous session on the next session. Hence, they do not need to be considered for aggregation either. In contrast, while symmetry-breaking constraints can be safely ignored during aggregation (as they are only relevant in the current instance), they might need to be reformulated. This is because they eliminate solutions and, thus, can only be kept as-is if the remaining solutions are compatible with those left by symmetry breaking constraints of previous instances and all aggregations. In general, symmetry-breaking constraints need to be carefully designed on a model-by-model basis. This is outside the scope of this paper and remains future work. The symmetry-breaking constraints for our DVRP model are compatible with the aggregations since the end nodes point at the next start nodes, except for the last end node, which points to the first start node. This is the exact pattern our circuit aggregation will enforce as well.

### 4.3 Phase Three: Garbage Collection

Our method aims to determine *objects* that are garbage for removal of any data or variables associated to these objects. Without this, instances would constantly grow with time. As shown for the circuit constraint above, some realized (and aggregated) objects cannot be removed. Therefore, we need to determine what can be safely removed and what cannot. We will say that realized objects, data value parameters, and variables are *garbage* if they can be removed without creating an inconsistency, and are *non-garbage* otherwise. Note that only realized objects, data value parameters, and variables can be garbage since, otherwise, their new values might change the solutions found by subsequent sessions.

The garbage collection phase for $P[\delta]$ is itself divided into three steps. The first step identifies the set of realized objects that are non-garbage for each constraint in $P[\delta]$ when considered *in isolation*, that is, the non-garbage objects local to each constraint. Note that these objects, together with the data value parameters and variables associated to them, are the only realized elements that have *not* been already aggregated during the previous phase (i.e., phase two). Formally, the identification of the non-garbage local to a given constraint in $P[\delta] = (\mathbf{X}[\delta], \mathbf{D}[\delta], \mathbf{C}[\delta])$ is defined in terms of the function $\mathcal{NG}^L \colon \mathbf{C}[\delta] \to \{G \mid G \subseteq \bigcup OS_\delta\}$, which returns the set of realized objects in every object set that the given constraint $c$ considers to be non-garbage and has, therefore, not been aggregated by $c$. For example, in our DVRP model, if $c$ is the objective constraint OBJ, then $\mathcal{NG}^L(c)$ will return the empty set, since all realized objects are garbage for OBJ. However, if $c$ is the circuit constraint PC5, then $\mathcal{NG}^L(c)$ will return any realized node $n$ whose predecessor node is not realized (e.g., an end node is the predecessor of the start node of another tour), that is, it returns $\{n \in \mathcal{R}(N) \mid succ_m = n \text{ and } m \in \mathcal{NR}(N)\}$. In the example for the circuit constraint provided in the previous section, this set would contain the realized node named $c$, but not the nodes named $d$ and $e$.

The second step in this phase collates the local non-garbage information to determine the realized objects that are non-garbage for the entire instance $P[\delta]$. To do this we define a global non-garbage function:

$$\mathcal{NG} \colon OS_\delta \to \{NG \mid NG \subseteq S,\ S \in OS_\delta\},\ \textbf{s.t. } \mathcal{NG}(S) \subseteq S,\ \forall S \in OS_\delta,$$

which maps each objects set $S \in OS_\delta$ to the objects in $S$ that are non-garbage:

$$\mathcal{NG}(S) = \mathcal{R}(S) \cap \bigcup\nolimits_{c \in \mathbf{C}[\delta]} \mathcal{NG}^L(c), \quad \forall S \in OS_\delta,$$

that is, the subset of all objects identified by any constraint $c$ in the instance as non-garbage, that are also realized objects of $S$. For ease of notation we use the complement to indicate the set of realized objects that are garbage $\mathcal{G}(S) = \mathcal{R}(S) \setminus \mathcal{NG}(S)$ and can therefore be removed.

The third and last step in this phase removes any garbage identified at the beginning of session $s_i$ to create instance $P[\delta_i]$, thus eliminating the garbage from all future instances as well. This is already partially achieved in phase two, by modifying the constraints to aggregate all realized data value parameters and variables. To complete the task, we must remove any objects identified as garbage from the input object sets. We do this by redefining each input object set $S$ as $\mathcal{NG}(S) + \mathcal{NR}(S)$, thus ensuring $S$ retains every realized object that has been identified as non-garbage ($\mathcal{NG}(S)$), plus every object that is not realized ($\mathcal{NR}(S)$). Note that $\mathcal{NR}(S)$ contains all new objects of $S$ and all old, non-realized ones. Note also that constructed object sets do not need to be redefined, as they are built from the input sets. For example, for our DVRP this step would require redefining the object set of tours $T = \mathcal{NG}(T) + \mathcal{NR}(T)$ and the object set of customer requests $C = \mathcal{NG}(C) + \mathcal{NR}(C)$. Once this is done, the definition of the constructed set of nodes $N = S(T) \cup E(T) \cup NC(C)$ automatically takes advantage of the (possibly smaller) $T$ and $C$ sets. We also extend the model to output the data for the new collections of objects, to use in the next session. This implicitly removes any data associated with a garbage object. Garbage variables are implicitly removed since the object set used to construct them no longer contains garbage objects. Therefore, once phase three finishes, $P[\delta_i]$ does not refer to objects, data value parameters, or variables identified as garbage.

## 5 Experimental Results

We present experimental results that demonstrate the effectiveness of our garbage collection approach. For this purpose, we took the standard MiniZinc [15] models of two optimization problems and transformed them (by hand), adding auxiliary functions, predicates and parameter definitions that implement the three phases of the approach. We use the MiniZinc 2.4.3 toolchain and the Gecode 6.1.1 solver. The iterative online optimization algorithm is implemented as a Python script that, for each session $s_i$, prepares the session data $\delta_i$ based on the previous session's result, and then calls MiniZinc to compile and solve the $P[\delta_i]$ instance.

### 5.1 Dynamic Vehicle Routing

Our first experiment is based on the DVRP model used as the running example, and the class 1 input data file `100-0-rc101-1`,[1] which provides coordinates for each node, service times for each customer request, the number of vehicles available (16), the start and end depots (always the same), and the start time of each time-window. We use this to compute the distance between any two nodes as their Euclidean distance, and to set the time at which a customer request arrives within a given time-window, as the start time of that time-window. No

[1] http://becool.info.ucl.ac.be/resources/benchmarks-dynamic-and-stochastic-vehicle-routing-problem-time-windows.

**Table 1.** Shows the session number, total number of customer requests so far, total number of tours added so far, (using garbage collection (GC) and aggregation:) best objective value found, compilation time, and runtime, number of garbage collected tours, and number of alive customers, (without GC and aggregation:) best objective value found, compilation time, and runtime.

| # | custs. | tours | with GC | | | | | without GC | | |
|---|---|---|---|---|---|---|---|---|---|---|
| | | | obj. | comp. (s) | run (s) | $\mathcal{G}(T)$ | alive $C$ | obj. | comp. (s) | run (s) |
| 1 | 2 | 16 | 64 | 0.18 | >45.00 | 0 | 2 | 64 | 0.21 | >45.00 |
| 2 | 3 | 31 | 128 | 0.46 | >45.00 | 14 | 3 | 128 | 3.49 | >45.00 |
| 3 | 10 | 47 | 572 | 0.55 | >45.00 | 29 | 8 | 572 | 38.01 | >45.00 |
| 4 | 18 | 63 | 861 | 0.76 | >45.00 | 45 | 15 | — | >45.00 | >45.00 |
| 5 | 24 | 79 | 1346 | 0.99 | >45.00 | 60 | 21 | — | >45.00 | >45.00 |
| 6 | 36 | 95 | 2010 | 1.61 | >45.00 | 76 | 32 | — | >45.00 | >45.00 |
| ⋮ | ⋮ | ⋮ | ⋮ | ⋮ | ⋮ | ⋮ | ⋮ | ⋮ | ⋮ | ⋮ |
| 14 | 91 | 223 | 5003 | 10.57 | >45.00 | 204 | 82 | — | >45.00 | >45.00 |
| 15 | 95 | 239 | 5352 | 13.34 | >45.00 | 220 | 86 | — | >45.00 | >45.00 |
| 16 | 96 | 255 | 5370 | 13.46 | >45.00 | 236 | 87 | — | >45.00 | >45.00 |
| 17 | 99 | 271 | 5432 | 13.52 | >45.00 | 252 | 90 | — | >45.00 | >45.00 |
| 18 | 99 | 287 | 5458 | 15.38 | >45.00 | 268 | 90 | — | >45.00 | >45.00 |
| 19 | 100 | 303 | 5483 | 15.61 | >45.00 | 284 | 91 | — | >45.00 | >45.00 |

other information in the data file was used. Note that whenever a tour ends, a new tour is added in that same session. We increased the current time $\tau$ by 50 time units each session, which is equivalent to 1 minute wall-clock time. Thus, the fleet gets updated instructions each minute. We also set the optimization runtime timeout to 45 s, to allow for potential overheads. We keep running until all 100 customers have been added.

The results in Table 1 show that our method significantly improves the compilation time; and without it, MiniZinc quickly gets overwhelmed with irrelevant tours and customers, slowing down the compilation. Further, using our method makes the problem suitable to run within a 45-s time-out. Note that empty tours in a session will be garbage collected in the next one.

### 5.2 Job-Shop Scheduling with Long Running Jobs

Our second experiment uses Job Shop Scheduling [10], a well studied, hard scheduling problem. We consider a rolling horizon version where jobs are spaced out by earliest start time. To highlight the advantages of automated garbage collection over the simple approach of [7], some of the jobs are given very long running times compared to other jobs. In the simple approach a job is garbage only if all tasks that arrived earlier are also garbage. These long running jobs prevent effective garbage collection under this policy, representing its worst case.

**Table 2.** Shows the session number, total number of jobs added so far, best objective value (makespan) found (all methods reported the same makespan), compilation time, and runtime without garbage collection, plus the same information and the garbage collected jobs for the simple garbage collection method of [7], and our proposed method.

| # | jobs | obj. | no GC | | simple GC | | | our GC | | |
|---|---|---|---|---|---|---|---|---|---|---|
| | | | comp. (s) | run (s) | comp. (s) | run (s) | $\mathcal{G}(J)$ | comp. (s) | run (s) | $\mathcal{G}(J)$ |
| 1 | 4 | 4278 | 0.09 | 0.11 | 0.09 | 0.11 | 0 | 0.08 | 0.10 | 0 |
| 2 | 8 | 4308 | 0.09 | 0.11 | 0.08 | 0.11 | 0 | 0.08 | 0.10 | 0 |
| ⋮ | ⋮ | ⋮ | ⋮ | ⋮ | ⋮ | ⋮ | ⋮ | ⋮ | ⋮ | ⋮ |
| 8 | 32 | 5428 | 0.46 | 0.50 | 0.47 | 0.51 | 0 | 0.27 | 0.30 | 7 |
| 9 | 36 | 5633 | 0.65 | >1.00 | 0.68 | >1.00 | 0 | 0.34 | >1.00 | 9 |
| 10 | 40 | 5874 | 0.88 | >1.00 | 0.90 | >1.00 | 0 | 0.23 | >1.00 | 18 |
| 11 | 44 | 8358 | >1.00 | >1.00 | >1.00 | >1.00 | 0 | 0.23 | 0.26 | 22 |
| 12 | 48 | 8388 | — | — | — | — | — | 0.17 | 0.20 | 30 |
| 13 | 52 | 8508 | — | — | — | — | — | 0.11 | 0.14 | 39 |
| ⋮ | ⋮ | ⋮ | ⋮ | ⋮ | ⋮ | ⋮ | ⋮ | ⋮ | ⋮ | ⋮ |
| 92 | 368 | 41028 | — | — | — | — | — | 0.19 | 0.22 | 350 |
| 93 | 372 | 41148 | — | — | — | — | — | 0.13 | 0.17 | 359 |
| ⋮ | ⋮ | ⋮ | ⋮ | ⋮ | ⋮ | ⋮ | ⋮ | ⋮ | ⋮ | ⋮ |

We run a similar experiment to that of [7], with the same model and data file `ft20` from the MiniZinc benchmarks,[2] and the same method to obtain an endless queue of jobs by repeatedly adding copies of the jobs in sequence. In each session we increase the current time $\tau$ by 408 scheduling time units, add the next 4 jobs in the queue, and set a time limit of 1 second. The first job (and every 40th subsequent job) is what we call a long-running job. We multiply the processing time, over all machines, of this job by 20.

Table 2 shows the benefit of using our method compared to no garbage collection and to the simple method proposed in [7]. By reasoning on objects individually (instead of finding the first non-realized job) we can remove any object that becomes realized. The simple method is prevented from garbage collecting them by the first job and, hence, becomes too slow.

As our experiments show, the overhead of the garbage collection steps is negligible compared to the overall compilation time, and compared to the performance gains in later sessions. The time complexity of the garbage collection steps depends on the complexity of the model, and is at most $O(n \log n)$ if the time complexity of the compilation without it is $O(n)$. Since garbage collection aims to minimize compilation time in future sessions, it will be quick and worthwhile in practice.

[2] https://github.com/MiniZinc/minizinc-benchmarks/tree/master/jobshop.

## 6 Related Work

Garbage collection is a well studied topic for programming languages and runtime environments [12], where it refers to the elimination of data that has no live references. In the online optimization context the data and past decisions always have references, but may still not be required for the current and future sessions, depending on which decisions are realized. Hence, determining what is garbage is more complex. As we have seen above, the model may require transformations such as aggregation in order to remove references to past decisions and data.

While online problems have been studied in great detail, almost all of this work concentrates on particular application problems (e.g., [11,16]). Any online problem has to tackle the difficulty that, without deletion, the problem continues to grow. While for particular application problems there may be simple rules that allow the modeler to define *garbage*, for a general definition, the interaction of realized decisions and earlier data with future decisions is complex.

Modeling support for online problem solving is still in its infancy. AIMMS has a notion of aggregation when using rolling horizon [17], which allows the effect of previous realized decisions to be accounted for in some parts of the model, such as the objective. The notion of realization used is much simpler, and automated aggregation of complex constraints like circuit is not considered.

Online MiniZinc [7] provides a framework that supports the automatic construction of the update model from an annotated offline version of the problem. It considers a simple form of garbage collection which only removes consecutive garbage objects until a non-garbage object is reached. The modeler is responsible for determining this directly, and aggregation is ignored. As a result, it is only usable when all aggregates are *true*. In contrast, the approach presented in this paper requires a complex model analysis, since we automate the understanding of how realized data and decisions can affect future decisions. In turn, our analysis requires a new view of modelling, where object sets are carefully used to represent dynamic data, and dependent object set creation is introduced.

Open global constraints [1,8] are a form of extensible constraint useful for online optimization problems whose size grows. The focus is on correct propagation of open constraints when not all information about them is available. They do not examine garbage collection since there is no notion of realization in the general open-world setting. The constraints simply grow as time progresses.

There is surprisingly little published work on aggregation of constraints. While the simplification of constraints when their variables become fixed is well understood, it is rarely documented. There are preprocessing methods that consider this, such as Boolean equi-propagation [14]. There has been work on eliminating variables during propagation [18], although we are not aware of a system that currently implements this. Aggregation, where information resulting from the simplification must be stored, does not seem to have been considered before.

## 7 Conclusion

Building online optimization systems has in the past been a rather complex process. Essentially, the modeler does not build a model of the problem, but an update model, which reasons about how to take information from the previous session and the new data to solve a new problem. We have previously [7] showed how to construct the update model automatically from the original model using annotations, but only introduced a very simple form of garbage collection: the modeler determines the latest object such that all previous objects are realized. In this paper, we now provide a comprehensive and automatic approach to garbage collection. The modeler specifies the rules for realization, and the garbage is automatically determined. We rely on uniquely named objects to ensure consistency of information across sessions. We tackle the key problem of aggregation, ignored in [7], where some constraints may need to be modified to record the effect of previously realized decisions.

## References

1. Barták, R.: Dynamic global constraints in backtracking based environments. Ann. Oper. Res. **118**, 101–119 (2003). https://doi.org/10.1023/A:1021805623454
2. Bent, R., Van Hentenryck, P.: Scenario-based planning for partially dynamic vehicle routing with stochastic customers. Oper. Res. **52**(6), 977–987 (2004)
3. Bent, R., Van Hentenryck, P.: Online stochastic optimization without distributions. In: ICAPS 2005, pp. 171–180 (2005)
4. Brown, K.N., Miguel, I.: Uncertainty and change. In: Handbook of Constraint Programming, pp. 731–760. Elsevier (2006). (Chap. 21)
5. Dantzig, G.B.: Linear programming under uncertainty. Manag. Sci. **1**(3/4), 197–206 (1955)
6. Dechter, R., Dechter, A.: Belief maintenance in dynamic constraint networks. In: AAAI 1988, pp. 37–42. AAAI Press, June 1988
7. Ek, A., Garcia de la Banda, M., Schutt, A., Stuckey, P.J., Tack, G.: Modelling and solving online optimisation problems. In: AAAI 2020, pp. 1477–1485. AAAI Press (2020)
8. Faltings, B., Macho-Gonzalez, S.: Open constraint programming. Artif. Intell. **161**(1–2), 181–208 (2005)
9. Gecode Team. Gecode: A generic constraint development environment (2020). http://www.gecode.org
10. Graham, R.: Bounds for certain multiprocessing anomalies. Bell Syst. Tech. J. **45**(9), 1563–1581 (1966)
11. Jaillet, P., Wagner, M.R.: Online vehicle routing problems: a survey. In: Golden, B., Raghavan, S., Wasil, E. (eds.) The Vehicle Routing Problem: Latest Advances and New Challenges. Operations Research/Computer Science Interfaces, vol. 43, pp. 221–237. Springer, Boston (2008). https://doi.org/10.1007/978-0-387-77778-8_10
12. Jones, R.E., Hosking, A.L., Moss, J.E.B.: The Garbage Collection Handbook: The Art of Automatic Memory Management. Chapman and Hall/CRC Applied Algorithms and Data Structures Series. CRC Press, Boca Raton (2011)

13. Mears, C., Garcia de la Banda, M., Wallace, M., Demoen, B.: A method for detecting symmetries in constraint models and its generalisation. Constraints **20**(2), 235–273 (2014). https://doi.org/10.1007/s10601-014-9175-5
14. Metodi, A., Codish, M., Stuckey, P.J.: Boolean equi-propagation for concise and efficient SAT encodings of combinatorial problems. J. Artif. Intell. Res. **46**, 303–341 (2013)
15. Nethercote, N., Stuckey, P.J., Becket, R., Brand, S., Duck, G.J., Tack, G.: MiniZinc: towards a standard CP modelling language. In: Bessière, C. (ed.) CP 2007. LNCS, vol. 4741, pp. 529–543. Springer, Heidelberg (2007). https://doi.org/10.1007/978-3-540-74970-7_38
16. Pruhs, K., Sgall, J., Torng, E.: Online scheduling. In: Handbook of Scheduling - Algorithms, Models, and Performance Analysis. Chapman and Hall/CRC (2004)
17. Roelofs, M., Bisschop, J.: Time-based modelling. In: AIMMS: The Language Reference, May 2, 2019 edn. (2019). www.aimms.com. (Chap. 33)
18. Schulte, C., Stuckey, P.J.: Dynamic variable elimination during propagation solving. In: Antoy, S., Albert, E. (eds.) Proceedings of the 10th International ACM SIGPLAN Conference on Principles and Practice of Declarative Programming, Valencia, Spain, 15–17 July 2008, pp. 247–257. ACM (2008)
19. Sethi, S., Soerger, G.: A theory of rolling horizon decision making. Ann. Oper. Res. **29**, 387–415 (1991). https://doi.org/10.1007/BF02283607
20. Van Hentenryck, P., Bent, R.: Online Stochastic Combinatorial Optimization. MIT Press, Cambridge (2009)
21. Verfaillie, G., Jussien, N.: Constraint solving in uncertain and dynamic environments: a survey. Constraints **10**(3), 253–281 (2005). https://doi.org/10.1007/s10601-005-2239-9

# Treewidth-Aware Quantifier Elimination and Expansion for QCSP

Johannes K. Fichte[1(✉)], Markus Hecher[2,3], and Maximilian F. I. Kieler[1]

[1] TU Dresden, Dresden, Germany
{johannes.fichte,maximilian.kieler}@tu-dresden.de
[2] TU Wien, Vienna, Austria
hecher@dbai.tuwien.ac.at
[3] University of Potsdam, Potsdam, Germany

**Abstract.** Research on the constraint satisfaction problem (CSP) is an active field both in theory and practice. An extension of CSP even considers universal and existential quantification of domain variables, known as QCSP, for which also solvers exist. The number of alternations between existential and universal quantifications is usually called quantifier rank. While QCSP is PSPACE-complete for bounded domain size, one can consider additional structural restrictions that make the problem tractable, for example, bounded treewidth. Chen [14] showed that one can solve QCSP instances of size $n$, quantifier rank $\ell$, bounded treewidth $k$, and bounded domain size $d$ in $\text{poly}(n) \cdot \text{tower}(\ell - 1, d^k)$, where tower are exponentials of height $\ell - 1$ with $d^k$ on top. We follow up on Chen's result and develop a treewidth-aware quantifier elimination technique that reduces the rank without an exponential blow-up in the instance size at the cost of exponentially increasing the treewidth. With this reduction at hand we show that one cannot significantly improve the blow-up of the treewidth assuming the exponential time hypothesis (ETH). Further, we lift a recently introduced technique for treewidth-decreasing quantifier expansion from the Boolean domain to QCSP.

## 1 Introduction

Constraint programming (CP) is a very popular solving paradigm for solving combinatorial problems that has applications in a variety of fields in computer science and human society [30,36,41]. CP allows for encoding a question by means of a set of objects (variables over given domains) and a collection of finite constraints, where a constraint is a relation between several variables limiting the values that these variables can take. Then, a solution to the considered question is an assignment to the variables (set of states) that satisfy the constraints. The

The work has been supported by the Austrian Science Fund (FWF), Grants Y698 and P32830, and the Vienna Science and Technology Fund, Grant WWTF ICT19-065. We would like to thank the anonymous reviewers for very detailed feedback and their suggestions. Special appreciation goes to Andreas Pfandler for early discussions.


H. Simonis (Ed.): CP 2020, LNCS 12333, pp. 248–266, 2020.
https://doi.org/10.1007/978-3-030-58475-7_15

resulting instance is called the constraint satisfaction problem (CSP) and CSAT asks for deciding whether a solution to such a problem exists.

CP usually asks whether there is an assignment and whether there is value to the variables that satisfy the constraints (existential quantification). An extension of CSPs considers existentially and universally quantified variables. A universally quantified variable ensures that a constraint is satisfied for that variable on all values of its domain. Since this allows for interleaving existentially and universally quantified variables, one often considers the number of alternations between the existential and universal quantifications, called quantifier rank. The resulting decision problem is referred to as QCSAT. While there are solvers [4,6,29], QCSAT is of very high complexity, i.e., PSPACE-complete if the quantifier rank is unbounded [14]. If one bounds the quantifier rank and the treewidth of the input instances the QCSAT problem is solvable in polynomial time. More precisely, Chen [14] has shown that one can solve QCSAT in time[1] $\text{poly}(n) \cdot \mathsf{tower}(\ell - 1, d^k)$ for instances of size $n$, quantifier rank $\ell$, bounded treewidth $k$, and bounded domain size $d$. There are also results involving extensions of treewidth [31].

Treewidth itself is a combinatorial invariant, which was originally introduced for graphs [7,9,11,48]. But treewidth also renders a large variety of NP-complete or #P-complete problems tractable when defining graph representations on the input instances and bounding the treewidth of these graph representations. For example, take the primal graph of an input instance, where we have as vertices the variables of the instance and an edge between two variables if these variables occur together in a clause or constraint, respectively. Then, the Boolean satisfiability problem (SAT) [49] and CSP [17,28] are polynomial-time tractable for instances of bounded treewidth on the primal graph. The technique also applies to other reasoning problems [43] and formalisms [20,32,47].

Practical research interests in treewidth waned with the success of CDCL-based solvers [1,38,44]. Reasons might be that those solvers implicitly take advantage of bounded treewidth [2]. However, recently, new practical applications for treewidth and problems that are higher in the polynomial hierarchy emerged. Particular examples include model counting of CNF formulas [24] and alternative computation models [25].

*Research Questions.* In this paper, we study treewidth and the QCSAT problem. Atserias and Oliva [3] showed that QCSAT for the Boolean domain remains intractable when parameterized by treewidth alone, which is established using a particular fragment of path decompositions. However, we want to go beyond this statement. We aim to extend insights into two directions. First, we study treewidth-aware reductions that eliminate or expand quantifiers and exponentially increase or decrease the treewidth, respectively. Second, we consider the effect of applying the reductions when assuming ETH. This allows for obtaining complexity lower bounds and show that results by Chen [14] cannot significantly be improved unless ETH fails. The exponential time hypothesis (ETH) [37] is a widely accepted standard hypothesis in the fields of exact and parameterized

[1] $\mathsf{tower}(\ell, x)$ is a tower of iterated exponentials of 2 of height $\ell$ with $x$ on top.

algorithms. ETH states that there is some real $s > 0$ such that we cannot decide satisfiability of a given 3-CNF formula $\phi$ in time $2^{s \cdot |\phi|} \cdot \|\phi\|^{\mathcal{O}(1)}$ [16, Ch.14]. The question to significantly improve the algorithm by Chen [14] has been answered negative when assuming ETH for a restricted version of QCSAT, considering only QCSPs over the Boolean domain and constraints in form of formulas in conjunctive normal form [23,45]. But the results do not allow for tight bounds for QCSAT.

*Our Contributions* are as follows:

1. We develop a treewidth-aware quantifier elimination reduction that reduces the rank without an exponential blow-up in the instance size at the cost of increasing the treewidth. Further, we lift a recently introduced reduction for treewidth-decreasing quantifier expansion from the Boolean domain to QCSAT. While this reduction costs an increase in the quantifier rank, the treewidth is exponentially decreased.
2. We consider increasing the treewidth and decreasing quantifier rank as well as decreasing the treewidth and increasing the quantifier rank. We show that under ETH our reduction cannot be improved significantly, i.e., the reductions are already ETH-tight with respect to treewidth. This is in contrast to other results [3,23,45], which do not provide tight bounds for arbitrary domains.
3. We investigate on the effect of applying the reductions when assuming ETH. In particular, quantifier elimination allows for establishing complexity lower bounds and show that results by Chen [14] cannot be significantly improved unless ETH fails. In turn, we obtain runtime bounds for using treewidth in QCSP solving under ETH.

## 2 Preliminaries

**Basics.** For an integer $x$, we define $\mathsf{tower} : \mathbb{N} \times \mathbb{N} \rightarrow \mathbb{N}$ by $\mathsf{tower}(0, x) := x$ and $\mathsf{tower}(\ell + 1, x) := 2^{\mathsf{tower}(\ell, x)}$ for all $\ell \in \mathbb{N}$. The domain $\mathcal{D}$ of a function $f : \mathcal{D} \rightarrow \mathcal{A}$ is given by $\mathsf{scope}(f)$. By $\log(\cdot)$ we mean the binary logarithm and we let $\log_b(\cdot)$ be the logarithm with base $b$. We assume familiarity with notions in computational complexity [46] and parameterized complexity [16,19,27]. For a given finite alphabet $\Sigma$, we call $I \in \Sigma^*$ an *instance* and $\|I\|$ denotes the *size of* $I$.

**Quantified Constraint Satisfaction Problems (QCSPs).** We define *Constraint Satisfaction Problems (CSPs)* over finite domains and their evaluation [21] as follows. A CSP $\mathcal{C}$ is a set of constraints and we denote by $\mathsf{var}(\mathcal{C})$ the set of *(constraint) variables* of $\mathcal{C}$. These constraint variables are over a (fixed) finite domain $\mathcal{D} = \{0, \ldots\}$ consisting of at least two values (at least "Boolean"). An *assignment* is a mapping $\iota : \mathsf{scope}(\iota) \rightarrow \mathcal{D}$ defined for a set $\mathsf{scope}(\iota)$ of variables. For a set $X$, let $\mathsf{ass}(X, \mathcal{D})$ be the *set of all assignments of the form* $X \rightarrow \mathcal{D}$. We say an assignment $\iota'$ *extends* $\iota$ (by $\mathsf{scope}(\iota') \setminus \mathsf{scope}(\iota)$) if $\mathsf{scope}(\iota') \supseteq \mathsf{scope}(\iota)$ and $\iota'(y) = \iota(y)$ for any $y \in \mathsf{scope}(\iota)$. Further, we let an *assignment* $\alpha|_X$ *restricted to* $X$ be the assignment with $\mathsf{scope}(\alpha|_X) = X \cap \mathsf{scope}(\alpha)$ and $(\alpha|_X)(y) := \alpha(y)$

for $y \in \mathsf{scope}(\alpha|_X)$. A *constraint* $C \in \mathcal{C}$ restricts certain variables $\mathsf{var}(C)$ of $\mathcal{C}$ and contains assignments to $\mathcal{D}$. More precisely, each $c \in C$ with $C \in \mathcal{C}$ is an *allowed assignment* $c : \mathsf{var}(C) \to \mathcal{D}$ of each variable $v \in \mathsf{var}(C)$ to $\mathcal{D}$. Formally, $\mathcal{C}$ is *satisfiable*, if there is an assignment $A : \mathsf{var}(\mathcal{C}) \to \mathcal{D}$, called *satisfying assignment*, such that for every $C \in \mathcal{C}$ there is an assignment $c \in C$ that equals $A|_{\mathsf{var}(C)}$.

Let $\ell \geq 0$ be an integer. A *quantified CSP (QCSP)* $Q$ is of the form $Q_1 V_1.Q_2 V_2. \cdots Q_\ell V_\ell.\mathcal{C}$ where $Q_i \in \{\forall, \exists\}$ for $1 \leq i \leq \ell$, $Q_\ell = \exists$, and $Q_j \neq Q_{j+1}$ for $1 \leq j \leq \ell - 1$; and where $V_i$ are disjoint, non-empty sets of constraint variables with $\bigcup_{i=1}^{\ell} V_i = \mathsf{var}(\mathcal{C})$; and $\mathcal{C}$ is a CSP. We call $\ell$ the *quantifier rank* of $Q$ and let $\mathsf{matrix}(Q) := \mathcal{C}$. The evaluation of QCSPs is defined as follows. Given a QCSP $Q$ and an assignment $\iota$, then $Q[\iota]$ is a QCSP that is obtained from $Q$, where every occurrence of any $x \in \mathsf{scope}(\iota)$ in $\mathsf{matrix}(Q)$ is replaced by $\iota(x)$, and variables that do not occur in the result are removed from preceding quantifiers accordingly. A QCSP $Q$ *evaluates to true*, or is valid, if $\ell = 0$ and the CSP $\mathsf{matrix}(Q)$ is satisfiable. Otherwise, if $Q_1 = \exists$, $Q$ evaluates to true if and only if there exists an assignment $\iota : V_1 \to \mathcal{D}$ such that $Q[\iota]$ evaluates to true. If $Q_1 = \forall$, then $Q[\iota]$ evaluates to true if for any assignment $\iota : V_1 \to \mathcal{D}$, $Q[\iota]$ evaluates to true. Given a QCSP $Q$, the *decision problem* QCSAT asks whether $Q$ evaluates to true.

For brevity, we denote constraints by a formula using equality $=$ between variables and elements of domain $\mathcal{D}$, negation $\neg$, which inverts (in-)equality expressions, as well as disjunction $\vee$, conjunction $\wedge$, and implication $\to$. For any Boolean variable $v \in \mathsf{var}(Q)$ the expression "$v$" is used as a shortcut for $v = 1$.

**Example 1.** *Consider the QCSP* $Q = \forall w, x.\exists y, z.\mathcal{C}$ *over domain* $\mathcal{D} = \{0, \ldots, 4\}$, *where* $\mathcal{C} := \{C_1, C_2, C_3\}$ *such that* $C_1 := z + 1 \leq w + y$, $C_2 := (w + x > 4) \vee (y \geq w + x)$, *and* $C_3 := y \leq w + x + 1$. *Given any assignment* $\iota$ *of* $\{w, x\}$, *we can construct an assignment* $\kappa$ *of* $Q$, *which sets* $\kappa(y) := \max(1, \min(4, w+x))$ *and* $\kappa(z) := w$. *Observe that then* $Q[\iota][\kappa]$ *simplifies to* $\forall w, x.\{(w+1 \leq w+1), (w+x \geq 4) \vee (w+x \geq w+x), (w+x \leq w+x+1)\}$, *which is valid over* $\mathcal{D}$. ∎

**Tree Decompositions (TDs).** For basic terminology on graphs and digraphs, we refer to standard texts [12,18]. For a tree $T = (N, A, r)$ with root $r$ and a node $t \in N$, we let $\mathsf{chldr}(t, T)$ be the set of all nodes $t'$, which have edge $(t, t') \in A$. Let $G = (V, E)$ be a graph. A *tree decomposition (TD)* of graph $G$ is a pair $\mathcal{T} = (T, \chi)$ where $T$ is a rooted tree, $r$ of $T$ is the root, and $\chi$ is a mapping that assigns to each node $t$ of $T$ a set $\chi(t) \subseteq V$, called a *bag*, such that the following conditions hold: (i) $V = \bigcup_{t \text{ of } T} \chi(t)$ and $E \subseteq \bigcup_{t \text{ of } T} \{\{u, v\} \mid u, v \in \chi(t)\}$; and (ii) *connectedness*: for each $s, t, u$, such that $t$ lies on the path from $s$ to $u$, we have $\chi(s) \cap \chi(u) \subseteq \chi(t)$. For a given vertex $x \in V$, we let $\mathsf{NodeI}(x) := \{t \mid t \text{ of } T, x \in \chi(t) \setminus (\bigcup_{t_i \in \mathsf{chldr}(t)} \chi(t_i))\}$ be the set of nodes, where $x$ is *introduced*. Then, $\mathsf{width}(\mathcal{T}) := \max_{t \text{ of } T} |\chi(t)| - 1$. The *treewidth* $\mathsf{tw}(G)$ of $G$ is the minimum $\mathsf{width}(\mathcal{T})$ over all tree decompositions $\mathcal{T}$ of $G$. For arbitrary but fixed $w \geq 1$, it is feasible in linear time to decide if a graph has treewidth at most $w$

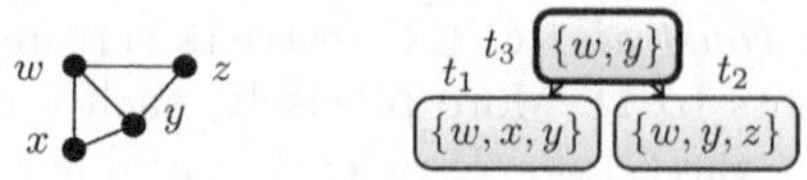


**Fig. 1.** Primal graph $P_Q$ of $Q$ from Example 1 (left) and a TD $\mathcal{T}$ of $P_Q$ (right).

and, if so, to compute a tree decomposition of width $w$ [10]. For a given tree decomposition $\mathcal{T} = (T, \chi)$ and an element $x \in \bigcup_{t \text{ of } T} \chi(t)$, we denote by $\mathcal{T}[x]$ the result $\mathcal{T}'$ of restricting $\mathcal{T}$ only to nodes, whose bags contain $x$. Observe that by the connectedness of $\mathcal{T}$, we have that $\mathcal{T}[x]$ is connected as well.

In order to apply tree decompositions for QCSPs, we need a graph representation of constraints. The *primal graph* $P_{\mathcal{C}}$ of a CSP $\mathcal{C}$ has the variables $\mathsf{var}(\mathcal{C})$ of $\mathcal{C}$ as vertices and an edge $\{x, y\}$ if there exists a constraint $C \in \mathcal{C}$ such that $x, y \in \mathsf{var}(C)$. For a QCSP $Q$, we define its primal graph as the primal graph of its matrix, i.e., $P_Q := P_{\mathsf{matrix}(Q)}$. Assume a tree decomposition $\mathcal{T} = (T, \chi)$ of $P_Q$. Then, we let $\mathcal{C}_t$ for a node $t$ of $T$ be defined by $\mathcal{C}_t := \{C \mid C \in \mathcal{C}, \mathsf{var}(C) \subseteq \chi(t)\}$.

**Example 2.** *Figure 1 illustrates the primal graph $P_Q$ of the QCSP from Example 1 and a tree decompositions of $P_Q$ of width 2, which coincides with the treewidth of $P_Q$ since, e.g., vertices $w$,$x$,$y$ are completely connected to each other [40].* ∎

## 3 Treewidth-Aware Quantifier Elimination

In this section, we present a reduction $\mathcal{R}\!\downarrow$ for eliminating the inner-most quantifier block while increasing the treewidth by a single exponent. Let $Q$ be a given QCSP of the form $Q := Q_1 V_1.Q_2 V_2. \cdots \exists V_\ell.\mathcal{C}$ over fixed domain $\mathcal{D}$.

### 3.1 Eliminating Quantifiers

In the reduction, we eliminate the last quantifier involving variables $V_\ell$. The overall process is guided along a tree decomposition $\mathcal{T} = (T, \chi)$ of $P_Q$ of width $k$ in order to ensure that the reduction is aware of the width and to increase treewidth only single-exponentially. More precisely, in $Q'$ we get rid of the inner-most quantifier $\exists V_\ell$ but at the cost of an increase of treewidth, where the treewidth of the primal graph of the resulting QCSP $Q' = \mathcal{R}\!\downarrow(Q, \mathcal{T})$ will be bounded by $\mathcal{O}(|\mathcal{D}|^k)$. The construction of our reduction is influenced by so-called *(non-serial) dynamic programming algorithms* that traverse tree decompositions [8,9]. Such algorithms compute for each node of the tree decomposition a table of assignments, which ensure that the subinstance given by the node evaluates to true. Then, these algorithms are typically designed in such a way that, whenever the table for the root node of the tree decomposition is not empty (i.e., in our case the table needs to contain at least one assignment), there is a solution to the problem. In our approach this is done similarly, but only for assignments $\alpha$ of variables $V_\ell$ of the inner-most quantifier block. Further, the reduction solves the inverse problem, where the reduced QCSP obtained by $\mathcal{R}\downarrow$ is valid if and only if $Q$ is invalid. However, if needed, one can invert the reduced QCSP accordingly, or use an adapted definition of QCSAT considering the inversion, cf. [15,35].

In the following, we present $\mathcal{R}\downarrow$ that simulates such an algorithm. To this end, we require the following variables. For each assignment $\alpha : V_\ell \cap \chi(t) \rightarrow \mathcal{D}$ of each tree decomposition node $t$ of $T$, we require $\mathcal{O}(|\mathcal{D}|^k)$ many (Boolean) variables of the form $\text{usat}_t^\alpha, \text{usat}_{<t}^\alpha$, and $\text{usat}_{>t}^\alpha$. Intuitively, if a variable $\text{usat}_t^\alpha$ evaluates to true for an assignment $\alpha$, we have that at least one constraint in $\mathcal{C}_t$ cannot be satisfied with $\alpha$. Then, we use $\text{usat}_{>t_i}^\alpha$ for a child node $t_i \in \mathsf{chldr}(t)$ of $t$ to indicate that no assignment $\beta : V_\ell \cap \chi(t_i) \rightarrow \mathcal{D}$ in node $t_i$, which is compatible with $\alpha$, can be extended to a satisfying assignment. However, the reason for that might lie even in a node below $t_i$. Finally, $\text{usat}_{<t}^\alpha$ evaluating to true implies that there is no assignment for at least one child node of $t$, that is compatible with $\alpha$ and can be extended to a satisfying assignment. These *unsatisfiability variables* are referred to by $\mathsf{VarUSatAss} := \{\text{usat}_t^\alpha, \text{usat}_{<t}^\alpha, \mid t \text{ of } T, \alpha \in \mathsf{ass}(V_\ell \cap \chi(t), \mathcal{D})\} \cup \{\text{usat}_{>t}^\alpha \mid t \text{ of } T, t \text{ is not the root of } T, \alpha \in \mathsf{ass}(V_\ell \cap \chi(t), \mathcal{D})\}$.

The reduction $\mathcal{R}\downarrow$ takes $Q$ and $\mathcal{T}$ and creates a fresh QCSAT instance

$$Q' := Q_2\ V_1.\ Q_3\ V_2.\ \cdots\ Q_{\ell-1}.\ V_{\ell-2}\ \exists V_{\ell-1}, \mathsf{VarUSatAss}.\ \mathcal{C}',$$

where $\mathcal{C}'$ is a CSP, whose constraints are denoted by the following formulas:

$$\text{usat}_t^\alpha \rightarrow \bigvee_{C \in \mathcal{C}_t} [\bigwedge_{\substack{c \in C \\ \alpha = c|_{V_\ell}}} \bigvee_{x \in \mathsf{var}(C) \setminus V_\ell} x \neq c(x)] \quad \text{for each } t \text{ of } T, \alpha \in \mathsf{ass}(V_\ell \cap \chi(t), \mathcal{D}) \tag{1}$$

$$\text{usat}_{>t_i}^\alpha \rightarrow \bigwedge_{\substack{\beta \in \mathsf{ass}(V_\ell \cap \chi(t_i), \mathcal{D}), \\ \beta|_{\chi(t)} = \alpha|_{\chi(t_i)}}} [\text{usat}_{<t_i}^\beta \vee \text{usat}_{t_i}^\beta] \quad \text{for each } t \text{ of } T, t_i \in \mathsf{chldr}(t), \alpha \in \mathsf{ass}(V_\ell \cap \chi(t), \mathcal{D}) \tag{2}$$

$$\text{usat}_{<t}^\alpha \rightarrow \bigvee_{t_i \in \mathsf{chldr}(t)} \text{usat}_{>t_i}^\alpha \quad \text{for each } t \text{ of } T, \alpha \in \mathsf{ass}(V_\ell \cap \chi(t), \mathcal{D}) \tag{3}$$

$$\bigwedge_{\alpha \in \mathsf{ass}(V_\ell \cap \chi(n), \mathcal{D})} [\text{usat}_{<n}^\alpha \vee \text{usat}_n^\alpha] \quad \text{for root node } n \text{ of } T \tag{4}$$

The reduction will be presented in two different blocks, namely Constraints (1) and Constraints (2)–(4), respectively. Constraints (1) ensure that unsatisfiability is computed accordingly. More precisely, with Constraints (1) we make sure that we have $\text{usat}_t^\alpha$ for node $t$ and assignment $\alpha$, if there is some constraint $C \in \mathcal{C}_t$, where each allowed assignment compatible with $\alpha$ fails. Note that these variables $\text{usat}_t^\alpha$ are used to separate evaluations of $\mathcal{C}_t$ for different assignments $\alpha$.

The second block takes care of defining variables of the form $\text{usat}_{>t}^\alpha$ and $\text{usat}_{<t}^\alpha$, and "guiding" the evaluation of constraints along tree decomposition $\mathcal{T}$. Constraints (2) ensure that we set $\text{usat}_{>t_i}^\alpha$ only for a child node $t_i$ of $t$ and an assignment $\alpha$ for node $t$, if there is no assignment $\beta$ for $t_i$ that is compatible with $\alpha$ and can be extended to a satisfying assignment. Intuitively, this achieves transport of unsatisfiability information for compatible assignments

**Table 1.** Detailed data on Constraints (1)–(4) consisting of the following: Number of instantiations, number of constraints per instantiation, as well as the size (number of allowed assignments) and arity (number of variables) per constraint.

| | #Instantiations | #Constraints | Size | Arity |
|---|---|---|---|---|
| (1) | $\mathcal{O}(\|T\| \cdot \|\mathcal{D}\|^{k+1})$ | $\mathcal{O}(\|C\| \cdot k)$ | $\mathcal{O}(\|\mathcal{D}\|)$ | 4 |
| (2) | $\mathcal{O}(\|T\| \cdot \|\mathcal{D}\|^{k+1} \cdot \max_{t \text{ of } T}\{\|\mathsf{chldr}(t)\|\})$ | $\mathcal{O}(\|\mathcal{D}\|^{k+1})$ | $2^3$ | 3 |
| (3) | $\mathcal{O}(\|T\| \cdot \|\mathcal{D}\|^{k+1})$ | $\max_{t \text{ of } T}\{\|\mathsf{chldr}(t)\|\}$ | $2^4$ | 4 |
| (4) | 1 | $\mathcal{O}(\|\mathcal{D}\|^{k+1})$ | $2^2$ | 2 |

between neighboring nodes. Then, Constraints (3) provide that finally for setting $\mathrm{usat}^{\alpha}_{<t}$ to 1, at least one $\mathrm{usat}^{\alpha}_{t_i}$ for any child node $t_i$ of $t$ is sufficient. Finally for the root $n$ of $T$, we require that no assignments $\alpha$ of $n$ leads to a satisfying one, which is ensured by Constraints (4). Table 1 lists details of constructed constraints.

*Remark: Reducing Treewidth Using Constraint Variables.* While it is easier to consider the assignment variables in VarUSatAss over the Boolean domain, during the construction of $\mathcal{R}{\downarrow}$ one can easily represent the same information by using variables over domain $\mathcal{D}$. Thereby, we form a group $g$ for $\lfloor\log(|\mathcal{D}|)\rfloor$ many assignments and variables in VarUSatAss are of the form $\mathrm{usat}^{g}_{t}$ over domain $\mathcal{D}$, which allows for referring to any of the $2^{\lfloor\log(|\mathcal{D}|)\rfloor}$ combinations of unsatisfiability for the assignments in $g$. Constraints (1) to (4) can be easily rewritten, where instead of $\mathrm{usat}^{\alpha}_{t}$ being 0 or 1, the corresponding bit of variable $\mathrm{usat}^{g}_{t}$ for group $g$ containing $\alpha$ is 0 or 1. The treewidth of the result $\mathcal{R}{\downarrow}(Q,\mathcal{T})$ is in $\mathcal{O}(|\mathcal{D}|^{k+1}/\lfloor\log(|\mathcal{D}|)\rfloor)$.

**Example 3.** *Recall QCSP $Q$ over $\mathcal{D}$ from Example 1 and TD $\mathcal{T} = (T, \chi)$ of $P_Q$ of Fig. 1. Next, we sketch instance $\mathcal{R}{\downarrow}(Q,\mathcal{T})$. Below, we denote constraints by formulas, as given in Example 1. For node $t_1$, we create constraints $\mathrm{usat}^{\alpha}_{t_1} \rightarrow ([w + x \leq 4] \wedge [\alpha(y) < w + x]) \vee (\alpha(y) > w + x + 1)$ of the Form (1) for each $\alpha \in \mathsf{ass}(\{y\}, \mathcal{D})$. Intuitively, this ensures that $\mathrm{usat}^{\alpha}_{t_1}$ can be set to 1 only if $\alpha$ cannot be extended to a satisfying assignment of $\mathcal{C}_{t_1}$. Further, since an empty disjunction evaluates to false, we create $\neg\mathrm{usat}^{\alpha}_{<t_1}$ of the Form (3) for each $\alpha \in \mathsf{ass}(\{y\}, \mathcal{D})$. Similar, for $t_2$, we create constraints of the Form (1), resulting in $\mathrm{usat}^{\alpha}_{t_2} \rightarrow (\alpha(z) + 1 > w + \alpha(y))$ for each $\alpha \in \mathsf{ass}(\{y, z\}, \mathcal{D})$ as well as constraints of the Form (3), namely $\neg\mathrm{usat}^{\alpha}_{<t_2}$ for each $\alpha \in \mathsf{ass}(\{y, z\}, \mathcal{D})$. Finally, root node $t_3$ has no constraints to check and hence constructs $\neg\mathrm{usat}^{\alpha}_{t_3}$ of Form (1) for $\alpha \in \mathsf{ass}(\{w\}, \mathcal{D})$. The root $t_3$ lifts information on unsatisfiability with the help of Constraints (2). To this end, Boolean variables of the Form of $\mathrm{usat}^{\alpha}_{>t_1}$ and $\mathrm{usat}^{\alpha}_{>t_2}$ for each assignment $\alpha \in \mathsf{ass}(\{y\}, \mathcal{D})$ are used, thereby requiring assignments $\beta$ (compatible with $\alpha$) for both child nodes $t_1$ and $t_2$. Then, for $\mathrm{usat}^{\alpha}_{<t_3}$ of such an assignment $\alpha$ evaluating to true, either $\mathrm{usat}^{\alpha}_{>t_1}$ or $\mathrm{usat}^{\alpha}_{>t_2}$ is required, as ensured by Constraints (3). Finally, Constraint (4) makes sure* *that unsatisfiability is given, i.e., for each such* $\alpha$, *we have that* $\mathrm{usat}^{\alpha}_{<t_3}$ *evaluates to true* ($\mathrm{usat}^{\alpha}_{t_3}$ *is forced to false by Constraints (1)).* ∎

### 3.2 Correctness, Runtime and Treewidth Expansion

**Theorem 4** (Correctness of Quantifier Elimination). *Given QCSP* $Q$ *over domain* $\mathcal{D}$ *and a tree decomposition* $\mathcal{T}$ *of the primal graph* $P_Q$. *Then, the reduction* $\mathcal{R}{\downarrow}(Q, \mathcal{T})$ *is correct: For any assignment* $\iota$ *over* $\mathsf{var}(Q) \setminus V_\ell$, *we have* $Q[\iota]$ *is valid, if and only if for* $Q' = \mathcal{R}{\downarrow}(Q, \mathcal{T})$, *we have* $Q'[\iota]$ *is invalid.*

*Proof (Sketch).* Let $Q' = \mathcal{R}{\downarrow}(Q, \mathcal{T})$. "$\Rightarrow$": Given any assignment $\iota : \mathsf{var}(Q) \setminus V_\ell$ such that $Q[\iota]$ evaluates to true. We show that for any assignment $\iota' : \mathsf{var}(Q) \cup \mathsf{VarUSatAss}$ extending $\iota$, we have that $Q'[\iota']$ evaluates to false. Assume towards a contradiction that $Q'[\iota']$ evaluates to true. By Constraints (4), we have $\iota'(\mathrm{usat}^{\alpha}_{<n}) = 1$ or $\iota'(\mathrm{usat}^{\alpha}_{n}) = 1$ for root node $n$ of $T$ and every assignment $\alpha : V_\ell \cap \chi(n) \rightarrow \mathcal{D}$. By construction of the reduction (Constraints (3) and (2)), and connectedness of $\mathcal{T}$, for each such $\alpha$, we have to find at least one node $t'$ (root $n$ or any descendant) of $n$, where there is an assignment $\beta$ with $\iota'(\mathrm{usat}^{\beta}_{t'}) = 1$. Then, by Constraints (1) and since $\iota'$ as well as $\alpha$ was an arbitrary extension of $\iota$, we conclude that $\iota$ does not satisfy at least one $C \in \mathcal{C}_t$ for any assignment extending $\iota$ by $V_\ell$, contradicting our assumption.

"$\Leftarrow$": The other direction works similarly by contraposition, where assuming that $Q[\iota]$ is invalid, we extend $\iota$ and construct an assignment $\iota'$ by greedily setting $\iota'(\mathrm{usat}^{\alpha}_{t}) = 1$ for nodes $t$ and assignments $\alpha$ such that no constraint of the Form (1) is unsatisfied. We repeat this for the other variables of $\mathsf{VarUSatAss}$ as well as Constraints (2), (3) and (4), and observe by connectedness of $\mathcal{T}$ and the construction of $\mathcal{R}{\downarrow}$ that then indeed $Q'[\iota']$ evaluates to true. □

**Theorem 5** (Runtime of Quantifier Elimination). *Given a QCSP* $Q$ *over fixed domain* $\mathcal{D}$, *where* $k$ *is the treewidth of the primal graph of* $Q$ *and* $\mathcal{C} = \mathsf{matrix}(Q)$. *Then, constructing a TD* $\mathcal{T}$ *and* $\mathcal{R}{\downarrow}(Q, \mathcal{T})$ *takes time* $\mathcal{O}(2^{k^3} \cdot |\mathsf{var}(\mathcal{C})| \cdot |\mathcal{D}|^{2k+2} \cdot \|\mathcal{C}\|)$.

*Proof.* First, we construct [10] a tree decomposition of the primal graph of $Q$ of width $k$ in time $2^{\mathcal{O}(k^3)} \cdot |\mathsf{var}(\mathcal{C})|$, consisting of at most $\mathcal{O}(2^{k^3} \cdot |\mathsf{var}(\mathcal{C})|)$ many nodes. Without loss of generality, we assume that for each node $t$ of $T$ we have at most two child nodes in $T$ as well as constant bag differences between two neighboring bags, since $\mathcal{T}$ can be modified accordingly in time $\mathcal{O}(k^2 \cdot 2^{k^3} \cdot |\mathsf{var}(\mathcal{C})|)$ [40, Lemma 13.1.3] without increasing the width $k$. Then, the number of variables in $\mathsf{VarUSatAss}$ is at most $\mathcal{O}(2^{k^3} \cdot |\mathsf{var}(\mathcal{C})| \cdot |\mathcal{D}|^{k+1})$ and the reduction constructs $\mathcal{O}(2^{k^3} \cdot |\mathsf{var}(\mathcal{C})| \cdot |\mathcal{D}|^{2k+2} \cdot \|\mathcal{C}\|)$ many constraints, cf. Table 1 for details. □

**Theorem 6** (Treewidth Expansion). *Given any QCSP* $Q$ *over fixed domain* $\mathcal{D}$. *Then, there is a TD* $\mathcal{T} = (T, \chi)$ *of primal graph* $P_Q$ *of* $Q$ *with* $k = \mathsf{width}(\mathcal{T}) = \mathsf{tw}(P_Q)$ *such that for* $Q' = \mathcal{R}{\downarrow}(Q, \mathcal{T})$ *the treewidth of* $P_{Q'}$ *is in* $\mathcal{O}(\frac{|\mathcal{D}|^{(k+1)}}{\lfloor \log(|\mathcal{D}|) \rfloor})$.

*Proof.* We take as $\mathcal{T}$ the TD as constructed in Theorem 5. Then, we construct a tree decomposition $\mathcal{T}' = (T, \chi')$ of the primal graph of $Q' = \mathcal{R}{\downarrow}(Q, \mathcal{T})$. For each tree decomposition node $t$ of $N$ we set its bag $\chi'(t) := [\chi(t) \setminus V_\ell] \cup \{\text{usat}^\alpha_{<t}, \text{usat}^\alpha_t \mid \alpha \in \mathsf{ass}(\chi(t) \cap V_\ell, \mathcal{D})\} \cup \{\text{usat}^\alpha_{>t_i}, \text{usat}^\beta_{<t_i}, \text{usat}^\beta_{t_i} \mid \alpha \in \mathsf{ass}(\chi(t) \cap V_\ell, \mathcal{D}), \beta \in \mathsf{ass}(\chi(t_i) \cap V_\ell, \mathcal{D}), t_i \in \mathsf{chldr}(t), \alpha|_{\chi(t_i)} = \beta|_{\chi(t)}\} \cup \{\text{usat}^\alpha_{<t}, \text{usat}^\alpha_{>t_i} \mid \alpha \in \mathsf{ass}(\chi(t) \cap V_\ell, \mathcal{D}), t_i \in \mathsf{chldr}(t)\}$. Observe that all the properties of TDs are satisfied. Indeed, connectedness holds since the only elements that are shared between two neighboring bags (and which are not in $\mathsf{var}(\mathcal{C}) \setminus V_\ell$) are in $\mathsf{VarUSatAss}$ due to Constraints (2), which is taken care of. Each bag $\chi'(t)$ contains $\mathcal{O}(k + |\mathcal{D}|^{(k+1)})$ variables, which can be decreased to $\mathcal{O}(\frac{|\mathcal{D}|^{(k+1)}}{\lfloor \log(|\mathcal{D}|) \rfloor})$ as remarked above. $\square$

## 4 Treewidth-Aware Quantifier Expansion

Having established quantifier elimination, we provide a reverse reduction for the QCSP formalism that exponentially decreases treewidth, but at the same time increases the quantifier rank by one. The overall procedure is inspired by the recent work on *quantified Boolean formulas (QBFs)* [23] and based on the technique developed in this work. However, the technique for decreasing treewidth is slightly extended and adapted for the QCSP formalism. In the following, we define our reduction that takes both an instance $Q$ of QCSAT of quantifier rank $\ell$ over fixed domain $\mathcal{D}$ and a tree decomposition $\mathcal{T}$ of the primal graph $P_Q$ of width $k$, and returns an instance of QCSAT of width $\mathcal{O}(\log_{|\mathcal{D}|}(k))$ and quantifier rank $\ell + 1$. The reduction is influenced and guided by the tree decomposition $\mathcal{T}$, which then gives rise to a tree decomposition $\mathcal{T}'$ of $P_{R(Q)}$ of width $\mathcal{O}(\log_{|\mathcal{D}|}(k))$. It is crucial that the constructed instance only uses in each bag of $\mathcal{T}'$ a constant number of elements of the original bags of $\mathcal{T}$. However, $\log_{|\mathcal{D}|}(k)$ many elements in a bag allow us to represent a "pointer" to address one of the $k$ many elements of each bag of $\mathcal{T}$. Consequently, the restriction to $\mathcal{O}(\log_{|\mathcal{D}|}(k))$ many elements in a bag enables constantly many such pointers. Still, values of these pointers require to be "synchronized", i.e., it is ensured that if two pointers from different, neighboring bags are referring to the same variable, that the corresponding pointer values coincide. This is then achieved with the help of an additional quantifier block, by expressing that for all pointer (targets), the target's values have to be aligned if two neighboring pointers refer to the same target.

### 4.1 Decreasing Treewidth by Quantifier Expansion

In the following, we present our reduction $\mathcal{R}{\uparrow}$. To this end, let $Q$ be a given QCSP of the form $Q := Q_1 V_1 . Q_2 V_2 . \cdots \exists V_\ell . \mathcal{C}$, where each $C \in \mathcal{C}$ uses exactly $s \geq 3$ many variables, i.e., we present the reduction for QCSP $Q$, where $s$ is an assumed constant such that $|\mathsf{var}(C)| = s$. For each $C \in \mathcal{C}$ we refer to the first, second, ..., and $s$-th variable of $C$ by $\mathsf{var}(C, 1)$, $\mathsf{var}(C, 2)$, ..., $\mathsf{var}(C, s)$, respectively. Further, in the following we assume a TD $\mathcal{T} = (T, \chi)$ of the primal graph $P_Q$.

We use the following sets of variables. The overall idea is to have that $Q$ is invalid if and only if $\mathcal{R}{\uparrow}(Q, \mathcal{T})$ is valid. To this end, we need variables $\mathsf{VarUSat} :=$

$\{\text{usat}_{<t}, \text{usat}_t \mid t \text{ of } T\}$ to guide the evaluation and information of unsatisfiability along the tree. For a set $V \subseteq \mathsf{var}(\mathcal{C})$ of variables, we denote by $\mathsf{VarI}(V) := \{x_t \mid x \in V, t \in \mathsf{NodeI}(x)\}$ the set of fresh variables (called *introduce variables*) over domain $\mathcal{D}$ generated for each original variable $x$ and node $t$ of $T$, where $x$ is introduced. Then, we denote by $\mathsf{VarP} := \{p_t^0, \ldots, p_t^{\lceil \log_{|\mathcal{D}|}(|\chi(t)|) \rceil - 1} \mid t \text{ of } T\}$ the set of fresh variables over $\mathcal{D}$, referred to by *pointer variables*, for representing a pointer for each node $t$ that will be used to address particular elements of the corresponding bags, and by $\mathsf{VarPV} := \{v_t \mid t \text{ of } T\}$ the set of fresh variables over domain $\mathcal{D}$, called *pointer values*, to assign domain values to the pointer targets. Overall, these variables serve the purpose of guiding the evaluation of $\mathcal{C}$ along tree decomposition $\mathcal{T}$. For checking constraints, we again need these pointers and pointer values, but $s$ more times since in order to check a constraint $C \in \mathcal{C}$, we require to refer to $s$ many bag elements of $\chi(t)$ for each node $t$ of $T$ at once. This results in the sets $\mathsf{VarCP} := \{p_{t,j}^0, \ldots, p_{t,j}^{\lceil \log_{|\mathcal{D}|}(|\chi(t)|) \rceil - 1} \mid t \text{ of } T, 1 \leq j \leq s\}$ and $\mathsf{VarCPV} := \{v_{t,j} \mid t \text{ of } T, 1 \leq j \leq s\}$ of fresh variables over domain $\mathcal{D}$. We call $\mathsf{VarCP}$ *check pointer variables*, and $\mathsf{VarCPV}$ is called *check pointer values*.

Before we discuss the reduction, we require for the pointers a vector representation ($|\mathcal{D}|$-ary representation) of the elements in a bag of $\mathcal{T}$, and a mapping that assigns bag elements to its corresponding $|\mathcal{D}|$-ary representation. In particular, we assume an arbitrary, but fixed total order $\prec$ of elements of a bag $\chi(t)$ of any given node $t$ of $T$. With $\prec$, we can then assign each element $x$ in $\chi(t)$ its unique (within the bag) induced ordinal number $o(t, x)$. This ordinal number $o(t, x)$ is expressed in a $|\mathcal{D}|$-ary number system (if $|\mathcal{D}| = 2$, we have binary). For that we need precisely $\lceil \log_{|\mathcal{D}|}(|\chi(t)|) \rceil$ many variables $\{p_t^0, \ldots, p_t^{\lceil \log_{|\mathcal{D}|}(|\chi(t)|) \rceil - 1}\} \subseteq \mathsf{VarP}$. We denote by $[\![x]\!]_t$ the (consistent) set of expressions $e$ of the form $p_t^i = d$ for a constant $d \in \mathcal{D}$ over those variables that encode (in $|\mathcal{D}|$-ary number system) the ordinal number $o(t, x)$ of $x \in \chi(t)$ in $t$, such that whenever $p_t^i = c$ is contained in the set $[\![x]\!]_t$, the $i$-th position in the unique representation of $o(t, x)$ is $c$. Analogously, for $1 \leq j \leq s$ we denote by $[\![x]\!]_{t,j}$ the consistent expressions over variables in $\{p_{t,j}^0, \ldots, p_{t,j}^{\lceil \log_{|\mathcal{D}|}(|\chi(t)|) \rceil - 1}\} \subseteq \mathsf{VarCP}$ encoding $o(t, x)$ of $x \in \chi(t)$.

**Example 7.** *Consider again QCSP $Q$ over domain $\mathcal{D}$ from Example 1 as well as TD $\mathcal{T} = (T, \chi)$ of $P_Q$ of Fig. 1. For encoding a pointer to any element of bag $\chi(t_1) = \{w, x, y\}$ over domain $\mathcal{D}$ we only need $\lceil \log_{|\mathcal{D}|}(|\chi(t_1)|) \rceil = 1$ variable $p_{t_1}^0$. We assume from now on, that the ordinal number $o(t, x)$ for each element $x$ of any bag and node $t$ of $\mathcal{T}$ is in* alphabetic *order. Consequently, we reach via $p_{t_1}^0 = 0$ variable $w$, $p_{t_1}^0 = 1$ refers to $x$ and $p_{t_1}^0 = 2$ targets at $y$.* ■

In the reduction, we need to distinguish whether the set $V_i$ of variables is universally or existentially quantified. Intuitively, since among the introduce variables we have copies $x_t$ of variables $x$, for each node $t$ in $\mathsf{NodeI}(x)$, we need to take special care of copies $x_t$ for universally quantified variables $x$. More precisely, we cannot place more than one of these copies $x_t$ under universal quantification, since ultimately these copies need to have the same value for a variable $x$. As a result, universal quantification requires to shift for each $x \in V_i$ all but one variable of $\mathsf{VarI}(\{x\})$ to the next existential quantifier block $Q_{i+1}$. To this end, we let for each universally quantified variable $x$, one representative variable in $\mathsf{VarI}(\{x\})$ that is not shifted be denoted by $\mathsf{rep}(x)$. Given a quantifier block $Q_i$, its variables $V_i$ and the variables $V'$ of the preceding quantifier block (if exists), we define the quantified introduce variables: $\mathsf{VarI}_\forall(V_i, V') := \{\mathsf{rep}(x) \mid x \in V_i\}$ for $Q_i = \forall$ and $\mathsf{VarI}_\exists(V_i, V') := \mathsf{VarI}(V_i \cup V') \setminus \mathsf{VarI}_\forall(V', \emptyset)$ for $Q_i = \exists$.

The reduction $\mathcal{R}\!\uparrow$ takes $Q$ and $\mathcal{T}$ and creates a fresh QCSAT instance $Q' :=$

$$Q_2\ \mathsf{VarI}_{Q_2}(V_1, \emptyset).\ Q_3\ \mathsf{VarI}_{Q_3}(V_2, V_1).\ \cdots \forall\ \mathsf{VarI}_\forall(V_\ell, V_{\ell-1}), \mathsf{VarP}.$$
$$\exists\ \mathsf{VarI}_\exists(\emptyset, V_\ell), \mathsf{VarPV}, \mathsf{VarCP}, \mathsf{VarCPV}, \mathsf{VarSat}.\ \mathcal{C}',$$

where $\mathcal{C}'$ is a CSP, whose constraints are denoted by the following formulas:

$$[\bigwedge_{e \in [\![x]\!]_t} e] \rightarrow x_t{=}v_t \qquad \text{for each } x_t \in \mathsf{VarI}(\mathsf{var}(\mathcal{C})) \tag{5}$$

$$[\bigwedge_{e \in [\![x]\!]_t} e \wedge \bigwedge_{e' \in [\![x]\!]_{t_i}} e'] \rightarrow v_t{=}v_{t_i} \qquad \text{for each } t \text{ of } T, t_i{\in}\mathsf{chldr}(t), x{\in}\chi(t){\cap}\chi(t_i) \tag{6}$$

$$[\bigwedge_{e \in [\![x]\!]_t} e \wedge \bigwedge_{e' \in [\![x]\!]_{t,j}} e'] \rightarrow v_t{=}v_{t,j} \qquad \text{for each } t \text{ of } T, x \in \chi(t), 1 \le j \le s \tag{7}$$

$$\mathrm{usat}_t \rightarrow \bigvee_{C \in \mathcal{C}_t} [\bigwedge_{\substack{1 \le i \le s,\\ x = \mathsf{var}(C,i),\\ e \in [\![x]\!]_{t,i},\\ c \in C}} e \wedge \bigvee_{\substack{1 \le j \le s,\\ x = \mathsf{var}(C,j)}} v_{t,j} {\neq} c(x)] \qquad \text{for each } t \text{ of } T \tag{8}$$

$$\mathrm{usat}_{\le t} \rightarrow \bigvee_{t_i \in \mathsf{chldr}(t)} \mathrm{usat}_{\le t_i} \vee \mathrm{usat}_t \qquad \text{for each } t \text{ of } T \tag{9}$$

$$\mathrm{usat}_{\le n} \qquad \text{for root node } n \text{ of } T \tag{10}$$

For each variable $x \in \mathsf{var}(\mathcal{C})$, Constraints (5) take care of the equivalence between the introduce variables in $\mathsf{VarI}(\{x\})$ and pointer values if the corresponding pointer targets $x$. More precisely, Constraints (5) ensure that whenever the pointer in a node $t \in \mathsf{NodeI}(x)$ targets an element $x \in \chi(t)$, pointer value $v_t$ coincides with variable $x_t$. The idea of this is similar to the "element_var" constraint [5].

The next block of constraints consisting of Constraints (6) as well as (7) is responsible for keeping both neighboring pointer variables as well as (check) pointer variables within a bag synchronized, respectively. In more details, Constraints (6) take care that whenever the pointer for a node $t$ targets an element $x \in \chi(t)$, and the pointer for a child node $t_i$ of $t$ targets also $x$, the pointer value variables $v_t$ and $v_{t_i}$ have to coincide. Further, Constraints (7) ensure equivalence between pointer value variables if a pointer targets the same element $x$ that a check pointer targets. However, these constraints can also be further strengthened if required, by only generating them for those $x \in \chi(t)$, where there exists a constraint $C \in \mathcal{C}_t$ with $x = \mathsf{var}(C, j)$, i.e., $x$ occurs in some constraint in $\mathcal{C}_t$.

**Table 2.** Detailed data on Constraints (5)–(10) consisting of the following: Number of instantiations, number of constraints per instantiation, as well as the size (number of allowed assignments) and arity (number of variables) per constraint.

| | #Instantiations | #Constraints | Size | Arity |
|---|---|---|---|---|
| (5) | $\mathcal{O}(\mathsf{VarI}(\mathsf{var}(\mathcal{C})))$ | 1 | $\mathcal{O}(\lvert\mathcal{D}\rvert^2 \cdot k)$ | $\mathcal{O}(\log_{\lvert\mathcal{D}\rvert}(k))$ |
| (6) | $\mathcal{O}(k \cdot \lVert T\rVert \cdot \max_{t \text{ of } T}\{\lvert\mathsf{chldr}(t)\rvert\})$ | 1 | $\mathcal{O}(\lvert\mathcal{D}\rvert^2 \cdot k)$ | $\mathcal{O}(\log_{\lvert\mathcal{D}\rvert}(k))$ |
| (7) | $\mathcal{O}(k \cdot \lVert T\rVert \cdot s)$ | 1 | $\mathcal{O}(\lvert\mathcal{D}\rvert^2 \cdot k)$ | $\mathcal{O}(\log_{\lvert\mathcal{D}\rvert}(k))$ |
| (8) | $\mathcal{O}(\lVert T\rVert)$ | $\mathcal{O}(\lVert\mathcal{C}\rVert \cdot s \cdot k)$ | $\mathcal{O}(\lvert\mathcal{D}\rvert)$ | 4 |
| (9) | $\mathcal{O}(\lVert T\rVert)$ | $\max_{t \text{ of } T}\{\lvert\mathsf{chldr}(t)\rvert\}$ | $2^5$ | 5 |
| (10) | 1 | 1 | 2 | 1 |

Then, we ensure by Constraints (8) with the help of check pointer and check pointer values that if $\text{usat}_t$ is set, then some constraint of $\mathcal{C}_t$ is unsatisfied. Concretely, in this case there has to be a constraint $C \in \mathcal{C}_t$ such that whenever the $i$-th check pointer of $t$ points to the $i$-th variable $x$ of constraint $C$ for every $i$, there has to be some allowed assignment $c$ of $C$ such that some check pointer value $v_{t,j}$ is different from $c(x)$. Finally, this information of unsatisfiability is passed towards the root by means of Constraints (9), and enforced by Constraint (10). Table 2 lists detailed data of the constraints constructed by the reduction $\mathcal{R}\!\uparrow$.

**Example 8.** *Recall QCSP $Q$ over $\mathcal{D}$ from Example 1 and TD $\mathcal{T} = (T, \chi)$ of $P_Q$ of Fig. 1. Instance $\mathcal{R}\!\uparrow(Q, \mathcal{T})$ equals to $\exists w_{t_1}, w_{t_2}, x_{t_1}.\forall y_{t_1}, z_{t_2}, p^0_{t_1}, p^0_{t_2}, p^0_{t_3}.\exists y_{t_2}, v_{t_1}, v_{t_2}, v_{t_3}, \quad p^0_{t_1,1}, p^0_{t_1,2}, p^0_{t_1,3}, p^0_{t_2,1}, p^0_{t_2,2}, p^0_{t_2,3}, p^0_{t_3,1}, p^0_{t_3,2}, p^0_{t_3,3}, v_{t_1,1}, v_{t_1,2}, v_{t_1,3}, v_{t_2,1}, v_{t_2,2}, v_{t_2,3}, v_{t_3,1}, v_{t_3,2}, v_{t_3,3}, \text{usat}_{t_1}, \text{usat}_{t_2}, \text{usat}_{t_3}, \text{usat}_{\leq t_1}, \text{usat}_{\leq t_2}, \text{usat}_{\leq t_3}.\mathcal{C}'$, such that the matrix $\mathcal{C}'$ contains the constraints discussed below.*

*For node $t_1$ we construct for each $w_{t_1}, x_{t_1}, y_{t_1}$ Constraints (5) as follows: $p^0_{t_1} = 0 \rightarrow w_{t_1} = v_{t_1}$, $p^0_{t_1} = 1 \rightarrow x_{t_1} = v_{t_1}$ and $p^0_{t_1} = 2 \rightarrow y_{t_1} = v_{t_1}$. Then, Constraints (7) synchronize pointers and check pointers, where for $1 \leq i \leq 3$, we have: $p^0_{t_1} = p^0_{t_1,i} \rightarrow v_{t_1} = v_{t_1,i}$. Constraints (8) ensure that $\text{usat}_{t_1}$ is set to 1 only if a constraint in $\mathcal{C}_{t_1}$ is violated. Concretely, $\text{usat}_{t_1} \rightarrow [p^0_{t_1,1} = 0 \wedge p^0_{t_1,2} = 1 \wedge p^0_{t_1,2} = 2] \wedge [(v_{t_1,1} \neq 0 \vee v_{t_1,2} \neq 0 \vee v_{t_1,3} \neq 0) \wedge \cdots]$ ensures that if the check pointer targets at $w, x$ and $y$, no allowed assignment of either $C_2$ or $C_3$ coincides with $v_{t_1,1}, v_{t_1,2}$ and $v_{t_1,3}$ respectively. Constraints (9) ensure that $\text{usat}_{<t_1} \rightarrow \text{usat}_{t_1}$.*

*For node $t_2$ the constructed constraints are similar. Then, for node $t_3$, there are no Constraints (5), since $t_3$ has no introduce variables, and Constraints (7) are constructed similar as for $t_1$, but for variables $w, y$ only. Constraints (6) ensure synchronization of values for $w$ and $y$ between child nodes, constructed as follows. Concretely, for $w$ we have $p^0_{t_3} = 0 \wedge p^0_{t_1} = 0 \rightarrow v_{t_3} = v_{t_1}$ and $p^0_{t_3} = 0 \wedge p^0_{t_2} = 0 \rightarrow v_{t_3} = v_{t_2}$. For $y$ we construct $p^0_{t_3} = 1 \wedge p^0_{t_1} = 2 \rightarrow v_{t_3} = v_{t_1}$ and $p^0_{t_3} = 1 \wedge p^0_{t_2} = 1 \rightarrow v_{t_3} = v_{t_2}$. Then, Constraints (8) corresponds to $\neg\text{usat}_{t_3}$ since $\mathcal{C}_{t_3} = \emptyset$ and (9) simplifies to $\text{usat}_{\leq t_3} \rightarrow \text{usat}_{\leq t_1} \vee \text{usat}_{\leq t_2}$. Finally, we have $\text{usat}_{\leq t_3}$ as Constraint (10). Observe that one can easily construct a TD $\mathcal{T}'$ of $\mathcal{R}\!\downarrow(Q, \mathcal{T})$ of* *width* $\mathcal{O}(\log_{|\mathcal{D}|}(\mathsf{width}(\mathcal{T})))$ *if the number of variables in* $\mathsf{VarI}(\mathsf{var}(\mathcal{C}))$ *is constant for each node in* $T$, *as discussed in the next section.* ■

*Remark: Reducing the Arity of Constraints.* The constraints above can be easily modified to contain at most $s$ many variables. This requires auxiliary variables, but does not affect satisfiability and increases treewidth by a constant (cf. [42]).

### 4.2 Correctness, Runtime and Treewidth Decrease

Assume a given QCSP formula $Q := Q_1V_1.Q_2V_2.\cdots\forall V_\ell.\mathcal{C}$ with at most $s \geq 3$ variables per constraint and let $\mathcal{T} = (T, \chi)$ be a TD of primal graph $P_Q$. Further, the obtained instance is addressed by $Q' = \mathcal{R}\!\uparrow\!(Q, \mathcal{T})$. Then, correctness relies on the following lemma, which establishes, that in a satisfying assignment of $Q'$ introduce variables $\mathsf{VarI}(\{x\})$ for every $x \in \mathsf{var}(\mathcal{C})$ get assigned equal values.

**Lemma 9** (Synchronization, $\star$[33]). *Let* $x_t, x_{t'} \in \mathsf{VarI}(\mathsf{var}(\mathcal{C}))$ *be any introduce variables for two nodes* $t, t' \in \mathsf{NodeI}(x)$. *Then, for any assignment* $\iota$ *assigning variables* $x_t, x_{t'}$ *such that* $\iota(x_t) \neq \iota(x_{t'})$, *we have that* $Q'[\iota]$ *is invalid. Further, let* $\iota$ *be any assignment that assigns for some* $1 \leq j \leq s$ *all variables in* $[\![x]\!]_{t,j} \cup \{v_{t,j}\}$ *such that assignment* $\iota$ *satisfies* $[\![x]\!]_{t,j}$. *Then, if* $\iota(x_{t'}) \neq \iota(v_{t,j})$, $Q'[\iota]$ *is invalid.*

**Theorem 10** (Correctness). *Let* $Q$ *be a QCSP over finite domain* $\mathcal{D}$, *where constraints have at most* $s \geq 3$ *variables, and let* $\mathcal{T} = (T, \chi)$ *be a tree decomposition of* $P_Q$. *Then,* $Q$ *is valid if and only if* $\mathcal{R}\!\uparrow\!(Q, \mathcal{T})$ *is invalid.*

*Proof (Sketch).* Proof by induction on quantifier rank $\ell$ that shows if $Q[\alpha]$ is valid under an assignment $\alpha$ to variables $V_1$ of $Q$, then $\mathcal{R}\!\uparrow\!(Q, \mathcal{T})[\alpha']$ is indeed invalid under any assignment $\alpha'$. By Lemma 9, there is a bijective correspondance between $\alpha$ and $\alpha'$, since all the introduce variables $\mathsf{VarI}(\mathsf{var}(\mathsf{matrix}(Q)))$ have to have the same value in $\alpha'$. It is therefore easy to see that in $\mathcal{R}\!\uparrow\!(Q, \mathcal{T})$ the introduce variables except $\mathsf{rep}(x)$ for any universally quantified variable $\mathsf{VarI}(\{x\})$ are required to be shifted to the next existential quantifier. Vice versa, one can show that if $Q$ is invalid, then $\mathcal{R}\!\uparrow\!(Q, \mathcal{T})[\alpha']$ is valid under any assignment $\alpha'$. □

**Theorem 11** (Runtime of Quantifier Expansion, $\star$[33]). *Given a QCSP* $Q$ *over domain* $\mathcal{D}$ *with* $s \geq 3$ *variables per constraint, where* $k$ *is the treewidth of the primal graph of* $Q$. *Then, constructing both a tree decomposition* $\mathcal{T}$ *and instance* $\mathcal{R}\!\uparrow\!(Q, \mathcal{T})$ *takes time* $\mathcal{O}(s \cdot 2^{k^3} \cdot |\mathcal{D}|^2 \cdot \|\mathsf{matrix}(Q)\|)$.

**Theorem 12** (Treewidth Decrease). *Given any QCSP* $Q$ *with at most* $s \geq 3$ *many variables per constraint over fixed domain* $\mathcal{D}$. *Then, there is a tree decomposition* $\mathcal{T} = (T, \chi)$ *of the primal graph* $P_Q$ *of* $Q$ *with* $k = \mathsf{width}(\mathcal{T}) = \mathsf{tw}(P_Q)$ *such that* $Q' = \mathcal{R}\!\uparrow\!(Q, \mathcal{T})$ *is a QCSP, where* $\mathsf{tw}(P_{Q'})$ *is in* $\mathcal{O}(s \cdot \lceil \log_{|\mathcal{D}|}(k) \rceil)$.

*Proof.* We take any TD $\mathcal{T} = (T, \chi)$ of $P_Q$ of width $k$, which has to exist since $k = \mathsf{tw}(P_Q)$. We assume that for each node $t$ of $T$ the number $|\mathsf{VarI}(\chi(t))|$ of introduced vertices is bounded by $\mathcal{O}(\log_{|\mathcal{D}|}(k))$. This can be assumed without loss of generality, since otherwise one can easily modify $\mathcal{T}$ by adding intermediate nodes between $t$ and the parent node of $t$. Further, as in Theorem 5, we assume at most two child nodes per node $t$ of $T$. From this we will construct a TD $\mathcal{T}' = (T, \chi')$ of the primal graph of $Q' = \mathcal{R}{\uparrow}(Q, \mathcal{T})$. For each TD node $t$ of $T$ with parent node $t^*$, where $t \in \mathsf{chldr}(t^*)$, (if exists) we set its bag $\chi'(t) := \{p \mid x \in \chi(t), p \in \mathsf{var}([\![x]\!]_t) \cup \mathsf{var}([\![x]\!]_{t^*})\} \cup \{p \mid x \in \chi(t), 1 \leq j \leq s, p \in \mathsf{var}([\![x]\!]_{t,j})\} \cup \{v_{t'}, v_{t,j} \mid t' \in \{t, t^*\}, 1 \leq j \leq s\} \cup \mathsf{VarI}(\chi(t)) \cup \{\text{usat}_{\leq t}, \text{usat}_t, \text{usat}_{\leq t_i} \mid t_i \in \mathsf{chldr}(t)\}$. Observe that all the properties of TDs are satisfied. Connectedness, in particular, holds since the only elements that are shared between two neighboring tree decompositions nodes are in $\mathsf{VarP}$ and $\mathsf{VarPV}$. Each bag $\chi'(t)$ contains for each $x \in \chi(t)$ at most $2 \cdot \lceil \log_{|\mathcal{D}|}(k+1) \rceil$ many variables $\mathsf{var}([\![x]\!]_t \cup [\![x]\!]_{t^*})$ and at most $s \cdot \lceil \log_{|\mathcal{D}|}(k+1) \rceil$ many variables $\mathsf{var}([\![x]\!]_{t,j})$ for $1 \leq j \leq s$. In total everything sums up to $\mathcal{O}(s \cdot \lceil \log_{|\mathcal{D}|}(k) \rceil)$ many elements per node. □

## 5 Consequences of Quantifier Elimination and Expansion

This section deals with final discussions on limitations of both reductions $\mathcal{R}{\downarrow}$ and $\mathcal{R}{\uparrow}$, and covers lower bounds for QCSAT. Before we show lower bounds for the QCSP formalism, we present an adaption of the *exponential time hypothesis (ETH)* for CSPs and treewidth, which was originally presented [37] for SAT and Boolean formulas. For the restricted case of Boolean QCSPs, called quantified Boolean formulas (QBFs), lower bounds under the ETH are known.

**Proposition 13** ([23]). *Given any QBF $Q$ of quantifier rank $\ell \geq 1$, where each constraint has at most 3 variables. Then, under ETH, $Q$ cannot be solved in time $\mathsf{tower}(\ell, o(k)) \cdot \text{poly}(\mathsf{matrix}(Q))$, where $k$ is the treewidth of graph $P_Q$.*

The following result presents lower bounds for CSPs over finite domain and treewidth. Thereby, using the reduction $\mathcal{R}{\downarrow}$ and existing results for QBFs as given in Proposition 13, we show that one can indeed generate CSPs of large domain, by transforming QBFs of quantifier rank $\ell$ into a CSP.

**Proposition 14** (Lower Bound for CSP). *Under ETH, an arbitrary CSP $\mathcal{C}$ over finite domain $\mathcal{D}$ can not be solved in time $|\mathcal{D}|^{o(k)} \cdot \text{poly}(\|\mathcal{C}\|)$, where $k$=$\mathsf{tw}(P_{\mathcal{C}})$.*

*Proof (Sketch).* We reduce from an arbitrary QCSP formula $Q'$ with $\mathcal{C}' = \mathsf{matrix}(Q')$ over quantifier rank $\ell$=2, where the treewidth of $P_{Q'}$ is $k'$, and the domain is Boolean (QBF). We construct $Q = \mathcal{R}{\downarrow}(Q', \mathcal{T}')$ using a TD $\mathcal{T}'$ of $P_{Q'}$. The treewidth $k$ of $P_Q$ is in $\mathcal{O}(2^{k'})$, but can be reduced to $k = f(k')$, where $f(k') := 2^{\mathcal{O}(k')}/\lfloor \log(|\mathcal{D}|) \rfloor$ if using variables $\mathsf{VarUSatAss}$ over some fixed domain $\mathcal{D}$. Consequently, if $Q$ could be solved in time $2^{o(f^{-1}(k))} \cdot \text{poly}(\|\mathcal{C}'\|)$ or in time $2^{\lfloor \log(|\mathcal{D}|) \rfloor \cdot \log(2^{o(k)})} \cdot \text{poly}(\|\mathcal{C}'\|) = |\mathcal{D}|^{o(k)} \cdot \text{poly}(\|\mathcal{C}\|)$, then also $Q'$ can be solved in time $2^{2^{o(k')}}$. This contradicts Proposition 13 for $Q'$ and establishes the claim for $\mathsf{matrix}(Q)$ for domain sizes up to $|\mathcal{D}|$ in $\mathcal{O}(2^{2^k})$. One can easily show the result for $|\mathcal{D}|$ in $\mathcal{O}(\mathsf{tower}(\ell, k))$ if reducing from QBF of quantifier rank $\ell > 2$. □

This result can be lifted to QCSP, by reducing from an arbitrary instance of QCSAT, but restricted to QBFs, to QCSAT, and doing a similar idea to the proof of Proposition 14, but for each tree decomposition bag. However, the result is then restricted to QCSPs $Q$, whose domain sizes $|\mathcal{D}|$ are bounded by $\mathcal{O}(2^k)$, where $k$ is the treewidth of $P_Q$. The reason for that is that the domain can be effectively used to reduce Boolean variables within a TD bag only if $\log(|\mathcal{D}|) \in \mathcal{O}(k)$. Consequently, we need a different technique. Indeed, for the full result using finite domains, we use reduction $\mathcal{R}\downarrow$ for quantifier elimination as follows.

**Theorem 15** (QCSP Lower Bound). *Given any QCSP $Q$ over finite domain $\mathcal{D}$ and quantifier rank $\ell \geq 1$, where each constraint has at most $s \geq 3$ many variables. Then, under ETH, $Q$ cannot be solved in time* $\mathsf{tower}(\ell - 1, |\mathsf{matrix}(Q)|^{o(k)}) \cdot \mathrm{poly}(\|\mathcal{C}\|)$, *where $k$ is the treewidth of the primal graph $P_Q$.*

*Proof.* We show the theorem by induction on the quantifier rank $\ell$. For the base case $\ell = 1$, the result follows from the Proposition 14 (ETH). For the induction step, we assume that the theorem holds for given $Q$ of quantifier rank $\ell \geq 1$. Let $Q'$ be any QCSP of quantifier rank $\ell + 1$ over fixed domain $\mathcal{D}$, where $k' = \mathsf{tw}(P_{Q'})$ and $C' = \mathsf{matrix}(Q')$. Then, let $\mathcal{T}'$ be a TD of $P_{Q'}$ of width $k'$, $Q = \mathcal{R}\downarrow(Q', \mathcal{T}')$ such that $C = \mathsf{matrix}(Q)$. By Theorem 6, $k$ in $|\mathcal{D}|^{\mathcal{O}(k')}$ is the treewidth of the primal graph of $C$. Since $k$ is in $|\mathcal{D}|^{\mathcal{O}(k')}$, by simplifications using properties of logarithms, $k$ is in $|\mathcal{D}|^{\mathcal{O}(k')} \cdot \log_{|\mathcal{D}|}(2) = \log_{|\mathcal{D}|}(2^{|\mathcal{D}|^{\mathcal{O}(k')}})$. By the induction hypothesis, we can *not* solve $Q$ in time $\mathsf{tower}(\ell - 1, |\mathcal{D}|^{o(k)}) \cdot \mathrm{poly}(\|\mathcal{C}\|) = \mathsf{tower}(\ell - 1, |\mathcal{D}|^{\log_{|\mathcal{D}|}(2^{|\mathcal{D}|^{o(k')}})}) \cdot \mathrm{poly}(\|\mathcal{C}\|) = \mathsf{tower}(\ell, |\mathcal{D}|^{o(k')}) \cdot \mathrm{poly}(\|\mathcal{C}\|)$. Towards a contradiction, assume that we can solve $Q'$ in time $\mathsf{tower}(\ell, |\mathcal{D}|^{o(\mathsf{tw}(P_{Q'}))}) \cdot \mathrm{poly}(\|\mathcal{C}'\|)$. But then since $Q$ can be obtained in polynomial time according to Theorem 5, by correctness of the reduction (Theorem 4), and since the solution of the inverse of $Q$ can be inverted in constant time, we can solve $Q$ in time $\mathsf{tower}(\ell, |\mathcal{D}|^{o(k')}) \cdot \mathrm{poly}(\|\mathcal{C}\|)$. This contradicts with the induction hypothesis. □

**Remark 16.** *The* QCSAT *lower bound above holds for the case of $s$=2 and $|\mathcal{D}|$≥3 by reducing from* 3-SAT over 3-COL [39] to CSAT with $s$=2 and $|\mathcal{D}|$=3.

With the lower bound of Theorem 15, we can show that the reduction $\mathcal{R}\downarrow$ as presented in Sect. 3 can probably not be significantly improved.

**Theorem 17** (ETH-Tightness: Quantifier Elimination). *Given any QCSP $Q$ over fixed domain $\mathcal{D}$ of quantifier rank $\ell$. Then, assuming any TD $\mathcal{T}$ of $P_Q$, under ETH, the treewidth increase of $\mathcal{R}\downarrow(Q, \mathcal{T})$ cannot be significantly improved.*

*Proof.* Let $\mathcal{T}$ be a TD of $P_Q$ of width $k = \mathsf{tw}(Q)$, $Q' = \mathcal{R}\downarrow(Q, \mathcal{T})$ such that $C' = \mathsf{matrix}(Q')$, and $k'$ be the treewidth of the primal graph of $C'$. By Theorem 6, $k'$ is in $\mathcal{O}(|\mathcal{D}|^{(k+1)} / \log(|\mathcal{D}|)$. This can be simplified using calculus (equations for logarithms) accordingly to show that $k'$ is in $\mathcal{O}(|\mathcal{D}|^{(k+1)} \cdot \log_{|\mathcal{D}|}(2)) =$

$\mathcal{O}(\log_{|\mathcal{D}|}(2^{|\mathcal{D}|^{(k+1)}}))$. As a result, since $Q'$ can be obtained in polynomial time according to Theorem 5, one can solve $Q'$ in time $\mathsf{tower}(\ell-1, |\mathcal{D}|^{\mathcal{O}(k)} \cdot \text{poly}(\|\mathcal{C}'\|)$. This is in line with the lower bound under ETH given by Theorem 15, which matches with the upper bound [14] and the treewidth $k'$ cannot be significantly improved. □

Notably, a similar result holds for $\mathcal{R}\!\uparrow$, since for given QCSP over domain $\mathcal{D}$ of treewidth $k$, one can reshape the expression of the treewidth of the primal graph of the QCSP obtained by $\mathcal{R}\!\uparrow$ (cf. Theorem 12) and show that $k$ is in $\mathcal{O}(\log_{|\mathcal{D}|}(k)) = \mathcal{O}(\log_{|\mathcal{D}|}(\log(|\mathcal{D}|^k)))$. However, since intuitively this method $\mathcal{R}\!\uparrow$ works by reducing treewidth, it can not reduce treewidth to a value smaller than 1. Therefore, one can only show that $\mathcal{R}\!\uparrow$ can not be significantly improved if $|\mathcal{D}|$ is bounded by $\mathcal{O}(k)$, since otherwise the logarithm is smaller than 1. Consequently, and in contrast to related work on QBFs [23], the reduction $\mathcal{R}\!\uparrow$ on treewidth decrease can not be used to establish the full result of Theorem 15. This is why for fixed, but large (non-binary) domains we indeed require $\mathcal{R}\!\downarrow$ to show the result.

## 6 Conclusion and Future Work

In this work, we present methods for quantifier elimination and expansion for the formalism of quantified constraint satisfaction problems (QCSPs) over fixed domains. Our reduction $\mathcal{R}\!\downarrow$ shows that for QCSPs of bounded treewidth, one can eliminate quantifiers and still avoid even in the worst-case an exponential blow-up in the size of the resulting QCSP. However, this comes at the price of an exponential blow-up in the treewidth of the resulting QCSP. While we hope that this can be avoided in several special cases, we also showed that under reasonable assumptions in computational complexity theory, namely if assuming the exponential time hypothesis (ETH), one can not expect to significantly improve this blow-up of treewidth. This new reduction $\mathcal{R}\!\downarrow$ for quantifier elimination allows for lifting a recently established lower bound result [23] for quantified Boolean formulas (QBFs) to QCSPs, which confirms that the upper bound of solving QCSPs by Chen [14] is ETH-tight. Further, we also provide the other direction, namely a reduction $\mathcal{R}\!\uparrow$ for quantifier expansion, where the treewidth is exponentially decreased, at the cost of increasing quantifier rank. This reduction $\mathcal{R}\!\uparrow$ is lifted from QBFs [23] and might serve well in reducing treewidth in practice. In particular, QCSP and QBF solvers based on treewidth and tree decompositions, such as the solver dynQBF [13] and others [22,26,34], could benefit from significantly reduced treewidth.

## References

1. Allouche, D., et al.: Tractability-preserving transformations of global cost functions. Artif. Intell. **238**, 166–189 (2016)

2. Atserias, A., Fichte, J.K., Thurley, M.: Clause-learning algorithms with many restarts and bounded-width resolution. J. Artif. Intell. Res. **40**, 353–373 (2011)
3. Atserias, A., Oliva, S.: Bounded-width QBF is PSPACE-complete. J. Comput. Syst. Sci. **80**(7), 1415–1429 (2014)
4. Bacchus, F., Stergiou, K.: Solution directed backjumping for QCSP. In: Bessière, C. (ed.) CP 2007. LNCS, vol. 4741, pp. 148–163. Springer, Heidelberg (2007). https://doi.org/10.1007/978-3-540-74970-7_13
5. Beldiceanu, N., Demassey, S.: Global Constraint Catalog: 5.117. element (2020). https://web.imt-atlantique.fr/x-info/sdemasse/gccat/Celement.html
6. Benedetti, M., Lallouet, A., Vautard, J.: QCSP made practical by virtue of restricted quantification. In: IJCAI, pp. 38–43 (2007)
7. Bertelè, U., Brioschi, F.: Contribution to nonserial dynamic programming. J. Math. Anal. Appl. **28**(2), 313–325 (1969)
8. Bertelè, U., Brioschi, F.: Nonserial Dynamic Programming. Academic Press Inc., Cambridge (1972)
9. Bertelè, U., Brioschi, F.: On Non-serial dynamic programming. J. Comb. Theory Ser. A **14**(2), 137–148 (1973)
10. Bodlaender, H.L.: A linear-time algorithm for finding tree-decompositions of small treewidth. SIAM J. Comput. **25**(6), 1305–1317 (1996)
11. Bodlaender, H.L.: A partial k-arboretum of graphs with bounded treewidth. Theoret. Comput. Sci. **209**(1–2), 1–45 (1998)
12. Bondy, J.A., Murty, U.: Graph Theory. Graduate Texts in Mathematics, vol. 244. Springer, Heidelberg (2008)
13. Charwat, G., Woltran, S.: Expansion-based QBF solving on tree decompositions. Fundamenta Informaticae **167**(1–2), 59–92 (2019)
14. Chen, H.: Quantified constraint satisfaction and bounded treewidth. In: Proceedings of the 16th European Conference on Artificial Intelligence (ECAI 2004), vol. IOS Press, pp. 161–170 (2004)
15. Creignou, N., Vollmer, H.: Boolean constraint satisfaction problems: when does Post's lattice help? In: Creignou, N., Kolaitis, P.G., Vollmer, H. (eds.) Complexity of Constraints. LNCS, vol. 5250, pp. 3–37. Springer, Heidelberg (2008). https://doi.org/10.1007/978-3-540-92800-3_2
16. Cygan, M., et al.: Parameterized Algorithms. Springer, Heidelberg (2015). https://doi.org/10.1007/978-3-319-21275-3
17. Dechter, R.: Tractable structures for constraint satisfaction problems. In: Handbook of Constraint Programming, vol. I, pp. 209–244. Elsevier (2006). (Chap. 7)
18. Diestel, R.: Graph Theory. Graduate Texts in Mathematics, vol. 173, 4th edn. Springer, Heidelberg (2012)
19. Downey, R.G., Fellows, M.R.: Fundamentals of Parameterized Complexity. TCS. Springer, London (2013). https://doi.org/10.1007/978-1-4471-5559-1
20. Dvořák, W., Pichler, R., Woltran, S.: Towards fixed-parameter tractable algorithms for abstract argumentation. Artif. Intell. **186**, 1–37 (2012)
21. Ferguson, A., O'Sullivan, B.: Quantified constraint satisfaction problems: from relaxations to explanations. In: Proceedings of the Twenty-Sixth International Joint Conference on Artificial Intelligence (IJCAI 2007), pp. 74–79. The AAAI Press (2007)
22. Fichte, J.K., Hecher, M., Morak, M., Woltran, S.: DynASP2.5: dynamic programming on tree decompositions in action. In: Lokshtanov, D., Nishimura, N. (eds.) Proceedings of the 12th International Symposium on Parameterized and Exact Computation (IPEC 2017). Dagstuhl Publishing (2017)

23. Fichte, J.K., Hecher, M., Pfandler, A.: Lower bounds for QBFs of bounded treewidth. In: Kobayashi, N. (ed.) Proceedings of the 35th Annual ACM/IEEE Symposium on Logic in Computer Science (LICS 2020), pp. 410–424. ACM (2020)
24. Fichte, J.K., Hecher, M., Thier, P., Woltran, S.: Exploiting database management systems and treewidth for counting. In: Komendantskaya, E., Liu, Y.A. (eds.) PADL 2020. LNCS, vol. 12007, pp. 151–167. Springer, Cham (2020). https://doi.org/10.1007/978-3-030-39197-3_10
25. Fichte, J.K., Hecher, M., Zisser, M.: An improved GPU-based SAT model counter. In: Schiex, T., de Givry, S. (eds.) CP 2019. LNCS, vol. 11802, pp. 491–509. Springer, Cham (2019). https://doi.org/10.1007/978-3-030-30048-7_29
26. Fichte, J.K., Hecher, M., Woltran, S., Zisser, M.: Weighted model counting on the GPU by exploiting small treewidth. In: Azar, Y., Bast, H., Herman, G. (eds.) Proceedings of the 26th Annual European Symposium on Algorithms (ESA 2018). LIPIcs, vol. 112, pp. 28:1–28:16. Dagstuhl Publishing (2018)
27. Flum, J., Grohe, M.: Parameterized Complexity Theory. TTCSAES. Springer, Heidelberg (2006). https://doi.org/10.1007/3-540-29953-X
28. Freuder, E.C.: A sufficient condition for backtrack-bounded search. J. ACM **32**(4), 755–761 (1985)
29. Gent, I.P., Nightingale, P., Stergiou, K.: QCSP-solve: a solver for quantified constraint satisfaction problems. In: IJCAI, pp. 138–143. Professional Book Center (2005)
30. Godet, A., Lorca, X., Hebrard, E., Simonin, G.: Using approximation within constraint programming to solve the parallel machine scheduling problem with additional unit resources. In: Conitzer, V., Sha, F. (eds.) Proceedings of the 34th AAAI Conference on Artificial Intelligence. The AAAI Press, New York (2020)
31. Gottlob, G., Greco, G., Scarcello, F.: The complexity of quantified constraint satisfaction problems under structural restrictions. In: IJCAI, pp. 150–155. Professional Book Center (2005)
32. Gottlob, G., Pichler, R., Wei, F.: Bounded treewidth as a key to tractability of knowledge representation and reasoning. Artif. Intell. **174**(1), 105–132 (2010)
33. Hecher, M., Fichte, J.K., Kieler, M.F.I.: Supplemental material of this submission, containing proofs of theorems marked with "⋆" (2020). Self-archived copy by the authors available online
34. Hecher, M., Thier, P., Woltran, S.: Taming high treewidth with abstraction, nested dynamic programming, and database technology. In: Pulina, L., Seidl, M. (eds.) SAT 2020. LNCS, vol. 12178, pp. 343–360. Springer, Cham (2020). https://doi.org/10.1007/978-3-030-51825-7_25
35. Hemaspaandra, E.: Dichotomy theorems for alternation-bounded quantified Boolean formulas. CoRR cs.CC/0406006 (2004). http://arxiv.org/abs/cs/0406006
36. Hooker, J.N., van Hoeve, W.-J.: Constraint programming and operations research. Constraints **23**(2), 172–195 (2017). https://doi.org/10.1007/s10601-017-9280-3
37. Impagliazzo, R., Paturi, R., Zane, F.: Which problems have strongly exponential complexity? J. Comput. Syst. Sci. **63**(4), 512–530 (2001)
38. Jégou, P., Terrioux, C.: Hybrid backtracking bounded by tree-decomposition of constraint networks. Artif. Intell. **146**(1), 43–75 (2003)
39. Karp, R.M.: Reducibility among combinatorial problems. In: Complexity of Computer Computations. The IBM Research Symposia Series, pp. 85–103. Plenum Press, New York (1972)
40. Kloks, T. (ed.): Treewidth: Computations and Approximations. LNCS, vol. 842. Springer, Heidelberg (1994). https://doi.org/10.1007/BFb0045375

41. Kuo, C., Ravi, S.S., Dao, T., Vrain, C., Davidson, I.: A framework for minimal clustering modification via constraint programming. In: Singh, S.P., Markovitch, S. (eds.) Proceedings of the 31st AAAI Conference on Artificial Intelligence, pp. 1389–1395. The AAAI Press, San Francisco (2017)
42. Lampis, M., Mitsou, V.: Treewidth with a quantifier alternation revisited. In: Proceedings of the 12th International Symposium on Parameterized and Exact Computation (IPEC 2017). LIPIcs, vol. 89, pp. 26:1–26:12. Dagstuhl Publishing (2017)
43. Ordyniak, S., Szeider, S.: Parameterized complexity results for exact Bayesian network structure learning. J. Artif. Intell. Res. **46**, 263–302 (2013)
44. Otten, L., Dechter, R.: Anytime AND/OR depth-first search for combinatorial optimization. AI Commun. **25**(3), 211–227 (2012)
45. Pan, G., Vardi, M.Y.: Fixed-parameter hierarchies inside PSPACE. In: LICS, pp. 27–36. IEEE Computer Society (2006)
46. Papadimitriou, C.H.: Computational Complexity. Addison-Wesley, Boston (1994)
47. Pichler, R., Rümmele, S., Woltran, S.: Counting and enumeration problems with bounded treewidth. In: Clarke, E.M., Voronkov, A. (eds.) LPAR 2010. LNCS (LNAI), vol. 6355, pp. 387–404. Springer, Heidelberg (2010). https://doi.org/10.1007/978-3-642-17511-4_22
48. Robertson, N., Seymour, P.: Graph minors. I. Excluding a forest. J. Comb. Theory Ser. B **35**(1), 39–61 (1983)
49. Samer, M., Szeider, S.: Algorithms for propositional model counting. J. Discrete Algorithms **8**(1), 50–64 (2010)

# A Time Leap Challenge for SAT-Solving

Johannes K. Fichte[1(✉)], Markus Hecher[2], and Stefan Szeider[3]

[1] School of Engineering Sciences, TU Dresden, Dresden, Germany
johannes.fichte@tu-dresden.de

[2] Database and Artificial Intelligence Group, TU Wien, Vienna, Austria
hecher@dbai.tuwien.ac.at

[3] Algorithms and Complexity Group, TU Wien, Vienna, Austria
sz@ac.tuwien.ac.at

**Abstract.** We compare the impact of hardware advancement and algorithm advancement for SAT-solving over the last two decades. In particular, we compare 20-year-old SAT-solvers on new computer hardware with modern SAT-solvers on 20-year-old hardware. Our findings show that the progress on the algorithmic side has at least as much impact as the progress on the hardware side.

**Keywords:** SAT-solving · Benchmarking · Old hardware · Hardware advancement · Algorithm advancement · Experimenting

## 1 Introduction

The last decades have brought enormous technological progress and innovation. Two main factors that are undoubtedly key to this development are (i) *hardware advancement* and (ii) *algorithm advancement*. Moore's Law, the prediction made by Gordon Moore in 1965 [54], that the number of components per integrated circuit doubles every year, has shown to be astonishingly accurate for several decades. Given such an exponential improvement on the hardware side, one is tempted to overlook the progress made on the algorithmic side.

This paper aims to compare the impact of hardware advancement and algorithm advancement based on a genuine problem, the propositional satisfiability problem (SAT). This problem is well-suited for such a comparison since it is one of the first problems for which progress in solving has been measured regularly through competitions [36]. Also, a standardized instance format has been established very early. By focusing on this problem, the comparison allows us to fathom the SAT and CP community's contribution to the overall progress.

Of course, the advancements in hardware and algorithms cannot be separated entirely. Targeted algorithm engineering can make use of new hardware features [10,13,22,37] and hardware development can be guided by the specific

The work has been supported by the Austrian Science Fund (FWF), Grants Y698, 32441, and 32830, and the Vienna Science and Technology Fund, Grant WWTF ICT19-065. Markus Hecher is also affiliated with the University of Potsdam, Germany.



H. Simonis (Ed.): CP 2020, LNCS 12333, pp. 267–285, 2020.
https://doi.org/10.1007/978-3-030-58475-7_16

demands of modern algorithms. This can quickly end up in comparing apples and oranges. Nevertheless, we think that by carefully setting up the experiment and choosing hardware and algorithms, it still allows us to draw some conclusions on the impact of the individual components.

We base the general setup of the comparison on a *Time Leap Challenge*, where virtual teams compete. Team SW uses new solvers on old hardware; Team HW uses old solvers on new hardware. The time between "old" and "new" spans about two decades. Which team can solve more instances? Depending on the outcome, one can compare the impact of hardware advancement and algorithm advancement. The idea for this time leap challenge for SAT-solvers was inspired by a thought experiment on algorithms in mixed-integer linear programming (MILP), suggested by Sebastian Stiller [69].

In the early 1990s, the dominant complete method for SAT-solving was the *DPLL Algorithm* (Davis-Putnam-Logemann-Loveland [15,16]), which combines backtracking search with Boolean constraint propagation [72]. However, in the late 1990s, the *CDCL Solvers* (Conflict-Driven Clause Learning) took over. They extended the basic DPLL framework with new methods, including clause learning [64], lazy data structures like watched literals [55], backjumping [55,64], and dynamic branching heuristics [55]; the combination of these methods resulted in a significant performance boost, often referred to as the "CDCL Revolution." Although the CDCL paradigm is still predominating today's SAT-solving, there have been several significant improvements made over the last two decades, including efficient preprocessing [17] and inprocessing [37], aggressive clause deletion [2], fast restarts [49], lightweight component caching [59], implication queue sorting [46], and new branching heuristics [48].

### 1.1 Experimental Setting

For our Time Leap Challenge, Team HW (old solvers on new hardware) is composed of the solvers `Grasp` (1996), `zChaff` (2001), and `siege` (2003) running on a computer from 2019 with an Intel Xeon Silver 4112 CPU at 2.60 GHz base frequency and 128GB RAM. Team SW (new solvers on old hardware) is composed of the solvers `MapleSat19` (2019), `CaDiCal` (2019), and `Glucose` (2016) running on a computer from 1999 with a Pentium III processor at 467 MHz frequency and 1.5 GB RAM. An essential question for setting up the experiment was the choice of a suitable set of benchmark instances. On the one hand, the instances should not be too challenging so that they are not entirely out of reach for old solvers or old hardware; on the other hand, the instances should still be challenging enough to provide interesting results. We settled on the benchmark set *set-asp-gauss* [33] that provides a reasonably good compromise, as it contains a large variety of instances, tailors adapted instance hardness, is free of duplicates, reproducible, and publicly available. We used a timeout of 900 seconds, which is the default for SAT competitions. Right in the beginning, we state a clear disclaimer. While a theoretical challenge is easy to design, a practical comparison can rarely be comprehensive and complete. About 20 years of evolution increases the practical search space by orders. There are many possibilities to

**Table 1.** Summary of experimental results

| | Grasp (1996) | zChaff (2001) | siege_v3 (2003) | Glucose (2016) | CaDiCal (2019) | Maple (2019) |
|---|---|---|---|---|---|---|
| | | | | Team SW | | |
| old HW (1999) | 73 | 48 | 37 | 106 | 98 | 77 |
| | Team HW | | | | | |
| new HW (2019) | 76 | 71 | 93 | 188 | 190 | 195 |

combine hardware, software, benchmarks, and solvers. Particularly, there might be solvers that are still available, but we missed during our research. Still, we provide a clear guideline on how we selected the teams and provide extensive details beyond. Our results are reproducible in the setting, and the conclusions provide a general idea. However, the ideas might not generalize to conclusions over other benchmark sets or solvers we might have missed. This is, however, a usual situation in many experiments with combinatorial solving as there is no good theoretical understanding of the practical effects [58]. Still, we aimed to put the concept of a time leap challenge from literature in popular science into a practical scientific context.

### 1.2 Results

Table 1 gives a summary of our results (we provide more details in Sect. 3). We see that both teams perform in a similar range with a slight advantage for Team SW.

### 1.3 Related Work

Knuth [42] provides an overview of various aspects of SAT-solving, including commented implementations of algorithms from several epochs of SAT-solving. His implementations assemble a DPLL solver (SAT10), a DPLL look-ahead solver (SAT11), and a CDCL solver (SAT13), as well as a preprocessor (SAT12). Since all these solvers are implemented uniformly, without special implementation or hardware tricks, they provide an excellent comparison of the algorithmic advancement of solver techniques. We therefore included, for comparison, the results of Knuth's solvers on the same benchmark set and hardware platform as the time leap challenge. Mitchell [53] provides an overview of techniques, implementations, and algorithmic advances of the year 2005 and looking back for 15 years. He already mentioned that the success of SAT-solving is due to three factors: improved algorithms, improved implementation techniques, and increased machine capacity. However, Mitchell's work does not provide evaluations on any actual practical effects at the time. Kohlhase [44] recently published work on collecting and preserving the comparability of old theorem provers to preserve cultural artifacts and history in Artificial Intelligence.[1] For an overview on the technique of CDCL-based solvers we refer the reader to introductory literature such as a chapter in the Handbook of Knowledge Representation [29], chapters on the history of modern SAT-solving [24], and CDCL-solvers [52] in the Handbook of Satisfiability [9]. Katebi, Sakallah, and Marques-Silva [41,63] considered various techniques of modern SAT-solvers under an empirical viewpoint. They designed experiments to evaluate factors and the aggregation of different SAT-enhancements that contribute to today's practical success of modern solvers. Works on targeted algorithm engineering for SAT-solvers are extensive. Just to name a few examples, there is work on exploiting features such as optimizing memory footprints for the architecture [10], on implementing cache-aware [13], on using huge pages [22], on how to benefit from parallel solving [34] or employing inprocessing. Inprocessing particularly takes advantage of modern hardware as one can execute much more instructions on a modern CPU than accessing bytes on memory [30,50]. Very recently, Audemard, Paulevé, and Simon [1] published a heritage system for SAT solvers. It allows for compiling, archiving, and running almost all released SAT solvers and is based on Docker, GitHub, and Zenodo. While they aim for archivability, our work provides an actual experiment incorporating soft- and hardware advances. We hope that their system allows for long term preservation and, if there is no major change in the computer architecture, that one can repeat our time leap challenge in another decade.

## 2 The Arena: Designing the Time Leap Challenge

To run a proper challenge, we design an arena by selecting from standard benchmark sets and several contestants out of a vast space of possibilities. We aim for the reasonable oldest hardware on which we can still run modern benchmark sets and solvers. In turn, this requires to set up a modern operating system on old hardware. To make it a time leap challenge, we are interested in solvers and hardware from similar generations, so a preferably small time frame from which both originate. The physical effort restricts us to consider only two time frames in the following. We take modern hardware and solvers from 2019 and old hardware from around 2000 and solvers from 2001/2002. Following academic ideas by Stallman [68], we focus on benchmark sets and solvers that are publicly available. Throughout the experimental work, we follow standard guidelines for benchmarking [45]. In the course of this section, we elaborate on various technical and organisational obstacles. Setting up a time leap challenge is also somewhat of an archaeological challenge.

In theory, a variety of competitions have been organized in the past. The competition results give a broad picture of benchmark instances and solvers. Old hardware and operating systems should still be widely available. In practice, neither open source, nor version control systems, nor public platforms to host

[1] The Theorem Prover Museum is available online at https://theoremprover-museum.github.io/.

software projects such as SourceForge[2], bitbucket, github, or gitlab, were popular in the community around the millennium. Publicly funded data libraries such as Zenodo [57] were also established much later. While the culture of storing text in libraries dates back to Alexandria and the first librarian Zenodotus in 280 BCE, searching for datasets and source codes from 20 years ago feels like digging through a burnt library. Enthusiasts maintained datasets and source codes from early competitions. Sometimes source codes were kept as a secret [28]. Some links redirect to grabbed domains, or people moved and with them, the webpages. Sometimes binaries show up from private collections or the Internet Archive [40]. However, it turned out that they do not run, as libraries on which they depend do not run on modern Linux or Unix distributions.

Below we report and explain details of the selection process.

*Instance Format.* Johnson and Trick suggested a uniform input format description in 1993, which is still used as the standard for SAT input instances [38]. The standardized input format and backward compatibility substantially simplified our selection process.

### 2.1 Selecting a Suitable Benchmark Set

Our focus on selecting a benchmark set is to consider a larger benchmark set, say of a cardinality ranging from 100 to 300. We are interested in a safe and stable choice of instances since benchmarks run a wide variety of experiments with preferably more than 10 solvers resulting in months of running time. Hence, we push to a reasonable state-of-the-art benchmark setting. We prefer instances that (i) are publicly available, (ii) contain a good selection of domains, including an industrial background, random, and combinatorial instances, and (iii) highlight differences for modern solvers. We summarize runtime and number of solved instances during our instance selection process in Table 2. For an initial selection, we ran instances only with the solver `Glucose` [4], which showed robust performance on many earlier experimental works that we carried out.

*Available Instances.* The first available benchmark instances *DIMACS-2* date back to 1992 and the 2nd DIMACS Challenge 1992–1993 on NP-hard problems, which also considered SAT as a problem [71]. The 241 instances are still well maintained and downloadable[3]. Note that the 1st SAT competition already took part in 1992 [11]. However, the instances are not publicly available. Over time researchers collected benchmarks such as *SATLIB* [31], which count more than 50,000 instances in total. The instances are still available on an old webpage by the collector.[4] A subset of these instances was also used for the SAT Competition 2002. However, those instances are not available from the SAT Competition website due to an abandoned domain. Instances from one of the annual SAT

[2] https://en.wikipedia.org/wiki/SourceForge.
[3] See: http://archive.dimacs.rutgers.edu/pub/challenge/sat/benchmarks/.
[4] See: https://www.cs.ubc.ca/~hoos/SATLIB/benchm.html.

**Table 2.** Runtime of a modern solver and modern hardware on selected benchmark sets. # refers to the number of solved instances, TO refers to the number of instances on which the solver timed out, ERR refers to the number of instances on which the solver found an input error, $t[h]$ refers to the total running time on the solved instances in hours, avg[s] refers to the average running time of an instance.

| Benchmark | Solver | # | TO | ERR | $t[h]$ | avg[s] |
|---|---|---|---|---|---|---|
| DIMACS2 | `Glucose` | 225 | 15 | 1 | 0.34 | 5.46 |
| SATLIB | `Glucose` | 43892 | 15 | 6399 | 4.45 | 0.36 |
| Set-asp-gauss | `Glucose` | 189 | 11 | 0 | 4.50 | 85.71 |

competitions from 2002 to 2019[5] follow stricter rules and detailed reports are available [36]. There are plenty of tracks, thousands of instances, and many of the more modern instances are enormous in size. A popular benchmark set with various instances from SAT competitions until 2013 and various fields is the benchmark set *set-asp-gauss* [33]. The set is a composition of representative benchmarks from a variety of sources. It has been widely used as a robust selection for tuning solvers in the past and was obtained by classifying the practical hardness of the instances from the SAT Competition 2009 and SAT Challenge 2012 and then selecting instances by sampling with the Gaussian probability distribution [33].

*Initial Evaluations.* In order to gather initial insights, we run all available solvers on our cluster. The hardware for the benchmark selection process consisted of a cluster of RHEL 7.7 Linux machines equipped with two Intel Xeon E5-2680v3 CPUs of 12 physical cores each running at 2.50 GHz, which we enforced by performance governors. The machines are equipped with 64 GB main memory of which 60.5 GB are freely available to programs. We compare wall clock time and number of timeouts. However, we avoid IO access on the CPU solvers whenever possible, i.e., we load instances into the RAM before we start solving. We run four solvers on one node at most, set a timeout of 900 seconds, and limit available RAM to 8 GB per instance and solver. We summarize our initial evaluation of the early benchmark sets in Table 2. The DIMACS-2 instances turned out to be very easy for modern solvers. For example, the solver Glucose solved almost all instances within less than one second, only five large instances (par32-X.cnf) of a parity learning problem remained unsolved within 900 s. The SATLIB instances are more challenging but still fairly easy for modern solvers. The SAT Competition 2002–2019 instances provide a broad selection. Since the results are still publicly available, we refrained from rerunning these sets. The runtime results on the benchmark set *set-asp-gauss* reveals that modern solvers can solve many instances. However, the instances are still challenging as the overall runtimes are

[5] The webpage http://www.satcompetition.org/ gives a broad overview on the results and details of the competitions since 2002.

reasonably long. Old solvers are still able to solve plenty of instances on modern hardware. The benchmark set consists of 200 instances in total.

*Decision.* After running the instances, we picked one existing benchmark set. Since the set *DIMACS-2* contains almost only easy instances, we rejected the set right away. While the *SATLIB* instances contain mainly easy instances, they are not very challenging for modern solvers. Further, the contained benchmarks have a strong bias towards handcrafted and random instances. The SAT 2002–2019 instances contain very interesting sets. However, some of the more modern instances are very large, and we figured that it is impossible to transfer and run the instances on old hardware. After reviewing the initial results and sampling memory requirements from earlier SAT competitions, we decided to use the benchmark set *set-asp-gauss* [33], which provides a reasonably good compromise. It contains a large variety of instances, tailors adapted instance hardness, is free of duplicates, reproducible, and publicly available.

### 2.2 Selecting Solvers

In the following section, we describe the selection process of SAT-solvers for our challenge. In order to foster reproducibility and favor open-source, we focus on publicly available solvers (binary or source code). Note that modern SAT-solving also includes various parallel algorithms. Due to the unavailability of wide parallel computation on old hardware, we restrict ourselves to sequential solvers. Further, we consider only solvers that are, vaguely speaking, descendants of the DPLL [15,16] algorithm, i.e., CDCL. These solvers are often referred to as solvers implementing complete and systematic search. However, restarts and deletion might affect completeness under certain conditions in practice [52]. To our knowledge, CDCL-based solvers with various additional techniques on top, which even extend the underlying proof system, are still the most prevailing paradigm for SAT-solvers. However, today, some solvers use strong proof techniques such as the division rule in cutting planes [20,27] or Gaussian Elimination [65,66].

*Researching for Solvers.* The 1st SAT Competition [11] and 2nd DIMACS Challenge [71] took place around 1992. However, no online resources on detailed solvers or source codes are available. The earliest public collection of solvers which is still available online[6], is the SATLIB Solver Collection [32]. The collection contains implementations on DPLL-based implementations as well as stochastic local search solvers. DPLL-based Implementations in the collection are `Grasp` [64], `NTAB` [14], `POSIT` [25], various versions of `REL_SAT` [6,39], which are also available on github[7], two versions of `SATO` [73], and four versions of `Satz` [47]. Further, we asked colleagues for the source code of old solvers and received an even older version of `Grasp` from 1996 [51]. The era of CDCL solvers

[6] See: https://www.cs.ubc.ca/~hoos/SATLIB/solvers.html.
[7] See: https://github.com/roberto-bayardo/relsat.

started in 2001 [55]. There, successful solvers such as BerkMin [28], siege [62], and zChaff [26] materialized. Siege [62] is publicly available with binaries in three versions from 2003 to 2004. We contacted colleagues on the source code of siege, but the author retired and the sources seem to be lost. For zChaff [26] even the source code is publicly available in four versions from 2001 to 2007. Binaries of BerkMin showed up in a backup of experiments on SAT-solvers from earlier works. We contacted the authors on source codes but received no answer. A famous solver in the SAT-solvers line is MiniSat, which is available online[8] in various versions [18,19,67]. The development of MiniSat started around 2003 [18] intending to create a compact, readable, and efficient solver for the community. The earliest version online is from 2005 and the most known and very popular version 2.2 from 2008. Another popular SAT-solver is Glucose [3], which was developed to aggressively remove clauses that are not helpful during clause learning of the CDCL procedure. This results in an incomplete algorithm as keeping learnt clauses is essential for completeness. We consider the version Glucose syrup 4.2.1 [4]. A very popular, successful and recent solver is Lingeling [7], which won several SAT competitions and the prize on the most innovative solver [5] in 2015. Two medalists of the SAT 2019 Race were CaDiCaL 1.0.3 [8] and a descendant of the solver MapleSAT [48], namely MapleLCMDistChronoBTDL-v3 (MapleSat19) [43].

*Testing the Solvers.* In order to benchmark a solver, we first need to compile it or run the binary with a modern operating system as there is otherwise no chance to get the solvers running on modern hardware. First, we considered all solvers from the SATLIB collection. We were able to compile and successfully run the solvers Grasp, Relsat, Satz, and SATO. However, we had to modify the source codes and build files so that they would compile with a modern compiler due to harder interpretations of language standards in modern compilers. Since the solvers were originally designed for 32bit Linux, we compiled the solvers on 32bit Linux and used them late on 64bit Linux by compatibility layers. While we were also successful in compiling solvers on 64bit systems, the 64bit binary would often solve fewer instances on the 64bit system or result in many segfaults. We suspect compatibility issues as either the developers of the old solvers could not expect certain datatypes on a future architecture or implemented sloppy memory management. All versions of the solver siege, which were available as a binary, still ran on a modern Linux using the 32bit compatibility mode. We were successful in building all versions of the solver zChaff; both on a 32bit as well as 64bit architecture. Unfortunately, the solver BerkMin does not run on modern or fairly recent Linux distribution. It turns out that the binary was compiled with an old gcc and linked to an old version of the glibc, which we discovered in an old Red Hat Enterprise Linux, but we were unable to integrate it into a modern Linux distribution. We found that all modern solvers were well maintained and still compiled on 32 and 64bit Linux distributions as well as a 64bit version of NetBSD.

[8] See: http://minisat.se/MiniSat.html.

*Final Teams.* In order to have a comparison on theoretical advances in SAT-solving between DPLL and CDCL from an abstract perspective and out of the hand of programmer, we picked the implementations by Donald Knuth [42]. The implementations represent particular time periods, more precisely, DPLL solver (`SAT10`), a DPLL look-ahead solver (`SAT11`), and a CDCL solver (`SAT13`), as well as a preprocessor (`SAT12`). We still tested the old solvers `Relsat`, `Satz`, and `SATO`, which resulted in less than 20 solved instances on our modern hardware for the best solver among them (`SATO`). Since it is theoretically well-known that CDCL can be significantly faster than DPLL [58,60], we already have the solvers by Knuth. There has already been work on the technological advances of various techniques between techniques in DPLL and CDCL solvers, we focus on the more modern CDCL solvers for both teams. However, since the solver `Grasp` decides a considerable number of instances and already implements conflict learning, we include `Grasp` into Team HW. Then, there are three solvers left for a team of solvers from about 20 years ago (Team HW), namely, `zChaff` (2001), `siege` (2003), and an early version of `MiniSat` (2005). We decided to include the earliest solver of `zChaff` (2001.2.17) into Team HW, since the numbers of solved instances did not differ much between the 2001 and 2004 versions on our reference hardware. We preferred to include version 3 of the solver `siege` (2003) as it solved about 12 instances more than version 1 (2001) on our modern reference hardware. We discarded `MiniSat` as the youngest of the older solvers. We picked `CaDiCaL` 1.0.3 [8] and `MapleLCMDistChronoBTDL-v3 (MapleSat19)` [43] for Team SW (new solvers on old hardware) due to their good performance in the SAT 2019 Race. `MapleSat19` won the SAT 2019 Race, and `CaDiCal` scored a second place. Since the slightly older solver `Glucose` syrup 4.2.1 [4] solved about ten instances more than the solver `Lingeling` 7d5db72 [7] on our modern reference hardware, we decided to pick `Glucose` for our Team SW.

### 2.3 Selecting the Environment: Operating System and Compiler

Since we are interested in comparing the team new solvers on old hardware and the team old solvers on new hardware, we think that it is only fair to also include advancements in kernel architecture, compilers, and operating systems into the consideration for new solvers. Anyway, it is not possible to obtain ancient Linux or Unix distributions due to missing source code mirrors and it is not possible to run such Linux or Unix distributions on modern hardware due to the lack of modern chipset drivers in ancient kernels. Due to long term support of hardware, we decided to favor Debian 10 codename buster (July 2019) [12] and try NetBSD 9 (Feb. 2020) [70] as operating systems. We ran the experiments on Linux kernel version 4.19.0-8-686-pae. We use gcc 8.3.0 on Debian and NetBSD. Our modern hardware at university was equipped with Linux Mint 19 codename Tara, kernel version 4.15.0-91, and gcc compiler version 7.5.0-3.

### 2.4 Selecting the Hardware

To have a wide variety of hardware, we started to gather old hardware from friends and colleagues. We collected ten systems over different generations, more precisely, systems containing a Pentium II (1998), a Pentium III (1999), an Ultra Sparc IIe (2001), a Pentium IV (2002), a Pentium IV Prescott (2004), a Core2 Duo (2007), an i5 Nehalem (2009), a Xeon Haswell (2013), a Xeon Skylake (2017), and an i7 Icelake (2019). A colleague prepared a SPARCstation II (1995) and SPARCstation Voyager (1995) for us.

*Technical Restrictions.* The selection of a benchmark set and operating systems restricted the space of possibilities on the potential old hardware. Preferably, we are interested in the oldest possible hardware and the youngest possible hardware. In more detail, modern Linux distributions such as Debian 10 still supports all x86-based (IA-32) i686 processors, including various AMD, Intel, and VIA processors. However, the i686 architectures limits experiments to Pentium II processors (1997) or later [35]. BSD distributions such as NetBSD 9 still supports the Sparc64 architecture, which in theory still allows the running of systems with processors SPARC64 (1995) and UltraSPARC IIe (1999). We were able run NetBSD 9 on a system with an Ultra Sparc IIe, namely, the Sun Netra X1 from about 2000/2001. Since for some solvers, we only had access to Linux or Solaris binaries and we were unable to setup Debian 10 or Solaris onto the Netra system in decent time due to a required setup via serial LOM interface and network boot, we discarded the Sun system from our final hardware selection. It is well known that modern operating systems and SAT-solvers are very memory-demanding [23] resulting in a requirement of having at least 1GB of total RAM inside the system. Since the L2 cache controllers of the Pentium II only allow the use of 512 MB of RAM and we could not get access to a system with a Pentium Pro processor, our oldest possible system (1999) was a Pentium III processor running at 467 MHz equipped with 1.5 GB RAM. Hence, we picked this system to run the solvers of Team SW. While the most modern CPU architecture we had access to was an i7 Icelake (2019), we decided to prefer the system running a Xeon Skylake due to the much larger caches, which are usually beneficial for SAT-solving. Still, the modern system with the Xeon Skylate was bought in 2019 for dedicated benchmarking, while the i7 was just a small-form-factor barebone desktop computer for which we feared that high permanent load over months might significantly degenerate performance due to overheating. The system for Team HW then contained two Intel Xeon Silver 4112 CPUs (Skylake architecture) of 2.60 GHz base-frequency equipped with 128 GB RAM. We ran the experiments at the maximum frequency of 3.00 GHz. Since the Netra X1 from 2000 was equipped with 2 GB and the NetBSD allowed to still run all source code based solvers, even the very modern ones, the Sun system serves as a point of reference.

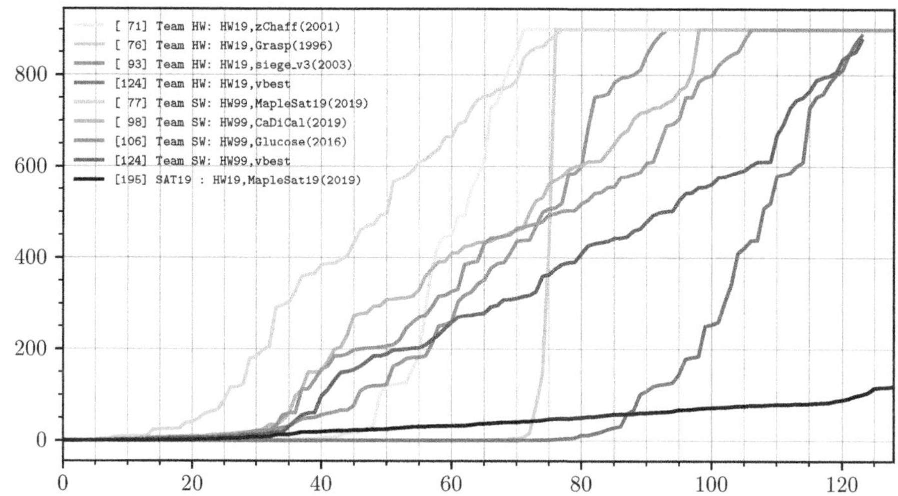


**Fig. 1.** Runtime for the SAT-solvers on all considered instances. The x-axis refers to the number of instances, and the y-axis depicts the runtime sorted in ascending order for each solver individually. vbest refers to the virtual best solver, i.e., we take the union over the solved instances for each team and consider the minimum for each instance. In the legend $[X]$ refers to a number of $X$ solved instances. `HW19` refers to the new hardware, and `HW99` refers to the old hardware. `SAT19` refers to a modern solver on modern hardware, which one can consider as a potential baseline.

### 2.5 The Final Stage: Experimental Setting and Limitations

We compare wall clock time and number of timeouts. However, we avoid IO access on the CPU solvers whenever possible, i.e., we load instances into the RAM if a network file system is involved and store uncompressed instances. We set a timeout of 900 s, and limited available RAM to 512 MB per instance and solver. We also tested for some solvers with resident set size restricted to 1 GB RAM and observed only a very small difference. Since Intel hardware around 2002 rarely had more than 512 MB RAM available, we went for the 512 MB setup. We follow standard guidelines for benchmarking [45]. Note that we do not validate for correctness of the solver outputs. We set and enforce resource limits by the tool *runsolver* [61].

## 3 The Trophies

Table 3 gives an overview on the number of solved instances for each solver and the two hardware generations. Figure 1 illustrates the runtime of the selected solvers and hardware as a cactus plot. Our results and gathered source codes are all publicly available [21]. Note that we report only on the two Intel-based hardware generations in this table. The results on the Ultra Sparc IIe system look very similar, usually, a few more instances were solved. Detailed data can be found in the supplemental material [21].

**Table 3.** Overview of the number of solved instances for the various solvers on our old and new hardware. HW99 represents the number of solved instances on the old hardware. HW19 represents the number of solved instances on the new hardware. vbest represents virtual best solvers, which are virtual solvers that we obtain by taking all instances that have been solved by the solvers considered in the group listed above.

| | Solver | Year/Generation | HW99 | HW19 |
|---|---|---|---|---|
| Team SW | MapleSat19 | 2019 | **77** | 195 |
| Team SW | CaDiCal | 2019 | **98** | 190 |
| Team SW | Glucose | 2016 | **106** | 188 |
| | vbest | | **124** | 198 |
| | sum | | 281 | 573 |
| | avg (%) | | 46.8 | 95.5 |
| Team HW | siege_v3 | 2003 | 37 | **93** |
| Team HW | zChaff | 2001 | 48 | **71** |
| Team HW | Grasp | 1996 | 73 | **76** |
| | vbest | | 87 | **124** |
| | sum | | 158 | 240 |
| | avg (%) | | 26.3 | 40.0 |
| Knuth | SAT13+12 | CDCL+P | 31 | 104 |
| Knuth | SAT13 | CDCL | 31 | 98 |
| Knuth | SAT11+12 | LH+P | 8 | 15 |
| Knuth | SAT11 | LH | 15 | 20 |
| Knuth | SAT10+12 | DPLL+P | 4 | 45 |
| Knuth | SAT10 | DPLL | 6 | 4 |
| Other Solvers | Lingeling | 2019 | **70** | 179 |
| Other Solvers | Lingeling-aqw-27d9fd4 | 2013 | **87** | 186 |
| Other Solvers | Lingeling-276 | 2011 | **83** | 177 |
| Other Solvers | MiniSat | 2008 | 84 | 178 |
| Other Solvers | siege_v4 | 2004 | 45 | 93 |
| Other Solvers | siege_v1 | 2003 | 33 | 81 |
| Other Solvers | sato | 2000 | 15 | 19 |
| Other Solvers | satz | 1998 | 7 | 9 |

### 3.1 Results

When we consider the number of solved instances on the hardware from 2019, MapleSat19 solves 195 instances. Recall that Team HW consists of the old solvers on modern hardware. It solves 93 instances (siege_v3), 76 instances (Grasp), and 71 instances (`zChaff`). On average, it solves about 80 instances (40% of the instances) at a standard deviation of about 12. However, the virtual best solver (`vbest`) for Team HW solves 124 instances, i.e., about 62% of the instances. The virtual best solver is the virtual solver that we obtain from taking the union over the solved instances by all three solvers and keeping the instance with best solved runtime. Team SW consists of the new solvers on old hardware. It solves 77 instances (`MapleSat19`), 98 instances (`CaDiCal`), and 106 instances (`Glucose`). On average, it solves about 94 instances (46.8% of the instances) with a standard deviation of 15. Its virtual best solver (`vbest`) solves 124 instances, i.e., about 62% of the instances. When considering the results on the solvers `MapleSat19`, `CaDiCal`, and `Glucose` on modern hardware, they solve 191 instances on average with a very low standard deviation of 3.6 instances. When considering the results on the solvers `siege_v3`, `zChaff`, and `Grasp`) on old hardware, they solve on average about 53 instances (26% of the instances) at a standard deviation of about 18.

### 3.2 Discussion of the Results

*Comparing the Teams.* The solver `MapleSat19`, which is the best solver from the 2019 SAT Race, solves as expected the highest number of instances on the new hardware. We are not surprised that neither Team SW nor Team HW or their virtual best solver gets anywhere close to this result. In view of Table 3 and Fig. 1, there are plenty of ways to compare the two teams. One can carry out (i) an individual comparison by the best (vbest), worst, or average solver, or even consider the individual solvers in direct comparison to each other, but one could also (ii) consider the virtual best solver for each team. If we choose Method (i) and individually compare the solvers, Team SW clearly wins for the measure best, worst, or average solver. We can also do one-by-one comparison and compare the solvers from each team individually with the solvers from the other team. Then, we take the number of solved instances for each solver X from Team SW against each solver Y from Team HW, and we give X a point if it solves more instances than Y or give a point to Y in the opposite case. Then, `Glucose` obtains 3 points (because it solves more instances than `siege_v3`, `zChaff`, and `Grasp`), `CaDiCal` obtains 3 points, and `MapleSat19` obtains 1 point, which totals 7 points for Team SW. In comparison, Team HW receives 0 points for `zChaff`, 0 points for `Grasp`, and 1 point for `siege_v3`, which totals 1 point. Hence, Team SW also wins. Nevertheless, if we consider the virtual best solvers, Team HW performs equally well as Team SW.

*Notable Observations.* The winner from the SAT Race 2019 (`MapleSat19/HW99`) solves less instances than the best solver (`siege_v3/HW19`) of Team HW. This seems surprising to us and we currently do not have a good explanation why `MapleSat19` solves so few instances on the old hardware, namely 21 instances less than `CaDiCal` and 29 instances less than `Glucose`. Since we observed a similar behavior with the latest implementation of `Lingeling` but not with `CaDiCal`, which also implements inprocessing techniques, we suspect that the advanced data structures in the solvers, the learning and restarting policy, and strong tuning towards modern hardware might be contributing factors. We found it interesting that the old solvers `siege_v3`, `zChaff`, and `Grasp` still solve a considerable number of instances on the new hardware. In particular, the solver `siege_v3` seems to benefit substantially from the new hardware, while `Grasp` gains almost no benefit from the new hardware. When we consider the implementations by Knuth, it is particularly remarkable that the DPLL solver with preprocessing on new hardware overtakes the CDCL solver with 45 solved instances. The CDCL implementation solves 31 instances with or without preprocessing on the old hardware.

### 3.3 Summary

When reviewing the results, we believe that our test setting reveals that both Team SW and HW perform in a similar range. If we compare individually, Team SW wins, which is also well visible in the cactus plot in Fig. 1. However, if we consider virtual best solvers, Team HW performs equally well. This leaves us with the conclusion that the last decades have brought enormous technological progress and innovation for SAT-solving, and the two main factors (i) *hardware advancement* and (ii) *algorithm advancement* both have a considerable influence.

## 4 Conclusion

We compare the impact of hardware and algorithm advancement on a genuine problem, namely, the propositional satisfiability problem (SAT). We describe in detail the decisions and challenges from a thought experiment to an actual experiment between old solvers and new solvers on new and old hardware with a time difference of about two decades. Our experiment's outcome confirms that modern algorithms have a strong influence on the performance of solvers, even when they run on old hardware. Nonetheless, solving significantly profits from technological advancement in hardware development and there is no clear winner between Team SW (new solvers on old hardware) and Team HW (old solvers on new hardware) in our time leap challenge. Overall, both teams perform in a similar range with a slight advantage for Team SW, which leads us to the conclusion that both hardware and software advances in science and industry have a mutual influence on modern solving. Hence, algorithm advancements are at least as important for the field of SAT-solving as hardware advancement. Further, algorithm engineering becomes of greater importance.

During our research, we noticed that long term reproducibility highly depends on available source code or static binaries with few dependencies. Further, it turned out helpful if the setup of a solver requires few additional system tools and few dependencies on external libraries. The dependencies within the operating system and source codes usually were not the problem as architectural dependencies would forbid to run the solvers. From our archaeological investigations, we suggest avoiding any external system for the setup for future long term experiments, i.e., tight dependencies on kernel versions or software containers such as Docker. Still, one uniform shared system for the entire community such as the SAT heritage project might prove helpful [1] if implemented also by competition organizers. Further, we think that public data libraries would be beneficial to understand long term advancements, not just source code repositories of private companies or university webpages.

One could post an open call and repeat the experiment with any solver. However, we believe that this would probably challenge developers of modern solvers to optimize their implementation for old hardware, which would result in a distorted picture for old solvers. Hence, we do not primarily intend to repeat the experiments in the near future [56].

We hope that our work stimulates research for others to also set up a time leap challenge in their fields such as for stochastic SAT-solvers, CSP-solvers, MaxSAT-solvers, and ILP-solvers.

**Acknowledgements.** We would like to thank several colleagues. Dave Mitchell and João Marques-Silva supported us with source codes from their old disks or mailboxes. Uwe Pretzsch and Siegmar Schöne helped to organize old hardware and Toni Pisjak maintained the modern benchmarking system, which we used.

## References

1. Audemard, G., Paulevé, L., Simon, L.: SAT heritage: a community-driven effort for archiving, building and running more than thousand SAT solvers. In: Pulina, L., Seidl, M. (eds.) SAT 2020. LNCS, vol. 12178, pp. 107–113. Springer, Cham (2020). https://doi.org/10.1007/978-3-030-51825-7_8
2. Audemard, G., Simon, L.: Predicting learnt clauses quality in modern SAT solvers. In: Boutilier, C. (ed.) IJCAI 2009, Proceedings of the 21st International Joint Conference on Artificial Intelligence, Pasadena, California, USA, 11–17 July 2009, pp. 399–404 (2009)
3. Audemard, G., Simon, L.: Glucose 2.1: aggressive - but reactive - clause database management, dynamic restarts. In: Berre, D.L., Gelder, A.V. (eds.) Proceedings of 3rd International Workshop of Pragmatics of SAT (PoS 2012), Trento, Italie (2012)
4. Audemard, G., Simon, L.: Glucose in the SAT Race 2019. In: Heule, M.J., Järvisalo, M., Suda, M. (eds.) Proceedings of SAT Race 2019: Solver and Benchmark Descriptions. Department of Computer Science Report Series, vol. B-2019-1, pp. 19–20. University of Helsinki (2019)
5. Balyo, T., Biere, A., Iser, M., Sinz, C.: SAT race 2015. Artif. Intell. **241**, 45–65 (2016). https://doi.org/10.1016/j.artint.2016.08.007
6. Bayardo, R.J., Schrag, R.C.: Using CSP look-back techniques to solve real-world sat instances. In: Kuipers, B., Webber, B. (eds.) Proceedings of the Fourteenth National Conference on Artificial Intelligence and Ninth Conference on Innovative Applications of Artificial Intelligence (AAAI 1997), pp. 203–208. The AAAI Press, Providence (1997)
7. Biere, A.: CaDiCaL, Lingeling, Plingeling, Treengeling, YalSAT entering the SAT competition 2017. In: Balyo, T., Heule, M., Järvisalo, M. (eds.) Proceedings of SAT Competition 2017 - Solver and Benchmark Descriptions. Department of Computer

Science Series of Publications B, vol. B-2017-1, pp. 14–15. University of Helsinki (2017)
8. Biere, A.: CaDiCaL simplified satisfiability solver (2019). http://fmv.jku.at/cadical/
9. Biere, A., Heule, M., van Maaren, H., Walsh, T. (eds.): Handbook of Satisfiability, Frontiers in Artificial Intelligence and Applications, vol. 185. IOS Press, Amsterdam (2009)
10. Bornebusch, F., Wille, R., Drechsler, R.: Towards lightweight satisfiability solvers for self-verification. In: Proceedings of the 7th International Symposium on Embedded Computing and System Design (ISED 2017), pp. 1–5, December 2017. https://doi.org/10.1109/ISED.2017.8303924
11. Büning, H.K., Buro, M.: Report on a SAT competition. Bull. Eur. Assoc. Theoret. Comput. Sci. **49**(1), 143–151 (1993)
12. Carter, J.: Debian 10 buster released (2019). https://www.debian.org/News/2019/20190706
13. Chu, G., Harwood, A., Stuckey, P.: Cache conscious data structures for Boolean satisfiability solvers. J. Satisf. Boolean Model. Comput. **6**, 99–120 (2009). https://doi.org/10.3233/SAT190064
14. Crawford, J.M., Auton, L.D.: Experimental results on the crossover point in satisfiability problems. In: Fikes, R., Lehnert, W. (eds.) Proceedings of the 11th National Conference on Artificial Intelligence (AAAI 1993), pp. 21–27. The AAAI Press, Washington, D.C. (1993)
15. Davis, M., Logemann, G., Loveland, D.: A machine program for theorem-proving. Commun. ACM **5**(7), 394–397 (1962). https://doi.org/10.1145/368273.368557
16. Davis, M., Putnam, H.: A computing procedure for quantification theory. J. ACM **7**(3), 201–215 (1960). https://doi.org/10.1145/321033.321034
17. Eén, N., Biere, A.: Effective preprocessing in SAT through variable and clause elimination. In: Bacchus, F., Walsh, T. (eds.) SAT 2005. LNCS, vol. 3569, pp. 61–75. Springer, Heidelberg (2005). https://doi.org/10.1007/11499107_5
18. Eén, N., Sörensson, N.: An extensible SAT-solver. In: Giunchiglia, E., Tacchella, A. (eds.) SAT 2003. LNCS, vol. 2919, pp. 502–518. Springer, Heidelberg (2004). https://doi.org/10.1007/978-3-540-24605-3_37
19. Eén, N., Sörensson, N.: MiniSAT (2008). http://minisat.se/
20. Elffers, J., Nordström, J.: Divide and conquer: towards faster pseudo-Boolean solving. In: Lang, J. (ed.) Proceedings of the Twenty-Seventh International Joint Conference on Artificial Intelligence, IJCAI-18, pp. 1291–1299. International Joint Conferences on Artificial Intelligence Organization, July 2018. https://doi.org/10.24963/ijcai.2018/180
21. Fichte, J.K., Hecher, M., Szeider, S.: Analyzed benchmarks and raw data on experiments for time leap challenge for SAT-solving (dataset). Zenodo, July 2020. https://doi.org/10.5281/zenodo.3948480
22. Fichte, J.K., Manthey, N., Stecklina, J., Schidler, A.: Towards faster reasoners by using transparent huge pages. CoRR abs/2004.14378 (2020)
23. Fichte, J.K., Manthey, N., Schidler, A., Stecklina, J.: Towards faster reasoners by using transparent huge pages. In: Simonis, H. (ed.) CP 2020. LNCS, vol. 12333, pp. 304–322. Springer, Heidelberg (2020)
24. Franco, J., Martin, J.: Chapter 1: A history of satisfiability. In: Biere, A., Heule, M., van Maaren, H., Walsh, T. (eds.) Handbook of Satisfiability. Frontiers in Artificial Intelligence and Applications, vol. 185, pp. 3–74. IOS Press, Amsterdam (2009). https://doi.org/10.3233/978-1-58603-929-5-3

25. Freeman, J.W.: Improvements to propositional satisfiability search algorithms. Ph.D. thesis, Department of Computer and Information Science, University of Pennsylvania, Philadelphia, PA, USA (1995)
26. Fu, Z., Mahajan, Y., Malik, S.: zchaff (2004). https://www.princeton.edu/~chaff/zchaff.html
27. Gocht, S., Nordström, J., Yehudayoff, A.: On division versus saturation in pseudo-Boolean solving. In: Proceedings of the Twenty-Eighth International Joint Conference on Artificial Intelligence, IJCAI-19, pp. 1711–1718. International Joint Conferences on Artificial Intelligence Organization, July 2019. https://doi.org/10.24963/ijcai.2019/237
28. Goldberg, E., Novikov, Y.: Berkmin (2003). http://eigold.tripod.com/BerkMin.html
29. Gomes, C.P., Kautz, H., Sabharwal, A., Selman, B.: Chapter 2: Satisfiability solvers. In: Frank van Harmelen, V.L., Porter, B. (eds.) Handbook of Knowledge Representation. Foundations of Artificial Intelligence, vol. 3, pp. 89–134. Elsevier Science Publishers, North-Holland (2008). https://doi.org/10.1016/S1574-6526(07)03002-7
30. Hennessy, J.L., Patterson, D.A.: Computer Architecture: A Quantitative Approach, 5th edn. Morgan Kaufmann, Burlington (2011)
31. Hoos, H.: Satlib - benchmark problems (2000). https://www.cs.ubc.ca/~hoos/SATLIB/benchm.html
32. Hoos, H.: Satlib - solvers (2000). https://www.cs.ubc.ca/~hoos/SATLIB/solvers.html
33. Hoos, H.H., Kaufmann, B., Schaub, T., Schneider, M.: Robust benchmark set selection for Boolean constraint solvers. In: Nicosia, G., Pardalos, P. (eds.) LION 2013. LNCS, vol. 7997, pp. 138–152. Springer, Heidelberg (2013). https://doi.org/10.1007/978-3-642-44973-4_16
34. Iser, M., Balyo, T., Sinz, C.: Memory efficient parallel sat solving with inprocessing. In: Keefer, R. (ed.) Proceedings of the IEEE 31st International Conference on Tools with Artificial Intelligence (ICTAI 2019), pp. 64–70. IEEE Computer Society (2019)
35. Jackson, I., Schwarz, C., Morris, D.A.: Debian GNU/Linux installation guide: 2.1. supported hardware (2019). https://www.debian.org/releases/stable/i386/ch02s01.en.html#idm272
36. Järvisalo, M., Berre, D.L., Roussel, O., Simon, L.: The international SAT solver competitions. AI Mag. **33**(1) (2012). http://www.aaai.org/ojs/index.php/aimagazine/article/view/2395
37. Järvisalo, M., Heule, M.J.H., Biere, A.: Inprocessing rules. In: Gramlich, B., Miller, D., Sattler, U. (eds.) IJCAR 2012. LNCS (LNAI), vol. 7364, pp. 355–370. Springer, Heidelberg (2012). https://doi.org/10.1007/978-3-642-31365-3_28
38. Johnson, D., Trick, M.: Satisfiability suggested format (1993). https://www.cs.ubc.ca/~hoos/SATLIB/Benchmarks/SAT/satformat.ps
39. Bayardo Jr., R.J., Pehoushek, J.D.: Counting models using connected components. In: Kautz, H.A., Porter, B. (eds.) Proceedings of the 17th Conference on Artificial Intelligence (AAAI 2000). The AAAI Press, Austin (2000)
40. Kahle, B.: Internet archive (2020). https://archive.org/
41. Katebi, H., Sakallah, K.A., Marques-Silva, J.P.: Empirical study of the anatomy of modern SAT solvers. In: Sakallah, K.A., Simon, L. (eds.) SAT 2011. LNCS, vol. 6695, pp. 343–356. Springer, Heidelberg (2011). https://doi.org/10.1007/978-3-642-21581-0_27

42. Knuth, D.E.: The Art of Computer Programming, Volume 4, Fascicle 6: Satisfiability. Addison-Wesley, Boston (2015)
43. Kochemazov, S., Zaikin, O., Kondratiev, V., Semenov, A.: MapleLCMDistChronoBT-DL, duplicate learnts heuristic-aided solvers at the SAT Race 2019. In: Heule, M.J., Järvisalo, M., Suda, M. (eds.) Proceedings of SAT Race 2019: Solver and Benchmark Descriptions. Department of Computer Science Report Series, vol. B-2019-1, pp. 24–24. University of Helsinki (2019)
44. Kohlhase, M.: The theorem prover museum - conserving the system heritage of automated reasoning. CoRR abs/1904.10414 (2019)
45. van der Kouwe, E., Andriesse, D., Bos, H., Giuffrida, C., Heiser, G.: Benchmarking crimes: an emerging threat in systems security. CoRR abs/1801.02381 (2018). http://arxiv.org/abs/1801.02381
46. Lewis, M.D.T., Schubert, T., Becker, B.W.: Speedup techniques utilized in modern SAT solvers. In: Bacchus, F., Walsh, T. (eds.) SAT 2005. LNCS, vol. 3569, pp. 437–443. Springer, Heidelberg (2005). https://doi.org/10.1007/11499107_36
47. Li, C.M., Anbulagan, A.: Heuristics based on unit propagation for satisfiability problems. In: Georgeff, M.P., Pollack, M.E. (eds.) Proceedings of the 15th International Joint Conference on Artificial Intelligence (IJCAI 1997), pp. 366–371. Morgan Kaufmann, Nagoya (1997)
48. Liang, J.H., Ganesh, V., Poupart, P., Czarnecki, K.: Learning rate based branching heuristic for SAT solvers. In: Creignou, N., Le Berre, D. (eds.) SAT 2016. LNCS, vol. 9710, pp. 123–140. Springer, Cham (2016). https://doi.org/10.1007/978-3-319-40970-2_9
49. Luby, M., Sinclair, A., Zuckerman, D.: Optimal speedup of Las Vegas algorithms. Inf. Process. Lett. **47**(4), 173–180 (1993)
50. Mahapatra, N.R., Venkatrao, B.: The processor-memory bottleneck: problems and solutions. XRDS **5**(3es), 2–9 (1999)
51. Marques-Silva, J.: Grasp (SAT solver). Pers. Commun. (2020)
52. Marques-Silva, J., Lynce, I., Malik, S.: Chapter 4: Conflict-driven clause learning SAT solvers. In: Biere, A., Heule, M., van Maaren, H., Walsh, T. (eds.) Handbook of Satisfiability. Frontiers in Artificial Intelligence and Applications, vol. 185, pp. 127–148. IOS Press, Amsterdam (2009)
53. Mitchell, D.G.: A SAT solver primer. In: Bulletin of the EATCS 85, The Logic in Computer Science Column, pp. 112–132 (2005)
54. Moore, G.E.: Cramming more components onto integrated circuits. Electronics **38**(8), 114 ff. (1965)
55. Moskewicz, M.W., Madigan, C.F., Zhao, Y., Zhang, L., Malik, S.: Chaff: engineering an efficient SAT solver. In: Rabaey, J. (ed.) Proceedings of the 38th Annual Design Automation Conference (DAC 2001), pp. 530–535. Association for Computing Machinery, New York (2001)
56. Munroe, R.: xkcd2268 (2019). https://xkcd.com/2268/
57. Nielsen, L.H.: Software citations now available in Zenodo (2019). https://blog.zenodo.org/2019/01/10/2019-01-10-asclepias/
58. Nordström, J.: On the interplay between proof complexity and sat solving. ACM SIGLOG News **2**(3), 19–44 (2015)
59. Pipatsrisawat, K., Darwiche, A.: A lightweight component caching scheme for satisfiability solvers. In: Marques-Silva, J., Sakallah, K.A. (eds.) SAT 2007. LNCS, vol. 4501, pp. 294–299. Springer, Heidelberg (2007). https://doi.org/10.1007/978-3-540-72788-0_28

60. Pipatsrisawat, K., Darwiche, A.: On the power of clause-learning SAT solvers with restarts. In: Gent, I.P. (ed.) CP 2009. LNCS, vol. 5732, pp. 654–668. Springer, Heidelberg (2009). https://doi.org/10.1007/978-3-642-04244-7_51
61. Roussel, O.: Controlling a solver execution with the runsolver tool. J. Satisf. Boolean Model. Comput. **7**, 139–144 (2011)
62. Ryan, L.: The siege SAT solver (2003). https://www2.cs.sfu.ca/research/groups/CL/software/siege/
63. Sakallah, K.A., Marques-Silva, J.: Anatomy and empirical evaluation of modern SAT solvers. Bull. EATCS **103**, 96–121 (2011). http://eatcs.org/beatcs/index.php/beatcs/article/view/138
64. Silva, J.P.M., Sakallah, K.A.: GRASP - a new search algorithm for satisfiability. In: Proceedings on the 7th International Conference on Computer-Aided Design (ICCAD 1996), pp. 220–227. Association for Computing Machinery, New York, November 1996
65. Soos, M.: Enhanced Gaussian elimination in DPLL-based SAT solvers. In: Berre, D.L., Gelder, A.V. (eds.) Proceedings of the Pragmatics of SAT (POS 2010) (2010)
66. Soos, M.: The CryptoMiniSat 5.5 set of solvers at the SAT competition 2018. In: Heule, M.J.H., Järvisalo, M., Suda, M. (eds.) Proceedings of SAT Competition 2018: Solver and Benchmark Descriptions. Department of Computer Science Series, vol. B-2018-1, pp. 17–18. University of Helsinki (2018)
67. Sörensson, N., Een, N.: Minisat v1.13 - a SAT solver with conflict clause minimization. Technical report, Chalmers University of Technology, Sweden (2005)
68. Stallman, R.: The GNU manifesto (1985). https://www.gnu.org/gnu/manifesto.en.html
69. Stiller, S.: Planet der Algorithmen. Albrecht Knaus Verlag, Munich (2015)
70. The NetBSD www team: The NetBSD project (2020). https://www.netbsd.org/
71. Trick, M., Chvatal, V., Cook, B., Johnson, D., McGeoch, C., Tarjan, B.: The 2nd DIMACS implementation challenge: 1992–1993 on NP hard problems: maximum clique, graph coloring, and satisfiability (1993). http://archive.dimacs.rutgers.edu/pub/challenge/sat/benchmarks/
72. Zabih, R., McAllester, D.A.: A rearrangement search strategy for determining propositional satisfiability. In: Shrobe, H.E., Mitchell, T.M., Smith, R.G. (eds.) Proceedings of the 7th National Conference on Artificial Intelligence, St. Paul, MN, USA, 21–26 August 1988, pp. 155–160. AAAI Press/The MIT Press (1988)
73. Zhang, H.: Sato: an efficient propositional prover. In: McCune, W. (ed.) Proceedings of the 14th International Conference on Automated Deduction (CADE 1997), Townsville, QLD, Australia, pp. 272–275, July 1997

# Breaking Symmetries with RootClique and LexTopSort

Johannes K. Fichte[1(✉)], Markus Hecher[2], and Stefan Szeider[3]

[1] School of Engineering Sciences, TU Dresden, Dresden, Germany
johannes.fichte@tu-dresden.de

[2] Database and Artificial Intelligence Group, TU Wien, Vienna, Austria
hecher@dbai.tuwien.ac.at

[3] Algorithms and Complexity Group, TU Wien, Vienna, Austria
sz@ac.tuwien.ac.at

**Abstract.** Bounded fractional hypertree width is the most general known structural property that guarantees polynomial-time solvability of the constraint satisfaction problem. Fichte et al. (CP 2018) presented a robust and scalable method for finding optimal fractional hypertree decompositions, based on an encoding to SAT Modulo Theory (SMT). In this paper, we provide an in-depth study of two powerful symmetry breaking predicates that allow us to further speed up the SMT-based decomposition: RootClique fixes the root of the decomposition tree; LexTopSort fixes the elimination ordering with respect to an underlying DAG. We perform an extensive empirical evaluation of both symmetry-breaking predicates with respect to the primal graph (which is known in advance) and the induced graph (which is generated during the search).

**Keywords:** Symmetry breaking · Hypergraphs · Elimination orderings · SAT Modulo Theory

## 1 Introduction

Bounded *fractional hypertree width*, introduced by Grohe and Marx [24,25], is the most general known purely structural restriction that guarantees polynomial-time tractability of the CSP. It generalizes all previously introduced structural restriction, including treewidth [11,19], spread-cut width [10] and hypertree width [22]. However, in order to utilize bounded fractional hypertree width of a CSP instance for solving it efficiently, one needs to have a fractional hypertree decomposition of the constraint hypergraph of the CSP instance available, witnessing the bounded fractional hypertree width. Computing such a decomposition of smallest width is again an NP-hard task [18]. Nevertheless, previous work by Fichte et al. [16] showed that a practically feasible SMT (SAT Modulo

The work has been supported by the Austrian Science Fund (FWF), Grants Y698, 32441, and 32830, and the Vienna Science and Technology Fund, Grant WWTF ICT19-065. Hecher is also affiliated with the University of Potsdam, Germany.


H. Simonis (Ed.): CP 2020, LNCS 12333, pp. 286–303, 2020.
https://doi.org/10.1007/978-3-030-58475-7_17

Theory) encoding exists, which supports the computation of optimal fractional hypertree decompositions of constraint hypergraphs with several hundred of vertices.

*Contribution.* In this paper, we introduce and study new symmetry breaking methods that speed up the SMT-approach for finding optimal fractional hypertree decompositions. Fractional hypertree decompositions are defined in terms of an (unrooted) decomposition tree, whose nodes are labeled with so-called bags of vertices of the decomposed hypergraph. However, the SMT encoding is based on a characterization of fractional hypertree width in terms of linear elimination orderings of the vertices of the decomposed constraint hypergraph. A hypergraph with $n$ vertices has $n!$ linear elimination orderings, where many of these correspond to the same decomposition tree. Consequently, that there is much room for *symmetry breaking* (SB) strategies. We take a closer look at two symmetry breaking methods: *RootClique* and *TopSort.*

RootClique is based on the observation, that we can pick any clique of the primal graph of the decomposed hypergraph, and assume that this clique appears in the bag of the decomposition tree's root; hence we call it a root clique. This symmetry breaking predicate allows us to restrict the considered linear orderings to only those where vertices of the root clique appear at the very end.

A linear ordering of the vertices gives rise to a decomposition DAG on the same vertex set, whose arcs correspond to the edges of the induced primal graph of the decomposed hypergraph, oriented according to the linear ordering. LexTopSort is based on the observation that from the many linear orderings that are all topological sort of the same decomposition DAG, it suffices to consider only one of them.

For both symmetry breaking predicates, we consider *static* and *dynamic* variants. The static variants operate on the *primal graph* of the given hypergraph. The dynamic variant operates on the *induced primal graph*, which is obtained from the primal graph during the search by adding fill-in edges according to the computed elimination ordering. Whereas the static variants have the advantage that the symmetry breaking constraints can be computed in a preprocessing phase before the decomposition process starts, it has the disadvantage of having fewer edges available and thus breaks fewer symmetries. Our experiments show whether advantage or disadvantage dominates.

A static version of RootClique was initially suggested for tree decompositions by Bodlaender et al. [7] and then ported to fractional hypertree decompositions by Fichte et al. [16]. For tree decompositions, it is reasonable to take a largest clique as the root clique, as it has the best chance to break the most symmetries. For a hypergraph, it is not clear what makes a clique well-suited for symmetry breaking. In addition to the size of the root clique, we consider several other criteria such as the size of the root clique including its neighborhood, the size of the root clique not counting twin vertices, or the number of hyperedges being incident with a vertex of the root clique. We also introduce a dynamic variant of RootClique, which requires a nontrivial SMT encoding.

For the other symmetry breaking predicate, LexTopSort, it is the other way around: a dynamic version has been suggested for hypertree decompositions by Schidler and Szeider [35], and we introduce and test a first static variant.

We provide an extensive experimental evaluation of all the discussed variants and combinations of RootClique and LexTopSort within the basic SMT encoding for fractional hypertree width *frasmt* by Fichte et al. [16]. For the experiments, we ran all the mentioned variants on the two leading SMT solvers *z3* [32] and *optimathsat* [36]. Overall, RootClique seems to show better results than LexTopSort. However, combining the two techniques at the same time even further improved the performance and the number of solved instances. Notably, it seems that using both solvers *z3* and *optimathsat* in combination, where we preferred the latter for instances of higher fractional hypertree width, the resulting portfolio is quite close to the virtual best solver of our experiments.

## 2 Preliminaries

**Hypergraphs.** A *hypergraph* is a pair $H = (V(H), E(H))$, consisting of a set $V(H)$ of *vertices* and a set $E(H)$ of *hyperedges*, each hyperedge is a subset of $V(H)$. For a hypergraph $H = (V, E)$ and a vertex $v \in V$, we write $E_H(v) = \{ e \in E \mid v \in e \}$ and $N_H(v) = (\bigcup E_H(v)) \setminus \{v\}$; the latter set is the *neighborhood* of $v$. If $u \in N_H(v)$ we say that $u$ and $v$ are *adjacent*. The *hypergraph* $H - v$ is defined by $H = (V \setminus \{v\}, \{e \setminus \{v\} \mid e \in E\})$. The *primal graph* (or 2-*section*) of a hypergraph $H = (V, E)$ is the graph $P(H) = (V, E_{P(H)})$ with $E_{P(H)} = \{ \{u, v\} \mid u \neq v$, there is some $e \in E$ such that $\{u, v\} \subseteq e \}$.

Consider a hypergraph $H = (V, E)$ and a set $S \subseteq V$. An *edge cover* of $S$ (with respect to $H$) is a set $F \subseteq E$ such that for every $v \in S$ there is some $e \in F$ with $v \in e$. A *fractional edge cover* of $S$ (with respect to $H$) is a mapping $\gamma : E \to [0, 1]$ such that for every $v \in S$ we have $\sum_{e \in E_H(v)} \gamma(e) \geq 1$. The *weight* of $\gamma$ is defined as $\sum_{e \in E} \gamma(e)$. The *fractional edge cover number* of $S$ (with respect to a hypergraph $H$), denoted $fn_H(S)$, is the minimum weight over all its fractional edge covers with respect to $H$.

A *tree decomposition* of a hypergraph $H = (V, E)$ is a pair $\mathcal{T} = (T, \chi)$ where $T = (V(T), E(T))$ is a tree and $\chi$ is a mapping that assigns each $t \in V(T)$ a set $\chi(t) \subseteq V$ (called the *bag* at $t$) such that the following properties hold:

- for each $v \in V$ there is some $t \in V(T)$ with $v \in \chi(t)$ ("$v$ is covered by $t$"),
- for each $e \in E$ there is some $t \in V(T)$ with $e \subseteq \chi(t)$ ("$e$ is covered by $t$"),
- for any three $t, t', t'' \in V(T)$ where $t'$ lies on the path between $t$ and $t''$, we have $\chi(t') \subseteq \chi(t) \cap \chi(t'')$ ("bags containing the same vertex are connected").

The width of a tree decomposition $\mathcal{T}$ of $H$ is the size of a largest bag of $\mathcal{T}$ minus 1. The treewidth $tw(H)$ of $H$ is the smallest width over all its tree decompositions.

**Hypertree Decompositions.** A *generalized hypertree decomposition* of $H$ is a triple $\mathcal{G} = (T, \chi, \lambda)$ where $(T, \chi)$ is a tree decomposition of $H$ and $\lambda$ is a mapping that assigns each $t \in V(T)$ an *edge cover* $\lambda(t)$ of $\chi(t)$. The *width* of $\mathcal{G}$ is the size of a largest edge cover $\lambda(t)$ over all $t \in V(T)$. A *hypertree decomposition* is a generalized hypertree decomposition that satisfies a certain additional property, which was added to make the computation of the decomposition tractable [22]. The *generalized hypertree width* $ghtw(H)$ of $H$ is the smallest width over all generalized hypertree decompositions of $H$. The *hypertree width* $htw(H)$ is the smallest width over all hypertree decompositions of $H$.

A *fractional hypertree decomposition* of $H$ is a triple $\mathcal{F} = (T, \chi, \gamma)$ where $(T, \chi)$ is a tree decomposition of $H$ and $\gamma$ is a mapping that assigns each $t \in V(T)$ a fractional edge cover $\lambda(t)$ of $\chi(t)$ with respect to $H$. The *width* of $\mathcal{F}$ is the largest weight of the fractional edge covers $\lambda(t)$ over all $t \in V(T)$. The fractional hypertree width $fhtw(H)$ of $H$ is the smallest width over all fractional hypertree decompositions of $H$.

To avoid trivial cases, we consider only hypergraphs $H = (V, E)$ with $E_H(v) \neq \emptyset$ for all $v \in V$. Consequently, every considered hypergraph $H$ has a (fractional) edge cover and $fhtw(H)$ is always defined. If $|V| = 1$ then $fhtw(H) = 1$.

Since an edge cover can be seen as the special case of a fractional edge cover, with weights restricted to $\{0, 1\}$, it follows that for every hypergraph $H$ we have $fhtw(H) \leq ghtw(H) \leq htw(H) \leq tw(P(H))$.

**Elimination Orderings.** The first SAT encoding of treewidth was suggested by Samer and Veith [34]. It uses an ordering-based characterization of treewidth which is also used by more recent SAT encodings of treewidth [3,6]. Later, ordering-based encodings where used for hypertree width [35], generalized hypertree width [5], and fractional hypertree width [16,28].

Let $H = (V, E)$ be a hypergraph with $n = |V|$ and $L = (v_1, \dots, v_n)$ a linear ordering of the vertices of $H$. We define the *hypergraph induced by* $L$ as $H_L^n = (V, E^n)$ where $E^n$ is obtained from $E$ by adding hyperedges successively as follows. We let $E^0 = E$, and for $1 \leq i \leq n$ we let $E^i = E^{i-1} \cup \{e_i\}$ where $e_i = \{\, v \in \{v_{i+1}, \dots, v_n\} \mid$ there is some $e \in E^{i-1}$ containing $v$ and $v_i \,\}$. We consider the binary relation $Arc_L = \{\, (v_i, v_j) \in V \times V \mid i < j$ and $v_i$ and $v_j$ are adjacent in $H_L^n \,\}$. We write $Arc_L(i) = \{v_i\} \cup \{\, v_j \mid (v_i, v_j) \in Arc_L \,\}$, hence $Arc_L(i) = \{v_i\} \cup e_i$. We refer to $P(H_L^n)$, the primal graph of the induced hypergraph $H_L^n$, as the *induced primal graph.*

The *fractional hypertree width of* $H$ *with respect to a linear ordering* $L$, denoted $fhtw_L(H)$, is the largest fractional edge cover number with respect to $H$ over all the sets $Arc_L(i)$, i.e.,

$$fhtw_L(H) = \max_{i=1}^{n} fn_H(Arc_L(i)).$$

**Theorem 1** ([16]). *The fractional hypertree width of a hypergraph $H$ equals the smallest fractional width over all its linear orderings:* $fhtw(H) = \min_L fhtw_L(H)$.

## 3 Symmetry Breaking for Elimination Orderings

In this section, we define all the symmetry breaking predicates, static and dynamic, and describe their encoding. Throughout this section, let $H = (V, E)$ be a fixed hypergraph.

As laid out in the proof of Theorem 1, one can translate back and forth between linear orderings and fractional hypertree decomposition, preserving the fractional width. The translation from the linear ordering into a decomposition is canonical and deterministic. The translation in the other direction, however, depends on several choices. Let us briefly describe the translation.

Let $\mathcal{F} = (T, \chi, \gamma)$ be a fractional hypertree decomposition of $H$. First, we choose a node $r$ of $T$ as the root and consider a rooted version $T_r$ of $T$. For each vertex $v \in V$, let $t = f(v)$ be the node of $T_r$ with $v \in \chi(t)$ that is closest to the root $r$. This consideration yields a partial ordering $\leq_{\mathcal{F}}$ of $V$, where $u \leq_{\mathcal{F}} v$ if and only if $f(u)$ is a descendant of $f(v)$ in $T_r$. The proof of Theorem 1 [16] shows that any linear ordering $L$ of $V$ that refines $\leq_{\mathcal{F}}$, has the same fractional width as $\mathcal{F}$. We observe that the linear orderings that refine $\leq_{\mathcal{F}}$ are exactly the *topological sorts* of the DAG $D_{\mathcal{F}} = (V, A)$ where $A = \{ (u, v) \mid u \neq v \text{ and } u \leq_{\mathcal{F}} v \}$. Any topological sort $L$ can be obtained from $D_{\mathcal{F}}$ by repeatedly deleting vertices without incoming arcs until all vertices have been deleted: $L$ is then the set of vertices arranged by their succession of deletion.

When we fix the root and the topological sort, we have determined the linear ordering uniquely. We will fix the root with the RootClique symmetry breaking predicate, and we will fix the topological sort with the LexTopSort symmetry breaking predicate.

### 3.1 RootClique

The static RootClique predicate is based on the well-known fact that, if $(T, \chi)$ is a tree decomposition of a graph $G$ and $K$ a clique in $G$, then there exists a node $t \in V(T)$ with $V(K) \subseteq \chi(t)$ (see, e.g., [8]). Hence, when we choose any clique $K$ in the primal graph $P(H)$, the static RootClique predicate requires that the root $r$ is chosen among the nodes for which $V(K) \subseteq \chi(r)$ holds. This is not a full symmetry breaking, since a clique may appear in different bags. Hence, we suggest several strategies for choosing a clique that suits this purpose. In particular, we choose a clique in $P(H)$, maximizing

1. the size of the clique,
2. the size of the clique including its neighborhood $N_H(K)$ in $P(H)$,
3. the size of the clique not counting *twin vertices*, which are any two vertices $u, v$ sharing the same neighborhood, i.e., where $N_H(u) = N_H(v)$, and
4. the number of hyperedges being incident with a vertex of the clique.

We also consider *k-hypercliques* which are cliques in $P(H)$ not intersecting with any hyperedge of $H$ in more than $k$ vertices. This concept was proposed by Fichte et al. [16], and was primarily intended to obtain lower bounds during the computation of fractional hypertree width. The dynamic RootClique predicate is based on the fact that each bag $\chi(t)$ is a clique in the *induced* primal graph, and conversely, every maximal clique in the induced primal graph corresponds to $\chi(t)$ for some node $t$. We uniquely determine the root of the decomposition tree by fixing the largest clique of the induced primal graph as the root clique.

Further details and the encodings are given after the next subsection.

### 3.2 LexTopSort

We start by explaining the dynamic variant of LexTopSort. For better integration with RootClique, we use the inverse variant of topological sorting, where vertices without outgoing arcs are deleted, proving the ordering "from right to left." When there are several vertices without outgoing arcs, we choose the lexicographically smallest vertex next. This choice uniquely determines the linear ordering, which corresponds to the reflected lexicographically smallest topological sort. We denote for two vertices $v_i, v_j$ by $\text{lex}(v_i, v_j)$ that $v_i$ precedes $v_j$ in the lexicographic ordering.

We enforce this restriction on the linear ordering $L = (v_1, \ldots, v_n)$ of $V$ with the following predicate: *For any* $1 \leq i < j \leq n$, *if* $\text{lex}(v_i, v_j)$, *then there must be some* $k \in \{i+1, \ldots, j\}$ *such that the induced primal graph contains the edge* $\{v_i, v_k\}$. In other words, when we delete the lexicographically larger vertex $v_j$ before we delete the lexicographically smaller vertex $v_i$, then $v_i$ must have a neighbor $v_k$ which has not been eliminated at that time, i.e., $v_i$ has an outgoing arc in $D_{\mathcal{F}}$ to a vertex $v_k$ that is still present when $v_j$ is deleted.

Since the encoding of dynamic LexTopSort is expensive, we propose a new relaxed static version, which does not break all symmetries but can be encoded in a significantly more compact way. The static version is obtained by a small but influential change in the symmetry breaking predicate, by using the primal graph, not the induced primal graph: *For any* $1 \leq i < j \leq n$, *if* $\text{lex}(v_i, v_j)$, *then there must be some* $k \in \{i+1, \ldots, j\}$ *such that the primal graph contains the edge* $\{v_i, v_k\}$.

### 3.3 Encodings for Symmetry Breaking

In this section, we describe how RootClique and LextTopSort can be encoded within the SMT encoding for fractional hypertree width due to Fichte et al. [16], which we briefly review. To this end, let $H = (V, E)$ be a given hypergraph with $V = \{v_1, \ldots, v_n\}$ and $w$ be a rational number. The encoding is an SMT formula that is satisfiable if and only if the hypergraph has a linear ordering $L$ of $V$ such that $\mathit{fhtw}_L(H) \leq w$. For computing the relation $\mathit{Arc}_L$, it uses Boolean *ordering variables* $o_{i,j}$ for $1 \leq i < j \leq n$ and Boolean *arc variables* $a_{i,j}$ for $1 \leq i, j \leq n$. Clauses are added that ensure that an ordering variable $o_{i,j}$ is true if and only if $i < j$ and $v_i$ precedes $v_j$ in $L$. In the following, we let $o^*(i,j)$ refer to $o(i,j)$ if

$i < j$ and $\neg o(j, i)$ otherwise. The arc variables are used to represent the relation $Arc_L$ for the ordering $L$ represented by the ordering variables, where $a(i, j)$ is true if and only if $(v_i, v_j) \in Arc_L$, i.e., if $v_j \in Arc_L(i)$. Finally, *weight variable* $w(i, e)$ for each $1 \leq i \leq n$ and $e \in E$ is used to represent the weight of $e$ in a fractional edge cover $\gamma_L(i)$ of the set $Arc_L(i)$, where $L$ is the ordering represented by the ordering variables.

**Static RootClique (***s-RQ***).** We encoded all the different variants for choosing cliques, and computed them by a solver in a separate solving step, executed before the actual decomposition. To this end, we use Boolean *clique variables* of the form $k(i)$ for each vertex $v_i \in V(H)$, where those vertices set to true form a clique $K$. All variants have in common, that a clique is computed as follows.

$[\neg k(i) \vee \neg k(j)]$ for any two vertices $v_i, v_j \in V(H)$ with $v_j \notin N_H(v_i)$.

Then, we considered different maximization constraints on top, resulting in different variants. In the following, we present variants for computing cliques that require adding different constraints to the constraint above.

*Largest Clique (LQ).* For obtaining a clique of size at least $\ell$, we add the following constraints.

$[\sum_{v_i \in V(H)} k(i) \geq \ell]$

*Largest Clique Including Neighbors (LQ+N).* We also considered maximizing a clique, where we additionally count the neighbors of the clique.

$[\sum_{v_i \in V(H)} k(i) + |N_H(v_i)| \geq \ell]$

*Largest Clique Excluding Twins (LQ-T).* The following variant excludes twin vertices, when computing a maximal clique.

$[\sum_{v_i \in V(H)} k(i) - (\sum_{v_i \in V(H)} |\{v_j \in V(H) \mid j > i, N_H(v_j) = N_H(v_i)\}|) \geq \ell]$

*Largest k-Hyperclique (k-Hy).* For k-hypercliques, we need the following additional constraints.

$[\neg k(i_1) \vee \cdots \vee \neg k(i_k)]$ for any $k$ vertices $v_{i_1}, \ldots, v_{i_k}$ of hyperedge $e \in E(H)$,
$[\sum_{v_i \in V(H)} k(i) \geq \ell]$

*Clique with Largest Number of Used Hyperedges (LuH).* This variant concerns only about maximizing the number of hyperedges that are adjacent to a clique.

$[\sum_{e \in E(H)} (\bigvee_{v_i \in e} k(i)) \geq \ell]$

Finally, after having computed a clique $K$, which can be obtained with any of the variants above, one can add the following constraints to the base encoding in order to actually break symmetries, statically guided by $K$. More precisely, the clique $K$ is ensured to be eliminated before the other vertices (and considered the root of the decomposition) such that each vertex of $K$ is eliminated in lexicographic order, i.e., according to $L$.

$[o^*(i,j)]$ for $v_j \in V(H) \setminus K$, $v_i \in K$,

$[o^*(i,j)]$ for $v_i, v_j \in K$, $v_i \neq v_j$, $\mathrm{lex}(v_i, v_j)$.

**Dynamic RootClique** (*d-RQ*)**.** While the main idea of this approach is similar to static RootClique, here we aim for a largest bag of the resulting decomposition to be the root node. However, this does not depend on the presence of a clique in the primal graph $P(H)$. Instead, we require such a clique in the induced primal graph of $H$. As a result, the SMT encoding for fractional hypertree decompositions based on elimination orderings is directly extended. For finding a largest bag, Boolean variables $B(i), b(i)$ for $1 \leq i \leq n$ and integer variables $d(i)$ for computing the degree of outgoing arcs in the induced primal graph of $H$ are used. Intuitively, $B(i)$ indicates that $v_i \in V(H)$ is the lexicographically largest vertex in $L$ that is contained in a largest bag. Consequently, smaller vertices of $i$ are in this largest bag, whose member elements are indicated by variables $b(j)$.

The following constraints model this construction, where the degree variables are computed and only one largest bag is allowed.

$[d(i) = \sum_{1 \leq j \leq n, j \neq i} a(i,j)]$ for $1 \leq i \leq n$,

$[\neg B(i) \vee \neg B(j)]$ for $1 \leq i < j \leq n$,

$[\bigvee_{v_i \in V(H)} B(i)]$.

We ensure that if for vertex $v_i$ there is a lex-smaller vertex $v_j$, where there is no arc from $v_i$ to $v_j$, $v_i$ cannot be the largest vertex in a largest bag. Further, for vertex $v_i$ with $B(i)$ it is not allowed that there is a larger bag (of larger degree) with a lexicographically larger vertex $v_j$.

$[\neg o^*(j,i) \vee a(i,j) \vee \neg B(i)]$ for $1 \leq i, j \leq n$, $i \neq j$,

$[\neg o^*(i,j) \vee B(j) \vee \neg B(i) \vee d(j) \leq d(i)]$ for $1 \leq i, j \leq n$, $i \neq j$.

For fixing the order of the elements within the bag and in relation to elements outside this bag, we compute the elements of this largest bag as follows.

$[\neg b(i) \vee \neg o^*(j,i) \vee \neg B(j)]$ for $1 \leq i, j \leq n$

$[\neg B(i) \vee \neg o^*(j,i) \vee b(j)]$ for $1 \leq i, j \leq n$.

Then, we fine-tune the symmetry breaking by setting the order within this (largest) bag and in relation to the other vertices. This is similar to symmetry breaking for RootClique, but depending on the elements of the largest bag.

$[\neg b(i) \vee b(j) \vee o^*(i,j)]$ for $1 \leq i < j \leq n$,

$[\neg b(i) \vee \neg b(j) \vee o^*(i,j)]$ for $1 \leq i < j \leq n$, $\mathrm{lex}(v_i, v_j)$.

**LexTopSort.** For encoding LexTopSort, we need the following additional SMT variables. Boolean variables $s(i, j)$ for each $1 \leq i, j \leq n$ with $i \neq j$ are used to represent that $v_i$ has $v_j$ as the lex-smallest vertex with an arc from $v_i$ to $v_j$ in the induced primal graph. For connected hypergraphs $H$, obviously, such a vertex $v_j$ has to exist for every vertex $v_i$, except for the smallest vertex in the ordering $L$. Being the smallest vertex $v_i$ is represented with Boolean variables $l(i)$ (for every $1 \leq i \leq n$).

For both static and dynamic LexTopSort, we need to encode that there is only one such smallest vertex. Further, we need to make sure that if for three vertices $v_i, v_j, v_k$ with $\text{lex}(v_i, v_j)$ either we have that $o^*(i, j)$ ($i$ is eliminated before $j$), or $v_k$ is not the lex-smallest vertex of $v_i$, or otherwise it is guaranteed that $j$ is eliminated before $k$. Intuitively, this ensures that either the succession of elimination coincides with $L$ or deleting the lexicographically larger vertex $v_j$ is allowed since $v_k$ is present in $D_{\mathcal{F}}$ when $v_j$ is deleted.

$$[\neg l(i) \vee \neg l(j)] \qquad \text{for } 1 \leq i < j \leq n,$$

$$[o^*(i,j) \vee \neg s(i,k) \vee o^*(j,k)] \qquad \text{for } 1 \leq i, j, k \leq n,\ \text{lex}(v_i, v_j).$$

Then, we add one of the following two blocks of constraints, depending on the static or dynamic variant of LexTopSort.

*Static LexTopSort (s-LT).* The static variant ensures that for every vertex $v_i$ that either $v_i$ is the smallest vertex or $v_i$ has a lex-smallest vertex. Then, for two neighbors $v_j, v_k$ of $v_i$, if $s(i, k)$, $v_j$ cannot be eliminated before $v_k$ in $D_{\mathcal{F}}$.

$$[\textstyle\bigvee_{\{v_i,v_j\} \in E(P(H))} s(i,j) \vee l(i)] \qquad \text{for } v_i \in V(H),$$

$$[\neg o^*(j,k) \vee \neg s(i,k)] \qquad \text{for } 1 \leq i, j, k \leq n,\ j \neq k, \{v_j, v_k\} \subseteq N_H(v_i).$$

*Dynamic LexTopSort (d-LT).* Conceptually, dynamic LexTopSort is similar to static LexTopSort, although the variants show major differences in runtime as we will see in Sect. 4.2. First, for every vertex $v_i$ either $v_i$ is the smallest vertex or there is a lex-smallest vertex for $v_i$. Then, if $v_i$ has $v_j$ as the lex-smallest vertex, we require an arc from $v_i$ to $v_j$ in the induced primal graph of $H$. Similar to static LexTopSort, if there are two candidates $v_j$ and $v_k$ for being the lex-smallest vertex of $v_i$, it is prohibited to take $v_k$ if $v_j$ is eliminated before $v_k$.

$$[\textstyle\bigvee_{i \neq j} s(i,j) \vee l(i)] \qquad \text{for } v_i \in V(H),$$

$$[\neg s(i,j) \vee a(i,j)] \qquad \text{for } 1 \leq i, j \leq n,\ i \neq j,$$

$$[\neg a(i,j) \vee \neg a(i,k) \vee \neg o(j,k) \vee \neg s(i,k)] \quad \text{for } 1 \leq i, j, k \leq n,\ i \neq j,\ i \neq k,\ j \neq k.$$

**Combining RootClique with LexTopSort.** For combining RootClique with LexTopSort, we have to take care that the ordering $L$ is in line with the vertices of the root clique being lexicographically first.

Further, for static RootClique, where we have a (static) clique $K$ prior the actual solving with the SMT encoding, we can easily fix the smallest vertex of the LexTopSort encoding as follows.

$$[l(i)] \qquad \text{for } v_i \in K, \text{ if for every } v_j \in K \text{ with } j \neq i, \text{ we have } \text{lex}(v_i, v_j).$$

# 4 Implementation and Experiments

We ported *frasmt* to *Python 3.8.* and implemented the different strategies for symmetry breaking. The source code of our solver *frasmt* is readily available at github.com/daajoe/frasmt and detailed results as well as analysis are online at Zenodo [12]. In our implementation, we support two SMT solvers, namely *z3 4.8.7* [32] as well as *optimathsat 1.6.4* [36]. Hence, we have two configurations: *frasmt_z3*, which uses the SMT solver *z3* and *frasmt_om*, which uses the SMT solver *optimathsat*. As it turns out, while one solver is overall better than the other, both solvers complement each other quite well. To demonstrate this finding, our results also show a portfolio variant *frasmt_z3+om* that uses both solvers, where for instances of fractional hypertree width larger than 4 solver *optimathsat* is invoked and below, solver *z3* is invoked. Further, for computing the cliques that are used in the static variant of RootClique, we applied a solver called *clingo 5.4.0* [20], which is an extension of SAT solvers and allows for incremental solving as well as optimization without manual cardinality constraints. For obtaining these cliques, we relied on the any-time algorithm of *clingo*. We allowed this solver to use up to 500 s, which showed almost the same results as using no internal time limit and –in the worst case– spending the total runtime on symmetry breaking only. Indeed, for symmetry breaking, we then used the best clique according to the optimization criteria of the clique variant that could be computed within these 500 s. However, we observed that it is indeed crucial to allow some time for symmetry breaking since the vanilla configuration of *frasmt_z3* is almost identical to *frasmt*, which spends only 10 s on limited approaches of symmetry breaking.

## 4.1 Benchmark Setup

We compared the different strategies of symmetry breaking with respect to the goal of finding the best variant. To this end, we configured the following setup.

*Measure and Resources.* In order to draw conclusions about the efficiency of the compared solvers, we mainly inspected wall clock times. We set a timeout of 7200 s and limited available RAM to 16 GB per instance. Resource limits were set and enforced by the tool *runsolver* [33].

*Benchmark Instances.* We considered a selection of 2191 instances collected by Fischl et al. [17] (publicly available at [15]) from various sources, consisting of hypergraphs that originate from CSP instances and conjunctive database queries. The instances and their original sources are summarized in Table 1. The instances contain up to 2993 vertices and 2958 hyperedges.

**Table 1.** The benchmark sets consisting of the type, the number of instances as well as the origin, of the instances we considered in our experiments. Note that some benchmark sets are overlapping and therefore the numbers do not add up to 2191.

| Benchmark set | Type | #Instances | Origin |
|---|---|---|---|
| `DaimlerChrysler` | Industrial | 15 | |
| `Grid2D` | Grid | 12 | |
| `ISCAS'89` | Competition | 24 | [23] |
| `MaxSAT` | MaxSAT | 35 | [5] |
| `csp_application` | XCSP | 1090 | [2] |
| `csp_random` | XCSP | 863 | [2] |
| `csp_other` | Misc | 82 | |
| `CQ` | Conjunctive Queries | 156 | [1,4,21,26,30,37] |
| $\sum$ | Hyperbench | 2191 | [17] |

*Benchmark Hardware.* Solvers were executed on a cluster of 12 nodes. Each node is equipped with two Intel Xeon E5-2650 CPUs consisting of 12 physical cores each at 2.2 GHz clock speed, 256 GB RAM and 1 TB hard disc drives (*not* an SSD) Seagate ST1000NM0033. The results were gathered on Ubuntu 16.04.1 LTS machines with disabled hyperthreading on kernel 4.4.0-139.

*Compared Solvers.* We mainly compare variants of *frasmt_z3* and *frasmt_om* to see the influence of symmetry breaking. The vanilla configuration *frasmt_z3* has the same features as the best reported configuration [16] of *frasmt*, where no extensive symmetry breaking is used. The results of *frasmt_z3* and *frasmt* are almost identical (small differences may occur due to *Python* version upgrade), which is why we refrained from further adding additional data to our plots and tables. We also considered the recent solver *triangulator* [29]. While *triangulator* overall is extremely fast on about half of the instances (about 1190), we observed that the solver quickly runs out of main memory on most of the other instances, which still persists if increasing main memory to 64GB. In consequence, we only report results for the more recent solver *triangulator-msc*, which uses *cplex* and is available at github.com/Laakeri/triangulator-msc. However, we follow other recent work on symmetry breaking [9] and stress that the main goal of our experiments is to demonstrate the *benefit* of our symmetry breaks, not to compare the speed of our approach to other algorithms with different techniques.

### 4.2 Benchmark Results

We discuss the following three aspects, where we first elaborate on the variants of RootClique. Then, we cover the performance of LexTopSort, including the combination with RootClique, followed by static vs. dynamic symmetry breaking.

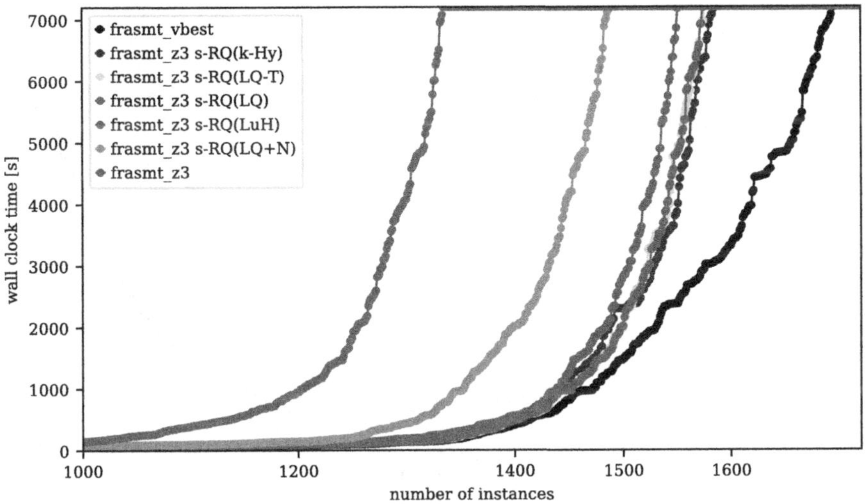


**Fig. 1.** A cactus plot of the static variants of RootClique, compared to the vanilla configuration *frasmt_z3*. The x-axis shows the number of instances and the y-axis depicts wall clock runtimes in seconds, which are sorted in ascending order, but for each solver configuration individually. Solver "*frasmt* _vbest" refers to the virtually best solver by taking for each instance the best result among all displayed solver (configurations). The legend is ordered from best to worst (from right to left in the plot).

*Computing Static RootCliques.* Figure 1 depicts a cactus plot of the variants *s-RQ* for static RootClique, as presented in the previous section. In this figure, the x-axis refers to the number of instances, where for each solver the runtime (y-axis) is sorted in ascending order. Therefore, this plot provides an overview of the variants over all instances. In this plot, we mainly focused on showing the variants for solver *z3*, which showed overall the best performance, since the results for *om*, while different compared to *z3*, draw a similar picture. Surprisingly, the k-hyperclique variant *(k-Hy)* shows the best results[1], which is, however, almost as good as the two variants *LQ-T* and *LQ* for computing largest cliques. While the variant on aiming for the largest clique without twin vertices *(LQ-T)* seems to have a slight advantage over going for the largest clique *(LQ)*, the differences are minor. The fact that k-hyperclique performs best was, however, surprising.

[1] For comparability with *frasmt*, we used $k = 6$ (the option reported best [16]).

On the other hand, if one also considers the variant *LuH* for preferring cliques with the largest number of used hyperedges, it seems that k-hypercliques might form a good compromise. Notably, if also considering LexTopSort, the situation changes slightly. It turns out that the variant *LQ-T* performs better than using the k-hyperclique, followed by *LQ*.

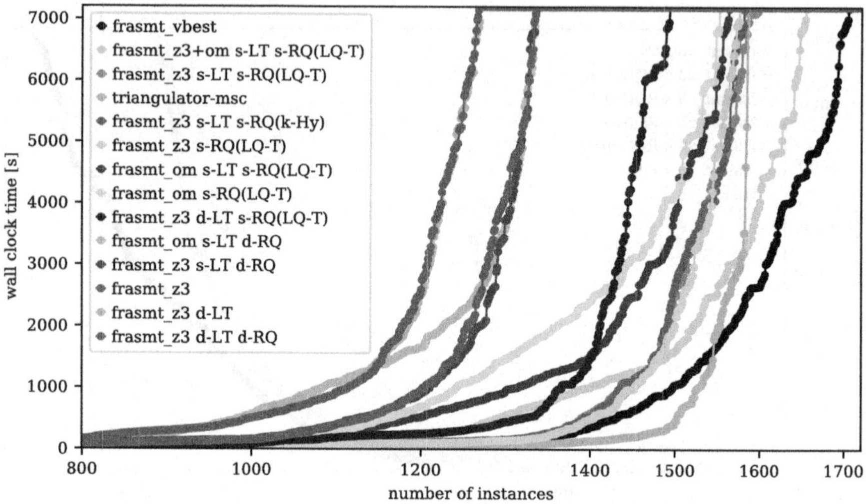


**Fig. 2.** A cactus plot showing (combinations of) both RootClique and LexTopSort symmetry breaking, where the x-axis refers to the number of instances, and the y-axis depicts wall clock runtimes in seconds. Runtimes are sorted in ascending order for each solver configuration individually. Solver "*frasmt* _vbest" refers to the virtually best solver by taking for each instance the best result among all displayed *frasmt* configurations. The legend is ordered from best to worst (from right to left in the plot).

*Combining RootClique with LexTopSort.* Before we discuss the combination of RootClique with LexTopSort, we briefly elaborate on the performance of the variants of LexTopSort without RootClique. It seems that especially dynamic LexTopSort worsens the picture. In more detail, both static and dynamic LexTopSort without RootClique show a rather bad performance, which is sometimes even worse than the vanilla configuration *frasmt_z3*. This observation is underlined by Fig. 2, which shows a cactus plot of variants of LexTopSort and combinations with variants of RootClique. The best variants use full static symmetry breaking only (*s-LT*, *s-RQ*). While the result for *frasmt_z3* suggests not much difference between full static symmetry breaking and static RootClique only (*s-RQ*), the results for *frasmt_om* reveal that indeed with static LexTopSort, one can further improve the results obtained by static RootClique only. This might also be emphasized due to the fact that the combination of solvers *z3* and *optimathsat* provides significant improvements compared to both single configurations. Notably, *triangulator-msc* is very fast, but overall *frasmt_z3* solves more instances. Table 2 reports on the number of solved instances and total runtimes for the larger benchmark sets, where timeouts count as 7200 s, and results are detailed and grouped by fractional hypertree width.

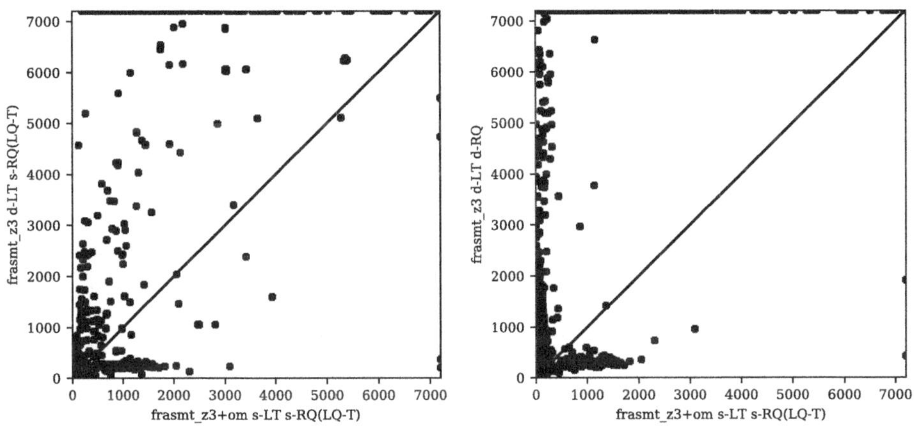


**Fig. 3.** Scatter plots comparing runtimes of instances (in seconds) one-by-one of our best configuration (x-axis) with the best dynamic symmetry breaking method (y-axis, left) and with the worst dynamic configuration (y-axis, right).

*Static vs. Dynamic Symmetry Breaking.* The results of the previous paragraphs seem to be rather bad for the dynamic variants of symmetry breaking. However, this is not too surprising if one considers that the encoding size of dynamic LexTopSort is in $\mathcal{O}(n^3)$ and that for RootClique the encoding is in $\mathcal{O}(n^2)$, where $n$ is the number of vertices contained in the hypergraph. Still, against all odds the cactus plot of Fig. 2 already depicts dynamic variants, whose curve is sometimes below other static variants and even below our best variant *frasmt_z3+om*. Further, Fig. 3 shows two scatter plots comparing runtimes instance-by-instance of the best variant of the previous paragraph (*frasmt_z3+om*, x-axis) with (y-axis) both the best variant (left) of dynamic symmetry breaking and the worst dynamic variant (right). Both dynamic variants show that, while *frasmt_z3+om* performs better on plenty of instances, there are still instances on the bottom left of the plots, where the dynamic variants are faster. Further, some instances on the bottom right of both plots cannot be solved by *frasmt_z3+om*, which are solved by dynamic variants.

**Table 2.** Detailed results on the number of solved instances grouped by fractional hypertree width of the solved instance. Runtimes are cumulated wall clock times in hours, where timeouts count as 7200 s.

| Set | Solver | Σ | *fhtw* range group | | | | time[h] |
|---|---|---|---|---|---|---|---|
| | | | max(*fhtw*) | 0-2 | 3-4 | >4 | |
| csp_other | frasmt_vbest | 46 | **7.0** | **28** | **5** | 13 | 103.57 |
| | frasmt_z3 s-RQ(LQ-T) | **44** | 6.0 | **28** | **5** | **11** | **104.62** |
| | frasmt_z3 s-LT s-RQ(LQ-T) | 43 | 6.0 | 27 | **5** | **11** | 106.09 |
| | frasmt_z3 s-LT s-RQ(k-Hy) | 43 | 6.0 | 27 | **5** | **11** | 106.69 |
| | frasmt_z3 | 43 | 6.0 | **28** | **5** | 10 | 110.17 |
| | frasmt_z3 s-LT d-RQ | 41 | 6.0 | 27 | **5** | 9 | 111.02 |
| | frasmt_z3+om s-LT s-RQ(LQ-T) | 40 | **7.0** | 27 | **5** | 8 | 112.63 |
| | frasmt_z3 d-LT s-RQ(LQ-T) | 37 | 6.0 | 26 | 4 | 7 | 118.04 |
| | frasmt_om s-LT s-RQ(LQ-T) | 37 | **7.0** | 26 | 3 | 8 | 118.76 |
| | frasmt_om s-RQ(LQ-T) | 36 | 6.0 | 26 | 3 | 7 | 119.69 |
| | frasmt_z3 d-LT | 36 | 6.0 | 26 | 4 | 6 | 124.27 |
| | frasmt_om s-LT d-RQ | 34 | 6.0 | 26 | 3 | 5 | 123.96 |
| | frasmt_z3 d-LT d-RQ | 34 | 6.0 | 25 | 4 | 5 | 125.89 |
| | triangulator-msc | 25 | $5.\bar{3}$ | 17 | 4 | 4 | 139.92 |
| csp_application | frasmt_vbest | 674 | **7.0** | **43** | 397 | 234 | 775.44 |
| | frasmt_z3 s-LT s-RQ(LQ-T) | **648** | **7.0** | **43** | **396** | 209 | 835.76 |
| | frasmt_z3 s-RQ(LQ-T) | 646 | **7.0** | **43** | **396** | 207 | **824.72** |
| | frasmt_z3 s-LT s-RQ(k-Hy) | 641 | **7.0** | **43** | **396** | 202 | 842.05 |
| | frasmt_z3+om s-LT s-RQ(LQ-T) | 640 | **7.0** | **43** | **396** | 201 | 827.50 |
| | frasmt_z3 | 559 | **7.0** | **43** | 370 | 146 | 974.08 |
| | frasmt_z3 s-LT d-RQ | 554 | **7.0** | **43** | 365 | 146 | 980.45 |
| | frasmt_om s-LT s-RQ(LQ-T) | 552 | **7.0** | 41 | 310 | 201 | 1003.93 |
| | triangulator-msc | 551 | **7.0** | **43** | 288 | **220** | 973.13 |
| | frasmt_z3 d-LT s-RQ(LQ-T) | 549 | **7.0** | 40 | 326 | 183 | 997.30 |
| | frasmt_om s-RQ(LQ-T) | 535 | **7.0** | 40 | 293 | 202 | 1040.55 |
| | frasmt_z3 d-LT | 498 | **7.0** | 40 | 312 | 146 | 1113.74 |
| | frasmt_z3 d-LT d-RQ | 498 | **7.0** | 40 | 312 | 146 | 1114.50 |
| | frasmt_om s-LT d-RQ | 482 | **7.0** | 38 | 293 | 151 | 1146.85 |
| csp_random | triangulator-msc | **860** | $6.\bar{6}$ | **54** | **39** | **767** | **203.21** |
| | frasmt_vbest | 835 | **9.0** | **54** | **39** | 742 | 365.57 |
| | frasmt_om s-RQ(LQ-T) | 830 | **9.0** | **54** | 38 | 738 | 444.29 |
| | frasmt_z3+om s-LT s-RQ(LQ-T) | 826 | **9.0** | **54** | **39** | 733 | 419.60 |
| | frasmt_om s-LT s-RQ(LQ-T) | 824 | **9.0** | **54** | 37 | 733 | 424.79 |
| | frasmt_z3 s-LT s-RQ(LQ-T) | 757 | **9.0** | **54** | **39** | 664 | 478.49 |
| | frasmt_z3 d-LT s-RQ(LQ-T) | 755 | **9.0** | **54** | **39** | 662 | 488.93 |
| | frasmt_z3 s-LT s-RQ(k-Hy) | 747 | **9.0** | **54** | **39** | 654 | 478.59 |
| | frasmt_z3 s-RQ(LQ-T) | 734 | **9.0** | **54** | **39** | 641 | 490.29 |
| | frasmt_om s-LT d-RQ | 670 | **9.0** | **54** | 31 | 585 | 691.12 |
| | frasmt_z3 s-LT d-RQ | 590 | **9.0** | **54** | 23 | 513 | 808.34 |
| | frasmt_z3 d-LT d-RQ | 585 | **9.0** | **54** | 19 | 512 | 830.63 |
| | frasmt_z3 d-LT | 584 | **9.0** | **54** | 19 | 511 | 830.89 |
| | frasmt_z3 | 582 | **9.0** | **54** | 18 | 510 | 824.62 |
| Σ | frasmt_vbest | 1706 | **9.0** | **271** | 446 | 989 | 1247.01 |
| | frasmt_z3+om s-LT s-RQ(LQ-T) | **1657** | **9.0** | 270 | **445** | 942 | 1362.19 |
| | frasmt_z3 s-LT s-RQ(LQ-T) | 1600 | **9.0** | 270 | **445** | 885 | 1423.34 |
| | triangulator-msc | 1585 | 7.0 | 259 | 336 | **990** | **1322.97** |
| | frasmt_z3 s-LT s-RQ(k-Hy) | 1583 | **9.0** | 270 | **445** | 868 | 1430.35 |
| | frasmt_z3 s-RQ(LQ-T) | 1576 | **9.0** | **271** | **445** | 860 | 1422.65 |
| | frasmt_om s-LT s-RQ(LQ-T) | 1564 | **9.0** | 267 | 355 | 942 | 1549.95 |
| | frasmt_om s-RQ(LQ-T) | 1553 | **9.0** | 266 | 339 | 948 | 1607.00 |
| | frasmt_z3 d-LT s-RQ(LQ-T) | 1493 | **9.0** | 266 | 374 | 853 | 1607.46 |
| | frasmt_om s-LT d-RQ | 1337 | **9.0** | 264 | 332 | 741 | 1966.01 |
| | frasmt_z3 s-LT d-RQ | 1336 | **9.0** | 270 | 398 | 668 | 1903.86 |
| | frasmt_z3 | 1335 | **9.0** | **271** | 398 | 666 | 1912.94 |
| | frasmt_z3 d-LT | 1269 | **9.0** | 266 | 340 | 663 | 2073.04 |
| | frasmt_z3 d-LT d-RQ | 1268 | **9.0** | 265 | 340 | 663 | 2075.15 |

## 5 Conclusion and Future Work

In this work, we analyzed different strategies for symmetry breaking in characterizations of fractional hypertree width. While we focused on this particular width parameter, the overall idea of our methods immediately apply to the computation of other width parameters, such as treewidth. Hence, we expect that our findings can be used to gain significant improvements for other ordering-based encodings [34]. Our two methods RootClique and LexTopSort for eliminating symmetries seem to be good companions, since a combination of both state-of-the-art SMT solvers *z3* [32] and *optimathsat* [36] reached the virtually best configuration. Still, we only considered the best variant of RootClique in this configuration, and it seems there is potential for algorithm selection involving machine-learning tools like autofolio [31]. Overall, we perceived static symmetry breaking strategies as superior to dynamic techniques. While this might not be too surprising, we observed that some instances could still be solved faster with dynamic techniques. We see future work in analyzing the impact of fractional hypertree width for practical solving (counting) similar to treewidth [13,14,27] and in the context of other measurements such as bag size and domain size.

## References


1. Arocena, P.C., Glavic, B., Ciucanu, R., Miller, R.J.: The iBench integration metadata generator. In: Li, C., Markl, V. (eds.) Proceedings of Very Large Data Bases (VLDB) Endowment, vol. 9:3, pp. 108–119. VLDB Endowment, November 2015. https://github.com/RJMillerLab/ibench
2. Audemard, G., Boussemart, F., Lecoutre, C., Piette, C.: XCSP3: an XML-based format designed to represent combinatorial constrained problems (2016). http://xcsp.org
3. Bannach, M., Berndt, S., Ehlers, T.: Jdrasil: a modular library for computing tree decompositions. In: Iliopoulos, C.S., Pissis, S.P., Puglisi, S.J., Raman, R. (eds.) 16th International Symposium on Experimental Algorithms, SEA 2017. LIPIcs, London, UK, 21–23 June 2017, vol. 75, pp. 28:1–28:21. Schloss Dagstuhl - Leibniz-Zentrum fuer Informatik (2017)
4. Benedikt, M., et al.: Benchmarking the chase. In: Geerts, F. (ed.) Proceedings of the 36th ACM SIGMOD-SIGACT-SIGAI Symposium on Principles of Database Systems (PODS 2017), pp. 37–52. Association for Computing Machinery, New York (2017). https://github.com/dbunibas/chasebench
5. Berg, J., Lodha, N., Järvisalo, M., Szeider, S.: MaxSAT benchmarks based on determining generalized hypertree-width. Technical report, MaxSAT Evaluation 2017 (2017)
6. Berg, J., Järvisalo, M.: SAT-based approaches to treewidth computation: an evaluation. In: 26th IEEE International Conference on Tools with Artificial Intelligence, ICTAI 2014, Limassol, Cyprus, 10–12 November 2014, pp. 328–335. IEEE Computer Society (2014)
7. Bodlaender, H.L., Fomin, F.V., Koster, A.M.C.A., Kratsch, D., Thilikos, D.M.: On exact algorithms for treewidth. In: Azar, Y., Erlebach, T. (eds.) ESA 2006. LNCS, vol. 4168, pp. 672–683. Springer, Heidelberg (2006). https://doi.org/10.1007/11841036_60

8. Bodlaender, H.L., Möhring, R.H.: The pathwidth and treewidth of cographs. SIAM J. Discret. Math. **6**(2), 181–188 (1993)
9. Codish, M., Miller, A., Prosser, P., Stuckey, P.J.: Constraints for symmetry breaking in graph representation. Constraints **24**(1), 1–24 (2018). https://doi.org/10.1007/s10601-018-9294-5
10. Cohen, D., Jeavons, P., Gyssens, M.: A unified theory of structural tractability for constraint satisfaction problems. J. Comput. Syst. Sci. **74**(5), 721–743 (2008)
11. Dechter, R.: Bucket elimination: a unifying framework for reasoning. Artif. Intell. **113**(1–2), 41–85 (1999)
12. Fichte, J.K., Hecher, M., Szeider, S.: Analyzed Benchmarks on Experiments for FraSMT v2.0.0 (Dataset). Zenodo, July 2020. https://doi.org/10.5281/zenodo.3950097
13. Fichte, J.K., Hecher, M., Thier, P., Woltran, S.: Exploiting database management systems and treewidth for counting. In: Komendantskaya, E., Liu, Y.A. (eds.) PADL 2020. LNCS, vol. 12007, pp. 151–167. Springer, Cham (2020). https://doi.org/10.1007/978-3-030-39197-3_10
14. Fichte, J.K., Hecher, M., Zisser, M.: An improved GPU-based SAT model counter. In: Schiex, T., de Givry, S. (eds.) CP 2019. LNCS, vol. 11802, pp. 491–509. Springer, Cham (2019). https://doi.org/10.1007/978-3-030-30048-7_29
15. Fichte, J.K., Hecher, M., Lodha, N., Szeider, S.: A Benchmark Collection of Hypergraphs. Zenodo, June 2018. https://doi.org/10.5281/zenodo.1289383
16. Fichte, J.K., Hecher, M., Lodha, N., Szeider, S.: An SMT approach to fractional hypertree width. In: Hooker, J. (ed.) CP 2018. LNCS, vol. 11008, pp. 109–127. Springer, Cham (2018). https://doi.org/10.1007/978-3-319-98334-9_8
17. Fischl, W., Gottlob, G., Longo, D.M., Pichler, R.: HyperBench: a benchmark of hypergraphs (2017). http://hyperbench.dbai.tuwien.ac.at
18. Fischl, W., Gottlob, G., Pichler, R.: General and fractional hypertree decompositions: hard and easy cases. In: den Bussche, J.V., Arenas, M. (eds.) Proceedings of the 37th ACM SIGMOD-SIGACT-SIGAI Symposium on Principles of Database Systems (PODS 2018), pp. 17–32. Association for Computing Machinery, New York, June 2018
19. Freuder, E.C.: A sufficient condition for backtrack-bounded search. J. ACM **29**(1), 24–32 (1982)
20. Gebser, M., Kaminski, R., Kaufmann, B., Schaub, T.: Multi-shot ASP solving with clingo. Theory Pract. Log. Program. **19**(1), 27–82 (2019)
21. Geerts, F., Mecca, G., Papotti, P., Santoro, D.: Mapping and cleaning. In: Cruz, I., Ferrari, E., Tao, Y. (eds.) Proceedings of the IEEE 30th International Conference on Data Engineering (ICDE 2014), pp. 232–243, March 2014
22. Gottlob, G., Leone, N., Scarcello, F.: Hypertree decompositions and tractable queries. J. Comput. Syst. Sci. **64**(3), 579–627 (2002)
23. Gottlob, G., Samer, M.: A backtracking-based algorithm for hypertree decomposition. J. Exp. Algorithmics **13**, 1:1.1–1:1.19 (2009)
24. Grohe, M., Marx, D.: Constraint solving via fractional edge covers. In: Proceedings of the of the 17th Annual ACM-SIAM Symposium on Discrete Algorithms (SODA 2006), pp. 289–298. ACM Press (2006)
25. Grohe, M., Marx, D.: Constraint solving via fractional edge covers. ACM Trans. Algorithms **11**(1), Article no. 4, 20 (2014)
26. Guo, Y., Pan, Z., Heflin, J.: LUBM: a benchmark for OWL knowledge base systems. Web Semant. Sci. Serv. Agents World Wide Web **3**(2), 158–182 (2005)

27. Hecher, M., Thier, P., Woltran, S.: Taming high treewidth with abstraction, nested dynamic programming, and database technology. In: Pulina, L., Seidl, M. (eds.) SAT 2020. LNCS, vol. 12178, pp. 343–360. Springer, Cham (2020). https://doi.org/10.1007/978-3-030-51825-7_25
28. Khamis, M.A., Ngo, H.Q., Rudra, A.: FAQ: questions asked frequently. In: Milo, T., Tan, W. (eds.) Proceedings of the 35th ACM SIGMOD-SIGACT-SIGAI Symposium on Principles of Database Systems, PODS 2016, San Francisco, CA, USA, 26 June–01 July 2016, pp. 13–28. Association for Computing Machinery, New York (2016)
29. Korhonen, T., Berg, J., Järvisalo, M.: Solving graph problems via potential maximal cliques: an experimental evaluation of the bouchitté-todinca algorithm. ACM J. Exp. Algorithmics **24**(1), 1.9:1–1.9:19 (2019)
30. Leis, V., Gubichev, A., Mirchev, A., Boncz, P., Kemper, A., Neumann, T.: How good are query optimizers, really? Proc. Very Large Data Bases (VLDB) Endow. **9**(3), 204–215 (2015)
31. Lindauer, M., Hoos, H.H., Hutter, F., Schaub, T.: AutoFolio: an automatically configured algorithm selector. J. Artif. Intell. Res. **53**, 745–778 (2015)
32. de Moura, L., Bjørner, N.: Z3: an efficient SMT solver. In: Ramakrishnan, C.R., Rehof, J. (eds.) TACAS 2008. LNCS, vol. 4963, pp. 337–340. Springer, Heidelberg (2008). https://doi.org/10.1007/978-3-540-78800-3_24
33. Roussel, O.: Controlling a solver execution with the runsolver tool. J. Satisf. Boolean Model. Comput. **7**, 139–144 (2011)
34. Samer, M., Veith, H.: Encoding treewidth into SAT. In: Kullmann, O. (ed.) SAT 2009. LNCS, vol. 5584, pp. 45–50. Springer, Heidelberg (2009). https://doi.org/10.1007/978-3-642-02777-2_6
35. Schidler, A., Szeider, S.: Computing optimal hypertree decompositions. In: Blelloch, G., Finocchi, I. (eds.) Proceedings of ALENEX 2020, the 22nd Workshop on Algorithm Engineering and Experiments, pp. 1–11. SIAM (2020)
36. Sebastiani, R., Trentin, P.: OptiMathSAT: a tool for optimization modulo theories. J. Autom. Reason. **64**(3), 423–460 (2020)
37. Transaction Processing Performance Council (TPC): TPC-H decision support benchmark. Technical report, TPC (2014). http://www.tpc.org/tpch/default.asp

# Towards Faster Reasoners by Using Transparent Huge Pages

Johannes K. Fichte[1](✉), Norbert Manthey[2], Julian Stecklina[2], and André Schidler[3]

[1] School of Engineering Sciences, TU Dresden, Dresden, Germany
johannes.fichte@tu-dresden.de
[2] Dresden, Germany
[3] Algorithms and Complexity Group, TU Wien, Vienna, Austria
aschidler@ac.tuwien.ac.at

**Abstract.** Various state-of-the-art automated reasoning (AR) tools are widely used as backend tools in research of knowledge representation and reasoning as well as in industrial applications. In testing and verification, those tools often run continuously or nightly. In this work, we present an approach to reduce the runtime of AR tools by 10% on average and up to 20% for long running tasks. Our improvement addresses the high memory usage that comes with the data structures used in AR tools, which are based on conflict driven no-good learning. We establish a general way to enable faster memory access by using the memory cache line of modern hardware more effectively. Therefore, we extend the standard C library (glibc) by dynamically allowing to use a memory management feature called huge pages. Huge pages allow to reduce the overhead that is required to translate memory addresses between the virtual memory of the operating system and the physical memory of the hardware. In that way, we can reduce runtime, which in turn decreases costs of running AR tools and applications with similar memory access patterns by linking the tool against this new glibc library when compiling it. In every day industrial applications, runtime savings allow to include more detailed verification tasks, getting better results of any-time optimization algorithms with a bound execution time, and save energy during nightly software builds. To back up the claimed speed-up, we present experimental results for tools that are commonly used in the AR community, including the domains ASP, hardware and software BMC, MaxSAT, and SAT.

The work has been supported by the Austrian Science Fund (FWF), Grants P32441 and W1255. We would like to thank the anonymous reviewers for very detailed feedback and their suggestions. Special thanks go to the reviewer who provided comments on CP solvers and propagators. The implementation is available at github:daajoe/thp_docker_build.


H. Simonis (Ed.): CP 2020, LNCS 12333, pp. 304–322, 2020.
https://doi.org/10.1007/978-3-030-58475-7_18

# 1 Introduction

Automated reasoning (AR) tools such as solvers for Answer Set Programming (ASP), Bounded Model Checking (BMC), and Boolean Satisfiability (SAT) are widely used to test and verify software and hardware in industry [14,20,27]. When improving AR tools, which usually run in industry each day for a few hours, one is also interested in speeding up existing implementations. A classical way to improve such tools is to consider their memory dependency when designing algorithms and data structures. While algorithm engineering often focuses on the source code of each of the solvers, we believe that an entirely overlooked direction is to take the interplay between hardware and software into account and speed up solving by modifying standard libraries (glibc) that programmers use to handle memory management. In the following, we outline how and why one can gain considerable improvements for many solvers by modifying glibc.

The underlying algorithms for AR tools are often based on a technique called conflict driven no-good learning (CDNL) with watch lists whose efficiency highly depends on their memory consumption [13]. On modern hardware, the accessible memory is virtual and handled by a physical memory management unit (MMU). The mapping between virtual and physical memory is stored in page tables by the MMU. In order to reduce access time, recently used mappings are stored in a *translation lookaside buffer (TLB)*. In practice, when using watch lists extensively, high memory consumption results in unpredictable memory accesses, many cache misses, and so called page translation failures (TLB misses). A common way to speed up a CDNL-solver is to use advanced data structures, which improve the memory access for frequently running operations in the solver, for example, data structures for learnt clauses, two watched literals, and linear lookup tables. Another way to obtain a speed-up is to respect the memory caches of the underlying hardware in the implementation. For example, one can considerably improve the performance of a naive SAT solver by modifying the implementation in such a way that the memory cache is used as much as possible (reducing cache misses) as implemented in Riss [29], using cache pre-fetching as done in the solver CryptoMiniSat [50], or implementing very sophisticated data structures with the goal to optimize the overall memory usage as in the solver Lingeling [9]. The hardware related line of optimization opens the research question whether one can obtain further speed-up by exploiting other cache layers that are available in modern CPUs. One such cache layer is used for address translation, i.e., the translation look aside buffer (TLB). As an extension, we want to investigate whether this kind of improvement is applicable to multiple AR tools.

### New Contribution

In this paper, we introduce a simple and transparent approach to effectively reduce the number of TLB misses in order to speed up the execution of modern memory dependent solvers, in particular, unit propagation, sometimes also called

Boolean constraint propagation. We employ a Linux memory management feature called *transparent huge pages (THP)*, which reduces the overhead of virtual memory translations by using larger virtual memory page sizes [52], effectively increasing the size of that cache. Our approach is based on modifying the standard C library (glibc), which is the default standard library in Linux systems [2]. Whenever a solver allocates memory, we make sure that we additionally give the operating system kernel advice about the use of the memory (madvise). This feature can then be used for a solver simply by recompiling it and statically linking it against our modified glibc. In that way, we obtain a significant speed-up on benchmarks in model checking of up to 15% and for most other solvers of up to 10%. The approach is based on a hardware feature and thus generalizable to other operating systems and CPU architectures supporting large page sizes.

Our advances summarize as follows:

1. We propose an easily accessible way to reduce the number of TLB misses in combinatorial memory-dependent solvers by patching the glibc in a way that our modifications can be activated or deactivated at runtime.
2. We provide a build system to easily patch glibc and statically link a solver against the patched version. Our system is based on a setup that uses OS-level virtualization (Docker) [31] and is available to all modern Linux systems. We already provide various pre-compiled state-of-the-art reasoning tools.
3. We carry out extensive benchmarks and present detailed results for various reasoning tools.

### Related Work

Chu, Hardwood, and Stuckey [17] as well as Hölldobler, Manthey, Saptawijaya [29] considered cache utilization in SAT solvers and illustrated how a resource-unaware SAT solver can be improved by utilizing the cache sensibly, resulting in reasonable speed-ups. The latter already hinted that using larger pages results in a speed-up of 10% for SAT solvers, and was a motivation for this work. The effect of huge pages has already been widely investigated in the field of operating systems, e.g., [47]. However, the focus was mostly on database systems, while an analysis of the effect for reasoning tools was not yet available. Recent research considered benchmarking system tools [38], selecting benchmarks to tune solvers [30], and treating input benchmarks for benchmarking [12]. These topics are orthogonal to our work. In contrast, we consider computational resources and memory management of solvers, in particular, its effect on the runtime. Bornebusch, Wille, and Drechsler [16] analyzed the memory footprint of SAT solvers and tried to improve them. However, they did not consider propagation and its data structures, which is reasonable from a complexity point of view due to large formula sizes.

## 2 Modern CDNL-Based Solvers and Memory

Before we present our advances, we give a brief explanation on how modern SAT solvers are implemented and introduce components and mechanisms that are relevant for memory access. As many reasoners are based on SAT technology, e.g., [13, 35, 53], core concepts are very similar for various reasoners. First, we define (propositional) formulas and their evaluation in the usual way and assume familiarity with standard notations, including satisfiability. For basic literature, we refer to introductory work [36]. We consider a universe $U$ of propositional variables. A *literal* is a variable or its negation and a *clause* is a finite set of literals. A (CNF) *formula* is a finite set of clauses. A *(partial) truth assignment* is a mapping $\tau : \text{var}(X) \rightarrow \{0, 1\}$ defined for a set $X \subseteq U$ of variables. For $x \in X$, we put $\tau(\neg x) = 1 - \tau(x)$. For a formula $F$, we abbreviate by $\text{var}(F)$ the variables that occur in $F$. We say that a truth assignment $\tau$ *satisfies* a clause $C$, if for at least one literal $\ell \in C$ we have $\tau(\ell) = 1$. We say that a truth assignment $\tau$ falsifies a clause $C$, if it assigns all its literals to 0. We call a clause $C$ *unit* if $\tau$ assigns all but one literal to 0. A truth assignment $\tau$ is *satisfying* if for each clause $C \in F$, the truth assignment $\tau$ satisfies $C$.

## 2.1 SAT Solvers

So far, there are two main contributing factors to advances in the efficiency of modern SAT solvers: (i) theoretical improvements in terms of more advanced algorithms and heuristics and (ii) algorithm engineering in terms of data structures. The core algorithm that drives search in modern SAT solvers is based on *conflict driven no-good learning (CDNL)* or also known as *conflict driven clause learning (CDCL)* [26, 42], which was widely extended by search heuristics [13, 34] and simplification techniques during search [33]. A key technique is *unit propagation*, which aims at finding clauses where all literals but one are already assigned and then setting the remaining literal to a value that satisfies the clause. Unit propagation is responsible for the vast majority of the overall runtime even in modern solvers [34]. Hence, algorithm engineering and efficient data structures are essential for practical solving, i.e., the *two-watched-literal* scheme for unit propagation [45] and fast lookup tables, which are also important for the used heuristics and learning techniques. The watched literal scheme reduces the number of steps in the algorithm and memory accesses, but decreases the efficiency of the memory access [29]. Still, this results in a considerable overall runtime improvement [34]. While lookup tables provide fast access to relevant clauses, they result in a much higher memory footprint and may yield unpredictable memory access [29].

## 2.2 CPUs, Virtual Memory, and Paging

Modern *operating systems (OSes)* provide the concept of *virtual memory* to applications. Thereby, the OS releases software developers from worrying about the actual physical memory layout and also allows for overcommitting resources. Virtual memory is managed at the granularity of *pages*. A page is a contiguous block of memory in the virtual address space. The OS can map a page to a *page frame*, which is a corresponding location in physical memory. On the Intel Architecture [32], page tables describe the mapping from pages to page frames

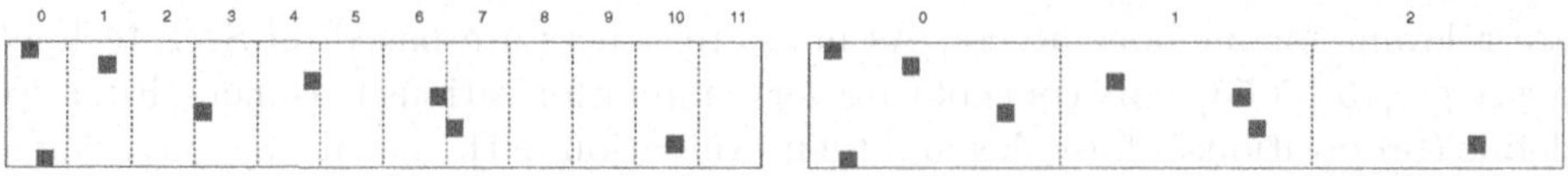


**Fig. 1.** This figure illustrates how pages cover a sequence of memory accesses for two different sizes of pages for a given amount of memory.

and thus virtual to physical addresses. On 64-bit Intel systems, these page tables are trees with a depth of commonly up to four levels for $2^{48}$-byte address spaces.

Walking these data structures to provide a translation for each memory access is infeasible, because it would add one page table read per level for each intended memory access. Instead, processors take advantage of spatial and temporal locality of memory accesses and cache translations in Translation Lookaside Buffers (TLBs). An Intel Skylake system has two levels of TLBs and the unified L2 TLB can hold 1536 entries [54]. Other recent CPUs have similar specifications. With 4 KiB pages, this translates to holding translations for 6 MiB of virtual memory in the TLB. A straight-forward way to increase the capacity of the TLB is for the processor architecture to allow for larger page sizes. On Intel 64-bit systems, in addition to 4 KiB pages, the system also supports 2 MiB and 1 GiB pages. While 1 GiB pages have (few) dedicated TLB entries, 2 MiB pages share the same entries as 4 KiB entries in the TLB on Skylake.

To benefit from large pages, the OS needs to make them accessible to applications [1] by constantly freeing memory (defragmenting) to obtain continuous blocks from which large pages can be allocated [46]. In more detail, fragmentation originates from applications that use 4 KiB pages for which short blocks of memory are frequently allocated and freed. If many applications using 4 KiB pages run in parallel, memory fragmentation is more likely. Then, it is harder to obtain free blocks of 2 MiB memory. To still be able to use larger pages, the OS tries to restructure the memory mapping for future 2 MiB requests.

Figure 1 illustrates the usage of memory with pages of different sizes. When using larger pages, fewer pages are required to cover the same area of memory. Hence, fewer TLB entries are occupied. In more detail, the black boxes in Fig. 1 illustrate a sequence of accesses (from top to bottom). While for larger pages (right) it is sufficient to memorize the translation for three pages, smaller pages require seven pages (left). In case the TLB can only hold four entries, the entry of Page 0 would be evicted before it can be re-used to access the same clause again. When using larger pages, fewer initial translations have to be done, and only three pages are required to perform all accesses.

### 2.3 Large Pages in Linux

Large page sizes are supported in Linux by a feature called *transparent huge pages (THP)*, which offers both implicit and explicit use of large page sizes and was introduced in Linux 2.6.38 [52]. If THP is enabled, memory does not have to be statically provisioned for applications to use large pages, which is a clear advantage over previous attempts involving large pages [48]. Instead, the system is continuously compacting memory to free up contiguous space to allocate large pages. The Linux kernel can then, depending on the system configuration, transparently allocate large pages for applications. If intended, a system administrator can still additionally provision large pages manually. THP can be globally enabled or configured as an opt-in feature. Both mechanisms degrade gracefully when no large pages are available and will instead back memory using the standard page size. When THP is configured for opt-in, an application can use the system call "madvise" with the "MADV_HUGEPAGE" flag to mark memory regions as eligible. If this is done for virtual memory regions that have not been backed by physical memory yet, e.g., directly after a "mmap" call, the kernel will try to allocate a large page on the first access to this memory. Otherwise, the kernel will occasionally scan virtual memory that is eligible for THP to create large pages. One downside of THP is that the kernel has to run scan and compact operations. Linux allows to configure this behavior to mitigate the impact by paying the cost for scanning and compacting at allocation time instead of doing it as a background job.

### 2.4 The Effect of THP

System workloads are known to speed up with huge pages. However, it may also reduce reproducibility, as huge pages have to be enabled in the kernel and globally for all applications on the system [48] and not every request might actually get a huge page. Hence, it is recommended to use small pages for benchmarking. Unfortunately, the StarExec cluster [51] has enabled THP by default for all executed programs making it incomparable to standard university computers, clusters, and industrial settings. The presented contribution is hence more targeted towards industrial use cases that want to solve a problem at hand as fast as possible. In case of virtualized machines, using small pages can result in almost 50% of the runtime being spent in address translation [39]. Using huge pages in both the guest and the host, which is 2M instead of 4k on the given architecture, reduces this value to about 4%. We expect similar savings for tools that are run in a virtualized environment, as virtualization typically uses huge pages internally. In the remainder of the paper, we investigate the actual effect on a bare metal setup for our experimental work.

## 3 TLB Misses in SAT Solvers

Typically, SAT solvers do not exhibit the memory access locality that caches or TLB are optimized for. While previous works considered caches [17,29], memory translation and the TLB have not been taken into account. Hence, we focus on memory accesses in the most time consuming part of SAT solvers: unit propagation, also called Boolean constraint propagation.

*UnitPropagate* (formula $F$, truth assignment $\tau$, literals $P$, watch lists $L$)

B1 **while** the list $P$ of literals to propagate is not empty //**compute closure**
B2 pick $p \in P$, and remove from $P$ //**typically DFS**
B3 access watch list $L_p$ of clauses such that $\neg p \in C$ //**propagate**
B4 **for all** clauses $C$ in $L_p$:
B5 **if** $C \neq \emptyset$, $x \in C$, and $x$ not falsified in assignment $\tau$ //**watchable literal**
B6 remove $C$ from $L_p$ //**maintain lists**
B7 add $C$ to watch list $L_{\neg x}$ for $\neg x$ //**maintain lists**
B8 **else if** $C = (x)$ unit, extend $P$ and $\tau$ with $x$ //**unit rule**
B9 **else if** $C$ is falsified, trigger **conflict analysis(**$\tau$, $C$**)** //**conflict**

Listing 1: Pseudo code for an implementation of unit propagation with the watched literal scheme. The state of the solver holds the formula $F$ as watch lists $L$, one list $L_x$ for each literal $x$, as well as a truth assignment $\tau$, and the list $P$ of literals to propagate. The result of the algorithm will either be an extension of the truth assignment, or a tuple truth assignment $\tau$ and a conflict clause that is falsified by truth assignment $\tau$.

Assume that a formula $F$ and a partial truth assignment $\tau$ is given and a two watched literal data structure is used [45]. Briefly, *unit propagation* works as outlined in Listing 1. Initially, for each clause $C \in F$, one selects two literals from $C$, which are not non-falsified by truth assignment $\tau$. Then, the truth assignment $\tau$ is extended by setting additional variables. Since assigning a literal $\ell \in C$ such that $\tau(\ell) = 1$ results in clause $C$ that is satisfied, which in turn allows to remove clause $C$ from the considered clauses right away, the only interesting case is if truth assignment $\tau$ sets literal $\ell \in C$ such that $\tau(\ell) = 0$. In that case, the clause $C$ might be falsified and be involved in a conflict or have unassigned literals, which can be used to imply the truth value of other literals. Then, UnitPropagate checks every clause $C$ that contains a literal which might be falsified during propagation. Therefore, the list $P$ of literals to propagate is traversed (Line B1), and each watch list $L_p$ for literal $p$ is processed (Line B3). Then, each clause $C$ in list $L_p$ contains $\neg p$, so that the new state of clause $C$ has to be evaluated by processing the other literals in clause $C$. Hence there are two cases: either (i) clause $C$ is satisfied by another literal, or (ii) clause $C$ contains another literal $x$ that is not yet falsified by the truth assignment $\tau$ (Line B5). Then, we watch literal $x \in C$ for being set to false instead of $\neg p$, and consequently have to update list $L_p$ (Line B6) and list $L_{\neg x}$ (Line B7). Otherwise, clause $C$ might be a unit clause (Line B8) or might be falsified by truth assignment $\tau$ (Line B9). In both cases, clause $C$ can remain in list $L_p$.

When considering the memory access pattern, the unit propagation algorithm has the following properties: the literals in list $P$ are not easily determined in advance. Hence, accesses in Line B3 to load the list are hard to predict. One could reduce the memory accesses in Lines B3 and B4 by pre-fetching data from memory in advance. This has been proposed in previous works [29]. Accessing the clauses in Line B4 are hard to predict, as the order of the clauses in list $L_p$ changes. In more detail, in Line B6 some clauses are removed and in Line B8 or B9 others are kept. To improve the access behavior in Line B5, Een and Sorensson [22] proposed for MiniSAT 2.1 an optimization to avoid the access of clause $C$ for the satisfied case. There, another literal of clause $C$ is stored in list $L_p$. Since their introduction, blocking literals are commonly used in most modern solvers. Accessing literal $x$ in Line B5 is also unpredictable, as literal order in clauses also changes. Typically, the two watched literals are the first two stored literals and they change whenever the clause is moved to another list.

Our hypothesis is that unit propagation is the major source of memory accesses, as most runtime is spent in unit propagation, and many different memory locations, i.e., clauses are accessed non-linearly during unit propagation. This drives us to the following hypothesis:

**Hypothesis 1.** *Accessing clauses during unit propagation as well as updating and accessing watch lists has a high impact on TLB misses.*

We support our hypothesis by the following observation. The two watched literals data structure allows to keep the number of overall accesses low, but has high memory footprint with additional data structures and lists. Further, the memory accesses for (i) clause to access next, (ii) literals of a clause to watch next, or (iii) list to place it are difficult to predict. Hence, Lines B3, B4, and B7 are prime candidates to access memory locations that have not been accessed recently, and hence, are not cached, nor served with current TLB entries.

*Analyzing Unit Propagation.* To back up Hypothesis 1 with data, we analyze the distribution of TLB misses in the SAT solvers MiniSat and Glucose[1]. When running MiniSat [22], we observe that 90% of TLB misses occur in unit propagation; thereof, about 10% when moving clauses to another (unpredictable) watch list and about 80% when accessing the first literal of the next watched clause. This data matches the assumption that Line B3 and B4 are responsible for most of the TLB misses. In addition, moving clauses to new watch lists contributes another 10%. When running Glucose version 4.2.1 [3], we can see similar results. 90% of the TLB misses happen in unit propagation. In the modern solver Glucose, unit propagation is split into (i) propagating binary clauses that contributed 5 % of all TLB misses, (ii) propagating during learned clause minimization [40] that contributes about 20 %, and (iii) propagating larger clauses and pushing them to watch lists that consume the majority of the TLB misses. Empirical observations for these two solvers confirm our hypothesis, unit propagation is the major source for TLB misses. The random memory accesses to check the next clause in the list for being unit, which can have an arbitrary memory location, as well as putting clauses into another watch list, are the major contributors. Unit propagation is responsible for a large fraction of the runtime of SAT solvers, which is actually spent in address translation. In Lingeling, Biere [8] places watch lists and its clauses closer to each other, to avoid TLB misses related to Line B4 (matching our observations in Sect. 5.1). Here, we present an orthogonal approach to avoid TLB misses, which allows to improve the implementation [8] further and can be applied to many other solvers.

[1] We use a sampling approach of CPU performance counters for TLB misses with the system tool perf.

*Unit Propagation Implementation Outlook.* We believe that additional data structure improvements along [17] are hardly feasible. Clauses would have to be even more compact and the changes require a huge effort for a single solver [7]. Changes to the underlying algorithms likely result in reduced performance and require tuning parameters again. On that account, we propose a general approach to THP, which can be easily used by many other tools.

## 4 Improving Unit Propagation with THP via Madvise

Modern Linux distributions provide native support for transparent huge pages. Usually, the systems allow a superuser to define the behavior via the configuration file "`/sys/kernel/mm/transparent_hugepage/enabled`" whose values "always" or "never" apply to all running processes. Because there might be applications running on the host that would suffer from larger pages, THP is usually disabled on physical systems and it is not advised to set the value to always. Fortunately, as described above, Linux also allows to use huge pages via the madvise system call. While this sounds fairly trivial, it requires (i) lots of manual adaption of the source code to mark memory regions as eligible and in turn makes the implementations of solvers (ii) fairly incomparable on an algorithmic level. When using "madvise" in combination with every malloc in a SAT solver, no speed-up is obtained, as most allocations are not aligned to the required large page size of 2M, and consequently are not backed by a huge page. In the following, we suggest an easily accessible way to reduce the number of TLB misses in combinatorial memory-dependent solvers.

### 4.1 Using More Huge Pages

In the previous section, we explained that using more huge pages seems to be a reasonable approach to speed up the memory access of modern solvers. This can be obtained by running a madvise system call to instruct the kernel to use transparent huge pages of 2M whenever the solver allocates memory. Then, we align all requested memory to 2M addresses and increase the size of the reservation accordingly, so that huge pages can actually be used. If we would not do so, two memory requests of the application can be in the middle of a 2M page, which results in not using a huge page. Compared to the system setting, this change results in using one more huge page per misaligned memory request.

### 4.2 Patching the Standard C Library (glibc)

In order to provide a transparent way to various solver developers, and offering a way for algorithms engineering to consider the effect of transparent huge pages on many AR tools, we want to avoid manual source code adaption as much as possible. To this end, we put our focus on the standard C system library glibc, which already provides standard functions to access the system memory. The library is used in Linux to compile most of the solvers. Instead of modifying the source code of various solvers, we implement the above mentioned ideas into glibc[2]. Whenever a certain runtime flag is activated, our modified glibc takes care of the above mentioned changes. The current approach uses a system environment variable (`GLIBC_THP_ALWAYS=1`) that can be specified before calling a program. This way we allow setting the flag for a specific program instead of all running programs. Globally enabling transparent huge pages for all running applications on the system is usually forbidden both in industry and academia by administrators due to a variety of potential side effects, which might slow down a variety of programs. If the flag is not set, we disable the use of huge pages. The additional cost, compared to glibc, is a single if-statement in combination with the madvise system call. We implemented our patches into glibc 2.23 [2] and enable the feature without any source code modification of the various solvers themselves. In that way it is entirely sufficient for the user of a solver to recompile the solver and link it against our modified glibc.

### 4.3 Huge Pages in a Solver

In order to use the feature, there are two ways to proceed: (i) *link* the solver *statically* against our modified glibc or (ii) patch the system glibc and then *dynamically link* the solver against the new glibc. We provide an easy and accessible way for the former, since patching the system glibc is usually considered problematic due to side effects and as it requires superuser permissions, which makes it very unlikely that actual users of the reasoning tools will use this feature. We introduce a virtual environment that allows for easy compiling of the solver, in order to avoid problematic setups of a new secondary glibc.

Our system is based on the OS-level virtualization Docker [31], which isolates running programs entirely from each other. Docker itself is available on all modern OSes and allows to deliver software in packages, which are called containers. A running container is entirely isolated and can bundle its own software. We use this to not interfere with the system glibc. But we do not publish only a Docker container, instead we provide the scripts to build containers in which the compilation then runs. The user just needs to install Docker and we provide the tooling to link a solver with THP support. Along, we give many exemplary scripts to highlight how to run the tools and various pre-compiled state-of-the-art solvers.

We would like to emphasize that in our approach Docker is only used to compile the solver, not to actually run the solver. Compiling within a Docker environment can be done on a local machine or a trusted machine where the user has privileged access (as Docker often requires certain additional privileges) or is simply allowed to run Docker containers. Then, the resulting binary is transferred to the runtime environment. Since we compile files statically inside the Docker container to the modified glibc, there is absolutely no need to install a patched glibc or a Docker environment on the actual runtime system.

[2] Our latest implementation is publicly available at github:daajoe/thp_docker_build and the glib patch at github:conp-solutions/thp. An upstream to the glibc library is in progress and we are in contact with the glibc maintainers.

**Table 1.** Overview on the speed-up between solving when using THP for SAT solvers. We distinguish by non-THP and THP by $\cdot_n$ and $\cdot_{thp}$, respectively. # counts the number of solved instances. $t$ contains the runtime in hours for the mutually solved instances, $s$ the saved runtime in %, i.e., $s = (t_n - t_{thp})/t_n \cdot 100$ factor. TLB represents the TLB load misses on the mutually solved instances. $r_{tlb}$ summarizes the % of TLB misses over the original TLB misses, i.e., $r_{tlb} = \text{TLB}_{thp}/\text{TLB}_n \cdot 100$. tlb[$s$] contains the TLB misses per second in relation to the physical bus speed of the CPU for 4 threads sharing the memory (9.6 GT/s /4).

| | Solver | $\#_n$ | $\#_{thp}$ | $t_n$ | $t_{thp}$ | $s$[%] | $\text{TLB}_n$ | $\text{TLB}_{thp}$ | $r_{tlb}$ | tlb[$s$]$_n$ | tlb[$s$]$_{tlb}$ |
|---|---|---|---|---|---|---|---|---|---|---|---|
| 1 | Winner19 | 194 | 197 | 7.38 | 6.13 | 16.97 | 3.4e+11 | 1.3e+10 | 3.94 | 5.35 | 0.25 |
| 2 | MergeSAT | 190 | 192 | 6.89 | 5.70 | 17.24 | 3.4e+11 | 1e+10 | 3.04 | 5.77 | 0.21 |
| 3 | CaDiCaL | 189 | 191 | 5.62 | 5.00 | 11.04 | 1.7e+11 | 5.6e+09 | 3.22 | 3.55 | 0.13 |
| 4 | Glucose | 189 | 191 | 4.50 | 3.66 | 18.72 | 2.6e+11 | 6.3e+09 | 2.44 | 6.61 | 0.20 |
| 5 | CryptoM | 183 | 185 | 6.54 | 5.75 | 12.06 | 2.5e+11 | 8.2e+09 | 3.21 | 4.51 | 0.16 |
| 6 | Lingeling | 177 | 178 | 6.29 | 5.97 | 5.08 | 5.6e+10 | 6.3e+08 | 1.12 | 1.04 | 0.01 |
| 7 | MiniSat | 169 | 173 | 6.99 | 5.96 | 14.75 | 2.8e+11 | 3.3e+09 | 1.20 | 4.56 | 0.06 |

## 5 Experimental Evaluation

We conducted a series of experiments using standard benchmark sets for various reasoning tools to analyze the effect of THP beyond pure SAT. All benchmark sets and our results are publicly available.

*Benchmarked Solvers and Instances.* In our experimental work, we present results for recent versions of publicly available SAT solvers: CaDiCaL, CryptoMiniSat 5, Glucose 4.2.1 [3], Lingeling [9], MapleLCMDistChronoBTDL (Winner2019) [37], MergeSAT [41], and MiniSat [22]. We selected the recommended benchmark for tool tuning [30] and compare MiniSat and MergeSAT from the above set of SAT tools again on a second hardware. For answer set programming (ASP), we used clasp [35] and a robust benchmark set, which was developed for solver optimization and provides a large variety of instances, with adapted instance hardness, and free of duplicates [30]. From software model checking (SWMC), we use CBMC [18], which uses a single call to a SAT solver. As SWMC benchmark, we use the benchmark provided when introducing the LLBMC tool [44]. As another group, we collected tools, which use incremental SAT solvers as a backend. For hardware model checking (HWMC), we use the bounded model checker aigbmc [11], with an unrolling limit of 100, and use the benchmark of the deep bound track of the HWMC Competition of 2017 [10]. For optimization, we use the MaxSAT solver Open-WBO [43], which uses the SAT solver Glucose as a backend. As MaxSAT benchmark, we picked the weighted partial maxsat formulas from 2014, to make sure the incremental interface is actually used. Finally, we consider muser-2 [6], which computes a minimal unsatisfiable subformula (MUS) from a CNF formula, and use the group MUS benchmark from the MUS competition 2011[3].

### 5.1 Evaluating Boolean Satisfiability (SAT) Solvers

For plain SAT solvers we carried out the following extensive study.

**Measure, Setup, and Resource Enforcements.** Our results were gathered on a cluster of RHEL 7.7 Linux machines with kernel 3.10.0-1062 with activated Meltdown and Spectre mitigations[4]. We evaluated the solvers on machines with two sockets equipped with Intel Xeon E5-2680v3 CPUs of 12 physical cores each at 2.50 GHz base frequency. We forced the performance governors to 2.5 Ghz [28] and disabled hyper-threading. To follow the hardware structure of the system, we manually grouped the solver resources as follows: on each socket the first two cores where restricted to controlling threads and processor cores 2, 3, 4, 5, 8 (virtual cores 2–6) and 9, 10, 11, 12, 13 (virtual cores 7–11) were assigned each to one running solver. The machines are equipped with 64 GB main memory of which 60.5 GB are freely available to programs. We compare wall clock time and number of timeouts. However, we avoid IO access on the CPU solvers whenever possible, i.e., we load instances into the RAM before we start solving. We run at most 4 solvers on one node, set a timeout of 900 s, and limited available RAM to 8 GB per instance and solver. We follow standard guidelines for benchmarking [38].

**SAT Results.** Table 1 gives an overview of the number of solved instances for each solver with and without THP. Note that we report in this table only for columns $\#_n$ and $\#_{thp}$ on individually solved instances. To obtain comparability for all other columns, we present data for instances that have been solved by both configurations. The results show that a solver with activated THP solves overall more instances than without THP. When considering runtime, the configurations that employ THP solve the considered instances faster. Runtime improvements range from a smaller improvement for Lingeling, which was about 5% faster (which translates to 0.32 h), up to more than 17% (i.e. one hour) faster for MergeSAT, MiniSat, and Winner2019. In terms of factor of saved runtime hours, we can see that the solvers that employ THP are up to almost 19% faster. As these values are gathered over the whole benchmark, these values are averages. For higher run-times, the speed-up is typically higher than for instances with a smaller runtime. The number of TLB misses that we observed reduce up to 2 orders, namely, $r_{tlb} = \text{TLB}_{thp}/\text{TLB}_n$ goes down to 1% for Lingeling and MiniSat. When we consider the number of solved instances, the value improves for all presented solvers when using THP, between 1 and 4 instances. As the presented

[3] MUS benchmarks are available at cril.univ-artois.fr/SAT11.

[4] Note that we initially also ran experiments with an earlier kernel 3.10.0-693. There, non-THP runtime results were comparable, but improvements were slightly smaller.

technique is a linear speed-up, we do not expect a major change in number of solved instances for a specific timeout, due to the heavy-tailed runtime behavior of SAT solvers. Three solvers show a slightly different pattern. For Lingeling, we reduced the TLB misses to 1.1% and obtained only a speed-up of 5.4%. For CaDiCaL, we reduced the TLB misses to 3.22% and obtained a speed-up of only 12.4%. For CryptoMiniSat, we reduced the TLB misses to 3.21% and obtained a speed-up of 13.7%. The last two columns illustrate the number of TLB misses per second (on average) in relation to the physical bus speed for 4 threads. Solvers with a higher initial value have a higher potential to benefit, which is reflected in our results.

**Discussion and Summary.** The results show that a solver with activated THP solves overall more instances than without THP. Throughout all solvers, we reduced the overall running time by 5% up to 19%. Our approach makes all solvers faster without spending a significant amount of time on optimizing the solver itself. Throughout our experiments, the number of TLB misses goes down significantly for all considered solvers. For Lingeling and MiniSat, the number of TLB misses even reduces to 1% of the original number of misses. Since unit propagation is a major source of memory accesses and a major cause of a high number of TLB misses and responsible of a large part of the solving time, the reduced TLB misses also yield a speed-up in the overall runtime. The observed results above confirm our hypothesis that improving on the number of TLB misses improves the runtime. We observe that three solvers have a somewhat different pattern, namely, CaDiCaL, CryptoMiniSat, Lingeling. While we obtain a speed-up, it is smaller than the order of the reduction of the TLB misses. If we compare the numbers to TLB misses per second and relate this to the physical bus speed that is shared by 4 threads – the value tlb[$s$] in Table 1, it suggests that the solvers CaDiCaL, CryptoMiniSat, Lingeling, and MiniSat are more optimized than the other solvers, where MiniSat also only maintains one watch list per literal, whereas the other solvers have a separate list for binary clauses, resulting in more memory fragmentation.

### 5.2 Evaluating Other Reasoners

We believe that the THP approach does not only boost SAT solvers, but other reasoners as well. On that account, we run additional experiments on tools that are either based on SAT solvers or implemented closely to the CDNL algorithm. To broaden the applicability, we also consider tools that use SAT solvers via their incremental interface [22]. As the SAT calls in these tools are shorter, we expect that the benefit of using THP is smaller.

**Hypothesis 2.** *When using incremental SAT solvers inside a reasoner, the benefit of THP is smaller.*

**Table 2.** Overview on the runtime of various reasoners with(out) THP evaluated on their respective competition benchmarks. $t_n$ and $t_{thp}$ represent the overall (PAR1) runtime of the reasoner in hours and $s$ represents the saved runtime in %.

| Category | Tool | $t_n$ | $t_{thp}$ | $s[\%]$ |
|---|---|---|---|---|
| SAT | MiniSat | 8.17 | 7.03 | 13.99 |
| SAT | MergeSAT | 7.94 | 6.90 | 13.13 |
| ASP | clasp | 3.66 | 3.29 | 10.18 |
| MaxSAT | open-wbo | 1.19 | 1.09 | 8.49 |
| MUS | muser2 | 4.18 | 3.97 | 5.16 |
| HWMC | aigbmc | 0.89 | 0.86 | 4.11 |
| SWMC | cbmc | 0.23 | 0.22 | 2.76 |

*Measure, Setup, and Resource Enforcements.* To support Hypothesis 2, we run a second analysis. To make sure the above results are not CPU and OS dependent, we used a second environment of the same architecture and repeat the results for MiniSat and MergeSAT. The computer has an Intel Core i5-2520M CPU running at 2.50 GHz, with an Ubuntu 16.04 and Linux 4.15, using 5 GB as memory limit and the 900 s as timeout per instance.

**Results.** Table 2 states the results for the considered tools and benchmarks. To measure the speed-up, we show only instances that have been solved by both variants. When using THP, we can usually solve a few more instances. First, we can see a similar improvement like in the previous setting where we used the same architecture, but different hardware. The improvement when using THP for tools with a single call is similarly high as presented above, i.e., SAT as well as ASP show improvements above 10%. Only cbmc from SWMC is an outlier, which might be related to its memory usage and the time it spends in other algorithms. For tools that use incremental SAT solvers as a backend, the improvements range from 4% from HWMC to 8% in MaxSAT. The low speed-up can be explained with their memory usage: over each benchmark, cbmc's memory usage is rather low, i.e., the median memory footprint is 8.8 MB. For all the other tools and categories, the memory footprint is higher, e.g., the median for ASP is 28.8 MB and for SAT 123.3 MB. The tools with incremental SAT backends also consume more memory than cbmc: MaxSAT 21.3 MB, HWMC 164 MB, and MUS 298.1 MB. As expected, tools with a higher memory footprint result in a higher speed-up due to transparent huge pages.

## 6 Conclusion and Future Work

*Summary.* Although reasoners solve NP-hard problems, they are used across the research community to solve many tasks in artificial intelligence. Reasoners are also employed in industry to verify properties, generate tests, or run similar tasks.

In this paper, we introduced a simple and transparent approach to effectively speed up memory access of reasoners by reducing the number of TLB misses, which in turn allows to significantly improve their runtime. Our approach is based on a modification of glibc, which is the C standard library of the GNU Project and widely used in Linux for C and C++ programs. A user of a reasoner can benefit from our improvement simply by recompiling his favorite reasoner and enabling the feature by an environment flag when invoking it. Our experiments confirmed that an application can save up to 25% runtime on certain instances and on average more than 10%. Since the tools based on SAT solvers are often also used for long running jobs, for example in systems biology or verification, we can save a significant amount of runtime, and hence operational cost. In that way, the number of solved instances might not always be the right measure to evaluate a reasoner, but the overall runtime should be taken into account as well.

*Other Applications.* We believe that our approach can also be very beneficial for other tools in automated reasoning, because there are many memory intensive applications that have simply not been tuned to reduce the number of TLB misses by using THP. One such domain might be computing (hyper-)graph parameters, which employ SAT [25] or SMT [23,24,49]. Another domain might be graph algorithms [15,19,21], which can have random memory access patterns if the underlying data structure is updated often.

*Potential Benefits for CP Solvers.* Our current work is tailored to algorithms that heavily use unit propagation and two watched literal data structures. While we expect similar behavior for algorithms that employ similar features and memory usage behavior, many CP solvers implement propagators that are heavily computation intensive and less memory dependent. So far, we have no evidence that the proposed techniques improve such CP solvers. Still, it could be helpful for some solvers that use lazy clause generation. This, however, requires more detailed experimental evaluations on a variety of benchmarks.

*Industrial Effects.* We believe that our approach can be very beneficial for industry in settings where AR tools are used for continuous builds, testing, and integration. In particular, we expect that the technique proves useful when there are monetary demands introduced by using computation platforms, which allow for easy scalability, but bill resources by allocation time, such as Amazon Web Services, Microsoft Azure, and Google Cloud [4,5].

*Future Work.* In the future, we are interested in the influence of THP on various other domains and in increasing the amount of tested benchmarks and reasoning tools. We hope that this opens up both theoretical and practical research on more general algorithm engineering techniques.

## References

1. Arcangeli, A.: Transparent hugepage support. In: KVM forum, vol. 9 (2010)
2. Arnold, R.S., et al.: The GNU C library (glibc) (2019). https://www.gnu.org/software/libc/
3. Audemard, G., Simon, L.: Glucose in the SAT race 2019. In: Heule, M.J., Järvisalo, M., Suda, M. (eds.) Proceedings of SAT Race 2019: Solver and Benchmark Descriptions. Department of Computer Science Report Series, vol. B-2019-1, pp. 19–20. University of Helsinki (2019)
4. AWS: Astera labs uses AWS to accelerate chip development (2020). https://aws.amazon.com/solutions/case-studies/astera-labs/?did=cr_card&trk=cr_card
5. AWS: Aws customer success story: Zalando (2020). https://aws.amazon.com/solutions/case-studies/zalando/
6. Belov, A., Marques-Silva, J.: MUSer2: an efficient MUS extractor. J. Satisf. Boolean Model. Comput. **8**(3/4), 123–128 (2012)
7. Biere, A.: Lingeling essentials, a tutorial on design and implementation aspects of the the SAT solver Lingeling. In: Berre, D.L. (ed.) POS 2014, Fifth Pragmatics of SAT Workshop. EPiC Series in Computing, vol. 27, p. 88. EasyChair (2014). https://doi.org/10.29007/jhd7. https://easychair.org/publications/paper/xJs
8. Biere, A.: Splatz, Lingeling, Plingeling, Treengeling, YalSAT entering the SAT competition 2016. In: Balyo, T., Heule, M., Järvisalo, M. (eds.) Proceedings of SAT Competition 2016 - Solver and Benchmark Descriptions. Department of Computer Science Series of Publications B, vol. B-2016-1, pp. 44–45. University of Helsinki (2016)
9. Biere, A.: CaDiCaL, Lingeling, Plingeling, Treengeling, YalSAT entering the SAT competition 2017. In: Balyo, T., Heule, M., Järvisalo, M. (eds.) Proceedings of SAT Competition 2017 - Solver and Benchmark Descriptions. Department of Computer Science Series of Publications B, vol. B-2017-1, pp. 14–15. University of Helsinki (2017)
10. Biere, A., van Dijk, T., Heljanko, K.: Hardware model checking competition 2017. In: Stewart, D., Weissenbacher, G. (eds.) Formal Methods in Computer-Aided Design, FMCAD 2017, Vienna, Austria, 02–06 October 2017, p. 9. IEEE (2017)
11. Biere, A., Heljanko, K., Wieringa, S.: AIGER 1.9 and beyond. Technical report 11/2, Institute for Formal Models and Verification, Johannes Kepler University, Altenbergerstr. 69, 4040 Linz, Austria (2011)
12. Biere, A., Heule, M.: The effect of scrambling CNFs. In: Berre, D.L., Järvisalo, M. (eds.) Proceedings of Pragmatics of SAT 2015 and 2018. EPiC Series in Computing, vol. 59, pp. 111–126. EasyChair (2019)
13. Biere, A., Heule, M., van Maaren, H., Walsh, T. (eds.): Handbook of Satisfiability, Frontiers in Artificial Intelligence and Applications, vol. 185. IOS Press, Amsterdam (2009)
14. Bjørner, N.: SMT in verification, modeling, and testing at Microsoft. In: Biere, A., Nahir, A., Vos, T. (eds.) HVC 2012. LNCS, vol. 7857, p. 3. Springer, Heidelberg (2013). https://doi.org/10.1007/978-3-642-39611-3_3
15. Bonnet, É., Sikora, F.: The PACE 2018 parameterized algorithms and computational experiments challenge: the third iteration. In: Paul, C., Pilipczuk, M. (eds.) Proceedings of the 13th International Symposium on Parameterized and Exact Computation (IPEC 2018). Leibniz International Proceedings in Informatics (LIPIcs), vol. 115, pp. 26:1–26:15. Dagstuhl Publishing, Helsinki (2019). https://doi.org/10.4230/LIPIcs.IPEC.2018.26

16. Bornebusch, F., Wille, R., Drechsler, R.: Towards lightweight satisfiability solvers for self-verification. In: Proceedings of the 7th International Symposium on Embedded Computing and System Design (ISED 2017), pp. 1–5, December 2017. https://doi.org/10.1109/ISED.2017.8303924
17. Chu, G., Harwood, A., Stuckey, P.: Cache conscious data structures for Boolean satisfiability solvers. J. Satisf. Boolean Model. Comput. **6**, 99–120 (2009). https://doi.org/10.3233/SAT19006
18. Clarke, E., Kroening, D., Lerda, F.: A tool for checking ANSI-C programs. In: Jensen, K., Podelski, A. (eds.) TACAS 2004. LNCS, vol. 2988, pp. 168–176. Springer, Heidelberg (2004). https://doi.org/10.1007/978-3-540-24730-2_15
19. Dell, H., Komusiewicz, C., Talmon, N., Weller, M.: The pace 2017 parameterized algorithms and computational experiments challenge: the second iteration. In: Lokshtanov, D., Nishimura, N. (eds.) Proceedings of the 12th International Symposium on Parameterized and Exact Computation (IPEC 2017), pp. 30:1–30:13. Leibniz International Proceedings in Informatics (LIPIcs), Dagstuhl Publishing (2017). https://doi.org/10.4230/LIPIcs.IPEC.2017.30
20. D'Silva, V., Kroening, D., Weissenbacher, G.: A survey of automated techniques for formal software verification. IEEE Trans. Comput. Aided Des. Integr. Circuits Syst. **27**(7), 1165–1178 (2008)
21. Dzulfikar, M.A., Fichte, J.K., Hecher, M.: The PACE 2019 parameterized algorithms and computational experiments challenge: the fourth iteration (invited paper). In: Jansen, B.M.P., Telle, J.A. (eds.) 14th International Symposium on Parameterized and Exact Computation (IPEC 2019). Leibniz International Proceedings in Informatics (LIPIcs), vol. 148, pp. 25:1–25:23. Schloss Dagstuhl-Leibniz-Zentrum fuer Informatik, Dagstuhl, Germany (2019). https://doi.org/10.4230/LIPIcs.IPEC.2019.25. https://drops.dagstuhl.de/opus/volltexte/2019/11486
22. Eén, N., Sörensson, N.: An extensible SAT-solver. In: Giunchiglia, E., Tacchella, A. (eds.) SAT 2003. LNCS, vol. 2919, pp. 502–518. Springer, Heidelberg (2004). https://doi.org/10.1007/978-3-540-24605-3_37
23. Fichte, J.K., Hecher, M., Lodha, N., Szeider, S.: An SMT approach to fractional hypertree width. In: Hooker, J. (ed.) CP 2018. LNCS, vol. 11008, pp. 109–127. Springer, Cham (2018). https://doi.org/10.1007/978-3-319-98334-9_8
24. Fichte, J.K., Hecher, M., Szeider, S.: Breaking symmetries with RootClique and LexTopsort. In: Simonis, H. (ed.) CP 2020. LNCS, vol. 12333, pp. 286–303. Springer, Cham (2020)
25. Fichte, J.K., Lodha, N., Szeider, S.: SAT-based local improvement for finding tree decompositions of small width. In: Gaspers, S., Walsh, T. (eds.) SAT 2017. LNCS, vol. 10491, pp. 401–411. Springer, Cham (2017). https://doi.org/10.1007/978-3-319-66263-3_25
26. Gomes, C.P., Selman, B., Crato, N.: Heavy-tailed distributions in combinatorial search. In: Smolka, G. (ed.) CP 1997. LNCS, vol. 1330, pp. 121–135. Springer, Heidelberg (1997). https://doi.org/10.1007/BFb0017434
27. Gupta, A., Ganai, M.K., Wang, C.: SAT-based verification methods and applications in hardware verification. In: Bernardo, M., Cimatti, A. (eds.) SFM 2006. LNCS, vol. 3965, pp. 108–143. Springer, Heidelberg (2006). https://doi.org/10.1007/11757283_5
28. Hackenberg, D., Schöne, R., Ilsche, T., Molka, D., Schuchart, J., Geyer, R.: An energy efficiency feature survey of the intel Haswell processor. In: Lalande, J.F., Moh, T. (eds.) Proceedings of the 17th International Conference on High Performance Computing & Simulation (HPCS 2019) (2019)

29. Hölldobler, S., Manthey, N., Saptawijaya, A.: Improving resource-unaware SAT solvers. In: Fermüller, C.G., Voronkov, A. (eds.) LPAR 2010. LNCS, vol. 6397, pp. 519–534. Springer, Heidelberg (2010). https://doi.org/10.1007/978-3-642-16242-8_37
30. Hoos, H.H., Kaufmann, B., Schaub, T., Schneider, M.: Robust benchmark set selection for Boolean constraint solvers. In: Nicosia, G., Pardalos, P. (eds.) LION 2013. LNCS, vol. 7997, pp. 138–152. Springer, Heidelberg (2013). https://doi.org/10.1007/978-3-642-44973-4_16
31. Hykes, S., et al.: Docker CE (2019). https://github.com/docker/docker-ce
32. Intel: Intel® 64 and IA-32 Architectures Software Developer's Manual (2019). Order Number: 325462–069US
33. Järvisalo, M., Heule, M.J.H., Biere, A.: Inprocessing rules. In: Gramlich, B., Miller, D., Sattler, U. (eds.) IJCAR 2012. LNCS (LNAI), vol. 7364, pp. 355–370. Springer, Heidelberg (2012). https://doi.org/10.1007/978-3-642-31365-3_28
34. Katebi, H., Sakallah, K.A., Marques-Silva, J.P.: Empirical study of the anatomy of modern SAT solvers. In: Sakallah, K.A., Simon, L. (eds.) SAT 2011. LNCS, vol. 6695, pp. 343–356. Springer, Heidelberg (2011). https://doi.org/10.1007/978-3-642-21581-0_27
35. Kaufmann, B., Gebser, M., Kaminski, R., Schaub, T.: clasp - a conflict-driven nogood learning answer set solver (2015). http://www.cs.uni-potsdam.de/clasp/
36. Kleine Büning, H., Lettman, T.: Propositional Logic: Deduction and Algorithms. Cambridge University Press, Cambridge (1999)
37. Kochemazov, S., Zaikin, O., Kondratiev, V., Semenov, A.: MapleLCMDistChronoBT-DL, duplicate learnts heuristic-aided solvers at the SAT Race 2019. In: Heule, M.J., Järvisalo, M., Suda, M. (eds.) Proceedings of SAT Race 2019: Solver and Benchmark Descriptions. Department of Computer Science Report Series, vol. B-2019-1, pp. 24–24. University of Helsinki (2019)
38. van der Kouwe, E., Andriesse, D., Bos, H., Giuffrida, C., Heiser, G.: Benchmarking crimes: an emerging threat in systems security. CoRR abs/1801.02381 (2018). http://arxiv.org/abs/1801.02381
39. Kwon, Y., Yu, H., Peter, S., Rossbach, C.J., Witchel, E.: Coordinated and efficient huge page management with ingens. In: Proceedings of the 12th USENIX Conference on Operating Systems Design and Implementation (OSDI 2016), pp. 705–721. USENIX Association, Savannah (2016)
40. Luo, M., Li, C.M., Xiao, F., Manyà, F., Lü, Z.: An effective learnt clause minimization approach for CDCL SAT solvers. In: Proceedings of the Twenty-Sixth International Joint Conference on Artificial Intelligence, IJCAI 2017, pp. 703–711 (2017). https://doi.org/10.24963/ijcai.2017/98
41. Manthey, N.: MergeSAT. In: Heule, M.J., Järvisalo, M., Suda, M. (eds.) Proceedings of SAT Race 2019: Solver and Benchmark Descriptions. Department of Computer Science Report Series, vol. B-2019-1, pp. 29–30. University of Helsinki, Helsinki (2019)
42. Marques-Silva, J., Sakallah, K.: GRASP: a search algorithm for propositional satisfiability. IEEE Trans. Comput. **48**(5), 506–521 (1999). https://doi.org/10.1109/12.769433
43. Martins, R., Manquinho, V., Lynce, I.: Open-WBO: a modular MaxSAT solver'. In: Sinz, C., Egly, U. (eds.) SAT 2014. LNCS, vol. 8561, pp. 438–445. Springer, Cham (2014). https://doi.org/10.1007/978-3-319-09284-3_33
44. Merz, F., Falke, S., Sinz, C.: LLBMC: bounded model checking of C and C++ programs using a compiler IR. In: Joshi, R., Müller, P., Podelski, A. (eds.) VSTTE

2012. LNCS, vol. 7152, pp. 146–161. Springer, Heidelberg (2012). https://doi.org/10.1007/978-3-642-27705-4_12
45. Moskewicz, M.W., Madigan, C.F., Zhao, Y., Zhang, L., Malik, S.: Chaff: engineering an efficient SAT solver. In: Rabaey, J. (ed.) Proceedings of the 38th Annual Design Automation Conference (DAC 2001), pp. 530–535. Association for Computing Machinery, New York (2001). https://doi.org/10.1145/378239.379017
46. Navarro, J., Iyer, S., Druschel, P., Cox, A.: Practical, transparent operating system support for superpages. SIGOPS Oper. Syst. Rev. **36**(SI), 89–104 (2003). https://doi.org/10.1145/844128.844138. This paper describes Super Page implementation in FreeBSD. It also has performance numbers, but really ancient ones. They roughly match the SAT solver performance improvements, though
47. Panwar, A., Prasad, A., Gopinath, K.: Making huge pages actually useful. In: Bianchini, R., Sarkar, V. (eds.) Proceedings of the 23rd ACM International Conference on Architectural Support for Programming Languages and Operating Systems (ASPLOS 2018), pp. 679–692. Association for Computing Machinery, New York, March 2018. https://doi.org/10.1145/3173162.3173203
48. Park, S., Kim, M., Yeom, H.Y.: GCMA: guaranteed contiguous memory allocator. IEEE Trans. Comput. **68**(3), 390–401 (2019). https://doi.org/10.1109/TC.2018.2869169
49. Schidler, A., Szeider, S.: Computing optimal hypertree decompositions. In: Blelloch, G., Finocchi, I. (eds.) Proceedings of the 2020 Symposium on Algorithm Engineering and Experiments (ALENEX 2020), Salt Lake City, UT, USA (2020). https://doi.org/10.1137/1.9781611976007.1
50. Soos, M.: Cryptominisat 5.7.1 (2020). https://github.com/msoos/cryptominisat
51. Stump, A., Sutcliffe, G., Tinelli, C.: StarExec: a cross-community infrastructure for logic solving. In: Demri, S., Kapur, D., Weidenbach, C. (eds.) IJCAR 2014. LNCS (LNAI), vol. 8562, pp. 367–373. Springer, Cham (2014). https://doi.org/10.1007/978-3-319-08587-6_28. Held as Part of the Vienna Summer of Logic, VSL 2014
52. Torvalds, L.: kernel.org: transparent hugepage support, May 2017. https://www.kernel.org/doc/Documentation/vm/transhuge.txt
53. Voronkov, A.: AVATAR: the architecture for first-order theorem provers. In: Biere, A., Bloem, R. (eds.) CAV 2014. LNCS, vol. 8559, pp. 696–710. Springer, Cham (2014). https://doi.org/10.1007/978-3-319-08867-9_46. Held as Part of the Vienna Summer of Logic (VSL)
54. Wikichip, C.: Skylake (client) - Microarchitectures - Intel (2020). https://en.wikichip.org/wiki/intel/microarchitectures/skylake_(client)

# The Argmax Constraint

Graeme Gange and Peter J. Stuckey(✉)

Monash University, Melbourne, Australia
{graeme.gange,peter.stuckey}@monash.edu

**Abstract.** The ARGMAX function returns the index of the (first copy of the) maximum value occuring in a list of values. ARGMAX is important in models where we choose a characteristic value based on a seperate criteria, and for modelling neural networks which make use of ARGMAX in their definition. The ARGMAX constraint has been studied for the special case of its use to find the index of the first `true` in a list of Booleans, since this is useful for encoding if-then-else chains. Here we examine the general ARGMAX constraint for the first time. We define an efficient domain consistent propagator and a decomposition for integers which implements domain consistency.

## 1 Introduction

The ARGMAX function returns the index of the maximum value occuring in a list of values. Importantly, if two or more entries in the list take the maximum value it returns the index of the *first* such value. While ARGMAX is a well understood mathematical operator, it not nearly as widely used as, for example, MAX. In that sense it has not been carefully considered previously. Recently, a propagation algorithm for ARGMAX was developed for the restricted case of a Boolean array with known upper bound [12]. However, the general case of non-fixed numeric arguments has never been considered in detail.

ARGMAX is important as a constraint since it allows us to embed optimization problems as subproblems of another problem, where we want have access to the solution that leads to the optimal result, rather than just the optimal value.

*Example 1.* ARGMAX is used to model selecting a value based on a separate criteria. Let $(x_i, y_i)$ be pairs of possible values to be selected $x_i$ and the worthiness criteria for that selection $y_i$, then a model to select the best value $v$ is $v = x_{argmax\ y}$. More concretely, in MiniZinc this is modelled as

```
array[1..n] of var int: x;
array[1..n] of var int: y;
var int: v = x[arg_max(y)];
```

ARGMAX is increasingly important for discrete optimization because of its use in machine learning models. Many forms of machine learning rely on ARGMAX to define the result of a machine learning algorithm. If we want to use these models inside a discrete optimization problem then we need to be able to model the ARGMAX behaviour.


H. Simonis (Ed.): CP 2020, LNCS 12333, pp. 323–337, 2020.
https://doi.org/10.1007/978-3-030-58475-7_19

*Example 2.* An important modern use of ARGMAX is in the output layer of classification style neural nets, where the last layer is typically an argmax returning the class which is most highly likely to be present. A simple MiniZinc model for such a neural net layer is

```
array[CLASS] of var float: likelihood; % likelihood from NN of each class
var CLASS: x = arg_max(likelihood);     % most likely class
```

This is particularly important given the trend towards embedding classifiers as part of optimization models (e.g. [2]). But also given recent interest in techniques for *explainable AI* (e.g. [1,4]). Given a classifier $K$, observations (or *hypotheses*) $H$ and output class $c$, an *abductive explanation* [7] is a (minimal) subset of $H$ which still guarantees output class $c$. For multi-class classification problems, this requires querying whether $H' \wedge c \neq$ ARGMAX(*likelehood*) is satisfiable for various subsets $H'$ of $H$. Hence the need to reason about classifiers.

In this paper we investigate the ARGMAX constraint from a constraint programming perspective. The contributions of this paper are:

- A complete characterization of the domain consistent propagation possible from an ARGMAX constraint;
- An efficient domain consistent propagator for ARGMAX;
- A new decomposition for ARGMAX over integers which maintains domain consistency; and
- Experimental results showing the benefit of the new propagator and decomposition

The remainder of this paper is organized as follows. In the next section we introduce notation. In Sect. 3 we examine the current default decomposition for ARGMAX in the MiniZinc [9] library, which is the only existing ARGMAX implementation we are aware of, and highlight its shortcomings. In Sect. 4 we define all the propagation rules that are possible for an ARGMAX constraint, and indeed prove that applying them results in a domain consistent propagator. In Sect. 5 we define a new decomposition for ARGMAX which is domain consistent (under some easy to satisfy conditions about the constraints that implement it). In Sect. 6 we discuss a simpler variant of ARGMAX. In Sect. 7 we provide experiments to demonstrate the effectiveness of the new propagator and decomposition, on both unit tests and complete examples that make use of ARGMAX. Finally in Sect. 8 we conclude.

## 2 Preliminaries

A constraint problem $\mathcal{P} = \langle \mathcal{V}, D_{init}, \mathcal{C}, o \rangle$ consists of a set of variables $\mathcal{V}$, an initial domain $D_{init}$, a set of propagators for constraints $\mathcal{C}$ and an objective $o$ w.l.o.g. to be minimized. A *domain* $D$ is a mapping from each $v \in \mathcal{V}$ variables to a set of integer values $D(v)$, which defined the possible values that $v$ can take. We will use *range notation* $l..u$ to represent the set of integer $R = \{d \in \mathbb{Z} | l \leq d \wedge d \leq u\}$ to define domains.

An assignment $\theta$ is a mapping from $\mathcal{V}$ to integers. We say $\theta \in D$ iff $\forall v \in \mathcal{V}, \theta(v) \in D(v)$. An assignment satisfies a constraint $c \in \mathcal{C}$ if $\theta(c)$ is true. A *solution* of problem $\mathcal{P} = \langle \mathcal{V}, D_{init}, \mathcal{C}, o \rangle$ is an assignment $\theta \in D_{init}$ that satisfies all constraints $c \in \mathcal{C}$. An *optimal solution* $\theta$ of $\mathcal{P}$ is a solution of $\mathcal{P}$ such that for all other solutions $\theta'$ of $\mathcal{P}$, $\theta(o) \leq \theta'(o)$.

We will be interested in discussing propagation behaviour so we introduce some finer grained notation for reasoning about the current state of a CP solver. The least value a variable $x$ can take in domain $D$ is $\mathsf{lb}_D(x) = \min D(x)$, similarly the greatest value variable $x$ can take in domain $D$ is $\mathsf{ub}_D(x) = \max D(x)$. The *atomic constraints* for problem $\mathcal{P}$ are defined as

$$\{\mathit{false}\} \cup \{\langle x = d \rangle \mid x \in \mathcal{V}, d \in D_{init}(x)\} \cup \{\langle x \geq d \rangle \mid x \in \mathcal{V}, d \in D_{init}(x) - \{\mathsf{lb}_{D_{init}}(x)\}$$

We treat atomic constraints as propositions that reason about the current domain $D$. For example, $\langle x = d \rangle$ is true in the current domain $D$ if $D(x) = \{d\}$, false if $d \notin D(x)$ and unknown otherwise. Similarly $\langle x \geq d \rangle$ is true if $\mathsf{lb}_D(x) \geq d$, false if $\mathsf{ub}_D(x) < d$ and unknown otherwise.

For real variables[1] we need to introduce the additional open atomic constraints $\{\langle x > d \rangle \mid d \in D_{init}(x) - \{lb_{D_{init}(x)}(x)\}$. For integers $\langle x > d \rangle$ is just shorthand for $\langle x \geq d + 1 \rangle$.

We use the notation $\langle x \neq d \rangle$ as shorthand for $\neg \langle x = d \rangle$, $\langle x \leq d \rangle$ as shorthand for $\neg \langle x > d \rangle$, and $\langle x < d \rangle$ as shorthand for $\neg \langle x \geq d \rangle$.

Finally, we introduce the notation $x \ll^b y$ where $x$ and $y$ are numeric and $b$ is Boolean, which is shorthand for $x \leq y$ if $b = 0$ and $x < y$ where $b = 1$. That is $b$ is an indicator of the *strictness* of the comparison. $x \gg^b y$ is similarly defined. Note that for integers $x \ll^b y$ is equivalent to $x \leq y - b$, and $x \gg^b y$ is equivalent to $x \geq y + b$. The importance of this notation is that given an array $[x_1, \ldots, x_n]$ if $x_i \ll^{i<j} x_j, i \neq j$ then position $i$ cannot be the argmax of the array since $x_j$ is either equal and earlier or greater.

## 3 Current Argmax Decomposition

The standard MiniZinc encoding (as of `2.4.3`) of $\mathtt{z} = \mathit{argmax}([\mathtt{x}_1, \ldots, \mathtt{x}_n])$ is reasonably straightforward: it introduces auxiliary variables $m_i$ and $p_i$ respectively indicating the value and position of the maximum over the first $i$ elements, with constraints:

$$\begin{aligned} m_1 &= x_1 \wedge p_1 = 1 \\ m_{i+1} &= \max(m_i, x_{i+1}) \\ p_{i+1} &= \textbf{if } m_i < x_{i+1} \textbf{ then } i + 1 \textbf{ else } p_i \\ z &= p_n \end{aligned}$$

This encoding is sound, but unsatisfactory. Assuming an appropriate encoding of the conditional [12] it can propagate forwards to $z$, but neither enforces domain consistency on $z$ nor effectively incorporates knowledge of $z$ back into the bounds.

---
[1] Implemented by floating point ranges.

*Example 3.* Consider $\mathtt{z} = argmax([\mathtt{x}_1, \mathtt{x}_2, \mathtt{x}_3, \mathtt{x}_4])$, with current domains $\{z \mapsto \{1,3,4\}, x_1 \in 1..3, x_2 \in 2..10, x_3 \in 3..5, x_4 \in 4..6\}$. Using the decomposition above, we find:

| $i$ | 1 | 2 | 3 | 4 |
|---|---|---|---|---|
| $D(m_i)$ | 1..3 | 2..10 | 3..10 | 4..10 |
| $D(p_i)$ | {1} | 1..2 | 1..3 | {1, 3, 4} |

At this point, no further pruning can be performed. However, we miss several inferences. First, since $\mathsf{lb}_D(x_4) > \mathsf{ub}_D(x_1)$, we can infer $1 \notin D(z)$. And for $x_2$ *not* to be the max, it can be at most 5. □

The underlying problem with the decomposition is that it doesnt propagate information backwards about $p_i$. We know that $p_4 \neq 2$ which means $p_2 \neq 2$ which would allow the pruning of $x_2$.

# 4 Propagation Behaviour of Argmax

A domain consistent propagator for ARGMAX is the strongest possible implementation we can hope for in a constraint programming solver. The constraint $z = argmax([x_1, \ldots, x_n])$ can correctly propagate as follows:

- If a value $j$ is known not to be the maximum index value $z$, then it cannot take a value greater than (or equal to if it occurs earlier than the index of the max $ubi$) to the maximum possible value of the maximum of $x_1, \ldots, x_n$, $ub$:

$$\begin{aligned} \forall j.ubi = argmax_{i \in D(z)} \mathsf{ub}_D(x_i) \wedge ub = \mathsf{ub}_D(x_{ubi}) \wedge \langle z \neq j \rangle \\ \Rightarrow \langle x_j \ll^{j<ubi} ub \rangle \end{aligned} \quad (1)$$

- If a variable $x_j$'s maximum value is less than the least possible maximum value of $x_1, \ldots, x_n$, $lb$, (or equal to and it occurs at a later position than where this value first occurs $lbi$) then it cannot be the argmax position $z$:

$$\begin{aligned} \forall j.lbi = argmax_{i \in 1..n} \mathsf{lb}_D(x_i) \wedge lb = \mathsf{lb}_D(x_{lbi}) \\ \wedge \langle x_j \ll^{j<lbi} lb \rangle \Rightarrow \langle z \neq j \rangle \end{aligned} \quad (2)$$

- If the index of the max $z$ is known to be $j$ then $x_j$ cannot take values that would be too low to be consistent with the least possible value $lb$ of the max of $x_1, \ldots, x_n$:

$$\begin{aligned} D(z) = \{j\} \wedge lbi = argmax_{i \in 1..n} \mathsf{lb}_D(x_i) \wedge lb = \mathsf{lb}_D(x_{lbi}) \\ \Rightarrow \langle x_j \gg^{j>lbi} lb \rangle \end{aligned} \quad (3)$$

*Example 4.* Consider Example 3. Figure 1 illustrates the application of the propagation rules. Equation (1) finds $ubi = 4, ub = 6$ and prunes any values which would set $z$ to a removed value (marked in green), inferring $x_2 < 6$. Equation (2) finds $lbi = 4, lb = 4$ and removes any indices which can no longer become $lbi$, discovering $z \neq 1$ since $x_1 < 4$.

Suppose instead that also $\mathcal{D}(z) = 1$ then Equation 3 will apply enforcing that $x_1 \geq 4$ and thus causing failure. □

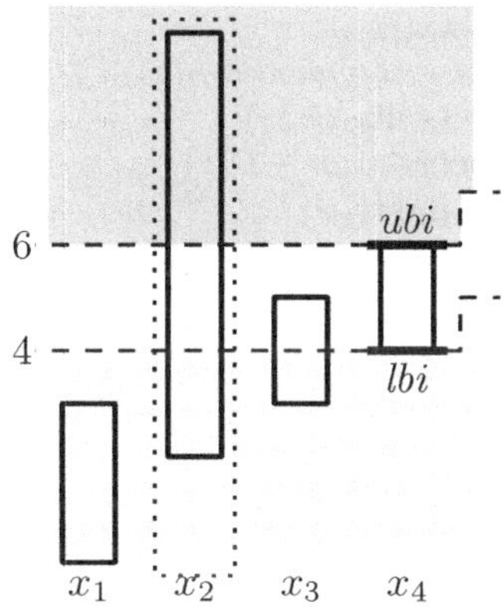


**Fig. 1.** Propagation of ARGMAX from domains in Example 3. 2 has already been removed from $\mathcal{D}(z)$, so cannot be chosen as $ubi$.

We can indeed show that these three rules define all the propagation behaviour we can correctly define for the constraint $z = argmax([x_1, \ldots, x_n])$.

**Theorem 1.** *Exhaustive application of the rules of Eqs. (1), (2) and (3) enforces domain consistency for* $z = argmax([x_1, \ldots, x_n])$.

*Proof.* Given domain $D$ which results from the exhaustive application of the rules of Eqs. (1), (2) and (3), we show that it provides a supporting solution for each value remaining in the domain.

Suppose $j \in D(z)$. We construct a solution $\theta \in D$ of the constraint where $z = j$. We claim that $\theta = \{z \mapsto j, x_j \mapsto \mathsf{ub}_D(x_j), x_i \mapsto \mathsf{lb}_D(x_i), i \neq j\}$ is a solution. Suppose to the contrary, then there exists $i \neq j$ where either $\mathsf{lb}_D(x_i) > \mathsf{ub}_D(x_j)$ or $\mathsf{lb}_D(x_i) = \mathsf{ub}_D(x_j) \wedge i < j$. Let $lbi$ and $lb$ be defined as in Eq. (2), then clearly $lb \geq \mathsf{lb}_D(x_i)$ and the conditions for the Eq. (2) hold, so $j \notin D(z)$. Contradiction.

Suppose $v \in D(x_j)$. We construct a solution $\theta \in D$ of the constraint where $x_j = v$. Let $ubi$ and $ub$ be defined as in (Eq. 1).

Consider the case where $j \neq ubi$. Consider the valuation $\theta = \{z \mapsto ubi, x_j \mapsto v, x_i \mapsto \mathsf{ub}_D(x_i), i \neq j\}$. We claim this is a solution. Suppose to the contrary then $v > \mathsf{ub}_D(x_{ubi})$ or $v = \mathsf{ub}_D(x_{ubi}) \wedge j < ubi$. For this to occur $j \notin D(z)$ otherwise its impossible by the definition of $ubi$ and $ub$. But then all the conditions of Eq. (1) apply and we would have imposed $x_j \ll^{j<ubi} ub$, meaning $v \notin D(x_j)$. Contradiction.

Consider the case where $j = ubi$. Consider the valuation $\theta = \{z \mapsto j, x_j \mapsto v, x_i \mapsto \mathsf{lb}_D(x_i)), i \neq j\}$. We claim this is a solution or we can construct an alternate. Note that $j = ubi \in D(z)$ by definition.

Suppose to the contrary $\theta$ is not a solution. Then there exists $i \neq j$ where either $\mathsf{lb}_D(x_i) > v$ or $\mathsf{lb}_D(x_i) = v \wedge i < j$. Let $lbi$ and $lb$ be defined as in Eq. (3), then clearly $lb \geq v$. There are two cases. If $D(z) = \{j\}$ then the conditions of Eq. (3) hold and we would have imposed $x_j \gg^{j>lbi} lb$, meaning $v \notin D(x_j)$. Contradiction.

```
ARGMAX([x_1, ..., x_n], z)
  ubi ← argmax(ub_D(x_i)|i ∈ D(z))
  ub ← ub_D(x_ubi)
  for(j ∈ 1..n − D(z))
    post(⟨x_j ≪^{j<ubi} ub⟩)
  lbi ← argmax(lb_D(x_i)|i ∈ 1..n)
  lb ← lb_D(x_lbi)
  for(j ∈ D(z))
    if(ub_D(x_j) ≪^{j<lbi} lb):
      post(⟨z ≠ j⟩)
  if(D(z) = {j})
    post(⟨x_j ≫^{lbi<j} lb_D(x_lbi)⟩)
```

**Fig. 2.** Basic propagator for $\mathbf{z} = argmax([\mathbf{x_1}, ..., \mathbf{x_n}])$.

Otherwise, there is some other value $j' \neq j, j' \in D(z)$. We construct $\theta' = \{z \mapsto j', x_j \mapsto v, x_{j'} \mapsto \mathsf{ub}_D(x'_j), x_i \mapsto \mathsf{lb}_D(x_i), i \notin \{j, j'\}\}$. Suppose to the contrary that $\theta'$ is not a solution, then there exists $i \neq j'$ where either $\mathsf{lb}_D(x_i) > \mathsf{ub}_D(x_{j'})$ or $\mathsf{lb}_D(x_i) = \mathsf{ub}_D(x_{j'}) \wedge i < j'$. Note that $v \leq lb$ so it cannot be the unique cause why $\mathsf{ub}_D(x_{j'})$ does not represent the max value assigned to an $x$ variable in $\theta'$. But this means the conditions of Eq. 2 apply for $j = j'$ which would mean we would propagate $z \neq j'$. Contradiction.

Since every value in the domain $D$ is supported by a solution $\theta$ of the constraint in $D$ the propagation enforces domain consistency. □

Theorem 1 shows that the propagation rules enforce domain consistency, but we need to enforce them efficiently. A straightforward propagator implementation is shown in Figure 2, more or less a direct implementation of rules (1)–(3). Note that **post**$(a)$ make the domain change given by the atomic constraint $a$. The propagator is clearly $O(n)$.

Table 1 implicitly defines an incremental propagator for the ARGMAX constraint. It shows how Eqs. (1–3) are applied on relevant domain change events. *lbi* and *ubi* are maintained between calls. Whenever the current *ubi* is (potentially) invalidated, we call filter-ub to compute an updated *ubi*, then apply Eq. (1) to tighten upper bonds. When the maximum lower bound increases, filter-z is called to apply Eq. (2) to prune any invalidated indices. Equation (2) is also applied upon $\langle x_i \leq k \rangle$, but only for the changed index. The complexity of the incremental propagator is still $O(n)$ since it may need to update *ubi* or filter multiple domains.

*Example 5.* Consider Example 3 where the initial domains are instead $\{z \in \{3, 4\}, x_1 \in 1..3, x_2 \in 2..5, x_3 \in 3..5, x_4 \in 4..6\}$ which is a fixpoint for the rules. The propagator state has $ubi = 4$, and $lbi = 4$. If we have new event $\langle x_2 \geq 3 \rangle$ then the test immediately fails and nothing happens. If we have new event $\langle x_1 \leq 2 \rangle$ similarly both test fails and nothing happens. On $\langle z \neq 4 \rangle$ we determine the new

**Table 1.** Propagation events for $\mathbf{z} = \mathbf{argmax}([\mathtt{x}_1, ..., \mathtt{x}_n])$.

| | |
|---|---|
| **on** $\langle \mathbf{z} \neq \mathbf{j} \rangle$:<br>**if**$(j = ubi)$:<br>  filter-ub$(\mathcal{D})$<br>**else**:<br>  post$(x_i \ll^{j<lbi} \mathsf{ub}_D(x_{ubi}))$<br>**if**$(\mathcal{D}(z) = \{j'\})$:<br>  filter-lb$(j', \mathcal{D})$ | filter-ub$(\mathcal{D})$:<br>  $ubi \leftarrow argmax(\mathsf{ub}_D(x_i) \mid i \in \mathcal{D}(z))$<br>  $ub \leftarrow \mathsf{ub}_D(x_{ubi})$<br>  **for**$(j \in 1..n - \mathcal{D}(z))$<br>    **post**$(\langle x_j \ll^{j<ubi} ub \rangle)$ |
| **on** $\langle \mathbf{x_j} \leq \mathbf{k} \rangle$:<br>**if**$(j = ubi)$:<br>  filter-ub$(\mathcal{D})$<br>**if**$(k \ll^{j<lbi} \mathsf{lb}_D(x_{lbi}))$:<br>  post$(z \neq j)$ | filter-z$(\mathcal{D})$:<br>  **for**$(j \in \mathcal{D}(z))$:<br>    **if**$(\mathsf{ub}_D(x_j) \ll^{j<lbi} lb)$:<br>      **post**$(\langle z \neq j \rangle)$ |
| **on** $\langle \mathbf{x_j} \geq \mathbf{k} \rangle$:<br>**if**$(\neg(k \ll^{j \leq lbi} \mathsf{lb}_D(x_{lbi})))$<br>  $lbi \leftarrow j$<br>  **if**$(\mathcal{D}(z) = \{j'\})$<br>    filter-lb$(j', \mathcal{D})$<br>  **else**<br>    filter-z$(\mathcal{D})$ | filter-lb$(j, \mathcal{D})$:<br>  **post**$(\langle x_j \gg^{lbi<j} \mathsf{lb}_D(x_{lbi}) \rangle)$<br>  $lbi \leftarrow j$ |

upper bound $ub = 5$ and $ubi = 3$ and post $\langle x_4 \leq 5 \rangle$ and since $D(z)$ is now a singleton we also post $\langle x_3 \geq 4 \rangle$. □

As mentioned above, $lbi$ and $ubi$ are perserved between propagator calls. In practice, we also persistently track the set $I_r = \{i \in 1..n - D(z) \mid \neg(ub(x_i) \ll^{i<lbi} lb(x_{lbi}))\}$. This is the set of indices which may be updated by FILTER-UB: they have been removed from $\mathcal{D}(z)$, but would otherwise be candidates for the maximum index. Then in FILTER-UB we iterate over $I_r$ directly, skipping irrelevant indices.

*Explaining Propagation.* In any lazy clause generation [10] solver (as we shall use in Sect. 7), a propagator for constraint $c$ must provide an *explanation* for any inference $\varphi$: a conjunction $E$ of atomic constraints such that $D \rightarrow E$, and $E \wedge c \rightarrow \varphi$. Fortunately, for ARGMAX the explanations are readily extracted from Eqs. (1–3). The explanation for Eq. (1) for $\langle x_j \ll^{j<ubi} ub \rangle$ is the most involved, as we need to justify *why* $ub(x_{ubi})$ takes its value. That is, for each index, why it does not support a greater upper bound. The explanation becomes:

$$\forall_{k \in \mathcal{D}(z)} \langle x_k \ll^{k<ubi} ub \rangle \wedge \forall_{k \notin \mathcal{D}(z)} \langle z \neq k \rangle \rightarrow \langle x_j \ll^{j<ubi} ub \rangle .$$

The explanation from Eq. (2) as to why $\langle z \neq j \rangle$ is that there was some other index $k$ such that $lb(x_k) \gg^{k>j} \mathsf{ub}_D(x_j)$. At the time of propagation, $lbi$ is always such an index, so our explanation is

$$\langle x_j \ll^{j<lbi} lb \rangle \wedge \langle x_{lbi} \geq lb \rangle \rightarrow \langle z \neq j \rangle .$$

Similarly, the explanation from Eq. (3) for $\langle x_j \gg^{lbi>j} c\rangle$ is always

$$\langle z = j\rangle \wedge \neg\langle x_{lbi} \ll^{lbi<j} c\rangle \rightarrow \langle x_j \gg^{lbi>j} c\rangle .$$

*Example 6.* Examining the propagation in Example 4, the explanation for $\langle x_2 \leq 5\rangle$ is $\langle x_1 < 6\rangle \wedge \langle x_3 \leq 6\rangle \wedge \langle x_4 \leq 6\rangle \wedge \langle z \neq 2\rangle \rightarrow \langle x_2 < 6\rangle$, while the explanation for $\langle z \neq 1\rangle$ is $\langle x_1 < 4\rangle \wedge \langle x_4 \geq 4\rangle \rightarrow \langle z \neq 1\rangle$. □

## 5 Decomposition

For the case of integer values, can produce a better decomposition of the ARGMAX propagator by reasoning lexicographically about the pairs $(x_j, -j)$. The maximum of these pairs defines the position and value of the max. Using this idea results in the decomposition below. We introduce auxiliary variables $q_j$ to represent the value of the pair $(x_j, -j)$ as a single integer $n \times x_j + n - j$. We introduce $M$ to represent the maximum value of the pairs. The decomposition is:

$$\begin{aligned} q_j &= n \times x_j + n - j, \quad 1 \leq j \leq n \\ M &= max([q_1, \ldots, q_n]) \\ z \neq j &\leftrightarrow M > q_j, \quad 1 \leq j \leq n \end{aligned}$$

Note that $q_i < q_j$ iff $x_i \ll^{i<j} x_j$. (We cannot do the same for real or floating point values, as we cannot readily define a view $T$ s.t. $T(x_i, i) < T(x_j, j) \equiv x_i \ll^{i<j} x_j$). Note that the $q_j$ variables can be defined by variable views [11] and hence dont exist as separate variables, thus the main cost of the decomposition is the global `max` constraint and the reified inequalities $b_j \leftrightarrow M > q_j$, where $b_j$ is equated to $\langle z \neq j\rangle$.

**Theorem 2.** *The decomposition defined above implements domain consistency for the constraint* $z =$ ARGMAX$([x_1, \ldots, x_n]$, *assuming bounds consistent propagation of* **max** *and the reified inequalities, and domain consistent propagation of* $q_j = n \times x_j + (u - j)$.

*Proof.* We show that it implements all the propagation behaviour discussed in Sect. 4. Note that $q_j = n \times x_j + (n - j)$ is often implemented by a view [11] which is equivalent to domain consistent propagation.

Examining Eq. (1) and give a domain $D$ where $ubi$ and $ub$ are defined as shown in the equation. Suppose the value $\mathsf{ub}_D(M)$ arises from position $i$. If $i \in D(z)$ then $ubi = i$ by definition and bounds consistency of the max and $\mathsf{ub}_D(M) = n \times ub - (n - ubi)$. The fact that $z \neq j$ will immediately assert $q_j < n \times ub + (n - ubi)$ which is equivalent to $x_j \leq ub - (j < ubi)$. This holds since the propagation of $q_j = n \times x_j + (n - j)$ is domain consistent. Now consider when $i \notin D(z)$. Then by bounds consistency of the max $\mathsf{ub}_D(M) = n \times \mathsf{ub}_D(x_i) - (n - i)$. But since $z \neq i$ the decomposition enforces $q_i < n \times \mathsf{ub}_D(x_i) - (n - i)$ or equivalently $x_i < \mathsf{ub}_D(x_i)$ by the domain consistent propagation of $q_j = n \times x_j + (n - j)$. This reduces the upper bound of $x_i$. This process will continue until the (new) index $i$ giving the max value for $M$ is in $D(z)$ otherwise its upper bound will also be reduced. So eventually we hit the first case above. Hence Eq. (1) is implemented by the decomposition.

Examining Eq. (2) and give a domain $D$ where $lbi$ and $lb$ are defined as shown in the equation. By bounds consistency of the max $\mathsf{lb}_D(M) = n \times lb + (n - lbi)$. The fact that $x_{lbi} \leq lb - (j < lbi)$ means that $q_{lbi} < n \times lb + (n - lbi)$ is known to be true (by domain consistency of $q_i = n \times x_j + (n - j)$), meaning that the constraint will immediately assert $z \neq j$. Hence Eq. (2) is implemented by the decomposition.

Examining Eq. (3) and give a domain $D$ where $lbi$ and $lb$ are defined as shown in the equation. By bounds consistency of the max $\mathsf{lb}_D(M) = n \times lb + (n - lbi)$. Making $z = j$ enforces $M \leq q_j$ which given the bound on $M$ bounds consistency enforces (for all $q_k$ and in particular) $q_j \geq n \times lb + (n - lbi)$ which is by domain propagation of $q_i = n \times x_j + (n - j)$ equivalent to $x_j \geq lb + (lbi < j)$. Hence Eq. (3) is implemented by the decomposition. □

The decomposition has an advantage over the propagator since the introduced variables $M$ and $q$ allow for concise explanations.

*Example 7.* Consider Example 3. Using the new decomposition we the definition of the new variables gives domains.

| $i$ | 1 | 2 | 3 | 4 | $M$ |
|---|---|---|---|---|---|
| $D(q_i)$ | 7..15 | 10..42 | 13..21 | 16..24 | 16..42 |

The fact that $q_1 < M$ must hold enforces $\langle z \neq 1 \rangle$. The explanation is simply $\langle q_1 \leq 15 \rangle \wedge \langle M \geq 16 \rangle \rightarrow \langle z \neq 1 \rangle$. Assuming the $q$ are defined by views this will be $\langle x_1 \leq 3 \rangle \wedge \langle M \geq 16 \rangle \rightarrow \langle z \neq 1 \rangle$. Note that this is more reusable than the explanation by the propagator which commits to a single variable $x_4$.

Then $\langle z \neq 2 \rangle$ enforces that $\langle q_2 < 42 \rangle$ which removes the top value from its domain, this then changes the domain of $x_2$ to 2..9 and $q_2$ to 10..38, the process repeats until the domain of $x_2$ becomes 2..5 and $q_2$ is 10..22 and $M$ is 16..24. The explanation chain is long, the final explanation for $\langle x_2 \leq 5 \rangle$ is $\langle z \neq 2 \rangle \wedge \langle M \leq 26 \rangle \rightarrow \langle x_2 \leq 5 \rangle$, which again is agnostic about the reason for the upper bound on $M$. □

## 6 Argmax Variant

There is a very similar constraint to ARGMAX in the global constraint catalog, called `max_index` [8]. This relation is defined as

$$max\_index(z, [x_1, \ldots, x_n]) \equiv x_z = max([x_1, \ldots, x_n])$$

so it simply requires $z$ to be *an* index where the maximum value occurs, not the first. This means it is not functional, and cannot thus be used for encoding conditional constructs like ARGMAX. The simple definition by decomposition above is not domain consistent.

*Example 8.* Consider the problem of Example 3 where the constraint is instead max_index(z, [$\texttt{x}_1, \texttt{x}_2, \texttt{x}_3, \texttt{x}_4$]). Then the decomposition determines $max(x) \in 4..10$ and propagates $z \neq 1$. It does not determine that $x_2 \leq 6$ which is a consequence of the constraint since if $x_2 = 7$ then $z$ could not be the correct index 2. □

We can define the domain consistent propagator for max_index by using rules analogous to rules (1)–(3).

- If a value $j$ is known not to be the maximum index value $z$, then it cannot take a value greater than the maximum possible value of the maximum of $x_1, \ldots, x_n$, $ub$:

$$\forall j.ub = \max_{i \in D(z)} \mathsf{ub}_D(x_i) \wedge \langle z \neq j \rangle \Rightarrow \langle x_j \leq ub \rangle \quad (4)$$

- If a variable $x_j$'s maximum value is less than the least possible maximum value of $x_1, \ldots, x_n$, $lb$, then it cannot be the argmax position $z$:

$$\forall j.lb = \max_{i \in 1..n} \mathsf{lb}_D(x_i) \wedge \langle x_j < lb \rangle \Rightarrow \langle z \neq j \rangle \quad (5)$$

- If the index of the max $z$ is known to be $j$ then $x_j$ cannot take values that would be too low to be consistent with the least possible value $lb$ of the max of $x_1, \ldots, x_n$:

$$D(z) = \{j\} \wedge lb = \max_{i \in 1..n} \mathsf{lb}_D(x_i) \Rightarrow \langle x_j \geq lb \rangle \quad (6)$$

We can straightforwardly modify the propagator of Fig. 1 and incremental update rules of Table 1 to implement this simpler variant.

## 7 Experimental Evaluation

We have implemented the ARGMAX propagator presented in Sect. 4 in geas [5], a lazy clause generation-based CP solver. We compare the standard MiniZinc decomposition DEC with the new propagator PROP and the new decomposition NEW presented in Sect. 5. Experiments were conducted on a Core i7 7820-HQ 2.9 Ghz with 32 Gb ram, running Ubuntu 18.04. All experiments were conducted with a 600 s timeout. All times are in seconds; a '—' entry indicates a timeout. Models and instances used in this section are available at https://gkgange.github.io/papers/2020/argmax.

### 7.1 Unit Tests

Let $\pi$ be an $n \times n$ matrix of values generated randomly in $[1, n]$. We construct a constraint system:

```
forall (i in 1..n) (x[i] = arg_max( [ pi[i, j] * x[j] | j in 1..n ] ));
alldifferent(x);
```

| $n$ | DEC | PROP | NEW |
|---|---|---|---|
| 30 | 4.1 | **0.07** | 0.11 |
| 40 | 57.16 | **0.14** | 0.25 |
| 50 | 362.71 | **0.86** | 1.32 |
| 60 | — | **3.88** | 5.72 |
| 70 | — | **5.96** | 9.89 |
| 80 | — | **21.82** | 31.67 |
| 90 | — | **43.93** | 82.46 |
| 100 | — | **152.32** | 224.06 |
| 110 | — | **258.95** | 345.13 |
| 120 | — | — | — |

**Fig. 3.** Unit-test performance for ARGMAX on integer variables, for increasing $n$.

The constraint system is vanishingly unlikely to be satisfiable (and indeed this occurs in none of our tests).

Results comparing the time for the three ARGMAX implementations to prove unsatisfiability are given in Fig. 3, for increasing values of $n$. The standard decomposition is evidently not competitive. For these problems, the faster propagation speed of the propagator is more essential than the search reduction from more expressive nogoods generated by the decomposition. Its often the case that nogood learning is of limited utility for random problems.

### 7.2 Wireless Signal Selection

We consider a modified facility location problem, modelling a WLAN planning task. The problem is to assign signal strengths to a set of access points $A$, maximizing connection quality for a set of clients $C$, subject to:

- Each client connects to the access point with maximum (observed) signal strength;
- Each client has a desired throughput; and
- Each access point has a maximum total throughput; all connected clients incur a penalty if their combined bandwidth demands are not met.

We generated instances by placing $n$ access points and $m$ clients randomly on a $70 \times 70$ grid, for $n \in \{10, 15, 20\}$ and $m \in \{40, 55, 70, 85\}$. Each client is assigned a random throughput chosen uniformly in [1, 100], and each access point has maximum throughput 300. The ARGMAX constraint is used for the first constraint.

The results shown in Fig. 4 given the average time on 10 instances of each size, and in parentheses the number of the 10 instances that succeeded. They clearly demonstrate that the new propagator and decomposition are substantially better than the current decomposition. Here the intermediate variable $M$ provides more reusable nogoods, meaning the new decomposition improves over the propagator.

| APs | clients | DEC | PROP | NEW |
|---|---|---|---|---|
| 10 | 40 | 22.81 (10) | 1.87 (10) | **1.40** (10) |
| 10 | 55 | 264.87 (9) | 91.47 (10) | **31.01** (10) |
| 10 | 70 | 570.91 (2) | 416.46 (7) | **173.07** (10) |
| 10 | 85 | — (0) | 564.23 (1) | **461.60** (6) |
| 15 | 40 | 49.32 (10) | 3.86 (10) | **2.00** (10) |
| 15 | 55 | 437.44 (5) | 185.84 (8) | **78.14** (10) |
| 15 | 70 | 583.80 (1) | 403.88 (4) | **310.15** (7) |
| 15 | 85 | — (0) | 384.45 (5) | **164.66** (10) |
| 20 | 40 | 78.58 (10) | 7.03 (10) | **5.81** (10) |
| 20 | 55 | 366.93 (7) | 96.25 (9) | **55.10** (10) |
| 20 | 70 | 545.76 (3) | 303.65 (7) | **138.06** (9) |
| 20 | 85 | — (0) | 574.85 (1) | **519.16** (2) |

**Fig. 4.** Comparison of methods on wireless signal selection problems of various sizes. Columns give average runtime over instances of each size (timeouts are counted as 600 s), and number of instances solved (out of 10).

### 7.3 Boosted Tree Explanation

For this experiment, we considered the SOYBEAN and CHESS (KING-ROOK VS. KING) classification tasks from the UCI Machine Learning Repository.[2] These tasks have 35 and 6 attributes respectively, and both have 18 possible output classes. We generated boosted-tree models using XGBOOST [3], consisting of 50 trees per class, and encoded the resulting classifiers into MiniZinc. We then took the first 300 instances from each data-set set, and derived cardinality-minimal explanations for the observed class label using GEAS' core-guided optimization capabilities [5] to emulate the implicit hitting-set method of [6].

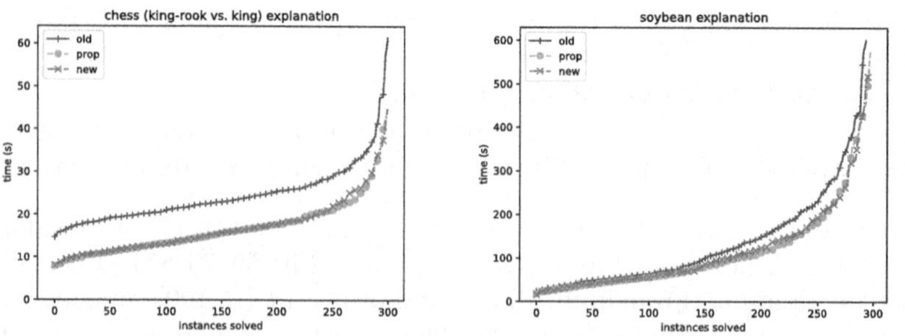


**Fig. 5.** Time to compute cardinality-minimal explanations for instances of the SOYBEAN and CHESS classification tasks.

These instances contain only a single ARGMAX constraint, at the head of a very large system of ELEMENT and LINEAR constraints. Here, the ARGMAX

[2] https://archive.ics.uci.edu/ml/.

| $n$ | B-DEC | B-PROP | NEW |
|---|---|---|---|
| 25 | **0.00** | **0.00** | **0.00** |
| 50 | **0.00** | **0.00** | **0.00** |
| 100 | **0.00** | **0.00** | 0.01 |
| 200 | 0.01 | **0.00** | 0.01 |
| 400 | 0.01 | **0.00** | 0.01 |
| 800 | 0.01 | **0.00** | 0.03 |
| 1600 | 0.01 | **0.01** | 0.05 |
| 3200 | 0.03 | **0.01** | 0.10 |
| 6400 | 0.05 | **0.02** | 0.25 |

(a) Unit test from [12]

| $n$ | B-DEC | B-PROP | NEW |
|---|---|---|---|
| 25 | **0.00** | **0.00** | **0.00** |
| 50 | 0.01 | **0.00** | 0.01 |
| 100 | 0.02 | **0.01** | 0.02 |
| 200 | 0.04 | **0.02** | **0.02** |
| 400 | 0.13 | 0.08 | **0.04** |
| 800 | 0.51 | 0.36 | **0.09** |
| 1600 | 18.69 | **2.05** | 7.53 |
| 3200 | 121.64 | **13.05** | 33.08 |
| 6400 | - | **59.09** | 127.88 |

(b) Alternate unit test.

**Fig. 6.** Performance of the new decomposition for the specialized Boolean case of ARGMAX.

propagation constitutes a smaller fraction of runtime – the model spends much more time in the score computations – but its *strength* is still important. On these instances, we see that the propagator and new decomposition perform similarly, both of which demonstrate advantage over the old decomposition.

### 7.4 Boolean Arg-Max

Given that the decomposition of Sect. 5 is domain consistent, it is interesting to consider how it compares to the specialised decomposition of [12] which is also domain consistent for the case of a Boolean array with at least one *true* value (Fig. 5).

Figure 6 compares the performance of our new decomposition with the decomposition (B-DEC) and propagator (B-PROP) presented in [12]. Figure 6(a) give results on the unit-test given in [12]. On this, the existing decomposition appears better. However, we note that all solvers prove this problem unsatisfiable without search – so overhead in setting up the constraints plays a major factor. (b) reports performance on the following constraint system:

```
array[1..n] of int: d1 = [ (23*i+2) mod n + 1 | i in 1..n];
array[1..n+1] of int: t1 = [ (37*i) mod n + 1 | i in 1..n+1];

array[1..n] of int: d2 = [ (17*i+3) mod n + 1 | i in 1..n];
array[1..n+1] of int: t2 = [ (31*i) mod n + 1 | i in 1..n+1];
  y = t1[ arg_max([ x = d1[i] | i in 1..n] ++ [true]) ];
  x = t2[ arg_max([ y = d2[i] | i in 1..n] ++ [true]) ];
```

This system is still unsatisfiable, but the unsatisfiability is not discovered at the root. In this case, the results are rather different. The new decomposition shows a $2 - 4\times$ speedup compared with the existing decomposition. However, the dedicated propagator still has a clear advantage on larger instances.

## 8 Conclusion

The ARGMAX constraint is a valuable if not well studied constraint. The Boolean version is important for `if-then-else-endif` constructs, and has recently been improved [12]. In this paper we define the domain consistent propagation possible for ARGMAX and then devise a propagator, and domain consistent decomposition for the integer case. We show that these markedly improve on the default decomposition. The decomposition is usually superior for integer problems in learning solvers which can take advantage of the reasoning about the introduced variable $M$. The propagator is superior in contexts where the new decomposition cannot be used (e.g. for floating point/rational domains), and presumably also for non-learning solvers which cannot take advantage of the richer language of learning of the decomposition.

## References

1. Arrieta, A.B., et al.: Explainable artificial intelligence (XAI): concepts, taxonomies, opportunities and challenges toward responsible AI. Inf. Fusion **58**, 82–115 (2020). https://doi.org/10.1016/j.inffus.2019.12.012. http://www.sciencedirect.com/science/article/pii/S1566253519308103
2. Bartolini, A., Lombardi, M., Milano, M., Benini, L.: Neuron constraints to model complex real-world problems. In: Lee, J. (ed.) CP 2011. LNCS, vol. 6876, pp. 115–129. Springer, Heidelberg (2011). https://doi.org/10.1007/978-3-642-23786-7_11
3. Chen, T., Guestrin, C.: XGBoost: a scalable tree boosting system. In: Krishnapuram, B., Shah, M., Smola, A.J., Aggarwal, C.C., Shen, D., Rastogi, R. (eds.) Proceedings of the 22nd ACM SIGKDD International Conference on Knowledge Discovery and Data Mining, San Francisco, CA, USA, 13–17 August 2016, pp. 785–794. ACM (2016). https://doi.org/10.1145/2939672.2939785
4. Explainable artificial intelligence: Darpa program (2020). http://darpa.mil/program/explainable-artificial-intelligence
5. Gange, G., Berg, J., Demirović, E., Stuckey, P.J.: Core-guided and core-boosted search for constraint programming. In: Hebrard, E., Musliu, N. (eds.) Proceedings of Seventeenth International Conference on Integration of Artificial Intelligence and Operations Research techniques in Constraint Programming (CPAIOR2020). Springer (2020, to appear)
6. Ignatiev, A., Morgado, A., Marques-Silva, J.: Propositional abduction with implicit hitting sets. In: Kaminka, G.A., et al. (eds.) ECAI 2016–22nd European Conference on Artificial Intelligence, The Hague, The Netherlands, 29 August–2 September 2016 - Including Prestigious Applications of Artificial Intelligence (PAIS 2016). Frontiers in Artificial Intelligence and Applications, vol. 285, pp. 1327–1335. IOS Press (2016). https://doi.org/10.3233/978-1-61499-672-9-1327
7. Ignatiev, A., Narodytska, N., Marques-Silva, J.: Abduction-based explanations for machine learning models. In: The Thirty-Third AAAI Conference on Artificial Intelligence, AAAI 2019, The Thirty-First Innovative Applications of Artificial Intelligence Conference, IAAI 2019, The Ninth AAAI Symposium on Educational Advances in Artificial Intelligence, EAAI 2019, Honolulu, Hawaii, USA, 27 January–1 February 2019, pp. 1511–1519. AAAI Press (2019). https://doi.org/10.1609/aaai.v33i01.33011511
8. Global constraint catalog: `max_index` (2020). https://web.imt-atlantique.fr/x-info/sdemasse/gccat/Cmax_index.html

9. Nethercote, N., Stuckey, P.J., Becket, R., Brand, S., Duck, G.J., Tack, G.: MiniZinc: towards a standard CP modelling language. In: Bessière, C. (ed.) CP 2007. LNCS, vol. 4741, pp. 529–543. Springer, Heidelberg (2007). https://doi.org/10.1007/978-3-540-74970-7_38
10. Ohrimenko, O., Stuckey, P., Codish, M.: Propagation via lazy clause generation. Constraints **14**(3), 357–391 (2009). https://doi.org/10.1007/s10601-008-9064-x
11. Schulte, C., Tack, G.: Views and iterators for generic constraint implementations. In: van Beek, P. (ed.) CP 2005. LNCS, vol. 3709, pp. 817–821. Springer, Heidelberg (2005). https://doi.org/10.1007/11564751_71
12. Stuckey, P.J., Tack, G.: Compiling conditional constraints. In: Schiex, T., de Givry, S. (eds.) CP 2019. LNCS, vol. 11802, pp. 384–400. Springer, Cham (2019). https://doi.org/10.1007/978-3-030-30048-7_23

# Certifying Solvers for Clique and Maximum Common (Connected) Subgraph Problems

Stephan Gocht[1,2], Ross McBride[3], Ciaran McCreesh[3](✉), Jakob Nordström[1,2], Patrick Prosser[3], and James Trimble[3]

[1] Lund University, Lund, Sweden
stephan.gocht@cs.lth.se
[2] University of Copenhagen, Copenhagen, Denmark
jn@di.ku.dk
[3] University of Glasgow, Glasgow, Scotland
{ciaran.mccreesh,patrick.prosser}@glasgow.ac.uk,
j.trimble.1@research.gla.ac.uk

**Abstract.** An algorithm is said to be *certifying* if it outputs, together with a solution to the problem it solves, a proof that this solution is correct. We explain how state of the art maximum clique, maximum weighted clique, maximal clique enumeration and maximum common (connected) induced subgraph algorithms can be turned into certifying solvers by using pseudo-Boolean models and cutting planes proofs, and demonstrate that this approach can also handle reductions between problems. The generality of our results suggests that this method is ready for widespread adoption in solvers for combinatorial graph problems.

## 1 Introduction

McConnell et al. [40] argue that all algorithm implementations should be *certifying*: that is, along with their output, they should produce an easily verified proof that the output is correct. Given the relative frequency of bugs in constraint programming (CP) solvers and in dedicated algorithms for hard combinatorial problems [7,12,25,42], it would be desirable to see certification becoming a social requirement for all new solvers—as has already happened in the Boolean satisfiability community through proof logging formats such as RUP [27], TraceCheck [5], DRAT [29,30,74], LRAT [13] and GRIT [14]. A proof log is a particular kind of certificate which records the steps taken by a solver in such a way that the correctness of each step can easily be checked, given that

The first and fourth authors were funded by the Swedish Research Council (VR) grant 2016-00782. The fourth author was also supported by the Independent Research Fund Denmark (DFF) grant 9040-00389B. The third, fifth and sixth authors were supported by the Engineering and Physical Sciences Research Council [grant numbers EP/P026842/1 and EP/M508056/1]. Some code development used resources provided by the Swedish National Infrastructure for Computing (SNIC).


H. Simonis (Ed.): CP 2020, LNCS 12333, pp. 338–357, 2020.
https://doi.org/10.1007/978-3-030-58475-7_20

all previous steps are known to be correct; the intent is that verifying a proof log should be very simple, even if solvers carry out complex reasoning.

Until recently, it was generally assumed that proof logging for more powerful CP-style solvers would require either very complicated (and hard to verify) certificates that must be aware of every kind of propagation performed by every constraint [72], or an exponential slowdown [22]. However, it has recently been shown that reasoning over pseudo-Boolean formulae can compactly express all-different reasoning [19], as well as all of the reasoning carried out by state-of-the-art subgraph isomorphism solvers [26], despite pseudo-Boolean reasoning not knowing anything about Hall sets, matchings, vertices, degrees, or paths.

The general idea behind this proof logging is that a constraint satisfaction problem (or other hard problem) is compiled to a pseudo-Boolean (PB) formula—that is, a 0–1 integer linear program. Then, either a witness of satisfiability is provided, or a proof showing that the PB formula implies $0 \geq 1$ is given. (Optimisation and enumeration problems are also supported.) The proofs of unsatisfiability demonstrated so far have consisted of a mix of "reverse unit propagation" (RUP) steps [19,24] to describe the backtracking search tree produced by the solver, and assistance in deriving any information used by propagators that is not immediately apparent to unit propagation (such as Hall sets and Hall violators). In this work, we report that this approach can be used in a more general way to obtain certifying algorithms (with proofs that can be checked by the VeriPB verifier [19]) for a range of other hard problems:

- We show how a wide variety of maximum clique algorithms from the literature can all be enhanced with proof logging, using very similar proof techniques. We also explain how to adapt this proof logging method to cover the inference used by a state of the art maximum weight clique solver. Finally, we discuss certification for all maximal clique enumeration algorithms.
- We also demonstrate proof logging for a state of the art CP-style maximum common induced subgraph algorithm, including for the connected variant of the problem.
- Finally, we look at a reduction from maximum common induced subgraph to maximum clique, which outperforms CP approaches on certain graph classes. We show that this reduction can be expressed inside the proof log, so we can take a PB model that was generated for the CP encoding, but then provide a proof from a clique algorithm—this is a bit like channelling [8], but for proofs. There are also clique-like algorithms with a propagator to enforce connectedness. Because the reduction can be viewed as a bijection, we can continue to express the connectedness constraint only on the CP encoding (where it is much easier to understand than on the clique encoding), but still validate clique-like proofs.

Our main conclusion is that proof logging using pseudo-Boolean reasoning is general and powerful enough to concisely describe the inference used in a wide range of combinatorial graph algorithms. Although the current implementation does not scale well enough to deal with the largest instances, it can already be used to provide, for the first time, fully verifiable proofs of correctness for some highly nontrivial medium-sized instances. Also, even if the overhead is currently too high to have proof logging switched on by default in production, it provides an excellent tool for debugging of nontrivial optimisation techniques during solver development. This is because incorrectly implemented steps are likely to lead to incorrect proofs, which can be detected even when the results produced by the solver happen to be correct. We believe this tool is mature enough for widespread adoption, and that requiring all solvers be able to output proofs would be a natural and desirable step to increase the confidence in the correctness of state-of-the-art solvers.

## 2 Clique Problems

A clique in a graph is a set of vertices, where every vertex in this set is adjacent to every other in this set. The problem of finding a maximum-sized clique in a graph is broadly applicable, and there are many dedicated solvers for the problem (which we will discuss below). However, as McCreesh et al. [42] note, at least some of these solvers are buggy—including the one [35] which was used as a sub-component by the winner of the 2019 PACE Implementation Challenge [28]. We therefore begin with a worked example, showing how a machine-verifiable proof could be constructed to demonstrate and prove the correctness of a solution for a simple maximum clique instance.

Consider the graph in Fig. 1. To prove that the maximum clique size of this graph is four, we have to show two things: that it has a clique of four vertices, and that there is no larger clique. To do so, we use the VeriPB proof verifier, which takes two files as its input: a pseudo-Boolean model in the standard OPB format [55], and a proof log which provides a verifiable solution to this model. Therefore, our first step is to encode the problem of finding a maximum clique in this graph as a pseudo-Boolean model. We have a 0–1 variable $x_i$ for each vertex $i$ in the graph, an objective which is to maximise the sum of the vertices taken, and for every non-adjacent pair of vertices, a constraint saying they cannot both be taken simultaneously. In OPB format, this looks like:

```
* #variable= 12 #constraint= 41
min: -1 x1 -1 x2 -1 x3 -1 x4 ...and so on... -1 x11 -1 x12 ;
1 ~x3 1 ~x1 >= 1 ;
1 ~x3 1 ~x2 >= 1 ;
1 ~x4 1 ~x1 >= 1 ;
* ...and a further 38 similar lines for the remaining non-edges
```

Here the first line is a special header comment, the second line specifies that the objective is to minimise $\sum_{i=1}^{12} -x_i$ (i.e. to maximise the number of vertices selected, $\sum_{i=1}^{12} x_i$, but OPB supports only minimisation), and subsequent lines specify constraints. An expression like `1 ~x3 1 ~x1 >= 1` corresponds to the linear inequality $1\overline{x}_3 + 1\overline{x}_1 \geq 1$, where the overline means negation, $\overline{x}_i = 1 - x_i$.

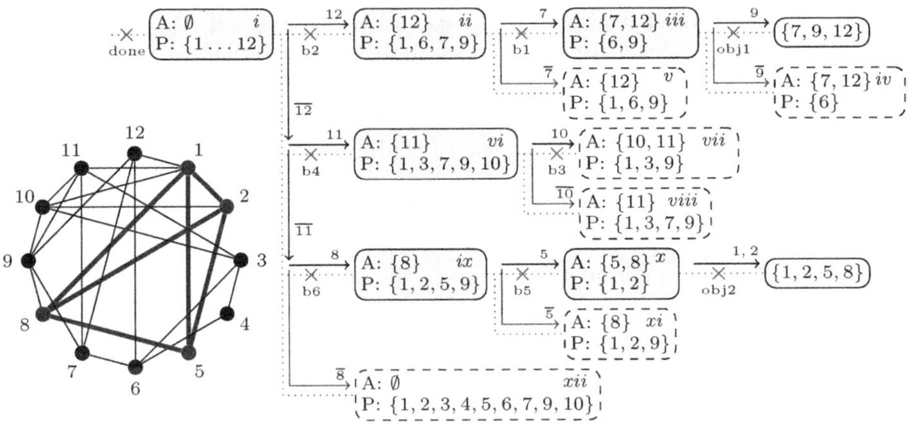


**Fig. 1.** On the left, a graph, with a 4-vertex clique highlighted. On the right, an illustration of the proof tree used in our worked example to show that this clique is maximum. The solid arrows show the solver's view of the search tree, and are labelled either with a vertex number being accepted or an overlined vertex number being rejected. Shaded boxes represent states in the search tree where we have accepted the vertices labelled "A" and can potentially accept the ones labelled "P", dashed boxes represent states that are eliminated by a bound, and clear boxes are candidate solutions. Roman numerals denote states discussed in the text. Dotted lines show the search tree used by the proof: the crosses with labels correspond to statements that justify a backtrack.

Note the simplicity of this encoding. This is important: the proof we will produce is expressed in terms of this encoding, and because this process is not formally verified, any errors in the encoding could potentially lead to a proof which "proves the wrong thing" being accepted.[1]

Now we move on to the proof. We could produce proofs of the decision problems for 4- and 5-cliques, but the VeriPB format also allows us to verify a branch-and-bound search directly. We now give such a proof, tracing a possible (and intentionally not very good) algorithm execution as we do so. The proof log must begin with a header, as follows (asterisk lines are comments):

```
pseudo-Boolean proof version 1.0
* load the objective function, and the 41 model constraints
f 41 0
```

Typically, maximum clique algorithms maintain two sets during search: a set of accepted vertices, $A$, which is always a growing clique, and a set of possible vertices $P$, each of which is adjacent to every vertex in $A$. Rather than a binary branching scheme, we will iterate over each vertex in $P$ in turn and first accept

[1] We are not aware of any obstacles for providing formal verification for this translation step. However, since this translation is so simple, in this paper we focus on the more challenging task of formally verifying the correctness of solvers' reasoning.

that vertex, then reject it and accept a second vertex instead. We will carry out a typical depth-first branch and bound search, with a rather ad hoc bound for illustration purposes. Our solver will begin in the state labelled *i* in Fig. 1, with no vertices accepted and every vertex being possible.

Suppose our solver first branches by deciding to accept vertex 12. Then by adjacency, only vertices 1, 6, 7 and 9 remain acceptable; we are in state *ii.* Suppose now we also accept vertex 7. This leaves vertices 6 and 9 possibly to be accepted; we are in state *iii.* We accept vertex 9, which is not adjacent to 6. We have found a maximal clique with three vertices. We therefore record this in the proof log, using an "o" rule. This rule tells the verifier to check that the solution we specified is in fact feasible, and then to create a new constraint, $\sum_{i=1}^{12} x_i \geq 4$, saying that any future solution must be better; this constraint also allows us to backtrack, which is marked as "obj1" in Fig. 1. We log this as:

```
o x7 x9 x12
```

We are now back to having accepted vertices 7 and 12, but now only 6 remains possible; this is state *iv.* Now that we have introduced a new constraint saying we must set at least four variables to true, it is obvious to a human that we are at a dead-end and must backtrack. We now have two options: we can explicitly justify why we can backtrack by deriving a new constraint manually, or we can rely upon some help from the proof verifier.

To derive the constraint manually, we would proceed as follows. If we sum the objective line, every non-adjacency constraint involving $x_7$ or $x_{12}$, and the non-adjacency constraint involving $x_6$ and $x_9$, we get $\overline{x}_2 + \overline{x}_3 + \overline{x}_4 + \overline{x}_5 + 6\overline{x}_7 + \overline{x}_8 + \overline{x}_{10} + 6\overline{x}_{12} \geq 7$. Now, for any variable $x_i$, we have an axiom $x_i \geq 0$. By also adding these axioms for each $i \in \{2, 3, 4, 5, 8, 10\}$, and normalising by using the fact that $x_i + \overline{x}_i = 1$, the sum reduces to $6\overline{x}_7 + 6\overline{x}_{12} \geq 1$, which we may then divide by 6 to get $\overline{x}_7 + \overline{x}_{12} \geq 1$ as desired. We *could* express these steps explicitly in the proof log—and we *could* also explain an algorithm a solver could use to know exactly which constraints to sum together and what constant to divide by. But fortunately, there is an easier approach. By using a "u" rule, we may tell the proof verifier to introduce a new constraint which is "obviously" true, given what it knows already. So, we may simply assert:

```
u 1 ~x12 1 ~x7 >= 1 ;
```

and the proof verifier will work out the rest. It is able to do this because this new constraint follows by *reverse unit propagation* (RUP) [19,24]: that is, if we add the negation of this constraint and unit propagate,[2] then contradiction follows without search. We may verify this: the negation of the constraint $\overline{x}_{12} + \overline{x}_7 \geq 1$ is $x_{12} + x_7 \geq 2$. From this, unit propagation infers that both $x_7$ and $x_{12}$ are 1. Then, using the non-adjacency constraints, all variables except $x_6$ and $x_9$

[2] In a PB setting, unit propagation is equivalent to achieving integer bounds consistency [9] on all constraints. This is identical to SAT unit propagation on clausal constraints, but is stronger in general.

will unit propagate to 0. Now, looking at the new objective constraint, we have to set at least four variables to 1, so $x_6$ and $x_9$ must both be 1. However, vertices 6 and 9 are non-adjacent, and so there is a constraint forbidding them both to be 1. Thus, RUP can derive a contradiction, and can safely add the asserted constraint—without our hypothetical solver authors having to perform any complicated bookkeeping. This new constraint is labelled "b1" in Fig. 1.

Our solver is now back in the state that it has accepted vertex 12, and it has vertices 1, 6, and 9 to choose from; this is state *v*. Observe that vertices 1 and 6 are non-adjacent, and so it is not possible to make a 4-clique using vertex 12 plus a subset of these vertices. We may therefore backtrack—again, this fact follows using RUP. We label this "b2" in the figure, and log it as:

```
u 1 ~x12 >= 1 ;
```

We are now back at the top of the search tree, having rejected vertex 12 entirely. Suppose we accept vertex 11 next: this leaves vertices 1, 3, 7, 9, and 10 as possibilities, state *vi*. Then suppose we accept vertex 10, leaving vertices 1, 3, and 9 as possibilities, state *vii*. Note that none of these vertices are adjacent, and so we may select at most one of these. To a human, it is now obvious that we may backtrack, but we must give the proof verifier a little help. Before we can use a RUP rule to backtrack, we must derive an at-most-one rule showing that $x_1 + x_3 + x_9 \leq 1$. We may do this as follows:

```
p 1 2 * 19 + 21 + 3 d
p 42 47 +
u 1 ~x11 1 ~x10 >= 1 ;
```

However, these two "p" lines are not easy to read, as expressed: some of the numbers are literal constants, some refer to constraints in the model file, and some refer to constraints we have generated earlier in the verification process. For this discussion, we will therefore take a few liberties with the proof format in our running example. Instead of writing "42" for the objective constraint (which got that number because it was the first introduced constraint, and there are 41 model constraints before it), we will write `obj1`. Similarly, rather than writing 19 to refer to the model constraint $\overline{x}_1 + \overline{x}_9 \geq 1$, we will write `nonadj1_9`. Finally, we will use the notation ⇝ `name` to give a name to the result of a rule that we will refer to later on in the proof, or to refer to a point in Fig. 1. After this, any remaining numbers are literal constants. Thus, the above snippet becomes:

```
* at most one [ x1 x3 x9 ]
p nonadj1_3 2 * nonadj1_9 + nonadj3_9 + 3 d        ⇝ tmp1
p obj1 tmp1 +
u 1 ~x11 1 ~x10 >= 1 ;                              ⇝ b3
```

and we may explain the two "p" rules more easily. In the cutting planes proof system for pseudo-Boolean formulae [11] upon which VeriPB is based, we can add together existing constraints, multiply existing constraints by a non-negative integer constant, and divide existing constraints by a positive integer constant. A "p" rule expresses these steps in reverse Polish notation. The first "p" rule multiplies the non-adjacency constraint $\overline{x}_1 + \overline{x}_3 \geq 1$ by 2 to get $2\overline{x}_1 + 2\overline{x}_3 \geq 2$, adds two more non-adjacency constraints to get $3\overline{x}_1 + 3\overline{x}_3 + 2\overline{x}_9 \geq 4$, and divides this by 3 to get the at-most-one constraint $\overline{x}_1 + \overline{x}_3 + \overline{x}_9 \geq 2$. The second "p" rule adds this to our objective constraint, $\sum_{i=1}^{12} x_i \geq 4$, to show that the remaining nine variables must sum to at least 3. This is now sufficient for reverse unit propagation to justify backtracking step "b3".

A very similar argument allows us to backtrack again: having accepted vertex 11, and rejected vertices 10 and 12, we may pick at most one of vertices 1, 3, and 7, plus possibly vertex 9 (state *viii*). We must help the verifier by generating another at-most-one constraint:

```
* at-most-one [ x1 x3 x7 ]
p nonadj1_3 2 * nonadj1_7 + nonadj3_7 + 3 d                    ⇝ tmp2
p obj1 tmp2 +
u 1 ~x11 >= 1 ;                                                 ⇝ b4
```

We are back at the top of search. Having rejected vertices 11 and 12, if we now branch accepting vertex 8 (state *ix*), and then vertex 5 (state *x*), the remaining possible vertices 1 and 2 can both be added to form a clique. We thus log this as a solution, which generates a new objective constraint $\sum_{i=1}^{12} x_i \geq 5$.

```
o x1 x2 x5 x8                                                   ⇝ obj2
u 1 ~x8 1 ~x5 >= 1 ;                                            ⇝ b5
```

Backtracking to the top of the search tree from state *xi* can be justified by observing that we may pick at most one of vertices 1 and 9:

```
p obj2 nonadj1_9 +
u 1 ~x8 >= 1 ;                                                  ⇝ b6
```

Finally, having rejected vertices 8, 11, and 12 at the top of search, we are in state *xii*, and the remaining nine vertices can be partitioned into independent sets to create three at-most-one constraints. To allow RUP to unset all nine vertices, we will combine these constraints incrementally, as follows.

```
* at-most-one [ x1 x3 x7 ] [ x2 x4 x9 ] [ x5 x6 x10 ]
p nonadj1_3 2 * nonadj1_7 + nonadj3_7 + 3 d                     ⇝ tmp3
p obj2 tmp3 +
p nonadj2_4 2 * nonadj2_9 + nonadj4_9 + 3 d                     ⇝ tmp4
p obj2 tmp3 + tmp4 +
p nonadj5_6 2 * nonadj5_10 + nonadj6_10 + 3 d                   ⇝ tmp5
p obj2 tmp3 + tmp4 + tmp5 +
```

The proof terminates by asserting that we have proved unsatisfiability—that is, there is nothing remaining that can beat the best solution we have found. This is done through a RUP check for contradiction, i.e. that $0 \geq 1$, followed by a "c" rule to terminate the proof.

```
u >= 1 ;                                                    ⇝ done
c done 0
```

Having produced this log, we may now hand it and the associated pseudo-Boolean model file to VeriPB, which will successfully verify it.

There is one other important detail that we have omitted from this proof: in practice, it is extremely helpful to the verifier if we delete temporary constraints when they are used, as well as intermediate backtracking constraints after we have backtracked further up the tree. (This is also crucial for the performance of proof logging for SAT [74].) VeriPB supports both deletion of numbered constraints, and a notion of "levels" which allow all constraints generated below a certain depth to be deleted simultaneously.

### 2.1 Maximum Clique Algorithms in General

The majority of maximum clique algorithms that are aimed at hard, dense graphs make use of backtracking search with branch and bound [4,33,35–37,39,44,51, 53,57,58,60,62,66–69,71]. The inference on adjacency performed by all of these algorithms is straightforward, with all of the cleverness being in branching and how bounds are computed [1]. We may therefore produce proof logs for all of these algorithms using only RUP, logging of solutions as they are found, and some additional help for the bounds.

*Colour Bounds.* If a graph can be coloured using $k$ colours (where adjacent vertices must be given different colours) then it cannot contain a clique of more than $k$ vertices. Producing an optimal colouring is hard (and typically harder in practice than finding a maximum clique), but various greedy methods exist, and have been used to give a dynamic bound during search inside clique algorithms. Suppose we have, after branching, our set of accepted vertices $A$, a set of undecided vertices $P$, and have already found a clique of $n$ vertices. If $c(P)$ is the number of colours used in some legal colouring of the subgraph induced by $P$, then if $|A| + c(P) \leq n$, we can immediately backtrack.

Using cutting planes, if we are given a colouring then it is easy to produce a proof that this bound is valid. By definition, for each pair of vertices in a given colour class, the PB model must have a constraint saying that both vertices cannot be taken simultaneously (because they do not have an edge between them). As we saw in the worked example, it is routine to combine these constraints into an at-most-one constraint, using a single sequence of arithmetic operations that mentions each pairwise constraint only once. We can then sum these new at-most-one constraints, add them to the objective constraint, and the rest of the work follows by unit propagation.

*Incremental Colour Bounds.* Producing a colouring can be relatively expensive. In order to reduce the number of colourings needed, many solvers reuse colourings. Suppose we have produced colour classes $C_1, \ldots, C_c$. Instead of making a single branching decision, we may branch on accepting each vertex in colour class

$C_c$ in turn first, followed by those in $C_{c-1}$, then $C_{c-2}$ and so on, stopping after we have visited only $n - |A| + 1$ colour classes. Ideally, in a proof log, we would not have to produce individual statements to justify not exploring each vertex in each remaining colour class. This is indeed possible: we derive an at-most-one constraint for colour class $C_1$, and remember its number $\ell_1$. We then add this constraint to the objective constraint. Next, we derive an at-most-one constraint for colour class $C_2$, add this to $\ell_1$, and remember its number $\ell_2$. Now we sum the objective constraint, $\ell_1$, and $\ell_2$. We continue until we reach a colour class which was used for branching—again, the worked example made use of this.

Other changes to the details of how colour bounds are produced has formed a substantial line of work in maximum clique algorithms [33,51,53,57,58,62,66–69]. However, proof logging is completely agnostic to this: we care only that we have a valid colouring, and do not need to understand any of the details of the algorithm that produced it.

*Stronger Bounds.* Even when a good colouring is found, colour bounds can be quite weak in practice. Some clique solvers identify subsets of $k$ colour classes which cannot form a clique of $k$ vertices. For example, San Segundo et al. [60] will find certain cases where there is a pair of colour classes $C_1$ and $C_2$, together with a vertex $v$, such that no triangle exists using $v$ and a vertex each from $C_1$ and $C_2$, and uses this to reduce the bound by one for vertex $v$. If such a case is identified, RUP is sufficient to justify it. Similarly, because pseudo-Boolean unit propagation is at least as strong as SAT unit propagation, bounds using MaxSAT reasoning on top of colour classes [35–37] are also easily justified.

*Algorithm Features Not Affecting Proof Logging.* Maximum clique algorithms have used a variety of different search orders [44]; as with the details of how colourings are produced, these details are irrelevant for proof logging. Similarly, bit-parallelism [59,61] has no effect on proof logging; thread-parallelism [16,43,45] remains to be seen, but since proof logging is largely I/O bound, it is likely that gains from multi-core parallelism will be lost on current hardware when logging. And finally, running a local search algorithm and "priming" the branch and bound algorithm with a strong initial incumbent [4,39,71] requires only that the new incumbent be logged before the search starts, regardless of how that incumbent was found.

*Implementation.* We implemented proof logging for the dedicated clique solver which is included in the Glasgow Subgraph Solver [48], and tested it on a system with dual Intel Xeon E5-2697A v4 CPUs, 512 GBytes RAM, and a conventional hard drive, running Ubuntu 18.04. Without proof logging, this solver is able to solve 59 of the 80 instances from the second DIMACS implementation challenge [32] in under 1,000 s. With proof logging enabled, we produced proof logs for 57 of the 59 instances, incurring a mean slowdown of 80.1; the final two instances were cancelled when their proof logs reached 1 TByte in size. We were then able to verify all 57 of these proofs, with verification being a mean of 10.1 times more expensive than writing the proofs. Note that the logging slowdown is to

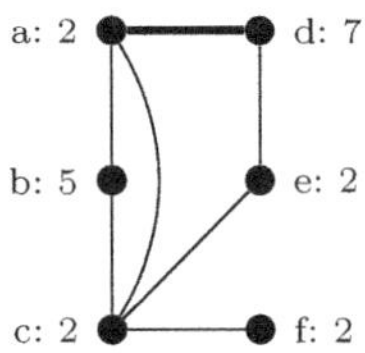


```
pseudo-Boolean proof version 1.0
f 8 0
o xa xd                                                   ⇝ obj
p nonadja_e 2 * nonadja_f + nonadje_f + 3 d 2 *           ⇝ cc1
p nonadjb_d 5 *                                           ⇝ cc2
p nonadjc_d 2 *                                           ⇝ cc3
p obj cc1 + cc2 + cc3 +                                   ⇝ done
c done 0
```

**Fig. 2.** On the left, a weighted graph, with a clique of weight ten from vertices $a$ and $d$ highlighted. On the right, a proof that there is no heavier clique.

be expected [26]: the original solver is able to carry out a full recursive call and bounds calculation in under a microsecond. If each such call requires 1 KByte of logged information then this already exceeds the 100 MBytes per second write capabilities of a hard disk by an order of magnitude.

### 2.2 Weighted Clique Algorithms

In the maximum weight clique problem, vertices have weights, and we are now looking to find the clique with the largest sum of the weights of its vertices, rather than the most vertices. A simple bound for this problem is to produce a colouring, and then sum the maximum weight of each colour class. Consider the example in Fig. 2, and the three colour classes $\{a, e, f\}$, $\{b, d\}$ and $\{c\}$. By looking only at the largest weight in each colour class, we obtain a bound of $2 + 7 + 2 = 11$. This bound may be justified in a cutting planes proof by generating the at-most-one constraints for each colour class as previously, and then multiplying each colour class by its maximum weight before summing them. However, better bounds can be produced by allowing a vertex to appear in multiple colour classes, and by splitting its weight among these colour classes. If we allow vertex $d$ to appear in the second colour class with weight 5 and in the third colour class with weight 2, then our bound is $2 + 5 + 2 = 9$. This technique originates with Babel [2], and is used in algorithms due to Tavares et al. [64,65], which are the current state of the art for many graph classes [46]. From a proof logging perspective, this splitting does not affect how we generate the bound, and so we may generate the proof shown on the right of Fig. 2.

*Implementation.* We implemented a simple certifying maximum weight clique algorithm using the Tavares et al. [64,65] bound in Python. With a timeout of 1,000 s, we were able to produce proof logs for 174 of the 289 benchmark instances from a recent collection [46]; all were verified successfully.

### 2.3 Maximal Clique Enumeration

Finally, in some applications we want to find every maximal clique (that is, one which cannot be made larger by adding vertices without removing vertices).

This problem also has a straightforward PB encoding: we express maximality by having a constraint for every vertex $v$ saying that either $x_v$ is selected, or at least one of its non-neighbours is.

The classic Bron-Kerbosch algorithm [6] uses a simple backtracking search, employing special data structures to minimise memory usage; it ensures maximality through a data structure called a *not*-set. We do not explain this data structure here, because it turns out to be equivalent in strength to unit propagation on the above PB model—indeed, to create a proof-logging Bron-Kerbosch algorithm, one needs only output a statement for every found solution, and a statement on every backtrack. More recent variations on this algorithm make use of different branching techniques [20, 49, 56, 70] and supporting data structures [15, 21, 56], but although these new techniques can make a huge difference to theoretical worst-case guarantees and to empirical runtimes, they do not require any changes to how proof logging is performed.

We are interested in proof logging for this problem because there are discrepancies in tables of published results for some common benchmark instances—for example, does the "celegensneural" instance from the Newman dataset have 856 [21], 1,386 [20], or some other number of maximal cliques? We implemented proof logging for Tomita et al.'s variant of the algorithm [70], and were able to confirm that 1,386 is the correct answer. We were also able to confirm the published values of Eppstein et al. [20] for all of the BioGRID instances, the listed DIMACS instances, and for the Newman instances with no more than 10,000 vertices. We were unable to produce certified results for larger sparse graphs, because the OPB encoding size is linear in the number of *non*-edges in the inputs.

## 3 Maximum Common Induced Subgraph Algorithms

The maximum common induced subgraph problem can be defined in various equivalent ways, but the most useful is that we are given two graphs, and must find an injective partial mapping from the first graph to the second, where adjacent vertices are mapped to adjacent vertices and non-adjacent vertices are mapped to non-adjacent vertices, mapping as many vertices as possible. The problem arises in applications including in chemistry and biology [18, 23, 54]. However, in many cases, the common subgraph is required to be *connected*: that is, if we take any two assigned vertices from the first graph, then we must be able to find a path from one to the other without using unassigned vertices.

McCreesh et al. [41] compared two approaches to the problem, one based upon CP [50, 73] and one based upon reduction to clique [3, 17, 34, 54], and found that the best approach varied depending upon the kinds of graph being used. Since then, improvements have come from two different lines of research: one based upon weakening subgraph isomorphism algorithms [31], and one called McSplit which replaces general algorithms and data structures used in CP with much faster domain-specific ones [38, 47]. We will discuss CP and McSplit, and then the clique reduction later, but first we must provide an appropriate PB encoding.

### 3.1 Pseudo-Boolean Encodings

To encode maximum common induced subgraph in PB form, we may adapt the subgraph isomorphism encoding of Gocht et al. [26]. For each vertex $f$ in the first graph $F$, and for each vertex $s$ in the second graph $S$, we have a variable $x_{f,s}$ which takes the value 1 if $f$ is mapped to $s$; we also have a variable $x_{f,\bot}$ if $f$ is unassigned. We then have exactly-one constraints over each set of $x_{f,-}$ variables, at-most-one constraints over each set of $x_{-,s}$ variables for injectivity, and induced adjacency constraints which are expressed using Gocht et al.'s second encoding,

$$x_{f,\bot} + \sum_{s\in \mathrm{V}(S)} x_{f,s} = 1 \qquad f \in \mathrm{V}(F)$$

$$\sum_{f\in \mathrm{V}(F)} x_{f,s} \leq 1 \qquad s \in \mathrm{V}(S)$$

$$\overline{x}_{f,s} + x_{g,\bot} + \sum_{t\in \mathrm{N}(s)} x_{q,t} \geq 1 \qquad f \in \mathrm{V}(F),\ q \in \mathrm{N}(f),\ s \in \mathrm{V}(S)$$

$$\overline{x}_{f,s} + x_{g,\bot} + \sum_{t\in \overline{\mathrm{N}}(s)} x_{q,t} \geq 1 \qquad f \in \mathrm{V}(F),\ q \in \overline{\mathrm{N}}(f),\ s \in \mathrm{V}(S)$$

and the objective is to maximise the sum of the non-$\bot$ variables.

For the connected version of the problem, expressing connectedness as a constraint is a little trickier. Our encoding is informed by two simple observations: a subgraph with $k$ vertices is connected if, for every pair of vertices in the subgraph, there is a walk of length no more than $k$ between them, and secondly, for $k > 1$, there is a walk of length $2k$ between two vertices $f$ and $g$ if and only if there is some vertex $h$ such that there are walks of length $k$ between $f$ and $h$ and also between $h$ and $g$.

Therefore, we first introduce auxiliary variables $x^1_{f,g}$ for every pair of vertices $f$ and $g$ in the first graph.[3] If $f$ and $g$ are non-adjacent, these variables are forced to false; otherwise we add constraints to specify that $x^1_{f,g}$ is true if and only if both $x_{f,\bot}$ and $x_{g,\bot}$ are false. In other words, $x^1_{f,g}$ is true precisely if $f$ and $g$ are adjacent and in the chosen subgraph. Writing $f \sim_F g$ and $f \not\sim_F g$ to mean vertices $f$ and $g$ are adjacent or not adjacent in the graph $F$ respectively, this is:

$$\overline{x}^1_{f,g} \geq 1 \qquad f, g \in \mathrm{V}(F), f \not\sim_F g$$

$$\overline{x}^1_{f,g} + \overline{x}_{f,\bot} \geq 1 \qquad f, g \in \mathrm{V}(F), f \sim_F g$$

$$\overline{x}^1_{f,g} + \overline{x}_{g,\bot} \geq 1 \qquad f, g \in \mathrm{V}(F), f \sim_F g$$

$$x^1_{f,g} + x_{f,\bot} + x_{g,\bot} \geq 1 \qquad f, g \in \mathrm{V}(F), f \sim_F g$$

Next, we introduce auxiliary variables $x^2_{f,g}$, which will tell us if there is a walk of length 2 between vertices $f$ and $g$. To do this, for each other vertex $h$,

[3] In all of what follows, these variables are equivalent under the exchange of $f$ and $g$, and so we may halve the number of variables needed by exchanging $f$ and $g$ if $f > g$. We do this in practice, but omit this in the description for clarity.

we have a variable $x^2_{f,h,g}$ which we constrain to be true if and only if there is a walk of length 1 from $f$ to $h$, and from $h$ to $g$. Now, $x^2_{f,g}$ may be constrained to be true if and only if either there is a walk of length 1 between $f$ and $g$, or at least one $x^2_{f,h,g}$ variable is true. We then repeat this process for walks of length 4, 8, and so on, until we reach a length $k$ which equals or exceeded the number of vertices in the first graph. For $k \in \{2, 4, 8, \ldots, 2^{\lceil \log |V(F)| \rceil}\}$:

$$
\begin{aligned}
x^{k/2}_{f,h} + \overline{x}^k_{f,h,g} &\geq 1 && f, g, h \in V(F), h \neq f, h \neq g \\
x^{k/2}_{h,g} + \overline{x}^k_{f,h,g} &\geq 1 && f, g, h \in V(F), h \neq f, h \neq g \\
x^k_{f,h,g} + \overline{x}^{k/2}_{f,h} + \overline{x}^{k/2}_{h,g} &\geq 1 && f, g, h \in V(F), h \neq f, h \neq g \\
\overline{x}^k_{f,g} + x^{k/2}_{f,g} + \sum_{h \in V(F) \setminus \{f,g\}} x^k_{f,h,g} &\geq 1 && f, g \in V(F) \\
x^k_{f,g} + \overline{x}^k_{f,h,g} &\geq 1 && f, g, h \in V(F), h \neq f, h \neq g \\
x^k_{f,g} + \overline{x}^{k/2}_{f,g} &\geq 1 && f, g \in V(F)
\end{aligned}
$$

Finally, to enforce connectedness, for each pair of vertices $f$ and $g$, we require that either $x_{f,\perp}$ or $x_{g,\perp}$ or $x^k_{f,g}$ is true.

$$
x_{f,\perp} + x_{g,\perp} + x^k_{f,g} \geq 1 \qquad f, g \in V(F), k = 2^{\lceil \log |V(F)| \rceil}
$$

An important property of this encoding is that all the auxiliary variables are *dependent*: that is, for every solution to the maximum common connected induced subgraph problem, there is exactly one feasible way of setting the auxiliary variables. In other words, the number of solutions to the PB encoding is exactly the same as the number of solutions to the real problem.

### 3.2 Proof Logging for Constraint Programming Algorithms

The McSplit algorithm [47] performs a CP-style backtracking search [50,73], looking to map as many vertices from the first graph as possible to distinct vertices in the second graph. We will therefore continue to use RUP to generate proofs. For adjacency and non-adjacency constraints, McSplit's reasoning is equivalent to unit propagation on our PB constraints, and so no help is needed. For the bound, McSplit performs "all different except ⊥" propagation, but with the number of occurrences of ⊥ constrained to beat the best solution found so far. Due to the special structure of the domains during search, it is able to do this in linear time, without needing the usual matching and components algorithm [52]. However, when it fails, it produces a sequence of Hall sets, and so we may reuse the justification technique described by Elffers et al. [19] with only a simple modification to cope with the objective function.

For the connected variant, McSplit uses a restricted branching scheme [47,73] rather than a conventional propagator: once at least one vertex is assigned a non-null value, it may only branch on vertices adjacent to a vertex already assigned a non-null value. If no such vertices exist, it backtracks. Interestingly, this requires no explicit support in proof logging: by carefully stepping through the auxiliary variables in the PB encoding, level by level, it can be seen that RUP will propagate all remaining variables to false when in this situation.

Therefore, implementing proof logging in McSplit requires four kinds of statement to be logged. Firstly, any new incumbent must be noted, as in the previous section. Secondly, all backtracks must be logged using a RUP rule. Thirdly, whenever the bound function detects that the current state may be pruned, we must derive a new constraint justifying this. And fourthly, it is extremely helpful to delete intermediate constraints using "level" statements. Again, this proof logging is completely agnostic to changes to the branching heuristic [38].

We implemented this proof logging inside the original McSplit implementation, and tested it for both the connected and non-connected variants of the problem on a commonly used set of benchmark instances [10,63]. We successfully verified McSplit's solutions to all 16,300 instances of no more than 25 vertices. Proof logging introduced a mean slowdown of 67.0 and 298.9 for non-connected and connected respectively, whilst verification was a further 13.4 and 21.6 times slower; again, writing to hard disk was by far the biggest bottleneck, as McSplit can make over five million recursive calls per second.

### 3.3 Maximum Common (Connected) Subgraph via Clique

An alternative approach to the maximum common subgraph problem is via a reduction to the maximum clique problem [3,34,54]. This reduction resembles the microstructure encoding of the CP representation, and is the best known approach on labelled graphs; we refer to McCreesh et al. [41] for a detailed explanation. From a proof logging perspective, one might expect that this encoding would require a whole new PB representation, or perhaps a large change to how proof logging is performed by a maximum clique algorithm. However, this is not the case: given the PB model for a maximum common subgraph problem from earlier in this section, we can derive the non-adjacency constraints needed for the clique model described in the previous section using only RUP, whilst the objective function needs no rewriting at all. Therefore, the only changes needed to a proof-logging clique algorithm is in the lookup of constraint identifiers.

McCreesh et al. [41] also show how a maximum clique algorithm can be adapted to deal with the connected variant of the problem, by embedding a propagator inside the clique algorithm. From a proofs perspective, we can work with the PB model and the clique reformulation, similar to channelling [8]—and since connectedness propagation requires no explicit proof logging with the original PB representation, it also requires no proof logging when performed inside a clique algorithm.

We therefore reimplemented McCreesh et al.'s clique common (connected) subgraph algorithm [41], and added proof logging support. Proof logging immediately caught a bug in our reimplementation that testing had failed to identify: we were only updating the incumbent when a maximal clique was found, which is correct in conventional clique algorithms but not for the connected variant, but this very rarely caused incorrect results. Once corrected, for both variants of the problem, we were able to verify all 11,400 instances of no more than 20 vertices from the same set of instances [10,63]. Proof logging introduced an average slowdown of 28.6 and 39.7 for non-connected and connected respectively, and verification was on average a further 11.3 and 73.1 times slower.

## 4 Conclusion

We have shown that pseudo-Boolean proof logging is sufficiently powerful and flexible to make certification possible for a wide range of graph solvers. Particularly of note is how proof logging is largely agnostic towards most changes to details in algorithm behaviour (such as search order, methods for calculating bounds, and underlying algorithms and data structures), and how it is able to deal with reformulation or changes of representation. This suggests that requiring certification should not be an undue burden on solver authors going forward. We also stress the simplicity of implementation: for every algorithm we considered, proof logging only required access to information that was already easily available inside the existing solvers. In particular, we do not need to implement any form of pseudo-Boolean constraint processing in order to generate these proofs, nor does a solver have to in any way understand or otherwise reason about the proofs it is producing. Furthermore, in each case, adding in support for proof logging was considerably easier than implementing the algorithm itself.

It is important to remember that proof logging does not prove that any algorithm or solver is correct. Instead, it provides a proof that a *claimed solution* is correct—and if a solution was produced using unsound reasoning, this will be caught, even if the solution is correct, or if it was produced by a correct algorithm being run on faulty hardware or with a buggy compiler. Additionally, this process does not verify that the encoding from a high level model to the PB representation is correct. To offset this, the encodings we use are deliberately simple, and when a more complex internal representation is used (such as in the clique model for maximum common subgraph), we can log the reformulation and verify the log in terms of the simpler model. This reformulation also suggests that for competitions, providing a standard encoding would not be a problem.

Although proof logging introduces considerable overheads (particularly when compared to the techniques used in the SAT community, which do not need to deal with powerful but highly efficient propagators), it can still be used to verify medium-sized instances involving tens of millions of inference steps. Given the abundance of buggy solver implementations that *usually* produce correct answers, we suggest that all authors of dedicated graph solvers should adopt proof logging from now on, and that competition organisers should strongly consider requiring proof logging support from entrants. For larger and harder instances, proof logging can be disabled, but because proof logging does not require intrusive changes to solver internals, this would still give us a large increase in confidence in the correctness of results compared to conventional testing.

## References


1. Atserias, A., Bonacina, I., de Rezende, S.F., Lauria, M., Nordström, J., Razborov, A.A.: Clique is hard on average for regular resolution. In: Proceedings of the 50th Annual ACM SIGACT Symposium on Theory of Computing, STOC 2018, Los Angeles, CA, USA, 25–29 June 2018, pp. 866–877 (2018)
2. Babel, L.: A fast algorithm for the maximum weight clique problem. Computing **52**(1), 31–38 (1994)
3. Balas, E., Yu, C.S.: Finding a maximum clique in an arbitrary graph. SIAM J. Comput. **15**(4), 1054–1068 (1986)
4. Batsyn, M., Goldengorin, B., Maslov, E., Pardalos, P.M.: Improvements to MCS algorithm for the maximum clique problem. J. Comb. Optim. **27**(2), 397–416 (2014)
5. Biere, A.: Tracecheck (2006). http://fmv.jku.at/tracecheck/
6. Bron, C., Kerbosch, J.: Finding all cliques of an undirected graph (algorithm 457). Commun. ACM **16**(9), 575–576 (1973)
7. Brummayer, R., Lonsing, F., Biere, A.: Automated testing and debugging of SAT and QBF solvers. In: Strichman, O., Szeider, S. (eds.) SAT 2010. LNCS, vol. 6175, pp. 44–57. Springer, Heidelberg (2010). https://doi.org/10.1007/978-3-642-14186-7_6
8. Cheng, B.M.W., Lee, J.H.M., Wu, J.C.K.: Speeding up constraint propagation by redundant modeling. In: Freuder, E.C. (ed.) CP 1996. LNCS, vol. 1118, pp. 91–103. Springer, Heidelberg (1996). https://doi.org/10.1007/3-540-61551-2_68
9. Choi, C.W., Harvey, W., Lee, J.H.M., Stuckey, P.J.: Finite domain bounds consistency revisited. In: Sattar, A., Kang, B. (eds.) AI 2006. LNCS (LNAI), vol. 4304, pp. 49–58. Springer, Heidelberg (2006). https://doi.org/10.1007/11941439_9
10. Conte, D., Foggia, P., Vento, M.: Challenging complexity of maximum common subgraph detection algorithms: a performance analysis of three algorithms on a wide database of graphs. J. Graph Algorithms Appl. **11**(1), 99–143 (2007)
11. Cook, W.J., Coullard, C.R., Turán, G.: On the complexity of cutting-plane proofs. Discret. Appl. Math. **18**(1), 25–38 (1987)
12. Cook, W.J., Koch, T., Steffy, D.E., Wolter, K.: A hybrid branch-and-bound approach for exact rational mixed-integer programming. Math. Program. Comput. **5**(3), 305–344 (2013)
13. Cruz-Filipe, L., Heule, M.J.H., Hunt, W.A., Kaufmann, M., Schneider-Kamp, P.: Efficient certified RAT verification. In: de Moura, L. (ed.) CADE 2017. LNCS (LNAI), vol. 10395, pp. 220–236. Springer, Cham (2017). https://doi.org/10.1007/978-3-319-63046-5_14
14. Cruz-Filipe, L., Marques-Silva, J., Schneider-Kamp, P.: Efficient certified resolution proof checking. In: Legay, A., Margaria, T. (eds.) TACAS 2017. LNCS, vol. 10205, pp. 118–135. Springer, Heidelberg (2017). https://doi.org/10.1007/978-3-662-54577-5_7
15. Dasari, N.S., Ranjan, D., Zubair, M.: pbitMCE: a bit-based approach for maximal clique enumeration on multicore processors. In: 20th IEEE International Conference on Parallel and Distributed Systems, ICPADS 2014, Hsinchu, Taiwan, 16–19 December 2014, pp. 478–485 (2014)
16. Depolli, M., Konc, J., Rozman, K., Trobec, R., Janezic, D.: Exact parallel maximum clique algorithm for general and protein graphs. J. Chem. Inf. Model. **53**(9), 2217–2228 (2013)
17. Durand, P.J., Pasari, R., Baker, J.W., Tsai, C.C.: An efficient algorithm for similarity analysis of molecules. Internet J. Chem. **2**(17), 1–16 (1999)

18. Ehrlich, H.C., Rarey, M.: Maximum common subgraph isomorphism algorithms and their applications in molecular science: a review. Wiley Interdiscip. Rev.: Comput. Mol. Sci. **1**(1), 68–79 (2011)
19. Elffers, J., Gocht, S., McCreesh, C., Nordström, J.: Justifying all differences using pseudo-boolean reasoning. In: The Thirty-Fourth AAAI Conference on Artificial Intelligence, AAAI 2020, The Thirty-Second Innovative Applications of Artificial Intelligence Conference, IAAI 2020, The Tenth AAAI Symposium on Educational Advances in Artificial Intelligence, EAAI 2020, New York, NY, USA, 7–12 February 2020, pp. 1486–1494 (2020)
20. Eppstein, D., Löffler, M., Strash, D.: Listing all maximal cliques in large sparse real-world graphs. ACM J. Exp. Algorithmics **18** (2013)
21. Eppstein, D., Strash, D.: Listing all maximal cliques in large sparse real-world graphs. In: Pardalos, P.M., Rebennack, S. (eds.) SEA 2011. LNCS, vol. 6630, pp. 364–375. Springer, Heidelberg (2011). https://doi.org/10.1007/978-3-642-20662-7_31
22. Gange, G., Stuckey, P.J.: Certifying optimality in constraint programming, February 2019. Presentation at KTH Royal Institute of Technology. Slides https://www.kth.se/polopoly_fs/1.879851.1550484700!/CertifiedCP.pdf
23. Gay, S., Fages, F., Martinez, T., Soliman, S., Solnon, C.: On the subgraph epimorphism problem. Discret. Appl. Math. **162**, 214–228 (2014)
24. Gelder, A.V.: Verifying RUP proofs of propositional unsatisfiability. In: International Symposium on Artificial Intelligence and Mathematics, ISAIM 2008, Fort Lauderdale, Florida, USA, 2–4 January 2008 (2008)
25. Gillard, X., Schaus, P., Deville, Y.: SolverCheck: declarative testing of constraints. In: Schiex, T., de Givry, S. (eds.) CP 2019. LNCS, vol. 11802, pp. 565–582. Springer, Cham (2019). https://doi.org/10.1007/978-3-030-30048-7_33
26. Gocht, S., McCreesh, C., Nordström, J.: Subgraph isomorphism meets cutting planes: solving with certified solutions. In: Proceedings of the Twenty-Ninth International Joint Conference on Artificial Intelligence, IJCAI 2020 [scheduled for July 2020, Yokohama, Japan, postponed due to the Corona pandemic], pp. 1134–1140 (2020)
27. Goldberg, E.I., Novikov, Y.: Verification of proofs of unsatisfiability for CNF formulas. In: 2003 Design, Automation and Test in Europe Conference and Exposition (DATE 2003), Munich, Germany, 3–7 March 2003, pp. 10886–10891. IEEE Computer Society (2003)
28. Hespe, D., Lamm, S., Schulz, C., Strash, D.: WeGotYouCovered: the winning solver from the PACE 2019 implementation challenge, vertex cover track. CoRR abs/1908.06795 (2019)
29. Heule, M., Hunt Jr., W.A., Wetzler, N.: Trimming while checking clausal proofs. In: Formal Methods in Computer-Aided Design, FMCAD 2013, Portland, OR, USA, 20–23 October 2013, pp. 181–188 (2013)
30. Heule, M.J.H., Hunt, W.A., Wetzler, N.: Verifying refutations with extended resolution. In: Bonacina, M.P. (ed.) CADE 2013. LNCS (LNAI), vol. 7898, pp. 345–359. Springer, Heidelberg (2013). https://doi.org/10.1007/978-3-642-38574-2_24
31. Hoffmann, R., McCreesh, C., Reilly, C.: Between subgraph isomorphism and maximum common subgraph. In: Proceedings of the Thirty-First AAAI Conference on Artificial Intelligence, San Francisco, California, USA, 4–9 February 2017, pp. 3907–3914 (2017)

32. Johnson, D.S., Trick, M.A.: Introduction to the second DIMACS challenge: cliques, coloring, and satisfiability. In: Cliques, Coloring, and Satisfiability, Proceedings of a DIMACS Workshop, New Brunswick, New Jersey, USA, 11–13 October 1993, pp. 1–10 (1993)
33. Konc, J., Janežič, D.: An improved branch and bound algorithm for the maximum clique problem. MATCH Commun. Math. Comput. Chem. **58**(3), 569–590 (2007)
34. Levi, G.: A note on the derivation of maximal common subgraphs of two directed or undirected graphs. CALCOLO **9**(4), 341–352 (1973). https://doi.org/10.1007/BF02575586
35. Li, C., Jiang, H., Manyà, F.: On minimization of the number of branches in branch-and-bound algorithms for the maximum clique problem. Comput. Oper. Res. **84**, 1–15 (2017)
36. Li, C.-M., Jiang, H., Xu, R.-C.: Incremental MaxSAT reasoning to reduce branches in a branch-and-bound algorithm for MaxClique. In: Dhaenens, C., Jourdan, L., Marmion, M.-E. (eds.) LION 2015. LNCS, vol. 8994, pp. 268–274. Springer, Cham (2015). https://doi.org/10.1007/978-3-319-19084-6_26
37. Li, C.M., Quan, Z.: An efficient branch-and-bound algorithm based on MaxSAT for the maximum clique problem. In: Proceedings of the Twenty-Fourth AAAI Conference on Artificial Intelligence, AAAI 2010, Atlanta, Georgia, USA, 11–15 July 2010 (2010)
38. Liu, Y., Li, C., Jiang, H., He, K.: A learning based branch and bound for maximum common subgraph related problems. In: The Thirty-Fourth AAAI Conference on Artificial Intelligence, AAAI 2020, The Thirty-Second Innovative Applications of Artificial Intelligence Conference, IAAI 2020, The Tenth AAAI Symposium on Educational Advances in Artificial Intelligence, EAAI 2020, New York, NY, USA, 7–12 February 2020, pp. 2392–2399 (2020)
39. Maslov, E., Batsyn, M., Pardalos, P.M.: Speeding up branch and bound algorithms for solving the maximum clique problem. J. Glob. Optim. **59**(1), 1–21 (2014)
40. McConnell, R.M., Mehlhorn, K., Näher, S., Schweitzer, P.: Certifying algorithms. Comput. Sci. Rev. **5**(2), 119–161 (2011)
41. McCreesh, C., Ndiaye, S.N., Prosser, P., Solnon, C.: Clique and constraint models for maximum common (connected) subgraph problems. In: Rueher, M. (ed.) CP 2016. LNCS, vol. 9892, pp. 350–368. Springer, Cham (2016). https://doi.org/10.1007/978-3-319-44953-1_23
42. McCreesh, C., Pettersson, W., Prosser, P.: Understanding the empirical hardness of random optimisation problems. In: Schiex, T., de Givry, S. (eds.) CP 2019. LNCS, vol. 11802, pp. 333–349. Springer, Cham (2019). https://doi.org/10.1007/978-3-030-30048-7_20
43. McCreesh, C., Prosser, P.: Multi-threading a state-of-the-art maximum clique algorithm. Algorithms **6**(4), 618–635 (2013)
44. McCreesh, C., Prosser, P.: Reducing the branching in a branch and bound algorithm for the maximum clique problem. In: O'Sullivan, B. (ed.) CP 2014. LNCS, vol. 8656, pp. 549–563. Springer, Cham (2014). https://doi.org/10.1007/978-3-319-10428-7_40
45. McCreesh, C., Prosser, P.: The shape of the search tree for the maximum clique problem and the implications for parallel branch and bound. ACM Trans. Parallel Comput. **2**(1), 8:1–8:27 (2015)
46. McCreesh, C., Prosser, P., Simpson, K., Trimble, J.: On maximum weight clique algorithms, and how they are evaluated. In: Beck, J.C. (ed.) CP 2017. LNCS, vol. 10416, pp. 206–225. Springer, Cham (2017). https://doi.org/10.1007/978-3-319-66158-2_14

47. McCreesh, C., Prosser, P., Trimble, J.: A partitioning algorithm for maximum common subgraph problems. In: Proceedings of the Twenty-Sixth International Joint Conference on Artificial Intelligence, IJCAI 2017, Melbourne, Australia, 19–25 August 2017, pp. 712–719 (2017)
48. McCreesh, C., Prosser, P., Trimble, J.: The Glasgow subgraph solver: using constraint programming to tackle hard subgraph isomorphism problem variants. In: Gadducci, F., Kehrer, T. (eds.) ICGT 2020. LNCS, vol. 12150, pp. 316–324. Springer, Cham (2020). https://doi.org/10.1007/978-3-030-51372-6_19
49. Naudé, K.A.: Refined pivot selection for maximal clique enumeration in graphs. Theor. Comput. Sci. **613**, 28–37 (2016)
50. Ndiaye, S.N., Solnon, C.: CP models for maximum common subgraph problems. In: Lee, J. (ed.) CP 2011. LNCS, vol. 6876, pp. 637–644. Springer, Heidelberg (2011). https://doi.org/10.1007/978-3-642-23786-7_48
51. Nikolaev, A., Batsyn, M., Segundo, P.S.: Reusing the same coloring in the child nodes of the search tree for the maximum clique problem. In: Dhaenens, C., Jourdan, L., Marmion, M.-E. (eds.) LION 2015. LNCS, vol. 8994, pp. 275–280. Springer, Cham (2015). https://doi.org/10.1007/978-3-319-19084-6_27
52. Petit, T., Régin, J.-C., Bessière, C.: Specific filtering algorithms for over-constrained problems. In: Walsh, T. (ed.) CP 2001. LNCS, vol. 2239, pp. 451–463. Springer, Heidelberg (2001). https://doi.org/10.1007/3-540-45578-7_31
53. Prosser, P.: Exact algorithms for maximum clique: a computational study. Algorithms **5**(4), 545–587 (2012)
54. Raymond, J.W., Willett, P.: Maximum common subgraph isomorphism algorithms for the matching of chemical structures. J. Comput. Aided Mol. Des. **16**(7), 521–533 (2002)
55. Roussel, O., Manquinho, V.M.: Input/output format and solver requirements for the competitions of pseudo-Boolean solvers, January 2016. Revision 2324. http://www.cril.univ-artois.fr/PB16/format.pdf
56. San Segundo, P., Furini, F., Artieda, J.: A new branch-and-bound algorithm for the maximum weighted clique problem. Comput. Oper. Res. **110**, 18–33 (2019)
57. San Segundo, P., Lopez, A., Batsyn, M.: Initial sorting of vertices in the maximum clique problem reviewed. In: Pardalos, P.M., Resende, M.G.C., Vogiatzis, C., Walteros, J.L. (eds.) LION 2014. LNCS, vol. 8426, pp. 111–120. Springer, Cham (2014). https://doi.org/10.1007/978-3-319-09584-4_12
58. San Segundo, P., Lopez, A., Batsyn, M., Nikolaev, A., Pardalos, P.M.: Improved initial vertex ordering for exact maximum clique search. Appl. Intell. **45**(3), 868–880 (2016)
59. San Segundo, P., Matía, F., Rodríguez-Losada, D., Hernando, M.: An improved bit parallel exact maximum clique algorithm. Optim. Lett. **7**(3), 467–479 (2013)
60. San Segundo, P., Nikolaev, A., Batsyn, M., Pardalos, P.M.: Improved infra-chromatic bound for exact maximum clique search. Informatica Lith. Acad. Sci. **27**(2), 463–487 (2016)
61. San Segundo, P., Rodríguez-Losada, D., Jiménez, A.: An exact bit-parallel algorithm for the maximum clique problem. Comput. Oper. Res. **38**(2), 571–581 (2011)
62. San Segundo, P., Tapia, C.: Relaxed approximate coloring in exact maximum clique search. Comput. Oper. Res. **44**, 185–192 (2014)
63. Santo, M.D., Foggia, P., Sansone, C., Vento, M.: A large database of graphs and its use for benchmarking graph isomorphism algorithms. Pattern Recogn. Lett. **24**(8), 1067–1079 (2003)
64. Tavares, A.W.: Algoritmos exatos para problema da clique maxima ponderada. Ph.D. thesis, Universidade federal do Ceará (2016)

65. Tavares, W.A., Neto, M.B.C., Rodrigues, C.D., Michelon, P.: Um algoritmo de branch and bound para o problema da clique máxima ponderada. In: Proceedings of XLVII SBPO, vol. 1 (2015)
66. Tomita, E., Kameda, T.: An efficient branch-and-bound algorithm for finding a maximum clique with computational experiments. J. Glob. Optim. **37**(1), 95–111 (2007)
67. Tomita, E., Seki, T.: An efficient branch-and-bound algorithm for finding a maximum clique. In: Calude, C.S., Dinneen, M.J., Vajnovszki, V. (eds.) DMTCS 2003. LNCS, vol. 2731, pp. 278–289. Springer, Heidelberg (2003). https://doi.org/10.1007/3-540-45066-1_22
68. Tomita, E., Sutani, Y., Higashi, T., Takahashi, S., Wakatsuki, M.: A simple and faster branch-and-bound algorithm for finding a maximum clique. In: Rahman, M.S., Fujita, S. (eds.) WALCOM 2010. LNCS, vol. 5942, pp. 191–203. Springer, Heidelberg (2010). https://doi.org/10.1007/978-3-642-11440-3_18
69. Tomita, E., Sutani, Y., Higashi, T., Wakatsuki, M.: A simple and faster branch-and-bound algorithm for finding a maximum clique with computational experiments. IEICE Trans. Inf. Syst. **96-D**(6), 1286–1298 (2013)
70. Tomita, E., Tanaka, A., Takahashi, H.: The worst-case time complexity for generating all maximal cliques and computational experiments. Theor. Comput. Sci. **363**(1), 28–42 (2006)
71. Tomita, E., Yoshida, K., Hatta, T., Nagao, A., Ito, H., Wakatsuki, M.: A much faster branch-and-bound algorithm for finding a maximum clique. In: Zhu, D., Bereg, S. (eds.) FAW 2016. LNCS, vol. 9711, pp. 215–226. Springer, Cham (2016). https://doi.org/10.1007/978-3-319-39817-4_21
72. Veksler, M., Strichman, O.: A proof-producing CSP solver. In: Proceedings of the Twenty-Fourth AAAI Conference on Artificial Intelligence, AAAI 2010, Atlanta, Georgia, USA, 11–15 July 2010 (2010)
73. Vismara, P., Valery, B.: Finding maximum common connected subgraphs using clique detection or constraint satisfaction algorithms. In: Le Thi, H.A., Bouvry, P., Pham Dinh, T. (eds.) MCO 2008. CCIS, vol. 14, pp. 358–368. Springer, Heidelberg (2008). https://doi.org/10.1007/978-3-540-87477-5_39
74. Wetzler, N., Heule, M.J.H., Hunt, W.A.: DRAT-trim: efficient checking and trimming using expressive clausal proofs. In: Sinz, C., Egly, U. (eds.) SAT 2014. LNCS, vol. 8561, pp. 422–429. Springer, Cham (2014). https://doi.org/10.1007/978-3-319-09284-3_31

# Phase Transition Behavior in Knowledge Compilation

Rahul Gupta[1], Subhajit Roy[1](✉), and Kuldeep S. Meel[2]

[1] Indian Institute of Technology Kanpur, Kanpur, India
[2] School of Computing, National University of Singapore, Singapore, Singapore

**Abstract.** The study of phase transition behaviour in SAT has led to deeper understanding and algorithmic improvements in modern SAT solvers. Motivated by these prior studies of phase transitions in SAT, we seek to study the behaviour of size and compile-time behaviour for random $k$-CNF formulas in the context of *knowledge compilation.*

We perform a rigorous empirical study and analysis of the size and runtime behavior for different knowledge compilation forms (and their corresponding compilation algorithms): d-DNNFs, SDDs and OBDDs across multiple tools and compilation algorithms. We employ instances generated from the random $k$-CNF model with varying generation parameters to empirically reason about the expected and median behavior of size and compilation-time for these languages. Our work is similar in spirit to the early work in the CSP community on phase transition behavior in SAT/CSP. We identify the interesting behavior with respect to different parameters: clause density and *solution density*, a novel control parameter that we identify for the study of phase transition behavior in the context of knowledge compilation. We summarize our empirical study in terms of two concrete conjectures; a rigorous study of these conjectures will possibly require new theoretical tools.

## 1 Introduction

Phase transition is concerned with a sudden change in the behavior of a property of interest of an object pertaining to variations of a parameter of interest. In the context of combinatorial problems, the phase transition behavior was first demonstrated in random graphs in the seminal work of Erdos and Renyi [18]. With the advent of SAT as a modeling language, the initial studies observed phase transition in the satisfiability of random $k$-CNF formulas and the seminal work of Mitchell, Selman, and Levesque [26] demonstrated empirical hardness around the phase transition region for modern SAT solvers. Theoretical investigations into determining the location of the phase transition region have led to several exciting results that yield insights into the algorithmic behavior of modern SAT heuristics [2]. In a significant theoretical breakthrough, the existence

The full version of this paper is available at http://arxiv.org/abs/2007.10400.


H. Simonis (Ed.): CP 2020, LNCS 12333, pp. 358–374, 2020.
https://doi.org/10.1007/978-3-030-58475-7_21

of the phase transition behavior for random $k$-CNF for large $k$ was theoretically proved; the question for small $k$ (>2) is still open [15].

The success of SAT solvers has led the development of tools and techniques for problems and sub-fields broadly classified under the umbrella of *Beyond NP*. This has led to an interest in the behavior of solvers through the lens of phase transition [16,17,31]. Motivated by the success of these studies in uncovering surprising insights into the solution space structure and regions of hardness for the modern SAT solvers, we turn our focus to another sub-field in *Beyond NP* that has found several practical applications: *Knowledge Compilation*.

Knowledge compilation broadly refers to the approaches that seek to compile propositional formulae into tractable representations; tractability is defined with respect to queries that can be performed in polynomial time over the compiled form. As the runtime of queries such as counting, sampling, equivalence is often polytime in the size of a representation, efficient compilation gains importance. Several compilations forms have been proposed showcasing the tradeoff between tractability and succinctness: OBDDs (Ordered Binary Decision Diagrams) [9], SDDs (Sentential Decision Diagrams) [13] and d-DNNFs (deterministic-Decomposable Negation Normal Forms) [11] and others. We refer the reader to [14] for a detailed survey on the size and tractability for several target languages. While every tractable language known so far has exponential size complexity in the worst case when the input is presented in CNF form, the size of compiled forms can often be only polynomially larger than the input CNF formula, which has highlighted the need for more detailed study of the size and runtime complexity of compilation procedures.

In this work, we undertake a rigorous empirical study of phase transition behavior in knowledge compilation. Our experimental setup employs instances generated from the random $k$-CNF model with varying generation parameters (number of variables and clauses as well as length of clauses). Our study is multidimensional, comparing and contrasting phase transition behaviors spanning:

- knowledge compilations: d-DNNFs, SDDs and OBDDs;
- properties of interest: size and compilation times;
- compilation algorithms and tools:
  - C2D [12], D4 [25], DSharp [28] for d-DNNF,
  - MiniC2D [29] and TheSDDPackage [33] for SDD,
  - CUDD with different variable ordering heursitics [19,32] for BDD.

A primary contribution of the seminal work of Mitchell et al. [26] was the establishment of clause density as a popular parameter of interest in the study of SAT solving. In a similar spirit, one of the key contributions of this work is the proposal of *solution density* as a new control parameter, along with clause density, for studying knowledge compilations. Solution density is defined as the ratio of the logarithm of the number of satisfying assignments to the number of variables. We show that while clause density is linked with solution density in expectation, the size and compilation-time varies significantly with varying solution density for a fixed clause density. We discover that for low clause densities, varying solution density has minimal effect on the size of compilations. In contrast, for high clause densities, solution density dictates the size as there is a minimal variation with clause density for a given fixed solution density.

Based on our experiments, we make two conjectures for compiled structures in a language $L$ that is a subset of DNNF:

1. Over a population of k-CNF formulas, $F_k(n, \lceil rn \rceil)$ of $k$-clauses over $n$ variables with a clause density $r$, for all integers $k \geq 2$, there exists a clause density $r$ that witnesses a phase transition on the size of the compiled structures in a language $L$; that is, there always exists a clause density $r_t$ s.t.:
   (a) for each pair $(r_1, r_2)$, such that $r_1 < r_2 < r_t$, the expected size of the compiled structure in $L$ for $F_k(n, r_1 n)$ is strictly smaller than that for clause density $r_2$;
   (b) for each pair $(r_1, r_2)$, such that $r_t < r_1 < r_2$, the expected size of the compiled structure in $L$ for $F_k(n, r_1 n)$ is strictly larger than that for clause density $r_2$.
2. Over a population of k-CNF formulas, $G_k(n, \lceil 2^{\alpha n} \rceil)$ of $k$-clauses over $n$ variables and having $\lceil 2^{\alpha n} \rceil$ solutions, for all integers $k \geq 2$, there exists a solution density $\alpha_k$ that witnesses a phase transition on the size of the compiled structures in a language $L$; that is, there always exists a solution density $\alpha_t$ s.t.:
   (a) for each pair $(\alpha_1, \alpha_2)$ such that $0 \leq \alpha_1 < \alpha_2 < \alpha_k$, the expected size of the compiled structure in $L$ for $G_k(n, 2^{\alpha_1 n})$ is strictly smaller than that for solution density $\alpha_2$;
   (b) for each pair $(\alpha_1, \alpha_2)$ such that $\alpha_k < \alpha_1 < \alpha_2 \leq 1$, the expected size of the compiled structure in $L$ for $G_k(n, 2^{\alpha_1 n})$ is strictly larger than that for solution density $\alpha_2$.

The study of phase-transition behavior for satisfiability of CNF formulae was instrumental in driving several breakthroughs in the design of new solvers and better understanding the problem structure [1,8]. We hope that our experimental study of phase transitions for knowledge compilations would lead to similar developments in the knowledge compilation. Our work is similar in spirit to the seminal work by the CSP community in empirical identification of phase transition phenomenon in the early 1990s and their summarization in the form of conjectures [10,21–23]. It is worth emphasizing that theoretical proofs of these conjectures were presented nearly 20 years since the first empirical studies [15], and the efforts to establishing these conjectures contributed to the development of several theoretical tools of widespread applicability [2]. We hope our empirical results will inspire similar efforts in the context of knowledge compilation.

The organization of the rest of the paper is as follows: Sect. 2 describes the notations and preliminaries, along with a survey of prior work. Section 3 describes the design of our experiments while Sects. 4 and 5 provide detailed observations with respect to clause density and solution density respectively. Due to space restrictions, this article is limited to experiments using the following tools: D4 for d-DNNF, TheSDDPackage for SDD, and CUDD (with SIFT variable reordering) for BDD. We, however, observed similar behaviors across all the other tools and summarise them in Sect. 6. Furthermore, throughout the article, we present only representative plots and an extended collection of corresponding plots is deferred to the full version of the paper.

## 2 Notations and Preliminaries

Let $X = \{x_1, x_2, \ldots, x_n\}$ be a set of propositional variables. A literal is a propositional variable ($x_i$) or its negation ($\neg x_i$). For a *formula* $F$ defined over $X$, a satisfying assignment or *witness* of $F$ is an assignment of truth values to the variables in $X$ such that $F$ evaluates to `true`. The total number of witnesses for a formula $F$ is denoted by $\#F$.

Let a $k$-clause be a disjunction of $k$ literals drawn over $X$ without repetition. Given $k \in \mathbb{N}$, $n \in \mathbb{N}$ and $r \in \mathbb{R}_{>0}$, let the random variable $F_k(n, \lceil rn \rceil)$ denote a Boolean formula consisting of the conjunction of $\lceil rn \rceil$ $k$-CNF clauses selected uniformly and independently from $\binom{n}{k}2^k$ possible $k$-clauses over $n$ variables. For a given $k$, $k$-CNF denotes the set of all possible boolean formulas $F_k(n, \lceil rn \rceil)$. For any given formula $F$, we denote the *clause density* ($r$) as the ratio of number of clauses to variables and the *solution density* ($\alpha$) as the ratio of logarithm of the number of satisfying assignments to the number of variables, i.e. $\alpha = \log(\#F)/n$. We denote the SAT phase transition clause density for random $k$-CNF by $r_k^p$. Given an $\alpha \in [0, 1]$, we define another random variable $G_k(n, \lceil 2^{\alpha n} \rceil)$ that denotes a randomly chosen boolean formula from $k$-CNF with $n$ variables and $\lceil 2^{\alpha n} \rceil$ satisfying assignments.

### 2.1 Target Compilation Languages

We briefly describe some of the prominent target compilation languages that we study—d-DNNFs, OBDDs and SDDs [14].

**Definition 1** [14]. *Let $V$ be the set of propositional variables. A formula in NNF is a rooted, directed acyclic graph (DAG) where each leaf node is labelled with true, false, $x$ or $\neg x$, $x \in V$; and each internal node is labelled with $\vee$ or $\wedge$ and can have arbitrarily many children.*

*Deterministic Decomposable Negation Normal Form* (d-DNNFs), a subset of NNF, satisfies *determinism* (operands of $\vee$ are mutually inconsistent) and *decomposition* (operands of $\wedge$ are expressed on a mutually disjoint set of variables).

*Ordered Binary Decision Diagrams* (OBDD) is a subset of d-DNNFs where the root node is a decision node and the order of decision variables is same for all paths from the root to a leaf; OBDDs are canonicalised by the variable ordering. A decision node is either a constant (`T`/`F`), or of the form $(X \wedge \alpha) \vee (\neg X \wedge \beta)$, on decision nodes $\alpha$ and $\beta$, and decision variables $X$.

*Sentential Decision Diagrams* (SDDs) are a subset of d-DNNFs that hold the properties of *structured decomposability* and *strongly deterministic decompositions*. The idea of structured decomposability is captured by the notion of *vtrees*; vtrees are binary trees whose leaves correspond to variables of the formula and internal nodes mark the decomposition into variables given by their left and right child. Vtrees seek to generalise variable ordering and precisely specify a decomposition scheme to be followed by the corresponding

SDD. Strongly deterministic decompositions seek to generalise Shannon's expansion in which decision is made on a single variable. For a formula $F$ where $\forall U, V \subseteq vars(F)$ such that $vars(F) = U \cup V$ and $U \cap V = \phi$, representing $F = (p_1(U) \wedge s_1(V)) \vee \ldots \vee (p_n(U) \wedge s_n(V))$ where $p_i$'s and $q_i$'s are boolean functions and $p_i(U) \wedge p_j(U) = \bot$ for $i \neq j$ captures the idea of strongly deterministic decompositions. Here, the decomposition of $vars(F)$ into $U$ and $V$ is the decomposition scheme.

We denote a target language by $\mathcal{L} \in \{$d-DNNF, OBDD, SDD$\}$. For a formula $F$ compiled to $\mathcal{L}$, let $\mathcal{N}_{\mathcal{L}}(F)$ denote the size (in the number of nodes) of the target representation and $\mathcal{T}_{\mathcal{L}}(F)$ denote the compilation time. The expected size for $k$-CNF with clause density $r$ and target language $\mathcal{L}$ can, therefore, be represented as $\mathsf{E}(\mathcal{N}_{\mathcal{L}}(F_k(n, rn)))$ and the expected time as $\mathsf{E}(\mathcal{T}_{\mathcal{L}}(F_k(n, rn)))$.

### 2.2 Related Work

Cheeseman et al. [10] studied the phase transition behavior for several CSP problems, noting an easy-hard-easy behavior with respect to an order parameter for these problems. For Hamiltonian circuits, graph coloring and k-SAT, the average connectivity of the graph (which translates to clause density for k-SAT) was found to be a good order parameter demonstrating phase transition behavior. Subsequently, exploring phase transition behaviors with various random generative models has drawn considerable attention [3,21–23,27,34].

Aguirre and Vardi [4] identified an "easy-hard-less hard" behaviour in OBDD compilations with clause density and discovered a phase transition from polynomial running time to exponential running time for SAT. Huang and Darwiche [24] proposed a new top-down algorithm for OBDD compilation based on the DPLL algorithm typically used for SAT solving and observed that the expected size of OBDD peaks around clause density equal to 2. Later, Gao, Yin, and Xu [20] conducted an experimental study for phase transitions in random $k$-CNF formulas taking d-DNNFs, OBDDs and DFAs (Deterministic Finite Automata) as the target compilation languages. They observe a phase transition behavior with respect to size of compilations and draw a conjecture stating that all subsets of DNNF show a "small-large-small" behavior with a unique peak. Our study is more comprehensive and we discuss in Sect. 4 several behaviors that were not explored by Gao et al.

Birnbaum and Lozinskii [6] presented a procedure called as Counting Davis-Putnam (CDP) for #SAT whose trace lies in FBDD, a stricter subset of d-DNNF language. They found that the median number of recursive calls of CDP reaches its peak when for clause density is 1.2 for random 3-CNF formulas. A different DPLL extension for solving #SAT, called Decomposing Davis-Putnam (DDP) leveraging component decomposition, was developed by Bayardo and Pehoushek [5]. The trace of DPP-model counter lies in d-DNNF language and the authors observed phase transition around the clause density of 1.5.

## 3 Design of Experiments

We aim to dive deeper into the size and compile-time behavior for d-DNNF, SDD and OBDD compilations owing to their wide range of applications. We measure the size of compilations by the number of nodes (edges also display a very similar behavior). To study the phase transition behavior, we identify a new parameter, *solution density*, the ratio of log of the number of satisfying assignments to the number of variables. While it is trivial to generate a random instance $F_k(n, rn)$ given $r$, a direct way to generate $G_k(n, 2^{\alpha n})$ given $\alpha$ is unknown. Therefore, to analyze the effect of varying $\alpha$ in compilations, we study the variation in size and compile-time against $\alpha$ for individual instances generated by varying $r$. Our experimental observations with varying clause density are given in Sect. 4 while those with solution density follow in Sect. 5.

As SAT is a decision problem, its phase transition is often characterized by the existence of a unique *cross-over point*, at the intersection of the curves representing probability of *true* (satisfiable) decision against clause density for different number of variables. However, in our case, we are dealing with a functional problem and therefore, we characterize phase transition with respect to the gradient of the function, i.e., when the underlying function achieves a local maxima. More concretely, we will be concerned with the size and runtime of the underlying knowledge compilation form and its corresponding compiler.

We carried out our experiments on a high performance compute cluster whose each node is an Intel E5-2690 v3 CPU with 24 cores and 96 GB of RAM. We utilize a single core per benchmark instance. Note that knowledge compilation is significantly harder and memory intensive than satisfiability, which restricts the scale of our experiments. Therefore, we have conducted experiments up to the largest number of variables for which we could gather all the needed statistics. We utilized more than 40,000 computational hours for our experimentation.

### 3.1 The Variables of Our Study

We conducted our experiments on a large number of random $k$-CNF formulas with a varying number of variables and clauses. For studying variations with clause density, we aggregated the results over 1000 instances of the random variable $F_k(n, rn)$ for each $r$ given a fixed $n$ and $k$. Here, $r$ is incremented with step size of at least 0.1 in a range containing 0 to $r_k^p$. For studying variations with solution density, we aggregated the results over at least 5 instances of the random variable $F_k(n, rn)$ for each $r$ given a fixed $n$ and $k$. Here, $r$ is incremented with step size of $1/n$ in a range from 0 to $r_k^p$. As clause density and solution density are linked in expectation, we vary $r$ finely for study with solution density to facilitate uniform distribution of instances with $\alpha$. Specifically, a representative subset of the following variations are a part of our study:

- number of variables($n$): 20–70 in general, upto an exponential number ($\lfloor 1.3^{37} \rfloor$) for small $r$
- length of clauses($k$): 2–7

- target languages: d-DNNFs, OBDDs, SDDs
- knowledge compilers: D4 [25], CUDD [19,32], TheSDDPackage [33]

## 4 Phase Transition with Clause Density

In this section, we present our observations for variations in size and compile times with d-DNNFs, SDDs, and OBDDs as target languages. We also dive deeper into the complexity of compilations. We present plots for a representative subset of the experimental results here; a more comprehensive set of plots appears in the full version of the paper.

### 4.1 Observing the Phase Transitions

**Size of Compilations.** Figure 1 shows the variations in the mean and median of the number of nodes for d-DNNF, SDD, and OBDD with respect to clause density. Figure 2 shows variations in mean number of nodes for d-DNNF on a *log* scale versus number of variables. The small-large-small pattern in the size of compilations is reminiscent of the empirical hardness of SAT near the phase transition, which is marked by an exponential increase in SAT solvers' runtimes near a specific (phase transition) clause density.

**Runtime of Compilation.** From an empirical usage perspective, the runtime of compilation assumes paramount importance. It is worth emphasizing that the size of the compiled form does not solely determine the runtime. In particular, the runtime of compilation broadly depends upon: (1) the search space itself, (2) the time spent in heuristics. In state-of-the-art knowledge compilers, newer heuristics increasingly attempt to prune the search space leading to reduced memory and runtime for subsequent exploration. However, they incur an increased overhead of applying heuristics. Therefore, to provide a holistic view of the hardness of compilations, Fig. 1 also shows the variation in average runtimes against clause density for d-DNNF, SDD, and OBDD. We observe that the location of phase transition appears slightly shifted (within $\pm 0.3$) compared to that for the size of compilations in the case of d-DNNF and SDD. However, in the case of OBDD, we have a starkly different behavior due to the different compilation procedures, even though both SDDs and OBDDs are compiled by repeatedly conjoining the clauses one by one using the polytime APPLY operation.

A plausible explanation is that while SDD compilation using TheSDDPackage involves dynamic vtree search with clause reordering, OBDD compilation using CUDD supports only dynamic variable ordering. This can lead to an easy-hard-less hard behavior for SDD compilations as clause reordering facilitates early pruning of intermediate SDDs [33]. On the other hand, OBDD compilation involves large intermediate OBDDs near phase transition clause density irrespective of the total number of clauses as clauses are selected randomly for the APPLY operation. This can lead to a sharp step transition in runtimes as the

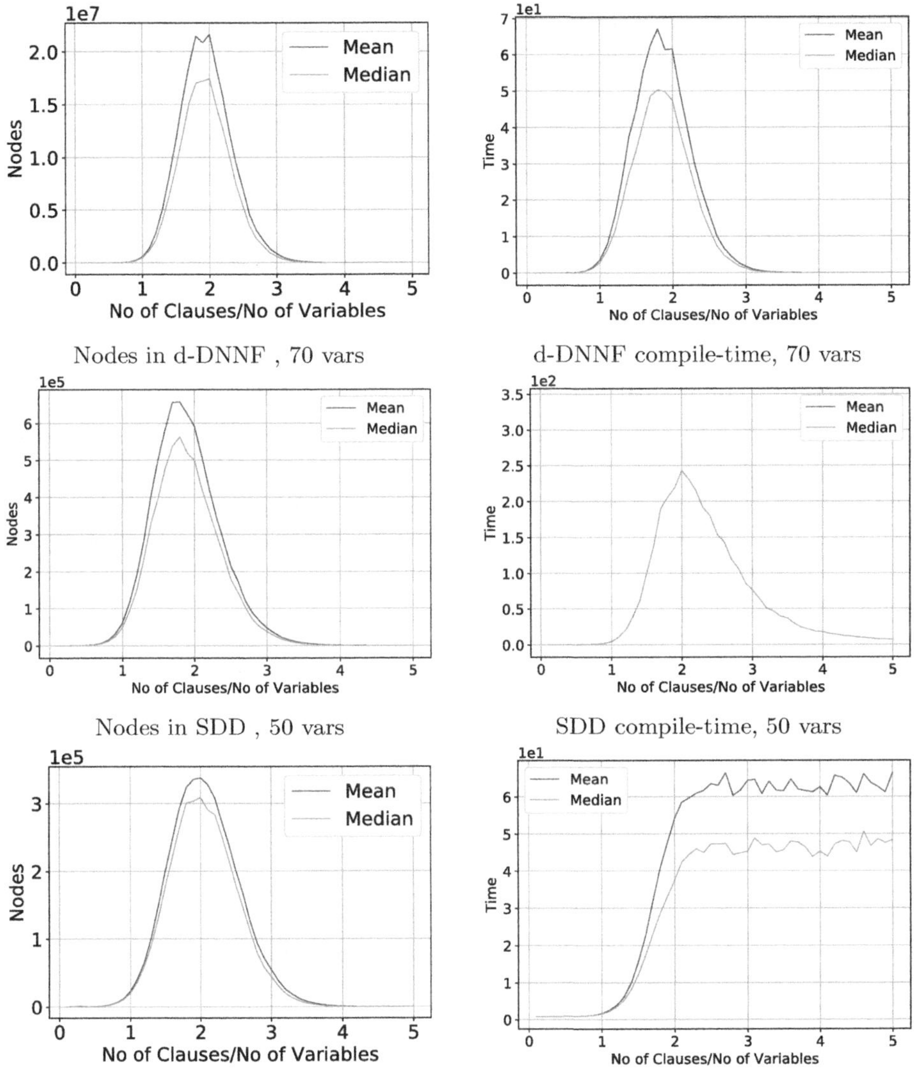


Nodes in d-DNNF , 70 vars

d-DNNF compile-time, 70 vars

Nodes in SDD , 50 vars

SDD compile-time, 50 vars

Nodes in OBDD, 50 vars

OBDD compile-time, 50 vars

**Fig. 1.** Mean number of nodes, compile-times for 3-CNF

addition of more clauses after phase transition prunes the OBDD, translating to reduced cost of successive APPLY operations.

We refer the reader to the full version for variation with respect to clause length. We sum up our observations via the following conjecture:

*Conjecture 1.* For every integer $k \geq 2$, given the number of variables $n$ and a target language $\mathcal{L}$ that is a subset of DNNF, there exists a positive real number $r_k$ such that
For each pair $(r_1, r_2)$, if $r_1 < r_2 < r_k$ then,

$$\mathsf{E}(\mathcal{N}_{\mathcal{L}}(F_k(n, r_1 n))) < \mathsf{E}(\mathcal{N}_{\mathcal{L}}(F_k(n, r_2 n)))$$

For each pair $(r_1, r_2)$, if $r_k < r_1 < r_2$ then,

$$\mathsf{E}(\mathcal{N}_{\mathcal{L}}(F_k(n, r_1 n))) > \mathsf{E}(\mathcal{N}_{\mathcal{L}}(F_k(n, r_2 n)))$$

### 4.2 Diving Deep into the Asymptotic Complexity

The size and runtime of compilations are exponential in the number of variables in the worst case. We are, however, interested in more precise relationship as size and runtime play crucial roles in many applications. While it is known that the exponent increases towards phase transition clause density [20], it is not clear if the exponent is linear or sublinear in the number of variables and whether this behavior changes with changing clause density.

Since, the size of OBDDs (and d-DNNFs and SDDs being more succinct [7]) is bounded by $\mathcal{O}(2^n)$, we plot $\log(\mathsf{E}(\mathcal{N}_{\text{d-DNNF}}(F_k(n, rn))))/n$ for different $n$ against varying $r$ in Fig. 2 and 3 for further investigation. From Fig. 2, we observe phase transition with respect to clause density even though the location of the phase transition point seems to shift slightly with different $n$. Next, we turn our attention to Fig. 3, wherein we observe that $\log(\mathsf{E}(\mathcal{N}_{\text{d-DNNF}}(F_k(n, rn))))/n$ decreases while showing signs of possible convergence with increasing $n$ for a given $r$ for all compilations in our study. In order to understand it better, we scale our experiments to a larger $n$ for small $r$ in Fig. 3b. However, even in this case, the answer remains unclear if $\log(\mathsf{E}(\mathcal{N}_{\text{d-DNNF}}(F_k(n, rn))))/n$ shall converge to a constant >0. Therefore, we can only say that for a constant, $c \geq 0$, $\lim_{n \to \infty} \frac{\log(\mathsf{E}(\mathcal{N}_{\mathcal{L}}(F_k(n, rn))))}{n} = c$.

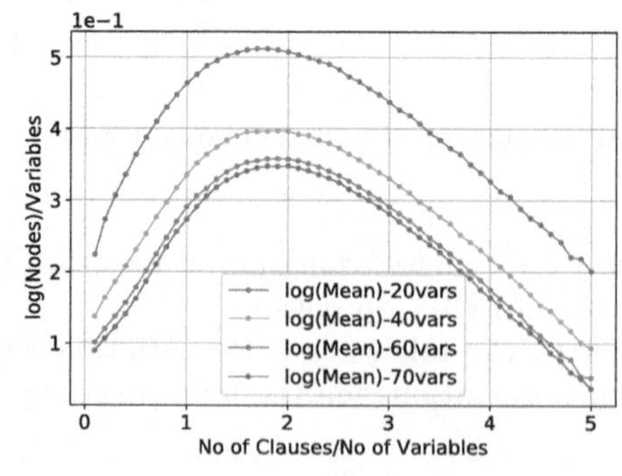


**Fig. 2.** $\log(\mathsf{E}(\mathcal{N}_{\mathcal{L}}(F_k(n, rn))))/n$ vs $r$ for different $n$ for $\mathcal{L} =$ d-DNNF

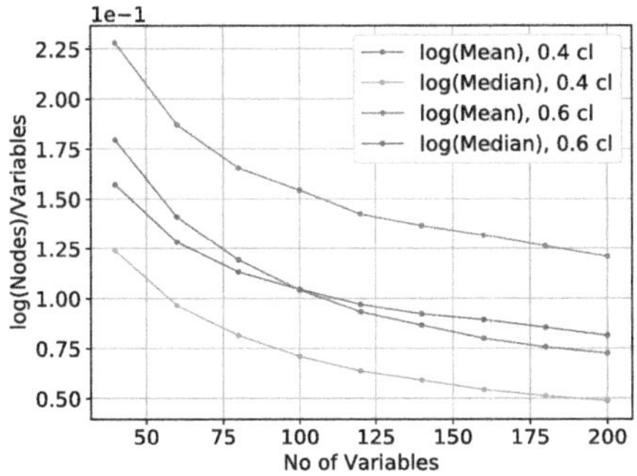


(a) clause densities 0.6 and 0.4

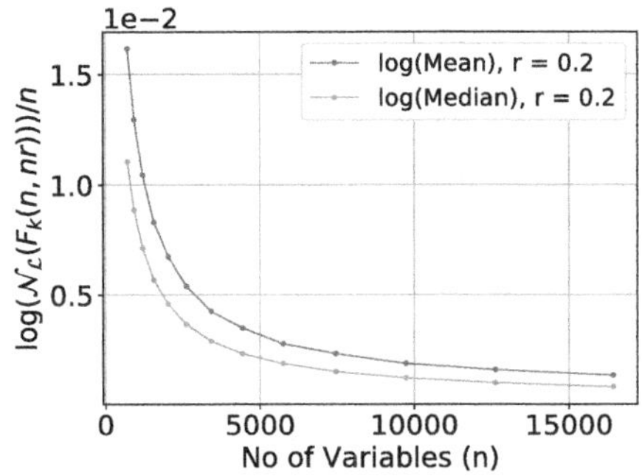


(b) clause density 0.2

**Fig. 3.** $\log(\mathcal{N}_{\mathcal{L}}(F_k(n, rn)))/n$ vs $n$ for $\mathcal{L}$ = d-DNNF

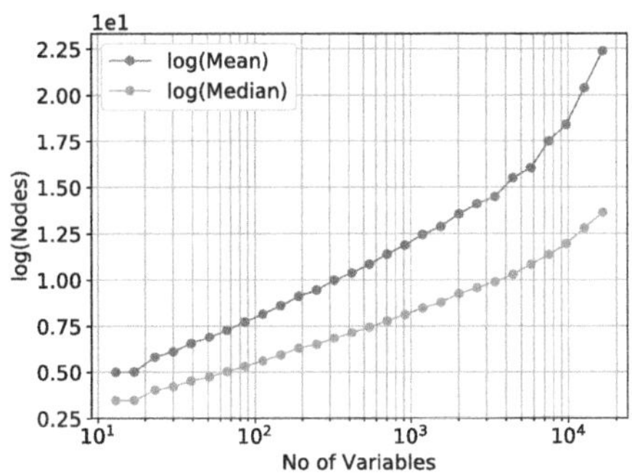


(a) $\log(\mathcal{N}_{\mathcal{L}}(F_k(n, rn)))$ vs $log(n)$

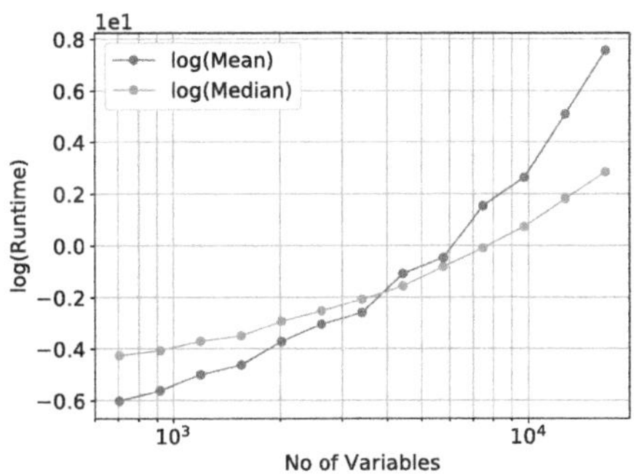


(b) $\log(\mathcal{T}_{\mathcal{L}}(F_k(n, rn)))$ vs $log(n)$

**Fig. 4.** log-log graph at $r = 0.2$ for d-DNNF

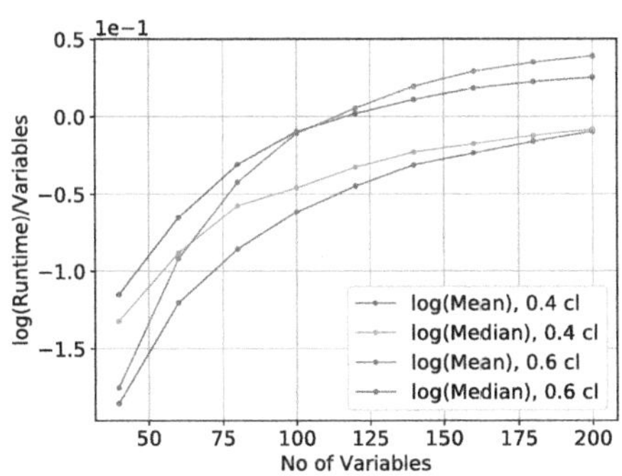


(a) clause densities 0.6 and 0.4

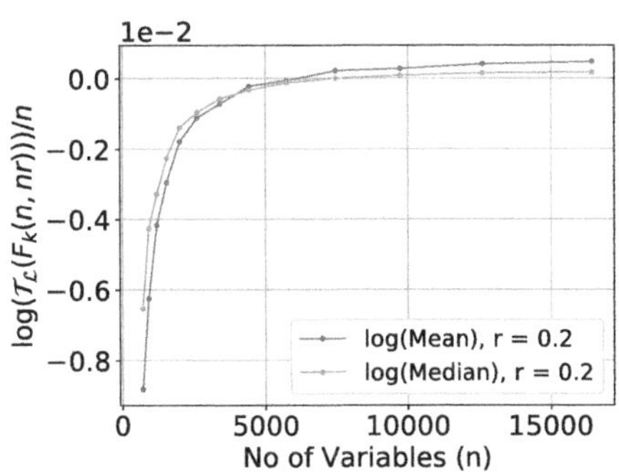


(b) clause density 0.2

**Fig. 5.** $\log(\mathcal{T}_{\text{d-DNNF}}(F_k(n, rn)))/n$ vs $n$ for d-DNNF

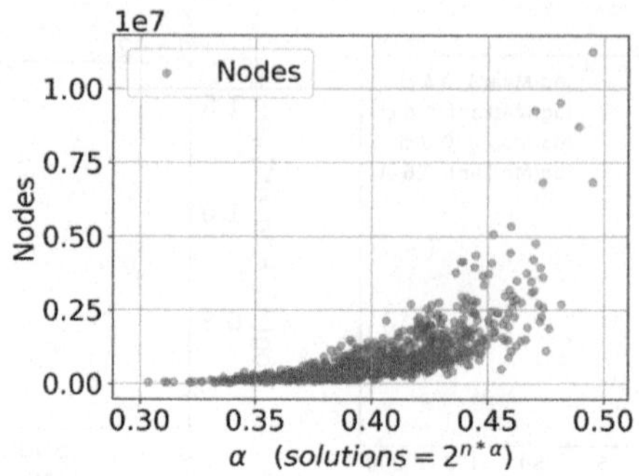


**Fig. 6.** Size vs solution density for individual instances with $r = 3.0$ for d-DNNF

While the question if $c = 0$ remains open for size, the case for runtime of compilation is a different story: Fig. 5 shows that $\frac{\mathsf{E}(\mathcal{T}_{\mathcal{L}}(F_k(n,rn))))}{n}$ increases with $n$. Knowing that the runtime in worst case is $poly(2^n)$ and extrapolating the observation, we conjecture that $\lim_{n\to\infty} \frac{\mathsf{E}(\mathcal{T}_{\mathcal{L}}(F_k(n,rn))))}{n} = c$ where $c > 0$ and depends on $r$. In other words, $\mathsf{E}(\mathcal{T}_{\mathcal{L}}(F_k(n, rn)))) = \theta(2^{cn})$ for state-of-the-art compilers.

**Polynomial to Exponential Size Phase Transition.** Gao et al. [20] had shown the existence of a polynomial to exponential phase transition for size of compilations around $r = 0.3$ by showing increase in slope for $r > 0.3$ and near constant slope for $r < 0.3$ on a $\log(\mathsf{E}(\mathcal{N}_{\text{d-DNNF}}(F_k(n, rn))))$-$\log(n)$ graph. Relationships of the form $y = ax^k$ appear as straight lines in a log-log graph. Our observations in Fig. 3 and 4a show that while the instances with $r = 0.2$ are indeed very easy compared to $r = 0.4$, the behavior is still exponential or quasi-polynomial for $r = 0.2$ which becomes dominant for large enough number of variables. In Fig. 4a, the behavior appears to change from a straight line to a line with increasing slope around $n = 7482$. On extrapolating the behavior for even smaller $r$, we can conjecture that $\forall\, r$, $\mathsf{E}(\mathcal{N}_{\text{d-DNNF}}(F_k(n, rn)))$ is at least quasi-polynomial in $n$.

## 5 Phase Transitions with Solution Density

Since knowledge compilations are a compact way of representing solutions, one can expect that they show variations in sizes with respect to the solution density as well. The solution density, however, is not independent of clause density for random $k$-CNF as given a clause density, expected solution density is fixed, and vice versa. Notwithstanding, we observe (Fig. 6) that solution density also appears to be a fundamental parameter given a fixed clause density, instances with different solution density have marked changes in their size of d-DNNF compilations. We, now, look at the size and compile-time behavior with respect to solution density.

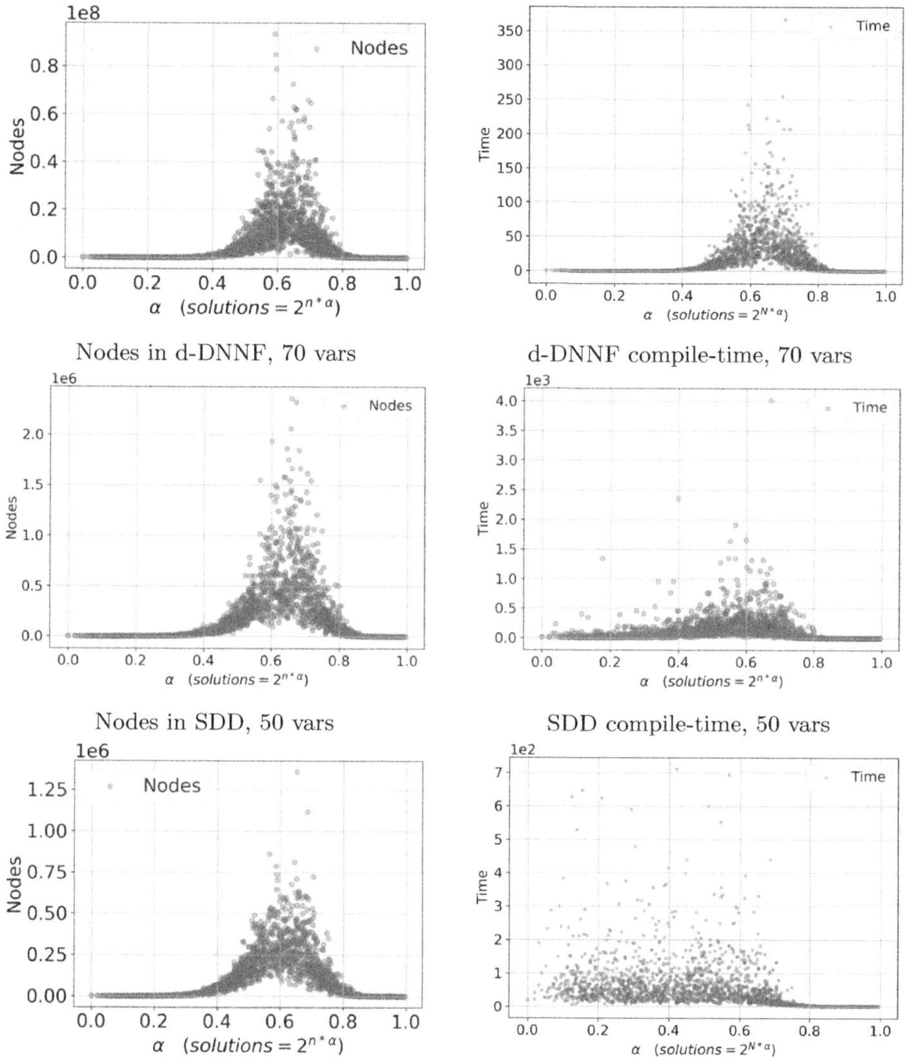


Nodes in d-DNNF, 70 vars

d-DNNF compile-time, 70 vars

Nodes in SDD, 50 vars

SDD compile-time, 50 vars

Nodes in OBDD, 50 vars

OBDD compile-time, 50 vars

**Fig. 7.** Number of Nodes and compile-times for individual instances of 3-CNF against solution density

## 5.1 Observing the Phase Transition

Figure 7 shows the small-large-small variation in size of compilations with respect to solution density. We can observe that there exists a region of critical solution density for each target language around which the size of instances are very large. The location of phase transition appears to depend upon the target compilation

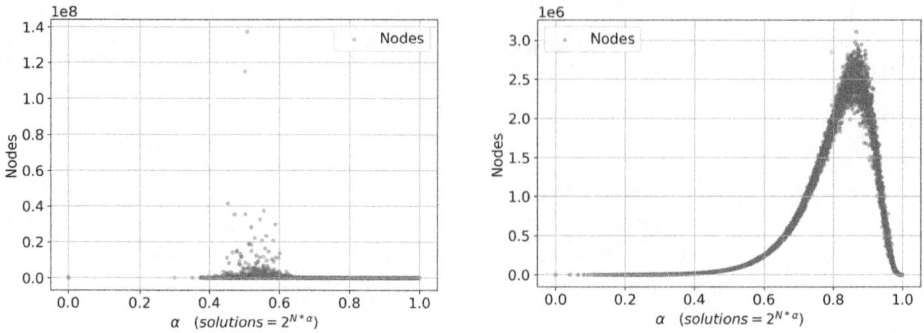


(a) 2-CNF and 600 vars (b) 7-CNF and 30 vars

**Fig. 8.** Nodes in d-DNNF vs solution density for different clause lengths($k$)

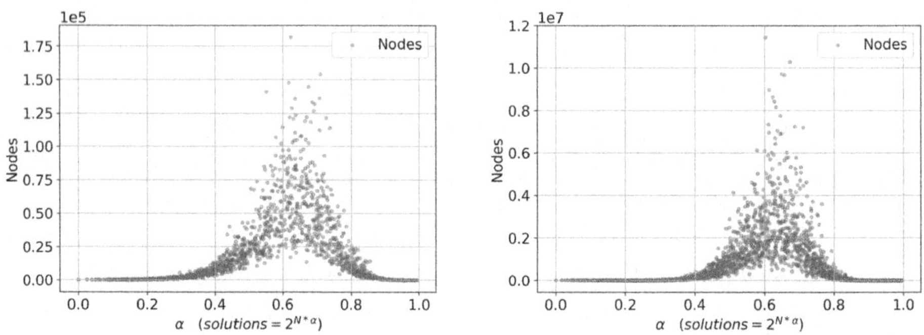


(a) 40 vars (b) 60 vars

**Fig. 9.** Nodes in d-DNNF vs solution density($\alpha$) for different number of variables

but it is difficult to comment upon the precise location and an extended study is required to generate independent instances for a given solution density.

We see that solution density is a major parameter affecting the size and runtime of compiled instances. In this context, we seek to understand the impact of other parameters on the phase transition location with respect to solution density. We focus on two such parameters: clause length and the number of variables.

*Impact of Clause Length.* Figure 8 shows that as we increase the clause length, the location of phase transition point with respect to solution density moves closer to 1. It is worth remarking that in the context of the satisfiability, the location of phase transition, albeit with respect to clause densityk, is known to depend on the clause length, so a similar behavior in the context of knowledge compilation is indeed not surprising.

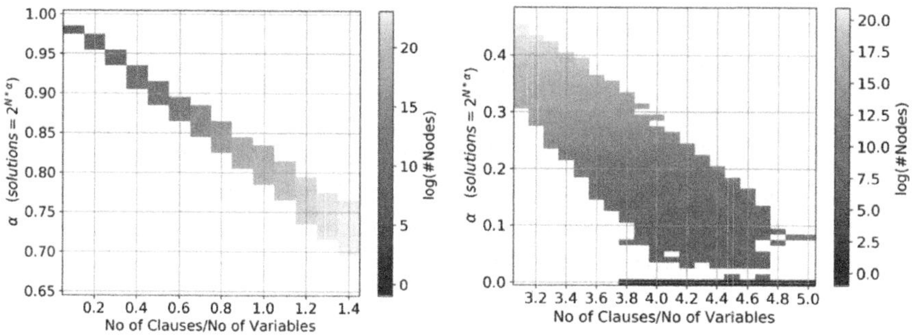


(a) $r \in [0.1, 1.4]$ (b) $r \in [3.1, 5.0]$

**Fig. 10.** log(Nodes) in d-DNNF in solution vs clause density grid for 3-CNF

*Size with Number of Variables.* From Fig. 9 we observe that the distribution of Nodes in d-DNNF compilation becomes sharper around the phase transition solution density with increasing number of variables.

*Runtime of Compilation.* From Fig. 7, we observe that the distribution of runtimes follows a similar easy-hard-easy pattern for d-dNNFs as well as SDDs but not OBDDs. The runtime behavior is similar to that of distribution of the number of nodes, as discussed in Sect. 4.

We sum up our observations in the following conjecture:

*Conjecture 2.* For every integer $k \geq 2$, given the number of variables $n$ and a target language $\mathcal{L}$ that is a subset of DNNF, there exists a positive real number $\alpha_k$ such that for each pair $(\alpha_1, \alpha_2)$, if $0 \leq \alpha_1 < \alpha_2 < \alpha_k$ then,

$$\mathsf{E}(\mathcal{N}_{\mathcal{L}}(G_k(n, 2^{\alpha_1 n}))) < \mathsf{E}(\mathcal{N}_{\mathcal{L}}(G_k(n, 2^{\alpha_2 n})))$$

For each pair $(\alpha_1, \alpha_2)$, if $\alpha_k < \alpha_1 < \alpha_2 \leq 1$ then,

$$\mathsf{E}(\mathcal{N}_{\mathcal{L}}(G_k(n, 2^{\alpha_1 n}))) > \mathsf{E}(\mathcal{N}_{\mathcal{L}}(G_k(n, 2^{\alpha_2 n})))$$

### 5.2 Combined Effect of Clause and Solution Density

We have seen that the density of solutions and clauses play a pivotal role, affecting the size and runtime. Also, given a clause density, the expected solution density is fixed and vice versa. This makes one wonder if there is a more complex relationship at play that affects the phase transition behavior of knowledge compilations. To investigate, we plot a heatmap (Fig. 10) on $\alpha \times r$ grid where the colours indicate the size of compilations. We employ random 3-CNF instances with 70 variables used for comparisons with clause density, as described earlier in Sect. 3. For each cell in the grid, we take an average of instances with the corresponding clause density that lie within the interval of solution density marked by the cell. Since the number of such instances can differ across the cells, we mark the average for a cell only if the number of instances is greater than 5 to minimize the variance to a feasible extent. From Fig. 10, we observe that for low clause densities, varying solution density has minimal effect on the size of compilations. In contrast, for high clause densities, solution density has a dominant effect on the size as there is a minimal variation with clause density. On the other hand, near phase transition, both the parameters play a significant role, and the precise relationship is murkier.

## 6 Effect of Different Tools

Heuristics and compilation algorithms play a crucial role in perceived hardness as elaborated in Sect. 4.1. We experimented with bottom up (TheSDDPackage) as well as top down (MiniC2D) compilation strategies for SDDs, predefined total variable ordering against dynamic ordering for OBDDs (CUDD) and different decomposition techniques for d-DNNFs (C2D, Dsharp and D4). Notably, target language for MiniC2D is decision-SDD and CUDD with predefined total variable order is $\text{OBDD}_{>}$ [14,30], which are less succinct than SDD and OBDD respectively. We state our representative observations for 3-CNF here. We observed that both MiniC2D and The SDD Package show maximum number of nodes around clause density 1.8 and solution density around 0.62. However, runtime for MiniC2D peaks around clause density 1.8 while TheSDDPackage peaks around clause density 2.0. In case of BDDs, we observed that disabling dynamic variable reordering shifts the peak (number of nodes) clause density from 2.0 to 1.5 and peak (number of nodes) solution density from 0.62 to 0.75. The observations for runtimes of OBDDs are much more involved due to reasons discussed in Sect. 4.1. For d-DNNFs, we observe that the peak (number of nodes) clause density stays around 1.8 and peak solution density stays around 0.62 irrespective of the hypergraph partitioning algorithm. Similar observations were recorded for runtime as well in case of d-DNNF. Summing up, we observe that while the precise behaviour of phase transition (for example, its location) can depend upon the heuristics employed in the process, the general behaviour persists irrespectively.

## 7 Conclusion

Our study provides evidence of phase transition behavior with respect to clause as well as solution density. While, both these parameters are linearly linked in expectation, it is interesting that varying the number of solutions on a fixed clause density leads to significant variation in the expected size of compilations. In terms of the complexity of compilations, we found the expected size is at least quasi-polynomial and expected runtime is exponential in the number of variables with varying clause density for state-of-the-art knowledge compilers. We believe that this paper opens up new directions for theoretical studies in an attempt to explain our empirical observations and conjectures.

**Acknowledgments.** This work was supported in part by National Research Foundation Singapore under its NRF Fellowship Programme[NRF-NRFFAI1-2019-0004] and AI Singapore Programme [AISG-RP-2018-005], and NUS ODPRT Grant [R-252-000-685-13]. The computational work for this article was performed on resources of the National Supercomputing Centre, Singapore https://www.nscc.sg. Any opinions, findings and conclusions or recommendations expressed in this material are those of the author(s) and do not reflect the views of National Research Foundation, Singapore.

## References

1. Achlioptas, D., Coja-Oghlan, A.: Algorithmic barriers from phase transitions. In: IEEE Symposium on Foundations of Computer Science, pp. 793–802, October 2008. https://doi.org/10.1109/FOCS.2008.11
2. Achlioptas, D.: Random satisfiability. In: Handbook of Satisfiability, pp. 245–270 (2009). https://doi.org/10.3233/978-1-58603-929-5-245
3. Achlioptas, D., Kirousis, L.M., Kranakis, E., Krizanc, D., Molloy, M.S.O., Stamatiou, Y.C.: Random constraint satisfaction: a more accurate picture. In: Smolka, G. (ed.) CP 1997. LNCS, vol. 1330, pp. 107–120. Springer, Heidelberg (1997). https://doi.org/10.1007/BFb0017433
4. Aguirre, A.S.M., Vardi, M.Y.: Random 3-SAT and BDDs: the plot thickens further. In: Walsh, T. (ed.) CP 2001. LNCS, vol. 2239, pp. 121–136. Springer, Heidelberg (2001). https://doi.org/10.1007/3-540-45578-7_9
5. Bayardo Jr., R.J., Pehoushek, J.D.: Counting models using connected components. In: AAAI, pp. 157–162 (2000)
6. Birnbaum, E., Lozinskii, E.L.: The good old Davis-Putnam procedure helps counting models. J. Artif. Int. Res. **10**(1), 457–477 (1999). http://dl.acm.org/citation.cfm?id=1622859.1622875
7. Bova, S.: SDDs are exponentially more succinct than OBDDs. In: Proceedings of AAAI, pp. 929–935. AAAI Press (2016). http://dl.acm.org/citation.cfm?id=3015812.3015951
8. Braunstein, A., Mézard, M., Zecchina, R.: Survey propagation: an algorithm for satisfiability. Random Struct. Algorithms **27**(2), 201–226 (2005). https://doi.org/10.1002/rsa.v27:2
9. Bryant, R.E.: Graph-based algorithms for Boolean function manipulation. IEEE Trans. Comput. **35**(8), 677–691 (1986). https://doi.org/10.1109/TC.1986.1676819
10. Cheeseman, P., Kanefsky, B., Taylor, W.M.: Where the really hard problems are. In: Proceedings of IJCAI, pp. 331–337 (1991). http://dl.acm.org/citation.cfm?id=1631171.1631221
11. Darwiche, A.: On the tractable counting of theory models and its application to truth maintenance and belief revision. J. Appl. Non-Classical Logics **11**, 11–34 (2001)
12. Darwiche, A.: New advances in compiling CNF to decomposable negation normal form. In: Proceedings of ECAI, pp. 318–322 (2004)
13. Darwiche, A.: SDD: a new canonical representation of propositional knowledge bases. In: Proceedings of IJCAI, pp. 819–826 (2011)
14. Darwiche, A., Marquis, P.: A knowledge compilation map. J. Artif. Int. Res. **17**, 229–264 (2002)
15. Ding, J., Sly, A., Sun, N.: Proof of the satisfiability conjecture for large k. In: Proceedings of STOC (2015). https://doi.org/10.1145/2746539.2746619

16. Dudek, J., Meel, K.S., Vardi, M.Y.: Combining the k-CNF and XOR phase-transitions. In: Proceedings of IJCAI, July 2016
17. Dudek, J., Meel, K.S., Vardi, M.Y.: The hard problems are almost everywhere for random CNF-XOR formulas. In: Proceedings of IJCAI, August 2017
18. Erdos, P., Renyi, A.: On the evolution of random graphs. In: Publication of the Mathematical Institute of the Hungarian Academy of Sciences, pp. 17–61 (1960)
19. Filippidis, I.: dd Python Package, Release 0.5.4 (2019). Mirrored on https://pypi.org/project/dd/. Accessed 04 Sept 2019
20. Gao, J., Yin, M., Xu, K.: Phase transitions in knowledge compilation: an experimental study. In: Sakallah, K.A., Simon, L. (eds.) SAT 2011. LNCS, vol. 6695, pp. 364–366. Springer, Heidelberg (2011). https://doi.org/10.1007/978-3-642-21581-0_31
21. Gent, I.P., MacIntyre, E., Prosser, P., Smith, B.M., Walsh, T.: Random constraint satisfaction: flaws and structure. Constraints Int. J. **6**(4), 345–372 (2001). https://doi.org/10.1023/A:1011454308633
22. Gent, I.P., Walsh, T.: The SAT phase transition. In: Proceedings of ECAI 1994, pp. 105–109 (1994)
23. Gent, I.P., Walsh, T.: Beyond NP: the QSAT phase transition, pp. 648–653. AAAI Press (1999)
24. Huang, J., Darwiche, A.: Using DPLL for efficient OBDD construction. In: Hoos, H.H., Mitchell, D.G. (eds.) SAT 2004. LNCS, vol. 3542, pp. 157–172. Springer, Heidelberg (2005). https://doi.org/10.1007/11527695_13
25. Jean-Marie Lagniez, P.M.: An improved decision-DNNF compiler. In: Proceedings of IJCAI, pp. 667–673 (2017)
26. Mitchell, D., Selman, B., Levesque, H.: Hard and easy distributions of SAT problems. In: AAAI, vol. 92, pp. 459–465 (1992)
27. Mora, T., Zdeborová, L.: Random subcubes as a toy model for constraint satisfaction problems. J. Stat. Phys. **131**, 1121–1138 (2008)
28. Muise, C., McIlraith, S.A., Beck, J.C., Hsu, E.: DSHARP: fast d-DNNF Compilation with sharpSAT. In: Proceedings of AAAI, pp. 356–361 (2016)
29. Oztok, U., Darwiche, A.: A top-down compiler for sentential decision diagrams. In: Proceedings of the 24th International Conference on Artificial Intelligence, IJCAI 2015, pp. 3141–3148. AAAI Press (2015)
30. Oztok, U., Darwiche, A.: An exhaustive DPLL algorithm for model counting. J. Artif. Int. Res. **62**(1), 1–32 (2018). https://doi.org/10.1613/jair.1.11201
31. Pote, Y., Joshi, S., Meel, K.S.: Phase transition behavior of cardinality and XOR constraints. In: Proceedings of IJCAI, August 2019
32. Somenzi, F.: CUDD Package, Release 3.0.0 (2004). Mirrored on https://sourceforge.net/projects/cudd-mirror/. Accessed 04 Sept 2019
33. The SDD package, ver 2.0 (2018). http://reasoning.cs.ucla.edu/sdd/. Accessed 5 June 2019
34. Williams, C., Hogg, T.: Exploiting the deep structure of constraint problems. Artif. Intell. **70**, 73–117 (1994)

# A Faster Exact Algorithm to Count X3SAT Solutions

Gordon Hoi[1], Sanjay Jain[1(✉)], and Frank Stephan[1,2]

[1] School of Computing, National University of Singapore,
13 Computing Drive, Block COM1, Singapore 117417, Republic of Singapore
e0013185@u.nus.edu, sanjay@comp.nus.edu.sg

[2] Department of Mathematics, National University of Singapore,
10 Lower Kent Ridge Road, Block S17, Singapore 119076, Republic of Singapore
fstephan@comp.nus.edu.sg

**Abstract.** The Exact Satisfiability problem, XSAT, is defined as the problem of finding a satisfying assignment to a formula in CNF such that there is exactly one literal in each clause assigned to be "1" and the other literals in the same clause are set to "0". If we restrict the length of each clause to be at most 3 literals, then it is known as the X3SAT problem. In this paper, we consider the problem of counting the number of satisfying assignments to the X3SAT problem, which is also known as #X3SAT.

The current state of the art exact algorithm to solve #X3SAT is given by Dahllöf, Jonsson and Beigel and runs in $O(1.1487^n)$ time, where $n$ is the number of variables in the formula. In this paper, we propose an exact algorithm for the #X3SAT problem that runs in $O(1.1120^n)$ time with very few branching cases to consider, by using a result from Monien and Preis to give us a bisection width for graphs with at most degree 3.

**Keywords:** #X3SAT · Counting models · Exponential time algorithms

## 1 Introduction

Given a propositional formula $\varphi$ in conjunctive normal form (CNF), a common question to ask would be if there is a satisfying assignment to $\varphi$. This is known as the satisfiability problem, or SAT. Many other variants of the satisfiability problem have also been explored. An important variant is the Exact Satisfiability problem, XSAT, where it asks if one can find a satisfying assignment such that exactly one of the literals in each clause is assigned the value "1" and all other

Sanjay Jain and Frank Stephan are supported in part by the Singapore Ministry of Education Tier 2 grant AcRF MOE2019-T2-2-121/R146-000-304-112. Further, Sanjay Jain is supported in part by NUS grant number C252-000-087-001. We thank the anonymous referees of CP2020 for several helpful comments. An extended technical report is available at https://arxiv.org/pdf/2007.07553.pdf.


H. Simonis (Ed.): CP 2020, LNCS 12333, pp. 375–391, 2020.
https://doi.org/10.1007/978-3-030-58475-7_22

literals in the same clause are assigned “0”. Another variant that has been heavily studied is the restriction of the number of literals allowed in each clause. In both SAT and XSAT, one allows arbitrary number of literals to be present in each clause. If we restrict the number of literals to be at most $k$ in each clause, then the above problems are now known as $k$SAT and X$k$SAT respectively. The most famous of these variants are 3SAT and X3SAT. All the mentioned problems, SAT, 3SAT, XSAT and X3SAT are known to be NP-complete [1–3,10,17].

Apart from decision problems and optimization problems, one can also work on counting the number of different models that solves the decision problem. For example, we can count the number of different satisfying assignments that solves SAT, and this is known as #SAT. The problem #3SAT, #XSAT and #X3SAT are defined similarly. Counting problems seem much harder than their decision counterparts. One may use the output of a counting algorithm to solve the decision problem. Another convincing example can be seen in that 2SAT is known to be in P [11] but #2SAT is #P-complete [18]. In fact, #SAT, #3SAT, #X3SAT and #XSAT are all known to be in #P-complete [18,19]. The problem of model counting has found wide applications in the field of AI such as the use of inference in Bayesian belief networks or probabilistic inference [15,16]. In this paper, we will focus on the #X3SAT problem.

Let $n$ denote the number of variables in the formula. Algorithms to solve #XSAT have seen numerous improvements [4,5,14,20] over the years. To date, the fastest #XSAT algorithm runs in $O(1.1995^n)$ time [21]. Of course, to solve the #X3SAT problem, one can rely on any of the mentioned algorithm that solves #XSAT to solve them directly. However, it is possible to exploit the structure of X3SAT and hence solve #X3SAT in a much faster manner. Dahllöf, Jonsson and Beigel gave an #X3SAT algorithm in $O(1.1487^n)$ time [5].

In this paper, we propose a faster and simpler algorithm to solve the #X3SAT problem in $O(1.1120^n)$ time. The novelty here lies in the use of a result by Monien and Preis [13] to help us to deal with a specific case. Also using a different way to analyze our algorithm allows us to tighten the analysis further.

## 2 Preliminaries

In this section, we will introduce some common definition needed by the algorithm and also the techniques needed to understand the analysis of the algorithm. The main design of our algorithm is a Davis Putnam Logemann Loveland (DPLL) [6,7] style algorithm, or also known as the branch and bound algorithm. Such algorithms are recursive in nature and have two kinds of rules associated with them: Simplification and Branching rules. Simplification rules help us to simplify a problem instance. Branching rules on the other hand, help us to solve a problem instance by recursively solving smaller instances of the problem. To illustrate the execution of the DPLL algorithm, a search tree is commonly used. We assign the root node of the search tree as the original problem. The subsequent child nodes are assigned whenever we invoke a branching rule. For more information, one may refer to [8]. Let $\mu$ denote our parameter of complexity.

To analyse the running time of the DPLL algorithm, one in fact just needs to bound the number of leaves generated in the search tree. This is due to the fact that the complexity of such algorithm is proportional to the number of leaves, modulo polynomial factors, i.e., $O(poly(|\varphi|, \mu) \times$ number of leaves in the search tree$)$ $= O^*($number of leaves in the search tree$)$, where the function $poly(|\varphi|, \mu)$ is some polynomial based on $|\varphi|$ and $\mu$, while $O^*(g(\mu))$ is the class of all functions $f$ bounded by some polynomial $p(\cdot)$ times $g(\mu)$.

Then we let $T(\mu)$ denote the maximum number of leaf nodes generated by the algorithm when we have $\mu$ as the parameter for the input problem. Since the search tree is only generated by applying a branching rule, it suffices to consider the number of leaf nodes generated by that rule (as simplification rules take only polynomial time). To do this, we employ techniques in [12]. Suppose a branching rule has $r \geq 2$ children, with $t_1, t_2, \ldots, t_r$ number of variables eliminated for these children. Then, any function $T(\mu)$ which satisfies $T(\mu) \geq T(\mu - t_1) + T(\mu - t_2) + \ldots T(\mu - t_r)$, with appropriate base cases, would satisfy the bounds for the branching rule. To solve the above linear recurrence, one can model this as $x^{-t_1} + x^{-t_2} + \ldots + x^{-t_r} = 1$. Let $\beta$ be the root of this recurrence, where $\beta \geq 1$. Then any $T(\mu) \geq \beta^{\mu}$ would satisfy the recurrence for this branching rule. In addition, we denote the branching factor $\tau(t_1, t_2, \ldots, t_r)$ as $\beta$. Tuple $(t_1, t_2, \ldots, t_r)$ is also known as the branching vector [8]. If there are $k$ branching rules in the DPLL algorithm, then the overall complexity of the algorithm can be seen as the largest branching factor among all $k$ branching rules; i.e. $c = max\{\beta_1, \beta_2, \ldots, \beta_k\}$, and therefore the time complexity of the algorithm is bounded above by $O^*(c^{\mu})$.

We will introduce some known results about branching factors. If $k < k'$, then we have that $\tau(k', j) < \tau(k, j)$, for all positive $k, j$. In other words, comparing two branching factors, if one eliminates more variable, then this will result in a a smaller branching factor. Suppose that $i + j = 2\alpha$, for some $\alpha$, then $\tau(\alpha, \alpha) \leq \tau(i, j)$. In other words, a more balanced tree will give a smaller branching factor.

Finally, suppose that we have a branching vector of $(u, v)$ for some branching rule. Suppose that for the first branch, we immediately do a follow up branching to get a branching vector of $(w, x)$, then we can apply branching vector addition to get a combined branching vector of $(u + w, u + x, v)$. This technique can sometimes help us to bring down the overall complexity of the algorithm further.

Finally, the correctness of DPLL algorithms usually follows from the fact that all cases have been covered. We now give a few definitions before moving onto the actual algorithm. We fix a formula $\varphi$:

**Definition 1.** Two clauses are called neighbours if they share at least a common variable. Two variables are called neighbours if they appear in some clause together. We say that a clause $C$ is a degree $k$ clause if $C$ has $k$ neighbours. Finally, a variable is a singleton if it appears only once in $\varphi$.

Suppose we have clauses $C_1 = (x \vee y \vee z)$, $C_2 = (x \vee a \vee b)$ and $C_3 = (y \vee a \vee c)$. Then $C_1$ is a neighbour to $C_2$ and $C_3$. In addition, all three are degree 2 clauses. Variables $a, b, y, z$ are neighbours of $x$, while $b, c, z$ are singletons.

**Definition 2.** We say that two variables, $x$ and $y$, are linked when we can deduce either $x = y$ or $x = \bar{y}$. When this happens, we can proceed to remove one of the linked variable, either $x$ or $y$, by replacing it with the other.

For example, in clause $(0 \vee x \vee y)$, we know that $x = \bar{y}$ to satisfy it. Thus, we can link $x$ with $\bar{y}$ and remove one of the variables, say $y$.

**Definition 3.** We denote the formula $\varphi[x = 1]$ obtained from $\varphi$ by assigning a value of 1 to the literal $x$. We denote the formula $\varphi[x = y]$ as obtained from $\varphi$ by substituting all instances of $x$ by $y$. Similarly, let $\delta$ be a subclause. We denote $\varphi[\delta = 0]$ as obtained from $\varphi$ by substituting all literals in $\delta$ to 0.

Suppose we have $\varphi = (x \vee y \vee z)$. Then if we assign $x = 1$, then $\varphi[x = 1]$ gives us $(1 \vee y \vee z)$. On the other hand, if we have $\varphi[y = x]$, then we have $(x \vee x \vee z)$. If $\delta = (y \vee z)$, then $\varphi[\delta = 0]$ gives us $(x \vee 0 \vee 0)$.

**Definition 4.** A sequence of degree 2 clauses $C_1, C_2, \ldots, C_k$, $k \geq 1$ is called a chain if for $1 \leq j \leq k - 1$, we have $C_j$ is a neighbour to $C_{j+1}$. Given any two clauses $C_e$ and $C_f$ that are at least degree 3, we say that they are connected via a chain if we have a chain $C_1, C_2, \ldots, C_k$ such that $C_1$ is a neighbour of $C_e$ (respectively $C_f$) and $C_k$ is a neighbour of $C_f$ (respectively $C_e$). Moreover, if we have a chain of degree 2 clauses $C_1, C_2, \ldots, C_k, C_1$, then we call this a cycle.

Suppose we have the following degree 3 clauses: $(a \vee b \vee c)$ and $(s \vee t \vee u)$, and the following chain: $(c \vee d \vee e)$, $(e \vee f \vee g)$, ..., $(q \vee r \vee s)$. Then note that the degree 3 clause $(a \vee b \vee c)$ is a neighbour to $(c \vee d \vee e)$ and $(s \vee t \vee u)$ is a neighbour to $(q \vee r \vee s)$. Therefore, we say that $(a \vee b \vee c)$ and $(s \vee t \vee u)$ are connected via a chain.[1]

**Definition 5.** A path $x_1, x_2, \ldots, x_i$ is a sequence of variables such that for each $j \in \{1, \ldots, i - 1\}$, the variables $x_j$ and $x_{j+1}$ are neighbours. A component is a maximal set of clauses such that any two variables, found in any clauses in the set has a path between each other. A formula is connected if any two variables have a path between each other. Else we say that the formula is disconnected, and consists of $k \geq 2$ components.

For example, let $\varphi = (x \vee y \vee z) \wedge (x \vee a \vee b) \wedge (e \vee c \vee d) \wedge (e \vee f \vee g)$. Then $\varphi$ is disconnected and is made up of two components, since $x$ has no path to $e$, while variables in the set $\{(x \vee y \vee z), (x \vee a \vee b)\}$ have a path to each other. Similarly, for $\{(e \vee c \vee d), (e \vee f \vee g)\}$. Therefore, $\{(x \vee y \vee z), (x \vee a \vee b)\}$ and $\{(e \vee c \vee d), (e \vee f \vee g)\}$ are two components.

**Definition 6.** Let $I$ be a set of variables of a fixed size. We say that $I$ is semi-isolated if there exists an $s \in I$ such that in any clause involving variables not in $I$, only $s$ from $I$ may appear.

[1] The definition of chains and cycles will be mainly used in Sect. 4.3 and Sect. 4.4.

For example consider the set $I = \{x, y, z, a, b\}$ and the clauses $(x \vee y \vee z)$, $(x \vee a \vee b)$, $(b \vee c \vee d)$, $(c \vee d \vee e)$. Since $b$ is the only variable in $I$ that appears in clauses involving variables not in $I$, $I$ is semi-isolated.

**Definition 7.** Suppose $G = (V, E)$ is a simple undirected graph. A *balanced bisection* is a mapping $\pi : V \to \{0, 1\}$ such that, for $V_i = \{v : \pi(v) = i\}$, $|V_0|$ and $|V_1|$ differ by at most one. Let $cut(\pi) = |\{(v, w) : (v, w) \in E, v \in V_0, w \in V_1\}|$. The bisection width of $G$ is the smallest $cut(\cdot)$ that can be obtained for a balanced bisection.

**Theorem 8 (see Monien and Preis [13]).** *For any $\varepsilon > 0$, there is a value $n(\varepsilon)$ such that the bisection width of any 3-regular graph $G = (V, E)$ with $|V| > n(\varepsilon)$ is at most $(\frac{1}{6} + \varepsilon)|V|$. This bisection can be found in polynomial time.*

The above result extends to all graphs $G$ with maximum degree of 3 [9].

## 3 Algorithm

Our algorithm takes in a total of 4 parameters: a formula $\varphi$, a cardinality vector $\mathbf{c}$, two sets $L$ and $R$.

The second parameter, a cardinality vector $\mathbf{c}$, maps literals to $\mathbb{N}$. The idea behind introducing this cardinality vector $\mathbf{c}$ is to help us to keep track of the number of models while applying simplification and branching rules. At the start, $\mathbf{c}(l) = 1$ for all literals in $\varphi$ and will be updated along the way whenever we link variables together or when we remove singletons. Since linking of variables is a common operation, we introduce a function to help us perform this procedure. The function *Link*(.), takes as inputs the cardinality vector and two literals involving different variables to link them[2]. It updates the information of the eliminated variable ($y$) onto the surviving variable ($x$) and after which, drops the entries of eliminated variable ($y$ and $\bar{y}$) in the cardinality vector $\mathbf{c}$. When we link $x$ and $y$ as $x = y$ (respectively, $x = \bar{y}$), then we call the function $Link(\mathbf{c}, x, y)$ (respectively, $Link(\mathbf{c}, x, \bar{y})$). We also use a function $MonienPreis(.)$ to give us partition based on Theorem 8.

**Function:** *Link*(.)
Input: A Cardinality Vector $\mathbf{c}$, literal $x$, literal $y$
Output: An updated Cardinality Vector $\mathbf{c'}$

- Update $\mathbf{c}(x) = \mathbf{c}(x) \times \mathbf{c}(y)$, and $\mathbf{c}(\bar{x}) = \mathbf{c}(\bar{x}) \times \mathbf{c}(\bar{y})$. After which, drop entries of $y$ and $\bar{y}$ from $\mathbf{c}$ and update it as $\mathbf{c'}$. Finally, return $\mathbf{c'}$

**Function:** $MonienPreis(.)$
Input: A graph $G_\varphi$ with maximum degree 3
Output: $L$ and $R$, the left and right partitions of minimum bisection width

For the third and fourth parameter, we have the sets of clauses $L$ and $R$. $L$ and $R$ will be used to store partitions of clauses after calling $MonienPreis(.)$,

[2] As seen in Definition 2.

based on the minimum bisection width. Initially, $L$ and $R$ are empty sets and will continue to be until we first come to Line 17 of the algorithm.[3]

We call our algorithm $CountX3SAT(\cdot)$. Whenever a literal $l$ is assigned a constant value, we drop both the entries $l$ and $\bar{l}$ from the cardinality vector and multiply the returning recursive call by $\mathbf{c}(l)$ if $l = 1$, or $\mathbf{c}(\bar{l})$ if $\bar{l} = 1$. In each recursive call, we ensure that the cardinality vector is updated to contain only entries where variables in the remaining formula have yet to be assigned a constant value. By doing so, we guarantee the following invariant: For any given $\varphi$, let $S_\varphi = \{h : h \text{ is an exact-satisfiable assignment for } \varphi\}$. Now for any given $\varphi$ and a cardinality vector $\mathbf{c}$, the output of $CountX3SAT(\varphi, \mathbf{c}, L, R)$ is given as $\sum_{h \in S_\varphi} \prod_{l: l \text{ is assigned true in } h} \mathbf{c}(l)$. Initial call to our algorithm would be $CountX3SAT(\varphi, \mathbf{c}, \emptyset, \emptyset)$, where the cardinality vector $\mathbf{c}$ has $\mathbf{c}(l) = 1$ for all literals at the start. The correctness of the algorithm follows from the fact that each step will maintain the invariant that $CountX3SAT(\varphi, \mathbf{c}, L, R)$ returns $\sum_{h \in S_\varphi} \prod_{l: l \text{ is assigned true in } h} \mathbf{c}(l)$, where if $\varphi$ is not exactly satisfiable, it returns 0. Note that in the algorithm below possibilities considered are exhaustive.

Algorithm: CountX3SAT(.)
Input: A formula $\varphi$, a cardinality vector $\mathbf{c}$, a set $L$, a set $R$
Output: $\sum_{h \in S_\varphi} \prod_{l: l \text{ is assigned true in } h} \mathbf{c}(l)$

1: If any clause is not exact satisfiable (by analyzing this clause itself) then return 0. If all clauses consist of constants evaluating to 1 or no clause is left then return 1.
2: If there is a clause $(1 \vee \delta)$, then let $\mathbf{c}'$ be the new cardinality vector by dropping the entries of the variables in $\delta$. Drop this clause from $\varphi$. Return $CountX3SAT(\varphi[\delta = 0], \mathbf{c}', L, R) \times \prod_{i \text{ is a literal in } \delta} \mathbf{c}(\bar{i})$
3: If there is a clause $C = (0 \vee \delta)$, then update $C = \delta$ in $\varphi$. Return $CountX3SAT(\varphi, \mathbf{c}, L, R)$.
4: If there is a single literal $x$ in a clause, then let $\mathbf{c}'$ be the new cardinality vector by dropping the entries $x$ and $\bar{x}$ from $\mathbf{c}$. Return $CountX3SAT(\varphi[x = 1], \mathbf{c}', L, R) \times \mathbf{c}(x)$.
5: If there is a 2-literal clause $(x \vee y)$, for some literals $x$ and $y$ with $x \neq y$ and $x \neq \bar{y}$, then $\mathbf{c}' = Link(\mathbf{c}, x, \bar{y})$. Return $CountX3SAT(\varphi[y = \bar{x}], \mathbf{c}', L, R)$.
6: If there is a clause $(x \vee \bar{x})$, for some variable $x$. Check if $x$ appears in other clauses. If yes, then drop this clause from $\varphi$ and return $CountX3SAT(\varphi, \mathbf{c}, L, R)$. If no, then let $\mathbf{c}'$ be the new cardinality vector by dropping $x$ and $\bar{x}$. Drop this clause from $\varphi$ and return $CountX3SAT(\varphi, \mathbf{c}', L, R) \times (\mathbf{c}(x) + \mathbf{c}(\bar{x}))$.
7: If there are $k \geq 2$ components in $\varphi$ and there are no edges between $L$ and $R$, then let $\varphi_1, \ldots, \varphi_k$ be the $k$ components of $\varphi$. Let $\mathbf{c}_i$ be the cardinality vector for $\varphi_i$ by only keeping the entries of the literals involving variables appearing in $\varphi_i$, and dropping the rest. Let $L = R = \emptyset$. Return $CountX3SAT(\varphi_1, \mathbf{c}_1, L, R) \times \ldots \times CountX3SAT(\varphi_k, \mathbf{c}_k, L, R)$.
8: If there exists a clause $(x \vee x \vee y)$, for some literals $x$ and $y$, then let $\mathbf{c}'$ be the new cardinality vector by dropping the entries $x$ and $\bar{x}$ from $\mathbf{c}$. Return $CountX3SAT(\varphi[x = 0], \mathbf{c}', L, R) \times \mathbf{c}(\bar{x})$

[3] More details about their role will be given in Sect. 4.3.

9: If there is a clause $(x \vee \bar{x} \vee y)$, then let $\mathbf{c}'$ be the new cardinality vector by removing the entries $y$ and $\bar{y}$. Return $CountX3SAT(\varphi[y = 0], \mathbf{c}', L, R) \times \mathbf{c}(\bar{y})$

10: If there exists a clause containing two singletons $x$ and $y$, then update $\mathbf{c}$ as: $\mathbf{c}(x) = \mathbf{c}(x) \times \mathbf{c}(\bar{y}) + \mathbf{c}(\bar{x}) \times \mathbf{c}(y)$, $\mathbf{c}(\bar{x}) = \mathbf{c}(\bar{x}) \times \mathbf{c}(\bar{y})$.
Let $\mathbf{c}'$ be the new cardinality vector by dropping the entries $y$ and $\bar{y}$ from $\mathbf{c}$. Drop $y$ from $\varphi$. Return $CountX3SAT(\varphi, \mathbf{c}', L, R)$.

11: There are two clauses $(x \vee y \vee z)$ and $(x \vee y \vee w)$, for some literals $x, y, z$ and $w$. Then in this case, let $\mathbf{c}' = Link(\mathbf{c}, z, w)$. Drop one of the clauses. Return $CountX3SAT(\varphi[w = z], \mathbf{c}', L, R)$.

12: There are two clauses $(x \vee y \vee z)$ and $(x \vee \bar{y} \vee w)$, for some literals $x, y, z$ and $w$. Then let $\mathbf{c}'$ be the new cardinality vector by dropping entries of $x$ and $\bar{x}$. Return $CountX3SAT(\varphi[x = 0], \mathbf{c}', L, R) \times \mathbf{c}(\bar{x})$.

13: There are two clauses $(x \vee y \vee z)$ and $(\bar{x} \vee \bar{y} \vee w)$, for some literals $x, y, z$ and $w$. Then $\mathbf{c}' = Link(\mathbf{c}, x, \bar{y})$. Return $CountX3SAT(\varphi[y = \bar{x}], \mathbf{c}', L, R)$.

14: If there exists a semi-isolated set $I$, with $3 \leq |I| \leq 20$, then let $x$ be the variable appearing in further clauses with variables not in $I$. Let $\mathbf{c}'$ be the new cardinality vector by updating the entries of $x$ and $\bar{x}$, dropping of entries of variables in $I - \{x\}$. Drop all the entries of $I - \{x\}$ from $\varphi$. Return $CountX3SAT(\varphi, \mathbf{c}', L, R)$.[4]

15: This rule is not analyzed for all cases, but only specific cases as mentioned in Sections 4.1 and 4.2 (more specifically this applies only when some variable appears in at least 3 clauses). If there exists a variable $x$ such that branching $x = 1$ and $x = 0$ allows us to either remove at least 7 variables on both branches, or at least 8 on one and 6 on the other, or at least 9 on one and 5 on the other, then branch $x$. Let $\mathbf{c}'$ be the new cardinality vector by dropping the entries $x$ and $\bar{x}$. Return $CountX3SAT(\varphi[x = 1], \mathbf{c}', L, R) \times \mathbf{c}(x) + CountX3SAT(\varphi[x = 0], \mathbf{c}', L, R) \times \mathbf{c}(\bar{x})$[5].

16: If there exists a variable $x$ appearing at least 3 times, then let $\mathbf{c}'$ be the new cardinality vector by dropping the entries $x$ and $\bar{x}$. Return $CountX3SAT(\varphi[x = 1], \mathbf{c}', L, R) \times \mathbf{c}(x) + CountX3SAT(\varphi[x = 0], \mathbf{c}', L, R) \times \mathbf{c}(\bar{x})$[5].

17: If there is a degree 3 clause in $\varphi$, then check if $\exists$ an edge between $L$ and $R$. If no, then construct $G_\varphi$ and let $(L', R') \leftarrow MonienPreis(G_\varphi)$. Then return $CountX3SAT(\varphi, \mathbf{c}, L', R')$. If $\exists$ an edge between $L$ and $R$, apply only the simplification rules (if any) as stated in Section 4.3. Choose an edge $e$ between $L$ and $R$. Then branch the variable $x_e$ represented by $e$. Let the cardinality vector $\mathbf{c}'$ be the new cardinality vector by dropping off entries $x_e$ and $\bar{x}_e$. Return $CountX3SAT(\varphi[x_e = 1], \mathbf{c}', L, R) \times \mathbf{c}(x_e) + CountX3SAT(\varphi[x_e = 0], \mathbf{c}', L, R) \times \mathbf{c}(\bar{x}_e)$[3].

18: If every clause in the formula is degree 2, choose any variable $x$ and we branch $x = 1$ and $x = 0$. Let $\mathbf{c}'$ be the new cardinality vector by dropping the entries $x$ and $\bar{x}$. Return $CountX3SAT(\varphi[x = 1], \mathbf{c}', L, R) \times \mathbf{c}(x) + CountX3SAT(\varphi[x = 0], \mathbf{c}', L, R) \times \mathbf{c}(\bar{x})$[5].

[4] More details on the updating of $\mathbf{c}'$ below in this section.

[5] More details on this branching rule is given in Section 4.

Note that every line in the algorithm has descending priority; Line 1 has higher priority than Line 2, Line 2 than Line 3 etc.

Line 1 of the algorithm is our stopping condition. If any clause is not exact satisfiable, immediately return 0. When no variables are left, then check if every clause is exactly satisfied. If yes, then return 1, else 0.

Line 2 of the algorithm deals with any clause that contains a constant 1. In this case, all the other literals in the clause must be assigned 0 and we can safely drop off this clause after that. Line 3 deals with any clause with a constant 0 in it. We can then safely drop the constant 0 from the clause. Line 4 deals with single-literal clauses. This literal must be assigned 1. Line 5 deals with two literal clauses when the two literals involve two different variables. Line 6 deals with two literal clauses when they come from the same variable, say $x$. Now if $x$ does not appear elsewhere, then either $x = 1$ or $x = 0$ will satisfy this clause. Thus as done in Line 6, multiplying $CountX3SAT(\varphi, \mathbf{c}', L, R)$ by the sum of $(\mathbf{c}(x) + \mathbf{c}(\bar{x}))$ would give us the correct value. Regardless of whether $x$ appears elsewhere or not, drop this clause.

After Line 6, we know that all clauses are of length 3. In Line 7, if the formula is disconnected, then we deal with each components separately. Line 7 has some relation with Line 17. If the algorithm is not currently processing Line 17, then basically we just call the algorithm on different components. The explicit relationship between Line 7 and Line 17 will be given in Sect. 4.3. In Line 8, we deal with a literal that appears twice in a clause. Then we can assign that literal as 0. In Line 9, we have a literal and its negation appearing in the same clause, then we assign the last literal to be 0. In Line 10, we deal with clauses having two singletons and we need to update the cardinality vector $\mathbf{c}$ before we are allowed to remove one. Suppose we have two singletons $x$ and $y$ and we wish to remove say $y$, then we need to update the entries of $\mathbf{c}(x)$ and $\mathbf{c}(\bar{x})$ to retain the information of $\mathbf{c}(y)$ and $\mathbf{c}(\bar{y})$. Note that in the updated $x$, when $x = 0$, this means that both the original $x$ and $y$ are 0. On the other hand, when we have $x = 1$ in the updated $x$, this means that we can either have $x = 1$ in the original $x$, or $y = 1$. Thus, this gives us the following update: $\mathbf{c}(x) = \mathbf{c}(x) \times \mathbf{c}(\bar{y}) + \mathbf{c}(\bar{x}) \times \mathbf{c}(y)$ when $x$ is assigned "1", and $\mathbf{c}(\bar{x}) = \mathbf{c}(\bar{x}) \times \mathbf{c}(\bar{y})$ when $x$ is assigned "0". After which, we can then safely remove the entries of $y$ and $\bar{y}$ from the cardinality vector $\mathbf{c}$.

In Lines 11, 12 and 13, we deal with two overlapping variables (in different permutation) between any two clauses. After which, any two clauses can only have at most only 1 overlapping variable between them. In Line 14, we deal with semi-isolated sets $I$ such that we can remove all but one of its variable. In Line 15, if we can find a variable $x$ such that by branching it, we can remove that amount of variables as stated, then we proceed to do so. The goal of introducing Line 14 and Line 15 is to help us out for Line 16, where we deal with variables that appear at least 3 times. Their relationship will be made clearer in the latersections. After which, all variables will appear at most 2 times and each clause must have at most degree 3. In Line 17, the remaining formula must consist of clauses of degree 2 and 3. Then we construct a graph $G_\varphi$, apply $MonienPreis(.)$ to it and choose a variable to branch, followed by applying simplification rules. We'll continue doing so until no degree 3 clauses exist. Lastly in Line 18, the formula will only consist of degree 2 clauses, and we will select any variable and branch $x = 1$ and $x = 0$. Hence, we have covered all cases in the algorithm.

Now, we give the details of Line 14. As $I$ is semi-isolated, let $x$ be the variable in $I$, such that $x$ appears in further clauses containing variables not in $I$. Note that when $x = 1$ or when $x = 0$, the formula becomes disconnected and clauses involving $I - \{x\}$ become a component of constant size. Therefore, we can use brute force (requiring constant time), to check which assignments to the $|I| - 1$ variables satisfy the clauses involving variables from $I$, and then correspondingly update $\mathbf{c}(x)$ and $\mathbf{c}(\bar{x})$, and drop all variables in $I - \{x\}$ from $\varphi$. We call such a process *contraction* of $I$ into $x$. Details given below.

**Updating of Cardinality Vector in Line 14 (Contracting Variables).** Let $S$ be the set of clauses which involve only variables in $I$. $\delta$ below denotes assignments to variables in $I - \{x\}$. For $i \in \{0, 1\}$, let

$Z_i = \{\delta$ : all clauses in $S$ are satisfied when variables in $I$ are set according to $\delta$ and $x = i\}$.

The following formulas update the cardinality vector for coordinate $x$ and $\bar{x}$, by considering the different possibilities of $\delta$ which make the clauses in $S$ satisfiable. This is done by summing over all such $\delta$ in $Z_i$ (for $i = x = 0$ and $i = x = 1$), the multiplicative factor formed by considering the cardinality vector values at the corresponding true literals in $\delta$. Here the literals $\ell$ in the formula range over literals involving the variables in $I - \{x\}$.

Let $\mathbf{c}(x) = \mathbf{c}(x) \times \sum_{\delta \in Z_1} \prod_{\ell \text{ is true in } \delta} \mathbf{c}(\ell)$.
Let $\mathbf{c}(\bar{x}) = \mathbf{c}(\bar{x}) \times \sum_{\delta \in Z_0} \prod_{\ell \text{ is true in } \delta} \mathbf{c}(\ell)$.

# 4 Analysis of the Branching Rules of the Algorithm

Note that Lines 1 to 14 are simplification rules and Lines 15 to 18 are branching rules. For Line 7, note that since the time of our algorithm is running in $O^*(c^n)$, for some $c$, then calling our algorithm onto different components will still give us $O^*(c^n)$. Therefore, we will analyse Lines 15 to 18 of the algorithm.

## 4.1 Line 15 of the Algorithm

The goal of introducing Lines 14 and 15 is to ultimately help us to simplify our cases when we deal with Line 16 of the algorithm. In Line 16, there can be some ugly overlapping cases which we don't have to worry after adding Lines 14 and 15 in the algorithm. The cases we are interested in are as follows.

(A) There exists a variable which appears in at least four clauses.

Suppose the variable is $x_0$, and the four clauses it appears in are $(x_0' \vee x_1 \vee x_2)$, $(x_0'' \vee x_3 \vee x_4)$, $(x_0''' \vee x_5 \vee x_6)$, $(x_0'''' \vee x_7 \vee x_8)$, where $x_0', x_0'', x_0''', x_0''''$ are either $x_0$ or $\bar{x_0}$. Note that $x_0, x_1, x_2, \ldots, x_8$ are literals involving different variables (by Lines 8,9,11,12,13). Note that setting literal $x_0'$ to 1 will correspondingly set both $x_1$ and $x_2$ to 0; when $x_0'$ is set to 0 correspondingly $x_1$ and $\bar{x_2}$ get linked. Similarly, when we set $x_0'', x_0''', x_0''''$. Thus, setting $x_0$ to 1 or 0 will give us removal of $i$ variables on one setting and $12 - i$ variables on the other setting, where $4 \leq i \leq 8$. Thus, including $x_0$, this gives us, in the worst case, a branching factor of $\tau(9, 5)$.

(B) There exists a variable which appears in exactly three clauses.

Suppose $x_0$ is a variable appearing in the three clauses $(x_0' \vee x_1 \vee x_2)$, $(x_0'' \vee x_3 \vee x_4)$, $(x_0''' \vee x_5 \vee x_6)$ where $x_0', x_0'', x_0'''$ are either $x_0$ or $\bar{x_0}$. Note that $x_0, x_1, x_2, \ldots, x_6$ are literals involving different variables. Let $I = \{x_0, v_1, v_2, \ldots, v_6\}$, where $v_i$ is the variable for the literal $x_i$.

(B.1) If $I$ is semi-isolated, or $I \cup \{u\}$ is semi-isolated for some variable $u$, then Line 14 takes care of this.

(B.2) If there are two other variables $u, w$ which may appear in any clause involving variables from $I$, then we can branch on one of the variables $u$ and then do contraction as in Line 14 for $I \cup \{w\}$ to $w$. Thus, we will have a branching factor of at least $\tau(8, 8)$.

(B.3) If there are at most two clauses $C1$ and $C2$ which involve variables from $I$ and from outside $I$ and these two together involve at least three variables from outside $I$, then consider the following cases.

Case 1: If both $C1$ and $C2$ have two variables from outside $I$. Then, let $C1$ have literal $x_i'$ and $C2$ have literal $x_j'$, where $x_i'$ is either $x_i$ or $\bar{x_i}$ and $x_j'$ is either $x_j$ or $\bar{x_j}$, and $i, j \in \{0, 1, \ldots, 6\}$. Now, one can branch on literal $x_i'$ being 1 or 0. In both cases, we can contract the remaining variables of $I$ into $x_j$ (using Line 14). Including the two literals set to 0 in $C1$ when $x_i'$ is 1, we get branching factor of $\tau(8, 6)$.

Case 2: $C1$ and $C2$ together have three variables from outside $I$. Without loss of generality assume $C1$ has one variable from outside $I$ and $C2$ has two variables from outside $I$. Then let $C1$ have literal $y$ which is outside $I$ and $C2$ have literal $x_j'$, where $x_j'$ is either $x_j$ or $\bar{x_j}$. Now, one can branch on literal $y$ being 1 or 0. In both cases, we can contract the variables of $I$ into $x_j$ (using Line 14). Including the literal $y$ we get branching factor of $\tau(7, 7)$.

(B.4) Case 2.3 and Case 2.4 in Lemma 10 for Line 16.

**Lemma 9.** *Branching the variable in Line 15 takes $O(1.1074^n)$ time. (The worst branching factor is $\tau(9, 5)$).*

### 4.2 Line 16 of the Algorithm

In this case, we deal with variables that appear exactly 3 times.

**Lemma 10.** *The time complexity of branching variables appearing 3 times is $O(1.1120^n)$.*

**Proof.** Suppose $x_0$ appears three times. Then we let the clauses that $x_0$ appear in be $(x_0' \lor x_1 \lor x_2)$, $(x_0'' \lor x_3 \lor x_4)$, $(x_0''' \lor x_5 \lor x_6)$, where the primed versions of $x_0$ denote either $x_0$ or $\bar{x_0}$.

Let $I = \{x_0, v_1, \ldots, v_6\}$, where $v_i$ is the variable in the literal $x_i$.

Note that when $x_0'$ is set to 1, then $x_1$ and $x_2$ are also set to 0. When $x_0'$ is set to 0 then $x_1$ and $x_2$ get linked. Similarly, for setting of $x_0''$ and $x_0'''$. Thus, setting of $x_0$ to 1 or 0 allows us to remove $i$ variables and $9 - i$ variables respectively among $v_1, \ldots, v_6$, where $3 \leq i \leq 6$ (the worst case for us thus happens with removal of 3 variables on one side and 6 on the other). We will show how to remove three further variables outside $I$ in the following cases (these may fall on either side of setting of $x_0$ to 1 or 0 above). Including $x_0$, we get the worst case branching factor of $\tau(10, 4)$.

Let the variables outside $I$ be called outside variables for this proof. Let a clause involving both variables from $I$ and outside $I$ be called a mixed clause. By Line 14 and 15 of the algorithm, there are at least 3 mixed clauses, and at least three outside variables which appear in mixed clauses.

Consider 3 mixed clauses $C1 = (x_i' \lor a_1 \lor a_2)$, $C2 = (x_j' \lor a_3 \lor a_4)$ and $C3 = (x_k' \lor a_5 \lor a_6)$, where $a_2, a_4, a_6$ are literals involving outside variables, and $x_i', x_j', x_k'$ are literals involving variables from $I$.

Case 1: It is possible to select the three mixed clauses such that $a_4$ involves a variable not appearing in $C1$ and $a_6$ involves a variable not appearing in $C1, C2$.

Note that this can always be done when there are at least four outside variables which appear in some mixed clauses.

In this case, $x_i'$ is set in at least one of the cases of $x_0$ being set to 1 or 0. Similarly for $x_j'$ and $x_k'$. In the case when $x_i'$ is set, one can either set $a_2$ or link it to $a_1$. In the case when $x_j'$ is set, one can either set $a_4$ or link it to $a_3$. In the case when $x_k'$ is set, one can either set $a_6$ or link it to $a_5$. Note that the above linkings are not cyclic as the variable for $a_4$ is different from that of $a_1$ and $a_2$. and the variable for $a_6$ is different from that of $a_1, a_2, a_3, a_4$. Thus, in total three outside variables are removed when $x_0$ is set to 1 and 0.

Case 2: Not Case 1. Here, the number of outside variables which appear in some mixed clause is exactly three. Choose some mixed clauses $C1, C2, C3$ such that exactly three outside variables are present in them. Suppose these variables are $a, b, c$. Suppose the number of outside variables in C1, C2, C3 is given by triple $(s_1, s_2, s_3)$ (without loss of generality assume $s_1 \leq s_2 \leq s_3$). We assume that the clauses chosen are so as to have the earlier case applicable below. That is, if all three variables $a, b, c$ appear in some mixed clause as only outside variable, then Case 2.1 is chosen; Otherwise, if at least 2 mixed clauses involving 2 outside variables are there and a mixed clause involving only one outside variable is there then Case 2.2. is chosen. Otherwise, if only one mixed clause involving two outside variable is there then Case 2.3 is chosen. Else, case 2.4 is chosen.

Case 2.1: $(s_1, s_2, s_3) = (1, 1, 1)$. This would fall in Case 1, as all three outside variables are different.

Case 2.2: $(s_1, s_2, s_3) = (1, 2, 2)$. As two variables cannot overlap in two different clauses, one can assume without loss of generality that the outside variables in $C1$ is $a$ or $b$, in $C2$ are $(a, b)$ and $C3$ are $(b, c)$. But then this falls in Case 1.

Case 2.3: $(s_1, s_2, s_3) = (1, 1, 2)$. For this not to fall in Case 1, we must have the same outside variable in $C1$ and $C2$. Suppose $a$ appears in $C1, C2$ and $b, c$ in $C3$. Furthermore, to not fall in Case 1, we must have that all other outside clauses must have $a$ only as the outside variable (they cannot have both $b, c$ as outside variable, as overlapping of two variables is not allowed). Thus, by branching on $a$, and then contracting, using Line 14, $I$ to $x_k$, will allow us to have a worst case branching factor $\tau(7, 7)$. Thus, this is covered under Line 15.

Case 2.4: $(s_1, s_2, s_3) = (2, 2, 2)$. Say $a, b$ are the outside variables in C1, $a, c$ are the outside variables in $C2$ and $b, c$ are the outside variables in $C3$. Furthermore, no other mixed clauses are there (as no two clauses can overlap in two literals).

Case 2.4.1: At least one of $a, b, c$ appears both as positive and negative literal in $C1, C2, C3$.

Suppose without loss of generality that $a$ appears as positive in $C1$ and negative in $C2$. Then, setting $a$ to be 1, allows us to set $b$ as well as contract all of $I$ to $c$ using Line 14. Setting $a$ to be 0, allows us to set $c$ as well as contract all of $I$ to $b$ using Line 14. Thus, we get a worst case branching factor of $\tau(9, 9)$.

Thus, this is covered under Line 15.

Case 2.4.2: None of $a, b, c$ appears both as positive and negative literal in $C1, C2, C3$. Without loss of generality assume $a, b, c$ all appear as positive literals in $C1, C2, C3$.

When, we set $x'_i = 1$, we have that $a = b = 0$ and we can contract rest of $I$ to $c$ using Line 14. This gives us removal of 9 variables. When we set $x'_i = 0$, we have that $a = \bar{b}$, and thus $c$ must be 0 (from $C2$ and $C3$), and thus we can contract rest of $I$ into $a$ using Line 14. Thus we get a worst case branching factor of $\tau(9, 9)$. Thus, this is covered under Line 15.

Therefore, the worst case time complexity is $O(\tau(10, 4)^n) \subseteq O(1.1120^n)$.

### 4.3 Line 17 of the Algorithm

We now deal with degree 3 clauses.

17: If there is a degree 3 clause in $\varphi$, then check if $\exists$ an edge between $L$ and $R$. If no, then construct $G_\varphi$ and let $(L', R') \leftarrow MonienPreis(G_\varphi)$. Then return $CountX3SAT(\varphi, \mathbf{c}, L', R')$. If $\exists$ an edge between $L$ and $R$, apply only the simplification rules (if any) as stated in this section (Section 4.3). Choose an edge $e$ between $L$ and $R$. Then branch the variable $x_e$ represented by $e$. Let the cardinality vector $\mathbf{c}'$ be the new cardinality vector by dropping off entries $x_e$ and $\bar{x}_e$. Return $CountX3SAT(\varphi[x_e = 1], \mathbf{c}', L, R) \times \mathbf{c}(x_e) + CountX3SAT(\varphi[x_e = 0], \mathbf{c}', L, R) \times \mathbf{c}(\bar{x}_e)$.

Now, we discuss Line 17 of the algorithm in detail. As long as a degree 3 clause exists in the formula, we repeat this process. First, we describe how to construct the graph $G_\varphi$.

**Construction.** We construct a graph $G_\varphi = (V, E)$, where $V = \{v_C : C$ is a degree 3 clause in $\varphi\}$. Given any vertices $v_{C'}$ and $v_{C''}$, we add an edge between them if any of the below conditions occur on clauses $C'$ and $C''$, where $C'$ and $C''$ are clauses with 3 neighbours:

1. If a common variable appears in both $C'$ and $C''$
2. $C'$ and $C''$ are connected by a chain of 2-degree clauses.

By construction, the graph $G_\varphi$ has maximum degree 3. Let $m_3$ denote the number of degree 3 clauses in $\varphi$. This gives us $|V| = m_3$. We can therefore apply the result by Monien and Preis, with the size of the bisection width $k \leq m_3(\frac{1}{6} + \varepsilon)$.

We construct the graph $G_\varphi$ when there are no edges between $L$ and $R$, and then apply $MonienPreis(.)$ to get our new partitions $L'$ and $R'$, which are sets of clauses. These partitions will remain connected until all edges between them are removed. In other words, the variables represented by them are branched. Now instead of bruteforcing all the variables in the bisection width at the same time, we branch them edge by edge. After each branching, we apply simplification rules before branching again. By our construction, we will not increase the degree of our clauses or variables (except temporarily due to linking; the corresponding clause will then be removed via Line 6). Therefore, we never need to resort to the earlier branching rules (Line 15 and 16) that deal with variables appearing at least 3 times again. In other words, once we come into Line 17, we will be repeating this branching rule in a recursive manner until all degree 3 clauses have been removed. Applying the simplification rules could mean that some variables have been removed directly or via linking, or some degree 3 clauses have now been dropped to a degree 2 clause etc. In other words, the clauses in the sets $L$ and $R$ have changed. Therefore, we need to update $L$ and $R$ correspondingly to reflect these changes before we repeat the branching again.

After branching the last variable between the two partitions, the formula becomes disconnected with two components and Line 7 handles this. Recall that in Line 7, we gave an additional condition to check for any edges between $L$ and $R$. During the course of applying simplification rules or branching the variables, it could be that additional components can be created before all the edges between $L$ and $R$ have been removed. Therefore, this condition to check for any edges between the partition is to ensure that Line 7 will not be called prematurely until all edges have been removed. We will now give in detail the choosing of the variable to branch below.

**Choosing of Variables to Branch.** Based on the construction earlier, an edge is added if any of the two possibilities mentioned above happen in the formula. Let $e$ be an edge in the bisection width. We choose a specific variable to branch in the different scenarios listed.

1. Case 1: The edge $e$ represents a variable sitting on two degree 3 clauses. Branch this variable.
2. Case 2: The edge $e$ represents a chain of 2 degree clauses. We alternate the branchings between the variables that appear in a degree 3 clause and a

degree 2 clause at both ends whenever Case 2 arises for symmetry reasons. For example, if we have degree 3 clause $(a \vee b \vee c)$ in the left partition connected to degree 3 clause $(s \vee t \vee u)$ in the right partition via a chain $(c, d, e), \ldots, (q, r, s)$, and it is left partition end turn, then we branch on variable $c$; if it is right partition end turn then we branch on variable $s$. These branchings will remove the whole chain, and convert the two degree 3 clauses into degree two or lower clause by compression as described below.

**Compression.** Suppose $C'$ and $C''$ are two degree 3 clauses connected via a chain $C_1, C_2, \ldots, C_k$, where $c$ is a common variable between $C'$ and $C_1$, and $s$ is a common variable between $C''$ and $C_k$. When $s$ is assigned either a value of 0 or 1, $C''$ drops to a clause of degree at most 2. $C_k$ becomes a 2-literal clause (in the worst case) and we can link the two remaining literals in it together and the clause is dropped. Therefore, the neighbouring clause $C_{k-1}$ has now become a degree 1 clause. By Line 10 of the algorithm, we can remove 1 singleton and $C_{k-1}$ drops to a 2-literal clause. Continuing the process of linking, dropping of clause and removing of singletons, the degree 3 clause at the end, $C'$, will drop to become a clause of at most degree 2 when $C_1$ is removed. Therefore, $C'$ and $C''$ will drop to a clause of at most degree 2.

With the Compression method, we now have the following. Let $C$ be a degree 3 clause. Since $C$ is a degree 3 clause, it has an edge to three other degree 3 clauses, say $E_1, E_2, E_3$. Choose any edge, say between $E_1$ and $C$. Now this edge can either represent a variable appearing in both $C$ and $E_1$, or a chain between $E_1$ and $C$ with variables at both ends appearing in $E_1$ and $C$. Therefore, assigning a value of 0 or 1 to this chosen variable represented by the edge will cause $C$ to drop to a clause of degree at most 2.

**Self-loop.** Note that such a case can arise, where a degree 3 clause can be connected via a degree 2 chain to itself. The idea to handle this is similar to Line 14 and by adopting the idea in Compression. Due to space constraints, details are omitted. More information is available at https://arxiv.org/abs/2007.07553.

Based on the choice of variables as mentioned above, we now give the time analysis for Line 17 of the algorithm. Note that the measure of complexity for our branching factors here is $m_3$, the number of degree 3 clauses.

**Lemma 11.** *The time complexity of dealing of branching variables in the bisection width is* $O(1.1092^n)$.

**Proof.** For $m_3$, the current number of degree 3 clauses, we have that each variable in a degree 3 clause occurs in exactly one further clause and that there are three variables per clause. Thus $3m_3 \le 2n$ and $m_3 \le \frac{2}{3}n$, where $n$ is the current number of variables. Note that the bisection width has size $k \le m_3(\frac{1}{6} + \varepsilon)$.

Once we remove the edges in the bisection width, the two sides (call them left (L) and right (R)) get disconnected, and thus each component can be solved independently. Here note that after the removal of all the edges in the bisection width, we have at most $m_3/2$ degree 3 clauses in each partition. As we ignore polynomial factors in counting the number of leaves, it suffices to concentrate on one (say left) partition. We consider two kinds of reductions: (i) a degree 3 clause on the left partition is removed or becomes of degree less than three due to a branching, and (ii) the degree 3 clauses on the right partition are not part of the left partition. The reduction due to (ii) is called bookkeeping reduction because we spread it out over the removal of all the edges in the bisection width. Note that after all the edges between $L$ and $R$ have been removed, $\frac{m_3}{2}$ many clauses are reduced due to the right partition not being connected to the left partition. As the number of edges in the bisection width is at most $\frac{m_3}{6}$, in the worst case, we can count at least $\frac{m_3}{2} \div \frac{m_3}{6} = 3$ degree 3 clauses for each edge in the bisection width that we remove. For the removal of degree 3 clauses in the left partition, we analyze as follows.

Let an edge be given between $L$ and $R$. We let the degree 3 clause $C = (a \vee b \vee c)$ be on the left partition, and the degree 3 clause $T = (s \vee t \vee u)$ be on the right partition. Then the edge can be represented by $c$, with $s = c$ or $s = \bar{c}$, or the edge is represented by a chain of degree 2 clauses, with the ends being $c$ and $s$. We branch the variable $c = 1$ and $c = 0$.

When $c = 0$, $C$ gets dropped to a degree 2 clause. Now this also means that the given edge gets removed (either directly or via Compression). Counting an additional 3 degree 3 clauses from the bookkeeping process, we remove a total of 4 degree 3 clauses here.

When $c = 1$, then $a = b = 0$. Since $C$ is a degree 3 clause, it is connected to 3 other degree 3 clauses. Now all 3 degree 3 clauses will either be removed, or will drop to a degree 2 clause (again either directly, or via Compression). Hence, this allows us to remove $1 + 3i + (3 - i)$ degree 3 clauses, where removing $C$ counts as 1, $i$ is the number of neighbours of $C$ in the right partition (bookkeeping) while $(3 - i)$ be the number of neighbours on the left. Since $i \in \{1, 2, 3\}$, the minimum number of degree 3 clauses we can remove here happens to be for $i = 1$, giving us 6 degree 3 clauses for this branch. This gives us a branching factor of $\tau(6, 4)$.

When we branch the variable $s = 1$ and $s = 0$, $C$ gets dropped to a degree 2 clause via Compression, and in both branches, the edge gets removed and we can count 3 additional clauses from the bookkeeping process. In both branches, we remove 4 degree 3 clauses. This gives us a branching factor of $\tau(4, 4)$. Since we are always doing alternate branching for Case 2 (branching at point $c$ and then at point $t$), we can apply branching vector addition on $(6, 4)$ to $(4, 4)$ on both branches to get a branching vector of $(8, 8, 10, 10)$.

Hence, Case 1 takes $O(\tau(6, 4)^{m_3})$ time, while Case 2 takes $O(\tau(8, 8, 10, 10)^{m_3})$ time. Since Case 2 is the bottleneck, this gives us $O(\tau(8, 8, 10, 10)^{m_3}) \subseteq O(\tau(8, 8, 10, 10)^{\frac{2}{3}n}) \subseteq O(1.1092^n)$, which absorbs all subexponential terms.

### 4.4 Line 18 of the Algorithm

In Line 18, the formula $\varphi$ is left with only degree 2 clauses in the formula. Now suppose that no simplification rules apply, then we know that the formula must consist of cycles of degree 2 because of Lines 2, 3, 5, 6 and 10 of the algorithm. Now if $\varphi$ consists of many components, with each being a cycle, then we can handle this by Line 7 of the algorithm. Therefore, $\varphi$ consists of a cycle.

Now, we choose any variable $x$ in this cycle and branch $x = 1$ and $x = 0$. Since all the clauses are of degree 2, we can repeatedly apply Line 10 and other simplification rules to solve the remaining variables (same idea as in Compression). Therefore, we would only need to branch one variable in this line. This, and repeatedly applying the simplification rules, will only take polynomial time.

Putting everything together, we have the following result.

**Theorem 12.** *The whole algorithm runs in* $O(1.1120^n)$ *time.*

## References

1. Byskov, J., Madsen, B., Skjernaa, B.: New algorithms for exact satisfiability. Theor. Comput. Sci. **332**, 513–541 (2005)
2. Cook, S.: The complexity of theorem proving procedures. In: Third Annual ACM Symposium on Theory of Computing (STOC 1971), pp. 151–158 (1971)
3. Dahllöf, V.: Exact algorithms for exact satisfiability problems. Linköping Studies in Science and Technology, Ph.D. dissertation no 1013 (2006)
4. Dahllöf, V., Jonsson, P.: An algorithm for counting maximum weighted independent sets and its applications. In: Proceedings of the 13th Annual ACM-SIAM Symposium on Discrete Algorithms (SODA 2002), pp. 292–298 (2002)
5. Dahllöf, V., Jonsson, P., Beigel, R.: Algorithms for four variants of the exact satisfiability problem. Theor. Comput. Sci. **320**(2–3), 373–394 (2004)
6. Davis, M., Logemann, G., Loveland, D.: A machine program for theorem proving. Commun. ACM **5**(7), 394–397 (1962)
7. Davis, M., Putnam, H.: A computing procedure for quantification theory. J. ACM **7**(3), 201–215 (1960)
8. Fomin, F.V., Kratsch, D.: Exact Exponential Algorithms. Texts in Theoretical Computer Science. An EATCS Series. Springer, Heidelberg (2010). https://doi.org/10.1007/978-3-642-16533-7
9. Gaspers, S., Sorkin, G.B.: Separate, measure and conquer: faster polynomial-space algorithms for Max 2-CSP and counting dominating sets. ACM Trans. Algorithms (TALG) **13**(4), 44:1–44:36 (2017)
10. Karp, R.M.: Reducibility among combinatorial problems. In: Miller, R.E., Thatcher, J.W., Bohlinger, J.D. (eds.) Complexity of Computer Computations. IRSS, pp. 85–103. Springer, Boston (1972). https://doi.org/10.1007/978-1-4684-2001-2_9
11. Krom, M.R.: The decision problem for a class of first-order formulas in which all disjunctions are binary. Zeitschrift für Mathematische Logik und Grundlagen der Mathematik **13**(1–2), 15–20 (1967)
12. Kullmann, O.: New methods for 3-SAT decision and worst-case analysis. Theor. Comput. Sci. **223**(1–2), 1–72 (1999)
13. Monien, B., Preis, R.: Upper bounds on the bisection width of 3- and 4-regular graphs. J. Discret. Algorithms **4**(3), 475–498 (2006)
14. Porschen, S.: On some weighted satisfiability and graph problems. In: Vojtáš, P., Bieliková, M., Charron-Bost, B., Sýkora, O. (eds.) SOFSEM 2005. LNCS, vol. 3381, pp. 278–287. Springer, Heidelberg (2005). https://doi.org/10.1007/978-3-540-30577-4_31
15. Roth, D.: On the hardness of approximate reasoning. Artif. Intell. **82**, 273–302 (1996)

16. Sang, T., Beame, P., Kautz, H.A.: Performing Bayesian inference by weighted model counting. In: AAAI, vol. 5, pp. 475–481 (2005)
17. Schaefer, T.J.: The complexity of satisfiability problems. In: Tenth Annual Symposium on Theory of Computing (STOC 1978), pp. 216–226 (1978)
18. Valiant, L.G.: The complexity of enumeration and reliability problems. SIAM J. Comput. **8**(3), 410–421 (1979)
19. Leslie, G.: Valiant the complexity of computing the permanent. Theor. Comput. Sci. **8**(2), 189–201 (1979)
20. Wahlström, M.: Algorithms, measures and upper bounds for satisfiability and related problems. Ph.D. thesis, Department of Computer and Information Science, Linköpings Universitet (2007)
21. Zhou, J., Weihua, S., Wang, J.: New worst-case upper bound for counting exact satisfiability. Int. J. Found. Comput. Sci. **25**(06), 667–678 (2014)

# Visualizations to Summarize Search Behavior

Ian S. Howell[1,2](✉), Berthe Y. Choueiry[1], and Hongfeng Yu[2]

[1] Constraint Systems Laboratory, University of Nebraska-Lincoln, Lincoln, NE 68508, USA
{ihowell,choueiry}@cse.unl.edu

[2] Visualization Laboratory, University of Nebraska-Lincoln, Lincoln, NE 68508, USA
yu@cse.unl.edu

**Abstract.** In this paper, we argue that metrics that assess the performance of backtrack search for solving a Constraint Satisfaction Problem should not be visualized and examined only at the end of search, but their evolution should be tracked throughout the search process in order to provide a more complete picture of the behavior of search. We describe a process that organizes search history by automatically recognizing qualitatively significant changes in the metrics that assess search performance. To this end, we introduce a criterion for quantifying change between two time instants and a summarization technique for organizing the history of search at controllable levels of abstraction. We validate our approach in the context of two algorithms for enforcing consistency: one that is activated by a surge of backtracking and the second that modifies the structure of the constraint graph. We also introduce a new visualization for exposing the behavior of variable ordering heuristics and validate its usefulness both as a standalone tool and when displayed alongside search history.

**Keywords:** Search · Visualization · Constraint Satisfaction

## 1 Introduction

In this paper, we propose a new perspective and visualization tools to understand and analyze the behavior of the backtrack-search procedure for solving Constraint Satisfaction Problems (CSPs). Backtrack search is commonly used for solving CSPs. However, its performance is unpredictable and can differ widely on similar instances. Further, maintaining a given consistency property during search has become a common practice [13,22,30,42,43] and new strategies for

This research is supported by NSF Grant No. RI-1619344 and NSF CAREER Award No. III-1652846. The experiments were completed utilizing the Holland Computing Center of the University of Nebraska, which receives support from the Nebraska Research Initiative.


H. Simonis (Ed.): CP 2020, LNCS 12333, pp. 392–409, 2020.
https://doi.org/10.1007/978-3-030-58475-7_23

dynamically switching between consistency algorithms are being investigated [1–3,9,15,16,18,25,29,41,46,48]. While consistency algorithms can dramatically reduce the size of the search space, their impact on the CPU cost of search can vary widely. This cost is currently poorly understood, hard to predict, and difficult to control.

To gain an understanding of the behavior of search and its performance, most previous work has focused on the visualization of various metrics at the *end* of search. In contrast, we claim that it is sometimes advantageous to *inspect the history* [20] or evolution of these metrics along the search process in order to detect bottlenecks that can be addressed by specific tools (e.g., modify the topological structure of the constraint network [43] or enforce a particular high-level consistency). As we demonstrate in this paper, it is sometimes not practical, or even feasible, to manually carry out such an analysis. For this reason, we advocate to automatically organize the history of search (i.e., its evolution over time) as a sequence of *regimes* [23] of qualitatively distinct behaviors detected by sampling the performance metrics during search. We claim that our visualizations are useful to the researcher designing new adaptive strategies and consistency algorithms as well as to the savvy engineer deploying constraint-programming solutions.

Building on the visualizations proposed by Woodward et al. [46] for describing search performance, we introduce the following contributions:

1. A criterion for computing the distance between two samples, quantifying the change in search behavior.
2. A clustering technique for automatically summarizing the history of search and its progress over time in a hierarchical structure that a human user can inspect at various levels of details.
3. A new visualization that reveals and summarizes the behavior of a dynamic variable ordering heuristic.

This paper is structured as follows. We first review background information and the relevant literature. Then, we describe our contributions. For each contribution, we highlight its usefulness with illustrative examples. Finally, we discuss the usefulness of our approach and conclude with directions for future research.

## 2 Background and Related Work

Constraint Satisfaction Problems (CSPs) are used to model many combinatorial decision problems of practical importance in Computer Science, Engineering, and Management. A CSP is defined by a tuple $(X, D, C)$, where $X$ is a set of *variables*, $D$ is the set of the variables' *domains*, and $C$ a set of *constraints* that restrict the combinations of values that the variables can take at the same time. We denote the number of variables $|X| = n$. A solution to a CSP is an assignment of values to variables such that all constraints are simultaneously satisfied. Determining whether or not a given CSP has a solution is known to be NP-complete. The *constraint network* of a CSP instance is a graph where the vertices represent the variables in the CSP and the edges represent the constraints and connect the variables in their scope.

To date, backtrack search (BT) is the only sound and complete algorithm for solving CSPs [7]. Backtracking suffers from *thrashing* when it repeatedly explores similar subtrees without making progress. Strategies for reducing thrashing include intelligent backtracking, ordering heuristics, learning no-goods, and enforcing a given *consistency property* after every variable instantiation. Enforcing consistency reduces the size of the search space by deleting, from the variables' domains, values that cannot appear in a consistent solution given the current search path (i.e., conditioning). In recent years, the research community has investigated *higher-level consistencies* (HLC) as inference techniques to prune larger portions of the search space at the cost of increased processing effort [6,13,17,33], leading to a trade-off between the search effort and the time for enforcing consistency. We claim that our visualizations are insightful tools to understand why and where to apply HLCs and to assess their effectiveness.

In this paper, we use by default dom/wdeg [10], a *dynamic* variable ordering heuristic (and dom/deg [5] where indicated). Below, we summarize the consistency properties that we enforce. We always maintain Generalized Arc Consistency (GAC) [26] during search. In addition to GAC, we enforce, as HLCs, Partition-One Arc-Consistency (POAC) [4], which filters the domains of the variables by performing singleton testing, and Partial Path Consistency (PPC) [8], which filters the constraints after triangulating the constraint graph. Both POAC and (strong) PPC are strictly stronger than GAC. In order to control their cost, we enforce the HLCs in a 'selective/adaptive' manner using the PREPEAK$^+$ strategy of Woodward et al. [46] for POAC (denoted PREPEAK$^+$-POAC) and the Directional Partial Path Consistency of Woodward [45] (denoted DPPC$^+$). PREPEAK$^+$ triggers POAC by watching the numbers of backtracks and follows geometric laws to increase or decrease the calls to POAC. DPPC$^+$ chooses a subset of the triangles of the triangulated constraint network on which it calls PPC in a manner to reduce the addition of new constraints while increasing propagation across the network. Finally, we use a $k$-way branching strategy for search. However, preliminary tests on Choco [12], on which we tested GAC and implemented POAC, show similar shapes of the visualizations.

Prior research on search visualization has appeared in the Constraint Programming literature. Most previous approaches, such as the ones discussed below, can be viewed as tools for *inspecting* and *debugging* search. The DiSCiPl project provides extensive visual functionalities to develop, test, and debug constraint logic programs such as displaying variables' states, effect of constraints and global constraints, event propagation at each node of the search tree, and identifying isomorphic subtrees [11,38]. Many useful methodologies from the DiSCiPl project are implemented in CP-Viz [39] and other works [37]. The OZ Explorer displays the search tree allowing the user to access detailed information about each tree node and to collapse and expand failing trees for closer examination [34]. This work is currently incorporated into Gecode's Gist [35]. The above approaches focus on exploring in detail the search tree to examine and inspect the impact of individual search decisions on a particular variable, constraint, variable domain, or even a subtree of the search. In contrast, our work proposes to *summarize* the evolution of search along a given dimension (i.e., projection), providing a qualitative, abstract view of a search tree that is too large to be explored. We believe that the above approaches are orthogonal and complementary to ours.

Ghoniem et al. [19] propose a variety of heatmap visualizations to illustrate the operation of constraint propagation across an instance. Their heatmaps illustrate the impact of the activity of a variable or of a constraint on other variables or constraints in terms of filtering the variables' domains or causing further constraint propagation. These visualizations are fine grained and inform the user of the effect of a particular decision. They have pedagogical value and can also be useful for debugging. They differ from our approach in that the user has the burden of identifying the animation frames of interests. Relatively short behaviors can go unnoticed and long behaviors can be wasteful of the user's time. In contrast, our regime-identification process automatically organizes search history into a sequence of 'interesting' episodes. We consider their contribution too to be orthogonal and complementary to our approach.

Tracing the search effort by *depth* of the search tree was first proposed for the number of constraint checks and values removed per search level (Epstein et al. [16]) and for the number of nodes visited (Simonis et al. [39] in CP-Viz, also used for solving a packing problem [40]). More recently, Woodward et al. [46] propose to focus on the number of *backtracks* per depth (BpD) to assess the severity of thrashing and on the number of *calls* to an HLC per depth (CpD) to explain the cost of enforcing an HLC, see Fig. 1. Further, they split the CpD into three line charts, showing in green the HLC calls that prune inconsistent subtrees and yield domain wipeouts (most effective), in blue those that filter the domains of future variables (effective), and in red those that do not filter the

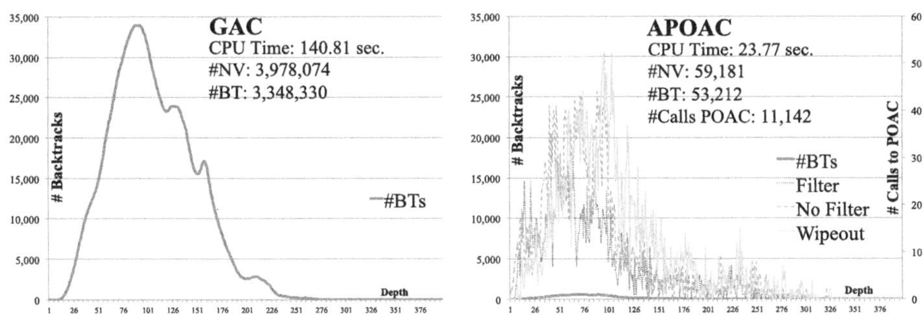


**Fig. 1.** From Woodward et al. [46]: Number of backtracks (BpD) and calls (CpD) to POAC per depth at the end of search for benchmark instance pseudo-aim-200-1-6-4 using dom/wdeg. Left: GAC. Right: APOAC, an adaptive POAC by Balafrej et al. [1]. The CPU time, #NV (number of nodes visited), and #BT (number of backtracks) are typically used to indicate the cost of search. (Color figure online)

domain of any future variable (wasteful). Finally, they superimpose the CpD to the BpD chart in order to reveal the effectiveness of the consistency enforced (and that of the adaptive strategy for enforcing an HLC, when applicable).

## 3 Tracking Search History

To the best of our knowledge, previous approaches to visualizing the various search 'metrics per depth' of the search tree (e.g., number of nodes visited, constraint checks, or backtracks) present the values reached at the *end* of search thus providing a 'blanket' summary where thrashing occurred and search invested the most resources. They fail to identify the problems that occur *during* search. We argue that these visualizations may mask important information, such as transient behaviors of search, that an algorithm developer or a savvy user need to examine and address in order to solve the problem. In this paper, we propose to track the progress of search by collecting samples of the BpD 'profile' during search, see Fig. 2.[1] In the examples discussed in this paper, we use a sampling rate of 100 ms. To this end, we propose a criterion for deciding whether two consecutive samples are 'equivalent' and pertain to the same qualitative *regime* [23] of search. The sequence of regimes yields a *history* [20] of the search.

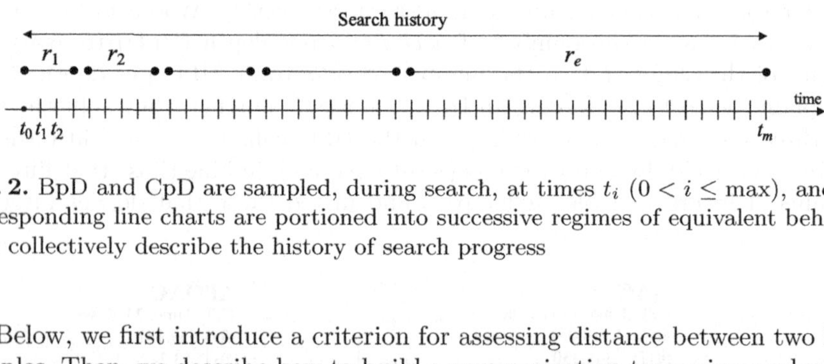


**Fig. 2.** BpD and CpD are sampled, during search, at times $t_i$ $(0 < i \leq \max)$, and the corresponding line charts are portioned into successive regimes of equivalent behavior that collectively describe the history of search progress

Below, we first introduce a criterion for assessing distance between two BpD samples. Then, we describe how to build a summarization tree using agglomerative hierarchical clustering and how the tree can be used to allow the human user to examine the search history at the desired level of detail. Finally, we provide two illustrative examples.

### 3.1 Distance Between Two BpD Samples

We adopt the following notations. $D = [1, d_{\max}]$ is the domain of the depth of the search tree for solving the CSP. In a $k$-way backtrack search, $d_{\max} = n$; in a binary-branching scheme, $d_{\max} \leq na$, where $a$ is the maximum domain size. We sample search, recording the cumulative number of backtracks per depth (BpD)

[1] Note that we can interrupt search at any time to conduct the analysis and need not wait until the end of search.

at each sampling instant. $\mathrm{BT}(d,t)$ designates the number of backtracks executed by search at depth $d \in D$ from the beginning of search until time $t$. The first regime starts at $t = 0$ and we set $\mathrm{BT}_{d\in D}(d,0) \equiv 0$. The last regime ends at $t_{\mathrm{m}}$ and corresponds to the last sample collected. We say that two time instants $t$ and $t'$ do not belong to the same qualitative regime if their corresponding BpD curves 'differ enough' according to some divergence or distance criterion. Below, we introduce one such criterion.

To measure the distance between BpDs at time stamps $t$ and $t'$, we adapt, to our context, the Kullback-Leibler (KL) divergence (also called *relative entropy*) [24]. Given two discrete probability distributions $P$ and $Q$ defined on the same probability space, $\chi$, the KL-divergence from $Q$ to $P$, which is always non-negative, is defined as: $D_{\mathrm{KL}}(P \,\|\, Q) = \sum_{x\in\chi} P(x) \log\left(\frac{P(x)}{Q(x)}\right)$.

In order to adapt the KL-divergence to our context, we introduce the probability that $\mathrm{BT}(d,t)$ take a given value as $p(d,t) = \frac{\mathrm{BT}(d,t)}{\sum_{d\in D}\mathrm{BT}(d,t)}$. In order to avoid having this probability be zero when $\mathrm{BT}(d,t) = 0$, we use Laplace Smoothing [27], approximating this probability with $\widehat{p}(d,t) = \frac{\mathrm{BT}(d,t)+\alpha}{\sum_{d\in D}\mathrm{BT}(d,t)+\alpha d_{\max}}$ and using $\alpha = 0.1$ in our experiments. Finally, we define the divergence $\mathrm{div}(t,t') = \mathrm{div}(t',t)$ between two BpD curves at time $t$ and $t'$ as

$$\mathrm{div(t,t')} = \max\left(\sum_{d\in D}\widehat{p}(d,t)\log\left(\frac{\widehat{p}(d,t)}{\widehat{p}(d,t')}\right), \sum_{d\in D}\widehat{p}(d,t')\log\left(\frac{\widehat{p}(d,t')}{\widehat{p}(d,t)}\right)\right) \quad (1)$$

Further, we define the distance between two regimes as the divergence between the two BpDs located at the *middle* sample of the corresponding regimes. Another alternative, a little more costly, is to use the largest divergence between any two samples in the regimes. Other options exist, preliminary evaluations did not exhibit a significant sensitivity to this choice.

### 3.2 Clustering Tree and Summarization Tree

We propose to summarize the history of the search following the idea of *agglomerative hierarchical clustering* described by Nielsen [28, Chapter 8], which yields a binary tree.

Starting from the sequence of BpD samples (collected during search and ordered along the time line), we form the leaf nodes of the *clustering tree* storing one sample per node. We place these nodes in a vector $Q$. Then,

1. We determine the divergence between every two consecutive nodes in $Q$ (those that store temporally *adjacent* samples).
2. We generate a parent for the two nodes with the smallest divergence value.
3. We replace, in $Q$, the two nodes with their parent.
4. We associate, to the parent, the concatenation of the sample sequences of the two children, ordered in increasing time.
5. We store, for the parent, the divergence value between its children.

6. We choose as representative of the new node the middle sample, breaking ties by using the earlier sample.

We iteratively apply the process until $|Q|= 1$. We call this tree the *clustering tree* and reuse it every time a new summarization is needed.

Finally, we identify the deepest nodes in the tree with a stored divergence value larger than or equal to a user-defined value $\delta$ and replace each such node by a new *regime node.* The regime node inherits from the node that it replaces, its representative middle sample, its divergence value, and the sequence of samples stored at the leaves of the corresponding subtree. We call the resulting structure a *$\delta$-summarization tree.* The dendrogram shown in Fig. 3 illustrates this process.

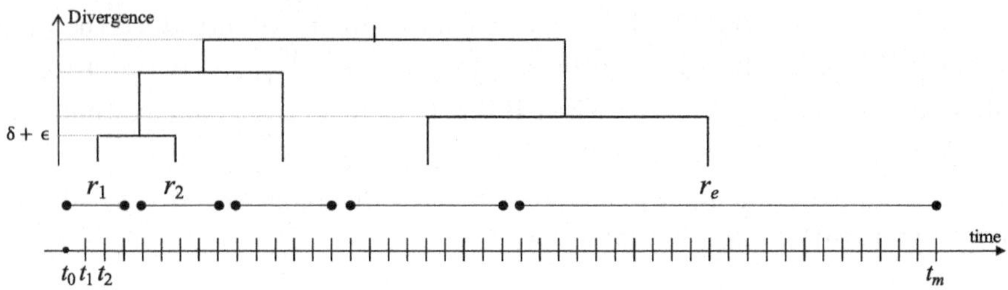


**Fig. 3.** Dendrogram of the summarization tree, showing the sequence of 'cluster' mergers and the divergence value at which each merger occurs (with $\epsilon \geq 0$)

The clustering tree stores samples at its leaves and does not use or depend on $\delta$. The summarization tree uses $\delta$ to 'trim' the *lower* levels of the clustering tree, replacing them with regimes for explanation, which can be done interactively and is computationally cheap. Thus, if the summarization is too detailed or too abstract, the user may adjust $\delta$ to quickly regenerate a new history from the clustering tree to reach a level of detail that is cognitively satisfying. Further, iteratively adding samples to the clustering tree as search proceeds can be done, online, in logarithmic time. Consequently, building and maintaining the clustering tree can efficiently be done online, during search. However, this direction remains to be investigated.

In addition to the data structures used for the hierarchical clustering [36], we introduce two data structures: first, $Q$, implemented as a vector of size $m$, where $m$ is the number of samples collected. Then, another vector to store the data of the non-leaf nodes of the clustering tree. This size of this vector is $m-1$ because the clustering tree is a full binary tree. The clustering tree can be built in $\mathcal{O}(m^2)$ time and $\mathcal{O}(m)$ space [14].

### 3.3 Illustrative Example: Explaining Adaptive Consistency

In this section, we compare the behavior of search for solving the SAT instance `aim-200-2-0-sat-1` [50]. GAC takes less than 10 min and generates 14 million

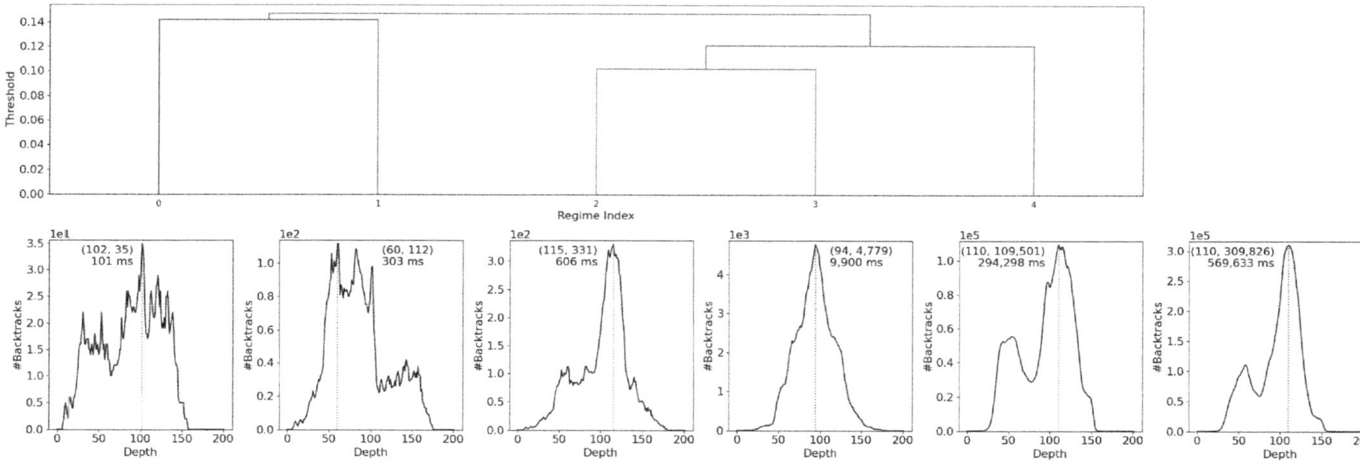


(a) aim-200-2-0-sat-1, GAC, dom/wdeg: Dendrogram (top), history with $\delta$=0.1 (bottom)

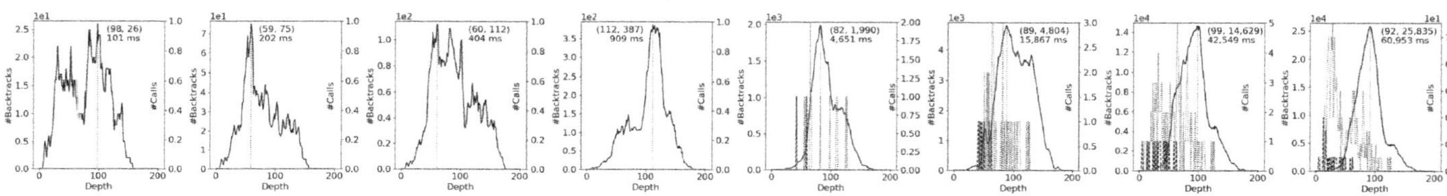


(b) aim-200-2-0-sat-1, PrePeak$^+$-POAC, dom/wdeg: History ($\delta$=0.05) superimposed CpD (green: wipeout; blue: filtering)

**Fig. 4.** aim-200-2-0-sat-1: Histories with GAC (top) and with PrePeak$^+$-POAC (bottom), using dom/wdeg. Showing the middle BpD sample in each regime and, for reference, the final sample collected. (Color figure online)

backtracks. PrePeak$^+$-POAC, on the other hand, solves the instance in little over a minute, netting only 1.2 million backtracks. To see why, we examine the corresponding search histories shown in Fig. 4a and 4b. We see that initially, both searches proceed exactly in the same manner. The behavior of PrePeak$^+$-POAC starts to diverge from that of GAC at the fifth regime of PrePeak$^+$-POAC (Fig. 4b) where PrePeak$^+$ reacts to the sharp increase of the number of backtracks around depth 100 and triggers POAC. POAC successfully prunes inconsistent subtrees (green calls) or decreases the size of search tree (blue calls). By examining the last three regimes of PrePeak$^+$ (Fig. 4b), we see how the HLC successfully reduces the severity of the backtracking, reducing the peak in height, width, and even in position. By investing in opportunistic and effective (even if costly) calls to POAC, we significantly reduce the search time.

### 3.4 Illustrative Example: Analyzing Structure

In this section, we show how the identification and visualization of the regimes of search provide insight into the structure of a problem instance and guide the choice of the appropriate type of consistency for solving it.

We consider the graph-coloring instance `mug100-25-3` [50] and try to solve it with GAC using dom/wdeg for dynamic variable ordering. Search fails to terminate within two hours. The inspection of the BpD of GAC (left of Fig. 5) at the end of the unsuccessful search reveals the presence of a 'dramatic' peak of the number of backtracks at depth 83 (i.e., relatively deep in the search tree). This behavior hints that search may have made a bad decision at a shallower level of search from which it could not recover.

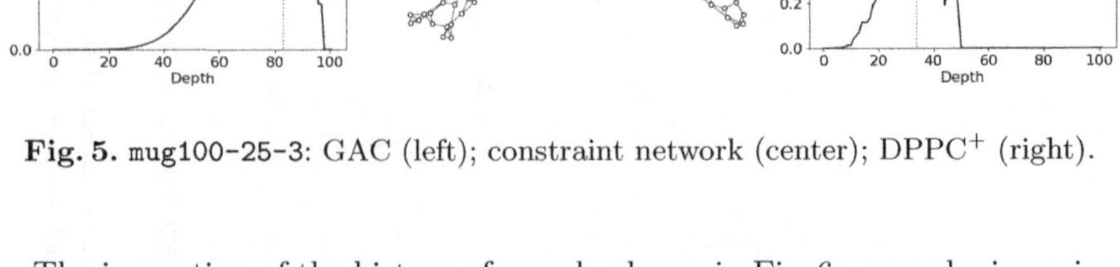


**Fig. 5.** `mug100-25-3`: GAC (left); constraint network (center); DPPC$^+$ (right).

The inspection of the history of search, shown in Fig. 6a, reveals six regimes. Search generates a first peak at depth 34 (Regimes 1 and 2) before generating a second peak at depth 86 (Regimes 3 and 4). Then, this second peak grows larger, dwarfs the first, shifts to depth 61, and settles at depth 83. Further, Regime 1 reveals a peak of magnitude 2834 backtracks at depth 34, which corresponds to instantiating about one third of the variables in the problem. Regime 2

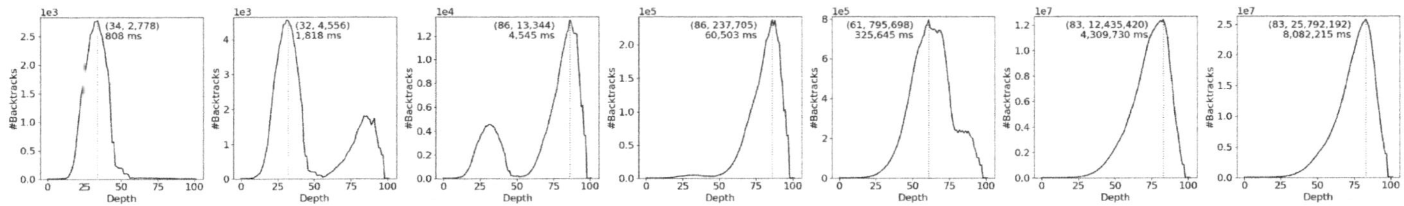


(a) History ($\delta$=0.1) of GAC, dom/wdeg shows six regimes

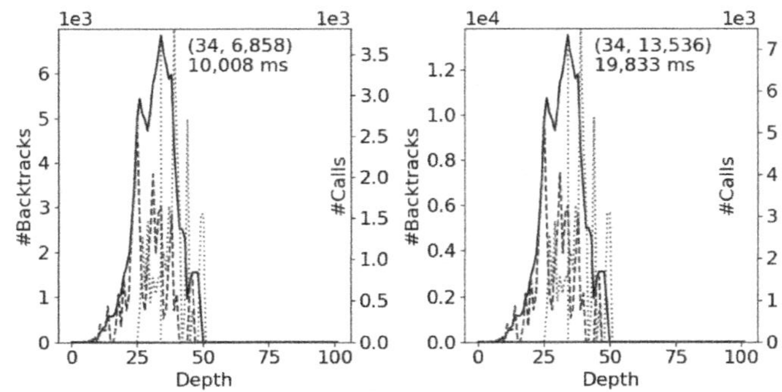


(b) History ($\delta$=0.1) of DPPC$^+$, dom/wdeg shows a unique regime

**Fig. 6.** Histories of `mug100-25-3` with GAC and DPPC$^+$. Showing the middle BpD sample in each regime and, for reference, the final sample collected.

shows that search has overcome the initial bottleneck in the instance occurring at depth 34 with a magnitude of 4556 backtracks but is struggling again with a second bottleneck at depth 86 with a magnitude of 4624 backtracks. Regime 3 shows that the severity of the first bottleneck is dwarfed by a dramatic increase of the second bottleneck, which reaches 237705 backtracks at depth 86. As time progresses, the peak moves to shallower levels, to depth 61. Although the peak keeps moving to deeper levels, search struggles and is unable to conclude. Another noteworthy information is the time scale at which the regimes are identified. The first three regimes take place within the first five seconds of search whereas the last three regimes occur at larger time scales. It seems presumptuous to expect a human user to identify and isolate behavior occurring at such small time scales.

The observation of the two bottlenecks prompts us to examine the constraint network of the problem instance looking for some structural explanation for the non-termination of search. Indeed, we expect the graph to exhibit some particular configuration such as two overlapping cliques. The examination of the constraint graph, shown at the center of Fig. 5, refutes our two-cliques hypothesis but reveals the existence of a few large cycles as well as many cycles of size three or more. Previous work argued that the presence of cycles can be detrimental to the effectiveness of constraint propagation and showed how triangulation of the constraint network allows us to remedy the situation [43,44,47]. With this insight, we consider enforcing, during search, Partial Path Consistency (PPC) instead of GAC because the algorithm for PPC operates on existing triangles and on triangulated cycles of the constraint network [8]. Because PPC is too expensive to enforce during search, Woodward proposed a computationally competitive algorithm to enforce a relaxed version of Directional Partial Path Consistency (DPPC$^+$) [45]. By enforcing a strong consistency along cycles, search is able to detect the insolvability of `mug100-25-3` and terminates in less than 20 s. The corresponding history of search, shown in Fig. 6b, exhibits a unique regime.

Figure 5 (right) shows the BpD of search while enforcing DPPC$^+$ on the `mug100-25-3` instance. This chart shows a peak of 13536 backtracks at depth 34. Table 1 compares the cost of search with GAC and with DPPC$^+$. The sign '>' indicates that search does not terminate within the allotted two hours.

**Table 1.** Performance of search (dom/wdeg) on `mug100-25-3`: GAC versus DPPC$^+$

| | GAC | DPPC$^+$ |
|---|---|---|
| CPU time (sec) | >8082.4 | 19.8 |
| #Nodes visited | 851130992 | 288976 |
| #Backtracks | 697346084 | 236356 |
| $\max_{d \in D}(\text{BT}(d, t_{max}))$ | 25953973 | 13536 |
| $\arg\max_{d \in D}(\text{BT}(d, t_{max}))$ | 83 | 34 |

Above, we showed how the visualizations and regimes of search behavior allow us to form a hypothesis about the issues encountered by search and attempt an alternative technique that is usually avoided because of its cost, culminating in successfully solving a difficult instance.

# 4 Visualizing Variable Ordering

Heuristics for variable ordering remain an active area of research in Constraint Programming. Currently, dom/wdeg [10] is considered the most effective general-purpose heuristic. When running experiments on benchmark problems, we encountered a puzzling case where search did not terminate when using dom/wdeg but safely terminated, determining inconsistency, when using dom/deg. We propose *variable-instantiation per depth* (VIpD) as a visualization to summarize the behavior of a variable ordering heuristic.

## 4.1 Variable-Instantiation per Depth (VIpD)

We denote by $I(v, d, t)$ the number of times variable $v$ is instantiated at a given depth $d$ from the beginning of search until time $t$. For a variable $v \in X$, we define the weighted depth:

$$d_w(v, t) = \frac{\sum_{d \in D} I(v, d, t) \cdot d}{\sum_{d \in D} I(v, d, t)} \tag{2}$$

We introduce the VIpD chart at time $t$ as a two-dimensional heatmap of $I(\cdot, \cdot, t)$ using, as $x$-axis, the depth of search and, as $y$-axis, the variables of the CSP listed in their increasing $d_w(\cdot, t)$ values (ties broken using a lexicographical ordering of the names of the variables).

## 4.2 Illustrative Example

We try to solve the instance `mug100-1-3` [50] of a graph coloring problem using GAC, but searches with dom/wdeg and with dom/deg fail to terminate after two hours of wall-clock time. In a next attempt, we maintain POAC during search; search with dom/wdeg did not terminate but, with dom/deg, it terminated after 47 min and determined that the instance is inconsistent. Table 2 compares the performance of POAC with dom/wdeg and dom/deg. The sign '>' indicates that search did not terminate within two hours.

**Table 2.** Performance of search on `mug100-1-3` using POAC

| POAC | dom/wdeg | dom/deg |
|---|---|---|
| CPU time (sec) | >8085.4 | 2836.5 |
| #Nodes visited | 37811011 | 25178511 |
| #Backtracks | 32031950 | 14955444 |
| $\max_{d \in D}(\mathrm{BT}(d, t_{max}))$ | 1042601 | 1393204 |
| $\arg\max_{d \in D}(\mathrm{BT}(d, t_{max}))$ | 54 | 57 |

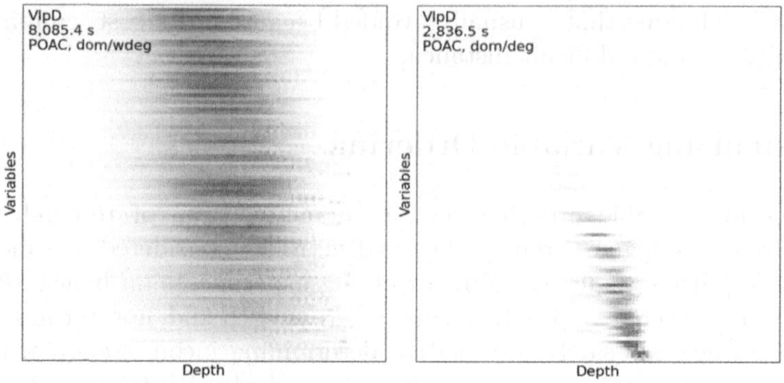


**Fig. 7.** Variable Instantiation per Depth (VIpD) on `mug100-1-3`. Left: POAC with dom/wdeg after two hours. Right: POAC with dom/deg after 47 min.

The comparison of the VIpD of POAC with dom/wdeg and that with dom/deg reveals an erratic behavior of dom/wdeg (Fig. 7). Indeed, dom/wdeg, which learns from domain wipeouts, is unable to determine the most conflicting variables on which it should focus. Instead, it continually instantiates a large number of variables over a wide range of depth levels in search. In contrast, dom/deg exhibits a more stable, less chaotic behavior. It focuses on the variables that yield inconsistency and is able to successfully terminate. The VIpD charts are also available as an Excel file where the user may analyze individual cells to examine their $I(\cdot, \cdot, t)$ values.

Comparing the histories alongside the VIpD charts (see Fig. 8), we see clearly that dom/wdeg is erratic and chaotic since the beginning of search whereas dom/deg is able to quickly focus on a relatively localized conflict to determine inconsistency of the problem instance. The peak of the BpD is sharper for dom/deg than for dom/wdeg and the VIpD shows a smaller set of variables is affected by the backtracking in dom/deg than in dom/wdeg. Future work could analyze this information to identify the source of conflict.

## 5 Discussion

We illustrated the usefulness of our approach in three case studies (summarized in Table 3). The first example (Sect. 3.3) compares two search histories to explain the operation and effectiveness of an adaptive technique (perhaps to a student or to a layperson). The same process can be used by a researcher or an application developer to explore the impact of, and adjust, the parameters of a new method. In the second example (Sect. 3.4), search fails because of a transient behavior at a tiny time scale detected by our automatic regime identification. Without this tool, the user may never notice the issue in order to explore effective solutions.

Beyond the scenarios and the metrics discussed above, we believe that building and comparing search histories are useful to explore, understand, and explain

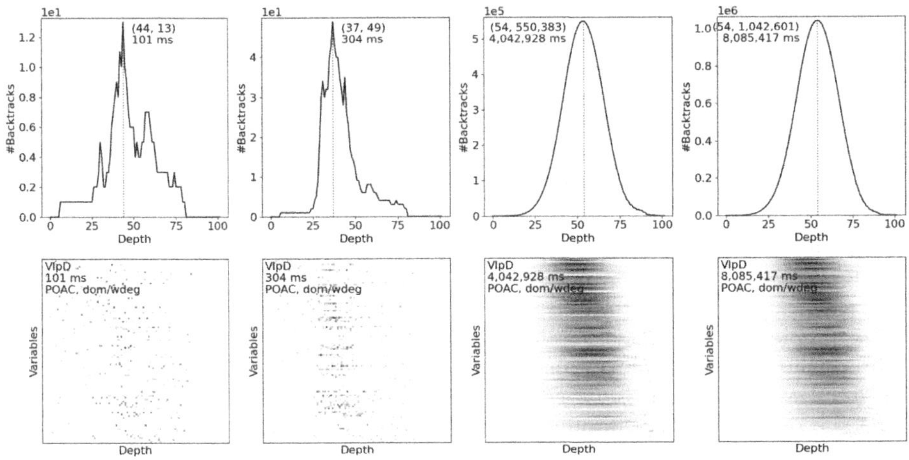


(a) POAC with dom/wdeg after two hours.

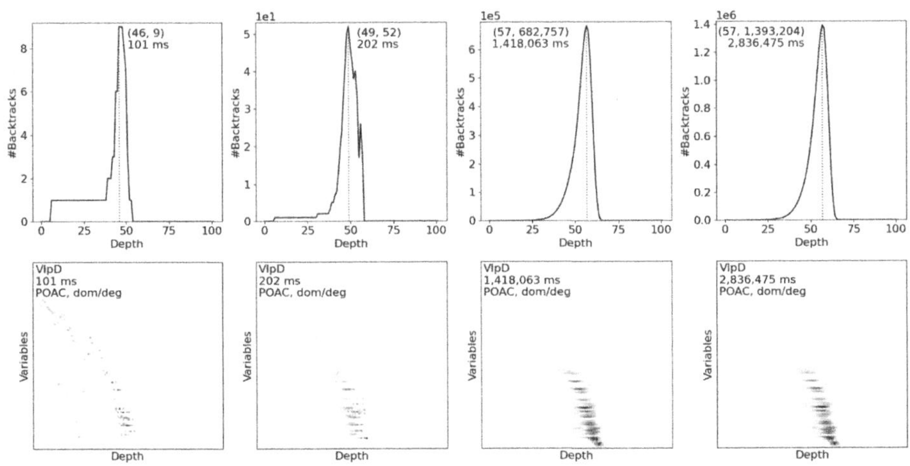


(b) POAC with dom/deg after 47 minutes.

**Fig. 8.** History and VIpD of POAC on `mug100-1-3`: dom/wdeg (top), dom/deg (bottom)

the impact, on the behavior and performance of search, of many advanced techniques such as no-good/clause learning, restart strategies, consistency algorithms, variable ordering heuristics, as well as the design of new such algorithms.

We strongly believe that our approach is beneficial in the context of existing constraint solvers such as Choco [12] and Gecode [35]. As stated above, preliminary tests on GAC and POAC [32] on Choco have shown similar BpD and CpD line-charts to those we see in $k$-way branching. Further, propagators in such solvers are often implemented on the individual constraints themselves, which

**Table 3.** Illustrative examples: the instances used and the corresponding search history

| | aim-200-2-0-sat-1 | | mug100-25-3 | | mug100-1-3 | |
|---|---|---|---|---|---|---|
| Consistency | GAC | PrePeak$^+$-POAC | GAC | DPPC$^+$ | POAC | |
| Displayed in... | Fig. 4a | Fig. 4b | Figs. 5, 6a | Figs. 5, 6b | Figs. 7, 8a | Figs. 7, 8b |
| CPU (ms) | 569633.2 | 60953.1 | 8082215.3 | 19833.0 | 8085417.6 | 2836475.7 |
| Threshold ($\delta$) | 0.1 | 0.05 | 0.1 | 0.1 | 0.1 | 0.1 |
| # Samples | 5639 | 604 | 80014 | 197 | 79, 345 | 28025 |
| # Regimes | 5 | 7 | 6 | 1 | 3 | 3 |
| History | H1 | H2 | H3 | H4 | H5 | H6 |

The histories with the durations of their regimes, listed chronologically:

H1: [202 ms, 101 ms, 606 ms, 17982 ms, 550742 ms]

H2: [101 ms, 202 ms, 101 ms, 1010 ms, 6370 ms, 16175 ms, 36994 ms]

H3: [1515 ms, 606 ms, 4848 ms, 107067 ms, 423119 ms, 7545060 ms]

H4: [19833 ms]

H5: [101 ms, 304 ms, 8085012 ms]

H6: [101 ms, 101 ms, 2836273 ms]

makes the accounting of the calls to various types of consistencies particularly well adapted.

The SAT community is another one that could benefit from our work. In SAT solvers, *inprocessing* (in the form of the application of the resolution rule) interleaves search and inference steps [21,49]. Resolution is allocated a fixed percentage of the CPU time (e.g., 10%) and sometimes its effectiveness is monitored for early termination. We believe that inference should be targeted at the 'areas' where search is struggling rather than following a predetermined and fixed effort allocation. We claim that understanding where search struggles and how that struggle changes can be used to identify where inference is best invested.

While our work does not generate verbal explanations, we claim that the graphical tools directly 'speak' to a user's intuitions.

## 6 Conclusions

In this paper, we presented a summarization technique and a visualization to allow the user to understand search behavior and performance on a given problem instance.

Currently, our approach provides a 'post-mortem' analysis of search, but our ultimate goal is to provide an 'in-vivo' analysis and allow the user to intervene in, and guide, the search process, trying alternatives and mixing strategies while observing their effects on the effectiveness of problem solving. A significantly more ambitious objective is to use the tools discussed in this paper as richer representations than mere numerical values to automatically guide search [31].

Future work includes the development of animation techniques based on the proposed visualizations.

## References


1. Balafrej, A., Bessiere, C., Bouyakhf, E., Trombettoni, G.: Adaptive singleton-based consistencies. In: Proceedings of the Twenty-Eighth AAAI Conference on Artificial Intelligence (AAAI 2014), pp. 2601–2607 (2014)
2. Balafrej, A., Bessiere, C., Coletta, R., Bouyakhf, E.H.: Adaptive parameterized consistency. In: Schulte, C. (ed.) CP 2013. LNCS, vol. 8124, pp. 143–158. Springer, Heidelberg (2013). https://doi.org/10.1007/978-3-642-40627-0_14
3. Balafrej, A., Bessière, C., Paparrizou, A.: Multi-armed bandits for adaptive constraint propagation. In: Proceedings of the Twenty-Fourth International Joint Conference on Artificial Intelligence (IJCAI 2015), pp. 290–296 (2015)
4. Bennaceur, H., Affane, M.-S.: Partition-k-AC: an efficient filtering technique combining domain partition and arc consistency. In: Walsh, T. (ed.) CP 2001. LNCS, vol. 2239, pp. 560–564. Springer, Heidelberg (2001). https://doi.org/10.1007/3-540-45578-7_39
5. Bessière, C., Régin, J.-C.: MAC and combined heuristics: two reasons to forsake FC (and CBJ?) on hard problems. In: Freuder, E.C. (ed.) CP 1996. LNCS, vol. 1118, pp. 61–75. Springer, Heidelberg (1996). https://doi.org/10.1007/3-540-61551-2_66
6. Bessière, C., Stergiou, K., Walsh, T.: Domain filtering consistencies for non-binary constraints. Artif. Intell. **172**, 800–822 (2008)
7. Bitner, J.R., Reingold, E.M.: Backtrack programming techniques. Commun. ACM **18**(11), 651–656 (1975). http://doi.acm.org/10.1145/361219.361224
8. Bliek, C., Sam-Haroud, D.: Path consistency for triangulated constraint graphs. In: Proceedings of the Sixteenth International Joint Conference on Artificial Intelligence (IJCAI 1999), pp. 456–461 (1999)
9. Borrett, J.E., Tsang, E.P., Walsh, N.R.: Adaptive constraint satisfaction: the quickest first principle. In: Proceedings of the Twelveth European Conference on Artificial Intelligence (ECAI 1996), pp. 160–164 (1996)
10. Boussemart, F., Hemery, F., Lecoutre, C., Sais, L.: Boosting systematic search by weighting constraints. In: Proceedings of the Seventeenth European Conference on Artificial Intelligence (ECAI 2006), pp. 146–150 (2004)
11. Carro, M., Hermenegildo, M.: Tools for constraint visualisation: the VIFID/TRIFID tool. In: Deransart, P., Hermenegildo, M.V., Małuszynski, J. (eds.) Analysis and Visualization Tools for Constraint Programming. LNCS, vol. 1870, pp. 253–272. Springer, Heidelberg (2000). https://doi.org/10.1007/10722311_11
12. Choco: An Open-Source Java Library for Constraint Programming. https://choco-solver.org
13. Debruyne, R., Bessière, C.: Some practicable filtering techniques for the constraint satisfaction problem. In: Proceedings of the Fifteenth International Joint Conference on Artificial Intelligence (IJCAI 1997), pp. 412–417 (1997)
14. Defays, D.: An efficient algorithm for a complete link method. Comput. J. **20**(4), 364–366 (1977)
15. Eén, N., Biere, A.: Effective preprocessing in SAT through variable and clause elimination. In: Bacchus, F., Walsh, T. (eds.) SAT 2005. LNCS, vol. 3569, pp. 61–75. Springer, Heidelberg (2005). https://doi.org/10.1007/11499107_5
16. Epstein, S.L., Freuder, E.C., Wallace, R.M., Li, X.: Learning propagation policies. In: Working Notes of the Second International Workshop on Constraint Propagation and Implementation, Held in Conjunction with CP-05, pp. 1–15 (2005)
17. Freuder, E.C., Elfe, C.D.: Neighborhood inverse consistency preprocessing. In: Proceedings of the Thirteenth National Conference on Artificial Intelligence (AAAI 1996), pp. 202–208 (1996)

18. Freuder, E.C., Wallace, R.J.: Selective relaxation for constraint satisfaction problems. In: Proceedings of the IEEE Third International Conference on Tools with Artificial Intelligence (ICTAI 1991), pp. 332–339 (1991)
19. Ghoniem, M., Cambazard, H., Fekete, J.D., Jussien, N.: Peeking in solver strategies using explanations visualization of dynamic graphs for constraint programming. In: Proceedings of the 2005 ACM Symposium on Software Visualization (SoftVis 2005), pp. 27–36 (2005)
20. Hayes, P.J.: The second Naive physics manifesto. In: Readings in Qualitative Reasoning About Physical Systems, pp. 46–63. Morgan Kaufmann (1990)
21. Järvisalo, M., Heule, M.J.H., Biere, A.: Inprocessing rules. In: Gramlich, B., Miller, D., Sattler, U. (eds.) IJCAR 2012. LNCS (LNAI), vol. 7364, pp. 355–370. Springer, Heidelberg (2012). https://doi.org/10.1007/978-3-642-31365-3_28
22. Karakashian, S., Woodward, R., Reeson, C., Choueiry, B.Y., Bessiere, C.: A first practical algorithm for high levels of relational consistency. In: Proceedings of the Twenty-Fourth AAAI Conference on Artificial Intelligence (AAAI 2010), pp. 101–107 (2010)
23. Kuipers, B.: Qualitative Reasoning - Modeling and Simulation with Incomplete Knowledge. MIT Press, Cambridge (1994)
24. Kullback, S., Leibler, R.A.: On information and sufficiency. Ann. Math. Stat. **22**(1), 79–86 (1951)
25. Lagerkvist, M.Z., Schulte, C.: Propagator groups. In: Gent, I.P. (ed.) CP 2009. LNCS, vol. 5732, pp. 524–538. Springer, Heidelberg (2009). https://doi.org/10.1007/978-3-642-04244-7_42
26. Mackworth, A.K.: On reading sketch maps. In: Proceedings of the Fifth International Joint Conference on Artificial Intelligence (IJCAI 1977), pp. 598–606 (1977)
27. Manning, C.D., Raghavan, P., Schütze, H.: Introduction to Information Retrieval. Cambridge University Press, Cambridge (2008)
28. Nielsen, F.: Introduction to HPC with MPI for Data Science. UTCS. Springer, Cham (2016). https://doi.org/10.1007/978-3-319-21903-5
29. Paparrizou, A., Stergiou, K.: Evaluating simple fully automated heuristics for adaptive constraint propagation. In: Proceedings of the IEEE Twenty-Fourth International Conference on Tools with Artificial Intelligence (ICTAI 2012), pp. 880–885 (2012)
30. Paparrizou, A., Stergiou, K.: On neighborhood singleton consistencies. In: Proceedings of the Twenty-Sixth International Joint Conference on Artificial Intelligence (IJCAI 2017), pp. 736–742 (2017)
31. Petit, T.: Personal communication (2020)
32. Phan, K.: Extending the Choco constraint-solver with high-level consistency with evaluation on the nonogram puzzle. Undergraduate thesis. Department of Computer Science and Engineering, University of Nebraska-Lincoln (2020)
33. Samaras, N., Stergiou, K.: Binary encodings of non-binary constraint satisfaction problems: algorithms and experimental results. JAIR **24**, 641–684 (2005)
34. Schulte, C.: Oz explorer: a visual constraint programming tool. In: Kuchen, H., Doaitse Swierstra, S. (eds.) PLILP 1996. LNCS, vol. 1140, pp. 477–478. Springer, Heidelberg (1996). https://doi.org/10.1007/3-540-61756-6_108
35. Schulte, C., Tack, G., Lagerkvist, M.Z.: Modeling and Programming with Gecode (2015). http://www.gecode.org/doc-latest/MPG.pdf
36. SciPy.org: scipy.cluster.hierarchy.linkage. https://docs.scipy.org/doc/scipy/reference/generated/scipy.cluster.hierarchy.linkage.html. Accessed 26 May 2020
37. Shishmarev, M., Mears, C., Tack, G., de la Banda, M.G.: Visual search tree profiling. Constraints **21**(1), 77–94 (2016)

38. Simonis, H., Aggoun, A.: Search-tree visualisation. In: Deransart, P., Hermenegildo, M.V., Małuszynski, J. (eds.) Analysis and Visualization Tools for Constraint Programming. LNCS, vol. 1870, pp. 191–208. Springer, Heidelberg (2000). https://doi.org/10.1007/10722311_8
39. Simonis, H., Davern, P., Feldman, J., Mehta, D., Quesada, L., Carlsson, M.: A generic visualization platform for CP. In: Cohen, D. (ed.) CP 2010. LNCS, vol. 6308, pp. 460–474. Springer, Heidelberg (2010). https://doi.org/10.1007/978-3-642-15396-9_37
40. Simonis, H., O'Sullivan, B.: Almost square packing. In: Achterberg, T., Beck, J.C. (eds.) CPAIOR 2011. LNCS, vol. 6697, pp. 196–209. Springer, Heidelberg (2011). https://doi.org/10.1007/978-3-642-21311-3_19
41. Stergiou, K.: Heuristics for dynamically adapting propagation. In: Proceedings of the Eighteenth European Conference on Artificial Intelligence (ECAI 2008), pp. 485–489 (2008)
42. Wallace, R.J.: SAC and neighbourhood SAC. AI Commun. **28**(2), 345–364 (2015)
43. Woodward, R., Karakashian, S., Choueiry, B.Y., Bessiere, C.: Solving difficult CSPs with relational neighborhood inverse consistency. In: Proceedings of the Twenty-Fifth AAAI Conference on Artificial Intelligence (AAAI 2011), pp. 112–119 (2011)
44. Woodward, R.J.: Relational neighborhood inverse consistency for constraint satisfaction: a structure-based approach for adjusting consistency and managing propagation. Master's thesis, Department of Computer Science and Engineering, University of Nebraska-Lincoln, Lincoln, NE, December 2011
45. Woodward, R.J.: Higher-level consistencies: where, when, and how much. Ph.D. thesis, University of Nebraska-Lincoln, December 2018
46. Woodward, R.J., Choueiry, B.Y., Bessiere, C.: A reactive strategy for high-level consistency during search. In: Proceedings of the Twenty-Seventh International Joint Conference on Artificial Intelligence (IJCAI 2018), pp. 1390–1397 (2018)
47. Woodward, R.J., Karakashian, S., Choueiry, B.Y., Bessiere, C.: Revisiting neighborhood inverse consistency on binary CSPs. In: Milano, M. (ed.) CP 2012. LNCS, pp. 688–703. Springer, Heidelberg (2012). https://doi.org/10.1007/978-3-642-33558-7_50
48. Woodward, R.J., Schneider, A., Choueiry, B.Y., Bessiere, C.: Adaptive parameterized consistency for non-binary CSPs by counting supports. In: O'Sullivan, B. (ed.) CP 2014. LNCS, vol. 8656, pp. 755–764. Springer, Cham (2014). https://doi.org/10.1007/978-3-319-10428-7_54
49. Wotzlaw, A., van der Grinten, A., Speckenmeyer, E.: Effectiveness of pre- and inprocessing for CDCL-based SAT solving. Technical report, Institut für Informatik, Universität zu Köln, Germany (2013). http://arxiv.org/abs/1310.4756
50. XCSP3: Benchmark Series in XCSP3 Format. http://xcsp.org/series. Accessed 24 May 2020

# Parallelization of TSP Solving in CP

Nicolas Isoart(✉) and Jean-Charles Régin

Université Côte d'Azur, CNRS, I3S, Nice, France
nicolas.isoart@gmail.com, jcregin@gmail.com

**Abstract.** The currently best CP method for solving the Travelling Salesman Problem is the Weighted Circuit Constraint associated with the LCFirst search strategy. The use of Embarrassingly Parallel Search (EPS) for this model is problematic because EPS decomposition is a depth-bounded process unlike the LCFirst search strategy which is depth-first. We present *Bound-Backtrack-and-Dive*, a method which solves this issue. First, we run a sequential solving of the problem with a bounded number of backtracks in order to extract key information from LCFirst, then we decompose with EPS using that information rather than LCFirst. The experimental results show that we obtain almost a linear gain on the number of cores and that *Bound-Backtrack-and-Dive* may considerably reduce the number of backtracks performed for some problems.

## 1 Introduction

In graph theory, the Travelling Salesman Problem (TSP) consists in searching for a single cycle covering a graph such that the sum of the cycle edge cost is minimal. This problem has been widely studied as there is a huge number of direct applications such as routing problems, school bus problems, *etc.* and indirect applications such as scheduling where cities are tasks and arcs are transition times. As the decision version of the TSP is NP-Complete, many methods have been tried to solve it (MIP, CP, SAT, *etc.*). Among them, a MIP method, named Concorde [1], is indisputably the state of the art for solving the pure TSP. Unfortunately, there are a lot of applications of the TSP where side constraints are involved. For example, precedence constraints between nodes or time windows (*i.e.* a period of time during which a node can be visited). In such cases, Concorde can no longer be used and CP with the Weighted Circuit Constraint (WCC) [2,3] and k-cutsets constraint [8] becomes a competitive approach for exact solving.

The WCC is composed of a Lagrangian relaxation of a 1-tree (*i.e.* a special minimum spanning tree plus one minimum cost edge) and a degree constraint. The Weighted Circuit and cutsets Constraint (WCCC) is the association of the WCC and k-cutsets constraint. In addition, it uses a search strategy integrating a graph interpretation of Last Conflict heuristics [5,10], named LCFirst [4]. This search strategy selects one node from the graph according to a heuristic and keeps branching on the node until there are no more candidates around it, no


H. Simonis (Ed.): CP 2020, LNCS 12333, pp. 410–426, 2020.
https://doi.org/10.1007/978-3-030-58475-7_24

matter if we backtrack or not. Hence, this search strategy learns from previous branching choices and tends to keep those that previously caused a fail. Results in [4] show that LCFirst clearly outperforms all other search strategies, it is the only one that really exploits the graph structure. So far, LCFirst beats all other search strategies by one or more orders of magnitude.

Over the last few years, machines with multiple cores have been commercialized. When the solving time is important, making the best use of your machine is also important. As a result, parallelism has been widely used to improve the solving of some problems. When the solving method uses branch-and-bound, Embarrassingly Parallel Search (EPS) [11,12] allows parallelism in problem solving in a non-intrusive way. The idea is to statically decompose the initial problem into a huge number of sub-problems that are consistent with propagation (*i.e.* running the propagation mechanism on them does not detect any inconsistency). These sub-problems are added to a queue, which is managed by a master. Then, each waiting worker takes a sub-problem from the queue and solves it. The process is repeated until all the sub-problems have been solved. The decomposition is done by selecting a subset of the variables and enumerating the consistent combinations of the values of these variables. More precisely, EPS decomposes step by step and at each step, sub-problems are decomposed into other sub-problems. The decomposition of a subproblem is done by calculating a set of variables to be assigned, then, generates all subproblems consistent with propagation assigning the set of variable to be affected. The decomposition is done when a specific number of sub-problems is obtained, therefore EPS repeats this step in a breadth-first way. Thus, we define an assignment limit in the enumeration mechanism, which we will increase iteratively until we obtain at least the required number of sub-problems.

Since LCFirst is a depth-based search strategy, it must solve a left branch completely before starting to solve a right branch to complete its learning. However, EPS involves a breadth-based decomposition method, thus it stays shallow in the search tree. The use of LCFirst during the decomposition is therefore difficult since one requires to go deep while the other stays shallow. Experiments show that the use of LCFirst during the decomposition gives results that are not robust. Sometimes the decomposition will be deep enough for LCFirst, but most of the time this is not the case and it leads to sub-problems such that the overall number of solving backtracks will be several times higher than a sequential execution.

The parallel solving of the TSP cannot be performed without LCFirst, since the results would be deteriorated by several orders of magnitude. Thus, we introduce *Bound-Backtrack-and-Dive*, a method used to approximate the information learned by LCFirst. It consists of running a sequential execution with a low number of backtrack allowed in order to build an ordered set containing the nodes selected by LCFirst in the search tree. The order is defined according to the number of times the nodes are backtracked and how deep they are. Then, the parallel execution of the TSP is started (decomposition step) such that the LCNodes are chosen accordingly to the previous defined order. Finally, the solving step is performed with the classical LCFirst algorithm. The idea comes from our analysis of LCFirst behavior, we notice that it quickly distinguishes some nodes and then refines its knowledge. We propose to study the information learned by LCFirst at the beginning of the search and to use it to simulate the main trends of LCFirst during the decomposition with EPS. We experimentally show that the study of the first $1,000$ backtracks of the search tree is enough to simulate LCFirst during the decomposition.

Indeed, without *Bound-Backtrack-and-Dive* we show that the decomposition is not robust and that it sometimes leads to low or no gains compared to a sequential execution. We also show that *Bound-Backtrack-and-Dive* allows us to obtain robust results on 4 cores, and then to improve solving times by a factor of 5.4 and backtracks by a factor of 2.3 compared to a sequential execution.

The decomposition of a problem such as TSP is not trivial and requires some modifications of the EPS decomposition mechanism for two reasons. First, the model of the TSP in CP contains a set-variable with the mandatory edges as lower bound and optional edges as upper bound. Decomposing with a set-variable is not as trivial as a classical Cartesian product because we have to be careful with the order while enumerating. Second, a new phenomenon appeared in the decomposition while experimenting: the increase of the assignment limit in the enumeration may not lead to an increase in the number of sub-problems. It turns out that the TSP search tree is extremely heterogeneous. Thus, we introduce an extended mechanism for coherent enumeration of set-variables and for stopping the decomposition in case of non-progression.

The article is organized as follows. We recall some definitions. Then, we give details on the modification of EPS decomposition mechanisms. Afterwards, we detail the difficulties of integrating EPS into TSP because of the combination of LCFirst and decomposition. Thus, we introduce the *Bound-Backtrack-and-Dive* algorithm to fix the issue of LCFirst during the decomposition. Before concluding, we give some experimental results.

## 2 Preliminaries

### 2.1 Graph Theory

A **directed graph** or **digraph** $G = (X, U)$ consists of a **vertex set** X and an **arc set** U, where every arc $(u, v)$ is an ordered pair of distinct vertices. We will denote by $X(G)$ the vertex set of $G$ and by $U(G)$ the arc set of $G$. The **cost** of an arc is a value associated with the arc. An **undirected graph** is a digraph such that for each arc $(u, v) \in U$, $(u, v) = (v, u)$. If $G_1 = (X_1, U_1)$ and $G_2 = (X_2, U_2)$ are graphs, both undirected or directed, $G_1$ is a **subgraph** of $G_2$ if $V_1 \subseteq V_2$ and $U_1 \subseteq U_2$. For each arc $(u, v) \in U$, $(u, v) \in \delta(u)$ and $(u, v) \in \delta(v)$. A **path** from node $v_1$ to node $v_k$ in $G$ is a list of nodes $[v_1, ..., v_k]$ such that $(v_i, v_{i+1})$ is an arc for $i \in [1..k-1]$. The path **contains** node $v_i$ for $i \in [1..k]$ and arc $(v_i, v_{i+1})$ for $i \in [1..k-1]$. The path is **simple** if all its nodes are distinct. The path is a **cycle** if $k > 1$ and $v_1 = v_k$. A cycle is **hamiltonian** if $[v_1, ..., v_{k-1}]$ is a simple path and contains every vertices of U. The **length** of a path $p$, denoted by $length(p)$, is the sum of the costs of the arcs contained in $p$. For a graph $G$, a solution to the **Travelling Salesman Problem (TSP)** in $G$ is a Hamiltonian cycle $HC \in G$ cycle minimizing $length(HC)$. An undirected graph $G$ is **connected** if there is a path between each pair of vertices, otherwise it is **disconnected**. A **tree** is a connected graph without a cycle. A tree $T = (X', U')$ is a **spanning tree** of $G = (X, U)$ if $X' = X$ and $U' \subseteq U$. A **Minimum Spanning Tree (MST)** is a spanning tree $T = (X', U')$ minimizing the sum of the costs of $U'$.

Without loss of generality, we will only consider undirected graphs.

### 2.2 Constraint Programming

A finite **constraint network** $\mathcal{N}$. is defined as a set of $n$ **variables** $X = \{x_1, \ldots, x_n\}$, a set of current **domains** $\mathcal{D} = \{D(x_1), \ldots, D(x_n)\}$ where $D(x_i)$ is the finite set of possible **values** for variable $x_i$, and a set $\mathcal{C}$ of **constraints** between variables. We introduce the particular notation $\mathcal{D}_0 = \{D_0(x_1), \ldots, D_0(x_n)\}$ to represent the set of initial domains of $\mathcal{N}$. on which constraint definitions were stated.

A **constraint** $C$ on the ordered set of variables $X(C) = (x_{i_1}, \ldots, x_{i_r})$ is a subset $T(C)$ of the Cartesian product $D_0(x_{i_1}) \times \cdots \times D_0(x_{i_r})$ that specifies the **allowed** combinations of values for the variables $x_{i_1}, \ldots, x_{i_r}$. An element of $D_0(x_{i_1}) \times \cdots \times D_0(x_{i_r})$ is called a **tuple on** $X(C)$. A value $a$ for a variable $x$ is often denoted by $(x, a)$. Let $C$ be a constraint. A tuple $\tau$ on $X(C)$ is **valid** if $\forall (x, a) \in \tau, a \in D(x)$. $C$ is **consistent** iff there exists a tuple $\tau$ of $T(C)$ which is valid. A value $a \in D(x)$ is **consistent with** $C$ iff $x \notin X(C)$ or there exists a valid tuple $\tau$ of $T(C)$ with $(x, a) \in \tau$. A constraint is **arc consistent** iff $\forall x_i \in X(C), D(x_i) \neq \emptyset$ and $\forall a \in D(x_i)$, $a$ is consistent with $C$. A **set-variable** $x_i$ is a particular variable such that its domain is a set of set defined by $D(x_i) = \{S | lb(x_i) \subseteq S \subseteq ub(x_i)\}$. The domain can also be represented by the sets $lb(x_i)$ and $ub(x_i)$ such that $v \in lb(x_i)$ iff $\forall S \in D(x_i), v \in S$ and $v \in ub(x_i)$ iff $\exists S \in D(x_i), v \in S$. We say that $lb(x_i)$ is the set of mandatory elements, and $ub(x_i) \setminus lb(x_i)$ is the set of optionals elements. Usually a set-variable is also associated with an interval representing its cardinality. A constraint is **bound consistent** iff $\forall x_i \in X(C), ub(x_i) \neq \emptyset$ and $\forall a \in lb(x_i)$.

### 2.3 TSP Model

The TSP has three strong constraints: connectivity, cycle covering, and cycle unicity. The WCC is a constraint mainly based on Lagrangian relaxation. Intuitively, it takes a lower bound of the TSP computable in a polynomial time and derives it until the optimal solution of the TSP is obtained. More precisely, this Lagrangian relaxation computes a succession of 1-trees. A 1-tree is a minimum spanning tree in $G = (X - \{x\}, U)$ such that $x \in X$ and $x$ is connected to the MST by its two closest neighbors. Thus, this lower bound respects the constraint of connectivity and unicity of the cycle. We still need to cover the whole graph with this cycle. If every node of the 1-tree has a degree 2 then the cycle covering constraint is respected, and therefore the 1-tree is an optimal solution of the TSP. From this observation, the idea of Held and Karp [6,7] was to use a Lagrangian relaxation integrating the degree constraint into the objective function. For each node $i$ of the graph a multiplier $\pi_i$ is associated. Suppose $d$ is the neighbor number of the node $i$ in the 1-tree. If $d > 2$ then the value of $\pi_i$ is increased. If $d < 2$ then the value of $\pi_i$ is decreased. At each iteration, the cost of the edges of the graph is modified such that $w'((i,j))$ is the new cost and $w'((i,j)) = w((i,j)) + \pi_i + \pi_j$.

In the WCC, the Held and Karp Lagrangian relaxation is coupled with a branch-and-bound and some filtering algorithms. In order to strenghten the results, we use the WCCC which is an extension of the WCC including the k-cutsets constraint. The k-cutsets constraint finds all sizes 2 and 3 cutsets in the graph and imposes that an even number of edges for each cutset is mandatory.

Then, we use a single undirected graph variable where all nodes are mandatory. The branch-and-bound considers only the edges, it consists in making a binary search where a left branch is an edge assignment and a right branch is an edge removal. In the context of a single undirected graph variable, we note $O$ the set of optional edges, $M$ the set of mandatory edges and $D$ the set of deleted edges such that $O \cup M \cup D = U$ and $O \Delta M \Delta D = U$.

### 2.4 LCFirst

Introduced by Fages et al. [4], LCFirst introduces the principle of Last Conflict [10] in the context of a single graph variable. It selects an edge $(u, v)$ with a search strategy and keeps one of its two extremities, say $u$. Then, next edge decisions will be made in the neighborhood of $u$ until there are no more optional edges around $u$, no matter if a fail is raised. Next, we will note $u$ as the LCNode.

Algorithm 1 is the LCFirst algorithm, an example is depicted in Fig. 1.

**Algorithm 1:** $LCFirst(G = (X, U), \gamma)$

```
LCFirst(G, γ):
    global LCNode;
    if LCNode ≠ nil then
        //set of optional edges in the neighborhood of the LCNode;
        set ← O(δ(LCNode));
        if set ≠ ∅ then
            //select an edge in set w.r.t heuristic search strategy;
            return select(set);
    //select an edge in all the optional edges w.r.t heuristic search strategy;
    (u, v) ← select(O(U));
    LCNode ← u;
    return (u, v);
```

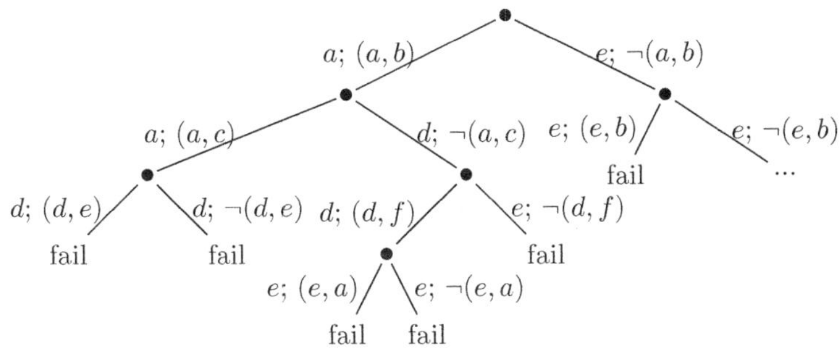


**Fig. 1.** An example of a search strategy with LCFirst. We note "LCNode; $(u, v)$" where $(u, v)$ is an assigned edge, $\neg(u, v)$ is a removed edge. We observe that when $(d, e)$ and $\neg(d, e)$ have caused a fail, $d$ is backtracked as the LCNode choice. We also observe that the choice of LCNode=$e$ is backtracked from the fail of $(e, a)$ to the first refutation because no choice could so far exhaust the neighborhood of $e$.

As the results show in [4], the use of LCFirst combined with selection heuristics outperforms all selection heuristics without a LCFirst policy. Thus, it is currently considered as the best search strategy for solving the TSP in CP.

However, the interaction of LCFirst with the TSP model is not obvious. As shown in [8], the use of k-cutsets improves the solving times. If a static search strategy (*i.e.* without LCFirst) is used as maxCost (*i.e.* select the edge by decreasing cost), the use of k-cutsets improves even more the solving times. Before k-cutsets, LCFirst maxCost and WCC was considered as one of the best models in CP. However, the integration of k-cutsets with WCC and LCFirst maxCost has the effect of degrading the results. On the other hand, using the search strategy LCFirst minDeltaDegree (*i.e.* selects the edge for which the sum of the endpoint degrees in the upper bound minus the sum of the endpoint degrees in the set-variable lower bound is minimal) combined with WCC and k-cutsets constraint allows to obtain a strong gain on solving times and to become the best CP model.

If not specified, the TSP model is the WCCC and the search strategy is LCFirst minDeltaDegree.

### 2.5 EPS

Embarrassingly Parallel Search (EPS) decomposes the initial problem into a huge number of sub-problems that are consistent with propagation (*i.e.* running the propagation mechanism on them does not detect any inconsistency), and adds them to a queue. Then, each waiting worker takes a sub-problem from the queue and solves it. The process stops when all sub-problems have been solved.

The main challenge of the decomposition is not to define equivalent problems, it is to avoid having some workers without work whereas some others are running during the solving step. In order to increase our chances to obtain well-balanced activity times for the workers[1], EPS decomposes the initial problem into a lot of sub-problems. It is usually considered that a good number of sub-problems per worker is between 30 to 300.

EPS must also avoid doing something in parallel that we would not have done sequentially. In order to reach that goal, EPS generates only sub-problems that are consistent with the propagation, that is such that if we run the propagation mechanism on them then there is no failure. This means that they are not known to be inconsistent, and would be considered by a sequential process. The generation of $q$ sub-problems is not straightforward because the number of sub-problems consistent with the propagation may not be related to the Cartesian product of some domains. A simple algorithm could be to perform a Breadth First Search (BFS) in the search tree until the desired number of sub-problems consistent with the propagation is reached. Unfortunately, it is not easy to perform a BFS efficiently mainly because BFS is not an incremental algorithm like Depth-First Search (DFS). Therefore, EPS uses a process resembling an iterative deepening depth-first search [9]: we consider a set $Y \subseteq X$ of variables: we only assign the variables of $Y$ and we stop the search when they are all assigned. In other words, we never try to assign a variable that is not in $Y$. This process is repeated until all assignments of $Y$ consistent with the propagation has been found. Each branch of a search tree computed by this search defines an assignment (*i.e.* a sub-problem). To generate $q$ sub-problems, we repeat the previous method by adding variables to $Y$ if necessary, until the number of sub-problems is greater than or equal to $q$. Note that the decomposition can be performed in parallel.

Optimization Problems, like the TSP, deserve a little more attention. EPS manages the value of the objective function as follows: when a worker takes a sub-problem, he also takes the best objective value that a worker has obtained so far, and when a worker solves a sub-problem, he communicates the best objective value found in order to update it for the next sub-problems. EPS do not use any additional communications. Note that it is not possible to use the objective value while decomposing the problem, because the objective value associated with the assignment of a subset of variable is not necessarily a valid bound for the problem and that assignment cannot necessarily be extended to a solution.

# 3 Adaptation of EPS Mechanism

## 3.1 Set-Variable

A set-variable is an efficient way to represent a set of classic variables and to break symmetries existing between these variables. When decomposing with EPS, we have two possibilities for the set-variables: we try to transform them into a set of classic variables and constraints (to break symmetries), or we adapt the usual algorithm to the set-variables. In the first case, it is important to break the symmetries. Indeed, a set-variable $sx$ involving the values $\{a, b, c, d\}$ and whose

[1] The activity time of a worker is the sum of the solving times of its sub-problems.

cardinality is equal to 3 has only one solution which implies the values $a$, $b$ and $c$, whereas if we replace $sx$ by 3 variables $x_1$, $x_2$ and $x_3$ we risk generating the solutions $x_1 = a$, $x_2 = b$, $x_3 = c$, and $x_1 = b$, $x_2 = a$, $x_3 = c$, *etc.* The introduction of an order between the variables avoids this concern. Nevertheless, when the cardinality is not fixed, this transformation is more delicate. It is therefore preferable to adapt the decomposition algorithm to the set-variable. When we consider a set $Y \subseteq X$ of variables, we pay attention to the set-variables. A classical variable is instantiated by a single value, whereas for a set-variable we will determine how many of its values should be instantiated at most. For instance, $sx$ can be instantiated with at most 1, 2 or 3 values. Its cardinality defines only the maximum because we search for partial assignments. In general, all but one set-variable of $Y$ will be potentially instantiated with their maximum possible values (*i.e.* the maximum cardinality).

### 3.2 Decomposition Issue

While experimenting on the TSP, results have shown some special cases such as decomposition. We observed in some cases that the increase of the enumeration limit on the mandatories leads to a reduction in the number of sub-problems generated. If suddenly many branches fail in the search tree, it may be possible that more problems have been removed than generated. For special cases such as this one, it is preferable to stop the decomposition as it may go into combinatorial. Thus, we have defined a stopping criterion other than the number of sub-problems generated. Now, if there is a decrease of the number of generated sub-problems greater than a given quantity, from our experiments we have empirically chosen 20%, we stop the decomposition in order to avoid this pathological case.

## 4 LCFirst and Decomposition

As explained above, the search strategy uses a graph interpretation of Last Conflict heuristic. Such a search strategy learns from left branches for the right branches. If the depth is bounded, as in the decomposition of EPS, it can happen that we do not reach the bottom of the search tree and therefore the right branches use different LCNode than the one they would have obtained if the left branch had been completely performed. This leads to different search tree and therefore different results (see Fig. 2).

In Fig. 2, with an enumeration limit set to 2, the grey area is not visited since the size of all assigned edges ($\{ab, ac\}$) is equal to 2. When solving sequentially, the grey area is visited since there is no enumeration limit whereas the grey area is not visited when decomposing the problem because an enumeration limit is set to 2. Thus, the LCNode value (yellow zone) for the branch $root - ab - \neg ac$ depends on the visit of the grey area. If it is visited then LCNode=$d$ otherwise LCNode=$a$. For the branch $root - \neg ab$ (orange zone) the LCNode value depends on the

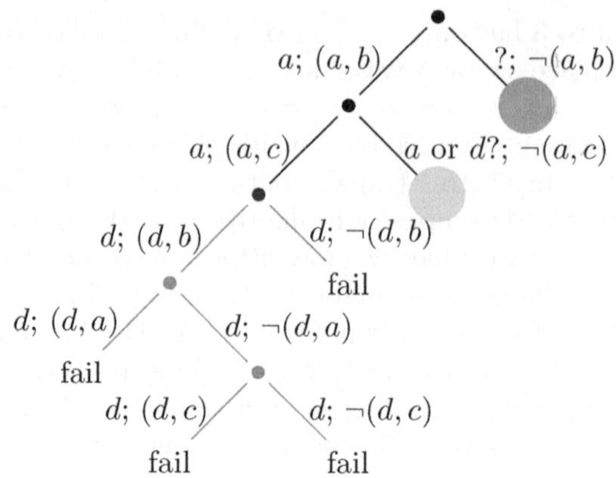


**Fig. 2.** Search for an assignment enumeration limit of 2, represented by a blue node when it is reached. The black area is the search tree with an enumeration limit. The gray area is the search tree when there is no assignment enumeration limit. The orange and yellow areas are where decomposition does not have full information about LCNode. (Color figure online)

result of the previous LCNode values (yellow zone). Hence, the decomposition can generate a very different search tree than a sequential execution.

One can legitimately ask the following question: how is the solving impacted if we get the wrong LCNode?

The impact can be very bad. For example, if we take the instance pcb442 of TSPLib [13], its sequential solving time is 17, 109 s for 3, 571, 219 backtracks, but its solving time on 4 cores with EPS is 21, 539 s for 11, 473, 274 backtracks. So, the parallel version is 26% slower and it loses a factor of 3.2 in backtracks whereas it involves 4 more cores. Later, the various experiments will show that this problem is not unusual. These results can be explained by the remarks made for Fig. 2. This shows that a few bad LCNode can significantly increase combinatorics. Disabling LCFirst from the decomposition strongly degrades the results. In practice, it leads to a search strategy that jumps in the graph and does not exploit the structure of the graph.

Thus, the classical EPS decomposition is not enough for a dynamic search strategy such as LCFirst.

## 5 Bound-Backtrack-and-Dive and Decomposition

To solve the problem of LCFirst during the decomposition, we propose to use a diving heuristic, named *Bound-Backtrack-and-Dive*. It consists in running a sequential solving of the problem with a bounded number of backtracks in order to build an ordered set of nodes, called $\gamma$, that we use to represent the impact of LCNode choice. Then, we run EPS with $\gamma$ as a parameter and we use it for LCNode selection during decomposition.

### 5.1 Computation of $\gamma$

For each $\gamma_i$ such that $i \in X$, a value representing the fail impact of branching around $i$ is associated. Thus, the higher $\gamma_i$ is, the more important the node is for LCFirst. More exactly, when a node $i$ is backtracked in the search tree, we update $\gamma_i$ with $\gamma_i \leftarrow \gamma_i + C/depth^2$ where $C$ is a constant such that $C > depth^2_{max}$. Therefore, a fail thrown at a shallow depth in the search tree will increase the value of $\gamma_{LCNode}$ much more than if the fail is thrown at a deep depth in the search tree. In order to represent this priority, a non-linear denominator has been chosen. The idea of prioritizing LCNodes causing shallow fails in the search tree allows us to narrow the search strategy to the difficult areas of the graph. Thus, we propose to calculate $\gamma$ with the above method for a fixed number of backtracks, when the limit is reached we stop the solving and return $\gamma$.

### 5.2 Decomposition

After computing $\gamma$, we decompose the initial problem with a modified LCFirst algorithm. When a node is emptied (*i.e.* there are no more selectable edges around LCNode) LCFirst will usually look for an edge in the graph using a heuristic, keep one end and empty it again. Here, we propose that when a node is emptied, instead of looking for an edge in the graph and keeping one extremity, we select the non-emptied node with the highest value in $\gamma$. Then, with a classical search strategy (*i.e.* minDeltaDeg), we select an edge around this node until we have emptied it. A possible implementation of modified LCFirst is described in Algorithm 2.

At the end of the decomposition, we drop $\gamma$ and go back to classical LCFirst. We mainly use $\gamma$ because the decomposition is done in breadth and LCFirst gets good LCNode by depth. Thus, while solving EPS sub-problems we can use depth-based methods (classical LCFirst) since we solve them. It allows us to learn locally about each sub-problem and provide better results.

## 6 Experiments

The algorithms have been implemented in Java 11 in a locally developed constraint programming solver. The experiments were performed on a Windows 10 laptop with an Intel Core i7-7820HQ CPU @ 2.90 GHz and 32 Go of RAM. The reference instances are from the TSPLib [13], a library of reference graphs for the TSP. All instances considered are symmetric graphs.

In order to represent as well as possible current machines, the parallel experiments were carried out on 4 cores. We present the results in tables. Each of them reports the solving time in seconds. If it is strictly greater than 48 hours, it is specified as *t.o.* The number of backtracks is denoted by #bk. The load balancing, noted l.b., is the sum of the solving times of each worker divided by the wall clock time. It allows us to evaluate the distribution quality of the done work, the closer the load balancing is to the number of workers the better.

**Algorithm 2:** $LCFirst(G = (X), \gamma)$

```
LCFirst(G,γ):
    global LCNode;
    if LCNode ≠ nil then
        set ← O(δ(LCNode));
        if set ≠ ∅ then
            return select(set);
    idMax ← nil;
    valMax ← −MAX_VALUE;
    for each γ_i ∈ γ do
        //search for the node with the highest γ_i having optional neighbors;
        if γ_i > valMax and O(δ(i)) ≠ ∅ then
            idMax ← i;
            valMax ← γ_i;
    if idMax ≠ nil then
        (u, v) ← select(O(δ(idMax));
        LCNode ← idMax;
    else
        (u, v) ← select(O(U));
        LCNode ← u;
    return (u, v);
```

We give the arithmetic mean and the geometric mean, the latter is less affected by large variations between the values. The *Bound-Backtrack-and-Dive* algorithm is noted BBD.

Initially, the goal is to make the overall same number of backtracks with and without EPS. In practice, it is very difficult to achieve it because of the dynamic strategies, the Lagrangian relaxation and the order of each sub-problems are taken. However, we show by experiments that the use of *Bound-Backtrack-and-Dive* allows to obtain a similar or even improving backtrack number. Because of load balancing problems in parallelism, backtracks are a good metric. Indeed, a problem can be solved with less backtrack and a higher solving time. For example, if 1,000 sub-problems are generated and each sub-problem takes 1 s to solve except one that takes 1,000 s to solve. If this problem is the last one to be handled, only one worker will work while the others have finished their jobs. It often happens because the TSP sub-problems can be extremely heterogeneous.

In Table 1, compared to a sequential solving, EPS improves the results on average with or without LCFirst during the decomposition. More precisely, the gain factor is better if we have LCFirst during the decomposition. It can be explained by the fact that we keep branching in the same areas of the graph. However, if we look at the ratio column for each of the runs with EPS, many instances do many more backtracks than their sequential runs. For example, pcb442 does 3,571,219bk with a sequential solving compared to 11,473,274bk

**Table 1.** Comparison of sequential solving, EPS without diving, EPS without diving and without LCFirst during decomposition. A ratio column compares the sequential method with each EPS method. Greater ratio are better.

| Instance | Sequential | | EPS noDiveLCDec | | | ratio | | EPS noDiveNoLCDec | | | ratio | |
|---|---|---|---|---|---|---|---|---|---|---|---|---|
| | time | #bk | time | #bk | l.b | time | #bk | time | #bk | l.b | time | #bk |
| pr136 | 23.4 | 19,147 | 11.1 | 25,748 | 3.1 | 2.1 | 0.7 | 9.0 | 22,650 | 3.7 | 2.6 | 0.8 |
| kroA150 | 3.8 | 2,493 | 4.8 | 9,796 | 3.5 | 0.8 | 0.3 | 4.7 | 9,654 | 3.2 | 0.8 | 0.3 |
| kroB150 | 212.1 | 176,773 | 82.4 | 183,964 | 2.9 | 2.6 | 1.0 | 91.1 | 190,276 | 2.7 | 2.3 | 0.9 |
| si175 | 461.0 | 401,428 | 277.7 | 220,874 | 1.0 | 1.7 | 1.8 | 504.7 | 424,430 | 1.0 | 0.9 | 0.9 |
| rat195 | 36.8 | 20,977 | 15.9 | 29,062 | 3.7 | 2.3 | 0.7 | 17.0 | 28,030 | 3.5 | 2.2 | 0.7 |
| kroA200 | 650.0 | 309,343 | 209.3 | 396,704 | 3.9 | 3.1 | 0.8 | 157.0 | 279,564 | 3.7 | 4.1 | 1.1 |
| kroB200 | 113.5 | 61,107 | 55.4 | 105,660 | 3.6 | 2.0 | 0.6 | 65.4 | 132,836 | 3.8 | 1.7 | 0.5 |
| tsp225 | 161.4 | 76,397 | 69.9 | 95,016 | 3.2 | 2.3 | 0.8 | 126.9 | 134,348 | 2.4 | 1.3 | 0.6 |
| gr229 | 247.0 | 165,826 | 77.4 | 167,368 | 3.7 | 3.2 | 1.0 | 128.4 | 139,272 | 1.9 | 1.9 | 1.2 |
| a280 | 9.3 | 2,829 | 8.8 | 9,442 | 3.2 | 1.1 | 0.3 | 6.5 | 6,408 | 2.8 | 1.4 | 0.4 |
| lin318 | 19.1 | 4,063 | 14.1 | 6,396 | 3.2 | 1.4 | 0.6 | 19.1 | 10,290 | 3.8 | 1.0 | 0.4 |
| pcb442 | 17,109.4 | 3,571,219 | 21,539.2 | 11,473,274 | 2.6 | 0.8 | 0.3 | 20,988.8 | 12,829,764 | 3.2 | 0.8 | 0.3 |
| gr431 | 1,532.2 | 211,585 | 852.0 | 258,832 | 2.3 | 1.8 | 0.8 | 657.7 | 242,232 | 3.1 | 2.3 | 0.9 |
| d493 | *t.o.* | 24,733,443 | 57,995.1 | 12,934,364 | 2.2 | >3.0 | >1.9 | 32,590.3 | 10,927,816 | 3.5 | >5.3 | >2.3 |
| ali535 | 123,620.0 | 13,119,783 | 85,733.0 | 10,014,704 | 1.0 | 1.4 | 1.3 | *t.o.* | *t.o.* | - | >0.7 | *t.o.* |
| mean | >21,133 | >2,858,427 | 11,130 | 2,395,414 | | | | >15,211 | >3,037,520 | | | |
| geo mean | >340 | >129,436 | 189 | 175,493 | | | | >207 | >190,302 | | | |

with EPS and LCFirst during the decomposition compared to 12,829,764bk with EPS and no LCFirst during the decomposition. Due to the large increase in the number of backtracks, this instance with these models takes more time to solve with EPS than in sequential. Conversely, other instances such as kroA200 work well with these models and EPS. While its sequential execution performs 309,343bk, EPS noDivLCDec performs 396,704bk and EPS noDiveNoLCDec performs 279,564bk. Thus, a gain of a factor of 3.1 and 4.1 respectively in solving time is observed. It can also happen that instances like d493 have a better behavior without LCFirst during the decomposition, here it gets a gain of 16% of backtracks. In short, EPS without diving works but is not robust.

In Table 2, we observe that the limit on the number of backtracks for diving has an impact on the overall solving time. First, for problems with a low solving time, the time spent diving can become significantly high compared to the overall solving time. For example, for a280 with limitBk = 1,000, we spend 3.7 s diving and 2.0 s decomposing and solving the problem. Increasing limitBk here increases the overall solving time significantly since the value of limitBk exceeds the number of backtracks performed sequentially. Thus, the instance is completely solved in sequential before being solved with EPS. The lin318 instance has the same issue, increasing the backtrack limit fully solves the problem in sequential during diving before doing the decomposition and solving, which is actually very fast (about 5 s) while diving takes 5 s for limitBk = 1,000, then about 16.5 s for limitBk = 5,000 and 10,000. Thus, it is better to use a very small limitBk for problems having very low solving times because the diving should not take a significant amount of time from the overall solving time. Then, for bigger problems,

**Table 2.** Benchmark of the results according to the limit of the number of backtrack allowed for diving.

| Instance | BBD limitBk = 1k | | | | BBD limitBk = 5k | | | | BBD limitBk = 10k | | | |
|---|---|---|---|---|---|---|---|---|---|---|---|---|
| | Div. | Decomp. and Solving | | | Div. | Decomp. and Solving | | | Div. | Decomp. and Solving | | |
| | time | time | #bk | l.b. | time | time | #bk | l.b. | time | time | #bk | l.b. |
| pr136 | 1.7 | 13.9 | 37,134 | 3.9 | 6.6 | 8.9 | 24,522 | 3.9 | 12.7 | 7.4 | 19,648 | 3.9 |
| kroA150 | 1.8 | 0.8 | 792 | 1.1 | 3.9 | 0.9 | 858 | 1.1 | 3.9 | 0.9 | 896 | 1.3 |
| kroB150 | 1.6 | 17.2 | 42,320 | 3.5 | 6.4 | 41.5 | 105,696 | 4.0 | 13.2 | 35.8 | 88,812 | 3.6 |
| si175 | 2.1 | 129.2 | 338,752 | 3.6 | 7.1 | 133.3 | 399,854 | 3.8 | 13.6 | 147.6 | 446,822 | 3.9 |
| rat195 | 2.4 | 12.1 | 21,468 | 3.8 | 9.5 | 10.4 | 17,726 | 3.9 | 18.0 | 12.4 | 20,826 | 3.9 |
| kroA200 | 2.4 | 103.4 | 180,692 | 3.9 | 9.7 | 63.5 | 105,470 | 3.7 | 19.4 | 86.4 | 150,470 | 3.8 |
| kroB200 | 2.4 | 21.0 | 37,548 | 3.8 | 9.6 | 29.7 | 53,248 | 3.9 | 19.3 | 24.3 | 43,962 | 4.0 |
| tsp225 | 2.7 | 39.1 | 65,572 | 4.0 | 12.1 | 31.6 | 53,876 | 3.9 | 23.2 | 30.5 | 50,210 | 3.8 |
| gr229 | 2.3 | 90.9 | 158,970 | 4.0 | 9.9 | 118.8 | 222,222 | 3.8 | 18.2 | 75.0 | 143,458 | 4.0 |
| a280 | 3.7 | 2.0 | 934 | 0.3 | 9.1 | 1.6 | 742 | 0.5 | 9.1 | 1.5 | 710 | 1.0 |
| lin318 | 5.0 | 5.0 | 1,284 | 0.3 | 16.4 | 5.3 | 1,648 | 0.1 | 16.7 | 5.5 | 1,298 | 0.1 |
| pcb442 | 7.5 | 2,541.7 | 1,761,286 | 3.7 | 36.7 | 3,906.3 | 2,263,926 | 3.0 | 70.5 | 1,643.5 | 1,043,384 | 3.3 |
| gr431 | 11.7 | 420.5 | 153,428 | 3.9 | 51.0 | 888.3 | 253,494 | 2.0 | 92.3 | 602.3 | 220,972 | 3.9 |
| d493 | 14.8 | 6,303.4 | 2,621,658 | 3.8 | 64.4 | 14,644.7 | 4,554,156 | 2.8 | 128.6 | 12,880.4 | 4,221,914 | 3.0 |
| ali535 | 16.7 | 55,186.4 | 20,216,560 | 3.5 | 87.4 | 35,869.5 | 6,517,890 | 1.7 | 173.5 | 37,523.1 | 7,574,870 | 1.9 |
| mean | | 4,326 | 1,709,227 | | | 3,717 | 971,689 | | | 3,538 | 935,217 | |
| geo mean | | 73 | 74,031 | | | 82 | 77,552 | | | 73 | 69,964 | |

limitBk also has an impact. Indeed, we notice that the best mean solving times are obtained with limitBk = 5,000 and the best geometric mean solving times are obtained with limitBk=1,000 and limitBk = 10,000. However, the geometric mean shows that the differences between them are actually quite small in terms of solving time and backtracks. We also observe that increasing limitBk increases on average the robustness of results. In Table 3, we notice that, contrary to Table 1, almost all backtracks ratios are positive, very often improving. It allows to obtain a factor up to 5.41 in mean solving time and 2.35 in mean number of backtracks. Thus, the ratios show that whatever the choice of the limit for *Bound-Backtrack-and-Dive*, the results are robust and allow to obtain a strong gain in solving time.

If we look at the results in more detail, we notice that limitBk = 1,000 performed particularly well for kroB150 (gain of a ratio 11.28 vs 4.43 vs 4.33 for resp. limit = {1,000, 5,000, 10,000}) and especially for d493 (gain of a ratio 27.35 vs 11.75 vs 13.28 for resp. limit = {1,000, 5,000, 10,000}). The advantage of limitBk = 1,000 is that the diving learns very quickly on some areas of the graph. With limitBk = 5,000 or limitBk = 10,000, the diving can learn on more areas of the graph. If these areas are equivalent, it is possible that the search strategy will focus less on specific areas. This can be an advantage as well as a drawback depending on the instance. So, without loss of generality, we choose limitBk = 1,000 for the next experiments.

In Table 4, we compare the impact of the number of sub-problems per worker (sppw) on the results. We can observe that no matter how many sub-problems are required for the decomposition, the load balancing remains good and similar

**Table 3.** Ratio of solving time and numbers of backtracks. It is calculated by respectively dividing the results of a sequential execution with the results of *Bound-Backtrack-and-Dive* execution.

| Instance | seq./limitBk = 1k | | seq./limitBk = 5k | | seq./limitBk = 10k | |
|---|---|---|---|---|---|---|
| | time | #bk | time | #bk | time | #bk |
| pr136 | 1.50 | 0.52 | 1.51 | 0.78 | 1.16 | 0.97 |
| kroA150 | 1.46 | 3.15 | 0.79 | 2.91 | 0.79 | 2.78 |
| kroB150 | 11.28 | 4.18 | 4.43 | 1.67 | 4.33 | 1.99 |
| si175 | 3.51 | 1.19 | 3.28 | 1.00 | 2.86 | 0.90 |
| rat195 | 2.54 | 0.98 | 1.85 | 1.18 | 1.21 | 1.01 |
| kroA200 | 6.14 | 1.71 | 8.88 | 2.93 | 6.14 | 2.06 |
| kroB200 | 4.85 | 1.63 | 2.89 | 1.15 | 2.60 | 1.39 |
| tsp225 | 3.86 | 1.17 | 3.69 | 1.42 | 3.01 | 1.52 |
| gr229 | 2.65 | 1.04 | 1.92 | 0.75 | 2.65 | 1.16 |
| a280 | 1.63 | 3.03 | 0.87 | 3.81 | 0.88 | 3.98 |
| lin318 | 1.91 | 3.16 | 0.88 | 2.47 | 0.86 | 3.13 |
| pcb442 | 6.71 | 2.03 | 4.34 | 1.58 | 9.98 | 3.42 |
| gr431 | 3.55 | 1.38 | 1.63 | 0.83 | 2.21 | 0.96 |
| d493 | 27.35 | 9.43 | 11.75 | 5.43 | 13.28 | 5.86 |
| ali535 | 2.24 | 0.65 | 3.44 | 2.01 | 3.28 | 1.73 |
| mean | 5.41 | 2.35 | 3.48 | 2.00 | 3.68 | 2.19 |
| geo mean | 3.68 | 1.75 | 2.55 | 1.67 | 2.55 | 1.85 |

(around 3). In fact, the average number of backtrack is quite close ($1,709,227$ vs $1,940,812$ vs $1,553,287$). The solving time is slightly better for sppw = 100, but again, d493 worked particularly well and greatly reduced the solving time. If we look instance by instance, we can see that the variations are relatively small, and despite the variations the results remain robust. Thus, no matter how many sub-problems are generated, the requested robustness and a strong improvement of the solving times is obtained.

To increase the number of workers, the overall number of sub-problems generated must be at least 50 sub-problems per worker. In Table 4, we show that the solving times are good for 400 sub-problems per workers and 4 workers, *i.e.* $1,600$ sub-problems. In EPS, communication times are negligible. Then, we observe results of the same order although not as good for up to 32 workers (32 workers $\times$ 50 sub-problems).

In Table 5, EPS without BBD allows to obtain a mean improvement of 1.9 in solving time and 0.8 in backtracks compared with a sequential execution. The average number of backtracks gain shows a lack of robustness, many instances do more backtracks in parallel than in sequential, and therefore does not exploit the set up resources as desired. Thus, times can be bad with a naive EPS application at the TSP. Finally, *Bound-Backtrack-and-Dive* allows to better decompose the

**Table 4.** Comparison of the number of sub-problems per worker (sppw) with limitBk= 1, 000.

| Instance | BBD sppw = 100 | | | | BBD sppw = 2 × 100 | | | | BBD sppw = 4 × 100 | | | |
|---|---|---|---|---|---|---|---|---|---|---|---|---|
| | Div. | Decomp and Solving | | | Div. | Decomp and Solving | | | Div. | Decomp and Solving | | |
| | time | time | #bk | l.b. | time | time | #bk | l.b. | time | time | #bk | l.b. |
| pr136 | 1.7 | 13.9 | 37,134 | 3.9 | 1.7 | 16.0 | 42,738 | 3.9 | 1.7 | 20.7 | 56,578 | 3.9 |
| kroA150 | 1.8 | 0.8 | 792 | 1.1 | 1.8 | 0.8 | 762 | 1.0 | 1.8 | 0.9 | 994 | 2.7 |
| kroB150 | 1.6 | 17.2 | 42,320 | 3.5 | 1.6 | 22.8 | 58,556 | 3.7 | 1.8 | 31.3 | 68,940 | 3.6 |
| si175 | 2.1 | 129.2 | 338,752 | 3.6 | 1.9 | 132.0 | 339,744 | 3.6 | 2.0 | 189.0 | 516,548 | 3.7 |
| rat195 | 2.4 | 12.1 | 21,468 | 3.8 | 2.5 | 15.5 | 27,384 | 3.9 | 2.5 | 16.7 | 29,318 | 3.8 |
| kroA200 | 2.4 | 103.4 | 180,692 | 3.9 | 2.4 | 100.1 | 182,132 | 3.9 | 2.4 | 120.9 | 225,610 | 4.0 |
| kroB200 | 2.4 | 21.0 | 37,548 | 3.8 | 2.3 | 22.2 | 40,436 | 3.9 | 2.3 | 22.8 | 43,442 | 3.8 |
| tsp225 | 2.7 | 39.1 | 65,572 | 4.0 | 2.7 | 34.6 | 56,854 | 3.7 | 2.8 | 41.0 | 65,522 | 3.9 |
| gr229 | 2.3 | 90.9 | 158,970 | 4.0 | 2.3 | 96.4 | 171,222 | 4.0 | 2.3 | 118.6 | 191,146 | 3.8 |
| a280 | 3.7 | 2.0 | 934 | 0.3 | 3.9 | 1.5 | 656 | 0.5 | 3.9 | 1.5 | 680 | 0.6 |
| lin318 | 5.0 | 5.0 | 1,284 | 0.3 | 5.2 | 5.2 | 1,336 | 0.2 | 5.2 | 5.5 | 1,398 | 0.1 |
| pcb442 | 7.5 | 2,541.7 | 1,761,286 | 3.7 | 7.5 | 1,356.1 | 883,406 | 3.6 | 7.6 | 4,402.6 | 1,698,740 | 1.8 |
| gr431 | 11.7 | 420.5 | 153,428 | 3.9 | 11.8 | 585.4 | 203,692 | 3.9 | 11.9 | 716.7 | 249,318 | 4.0 |
| d493 | 14.8 | 6,303.4 | 2,621,658 | 3.8 | 14.8 | 13,649.6 | 3,587,410 | 2.2 | 15.0 | 7,321.5 | 2,439,330 | 2.8 |
| ali535 | 16.7 | 55,186.4 | 20,216,560 | 3.5 | 20.1 | 65,758.0 | 23,515,850 | 3.9 | 19.9 | 57,045.1 | 17,711,738 | 3.4 |
| mean | | 4,326 | 1,709,227 | 3.1 | | 5,453 | 1,940,812 | 3.1 | | 4,670 | 1,553,287 | 3.0 |
| geo mean | | 73 | 74,031 | 2.4 | | 78 | 76,190 | 2.4 | | 92 | 87,072 | 2.3 |

**Table 5.** General Results. It shows the differences between sequential execution, naive EPS application and *Bound-Backtrack-and-Dive* with limitBk=1, 000 and sppw=100.

| | Sequentiel LCFirst | | noDiveLCDec | | ratio | | BBD | | | ratio | |
|---|---|---|---|---|---|---|---|---|---|---|---|
| | | | | | | | Div. | Decomp. and Solv. | | | |
| Instance | time | #bk | time | #bk | time | #bk | time | time | #bk | time | #bk |
| pr136 | 23.4 | 19,147 | 11.1 | 25,748 | 2.1 | 0.7 | 1.7 | 13.9 | 37,134 | 1.5 | 0.5 |
| kroA150 | 3.8 | 2,493 | 4.8 | 9,796 | 0.8 | 0.3 | 1.8 | 0.8 | 792 | 1.5 | 3.1 |
| kroB150 | 212.1 | 176,773 | 82.4 | 183,964 | 2.6 | 1.0 | 1.6 | 17.2 | 42,320 | 11.3 | 4.2 |
| si175 | 461.0 | 401,428 | 277.7 | 220,874 | 1.7 | 1.8 | 2.1 | 129.2 | 338,752 | 3.5 | 1.2 |
| rat195 | 36.8 | 20,977 | 15.9 | 29,062 | 2.3 | 0.7 | 2.4 | 12.1 | 21,468 | 2.5 | 1.0 |
| kroA200 | 650.0 | 309,343 | 209.3 | 396,704 | 3.1 | 0.8 | 2.4 | 103.4 | 180,692 | 6.1 | 1.7 |
| kroB200 | 113.5 | 61,107 | 55.4 | 105,660 | 2.0 | 0.6 | 2.4 | 21.0 | 37,548 | 4.9 | 1.6 |
| tsp225 | 161.4 | 76,397 | 69.9 | 95,016 | 2.3 | 0.8 | 2.7 | 39.1 | 65,572 | 3.9 | 1.2 |
| gr229 | 247.0 | 165,826 | 77.4 | 167,368 | 3.2 | 1.0 | 2.3 | 90.9 | 158,970 | 2.7 | 1.0 |
| a280 | 9.3 | 2,829 | 8.8 | 9,442 | 1.1 | 0.3 | 3.7 | 2.0 | 934 | 1.6 | 3.0 |
| lin318 | 19.1 | 4,063 | 14.1 | 6,396 | 1.4 | 0.6 | 5.0 | 5.0 | 1,284 | 1.9 | 3.2 |
| pcb442 | 17,109.4 | 3,571,219 | 21,539.2 | 11,473,274 | 0.8 | 0.3 | 7.5 | 2,541.7 | 1,761,286 | 6.7 | 2.0 |
| gr431 | 1,532.2 | 211,585 | 852.0 | 258,832 | 1.8 | 0.8 | 11.7 | 420.5 | 153,428 | 3.5 | 1.4 |
| d493 | 172,800.0 | 24,733,443 | 57,995.1 | 12,934,364 | >3.0 | >1.9 | 14.8 | 6,303.4 | 2,621,658 | >23.7 | >9.4 |
| ali535 | 123,620.0 | 13,119,783 | 85,733.0 | 10,014, 704 | 1.4 | 1.3 | 16.7 | 55,186.4 | 20,216,560 | 2.2 | 0.6 |
| mean | >21,133 | >2,858,427 | 11,130 | 2,395,414 | >1.9 | >0.8 | | 4,326 | 1,709,227 | >5.4 | >2.3 |
| geo mean | >339 | >129,436 | 189 | 175,493 | >1.7 | >0.7 | | 73 | 74,031 | >3.7 | >1.7 |

TSP by simulating LCFirst. Indeed, EPS with BBD obtain a mean improvement of 5.4 in solving time and 2.3 in backtracks, so all instances have an improved robust result.

## 7 Conclusion

We have shown that the application of EPS to the TSP is not trivial. Indeed, EPS decomposition is breadth-based whereas TSP embeds LCFirst, a depth-based search strategy, so the two methods are incompatible. In order to combine the two approaches, we introduced *Bound-Backtrack-and-Dive*, a diving algorithm, which consists in a first step of performing a sequential execution with a bounded number of backtrack in order to study the behavior of LCFirst. Then, run EPS, simulate LCFirst during the decomposition using our preliminary study and finally solve with a classical LCFirst the generated sub-problems in parallel.

Experimental results show that the application of *Bound-Backtrack-and-Dive* allows to obtain robust results. Thus, the efficiency of parallelism applied to TSP with *Bound-Backtrack-and-Dive* allows a mean gain of a factor 5.4 in solving times and 2.3 in number of backtracks with 4 cores.

We think that this method can sometimes allow us to avoid dynamic learning strategies when it is an issue, here for the application of parallelism, and obtain a great improvement of solving times. We hope that similar results will be obtained for other learning search strategies.

## References

1. Applegate, D.L., Bixby, R.E., Chvatal, V., Cook, W.J.: The Traveling Salesman-problem: A Computational Study. Princeton University Press, Princeton (2006)
2. Benchimol, P., Régin, J.-C., Rousseau, L.-M., Rueher, M., van Hoeve, W.-J.: Improving the held and karp approach with constraint programming. In: Lodi, A., Milano, M., Toth, P. (eds.) CPAIOR 2010. LNCS, vol. 6140, pp. 40–44. Springer, Heidelberg (2010). https://doi.org/10.1007/978-3-642-13520-0_6
3. Benchimol, P., Van Hoeve, W.J., Régin, J.C., Rousseau, L.M., Rueher, M.: Improved filtering for weighted circuit constraints. Constraints **17**(3), 205–233 (2012). https://hal.archives-ouvertes.fr/hal-01344070
4. Fages, J.G., Lorca, X., Rousseau, L.M.: The salesman and the tree: the importance of search in CP. Constraints **21**(2), 145–162 (2016)
5. Haralick, R., Elliott, G.: Increasing tree search efficiency for constraint satisfaction problems. Artif. Intell. **14**, 263–313 (1979)
6. Held, M., Karp, R.M.: The traveling-salesman problem and minimum spanning trees. Oper. Res. **18**(6), 1138–1162 (1970)
7. Held, M., Karp, R.M.: The traveling-salesman problem and minimum spanning trees: part ii. Math Program. **1**(1), 6–25 (1971)
8. Isoart, N., Régin, J.-C.: Integration of structural constraints into TSP models. In: Schiex, T., de Givry, S. (eds.) CP 2019. LNCS, vol. 11802, pp. 284–299. Springer, Cham (2019). https://doi.org/10.1007/978-3-030-30048-7_17
9. Korf, R.E.: Depth-first iterative-deepening: an optimal admissible tree search. Artif. Intell. **27**(1), 97–109 (1985). https://doi.org/10.1016/0004-3702(85)90084-0
10. Lecoutre, C., Saïs, L., Tabary, S., Vidal, V.: Reasoning from last conflict(s) in constraint programming. Artif. Intell. **173**(18), 1592–1614 (2009)
11. Malapert, A., Régin, J., Rezgui, M.: Embarrassingly parallel search in constraint programming. J. Artif. Intell. Res. (JAIR) **57**, 421–464 (2016)

12. Régin, J.-C., Rezgui, M., Malapert, A.: Embarrassingly parallel search. In: Schulte, C. (ed.) CP 2013. LNCS, vol. 8124, pp. 596–610. Springer, Heidelberg (2013). https://doi.org/10.1007/978-3-642-40627-0_45
13. Reinelt, G.: TSPLIB–a traveling salesman problem library. ORSA J. Comput. **3**(4), 376–384 (1991)

# Using Resolution Proofs to Analyse CDCL Solvers

Janne I. Kokkala[1,2(✉)] and Jakob Nordström[1,2(✉)]

[1] Lund University, Lund, Sweden
janne.kokkala@cs.lth.se
[2] University of Copenhagen, Copenhagen, Denmark
jn@di.ku.dk

**Abstract.** We propose that CDCL SAT solver heuristics such as restarts and clause database management can be analysed by studying the resolution proofs produced by the solvers, and by trimming these proofs to extract the clauses actually used to reach the final conclusion.

We find that for non-adaptive Luby restarts higher frequency makes both untrimmed and trimmed proofs smaller, while adaptive restarts based on literal block distance (LBD) decrease proof size further mainly for untrimmed proofs. This seems to indicate that restarts improve the reasoning power of solvers, but that making restarts adaptive mainly helps to avoid useless work that is not needed to reach the end result.

For clause database management we find that switching off clause erasures often, though not always, leads to smaller untrimmed proofs, but has no significant effect on trimmed proofs. With respect to quality measures for learned clauses, activity in conflict analysis is a fairly good predictor in general for a clause ending up also in the trimmed proof, whereas for the very best clauses the LBD score gives stronger correlation. This gives more rigorous support for the currently popular heuristic of prioritizing clauses with very good LBD scores but sorting the rest of the clauses with respect to activity when deciding which clauses to erase. We remark that for these conclusions, it is crucial to use the actual proof found by the solver rather than the one reconstructed from the *DRAT* proof log.

## 1 Introduction

Boolean satisfiability (SAT) solving is one of the most striking success stories of computer science, but also one of its most puzzling mysteries. Though modern *conflict-driven clause learning (CDCL)* SAT solvers [29,31][1] are used on an every-day basis to solve real-world instances with hundreds of thousands or even millions of variables, there is still a very poor understanding of how they can perform so well on problems that are, after all, widely conjectured to be exponentially hard in the worst case [14,21].

[1] A similar idea in the context of constraint satisfaction problems (CSPs) was independently developed in [5].


H. Simonis (Ed.): CP 2020, LNCS 12333, pp. 427–444, 2020.
https://doi.org/10.1007/978-3-030-58475-7_25

The most important difference between CDCL and classic DPLL backtrack search [15] is in how conflicts guide the search by generating new learned clauses [29] and informing branching decisions [31], and this accounts for most of the performance gain of CDCL over DPLL [24]. Further improvements have been obtained through careful implementation of the basic CDCL algorithm with highly optimized data structures, as well as through the use of sophisticated heuristics such as *activity* [17] or *literal block distance (LBD)* [2] to identify useful clauses, *phase saving* [35] to guide variable decisions, *adaptive restarts* [3,10] to speed up the search, et cetera.

Unfortunately, our scientific understanding of the performance of these heuristics is still very limited. A natural approach to gain insights would seem to be to collect real-world benchmarks and run experiments with different heuristics to study how they contribute to overall performance. This has been done in [24,27], and there are also in-depth studies focusing specifically on, e.g., variable decisions [11] and restart schemes [12,20], but it has been hard to reach clear-cut conclusions from the diverse set of formulas found in real-world benchmark sets. Another approach has been to run experiments on crafted benchmarks [18,23], where detailed knowledge of the theoretical properties of the formulas makes it possible to draw conclusions about solver performance, but although this can uncover intriguing findings, it is not clear to what extent the conclusions are relevant in a real-world setting.

**Our Contributions.** In this paper, we investigate whether the proofs generated by SAT solvers can shed light on the effectiveness of solver heuristics. When a CDCL solver decides that a formula is unsatisfiable, it does so, in effect, by deriving a proof of contradiction in the *resolution proof system* [7].[2] Once the solver has terminated, such a proof can be trimmed to keep only the subset of clauses needed to reach this conclusion. We study such untrimmed and trimmed proofs obtained from a selection of benchmarks from the SAT competitions [36] in order to gain insights into solver performance, focusing on restarts and clause database management and how they affect the solver reasoning.

It is well-known that frequent restarts are crucial for the performance of CDCL solvers, but it has remained stubbornly open whether such restarts are just a helpful heuristic or whether they fundamentally increase the theoretical reasoning power. This question cannot be settled by experiments, but we give some empirical evidence that the latter alternative might apply by showing that solvers not only run faster with frequent restarts but also reason more efficiently.

In more detail, we study adaptive restarts as in *Glucose* [3,19] and compare to the non-adaptive Luby restarts in *MiniSat* [17,30], but with different multiplicative constants to get non-adaptive restart frequencies in the full range from the most frequent to least frequent adaptive restarts encountered for our benchmarks. For the non-adaptive policy we find that higher restart frequency correlates with smaller proof size for both untrimmed and trimmed proofs. Adaptive restarts yield smaller untrimmed proofs than all non-adaptive restart frequencies, so the effect of adaptiveness is not only about the frequency but also the

[2] Note, though, that this is not quite true for some pre- and inprocessing techniques.

exact timing of the restarts. The improvements from adaptive restarts are not as clear for the trimmed proofs, however. Our interpretation of this is that more frequent restarts improve the reasoning power of solvers, but that adaptive restarts mainly help to abort useless stages of the search process earlier.

When managing the clauses learned during search, there is a tension between on the one hand keeping as many clauses as possible, since they prune the search space and thus make the reasoning stronger, and on the other hand getting rid of them, since as the number of clauses grows the solver has to spend increasing time on handling them, which makes the reasoning slower. Conventional wisdom dictates that solvers should aggressively minimize memory usage, erasing an ever increasing fraction of learned clauses as the running time increases, but there is little scientific understanding of how this affects the quality of the reasoning performed, or of how to assess which clauses should be kept or thrown away.

When we experiment with switching off erasures completely, so that the solver keeps all learned clauses, we see that this most often leads to smaller untrimmed proofs, but far from always. That is, there exist formulas for which, perhaps somewhat counter-intuitively, clause erasures not only make the solver reason faster, but also better. Even more interestingly, even when the untrimmed proofs get smaller, we do not observe any significant effect on the trimmed proofs. This suggests that the core reasoning needed to decide the formula does not get stronger with more clauses in memory, only that these extra clauses help the solver to "focus" and avoid work that turns out to be useless with hindsight.

Regarding which learned clauses are more or less useful for the solver, it is not obvious how to answer this question, since it is unclear how to measure "usefulness". One approach is to fix a non-adaptive strategy for how many clauses should be removed at clause database reduction, and then decide which clauses to erase based on literal block distance (LBD) score or activity, as in *Glucose* and *MiniSat*, respectively. We find that both untrimmed and trimmed proof size is smaller for LBD-based erasure than activity-based erasure, and that (as a control) both are clearly better than randomly choosing which clauses to erase.

Another approach, following [4,25], is to consider learned clauses in the untrimmed proof "useful" if they remain in the proof after trimming. We find that very good LBD scores strongly correlate with appearing in the trimmed proof, but that clause activity is a better predictor over a wider range of values for which learned clauses survive the trimming process. This provides more rigorous evidence for the empirical claim in [34] that the clause database reduction policy should prioritize top LBD scores but gives more weight to clausal activity for clauses with worse LBD scores, a claim that is also supported by the experiments in [22].

A relevant observation in this context is that the conclusions in the last paragraph above rely on using the actual proof found by the solver. It is also possible to reconstruct a resolution proof from the *DRAT* proof logs used in the SAT competitions by applying *DRAT-trim* [41], but we find that such proofs can look quite different from the ones constructed by the solver during search, and so provide less insight into how the solver actually reasoned.

One obvious criticism of this approach is that our notion of usefulness of clauses is narrow—it might well be the case that learned clauses can be helpful for the solver in other ways than by appearing in a final, trimmed proof (as also noted in [4]). Furthermore, even if a clause appears in the trimmed proof, it might be that this appearance comes very soon after the clause was learned, and that after this the clause can safely be thrown away. A more refined approach here is to ask how likely it is at any given point in time that a given clause will be used in the future, a question that was approached in [39] using machine learning techniques. While these are valid points, we nevertheless hope that usage in the trimmed proof can serve as *one* relevant measure providing insights, even though there is certainly room for other measures providing additional information.

Another possible concern is that since we are looking at resolution proofs, we have to limit our attention to only unsatisfiable formulas. Since SAT solvers should work well on both satisfiable and unsatisfiable instances, it could be that we are missing out on important observations by studying only one of these categories of benchmarks. This is also true, but we consider this to be less of a concern. It is in fact possible to come up with a notion of "proofs" also for satisfiable formulas—namely, the learned clauses that guided the solver to the satisfying assignment found, together with all other clauses used to derive these guiding clauses—but we have to leave as interesting future work the task of studying such proofs for satisfiable formulas, and investigating which of our conclusions hold also in this setting and what new observations can be made.

**Related Work.** A very thorough study of untrimmed and trimmed proofs was performed in [37], where *Glucose* was used to examine the proportion of useful learned clauses across different instances, the effect of shuffling on the number of useful clauses in the input formula, and the correlation of proof size with maximal clause size. Interestingly, usages of older clauses were reported to be more likely to appear in trimmed proofs, but since these experiments were performed only with clause erasures switched on, it was pointed out that this might be due to the solver erasing bad clauses early. *Glucose* was also compared to a solver with *MiniSat*-style policies for restarts and database management, but these experiments did not try to isolate the effects of different heuristics. Furthermore, clause features such as size and LBD score were studied, and LBD was observed to be a better predictor of usefulness than size, but the method used did not allow for an analysis of more dynamic features such as activity.

An analogous idea of trimming appeared in [28], where a dependency graph containing both learned clauses and decided and propagated variables was constructed, and then pruned to contain only the clauses and propagations useful for reaching the final conflict (or the satisfying assignment in the case of satisfiable instances). This was used to study decision heuristics, but the same approach could be harnessed to define a broader notion of clause usefulness, giving credit for useful propagations even if the clause does not appear in conflict analysis.

**Outline of This Paper.** We start in Sect. 2 by discussing how resolution proofs can be extracted from CDCL solvers. In Sect. 3 we describe our experimental set-up, and in Sect. 4 we present our detailed results. Some final remarks, including suggestions for future research, are made in Sect. 5.

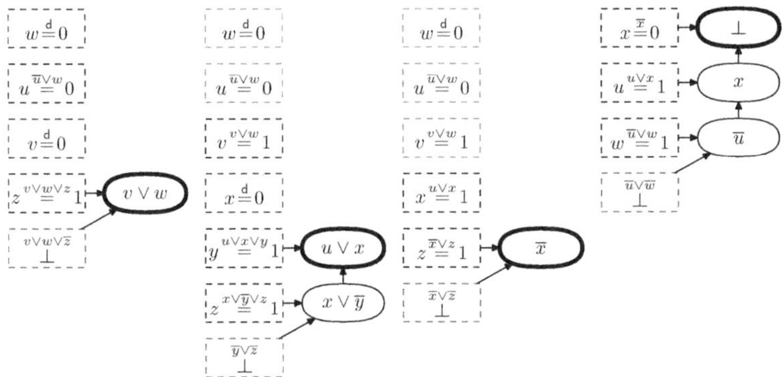

(a) Example CDCL execution (trail grows vertically and shifts right after conflict).

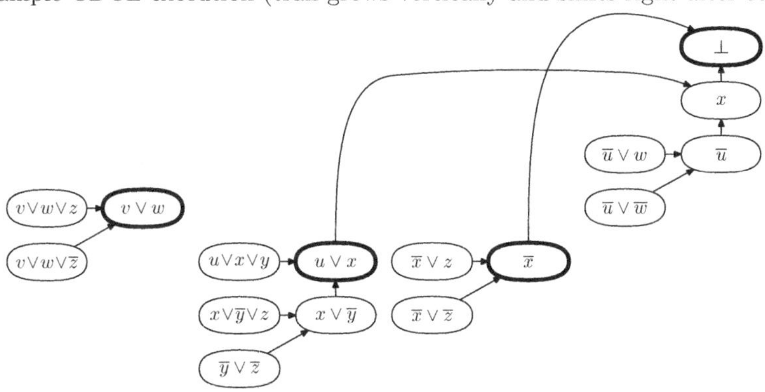


(b) Extracted resolution proof (with long arrows from used learned clauses).

**Fig. 1.** CDCL execution on formula in (1) and corresponding resolution proof.

## 2 CDCL SAT Solvers and Resolution Proofs

In this section we discuss briefly, and mostly by example, how resolution proofs can be obtained from CDCL solvers. We refer the reader to, e.g., [13,33] for more details on connections between the theory and practice of SAT solving.

The solver input is a formula in *conjunctive normal form (CNF)* such as

$$\begin{aligned}&(u \vee x \vee y) \wedge (v \vee w \vee z) \wedge (x \vee \overline{y} \vee z) \wedge (v \vee w \vee \overline{z}) \wedge \\ &(\overline{x} \vee z) \wedge (\overline{y} \vee \overline{z}) \wedge (\overline{x} \vee \overline{z}) \wedge (\overline{u} \vee w) \wedge (\overline{u} \vee \overline{w})\end{aligned} \tag{1}$$

(or, in words, a conjunction of clauses, where every clause is a disjunction of literals, i.e., unnegated or negated variables, with negation denoted by overline). A possible CDCL run for this formula is illustrated in Fig. 1a. At all times, the solver maintains a *trail* of variable assignments, and we show how this trail grows and shrinks vertically as time flows from left to right.

The solver starts by deciding the variable $w$ to false, or 0, which makes the clause $\overline{u} \vee w$ *unit propagate* $u = 0$, since all of this clause except the literal $\overline{u}$

has been falsified. The solver marks the *reason clause* for this propagation on the trail (stacked on the top of the assignment in our illustration). No further propagations can be made, so $w = 0$ and $u = 0$ are all the assignments at *decision level* 1 of the trail. To move things forward the solver has to make a second decision, in our example $v = 0$. Then the clause $v \lor w \lor z$ propagates $z$ to true, or 1, which leads to a *conflict* since the clause $v \lor w \lor \overline{z}$ is now falsified. At the time of conflict, decision level 2 contains $v = 0$ and $z = 1$. During *conflict analysis* the solver *learns* a new clause by applying the *resolution rule*—which resolves two clauses of the form $C \lor z$ and $D \lor \overline{z}$ over the variable $z$ to derive $C \lor D$—to the conflict clause and the reason clauses. In this case, $v \lor w \lor z$ and $v \lor w \lor \overline{z}$ are resolved to yield $v \lor w$, after which the analysis stops (since this is a *unique implication point (UIP)* clause with a single literal from the last decision level).

After learning $v \lor w$, the solver *backjumps* to the *assertion level*, which is the second highest decision level of any literal in the learned clause, by undoing all decisions and propagations at later levels, in our example leaving only $w = u = 0$ at assertion level 1. This causes unit propagation on the learned clause, flipping the value of $v$ (called the *asserting literal*). A new decision $x = 0$ followed by a couple of propagations lead to a second conflict where $u \lor x$ is learned, after which a third conflict results in the learned *unit clause* $\overline{x}$. Unit clauses cause backjumps to decision level 0 (incidentally, this has exactly the same effect as making a *restart*). In our example, this triggers a fourth conflict, and since no decisions have been made the solver can conclude that the formula is unsatisfiable. If, however, we would let the solver run a final conflict analysis, applying the resolution rule to the reasons propagating to the conflict, this would derive the empty clause $\perp$ containing no literals, as shown on the far right in Fig. 1a.

To obtain a resolution proof of unsatisfiability for (1) from Fig. 1a, we start with the final (imagined) derivation of $\perp$, and then go back in time, including the conflict analyses for all clauses used in this derivation, and then the conflict analyses for these clauses, et cetera, leading to the proof visualized in Fig. 1b. During this process, learned clauses that are not needed can be trimmed away. In our example, we see that the first conflict analysis was not needed to decide unsatisfiability. In this way, we obtain untrimmed and trimmed resolution proofs from CDCL executions. Our simplified description ignores aspects like *clause minimization* [40], but such steps also correspond to resolution derivations. Some *preprocessing* steps are not captured by resolution, however, and therefore we analyse CDCL executions on formulas as output by the preprocessor.

In complexity theory, the *proof size* is defined to be the number of clauses in a resolution proof, which in Fig. 1b is 16 for the untrimmed and 13 for the trimmed proof. In this paper, we will be slightly more relaxed and count just the number of learned clauses, so that the untrimmed and trimmed proofs have sizes 4 and 3, respectively. We have verified that this choice does not affect the analysis of our experiments. Conveniently, this means that the size of the untrimmed proof is just the total number of conflicts encountered during execution.

*Clause size* is defined to be the number of literals in a clause, so that $v \vee w \vee \overline{z}$ has size 3. The *literal block distance (LBD)* of a clause with respect to the current trail is the number of different decision levels represented in the clause. At the time of the first conflict in Fig. 1a, $w$ is assigned at level 1 and $v$ and $z$ at level 2, so the LBD score of $v \vee w \vee \overline{z}$ is 2. The clauses active in the first conflict analysis are $v \vee w \vee \overline{z}$ and $v \vee w \vee z$, and in the second conflict analysis the clauses that take part are $\overline{y} \vee \overline{z}$, $x \vee \overline{y} \vee z$, and $u \vee x \vee y$. Such clauses get their *clause activity* increased by 1, and a mild exponential smoothing is applied to the activity score to give greater weight to the recent history of conflicts.

## 3 Experimental Set-Up

Let us now describe our CDCL solver configuration and choice of benchmarks.

**Solver Configuration** We use version 3.0 of *Glucose* [19] (which serves as a basis also for many other modern CDCL solvers), but enhanced to output resolution proofs and to vary restart and clause database management policies.

For ***restarts***, we compare the following policies:

**Adaptive restarts.** The default in *Glucose*, where, essentially, restarts are triggered when the average LBD score of recently learned clauses becomes bad compared to the overall average.

**Luby restarts.** As proposed in [20] and used in *MiniSat*, the solver restarts after a predetermined number of conflicts as specified by the *Luby sequence* $1, 1, 2, 1, 1, 2, 4, 1, 1, 2, 1, 1, 2, 4, 8, \ldots$, multiplied by some constant.

**Fixed-interval restarts.** Restart after a constant number of conflicts.

We study how adaptive restarts affects proof size compared to restarting at predetermined points in time. To investigate whether the effect of adaptivity is mainly to adjust the overall restart frequency or to trigger restarts at specific points in time, we compare to Luby restarts with factors that give similar restart frequencies. Fixed-interval restarts are considered as a theoretically interesting extreme case, though in practice this is too inefficient in terms of running time.

We use Luby sequences with factors 1, 10, 20, 50, 100, and 200. In preliminary experiments with default *Glucose*, for around 95% of unsatisfiable SAT competition benchmarks the total number of restarts are below what would be obtained with Luby restarts with factor 200. We also compare adaptive restarts to the "virtual best Luby solver", picking the best Luby-restarting solver for each benchmark, and the "virtual closest Luby solver" with closest average restart frequency for this particular benchmark. Finally, we run experiments with solvers that restart every 20 conflicts, every 10 conflicts, and every conflict. For all these experiments we use the default *Glucose* clause database management policy.

Concerning ***learned clause deletion***, we investigate how untrimmed and trimmed proof size is affected when the clause database reduction is completely switched off, so that all learned clauses are kept. We run these experiments both for adaptive restarts and for Luby restarts with factor 100 (the *MiniSat* default).

We also consider how the solver chooses which clauses to erase when database reduction is switched on, something we refer to as ***clause assessment***. A first rough description of how the CDCL clause database is managed is as follows. When the solver reaches a certain number of conflicts, a method `reduceDB` is called that sorts the clauses in the database according to some clause assessment criterion, after which the worst half of the clauses are removed (but binary clauses, i.e., clauses of size 2, are never removed). The number of conflicts until the next database reduction is then increased by some constant, meaning that the number of learned clauses in memory after $N$ conflicts will be proportional to $\sqrt{N}$. Glossing over some low-level details (due to space constraints), *Glucose* refines the above model in the following way:

- Clauses with LBD score 1 or 2, so-called *glue clauses*, are never deleted.
- When a clause appears in conflict analysis the LBD score is recomputed, and decreased scores protect from deletion at the next database reduction.
- If many clauses with good LBD scores have been learned, the next clause reduction will be delayed, meaning that more clauses will be kept in memory.

It follows from this that there is a strong feedback loop between the LBD scores and how many learned clauses are kept in memory. As we report in this paper, it is also the case that more clauses in memory tends to yield more efficient reasoning (measured in terms of proof size, not time). If we want to compare different ways of assessing the quality of clauses, we have to break this feedback loop in order to get a fair comparison, since otherwise clause assessment based on LBD might look good just because it leads to more clauses being kept. The problem is, however, that the aggressive clause deletion policy in *Glucose* works well only because the solver keeps more clauses when the LBD scores are good [38].

Therefore, in our clause assessment experiments we use a non-adaptive database reduction strategy that yields clause database sizes that are reasonably close to standard *Glucose*, so that the comparisons will be meaningful, but (almost) never smaller, so that our experiments will not be biased by deleting clauses more aggressively than *Glucose* would do. After some experimentation, the best solution we found was to make each `reduceDB` call erase only 30% of the clauses and to increase the database reduction interval by 4600. This leads to database sizes that are larger than default *Glucose* except for 5% of our benchmarks.

We now have database reduction policy that always erases the same number of clauses regardless of how these clauses are chosen, so that we can study the effect of different clause assessment policies in isolation. Or at least almost: one final problem is that every time a unit clause is learned, all clauses implied by that unit are erased, meaning that the number of clauses in memory shrink, potentially quite significantly. Thus, if a particular clause assessment policy is successful in the sense of leading to more unit clauses being learned, this will make the solver manage memory more aggressively. In contrast to the LBD feedback loop discussed above, we see no way of countering this effect, since not erasing satisfied clauses immediately would also lead to unpredictable effects on the database size (which we cannot explain in detail due to space constraints).

In our clause assessment experiments, we always keep binary clauses and remove the worst 30% of the other clauses sorted according to the following criteria (except for the first default configuration):

**LBD+bumps.** Default policy in *Glucose* with database size being bumped if the LBD scores of clauses are good enough (evaluated for comparison).
**LBD.** Simulation of Glucose but with non-adaptive database size policy, prioritizing (a) first glue clauses (LBD score 1 or 2), (b) then clauses with updated LBD, and (c) finally other clauses sorted by LBD (breaking ties by activity).
**Activity.** Activity in conflicts, with higher activity being better (as in *MiniSat*).
**Size.** Clause size, with smaller clauses being better.
**Random.** Random choice of which clauses to erase.

For these experiments we use Luby restarts (with factor 100) to avoid feedback between clause assessment and restarts (except for the default configuration).

**Benchmark Selection and Analysis.** We ran three separate sets of experiments to measure the effects of restarts, clause deletion, and clause assessment as described above. To select benchmarks for these experiments, we first randomly sampled 200 unsatisfiable instances from the SAT competitions and races 2015–2019 [36] and ran them through the preprocessor of *Glucose* (since the extracted proofs are for CDCL search after preprocessing). Since we want to analyse proofs we cannot deal with time-outs, and so have to select benchmarks solvable by all solver configurations. We therefore ran each solver configuration for all 200 benchmarks, and let each configuration select the 150 instances that were solvest fastest. The final collection of benchmarks for each set of experiments was chosen as the intersection of the sets of benchmarks selected by each solver configuration. Just to give a sense of the computational effort involved, for standard *Glucose* this approach led to running times of around $6,000$ s or less. For the restart experiments, we had running times of up to $13,000$ s, for experiments with clause deletion switched off up to $55,000$ s, and for the clause assessment the control experiment with random erasures resulted in times of up to $173,000$ s (2 days). One instance was solved in only 9 conflicts, before any restarts or clause erasures, so it was ignored in the analysis. The number of benchmarks in the final experiments was between 120 and 133.

For each solver configuration and each instance, we collected data about trimmed and untrimmed proof size (where, as mentioned before, the latter is the total number of conflicts during execution), and compared different solver configurations for both trimmed and untrimmed proofs. In order not to give undue weight to the very hardest benchmarks, we consider logarithms of proof sizes. For two different solvers, we use the standard paired t-test to find a 99% confidence interval for the mean of the difference of the logarithms. This confidence interval can be transformed back to a confidence interval for the geometric mean of the ratio of the proof sizes. It is important to note that since we perform multiple experiments and tests, the 99% confidence level cannot be regarded as a proper measure of statistical significance, but the confidence intervals still provide a useful way of understanding the magnitude of the differences.

**Features of Useful Clauses.** For the experiments with clause deletion switched off, we compare untrimmed and trimmed proofs to see whether different properties of learned clauses can predict whether they will be useful or not, i.e., remain in the final, trimmed, proof. In order to obtain results that could be useful for future solver development, we focus on clause features that the solver could know during execution, rather than on information that can be computed only with hindsight. We consider *static features*, which are determined when the clause is learned, and *dynamic features*, which can change while the solver is running. Since dynamic features can vary over time, what we measure are features of *clause usages* in derivations rather than of the clauses themselves. If a clause in the database is used several times, every usage gives rise to a new data point.

Because features are often used to assess clauses relative to other clauses in the database, and because clauses that are never used by the solver would not appear in our analysis of usages, we also consider the percentile ranks of features in the database at the time of usage. The percentile rank also changes over time when the distribution of features of learnt clauses changes. When collecting data for the percentile ranks, ties are broken randomly.

We collect the following static features computed when the clause is learned:

**Size.** Size of the learned clause.
**Initial LBD.** Clause LBD score (with respect to the trail when learned).
**Decision level.** Decision level of conflict when learned.
**Backjump length.** Difference of conflict level and assertion level.
**Conflicts since restart.** Number of conflicts since the latest restart.

We also consider the following dynamic features:

**Dynamic LBD.** When a clause is learned, its dynamic LBD is set to the initial LBD score. Whenever a clause is used as a reason during conflict analysis, a new candidate LBD score is computed based on the current trail, and the dynamic LBD is updated if the score decreased by at least 2.
**Activity.** Conceptually speaking, the initial activity of a newly learned clause is 1; it is increased by 1 every time the clause appears in conflict analysis; and all clausal activities are multiplied by a factor $\alpha = 0.999$ after every conflict.

Since solvers will not remove unit or binary clauses, we focus on features for clauses of size at least 3, and the percentile ranks are also computed among these clauses. The reason that clause deletion is switched off is that we do not want the choice of which clauses to erase to bias which clauses seem useful. For the same reason, we use non-adaptive Luby restarts (with factor 100).

By the nature of how CDCL solvers work, we expect some features to correlate strongly with clause usage for trivial reasons. For example, small clauses are more likely to propagate and thus to appear more often in conflict analysis, and will also tend to have low LBD scores. A clause that currently has high activity has been used a lot in conflict analysis, meaning that all other things being equal it is also more likely to show up in the trimmed proof. Such correlations may not say too much about whether the clauses actually contribute to terminating the search, as they could also have many usages that do not appear in the trimmed proof. To measure the predictive power of a feature, we focus on the *conditional probability* that a usage of a clause appears also in the trimmed proof, conditioning on the value of the feature. For a completely uninformative predictor, this would simply be the ratio of all clause usages in the trimmed proof versus the untrimmed proof. If the conditional probabilities for some values of a feature differ from that, it suggests that the feature can be a predictor for usefulness.

We combine data from multiple benchmarks by summing the absolute counts of usages over all benchmarks. In general, this approach may make a few benchmarks with large proofs dominate. To check whether we have this problem, we performed the same analysis on random subsets of the selected instances. The results were similar, so the analysis appears robust to the effect of single large instances.

# 4 Results

The results of our experiments, and our analysis of them, are as follows. For full data and source code, see https://doi.org/10.5281/zenodo.3951538.

## 4.1 Proof Sizes

In Tables 1–3, we present the experimental data for some selected pairwise comparisons of solver configurations. For each pair of solvers and type of proof (untrimmed/trimmed), we calculate the ratio of sizes of the proofs provided by the solvers for each instance. In the tables, we show the quartiles of these ratios in the data and the geometric mean with the 99% confidence interval, computed independently for untrimmed and trimmed proofs.

Comparing different restart frequencies (see Table 1), we find that smaller Luby factors (i.e., faster restarts) tend to give shorter untrimmed and trimmed proofs on average. Restarting at every 20 conflicts gives shorter proofs than Luby restarts with factor 20, but for factor 1 there is no clear difference between Luby and fixed-interval restarts. For fixed-interval restarts every 1, 10, and 20 conflicts, more frequent restarts seem to increase the proof sizes, but the difference is not statistically significant. We interpret this as evidence that up to a certain limit, more frequent restarts generally improve the reasoning power of the solver. Adaptive restarts appear to be better than even the most frequent Luby restart policy, though, yielding clearly smaller untrimmed proofs, and perhaps also slightly smaller trimmed proofs. This seems to indicate that the advantage of adaptive restarts comes mainly from recognizing when the solver is doing useless work and not as much from finding better proofs. One could ask whether adaptive restarts work by simply selecting the best restart frequency for each instance. However, our data gives evidence to the contrary, as choosing the Luby solver with the closest average restart frequency for each instance would perform worse.

**Table 1.** Comparison of restart policies. Values larger than 1 mean that the first solver generates larger proofs.

| | | Untrimmed proof | | | | Trimmed proof | | | |
|---|---|---|---|---|---|---|---|---|---|
| Solvers | | quartiles | | | geom. mean | quartiles | | | geom. mean |
| Luby-200 | Luby-100 | 0.93 | 1.03 | 1.19 | $1.06 \pm 0.11$ | 0.91 | 1.01 | 1.17 | $1.04 \pm 0.08$ |
| Luby-100 | Luby-50 | 0.92 | 1.04 | 1.19 | $1.08 \pm 0.12$ | 0.95 | 1.06 | 1.26 | $1.11 \pm 0.09$ |
| Luby-50 | Luby-20 | 0.89 | 0.99 | 1.17 | $0.97 \pm 0.11$ | 0.92 | 1.01 | 1.12 | $1.02 \pm 0.11$ |
| Luby-20 | Luby-10 | 0.95 | 1.03 | 1.19 | $1.07 \pm 0.09$ | 0.95 | 1.04 | 1.19 | $1.06 \pm 0.10$ |
| Luby-10 | Luby-1 | 0.89 | 1.01 | 1.16 | $1.03 \pm 0.09$ | 0.94 | 1.03 | 1.15 | $1.06 \pm 0.06$ |
| Luby-20 | Fixed-20 | 0.78 | 1.13 | 1.56 | $1.13 \pm 0.17$ | 0.92 | 1.18 | 1.61 | $1.23 \pm 0.15$ |
| Luby-1 | Fixed-1 | 0.72 | 0.96 | 1.19 | $0.93 \pm 0.12$ | 0.91 | 1.10 | 1.28 | $1.03 \pm 0.11$ |
| Fixed-20 | Fixed-10 | 0.88 | 1.02 | 1.14 | $0.96 \pm 0.09$ | 0.90 | 1.01 | 1.10 | $0.98 \pm 0.07$ |
| Fixed-10 | Fixed-1 | 0.81 | 0.98 | 1.13 | $0.95 \pm 0.10$ | 0.86 | 0.99 | 1.09 | $0.96 \pm 0.09$ |
| adaptive | Luby-20 | 0.68 | 0.84 | 0.97 | $0.79 \pm 0.09$ | 0.80 | 0.95 | 1.07 | $0.92 \pm 0.09$ |
| adaptive | Luby Closest | 0.63 | 0.83 | 1.02 | $0.75 \pm 0.08$ | 0.76 | 0.91 | 1.04 | $0.83 \pm 0.09$ |
| adaptive | Luby VBS | 0.89 | 1.06 | 1.24 | $1.10 \pm 0.13$ | 0.99 | 1.13 | 1.33 | $1.19 \pm 0.11$ |
| adaptive | Fixed-20 | 0.57 | 0.90 | 1.35 | $0.89 \pm 0.14$ | 0.85 | 1.06 | 1.58 | $1.13 \pm 0.14$ |

**Table 2.** Effect of turning clause erasures off with adaptive restarts and Luby-100 restarts. Values larger than 1 mean that the first solver generates larger proofs.

| | | Untrimmed proof | | | | Trimmed proof | | | |
|---|---|---|---|---|---|---|---|---|---|
| Solver erasures/restarts | | quartiles | | | geom. mean | quartiles | | | geom. mean |
| Off/adaptive | On/adaptive | 0.70 | 0.81 | 0.96 | $0.77 \pm 0.07$ | 0.86 | 1.00 | 1.15 | $0.98 \pm 0.06$ |
| Off/Luby | On/Luby | 0.67 | 0.80 | 0.99 | $0.77 \pm 0.07$ | 0.82 | 0.98 | 1.18 | $0.98 \pm 0.08$ |

Turning clause erasures completely off decreases the untrimmed proof size in 83% of the instances, and by 23% on average, but there appears to be no measurable average difference for the trimmed proofs. The results are similar when using Luby restarts with factor 100 instead of adaptive restarts (see Table 2).

Comparing the two popular policies LBD and activity for clause assessment, we see (as shown in Table 3) that using LBD gives significantly smaller proofs. We also find that LBD is better than clause size, which is evidence that LBD contains more useful information than just size despite being strongly correlated with it. Clause size, in turn, seems to be slightly better than activity, but the difference is not statistically significant. Choosing which clauses to erase at random is clearly worse than all other policies, but even so it is not a completely hopeless approach, as it yields shorter proofs than LBD for 10–15% of the instances.[3]

To verify that our solver with fixed database size updates and LBD clause assessment is reasonably close to the actual behaviour of *Glucose* with adaptive

[3] For one of the selected benchmarks the solver with random clause erasures produced a proof too large to analyse with our tool chain, so this data point is missing. However, it would not make any significant difference.

**Table 3.** Comparison of clause assessment policies. Values larger than 1 mean that the first solver generates larger proofs.

| | | Untrimmed proof | | | | Trimmed proof | | | |
|---|---|---|---|---|---|---|---|---|---|
| Solvers | | quartiles | | | geom. mean | quartiles | | | geom. mean |
| LBD | activity | 0.74 | 0.92 | 1.02 | $0.84 \pm 0.09$ | 0.74 | 0.88 | 1.00 | $0.84 \pm 0.09$ |
| LBD | size | 0.79 | 0.94 | 1.01 | $0.91 \pm 0.07$ | 0.78 | 0.91 | 1.00 | $0.87 \pm 0.07$ |
| LBD | random | 0.66 | 0.80 | 0.97 | $0.73 \pm 0.07$ | 0.61 | 0.74 | 0.88 | $0.66 \pm 0.08$ |
| size | activity | 0.83 | 1.01 | 1.23 | $0.93 \pm 0.10$ | 0.85 | 1.00 | 1.14 | $0.96 \pm 0.09$ |
| activity | random | 0.81 | 0.91 | 1.00 | $0.86 \pm 0.08$ | 0.73 | 0.86 | 0.97 | $0.79 \pm 0.09$ |
| LBD+bumps | LBD | 1.00 | 1.17 | 1.37 | $1.21 \pm 0.12$ | 0.94 | 1.08 | 1.23 | $1.10 \pm 0.10$ |
| *Glucose* | LBD | 0.83 | 1.00 | 1.20 | $0.97 \pm 0.12$ | 0.81 | 1.00 | 1.13 | $0.95 \pm 0.12$ |

database size and adaptive restarts, we also compare the proof sizes for these two solvers. There is no statistically significant difference for the proof sizes, and 50% of proof sizes obtained from *Glucose* are within 19% of our LBD model, so we believe that this clause assessment experiment is relevant in practice.

**Clause Features.** We estimate the conditional probability that a clause usage in the untrimmed proof appears also in the trimmed proof by dividing the sampled frequency distribution of a feature in the trimmed proofs by the frequency distribution in the untrimmed proofs. In Fig. 2, we visualize the computed conditional probabilities for some features. In addition, the plots contain a dashed line that shows the ratio of all clause usages in the trimmed proof versus the untrimmed proof, which is what the graph for an uninformative, completely uncorrelated, predictor would look like. Similar values can also be computed for the *DRAT-trim* proof instead of the trimmed solver proof (although they cannot be interpreted as conditional probabilities since *DRAT-trim* usages are not a subset of solver usages); these are shown in the same plots for comparison. To indicate which values are relevant, plots also show the frequency distribution of all solver usages, transformed for the logarithmic x-axis so that area under the curve corresponds roughly to a probability measure (but with arbitrary scaling).

For dynamic LBD, glue clauses (with LBD scores at most 2) occur in the trimmed proofs more than average, and the top 5% of clauses have clearly larger probability of appearing in the trimmed proof than the rest. In the plot, there is also a peak around the value 250; however, as the solver usage distribution line shows there are not many usages with these values, so this is likely to be an effect of small sample size. Initial LBD and size are somewhat similar, but dynamic LBD is a better predictor for the top clauses than either of them.

Clauses with very small activity score are sometimes used by the solver, but they tend to be less common in trimmed proofs. Higher values indicate usefulness, except that clauses with very high activity scores (above 30) tend not to be useful; it appears that the solver uses some clauses a lot that are not needed in the final proof. Curiously, activity scores just below small integer values

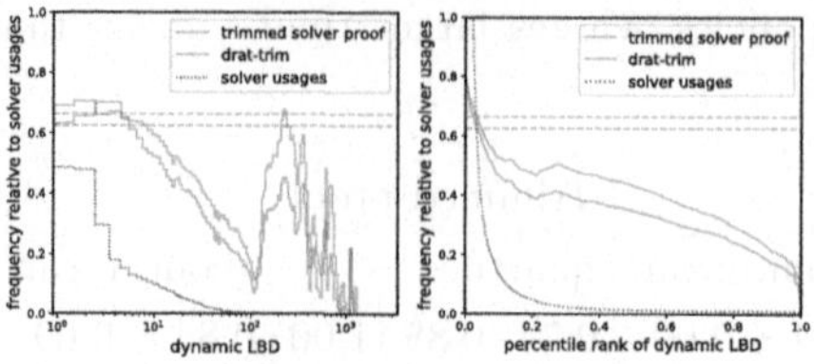


(a) Dynamic LBD.

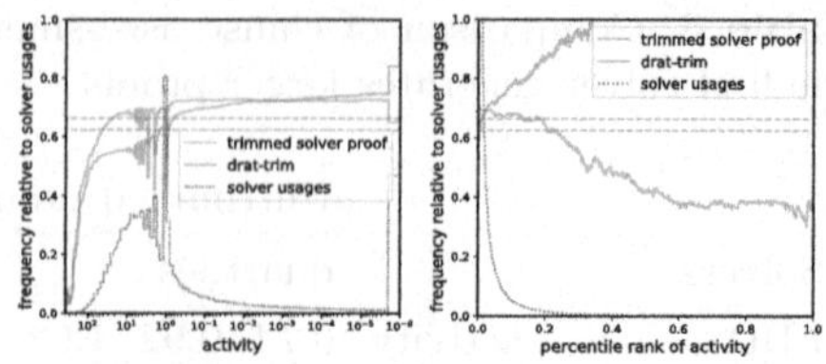


(b) Activity. Note the reversed x-axis in the absolute value plot.

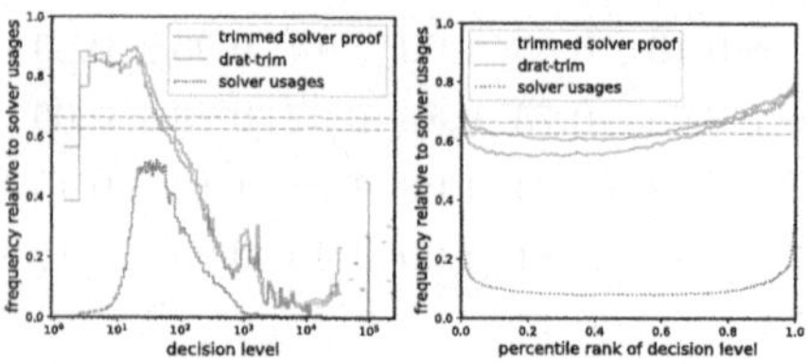


(c) Decision level when clause was learned.

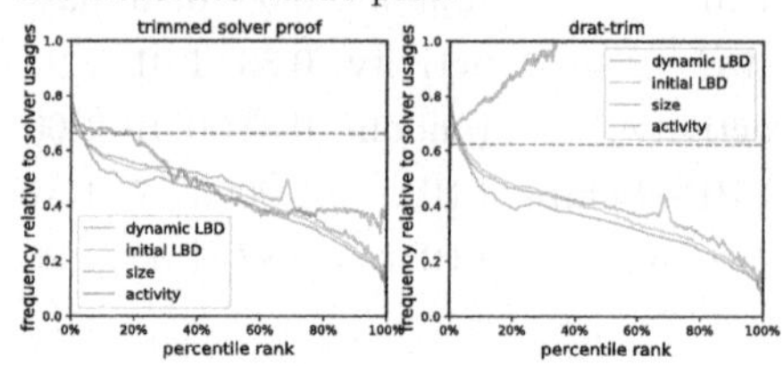


(d) Comparison of usefulness as a function of percentile rank.

**Fig. 2.** Sampled conditional probabilities that usages of clauses in the untrimmed proof appears also in the trimmed solver proof, and the analogous ratio for the *DRAT-trim* proof.

are less common in trimmed proofs. These come mostly from clauses that have been used recently and where the activity has not had much time to decay. One possible explanation is that being used many times in short succession may not indicate usefulness, but clauses that are used many times throughout a longer time interval are better.

A comparison of the computed conditional probabilities for percentile ranks of dynamic LBD, initial LBD, size, and activity is shown in Fig. 2d. When comparing the predictive power of the most popular measures, i.e., dynamic LBD and activity, it seems that LBD is a good predictor for the very best clauses, but that activity is relevant for a wider range of values. If we would use the *DRAT-trim* frequency distribution instead, we would not see as clear a difference between dynamic LBD and initial LBD or size. Also, it is clear that low-activity clauses are used by *DRAT-trim* much more often than by the solver.

Measuring the time elapsed from the most recent restart to when a clause is learned does not seem to provide any predictive power. Clauses that cause a backjump of only one decision level seem to appear often in conflict analysis, but tend to be less useful than clauses yielding longer backjumps. The data for the decision level at which a clause is learned seem contradictory: usages with small absolute value are more likely to appear in the trimmed proof, but so are usages with high percentile rank values. Such behaviour could potentially be caused by the distribution of the feature in the database changing over time, but understanding this in detail will require further research.

## 5 Concluding Remarks

The main philosophy underlying this paper is that in order to gain a better understanding of how CDCL SAT solvers work, it is fruitful to investigate the reasoning that they perform. Since CDCL solvers are search algorithms for resolution proofs when run on unsatisfiable formulas, we can study what kind of proofs they find, and what parts of these proofs are essential for establishing that the formulas are indeed unsatisfiable.

Using this method of analysis, we find that more frequent Luby-style restarts help solvers to produce shorter proofs (even if all too frequent restarts cause too much of a penalty in running time). Making restarts adaptive can significantly decrease proof size further, but mainly for the untrimmed proofs containing all derivations rather than for the trimmed proofs containing only essential clauses. This indicates that adaptive restarts are often successful in helping the solver avoid unnecessary work. When assessing whether a learned clause is likely to be useful, as measured by the probability of the clause appearing in the final, trimmed proof, we find that very good literal block distance (LBD) score is a strong predictor, but that clausal activity appears to be more relevant over a larger range of values. This supports the currently popular approach of prioritizing clauses with low LBD scores but sorting other clauses with respect to activity [34].

We consider our paper, and previous works in a similar spirit such as [28,37], to be only first steps, and see ample scope for future research in this direction. In particular, it would be very interesting to extend our method to satisfiable formulas, by looking at the "proofs" obtained by concatenating the conflict analyses for the learned clauses guiding the solver to the satisfying assignment.

In addition to the heuristics for restarts and clause database management studied in this work, it would be relevant to investigate variable decision heuristics such as VSIDS and phase saving, building on and extending [28]. An arguably even more urgent task is to gain a better understanding of relatively new techniques such as *learned clause minimization* [26] and *chronological backtracking* [32], which have played an important role in the SAT competitions [36] in recent years.

Our data analysis is relatively simple, and there should be room for using more advanced tools. A tempting idea is to combine our approach with the machine learning techniques in [39] (but, importantly, applied on the actual proof found by the solver rather than the one reconstructed by *DRAT-trim*). Also, it would be interesting to study more properties of proofs such as space complexity, and whether theoretical time-space trade-offs as in [1,6,8,9] could show up also in practice, in view of the aggressive memory management in modern solvers.

Finally, it is intriguing that some of our results are quite different from those in [18]. As an example, that paper found that activity-based clause assessment when choosing which clauses to erase is almost equally bad as random, whereas in our work it is clearly better. A natural question is how much of this discrepancy might be due to that we use "applied" SAT competition benchmarks, whereas only crafted, combinatorial formulas were considered in [18].

**Acknowledgements.** We wish to thank Jo Devriendt and Stephan Gocht for many interesting discussions, and helpful suggestions, throughout the project. We are also grateful to the anonymous reviewers, who helped improve the exposition considerably.

The computational experiments used resources provided by the Swedish National Infrastructure for Computing (SNIC) at the High Performance Computing Center North (HPC2N) at Umeå University. The authors were supported by the Swedish Research Council (VR) grant 2016-00782, and the second author also received funding from the Independent Research Fund Denmark (DFF) grant 9040-00389B.

## References

1. Alwen, J., de Rezende, S.F., Nordström, J., Vinyals, M.: Cumulative space in black-white pebbling and resolution. In: Proceedings of the 8th Innovations in Theoretical Computer Science Conference (ITCS 2017). Leibniz International Proceedings in Informatics (LIPIcs), vol. 67, pp. 38:1–38:21, January 2017
2. Audemard, G., Simon, L.: Predicting learnt clauses quality in modern SAT solvers. In: Proceedings of the 21st International Joint Conference on Artificial Intelligence (IJCAI 2009), pp. 399–404, July 2009
3. Audemard, G., Simon, L.: Refining restarts strategies for SAT and UNSAT. In: Milano, M. (ed.) CP 2012. LNCS, pp. 118–126. Springer, Heidelberg (2012). https://doi.org/10.1007/978-3-642-33558-7_11
4. Audemard, G., Simon, L.: Lazy clause exchange policy for parallel SAT solvers. In: Sinz, C., Egly, U. (eds.) SAT 2014. LNCS, vol. 8561, pp. 197–205. Springer, Cham (2014). https://doi.org/10.1007/978-3-319-09284-3_15
5. Bayardo Jr., R.J., Schrag, R.: Using CSP look-back techniques to solve real-world SAT instances. In: Proceedings of the 14th National Conference on Artificial Intelligence (AAAI 1997), pp. 203–208, July 1997
6. Beame, P., Beck, C., Impagliazzo, R.: Time-space tradeoffs in resolution: superpolynomial lower bounds for superlinear space. SIAM J. Comput. **45**(4), 1612–1645 (2016). Preliminary version in STOC '12
7. Beame, P., Kautz, H., Sabharwal, A.: Towards understanding and harnessing the potential of clause learning. J. Artif. Intell. Res. **22**, 319–351 (2004). Preliminary version in IJCAI '03
8. Beck, C., Nordström, J., Tang, B.: Some trade-off results for polynomial calculus. In: Proceedings of the 45th Annual ACM Symposium on Theory of Computing (STOC 2013), pp. 813–822, May 2013
9. Ben-Sasson, E., Nordström, J.: Understanding space in proof complexity: separations and trade-offs via substitutions. In: Proceedings of the 2nd Symposium on Innovations in Computer Science (ICS 2011), pp. 401–416, January 2011
10. Biere, A.: Adaptive restart strategies for conflict driven SAT solvers. In: Kleine Büning, H., Zhao, X. (eds.) SAT 2008. LNCS, vol. 4996, pp. 28–33. Springer, Heidelberg (2008). https://doi.org/10.1007/978-3-540-79719-7_4
11. Biere, A., Fröhlich, A.: Evaluating CDCL variable scoring schemes. In: Heule, M., Weaver, S. (eds.) SAT 2015. LNCS, vol. 9340, pp. 405–422. Springer, Cham (2015). https://doi.org/10.1007/978-3-319-24318-4_29
12. Biere, A., Fröhlich, A.: Evaluating CDCL restart schemes. In: Proceedings of Pragmatics of SAT 2015 and 2018. EPiC Series in Computing, vol. 59, pp. 1–17, March 2019. https://easychair.org/publications/paper/RdBL

13. Buss, S., Nordström, J.: Proof complexity and SAT solving (2020). Chapter to appear in the 2nd edition of Handbook of Satisfiability. Draft version available at http://www.csc.kth.se/~jakobn/research/
14. Calabro, C., Impagliazzo, R., Paturi, R.: The complexity of satisfiability of small depth circuits. In: Chen, J., Fomin, F.V. (eds.) IWPEC 2009. LNCS, vol. 5917, pp. 75–85. Springer, Heidelberg (2009). https://doi.org/10.1007/978-3-642-11269-0_6
15. Davis, M., Logemann, G., Loveland, D.: A machine program for theorem proving. Commun. ACM **5**(7), 394–397 (1962)
16. Eén, N., Sörensson, N.: An extensible SAT-solver. In: Giunchiglia, E., Tacchella, A. (eds.) SAT 2003. LNCS, vol. 2919, pp. 502–518. Springer, Heidelberg (2004). https://doi.org/10.1007/978-3-540-24605-3_37
17. Eén, N., Sörensson, N.: An extensible SAT-solver [extended version 1.2] (2004). http://minisat.se/downloads/MiniSat.pdf. Updated version of [16]
18. Elffers, J., Giráldez-Cru, J., Gocht, S., Nordström, J., Simon, L.: Seeking practical CDCL insights from theoretical SAT benchmarks. In: Proceedings of the 27th International Joint Conference on Artificial Intelligence (IJCAI 2018), pp. 1300–1308, July 2018
19. The Glucose SAT solver. http://www.labri.fr/perso/lsimon/glucose/
20. Huang, J.: The effect of restarts on the efficiency of clause learning. In: Proceedings of the 20th International Joint Conference on Artificial Intelligence (IJCAI 2007), pp. 2318–2323, January 2007
21. Impagliazzo, R., Paturi, R.: On the complexity of $k$-SAT. J. Comput. Syst. Sci. **62**(2), 367–375 (2001). Preliminary version in CCC '99
22. Jamali, S., Mitchell, D.: Centrality-based improvements to CDCL heuristics. In: Beyersdorff, O., Wintersteiger, C.M. (eds.) SAT 2018. LNCS, vol. 10929, pp. 122–131. Springer, Cham (2018). https://doi.org/10.1007/978-3-319-94144-8_8
23. Järvisalo, M., Matsliah, A., Nordström, J., Živný, S.: Relating proof complexity measures and practical hardness of SAT. In: Milano, M. (ed.) CP 2012. LNCS, pp. 316–331. Springer, Heidelberg (2012). https://doi.org/10.1007/978-3-642-33558-7_25
24. Katebi, H., Sakallah, K.A., Marques-Silva, J.P.: Empirical study of the anatomy of modern SAT solvers. In: Sakallah, K.A., Simon, L. (eds.) SAT 2011. LNCS, vol. 6695, pp. 343–356. Springer, Heidelberg (2011). https://doi.org/10.1007/978-3-642-21581-0_27
25. Katsirelos, G., Sabharwal, A., Samulowitz, H., Simon, L.: Resolution and parallelizability: barriers to the efficient parallelization of SAT solvers. In: Proceedings of the 27th AAAI Conference on Artificial Intelligence (AAAI 2013), pp. 481–488, July 2013
26. Luo, M., Li, C.M., Xiao, F., Manyá, F., Lü, Z.: An effective learnt clause minimization approach for CDCL SAT solvers. In: Proceedings of the 26th International Joint Conference on Artificial Intelligence (IJCAI 2017), pp. 703–711, August 2017
27. Lynce, I., Marques-Silva, J.P.: Building state-of-the-art SAT solvers. In: Proceedings of the 15th European Conference on Artificial Intelligence (ECAI 2002), pp. 166–170. IOS Press, May 2002
28. Malik, S., Ying, V.A.: On the efficiency of the VSIDS decision heuristic, August 2016. presentation at the workshop Theoretical Foundations of SAT Solving. Slides http://www.fields.utoronto.ca/sites/default/files/talk-attachments/SharadMalik-Fields2016.pdf
29. Marques-Silva, J.P., Sakallah, K.A.: GRASP: a search algorithm for propositional satisfiability. IEEE Trans. Comput. **48**(5), 506–521 (1999). Preliminary version in ICCAD '96

30. The MiniSat page. http://minisat.se/
31. Moskewicz, M.W., Madigan, C.F., Zhao, Y., Zhang, L., Malik, S.: Chaff: engineering an efficient SAT solver. In: Proceedings of the 38th Design Automation Conference (DAC 2001), pp. 530–535, June 2001
32. Nadel, A., Ryvchin, V.: Chronological backtracking. In: Beyersdorff, O., Wintersteiger, C.M. (eds.) SAT 2018. LNCS, vol. 10929, pp. 111–121. Springer, Cham (2018). https://doi.org/10.1007/978-3-319-94144-8_7
33. Nordström, J.: On the interplay between proof complexity and SAT solving. ACM SIGLOG News **2**(3), 19–44 (2015)
34. Oh, C.: Improving SAT solvers by exploiting empirical characteristics of CDCL. Ph.D. thesis, New York University (2016). https://cs.nyu.edu/media/publications/oh_chanseok.pdf
35. Pipatsrisawat, K., Darwiche, A.: A lightweight component caching scheme for satisfiability solvers. In: Marques-Silva, J., Sakallah, K.A. (eds.) SAT 2007. LNCS, vol. 4501, pp. 294–299. Springer, Heidelberg (2007). https://doi.org/10.1007/978-3-540-72788-0_28
36. The international SAT Competitions web page. http://www.satcompetition.org
37. Simon, L.: Post mortem analysis of SAT solver proofs. In: Proceedings of the 5th Pragmatics of SAT Workshop. EPiC Series in Computing, vol. 27, pp. 26–40, July 2014. https://easychair.org/publications/paper/N3GD
38. Simon, L.: Personal communication (2018)
39. Soos, M., Kulkarni, R., Meel, K.S.: CrystalBall: gazing in the black box of SAT solving. In: Janota, M., Lynce, I. (eds.) SAT 2019. LNCS, vol. 11628, pp. 371–387. Springer, Cham (2019). https://doi.org/10.1007/978-3-030-24258-9_26
40. Sörensson, N., Biere, A.: Minimizing learned clauses. In: Kullmann, O. (ed.) SAT 2009. LNCS, vol. 5584, pp. 237–243. Springer, Heidelberg (2009). https://doi.org/10.1007/978-3-642-02777-2_23
41. Wetzler, N., Heule, M.J.H., Hunt, W.A.: DRAT-trim: efficient checking and trimming using expressive clausal Proofs. In: Sinz, C., Egly, U. (eds.) SAT 2014. LNCS, vol. 8561, pp. 422–429. Springer, Cham (2014). https://doi.org/10.1007/978-3-319-09284-3_31

# Core-Guided Model Reformulation

Kevin Leo(✉), Graeme Gange(✉), Maria Garcia de la Banda(✉), and Mark Wallace(✉)

Faculty of Information Technology, Monash University, Clayton, Australia
{kevin.leo,graeme.gange,maria.garciadelabanda,mark.wallace}@monash.edu

**Abstract.** Constraint propagation and SAT solvers often underperform when dealing with optimisation problems that have an *additive* (or separable) objective function. The core-guided search introduced by MaxSAT solvers can overcome this weakness by detecting and exploiting *cores*: subsets of the objective components that cannot collectively take their lower bounds. This paper shows how to use the information collected during core-guided search, to reformulate the objective function for an entire *class of problems* (those captured by the problem *model*). The resulting (currently manual) method is examined on several case studies, with very promising results.

## 1 Introduction

Modern approaches for solving combinatorial optimisation problems first specify a *model* that formally describes the problem's parameters, variables, constraints and objective function. All parameters are later instantiated with input data, describing an *instance* of the problem. Each instance is then compiled to the format required by the selected *solver*, which explores the model's search space to find high quality solutions.

The economic impact of combinatorial optimisation problems has fuelled the design of powerful modelling languages, such as AMPL [9], OPL [23], Essence [10] and MiniZinc [18], and powerful solvers within the Mixed Integer Programming (MIP), Constraint Programming (CP), and MaxSAT solving paradigms. However, while there have been many advances in the variety and quality of solvers available, advances in technology that helps users improve their models have been scarce. This is unfortunate since, while the way in which a problem is modelled can significantly affect its solving time, designing good models is still very challenging, even for expert users. As a result, users must follow a time consuming, iterative, modify-and-test approach that can still yield poor results.

This paper aims at helping users identify model improvements, by taking advantage of some of the great advances achieved by Lazy Clause Generation (LCG) [8,19] and MaxSAT solvers. LCG solvers, such as Chuffed [6], Geas [11] and ORTools [20], combine the strengths of the CP and SAT solving paradigms. This allows LCG solvers to infer *nogoods* (i.e., reasons for failure), and use them to avoid repeatedly exploring infeasible subproblems. Previous work [24] showed that the nogoods inferred by LCG solvers for a model's instances can be used to identify (a) existing constraints that may

Partly funded by Australian Research Council grant DP180100151.


H. Simonis (Ed.): CP 2020, LNCS 12333, pp. 445–461, 2020.
https://doi.org/10.1007/978-3-030-58475-7_26

be strengthened, and (b) new redundant constraints on the existing model variables that are likely to increase performance.

Our work complements this line of research by providing a (currently manual) method that uses the information inferred by the core-guided search of MaxSAT solvers, to improve a common class of optimisation models: those with an *additive* (or separable) objective function, i.e. a sum of components, each containing a single variable. In particular, we show how the cores found by a core-guided solver can help identify components in the model (rather than in the instance) that can yield better bounds when grouped. We then show how to use these components to reformulate the model itself, by adding *new variables* to the model. Note that most previous works on model reformulation, have only used variables appearing in the original model. One of the few exceptions is [5], which introduced new variables to achieve a lower computational complexity in handling the *SEQUENCE* constraint. While adding new variables to a model is very unusual and challenging, our experimental results indicate that, if done appropriately, it can significantly speed up LCG (and sometimes CP) solvers. In addition, we show how the first two steps of the method can be automated. Automating the last step is a challenging and important future goal.

## 2 Background

**Constraint Optimisation Problems:** A constraint optimisation problem $P$ is a tuple $(C, D, f)$, where $C$ is a set of constraints, $D$ a *domain* mapping each variable $x$ appearing in $C$ to set of values $D(x)$, and $f$ an objective function. $C$ is logically interpreted as the conjunction of its elements, and $D(x)$ as the conjunction of unary constraints on $x$. A *literal* of $P$ is a unary constraint whose variable appears in $C$. To solve problem $P \equiv (C, D, f)$, a CP solver first applies constraint propagation to reduce domain $D$ to $D'$, by executing the propagator associated with the constraints in $C$ until reaching a fixpoint. If $D'$ is equivalent to *false* ($D'(x)$ is empty for some variable $x$), we say $P$ fails. If $D'$ is not equivalent to *false* and fixes all variables, we have found a solution to $P$. Otherwise, the solver splits $P$ into $n$ subproblems $P_i \equiv (C \wedge c_i, D', f), 1 \leq i \leq n$, where $C \wedge D' \Rightarrow (c_1 \vee c_2 \vee \ldots \vee c_n)$ and $c_i$ are literals (the *decisions*), and then iteratively searches these subproblems.

The search proceeds propagating and making decisions until either (1) a solution is found, or (2) a failure is detected. In case (1) the search computes the value of $f$, constraints the next value of $f$ to be better (greater or smaller, depending on $f$) and continues the search for this better value (the traditional *branch-and-bound*). In case (2) the search usually backtracks to a previous point to make a different decision.

**Lazy Clause Generation:** LCG solvers [8,19] extend CP solvers by instrumenting their propagators to explain domain changes in terms of *equality* ($x = d$ for $d \in D(x)$), *disequality* ($x \neq d$) or *inequality* ($x \geq d$ or $x \leq d$) literals. An *explanation* for literal $\ell$ is $S \rightarrow \ell$, where $S$ is a set of literals (interpreted as a conjunction). For example, the explanation for the propagator of constraint $x \neq y$, which infers literal $y \neq 5$ given literal $x = 5$, is $\{x = 5\} \rightarrow y \neq 5$. Each literal inferred when solving problem $P$ is recorded with its explanation, forming an *implication graph*. If failure is detected for subproblem $P'$, LCG solvers use this graph to compute a clause $L$ (or *nogood*): a disjunction of literals that holds under any solution of $P$ but is inconsistent under $P'$.[1] $L$ is then added to $P$'s constraints, to avoid failing for the same reasons.

**Core-Guided Optimisation:** CP solvers often underperform when proving optimality for additive objectives. This is because the lower bound of any objective component, say $oc_i$ of variable $x_i$ for minimising function $f \equiv oc_1 + \cdots + oc_n$, can often be achieved by sacrificing others, and $f$'s lower bound is inferred only from the bounds of its components. Core-guided solvers overcome this weakness by first fixing all components to their lower bounds, and then searching for a solution. If this succeeds, an optimum has been found. Otherwise, they return a *core*: a (hopefully small) subset of components that cannot collectively take their lower bounds. They then update $f$'s bound *without* committing to *which* core component incurs the cost, and adjust the lower bounds of the components in the core. Finally they re-solve, repeating this process until a solution is found. Different core-guided solvers differ mostly in how the interaction between cores, component bounds, and the objective is handled. We assume such solvers return either an empty set, indicating the current subproblem is satisfiable, or a set $S$ of literals of the form $x_i \geq k$ where variable $x_i$ appears in the objective, indicating at least one literal in $S$ must hold. Extending LCG solvers to support this interface is straightforward.

This paper uses the LCG, core-guided solver GEAS [11]. Its core-guided approach is based on the OLL [1] method, which progressively reformulates the objective to use the discovered cores: upon finding core $S$, OLL introduces a new variable $p = \sum S$ (with lower bound increased by at least 1), and rewrites the objective in terms of $p$. GEAS improves the basic OLL with stratification [2, 17] (extracting cores on high-coefficient terms first), weight-aware core extraction [4] (delaying the introduction of new variables until no cores can be found), and the 'hardening rule' [2] (upper-bound propagation on new variables).

## 3 Motivation for Core-Guided Model Reformulation

The reformulation of the objective performed by GEAS when solving an instance, can significantly reduce the search and, thus, the solving time for both LCG and branch-and-bound solvers. This is somewhat counter-intuitive, as the reformulation introduces new instance variables. The reasons for such reduction are twofold. First, LCG solvers can use the new variables to learn nogoods that shorten their optimality proof. This shows the importance of what we call the *language of learning*. Second, branch-and-bound solvers can use the bounds on the new variables to detect failed subproblems earlier.

This paper aims to achieve similar improvements to those achieved by core-guided solvers, but at the model (rather than instance) level. Thus, performance can be improved for multiple instances (rather than only for the one being executed), and also for non-core-guided solvers (either because they are not available, or are not as fast for the instance in question). Let us demonstrate via two (extreme) examples the radical performance improvements made possible by reformulating the objective to use variables whose bounds are detected by core-guided search.

[1] Note our *nogoods* denote a positive (implied) clause. In other works they denote the (conjunctive) negation of its literals.

*Example 1.* Consider an optimisation problem with $n$ pairs of variables, $x_i, y_i : i \in 1..n$, where each variable has $0..m$ domain, $\forall i \in 1..n : x_i + y_i \geq k$, and the objective is to minimise the sum of the variables ($\sum_{i=1}^{n} x_i + y_i$). With a CP or LCG solver, propagation ensures assignments to $x_i$ and $y_i$ are mutually consistent and, if the lower values in the domains are tried first by the search, the first solution to be found will be optimal. Given a direct model for this problem, the first row of the table below shows, for $n = 10$ and $m = 5$, the number of search steps required by CP solver Gecode [12] to find the first solution and prove its optimality, for several values of $k$.

| Search steps with: | $k = 2$ | $k = 3$ | $k = 4$ |
|---|---|---|---|
| original objective | 21 | 73,955 | 11,163,595 |
| reformulated objective | 21 | 21 | 21 |

The last row shows the number of search steps required after (a) adding to the model new variables $xy_i$ and constraints $xy_i = x_i + y_i, xy_i \geq k$ for each $i \in 1..n$, and (b) reformulating its objective as $\sum_{i=1}^{n} xy_i$. With this model reformulation, once Gecode finds the first solution and starts searching for a better (smaller) objective, the bounds on the $xy_i$ variables allow it to immediately realise that no higher value is possible for any $x_i$ or $y_i$. Thus, the search to prove optimality efficiently finishes right after the first solution is found, regardless of the value of $k$. □

*Example 2.* Consider now a problem with $n$ decision variables $x_1 \ldots x_n$, with domain $0..1$. Each triple of variables $(x_i, x_j, x_k) : i > j > k$ has a target values $(a_i, a_j, a_k)$ and the triple incurs a benefit of 1 if $x_i = a_i, x_j = a_j$ and $x_k = a_k$. The objective is the sum of these benefits. Given a direct model for this problem, the first row of the table below gives the average CPU time it took the Gecode, Chuffed and Gurobi [13] solvers to find an optimal solution and prove optimality for 10 randomly generated instances.

| CPU time (secs) | Gecode 6.1.1 | Chuffed 0.10.4 | Gurobi 7.5.2 |
|---|---|---|---|
| original objective | 57 | 236 | 232 |
| reformulated objective | 4 | 3 | 152 |

The last row shows the time required after (a) adding to the model a new variable $xa_i$ for each decision variable $x_i$, and (b) reformulating its objective as $\sum_i xa_i$. To define $xa_i$, let $tv_1$ be the first variable in the $t^{th}$ triple; $ta_1$ be its first matching value; $val_t$ be the value of this triple in an assignment; and $xa_i(b)$ be $\sum_{t:tv_1=x_i \wedge ta_1=b} val_t$. Then $xa_i = max(xa_i(1), xa_i(0))$. □

## 4 Methodology

The examples presented in the previous section clearly show the potential benefits of introducing new variables to a model. However, picking an effective reformulation is very difficult, since the number of possible new variables is huge. This is true even for our reduced scope, where these new variables must be formed from any (iterative) combination of the variables in the objective. The challenge then is to chose those variables that will achieve good improvements in a large number of instances.

Our core-guided reformulation method is designed to address this challenge. To achieve this, it takes two inputs: an optimisation model whose objective function is additive, and a set of input data files. It then performs the following main steps:

1. Use a core-guided solver to find, for each model instance, cores that are candidates for new variables;
2. Select some of these candidates, based on their likely reduction in solving time;
3. Modify the model to add new variables for the selected cores, constrain them, and use them to reformulate the objective, without changing the optimal solutions.

The rest of this section discusses the above three steps in more detail, using the Resource-Constrained Project Scheduling Problem with Weighted Earliness and Tardiness cost to illustrate the method. This problem tries to schedule tasks that have a given duration and desired start time, subject to precedence constraints and cumulative resource restrictions. The objective is to find an optimal schedule that minimises the weighted cost of the earliness and tardiness of any task that is not completed by their desired deadline.

The model we use (`rcpsp-wet` in the MiniZinc benchmarks) has the objective:

```
objective = sum (i in Tasks) (
  deadline[i,2]*max(0,deadline[i,1]-s[i]) + % earliness cost
  deadline[i,3]*max(0,s[i]-deadline[i,1])); % tardiness cost
```

that is, the sum of the earliness and tardiness costs for every task `i` in input set of `Tasks`, where parameter `deadline[i,1]` gives the desired start time for $i$, parameters `deadline[i,2]` and `deadline[i,3]` give the cost per time unit for task `i` to start before or after its desired time, respectively, and variable `s[i]` represents the start time for task `i`.

## 4.1 Step 1: Finding Core Candidates

*Step 1.1 Solver Instrumentation:* As we will see below, we currently find new variables by manually interpreting the cores found by the solver. Therefore, the solver needs to be instrumented to output them in human readable form. The GEAS solver, which connects to the MiniZinc system and is the core-guided solver we use, already produces verbose output for debugging purposes. This includes, for each iteration in which the objective is modified, the value of the objective function at the end of each iteration, and all cores found together with their individual impact on the lower bound (for a minimising objective). While our current manual method simply uses this output, any future automation of the method will require a formal protocol for communicating with the core-guided solver, similar to that used in the profiling of CP solvers [21].

*Step 1.2 Collect the Cores:* To collect the cores for a given problem, we run GEAS in core-opt mode on a subset of the model instances we have, and record its verbose output. Currently, the subset selected corresponds to at most 2 small instances, as this will simplify the remaining manual steps. For the case of `rcpsp-wet`, we used the instances obtained by instantiating the model with data files `j30_1_3.dzn` and `j30_43_10.dzn`. Once this step is automated, better results will be obtained by using a large and diverse set of instances. Note that we disabled GEAS core hardening for this step, as we do not want any literals (including those made false by hardening) to be omitted from the reported cores.

*Step 1.3 Rename the Cores:* Solvers express cores in terms of the variables created when compiling the instance, or those created by the solver itself. These names are generic, making them difficult for humans to interpret. For example, the following shows an extract from the verbose output created by GEAS when solving the `j30_1_3.dzn` instance:

```
Found core of size 2, new lb: 5
CORE: X_INTRODUCED_261_ >= 7,
      X_INTRODUCED_217_ >= 1
```

indicating that the core has 2 literals, resulted in a new lower bound of 5 for the objective function, and contains instance variables `X_INTRODUCED_261_` and `X_INTRODUCED_217_`. Typically, solver writers who want to interpret such names, must examine the compiled instance output to see what these variables might refer to. In [24], the authors used a source map produced by the MINIZINC compiler, to map instance variables back to variables and expressions in the original model. Herein, we use the same method to link back the core variables, which for the above core results in:

```
Found core of size 2, new lb: 5
CORE: 'max(0, deadline[16, 1] - s[16])' >= 7,
      'max(0, s[25] - deadline[25, 1])' >= 1
```

The variables can now be easily recognised (from the objective) as the earliness of task 16, and the tardiness of task 25.

*Step 1.4 Collect New Variable Core Candidates:* Cores containing more than one literal are candidates for new variables to be introduced in the model (singleton cores contain a variable that already exists in the model and are, therefore, not useful for our method). We collect all such cores by performing the previous two steps for the selected subset of model instances, and recording the results.

## 4.2 Step 2: Selecting Good Candidates

*Step 2.1 Find Patterns Among the Cores:* Once all candidate cores are collected, the next main step in our method involves interpreting these cores to determine subsets that are likely to reduce solving time for many instances. To achieve this, we first try to find *patterns* among the different cores found. The following details three of the patterns we have often found in our experiments. Importantly, we focus on finding patterns for the most effective cores, i.e., those with greater impact on the objective function value and its lower bound. In GEAS, these are often the cores found early in the search.

*Identical Up to Renaming:* Many of the cores collected differ only in the name of the parameters present in the core's variables, and their bounds. For example, core:

```
Found core of size 2, new lb: 10
CORE: 'max(0, s[14] - deadline[14, 1])' >= 4,
      'max(0, deadline[8, 1] - s[8])' >= 1
```

and the one in Step 1.3, have two variables with pattern `max(0,deadline[i,1]-s[i])`, `max(0,s[j]-deadline[j,1])`, where `i` represents tasks 16 in the first core and 8 in the second, and `j` represents tasks 25 and 14. Note that we always ignore the literals' bound (e.g., 4 and 1 in the above core). While we currently find these patterns manually, the method described in [24] for finding nogood patterns across instances, can be easily adapted to cores.

*Simple Ordering:* A simple but common pattern consists of pairs or triplets of variables that appear "near" each other in some ordering in the model. For example, variables representing the state of some object at time points $t$ and $t+1$, or variables representing two tasks where one is a successor of the other, as task 14 is of task 8 in the above core for `rcpsp-wet`.

*Element Constraints:* Sometimes cores have literals assigning all (or most of) the possible values of a variable, e.g., $1 * (x = 1) + 5 * (x = 2) + 6 * (x = 3) \ldots$ This often occurs when the variable's contribution to the objective is non-linear. These cores reconstruct an `element` global constraint (see, for example, the reformulation for the `jp-encoding` model in Sect. 5.1).

*Step 2.2 Interpret the Patterns:* We now look for reasons for the patterns to appear, that is, for the associated variables to appear often together in effective cores. This usually requires in-depth knowledge regarding the relationship between these variables. For example, for the pattern `max(0,deadline[i,1]-s[i])`, `max(0,s[j]-deadline[j,1])` mentioned above, we must understand what connects the earliness of the tasks represented by `i` (16 or 8) to the tardiness of those represented by `j` (25 or 14, respectively). Visualising the input data using a variation of a Gantt chart helped us realise, for example, that task `j` is often the direct successor of task `i`, they overlap in time, and have the highest earliness and tardiness costs. In other cores `j` is often a non-direct successor of `i`, and the penalty for scheduling the chain of tasks is also very high due to overlaps.

### 4.3 Step 3: Reformulating the Model

*Step 3.1 Reformulate the Objective:* Once the patterns are interpreted, we reformulate the objective using this information. The aim is to group objective components that are expected to form effective cores. We have observed that patterns often suggest an *ordering* of components that places them in cores together. For example, for `rcpsp-wet` we can sort the earliness/tardiness components of direct successor tasks that overlap, based on the cost of enforcing their precedence (we call this ordering *direct*), leaving the remaining components unchanged. Alternatively, we can use an ordering obtained by simply sorting the earliness/tardiness components based on the desired start time of their task (we call this ordering *start*).

For any such ordering of the components in the objective, our method recursively creates new variables for each disjoint pair of adjacent components, and replaces them in the objective function with the new variable. We achieve this by using the following group function, which we have implemented in MINIZINC and added to its library:

$$\begin{aligned}
&\mathsf{group}([x]) = x\\
&\mathsf{group}([x_1,\ldots,x_{2n}]) = \mathsf{group}([z_1,\ldots,z_n])\\
&\qquad \texttt{s.t. } z_i \geq x_{2i-1} + x_{2i},\ \forall.\, i \in 1\ldots n\\
&\mathsf{group}([x_1,\ldots,x_{2n},x_{2n+1}]) = \mathsf{group}([z_1,\ldots,z_n,x_{2n+1}])\\
&\qquad \texttt{s.t. } z_i \geq x_{2i-1} + x_{2i},\ \forall.\, i \in 1\ldots n
\end{aligned}$$

The function receives as its argument a list with the objective components in the given ordering $[x_1,\ldots,x_m]$, and creates a new variable $z_i$ for each pair of adjacent components $x_{2i-1} + x_{2i}$. It then recursively calls itself with a list of the new variables in the order they were created as input, while appropriately dealing with the case of $m$ being odd or even. The recursion ends when the list contains a single component $x$, simply returning $x$ as the new objective to minimize. Note that the best performance for GEAS occurs when $z_i$ is bound from below. However, other solvers may perform better when $z_i$ is defined as $z_i = x_{2i-1} + x_{2i}$. Therefore, in practice we use the bounding strategy that is best for a given solver. Note also that the group function uses recursion to combine more distant components. We do this to compensate for the locality brought by the fact we currently only consider the early cores of a very small selected subset of small instances, since these are the ones that are easier to interpret by hand. Once better support for the interpretation step is achieved, this should be reconsidered.

Our method adds to the model both the function that produces the ordering and the group function which uses it (in the case of the group function, this is done by adding it to the MINIZINC library, but it has the same effect). As a result, the time needed to compute this ordering is an overhead to the execution of the instance. Therefore, care needs to be taken when defining orderings that might be too time consuming to compute. The same can be said for the group function, although in this case reducing the time overhead might not be as important as reducing the number of variables created. We therefore experimented with a version of the group function that only performs $k$ iterations, with the aim of introducing fewer new variables. The resulting models did not yield noticeable improvements in our experiments (data not shown).

*Step 3.2 Add Bounds for New Variables:* The reformulation of the objective can improve the solving time of any clause-learning solver (such as LCG and MaxSAT). This is because it introduces new variables that can be used by these solvers to learn new clauses and, thus, reduce the search space. However, the reformulation would not usually help CP solvers, as they will be unable to infer tighter bounds on the introduced variables.

To counter this, we modify group to add a bound to the new $z_i$ variables it creates. For example, for `rcpsp-wet`, if the first iteration creates the following variable:

```
new_var = deadline[i,2]*max(0,deadline[i,1]-s[i]) +
          deadline[j,3]*max(0,s[j]-deadline[j,1])
```

we also add to the model the constraint:

```
new_var >= min(deadline[i,2],deadline[j,3])*
           (deadline[i,1]+d[i]-deadline[j,1]);
```

ensuring `new_var` is greater or equal than the minimum cost to enforce the precedence, that is, the minimum of the earliness cost of task `i` and the tardiness cost of task `j`, multiplied by their overlap (`d[i]` is a parameter of the model representing `i`'s duration).

## 4.4 Automating the Method

The initial stages of our method (all substeps in step 1: finding core candidates) are automatic, thanks to the use of core-guided optimisation to generate the cores and their information, and the use of existing MiniZinc infrastructure to collect and rename the cores, and identify the core candidates. While substep 2.1 (finding patterns among the cores) is currently done manually, it can be automated using similar technology to that used by [24] to identify patterns among nogoods.

The most difficult manual stages to automate are the analytical ones: pattern interpretation (step 2.2), and designing the reformulation (step 3). As these rely on insights regarding the underlying model structure, full automation is quite challenging. However, it is possible to automate certain processes to make these stages easier. For example, interpreting the patterns requires understanding *why* the cores represented by the pattern hold. Since each such core typically only relates to a small fragment of the model, identifying this fragment can often immediately reveal the meaning of the core. And *this* identification is something that can be automated: given a model $M$ and core $C$, we know $M \wedge \neg C$ is unsatisfiable. Thus, we can use tools such as FindMUS [16] to identify a minimal unsatisfiable subset of $M$ that causes the failure.

*Example 3.* Consider the following core which was part of the output created by Geas for the `rcpsp-wet` model, with the instance obtained from data file `j30_1_3.dzn`.

```
CORE: 'max(0, s[27] - deadline[27, 1])' >= 3,
      'max(0, deadline[17, 1] - s[17])' >= 1
```

We update the model with name annotations that explain what each constraint means.

```
constraint forall ( i in Tasks, j in suc[i] ) (
  ( s[i] + d[i] <= s[j] )
    :: "Task \(i) must finish before task \(j) starts" );
```

We then add the negation of the core to the MiniZinc model as follows:

```
constraint :: "Core 5"
           not (   max(0, s[27] - deadline[27, 1]) >= 3
                \/ max(0, deadline[17, 1] - s[17]) >= 1);
```

With just these modifications, FINDMUS is able to output the following:

```
MUS:  Core 5
      Task 17 must finish before task 24 starts
      Task 24 must finish before task 27 starts
```

Note that, while the core only mentions tasks 17 and 27, FINDMUS is able to identify the chain of precedence constraints required for explaining the negated core. This makes the task of interpreting the cores much easier.

Automating the model reformulation (step 3) is more involved, as it requires the design of higher-level groupings using the *reasons* for the cores found in step 2.2. However, as we will see in the next section, the reformulations we produce are usually structurally simple: either an ordering or hierarchical clustering, based on proximity with respect to numeric parameters (i.e. `jp-encoding` and `seat-moving`), or constraint structures (i.e. `spot5` and `rcpcsp-wet`). It is thus possible (although non-trivial) to use structural analysis methods, such as [15], to do this, since they are able to identify subsets of the model constraints that correspond to pre-defined constraint structures.

# 5 Experimental Evaluation

This section illustrates how to apply our core-guided model reformulation method to five models, and experimentally evaluates the efficiency of the reformulations obtained for different orderings.

## 5.1 Models and Their Reformulation

To evaluate the effectiveness of our method, we require models of optimisation problems with an additive objective function, and for which core-guided solvers can obtain better results than branch and bounds ones (otherwise, the method has no chance of success). Therefore, we selected the top five models in the MINIZINC annual competition [22], for which core-guided GEAS performs drastically better on at least one instance, than branch-and-bound GEAS. This yielded the `rcpsp-wet` model used above to illustrate our method, and the four models described below.

For each model, we selected 1–2 instances to analyze (typically, the smallest instance to identify cores, and a moderate one to check that the identified patterns reoccur). After modification, we evaluated the reformulated model over all instances from the challenge. The following describes how our core-guided reformulation method was applied to the other four selected models.

**The** `seat-moving` **Model:** Given a set of seats and the people sitting in them (some may be empty), the problem is to find the minimum number of moves and time-steps needed to reach a target seating plan. Some people can swap seats with anyone in one move; the rest must first move to an empty seat to make way. The objective is defined as:

```
cost = sum(i in 1..MAX_STEP-1,p in 1..P)
           (person[i,p]!=person[i+1,p]);
objective = cost + step*MAX_STEP*P;
```

where variable `cost` counts, for each time-step `i` and person `p`, the seats where `p` is at time `i` but not at `i+1` (note that boolean `person[i,p]!=person[i+1,p]` is coerced to an integer), and variable `step` is the number of time-steps needed. Therefore, the objective sums all moves performed in every step by any person,

Studying the early cores GEAS finds for instances `sm-10-12-00` and `sm-10-20-00`, we realised they contain the moves of a single person, rather than those of a time-step, which is how they appear in the sum that defines variable `cost`. Thus, we grouped the components using a *reverse* ordering that simply reverses the order of the sum indices:

```
cost = group(p in 1..P, i in 1..MAX_STEP-1)
         (person[i,p]!=person[i+1,p]);
```

In addition, we also added a (very weak) bound on the first set of new variables (i.e., those created in the first iteration of the group function), that ensures the number of moves for people not starting in their target seat is $\geq 1$.

**The** `jp-encoding` **Model:** The Japanese Encoding problem tries to find the most likely encoding used for each byte in a byte stream of encoded Japanese text, where multiple encodings may be used. The model considers the ASCII, EUC-JP, SJIS and UTF-8 encodings, each with a scoring table that maps each byte to its penalty score (based on likelihood) for that encoding, plus an "unknown" encoding with a large penalty. The objective to minimise is defined in the model as:

```
objective = 1000*n_unknown + sum(i in 1..len) (
   (encoding[i]==e_euc_jp)*eucjp_score[stream[i]+1]
 + (encoding[i]==e_sjis)*jis_score[stream[i]+1]
 + (encoding[i]==e_utf8)*utf8_score[stream[i]+1]);
```

that is, it sums the penalties (given by parameter tables `eujp_score`, `sjis_score` and `utf8_score`) for the encoding chosen by variable `encoding[i]` for each byte (represented by parameter `stream[i]`) in position `i` of the input stream. Note that the penalty is 0 for the ASCII encoding, and 1000 for an unknown encoding (in this case variable `n_unknown` has been incremented by 1).

Studying the early cores found by GEAS on `data200.dzn`, we realised they refer to all possible encodings of the byte in a given position (it must get *some* encoding). We thus used the `element` global constraint to create new variables `encoding_cost(i)`, representing the encoding penalty of position `i`:

```
function var int: encoding_cost(int: i) =
  array1d(0..4,[0,eucjp_score[stream[i]+1],
                sjis_score[stream[i]+1],
                utf8_score[stream[i]+1],1000])[encoding[i]];
objective = sum(i in 1..len)(encoding_cost(i));
```

We then re-applied our method to the reformulated model and realised that the new cores were *local*, i.e., involved `encoding_cost(i)` and either `encoding_cost(i+1)` or `encoding_cost(i+2)`. Thus, we defined a *local* ordering that simply sorted the encodings by position (which is the same as the original order in the model):

```
objective = group(i in 1..len)(encoding_cost(i));
```

**The** `rel2onto` **model:** The Relational-To-Ontology Mapping problem takes as input (a) an *alignment* graph formed by cnodes, corresponding to the ontology's classes, dnodes corresponding to the data properties of the classes, and weighted edges, (b) the set of attributes of a relational database, and (c) the set of dnodes each attribute might be matched to with a given cost. The problem is to find a single match for all attributes and a Steiner Tree for the alignment graph, such that the matched dnode of every attribute is in the tree, and the cost of the edges in the Steiner Tree and of the matched attributes is minimised. The objective to minimise is defined in the model as:

```
w = sum(i in edges)(es[i]) * ws[i]);
wm = sum(a in atts)(match_costs[a, match[a]]);
objective = w + wm;
```

that is, the sum `w` of the weight (given by parameter `ws[i]`) of every edge `i` in the graph that appears in the Steiner tree (true if variable `es[i]` holds), plus the sum `wm` of the cost for every attribute `a` of the match (given by variable `match[a]`).

When studying the cores inferred by GEAS for `5_5.dzn` and `5_28.dzn`, we discovered many cores involve two variables `es[i]` and `es[j]`, where `i` and `j` are edges of the alignment graph. Moreover, the cores were generated for attributes that could only be associated with two possible dnodes, indicating that, since each attribute must be in the matching, some edge adjacent to that attribute must be in the Steiner tree. Based on this, we constructed an *adjacent* ordering that groups edges associated with a given attribute. As this did not perform as well as hoped, we looked deeper into the cores and realised that, when an attribute's matches overlapped with a previous one, the solver would *merge* the old variable with the new adjacent edges. Thus, we encoded a similar iterative *merging* strategy: starting with each edge in a singleton partition, we repeatedly select a new attribute $a$, and introduce a fresh variable for the sum of all partitions containing edges adjacent to $a$. Once all attributes are processed, the objective sums the resulting cost variables.

**The** `spot5` **Model:** The SPOT5 earth observation satellite management problem [3] tries to find a subset of a given set of photographs to take, given many different constraints, including minimum distance, non-overlapping, and recording capacity. The model encodes these constraints as a set of binary and ternary `table` constraints. The objective to minimise is defined in the model as:

```
objective = sum(j in 1..num_variables)(costs[j]*(p[j]=0));
```

which sums the cost (given by parameter `costs[j]`) of each given photograph `j` (`j in 1..num_variables`) that is *not* taken (given by variable `p[j]` having value 0).

While the table encoding of constraints makes interpretation difficult, we did observe for `54.dzn` that the early cores have two variables connected by some binary table, constraining one of them to be 0. This ensures at most one of two photos is taken. Later cores also contain 2 or 3 variables, connecting a new variable to existing reformu-

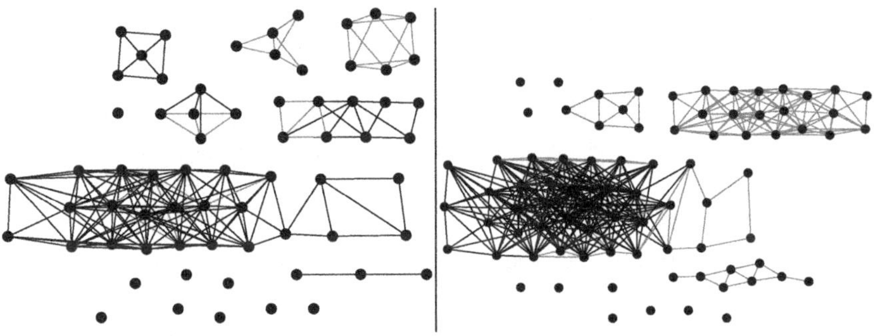

**Fig. 1.** Constraint structure of two `spot5` instances (left|right). Nodes represent variables; edges binary constraints (red if non-zero cost). (Color figure online)

lated costs, or grouping some new variables together. Moreover, the variables in these expanded cores formed cliques connected by non-zero cost tables.

Our first reformulation used a *merging* strategy that, given a set of (initially singleton) clusters, iteratively merged the two with greatest inter-cluster cost (creating a new cost variable). This yielded effective reformulations but was too slow to compute. Visualising the constraint graph of two instances, Fig. 1, we realised that they form almost (but not always) *interval graphs*: those where vertices are real intervals and edges their intersection [14].[2] If they form an interval graph, there must be an ordering where members of any maximal clique appear sequentially. This *interval* ordering would be a good candidate for grouping, as it keeps related vertices nearby. For interval graphs, the ordering can be obtained using lexicographic breadth-first searches (LexBFSs) [7]. We used this procedure, expecting suitable orderings even for constraints that do not form interval graphs. The bounds for group $G$ are computed by a greedy vertex cover of the subgraph of non-zero cost tables containing only leaves of $G$.

### 5.2 Experimental Results

Each of the sub-figures in Fig. 2 shows the results obtained for the problem shown in the caption. For each problem, the results are grouped by the data files used to create the instances, as given in the $x$-axis. The results for each data file are divided into 3 sets of bars separated by spaces. Each set of bars corresponds to the results given by one solver when executing the instances obtained by adding the data file to each of the reformulations named in the caption in the given order. Note the captions always start with the *original* model and a *naïve* grouping of the components in the order they appear in the original objective. The three solvers used are Gecode (set of bars on the left), Chuffed (middle) and the branch-and-bound version of GEAS (right). The values shown per instance are as follows: (a) the solving time as the height of the bar w.r.t. left $y$-axis in logarithmic scale, and with a 300 s timeout; (b) the objective value as a black dot on the bar w.r.t. right $y$-axis, and scaled to the range $[0, 1]$, and (c) the baseline for

[2] Vertices in the model correspond to observations made along the trajectory of a satellite (have an underlying ordering); edges correspond to observations that are close enough to interfere.

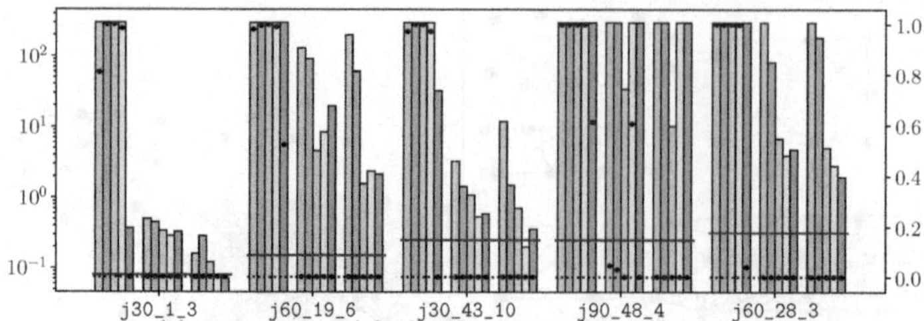


(a) `rcpsp-wet`: *original, naïve, start, direct, direct-bound*

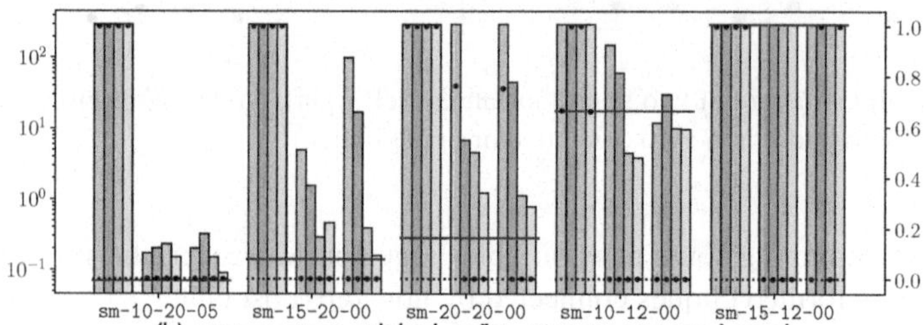


(b) `seat-moving`: *original, naïve, reverse, reverse-bound*

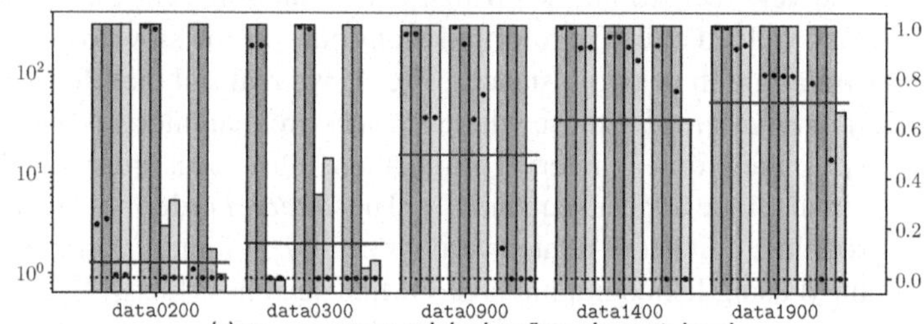


(c) `jp-encoding`: *original, naïve, element, local*

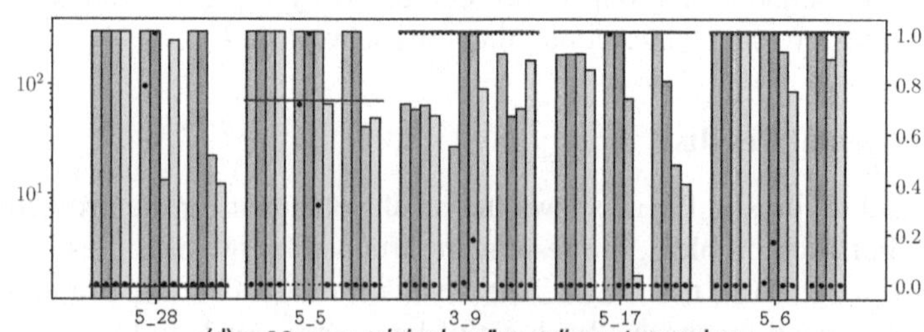


(d) `rel2onto`: *original, naïve, adjacent, merging*

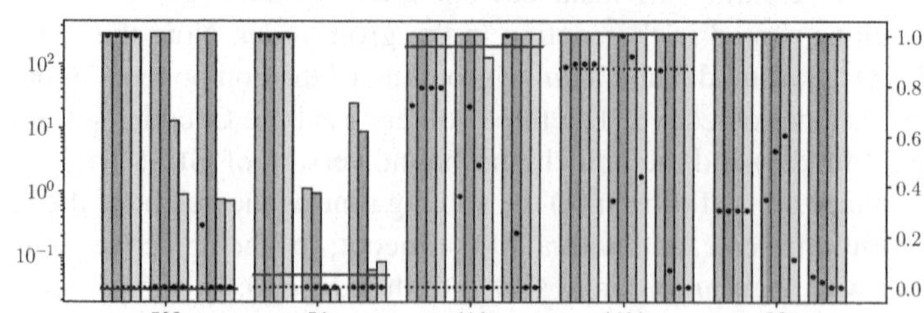


(e) `spot5`: *original, naïve, interval, interval-bound*

**Fig. 2.** Results for our five models

each instance (core-guided GEAS on the *original* model) as the horizontal blue line for time, and the black dotted line for the objective value. Note that we also obtained results for reformulations with random orderings. These results were consistently worse, and are omitted to improve legibility.

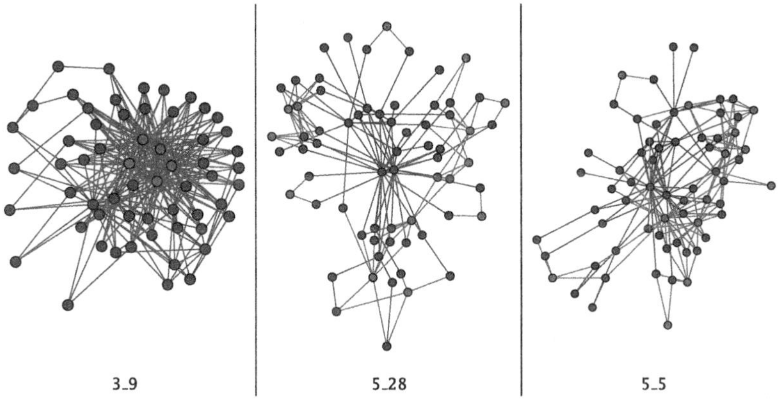


**Fig. 3.** Visualising three `rel2onto` instances: black edges for ontology structure; magenta for possible attribute/concept matchings.

For `rcpsp-wet`, *start*, *direct* and *direct-bound* achieve very good improvements for both LCG solvers (given the logarithmic scale for time), with *direct* able to improve the time of the largest instance, and *direct-bound* often performing best. As expected, Gecode does not benefit from *start* or *direct*, but drastically improves with *direct-bound* on two instances and gets a much better objective bound on the remaining three.

For `seat-moving`, both *reverse* and *reverse-bound* significantly improve the performance of the two LCG solvers for most instances. Instance `sm-10-12-00` is interesting, as the LCS solvers often outperforms the baseline time. The *reverse-bound* reformulation did not improve Gecode (the bound was too weak).

For `jp-encoding` *element* and *local* both yield good performance, with *element* often performing much better for Chuffed, and *local* performing outstandingly for GEAS (often better than the baseline). Interestingly, *naïve* performs badly for Chuffed but, for GEAS, despite never proving optimality, discovers similar or better bounds than *element*. This is because the terms in the implied element constraint are grouped together in the original objective, resulting in similar reformulations.

For `rel2onto`, *adjacent* and *merging* make Chuffed worse for `3_9`, whose structure is quite different from the two instances we used for our cores, as shown in Fig. 3). However, they significantly improve Chuffed and GEAS on most other instances except Interestingly, core-guided GEAS on the original model times out for three instances, but branch-and-bound LCG solvers perform much better with the reformulations.

For `spot5`, *interval* and *interval-bound* perform significantly better than *original* or *naïve* for both LCG solvers, while *interval-bound* did not improve Gecode.

## 6 Conclusions and Future Work

From the above results we conclude the following. First, non-bound core-guided reformulations are often enough to achieve excellent improvements for LCG solvers, as bounds can be learned for the new variables. Second, tight bounds (as for `rcpsp-wet`) can drastically improve CP solvers. Finally, our reformulations of simple models yield great results, but we believe the insights of model owners should enable even better groupings and tighter bounds for all models, as in-depth knowledge is key. Given the scarcity of core-guided (and LCG) solvers, a key contribution is to show modellers the importance of appropriately grouping the objective components and tightly bounding them. In particular, we show that significant speedups can be obtained by simply using our grouping function on orderings of the objective components that are based on "relatedness". While these orders can be tried speculatively without using core-guided optimization, its use can help to quickly identify where to look for "usefully related" terms (i.e., orderings), and for candidates for analytic bounds to add. We also show how parts of the process can be either automated or supported by automation. We are particularly excited by the idea of using an MUS enumeration tool to identify the *reasons* behind the cores. We are following this approach in our future work, where we aim to further automate our method as much as possible.

## References

1. Andres, B., Kaufmann, B., Matheis, O., Schaub, T.: Unsatisfiability-based optimization in clasp. In: Proceedings of ICLP Technical Communications. LIPIcs, vol. 17, pp. 211–221. Schloss Dagstuhl - Leibniz-Zentrum fuer Informatik (2012)
2. Ansótegui, C., Bonet, M.L., Gabàs, J., Levy, J.: Improving SAT-based weighted MaxSAT solvers. In: Milano, M. (ed.) CP 2012. LNCS, pp. 86–101. Springer, Heidelberg (2012). https://doi.org/10.1007/978-3-642-33558-7_9
3. Bensana, E., Lemaître, M., Verfaillie, G.: Earth observation satellite management. Constraints **4**(3), 293–299 (1999). https://doi.org/10.1023/A:1026488509554
4. Berg, J., Järvisalo, M.: Weight-aware core extraction in SAT-based MaxSAT solving. In: Beck, J.C. (ed.) CP 2017. LNCS, vol. 10416, pp. 652–670. Springer, Cham (2017). https://doi.org/10.1007/978-3-319-66158-2_42
5. Brand, S., Narodytska, N., Quimper, C.-G., Stuckey, P., Walsh, T.: Encodings of the SEQUENCE constraint. In: Bessière, C. (ed.) CP 2007. LNCS, vol. 4741, pp. 210–224. Springer, Heidelberg (2007). https://doi.org/10.1007/978-3-540-74970-7_17
6. Chu, G.: Improving combinatorial optimization. Ph.D. thesis, University of Melbourne, Australia (2011). http://hdl.handle.net/11343/36679
7. Corneil, D.G.: Lexicographic breadth first search – a survey. In: Hromkovič, J., Nagl, M., Westfechtel, B. (eds.) WG 2004. LNCS, vol. 3353, pp. 1–19. Springer, Heidelberg (2004). https://doi.org/10.1007/978-3-540-30559-0_1
8. Feydy, T., Stuckey, P.J.: Lazy clause generation reengineered. In: Gent, I.P. (ed.) CP 2009. LNCS, vol. 5732, pp. 352–366. Springer, Heidelberg (2009). https://doi.org/10.1007/978-3-642-04244-7_29
9. Fourer, R., Gay, D.M., Kernighan, B.W.: AMPL: a mathematical programming language. AT & T Bell Laboratories Murray Hill, NJ 07974 (1987)
10. Frisch, A.M., Grum, M., Jefferson, C., Martínez, B., Miguel, H.I.: The design of ESSENCE: a constraint language for specifying combinatorial problems. In: IJCAI-07, pp. 80–87 (2007)

11. Gange, G., Berg, J., Demirović, E., Stuckey, P.J.: Core-guided and core-boosted search for CP. In: Hebrard, E., Musliu, N. (eds.) Proceedings of Seventeenth International Conference on Integration of Artificial Intelligence and Operations Research techniques in Constraint Programming (CPAIOR2020). Springer, Heidelberg (2020, to appear)
12. Gecode Team: Gecode: Generic Constraint Development Environment (2006). http://www.gecode.org
13. Gurobi Optimization Inc.: Gurobi Optimizer Reference Manual Version 7.5. Houston, Texas: Gurobi Optimization (2017)
14. Lekkeikerker, C., Boland, J.: Representation of a finite graph by a set of intervals on the real line. Fundamenta Mathematicae **51**(1), 45–64 (1962)
15. Leo, K., Mears, C., Tack, G., Garcia de la Banda, M.: Globalizing constraint models. In: Schulte, C. (ed.) CP 2013. LNCS, vol. 8124, pp. 432–447. Springer, Heidelberg (2013). https://doi.org/10.1007/978-3-642-40627-0_34
16. Leo, K., Tack, G.: Debugging unsatisfiable constraint models. In: Salvagnin, D., Lombardi, M. (eds.) CPAIOR 2017. LNCS, vol. 10335, pp. 77–93. Springer, Cham (2017). https://doi.org/10.1007/978-3-319-59776-8_7
17. Marques-Silva, J., Argelich, J., Graça, A., Lynce, I.: Boolean lexicographic optimization: algorithms & applications. Ann. Math. Artif. Intell. **62**(3–4), 317–343 (2011)
18. Nethercote, N., Stuckey, P.J., Becket, R., Brand, S., Duck, G.J., Tack, G.: MiniZinc: towards a standard CP modelling language. In: Bessière, C. (ed.) CP 2007. LNCS, vol. 4741, pp. 529–543. Springer, Heidelberg (2007). https://doi.org/10.1007/978-3-540-74970-7_38
19. Ohrimenko, O., Stuckey, P.J., Codish, M.: Propagation = lazy clause generation. In: Bessière, C. (ed.) CP 2007. LNCS, vol. 4741, pp. 544–558. Springer, Heidelberg (2007). https://doi.org/10.1007/978-3-540-74970-7_39
20. Perron, L., Furnon, V.: OR-Tools (2019). https://developers.google.com/optimization/
21. Shishmarev, M., Mears, C., Tack, G., Garcia de la Banda, M.: Visual search tree profiling. Constraints 1–18 (2015)
22. Stuckey, P., Becket, R., Fischer, J.: Philosophy of the MiniZinc challenge. Constraints **15**(3), 307–316 (2010). https://doi.org/10.1007/s10601-010-9093-0
23. Van Hentenryck, P.: The OPL Optimization Programming Language. MIT Press, Cambridge (1999)
24. Zeighami, K., Leo, K., Tack, G., de la Banda, M.G.: Towards semi-automatic learning-based model transformation. In: Hooker, J.N. (ed.) CP 2018. LNCS, vol. 11008, pp. 403–419. Springer, Heidelberg (2018). https://doi.org/10.1007/978-3-319-98334-9_27

# Filtering Rules for Flow Time Minimization in a Parallel Machine Scheduling Problem

Margaux Nattaf[1](✉) and Arnaud Malapert[2]

[1] Institute of Engineering, Univ. Grenoble Alpes, CNRS, Grenoble INP, G-SCOP, 38000 Grenoble, France
margaux.nattaf@grenoble-inp.fr
[2] Université Côte d'Azur, CNRS, I3S, Nice, France
arnaud.malapert@univ-cotedazur.fr

**Abstract.** This paper studies the scheduling of jobs of different families on parallel machines with qualification constraints. Originating from semi-conductor manufacturing, this constraint imposes a time threshold between the execution of two jobs of the same family. Otherwise, the machine becomes disqualified for this family. The goal is to minimize both the flow time and the number of disqualifications. Recently, an efficient constraint programming model has been proposed. However, when priority is given to the flow time objective, the efficiency of the model can be improved.

This paper uses a polynomial-time algorithm which minimize the flow time for a single machine relaxation where disqualifications are not considered. Using this algorithm one can derived filtering rules on different variables of the model. Experimental results are presented showing the effectiveness of these rules. They improve the competitiveness with the mixed integer linear program of the literature.

**Keywords:** Parallel machine scheduling · Job families · Flow time · Machine disqualification · Filtering algorithm · Cost-based filtering

## 1 Introduction

This paper considers the scheduling of job families on parallel machines with time constraints on machine qualifications. In this problem, each job belongs to a family and a family can only be executed on a subset of qualified machines. In addition, machines can lose their qualifications during the schedule. Indeed, if no job of a family is scheduled on a machine during a given amount of time, the machine lose its qualification for this family. The goal is to minimize the sum of job completion times, i.e. the flow time, while maximizing the number of qualifications at the end of the schedule.

This problem, called scheduling Problem with Time Constraints (PTC), is introduced in [11]. It comes from the semiconductor industries. Its goal is to


H. Simonis (Ed.): CP 2020, LNCS 12333, pp. 462–477, 2020.
https://doi.org/10.1007/978-3-030-58475-7_27

introduce constraints coming from Advanced Process Control (APC) into a scheduling problem. APC's systems are used to control processes and equipment to reduce variability and increase equipment efficiency. In PTC, qualification constraints and objective come from APC and more precisely from what is called Run to Run control. More details about the industrial problem can be found in [10].

Several solution methods has been defined for PTC [7,10,11]. In particular, the authors of [7] present two pre-existing models: a Mixed Integer Linear Program (MILP) and a Constraint Programming (CP) model. Furthermore, they define a new CP model taking advantage of advanced CP features to model machine disqualifications. However, the paper shows that when the priority objective is the flow time, the performance of the CP model can be improved.

The objective of this paper is to improve the performances of the CP model for the flow time objective. To do so, a relaxed version of PTC where qualification constraints are removed is considered. For this relaxation, the results of Mason and Anderson [9] are adapted to define an algorithm to optimally sequence jobs on one machine in polynomial time. This algorithm is then used to define several filtering algorithms for PTC.

Although, the main result of this paper concerns the filtering algorithms for PTC, there is also two more general results incident to this work. First, those algorithms can be directly applied to any problem having a flow time objective and which can be relaxed to a parallel machine scheduling problem with sequence-independent family setup times. Secondly, the approach is related to cost-based domain filtering [3], a general approach to define filtering algorithms for optimization problems.

The paper is organized as follows. Section 2 gives a formal description of the problem the CP model for PTC. Section 3 presents the relaxed problem and the optimal machine flow time computation of the relaxed problem. Section 4 shows how this flow time is used to define filtering rules and algorithms for PTC. Finally, Sect. 5 shows the performance of the filtering algorithms and compares our results to the literature.

## 2 Problem Description and Modeling

In this section, a formal description of PTC is given. Then, a part of the CP model of [7] is presented. The part of the model presented is the part that is useful to present our cost based filtering rules and correspond to the modeling of the relaxation. Indeed, as we are interested in the flow time objective, the machine qualification modeling is not presented in this paper.

### 2.1 PTC Description

Formally, the problem takes as input a set of jobs, $\mathcal{N} = \{1, \dots, N\}$, a set of families $\mathcal{F} = \{1, \dots, F\}$ and a set of machines, $\mathcal{M} = \{1, \dots, M\}$. Each job $j$ belongs to a family and the family associated with $j$ is denoted by $f(j)$. For each family $f$, only a subset of the machines $\mathcal{M}_f \subseteq \mathcal{M}$, is able to process a job of $f$. A machine $m$ is said to be qualified to process a family $f$ if $m \in \mathcal{M}_f$. Each family $f$ is associated with the following parameters:

- $n_f$ denotes the number of jobs in the family. Note that $\sum_{f \in \mathcal{F}} n_f = N$.
- $p_f$ corresponds to the processing time of jobs in $f$.
- $s_f$ is the setup time required to switch the production from a job belonging to a family $f' \neq f$ to the execution of a job of $f$. Note that this setup time is independent of $f'$, so it is sequence -independent. In addition, no setup time is required neither between the execution of two jobs of the same family nor at the beginning of the schedule, i.e. at time 0.
- $\gamma_f$ is the threshold value for the time interval between the execution of two jobs of $f$ on the same machine. Note that this time interval is computed on a start-to-start basis, i.e. the threshold is counted from the start of a job of family $f$ to the start of the next job of $f$ on machine $m$. Then, if there is a time interval $]t, t + \gamma_f]$ without any job of $f$ on a machine, the machine lose its qualification for $f$.

The objective is to minimize both the sum of job completion times, i.e. the flow time, and the number of qualification looses or disqualifications. An example of PTC together with two feasible solutions is now presented.

*Example 1.* Consider the instance with $N = 10$, $M = 2$ and $F = 3$ given in Table 1(a). Figure 1 shows two feasible solutions. The first solution, described by Fig. 1(b), is optimal in terms of flow time. For this solution, the flow time is equal to $1 + 2 + 9 + 15 + 21 + 1 + 2 + 12 + 21 + 30 = 114$ and the number of qualification losses is 3. Indeed, machine 1 ($m_1$) loses its qualification for $f_3$ at time 22 since there is no job of $f_3$ starting in interval $]1, 22]$ which is of size $\gamma_3 = 21$. The same goes for $m_2$ and $f_3$ at time 22 and for $m_2$ and $f_2$ at time 26. The second solution, described by Fig. 1(c), is optimal in terms of number of disqualifications. Indeed, in this solution, none of the machines loses their qualifications. However, the flow time is equal to $1 + 2 + 9 + 17 + 19 + 9 + 18 + 20 + 27 + 37 = 159$.

### 2.2 CP Model

In the following section, the part of the CP model of [7] which is useful for this work is recalled. This part corresponds to a Parallel Machine Scheduling Problem (PMSP) with family setup times. New auxiliary variables used by our cost based filtering rules are also introduced. These variables are written in bold in the variable description. To model the PMSP with family setup times, (optional) interval variables are used [5,6]. To each interval variable $J$, a start time $st(J)$, an end time $et(J)$, a duration $d(J)$ and an execution status $x(J)$ is associated. The execution status $x(J)$ is equal to 1 if and only if $J$ is present in the solution and 0 otherwise.

| $f$ | $n_f$ | $p_f$ | $s_f$ | $\gamma_f$ | $\mathcal{M}_f$ |
|---|---|---|---|---|---|
| 1 | 3 | 9 | 1 | 25 | $\{2\}$ |
| 2 | 3 | 6 | 1 | 26 | $\{1,2\}$ |
| 3 | 4 | 1 | 1 | 21 | $\{1,2\}$ |

(a) Instance with $N = 10$, $M = 2$ and $F = 3$

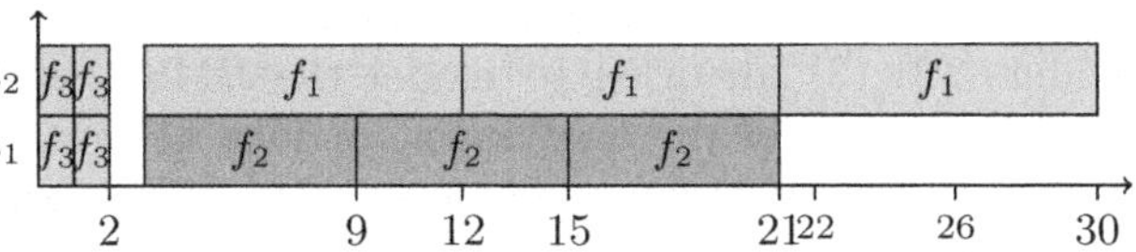


(b) An optimal solution for the flow time objective

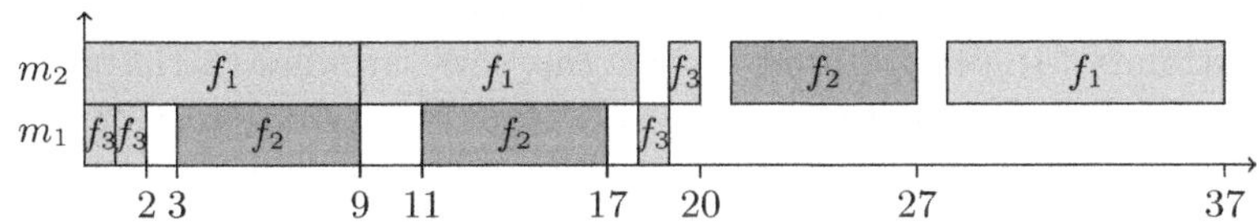


(c) An optimal solution for qualification losses

**Fig. 1.** Two solution examples for PTC.

The following set of variables is used:

- $jobs_j$, $\forall j \in \mathcal{N}$: Interval variable modeling the execution of job $j$;
- $altJ_{j,m}$, $\forall (j,m) \in \mathcal{N} \times \mathcal{M}_{f(j)}$: Optional interval variable modeling the assignment of job $j$ to machine $m$;
- $\mathbf{flowtime_m}$ and $flowtime$: Integer variables modeling respectively the flow time on machine $m$ and the global flow time;
- $\mathbf{nbJobs_{f,m}}, \forall (f,m) \in \mathcal{F} \times \mathcal{M}_f$: Integer variable modeling the number of jobs of family $f$ scheduled on $m$;
- $\mathbf{nbJobs_m}, \forall m \in \mathcal{M}$: Integer variable modeling the number of jobs scheduled on $m$.

To model the PMSP with setup time, the following sets of constraints is used:

$$flowTime = \sum_{j \in \mathcal{N}} et(jobs_j) \quad (1)$$

$$alternative\left(jobs_j, \left\{altJ_{j,m} | m \in \mathcal{M}_{f(j)}\right\}\right) \qquad \forall j \in \mathcal{N} \quad (2)$$

$$noOverlap\left(\left\{altJ_{j,m} | \forall j \text{ s.t. } m \in \mathcal{M}_{f(j)}\right\}, S\right) \qquad \forall m \in \mathcal{M} \quad (3)$$

$$flowtime = \sum_{m \in \mathcal{M}} flowtime_m \quad (4)$$

$$flowtime_m = \sum_{j \in \mathcal{N}} et(altJ_{j,m}) \qquad \forall m \in \mathcal{M} \quad (5)$$

$$nbJobs_{f,m} = \sum_{j \in \mathcal{N}; f(j)=f} x(altJ_{j,m}) \qquad \forall (f,m) \in \mathcal{F} \times \mathcal{M}_f \quad (6)$$

$$\sum_{m \in \mathcal{M}} nbJobs_{f,m} = n_f \qquad \forall f \in \mathcal{F} \quad (7)$$

$$\sum_{f \in \mathcal{F}} nbJobs_{f,m} = nbJobs_m \qquad \forall m \in \mathcal{M} \quad (8)$$

$$\sum_{m \in \mathcal{M}} nbJobs_m = N \quad (9)$$

Constraint (1) is used to compute the flow time of the schedule. Constraints (2)–(3) are used to model the PMSP with family setup time. Constraints (2) model the assignment of jobs to machine. Constraints (3) ensure that jobs do not overlap and enforce setup times. Note that $S$ denotes the setup time matrix: $(S_{f,f'})$ is equal to 0 if $f = f'$ and to $s_f$ otherwise. A complete description of *alternative* and *noOverlap* constraints can be found in [5,6].

In [7], additional constraints are used to make the model stronger, e.g. ordering constraints, cumulative relaxation. They are not presented in this paper.

Constraints (4)–(9) are used to link the new variables to the model. Constraints (4) and (5) ensure machine flow time computation. Constraints (6) compute the number of jobs of family $f$ executed on machine $m$. Constraints (7) make sure the right number of jobs of family $f$ is executed. Constraints (8) and (9) are equivalent to constraints (6) and (7) but for the total number of jobs scheduled on machine $m$. The bi-objective optimization is a lexicographical optimization or its linearization [2].

# 3 Relaxation Description and Sequencing

## 3.1 $\mathcal{R}$-PTC Description

The relaxation of PTC ($\mathcal{R}$-PTC) is a parallel machine scheduling problem with sequence-independent family setup time without the qualification constraints (parameter $\gamma_f$). The objective is then to minimize the flow time. In this section, it is assumed that a total the assignment of jobs to machines is already done and the objective is to sequence jobs so the flow time is minimal. Therefore, since the sequencing of jobs on $M$ machines can be seen as $M$ one machine problems, this section presents how jobs can be sequenced optimally on one machine. In Sect. 4, the cost-based filtering rules handle partial assignments of jobs to the machines.

## 3.2 Optimal Sequencing for $\mathcal{R}$-PTC

The results presented in this section were first described in [8]. They are adapted from Mason and Anderson [9] who considers an initial setup at the beginning of the schedule. The results are just summarized in this paper.

First, a solution can be represented as a sequence $S$ representing an ordered set of $n$ jobs. Considering job families instead of individual jobs, $S$ can be seen as a series of blocks, where a block is a maximal consecutive sub-sequence of jobs in $S$ from the same family (see Fig. 2). Let $B_i$ be the $i$-th block of the sequence, $S = \{B_1, B_2, \ldots, B_r\}$. Hence, successive blocks contain jobs from different families. Therefore, there will be a setup time before each block (except the first one).

The idea of the algorithm is to adapt the Shortest Processing Time ($SPT$) rule [12] for blocks instead of individual jobs. To this end, blocks are considered as individual jobs with processing time $P_i = s_{f_i} + |B_i| \cdot p_{f_i}$ and weight $W_i = |B_i|$ where $f_i$ denotes the family of jobs in $B_i$ (which is the same for all jobs in $B_i$).

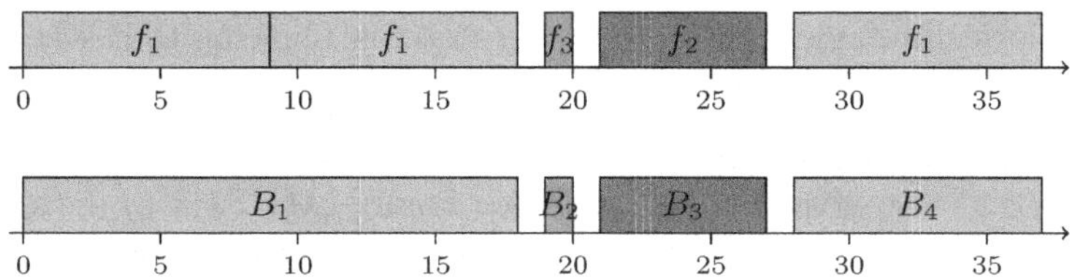


**Fig. 2.** Block representation of a solution.

The first theorem of this section states that there always exists an optimal solution $S$ containing exactly $|\mathcal{F}|$ blocks and that each block $B_i$ contains all jobs of the family $f_i$. That is, all jobs of a family are scheduled consecutively.

**Theorem 1.** *Let $\mathcal{I}$ be an instance of the problem. There exists an optimal solution $S^* = \{B_1, \ldots, B_{|\mathcal{F}|}\}$ such that $|B_i| = n_{f_i}$ where $f_i$ is the family of jobs in $B_i$.*

*Proof (Sketch).* For a complete proof of the theorem, see [8].
Consider an optimal solution $S = \{B_1, \ldots, B_u, \ldots, B_v, \ldots, B_r\}$ with two blocks $B_u$ and $B_v$ $(u < v)$, containing jobs of the same family $f_u = f_v = f$. Then, moving the first job of $B_v$ at the end of block $B_u$ can only improve the solution.

Indeed, let us define $P$ and $W$ as: $P = \sum_{i=u+1}^{v-1} P_i + s_f$ and $W = \sum_{i=u+1}^{v-1} |B_i|$. In addition, let $S'$ be the sequence formed by moving the first job of $B_v$, say job $j_v$, at the end of block $B_u$. The difference on the flow time between $S$ and $S'$, is as follows:

$$FT_{S'} - FT_S = \begin{cases} W \cdot p_f - P & \text{if } |B_v| = 1 \\ W \cdot p_f - P - \sum_{i=v+1}^{r} |B_i| \cdot s_f & \text{if } |B_v| > 1 \end{cases}$$

Using Lemma 1 of [8] stating that $P/W \geq p_f$, then $FT_{S'} - FT_S < 0$ and the flow time is improved in $S'$. Hence, moving the first job of $B_v$ at the end of block $B_u$ leads to a solution $S'$ at least as good as $S$.

Therefore, a block $B_i$ contains all jobs of family $f_i$. Indeed, if not, applying the previous operation leads to a better solution. Hence, $|B_i| = n_{f_i}$ and there are exactly $F$ blocks in the optimal solution, i.e. one block per family.

At this point, the number of block and theirs contents are defined. The next step is to order them. To this end, the concept of weighted processing time is also adapted to blocks as follows.

**Definition 1.** *The Mean Processing Time (MPT) of a block $B_i$ is defined as $MPT(B_i) = P_i/W_i$.*

One may think that, in an optimal solution, jobs are ordered by $SMPT$ (Shortest Mean Processing Time). However, this is not always true since no setup time is required at time 0. Indeed, the definition of block processing time always considers a setup time before the block. In our case, this is not true for the first block. Example 2 gives a counter-example showing that scheduling all blocks according to the SMPT rule is not optimal.

*Example 2 (Counter-example – Fig. 3).* Consider the instance given by Table 1(a) that also gives the MPT of each family ($MPT_f = p_f + (s_f \div n_f)$). Figure 3(b) shows the SMPT rule may lead to sub-optimal solutions when no setup time is required at time 0. Indeed, following the SMPT rule, jobs of family 1 have to be scheduled before jobs of family 2. This leads to a flow time of 198. However, schedule jobs of family 2 before jobs of family 1 leads to a better flow time, i.e. 181.

| $f$ | $n_f$ | $p_f$ | $s_f$ | $MPT_f$ |
|---|---|---|---|---|
| 1 | 2 | 11 | 2 | 12 |
| 2 | 3 | 12 | 9 | 15 |

(a) Instance with $N = 5$ and $F = 2$.

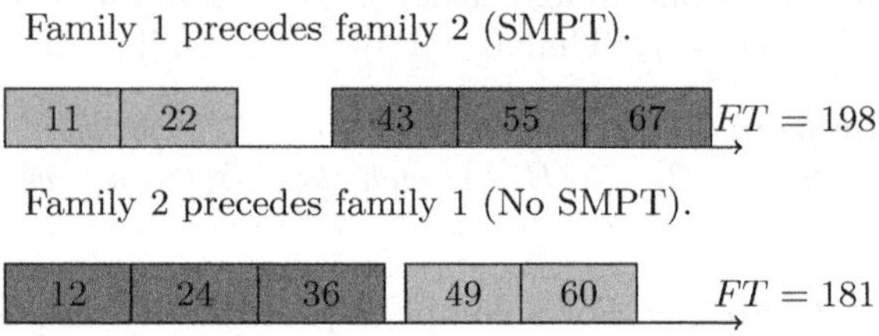


(b) Scheduling jobs of a family before those of the other one. Numbers are completion times of the jobs.

**Fig. 3.** Scheduling all blocks according to the SMPT rule is not optimal.

Actually, the only reason why the *SMPT* rule does not lead to an optimal solution is that no setup time is required at time 0. Therefore, in an optimal solution, all blocks except the first one are scheduled according to the *SMPT* rule. That is the statement of Theorem 2 for which a proof is given in [8].

**Theorem 2.** *In an optimal sequence of the problem, the blocks* 2 *to* $|\mathcal{F}|$ *are ordered by SMPT (Shortest Mean Processing Time). That is, if* $1 < i < j$ *then* $MPT(B_i) \leq MPT(B_j)$.

The remaining of this section explains how these results are used to define a polynomial time algorithm for sequencing jobs on a machine so the flow time is minimized. This algorithm is called `sequencing` in the remaining of the paper.

Theorem 1 states that there exists an optimal solution $S$ containing exactly $|\mathcal{F}|$ blocks and that each block $B_i$ contains all jobs of family $f_i$. Theorem 2 states that the blocks $B_2$ to $B_{|\mathcal{F}|}$ are ordered by *SMPT*. Finally, one only needs to determine which family is processed in the very first block.

Algorithm `sequencing` takes as input the set of jobs and starts by grouping them in blocks and sorting them in SMPT order. The algorithm then computes the flow time of this schedule. Each block is then successively moved to the first position (see Fig. 4) and the new flow time is computed. The solution returned by the algorithm is therefore the one achieving the best flow time.

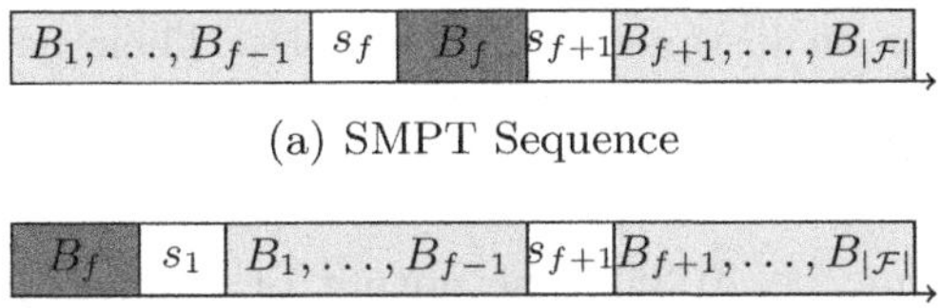


(a) SMPT Sequence

(b) $B_f$ is moved in the first position.

**Fig. 4.** SMPT sequence and move operation.

The complexity for ordering the families in $SMPT$ order is $O(F \log F)$. The complexity of moving each block to the first position and computing the corresponding flow time is $O(F)$. Indeed, there is no need to re-compute the entire flow time. The difference of flow time coming from the *Move operation* can be computed in $O(1)$. Hence, the complexity of `sequencing` is $O(F \log F)$.

## 4 Filtering Rules and Algorithms

This section is dedicated to the cost-based filtering rules and algorithms derived from the results of Sect. 3. They are separated into three parts, each one corresponding to the filtering of one variable: $flowtime_m$ (Sect. 4.1); $nbJobs_{f,m}$ (Sect. 4.2); $nbJobs_m$ (Sect. 4.3). Note that the two last variables constrained by the sum constraint 8.

During the solving of the problem, jobs are divided for each machine $m$ into three categories based on the interval variables $altJ_{j,m}$: each job either is, or can be, or is not, assigned to the machine. Note that the time windows of the interval variables are not considered in the relaxation. For a machine $m$, let $\mathcal{N}_m^A$ be the set of jobs for which the assignment to machine $m$ is decided. In the following of this section, an instance $\mathcal{I}$ is always considered with the set $\mathcal{N}^A = \cup_{m \in \mathcal{M}} \mathcal{N}_m^A$ of jobs already assigned to a machine. Thus, an instance is denoted by $(\mathcal{I}, \mathcal{N}^A)$.

Some notations are now introduced. For a variable $x$, $\underline{x}$ (resp. $\overline{x}$) represents the lower (resp. the upper) bound on the variable $x$. Furthermore, let $FT^*(\mathcal{X})$ be the flow time of the solution returned by the algorithm `sequencing` applied on the set of jobs $\mathcal{X}$.

### 4.1 Increasing the Machine Flow Time

The first rule updates the lower bound on the $flowtime_m$ variable and follows directly from Sect. 3. The complexity of this rule is $O(M \cdot F \cdot \log F)$.

**Rule 1.** $\forall m \in \mathcal{M},\ flowtime_m \geq FT^*(\mathcal{N}_m^A)$

*Proof.* It is sufficient to notice that, for a machine $m$, `sequencing` gives a lower bound on the flow time. In particular, $FT^*(\mathcal{N}_m^A)$ is a lower bound on the flow time of $m$ for the instance $(\mathcal{I}, \mathcal{N}^A)$.

*Example 3.* Consider an instance with 3 families. Their parameters are given by Table 1(a). A specific machine $m$ is considered and set $\mathcal{N}_m^A$ is composed of one job of each family. This instance and $\mathcal{N}_m^A$ is used in all the example of this section.

Suppose $flowtime_m \in [0, 35]$. The output of `sequencing` is given on the top of Fig. 5(b). Thus, the lower bound on the flow time can be updated to $2 + 7 + 13 = 22$, i.e. $flowtime_m \in [22, 35]$.

Suppose now that an extra job of family $f_2$ is added to $\mathcal{N}_m^A$ (on the bottom of Fig. 5(b)). Thus, $FT^*(\mathcal{N}_m^A) = 2+8+11+16 = 37$ and $37 > \overline{flowtime}_m = 35$. Thus, the assignment defined by $\mathcal{N}_m^A$ is infeasible.

| $f$ | $n_f$ | $p_f$ | $s_f$ |
|---|---|---|---|
| 1 | 3 | 2 | 5 |
| 2 | 3 | 3 | 3 |
| 3 | 4 | 4 | 1 |

(a) Instance data.

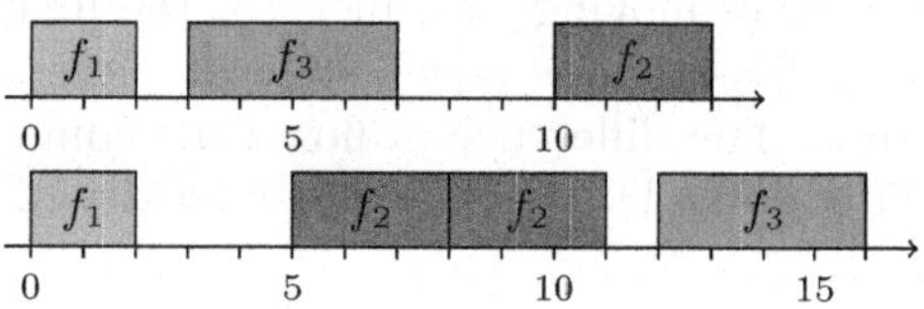


(b) Illustration of Rule 1 and Rule 3.

**Fig. 5.** Illustration of $flowtime_m$ filtering (Rule 1).

Another rule can be defined to filter $flowtime_m$. This rule is stronger than Rule 1 and is based on $\mathcal{N}_m^A$ and $\underline{nbJobs}_m$. Indeed, $\underline{nbJobs}_m$ denotes the minimum number of jobs that has to be assigned to $m$. Thus, if one can find the $\underline{nbJobs}_m$ jobs (including the one of $\mathcal{N}_m^A$) that leads to the minimum flow time, it will give a lower bound on the flow time of machine $m$.

Actually, it may be difficult to know exactly which jobs will contribute the least to the flow time. However, considering jobs in $SPT$ order and with 0 setup time gives a valid lower bound on $flowtime_m$. First, an example illustrating the filtering rule is presented and then, the rule is formally given.

*Example 4.* Consider the instance described in Example 3. Suppose $\mathcal{N}_m^A$ is composed of one job of each family and $flowtime_m \in [22, 60]$. Suppose also $\underline{nbJobs}_m = 6$. Thus, 3 extra jobs need to be assigned to $m$.

Families in $SPT$ order are $\{f_1, f_2, f_3\}$ and the remaining number of jobs in each family is 2, 2, 3. Hence, the 3 extra jobs are: 2 jobs of $f_1$ and 1 job of $f_2$.

To make sure the lower bound on the flow time is valid, those jobs are sequenced on $m$ with no setup time. In Fig. 6, $f_j$ denotes "classical" jobs of family $f_j$ while $f_j'$ denotes jobs of family $f_j$ with no setup time.

Figure 6 shows the results of `sequencing` on the set of jobs composed of $\mathcal{N}_m^A$ plus the 3 extra jobs, i.e. $\underline{nbJobs}_m = 6$. Here, $FT^* = 2 + 4 + 6 + 9 + 14 + 20 = 55$. Thus, the lower bound on $flowtime_m$ can be updated and $flowtime_m \in [55, 60]$.

Note that, because Rule 1 does not take $\underline{nbJobs}_m$ into account, it gives a lower bound of 22 in this case.

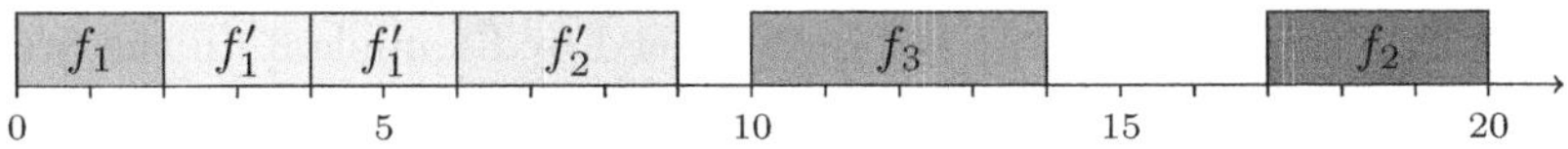


**Fig. 6.** Illustration of $flowtime_m$ filtering (Rule 2).

Let $\mathcal{N}_m^O$ denotes the set composed of the first $\underline{nbJobs}_m - |\mathcal{N}_m^A|$ remaining jobs in $SPT$ order with setup time equal to 0.

**Rule 2.** $\forall m \in \mathcal{M},\ flowtime_m \geq FT^*(\mathcal{N}_m^A \cup \mathcal{N}_m^O)$

*Proof.* First note that if $\mathcal{N}_m^O = \emptyset$, Rule 1 gives the result. Thus, suppose that $|\mathcal{N}_m^O| \geq 1$. By contradiction, suppose $\exists m \in \mathcal{M}$ such that $flowtime_m < FT^*(\mathcal{N}_m^A \cup \mathcal{N}_m^O)$. Thus, there exists another set of jobs $\overline{N_m^O}$ with:

$$FT^*(\mathcal{N}_m^A \cup \overline{N_m^O}) < FT^*(\mathcal{N}_m^A \cup \mathcal{N}_m^O) \tag{10}$$

First, note that w.l.o.g. $|\overline{N_m^O}| = |\mathcal{N}_m^O|$. Indeed, if $|\overline{N_m^O}| > |\mathcal{N}_m^O|$, we can remove jobs from $|\overline{N_m^O}|$ without increasing the flow time. Furthermore, w.l.o.g. we can consider that $\forall j \in \overline{N_m^O},\ s_{f_j} = 0$. Indeed, since setup times can only increase the flow time, thus inequality (10) is still verified.

Let $\overline{S} = \{\overline{j}_1, \cdots, \overline{j}_{nbJobs_m}\}$ be the sequence returned by the `sequencing` algorithm on $\mathcal{N}_m^A \cup \overline{N_m^O}$. Let also $\overline{j}_i$ be the job in $\overline{N_m^O} \backslash \mathcal{N}_m^O$ with the $SPT$. Finally, let $j_k$ be the job with the $SPT$ in $\in \mathcal{N}_m^O \setminus \overline{N_m^O}$. Thus, since $p_{f_{\overline{j}_i}} > p_{f_{j_k}}$, sequence $\overline{S}' = \{\overline{j}_1, \cdots, \overline{j}_{i-1}, j_k, \overline{j}_{i+1}, \cdots \overline{j}_{nbJobs_m}\}$ has a smaller flow time than $\overline{S}$.

Repeated applications of this operation yield to a contradiction with Eq. (10).

The complexity of Rule 2 is $O(M \cdot F \cdot \log F)$. Indeed, sorting families in $SPT$ order can be done in $O(F \cdot logF)$. Creating the set $\mathcal{N}_m^O$ is done in $O(F)$ and `sequencing` is applied in $O(F \cdot logF)$ which gives a total complexity of $O(F \cdot logF + M \cdot (F + F \cdot \log F))$.

### 4.2 Reducing the Maximum Number of Jobs of a Family

The idea of the filtering rule presented in this section is as follows. For a family $f$, $\overline{nbJobs}_{f,m}$ define the maximum number of jobs of family $f$ that can be scheduled on $m$. Thus, if adding those $\overline{nbJobs}_{f,m}$ to $\mathcal{N}_m^A$ leads to an infeasibility, $\overline{nbJobs}_{f,m}$ can be decreased by at least 1. Let denote by $\mathcal{N}_m^f$ the set composed of $\overline{nbJobs}_{f,m}$ jobs of family $f$ minus those already present in $\mathcal{N}_m^A$.

**Rule 3.** *If* $\exists (f, m) \in \mathcal{F} \times \mathcal{M}$ *such that* $FT^*(\mathcal{N}_m^A \cup \mathcal{N}_m^f) > \overline{flowtime}_m$, *then* $nbJobs_{f,m} \leq \overline{nbJobs}_{f,m} - 1$

*Proof.* Suppose that for a family $f$ and a machine $m$, we have a valid assignment such that $FT^*(\mathcal{N}_m^A \cup \mathcal{N}_m^f) > \overline{flowtime}_m$ and $nbJobs_{f,m} = \overline{nbJobs}_{f,m}$. By Rule 1, the assignment is infeasible which is a contradiction.

*Example 5.* Let us consider the instance defined by Example 3. In the first part of this example, $\mathcal{N}_m^A$ is composed of one job of each family and $flowtime_m \in [22, 35]$. Suppose that $\overline{nbJobs_{f_2,m}} = 2$. Thus, $\mathcal{N}_m^A \cup \mathcal{N}_m^f$ contains one job of family $f_1$ and $f_3$ and two jobs of family $f_2$. The bottom part of Fig. 5(b) shows that $FT^*(\mathcal{N}_m^A \cup \mathcal{N}_m^f) = 37 > \overline{flowtime_m} = 35$. Thus, $\overline{nbJobs_{f_2,m}} < 2$.

The complexity of Rule 3 is $O(M \cdot F \cdot \log F + M \cdot F^2 \cdot \log |\mathcal{D}_{nbJobs_{f,m}}^{MAX}|)$ with $|\mathcal{D}_{nbJobs_{f,m}}^{MAX}|$ the maximum size of the domain of variables $nbJobs_{f,m}$. Indeed, the first part corresponds to the complexity for applying the `sequencing` algorithm on all machine. This algorithm only need to be applied once since each time we remove jobs from $\mathcal{N}_m^f$, $FT^*$ can be updated in $O(F)$. Indeed, only the position of the family in the sequence has to be updated. Thus, the second part corresponds to the updating of $FT^*$ for each family and each machine. By proceeding by dichotomy, this update has to be done at most $\log |\mathcal{D}_{nbJobs_{f,m}}^{MAX}|$ times. Thus, the complexity of Rule 3 is $O(M \cdot F \cdot (\log F + F \cdot \log |\mathcal{D}_{nbJobs_{f,m}}^{MAX}|))$.

### 4.3 Reducing the Maximum Number of Jobs on a Machine

The idea behind Rule 2 can be used to reduce the maximum number of jobs on machine $m$. Indeed, for a machine $m$, $\overline{nbJobs_m}$ is the maximum number of jobs that can be scheduled to $m$. Thus, if it is not possible to schedule $\overline{nbJobs_m}$ on $m$ without exceeding the flow time, then $\overline{nbJobs_m}$ can be decreased.

The extra jobs that will be assigned on $m$ must be decided. Note that those jobs must give a lower bound on the flow time for $\overline{nbJobs_m}$ jobs with the pre-assignment defined by $\mathcal{N}_m^A$. Thus, jobs can be considered in $SPT$ order with no setup time. Before giving the exact filtering rule, an example is described.

*Example 6.* Consider the instance described in Example 3. In the first part of this example, $\mathcal{N}_m^A$ is composed of one job of each family and $flowtime_m \in [22, 60]$. Suppose $\overline{nbJobs_m} = 7$. Thus, the 4 extra jobs assigned to $m$ are: 2 jobs of $f_1$ and 2 jobs of $f_2$. Figure 7 shows the results of `sequencing` on the set of jobs composed of $\mathcal{N}_m^A$ plus the 4 extra jobs. Here, $FT^* = 2 + 4 + 6 + 11 + 14 + 17 + 23 = 77$ which is greater than $\overline{flowtime_m} = 60$. Thus, $\overline{nbJobs_m}$ cannot be equal to 7 and can be filtered.

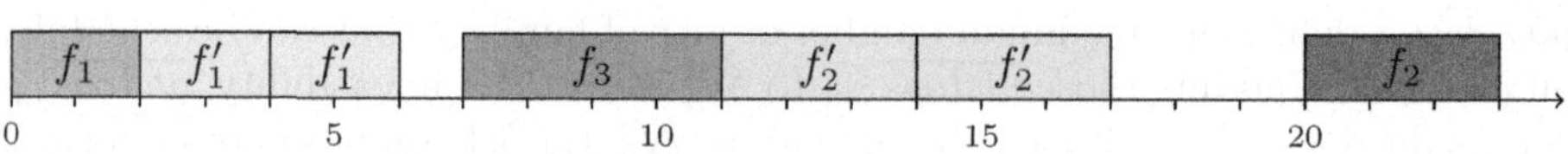


**Fig. 7.** Illustration of Rule 4.

Let $\mathcal{N}_m^{NS}$ denotes the set composed of the first $\overline{nbJobs_m} - |\mathcal{N}_m^A|$ jobs in $SPT$ order with setup time equal to 0.

**Rule 4.** *If* $\exists m \in \mathcal{M}$ *such that* $FT^*(\mathcal{N}_m^A \cup \mathcal{N}_m^{NS}) > \overline{flowtime_m}$*, then* $nbJobs_m \leq \overline{nbJobs_m} - 1$

*Proof.* The arguments are similar to those for Rule 2.

The complexity of Rule 4 is $O(M \cdot F \cdot \log F + M \cdot F^2 \cdot |\mathcal{D}_{nbJobs_m}^{MAX}|)$ with $|\mathcal{D}_{nbJobs_m}^{MAX}|$ the maximum size of the domain of variables $nbJobs_m$. Indeed, as for Rule 3, the algorithm `sequencing` only needs to be applied once and then can be updated in $O(F)$. For each machine and each family, this update has to be done at most $|\mathcal{D}_{nbJobs_m}^{MAX}|$ times. Indeed, proceeding by dichotomy here implies that $FT^*$ cannot be updated but has to be re-computed each time. Thus, the complexity of Rule 4 is $O(M \cdot F \cdot (\log F + F \cdot |\mathcal{D}_{nbJobs_m}^{MAX}|))$.

## 5 Experimental Results

This section starts with the presentation of the general framework of the experiments in Sect. 5.1. Following the framework, the filtering rules are evaluated in Sect. 5.2. Then, the model is compared to those of the literature in Sect. 5.3. Last, a brief sensitivity analysis is given in Sect. 5.4.

### 5.1 Framework

The experiment framework is defined so the following questions are addressed.

*Q1.* Which filtering rule is efficient? Are filtering rules complementary?
*Q2.* Which model of the literature is the most efficient?
*Q3.* What is the impact of the heuristics? Of the bi-objective aggregation?

To address these questions, the following metrics are analyzed: number of feasible solutions, number of proven optimum, upper bound quality; solving times; number of fails (for CP only).

The benchmark instances used to perform our experiments are extracted from [10]. In this paper, 19 instance sets are generated with different number of jobs ($N$), machines ($M$), family ($F$) and qualification schemes. Each of the instance sets is a group of 30 instances. There is a total of 570 feasible instances with $N = 20$ (180), $N = 30$ (180), $N = 40$ (30), $N = 50$ (30), $N = 60$ (60), $N = 70$ (90).

The naming scheme for the different solving algorithms is described in Table 1. The first letter represents the model where ILP model, $\text{CP}_O$ model and $\text{CP}_N$ model denotes respectively the ILP model and CP model of [7], and the CP model detailed in Sect. 2.2. The models are implemented using IBM ILOG CPLEX Optimization Studio 12.10 [4]. That is CPLEX for the ILP model and CP Optimizer for CP models. The second letter indicates whether two heuristics are executed to find solutions which are used as a basis for the models. These heuristics are called *Scheduling Centric Heuristic* and *Qualification Centric Heuristic* [10]. The goal of the first heuristic is to minimize the flow time while the second one tries to minimize the number of disqualifications. The third letter indicates the filtering rules that are activated for the $CP_N$ model. Rule 2 is used for the letter L because it has been shown more efficient than Rule 1 in preliminary experiments. The fourth letter indicates the bi-objective aggregation method: lexicographic optimization; linearization of lexicographic optimization. The last letter indicates the objective priority. Here, the priority is given to the flow time in all experiments because the cost-based filtering rules concern the flow time objective.

**Table 1.** Algorithms encoding.

| Model | | Heuristic | | Filtering rule | | | | Bi-objective | | Priority | |
|---|---|---|---|---|---|---|---|---|---|---|---|
| I | ILP model | _ | None | L | Rule 2 | _ | None | S | Weighted sum | F | Flow time |
| O | $CP_O$ model | H | All | F | Rule 3 | A | All | L | Lexicographic | Q | Disqualifications |
| N | $CP_N$ model | | | M | Rule 4 | | | | | | |

All the experiments were led on a Dell computer with 256 GB of RAM and 4 Intel E7-4870 2.40 GHz processors running CentOS Linux release 7 (each processor has 10 cores). The time limit for each run is 300 s.

### 5.2 Evaluation of the Filtering Rules

In this section, the efficiency and complementary nature of the proposed filtering rules are investigated. In other words, the algorithms N_*LF are studied. To this end, the heuristics are not used to initialize the solvers. The lexicographic optimization is used since it has been shown more efficient in preliminary experiments.

First, all algorithms find feasible solutions for more than 99% of the instances. Then, for each algorithm, the number of instances solved optimally is drawn as a function of the time in Fig. 8(a). The leftmost curve is the fastest whereas the topmost curve proves the more optima. Clearly, compared to N__LF, the filtering rules accelerates the proof and allow the optimal solving of around eighty more instances. One can notice that the advanced filtering rules (N_FLF, N_MLF, N_ALF), also slightly improves the proof compared to the simple update of the flow time lower bound (N_LLF). Here, the advanced filtering rules are indistinguishable.

Table 1(b) ranks the filtering rules according to a scoring procedure based on the Borda count voting system [1]. In this procedure, each benchmark instance is treated like a voter who ranks the algorithms. Each algorithm scores points based on their fractional ranking: the algorithms compare equal receive the same ranking number, which is the mean of what they would have under ordinal rankings. Here, the rank only depends on the answer: the solution status (optimum, satisfiable, and unknown); and then the objective value. So, the lower is the score, the better is the algorithm. Once again, advanced rules are really close, and slightly above the simple lower bound update. They clearly outperform the algorithm without cost-based filtering.

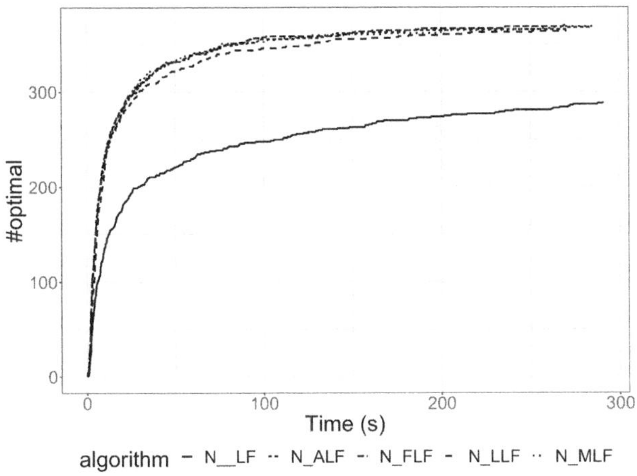


(a) Cumulative number of optima.

| Algorithm | Score |
|---|---|
| N_MLF | 1612 |
| N_ALF | 1644 |
| N_FLF | 1650 |
| N_LLF | 1712 |
| N__LF | 1932 |

(b) Borda ranking.

**Fig. 8.** Evaluation of filtering rules.

### 5.3 Comparison to the Literature

In this section, the best possible algorithms using the ILP model, $CP_O$ model, and $CP_N$ model are compared. Here, the heuristics are used to initialize the solvers, and thus the algorithms IH_SF, OH_LF, NHALF are studied.

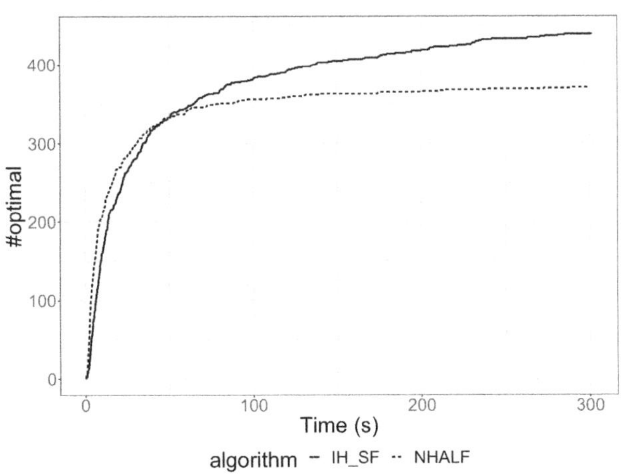


(a) Cumulative number of optima.

| Algorithm | Score |
|---|---|
| IH_SF | 942.5 |
| NHALF | 951.5 |
| OH_LF | 1526.0 |

(b) Borda ranking.

**Fig. 9.** Comparison of the models.

First, the models exhibit very different behaviors in terms of satisfiability and optimality. Then, for each model, the number of instances solved optimally is drawn as a function of the time in Fig. 9(a). The $CP_O$ model is not visible because it does not prove any optimum. The $CP_N$ model is faster than the ILP model, but proves less optimum (around seventy).

Table 1(b) ranks the models rules according to the Borda count voting system (see Sect. 5.2). It confirms the poor efficiency of the $CP_O$ model, but brings closer the performance of $CP_N$ model and ILP model. Table 1(a) explains these close scores. ILP model proves $67 = 70 - 3$ more optima than $CP_N$ model, but $CP_N$ model finds feasible solutions for 32 more instances.

To conclude, the $CP_N$ model is competitive with the ILP model and they offer orthogonal performance since the ILP model is more efficient for proving the optimality whereas the $CP_N$ model is more efficient for quickly finding feasible solutions.

### 5.4 Sensitivity Analysis

Here, the impact of the heuristics and of the bi-objective optimization on the efficiency of the models is analyzed. The $CP_O$ model is excluded since it is clearly dominated by the two other models.

The heuristics have a little impact on the $CP_N$ model (see Table 1(a)). The solution status stay identical 98.5% of the time and the solving times remain in the same order of magnitude. However, Table 1(b) shows the significant impact of the heuristics on the ILP model where the answer for 50 instances becomes satisfiable. For both models, the heuristics do not help to prove more optima.

Table 1(c) shows the significant impact of using the lexicographic optimization instead of the weighted sum method on the $CP_O$ model. Indeed, 69 instances with unknown status when using the weighted sum method become satisfiable using the lexicographic optimization. Note that the lexicographic optimization is not available for the ILP model.

| NHALF | IH_SF | | |
|---|---|---|---|
| | OPT | SAT | UNK |
| OPT | 369 | 3 | 0 |
| SAT | 70 | 94 | 32 |
| UNK | 1 | 0 | 1 |

(a) Two best algorithms.

| I__SF | IH_SF | | |
|---|---|---|---|
| | OPT | SAT | UNK |
| OPT | 433 | 3 | 0 |
| SAT | 6 | 44 | 0 |
| UNK | 1 | 50 | 33 |

(b) Heuristics impact.

| N_ALF | N_ASF | | |
|---|---|---|---|
| | OPT | SAT | UNK |
| OPT | 360 | 8 | 1 |
| SAT | 4 | 125 | 69 |
| UNK | 0 | 0 | 3 |

(c) Aggregation impact.

**Fig. 10.** Contingency table of the solution status between pairs of algorithms.

## 6 Conclusion

In this paper, cost-based domain filtering has been successfully applied to an efficient constraint programming model for scheduling problems with setup on parallel machines. The filtering rules derive from a polynomial-time algorithm which minimize the flow time for a single machine relaxation. Experimental results have shown the rules efficiency and the competitiveness of the overall algorithm. The first perspective is to tackle larger industrial instances since CP relies on its ability at finding feasible solutions. The second perspective is to pay more attention to the propagation of the flow time based on what has been done for the makespan.

## References


1. Brams, S.J., Fishburn, P.C.: Voting procedures. In: Arrow, K.J., Sen, A.K., Suzumura, K. (eds.) Handbook of Social Choice and Welfare, volume 1 of Handbook of Social Choice and Welfare, Chapter 4, pp. 173–236. Elsevier (2002). https://ideas.repec.org/h/eee/socchp/1-04.html
2. Ehrgott, M.: Multicriteria Optimization. Springer, Heidelberg (2000)
3. Focacci, F., Lodi, A., Milano, M.: Cost-based domain filtering. In: Jaffar, J. (ed.) Principles and Practice of Constraint Programming - CP 1999, pp. 189–203. Springer, Heidelberg (1999). ISBN 978-3-540-48085-3
4. IBM. IBM ILOG CPLEX Optimization Studio (2020). https://www.ibm.com/products/ilog-cplex-optimization-studio
5. Laborie, P., Rogerie, J.: Reasoning with conditional time-intervals. In: Proceedings of the Twenty-First International Florida Artificial Intelligence Research Society Conference, 15–17 May 2008, Coconut Grove, Florida, USA, pp. 555–560 (2008). http://www.aaai.org/Library/FLAIRS/2008/flairs08-126.php
6. Laborie, P., Rogerie, J., Shaw, P., Vilím, P.: Reasoning with conditional time-intervals. Part II: an algebraical model for resources. In: Proceedings of the Twenty-Second International Florida Artificial Intelligence Research Society Conference, 19–21 May 2009, Sanibel Island, Florida, USA (2009). http://aaai.org/ocs/index.php/FLAIRS/2009/paper/view/60
7. Malapert, A., Nattaf, M.: A new cp-approach for a parallel machine scheduling problem with time constraints on machine qualifications. In: Rousseau, L.-M., Stergiou, K. (eds.) Integration of Constraint Programming, Artificial Intelligence, and Operations Research, pp. 426–442. Springer, Heidelberg (2019). https://doi.org/10.1007/978-3-030-19212-9_28. ISBN 978-3-030-19212-9
8. Malapert, A., Nattaf, M.: Minimizing flow time on a single machine with job families and setup times. Publication Submitted and Accepted at the Project Management and Scheduling Conference (PMS 2020), May 2020. https://hal.archives-ouvertes.fr/hal-02530703
9. Mason, A.J., Anderson, E.J.: Minimizing flow time on a single machine with job classes and setuptimes.*Naval Res. Logist. (NRL)***38**(3), 333–350 (1991). https://doi.org/10.1002/1520-6750(199106)38:3⟨333::AID-NAV3220380305⟩3.0.CO;2-0. https://doi.org/10.1002/1520-6750%28199106%2938%3A3%3C333%3A%3AAID-NAV3220380305%3E3.0.CO%3B2-0
10. Nattaf, M., Dauzère-Pérès, S., Yugma, C., Wu, C.-H.: Parallel machine scheduling with time constraints on machine qualifications.*Comput. Oper. Res.***107**, 61–76 (2019). https://doi.org/10.1016/j.cor.2019.03.004
11. Obeid, A., Dauzère-Pérès, S., Yugma, C.: Scheduling job families on non-identical parallel machines with time constraints. Ann. Oper. Res. **213**(1), 221–234 (2012). https://doi.org/10.1007/s10479-012-1107-4
12. Smith, W.E.: Various optimizers for single-stage production. Naval Res. Logistics Q. **3**(1–2), 59–66 (1956). https://doi.org/10.1002/nav.3800030106

# MaxSAT-Based Postprocessing for Treedepth

Vaidyanathan Peruvemba Ramaswamy(✉) and Stefan Szeider

Algorithms and Complexity Group, TU Wien, Vienna, Austria
{vaidyanathan,sz}@ac.tuwien.ac.at

**Abstract.** Treedepth is an increasingly popular graph invariant. Many NP-hard combinatorial problems can be solved efficiently on graphs of bounded treedepth. Since the exact computation of treedepth is itself NP-hard, recent research has focused on the development of heuristics that compute good upper bounds on the treedepth.

In this paper, we introduce a novel MaxSAT-based approach for improving a heuristically obtained treedepth decomposition. At the core of our approach is an efficient MaxSAT encoding of a weighted generalization of treedepth arising naturally due to subtree contractions. The encoding is applied locally to the given treedepth decomposition to reduce its depth, in conjunction with the collapsing of subtrees. We show the local improvement method's correctness and provide an extensive experimental evaluation with some encouraging results.

**Keywords:** Tree-depth · Elimination-tree height · SAT encoding · MaxSAT · Computational experiments

## 1 Introduction

The treedepth [29,30] of a connected graph $G$ is the smallest integer $k$ such that $G$ is a subgraph of the transitive closure $[T]$ of a tree $T$ of height $k$. The transitive closure $[T]$ is obtained from $T$ by adding all edges $uv$ whenever $u$ is an ancestor of $v$ in $T$ (see Fig. 1 for an example). We call $T$ a *treedepth decomposition* of $G$ of depth $k$. The notion of treedepth was first investigated employing *elimination trees* (*e-trees*) and *elimination height* [18,20,22,34]. 1-*partition trees* [16] and *separation game* [23] are some other names used in literature for alternative approaches to treedepth.

Treedepth has algorithmic applications for several problems where treewidth cannot be used [8,10,15]. It admits algorithms for these problems whose running times are exponential in the treedepth but polynomial (of constant order) in the input size. These results request methods for computing treedepth decompositions of small (ideally minimal) depth, which is generally an NP-hard task [36].

The authors acknowledge the support by the FWF (projects P32441 and W1255) and by the WWTF (project ICT19-065).


H. Simonis (Ed.): CP 2020, LNCS 12333, pp. 478–495, 2020.
https://doi.org/10.1007/978-3-030-58475-7_28

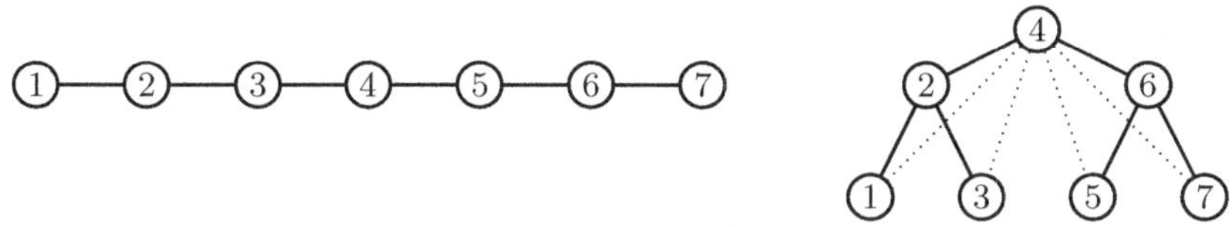


**Fig. 1.** Left: graph $P_7$. Right: treedepth decomposition of $P_7$ of depth 3.

Exact algorithms for computing the treedepth of graphs have been suggested in theoretical work [6,38]. Until recently, only very few practical implementations of algorithms that compute treedepth decompositions have been reported in the literature. Villaamil [42] discussed several heuristic methods based on minimal separators, and Ganian et al. [9] suggested two exact methods based on SAT-encodings. In general, exact methods are limited to small graphs (up to around 50 vertices [9]), whereas heuristic methods apply to large graphs but can get stuck at suboptimal solutions.

*Contribution.* In this paper, we propose the novel MaxSAT-based algorithm TD-SLIM that provides a crossover between exact and heuristic methods, taking the best of two worlds. The basic idea is to take a solution computed by a heuristic method and apply an exact (MaxSAT-based) method locally to parts of the solution, chosen small enough to admit a feasible encoding size. Although the basic idea sounds compelling and reasonably simple, its realization requires several conceptual contributions and novel results.

At the heart of our approach, local parts of the treedepth decomposition must reflect certain properties of the global decomposition. That way, an improved local decomposition can be fit back into the global one. This additional information gives rise to the more general decomposition problem, namely the *treedepth decomposition of weighted graphs with ancestry constraints*, for which we present a partition-based characterization, which leads to an efficient MaxSAT encoding. As the weights can become large, a distinctive feature of our encoding is that its size remains independent of the weights appearing in the instance.

We establish the correctness of our local-improvement method and provide an experimental evaluation on various benchmark graphs. Thereby, we compare different parameter settings and configurations and the effect of replacing several SAT calls by one MaxSAT call.

Our findings are significant, as they show that an improvement in the initial decomposition is feasible in practically all cases. The best configuration could, on average, almost reduce the depth of the initial treedepth decomposition by a half (52%) for a simple heuristic (DFS) and by a third (29%) when starting from a more elaborate heuristic (Sep). Somewhat surprisingly, it turned out that on smaller instances, that admit a single SAT encoding, the SAT-based local improvement method outperforms a single SAT call, achieving a similar depth in a tenth of the time.

## 2 Related Work

The idea of using SAT encodings for computing a decompositional graph invariant is due to Samer and Veith [39]. They proposed an *ordering-based* encoding for treewidth, which provides the basis for several subsequent improved encodings [1,2]. Heule and Szeider [13] proposed a SAT encoding for clique-width, thereby introducing a first *partition-based* encoding. These two general approaches (ordering-based and partition-based encodings) have been worked out and compared for several other decompositional graph invariants (also known as width parameters) [9,25]. Recently, several papers have proposed SMT-encodings for decompositional hypergraph parameters [4,40].

All the encodings mentioned above suffer from the limitation that the encoding size is at least cubic in the size of the input graph or hypergraph, which limits the practical applicability of these methods to graphs or hypergraph whose size is in the order of several hundred vertices. Lodha et al. [24,26] introduced the SAT-based local improvement method (SLIM), which extends the applicability of SAT-encodings to larger inputs. So far, there have been three concrete approaches that use SLIM, one for computing branchwidth (in the papers cited above [24,26]), one for computing treewidth [5], and very recently one for treewidth-bounded Bayesian network structure learning [33]. SLIM is a meta-heuristic that, similarly to Large Neighborhood Search [35], tries to improve a current solution by exploring its neighborhood of potentially better solutions. As a distinctive feature, SLIM defines neighboring solutions in a structurally highly constrained way and uses a complete method (SAT) to identify a better solution.

The comprehensive thesis by Villaamil [42] discusses four heuristics for computing treedepth. The first heuristic is based on computing a depth-first search (DFS) spanning tree of the given graph, which happens to be a valid treedepth decomposition. The remaining three heuristics are all based on finding minimal separators. Two of them are greedy local-search techniques while the third one makes use of a spectral algorithm to compute the separators, providing better decompositions at the expense of longer running times [37]. Several algorithms have been proposed for minimizing the height (among other metrics) of e-trees in the context of matrix factorizations [11,19,21,28]. More specifically, treedepth has been studied in the area of CSP under the name pseudo-tree height [3,7,17,19]. However, only few papers focus on minimizing the treedepth alone, they usually minimize a secondary measure such as fill-in. Very recently, due to the PACE Challenge 2020[1], the implementations of several new heuristics for computing treedepth decompositions became available.

## 3 Preliminaries

All considered graphs are finite, simple, and undirected. Let $G$ be a graph. $V(G)$ and $E(G)$ denote the vertex set and the edge set of $G$, respectively. The size

[1] https://pacechallenge.org/2020.

of the graph, denoted by $|G|$, is the number of vertices, i.e., $|V(G)|$. We denote an edge between vertices $u$ and $v$ by $uv$ (or equivalently by $vu$). The subgraph of $G$ induced by a set $S \subseteq V(G)$, denoted by $G[S]$, has as vertex set $S$ and as edge set $\{ uv \in E(G) \mid u, v \in S \}$. As a shorthand, we sometimes use $G[H]$ to represent $G[V(H)]$, where $G, H$ are graphs. For a set $S \subseteq V(G)$ we define $G - S = G[V(G) \setminus S]$.

For a rooted tree $T$ and a vertex $v \in V(T)$, we let $T_v$ denote the *subtree of $T$ rooted at $v$*. The vertex $v$ is the *parent* of vertex $u$ if $v$ is the first vertex (after $u$) on the path from $u$ to the root of the tree. A vertex $v$ is an *ancestor* of vertex $u$ if $v$ lies on the path from $u$ to the root of the tree and $v \neq u$. We use $r(T)$ to denote the root of a tree $T$. The *height of a vertex* $v$ in a rooted forest $T$, denoted by $height_T(v)$, is 1 plus the length of a longest path from $v$ to any leaf not passing through any ancestor (the height of a leaf is 1). The *height of a rooted tree* $T$ is then $height_T(r(T))$, i.e., the height of the root. Naturally, the height of a rooted forest is the maximum height of its constituent trees. The *depth of a vertex* $v$ in a tree $T$, denoted by $depth_T(v)$, is the length of the path from $r(T)$ to $v$ (the depth of $r(T)$ is 1). The *transitive closure* $[T]$ of a rooted forest $T$ is the undirected graph with vertex set $V(T)$ having an edge between two vertices if and only if one is an ancestor of the other in $T$.

The *treedepth* of an undirected graph $G$, denoted by $td(G)$, is the smallest integer $k$ such that there is a rooted forest $T$ with vertex set $V(G)$ of height $k$ for which $G$ is a subgraph of $[T]$. A forest $T$ for which $G$ is a subgraph of $[T]$ is also called a *treedepth decomposition*, whose *depth* is equal to the height of the forest. Informally, a graph has treedepth at most $k$ if it can be embedded in the closure of a forest of height $k$. If $G$ is connected, then it can be embedded in the closure of a tree instead of a forest.

Alternatively, letting $C(G)$ denote the set of connected components of a graph $G$, the treedepth of $G$ can be defined recursively as follows [29]:

$$td(G) = \begin{cases} 1 & \text{if } |V(G)| = 1; \\ \max_{G' \in C(G)} td(G') & \text{if } |C(G)| > 1; \\ 1 + \min_{v \in V(G)} td(G - v) & \text{otherwise.} \end{cases}$$

Based on this definition, it is clear that when we want to compute a treedepth decomposition $T$ of a graph $G$, we can assume without loss of generality, that $G$ is connected, and that the decomposition $T$ is a rooted tree.

## 4 Local Improvement

In this section, we lay out the theoretical foundations of our approach. Due to space constraints, we have omitted some proofs. For this section, let us fix a connected graph $G$ for which we would like to obtain a treedepth decomposition. We assume that $G$ is too large to admit a practically feasible SAT encoding. Hence, we can use a heuristic method (i.e., a *global solver*) to obtain a possibly suboptimal treedepth decomposition $T$. Now we would like to use a SAT-based encoding (i.e., a *local solver*) to possibly improve the decomposition $T$ and reduce its depth. We refer to this approach as the *SAT-based Local Improvement Method for Treedepth* (TD-SLIM).

In our description of the procedure, we use a *stack* data-structure, which is essentially an ordered set allowing constant-time access, insertion, and deletion at the last position. We denote an empty stack by $\emptyset$. Inserting a new element at the last position is called *pushing*. Retrieving and deleting the last element is called *popping*. Finally, we denote the last element of a stack $S$ by $last(S)$.

A first idea is to select subtrees $T_v$ at nodes $v$ of $T$ that are located low enough in $T$ such that $G[T_v]$ is small enough to admit a SAT encoding but $G[T_u]$ is not, where $u$ is the parent of $v$. This way, we could try all such $v$ and successively replace subtrees $T_v$ with new subtrees $T'_v$ of possibly lower height as long as $[T'_v]$ contains all the edges of the induced graph $G[T_v]$. All other edges of $G$, in particular those with one end in $T_v$ and one end outside $T_v$, still remain included in the overall closure. This simple procedure, however, can only improve those parts of $T$ that are close to the leaves, therefore, leaving large parts of $T$ untouched.

In order to overcome this limitation, we use an operation that contracts an entire subtree $T_v$ of $T$ into the single node $v$, and we label $v$ with the depth of $T_v$. With such contractions, we can successively eliminate lower parts of $T$, so that eventually all parts of $T$ become accessible to a potential local improvement. This requires the local solver not only to deal with weighted vertices, but now, the edges outside the induced graph $G[T_v]$ are "not safe" anymore, and one must take special care for that. We accomplish this by labeling $v$ with a set $A(v)$ of *ancestors*, which are all the vertices $u \in V(G) \setminus V(T_v)$, which are adjacent in $G$ with a vertex $w \in V(T_v) \setminus \{v\}$. Since $T$ is a treedepth decomposition, all ancestors of $v$ lie on the path between the root of $T$ and $v$. We add two more labels to $v$: $d(v)$ holding the weighted depth of $T_v$ (which we formally define in the next subsection) and $S(v)$ which is a stack maintaining the sequence of subtrees rooted at $v$ that were contracted, i.e., after contracting $T_v$, we push $T_v$ onto $S(v)$ so that we can reverse the contraction later. We refer to the process of storing or associating a decomposition with a particular vertex as "tagging."

For uniformity, we assume that already at the beginning, for all vertices $v$ of $G$, we have the trivial labels $A(v) = \emptyset$, $S(v) = \emptyset$, and $d(v) = 0$ and consider $G$ as a (trivially) *labeled graph*. We frequently deal with pairs of the form $(G, T)$ where $G$ is a labeled graph, and $T$ is a treedepth decomposition of $G$.

### 4.1 Treedepth of Labeled Graphs

The above considerations lead us to the following recursive definition of treedepth for labeled graphs. Let $G = (V, E, d, A, S)$ be a labeled graph. Importantly, when we delete a vertex and obtain $G - v$, the vertex $v$ is also deleted from all the ancestor sets $A(u)$ for all $u \in V(G - v)$.

$$td(G) = \begin{cases} 1 + d(v) & \text{if } |V(G)| = 1; \\ \max_{G' \in C(G)} td(G') & \text{if } |C(G)| > 1; \\ 1 + \min_{v \in V(G) \text{ with } A(v)=\emptyset} \max(td(G-v), d(v)) & \text{otherwise.} \end{cases}$$

Now, $T$ is a treedepth decomposition of the labeled graph $G$ of weighted depth $D$ if all the following properties hold:

T1 $T$ is a treedepth decomposition of the unlabeled graph $(V, E)$,
T2 for every $v \in V$, $d(v) + depth_T(v) \leq D$,
T3 for every $v \in V$ and $u \in A(v)$, $u$ is an ancestor of $v$ in $T$.

Given a labeled graph $G$, the *weighted depth of a rooted tree* $T$, denoted by $depth(T)$, with $V(T) \subseteq V(G)$ is $\max_{v \in V(T)} depth_T(v) + d(v)$. Since our definitions extend smoothly to the unlabeled case by means of trivial labels, we may sometimes refer to "weighted depth" simply as "depth."

### 4.2 Contracting Subtrees

For a labeled graph $G$ and a vertex $v \in V(G)$, we denote the operation of *contracting the subtree* $T_v$ in the decomposition $T$ of graph $G$ by $(G, T) \uparrow v = (G', T')$, yielding a new graph $G'$ and a decomposition $T'$. $G'$ is obtained by identifying the vertices $V(T_v)$ in $G$ and updating the labels as follows: we set $d(v) = depth(T_v)$, we push $T_v$ onto $S(v)$, we add to the set $A(v)$ the set of vertices $u \in V(G) \setminus V(T_v)$ which are adjacent to some $w \in V(T_v) \setminus \{v\}$, and we tag the newly added elements in $A(v)$ with the decomposition $T_v$. $T'$ is obtained by deleting the vertices $V(T_v) \setminus \{v\}$ from the decomposition $T$. It is easy to verify that $T'$ is a treedepth decomposition of $G'$.

### 4.3 Expanding Subtrees

Let $(G', T')$ be a pair consisting of a labeled graph and the corresponding decomposition obtained from $(G, T)$ by a sequence of contractions. Let $v \in V(G')$ be a vertex with a nontrivial label $S(v)$; $v$ is not necessarily a leaf of $T'$. From $(G', T')$ we obtain a labeled graph $G^*$ and a decomposition $T^*$ by the operation of *expanding the vertex* $v$ *in the decomposition* $T'$ *of graph* $G'$ denoted as $(G', T') \downarrow v = (G^*, T^*)$. $G^*$ is obtained as follows: we delete from $A(v)$ the elements tagged with $T_v$, we pop $T_v$ from $S(v)$, we set $d(v) = depth(last(S(v)))$, and we add to $G'$ all the vertices from $V(T_v) \setminus \{v\}$ and all the edges $uw \in E(G)$ with $u \in V(T_v) \setminus \{v\}$, $w \in V(G) \setminus V(T_v)$. Although the tree $T'$ can be very different from the original $T$ we started with, we can still extend it with $T_v$, just adding it as a subtree of $v$ (possibly next to some existing subtree in $T'$), obtaining a new tree $T^*$.

**Lemma 1.** *If* $(G', T') \downarrow v = (G^*, T^*)$ *is obtained by the process described above, then* $T^*$ *is a treedepth decomposition of* $G^*$.

### 4.4 Improving a Subtree

Let $G = (V, E, d, A, S)$ be a labeled graph and $T$ a treedepth decomposition of $G$. Let $v \in V(G)$ and $T_v$ a treedepth decomposition of the induced labeled graph $G[T_v]$. Let $T''$ be a different treedepth decomposition of $G[T_v]$ of weighted depth not exceeding that of $T_v$ (note that the root of $T''$ need not necessarily be $v$). Finally, let $T'$ be the tree obtained from $T$ by replacing $T_v$ by $T''$.

**Lemma 2.** *$T'$ is a treedepth decomposition of $G$ of weighted depth not larger than the weighted depth of $T$.*

### 4.5 The Improvement Procedure

We now provide a high-level overview of the *local improvement procedure* (see Algorithm 1). The procedure takes as input a (possibly labeled) graph $G$, a starting treedepth decomposition $T$ of $G$, and returns a treedepth decomposition $T'$ of $G$ such that $depth(T') \leq depth(T)$. It requires three parameters:

- *budget* $\beta$ which indicates a conservative estimate of the maximum size of instances that is practically feasible for a SAT-solver to solve reasonably quickly (within a few seconds),
- *timeout* $\tau$ (in seconds) for each individual SAT (or MaxSAT) call,
- *contraction size* $\kappa$ which denotes the maximum size of subtrees that can be contracted.

**Input** : Graph $G$, decomposition $T$ of $G$, budget $\beta$, timeout $\tau$ (in seconds), contraction size $\kappa$
**Output:** Decomposition $T'$ of $G$ such that $td(T') \leq td(T)$

**begin**
  $(G', T') \longleftarrow (G, T)$
  `// Contraction Phase`
  **repeat**
    $v \longleftarrow$ `GetNode(`$T'$, $\beta$`)`
    Improve $T'_v$ (either by using SAT-solver or Lemma 4)
    **while** *contraction condition* **do**
      $u \longleftarrow$ `GetNode(`$T'$, $\kappa$`)`
      $(G', T') \longleftarrow (G', T') \uparrow u$ `// contract at` $u$
    **end**
  **until** $r(T) \in T'_v$
  `// Expansion Phase`
  **while** *there exists $u$ with nontrivial $S(u)$* **do**
    $(G', T') \longleftarrow (G', T') \downarrow u$ `// expand at` $u$
  **end**
  **return** $T'$
**end**

**Algorithm 1.** Pseudocode for TD-SLIM

One individual pass of the procedure consists of a *contraction phase* during which only improvement and contraction operations occur and an *expansion phase* during which the contracted treedepth decomposition is expanded to obtain a treedepth decomposition for the original graph $G$. The contraction phase terminates when we encounter, as the instance to be improved, a local instance containing the root of the starting decomposition $T$. The *contraction condition* on Line 6 determines how many contractions to perform and is only relevant when $\kappa < \beta$. We stop contracting when the size (or depth) of the contracted subinstance falls below a threshold (see partial contraction strategy in Sect. 6.3). The local improvement procedure itself can be repeated any number of times and the potentially improved treedepth decomposition $T'$ returned by one iteration can be used as the starting decomposition for the next iteration. Each individual iteration requires polynomial time in addition to the time required for the SAT (or MaxSAT) call.

At the heart of the contraction phase of the improvement procedure is the `GetNode` subroutine, which is responsible for finding nontrivial sub-instances which can be improved and then contracted. This subroutine takes as input a treedepth decomposition $T$ and a budget $\beta$. It first tries to find a nonleaf vertex $v \in V(T)$ such that $|T_v| \leq \beta$. If no such vertex exists, it instead returns a vertex $u \in V(T)$ such that $height_T(u)$ is 2 and $u$ has at least $\beta$ children, the existence of which is proven by the following lemma.

**Lemma 3.** *Given an integer $k$ and a rooted tree $T$ such that $|T| \geq 2$, at least one of the following conditions is true:*

1. *There exists a nonleaf vertex $v \in V(T)$ such that $|T_v| \leq k$ and $|T_p| > k$ where $p$ is the parent of $v$ in $T$.*
2. *There exists a vertex $u \in V(T)$ such that $height_T(u)$ is 2 and $u$ has at least $k$ children.*

*Proof.* Let us assume for the sake of contradiction that both conditions are false. Let $v$ be a deepest leaf in $T$ and let $u$ be the parent of $v$ in $T$. Note that $height_T(u) = 2$, otherwise $v$ would not be a deepest leaf. Since by our assumption, the second condition is false, $p$ can have at most $k - 1$ children and since its height is 2, $p$'s children are its only descendants. Hence $|T_p| \leq k$, implying that the first condition is true, thus contradicting our assumption. □

When we are unable to find a nonleaf vertex $v$ such that $|T_v| \leq \beta$, it means that there are no subinstances remaining which can be improved and hence the contraction phase of the algorithm would have to terminate abruptly. It is also worth noting that any improvement in the depth of the final decomposition $T'$, as compared the initial decomposition $T$, can be traced back to the contraction phase. Thus, an early termination of this phase means a narrower scope for improvement. But as can be seen in Lemma 3, whenever we are unable to find a reasonably-sized subinstance, we can use this fact—it must be due to a high-degree parent—to our advantage by tackling this case separately.

**Lemma 4.** *Given a labeled graph $G$ which is a star on $n$ vertices, the treedepth of $G$ can be determined in time $\mathcal{O}(n \log n)$.*

# 5 MaxSAT Encoding

Ganian et al. [9], introduced and compared two SAT-encodings for treedepth, one explicitly guessing the tree-structure of a treedepth decomposition and one using a novel partition-based characterization of treedepth, the latter outperforming the former significantly. In both cases, given an unlabeled graph $G$ and an integer $k$, a CNF formula $F(G,k)$ is produced, which is satisfiable if and only if the treedepth of $G$ is at most $k$. By trying out different values of $k$ one can determine the exact treedepth of $G$.

In this section, we build upon Ganian et al.'s [9] partition-based encoding and describe extensions required to employ the SAT encoding for *labeled graphs* thereby addressing the problem of computing treedepth decomposition of weighted graphs with ancestry constraints. We further explain how the encoding can be lifted to MaxSAT, yielding a significant speedup.

## 5.1 Partition-Based Formulation

A *weak partition* of a set $S$ is a set $P$ of nonempty disjoint subsets of $S$. The elements of $P$ are called *equivalence classes*. Let $P, P'$ be two partitions of $S$, then $P'$ is a *refinement* of $P$ if any two elements $x, y \in S$ that are in the same equivalence class of $P'$ are also in the same equivalence class of $P$. Given a set $S$, a *derivation* $\mathcal{P}$ of *length* $\ell$ is a sequence $(P_1, P_2, \dots, P_\ell)$ of weak partitions of $S$. $P_i$ is called the *i-th level* of $\mathcal{P}$. For some $2 \le i \le \ell$, we say that a set $c \in P_{i-1}$ is a *child* of a set $p \in P_i$ if $c \subseteq p$. We denote by $\mathsf{c}^i_{\mathcal{P}}(p)$ the set of all children of $p$ at level $i$. Further, $\chi^i_{\mathcal{P}}(p)$ denotes the set $p \setminus \bigcup_{c \in \mathsf{c}^i_{\mathcal{P}}(p)} c$. Finally, the shorthand $\bigcup P_i$ denotes the set $\bigcup_{p \in P_i} p$. Given a labeled graph $G$, a *derivation* $\mathcal{P}$ *of* $G$ is a sequence $(P_1, \dots, P_\ell)$ of weak partitions of the set $V(G)$ satisfying the following properties:

- D1 $P_1 = \emptyset$ and $P_\ell = \{V(G)\}$;
- D2 for every $1 \le i \le \ell - 1$, $P_i$ is a refinement of $P_{i+1}$;
- D3 for every $1 \le i \le \ell$ and $p \in P_i$, $|\chi^i_{\mathcal{P}}(p)| \le 1$;
- D4 for every edge $uv \in E(G)$, there is a $p \in P_i$ for some $1 \le i \le \ell$ such that $u, v \in p$ and $\chi^i_{\mathcal{P}}(p) \cap \{u, v\} \ne \emptyset$;
- D5 for every $v \in V(G)$ and $1 \le i \le \ell$, if $v \in \bigcup P_i$ then $d(v) + 2 \le i$; and
- D6 for every $v \in V(G)$ and $u \in A(v)$, there is a $p \in P_i$ for some $1 \le i \le \ell$ such that $u, v \in p$ and $u \in \chi^i_{\mathcal{P}}(p)$; together with D3 that implies $\chi^i_{\mathcal{P}}(p) = \{u\}$.

**Theorem 1.** *Let $G = (V, E, d, A, S)$ be a labeled graph and $D$ an integer. $G$ has a treedepth decomposition of weighted depth at most $D$ if and only if $G$ has a derivation of length at most $D + 1$.*

*Proof.* Let $T$ be a treedepth decomposition of $G$ of weighted depth $D$. Let $\mathcal{P}$ be the derivation consisting of weak partitions $(P_1, \dots, P_{D+1})$ where $P_1 = \emptyset$, $P_i = \{\, V(T_u) \mid u \in V(T) \text{ and } \mathit{depth}_T(u) = D - i + 2 \,\}$ for every $2 \le i \le D + 1$. It is easy to see that $\mathcal{P}$ is a derivation of the unlabeled graph $(V, E)$ and the length of $\mathcal{P}$ is $D+1$. For any vertex $v \in V(G)$, let $2 \leq i \leq D+1$ such that $v \in \bigcup P_i$, thus, by construction of $\mathcal{P}$, we get $depth_T(v) \geq D - i + 2$. Since $T$ is a treedepth decomposition of the labeled graph $G$, it satisfies property T2, meaning $D \geq d(v) + depth_T(v) \geq d(v) + D - i + 2$ which implies $i \geq d(v) + 2$, which is precisely property D5. To show property D6, let $v \in V(G)$ and $u \in A(v)$, let $k = D - depth_T(u) + 2$. We observe that there exists $p \in P_k$ such that $u \in p$, and since $u$ is an ancestor of $v$ (from property T3), $v \in T_u$ therefore $v \in p$. We further observe that $u \notin \bigcup P_{k-1}$ meaning $u \in \chi^k_{\mathcal{P}}(p)$, hence property D6 holds.

Towards showing the converse, let $\mathcal{P}$ be a derivation of length $D+1$ of the labeled graph $G$. Note that w.l.o.g we can assume $\chi^i_{\mathcal{P}}(p) \neq \emptyset$ for every $1 \leq i \leq D+1$ and $p \in P_i$, because otherwise we could replace $p$ with all its children without increasing the length of the derivation and retaining all the properties. For every $v \in V(G)$ there exists exactly one $1 \leq i \leq D+1$ and $p \in P_i$ such that $\chi^i_{\mathcal{P}}(p) = \{v\}$; in that case we say that the set $p$ *introduces* $v$. Now we construct the treedepth decomposition $T$ with vertex set $V(G)$ by adding an edge between $u, v \in V(G)$ if the set introducing $u$ is a child of the set introducing $v$ or vice versa. It can be seen that $T$ is a treedepth decomposition of depth at most $D$ of the unlabeled graph $(V, E)$.

Towards showing T2, let $v \in V(G)$ and $1 < i \leq D+1$ such that $v \in \bigcup P_i$ and $v \notin \bigcup P_{i-1}$; in other words, $i$ is the smallest index such that $v \in \bigcup P_i$. By construction of $T$, this implies that $v$ is introduced in the $i$-th layer, meaning $depth_T(v) = D - i + 2$. Combining this with property D5, we get $d(v) + depth_T(v) \leq D$, thus satisfying property T2. Now, to show T3, since $\mathcal{P}$ satisfies property D6, let $2 \leq i \leq D+1$ such that $u, v \in p$ for some $p \in P_i$ and $\chi^i_{\mathcal{P}}(p) = \{u\}$. Thus $p$ introduces $u$. Further, $v \in \mathsf{c}^i_{\mathcal{P}}(p)$ since $v \notin \chi^i_{\mathcal{P}}(p)$, meaning $u$ is an ancestor of $v$ in $T$, therefore satisfying property T3. Hence $T$ is indeed a treedepth decomposition of the labeled graph $G$. □

We note that the treedepth of a labeled graph $G = (V, E, d, A, S)$ can be as large as $|V| + \max_{v \in V} d(v)$. Since the above proof explicitly encodes the weight labels $d(v)$ in the derivation, the treedepth $D$ affects the number of layers in the derivation, which in turn affects the size of the encoding. Thus, for graphs with large $d(v)$, the encoding size is also large. A neat observation can however remedy this: the derivation of $G$ does not need to have any more than $|V| + 1$ layers.

**Observation 1.** *Let $G = (V, E, d, A, S)$ be a labeled graph and $D > |V|$ be an integer. Let $G' = (V, E, d', A, S)$ where $d'(v) := \max(0, d(v) - (D - |V|))$ for $v \in V(G)$. $G$ has a derivation of length at most $D+1$ if and only if $G'$ has a derivation of length at most $|V| + 1$.*

### 5.2 Encoding of a Derivation

We now tersely describe the encoding of the formulation discussed in the previous subsection.

**Theorem 2.** *Given a labeled graph $G$ and an integer $D$, one can construct in polynomial time, a CNF formula $F(G, D)$ that is satisfiable if and only if $G$ has a derivation of length at most $D$.*

The remainder of this section describes the clauses constituting the formula $F$. We have a set variable $s(u, v, i)$, for every $u, v \in V(G)$ with $u \leq v$ and every $i$ with $1 \leq i \leq D$. The variable $s(u, v, i)$ indicates whether vertices $u$ and $v$ appear in the same equivalence class of $P_i$, and $s(u, u, i)$ indicates whether $u$ appears in some equivalence class of $P_i$. The following clauses ensure D1 and D2:

$$\neg s(u, v, 1) \wedge s(u, v, D) \qquad \text{for } u, v \in V(G), u \leq v, \text{ and}$$
$$\neg s(u, v, i) \vee s(u, v, i+1) \qquad \text{for } u, v \in V(G), u \leq v, 1 \leq i \leq D.$$

The following clauses ensure D3 and D4:

$$\neg s(u, v, i) \vee s(u, u, i-1) \vee s(v, v, i-1) \qquad \text{for } u, v \in V(G), u < v, 2 \leq i \leq D,$$
$$\neg s(u, u, i) \vee \neg s(v, v, i) \vee s(u, u, i-1) \vee s(u, v, i)$$
$$\neg s(u, u, i) \vee \neg s(v, v, i) \vee s(v, v, i-1) \vee s(u, v, i)$$
$$\text{for } uv \in E, u < v, 2 \leq i \leq D.$$

The following clauses ensure the semantics of the set variables of the form $s(u, u, i)$ as well as the transitivity of the set variables:

$$(\neg s(u, v, i) \vee s(u, u, i)) \wedge (\neg s(u, v, i) \vee s(v, v, i))$$
$$\text{for } u, v \in V(G), u < v, 2 \leq i \leq D,$$
$$(\neg s(u, v, i) \vee \neg s(u, w, i) \vee s(v, w, i))$$
$$\wedge(\neg s(u, v, i) \vee \neg s(v, w, i) \vee s(u, w, i))$$
$$\wedge(\neg s(u, w, i) \vee \neg s(v, w, i) \vee s(u, v, i))$$
$$\text{for } u, v, w \in V(G), u < v < w, 1 \leq i \leq D.$$

Finally, the following clauses ensure D5 and D6:

$$\neg s(u, u, i) \quad \text{for } u \in V(G), 2 \leq i \leq D, \text{if } |V(G)| + d(u) \geq D \text{ and } i < d(u) + 2,$$
$$\neg s(v, v, i) \vee s(u, u, i) \qquad \text{for } u \in V(G), v \in A(u) \text{ and } 2 \leq i \leq D.$$

This concludes the description of the SAT encoding. Note that, due to Observation 1, without loss of generality, we may assume that $D \leq |V| + 1$.

We now extend the above encoding to a Partial MaxSAT formulation $F'(G, D)$ containing soft clauses. An optimal solution for $F'(G, D)$ satisfies $\mu$ soft clauses if and only if $G$ has treedepth $D - \mu + 1$. First, all the clauses from the SAT encoding are added as hard clauses into $F'$. Then we introduce a *free layer* variable $f_i$ for $1 \leq i \leq D$, which is false if some vertex appears in the $i$-th layer, i.e., $u \in \bigcup P_i$ for some $u \in V(G)$. Further, if the $i$-th layer is free then all the lower layers must also be free. These conditions are encoded via the following hard clauses:

$$\neg f_i \vee \neg s(u, u, i) \qquad \text{for } u \in V(G) \text{ and } 2 \le i \le D,$$
$$\neg f_i \vee f_{i-1} \qquad \text{for } 2 \le i \le D.$$

Additionally, we need to take special care in the case of labeled graphs, as the depth labels could mean that even though no vertices appear in a layer, they are still occupied by a subtree represented by the depth labels. In other words, a vertex $v$ must not only "occupy" its own layer but also $d(v)$ many layers below its own layer. The following hard clause captures this condition:

$$\neg s(u, u, i) \vee \neg f_{i-j} \qquad \text{for } u \in V(G), 2 \le i \le D \text{ and } 1 \le j \le \min(d(u), i).$$

Finally, we introduce a soft unit clause $f_i$ for $1 \le i \le D$. This sets the objective of the MaxSAT solver to maximize the number of free layers, which consequently, minimizes the depth of the decomposition. This concludes the description of the MaxSAT encoding.

## 6 Experimental Evaluation

### 6.1 Experimental Setup

We ran all our experiments on a 10-core Intel Xeon E5-2640 v4, 2.40 GHz CPU, with each process having access to 8 GB RAM. We used Glucose 4.0[2] as the SAT-solver and UWrMaxSat as the MaxSAT-solver, both with standard settings. We use UWrMaxSat primarily due to its anytime nature (i.e., it can be terminated anytime, and it outputs the current best possibly suboptimal result). We also tried other solvers like RC2 and Loandra from the 2019 MaxSAT Evaluation contest, but UWrMaxSat worked better for our use case. The details of all the MaxSAT solvers can be found on the 2019 MaxSAT Evaluation webpage[3].

We implemented the local improvement algorithm in Python 3.6.9, using the Networkx 2.4 graph library [12] and the PySAT 0.1.5 library [14] for the MaxSAT solver RC2. The source code of our implementation is available online [32]. Our experiments aim to demonstrate the benefit of applying local improvement to any external heuristic, and not to provide a comparison between our approach coupled with the two considered heuristics and other standalone algorithms.

### 6.2 Instances

We tested our implementation on subsets of the public benchmark instances used by the PACE Challenge 2020, from both the Exact Track (smaller instances on average) and the Heuristic Track. We formed two datasets as follows: (i) *Dataset A* consists of all the instances on which the heuristic algorithms were able to compute a solution within 2 h. This yielded 140 instances in the range $|V| \in [10, 4941]$, $|E| \in [15, 86528]$. This dataset is meant to serve as a

[2] https://www.labri.fr/perso/lsimon/glucose/.

[3] https://maxsat-evaluations.github.io/2019/descriptions.html.

comprehensive dataset with a large variance in the graph sizes. (ii) *Dataset B* consists of 30 instances from the Exact Track (27–85). This dataset represents the set of instances that we expect to lie in the practically feasible zone of SAT-solvers (or MaxSAT solvers). The resulting graphs lie in the range $|V| \in [30, 72]$, $|E| \in [48, 525]$.

### 6.3 Experiment 1

We evaluate the quality of the solution in terms of the improvement in depth, where *absolute improvement* (or simply *improvement*) refers to the difference between the starting heuristic depth and the final reported depth, and *relative improvement* (RI) refers to the absolute improvement expressed as a percentage of the starting heuristic depth. Our implementation of the algorithm can be configured using the following parameters:

- budget $\beta$, can either be a single value or a sequence of budget values (we denote by *Multibudget* the sequence $(5+(5i \bmod 40))_{i\geq 0}$ where the next budget value is used when the current value fails to provide any improvement),
- timeout $\tau$,
- contraction ratio $\gamma$ which determines the contraction size $\kappa = \gamma\beta$,
- partial contraction strategy which determines when we switch from contraction to improvement—either when the local instance's size has been reduced by half or when the local instance's unweighted depth has been reduced by 2,
- target solver, i.e., SAT or MaxSAT,
- random seed, and
- global timeout.

Since the number of possible parameter configurations is huge, we first ran a preliminary experiment on a small number of instances with a large number of parameter configurations to narrow down the better performing configurations. We gathered a smaller set of configurations from this initial run, which we then tested rigorously on Dataset A. We tried $\beta \in \{20, \text{Multibudget}\}$, $\tau = 20$, $\gamma \in \{0.5, 1\}$, partial contraction by depth, target solver MaxSAT, and random seed $\in \{1, 2, 3\}$. For the starting decomposition, we compared two heuristics proposed by Villaamil [42]—a randomized variant of the DFS heuristic and one of the separator-based heuristics (denoted by Sep). Given a treedepth heuristic algorithm X, we denote by TD-SLIM(X) the algorithm obtained by running the local improvement procedure on top of the heuristic solution provided by algorithm X. The implementation of Sep was kindly provided to us by Oelschlägel [31]. We precomputed the heuristic solutions for Dataset A with a 2-h timeout and then ran the improvement procedure for 30 min.

We use three solvers or configurations to present the performance data and convey the likelihood of the different results:

- *Virtual Best Solver* (VBS): the hypothetical solver which, for each instance, knows the configuration that yields the best improvement,

- *Single Best Solver* (SBS): the solver with the configuration that resulted in the best improvement on average across all instances,
- *Average Solver* (AS): the hypothetical solver representing the average performance across all the configurations for a particular instance.

The average relative improvements (including the cases with no improvement) for AS, SBS, and VBS were 45.9%, 52.2%, and 53.0% when starting from the DFS heuristic and 21.9%, 29.2%, and 30.2% when starting from the Sep heuristic, respectively. Table 1 shows the instances from Dataset A with the best relative improvement. The parameter combination to achieve the best average relative improvement across all the 140 instances across both heuristics (i.e., SBS) was (Multibudget, MaxSAT, $\gamma = 0.5$, partial contraction by depth). In the experiment, we observed a rather robust performance over all considered configurations.

**Table 1.** Top 15 instances from Experiment 1 sorted by best relative improvement among both heuristics. *Start* and *Final* refer to the starting heuristic depth and the final reported depth, respectively *Imp.* refers to the absolute improvement.

| Instance | $\|V\|$ | $\|E\|$ | DFS heuristic | | | | Sep heuristic | | | |
|---|---|---|---|---|---|---|---|---|---|---|
| | | | Start | Final | Imp. | RI (%) | Start | Final | Imp. | RI (%) |
| heur_055 | 590 | 668 | 92 | 18 | 74 | 80.43 | 27 | **17** | 10 | 37.04 |
| heur_033 | 255 | 507 | 129 | **34** | 95 | 73.64 | 155 | **34** | 121 | 78.06 |
| heur_071 | 1023 | 2043 | 486 | **108** | 378 | 77.78 | 657 | 418 | 239 | 36.38 |
| exact_165 | 176 | 186 | 49 | 11 | 38 | 77.55 | 17 | **10** | 7 | 41.18 |
| exact_193 | 449 | 2213 | 128 | **29** | 99 | 77.34 | 151 | 75 | 76 | 50.33 |
| heur_021 | 195 | 340 | 100 | 23 | 77 | 77.00 | 43 | **22** | 21 | 48.84 |
| exact_173 | 198 | 692 | 69 | **16** | 53 | 76.81 | 57 | 50 | 7 | 12.28 |
| exact_169 | 181 | 253 | 77 | **18** | 59 | 76.62 | 44 | 20 | 24 | 54.55 |
| exact_195 | 451 | 587 | 131 | 31 | 100 | 76.34 | 74 | **23** | 51 | 68.92 |
| exact_157 | 163 | 195 | 49 | **12** | 37 | 75.51 | 27 | 13 | 14 | 51.85 |
| exact_103 | 92 | 131 | 57 | **14** | 43 | 75.44 | 24 | 16 | 8 | 33.33 |
| heur_025 | 212 | 257 | 67 | **17** | 50 | 74.63 | 36 | 19 | 17 | 47.22 |
| exact_177 | 204 | 248 | 34 | **9** | 25 | 73.53 | 16 | 11 | 5 | 31.25 |
| exact_107 | 95 | 121 | 41 | **11** | 30 | 73.17 | 23 | 13 | 10 | 43.48 |
| exact_185 | 276 | 1187 | 63 | **20** | 43 | 68.25 | 85 | 23 | 62 | 72.94 |

Out of the 140 instances solved by both heuristics, TD-SLIM(DFS) provided a strictly lower depth than TD-SLIM(Sep) for 48 instances and strictly higher depth for 38 instances. In the remaining 54 instances, both TD-SLIM(DFS) and TD-SLIM(Sep) reached the same depth value. Thus, TD-SLIM is often capable of achieving a comparable or even lower depth despite starting from a significantly

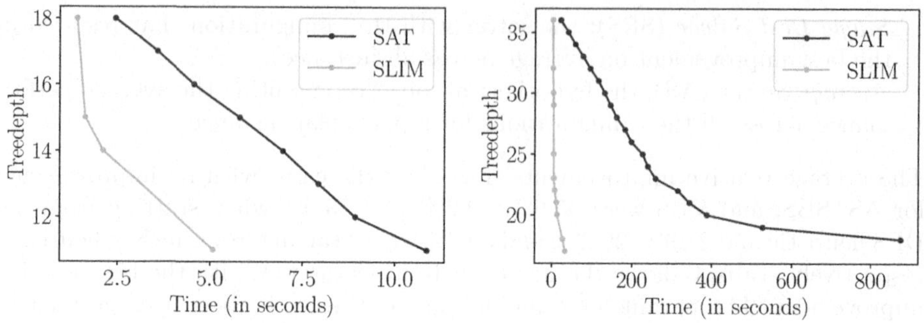


(a) exact_031.gr

(b) dolphin.gml

**Fig. 2.** TD-SLIM vs SAT

worse heuristic decomposition, but there are also cases that show that TD-SLIM is capable of utilizing and building upon a better starting decomposition. Comparing these depths with the lowest known depths from the PACE challenge, TD-SLIM(DFS) matches the lowest depth on 53 instances and is off by one on 20 instances. TD-SLIM(Sep) matches the lowest depth on 41 instances and is off by one on 17 instances.

Very recently, the results for the PACE challenge were announced and the implementations of the solvers were made available. Consequently, we tested TD-SLIM on top of the winning heuristic ExTREEm [41]. We ran the heuristic for 15 min and then TD-SLIM on top of the solution provided by the heuristic for 15 min, using the same total time limit of 30 min as used in the PACE challenge. Somewhat surprisingly, we observed that for 6 instances, TD-SLIM(ExTREEm) was able to compute better decompositions than simply running ExTREEm for 30 min. For 3 of these instances, TD-SLIM(ExTREEm) was even able to find a better decomposition than all the 55 participating heuristic solvers.

### 6.4 Experiment 2

We tested the effectiveness of TD-SLIM over a one-shot SAT or MaxSAT encoding where the entire instance is passed to the solver. To give the one-shot SAT or MaxSAT a massive advantage, we chose Dataset B, which contains much smaller instances, and we set the global timeout for TD-SLIM to 30s, whereas for SAT and MaxSAT we chose 300s. For this experiment, we used DFS as the starting heuristic. We observed that TD-SLIM performs comparably to SAT in terms of the final depth achieved, computing the best upper bound for 18 instances while SAT arrives at the best upper bound for 22 instances. Nevertheless, when it comes to slightly larger instances, TD-SLIM even outperforms SAT in 7 instances despite only having a tenth of the time. We also noticed that MaxSAT tends to perform significantly worse than SAT. We suspect this might be the case because the MaxSAT solver spends time improving the lower bounds, which is of little use in this case.

As a part of Experiment 2, we also compared the trajectory of improvement over time of TD-SLIM and SAT. We use graphs 'exact_031.gr' and 'dolphin.gml' (from [27]) as typical examples. As can be seen in Fig. 2a and Fig. 2b, TD-SLIM is much faster than SAT. Another interesting observation is that TD-SLIM is able to achieve depth values which were previously not possible, e.g., for the graph 'B10Cage', TD-SLIM improved the previously known upper bound from 23 [9] to 22 with a runtime of around 50s.

## 7 Concluding Remarks

Our (Max)SAT-based local improvement approach to treedepth provides a compelling showcase for demonstrating how (Max)SAT encodings can be scaled to large inputs, thus widening the scope of potential applications for exact constraint-based methods. Our experiments show that in many cases, our approach allows a significant improvement over heuristically obtained treedepth decompositions. We observed that TD-SLIM is able to improve even over strong heuristics like the winning heuristic from the PACE challenge. Rather unexpected is the finding that on smaller instances, the local improvement method significantly outperforms a one-shot (Max)SAT encoding.

In future work, we plan to systematically study the effect of postprocessing for different types of heuristics, as have been made available by the PACE challenge. Other topics for future work include the utilization of incremental solving techniques for the SAT-based optimization of treedepth, and its comparison to the MaxSAT approach, as well as the development of symmetry breaking methods for further speeding up the encoding.

## References

1. Bannach, M., Berndt, S., Ehlers, T.: Jdrasil: a modular library for computing tree decompositions. In: Iliopoulos, C.S., Pissis, S.P., Puglisi, S.J., Raman, R. (eds.) 16th International Symposium on Experimental Algorithms, SEA 2017, London, UK, 21–23 June 2017, vol. 75, pp. 28:1–28:21. LIPIcs, Schloss Dagstuhl - Leibniz-Zentrum fuer Informatik (2017)
2. Berg, J., Järvisalo, M.: SAT-based approaches to treewidth computation: an evaluation. In: 26th IEEE International Conference on Tools with Artificial Intelligence, ICTAI 2014, Limassol, Cyprus, 10–12 November 2014, pp. 328–335. IEEE Computer Society (2014)
3. Dechter, R., Mateescu, R.: AND/OR search spaces for graphical models. Artif. Intell. **171**(2–3), 73–106 (2007)
4. Fichte, J.K., Hecher, M., Lodha, N., Szeider, S.: An SMT approach to fractional hypertree width. In: Hooker, J. (ed.) CP 2018. LNCS, vol. 11008, pp. 109–127. Springer, Cham (2018). https://doi.org/10.1007/978-3-319-98334-9_8

5. Fichte, J.K., Lodha, N., Szeider, S.: SAT-based local improvement for finding tree decompositions of small width. In: Gaspers, S., Walsh, T. (eds.) SAT 2017. LNCS, vol. 10491, pp. 401–411. Springer, Cham (2017). https://doi.org/10.1007/978-3-319-66263-3_25
6. Fomin, F.V., Giannopoulou, A.C., Pilipczuk, M.: Computing tree-depth faster than $2^n$. Algorithmica **73**(1), 202–216 (2015)
7. Freuder, E.C., Quinn, M.J.: Taking advantage of stable sets of variables in constraint satisfaction problems. In: IJCAI, vol. 85, pp. 1076–1078. Citeseer (1985)
8. Gajarský, J., Hlinený, P.: Faster deciding MSO properties of trees of fixed height, and some consequences. In: D'Souza, D., Kavitha, T., Radhakrishnan, J. (eds.) IARCS Annual Conference on Foundations of Software Technology and Theoretical Computer Science, FSTTCS 2012, Hyderabad, India, 15–17 December 2012, vol. 18, pp. 112–123. LIPIcs, Schloss Dagstuhl - Leibniz-Zentrum fuer Informatik (2012)
9. Ganian, R., Lodha, N., Ordyniak, S., Szeider, S.: SAT-encodings for treecut width and treedepth. In: Kobourov, S.G., Meyerhenke, H. (eds.) Proceedings of ALENEX 2019, the 21st Workshop on Algorithm Engineering and Experiments, pp. 117–129. SIAM (2019)
10. Gutin, G., Jones, M., Wahlström, M.: Structural parameterizations of the mixed Chinese postman problem. In: Bansal, N., Finocchi, I. (eds.) ESA 2015. LNCS, vol. 9294, pp. 668–679. Springer, Heidelberg (2015). https://doi.org/10.1007/978-3-662-48350-3_56
11. Hafsteinsson, H.: Parallel sparse Cholesky factorization. Cornell University, Technical report (1988)
12. Hagberg, A.A., Schult, D.A., Swart, P.J.: Exploring network structure, dynamics, and function using NetworkX. In: Proceedings of the 7th Python in Science Conference (SciPy 2008), Pasadena, CA, USA, pp. 11–15, August 2008
13. Heule, M., Szeider, S.: A SAT approach to clique-width. ACM Trans. Comput. Log. **16**(3), 24 (2015). https://doi.org/10.1145/2736696
14. Ignatiev, A., Morgado, A., Marques-Silva, J.: PySAT: a Python toolkit for prototyping with SAT Oracles. In: Beyersdorff, O., Wintersteiger, C.M. (eds.) SAT 2018. LNCS, vol. 10929, pp. 428–437. Springer, Cham (2018). https://doi.org/10.1007/978-3-319-94144-8_26
15. Iwata, Y., Ogasawara, T., Ohsaka, N.: On the power of tree-depth for fully polynomial FPT algorithms. In: Niedermeier, R., Vallée, B. (eds.) 35th Symposium on Theoretical Aspects of Computer Science, STACS 2018, 28 February–3 March 2018, Caen, France, vol. 96, pp. 41:1–41:14. LIPIcs, Schloss Dagstuhl - Leibniz-Zentrum fuer Informatik (2018)
16. Iyer, A.V., Ratliff, H.D., Vijayan, G.: On a node ranking problem of trees and graphs. Technical reporrt, Georgia Inst of Tech Atlanta Production and Distribution Research Center (1986)
17. Jégou, P., Terrioux, C.: Hybrid backtracking bounded by tree-decomposition of constraint networks. Artif. Intell. **146**(1), 43–75 (2003)
18. Jess, J.A.G., Kees, H.G.M.: A data structure for parallel L/U decomposition. IEEE Trans. Comput. **3**, 231–239 (1982)
19. Kayaaslan, E., Uçar, B.: Reducing elimination tree height for parallel LU factorization of sparse unsymmetric matrices. In: 2014 21st International Conference on High Performance Computing (HiPC), pp. 1–10. IEEE (2014)
20. Kees, H.G.M.: The organization of circuit analysis on array architectures. Ph.D. thesis, Citeseer (1982)
21. Liu, J.W.: Reordering sparse matrices for parallel elimination. Parallel Comput. **11**(1), 73–91 (1989)

22. Liu, J.W.: The role of elimination trees in sparse factorization. SIAM J. Mat. Anal. Appl. **11**(1), 134–172 (1990)
23. Llewellyn, D.C., Tovey, C., Trick, M.: Local optimization on graphs. Discrete Appl. Math. **23**(2), 157–178 (1989)
24. Lodha, N., Ordyniak, S., Szeider, S.: A SAT approach to branchwidth. In: Creignou, N., Le Berre, D. (eds.) SAT 2016. LNCS, vol. 9710, pp. 179–195. Springer, Cham (2016). https://doi.org/10.1007/978-3-319-40970-2_12
25. Lodha, N., Ordyniak, S., Szeider, S.: SAT-encodings for special treewidth and pathwidth. In: Gaspers, S., Walsh, T. (eds.) SAT 2017. LNCS, vol. 10491, pp. 429–445. Springer, Cham (2017). https://doi.org/10.1007/978-3-319-66263-3_27
26. Lodha, N., Ordyniak, S., Szeider, S.: A SAT approach to branchwidth. ACM Trans. Comput. Log. **20**(3), 15:1–15:24 (2019)
27. Lusseau, D., Schneider, K., Boisseau, O.J., Haase, P., Slooten, E., Dawson, S.M.: The bottlenose dolphin community of doubtful sound features a large proportion of long-lasting associations. Behav. Ecol. Sociobiol. **54**(4), 396–405 (2003)
28. Manne, F.: Reducing the height of an elimination tree through local recorderings. University of Bergen, Department of Informatics (1991)
29. Nesetril, J., de Mendez, P.O.: Tree-depth, subgraph coloring and homomorphism bounds. Eur. J. Comb. **27**(6), 1022–1041 (2006)
30. Nešetřil, J., de Mendez, P.O.: Sparsity - Graphs, Structures, and Algorithms. Algorithms and Combinatorics, vol. 28. Springer, Heidelberg (2012). https://doi.org/10.1007/978-3-642-27875-4
31. Oelschlägel, T.: Treewidth from Treedepth. Ph.D. thesis, RWTH Aachen University (2014)
32. Peruvemba Ramaswamy, V., Szeider, S.: aditya95sriram/td-slim: public release, July 2020. Zenodo. https://doi.org/10.5281/zenodo.3946663
33. Peruvemba Ramaswamy, V., Szeider, S.: Turbocharging treewidth-bounded Bayesian network structure learning (2020). https://arxiv.org/abs/2006.13843
34. Pieck, J.: Formele definitie van een e-tree. Eindhoven University of Technology: Department of Mathematics: Memorandum 8006 (1980)
35. Pisinger, D., Ropke, S.: Large neighborhood search. In: Gendreau, M., Potvin, J.Y. (eds.) Handbook of Metaheuristics, pp. 399–419. Springer, Boston (2010). https://doi.org/10.1007/978-1-4419-1665-5_13
36. Pothen, A.: The complexity of optimal elimination trees. Technical report (1988)
37. Pothen, A., Simon, H.D., Wang, L., Barnard, S.T.: Towards a fast implementation of spectral nested dissection. In: Supercomputing 1992: Proceedings of the 1992 ACM/IEEE Conference on Supercomputing, pp. 42–51. IEEE (1992)
38. Reidl, F., Rossmanith, P., Villaamil, F.S., Sikdar, S.: A faster parameterized algorithm for treedepth. In: Esparza, J., Fraigniaud, P., Husfeldt, T., Koutsoupias, E. (eds.) ICALP 2014. LNCS, vol. 8572, pp. 931–942. Springer, Heidelberg (2014). https://doi.org/10.1007/978-3-662-43948-7_77
39. Samer, M., Veith, H.: Encoding treewidth into SAT. In: Kullmann, O. (ed.) SAT 2009. LNCS, vol. 5584, pp. 45–50. Springer, Heidelberg (2009). https://doi.org/10.1007/978-3-642-02777-2_6
40. Schidler, A., Szeider, S.: Computing optimal hypertree decompositions. In: Blelloch, G., Finocchi, I. (eds.) Proceedings of ALENEX 2020, the 22nd Workshop on Algorithm Engineering and Experiments, pp. 1–11. SIAM (2020)
41. Swat, S.: swacisko/pace-2020: first release of ExTREEm, June 2020. Zenodo. https://doi.org/10.5281/zenodo.3873126
42. Villaamil, F.S.: About treedepth and related notions. Ph.D. thesis, RWTH Aachen University (2017)

# Perturbing Branching Heuristics in Constraint Solving

Anastasia Paparrizou(✉) and Hugues Wattez

CRIL, University of Artois & CNRS, Lens, France
{paparrizou,wattez}@cril.fr

**Abstract.** Variable ordering heuristics are one of the key settings for an efficient constraint solver. During the last two decades, a considerable effort has been spent for designing dynamic heuristics that iteratively change the order of variables as search progresses. At the same time, restart and randomization methods have been devised for alleviating heavy-tailed phenomena that typically arise in backtrack search. Despite restart methods are now well-understood, choosing *how* and *when* to randomize a given heuristic remains an open issue in the design of modern solvers. In this paper, we present several conceptually simple perturbation strategies for incorporating random choices in constraint solving with restarts. The amount of perturbation is controlled and learned in a bandit-driven framework under various stationary and non-stationary exploration policies, during successive restarts. Our experimental evaluation shows significant performance improvements for the perturbed heuristics compared to their classic counterpart, establishing the need for incorporating perturbation in modern constraint solvers.

## 1 Introduction

For decades now, researchers in Constraint Programming (CP) have put a tremendous effort in designing constraint solvers and advancing their internal components. Many mechanisms have been combined, leading to a technology that is now widely used to solve combinatorial problems. A constraint solver is typically composed of a backtracking search algorithm, a branching heuristic for guiding search, a filtering procedure for pruning the search space, and no-good recording.

Since the very beginning, the order in which variables are selected (assigned) by the branching heuristic holds a central place. It is referred to as the variable ordering heuristic. Choosing the appropriate variable ordering heuristic for solving a given constraint satisfaction problem is quite important since the solving time may vary by orders of magnitude from one heuristic to the other. Recent heuristics are more stable [15,36].

Backtrack search is also vulnerable to unstable behavior because of its inherent combinatorial nature. In the early '00s, the exponential time differences have been investigated under the phase-transition phenomena [19] and the heavy-tailed distributions of solving time [12]. We can decrease such undesired differences by introducing restart policies, randomization [6] and no-goods recording


H. Simonis (Ed.): CP 2020, LNCS 12333, pp. 496–513, 2020.
https://doi.org/10.1007/978-3-030-58475-7_29

during search. While restarts and no-goods are well established in CP solvers [11,18,24,27], randomization remains limited to ad-hoc techniques that have been found to work well in practice.

In this work, we use randomization to perturb the variable selection process. These perturbations are designed to keep a good and controlled balance between exploitation and exploration. We introduce conceptually simple *perturbation strategies* for incorporating random choices in constraint solving with restarts. Most of strategies that we present are adaptive, meaning that the amount of perturbation is learned during successive restarts. We exploit the restart mechanism that exists in all modern solvers to control the application of random choices. We deploy a reinforcement learning technique that determines at each run (i.e., at the beginning of the restart), if it will apply the standard heuristic, embedded in the constraint solver, or a procedure that makes random branching choices. This is a sequential decision problem and as such, it can be modeled as a multi-armed bandit problem (MAB) [3], precisely, as a double-armed. In reinforcement learning, the proportion between exploration and exploitation is specified by various policies. We tried several of them, such as `Epsilon-greedy` [31], `EXP3` [4], `Thompson Sampling` [32], the upper confidence bound `UCB1` [4] and `MOSS` [2]. The learning comes from the feedback taken after each run that reflects the efficiency of the run under a given choice, referred to as a reward function. We also propose a static strategy that perturbs a given heuristic with a fixed probability, found empirically. We evaluate the static and adaptive strategies for several well known heuristics, showing significant performance improvements in favor of the perturbed solver independently of the underlying heuristic used. A perturbed strategy always outperforms its baseline counterpart both in time and number of solved instances. We have also run experiments allowing the use of no-goods, an integral component of solvers nowadays, showing that perturbations still dominate the standard setting, as the no-goods obtained during random runs do not disorientate search.

Many useful observations are derived from this study. The more inefficient a heuristic is, the more effective the perturbation. A perturbed solver can compensate for a potentially bad heuristic choice done by the user, as it permits to automatically improve its performance by visiting unknown parts of the search space. This is due to the random runs, during which the heuristic acquires extra knowledge, other than what obtained when running alone. We show that adaptive strategies always outperform the static ones, as they can adjust their behavior to the instance to be solved and to the heuristic setting. Overall, the results show the benefits of establishing perturbation in CP solvers for improving their overall performance whatever their default setting is.

## 2 Related Work

Introducing a touch of randomization for better diversifying the search of local and complete procedures has been shown to be quite effective for both SAT (Satisfiability Testing) and CP (Constraint Programming). A stochastic local search requires the right setting of the "noise" parameter so as to optimize its performance. This parameter determines the likelihood of escaping from local minima by making non-optimal moves. In GSAT [30], it is referred as the random walk parameter and in walkSAT [29], simply as the "noise". Large Neighborhood Search uses randomization to perform jumps in the search space while freezing a fragment of the best solution obtained so far [26].

In complete CP solvers, the first evidence that diversification can boost search dates back to Harvey and Ginsberg research [17]. Harvey and Ginsberg proposed a deterministic backtracking search algorithm that differs from the heuristic path by a small number of decision points, or "discrepancies". Then, Gomes et al. [12,13] showed that a controlled form of randomization eliminates the phenomenon of "heavy-tailed cost distribution" (i.e., a non-negligible probability that a problem instance requires exponentially more time to be solved than any previously solved instances). Randomization was applied as a tie-breaking step: if several choices are ranked equally, choose among them at random. However, if the heuristic function is powerful, it rarely assigns more than one choice the highest score. Hence, the authors introduced a "heuristic equivalence" parameter in order to expand the choice set for random tie-breaking.

More recently, Grimes and Wallace [14] proposed a way to improve the classical `dom/wdeg` heuristic (based on constraint weighting) by using random probing, namely a pre-processing sampling procedure. The main idea is to generate the weights of the variables with numerous but very short runs (i.e., restarts) prior search, in order to make better branching decisions at the beginning of the search. Once the weights are initialized, a complete search is performed during which weights either remain frozen or continue updating.

## 3 Preliminaries

A *Constraint Network* $P$ consists in a finite set of variables $\texttt{vars}(P)$, and a finite set of constraints $\texttt{ctrs}(P)$. We use $n$ to denote the number of variables. Each variable $x$ takes values from a finite domain, denoted by $\texttt{dom}(x)$. Each constraint $c$ represents a mathematical relation over a set of variables, called the *scope* of $c$. A *solution* to $P$ is the assignment of a value to each variable in $\texttt{vars}(P)$ such that all constraints in $\texttt{ctrs}(P)$ are satisfied. A constraint network is *satisfiable* if it admits at least one solution, and the corresponding *Constraint Satisfaction Problem (CSP)* is to determine whether a given constraint network is satisfiable, or not. A classical procedure for solving this NP-complete problem is to perform a backtrack search on the space of partial solutions, and to enforce a property called *generalized arc consistency* [23] on each decision node, called *Maintaining Arc Consistency* (MAC) [28]. The MAC procedure selects the next variable to assign according to a *variable ordering heuristic*, denoted $H$. Then, the selected variable is assigned to a value according to its value ordering heuristic, which is usually the lexicographic order over $\texttt{dom}(x)$.

As mentioned in Sect. 2, backtrack search algorithms that rely on deterministic variable ordering heuristics have been shown to exhibit heavy-tailed behavior on both random and real-world CSP instances [12]. This issue can be alleviated using *randomization* and *restart* strategies, which incorporate some random choices in the search process, and iteratively restart the computation from the beginning, with a different variable ordering [6]. Since our randomization method will be discussed in Sect. 4, we focus here on restart strategies.

Conceptually, a restart strategy is a mapping $\texttt{res} : \mathbb{N} \rightarrow \mathbb{N}$, where $\texttt{res}(t)$ is the maximal number of "steps" which can be performed by the backtracking search algorithm at run, or *trial*, $t$. A constraint solver, equipped with the MAC procedure and a restart strategy $\texttt{res}$, builds a sequence of search trees $\langle \mathcal{T}^{(1)}, \mathcal{T}^{(2)}, \ldots \rangle$, where $\mathcal{T}^{(t)}$ is the search tree explored by MAC at run $t$. After each run, the solver can memorize some relevant information about the sequence $\langle \mathcal{T}^{(1)}, \mathcal{T}^{(2)}, \ldots, \mathcal{T}^{(t-1)} \rangle$, like the number of constraint checks in the previous runs, the no-goods that have appeared frequently in the search trees explored so far [20]. The *cutoff*, $\texttt{res}(t)$, which is the number of allowed steps, may be defined by the number of nodes, the number of wrong decisions [8], the number of seconds, or any other relevant measure. In a *fixed* cutoff restart strategy, the number $T$ of trials is fixed in advance, and $\texttt{res}(t)$ is constant for each trial $t$, excepted for the $T$th trial which allows an unlimited number of steps (in order to maintain a complete algorithm). This strategy is known to be effective in practice [13], but a good cutoff value $\texttt{res}(t)$ has to be found by trial and error. Alternatively, in a *dynamic* cutoff restart strategy, the number $T$ of trials is unknown, but $\texttt{res}$ increases geometrically, which guarantees that the whole space of partial solutions is explored after $O(n)$ runs [33]. A commonly used cutoff strategy is driven by the Luby sequence [22].

## 4 Perturbation Strategies

As indicated in Sect. 3, the process of constraint solving with a restart policy may be viewed as a sequence $\langle 1, 2, \cdots, T \rangle$ of *runs*. For the aforementioned restart functions, the sequence of runs is finite, but the horizon $T$ is not necessarily known in advance. During each run $t$, the solver calls the MAC algorithm for building a search tree $\mathcal{T}_t$, whose size is determined by the cutoff of the restart policy. If the solver has only access to a single variable ordering heuristic, say $H$, it will run MAC with $H$ after each restart. Yet, if the solver is also allowed to *randomize* its variable orderings, it is faced with a fundamental choice at each run $t$: either call MAC with the heuristic $H$ in order to "exploit" previous computations made with this heuristic, or call MAC with a random variable ordering $U$ so as to "explore" new search trees, and potentially better variable orderings. Here, $U$ is any variable ordering drawn at random according to a uniform distribution over the permutation group of $\texttt{vars}(P)$. We need to highlight here, that the intermediate random runs of $U$ perturb the involved classic heuristic $H$ by updating its parameters, which ultimately affects the behavior/performance of $H$. In other words, the subsequent heuristic runs, will not produce the same orderings as in the traditional solving process, allowing thus the solver to (potentially) tackle instances that neither $H$ nor $U$ would solve stand-alone.

**Algorithm 1:** Bandit-Driven Perturbations

**Input:** constraint network $P$, heuristic $H$, policy $B$

INITARMS$_B(H, U)$ // *Initialize the arms and the bandit policy*
**for** *each run* $t = 1, \cdots, T$ **do**
  $a_t \leftarrow$ SELECTARM$_B()$ // *Select an arm according to the bandit policy*
  $r_t(a_t) \leftarrow$ MAC$(P, a_t)$ // *Execute the solver and compute the reward*
  UPDATEARMS$_B(r_t(a_t))$ // *Update the bandit policy*

The task of incorporating perturbations into constraint solving with restarts can be viewed as a *double-armed* bandit problem: during each run $t$, we have to decide whether the MAC algorithm should be called using $H$ (exploitation arm) or $U$ (exploration arm). Once MAC has built a search tree $\mathcal{T}_t$, the performance of the chosen arm can be assessed using a reward function defined according to $\mathcal{T}_t$. The overall goal is to find a policy mapping each run $t$ to a probability distribution over $\{H, U\}$ so as to maximize cumulative rewards.

Multi-armed bandit algorithms have recently been exploited in CP in different contexts, i.e. for guiding search [21], for learning the right level of propagation [5] or the right variable ordering heuristic [34,35,37]. In the framework of Xia and Yap [37], a single search tree is explored (i.e., no restarts), and the bandit algorithm is called at each node of the tree to decide which heuristic to select. The trial is associated with explored subtrees, while in our approach, trials are mapped to runs using a restart mechanism. Our framework makes use of restarts in the same way as the ones of [34,35], as it was shown in [35] that such a framework offers greater improvements compared to the one of [37]. In our case, we utilise a double-armed framework in order to construct our bandit-driven perturbation given by Algorithm 1. The algorithm, takes as input a constraint network $P$, a variable ordering heuristic $H$, and a bandit policy $B$. As indicated above, the bandit policy has access to two arms, $H$ and $U$, where $U$ is the random variable ordering generated on the fly, during execution. The three main procedures used by the bandit policy are INITARMS$_B$ for initializing the distribution over $\{H, U\}$ according to policy $B$, SELECTARM$_B$ for choosing the arm $a_t \in \{H, U\}$ that will be used to guide the search all along the $t$th run, and UPDATEARMS$_B$ for updating the distribution over $\{H, U\}$ according to the observed reward $r_t(a_t)$ at the end of the $t$th run.

### 4.1 Reward Function

The feedback $r_t(a_t)$ supplied at the end of each run captures the performance of the MAC algorithm, when it is called using $a_t \in \{H, U\}$. To this end, the reward function $r_t$ maps the search tree $\mathcal{T}_t$ built by MAC$(a_t)$ to a numeric value in $[0, 1]$ that reflects the quality of backtracking search when guided by $a_t$.

As a reward function, we introduce the measure of the *explored sub-tree* denoted as `esb`. `esb` is given by the number of visited nodes during a run,

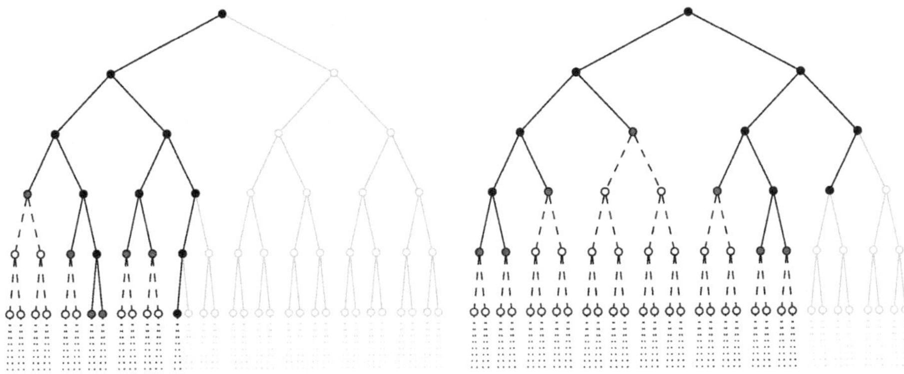

(a) $\texttt{esb}(\mathcal{T}_{\text{left}}) = \frac{\log(16)}{\log(2^6)} \approx 0.67$

(b) $\texttt{esb}(\mathcal{T}_{\text{right}}) = \frac{\log(16)}{\log(2^5)} \approx 0.80$

**Fig. 1.** Comparison of two runs with a restart cutoff fixed to 16 nodes.

divided by the size of the complete sub-tree defined over the variables selected during this run. The later is simply the size of the Cartesian product of the domains of the variables selected during the run. As selected variables, we consider those ones that have been chosen at least once by the arm in question. `esb` represents the search space covered by the solver, under a certain setting of a run, compared to the total possible space on the selected variables. The intuition is that an exploration that discovers failures deeply in the tree (meaning that, many variables are instantiated) will be penalized (due to the big denominator) against an exploration that discovers failures at the top branches.

In formal terms, given a search tree $\mathcal{T}$ generated by the MAC algorithm, let $\texttt{vars}(\mathcal{T})$ be the set of variables that have been selected at least once during exploration of $\mathcal{T}$ and $\texttt{nodes}(\mathcal{T})$ the number of visited nodes. Then,

$$r_t(a_t) = \texttt{esb}(\mathcal{T}_t) = \frac{\log(\texttt{nodes}(\mathcal{T}_t))}{\log\left(\prod_{x \in \texttt{vars}(\mathcal{T}_t)} |\texttt{dom}(x)|\right)}$$

A logarithmic scaling is needed to obtain a better discrimination between the arms as the numerator is usually significantly smaller than the denominator of the fraction. The reward values belong to [0, 1]. The higher the ratio/reward is, the better the performance of $a_t$ is.

Figure 1, shows a motivating example for the reward function. It displays an example of tree explorations done by two different runs. For simplicity, domains are binary and at each level the variable to be instantiated is fixed per run (namely left and right branches are on the same variable). Empty nodes represent non-visited or pruned nodes, solid black nodes are the visited ones, solid red ones denote failures. Below red nodes, there are the pruned sub-trees in dashed style while the non-visited sub-trees are slightly transparent. For both runs we consider the same number of node visits, i.e. 16. At the left run (Fig. 1a), the solver goes until level 6 (selecting 6 variables) while the solver at the right run (Fig. 1b) goes until level 5. The left run will take the score of 0.67 and the right one 0.80. Our bandit will prefer the arm that produced the right tree, namely a search that goes faster at the right branches than deeper in the tree (which implies that more search is required). Early failures is a desired effect that many heuristics consider explicitly on implicitly in order to explore smaller search spaces until the solution or unsatisfiability.

### 4.2 Perturbation Rate

As indicated in Algorithm 1, the perturbation framework is conceptually simple: based on a restart mechanism, the solver performs each run by first selecting an arm in $\{H, U\}$, next, observing a reward for that arm, and then, updating its bandit policy according to the observed reward. This simple framework, allows us to use a variety of computationally efficient bandit policies to adapt/control the amount of perturbation applied during search. From this perspective, we have opted for five well-studied "adaptive" policies for the double-armed bandit problem, and one "static" (or stationary) policy which serves as reference for our perturbation methods.

***$\epsilon$-Greedy.*** Arguably, this is the simplest adaptive bandit policy that interleaves exploitation and exploration using a parameter $\epsilon$. The policy maintains the empirical means $\hat{r}$ of observed rewards for $H$ and $U$. Initially, both $\hat{r}_1(H)$ and $\hat{r}_1(U)$ are set to 0. On each run $t$, the function $\text{SELECTARM}_{\epsilon G}$ returns with probability $(1-\epsilon)$ the arm $a_t$ that maximizes $\hat{r}_t(a_t)$, and returns with probability $\epsilon$ any arm $a_t$ drawn uniformly at random. Finally, based on the observed reward $r_t(a_t)$, the procedure $\text{UPDATEARMS}_{\epsilon G}$ updates the empirical mean of $a_t$ according to

$$\hat{r}_{t+1}(a_t) = \frac{t}{t+1}\hat{r}_t(a_t) + \frac{1}{t+1}r_t(a_t)$$

***EXP3.*** The EXPonentially weighted forecaster for EXPloration and EXPloitation (EXP3) is the baseline bandit policy operating in "non-stochastic" environments, for which no statistical assumption is made about the reward functions [4]. Several variants of EXP3 have been proposed in the literature, but we use here the simplest version defined in [10]. Here, the procedure $\text{INITARMS}_{\text{EXP3}}$ sets the initial distribution $\pi_1$ of arms to the uniform distribution $(1/2, 1/2)$. During each trial $t$, the procedure $\text{SELECTARM}_{\text{EXP3}}$ simply draws an arm $a_t$ according to the distribution $\pi_t$. Based on the observed reward $r_t(a_t)$, the procedure $\text{UPDATEARMS}_{\text{EXP3}}$ updates the distribution $\pi_t$ according to the multiplicative weight-update rule:

$$\pi_{t+1}(a) = \frac{\exp(\eta_t R_t(a))}{\exp(\eta_t R_t(H)) + \exp(\eta_t R_t(U))}$$

$\eta_t$ corresponds to the learning rate (usually set to $\frac{1}{\sqrt{t}}$),

$$R_t(a) = \sum_{s=1}^{t} \frac{r_s(a)}{\pi_s(a)} \mathbb{1}_{a \sim \pi_s}$$

and $\mathbb{1}_{a \sim \pi_s}$ indicates whether $a$ was the arm picked at trial $s$, or not.

*UCB1.* Upper Confidence Bound (UCB) policies are commonly used in "stochastic" environments, where it is assumed that the reward value $r_t(a)$ of each arm $a$ is drawn according to a fixed, but unknown, probability distribution. `UCB1` is the simplest policy in the Upper Confidence Bound family [3]. In the setting of our framework, this algorithm maintains two 2-dimensional vectors, namely, $n_t(a)$ is the number of times the policy has selected arm $a$ on the first $t$ runs, and $\hat{r}_t(a)$ is the empirical mean of $r_t(a)$ during the $n_t(a)$ steps. $\text{INITARMS}_{\texttt{UCB1}}$ sets both vectors to zero and, at each run $t$, $\text{SELECTARM}_{\texttt{UCB1}}$ selects the arm $a_t$ that maximizes

$$\hat{r}_t(a) + \sqrt{\frac{2\ln(t)}{n_t(a)}}$$

Finally, $\text{UPDATEARMS}_{\texttt{UCB1}}$ updates the vectors $n_t$ and $\hat{r}_t$ according to $a_t$ and $r_t(a_t)$, respectively.

*MOSS.* The Minimax Optimal Strategy in the Stochastic case (MOSS) algorithm is an instance of the UCB family. The only difference with `UCB1` lies in the confidence level which not only takes into account the number of plays of individual arms, but also the number of arms (2) and the number of runs ($t$). Specifically, $\text{SELECTARM}_{\texttt{MOSS}}$ chooses the arm $a_t$ that maximizes

$$\hat{r}_t(a) + \sqrt{\frac{4}{n_t(a)} \ln^+\left(\frac{t}{2n_t(a)}\right)}$$

where $\ln^+(x) = \ln\max\{1, x\}$.

*TS.* The Thompson Sampling algorithm is another well-known policy used in stochastic environments [1,32]. In essence, the TS algorithm maintains a beta distribution for the rewards of each arm. $\text{INITARMS}_{\texttt{TS}}$ sets $\alpha_1(a)$ and $\beta_1(a)$ to 1 for $a \in \{H, U\}$. On each run $t$, $\text{SELECTARM}_{\texttt{TS}}$ selects the arm $a_t$ that maximizes $\text{Beta}(\alpha_t(a), \beta_t(a))$, and $\text{UPDATEARMS}_{\texttt{TS}}$ uses $r_t(a_t)$ to update the beta distribution as follows:

$$\begin{aligned}\alpha_{t+1}(a) &= \alpha_t(a) + \mathbb{1}_{a=a_t} r_t(a_t)\\ \beta_{t+1}(a) &= \beta_t(a) + \mathbb{1}_{a=a_t}(1 - r_t(a_t))\end{aligned}$$

*SP.* Finally, in addition to the aforementioned adaptive bandit policies which learn a distribution on $\{H, U\}$ according to observed rewards, we shall consider the following Static Policy (SP): on each round $t$, $\text{SELECTARM}_{\texttt{SP}}$ chooses $H$ with probability $(1 - \epsilon)$, and $U$ with probability $\epsilon$. Although this policy shares some similarities with the $\epsilon$-greedy algorithm, there is one important difference: the distribution over $\{H, U\}$ is fixed in advance, and hence, SP does not take into account the empirical means of observed rewards. In other words, $\text{UPDATEARMS}_{\texttt{SP}}$ is a dummy procedure that always returns $(1 - \epsilon, \epsilon)$. This stationary policy will serve as reference for the adaptive policies in the experiments.

## 5 Experimental Evaluation

We have conducted experiments on a large dataset to demonstrate the performance of the proposed perturbations. The set includes all instances (612 in total) from the 2017's XCSP3 competition[1] coming from 60 different problem classes. The experiments have been launched on an 2.66 GHz Intel Xeon and 32 GB RAM nodes. We have used the `AbsCon`[2] solver in which we integrated our perturbation strategies and the strategies of [13] and [14]. We used 2-way branching, generalized arc consistency as the level of consistency, Luby progression based on node visits as restart policy (the constant is fixed to 100 in `AbsCon`) and the timeout set to 1,200 s. We have chosen a big variety of variable ordering heuristics, including recent, efficient and state-of-the-art ones: `dom` [16], `dom/ddeg` [7], `activity` [25], `dom/wdeg` [9], `CHS` [15], $\texttt{wdeg}^{\texttt{ca.cd}}$ [36] and finally `rand` which chooses uniformly randomly a variable order. Among these, `dom` and `dom/ddeg` do not record/learn anything between two runs (`dom` and `ddeg` are re-initialized at the root), while all the others learn during each (random) run and maintain this knowledge all along the solving process, which might change/improve their behavior. We have run all these original heuristics separately for a baseline comparison. Note that in our first experiments no-goods recording are switched off in the solver to avoid biasing the results of heuristics and strategies.

Regarding our perturbation strategies, we denote by `SP` the static perturbation and by `e-greedy`, `UCB1`, `MOSS`, `TS` and `EXP3` the various adaptive perturbation (`AP`) strategies. Epsilon of `SP` and `e-greedy` are fixed to 0.1. This value has been fixed offline after a linear search of the best value. Apart from comparing to the default solver settings (i.e., `original` heuristics), we compare to three other perturbation strategies from the bibliography. The one is the `sampling` algorithm of [14] that corresponds to the sampling pre-processing step which is fixed to 40 restarts with a cutoff of $n$ nodes corresponding to the number of variables of each instance. When the probing phase finishes, we continue updating the variable scoring as it produces better results. The second is the `equiv-30` that corresponds to the criterion of equivalence of [13]. This equivalence parameter is set to 30% as authors proposed. Last, we compare to the standard tie-breaking denoted `equiv-0`, where a random choice is done among the top ranked variables scored equally by the underlying heuristic. Note that there are no ties, `equiv-0` has no effect on the heuristic, as opposed to `equiv-30`.

Table 1 displays the results of the aforementioned settings and strategies on the XCSP'17 competition dataset. The comparison is given on the number of solved instances (*#inst*), within 1,200 s, the cumulative CPU time (*time*) computed from instances solved by at least one method and the percentage of perturbation (*%perturbation*) which is the mean perturbation of the solved instances, computed by the number of runs with the arm $U$ divided by the total of runs. Each time a setting has not solved an instance that another setting solved, it is penalized by the timeout time. Numbers in bold indicate that a strategy

[1] See http://www.cril.univ-artois.fr/XCSP17.
[2] See http://www.cril.fr/~lecoutre/#/softwares.

outperformed her corresponding default setting of the solver (i.e., `original`). Underlined numbers show the winning strategy. `dom` and `dom/ddeg`, the unaffected heuristics, appear at the top of the table as they cannot be perturbed by randomized runs. After each run their parameters are reinitialized and not accumulated as for the rest of the heuristics (perturbed ones). Hence, for `dom` and `dom/ddeg`, any additional instance that perturbation strategies are able to solve comes from an intervening run of $U$.

**Table 1.** Comparison of `original`, `sampling`, `equiv-30` and the proposed perturbed strategies for the XCSP'17 dataset.

| | | original | sampling | equiv-0 | equiv-30 | SP | | AP | | |
|---|---|---|---|---|---|---|---|---|---|---|
| | | | | | | | e-greedy | UCB1 | MOSS | TS | EXP3 |
| dom | #inst | 287 | **321** | **312** | **308** | **315** | **314** | **323** | **322** | **318** | **323** |
| | time (359) | 101,589 | **58,496** | **74,064** | **75,992** | **72,460** | **75,474** | **59,527** | **61,171** | **63,856** | **61,842** |
| | %perturb. | 0 | - | - | - | 10 | 7.7 | 32.8 | 21.0 | 24.9 | 44.1 |
| dom/ddeg | #inst | 307 | **321** | **319** | **324** | **343** | **337** | **345** | **346** | **342** | **343** |
| | time (365) | 85,131 | **61,071** | **67,537** | **66,005** | **46,179** | **51,371** | **41,404** | **43,280** | **44,981** | **40,260** |
| | %perturb. | 0 | - | - | - | 10 | 8.1 | 34.3 | 21.4 | 26.5 | 45.3 |
| activity | #inst | 342 | 311 | 334 | 329 | **356** | **356** | **352** | **353** | **350** | **351** |
| | time (372) | 52,989 | 82,041 | 60,304 | 64,619 | **36,688** | **36,125** | **39,552** | **37,463** | **40,306** | **39,371** |
| | %perturb. | 0 | - | - | - | 10 | 7.6 | 33.5 | 20.7 | 26.0 | 44.4 |
| dom/wdeg | #inst | 347 | 342 | **349** | **349** | **358** | **346** | **358** | **354** | **356** | **363** |
| | time (381) | 55,599 | 56,038 | **54,276** | **53,263** | **45,615** | 58,888 | **43,726** | **49,052** | **46,896** | **39,277** |
| | %perturb. | 0 | - | - | - | 10 | 11.6 | 34.3 | 23.9 | 28.6 | 45.2 |
| $\text{wdeg}^{\text{ca.cd}}$ | #inst | 366 | 354 | **368** | 361 | **370** | **368** | **371** | **372** | **367** | **369** |
| | time (389) | 42,565 | 52,344 | 43,944 | 49,717 | **38,966** | **39,292** | **41,745** | **40,433** | 44,941 | 42,661 |
| | %perturb. | 0 | - | - | - | 10 | 7.8 | 32.1 | 19.9 | 24.8 | 42.9 |
| CHS | #inst | 370 | 343 | **371** | 361 | **371** | **372** | 367 | **373** | 367 | 367 |
| | time (389) | 41,462 | 64,779 | 42,025 | 56,639 | **37,699** | **38,158** | 43,869 | **38,190** | 46,597 | 42,444 |
| | %perturb. | 0 | - | - | - | 10 | 7.3 | 32.6 | 19.9 | 25.7 | 44.5 |
| rand | #inst | 291 | | | | | | | | | |
| | time (291) | 12,921 | | | | | | | | | |
| | %perturb. | 100 | | | | | | | | | |

The existing perturbation techniques, `sampling`, `equiv-30` and `equiv-0`, solve more instances than their respective baseline heuristic for the case of `dom` and `dom/ddeg`. However, this never happens for `sampling` and `equiv-30` on the more sophisticated heuristics (except from `dom/wdeg`), where we see that the perturbation they apply disorientates totally the search, solving constantly less instances than the original heuristics (e.g., `sampling` missed 31 instances for `activity`). `equiv-0` just marginally outperforms pure heuristics by one or two instances while it is far inferior to `AP`s (e.g., it missed 22 instances compared to `e-greedy` on `activity`). `equiv-30` is superior to `sampling` on the perturbed heuristics but still far inferior to all proposed strategies. Among the proposed perturbation strategies, we observe that both `SP` and `AP` strategies constantly outperform the default setting of the solver, while for many heuristics they have close performance (e.g., `activity`). `MOSS` is the best strategy in terms of solved instances and time results, except for `activity` and `dom/wdeg` heuristics, where `SP` (`e-greedy` too) and `EXP3` dominate respectively. `UCB1` and `MOSS` are the best strategies for `dom` and `dom/ddeg`, with only one instance of difference, showing that a perturbation rate between 20% and 30% is the best choice. On the other side, `TS` with a similar rate seems not to select that well the right arm for all runs. Similarly for `equiv-30`, which also applies a randomization of 30% on the top ranked variables, it seems that the way it is applied (i.e., at every decision) is not that efficient. `SP` and `e-greedy`, despite winning their original counterparts are less competitive due to their low perturbation rate. `EXP3` is also a good candidate policy for `dom` and `dom/ddeg`.

`rand` solved the less instances in total, i.e. 291, in 12,921 seconds, which means that many of them correspond to quite easy instances. As `rand` is the "bad" arm, a small participation of 10% in `SP` is just enough to be beneficial for the heuristics that are by themselves efficient (e.g., $\texttt{wdeg}^{\texttt{ca.cd}}$, `CHS`), but as `SP` is not adaptive, it is rarely better than the `AP` strategies; it cannot adjust its behavior to heuristics that require more perturbation to improve (e.g., `dom`). `AP` strategies, being adaptive, allow usually much more exploration of the $U$ arm that makes them win several instances. An exception is the `CHS` heuristic, where a perturbation over 20% might be harmful (i.e., in `UCB1`, `TS` and `EXP3` policies). Each bandit policy follows a general trend that can vary between heuristics (e.g., `UCB1` is around 30%, `MOSS` around 20% and `e-greedy` with `EPX3` represent the two extremities). `EXP3` sets the initial distribution $\pi_1$ of arms to $(1/2, 1/2)$ such that both arms have equal chances at the beginning. As it is a non-stochastic bandit it needs more exploration than stochastic ones need to converge. Although one would expect that this could deteriorate the solver, it seems that the bandit utilizes the right arm at each run since it is efficient both in time and solved instances. The high perturbation rates come from the mean, that smoothens the high variance between instances. Also, many of the instances are solved fast and `EXP3` favorizes a lot $U$, being the best arm at the early (short) runs, while the long runs at the end are done by $H$, which explains its good overall performance. `TS` despite it is better than original heuristics and existing perturbation methods, it is usually inferior to `SP` by small differences.

We distinguish the `MOSS` policy, which apart from being the best policy for many heuristics, it never deteriorates the solver for any heuristic (as happens for some strategies on `CHS`). It applies the exploration when and where needed (more at the early runs) and converges faster to the best heuristic. Regarding the time performance, all perturbation strategies are faster than their baseline heuristic, even for the most efficient heuristics (i.e., `CHS` and $\texttt{wdeg}^{\texttt{ca.cd}}$).

Figure 2, visualizes in a cactus plot the performance of the best `AP` strategy, namely `MOSS`, for all heuristics compared to their corresponding `original` heuristic. On x-axis we see the instances solved as time progresses. y-axis displays the allowed time given to the solver. Dashed lines display the performance of `original` heuristics and solid lines the perturbed solver. The closer a line is to the right bottom corner the more instances has solved in less time. In general, `MOSS` policy perturbations appear always at the right side of their respective default heuristic. We observe that for the less efficient heuristics, as `dom` and `dom/ddeg`, the performance gap between the `original` and the respective perturbed version is big even for easy instance and increases significantly as time passes, corresponding to more difficult instances. It worths noticing that, `dom/ddeg` after being perturbed, becomes better than `activity` (solved 4 more instances), that originally was much more efficient than `dom/ddeg`. $\texttt{dom}_{\texttt{MOSS}}$ solved in total 35 (resp. 39) more instances than `dom` (resp. `dom/ddeg`). For `activity` and `dom/wdeg`, the curves are closer, though the gap is still significant especially for harder instances. Even for the most efficient heuristics proposed the last two years, namely `CHS` and $\texttt{wdeg}^{\texttt{ca.cd}}$, `MOSS` is almost all the time the best setting for the solver.

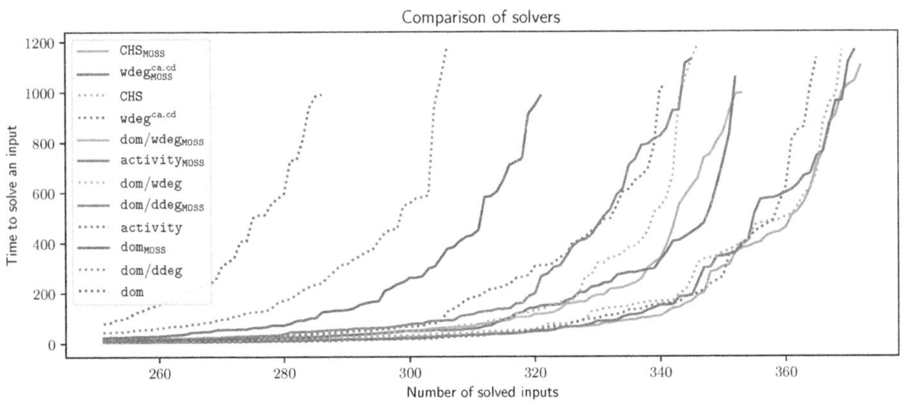


**Fig. 2.** Comparison of `original` and `MOSS` strategies while increasing the allowed time.

In the following, we do not present results for all adaptive strategies, but only for the most stable and efficient ones. We omit presenting results for `EXP3` because it failed for `CHS`, which is one of the two best heuristics in CP, and for `TS`, because it is never the winning strategy.

Table 2 gives complementary information derived from Table 1 for the best proposed strategies. We have calculated the number of instances solved exclusively by `rand` (i.e., instances that the heuristic alone could not solve), denoted by *#random*, and the number of instances, *#perturb*, that the solver solved due to the perturbation `rand` caused to the corresponding heuristic. Instances in *#perturb* are those that are solved neither by the `original` heuristic nor the `rand` heuristic, making thus the perturbed solver outperform even its corresponding virtual best solver. As expected, `dom` and `dom/ddeg` do not win instances due to perturbation, but only due to a good ordering during a random run (more than 30 instances for each policy). For other heuristics, we see a more balanced distribution of instances in *#random* and *#perturb*. We observe that the most robust heuristics (CHS and $\texttt{wdeg}^{ca.cd}$) solve extra instances mainly by perturbation (for MOSS: 10 and 7 respectively) rather than by calling rand (3 and 3). Indeed, both are so efficient that the majority of solved instances by rand are also solved by them, which explains why their perturbations'gain' fewer instances than the other heuristics do. In contrast, activity and dom/wdeg, that are less efficient than CHS and $\texttt{wdeg}^{ca.cd}$, gain more instances in total, most of which are gained by rand.

**Table 2.** Won instances by perturbation or random runs for SP, e-greedy, UCB1 and MOSS with nodes as cutoff for the XCSP'17 dataset

| | | dom | dom/ddeg | activity | dom/wdeg | $\texttt{wdeg}^{ca.cd}$ | CHS |
|---|---|---|---|---|---|---|---|
| SP | #perturb. | 0 | 0 | 8 | 8 | 5 | 4 |
| | #random | 28 | 36 | 12 | 10 | 3 | 3 |
| e-greedy | #perturb. | 0 | 0 | 8 | 5 | 5 | 6 |
| | #random | 29 | 31 | 10 | 6 | 3 | 2 |
| UCB1 | #perturb. | 0 | 0 | 5 | 7 | 8 | 7 |
| | #random | 38 | 41 | 13 | 10 | 3 | 2 |
| MOSS | #perturb. | 0 | 0 | 6 | 7 | 7 | 10 |
| | #random | 37 | 41 | 12 | 8 | 3 | 3 |

As the results in Table 1 are quite condensed, in Table 3 we show how strategies operate and adjust for certain problem classes. For each class and each heuristic, we present the number of solved instances, the total time and the perturbation rate. For *CoveringArray*, MOSS is the best strategy to apply the appropriate perturbation rate, independently of the heuristic chosen, compared to SP and e-greedy whose rate is too low. For *SuperSolutions* with dom as heuristic method, we see that e-greedy and MOSS are both winners despite their totally different rates. Though, it is notable that MOSS is twice faster than e-greedy on the same instances. SP with a rate close to e-greedy wins the half instances compared to it (and MOSS). Such observations are clear evidences, that not only the amount of randomization counts but also the when it appears. Policies as MOSS learn to discriminate on which run to apply $H$ or $U$. Recall that, compared to other policies, MOSS considers in $\text{SELECTARM}_{\texttt{MOSS}}$ more parameters, as the number or arms and the number of runs. In *KnightTour*, all strategies are efficient, but APs are always better than SP in terms of time. For *Blackhole* and *LatinSquare*, perturbation is not fruitful and thus, both APs converge to very low rates even when instances are easy (just few seconds per instance). Notice that, in general dom/wdeg is helped a lot by perturbation, as in all classes rates are higher compared to other heuristics (double percentages). Also, the percentage of perturbation varies a lot depending on the heuristic and the problem class, which is the reason of the success of APs.

**Table 3.** Comparison of `original`, `SP`, `e-greedy` and `MOSS` strategies on a subset of families from the XCSP'17 dataset.

| | | original | SP | e-greedy | MOSS |
|---|---|---|---|---|---|
| | dom/ddeg | 2 (4,803s, 0%) | **3 (4,533s, 10%)** | 2 (4,803s, 3%) | <u>**4 (2,952s, 17%)**</u> |
| *CoveringArray* | dom/wdeg | 4 (2,669s, 0%) | 4 (3,450s, 10%) | 3 (3,605s, 31%) | <u>**5 (2,156s, 34%)**</u> |
| | CHS | 4 (2,483s, 0%) | **4 (2,424s, 10%)** | **5 (1,705s, 3%)** | <u>**6 (1,491s, 17%)**</u> |
| | dom | 2 (8,404s, 0%) | **4 (6,972s, 10%)** | **6 (7,471s, 6%)** | <u>**6 (3,912s, 24%)**</u> |
| *SuperSolutions* | $\texttt{wdeg}^{\texttt{ca.cd}}$ | 5 (4,968s, 0%) | <u>**7 (3,101s, 10%)**</u> | **6 (4,467s, 9%)** | **7 (3,821s, 23%)** |
| | CHS | 7 (3,507s, 0%) | 6 (3,788s, 10%) | 6 (3,812s, 6%) | <u>**8 (1,689s, 20%)**</u> |
| | dom/ddeg | 5 (7,220s, 0%) | **9 (3,276s, 10%)** | <u>**9 (2,623s, 34%)**</u> | **9 (3,092s, 37%)** |
| *KnightTour* | activity | 7 (4,891s, 0%) | **9 (3,530s, 10%)** | **9 (2,823s, 26%)** | <u>**9 (2,513s, 30%)**</u> |
| | dom/wdeg | 5 (7,218s, 0%) | **9 (4,302s, 10%)** | **7 (5,509s, 25%)** | <u>**9 (3,286s, 39%)**</u> |
| | dom | 6 (2,609s, 0%) | **6 (2,489s, 10%)** | **6 (2,503s, 4%)** | <u>**6 (2,423s, 16%)**</u> |
| *Blackhole* | dom/ddeg | 8 (327s, 0%) | **8 (44s, 10%)** | **8 (55s, 4%)** | <u>**8 (25s, 20%)**</u> |
| | activity | 8 (897s, 0%) | 7 (1,618s, 10%) | 8 (908s, 6%) | <u>**8 (816s, 17%)**</u> |
| | dom | 8 (5,841s, 0%) | **8 (5,840s, 10%)** | 8 (5,928s, 2%) | <u>**8 (5,813s, 13%)**</u> |
| *LatinSquare* | dom/wdeg | 10 (2,481s, 0%) | <u>**12 (1,143s, 10%)**</u> | 10 (3,262s, 44%) | 10 (3,351s, 40%) |
| | CHS | <u>11 (3,031s, 0%)</u> | 9 (3,947s, 10%) | 9 (4,469s, 3%) | 8 (5,866s, 13%) |

As modern solvers exploit no-goods to improve their overall performance, we repeated our experiments by activating no-goods in order to examine the robustness of the proposed strategies and the interaction between no-goods and perturbation. Table 4 displays the results `sampling`, `equiv-0`, `equiv-30`, `SP` and the best `AP` strategies, namely `e-greedy`, `UCB1` and `MOSS`. As seen in Table 1, `sampling` and `equiv-30` make some improvements on the less efficient heuristics as `dom` and `dom/ddeg`, but are still inferior to `SP` and `MOSS`, while they are inefficient on all other heuristics. Surprisingly, `equiv-0`, despite being still more efficient than `sampling` and `equiv-30`, seems to interact badly with the presence of no-goods, as it can no longer improve the solver for any heuristic (just marginally `dom/ddeg`). `SP`, despite being static, it remains efficient apart from the case of `CHS`. Regarding the `AP` strategies, `e-greedy` (resp. `UCB1`) wins almost always the underlying heuristic except from the case of `CHS` (resp. $\texttt{wdeg}^{\texttt{ca.cd}}$). `MOSS` is again the most stable strategy, being able to improve all heuristics it perturbed. Note that the presence of no-goods has improved the performance of both the default and the perturbed solver. Therefore, there are slightly smaller differences between them compared to Table 1. The proposed perturbation strategies are robust to this fundamental parameter for solvers compared to existing strategies and adapt their behavior, especially `MOSS`.

**Table 4.** Comparison of `original`, `sampling`, `equiv-30`, `SP`, `e-greedy` and `MOSS` with nodes as cutoff and no-goods activated.

| | | original | sampling | equiv-0 | equiv-30 | SP | | AP | |
|---|---|---|---|---|---|---|---|---|---|
| | | | | | | | e-greedy | UCB1 | MOSS |
| dom | #inst | 309 | **321** | 306 | **316** | **338** | **334** | **342** | **340** |
| | time (359) | 73,367 | **57,279** | 74,364 | **68,446** | **43,327** | **51,168** | **42,219** | **44,027** |
| | %perturb. | 0 | - | - | - | 10 | 8.6 | 33.9 | 21.4 |
| dom/ddeg | #inst | 316 | **322** | **320** | **321** | **347** | **346** | **352** | **347** |
| | time (367) | 71,191 | **62,375** | **68,703** | **72,550** | **40,042** | **41,530** | **30,770** | **38,995** |
| | %perturb. | 0 | - | - | - | 10 | 8.8 | 34.9 | 22.5 |
| activity | #inst | 352 | 319 | 349 | 346 | **356** | **357** | **356** | **355** |
| | time (373) | 40,601 | 74,680 | 45,590 | 46,245 | **32,328** | **33,161** | **34,228** | **35,854** |
| | %perturb. | 0 | - | - | - | 10 | 7.8 | 34.1 | 20.7 |
| dom/wdeg | #inst | 352 | 340 | 349 | 344 | **359** | **356** | **364** | **364** |
| | c.time (377) | 41,882 | 53,238 | 48,073 | 53,841 | **34,253** | **40,100** | **29,559** | **30,995** |
| | %perturb. | 0 | - | - | - | 10 | 11.9 | 34.7 | 23.9 |
| $wdeg^{ca.cd}$ | #inst | 373 | 356 | 373 | 361 | 373 | **375** | 371 | **377** |
| | time (388) | 33,887 | 50,009 | 33,053 | 53,587 | **33,412** | **32,359** | 33,772 | **28,614** |
| | %perturb. | 0 | - | - | - | 10 | 7.8 | 32.8 | 19.8 |
| CHS | #inst | 375 | 348 | 372 | 366 | 373 | 373 | **375** | **376** |
| | time (387) | 30,990 | 58,121 | 36,829 | 47,159 | **30,658** | 31,547 | 31,020 | **30,585** |
| | %perturb. | 0 | - | - | - | 10 | 7.1 | 33.0 | 19.8 |
| rand | #inst | 293 | | | | | | | |
| | time (293) | 16,992 | | | | | | | |
| | %perturb. | 100 | | | | | | | |

## 6 Conclusion

We presented several strategies that significantly improve the performance and robustness of the solver by perturbing the default branching heuristic. It is the first time an approach tries to learn how and when to apply randomization in an on-line and parameter-free fashion. Controlled random search runs help variable ordering heuristics to acquire extra knowledge from parts of the search space that they were not dedicated to explore. We summarize the benefits of our approach which are manifold:

- Our perturbation techniques constantly improve the performance of the solver independently of the heuristic used as baseline in the solver. A perturbed strategy always outperforms its baseline counterpart both in time and solved instances.
- The presence of no-goods does not impact the efficacy of the perturbed solver. The produced no-goods are still fruitful.
- Perturbed heuristics can compensate for a wrong heuristic choice done by the user. Thanks to perturbation, the performance of the solver with a bad initial heuristic can reach or even outperform the performance of the solver with a

better baseline heuristic. This is a step towards autonomous and adaptive solving, where the solver learns and adjusts its behavior to the instance being solved.
- Our approach is generic and easy to embed in any solver that exploits restarts.

**Acknowledgments.** The authors would like to thank Frederic Koriche for his valuable advices on the Machine Learning aspects of the paper as well as the anonymous reviewers for their constructive remarks. This work has been partially supported by the project *Emergence* 2020 BAUTOM of INS2I and the project CPER Data from the region "Hauts-de-France".

## References

1. Agrawal, S., Goyal, N.: Near-optimal regret bounds for Thompson Sampling. J. ACM **64**(5), 30:1–30:24 (2017)
2. Audibert, J.Y., Bubeck, S.: Minimax policies for adversarial and stochastic bandits. In: COLT, Montreal, Canada, pp. 217–226 (2009)
3. Auer, P., Cesa-Bianchi, N., Fischer, P.: Finite-time analysis of the multiarmed bandit problem. Mach. Learn. **47**(2), 235–256 (2002)
4. Auer, P., Cesa-Bianchi, N., Freund, Y., Schapire, R.: The nonstochastic multiarmed bandit problem. SIAM J. Comput. **32**(1), 48–77 (2002)
5. Balafrej, A., Bessiere, C., Paparrizou, A.: Multi-armed bandits for adaptive constraint propagation. In: Proceedings of IJCAI 2015, pp. 290–296 (2015)
6. van Beek, P.: Backtracking search algorithms. In: Handbook of Constraint Programming, Chap. 4, pp. 85–134. Elsevier (2006)
7. Bessière, C., Régin, J.-C.: MAC and combined heuristics: two reasons to forsake FC (and CBJ?) on hard problems. In: Freuder, E.C. (ed.) CP 1996. LNCS, vol. 1118, pp. 61–75. Springer, Heidelberg (1996). https://doi.org/10.1007/3-540-61551-2_66
8. Bessiere, C., Zanuttini, B., Fernandez, C.: Measuring search trees. In: Proceedings of ECAI 2004 Workshop on Modelling and Solving Problems with Constraints, pp. 31–40 (2004)
9. Boussemart, F., Hemery, F., Lecoutre, C., Sais, L.: Boosting systematic search by weighting constraints. In: Proceedings of ECAI 2004, pp. 146–150 (2004)
10. Bubeck, S., Cesa-Bianchi, N.: Regret Analysis of Stochastic and Nonstochastic Multi-armed Bandit Problems. Foundations and Trends in Machine Learning. Now Publishers (2012)
11. Gecode Team: Gecode: generic constraint development environment (2006). http://www.gecode.org
12. Gomes, C., Selman, B., Crato, N., Kautz, H.: Heavy-tailed phenomena in satisfiability and constraint satisfaction problems. J. Autom. Reason. **24**(1), 67–100 (2000)
13. Gomes, C.P., Selman, B., Kautz, H.: Boosting combinatorial search through randomization. In: Proceedings of AAAI 1998, pp. 431–437 (1998)
14. Grimes, D., Wallace, R.J.: Sampling strategies and variable selection in weighted degree heuristics. In: Bessière, C. (ed.) CP 2007. LNCS, vol. 4741, pp. 831–838. Springer, Heidelberg (2007). https://doi.org/10.1007/978-3-540-74970-7_61
15. Habet, D., Terrioux, C.: Conflict history based branching heuristic for CSP solving. In: Proceedings of the 8th International Workshop on Combinations of Intelligent Methods and Applications (CIMA), Volos, Greece, November 2018

16. Haralick, R., Elliott, G.: Increasing tree search efficiency for constraint satisfaction problems. Artif. Intell. **14**, 263–313 (1980)
17. Harvey, W.D., Ginsberg, M.L.: Limited discrepancy search. In: Proceedings of the 14th International Joint Conference on Artificial Intelligence, IJCAI 1995, vol. 1, pp. 607–613 (1995)
18. Hebrard, E.: Mistral, a constraint satisfaction library. Proc. Third Int. CSP Solver Competition **3**(3), 31–39 (2008)
19. Hogg, T., Huberman, B.A., Williams, C.P.: Phase transitions and the search problem. Artif. Intell. **81**(1), 1–15 (1996)
20. Lecoutre, C., Sais, L., Tabary, S., Vidal, V.: Recording and minimizing nogoods from restarts. J. Satisfiability, Boolean Model. Comput. (JSAT) **1**, 147–167 (2007)
21. Loth, M., Sebag, M., Hamadi, Y., Schoenauer, M.: Bandit-based search for constraint programming. In: Schulte, C. (ed.) CP 2013. LNCS, vol. 8124, pp. 464–480. Springer, Heidelberg (2013). https://doi.org/10.1007/978-3-642-40627-0_36
22. Luby, M., Sinclair, A., Zuckerman, D.: Optimal speedup of Las Vegas algorithms. Inf. Process. Lett. **47**(4), 173–180 (1993)
23. Mackworth, A.: On reading sketch maps. In: Proceedings of IJCAI 1977, pp. 598–606 (1977)
24. Merchez, S., Lecoutre, C., Boussemart, F.: AbsCon: a prototype to solve CSPs with abstraction. In: Proceedings of CP 2001, pp. 730–744 (2001)
25. Michel, L., Van Hentenryck, P.: Activity-based search for black-box constraint programming solvers. In: Beldiceanu, N., Jussien, N., Pinson, É. (eds.) CPAIOR 2012. LNCS, vol. 7298, pp. 228–243. Springer, Heidelberg (2012). https://doi.org/10.1007/978-3-642-29828-8_15
26. Pisinger, D., Ropke, S.: Large neighborhood search. In: Gendreau, M., Potvin, J.Y. (eds.) Handbook of Metaheuristics. International Series in Operations Research & Management Science, vol. 146, pp. 399–419. Springer, Boston (2010). https://doi.org/10.1007/978-1-4419-1665-5_13
27. Prud'homme, C., Fages, J.G., Lorca, X.: Choco Solver Documentation. TASC, INRIA Rennes, LINA CNRS UMR 6241, COSLING S.A.S. (2016). http://www.choco-solver.org
28. Sabin, D., Freuder, E.C.: Contradicting conventional wisdom in constraint satisfaction. In: Borning, A. (ed.) PPCP 1994. LNCS, vol. 874, pp. 10–20. Springer, Heidelberg (1994). https://doi.org/10.1007/3-540-58601-6_86
29. Selman, B., Kautz, H.A., Cohen, B.: Noise strategies for improving local search. In: Proceedings of AAAI 1994, pp. 337–343 (1994)
30. Selman, B., Levesque, H., Mitchell, D.: A new method for solving hard satisfiability problems. In: Proceedings of the Tenth National Conference on Artificial Intelligence, AAAI 1992, pp. 440–446. AAAI Press (1992). http://dl.acm.org/citation.cfm?id=1867135.1867203
31. Sutton, R., Barto, A.G.: Reinforcement learning: an introduction **9**, 1054 (1998)
32. Thompson, W.R.: On the likelihood that one unknown probability exceeds another in view of the evidence of two samples. Biometrika **25**(3–4), 285–294 (1933)
33. Walsh, T.: Search in a small world. In: Proceedings of IJCAI 1999, pp. 1172–1177 (1999)
34. Wattez, H., Koriche, F., Lecoutre, C., Paparrizou, A., Tabary, S.: Heuristiques de recherche: un bandit pour les gouverner toutes. In: 15es Journées Francophones de Programmation par Contraintes - JFPC 2019 (2019). https://hal.archives-ouvertes.fr/hal-02414288

35. Wattez, H., Koriche, F., Lecoutre, C., Paparrizou, A., Tabary, S.: Learning variable ordering heuristics with multi-armed bandits and restarts. In: Proceedings of ECAI 2020 (2020). (To appear)
36. Wattez, H., Lecoutre, C., Paparrizou, A., Tabary, S.: Refining constraint weighting. In: Proceedings of ICTAI 2019, pp. 71–77 (2019)
37. Xia, W., Yap, R.H.C.: Learning robust search strategies using a bandit-based approach. In: Proceedings of AAAI 2018, pp. 6657–6665 (2018)

# Finding the Hardest Formulas for Resolution

Tomáš Peitl[1] and Stefan Szeider[2(✉)]

[1] Friedrich Schiller University Jena, Jena, Germany
tomas.peitl@uni-jena.de
[2] TU Wien, Vienna, Austria
sz@tuwien.ac.at

**Abstract.** A CNF formula is harder than another CNF formula with the same number of clauses if it requires a longer resolution proof. The *resolution hardness numbers* give for $m = 1, 2, \ldots$ the length of a shortest proof of a hardest formula on $m$ clauses. We compute the first ten resolution hardness numbers, along with the corresponding hardest formulas. We achieve this by a candidate filtering and symmetry breaking search scheme for limiting the number of potential candidates for formulas and an efficient SAT encoding for computing a shortest resolution proof of a given candidate formula.

**Keywords:** Resolution proofs · Minimal unsatisfiability · SAT encodings · Graph symmetries · Computational experiments

## 1 Introduction

Resolution is a fundamental proof system that can be used to certify the unsatisfiability of a propositional formula in conjunctive normal form (CNF). What makes resolution particularly interesting is that the length of a shortest resolution proof of a given CNF formula (called the *resolution complexity* of the formula) provides an unconditional lower bound on the running time of modern SAT solvers [17]. Since we know that there are classes of unsatisfiable CNF formulas (such as the formulas based on the Pigeon Hole Principle) with exponential resolution complexity [6], we have an exponential lower bound on the runtime. It is a natural question to ask: *which formulas are the hardest for resolution?* i.e., which formulas have the highest resolution complexity? This is a quite intriguing and hard question, which has been approached mainly in an asymptotic way by propositional proof complexity [22].

We address this question by following a recent trend in tackling combinatorial problems using SAT and CSP methods [2,7,8]. For small values of $n$ and $m$, we compute all the formulas (modulo isomorphisms) with $n$ variables and $m$ clauses

The authors acknowledge the support by the FWF (projects P32441 and J-4361) and by the WWTF (project ICT19-065).


H. Simonis (Ed.): CP 2020, LNCS 12333, pp. 514–530, 2020.
https://doi.org/10.1007/978-3-030-58475-7_30

that are the hardest formulas for resolution. With these results, we can compute the first few *resolution hardness numbers* $(h_m)_{m\geq 1}$, where $h_m$ gives the highest resolution complexity of a CNF formula with $m$ clauses.

We obtain our results by the combination of two techniques:

1. A *candidate filtering and symmetry breaking search scheme* for limiting the number of potential candidate formulas with $m$ variables whose resolution complexity is $h_m$.
2. An *efficient SAT encoding* for computing the resolution complexity of a given candidate formula.

In our search scheme, we reduce the candidate formulas to a certain class of minimally unsatisfiable (MU) formulas that obey additional degree constraints. We model these formulas by graphs of a particular kind. We generate these graphs modulo symmetries by a special adaptation of the Nauty graph symmetry package.

This still leaves us with a large number of formulas whose resolution complexity we must determine algorithmically. For this task, we devised an efficient SAT encoding that produces for a given candidate formula $F$ and an integer $s$, a CNF formula $\mathtt{short}_s(F)$, which is satisfiable if and only if $F$ admits a resolution proof of length $\leq s$. We determine the resolution complexity of $F$ by feeding $\mathtt{short}_s(F)$ to a SAT solver with various choices of $s$. While a SAT encoding for this problem has been proposed before [14], and we do take some inspiration from it, we make crucial adaptations tailored towards minimally unsatisfiable formulas. Furthermore, we introduce a symmetry-breaking scheme that fully breaks all symmetries resulting from permutations of the sequence of clauses.

In addition to the values of the resolution hardness numbers, we can draw a more detailed map of the hardest formulas with a particular number $n$ of variables and a particular number $m$ of clauses.

Our theoretical results reveal the significance of *regular saturated minimally unsatisfiable (RSMU)* formulas, which are unsatisfiable formulas that (i) become satisfiable by adding any further literal to any clause, and (ii) where each literal appears in at least two clauses. As a by-product of our computations, we obtain a catalog of RSMU formulas with a small number of variables and clauses, which may be of independent interest in the research on minimal unsatisfiability. For instance, the computed formulas' structure can possibly be used to come up with infinite sequences of hard formulas, which can lead to tighter general bounds.

An alternative but not very interesting object of study would be $v_n$, the highest resolution complexity of formulas with $n$ *variables*. It is not hard to see that every unsatisfiable formula on $n$ variables has a resolution refutation of length $\leq 2^{n+1} - 1$ and that indeed $v_n = 2^{n+1} - 1$, witnessed by the formula which contains all possible clauses of width $n$.

## 2 Preliminaries

*Formulas.* We consider propositional formulas in conjunctive normal form (CNF) represented as sets of clauses. We assume an infinite set *var* of (propositional) *variables*. A *literal* $\ell$ is a variable $x$ or a negated variable $\neg x$; we write $lit := \{ x, \neg x \mid x \in var \}$. For a literal $\ell$ we put $\overline{\ell} := \neg x$ if $\ell = x$, and $\overline{\ell} := x$ if $\ell = \neg x$. For a set of $C$ literals we put $\overline{C} := \{ \overline{\ell} \mid \ell \in C \}$. $C$ is *tautological* if $C \cap \overline{C} \neq \emptyset$. A finite non-tautological set of literals is a *clause*; a finite set of clauses is a (CNF) *formula*. The empty clause is denoted by $\square$. We write $\mathrm{CNF}(n, m)$ for the class of all CNF formulas on $n$ variables and $m$ clauses, and $\mathrm{CNF}(m) = \bigcup_{n=0}^{\infty} \mathrm{CNF}(n, m)$. For a clause $C$, we put $var(C) = \{ var(\ell) \mid \ell \in C \}$, and for a formula $F$, $var(F) = \bigcup_{C \in F} var(C)$. Similarly, we put $lit(F) := var(F) \cup \overline{var(F)}$. A formula $F$ is *satisfiable* if there is a mapping $\tau : var(F) \to \{0, 1\}$ such that every clause of $F$ contains either a literal $x$ with $\tau(x) = 1$ or a literal $\neg x$ with $\tau(x) = 0$, and *unsatisfiable* otherwise. A formula is *minimally unsatisfiable (MU)* if it is unsatisfiable, but every proper subset is satisfiable.

*Resolution Proofs.* If $C_1 \cap \overline{C_2} = \{\ell\}$ for clauses $C_1, C_2$ and a literal $\ell$, then the *resolution rule* allows the derivation of the clause $D = (C_1 \cup C_2) \setminus \{\ell, \overline{\ell}\}$; $D$ is the *resolvent* of the *premises* $C_1$ and $C_2$, and we say that $D$ is obtained by *resolving on* $\ell$. Let $F$ be a formula and $C$ a clause. A sequence $P = L_1, \dots, L_s$ of clauses (proof *lines*) is a *resolution derivation of* $L_s$ *from* $F$ if for each $i \in \{1, \dots, s\}$ at least one of the following holds.

1. $L_i \in F$ ("$L_i$ is an axiom");
2. $L_i$ is the resolvent of $L_j$ and $L_{j'}$ for some $1 \leq j < j' < i$ ("$L_i$ is obtained by resolution").

We write $|P| := s$ and call $s$ the *length* of $P$. If $L_s$ is the empty clause, then $P$ is a *resolution refutation* or *resolution proof* of $F$. A line $L_i$ in a resolution derivation may have different possible "histories;" i.e., $L_i$ may be the resolvent of more than one pair of clauses preceding $L_i$, or $L_i$ may be both an axiom and obtained from preceding clauses by resolution, etc. In the sequel, however, we assume that an arbitrary but fixed history is associated with each considered resolution derivation.

It is well known that resolution is a complete proof system for unsatisfiable formulas; i.e., a formula $F$ is unsatisfiable if and only if there exists a resolution refutation of it. The *resolution complexity* or *resolution hardness* $h(F)$ of an unsatisfiable formula $F$ is the length of a shortest resolution refutation of $F$ (for satisfiable formulas we put $h(F) := -\infty$). For a nonempty set $\mathcal{C}$ of formulas, we put $h(\mathcal{C}) = \max_{F \in \mathcal{C}} h(F)$.

*Isomorphisms of Formulas.* Two formulas $F$ and $F'$ are *isomorphic* if there exists a bijection $\varphi : lit(F) \to lit(F')$ such that for each literal $\ell \in lit(F)$ we have $\overline{\varphi(\ell)} = \varphi(\overline{\ell})$ and for each $C \subseteq lit(F)$ we have $C \in F$ if and only if $\varphi(C) \in F'$. For instance the formulas $F = \{\{x, \overline{y}\}, \{\overline{x}, y\}, \{\overline{y}\}\}$, and $F' = \{\{\overline{z}, w\}, \{z, w\}, \{\overline{w}\}\}$ are isomorphic.

Obviously, two isomorphic formulas have the same properties concerning satisfiability, minimal unsatisfiability, and resolution proof length. For a set $\mathcal{C}$ of formulas, we define $\mathrm{Iso}(\mathcal{C})$ to be an inclusion-maximal subset of $\mathcal{C}$ such that no two elements of Iso($\mathcal{C}$) are isomorphic. In other words, Iso($\mathcal{C}$) contains exactly one representative from each isomorphism class.

A *2-graph* is an undirected graph $G = (V, E)$ together with a partition of its vertex set into two subsets $V = V_1 \uplus V_2$. Two 2-graphs $G = (V_1 \uplus V_2, E)$ and $G' = (V_1' \uplus V_2', E')$ are *isomorphic* if there exists a bijection $\varphi : V_1 \uplus V_2 \to V_1' \uplus V_2'$ such that $v \in V_i$ if and only if $\varphi(v) \in V_i'$, $i = 1, 2$, and $\{u, v\} \in E$ if and only if $\{\varphi(u), \varphi(v)\} \in E'$.

The *clause-literal graph* of a formula $F$ is the 2-graph $G(F) = (V_1 \uplus V_2, E)$ with $V_1 = lit(F)$, $V_2 = F$, and $E = \{\, \{x, \overline{x}\} \mid x \in var(F) \,\} \cup \{\, \{C, \ell\} \mid C \in F, \ell \in C \,\}$.

The following statement is easy to verify.

**Proposition 1.** *Two formulas are isomorphic if and only if their clause-literal graphs are isomorphic.*

## 3 Theoretical Framework

We define the *m-th resolution hardness number* as the highest resolution complexity among formulas with $m$ clauses:

$$h_m = \max_{F \in \mathrm{CNF}(m)} h(F) = h(\mathrm{CNF}(m)).$$

In this section, we discuss various properties such that it suffices to consider only formulas with these properties for computing $h_m$.

Let $H(n,m) = \{\, F \in \mathrm{CNF}(n,m) \mid h(F) = h_m \,\}$ and $H(m) = \bigcup_{n=0}^{\infty} H(n,m)$; thus $h(\mathrm{CNF}(n,m)) = h(H(n,m))$, and $h(\mathrm{CNF}(m)) = h(H(m))$.

**Lemma 1.** *All formulas in $H(m)$ are minimally unsatisfiable.*

*Proof.* Suppose to the contrary, that there exists some $F \in H(m)$ which is not minimally unsatisfiable and choose any minimally unsatisfiable subset $F' \subsetneq F$ and let $d = m - |F'| \geq 1$. Pick a clause $C \in F'$ and take new variables $x_1, \ldots, x_d \notin var(F)$. We obtain a minimally unsatisfiable formula $F''$ from $F'$ by replacing $C$ by the clauses $C \cup \{x_1, \ldots, x_d\}$, $\{\overline{x_1}\}, \ldots, \{\overline{x_d}\}$. From a shortest resolution proof $P''$ of $F''$ we obtain a resolution proof $P'$ of $F'$. By construction, $|P'| + d = |P''|$, hence $h(F) \leq h(F') + d \leq h(F'')$, which is a contradiction to $h(F'') \leq h_m = h(F)$, a contradiction. □

A formula is *saturated minimally unsatisfiable* if it is unsatisfiable and adding a literal to any of its clauses makes it satisfiable. Every saturated minimally unsatisfiable formula is minimally unsatisfiable, since adding a pure literal to a clause has the same effect as deleting the clause.

**Lemma 2.** *$H(m)$ contains a saturated minimally unsatisfiable formula.*

*Proof.* Let $F$ be an arbitrary formula in $H(m)$. By Lemma 1, $F$ is minimally unsatisfiable. Assume now that $F$ is not saturated, and we can add to some clause $C$ of $F$ a literal $\ell$, obtaining a minimally unsatisfiable formula $F'$. We claim that $h(F') \geq h(F) = h_m$. Take a shortest proof $P$ of $F'$. Delete $\ell$ from the axiom $C \cup \{\ell\}$ in $P$ and propagate this deletion through $P$ to other clauses. This way, we obtain a sequence $P'$ of clauses, which contains as a subsequence a resolution proof of $F$. Hence indeed $h(F') \geq h(F) = h_m$, and so $F' \in H(m)$. □

A literal $\ell$ is called *$r$-singular* in a formula $F$ if there is exactly one clause in $F$ that contains $\ell$, and there are exactly $r$ clauses in $F$ that contain $\overline{\ell}$. A literal is *singular* in $F$ if it is $r$-singular for some $r \geq 0$ [21]. We also say a literal is $\geq r$-singular if it is $r'$-singular for some $r' \geq r$.

We denote by $\mathrm{MU}(n, m)$ the class of minimally unsatisfiable formulas with $n$ variables and $m$ clauses, and by $\mathrm{SMU}(n, m) \subseteq \mathrm{MU}(n, m)$ the subclass consisting of saturated formulas. $\mathrm{RSMU}(n, m)$ denotes the subclass of $\mathrm{SMU}(n, m)$ containing only formulas without singular variables. We call such formulas *regular*. We also use the shorthand $\mathrm{SSMU}(n, m) = \mathrm{SMU}(n, m) \setminus \mathrm{RSMU}(n, m)$.

Consider a formula $F$ and a variable $x$ of $F$. Let $\mathrm{DP}_x(F)$ denote the formula obtained from $F$ after adding all possible resolvents that can be obtained from clauses in $F$ by resolving on $x$ and removing all clauses in which $x$ occurs [21]. We say that $\mathrm{DP}_x(F)$ is obtained from $F$ by *Davis-Putnam reduction* or short *DP-reduction* on $x$ [3]. We will mainly use DP-reduction in the opposite direction, starting with a formula $F$ and generating a formula $F'$ such that $F = \mathrm{DP}_x(F')$. We then say that $F'$ has been obtained from $F$ by *DP-lifting*.

The following result by Kullmann and Zhao [12, Lemma 12] establishes an important link between DP-reduction on a singular variable and saturated minimal unsatisfiability.

**Lemma 3 (Kullmann and Zhao** [12]**).** *Let $F$ be a formula and $x$ an $r$-singular literal of $F$ such that $C_0$ is the only clause of $F$ containing $x$ and $C_1, \dots, C_r$ are the only clauses of $F$ containing $\overline{x}$. Then $F \in \mathrm{SMU}(n, m)$ if and only if the following three conditions hold: (i) $\mathrm{DP}_x(F) \in \mathrm{SMU}(n-1, m-1)$, (ii) $C_0 \setminus \{x\} = \bigcap_{i=1}^{r} C_i \setminus \{\overline{x}\}$, and (iii) for every $C' \in F \setminus \{C_0, \dots, C_r\}$ there is some literal $\ell \in C_0 \setminus \{x\}$ which does not belong to $C'$.*

The next lemma, a direct consequence of the preceding one, states that in the context of saturated minimally unsatisfiable formulas, DP-lifting is uniquely determined by a subset of the lifted formula.

**Lemma 4.** *Let $n \geq 2$ and let $F' \in \mathrm{SMU}(n-1, m-1)$. Then each formula $F \in \mathrm{SSMU}(n, m)$ which can be obtained from $F'$ by DP-lifting on a singular literal $x$ of $F$, can be generated by selecting $r$ clauses $C'_1, \dots, C'_r \in F'$ such that $\bigcap_{i=1}^{r} C'_i \not\subseteq C$ for any $C \in F \setminus \{C'_1, \dots, C'_r\}$, and replacing them by the $r+1$ clauses $C_0, \dots, C_r$ where $C_0 = \bigcap_{i=1}^{r} C'_i \cup \{x\}$ and $C_i = C'_i \cup \{\overline{x}\}$.*

The next lemma is useful when we know $h_{m-1}$, have a lower bound on $h_m$, and want to show that a formula $F$ containing singular literals does not require longer proofs than our current bound on $h_m$, without laboriously computing a shortest proof of $F$.

**Lemma 5.** *Let $F \in \mathrm{MU}(n, m)$ with an $r$-singular variable. Then $h(F) \leq h_{m-1} + r + 1$.*

*Proof.* We perform DP-reduction on the $r$-singular variable using $r+1$ axioms, then refute the resulting formula on $m-1$ remaining clauses. □

The *deficiency* $\delta(F)$ of a formula $F$ is defined as $|F| - |var(F)|$. By a lemma attributed to Tarsi [1], all minimally unsatisfiable formulas have a positive deficiency. That means that a minimally unsatisfiable formula with a fixed number of clauses cannot have *too many* variables. It is easy to see that it cannot have *too few* variables either: each clause must be falsified by an assignment that satisfies every other clause, whence we infer that the number of assignments bounds the number of clauses. Putting the two inequalities together yields Lemma 6.

**Lemma 6 (Aharoni and Linial [1]).** *Let $F$ be a minimally unsatisfiable formula. Then $\log_2 |F| \leq |var(F)| < |F|$.*

The structure of saturated minimally unsatisfiable formulas of deficiencies 1 and 2 is well understood [11]. In particular, it is known that for $m > 1$, each $F \in \mathrm{SMU}(m-1, m)$ has a 1-singular literal. It is also known that $|\mathrm{Iso}(\mathrm{RSMU}(m-2, m))| = 1$ for $m \geq 4$ [10] (otherwise, there are no minimally unsatisfiable formulas of deficiency 2). We pick the unique representative $\mathcal{F}^2_m$ for $\mathrm{Iso}(\mathrm{RSMU}(m-2, m))$, which consists of the clauses $\{\overline{x_1}, x_2\}, \ldots, \{\overline{x_{n-1}}, x_n\}, \{\overline{x_n}, x_1\}$, $\{x_1, \ldots, x_n\}$, and $\{\overline{x_1}, \ldots, \overline{x_n}\}$, $n = m-2$.

Due to their simple structure, we can determine the resolution hardness of $\mathrm{SMU}(m-1, m)$ and $\mathrm{RSMU}(m-2, m)$ formulas without any computation.

**Proposition 2.** *For every $m \geq 1$, $h(\mathrm{SMU}(m-1, m)) = 2m - 1$.*

*Proof.* Apart from the formula $\{\square\}$, every formula from $\mathrm{SMU}(m-1, m)$ contains a 1-singular variable [4, Theorem 12], so the statement follows by induction from Lemma 5 and the fact that $2m-1$ is the shortest possible proof length. □

**Proposition 3.** *For every $m \geq 4$, $h(\mathrm{RSMU}(m-2, m)) = h(\mathcal{F}^2_m) = 3m - 5$.*

*Proof.* $\mathcal{F}^2_m$ consists of binary strict Horn clauses (*BSH*—one negative and one positive literal) and the full positive and full negative clause. Resolving any pair of BSH clauses produces a BSH clause again. Resolving a BSH clause with a positive (negative) clause produces a positive (negative) clause, which is at most one shorter. Hence, to get to a positive (negative) unit clause, one must shorten the full positive (negative) clause at least $n - 1 = m - 3$ times. In total, we have $m$ axioms plus $2(m-3)$ shortening steps plus a final resolution step, altogether $3m - 5$ proof lines. It is easy to see that such proof exists for every $m$. □

Propositions 2 and 3, together with Lemma 6, give us a lower bound for $h_m$.

**Corollary 1.** *For* $m \leq 3$, $h_m = 2m - 1$. *For* $m \geq 4$, $h_m \geq 3m - 5$.

*Proof.* For $m \leq 3$ Lemma 6 rules out formulas with deficiency higher than 1, showing $h_1 = 1$, $h_2 = 3$, and $h_3 = 5$. The rest is a direct consequence of Proposition 3. □

For other formulas, we will need to generate the formulas and compute their shortest proofs. Our general approach for computing $\mathrm{Iso}(H(m))$ and in turn $h_m$ is to compute the sets $\mathrm{Iso}(\mathrm{SMU}(n, m))$ for $n = \lceil \log_2(m) \rceil, \ldots, m-1$, and test for each $F \in \mathrm{Iso}(\mathrm{SMU}(n, m))$ its resolution hardness $h(F)$ using the SAT encoding, which we describe in Sects. 4 and 5.

We split the computation of $\mathrm{Iso}(\mathrm{SMU}(n, m))$ into two parts. We first generate $\mathrm{Iso}(\mathrm{RSMU}(n, m))$ for $\lceil \log_2(m) \rceil \leq n < m$. Due to Proposition 1, we can do this by enumerating non-isomorphic 2-graphs, which correspond to clause-literal graphs of formulas in $\mathrm{RSMU}(n, m)$. We can limit ourselves to 2-graphs $G = (V_1 \uplus V_2, E)$ where $|V_1| = 2n$ and $|V_1| = m$, and where every vertex in $V_1$ has exactly one neighbor in $V_1$ and at least two neighbors in $V_2$. We use a tailor-made adaptation of the graph symmetry package Nauty [13] to enumerate such graphs; further details can be found in Sect. 6.

If $n$ and $m$ are such that $2^{n-1} < m-1$, we know there cannot be any formulas in $\mathrm{SSMU}(n, m)$ because singular DP-reduction would turn them into minimally unsatisfiable formulas on $n-1$ variables with $2^{n-1} < m-1$ clauses, but no such formulas exist. Hence, in those cases $\mathrm{RSMU}(n, m) = \mathrm{SMU}(n, m)$, and we already have $\mathrm{Iso}(\mathrm{SMU}(n, m))$. From these starting points, we repeatedly apply Lemma 4 to every formula in $\mathrm{Iso}(\mathrm{SMU}(n, m))$ to obtain $\mathrm{Iso}(\mathrm{SSMU}(n+1, m+1))$. Together with $\mathrm{Iso}(\mathrm{RSMU}(n+1, m+1))$ we then obtain $\mathrm{Iso}(\mathrm{SMU}(n+1, m+1))$.

The rationale for splitting the computation of $\mathrm{SMU}(n, m)$ into two pieces is the following. Enumerating non-isomorphic clause-literal graphs by Nauty for given parameters $n$ and $m$ is the hardest part of our computation. We often need to enumerate a significantly larger set than $\mathrm{SMU}(n, m)$. Therefore, we need to prune the enumeration phase as much as possible. When focusing on regular formulas, we can introduce additional bounds for Nauty, which significantly speed up the search. Applying then Lemma 4 inductively to the rather small set $\mathrm{SMU}(n, m)$ is computationally affordable (as long as the set $\mathrm{SMU}(n, m)$ remains reasonably small).

Since we are interested only in saturated minimally unsatisfiable formulas, we need to filter the graphs that we generate, which requires multiple calls to a SAT solver for every graph generated. Re-initializing the SAT solver with different formulas for different tests is expensive. Therefore it is desirable to bundle as many SAT calls together either by adding clauses incrementally or by using assumptions. While incrementally testing minimal unsatisfiability without solving multiple different formulas is relatively straightforward (via clause selector variables), it is not immediately clear how to do the same for saturation. We devised an algorithm that decides saturated minimal unsatisfiability using assumption-based calls to a SAT solver without the need to solve multiple different formulas. As a bonus, the formula for the saturation test contains all the clauses of the formula used for the minimality test, so both tests can proceed incrementally. The following lemma is the basis for our algorithm (a satisfiability test precedes the saturation test in our implementation, so it is safe to assume that the formula tested is unsatisfiable).

**Lemma 7.** *Let $F$ be an unsatisfiable formula, $C$ a clause of $F$, and $x \notin C$ a literal. The formula $E = F \cup (C \cup \{x\}) \setminus C$, where the clause $C$ was extended with the literal $x$, is unsatisfiable if and only if the formula $G = F \cup \{\{x\}\} \setminus C$ is unsatisfiable.*

*Proof.* If $E$ is unsatisfiable, so is $G$ because every clause in $G$ is a subset of some clause in $E$. Conversely, assume that $E$ is satisfiable with the assignment $\tau$. Because $\tau$ satisfies $F \setminus C$, and $F$ is unsatisfiable, $\tau$ must falsify $C$, and so it must satisfy $x$. Hence, it satisfies $G$. □

Lemma 7 gives rise to the following algorithm: for all $C \in F$ and every literal $x \notin C$, check whether $F \cup \{x\} \setminus C$ is unsatisfiable. If so, the clause $C$ can be extended with $x$ preserving unsatisfiability, meaning that $F$ is not saturated. This can be implemented in an assumption-based fashion with a single formula by augmenting $F$ with all possible unit clauses, adding selector variables to every clause, and turning clauses on and off as necessary using assumptions.

## 4 Encoding for Shortest Resolution Proofs

This section gives the details of our SAT encoding computing the shortest resolution proof of an input formula. We aim to encode the following question.

Given a formula $F$ with the clauses (*axioms*) $A_1, \ldots, A_m$ and $var(F) = V = \{x_1, \ldots, x_n\}$, does there exist a resolution refutation of $F$ of length at most $s$, i.e., does there exist a sequence $P = L_1, \ldots, L_s$ of $s$ lines (clauses), such that each $L_i$ is either some axiom $A_j$ or a resolvent of two previous $L_{i'}, L_{i''}$, $i', i'' < i$, and $L_s$ is empty. We denote this problem by $\texttt{SHORT}(F, s)$.

It is easy to see that $\texttt{SHORT}(F, s)$ is coNP-hard ($s$ given in binary): since each unsatisfiable formula $F$ with $n$ variables has a resolution refutation of length at most $2^{n+1} - 1$, we have $\texttt{UNSAT}(F) = \texttt{SHORT}(F, 2^{n+1} - 1)$. Therefore, using a SAT-based approach is indeed justified. On the other hand, membership in NEXPTIME can easily be seen as well: guess a refutation of length $s$ and verify that it is correct. The precise complexity of $\texttt{SHORT}(F, s)$ is an open problem—our intuition, based on our inability to construct a deterministic single-exponential-time algorithm for $\texttt{SHORT}(F, s)$, is that it might be NEXPTIME-complete.

The basic idea of our encoding is to have variables $\text{pos}[i, v]$ and $\text{neg}[i, v]$ that determine whether $v$ and $\overline{v}$ occur in $L_i$, and variables $\text{arc}[i, j]$, which hold the information about the structure of the resolution steps in the proof. Together, these variables fully determine a candidate resolution proof sequence $P$. We additionally use auxiliary variables to express certain constraints more succinctly. Table 1 lists the core variables used by the encoding.

We drew inspiration from a similar encoding proposed by Marques-Silva and Mencía [14] (henceforth referred to as MSM), but took several departures, afforded by the fact that we focus on minimally unsatisfiable formulas. One of the strongest points of MSM, enumerating *minimal correction subsets* (MCSes, i.e., inclusion-minimal sets of clauses whose deletion renders the formula satisfiable) in a preprocessing step, becomes trivial for minimally unsatisfiable formulas: the MCSes are precisely all singletons by definition of minimal unsatisfiability. Instead, we require that every axiom of the input formula is used in the proof.

On the other hand, we extend the encoding with powerful symmetry breaking predicates. These predicates, explained in detail in Sect. 5, completely break all symmetries resulting from different permutations of the same sequence of clauses, and as such, they constitute a valuable theoretical contribution. Moreover, thanks to this additional symmetry breaking, we were able to compute shortest proofs of several formulas, for which MSM failed to produce an answer in hours of running time. This symmetry breaking uses further auxiliary variables, which are introduced in Sect. 5.

Another novelty of our encoding is the capacity to reject a partially constructed proof early based on a counting argument involving the number of times a clause is used in resolution steps. We give the details at the end of this section.

**Table 1.** Variables used by the shortest-proof encoding. The symbol $v$ is understood to range over $V$, while the symbols $i, j$ range over the set $\{1, \dots, s\}$ with $i < j$, except for ax$[i, j]$, where $j$ ranges over $\{1, \dots, m\}$ instead. The *-marked terms are asymptotically dominating ($m \leq s$).

| Variable | Meaning | How many |
|---|---|---|
| pos$[i, v]$ | $v \in L_i$ | $O(ns)$ |
| neg$[i, v]$ | $\overline{v} \in L_i$ | $O(ns)$ |
| piv$[i, v]$ | $v$ is the pivot variable for the resolvent $L_i$ | $O(ns)$ |
| ax$[i, j]$ | $L_i = A_j$ | $O(ms)$ |
| isax$[i]$ | $\exists j : L_i = A_j$ | $O(s)$ |
| arc$[i, j]$ | $L_i$ is a premise of $L_j$ | $O(s^2)$ |
| upos$[i, v]$ | $v$ occurs in at least one premise of $L_i$ | $O(ns)$ |
| uneg$[i, v]$ | $\overline{v}$ occurs in at least one premise of $L_i$ | $O(ns)$ |
| poscarry$[i, j, v]$ | $v \in L_i$ and $L_i$ is a premise of $L_j$ | $O(ns^2)^*$ |
| negcarry$[i, j, v]$ | $\overline{v} \in L_i$ and $L_i$ is a premise of $L_j$ | $O(ns^2)^*$ |

In the following subsections, we list the clauses of the encoding, using complex Boolean expressions where convenient, and implicitly assuming that those are translated into a logically equivalent CNF in the natural way. Sometimes we write pos|neg to save space, meaning that the surrounding expression should be interpreted twice, with `pos` and `neg` substituted. We will also use cardinality constraints of the form $\sum_{x \in X} x \leq k$, which can be encoded using an arbitrary CNF cardinality-constraint encoding. We use the sequential counter [19], which seemed to perform best in our tests.

*Definitions.* First, we list all the clauses that provide definitions for the variables ax, isax, {pos|neg}carry, the union of premises via u{pos|neg}, and piv.

$$\bigwedge_{\substack{1\leq i\leq s\\1\leq j\leq m}} \mathrm{ax}[i,j] \rightarrow \left( \bigwedge_{v\in A_j} \mathrm{pos}[v,i] \bigwedge_{\overline{v}\in A_j} \mathrm{neg}[v,i] \bigwedge_{v\notin var(A_j)} \overline{\mathrm{pos}[v,i]} \wedge \overline{\mathrm{neg}[v,i]} \right),$$

$$\bigwedge_{1\leq i\leq s} \left( \mathrm{isax}[i] = \bigvee_{1\leq j\leq m} \mathrm{ax}[i,j] \right),$$

$$\bigwedge_{1\leq i,j\leq s; v\in V} \{\mathrm{pos}|\mathrm{neg}\}\mathrm{carry}[i,j,v] = \{\mathrm{pos}|\mathrm{neg}\}[i,v] \wedge \mathrm{arc}[i,j],$$

$$\bigwedge_{1\leq j\leq s; v\in V} \left( \mathrm{u}\{\mathrm{pos}|\mathrm{neg}\}[j,v] = \bigvee_{1\leq i<j} \{\mathrm{pos}|\mathrm{neg}\}\mathrm{carry}[i,j,v] \right),$$

$$\bigwedge_{1\leq i\leq s; v\in V} \mathrm{piv}[i,v] = \mathrm{upos}[i,v] \wedge \mathrm{uneg}[i,v].$$

*Essential Constraints.* The final clause is empty: $\bigwedge_{v\in V} \overline{\mathrm{pos}[i,v]} \wedge \overline{\mathrm{neg}[i,v]}$.

Axioms have no incoming arcs: $\bigwedge_{1\leq i<j\leq s} \mathrm{isax}[j] \rightarrow \overline{\mathrm{arc}[i,j]}$.

Clauses are non-tautological: $\bigwedge_{\substack{1\leq i\leq s\\ v\in V}} \overline{\mathrm{pos}[i,v]} \vee \overline{\mathrm{neg}[i,v]}$.

Non-pivot literals are retained after resolution.

$$\bigwedge_{1\leq i\leq s; v\in V} \overline{\mathrm{piv}[i,v]} \wedge \mathrm{u}\{\mathrm{pos}|\mathrm{neg}\}[i,v] \rightarrow \{\mathrm{pos}|\mathrm{neg}\}[i,v]$$

No new literals are introduced into resolvents.

$$\bigwedge_{1\leq i\leq s; v\in V} \overline{\mathrm{isax}[i]} \wedge \{\mathrm{pos}|\mathrm{neg}\}[i,v] \rightarrow \mathrm{u}\{\mathrm{pos}|\mathrm{neg}\}[i,v]$$

Every resolvent has a pivot: $\bigwedge_{1\leq i\leq s}\left(\overline{\text{isax}[i]} \rightarrow \bigvee_{v\in V} \text{piv}[i,v]\right)$, and the pivot is unique: $\bigwedge_{\substack{1\leq i\leq s\\ v\neq v'\in V}} \overline{\text{piv}[i,v]} \vee \overline{\text{piv}[i,v']}$. Every clause has exactly two premises[1].

$$\bigwedge_{3\leq j\leq s} \sum_{1\leq i<j} \text{arc}[i,j] = 2$$

*Redundant Constraints.* If we search for the proof by iteratively incrementing the bound $s$, we know that every clause must be used: $\bigwedge_{1\leq i<s} \bigvee_{i<j\leq s} \text{arc}[i,j]$. Axioms, do not have pivots: $\bigwedge_{\substack{1<i\leq s\\ v\in V}} \text{isax}[i] \rightarrow \overline{\text{piv}[i,v]}$. We require that the axioms are placed at the beginning of the proof $\bigwedge_{1\leq i<s} \text{isax}[i{+}1] \rightarrow \text{isax}[i]$, and in the same order as they appear in the original formula $\bigwedge_{1\leq i\leq s; 1\leq j_1\leq j_2\leq m} \overline{\text{ax}[i,j_2]} \vee \overline{\text{ax}[i+1,j_1]}$. Hence, $A_j$ can appear no later than as $L_j$, expressed by the unit clauses $\overline{\text{ax}[i,j]}$ for $1 < i \leq s$ and $1 \leq j \leq \min(i-1, m)$. When considering only MU formulas, we can omit the above and directly place all axioms at the start in a fixed order: $\bigwedge_{i=1}^{m} \text{ax}[i,i] \bigwedge_{i=m+1}^{s} \overline{\text{isax}[i]}$.

*Counting the In- and Out-Degrees.* For every sequence of clauses $P = L_1, \ldots, L_s$ that constitutes a resolution proof we can define a *directed acyclic graph (DAG)* $G(P)$ whose vertices are the clauses of $P$ and which has the arcs $(L_i, L_k)$ and $(L_j, L_k)$ if $L_k$ is a resolvent of $L_i$ and $L_j$. Using the redundant constraints from above and assuming minimal unsatisfiability of $F$, we will show how one can place an additional redundant constraint on the proof DAG structure. This feature is based on the simple identity $\sum_{L\in G} d_{\text{out}}(L) = \sum_{L\in G} d_{\text{in}}(L)$, which holds in every directed graph $G$. In a proof DAG $G(P)$ of size $s$, axioms have in-degree 0 and resolvents have in-degree 2, so $\sum_{L\in G(P)} d_{\text{in}}(L) = 2(s-m)$. At the same time, every clause except for the last has out-degree at least 1. Therefore, at any time of the search, with $d'_{\text{out}}(L_i) \geq 0$ outgoing arcs already added to $L_i$ in the partial DAG, it must hold that

$$\sum_{i=1}^{s-1} \max(d'_{\text{out}}(L_i), 1) \leq 2(s-m) \iff \sum_{i=1}^{s-1} \max(d'_{\text{out}}(L_i) - 1, 0) \leq s - 2m + 1.$$

We encode the latter inequality as a cardinality constraint. To capture the value of $\max(d'_{\text{out}}(L_i) - 1, 0)$, we introduce the notion of an *extra* arc: for a clause $L_i \in P$ with multiple outgoing arcs to clauses $L_{j_1}, \ldots, L_{j_k}$, $j_1 < \cdots < j_k$, we say that the arcs to $L_{j_2}, \ldots, L_{j_k}$ are extra.[2] Hence, $\max(d_{\text{out}}(L_i) - 1, 0)$ is precisely the number of extra outgoing arcs from $L_i$. We define the variables $\text{exarc}[i,j]$ whose meaning is that $\text{arc}[i,j]$ is an extra arc and enforce the cardinality constraint on them.

[1] It would be enough to specify the at-most-two constraint here: the presence of at least two premises for resolvents is already enforced by the existence of a pivot and because the clauses are non-tautological: the pivot appears in both polarities in the union of premises (upos and uneg), which could not happen with one premise. Nevertheless, including both constraints appears to improve performance.

[2] This includes symmetry breaking: the single non-extra arc is the first one.

$$\bigwedge_{1\le i<j<k\le s} \mathrm{arc}[i,j]\wedge\mathrm{arc}[i,k] \rightarrow \mathrm{exarc}[i,k]; \qquad \sum_{1\le i<j\le s} \mathrm{exarc}[i,j] \le s-2m+1.$$

Since the cardinality constraint is unsatisfiable if the right-hand side is negative, we additionally get that $s \geq 2m-1$; indeed, a shorter proof could not use all $m$ axioms. In other words, any proof of an MU formula must be "read-at-least-once."

## 5 Symmetry Breaking

Consider the proof DAG $G(P)$ of a resolution proof $P$. Any proof $P$ is simply a topological sort of its DAG $G(P)$. If two sequences $P_1$ and $P_2$ share the same DAG $G(P_1) = G(P_2) = G$, then $P_1$ and $P_2$ are essentially the same proof. Our aim now is to make sure that for each candidate proof DAG $G$, exactly one topological sort is accepted by the encoding.

A directed acyclic graph can be topologically sorted by repeatedly picking and deleting from $G$ a source vertex, i.e., one with no incoming arcs, as the next vertex in the resulting topologically sorted sequence. In the event that several sources are available, any one can be picked, which is why a given DAG, in general, has many topological sorts. We define a *canonical* topological sort of a given DAG $G$ in the following way. Let $\leq^*$ be an arbitrary total order on the vertices of $G$. The *canonical topological sort of* $G$ is the topological sort that results from always picking the greatest source vertex under $\leq^*$. The idea for this symmetry breaking is due to Schidler and Szeider [18] who introduced it in a different context; Fichte et al. [5] further studied this technique under the name *LexTopSort*.

To verify that a given sequence $P$ is the canonical topological sort of $G(P)$, we need to check that for every pair of vertices $L_i, L_j$, $i < j$, if $L_j$ was a source already at the time when $L_i$ was inserted, then $L_j \leq^* L_i$. We can check whether $L_j$ was a source *simultaneously* with $L_i$ by checking that there is no arc $(L_k, L_j)$ with $i \leq k$. This is the role of the variables $\mathrm{sim}[i,j]$.

We also need to reason about the order $\leq^*$ on clauses. We define the following order on the literals $x_1 < \overline{x_1} < \cdots < x_n < \overline{x_n}$, and order clauses of the proof lexicographically based on this order: $L_i <^* L_j$ if there is a literal $l \in L_j$ such that $l \notin L_i$ and $\{l'\} \cap L_i = \{l'\} \cap L_j$ for all $l' < l$. We represent $\leq^*$ using the variables $\mathrm{equal}[i,j,l]$, which say that the clauses $L_i$ and $L_j$ are equal up to position $l$ in the ordering of the literals, and the corresponding constraints below.

$$\bigwedge_{1\le i<j\le s} \mathrm{equal}[i,j,x_1] = (\mathrm{pos}[i,x_1] \iff \mathrm{pos}[j,x_1])$$

$$\bigwedge_{1\le i<j\le s;\ 1<k\le n} \mathrm{equal}[i,j,x_k] = \mathrm{equal}[i,j,\overline{x_{k-1}}] \wedge (\mathrm{pos}[i,x_k] \iff \mathrm{pos}[j,x_k])$$

$$\bigwedge_{1\le i<j\le s;\ 1\le k\le n} \mathrm{equal}[i,j,\overline{x_k}] = \mathrm{equal}[i,j,x_k] \wedge (\mathrm{neg}[i,x_k] \iff \mathrm{neg}[j,x_k])$$

Definition of sim: for $1 \leq i < s$, $\mathrm{sim}[i, i+1] = \overline{\mathrm{arc}[i, i+1]}$, and

$$\bigwedge_{1 \leq i < j-1 \leq s} \mathrm{sim}[i, j] = \mathrm{sim}[i+1, j] \wedge \overline{\mathrm{arc}[i, j]}.$$

The following constraint enforces that the sequence is the canonical topological sort (for resolvents only, the order of axioms is handled differently—see Sect. 4).

$$\bigwedge_{1 \leq i < j \leq s} \left(\mathrm{sim}[i, j] \wedge \overline{\mathrm{ax}[i]}\right) \rightarrow (\mathrm{pos}[i, x_1] \geq \mathrm{pos}[j, x_1])$$

$$\bigwedge_{\substack{1 \leq i < j \leq s \\ 1 \leq k \leq n}} \left(\mathrm{sim}[i, j] \wedge \overline{\mathrm{ax}[i]} \wedge \mathrm{equal}[i, j, x_k]\right) \rightarrow (\mathrm{neg}[i, x_k] \geq \mathrm{neg}[j, x_k])$$

$$\bigwedge_{\substack{1 \leq i < j \leq s \\ 1 \leq k < n}} \left(\mathrm{sim}[i, j] \wedge \overline{\mathrm{ax}[i]} \wedge \mathrm{equal}[i, j, \overline{x_k}]\right) \rightarrow (\mathrm{pos}[i, x_{k+1}] \geq \mathrm{pos}[j, x_{k+1}])$$

$$\bigwedge_{1 \leq i < j \leq s} \overline{\mathrm{equal}[i, j, \overline{x_n}]}$$

The following theorem summarizes the properties of our encoding.

**Theorem 1.** *Let $F$ be a propositional formula on $n$ variables and $m$ clauses and let $\texttt{short}_s(F)$ be the formula defined above. Then the following statements hold:*

1. *the size of $\texttt{short}_s(F)$ is polynomial in $\max(n, m, s)$ ($s$ can be exponential in the input length);*
2. *$\texttt{short}_s(F)$ is satisfiable if and only if $F$ has a resolution refutation of length $s$ in which every clause is used to derive the empty clause;*
3. *any model of $\texttt{short}_s(F)$ can naturally be interpreted as a sequence of clauses $P$ that constitutes a valid resolution proof of $F$;*
4. *$P$ is the canonical topological sort of $G(P)$.*

Theorem 1 gives rise to a simple algorithm. Start with $s = 1$, and increment $s$ by one while $\texttt{short}_s(F)$ is unsatisfiable. As soon as $\texttt{short}_s(F)$ becomes satisfiable, $s$ is the length of a shortest resolution refutation of $F$, and the refutation itself can be extracted from a model of $\texttt{short}_s(F)$. An improvement is possible for MU formulas, by starting not at $s = 1$, but $s = 2m - 1$, as described in Sect. 4.

## 6 Experiments

In this section, we describe how we performed our computations. We will refer to formulas and graphs interchangeably throughout this section, saying for instance that a graph is minimally unsatisfiable. In such cases, it is understood that we are using the correspondence between formulas and graphs sketched in Sect. 3, and implicitly mean the corresponding object.

To generate Iso(RSMU$(n, m)$), we run a modified version of the `genbg` utility[3] from the graph automorphism package `Nauty` [13], which enumerates isomorph-free 2-graphs. The modification is that the graphs generated are not bipartite as in `genbg`, but $V_1$ induces a *matching*, i.e., the graph is a clause-literal graph as defined in Sect. 2. We run `genbg` with the parameters `-cAtd3:2 2n m`, meaning we are interested in connected (`-c`) triangle-free (`-t`) 2-graphs $G = (V_1 \uplus V_2, E)$, such that $V_1$ has $2n$ vertices (the literals) whose minimum degree is 3 (every literal should occur twice, plus the edge between the two literals of a variable), and $V_2$ has $m$ vertices (the clauses) with minimum degree 2 (a unit clause would imply a singular literal, so we can skip such graphs), and such that no two vertices $V_2$ have neighborhoods that are subsets of one another (`-A`). This gives us a set $S$ of graphs that contains Iso(RSMU$(n, m)$), and such that all graphs in $S$ represent formulas without tautological clauses (triangle-freeness), without singular literals (degree bounds), and without subsumed clauses (`-A`). Hence it remains to filter the output of `genbg` for saturated minimal unsatisfiability. For that purpose we use CryptoMiniSAT [20] via its `C` API, in essentially the natural way of testing for saturated minimal unsatisfiability, however, with some technical optimizations worth mentioning.

It turns out that for small values of $n$ and $m$, the vast majority of graphs generated are satisfiable[4]. This, in turn, means that the vast majority of time is spent checking satisfiability, which happens to be relatively expensive due to the need to compute an explicit representation of the clauses and to re-initialize the solver for every graph from scratch. Upon realizing this limitation, we implemented a carefully tuned and highly cache-performant brute-force satisfiability test that uses Nauty's data structures directly and simply checks every assignment, and observed an orders-of-magnitude speed-up.

To check minimal unsatisfiability of a formula $F$, we construct $F' = \{\, C \cup \{\overline{x_C}\} \mid C \in F \,\}$ with the fresh selector variables $x_C$, and solve all formulas $F'[\tau_C]$, where $\tau_C$ sets $x_C$ to 0 and all other $x_{C'}$ to 1, via assumptions.

If $F$ passes the minimal unsatisfiability test, we build $F'' = F' \cup \{\, \{\ell, x_\ell\} \mid \ell \in \mathit{lit}(F) \,\}$ by adding the extra clauses to $F'$, which is already loaded in the SAT solver, and test for saturation by solving the formulas $F''[\tau_{C,\ell}]$, where $\tau_{C,\ell}$ sets $x_C$ and $x_\ell$ to 0 (turns off the clause $C$ and turns on the unit clause $\{\ell\}$) and all other auxiliary variables to 1, again via assumptions.

Once we have generated Iso(RSMU$(n, m)$), which is equal to Iso(SMU$(n, m)$) for values of $n, m$ where MU$(n-1, m-1)$ is empty, we use Lemma 4 to compute SMU$(n+1, m+1)$. Whenever we have computed SMU$(n+1, m+1)$, we simply run our encoding on every formula, incrementally increasing the proof length bound $s$, and compute all shortest proofs.

---

[3] We gratefully acknowledge the help of Brendan McKay, author of Nauty, who provided a modification of `genbg` for our purpose.

[4] As an example, for $n = 5$ and $m = 9$, a total of 9356316116 out of the 9360503942 generated graphs were satisfiable. The proportion was similar for other parameters.

We implemented the encoding and the iterative search for a shortest proof in Python using the PySAT framework [9]. We concluded from our initial tests that among the SAT solvers available in PySAT, CaDiCaL[5] performed best, and we decided to stick with it.

**Table 2.** Values of $h(H(n,m))$, and in parenthesis the number of formulas in RSMU$(n,m)$, up to isomorphism, that require resolution proofs of length $h(H(n,m))$. For all $3 \leq n \leq 9$ and $n+2 \leq m \leq 10$, we found $H(n,m) \subseteq$ RSMU$(n,m)$, except for $H(7,10)$ which also contains 19 singular formulas up to isomorphism. By Proposition 2, $h(H(m-1,m)) = 2m-1$ and so, no computation is necessary. By Lemma 6, there are no minimally unsatisfiable formulas in the areas marked by a hyphen.

| $n \backslash m$ | 1 | 2 | 3 | 4 | 5 | 6 | 7 | 8 | 9 | 10 |
|---|---|---|---|---|---|---|---|---|---|---|
| 0 | **1** | – | – | – | – | – | – | – | – | – |
| 1 | – | **3** | – | – | – | – | – | – | – | – |
| 2 | – | – | **5** | **7**(1) | – | – | – | – | – | – |
| 3 | – | – | – | **7** | **10**(1) | 11(3) | 13(1) | 15(1) | – | – |
| 4 | – | – | – | – | 9 | **13**(1) | 15(1) | **19**(1) | 20(1) | 21(5) |
| 5 | – | – | – | – | – | 11 | **16**(1) | 18(3) | **22**(1) | 25(1) |
| 6 | – | – | – | – | – | – | 13 | **19**(1) | **22**(3) | **26**(3) |
| 7 | – | – | – | – | – | – | – | 15 | **22**(1) | 25(5) |
| 8 | – | – | – | – | – | – | – | – | 17 | 25(1) |
| 9 | – | – | – | – | – | – | – | – | – | 19 |

Table 2 lists the length of the longest shortest proof required by an SMU$(n,m)$ formula, and, by taking the maximum in each column, also values of $h_m$. In particular, we obtain the first ten resolution hardness numbers:

$$(h_m)_{m \geq 1} = 1, 3, 5, 7, 10, 13, 16, 19, 22, 26, \ldots$$

We observe the interesting phenomenon that all computed formulas attaining the maximum hardness $h_m$ are regular.

It is known that every MU$(n,m)$ formula has a proof of length at most $2^{m-n-1}n + m$ [11, Section 11.3], and, along with the existence of formulas which require exponentially long proofs, this implies that maximum hardness cannot forever be attained by formulas of any bounded deficiency $m - n$. Our computations reveal that $m = 10$ is the tipping point where formulas of deficiency 2 "drop out of the race," as there is no longer a hardest formula of deficiency 2, see Table 2. Up to isomorphism, there are exactly three hardest formulas for $m = 10$, all of which are of deficiency 4. Figure 1 shows the clause-literal graphs of these three formulas.

Our encoding [16] and our catalog of SMU formulas [15] are publicly available.

[5] https://fmv.jku.at/cadical.

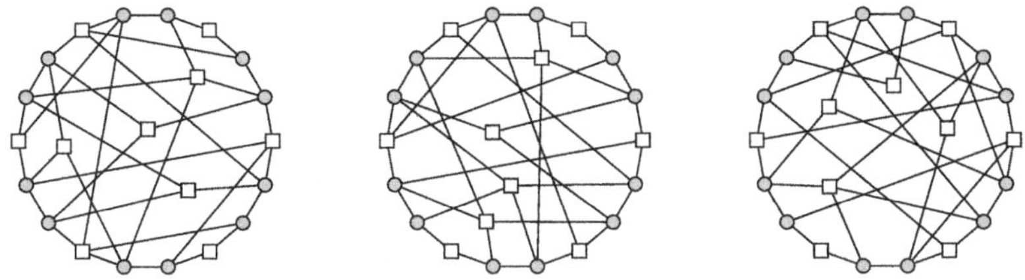

**Fig. 1.** Clause-literal graphs of the three hardest formulas with 10 clauses.

## 7 Conclusion

We conducted an extensive computational investigation into resolution hardness. First, we developed theoretical foundations that allowed us to pinpoint classes of formulas of maximum resolution hardness. Then, using a tight graph representation of formulas and carefully tuned generation procedures, we computed all candidates for hardest formulas for up to ten clauses. With this information, and using a SAT encoding for the computation of shortest resolution proofs targeted towards minimally unsatisfiable formulas and with powerful novel symmetry breaking, we calculated the first ten resolution hardness numbers. Our results indicate that regular saturated minimally unsatisfiable formulas achieve the highest hardness. It remains as an interesting theoretical question whether the hardest formulas are always regular.

## References

1. Aharoni, R., Linial, N.: Minimal non-two-colorable hypergraphs and minimal unsatisfiable formulas. J. Combin. Theory Ser. A **43**, 196–204 (1986)
2. Codish, M., Miller, A., Prosser, P., Stuckey, P.J.: Constraints for symmetry breaking in graph representation. Constraints An Int. J. **24**(1), 1–24 (2019). https://doi.org/10.1007/s10601-018-9294-5
3. Davis, M., Putnam, H.: A computing procedure for quantification theory. J. ACM **7**(3), 201–215 (1960)
4. Davydov, G., Davydova, I., Kleine Büning, H.: An efficient algorithm for the minimal unsatisfiability problem for a subclass of CNF. Ann. Math. Artif. Intell. **23**, 229–245 (1998)
5. Fichte, J.K., Hecher, M., Szeider, S.: Breaking symmetries with RootClique and LexTopSort. In: Simonis, H. (ed.) Proceedings of CP 2020, the 26rd International Conference on Principles and Practice of Constraint Programming. LNCS, vol. 12333, pp. 286–303. Springer, Heidelberg (2020). https://doi.org/10.1007/978-3-030-58475-7_17
6. Haken, A.: The intractability of resolution. Theor. Comput. Sci. **39**, 297–308 (1985)
7. Heule, M.J.H.: Schur number five. In: Proceedings of the Thirty-Second AAAI Conference on Artificial Intelligence, (AAAI 2018), the 30th innovative Applications of Artificial Intelligence (IAAI 2018), and the 8th AAAI Symposium on Educational Advances in Artificial Intelligence (EAAI 2018), New Orleans, Louisiana, USA, 2–7 February 2018, pp. 6598–6606 (2018)

8. Heule, M.J.H., Kullmann, O., Marek, V.W.: Solving and verifying the Boolean pythagorean triples problem via cube-and-conquer. In: Creignou, N., Le Berre, D. (eds.) SAT 2016. LNCS, vol. 9710, pp. 228–245. Springer, Cham (2016). https://doi.org/10.1007/978-3-319-40970-2_15
9. Ignatiev, A., Morgado, A., Marques-Silva, J.: PySAT: a Python toolkit for prototyping with SAT Oracles. In: Beyersdorff, O., Wintersteiger, C.M. (eds.) SAT 2018. LNCS, vol. 10929, pp. 428–437. Springer, Cham (2018). https://doi.org/10.1007/978-3-319-94144-8_26
10. Kleine Büning, H.: On subclasses of minimal unsatisfiable formulas. Discr. Appl. Math. **107**(1–3), 83–98 (2000)
11. Kleine Büning, H., Kullmann, O.: Minimal unsatisfiability and autarkies. In: Biere, A., Heule, M.J.H., van Maaren, H., Walsh, T. (eds.) Handbook of Satisfiability, Frontiers in Artificial Intelligence and Applications, chap. 11, vol. 185, pp. 339–401. IOS Press (2009)
12. Kullmann, O., Zhao, X.: On Davis-Putnam reductions for minimally unsatisfiable clause-sets. Theor. Comput. Sci. **492**, 70–87 (2013)
13. McKay, B.D., Piperno, A.: Practical graph isomorphism, II. J. Symbolic Comput. **60**(0), 94 – 112 (2014). https://doi.org/10.1016/j.jsc.2013.09.003, http://www.sciencedirect.com/science/article/pii/S0747717113001193
14. Mencía, C., Marques-Silva, J.: Computing shortest resolution proofs. In: Moura Oliveira, P., Novais, P., Reis, L.P. (eds.) EPIA 2019. LNCS (LNAI), vol. 11805, pp. 539–551. Springer, Cham (2019). https://doi.org/10.1007/978-3-030-30244-3_45
15. Peitl, T., Szeider, S.: Saturated minimally unsatisfiable formulas on up to ten clauses (2020). https://doi.org/10.5281/zenodo.3951546
16. Peitl, T., Szeider, S.: short.py: encoding for the shortest resolution proof (2020). https://doi.org/10.5281/zenodo.3952168
17. Pipatsrisawat, K., Darwiche, A.: On the power of clause-learning SAT solvers with restarts. In: Gent, I.P. (ed.) CP 2009. LNCS, vol. 5732, pp. 654–668. Springer, Heidelberg (2009). https://doi.org/10.1007/978-3-642-04244-7_51
18. Schidler, A., Szeider, S.: Computing optimal hypertree decompositions. In: Blelloch, G., Finocchi, I. (eds.) Proceedings of ALENEX 2020, the 22nd Workshop on Algorithm Engineering and Experiments. pp. 1–11. SIAM (2020)
19. Sinz, C.: Towards an optimal CNF encoding of Boolean cardinality constraints. In: van Beek, P. (ed.) CP 2005. LNCS, vol. 3709, pp. 827–831. Springer, Heidelberg (2005). https://doi.org/10.1007/11564751_73
20. Soos, M., Nohl, K., Castelluccia, C.: Extending SAT solvers to cryptographic problems. In: Kullmann, O. (ed.) SAT 2009. LNCS, vol. 5584, pp. 244–257. Springer, Heidelberg (2009). https://doi.org/10.1007/978-3-642-02777-2_24
21. Szeider, S.: Minimal unsatisfiable formulas with bounded clause-variable difference are fixed-parameter tractable. J. Comput. Syst. Sci. **69**(4), 656–674 (2004)
22. Urquhart, A.: The complexity of propositional proofs. Bull. Symbolic Logic **1**(4), 425–467 (1995)

# HADDOCK: A Language and Architecture for Decision Diagram Compilation

Rebecca Gentzel[1], Laurent Michel[1(✉)], and W.-J. van Hoeve[2]

[1] University of Connecticut, Storrs, CT 06269, USA
rebecca.gentzel@uconn.edu, ldm@engr.uconn.edu
[2] Carnegie Mellon University, Pittsburgh, PA 15213, USA
vanhoeve@andrew.cmu.edu

**Abstract.** Multi-valued decision diagrams (MDDs) were introduced into constraint programming over a decade ago as a powerful alternative to domain propagation. While effective MDD-propagation algorithms have been proposed for various constraints, to date no system exists that can generically compile and combine MDD propagation for arbitrary constraints. To fill this need, we introduce HADDOCK, a declarative language and architecture for MDD compilation. HADDOCK supports the specification, implementation, and composition of a broad range of MDD propagators that delivers the strength one expects from MDDs at a fraction of the development effort and with comparable performance. This paper describes the language, the framework architecture, outlines its performance credentials and demonstrates how to specify and implement novel MDD propagators.

## 1 Introduction

Binary decision diagrams (BDDs) were first introduced to represent Boolean functions in the context of switching circuit verification [1,26]. They became widely popular within various branches of computer science after Bryant proposed effective algorithms to compile BDDs with a fixed variable ordering [7,8]. Since then, many variants of decision diagrams have been developed, including multi-valued decision diagrams (MDDs) for representing functions with discrete (multi-valued) variables [30]. In the context of constraint programming, decision diagrams can be interpreted as a compact graphical representation of the solution set to a given discrete structure, typically represented by a (global) constraint. For example, decision diagrams were used to develop propagation algorithms for constraints over sets [13,20,25], $n$-ary table constraints [10], and REGULAR constraints [9]. All these works use MDDs as an efficient data structure to perform traditional domain propagation.

Andersen et al. [2] were the first to recognize that instead of propagating domains, it is possible to *propagate MDDs*: in addition to a domain store, the constraint solver maintains an MDD store on which constraints perform filtering.


H. Simonis (Ed.): CP 2020, LNCS 12333, pp. 531–547, 2020.
https://doi.org/10.1007/978-3-030-58475-7_31

The key contribution of [2] is to introduce *relaxed* decision diagrams that are of polynomial size by explicitly limiting their width (the maximum number of nodes per layer). When the width is limited to one, MDD propagation defaults to domain propagation. Larger widths can lead to increasingly more effective MDD propagation, as shown in a range of papers that study the compilation and application of MDD-based constraint programming [4,11,16–19,22,24].

MDD propagation methods are all based on the same underlying principles for compiling, refining, and filtering the decision diagram, as summarized in [22]. Yet, their implementations are often dedicated to their specific task or purpose [12,14,15,28,29]. To date, no system exists that allows users or developers to easily define and combine MDD propagators in constraint programming models. In this work, we take on this task and present HADDOCK,[1] a language and architecture that uses an MDD specification to automatically compile and integrate decision diagrams into a CP solver.

**Contributions.** Our primary contribution is the introduction of a *specification language* and associated implementation architecture that not only allows the automatic compilation of the diagram, but also generates the rules for refining (splitting) and filtering (propagating) MDD abstractions. This concretizes and generalizes the framework for MDD-based constraint programming proposed in [22]. While that framework allows one to describe MDD propagators based on transitions between states, it does not provide concrete functionality for generic propagation, refinement, or composition. Another related compilation framework is proposed in [5,6], which uses a dynamic programming formulation as input to compile the diagram. We will apply similar concepts, but we note that a dynamic programming model alone is not sufficient, e.g., for describing some of the MDD propagation rules or the integration into a CP solver. Instead, we adopt the formalism of *labeled transition systems* as an abstraction of MDDs.

We have implemented HADDOCK in the C++ version of MiniCP [27].[2] So far, HADDOCK contains MDD specifications for ALLDIFF, AMONG, GCC, SEQUENCE, and (weighted) SUM constraints, as well as some problem-specific MDD propagators. HADDOCK allows the user to declare multiple MDDs within a CP model, each associated with a suitable set of constraints, and automatically integrates the MDD propagators into the constraint propagation fixpoint computation. HADDOCK offers comparable performance for MDD propagation at a fraction of the development effort needed for dedicated implementations.

**Motivating Example.** We will use the AMONG constraint as the running example for this paper. Recall that AMONG$(x, lb, ub, S)$ is defined as

$$lb \leq \sum_{i=0}^{n-1} (x_i \in S) \leq ub \tag{1}$$

[1] HADDOCK stands for 'Handling Automatically Decision Diagrams Over Constraint Kernels'. It also refers to a saltwater fish, as well as to a fictional character from the Tintin comic series by Hergé, both of which are however irrelevant to this paper.

[2] Code: https://bitbucket.org/ldmbouge/minicpp/commits/tag/HADDOCK.

```
int main() {
  int width = 64, H = 40;
  int L1 =  0, U1 = 6,  N1 = 8;
  int L2 = 22, U2 = 30, N2 = 30;
  int L3 =  4, U3 = 5,  N3 = 7;
  auto cp   = makeSolver();
  auto vars = boolVarArray(cp, H);
  auto mdd = new MDD(cp, width);
  for (int i=0; i<H-N1+1; i++)
    amongMDD(mdd,
          vars.sub(i,i+N1),L1,U1,{1});
  for (int i=0; i<H-N2+1; i++)
    amongMDD(mdd,
          vars.sub(i,i+N2),L2,U2,{1});
  for (int i=0; i<H-N3+1; i+=7)
    amongMDD(mdd,
          vars.sub(i,i+N3),L3,U3,{1});
  cp->post(mdd);
  ...
}
```

```
void amongMDD(MDDSpec& mdd,
              vector x,int lb,int ub,
              set<int> S)  {
  mdd.add(x);
  auto c = mdd.makeConstraint(x,"amongMDD");
  int Ld = mdd.addState(c,0),Ud = mdd.addState(c,0);
  int Lu = mdd.addState(c,0),Uu = mdd.addState(c,0);
  mdd.arcExist(cs,[=](p, c, x, v) -> bool {
       return (p[Ld] + v∈S + c[Lu] <= ub) &&
              (p[Ud] + v∈S + c[Uu] >= lb);
  });
  mdd.forward(Ld,[=](o,p,x,v){o[Ld]=p[Ld]+v∈S;});
  mdd.forward(Ud,[=](o,p,x,v){o[Ud]=p[Ud]+v∈S;});
  mdd.reverse(Lu,[=](o,c,x,v){o[Lu]=c[Lu]+v∈S;});
  mdd.reverse(Uu,[=](o,c,x,v){o[Uu]=c[Uu]+v∈S;});
  mdd.relax(Ld,[=](o,l,r){ o[Ld]=min(l[Ld],r[Ld]);});
  mdd.relax(Ud,[=](o,l,r){ o[Ud]=max(l[Ud],r[Ud]);});
  mdd.relax(Lu,[=](o,l,r){ o[Lu]=min(l[Lu],r[Lu]);});
  mdd.relax(Uu,[=](o,l,r){ o[Uu]=max(l[Uu],r[Uu]);});
}
```

**Fig. 1.** Pseudo-C++ code to create a nurse rostering model using an MDD (left) and pseudo-C++ HADDOCK code for creating an AMONG MDD propagator (right).

for an array of $n$ variables $x$, a lower bound $lb$, an upper bound $ub$, and a set of values $S$. In [22], an MDD propagator for AMONG, establishing MDD consistency in polynomial time, was proposed and implemented. The pseudo-C++ HADDOCK fragment shown on the right of Fig. 1 generates code with equivalent MDD propagation behavior. The `amongMDD` function takes a reference to an `MDDSpec` object *mdd* that *accumulates* all the specifications. Its other arguments are the array of variables $x$, the bounds $lb$ and $ub$, and the set of values $S$. Line 4 tracks the array of variables and line 5 creates a descriptor for the four properties specified in lines 6–7 and to be held in a state. The remainder of the code relies on closures to define the arc existence condition as well as the transition functions and relaxations for the four properties. The method `forward` is used to add top-down transitions while `reverse` adds bottom-up transitions.

With 18 lines of code, a developer specifies a *reusable* factory that models instances of AMONG. Multiple calls to this factory results in a *composition* of specifications to model their conjunction. An example of this is shown on the left of Fig. 1 for the nurse rostering problem. It creates the traditional decision variables (line 8), an MDD propagator (line 9), and multiple `amongMDD` constraints for various shifts of windows of length $N1$, $N2$, and $N3$. A call to `x.sub(a,b)` returns the sub-sequence of variables $[x_a, x_{a+1}, ..., x_b]$. Finally, it posts the MDD representing the conjunction of these constraints on line 18. The search is omitted for brevity's sake. The remainder of the paper explains this MDD specification language, its semantics, and how to mechanically derive propagators. Section 2 formally describes MDDs in terms of labeled transition systems. Section 3 introduces the different elements of our description language using states and transitions. We formally introduce the resulting MDD language in Sect. 4. Section 5 describes the implementation details of HADDOCK, followed by an experimental evaluation Sect. 6. We conclude in Sect. 7.

## 2 Decision Diagrams as Labeled Transition Systems

MDDs are commonly defined as layered directed acyclic graphs [5,30]. For our purposes, however, it is more convenient to formalize an MDD using a labeled transition system (LTS) as abstraction. Namely, an LTS allows us to describe the *rules* that govern an MDD rather than the MDD itself, and it provides a computational device to compile MDDs. Furthermore, an LTS abstraction can more clearly express the steps of generic MDD compilation, such as the computation of intermediate states, than the concrete acyclic directed graph.

We first recall the definition of labeled transition systems [23]:

**Definition 1.** *A* labeled transition system *is a triplet* $\langle \mathcal{S}, \rightarrow, \Lambda \rangle$ *where* $\mathcal{S}$ *is a set of states,* $\rightarrow$ *is a relation of labeled transitions between states from* $\mathcal{S}$ *and* $\Lambda$ *is a set of labels used to tag transitions.*

A transition from state $S_0$ to state $S_1$ (both in $\mathcal{S}$) belonging to the relation $\rightarrow$ is denoted $S_0 \xrightarrow{\ell} S_1$ with $\ell \in \Lambda$. A start state $S_\perp$ has no predecessors according to the transition relation $\rightarrow$ while an end state $S_\top$ has no successors.

Next we define an LTS to represent a single constraint (without loss of generality) from which we will derive our MDD definition.

**Definition 2.** *Given a constraint* $c(x)$ *of arity* $k$ *over an ordered set of variables* $x = x_0, \ldots, x_{k-1}$ *with domains* $D(x_0), \ldots, D(x_{k-1})$*, we define the associated labeled transition system as* $L(c, x) = \langle \mathcal{S}, \rightarrow, \Lambda \rangle$ *in which* $\mathcal{S}$*,* $\Lambda$ *and* $\rightarrow$ *are defined as follows:*

- *the state set* $\mathcal{S}$ *is stratified in* $k+1$ *layers* $\mathcal{L}_0$ *through* $\mathcal{L}_k$ *with transitions from* $\rightarrow$ *connecting states between layers* $i$ *and* $i+1$ *exclusively;*
- *the transition label set* $\Lambda$ *is defined as* $\bigcup_{i \in 0..k-1} D(x_i)$*;*
- *a transition between two states* $a \in \mathcal{L}_i$ *and* $b \in \mathcal{L}_{i+1}$ *carries a label* $v \in D(x_i)$*;*
- *the layer* $\mathcal{L}_0$ *consists of a single* source *state* $S_\perp$*;*
- *the layer* $\mathcal{L}_k$ *consists of a single* sink *state* $S_\top$*.*

By interpreting states as nodes and transitions as arcs, $L(c, x)$ represents a directed acyclic graph. In particular, a path from $S_\perp$ to $S_\top$ corresponds to a complete variable assignment. While Definition 2 directly links variable assignments to transitions between layers in the LTS, the constraint $c$ is implicitly represented in the states $\mathcal{S}$ and the transition function. We next define MDDs as labeled transition systems with specific properties:

**Definition 3.** *Given a constraint* $c(x)$ *over an ordered set of variables* $x$*, a* relaxed MDD *with respect to* $c(x)$ *is an LTS* $L(c, x)$ *such that each state in* $\mathcal{S}$ *lies on at least one* $S_\perp$*-*$S_\top$ *path and each feasible solution to* $c$ *corresponds to an* $S_\perp$*-*$S_\top$ *path. An* exact MDD *with respect to* $c(x)$ *is a relaxed MDD in which additionally every path from* $S_\perp$ *to* $S_\top$ *corresponds to a feasible solution to* $c$*.*

In order to compile an MDD, constraints must be appropriately defined in terms of states $\mathcal{S}$ and the transition function $\rightarrow$. We must furthermore specify a suitable state relaxation function to merge non-identical states.

*Example 1.* Consider the constraint AMONG($\{x_0, x_1, x_2, x_3\}, l = 1, u = 2, S = \{1\}$) where each variable has domain $\{0, 1\}$. Similar to [22], we define a state $s$ by four properties, as a tuple $(L^{\downarrow}(s), U^{\downarrow}(s), L^{\uparrow}(s), U^{\uparrow}(s))$, where

$L^{\downarrow}(s)$ represents the minimum occurrence of values in $S$ along any $S_{\perp}$-$s$ path,
$U^{\downarrow}(s)$ represents the maximum occurrence of values in $S$ along any $S_{\perp}$-$s$ path,
$L^{\uparrow}(s)$ represents the minimum occurrence of values in $S$ along any $s$-$S_{\top}$ path,
$U^{\uparrow}(s)$ represents the maximum occurrence of values in $S$ along any $s$-$S_{\top}$ path.

We initialize layer $\mathcal{L}_0$ with the state $S_{\perp} = (0, 0, \infty, \infty)$ and layer $\mathcal{L}_n$ with the state $S_{\top} = (\infty, \infty, 0, 0)$.

We next consider the transition relation $s \xrightarrow{v} s'$ from state $s \in \mathcal{L}_i$ via an arc labeled with $v \in D(x_i)$ to a state $s' \in \mathcal{L}_{i+1}$. This transition affects the properties $L^{\downarrow}(s)$ and $U^{\downarrow}(s)$ as

$$L^{\downarrow}(s') = L^{\downarrow}(s) + (v \in S)$$
$$U^{\downarrow}(s') = U^{\downarrow}(s) + (v \in S).$$

We call such properties *forward* properties since they follow the orientation of the transition. The properties $L^{\uparrow}(s)$ and $U^{\uparrow}(s)$, however, are updated reversely along the transition:

$$L^{\uparrow}(s) = L^{\uparrow}(s') + (v \in S)$$
$$U^{\uparrow}(s) = U^{\uparrow}(s') + (v \in S).$$

We therefore call such properties *reverse* properties. Lastly, we use the state properties to define an existence rule, i.e., transition $s \xrightarrow{v} s'$ exists if

$$(L^{\downarrow}(s) + (v \in S) + L^{\uparrow}(s') \le u) \wedge (U^{\downarrow}(s) + (v \in S) + U^{\uparrow}(s') \ge l).$$

Figure 2(a) depicts an LTS that represents an exact MDD for our constraint. The values of the properties are indicated inside the nodes representing the states. One can inspect that each path from $S_{\perp}$ to $S_{\top}$ corresponds to a solution to the constraint, and vice-versa. The width of the exact MDD is three.

To define a relaxed MDD, we specify a *merge* operator that takes two states $a, b \in \mathcal{L}_i$ and merges them into a single state $s$. For AMONG, we define it as:

$$L^{\downarrow}(s) = \min(L^{\downarrow}(a), L^{\downarrow}(b)) \qquad L^{\uparrow}(s) = \min(L^{\uparrow}(a), L^{\uparrow}(b))$$
$$U^{\downarrow}(s) = \max(U^{\downarrow}(a), U^{\downarrow}(b)) \qquad U^{\uparrow}(s) = \max(U^{\uparrow}(a), U^{\uparrow}(b)).$$

Observe that this merging operator *relaxes* the state computation. It therefore may introduce non-solutions to the MDD but will never lead to the removal of solutions. Figure 2(b) depicts a relaxed MDD, where the width is limited to a maximum of two. Each solution to the constraint is present as a path from $S_{\perp}$ to $S_{\top}$, but it also contains non-solutions, e.g., $(0, 0, 0, 0)$. □

The properties of a state $s$ in the LTS represent information over the collection of paths from $S_{\perp}$ to $s$ (i.e., prefixes) and from $s$ to $S_{\top}$ (i.e., suffixes). Note that producing a suitable LTS specification for any given constraint $c$ will yield a different set of state properties, transition definitions, existence conditions, and merging rules. The following sections explain how to do this systematically.

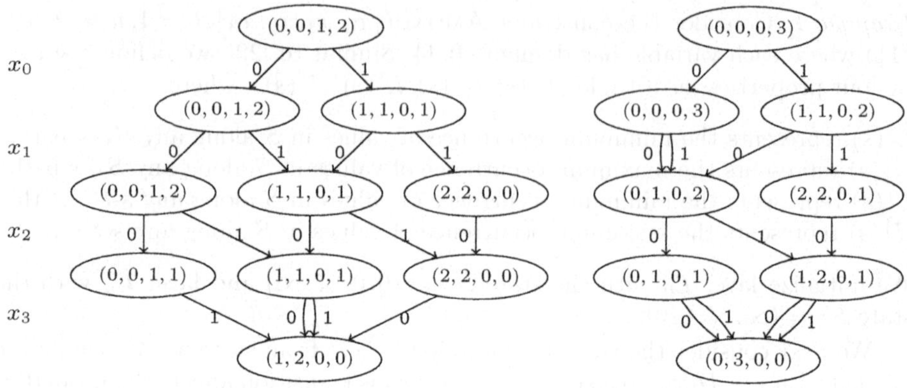

a. Exact decision diagram b. Relaxed decision diagram

**Fig. 2.** Exact (a) and relaxed (b) MDDs for the constraint AMONG($\{x_0, x_1, x_2, x_3\}$, $l = 1, u = 2, S = \{1\}$), where all variable domains are $\{0, 1\}$. Each state $s$ depicts its tuple of properties $(L^{\downarrow}(s), U^{\downarrow}(s), L^{\uparrow}(s), U^{\uparrow}(s))$ as defined in Example 1.

## 3 States and Transition Functions

This section describes the core elements of the LTS representation in HADDOCK in terms of states and transitions.

### 3.1 States

For exposition purposes, we assume that states are defined by properties that are integer-valued, but we note that these can represent Booleans and even richer types such as floating points or sets.

**Definition 4.** *A* state *is a tuple* $s = \langle P_{i_0}, P_{i_1}, \ldots, P_{i_{n-1}} \rangle$ *with* $n$ *properties denoted* $P_{i_k}$ *where each* $P_{i_k} \in \mathbb{Z}$.

It is important to realize that our LTS, and resulting MDD, can carry properties for multiple constraints. The identifiers (names) created for a constraint $c$ are collected in a set $\mathcal{P}(c) = \{i_0, \ldots, i_{n-1}\}$. These identifiers are unique across constraints, i.e., two constraints are guaranteed to use different property names.

### 3.2 Forward and Reverse Transition Rules

As shown in Example 1, a transition $s \xrightarrow{\ell} s'$ can be processed following the orientation of the transition (forward), or reversely. A state property may be subject to transition rules that are forward, reverse, or both. We therefore define the set of property indices $\Delta^{\downarrow} \subseteq \{0, \ldots, n-1\}$ and $\Delta^{\uparrow} \subseteq \{0, \ldots, n-1\}$ to indicate the properties that will be processed by the forward and reverse transition rules, respectively. We start by defining how individual properties are updated by a transition, by means of forward and reverse *property transition rules*:

**Definition 5.** *A* forward property transition rule *for property* $P_i$ *is a function* $T_i^{\downarrow} : \mathcal{S} \times \mathbb{Z} \times \mathcal{D} \to \mathbb{Z}$. *It takes as input a source state* $s \in \mathcal{L}_j$, *the layer index* $j$ *(i.e., variable* $x_j$*), and a domain value* $v \in D(x_j)$ *to produce a value for property* $P_i$ *in the destination state* $s' \in \mathcal{L}_{j+1}$.

**Definition 6.** *A* reverse property transition rule *for property* $P_i$ *is a function* $T_i^{\uparrow} : \mathcal{S} \times \mathbb{Z} \times \mathcal{D} \to \mathbb{Z}$. *It takes as input a state* $s \in \mathcal{L}_{j+1}$, *the layer index* $j$ *(i.e., variable* $x_j$*), and a domain value* $v \in D(x_j)$ *to produce a value for property* $P_i$ *in the destination state* $s' \in \mathcal{L}_j$.

We next use the individual property transition rules to define forward and reverse *state transition rules*:

**Definition 7.** *A* forward state transition rule *is a function* $T^{\downarrow} : \mathcal{S} \times \mathcal{S} \times \mathbb{Z} \times \mathcal{D} \to \mathcal{S}$. *Given a source state* $s = \langle P_0, \ldots, P_{n-1} \rangle \in \mathcal{L}_j$, *a destination state* $d = \langle Q_0, \ldots, Q_{n-1} \rangle \in \mathcal{L}_{j+1}$, *the layer index* $j$ *(i.e., variable* $x_j$*), a domain value* $v \in D(x_j)$ *and the set of forward property indices* $\Delta^{\downarrow}$, *it computes the forward property transitions of the successor state* $s'$ *in* $\mathcal{L}_j$. *The function is defined as follows:*

$$\forall i \in \{0, \ldots, n-1\} \; : \; Q_i' = \begin{cases} T_i^{\downarrow}(s, j, v) & \text{if } i \in \Delta^{\downarrow} \\ Q_i & \text{if } i \notin \Delta^{\downarrow} \end{cases}$$

*and, finally*

$$T^{\downarrow}(s, d, j, v) = \langle Q_0', \ldots, Q_{n-1}' \rangle.$$

**Definition 8.** *A* reverse state transition rule *is a function* $T^{\uparrow} : \mathcal{S} \times \mathcal{S} \times \mathbb{Z} \times \mathcal{D} \to \mathcal{S}$. *Given a source state* $s = \langle P_0, \ldots, P_{n-1} \rangle \in \mathcal{L}_{j+1}$, *a destination state* $d = \langle Q_0, \ldots, Q_{n-1} \rangle \in \mathcal{L}_j$, *the layer index* $j$ *(i.e., variable* $x_j$*), a domain value* $v \in D(x_j)$ *and the set of reverse property indices* $\Delta^{\uparrow}$, *it computes the reverse property transitions of the successor state* $s'$ *in* $\mathcal{L}_j$. *The function is defined as follows:*

$$\forall i \in \{0, \ldots, n-1\} \; : \; Q_i' = \begin{cases} T_i^{\uparrow}(s, j, v) & \text{if } i \in \Delta^{\uparrow} \\ Q_i & \text{if } i \notin \Delta^{\uparrow} \end{cases}$$

*and, finally*

$$T^{\uparrow}(s, d, j, v) = \langle Q_0', \ldots, Q_{n-1}' \rangle.$$

In the definitions of the state transition rules, the destination state $d$ and the source state $s$ provide values for properties not listed in $\Delta^{\downarrow}$ and $\Delta^{\uparrow}$, respectively.

### 3.3 Relaxation Functions

By design, a state $s$ is generally subject to multiple forward and reverse state transitions. These are aggregated, or merged, via pointwise *property relaxation functions*. Relaxation functions are also applied to merge non-identical states when a layer exceeds the given maximum MDD width.

**Definition 9.** *A* property relaxation function *for property* $P_i$ *takes the form* $R_i : \mathcal{S} \times \mathcal{S} \to \mathbb{Z}$. *It takes as input states* $s^{\ell} = \langle P_0^{\ell}, \ldots, P_{n-1}^{\ell} \rangle$ *and* $s^r = \langle P_0^r, \ldots, P_{n-1}^r \rangle$ *and returns* $P_i' = R_i(s^{\ell}, s^r)$ *as the merged property.*

The individual property relaxations are used to define *state relaxation functions*:

**Definition 10.** *A* state relaxation function *takes the form* $R : \mathcal{S} \times \mathcal{S} \to \mathcal{S}$. *Given states* $s^\ell$ *and* $s^r$ *it computes* $\langle R_0(s^\ell, s^r), \ldots, R_{n-1}(s^\ell, s^r)\rangle$.

### 3.4 Transition Existence Functions

A critical component of MDD propagation is *arc filtering*, i.e., the removal of arcs that do not belong to any path corresponding to a solution to the constraint. Arc filtering is performed by applying *transition existence functions*, which rely on the source and destination states, as well as the transition label:

**Definition 11.** *A* transition existence function *takes the form* $E_t : \mathcal{S} \times \mathcal{S} \times \mathbb{Z} \times \mathcal{D} \to \mathbb{B}$. *Given a parent state* $p$*, a child state* $c$*, a layer* $j$ *for state* $c$*, and a transition label* $v$*, the function returns the existence of path* $S_\top \rightsquigarrow p \xrightarrow{v} c \rightsquigarrow S_\bot$ *whose labels correspond to a solution to the constraint.*

### 3.5 State Functions

State transitions are defined *pointwise* in terms of forward and reverse property transitions. This does not provide for state-wide reasoning involving multiple state properties. In some cases, for example the MDD propagation for SEQUENCE constraints [4], it is desirable to process multiple state properties simultaneously. *State update functions* provide this capability:

**Definition 12.** *A* state update function *takes the form* $U : \mathcal{S} \to \mathcal{S}$. *It is a state transformation function that updates property* $P_i$ *based on properties* $P_0, \ldots, P_{i-1}, P_{i+1}, \ldots, P_{n-1}$ *for* $i = 0, \ldots, n-1$.

Analogous to transition existence functions, we also define *state existence functions*:

**Definition 13.** *A* state existence function *takes the form* $E_s : \mathcal{S} \to \mathbb{B}$. *Given a state* $p$ *the function returns the existence of paths* $S_\top \rightsquigarrow p$ *and* $p \rightsquigarrow S_\bot$ *whose labels correspond to a solution to the constraint.*

*Example 2.* Continuing Example 1, we recognize that the state properties, the forward and reverse transition functions, the relaxation functions, and the transition existence function all follow the specifications in this section. As an example of a state existence function for AMONG, we could define for a state $s$:

$$E_s(s) = (L^\downarrow(s) + L^\uparrow(s) \leq u) \wedge (U^\downarrow(s) + U^\uparrow(s) \geq l).$$

□

## 4 MDD Language

It is, perhaps, valuable to cast an LTS as a *virtual machine formalism* in which to capture the computational aspects of an MDD. An LTS faithfully models state derivation via tentative variable assignments as well as state relaxations with transitions. Yet, it remains a computational artifact that is delicate to describe and harder to maintain. This section introduces an *MDD language* to express the set of source-to-sink paths forming the MDD that an LTS computes. To a large extent, MDD languages are to LTS what regular expressions are to DFAs. Namely, an MDD language enables a developer to elevate the discourse and capture the language to be recognized in a fairly compact and declarative fashion. It leaves to a "compiler" the delicate task of producing the state machine. Figure 1 provides an illustration of an MDD language given in pseudo-C++. The purpose of this section is to describe the language more precisely.

**Definition 14.** *Given a constraint $c(x_0, \ldots, x_{k-1})$ over an ordered set of variables $X = \{x_0, \ldots, x_{k-1}\}$ with domains $D(x_0), \ldots, D(x_{k-1})$ the* MDD language *for c is a tuple $\mathcal{M}_c = \langle X, \mathcal{P}, S_\perp, S_\top, T^\downarrow, T^\uparrow, U, E_t, E_s, R\rangle$ where $\mathcal{P}$ is the set of properties used to model states, $S_\perp$ is the source state, $S_\top$ is the sink state, $T^\downarrow$ is the forward state transition rule, $T^\uparrow$ is the reverse state transition rule, $U$ is the state update function, $E_t$ is the transition existence function, and $E_s$ is the state existence function.*

We note that with the exception of $X$, all elements in the tuple $\mathcal{M}_c$ depend on the constraint $c$.

An MDD language is an abstraction to describe the states and rules that define a labeled transition system $\langle \mathcal{S}, \rightarrow, \Lambda\rangle$ for a given constraint. The LTS can be interpreted as a computational device that recognizes acceptable paths forming the MDD for the constraint. This is formalized in the following theorem.

**Theorem 1.** *Let $c(x_0, \ldots, x_{k-1})$ be a constraint over an ordered set of variables $X = \{x_0, \ldots, x_{k-1}\}$ with domains $D(x_0), \ldots, D(x_{k-1})$. An MDD language $\mathcal{M}_c$ is sufficient to define an exact or relaxed MDD for c.*

*Proof.* The LTS defines states by the properties in $\mathcal{P}$, and relies on $S_\perp$ and $S_\top$ for initialization. It produces new states by using the forward and reverse state transition rules $T^\downarrow, T^\uparrow$, the state update rule $U$, and the values in $D(x_j)$ for layer $j$. The existence of each transition and state is given by $E_t$ and $E_s$.

We use the LTS to compute an MDD by organizing the transitions into $k+1$ layers where $S_\perp$ initializes layer 0, $S_\top$ initializes layer $k$, and the transition values for layer $j$ correspond to the domain values of $x_j$. All transitions out of states in layer $k-1$ are directed to $S_\top$ via the relaxation operator $R$. This process may create paths not connected to $S_\perp$ or $S_\top$; we remove such paths from the LTS.

If no maximum width per layer is imposed, the process will create an exact MDD, which follows from the definition of the properties, state transition rules, state update rules, and transition and state existence functions, i.e., each path from $S_\perp$ to $S_\top$ corresponds to a solution to $c$.

In presence of a maximum width per layer, we can arbitrarily merge nodes using the relaxation operator $R$, which by definition computes a relaxation of the merged states. Hence this process produces a relaxed MDD. □

Since we use MDDs to represent and propagate multiple constraints simultaneously, we need a formalism that supports this functionality. Indeed, MDD languages can be combined by defining an appropriate conjunction operator, and the result is again an MDD language, as shown in the following theorem:

**Theorem 2.** *Let $\mathcal{M}_1$ and $\mathcal{M}_2$ be MDD languages. There exists a conjunction operator $\wedge$ for MDD languages such that $\mathcal{M}_1 \wedge \mathcal{M}_2$ is also an MDD language, and represents the set of paths that are common to $\mathcal{M}_1$ and $\mathcal{M}_2$.*

The proof relies on the definition of the conjunction operator, which is detailed in the Appendix[3]. It essentially concatenates the tuples that define $\mathcal{M}_1$ and $\mathcal{M}_2$. For example, if $\mathcal{M}_1$ and $\mathcal{M}_2$ are defined on overlapping sets of variables $X_1$ and $X_2$, the conjunction is defined on $X_1 \cup X_2$ (e.g., ordered first by $X_1$ and then by $X_2$). The conjunction operator forms the basis of the implementation in HADDOCK for composing multiple constraints into an MDD.

Observe that Theorem 1 leaves open many design choices for an actual implementation of the LTS and associated MDD. Section 5 explains how, operationally, HADDOCK uses MDD languages to produce the actual LTS and MDD.

## 5 Implementation

The purpose of an MDD specification is to mechanize the construction and the propagation of an actual MDD. Sub-sect. 5.1 first considers the creation (posting) of an MDD. Sub-sect. 5.2 explores the actual propagation of *events* occurring through the course of a CP-style search.

### 5.1 Posting

HADDOCK embraces an incremental refinement scheme to construct an MDD [17]. It starts by creating a width-one MDD connecting $S_\perp$ to $S_\top$ which is then refined via *node splitting* until each layer $\mathcal{L}_i$ for $i \in 1..k-1$ reaches at most its target width $w$ and the MDD settles into an MDD propagation fixpoint.

*Initialization.* Given an MDD language $\langle X, \mathcal{P}, S_\perp, S_\top, T^\downarrow, T^\uparrow, U, E_t, E_s, R\rangle$, the source state $S_\perp$ and sink state $S_\top$ have their properties initialized to values suitable for an empty prefix and suffix, respectively, for the constraint $c$ the MDD models. Specifically, each property in $P_i \in \mathcal{P}$ subject to a forward property transition rule $T_i^\downarrow$ should be initialized in $S_\perp$ while each property $P_i$ subject to a reverse property transition rule $T_i^\uparrow$ should be initialized in $S_\top$.

*Example 3.* Recall that for AMONG, $\mathcal{P} = \{L^\downarrow, U^\downarrow, L^\uparrow, U^\uparrow\}$, and, therefore, let $S_\perp = [L^\downarrow \mapsto 0, U^\downarrow \mapsto 0]$ and $S_\top = [L^\uparrow \mapsto 0, U^\uparrow \mapsto 0]$. Namely, the forward properties are initialized to 0 in the source state and the reverse properties are initialized to 0 in the sink state. □

[3] Curious readers can find it in the online supplement.

*Construction.* The construction of an MDD proceeds in phases. First, a two pass (forward and reverse) process creates a width one MDD. The second phase *widens* the MDD up to width $w$ through node splitting.

**First Phase.** The forward pass loops through layers $1..k-1$ to generate states. Assume that the algorithm operates on layer $\mathcal{L}_j$, i.e., all layers $\mathcal{L}_0 \cdots \mathcal{L}_{j-1}$ are already constructed (each layer is a singleton in this phase). It picks the one state $s \in \mathcal{L}_{j-1}$, and, for each value $v \in D(x_{j-1})$, it computes the successor state $s'$ according to $s \xrightarrow{v} s'$, i.e., it evaluates

$$s_v = T^{\downarrow}(s, S_{\top}, j, v).$$

Note how it relies on $S_{\top}$ as an initial approximation for the reverse-defined properties. Finally, it uses the state update function $U$, the relaxation function $R$ and the domain of $D(x_{j-1}) = \{v_0, v_1, v_2, \ldots, v_\ell\}$ to deliver

$$s' = U(R(\cdots R(R(s_{v_0}, s_{v_1}), s_{v_2}) \cdots, s_{v_\ell}))$$

and define $\mathcal{L}_j = \{s'\}$. The reverse pass iterates backward through the layers. At iteration $j$, it uses the singleton state $s' \in \mathcal{L}_j$, its one child $c \in \mathcal{L}_{j+1}$, and a value $v \in D(x_j)$ to compute $s_v = T^{\uparrow}(c, s', j, v)$. The computation integrates the down state $s'$ obtained from the forward pass. Once again, an application of the relaxation yields the now final value for $s'$ via

$$s' = U(R(\cdots R(R(s_{v_0}, s_{v_1}), s_{v_2}) \cdots, s_{v_\ell})) \text{ with } \mathcal{L}_j = \{s'\}.$$

**Second Phase.** The purpose is to widen each layer of the MDD to its final width. The *splitting* algorithm is applied to relaxed nodes in a layer $\mathcal{L}_j$ for which $|\mathcal{L}_j| < w$. Let $s$ be such a state and let $\delta^-(s)$ denote the set of its parent states in layer $\mathcal{L}_{j-1}$ and $\delta^+(s)$ denote the set of its child states in layer $\mathcal{L}_{j+1}$. One can compute, for each existing transition $p \xrightarrow{v} s$ with $p \in \delta^-(s)$, the true state one would reach from $p$ according to value $v$, i.e.,

$$s_p = T^{\downarrow}(p, s, j-1, v) \;\; : \forall p \in \delta^-(s) \;\text{ and } v \text{ on } p \xrightarrow{v} s.$$

Several parents $p$ might yield the same $s_p$, which conveys that multiple transitions lead to the same refined state. One can compute

$$\mathit{refine}(s) = \{s_p \mid \forall p \in \delta^-(s) \;\text{ and } v \text{ on } p \xrightarrow{v} s\}$$

as the set of refined states meant to replace $s$ and endow each $s' \in \mathit{refine}(s)$ with a transition from its parents and the same child states as $s$, i.e, $\delta^+(s)$. Some transitions to states in $\delta^+(s)$ may be invalid according to $E_t$ and are not added. Any element of *refine*$(s)$ that ends up childless as a result must be deleted. Such deletion can trickle back up through multiple layers and bring their width below the desired value. In such a situation one may wish to *reboot* the splitting to an earlier layer rather than continuing on the current one. Pragmatically, it may be useful to bound how far one can "reboot", which we investigate in Sect. 6.

It is desirable to order *refine*$(s)$ as its cardinality, together with that of $\mathcal{L}_j \backslash \{s\}$ may exceed $w$ in which case not all states in *refine*$(s)$ can be adopted "as is" and some merging is required. Namely, given a state ordering relation $\preccurlyeq$ between states, the ordering of *refine*$(s)$ is $s'_0 \preccurlyeq s'_1 \preccurlyeq \ldots \preccurlyeq s'_{m-1}$ ($|\mathit{refine}(s)| = m$). This ordering can induce up to $\ell$ equivalence classes of states deemed similar enough to warrant the use of a single state to represent each class. Naturally, $\ell = w - |\mathcal{L}_j| + 1$, i.e., $\ell$ is an upper bound on the number of classes based on the number of *unused slots* in layer $j$. The new layer $\mathcal{L}_j$ becomes[4]

$$\mathcal{L}'_j = \mathcal{L}_j \setminus \{s\} \cup \left( \bigcup_{C_k \in \{C_0,\ldots,C_{l-1}\}} U(R(C_k)) \right)$$

with $C_0 \cdots C_{l-1}$ as the $l \leq \ell$ classes formed from $s'_0 \preccurlyeq \ldots \preccurlyeq s'_{m-1}$. Layer $j$'s width satisfies $|\mathcal{L}'_j| \leq w$. Changes to the topology of the MDD (state addition, deletion and transitions) are *events* that require a fixpoint propagation to update the values of properties held in the affected states, which is discussed next.

### 5.2 Propagation

*MDD Events.* Propagation occurs in response to *events* affecting the MDD or its connection to finite domain variables. Solver events like $del(v, x_j)$ reporting the loss of a value $v$, i.e., $v \notin D(x_j)$ for a variable $x_j$ appearing in the MDD induce MDD events to convey the deletion of all transitions between layers $j$ and $j+1$ labeled by $v$. MDD events are handled *within* the MDD and are detailed below:

$del(s)$: loss of state $s$ in a layer
$add(s)$: addition of state $s$ in a layer
$del(p \xrightarrow{v} c)$: loss of a transition from $p$ to $c$
$add(p \xrightarrow{v} c)$: addition of a transition from $p$ to $c$
$state(i, s)$: properties $i \subseteq \mathcal{P}(c)$ appearing in state $s$ have changed

The first four events change the topology of the MDD and affect one or more states. The loss of $p \xrightarrow{v} c$ means that the properties of $p$ and $c$ may now be outdated and should be refreshed as that transition might have contributed to a relaxation. For any event $e$, let $\mathit{affected}^{\downarrow}(e)$ and $\mathit{affected}^{\uparrow}(e)$ be the set of states below and above the state, respectively, or the transition in $e$ and affected by it.

*Overall Propagation.* Any state $s$ of an MDD depends on its parent states $\delta^-(s)$ and its child states $\delta^+(s)$. The layered and acyclic structure of the MDD graph provides a natural strategy for updating states when changes occur. Since the MDD is Berge acyclic [3], processing the updates through forward and reverse passes that consider changes in reverse topological order and topological order, respectively, is sensible. Passes can be repeated until a fixpoint is reached.

[4] We take some liberty with notation and refer to the relaxation $R$ over a set of states.

To this end, one needs an event list $E$ and two priority queues $Q^{\downarrow}$ and $Q^{\uparrow}$ to hold onto states. The priority is simply the layer index of the state. Processing an event $e \in E$ schedules $\mathit{affected}^{\downarrow}(e)$ in $Q^{\downarrow}$ and $\mathit{affected}^{\uparrow}(e)$ in $Q^{\uparrow}$ accordingly. Eventually, one processes the states in $Q^{\uparrow}$ in reverse priority order, followed by the states in $Q^{\downarrow}$ in priority order. This propagation may delete transitions as well as states, which induces additional rounds of splitting (as described in the previous sub-section). Pragmatically, it may be desirable to *bound* the number of passes in which splitting occurs to curtail the computational efforts.

*Event Propagation.* A comprehensive treatment of events is not possible within page limitations. Consider as an example the propagation rule for the event $e = \mathit{state}(i, s)$ concerning a change to the properties in $i$ for state $s$ in layer $\mathcal{L}_j$. Processing $e$ means scheduling $\delta^-(s)$ in $Q^{\uparrow}$ and $\delta^+(s)$ in $Q^{\downarrow}$. When a child $c$ of $s$ is pulled from $Q^{\downarrow}$ (the case with $Q^{\uparrow}$ is similar), state $c$ must be updated since $s$ has changed and the relation $s \xrightarrow{v} c$ connects them. This can be done with

$$c_p = T^{\downarrow}(p, c, j, v_p) \ \forall p \in \delta^-(c), \text{with } p \xrightarrow{v_p} c$$

and computing the relaxation of states $c_p$ for all parents $p$ to yield

$$c' = U(R(\{c_p | p \in \delta^-(c), \text{with } p \xrightarrow{v_p} c\})).$$

If $E_s(c')$ does not hold, $c'$ itself is no longer sound and is deleted. If $E_s(c')$ holds and $c' \neq c$, $c$ is changing. Therefore, every transition $c \xrightarrow{v} d$ must be tested with $E_t$ and invalid transitions removed. Such removal should trigger the scheduling of affected nodes. Since $c$ is now changed to $c'$, an event $\mathit{state}(\psi, c')$ should be processed ($\psi$ refers to the subset of properties whose value differ in $c$ and $c'$).

## 6 Empirical Evaluation

HADDOCK is part of `MiniC++`, a C++ implementation of the MiniCP specification [27]. All benchmarks were executed on a Macbook Pro with an i7-5557U processor and 16GB. While a generic implementation is unlikely to match dedicated implementations, HADDOCK let developers produce MDD-based propagators for global constraints with minimal effort. We demonstrate below 1) how HADDOCK compares to an existing MDD implementation, 2) its new modeling capabilities, and 3) the performance impact of the reboot depth parameter.

*Experiment 1: Comparison to State-of-the-Art.* First, we compare HADDOCK to the 'Dedicated' AMONG MDD propagator developed in [21,22] and a classic finite-domain model (written in MiniCP) that uses a cumulative-sums encoding for SEQUENCE constraints as a reasonable baseline for domain propagation. We evaluate these methods on the nurse rostering benchmark problems from [22]. Those problems ask to schedule work shifts for a nurse, subject to a collection of AMONG (or SEQUENCE) constraints. There are three classes of instances with

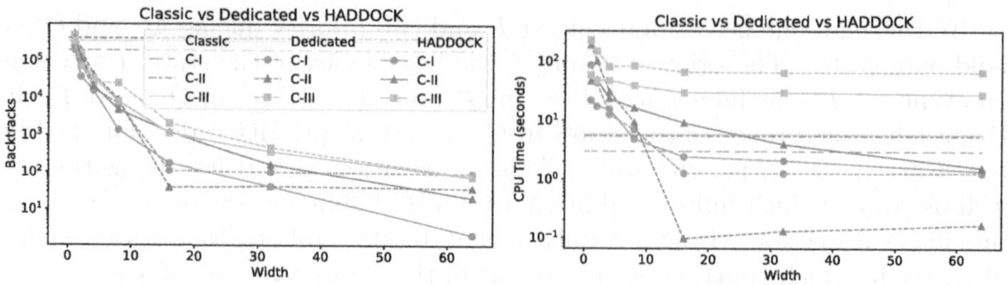


**Fig. 3.** Backtracks (left) and CPU time (right) for finding all solutions for `amongNurse`.

different lower and upper bounds on the number of work days in a sequence. Class C-I requires at most 6 out of 8 consecutive work days and at least 22 out of 30 consecutive work days. C-II uses 6 out of 9 and 20 out of 30 while C-III uses 7 out of 9 and 22 out of 30. Each instance also requires 4 or 5 work days each week and uses a horizon of 40 days. Runtimes reported in [21] were scaled to account for processor differences. We used the highest Geekbench 5 scores reported for our machine CPU (825) and an average score of 330 (standard deviation 10) for an Intel Xeon E5345 that matches the 2.3 GHz characterization in [21]. This yields a scaling by $\frac{330}{825} = 0.4$. Important differences between HADDOCK and [21] are that the implementation of [21] does not compute an MDD propagation fixpoint and adopts a dedicated splitting heuristic that differs from ours.

Figure 3 compares the search tree size (left) and runtimes (right) between the three approaches for finding *all solutions* (each model uses the same lexicographic search strategy). 'Classic' refers to the conventional finite-domain model, 'Dedicated' refers to [21], and 'Haddock' is our implementation with a reboot depth of 0 for clearer comparison. Observe that HADDOCK starts, at width 1, with the exact same number of backtracks as Classic (a bit better than Dedicated, which is attributed to the lack of a fixpoint within [21]). Second, the number of backtracks of HADDOCK are roughly comparable to [21]. Third, the runtimes are within a small factor of each other with HADDOCK in front for class C-III and behind (by a factor in the 1×–10× range) on C-II. Those results are very promising given the generic nature of HADDOCK.

*Experiment 2: Modeling Capabilities.* We next demonstrate the modeling capabilities of HADDOCK on the classic *All-Interval Series* problem (#007 on CSPLIB). Given an integer $n$, the problem asks to find a vector $x = (x_0, \ldots, x_{n-1})$ such that $x$ is a permutation of $\{0, \ldots, n-1\}$, and the interval vector $y = (|x_1 - x_0|, |x_2 - x_1|, \ldots, |x_{n-1} - x_{n-2}|)$ is a permutation of $\{1, \ldots, n-1\}$.

To model this problem using MDDs, we defined the ABSDIFFMDD$(z_0, z_1, z_2)$ constraint in HADDOCK, representing the relation $|z_0 - z_1| = z_2$. It maintains the set of values taken by each of the variables and defines explicit arc existence functions based on this relationship. We then introduce constraints ABSDIFFMDD$(x_i, x_{i+1}, y_i)$, for $i = 0, \ldots, n-2$, which gives us a natural variable ordering, interleaving vectors $x$ and $y$. Lastly, we define ALLDIFFMDD$(x)$

**Table 1.** MDD propagation on the All-Interval Series problem ($n = 11$).

| Reboot | Width 1 | Width 2 | Width 4 | Width 8 | Width 16 | Width 32 | Width 64 |
|---|---|---|---|---|---|---|---|
| 0 | 10,062 | 9,459 | 7,815 | 6,196 | 4,027 | 2,368 | 1,511 |
| 2 | 10,062 | 9,332 | 7,522 | 6,191 | 3,796 | 2,264 | 1,470 |
| 4 | 10,062 | 9,069 | 6,972 | 5,279 | 3,282 | 2,100 | 1,357 |
| 8 | 10,062 | 5,698 | 3,277 | 2,283 | 1,626 | 1,839 | 316 |
| MAX | 10,062 | 1,155 | 140 | 40 | 34 | 23 | 6 |

and ALLDIFFMDD($y$) constraints to model the permutations. The entire ABSDIFFMDD language requires less than 120 lines for declaring its states, transitions, and relaxation rules. Likewise, the ALLDIFFMDD language requires only 70 lines. For these constraints, HADDOCK applies a generic state ordering based on the number of parent states (smallest first) during the node splitting process.

We apply a lexicographic search on the $x$ variables to compare the performance over different settings. We report results for $n = 11$ (finding all solutions) in Table 1. To evaluate the impact of MDD propagation, consider the base case in which the reboot depth is set to 0. The table shows how increasing the width from 1 (domain propagation) to larger widths reduces the search tree size, yielding a factor 10 for maximum width 64. It is encouraging to see how MDDs can be used like any other global constraints within a CP solver and deliver huge reductions in the size of the search tree. We do note, however, that the reduction in search tree size has no positive impact on the solving time. It remains an area for future work to exploit better split ordering and an optimized implementation.

*Experiment 3: Reboot Depth Sensitivity.* We next investigate the impact of varying the reboot depth. Consider Table 1 once again. Rows 2 through 5 convey the impact of reboots on the search tree size at all considered widths. Recall that HADDOCK only relies on a syntactic value (the number of parents) to rank states. Also recall that rebooting considers abandoning splitting at layer $k$ when it has induced state deletion in prior layers and return to those. The `reboot` parameter controls how far back HADDOCK can jump when returning to earlier layers. In this situation, rebooting can have a *dramatic* impact on the size of the search at any width. The most extreme parameter settings of width = 64 and unbounded reboots yield a search tree with only 6 nodes, i.e., 4 orders of magnitude compared to the baseline in the upper left corner of the table.

## 7 Conclusion

This paper introduced HADDOCK, a system for Handling Automatically Decision Diagrams over Constraints Kernels. It described the language and framed its implementation in terms of a compilation down to an LTS form. The resulting system is generic and capable of capturing and automatically deriving implementation for several MDDs that were proposed before and for which only dedicated implementations exist. The empirical evaluation showed that this very first implementation of a generic MDD engine exhibits search tree reductions that are qualitatively on par with prior work and achieves comparable runtime.

**Acknowledgments.** Willem-Jan van Hoeve was partially supported by ONR Grant No. N00014-18-1-2129 and NSF Award #1918102. L. Michel and R. Gentzel were partially supported by Comcast under Grant #15228, Synchrony under #790057267 and ONR/NIUVT under #N000014-20-1-2040.

## References

1. Akers, S.B.: Binary decision diagrams. IEEE Trans. Comput. **C–27**, 509–516 (1978)
2. Andersen, H.R., Hadzic, T., Hooker, J.N., Tiedemann, P.: A constraint store based on multivalued decision diagrams. In: Bessière, C. (ed.) CP 2007. LNCS, vol. 4741, pp. 118–132. Springer, Heidelberg (2007). https://doi.org/10.1007/978-3-540-74970-7_11
3. Berge, C.: Hypergraphs - Combinatorics of Finite Sets. North-Holland Mathematical Library, vol. 45. North-Holland, Amsterdam (1989)
4. Bergman, D., Cire, A.A., van Hoeve, W.-J.: MDD propagation for sequence constraints. JAIR **50**, 697–722 (2014)
5. Bergman, D., Cire, A.A., van Hoeve, W.-J., Hooker, J.N.: Decision Diagrams for Optimization. Springer, Heidelberg (2016). https://doi.org/10.1007/978-3-319-42849-9
6. Bergman, D., Cire, A.A., van Hoeve, W.-J., Hooker, J.N.: Discrete optimization with decision diagrams. INFORMS J. Comput. **28**(1), 47–66 (2016)
7. Bryant, R.E.: Graph-based algorithms for boolean function manipulation. IEEE Trans. Comput. **C–35**, 677–691 (1986)
8. Bryant, R.E.: Symbolic boolean manipulation with ordered binary decision diagrams. ACM Comput. Surv. **24**, 293–318 (1992)
9. Cheng, K.C.K., Xia, W., Yap, R.H.C.: Space-time tradeoffs for the regular constraint. In: Milano, M. (ed.) CP 2012. LNCS, vol. 7514, pp. 223–237. Springer, Heidelberg (2012). https://doi.org/10.1007/978-3-642-33558-7_18
10. Cheng, K.C.K., Yap, R.H.C.: Maintaining generalized arc consistency on ad-hoc n-ary Boolean constraints. In Brewka, G., et al. (eds.) Proceedings of ECAI, pp. 78–82. IOS Press (2006)
11. Cire, A.A., van Hoeve, W.-J.: Multivalued decision diagrams for sequencing problems. Oper. Res. **61**(6), 1411–1428 (2013)
12. de Uña, D., Gange, G., Schachte, P., Stuckey, P.J.: Compiling CP subproblems to MDDs and d-DNNFs. Constraints **24**(1), 56–93 (2019). https://doi.org/10.1007/s10601-018-9297-2
13. Gange, G., Lagoon, V., Stuckey, P.J.: Fast set bounds propagation using BDDs. In: Ghallab, M. et al. (eds.) Proceedings of ECAI, pp. 505–509. IOS Press (2008)
14. Gange, G., Stuckey, P.J., Van Hentenryck, P.: Explaining propagators for edge-valued decision diagrams. In: Schulte, C. (ed.) CP 2013. LNCS, vol. 8124, pp. 340–355. Springer, Heidelberg (2013). https://doi.org/10.1007/978-3-642-40627-0_28

15. Gange, G., Stuckey, P.J., Szymanek, R.: MDD propagators with explanation. Constraints **16**(4), 407–429 (2011)
16. Hadžić, T., Hooker, J.N.: Cost-bounded binary decision diagrams for 0–1 programming. Technical report, Carnegie Mellon University (2007)
17. Hadzic, T., Hooker, J.N., O'Sullivan, B., Tiedemann, P.: Approximate compilation of constraints into multivalued decision diagrams. In: Stuckey, P.J. (ed.) CP 2008. LNCS, vol. 5202, pp. 448–462. Springer, Heidelberg (2008). https://doi.org/10.1007/978-3-540-85958-1_30
18. Hadzic, T., Hooker, J.N., Tiedemann, P.: Propagating separable equalities in an MDD store. In: Perron, L., Trick, M.A. (eds.) CPAIOR 2008. LNCS, vol. 5015, pp. 318–322. Springer, Heidelberg (2008). https://doi.org/10.1007/978-3-540-68155-7_30
19. Hadzic, T., O'Mahony, E., O'Sullivan, B., Sellmann, M.: Enhanced inference for the market split problem. In: Proceedings of ICTAI, pp. 716–723. IEEE Computer Society (2009)
20. Hawkins, P., Lagoon, V., Stuckey, P.J.: Solving set constraint satisfaction problems using ROBDDs. JAIR **24**(1), 109–156 (2005)
21. Hoda, S.: Essays on equilibrium computation, MDD-based constraint programming and scheduling. Ph.D. thesis, Carnegie Mellon University (2010)
22. Hoda, S., van Hoeve, W.-J., Hooker, J.N.: A systematic approach to MDD-based constraint programming. In: Cohen, D. (ed.) CP 2010. LNCS, vol. 6308, pp. 266–280. Springer, Heidelberg (2010). https://doi.org/10.1007/978-3-642-15396-9_23
23. Keller, R.M.: Formal verification of parallel programs. Commun. ACM **19**(7), 371–384 (1976)
24. Kinable, J., Cire, A.A., van Hoeve, W.-J.: Hybrid optimization methods for time-dependent sequencing problems. Eur. J. Oper. Res. **259**(3), 887–897 (2017)
25. Lagoon, V., Stuckey, P.J.: Set domain propagation using ROBDDs. In: Wallace, M. (ed.) CP 2004. LNCS, vol. 3258, pp. 347–361. Springer, Heidelberg (2004). https://doi.org/10.1007/978-3-540-30201-8_27
26. Lee, C.-Y.: Representation of switching circuits by binary-decision programs. Bell Syst. Tech. J. **38**(4), 985–999 (1959)
27. Michel, L., Schaus, P., Van Hentenryck, P.: MiniCP: a lightweight solver for constraint programming (2018). https://minicp.bitbucket.io
28. Perez, G., Régin, J.-C.: Efficient operations on MDDs for building constraint programming models. In: Proceedings of IJCAI, pp. 374–380 (2015)
29. Perez, G., Régin, J.-C.: Soft and cost MDD propagators. In: Proceedings of AAAI, pp. 3922–3928. AAAI Press (2017)
30. Wegener, I.: Branching Programs and Binary Decision Diagrams: Theory and Applications. SIAM Monographs on Discrete Mathematics and Applications. Society for Industrial and Applied Mathematics, Philadelphia (2000)

# Towards a Generic Interval Solver for Differential-Algebraic CSP

Simon Rohou[1](✉), Abderahmane Bedouhene[2], Gilles Chabert[3], Alexandre Goldsztejn[4], Luc Jaulin[1], Bertrand Neveu[2], Victor Reyes[5], and Gilles Trombettoni[5]

[1] Lab-STICC, ENSTA Bretagne, CNRS, Brest, France
Simon.rohou@ensta-bretagne.fr
[2] LIGM, Ecole des Ponts, Univ. Gustave Eiffel, CNRS, Marne-la-Vallée, France
[3] IRT Jules Verne, LS2N, Nantes, France
[4] LS2N, CNRS, Nantes, France
[5] LIRMM, University of Montpellier, CNRS, Montpellier, France

**Abstract.** In this paper, we propose an interval constraint programming approach that can handle the *differential-algebraic CSP* (DACSP), where an instance is composed of real and functional variables (also called dynamic variables or trajectories) together, and differential and/or "static" numerical constraints among those variables. Differential-Algebraic CSP systems can model numerous real-life problems occurring in physics, biology or robotics. We introduce a solver, built upon the `Tubex` and IBEX interval libraries, that can rigorously approximate the set of solutions of a DACSP system. The solver achieves temporal slicing and a tree search by splitting trajectories domains. Our approach provides a significant step towards a generic interval CP solver for DACSP that has the potential to handle a large variety of constraints. First experiments highlight that this solver can tackle interval Initial Value Problems (IVP), Boundary Value Problems (BVP) and integro-differential equations.

## 1 Introduction

Differential-Algebraic CSP (DACSP) systems include real variables, functional variables (also called trajectories or dynamical variables) describing the dynamics of the system, and differential and/or "static" numerical constraints among those variables. Because they are at the core of so many applications, such as biological systems, mechanism dynamics, astronomy, robotics, control, a lot of work has been dedicated to solving specific subclasses of the DACSP. Most of the approaches follow a probabilistic approach and are limited to linear constraints with Gaussian errors [17,35].

There are a number of advantages to using interval methods for handling dynamical and/or static systems. They can manage nonlinear constraints and approximate the solutions rigorously, whatever the uncertainties on the parameters (*e.g.*, uncertainties related to measurements or to inaccurate


H. Simonis (Ed.): CP 2020, LNCS 12333, pp. 548–565, 2020.
https://doi.org/10.1007/978-3-030-58475-7_32

physical models). Bounded intervals are used to characterize these uncertainties, as well as the errors due to operations over floating-point numbers.

With Jaulin *et al.* works [15,16], a significant step has been made towards a declarative constraint programming approach for dynamical systems. Contrary to dominant constraint approaches dealing with differential equations such as [8,30,37], the trajectories are viewed as variables of the CSP. They consider trajectories as variables, differential equations as constraints and *tubes* as domains. The solution and the domain are given by a tube representing a set of possible trajectories (see Fig. 1). This increases the level of abstraction and simplifies the formalization of the problem. For estimating a trajectory, a set of *contractors*, similar to *propagators* in CP, are applied to filter (contract) the bounds of the domain/tube. A benefit of the contractor programming approach [3] is the variety of dynamical systems that can be handled. More recently, the `Tubex` library [32] has provided data structures for representing tubes and a catalogue of contractors, similarly to the catalogue of propagators available in CP solvers [29]. The user defines a sequence of contractors for modeling his problem. The set of contractors is applied iteratively until a quasi fixpoint in terms of contraction is reached. For instance, the resulting framework has been recently applied to actual data for autonomous robot localization. [31] describes the problem of localizing an underwater robot. Its evolution is depicted by an Ordinary Differential Equation (ODE) and bounded measurements, but the initial condition (position) of the system is unknown. During the mission, its sonar detects indistinguishable rocks on the seabed, that all look alike but are known to belong to a discrete point map embedded beforehand. The map of rocks positions is also modeled as a constraint. The constraint propagation approach is able to merge all data coming from the robot evolution and rocks observations. The identity of the rocks is finally associated to items in the map, and the trajectory of the robot is accurately estimated. This application highlights how a complex problem involving discrete, continuous and differential constraints can be solved easily following a CP/contractor approach.

Contractor programming is relevant when the problem is defined by heterogeneous constraints, provided they are redundant and numerous enough to enable the contraction phase alone to solve the problem.

*Contributions.* In this paper, we introduce a generic solver using the `Tubex` library that can handle a DACSP instance made of differential constraints and/or static continuous constraints. The numerical constraints relate real variables that either represent states of the trajectories at given instants or are independent from the dynamics (*e.g.*, the rocks positions in the previous example). Compared to the `Tubex` approach alone, the solver is endowed with an operator that can perform a choice point by splitting (bisecting) a tube in two at a chosen time. Thus, a tree search can accurately estimate distinct trajectories (for problems with several solutions) and can better handle several hard problems. To our knowledge, no other tool for solving dynamical systems performs a tree search.

Our solver can manage most of the contractors available in the literature. On the one hand, several contractors coming from CP, *e.g.*, 3B [22] and CID [36], are included naturally to improve the pruning of domains of functional variables, *i.e.*, tubes. On the other hand, our generic framework enables to wrap existing solvers dedicated to specific types of (differential) constraints into contractors. These contractors efficiently reduce domains by taking into account space and time dependencies. In particular, we show in the experiments the benefits of using the contractor `CtcVnode` built upon the state-of-the-art VNODE guaranteed integration IVP solver [27,28].

This generic CP framework allows separating the problem description, which includes here all combinations of differential and algebraic problems, from its resolution. We also wanted to solve difficult standard problems coming from numerical analysis, *e.g.*, BVP for integro-differential equations, that are out of the scope of previous CP/ODE frameworks and remain today difficult to solve in the numerical analysis framework.

*Related Work.* VNODE [28], CAPD [18], COSY [30] and `DynIbex` [4] are state of the art interval analysis solvers dedicated to IVPs.[1] They are fully relevant for determining the guaranteed solution of a system at a final time, that has crucial applications such as the position determination of celestial bodies in astronomy [39] or to characterize chaotic attractors [37]. They use different algorithms to reliably simulate the initial information over time. In particular, the VNODE tool used in our solver combines a high-order interval Taylor form to integrate the state from an instant to a next one, and a step limiting the wrapping effect implied by interval calculation: it encloses the solution at the discrete times by an envelope sharper than a box, such as rotated boxes, zonotopes or polygons [23].

A BVP is generally defined by an ODE, but the trajectory is not entirely determined at the initial or final times, which prevents a solving algorithm from propagating (integrating) the information from the known state to the rest of the trajectory. Instead, static constraints are defined on specific states (at specified instants). To deal with BVPs, the shooting method [12] is a sampling-and-optimization algorithm that runs several integration processes from the initial state while trying to minimize the distance to a solution satisfying the ODE. This makes the approach incomplete and unable to determine several solutions, if any. Instead, our rigorous and deterministic solver can explore the whole search space and isolate distinct solutions.

The constraint programming community has contributed to dynamical systems. Janssen *et al.* propose a strategy [8] dedicated to interval IVPs that has similarities with VNODE or CAPD. The integration step between two consecutive states (instants) used a box consistency algorithm [1] to better contract the output state. A first attempt to create a constraint language extended to ODEs was made a few years later [6,7]. The language considers the entire ODE as a whole, but the unknown function has a special status that prevents direct manipulations: any constraint involving the unknown trajectory is managed by

[1] An Initial Value Problem is composed of an ODE and an initial condition. Numerical integration propagates the initial value through the whole trajectory by integrating the evolution function of the ODE.

a specific operator. The fact that the function is bounded in an interval requires the introduction of an ad-hoc constraint called minimum/maximum restriction. The link with real variables is also achieved via an ad-hoc constraint called value restriction. This language does not have the level of genericity targeted by our solver, where the concept of trajectories appears at the same level as other variables and ODEs are syntactic constructions among others. In our solver, we will rather use the $\mathcal{C}_{\text{eval}}$ contractor [33] to handle value restriction constraints.

Although [11] improves [6] in both modeling possibilities and solving efficiency, it is still restricted to ODE constraints relating solutions values at specified times. Finally, our model can accept various differential constraints (like an integro-differential equation) and static constraints such as distance relations between states at different instants, as long as the corresponding contractors exist.

*Outline.* Section 2 introduces the notations used in the paper and the background useful to understand the following sections. Sections 3 and 4 detail our DACSP solver and its different procedures and parameters: contractors, choice point heuristics, *etc.* Section 5 reports first experimental results obtained on interval IVPs, but also integro-differential equations and BVPs.

## 2 Background and Notations

We first provide some background about intervals, inclusion functions and contraction. We then present several differential constraints and interval techniques adapted to dynamical systems.

### 2.1 Intervals

Contrary to numerical analysis methods that work with single values, interval methods can manage sets of values enclosed in intervals. Interval methods are known to be particularly useful for handling nonlinear constraint systems.

**Definition 1 (Interval, box, box width, box hull)**
*An interval* $[x_i] = [\underline{x_i}, \overline{x_i}]$ *defines the set of reals* $x_i$ *such that* $\underline{x_i} \leq x_i \leq \overline{x_i}$. $\mathbb{IR}$ *denotes the set of all intervals. A box* $[\boldsymbol{x}]$ *denotes a Cartesian product of intervals* $[\boldsymbol{x}] = [x_1] \times ... \times [x_n]$. *The size, width or diameter of a box* $[\boldsymbol{x}]$ *is given by* $w([\boldsymbol{x}]) \equiv \max_i(w([x_i]))$ *where* $w([x_i]) \equiv \overline{x_i} - \underline{x_i}$. *The hull of boxes approximates the union operator. It returns the smallest box enclosing all the boxes hulled.*

*Interval arithmetic* [26] has been defined to extend to $\mathbb{IR}$ the usual mathematical operators over $\mathbb{R}$ For instance, the interval sum is defined by $[x_1] + [x_2] = [\underline{x_1} + \underline{x_2}, \overline{x_1} + \overline{x_2}]$. When a function $\boldsymbol{f}$ is a composition of elementary functions, an *inclusion function* $[\boldsymbol{f}]$ of $\boldsymbol{f}$ must be defined to ensure a conservative image computation. There are several inclusion functions. The *natural* inclusion function of a real function $f$ corresponds to the mapping of $f$ to intervals using interval arithmetic. For instance, the natural inclusion function $[f]_N$ of $f(x) = x(x+1)$

in the domain $[x] = [0,1]$ computes $[f]_N([0,1]) = [0,1] \cdot [1,2] = [0,2]$. Another inclusion function is based on an interval Taylor form [14].

Interval arithmetics can be used for solving the *numerical CSP* (NCSP), *i.e.* finding solutions to an NCSP instance $P = (\boldsymbol{x}, [\boldsymbol{x}], \boldsymbol{c})$, where $\boldsymbol{x}$ is an $n$-set of variables taking their real values in the domain $[\boldsymbol{x}]$ and $\boldsymbol{c}$ is an $m$-set of numerical constraints using operators like $+$, $-$, $\times$, $a^b$, exp, log, sin, *etc.* NCSP solvers, like GlobSol [19], Gloptlab [9], RealPaver [13] or IBEX [2] to name a few, follow a Branch and Contract method to solve an NCSP. The branching operation subdivides the search space by recursively bisecting variable intervals into two subintervals and exploring both sub-boxes independently. The combinatorial nature of this tree search is not always observed thanks to the *contraction* (filtering) operations applied at each node of the search tree. Informally, a contraction applied to an NCSP instance can reduce the domain without losing any solution. A contractor used in this paper is the well-known `HC4-revise` [1,25], also called *forward-backward.* This contractor handles a single numerical constraint and obtains a (generally non optimal [5]) contracted box including all the solutions of that constraint. To contract a box w.r.t. an NCSP instance, the HC4 algorithm performs a (generalized) AC3-like propagation loop applying iteratively the HC4-Revise procedure on each constraint individually until a quasi fixpoint is obtained in terms of contraction.

3B-consistency [22] and CID-consistency [36] are two other stronger consistencies, enforced on an NCSP, that are exploited by our solver. The corresponding 3B and CID algorithms should call their Var3B or VarCID procedure on all the NCSP variables for enforcing the 3B or CID consistency. In practice however the algorithms implemented apply these procedures on a subset of the variables to get a better tradeoff between contraction and performance. VarCID splits a variable interval in $k$ subintervals, and runs a contractor, such as HC4, on the corresponding sub-boxes. The $k$ sub-boxes contracted are finally hulled. Var3B is somehow a dual operator that tries to remove subintervals at the bounds of a variable interval. If a contraction, like HC4, applied to a sub-box, where the interval is replaced by a subinterval at a bound, leads to an empty domain, then it proves that the subinterval contains no solution and can be removed safely from the variable domain.

### 2.2 Trajectories and Tubes

A trajectory, denoted $\boldsymbol{x}(\cdot) = (x_1(\cdot), .., x_n(\cdot))$, is a function from $[t_0, t_f] \subset \mathbb{R}$ to $\mathbb{R}^n$. The input (argument) of $\boldsymbol{x}(\cdot)$ is named *time* in this article (and denoted $\cdot$ or $t$) while the output (image) is called *state.*

A tube $[\boldsymbol{x}](\cdot)$ is the interval counterpart of a trajectory and is defined as an envelope over the same temporal domain $[t_0, t_f]$. The concept appeared in [10,20] in the context of ellipsoidal estimations. In our solver, it is used as a domain on which we apply operations of contractions and bisections. We employ them as intervals of trajectories, which is consistent with the aforementioned tools.

Hence, we will use the definition given in [21] where a tube $[\boldsymbol{x}](\cdot) : [t_0, t_f] \rightarrow \mathbb{IR}^n$ is an interval of two trajectories $[\underline{\boldsymbol{x}}(\cdot), \overline{\boldsymbol{x}}(\cdot)]$ such that $\forall t \in [t_0, t_f]$, $\underline{\boldsymbol{x}}(t) \leqslant$

$\overline{x}(t)$. We also consider empty tubes that depict an absence of solutions. A trajectory $\boldsymbol{x}(\cdot)$ belongs to the tube $[\boldsymbol{x}](\cdot)$ if $\forall t \in [t_0, t_f]$, $\boldsymbol{x}(t) \in [\boldsymbol{x}](t)$. Figure 1 illustrates a one-dimensional tube enclosing a trajectory $x(\cdot)$.

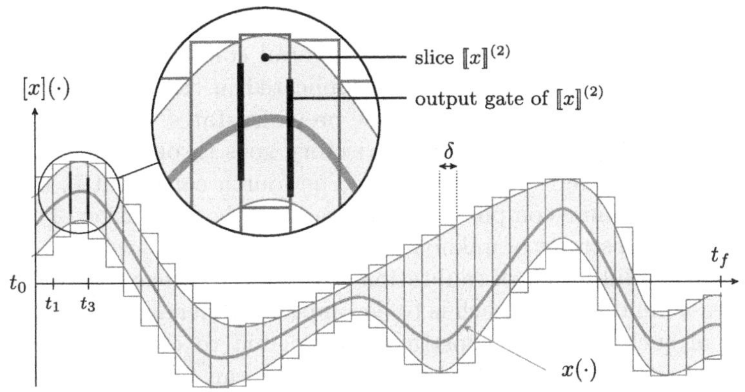


**Fig. 1.** A one-dimensional tube $[x](\cdot)$, interval of two functions $[\underline{x}(\cdot), \overline{x}(\cdot)]$, enclosing a random trajectory $x(\cdot)$ depicted in orange. The tube is numerically represented by a set of $\delta$-width slices illustrated by blue boxes. (Color figure online)

Our choice is to represent numerically a tube as a set of boxes corresponding to temporal slices. More precisely, an $n$-dimensional tube $[\boldsymbol{x}](\cdot)$ with a sampling time $\delta > 0$ is implemented as a box-valued function which is constant for all $t$ inside intervals $[k\delta, k\delta + \delta]$, $k \in \mathbb{N}$. The box $[k\delta, k\delta + \delta] \times [\boldsymbol{x}](t_k)$, with $t_k \in [k\delta, k\delta + \delta]$, is called the $k^{th}$ *slice* of the tube $[\boldsymbol{x}](\cdot)$ and is denoted by $[\![\boldsymbol{x}]\!]^{(k)}$. This implementation takes rigorously into account floating-point precision when building a tube: computations involving $[\boldsymbol{x}](\cdot)$ will be based on its slices, thus giving a reliable outer approximation of the solution set. The slices may be of same width as depicted in Fig. 1, but the tube can also be implemented with a customized temporal *slicing*. Finally, we endow the definition of a slice $[\![\boldsymbol{x}]\!]^{(k)}$ with the *slice (box) envelope* (blue painted in Fig. 1) and two input/output *gates* $[\boldsymbol{x}](t_k)$ and $[\boldsymbol{x}](t_{k+1})$ (black painted) that are intervals of $\mathbb{IR}^n$ through which trajectories are entering/leaving the slice.

Once a tube is defined, it can be handled in the same way as an interval. We can for instance use arithmetic operations as well as function evaluations. If $f$ is an elementary function such as sin, cos or exp, we define $f([x](\cdot))$ as the smallest tube containing all feasible values: $f([x](\cdot)) = \left[\, \{f(x(\cdot)) \mid x(\cdot) \in [x](\cdot)\} \,\right]$.

### 2.3 Dynamical Systems and Differential-Algebraic CSP

A differential constraint relates one or several trajectories and/or real variables. Numerous types of differential constraints can be considered in our approach, including:

1. $\dot{\boldsymbol{x}}(\cdot) = \boldsymbol{f}(\boldsymbol{x}(\cdot))$ (ODE)
2. $\dot{\boldsymbol{x}}(t) = \boldsymbol{f}(\boldsymbol{x}(t)) + \int_{t_0}^{t} \boldsymbol{x}(\tau)d\tau$ (integro-differential equation)
3. $\boldsymbol{x}(t_k) = \boldsymbol{y}$, $\dot{\boldsymbol{x}}(\cdot) = \boldsymbol{v}(\cdot)$ (evaluation constraint)
4. $\forall t \in [t]$, $\boldsymbol{x}(t) \notin [\boldsymbol{y}]$
5. $\boldsymbol{x}(t) = \boldsymbol{y}(t+\delta)$ (delay constraint)

The first one is the most widespread differential constraint (see Definition 2). The second constraint is a little more complicated in that the state at a given time depends on the sum (integral) of the previous states. The third evaluation constraint $\boldsymbol{y} = \boldsymbol{x}(t_k)$ states that the trajectory goes through an uncertain real value in $[\boldsymbol{y}]$ at an uncertain time in $[t_k]$. The fourth constraint is the complementary, although more complicated, constraint of the evaluation. The fifth one imposes a delay constraint of unknown real value $\delta$ between two trajectories and is particularly useful for clock calibration purposes in autonomous systems [38].

The idea behind our approach is to decompose a differential-algebraic system into a set of such primitive constraints associated to contractors (similar to propagators in CSP solvers [29]) that belong, or can be added to, the `Tubex` library. We formally define below the following differential constraints used in the experimental part.

**Definition 2 (ODE and integro-differential equation)**
*Consider* $\boldsymbol{x}(\cdot) : [t_0, t_f] \to \mathbb{R}^n$*, its derivative* $\dot{\boldsymbol{x}}(\cdot) : [t_0, t_f] \to \mathbb{R}^n$*, and an* evolution function $\boldsymbol{f} : \mathbb{R}^n \to \mathbb{R}^n$*, possibly non-linear. An ODE*[2] *is defined by:*

$$\dot{\boldsymbol{x}}(\cdot) = \boldsymbol{f}(\boldsymbol{x}(\cdot))$$

*An integro-differential equation is defined by:*

$$\dot{\boldsymbol{x}}(t) = \boldsymbol{f}(\boldsymbol{x}(t)) + \int_{t_0}^{t} \boldsymbol{x}(\tau)d\tau.$$

The ODEs considered are *explicit, i.e.* the evolution function $\boldsymbol{f}$ computes $\dot{\boldsymbol{x}}$ directly. These differential constraints can define dynamical systems.

**Definition 3 (IVP, interval IVP, BVP)**
*The initial value problem (IVP) is defined by an ODE* $\dot{\boldsymbol{x}}(\cdot) = \boldsymbol{f}(\boldsymbol{x}(\cdot))$ *and an initial condition* $\boldsymbol{x}(t_0) = \boldsymbol{x}_0$*, where* $\boldsymbol{x}_0$ *is a constant in* $\mathbb{R}^n$*. In an interval IVP, the initial condition is bounded by an interval, i.e.* $\boldsymbol{x}(t_0) \in [\boldsymbol{x}_0]$*. A boundary value problem (BVP) is defined by an ODE and a numerical constraint* $\mathbf{c}(\boldsymbol{x}(t_1)..\boldsymbol{x}(t_n)) = \mathbf{0}$*, where* $\mathbf{c} : \mathbb{R}^n \to \mathbb{R}^n$ *and* $\forall i \in \{1..n\}$*,* $t_i \in [t_0, t_f]$*.*

A BVP generalizes an IVP in that no initial condition fully determines the trajectory at a unique instant. Instead, $n$ algebraic constraints relate several states

[2] Note that a high-order problem can be transformed automatically into a first-order ODE shown in Definition 2 by introducing auxiliary variables. Also note that non autonomous ODEs of the form $\dot{\boldsymbol{x}}(t) = \boldsymbol{f}(\boldsymbol{x}(t), t)$ can also be transformed into autonomous ODEs $\dot{\boldsymbol{x}}(t) = \boldsymbol{f}(\boldsymbol{x}(t))$ whose derivative depends only on the state.

at times $t_1..t_n$ and enable the trajectory determination. Note that a condition known at any instant is equivalent to knowing the state at time $t_0$. Indeed, numerical integration can propagate the information forward or backward in time indifferently. We are now in position to define the DACSP.

**Definition 4 (Differential-algebraic CSP – DACSP)**
*A DACSP network is defined by a quintuplet* $(\boldsymbol{x}(\cdot), [\boldsymbol{x}](\cdot), \boldsymbol{y}, [\boldsymbol{y}], \boldsymbol{c})$, *where* $\boldsymbol{x}(\cdot)$ *is a trajectory variable of domain* $[\boldsymbol{x}](\cdot)$ *(a tube* $\mathbb{R} \rightarrow \mathbb{IR}^{n_1}$*, as defined in the previous section),* $\boldsymbol{y} \in \mathbb{R}^{n_2}$ *denotes the static numerical variables with a domain/box* $[\boldsymbol{y}]$ *and* $\boldsymbol{c}$ *denotes the set of static or differential constraints. Solving a DACSP instance consists in finding the set of values in* $[\boldsymbol{x}](\cdot)$ *and* $[\boldsymbol{y}]$ *satisfying* $\boldsymbol{c}$.

## 3 A Generic Solver for Differential-Algebraic CSP

In Algorithm 1, we give a description of a first generic solver for DACSP. The solver works on a network $P = (\boldsymbol{x}(\cdot), [\boldsymbol{x}](\cdot), \boldsymbol{y}, [\boldsymbol{y}], \boldsymbol{c})$ and returns a set of trajectories satisfying $\boldsymbol{c}$. The input tube $[\boldsymbol{x}](\cdot)$ is defined generally with one single slice $[t_0, t_f] \times [-\infty, \infty]^n$. We attempt to tackle a wide class of problems with possibly different behaviors. This may impair the effectiveness of a unique generic algorithm. In practice however, the user may already have an intuition of some instants from which things should propagate. As a consequence, in addition to $P$, we allow the user to provide a set of *special times*, *i.e.* elements of the temporal domain that involve states of the trajectory $\boldsymbol{x}(\cdot)$ and other static constraints. It allows this first solver to perform contraction more incrementally.

The solver works in two main phases: a so-called *slicing* step splitting the temporal domain into time slices, followed by a tree search subdividing the vectorial tube $[\boldsymbol{x}](\cdot)$. The last `TubeMerge` function compensates a potential clustering effect and merges together pairs of solution tubes that intersect along the temporal domain (on all the slices) in all the $(\boldsymbol{x})$ dimensions. It is necessary when bisection is not used for identifying different solutions, but helps the solver to compute accurate trajectories.

The main precision parameter of the solver is *maxTubeDiam*, a size expressed as the maximum width over all the slices envelopes of the tube. The solver can indifferently compute "thin" trajectories of theoretical null volume (*e.g.*, when dealing with pure IVPs) or "thick" trajectories (*i.e.*, continua of trajectories, *e.g.* when dealing with interval IVPs). In the latter case, the user has to tune this precision parameter to get a good approximation of thick trajectories. A second user-defined *#slicesMax* parameter is a maximum number of slices created during the solving, especially during the slicing phase. A large slice number leads to a better trajectory accuracy at the cost of worse performance. Note that the CPU time generally grows linearly in the number of slices.

The first slicing phase is performed by the first `do..while` loop interleaving contraction of $P$ and time slicing ("discretization"). The latter splits several slices of $[\boldsymbol{x}](\cdot)$ into two slices of equal temporal size. Three main slicing policies have been tested:

**Algorithm** DACSP-Solver *($P = (\boldsymbol{x}(\cdot), [\boldsymbol{x}](\cdot), \boldsymbol{y}, [\boldsymbol{y}], \boldsymbol{c})$, specialTimes, maxTubeDiam, #slicesMax,* Integration, *$\boldsymbol{\varepsilon} = (\varepsilon_{fpt}, \varepsilon_{integr}, \varepsilon_{3B})$, slicingPolicy, bisectionPolicy)*

- **do**
  - /* Slicing loop: */
  - $([\boldsymbol{x}](\cdot), [\boldsymbol{y}]) \leftarrow$ Contraction(*P*, *specialTimes*, Integration, $\boldsymbol{\varepsilon}$, *false*)
  - $[\boldsymbol{x}](\cdot) \leftarrow$ Slicing($[\boldsymbol{x}](\cdot)$, *slicingPolicy*)
- **while** *MaxDiam*($[\boldsymbol{x}](\cdot)$) > *maxTubeDiam* **and** *#Slices(tube)* < *#slicesMax*
- **if** *MaxDiam*($[\boldsymbol{x}](\cdot)$) $\leq$ *maxTubeDiam* **then** return $[\boldsymbol{x}](\cdot)$
- $L \leftarrow \{ ([\boldsymbol{x}](\cdot), null) \}$
- **while** $L \neq \emptyset$ */* Depth-first tree search */* **do**
  - $([\boldsymbol{x}](\cdot), gate) \leftarrow$ Pop($L$)
  - $([\boldsymbol{x}](\cdot), [\boldsymbol{y}]) \leftarrow$ Contraction(*P*, *specialTimes*, Integration, $\boldsymbol{\varepsilon}$, *true*, *gate*)
  - **if** *MaxDiam*($[\boldsymbol{x}](\cdot)$) $\leq$ *maxTubeDiam* **then**
    - *solutionsList* $\leftarrow$ *solutionsList* $\cup \{[\boldsymbol{x}](\cdot)\}$
  - **else**
    - $([\boldsymbol{x}_1](\cdot), [\boldsymbol{x}_2](\cdot), gate) \leftarrow$ Bisect ($[\boldsymbol{x}](\cdot)$, *bisectionPolicy*)
    - $L \leftarrow \{([\boldsymbol{x}_1](\cdot), gate)\} \cup \{([\boldsymbol{x}_2](\cdot), gate)\} \cup L$
- *solutionsList* $\leftarrow$ TubeMerge(*solutionsList*, $[\boldsymbol{x}](\cdot)$)
- return *solutionsList*

**Algorithm 1:** The DACSP solver.

- (*all*) Split *all* the slices in two.
- (*median*) Compute for all the bounded slices a $dx$ difference between the middle points of 2 consecutive gates, maximum over all the dimensions, *i.e.* $dx = \max_i |m([x_i^k]) - m([x_i^{k-1}])|$, where $m([x_i])$ denotes the middle of $[x_i]$. Split half of the slices with the largest $dx$ and all the unbounded slices.
- (*average*) Split the slices having a $dx$ greater than the average value and all the unbounded slices.

If the loop terminates because the number of slices reaches *#slicesMax*, the tree search will be in charge to get the *maxTubeDiam* precision. This slicing phase seems to contradict the principles of most numerical algorithms that decide to subdivide a given time step adaptively. However, the `Integration` procedure called by the `Contraction` method carries out these adaptive time discretization steps, so that both mechanisms, *i.e.* integration and slicing phases, perform time discretization in a complementary manner.

The second phase performed by the second `while` loop is combinatorial. It implements a tree search branching on the domains of the trajectory variables, *i.e.* tubes. Although depth-first search is well-known in the CP community, to our knowledge, no prior work proposed to make choice points on tubes, defined as follows.

**Definition 5 (Tube bisection)**
*Let $[\boldsymbol{x}](\cdot)$ be a tube of a trajectory $\boldsymbol{x}(\cdot)$ defined over $[t_0, t_f]$.*

*Let $t_k$ be an instant in $[t_0, t_f]$, $i$ a dimension in $\{1..n\}$, and $[x_i]$ the interval value of $[x_i](\cdot)$ at $t_k$. Let $mid(x_i)$ be $\frac{\overline{x_i}+\underline{x_i}}{2}$.*
*The tube bisection $(t_k, i)$ of $[\boldsymbol{x}](\cdot)$ produces two tubes $[\boldsymbol{x}^L](\cdot)$ and $[\boldsymbol{x}^R](\cdot)$ equal to $[\boldsymbol{x}](\cdot)$ except at time $t_k$, where $[x_i^L] = [\underline{x_i}, mid(x_i)]$ and $[x_i^R] = [mid(x_i), \overline{x_i}]$.*

In practice, a bisection $(t_k, i)$ is applied only to a gate of the tube. Two heuristics are proposed to the user for selecting the instant $t_k$. The first one picks randomly one instant among the "special times" specified by the user. The second one selects the $t_k$ having the largest box $[\boldsymbol{x}](t_k)$. The dimension $i \in \{1..n\}$, on which the bisection is performed, is decided according to the largest component $[x_i]$.

Note that the DACSP solver is sound because no operator used in Algorithm 1 can eliminate a solution: `Contraction`, `Slicing`, `Bisect`, `TubeMerge`.

## 4 Contractions in the Solver

The `Contraction` function consists of a simple propagation loop that calls the contractors corresponding to constraints in $\boldsymbol{c}$ until the relative gain in contraction volume is less than $\varepsilon_{fp}$. Contractors can be of any type: HC4-Revise for a numerical constraint or the "map" contractor mentioned in introduction for the robotic application. The propagation loop is followed by a call to a contractor `Dyn3B` enforcing a strong consistency on the tube (see Algorithm 3).

Let us detail in Algorithm 2 an important contractor, called `ExplicitDE`, that carries out tube contractions based on ODEs or integro-differential equations. The procedure is mainly responsible for launching integration steps forward and backward in time through the tube. The actual integration method used is a parameter of Algorithm 2. Note that a guaranteed integration algorithm inferring a new information, like a value of the state known at a specific instant, is incremental in that it may contract only a subset of a tube if no more contraction is obtained at a given gate. The tube contraction is not incremental in the first slicing phase or at the top of the search tree ($gateBis = null$) because the `Slicing` procedure can subdivide numerous slices everywhere in the tube (parameter $isIncremental$ set to false). Therefore integration is run from $t_0$ to $t_f$ (forward) and from $t_f$ to $t_0$ (backward). Conversely, during the tree search, integration is triggered by a tube bisection or a domain modification at a special time (whose state is related with a static variable). That is why incremental integrations start from each of these instants.[3] Our solver is endowed with two possible `Integration` procedures. The first one is an "internal" generic integration algorithm that will be incorporated in the Tubex library. Its signature is close to the procedure `Integration` shown in Algorithm 2. It can be triggered from any specified time in the tube, forward or backward, and with the possibility of running the simulation incrementally, *i.e.* stopping it if no sufficient contraction volume gain has been obtained in a gate box. This procedure

[3] The actual code is a little bit more complicated. An instant is skipped if it is handled by the previous integration step.

```
Algorithm ExplicitDE(f, [x](·), specialTimes, Integration, ε,
isIncremental, gateBis)
    if gateBis = null or not isIncremental then
        [x](·) ← Integration(f, [x](·), t_0, FORWARD, ε, false)
        [x](·) ← Integration(f, [x](·), t_f, BACKWARD, ε, false)
    else
        gates ← Sort({gateBis} ∪ specialTimes)
        forall the gate ∈ gates do
            // forward and incremental simulation:
            [x](·) ← Integration(f, [x](·), gate, FORWARD, ε, true)
        forall the gate ∈ gates, in reverse order do
            // backward and incremental simulation:
            [x](·) ← Integration(f, [x](·), gate, BACKWARD, ε, true)
```

**Algorithm 2:** A generic contractor for ODE and integro-diff equation

`Integration` is generic in that it can accept an evolution function $\boldsymbol{f}$ describing either an ODE or an integro-differential equation (see Definition 2). It can also be specialized by a "slice integration contractor" called at each time step of the simulation. Two slice contractors are highlighted in this paper. The first one, called `DynBasic` hereafter, wraps at the slice level two simple contractors available since the very first version of `Tubex`: `CtcDeriv` and an evaluation of the evolution function $\boldsymbol{f}$ called iteratively. `CtcDeriv` (denoted $\mathcal{C}_{\frac{d}{dt}}$ in the literature [34]) is a tube contractor treating the constraint $\dot{\boldsymbol{x}}(\cdot) = \boldsymbol{v}(\cdot)$, where $\boldsymbol{x}(\cdot)$ and $\boldsymbol{v}(\cdot)$ are two trajectories and $\boldsymbol{v}(\cdot)$ is the derivative of $\boldsymbol{x}(\cdot)$ over time. The fundamental theorem of calculus that relates differentiation and integration, is used by `CtcDeriv` for contracting the tube $[\boldsymbol{x}](\cdot)$. The second slice contractor, called `DynCIDGuess` hereafter, generates for each integration step a "slice" contractor graph, where the variables correspond to the two gates and the slice envelope (see Sect. 2.1). Based on the input gate box and the envelope, `DynCIDGuess` can improve the output gate box using sophisticated singleton consistencies based on 3B and CID (see Sect. 2.1). This contractor will be detailed in another article. This generic integration contractor starts by calling a Picard operator that allows one to set non-infinite initial bounds on some tube $[\boldsymbol{x}](\cdot)$, which is required for engaging contraction, and can create new slices adaptively [27].

A second `Integration` procedure wraps directly the state-of-the-art VNODE [28] guaranteed integration solver into a `CtcVnode` contractor. During the slicing phase, it calls VNODE simulations forward and backward from the smallest gate. After each bisection, it calls VNODE simulations forward and backward starting from the bisected gate. To make `CtcVnode` a contractor, the results obtained by the VNODE simulations are intersected with the current tube. VNODE performs its own slicing, especially in the first iterations, and the slices produced are added to those from the slicing phase. Finally, as we will see in the experiments, the contractor `CtcVnode`, and a slice integration contractor as `DynBasic` or `DynCIDGuess`, can be called successively inside the contraction loop performed by the `Contraction` procedure.

Another new and useful dynamic contractor is the `Dyn3B` contractor described in Algorithm 3. This is a dynamic adaption of the 3B algorithm described in Sect. 2.1. It selects iteratively the instant (gate) $t_k$ with the largest interval (the tube is thus not contracted at all instants) and applies a `VarDyn3B` shaving procedure to all the $[x_i]$ intervals at $t_k$. `VarDyn3B` is a straightforward adaptation of the standard `Var3B` shaving procedure (see Sect. 2.1) to tubes. Subintervals at the bounds of $[x_i]$ can be safely eliminated if an integration starting from the corresponding sub-tube leads to an empty domain. This integration procedure can be achieved by `DynBasic`, or `CtcVnode` followed by `DynBasic`.

**Algorithm** `Dyn3B` *(P, ε)*
**do**
  *volumeSave* ← volume($[\boldsymbol{x}](\cdot)$)
  $t_k$ ← `SelectGate`($[\boldsymbol{x}](\cdot)$)
  **forall the** $i \in \{1..n\}$ **do**
    $[\boldsymbol{x}](\cdot)$ ← `VarDyn3B`($P$, $t_k$, $i$)
**while** *VolumeGain(*$[\boldsymbol{x}](\cdot)$, *volumeSave)* $> \varepsilon_{3B}$

**Algorithm 3:** The Dynamic 3B algorithm

## 5 Experiments

The goal of this section is to highlight that the DACSP model, the contractors available via `Tubex` and our DACSP solver can handle a large variety of systems that no competitor or a few ones can deal with. All the results have been obtained on a CPU computer using an `x86-64` processor (1.6 GHz).

### 5.1 BVP for Integro-Differential Equation

Let us illustrate the versatility of our DACSP solver on the following problem. It combines an integro-differential equation defined on the domain $[0, 1]$ and a constraint between the initial and final values, as follows:

$$\begin{cases} \dot{x}(t) = 1 - 2x(t) - 5\int_0^t x(\tau)d\tau; \ t \in [0, 1] \\ x(0)^2 + x(1)^2 = 1 \end{cases} \tag{1}$$

Our solver can find both solutions in 8.35 seconds and needs to resort to 66 bisections (and a search tree depth of 25) to isolate them at a good accuracy. For both solutions, Table 1 reports some details. Note that only our generic `Integration` algorithm can be used in the solver for this particular problem

**Table 1.** Solutions obtained on the integro-differential based system. The table reports the diameters of the initial and final gates, the tube volume and the slices number.

| Solution | Diam. of gate $t_0 = 0$ | Diam. of gate $t_f = 1$ | Tube volume | slices |
|---|---|---|---|---|
| 1 | 0.015 | 0.030 | 0.018 | 400 |
| 2 | 0.034 | 0.022 | 0.024 | 400 |

since there is no ODE, contrarily to the following DACSP systems. It has been run with $maxTubeDiam = 0.02$ and $\#slicesMax = 400$.

For the next two DACSP categories tested, we show the best combination of the `CtcVnode` (refered by `vnode` in the tables), `DynBasic` (`basic`), `DynCIDGuess` (`CIDG`) and `Dyn3B` (`3Bvnode` or `3Bbasic`) contractors.

### 5.2 BVPs and Cruz & Barahona System

We show in Fig. 2 five BVPs and the Cruz system close to a BVP because no state is fully determined at a given instant. However, note that this system has a thick tube solution. The results are presented in Table 2.

$$\begin{cases} \dot{x}(\cdot) = x(\cdot) \\ x(0)^2 + x(1)^2 = 1 \\ [t_0, t_f] = [0, 1] \\ maxTubeDiam = 0.0005 \end{cases} \quad (2) \qquad \begin{cases} \ddot{x}(\cdot) = -x(\cdot) \\ x(0) = 0; x(\pi/2) = 2 \\ [t_0, t_f] = [0, \pi/2] \\ maxTubeDiam = 0.0005 \end{cases} \quad (3)$$

$$\begin{cases} \ddot{x}(\cdot) = 5\dot{x}(\cdot) \\ x(0) = 1;\ x(1) = 0 \\ \dot{x}(0) \in [-10, 10];\ \dot{x}(1) \in [-10, 10] \\ [t_0, t_f] = [0, 1] \\ maxTubeDiam = 0.02 \end{cases} \quad (4) \qquad \begin{cases} \ddot{x}(\cdot) = -10(\dot{x}(\cdot) + x(\cdot)^2) \\ x(0) = 0; x(1) = 0.5 \\ \dot{x}(0) \in [-20, 20]; \dot{x}(1) \in [-20, 20] \\ [t_0, t_f] = [0, 1] \\ maxTubeDiam = 0.05 \end{cases} \quad (5)$$

$$\begin{cases} \ddot{x}(\cdot) = -\exp(x(\cdot)) \\ x(0) = 0;\ x(1) = 0 \\ \dot{x}(0) \in [-20, 20] \\ \dot{x}(1) \in [-20, 20] \\ [t_0, t_f] = [0, 1] \\ maxTubeDiam = 0.05 \end{cases} \quad (6) \qquad \begin{cases} \dot{x}_1(\cdot) = -0.7x_1(\cdot) \\ \dot{x}_2(\cdot) = 0.7x_1(\cdot) - (\ln(2)/5)x_2(\cdot) \\ x_1(0) = 1.25 \\ x_2 \in [1.1, 1.3] \text{ during } [1, 3] \\ [t_0, t_f] = [0, 6] \\ maxTubeDiam = 0.04 \end{cases} \quad (7)$$

**Fig. 2.** Five BVPs and the Cruz system. (2) A one-dimensional problem with an algebraic constraint between the initial and final states; (3) Classical linear example cited in Wikipedia; (4) and (5) denote resp. Systems 2 and 23 in the BVPSolve benchmark [24]; (6) the Bratu system, the only one with two solutions in the BVPSolve benchmark, (7) the Cruz system with a partial information in the middle of the temporal domain.

### 5.3 Interval IVPs

Although solving interval IVPs is not the primary purpose of the DACSP solver, we present results obtained on three interval IVPs (defined in Fig. 3, results provided in Table 3).

$$\begin{cases} \dot{x} = -x^2 \\ x(0) \in [0.1, 0.4] \\ [t_0, t_f] = [0, 5] \\ eps = 0.2 \end{cases} \quad (8) \qquad \begin{cases} \dot{x}_1 = -x_1 - 2x_2 \\ \dot{x}_2 = -3x_1 - 2x_2 \\ x_1(0) \in [5.9, 6.1] \\ x_2(0) \in [3.9, 4.1] \\ [t_0, t_f] = [0, 1] \\ eps = 0.5 \end{cases} \quad (9) \qquad \begin{cases} \dot{x}_1 = -x_2 + 0.1\, x_1\, (1 - x_1^2 - x_2^2) \\ \dot{x}_2 = x_1 + 0.1\, x_2\, (1 - x_1^2 - x_2^2) \\ x_1(0) \in [0.7, 1.3] \\ x_2(0) = 0.0 \\ [t_0, t_f] = [0, 5] \\ eps = 0.15 \end{cases} \quad (10)$$

**Fig. 3.** Three interval IVPs tested. (8) and (9) were introduced in [8]. (10) describes a limit cycle and is particularly sensitive to the wrapping effect caused by interval computation.

**Table 2.** Solutions obtained on BVP systems. For each system, strategy and solution ($s_1$ and/or $s_2$), we report the diameters of the two unknown states in $t_0$ and $t_f$ (most of the systems tested are 2-dimensional, but 2 of the 4 bounds are provided as initial conditions), the volume of the solution tubes, the number of slices, the computational time and the number of choice points required (#bis.).

| Sys. | Best strategy | #sol | $t_0$ diam. | $t_f$ diam. | Tube vol. | #slices | Time | #bis. |
|---|---|---|---|---|---|---|---|---|
| (2) | vnode+basic | $s_1$ | 2e−8 | 5e−8 | 2.e−4 | 5,000 | 7.63 s | 1 |
| | | $s_2$ | 2e−8 | 5e−8 | 2.e−4 | 5,000 | | |
| (3) | vnode+basic | $s_1$ | 7e−15 | 7e−15 | 6e−4 | 12,288 | 12.7 s | 0 |
| (4) | vnode+CIDG+3Bvnode | $s_1$ | 2e−9 | 4e−7 | 5e−3 | 1,216 | 3.05 s | 0 |
| (5) | vnode+CIDG+3Bbasic | $s_1$ | 3.e−2 | 2.e−4 | 0.012 | 5,000 | 81 s | 6 |
| (6) | vnode+basic | $s_1$ | 3.e−6 | 2.e−6 | 7.e−4 | 2,000 | 75 s | 62 |
| | | $s_2$ | 5.e−3 | 5.e−3 | 0.025 | 2,000 | | |
| (7) | CIDG | $s_1$ | 0.0644 | 0.0282 | 0.2637 | 10,000 | 6.85 s | 1 |

For the 2 examples from [8], the exact solution is known, so we also report the relative error (column Gap in Table 3) on interval width of the final gates.

### 5.4 Discussion on Experiments

Note first that the `VNODE` solver alone (outside the DACSP solver) cannot cope with BVPs and is not efficient on the interval IVPs selected. `CtcVnode` (inside the DACSP solver) often provides a very good accuracy on the gates. The good performance is probably due to the high-order interval Taylor form used (order 11 has been set for the experiments). However, `CtcVnode` does generally not obtain good contraction on the whole tube (slice envelopes). Additional work is required to envisage obtaining a better tube volume accuracy using `CtcVnode`. `DynCIDGuess` alone is generally not efficient, except on Systems (7) and (8),

**Table 3.** Solutions obtained on interval IVPs systems. For each system, we report the best strategy, the diameters of the state at $t_f$, the volume of the solution tube, the number of slices, the computational time and the number of bisections.

| Sys. | Strategy | $t_f$ diam. | Tube vol. | Gap | #slices | Time | #bis. |
|---|---|---|---|---|---|---|---|
| (8) | CIDG | 0.06668 | 0.6934 | 0.02% | 40,000 | 9.35 s | 1 |
| (9) | vnode+CIDG | (0.544; 0.544) | 0.700 | (0.01%; 0.01%) | 2,000 | 3.93 s | 1 |
| (10) | vnode+basic | (0.0695; 0.2273) | 2.54 | | 1,000 | 13.3 s | 7 |

because it requires too many slices to reach the precision (recall that the CPU time generally grows linearly in the number of slices). Finally, the best option for the `Integration` procedure is generally to call first `CtcVnode` and then `DynBasic` or `DynCIDGuess`. A final call to `Dyn3B` is useful for Systems (4) (using `CtcVnode` as subcontractor) and (5) (using `DynBasic`).

Overall, different solver strategies provide the best results on the different systems tested. All the devices offered by the solver can be useful on different instances: contractors, slicing, choice points. When the best strategy includes a number of bisections, this means that the $\#slicesMax$ has been reached and the solver resorts to choice points to better approximate the solution tube. An issue for future work is to better study the interplay between slicing and bisection in order to obtain a more generic DACSP solver that can work without the $\#slicesMax$ parameter.

## 6 Conclusion and Future Work

We have presented a new generic solver that can handle together differential and static numerical constraints. The originality of the approach lies both in the underlying model considering trajectories as variables and in a novel backtracking mechanism applicable to DACSP. Our DACSP solver is endowed with an exploration operator that enables to bisect a tube at a chosen time. This allows the DACSP solver to better handle hard DACSP systems and accurately estimate distinct trajectories of problems having several solutions. We have shown on first experimental results that our solver is versatile enough to solve DACSP instances for which no or a few algorithmic solutions currently exist. We have also demonstrated the benefits of wrapping the state-of-the-art VNODE in a `CtcVnode` contractor implemented in `Tubex`.

With regard to future work, we will first try to limit the number of user-defined parameters, in particular remove $\#slicesMax$. Also, we want to propose ideally only one combination of contractors in the DACSP solver for every DACSP subclass. Second, we will study a more general search tree branching static and dynamical variables domains indifferently, though we need to explore new ideas on large-scale problems. Finally, in the current solver, the propagation between functional and real variables domains is somewhat naive and is partly ensured by the "special times" specified by the user (see Algorithm 2). The quite recent Tubex 3.0 accepts bi-level slice/tube variables, which will enable a fully incremental contraction achieved by a propagation engine.

Supplementary materials including the sources of the solver and the experiments are available on http://simon-rohou.fr/research/dacsp-solve/.

**Acknowledgements.** This work was supported by the French Agence Nationale de la Recherche (ANR) [grant number ANR-16-CE33-0024].

## References

1. Benhamou, F., Goualard, F., Granvilliers, L., Puget, J.-F.: Revising hull and box consistency. In: Proceedings of International Conference on Logic Programming (ICLP), pp. 230–244 (1999)
2. Chabert, G.: IBEX - an Interval-Based EXplorer (2020). http://www.ibex-lib.org/
3. Chabert, G., Jaulin, L.: Contractor programming. Artif. Intell. **173**, 1079–1100 (2009)
4. Chapoutot, A., Alexandre dit Sandretto, J., Mullier, O.: Dynibex (2015). http://perso.ensta-paristech.fr/~chapoutot/dynibex/
5. Collavizza, H., Delobel, F., Rueher, M.: Comparing partial consistencies. Reliable Comput. **5**(3), 213–228 (1999)
6. Cruz, J., Barahona, P.: Constraint satisfaction differential problems. In: Rossi, F. (ed.) CP 2003. LNCS, vol. 2833, pp. 259–273. Springer, Heidelberg (2003). https://doi.org/10.1007/978-3-540-45193-8_18
7. Cruz, J., Barahona, P.: Constraint reasoning with differential equations. Appl. Numer. Anal. Comput. Math. **1**(1), 140–154 (2004)
8. Deville, Y., Janssen, M., Van Hentenryck, P.: Consistency techniques in ordinary differential equations. In: Maher, M., Puget, J.-F. (eds.) CP 1998. LNCS, vol. 1520, pp. 162–176. Springer, Heidelberg (1998). https://doi.org/10.1007/3-540-49481-2_13
9. Domes, F.: GLOPTLAB: a configurable framework for the rigorous global solution of quadratic constraint satisfaction problems. Optim. Methods Softw. **24**, 727–747 (2009)
10. Filippova, T.F., Kurzhanski, A.B., Sugimoto, K., Vályi, I.: Ellipsoidal state estimation for uncertain dynamical systems. In: Milanese, M., Norton, J., Piet-Lahanier, H., Walter, É. (eds.) Bounding Approaches to System Identification, pp. 213–238. Springer, Boston (1996). https://doi.org/10.1007/978-1-4757-9545-5_14
11. Goldsztejn, A., Mullier, O., Eveillard, D., Hosobe, H.: Including ordinary differential equations based constraints in the standard CP framework. In: Cohen, D. (ed.) CP 2010. LNCS, vol. 6308, pp. 221–235. Springer, Heidelberg (2010). https://doi.org/10.1007/978-3-642-15396-9_20
12. Goodman, T.R., Lance, G.N.: The numerical solution of two-point boundary value problems. Math. Tables Other Aids Comput. **10**, 82–86 (1956)
13. Granvilliers, L., Benhamou, F.: RealPaver: an interval solver using constraint satisfaction techniques. ACM Trans. Math. Softw. - TOMS **32**, 138–156 (2006)
14. Hansen, E.R.: Global Optimization using Interval Analysis. Marcel Dekker, New York (1992)
15. Jaulin, L.: Nonlinear bounded-error state estimation of continuous-time systems. Automatica **38**, 1079–1082 (2002)

16. Jaulin, L.: Range-only SLAM with indistinguishable landmarks: a constraint programming approach. Constraints **21**(4), 557–576 (2016)
17. Kalman, R.E.: Contributions to the theory of optimal control. Bol. Soc. Mat. Mex. **5**, 102–119 (1960)
18. Kapela, T., Mrozek, M., Pilarczyk, P., Wilczak, D., Zgliczynski, P.: CAPD - a rigorous toolbox for computer assisted proofs in dynamics (2010). http://capd.ii.uj.edu.pl/
19. Kearfott, R.B.: GlobSol: history, composition, and advice on use. In: Bliek, C., Jermann, C., Neumaier, A. (eds.) COCOS 2002. LNCS, vol. 2861, pp. 17–31. Springer, Heidelberg (2003). https://doi.org/10.1007/978-3-540-39901-8_2
20. Kurzhanski, A.B., Filippova, T.F.: On the theory of trajectory tubes - a mathematical formalism for uncertain dynamics, viability and control. In: Kurzhanski, A.B. (eds.) Advances in Nonlinear Dynamics and Control: A Report from Russia, pp. 122–188, Birkhäuser, Boston (1993). https://doi.org/10.1007/978-1-4612-0349-0_4
21. Le Bars, F., Sliwka, J., Jaulin, L., Reynet, O.: Set-membership state estimation with fleeting data. Automatica **48**(2), 381–387 (2012)
22. Lhomme, O.: Consistency techniques for numeric CSPs. In: IJCAI, pp. 232–238 (1993)
23. Lohner, R.: Enclosing the solutions of ordinary initial and boundary value problems. In: Kaucher, E., Kulisch, U., Ullrich, Ch. (eds.) Computer Arithmetic: Scientific Computation and Programming Languages, pp. 255–286. BG Teubner, Stuttgart (1987)
24. Mazzia, F., Cash, J.R., Soetaert, K.: Solving boundary value problems in the open source software R: package bvpSolve. Opuscula Math. **34**(2), 387–403 (2014)
25. Messine, F.: Méthodes d'optimisation globale basées sur l'analyse d'intervalle pour la résolution des problèmes avec contraintes. Ph.D. thesis, LIMA-IRIT-ENSEEIHT-INPT, Toulouse (1997)
26. Moore, R.E.: Interval Analysis, vol. 4. Prentice-Hall, Englewood Cliffs (1966)
27. Nedialkov, N., Jackson, K., Corliss, G.: Validated solutions of initial value problem for ordinary differential equations. Appl. Math. Appl. **105**(1), 21–68 (1999)
28. Nedialkov, N.S.: VNODE-LP, a validated solver for initial value problems in ordinary differential equations. Technical report CAS-06-06-NN, Department of Computing and Software, McMaster University, Hamilton, Ontario, Canada (2006)
29. Prud'homme, C., Fages, J.-G., Lorca, X.: Choco documentation (2014). http://www.choco-solver.org
30. Revol, N., Makino, K., Berz, M.: Taylor models and floating-point arithmetic: proof that arithmetic operations are validated in COSY. J. Logic Algebraic Program. **64**, 135–154 (2005)
31. Rohou, S., Desrochers, B., Jaulin, L.: Set-membership state estimation by solving data association. In IEEE International Conference on Robotics and Automation (2020)
32. Rohou, S., et al.: The Tubex library - Constraint-programming for robotics (2020). http://simon-rohou.fr/research/tubex-lib/
33. Rohou, S., Jaulin, L., Mihaylova, L., Le Bars, F., Veres, S.M.: Reliable nonlinear state estimation involving time uncertainties. Automatica **93**, 379–388 (2018)
34. Rohou, S., Jaulin, L., Mihaylova, L., Le Bars, F., Veres, S.M.: Guaranteed computation of robot trajectories. Robot. Auton. Syst. **93**, 76–84 (2017)
35. Thrun, S., Bugard, W., Fox, D.: Probabilistic Robotics. MIT Press, Cambridge (2005)

36. Trombettoni, G., Chabert, G.: Constructive interval disjunction. In: Bessière, C. (ed.) CP 2007. LNCS, vol. 4741, pp. 635–650. Springer, Heidelberg (2007). https://doi.org/10.1007/978-3-540-74970-7_45
37. Tucker, W.: A rigorous ODE solver and Smale's 14th problem. Found. Comput. Math. **2**(1), 53–117 (2002)
38. Voges, R., Wagner, B.: Timestamp offset calibration for an IMU-camera system under interval uncertainty. In: 2018 IEEE/RSJ International Conference on Intelligent Robots and Systems (IROS), pp. 377–384 (2018)
39. Wilczak, D., Zgliczynski, P.: Heteroclinic connections between periodic orbits in planar restricted circular three-body problem-a computer assisted proof. Commun. Math. Phys. **234**(1), 37–75 (2003)

# *abstractXOR*: A global constraint dedicated to differential cryptanalysis

Loïc Rouquette[1,2] and Christine Solnon[1(✉)]

[1] CITI, INRIA, INSA Lyon, 69621 Villeurbanne, France
{loic.rouquette,christine.solnon}@insa-lyon.fr
[2] LIRIS, UMR5201 CNRS, 69621 Villeurbanne, France

**Abstract.** Constraint Programming models have been recently proposed to solve cryptanalysis problems for symmetric block ciphers such as AES. These models are more efficient than dedicated approaches but their design is difficult: straightforward models do not scale well and it is necessary to add advanced constraints derived from cryptographic properties. We introduce a global constraint which simplifies the modelling step and improves efficiency. We study its complexity, introduce propagators and experimentally evaluate them on two cryptanalysis problems (single-key and related-key) for two block ciphers (AES and Midori).

## 1 Motivations

Symmetric bloc ciphers use a secret key $K$ to cipher an input text $X_0$ into a cipher text $X_r$ in such a way that $X_r$ can be deciphered back into $X_0$ with the same key $K$. Differential cryptanalysis aims at evaluating if we can guess $K$ by studying difference propagation during ciphering [6]. These differences are obtained by applying a XOR (bitwise exclusive or, denoted $\oplus$) between two input texts. In the related-key attack [5], differences are also introduced in keys. For mounting these attacks, we must compute *Maximum Differential Characteristics (MDCs)*, *i.e.*, most probable differences.

A widely used symmetric block cipher is AES [10]. However, AES is rather time consuming, and lighter ciphers must be designed for devices with limited computational resources. Each time a new cipher is designed, we must compute MDCs to evaluate its robustness with respect to differential attacks. In this section, we illustrate MDCs on Midori128 [2], which is a lighweight cipher simpler to explain than AES. However, all models can be extended to AES and to other existing symmetric block ciphers, and we experimentally evaluate our approach on both Midori and AES.

*Midori128.* The ciphering iterates $r$ *rounds* and each round is composed of four operations: *SubBytes* replaces every byte with another byte according to a given lookup table; *ShuffleCells* moves bytes; and *MixColumns* and *AddKey* perform XORs. For each round $i \in [0, r-1]$, $X_i$ denotes the text state at the beginning


H. Simonis (Ed.): CP 2020, LNCS 12333, pp. 566–584, 2020.
https://doi.org/10.1007/978-3-030-58475-7_33

**Maximise** $\sum_{i\in[0,r-1],b\in[0,15]} P_i[b]$ **so that** $\forall i \in [0, r-1], \forall b \in [0, 15]$:
$(C_1)$ $(\delta X_i[b], \delta S_i[b], P_i[b]) \in$ ***subBytesTable***$_b$
$(C_2)$ $\delta Y_i[f(b)] = \delta S_i[b]$ **where** $f$ **is a given permutation from** $[0, 15]$ **to** $[0, 15]$
$(C_3)$ $\delta Z_i[b] \oplus \delta Y_i[(b+4)\%16] \oplus \delta Y_i[(b+8)\%16] \oplus \delta Y_i[(b+12)\%16] = 0$
$(C_4)$ $\delta Z_i[b] \oplus \delta K[b] \oplus \delta X_{i+1}[b] = 0$

**Fig. 1.** MDC problem for Midori128.

of round $i$, and $S_i$, $Y_i$, and $Z_i$ denote intermediate text states after each operation: $S_i$ is the result of applying *SubBytes* on $X_i$; $Y_i$ is the result of applying *ShuffleCells* on $S_i$; $Z_i$ is the result of applying *MixColumns* on $Y_i$; and $X_{i+1}$ is the result of applying *AddKey* on $Z_i$ and $K$.

The goal of the MDC problem is to compute differences. We denote $\delta K$ the differences in the key (*i.e.*, $\delta K$ is the result of applying a XOR between two keys), and $\delta X_i$ (resp. $\delta S_i$, $\delta Y_i$, and $\delta Z_i$) the differences in the text at the beginning of round $i$ (resp. after applying *SubBytes*, *ShuffleCells*, and *MixColumns*). For each $A \in \{K, X_i, S_i, Y_i, Z_i, X_r : i \in [0, r-1]\}$, $\delta A$ is a sequence of 16 bytes (where each byte is a sequence of 8 bits) and, given a byte position $b \in [0, 15]$, $\delta A[b]$ denotes the byte at position $b$ in $\delta A$. $\delta A[b]$ is called a *differential variable*, and $\delta$ denotes the set of all differential variables. The domain of each differential variable $\delta A[b] \in \delta$ is $D(\delta A[b]) = [0, 255]$.

The goal is to find the most probable assignment of differential variables. For all operations but *SubBytes*, differences are deterministically computed, *i.e.*, we can compute $\delta X_{i+1}$ given $\delta S_i$ and $\delta K$. In this case, the probability of observing $\delta X_{i+1}$ given $\delta S_i$ and $\delta K$ is equal to 1. However, this is not the case for *SubBytes*: when $\delta X_i[b] \in [1, 255]$, there are several possible values for $\delta S_i[b]$. The only case where we can deterministically compute $\delta S_i[b]$ given $\delta X_i[b]$ is when $\delta X_i[b] = 0$: in this case, $\delta S_i[b] = 0$. The table $subBytesTable_b$ contains all triples $(\delta_{in}, \delta_{out}, p)$ such that $p$ is the $\log_2$ probability that $\delta S_i[b] = \delta_{out}$ when $\delta X_i[b] = \delta_{in}$. This table depends on the position $b$ of the byte in $\delta S_i$. We introduce a variable $P_i[b]$ which corresponds to this $\log_2$ probability and whose domain is $D(P_i[b]) = \{-6, -5, -4, -3, -2, 0\}$.

Figure 1 describes the MDC problem for Midori128. The goal is to maximise the sum of all $\log_2$ probabilities $P_i[b]$. Constraints $(C_1)$ to $(C_4)$ correspond to the 4 operations applied at each round: $(C_1)$ is the table constraint corresponding to *SubBytes*; $(C_2)$ corresponds to *ShuffleCells*, which moves bytes from position $b$ in $\delta S_i$ to position $f(b)$ in $\delta Y_i$; $(C_3)$ and $(C_4)$ correspond to *MixColumns* and *AddKey*, respectively, and only involve XOR operations.

*Two Step Solving Process.* Most differential variables are equal to 0 in MDCs. Indeed, when $\delta S_i[b] = \delta X_i[b] = 0$, the $\log_2$ probability $P_i[b]$ is equal to 0 whereas in all other cases it is smaller than or equal to $-2$ for Midori, and $-6$ for AES. Hence, the MDC problem is usually solved in two steps [16]: in Step1, we search

$(A_0)$ $n = \sum_{i\in[0,r-1],b\in[0,15]} \Delta X_i[b]$
$\forall i \in [0, r-1], \forall b \in [0, 15]$:
$(A_1)$ $\Delta X_i[b] = \Delta S_i[b]$
$(A_2)$ $\Delta Y_i[b] = \Delta S_i[f(b)]$ where $f$ is a given permutation from $[0, 15]$ to $[0, 15]$
$(A_3)$ $\Delta Z_i[b] \circ \Delta Y_i[(b+4)\%16] \circ \Delta Y_i[(b+8)\%16] \circ \Delta Y_i[(b+12)\%16] = 0$
$(A_4)$ $\Delta Z_i[b] \circ \Delta K[b] \circ \Delta X_{i+1}[b] = 0$

**Fig. 2.** Step1 problem for Midori128

for difference positions, whereas in Step2 we search for the exact values of the differential variables.

More precisely, at Step1, the set of variables is $\Delta = \{\Delta_j : \delta_j \in \delta\}$. Each variable $\Delta_j \in \Delta$ has a binary domain $D(\Delta_j) = \{0, 1\}$ and indicates if there is a difference or not in $\delta_j$, *i.e.*, $\Delta_j = 0 \Leftrightarrow \delta_j = 0$ and $\Delta_j = 1 \Leftrightarrow \delta_j \in [1, 255]$. The Step1 problem is defined in Fig. 2 for Midori128. It is very similar to the problem of Fig. 1. The main difference is that $\log_2$ probability variables ($P_i[b]$) are removed, and the objective function and constraint $(C_1)$ are replaced with constraint $(A_0)$ which ensures that the number of $\Delta X_i[b]$ variables assigned to 1 is equal to a given value $n$. Constraint $(A_1)$ comes from the fact that $\delta X_i[b] = 0$ iff $\delta S_i[b] = 0$. Finally, XOR constraints $(C_3)$ and $(C_4)$ are replaced with abstract XOR constraints $(A_3)$ and $(A_4)$: an abstract XOR constraint $\Delta_1 \circ \ldots \circ \Delta_l = 0$ is satisfied iff, for each $\Delta_j$ assigned to 1 there exists an integer value in $[1, 255]$ such that the XOR of all these values is equal to 0. This constraint may be encoded by $\sum_{j=1}^{l} \Delta_j \neq 1$. Indeed, when all $\Delta_j$ are assigned to 0, the abstract XOR is trivially satisfied; when exactly one $\Delta_j$ is assigned to 1, it is trivially violated; otherwise, it is satisfied because it is always possible to find $k \geq 2$ values in $[1, 255]$ such that the result of XORing them is equal to 0. We refer to this encoding of an abstract XOR constraint as the $\text{sum}_{\neq 1}$ encoding.

Given a Step1 solution $s$, we define the Step2 model obtained from the model of Fig. 1 by adding the following constraint for each variable $\Delta_j \in \Delta$: if $\Delta_j$ is assigned to 0 in $s$ then $\delta_j = 0$, else $\delta_j \neq 0$. This model is much easier to solve than the original one as many $\Delta_j$ variables are assigned to 0 in $s$. However, some Step1 solutions may lead to inconsistent Step2 problems. These Step1 solutions are said to be *Step2-inconsistent*. These inconsistencies mainly come from the fact that XORs are poorly abstracted at Step1: every abstract XOR constraint ensures that there exist integer values whose XOR is equal to 0, but this is ensured for each constraint separately so that several abstract XORs can be satisfied at Step1 while they are Step2-inconsistent when considering them all together. For example, the two abstract XOR constraints $\Delta_1 \circ \Delta_2 = 0$ and $\Delta_1 \circ \Delta_2 \circ \Delta_3 = 0$ are satisfied when $\Delta_1$, $\Delta_2$, and $\Delta_3$ are assigned to 1, but this assignment is Step2-inconsistent because $(\delta_1 \oplus \delta_2 = 0) \Rightarrow (\delta_1 = \delta_2) \Rightarrow (\delta_3 = 0)$.

Finally, to compute MDCs, we iteratively search for all Step1 solutions with increasing values of $n$, and for each Step1 solution we solve the associated Step2 problem, until some conditions are reached (see [12] for details).

*Existing Approaches to Compute MDCs.* Two main dedicated approaches have been proposed to solve the Step1 problem for AES: a graph traversal approach [11], and a Branch & Bound approach [7]. Both approaches do not scale well and they are not able to solve all AES instances within a reasonable amount of time.

An appealing alternative to dedicated approaches is to use generic solvers such as Integer Linear Programming (ILP), Boolean satisfiability (SAT) or Constraint Programming (CP). ILP has been used to compute MDCs for block ciphers such as SIMON, PRESENT or LBlock [26]. However, it is difficult to model the *SubBytes* operation (modelled by constraint ($C_1$) in Fig. 1) by means of linear inequalities, and ILP does not scale well to solve Step2.

SAT has also been used to compute MDCs for ciphers such as ARX [21] or Simon [17]. CryptoMiniSat [23] introduces XOR-clauses and uses Gaussian elimination to efficiently propagate them. These XOR-clauses can be used to model XOR constraints ($C_3$) and ($C_4$) in Step2. However, they cannot be used to model abstract XOR constraints ($A_3$) and ($A_4$) in Step1. Indeed, if $1 \oplus 1 = 0$ at a bitwise level (during Step2), this is no longer true during Step1 because the XOR of two bytes different from 0 may be equal to 0. Similarly to ILP, non linear operations such as *SubBytes* are not straightforward to model by means of clauses. In [18], Lafitte shows how to encode a relation associated with a non linear operation into a set of clauses and, in [24], Sun et al. show how to reduce the number of clauses by using the same approach as in [1]. However, the resulting SAT model does not scale well and cannot solve Step2 for AES, for example.

CP has been used to compute MDCs for AES [14,15], Midori [13], and SKINNY [25]. These CP models are very efficient. However, if efficient Step2 models are easily derived from problem definitions (such as Fig. 1 for Midori128), efficient Step1 models are much harder to design. Indeed, a basic model is derived from Fig. 2 by replacing every abstract XOR with its $\text{sum}_{\neq 1}$ encoding. However, this model has a lot of Step2-inconsistent solutions. To reduce the number of Step2-inconsistent solutions, it is necessary to add constraints derived from advanced cryptographic properties.

*Contributions and Overview of the Paper.* Our goal is to ease the design of CP models for computing MDCs, while ensuring an efficient solving process. To this aim, we introduce a global constraint which propagates XORs in a global way in order to reduce the number of Step2-inconsistent solutions.

In Sect. 2, we introduce notations and preliminary definitions. In Sect. 3, we introduce the *abstractXOR* constraint which ensures that a set of abstract XOR constraints is Step2-consistent. We show that deciding of *abstractXOR* feasibility is $\mathcal{NP}$-complete when differential variables are constrained to belong to $[0, 255]$ whereas it is polynomial when the domain of differential variables is not upper bounded. Hence, we relax *abstractXOR* by removing this upper bound. In Sect. 4, we introduce two propagators for *abstractXOR*: the first one simply ensures feasibility, and the second one ensures Generalised Arc Consistency (GAC).

In Sect. 5, we experimentally evaluate them on two MDC problems (related-key and single-key) for Midori and AES.

## 2 Notations and Preliminary Definitions

Given two integer values $a$ and $b$, $[a, b]$ denotes the set of all integer values ranging from $a$ to $b$. $\mathbb{N}^+$ denotes the set of all natural numbers (excluding 0).

$\Delta$ denotes a set of variables such that the domain of each variable $\Delta_j \in \Delta$ is $D(\Delta_j) \subseteq \{0, 1\}$. $\Delta_j$ is assigned iff $\#D(\Delta_j) = 1$, and an assignment is complete if all variables of $\Delta$ are assigned. $\Delta^0$ denotes the set of variables assigned to 0 and $\Delta \setminus \Delta^0$ denotes the set of variables that are either assigned to 1 or not yet assigned.

$C$ denotes a set of abstract XOR constraints defined on $\Delta$. $C_{\downarrow\Delta^0}$ denotes the set of XOR constraints obtained from $C$ by (i) replacing each $\Delta_j \in \Delta^0$ with 0, (ii) replacing each $\Delta_j \in \Delta \setminus \Delta^0$ with an integer variable $\delta_j$ whose domain is $D(\delta_j) = \mathbb{N}^+$, and (iii) replacing each abstract XOR $\circ$ with the bitwise XOR $\oplus$. Examples are displayed in Fig. 3 (equations of $C_{\downarrow\Delta^0}$ are simplified by replacing $\delta_j \oplus 0$ with $\delta_j$).

$C_{\downarrow\Delta^0}$ is represented by a matrix $M$ which contains one row for each equation and one column for each variable in $\Delta \setminus \Delta^0$: $M[i, j] = 1$ if $\delta_j$ occurs in the $i^{\text{th}}$ equation of $C_{\downarrow\Delta^0}$; otherwise, $M[i, j] = 0$. We denote $n$ and $m$ the numbers of rows and columns of $M$. For each row $i \in [1, n]$, we define $nonZero_i = \{j \in [1, m] : M[i, j] = 1\}$, $pivot_i = \min nonZero_i$, and $nonPivot_i = nonZero_i \setminus \{pivot_i\}$. $M$ is in *reduced row-echelon (RRE)* form iff, for every row $i \in [1, n]$, there is exactly one non-zero cell in column $pivot_i$, *i.e.*, $\sum_{i'=1}^{n} M[i', pivot_i] = 1$ (see examples in Fig. 3).

$$\Delta = \{\Delta_1, \Delta_2, \Delta_3, \Delta_4, \Delta_5, \Delta_6, \Delta_7\}$$
$$C = \{\Delta_1 \circ \Delta_4 \circ \Delta_6 \circ \Delta_7 = 0, \quad \Delta_2 \circ \Delta_4 \circ \Delta_5 \circ \Delta_7 = 0, \quad \Delta_3 \circ \Delta_5 \circ \Delta_6 = 0\}$$

**Example 1:** $\Delta^0 = \{\Delta_7\}$

$C_{\downarrow\Delta^0}$:
$$\delta_1 \oplus \delta_4 \oplus \delta_6 = 0$$
$$\delta_2 \oplus \delta_4 \oplus \delta_5 = 0$$
$$\delta_3 \oplus \delta_5 \oplus \delta_6 = 0$$

$M$:

| $\delta_1$ | $\delta_2$ | $\delta_3$ | $\delta_4$ | $\delta_5$ | $\delta_6$ |
|---|---|---|---|---|---|
| 1 | 0 | 0 | 1 | 0 | 1 |
| 0 | 1 | 0 | 1 | 1 | 0 |
| 0 | 0 | 1 | 0 | 1 | 1 |

**Example 2:** $\Delta^0 = \{\Delta_2, \Delta_3, \Delta_7\}$

$C_{\downarrow\Delta^0}$:
$$\delta_1 \oplus \delta_4 \oplus \delta_6 = 0$$
$$\delta_4 \oplus \delta_5 = 0$$
$$\delta_5 \oplus \delta_6 = 0$$

$M$:

| $\delta_1$ | $\delta_4$ | $\delta_5$ | $\delta_6$ |
|---|---|---|---|
| 1 | 1 | 0 | 1 |
| 0 | 1 | 1 | 0 |
| 0 | 0 | 1 | 1 |

**Fig. 3.** Top: A set $\Delta$ of variables and a set $C$ of abstract XOR constraints. Bottom: $C_{\downarrow\Delta^0}$ and $M$ when $\Delta_7$ is assigned to 0 (Ex. 1, on the left) and when $\Delta_2$, $\Delta_3$, and $\Delta_7$ are assigned to 0 (Ex. 2, on the right). In Ex. 1, $M$ is in RRE form and $nonZero_1 = \{1, 4, 6\}$, $pivot_1 = 1$, and $nonPivot_1 = \{4, 6\}$. In Ex. 2, $M$ is not in RRE form because the pivot columns of rows 2 and 3 have two non-zero cells.

## 3 Definition and Complexity of *abstractXOR*

When computing MDCs in a two-step process, we aim at minimising as much as possible the number of Step1 solutions which are Step2-inconsistent. As many Step2-inconsistencies come from the fact that XOR constraints are poorly abstracted at Step1, we introduce a global constraint to obtain a tighter Step1 model.

**Definition 1.** *Given an integer value $u > 0$, the constraint abstractXOR$_{u,C}(\Delta)$ is satisfied by a complete assignment iff $C_{\downarrow\Delta^0} \cup \{\delta_j \leq u : \Delta_j \in \Delta \setminus \Delta^0\}$ is consistent.*

Let us consider Ex. 1 of Fig. 3. If $u = 3$, then *abstractXOR*$_{u,C}(\Delta)$ is satisfied because there exists a solution of $C_{\downarrow\Delta^0}$ such that every $\delta_j$ belongs to $[1, 3]$ (*e.g.*, $\delta_1 = \delta_5 = 1$, $\delta_2 = \delta_6 = 2$, and $\delta_3 = \delta_4 = 3$). However, if $u = 2$, then *abstractXOR*$_{u,C}(\Delta)$ is not satisfied because $C_{\downarrow\Delta^0}$ has no solution when every $\delta_j$ must belong to $[1, 2]$.

In Ex. 2, *abstractXOR*$_{u,C}(\Delta)$ is not satisfied because $(\delta_4 \oplus \delta_5 = 0 \wedge \delta_5 \oplus \delta_6 = 0) \Rightarrow (\delta_4 = \delta_5 = \delta_6) \Rightarrow (\delta_4 \oplus \delta_6 = 0)$. Therefore, $\delta_1$ must be equal to 0, which is impossible as $\delta_1$ must belong to $[1, u]$.

*abstractXOR* allows us to easily model Step1 problems. For example, for Midori128, we replace constraints $(A_3)$ and $(A_4)$ with *abstractXOR*$_{255,C}(\Delta)$ where $C = \{(A_3), (A_4)\}$, and $\Delta$ contains all variables involved in $(A_3)$ or $(A_4)$. The resulting model has less Step2-inconsistent solutions than the basic model obtained by replacing $(A_3)$ and $(A_4)$ with sum$_{\neq 1}$ constraints: *abstractXOR* ensures the consistency of $(C_3) \wedge (C_4)$ at Step2, whereas the basic model only ensures the feasibility of each XOR separately.

However, checking the feasibility of *abstractXOR* is intractable.

**Theorem 1.** *Deciding if a complete assignment satisfies abstractXOR$_{u,C}(\Delta)$ is an $\mathcal{NP}$-complete problem.*

*Proof.* To decide whether *abstractXOR*$_{u,C}(\Delta)$ is satisfied by a complete assignment, we must decide whether $C_{\downarrow\Delta^0}$ is consistent when all $\delta_j$ variables occurring in $C_{\downarrow\Delta^0}$ are constrained to belong to $[1, u]$. This problem trivially belongs to $\mathcal{NP}$ as we can decide in polynomial time whether a given assignment of all $\delta_j$ variables satisfies $C_{\downarrow\Delta^0}$. To show that it is $\mathcal{NP}$-complete, we give the intuition of a reduction from the graph colouring problem, which aims at deciding if we can assign a colour $c_i \in [1, u]$ to each vertex $i$ of a graph so that $c_i \neq c_j$ for each edge $(i, j)$. Given a graph $G$, we associate a variable $\delta_i$ (resp. $\delta_{ij}$) with every vertex $i$ (resp. edge $(i, j)$) of $G$, and we define the XOR constraints: $C = \{\delta_i \oplus \delta_j \oplus \delta_{ij} = 0 : (i, j) \text{ is an edge of } G\}$. If each variable must belong to $[1, u]$, then each XOR constraint associated with an edge $(i, j)$ ensures that $\delta_i \neq \delta_j$ (because $\delta_i = \delta_j \Leftrightarrow \delta_{ij} = 0$). Hence, we can show that every solution of $C$ corresponds to a valid colouring of $G$, and vice-versa.

Now, let us show that we can decide if *abstractXOR* is satisfied in polynomial time when $\delta_j$ variables are not upper bounded. In this case, we have to decide

**Algorithm 1:** RRE form of an $n \times m$ matrix $M$

1 $i \leftarrow 1$
2 **while** $i \leq n$ **do**
  /* *every row* $i' \in [1, i-1]$ *is in RRE form, i.e.,* $\sum_{i''=1}^{n} M[i'', pivot_{i'}] = 1$ */
3   **if** $nonZero_i = \emptyset$ **then** remove row $i$ and decrement $n$;
4   **else**
5     **for** *each row* $i' \in [1, n]$ *such that* $i' \neq i$ *and* $M[i', pivot_i] = 1$ **do**
6       **for** *each column* $j' \in [1, m]$ **do** $M[i', j'] \leftarrow M[i', j'] \oplus M[i, j']$ ;
7     $i \leftarrow i + 1$

if $C_{\downarrow\Delta^o}$ is consistent. We first show how to put the matrix $M$ associated with $C_{\downarrow\Delta^o}$ in RRE form. This is done by Algorithm 1, which uses a principle similar to Gaussian elimination of linear equations. Algorithm 1 does not change the set of solutions because it only removes empty rows (line 3), or replaces a row $i'$ with the result of XORing it with another row $i$ (line 6). To show that Algorithm 1 puts $M$ in RRE form, we show that the comment after line 2 is an invariant property of the loop lines 2–7. This property is trivially satisfied at the first iteration when $i = 1$ and, if it is satisfied at some iteration, then it is satisfied at the next iteration: if row $i$ is empty (line 3) then it is removed and $i$ is not incremented so that the property is still satisfied; otherwise (lines 4–7), every row $i' \neq i$ which contains a non-zero cell on column $pivot_i$ is XORed with row $i$ so that column $pivot_i$ only contains one non-zero cell on row $i$ just after lines 5–6. The complexity of this algorithm is $\mathcal{O}(mn^2)$.

We use Property *atLeast2* (defined below) to decide if $C_{\downarrow\Delta^o}$ is consistent.

**Definition 2.** *A matrix $M$ in RRE form satisfies Prop.* atLeast2 *if each row has at least two non-zero cells,* i.e., $\forall i \in [1, n], \#nonZero_i \geq 2$.

**Theorem 2.** *$C_{\downarrow\Delta^o}$ is consistent iff its associated matrix $M$ in RRE form satisfies Prop.* atLeast2.

*Proof.* If $M$ does not satisfy Prop. *atLeast2*, then it contains a row with exactly one non-zero cell, *i.e.*, there exists an equation of the form $\delta_j = 0$. In this case $C_{\downarrow\Delta^o}$ is inconsistent as $D(\delta_j) = \mathbb{N}^+$.

If $M$ satisfies Prop. *atLeast2*, then we can always build a solution for $C_{\downarrow\Delta^o}$. The idea is to first assign values to variables associated with non-pivot columns, and then compute values of variables associated with pivot columns by XORing the corresponding non-pivot variables. To ensure that values computed for pivot variables are always different from 0, we have to choose carefully the values of non-pivot variables. More precisely, non-pivot variables are assigned one after the other. When choosing a value for a non-pivot variable $\delta_j$, for each row $i$ such that $j \in nonPivot_i$, if all variables of $nonPivot_i$ but $\delta_j$ are already assigned, then we must choose a value different from the result of the XOR of these assigned variables. As the domains of $\delta_j$ variables are not upper bounded, we can always build a solution.

**Algorithm 2:** Check Prop. *atLeast2* of an $n \times m$ matrix $M$ in RRE form

1 **for** *each row* $i \in [1, n]$ *such that* $nonPivot_i = \emptyset$ **do**
2     **if** $D(\Delta_{pivot_i}) = \{1\}$ **then return** failure;
3     **else**
4         remove 1 from $D(\Delta_{pivot_i})$
5         remove row $i$ and decrement $n$
6         remove column $pivot_i$ and decrement $m$
7 **return** success

A consequence of Theorem 2 is that we can decide in polynomial time if a complete assignment satisfies $abstractXOR_{\infty,C}(\Delta)$. Indeed, this amounts to deciding whether $C_{\downarrow\Delta^0}$ is consistent. This can be done by using Algorithm 1 to put the matrix $M$ associated with $C_{\downarrow\Delta^0}$ in RRE form, and then checking that Prop. *atLeast2* is satisfied.

# 4 Propagation of *abstractXOR*

As deciding of the satisfaction of $abstractXOR_{u,C}(\Delta)$ is polynomial when $u = \infty$, we consider that $u = \infty$ from now on. In this section, we introduce an algorithm that checks feasibility (Sect. 4.1), an algorithm that ensures Generalised Arc Consistency (Sect. 4.2), and we discuss implementation and complexity issues (Sect. 4.3).

## 4.1 Checking Feasibility of *abstractXOR*

Before starting the search, we build the matrix $M$ associated with $C_{\downarrow\Delta^0}$ and use Algorithm 1 to put it in RRE form. During the search, we maintain $M$ in RRE form: each time a variable $\Delta_j \in \Delta$ is assigned to 0, we remove column $j$ from $M$ and, if $j$ is the pivot column of a row $i$, we execute lines 3–6 of Algorithm 1.

Once $M$ is in RRE form, we check feasibility by exploiting Theorem 2, as shown in Algorithm 2: for each row $i$ with only one non-zero cell, if $\Delta_{pivot_i}$ is assigned to 1 we trigger failure, otherwise we assign 0 to $\Delta_{pivot_i}$ and remove row $i$ and column $pivot_i$ from $M$.

**Theorem 3.** *Algorithm 2 returns success iff* $\text{abstractXOR}_{\infty,C}(\Delta)$ *can be satisfied.*

*Proof.* Algorithm 2 returns failure (line 2) when there is a row $i$ with a single non-zero cell and the corresponding variable $\Delta_{pivot_i}$ is assigned to 1. This row corresponds to the equation $\delta_{pivot_i} = 0$ which cannot be satisfied when $\Delta_{pivot_i} = 1$. When Algorithm 2 returns success (line 7), $M$ satisfies Prop. *atLeast2* and, for each column $j \in [1, m]$, the variable $\Delta_j$ can still be assigned to 1. In this case, Theorem 2 ensures that the constraint can be satisfied by assigning 1 to each non-assigned variable.

Algorithm 2 does not only check feasibility, but also filters domains: it removes 1 from the domain of every variable $\Delta_{pivot_i}$ associated with a row $i$ such that $nonPivot_i = \emptyset$ (line 4). This does not remove solutions as this row corresponds to the equation $\delta_{pivot_i} = 0$ which is satisfied iff $\Delta_{pivot_i} = 0$.

### 4.2 Ensuring the Generalized Arc Consistency of *abstractXOR*

To ensure GAC, we must ensure that for each variable $\Delta_j \in \Delta$ and each value $v \in D(\Delta_j)$, the couple $(\Delta_j, v)$ has a support, *i.e.*, there exists a consistent assignment which assigns $v$ to $\Delta_j$ and a value $v' \in D(\Delta_{j'})$ to every other variable $\Delta_{j'} \in \Delta \setminus \{\Delta_j\}$.

By maintaining $M$ in RRE form and ensuring Prop. *atLeast2*, we ensure that $(\Delta_j, 1)$ has a support for each variable $\Delta_j \in \Delta$ such that $1 \in D(\Delta_j)$. Also $(\Delta_j, 0)$ has a support for every variable $\Delta_j \in \Delta$ assigned to 0. However, when $\Delta_j$ is not assigned, $(\Delta_j, 0)$ may not have a support. This occurs when there exist $\Delta_j, \Delta_{j'} \in \Delta \setminus \Delta^0$ such that $D(\Delta_j) = \{0,1\} \wedge D(\Delta_{j'}) = \{1\} \wedge C_{\downarrow \Delta^0} \Rightarrow (\delta_j = \delta_{j'})$. In this case, the couple $(\Delta_j, 0)$ has no support because $C_{\downarrow \Delta^0} \wedge (\delta_j = 0) \wedge (\delta_{j'} = 1)$ is inconsistent.

Hence, to ensure GAC we need to identify cases where the equality of two variables is a logical consequence of $C_{\downarrow \Delta^0}$. This is done by the following theorem.

**Theorem 4.** *For each pair of variables* $\{\Delta_j, \Delta_{j'}\} \subseteq \Delta \setminus \Delta^0$, $C_{\downarrow \Delta^0} \Rightarrow (\delta_j = \delta_{j'})$ *iff one of the following cases holds in the matrix* $M$ *in RRE form associated with* $C_{\downarrow \Delta^0}$*:*

**Case 1:** $\exists i \in [1, n], nonZero_i = \{j, j'\}$
**Case 2:** $\exists i, i' \in [1, n], (pivot_i = j) \wedge (pivot_{i'} = j') \wedge (nonPivot_i = nonPivot_{i'})$

*Proof.* Case 1 occurs when $M$ contains a row $i$ with exactly two non-zero cells, and this row corresponds to the equation $\delta_j = \delta_{j'}$. Case 2 occurs when $M$ contains 2 rows $i$ and $i'$ such that $nonPivot_i = nonPivot_{i'}$. These rows imply that $\delta_{pivot_i} = \delta_{pivot_{i'}}$ because both $\delta_{pivot_i}$ and $\delta_{pivot_{i'}}$ are equal to the result of XORing a same set of variables.

There is no other case where $C_{\downarrow \Delta^0} \Rightarrow (\delta_j = \delta_{j'})$ because, when $M$ is in RRE form, every row $i$ has a different pivot column $pivot_i$. Therefore, every equation in $C_{\downarrow \Delta^0}$ contains a different pivot variable $\delta_{pivot_i}$. Hence, $\delta_j$ and $\delta_{j'}$ are constrained to be equal either because they occur in a same equation without any other variable, or because they are the pivot variables of two different equations which share the same non-pivot variables.

Let us illustrate these two cases on the example displayed in Fig. 3:

- If $\Delta^0 = \{\Delta_5, \Delta_7\}$ then $C_{\downarrow \Delta^0}$ contains the equation $\delta_2 \oplus \delta_4 = 0$, and if $D(\Delta_4) = \{1\}$ and $D(\Delta_2) = \{0, 1\}$, then $(\Delta_2, 0)$ has no support.

**Algorithm 3:** Propagation of the assignment of 1 to a variable $\Delta_j$

1 let $Q$ be an empty queue; enqueue $\Delta_j$ in $Q$
2 **while** *$Q$ is not empty* **do**
3     dequeue a variable $\Delta_j$ from $Q$ and remove 0 from $D(\Delta_j)$
4     **if** *$j$ is the pivot column of a row $i$* **then**
5         **if** *$nonPivot_i = \{j'\}$ and $D(\Delta_{j'}) = \{0, 1\}$* **then** enqueue $\Delta_{j'}$ in $Q$ ;
6         **else**
7             **for** *each $i' \in [1, n]$ such that $nonPivot_i = nonPivot_{i'}$* **do**
8                 **if** $D(\Delta_{pivot_{i'}}) = \{0, 1\}$ **then** enqueue $\Delta_{pivot_{i'}}$ in $Q$;
9     **else**
10         **for** *each $i \in [1, n]$ such that $nonPivot_i = \{j\}$* **do**
11             **if** $D(\Delta_{pivot_{i'}}) = \{0, 1\}$ **then** enqueue $\Delta_{pivot_{i'}}$ in $Q$;

- If $\Delta^0 = \{\Delta_3, \Delta_5, \Delta_6\}$, then $C_{\downarrow\Delta^0}$ is equal to: $\delta_1 \oplus \delta_4 \oplus \delta_7 = 0$, $\delta_2 \oplus \delta_4 \oplus \delta_7 = 0$ and $M$ is equal to:

| $\delta_1$ | $\delta_2$ | $\delta_4$ | $\delta_7$ |
|---|---|---|---|
| 1 | 0 | 1 | 1 |
| 0 | 1 | 1 | 1 |

  This implies that the pivot variables $\delta_1$ and $\delta_2$ are both equal to $\delta_4 \oplus \delta_7$, and if $D(\Delta_1) = \{1\}$ and $D(\Delta_2) = \{0, 1\}$, then $(\Delta_2, 0)$ has no support.

To maintain GAC during the search, we call Algorithm 3 each time a variable must be assigned to 1. This algorithm uses a queue $Q$ of variables that must be assigned to 1. At each iteration of the loop lines 2–11, a variable $\Delta_j$ is dequeued from $Q$, and it is assigned to 1. This assignment is propagated on every variable $\Delta_{j'}$ such that $C_{\downarrow\Delta^0} \Rightarrow (\delta_j = \delta_{j'})$. We exploit Theorem 4 to identify these variables:

- Case 1 has two sub-cases: if $j$ is the pivot column of a row $i$, we simply check if $nonPivot_i$ is reduced to a singleton (line 5); otherwise, we have to search for every row $i$ such that $nonPivot_i$ only contains $j$ (lines 10–11).
- Case 2 only holds when $j$ is the pivot column of a row $i$, and we have to search for every row $i'$ such that $nonPivot_i = nonPivot_{i'}$ (lines 7–8).

Also, each time a variable is assigned to 0, we proceed as explained in Sect. 4.1 to check feasibility. Then, for each line which has been modified when executing lines 3–6 of Algorithm 1, we check if cases 1 or 2 of Theorem 4 hold and imply that $\delta_j = \delta_{j'}$ with $D(\Delta_j) = \{0, 1\}$ and $D(\Delta_{j'}) = \{1\}$: in this case, we call Algorithm 3 to propagate the assignment of 1 to $\Delta_j$.

### 4.3 Implementation and Complexities

*Sparse Sets.* Our propagators mainly involve traversing non-zero cells of rows and columns of $M$. As $M$ is very sparse, we represent its rows and columns with sparse sets [19]: each sparse set contains the non-zero cells of a row or a column. This allows us to visit every non-zero cell of a column (resp. row) in linear time with respect to the number of non-zero cells instead of $\mathcal{O}(m)$ (resp. $\mathcal{O}(n)$), and to decide in constant time if an element belongs to a set. Sparse sets also allow to restore sets in constant time when backtracking, provided that we only remove elements at each choice point. Unfortunately, this is not the case here as new non-zero cells may appear when XORing lines. Hence, when backtracking, we undo all operations done before the recursive call.

*Time Complexity of the Propagators.* Let $n_1$ (resp. $m_1$) be the maximum number of non-zero cells in a row (resp. a column) of $M$. When using sparse sets, the complexity of putting $M$ in RRE form, as described by Algorithm 1, becomes $\mathcal{O}(nn_1m_1)$.

The complexity of the propagation of the assignment of a variable to 0 is $\mathcal{O}(n_1m_1)$. Indeed, when a variable $\Delta_j$ is assigned to 0, we have to (i) remove column $j$, (ii) execute lines 3–7 of Algorithm 1 if $j$ is a pivot column, and (iii) run Algorithm 2. The complexity of this depends on whether $j$ is a pivot column or not:

- if $j$ is a pivot column, then (i) is achieved in $\mathcal{O}(1)$ as column $j$ only contains one non-zero cell; (ii) is achieved in $\mathcal{O}(n_1m_1)$; and (iii) is achieved in $\mathcal{O}(n_1)$ provided that we keep track of the rows that have been modified at step (ii);
- if $j$ is not a pivot column, then (i) is achieved in $\mathcal{O}(n_1)$ and (iii) is achieved in $\mathcal{O}(n_1)$ provided that we keep track of the rows that have been modified at step (i).

The complexity of the propagation of the assignment of a variable to 1 by Algorithm 3 is $\mathcal{O}(mn_1m_1)$. Indeed, in the worst case, this implies to assign 1 to every other variable. Hence, the loop lines 2–11 is performed $\mathcal{O}(m)$ times. The loop lines 7–8 is iterated $\mathcal{O}(n_1)$ times (we traverse non-zero cells of the column of a variable in $nonPivot_i$ to identify the rows $i'$ for which we have to check if $nonPivot_i = nonPivot_{i'}$), and we decide if $nonPivot_i = nonPivot_{i'}$ in $\mathcal{O}(m_1)$. The loop lines 10–11 is iterated $\mathcal{O}(n_1)$ times as we only have to consider the non-zero cells of column $j$.

*Implementation.* Our global constraint has been implemented in Java and integrated in Choco 4 [22]. As its propagators are rather expensive, we give a low priority to *abstractXOR* so that, at each node of the search tree, Choco propagates all other constraints before propagating *abstractXOR*.

## 5 Experimental Evaluation

In this section, we experimentally evaluate the interest of *abstractXOR*. We first consider the related-key MDC problem, where differences can be injected both in the key and the input text, and we report results for Midori in Sect. 5.1 and for AES in Sect. 5.2. In Sect. 5.3, we consider the single-key MDC problem, where differences are injected only in the input text. All experiments have been done on a single core of an Intel Xeon E3-1270v3 (3.50 GHz) with 32 GB RAM.

### 5.1 Related-Key MDC for Midori

*Description of the Problem.* The related-key MDC for Midori is described in Sect. 1 for the case where the input text $X_0$ is a sequence of 128 bits (denoted Midori128). Midori is also defined for 64 bit texts (denoted Midori64). In this case, *SubBytes* and *subBytesTable* are defined for 4 bit sequences instead of 8 bit sequences. Also, a key schedule is used to compute a new subkey at each round (see [2] for details).

We consider different values for $r$, ranging from 3 to the number of rounds defined in [2], *i.e.*, 16 (resp. 20) for Midori64 (resp. Midori128). For each value of $r$, the constant $n$ used in constraint $(A_0)$ of Fig. 2 is set to the smallest value for which there exists a solution, as this is the most difficult instance: instances with smaller values of $n$ are often trivially inconsistent, whereas instances with larger values are useless.

We report results on two problems: $Enum_1$ aims at enumerating all solutions of the Step1 problem described in Fig. 2 for Midori128; $Opt_{1+2}$ aims at finding the MDC whose probability is maximal as described in Fig. 1 for Midori128.

*Models for* $Enum_1$. We consider two models. The first one, denoted $Enum_1$ *Global*, is derived from Fig. 2 in a straightforward way, by replacing $(A_3)$ and $(A_4)$ with $abstractXOR_{\infty,C}(\Delta)$ where $C = \{(A_3), (A_4)\}$ and $\Delta$ contains all variables occurring in $(A_3)$ or $(A_4)$. It is implemented in Java with Choco 4 [22], and we consider two propagators: $Global_{Feas}$ only checks feasibility, as described in Sect. 4.1, and $Global_{GAC}$ ensures GAC, as described in Sect. 4.2. In both cases, we use the minDom/wdeg variable ordering heuristic [8].

The second model, denoted $Enum_1$ *Advanced*, is introduced in [12] (and is more efficient than the model of [13]). It is obtained from Fig. 2 by replacing $(A_3)$ and $(A_4)$ with their $\text{sum}_{\neq 1}$ encodings, and by adding a constraint derived from a property of *MixColumns* called the *Maximum Distance Separable (MDS)* property. It further adds new variables and constraints to remove Step2-inconsistent solutions by reasoning on equality relations between $\Delta_j$ variables. This model is much more difficult to design than *Global*. For this model, we report results obtained by Picat-SAT [28], which encodes the problem into a SAT formula and then uses the SAT solver Lingeling [4]. We made experiments with other CP solvers (such as Choco, Gecode or Chuffed, for example), and we only report results obtained with Picat-SAT because it scales much better.

*Models for* $Opt_{1+2}$. The problem described in Fig. 1 cannot be solved within a reasonable amount of time (even for the smallest value of $r$) without decomposing it into two steps, as described in Sect. 1. We consider two models for this two step process. $Opt_{1+2}$ *Global* simply merges $Enum_1$ *Global* with the model of Fig. 1, and adds a constraint which relates $\delta_j$ and $\Delta_j$ variables, *i.e.*, $\delta_j = 0 \Leftrightarrow \Delta_j = 0$. Also, we add a variable ordering heuristic to assign $\Delta_j$ variables before $\delta_j$ variables. This model is implemented in Choco 4.

$Opt_{1+2}$ *Advanced* uses $Enum_1$ *Advanced* to search for Step1 solutions. However, we do not merge this model with the Step2 model of Fig. 1 and use a single solver to solve the two steps because CP solvers like Choco cannot efficiently solve $Enum_1$ *Advanced* whereas SAT solvers like Lingeling cannot efficiently solve Step2 [14]. Hence, $Opt_{1+2}$ *Advanced* uses Picat-SAT to solve $Enum_1$ *Advanced*, and each time a Step1 solution $s$ is found, it uses Choco with the model of Fig. 1 to search for the best Step2 solution associated with $s$. This process is stopped either when there is no more Step1 solution, or when an optimal Step2 solution is found (*i.e.*, a solution such that all $P_i[b]$ variables are assigned to $-2$ as this is the largest possible value).

*Results.* On the top row of Fig. 4, we display the number of choice points needed to enumerate all Step1 solutions. $Global_{GAC}$ explores less choice points than $Global_{Feas}$, though the difference is very small for Midori64 when $r \geq 12$.

In the middle row of Fig. 4, we display the CPU time spent to enumerate all Step1 solutions. For Midori64, the two *Global* variants have very similar times whereas, for Midori128, $Global_{GAC}$ is faster than $Global_{Feas}$. *Advanced* is much slower than *Global*.

In the bottom row of Fig. 4, we display the CPU time needed to solve the full MDC problem. For Midori64, $Global_{Feas}$ and $Global_{GAC}$ have very similar results, and are much faster than *Advanced*. For Midori128, $Global_{GAC}$ is faster than $Global_{Feas}$, which is faster than *Advanced*, especially when $r$ increases.

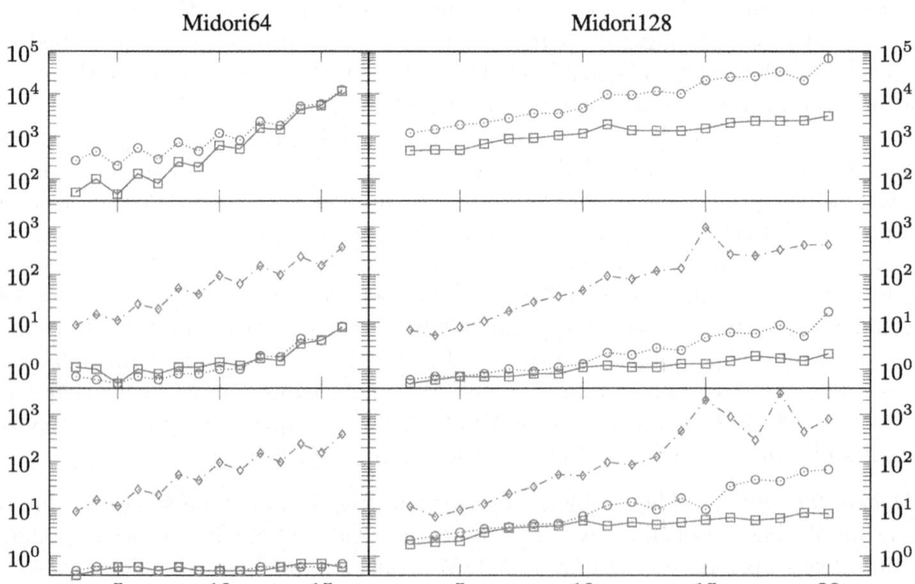


**Fig. 4.** Comparison of $Global_{Feas}$ (···○···), $Global_{GAC}$ (—□—), and Advanced (-·◇·-) for Midori. The x-axis gives the number of rounds $r$, and the y-axis the number of choice points for $Enum_1$ (up), and the run time for $Enum_1$ (Middle) and for $Opt_{1+2}$ (bottom). Times are in seconds.

### 5.2 Related-Key MDC for AES

*Description of the Problem.* Like Midori, AES iterates $r$ rounds, and each round is composed of four operations. However, AES computes a new sub-key at each round according to a key schedule which combines XOR and *SubBytes* operations. Also, the *MixColumns* operation is different and it combines XORs with a finite field multiplication by constant coefficients. This multiplication is easily modelled at Step2 using table constraints. However, it cannot be modelled at Step1 and constraint ($A_3$) is replaced with the following constraints which are derived from the MDS property of *MixColumns* (see [14] for more details):

$$\forall i \in [0, r-2], \forall k \in [0, 3], \sum_{j=0}^{3} (\Delta Z_i[k+4j] + \Delta Y_i[k+4j]) \in \{0, 5, 6, 7, 8\}$$

$$\forall i_1, i_2 \in [0, r-2], \forall k_1, k_2 \in [0, 3], \sum_{j=0}^{3} (x_{i_1 i_2 k_1 k_2 j} + y_{i_1 i_2 k_1 k_2 j}) \in \{0, 5, 6, 7, 8\}$$

where $x_{i_1 i_2 k_1 k_2 j}$ and $y_{i_1 i_2 k_1 k_2 j}$ are binary variables which are constrained as follows:

$$x_{i_1 i_2 k_1 k_2 j} = 1 \Leftrightarrow \delta Z_{i_1}[k_1 + 4j] \oplus \delta Z_{i_2}[k_2 + 4j] \neq 0$$
$$y_{i_1 i_2 k_1 k_2 j} = 1 \Leftrightarrow \delta Y_{i_1}[k_1 + 4j] \oplus \delta Y_{i_2}[k_2 + 4j] \neq 0$$

There exist three variants of AES, denoted AES$l$, where $l \in \{128, 192, 256\}$ corresponds to the number of bits in the key. The key schedule depends on $l$ whereas all other operations do not depend on $l$. For each key size $l$, we consider different values for the number of rounds $r$, ranging from 3 to an upper bound which depends on MDC probabilities: when increasing $r$, the probability decreases and it is useless to compute MDCs whenever the $\log_2$ probability becomes smaller than $-128$.

Like in Sect. 5.1, we consider two problems: $Enum_1$ aims at enumerating all Step1 solutions, and $Opt_{1+2}$ aims at finding the optimal MDC.

*Models for* $Enum_1$. We consider two CP models. *Global* is derived in a straightforward way from the definition of AES and the MDS property by replacing all XOR equations with an *abstractXOR* global constraint. It is implemented with Choco 4, and we consider two propagators (ensuring feasibility and GAC, respectively).

*Advanced* is the model introduced in [14] (which is more efficient than the ones of [15] and [20]). It uses a preprocessing step to infer new XOR equations from the key schedule, and it adds new variables and constraints to remove Step2-inconsistent solutions by reasoning on equality relations between $\Delta_j$ variables. This model is much more difficult to design than *Global*. It is implemented with Picat-SAT.

*Models for* $Opt_{1+2}$. Like in Sect. 5.1, *Global* solves the two steps with a single model implemented with Choco 4 whereas *Advanced* enumerates Step1 solutions with Picat-SAT and searches for optimal MDCs with Choco 4.

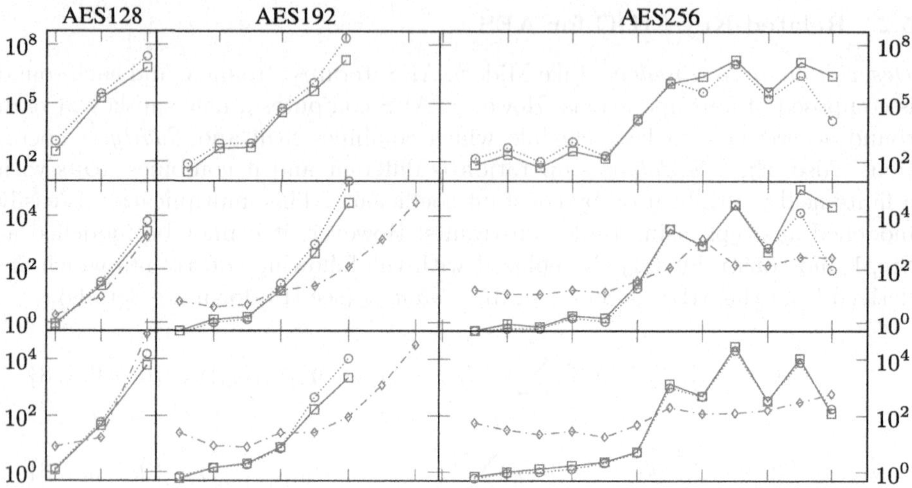


**Fig. 5.** Comparison of $Global_{Feas}$ (···○···), $Global_{GAC}$ (—□—), and *Advanced* (-·◇·-) for AES. The x-axis gives the number of rounds $r$, and the y-axis the number of choice points for $Enum_1$ (up), and the run time for $Enum_1$ (Middle) and $Opt_{1+2}$ (bottom). Times are in seconds.

*Results.* On the top row of Fig. 5, we display the number of choice points needed to enumerate all Step1 solutions. In most cases, $Global_{Feas}$ explores slightly more choice points than $Global_{GAC}$. However, for 5 instances of AES256, $Global_{Feas}$ explores slightly less choice points than $Global_{GAC}$. This is a bit surprising (as ensuring GAC is stronger than ensuring feasibility) but not impossible as filtering has an impact on the variable ordering heuristic.

In the middle row of Fig. 5, we display the time spent to enumerate all Step1 solutions. In many cases, $Global_{GAC}$ is faster than $Global_{Feas}$, but the difference is often rather small. *Advanced* is slower than *Global* when $r \leq 3$ (resp. 6 and 8) for AES128 (resp. 192 and 256), but it has better scale-up properties and it becomes faster for larger values of $r$. In particular, *Advanced* is able to solve AES192 when $r = 9$ (resp. $r = 10$) in 1,326 s (resp. 31,611 s) whereas *Global* is not able to complete the run within a time limit of 200,000 s.

In the bottom row of Fig. 5, we display the time needed to solve the full MDC problem. The performance of the three approaches are rather similar to the one in the middle row. However, for many instances the fact that the two steps are solved within a single model improves the solution process. This is the case, for example, for AES128 when $r = 5$. In this case, there are 103 Step1 solutions. If *Advanced* is more efficient than $Global_{GAC}$ to enumerate these solutions (1,694 s for *Advanced* instead of 2,656 s for $Global_{GAC}$), *Advanced* needs much more time to find the optimal MDC (76,103 s instead of 6,096 s).

### 5.3 Experimental Results for the Single-Key Problem

In the single-key differential attack, differences are introduced only in the initial text $X_0$, and no difference is introduced in the key, *i.e.*, $\delta K = 0$. Like for related-key, we consider two problems: $Enum_1$ (to enumerate all Step1 solutions), and $Opt_{1+2}$ (to find the optimal MDC). We also consider two block ciphers, *i.e.*, Midori and AES. In all cases, we consider *Global* and *Advanced* models, and these models are obtained from related-key models by assigning 0 to all variables associated with the key.

CPU times are reported in Table 1. For AES, the problem is the same whatever the length of the key (128, 192, or 256), as there is no difference in the key. For Midori, $Enum_1$ is the same whatever the length of the initial text (64 or 128) as bit sequences are abstracted by Boolean values. However, $Opt_{1+2}$ is different for Midori64 and Midori128. Surprisingly, single-key problems are much harder to solve than related-key ones, though the size of the search space is smaller (as all variables associated with the key are assigned to 0). This comes from the fact that the number of differences (defined by the constant $n$ in Fig. 2) is strongly increased: $n$ is increased from 3 (resp. 4 and 5) to 7 (resp. 16 and 23) when $r = 3$ (resp. 4 and 5) for Midori, and from 5 (resp. 12) to 9 (resp. 25) when $r = 3$ (resp. 4) for AES.

*Results for Midori.* *Advanced* finds much more Step1 solutions than *Global*: it finds 64 (resp. 4,908) solutions when $r = 3$ (resp. 4), whereas *Global* finds 16 (resp. 68) solutions. Every solution found by *Advanced* and not by *Global* is Step2 inconsistent and *Advanced* spends a lot of time to enumerate these useless solutions. Hence, *Advanced* is not able to solve Midori within one hour when $r > 3$. When $r = 4$, *Advanced* is able to solve $Enum_1$ in 59,036 s, but it is not able to solve $Opt_{1+2}$ within a reasonable amount of time because most Step1 solutions are Step2 inconsistent.

*Global* is able to solve up to $r = 5$ (resp. $r = 4$) for Midori64 (resp. Midori128). Step2 is much harder for Midori128 than for Midori64 because differential variables associated with the text take their values in $[0, 255]$ for Midori128 and in $[0, 16]$ for Midori64. Ensuring GAC often pays off and $Global_{GAC}$ is faster than $Global_{Feas}$, except for $Opt_{1+2}$/Midori128/$r = 3$.

*Results for AES.* When $r = 3$, both $Enum_1$ and $Opt_{1+2}$ are quickly solved, and *Global* is an order faster than *Advanced*. When $r = 4$, there is a huge number of Step1 solutions (we have enumerated 1,715,652 solutions within a 24 h time limit with $Global_{GAC}$, and all these solutions are Step2 consistent). Hence, *Global* fails at enumerating all Step1 solutions within a reasonable amount of time. However, when merging Step1 and Step2 models to solve $Opt_{1+2}$, we find an optimal solution in less than 2 s (the optimality proof is trivial because all $P_i[b]$ variables are assigned to the largest possible value).

When $r = 4$, the probability of the optimal MDC is equal to $2^{-150}$, which is smaller than $2^{-128}$. Hence, this MDC is useless to mount attacks. However, the fact that *Global* is able to enumerate a huge number of Step1 solutions in

**Table 1.** Single-Key results: Time (in seconds) needed by $Global_{Feas}$ ($G_{Feas}$), $Global_{GAC}$ ($G_{GAC}$), and *Advanced* (*Adv*) for Midori (left) and AES (right). We report '-' when time exceeds 3600 s.

| | $Enum_1$ | | | $Opt_{1+2}$ | | | | | |
|---|---|---|---|---|---|---|---|---|---|
| | Midori 64 and 128 | | | Midori64 | | | Midori128 | | |
| r | $G_{Feas}$ | $G_{GAC}$ | Adv | $G_{Feas}$ | $G_{GAC}$ | Adv | $G_{Feas}$ | $G_{GAC}$ | Adv |
| 3 | 0.6 | 0.6 | 8.7 | 0.7 | 0.6 | 11.1 | 1.3 | 8.1 | 12.5 |
| 4 | 22.4 | 8.0 | - | 17.6 | 11.1 | - | 434.9 | 290.5 | - |
| 5 | 2897.8 | 686.8 | - | 2608.1 | 689.7 | - | - | - | - |

| | AES 128, 192, and 256 | | | | | |
|---|---|---|---|---|---|---|
| | $Enum_1$ | | | $Opt_{1+2}$ | | |
| | $G_{Feas}$ | $G_{GAC}$ | Adv | $G_{Feas}$ | $G_{GAC}$ | Adv |
| 3 | 1.3 | 0.7 | 7.5 | 1.1 | 1.0 | 55.0 |
| 4 | - | - | - | 1.2 | 1.4 | - |

a reasonable amount of time opens new perspectives: we can search for a set of MDCs that share the same values in the initial text $\delta X_0$ and in the cipher text $\delta X_r$, and combine these MDCs to find better differentials.

## 6 Conclusion

We have introduced a new global constraint which eases the design of models for computing MDCs: these models are straightforwardly derived from problem definitions. This global constraint allows us to compute MDCs much faster than advanced models (which are much more difficult to design and which combine SAT and CP solvers) for single-key and related-key Midori, and for single-key AES. However, for related-key AES, it fails at solving the two largest instances of AES192 within a reasonable amount of time, and SAT has better scale-up properties for enumerating Step1 solutions. As pointed out in [14], clause learning is a key ingredient for solving this problem, and further work will aim at improving scale-up properties of Choco on this problem by adding clause learning to Choco.

We believe our new global constraint opens promising perspectives for cryptographs, and we aim at using it to solve new differential cryptanalysis problems such as those studied in [9] or [27], and new symmetric block ciphers such as Skinny [3].

**Acknowledgement.** This work has been funded by ANR DeCrypt (ANR-18-CE39-0007). We thank Charles Prud'homme for answering our numerous questions on Choco.

## References

1. Abdelkhalek, A., Sasaki, Y., Todo, Y., Tolba, M., Youssef, A.: MILP modeling for (large) s-boxes to optimize probability of differential characteristics. IACR Trans. Symmetric Cryptol. **2017**(4), 99–129 (2017)
2. Banik, S., et al.: Midori: a block cipher for low energy. In: Iwata, T., Cheon, J.H. (eds.) ASIACRYPT 2015. LNCS, vol. 9453, pp. 411–436. Springer, Heidelberg (2015). https://doi.org/10.1007/978-3-662-48800-3_17
3. Beierle, C., et al.: The SKINNY family of block ciphers and its low-latency ariant MANTIS. In: Robshaw, M., Katz, J. (eds.) CRYPTO 2016. LNCS, vol. 9815, pp. 123–153. Springer, Heidelberg (2016). https://doi.org/10.1007/978-3-662-53008-5_5

4. Biere, A.: Yet another local search solver and lingeling and friends entering the sat competition 2014, pp. 39–40, January 2014
5. Biham, E.: New types of cryptoanalytic attacks using related keys (extended abstract). In: EUROCRYPT, LNCS, vol. 765, pp. 398–409. Springer (1993)
6. Biham, E., Shamir, A.: Differential cryptanalysis of feal and N-Hash. In: Davies, D.W. (ed.) EUROCRYPT 1991. LNCS, vol. 547, pp. 1–16. Springer, Heidelberg (1991). https://doi.org/10.1007/3-540-46416-6_1
7. Biryukov, A., Nikolić, I.: Automatic search for related-key differential characteristics in byte-oriented block ciphers: application to AES, Camellia, Khazad and others. In: Gilbert, H. (ed.) EUROCRYPT 2010. LNCS, vol. 6110, pp. 322–344. Springer, Heidelberg (2010). https://doi.org/10.1007/978-3-642-13190-5_17
8. Boussemart, F., Hemery, F., Lecoutre, C., Sais, L.: Boosting systematic search by weighting constraints. In: Proceedings of the 16th Eureopean Conference on Artificial Intelligence, ECAI 2004, pp. 146–150. IOS Press (2004)
9. Cid, C., Huang, T., Peyrin, T., Sasaki, Yu., Song, L.: Boomerang connectivity table: a new cryptanalysis tool. In: Nielsen, J.B., Rijmen, V. (eds.) EUROCRYPT 2018. LNCS, vol. 10821, pp. 683–714. Springer, Cham (2018). https://doi.org/10.1007/978-3-319-78375-8_22
10. FIPS 197: Advanced Encryption Standard. Federal Information Processing Standards Publication 197 (2001). u.S. Department of Commerce/N.I.S.T
11. Fouque, P.-A., Jean, J., Peyrin, T.: Structural evaluation of AES and chosen-key distinguisher of 9-round AES-128. In: Canetti, R., Garay, J.A. (eds.) CRYPTO 2013. LNCS, vol. 8042, pp. 183–203. Springer, Heidelberg (2013). https://doi.org/10.1007/978-3-642-40041-4_11
12. Gérault, D.: Security Analysis of Contactless Communication Protocols. Ph.D. thesis, Université Clermont Auvergne (2018)
13. Gérault, D., Lafourcade, P.: Related-key cryptanalysis of Mmidori. In: Dunkelman, O., Sanadhya, S.K. (eds.) INDOCRYPT 2016. LNCS, vol. 10095, pp. 287–304. Springer, Cham (2016). https://doi.org/10.1007/978-3-319-49890-4_16
14. Gerault, D., Lafourcade, P., Minier, M., Solnon, C.: Computing AES related-key differential characteristics with constraint programming. Artif. Intell. **278**, 103183 (2020)
15. Gerault, D., Minier, M., Solnon, C.: Constraint programming models for chosen key differential cryptanalysis. In: Rueher, M. (ed.) CP 2016. LNCS, vol. 9892, pp. 584–601. Springer, Cham (2016). https://doi.org/10.1007/978-3-319-44953-1_37
16. Knudsen, L.R.: Truncated and higher order differentials. In: Preneel, B. (ed.) FSE 1994. LNCS, vol. 1008, pp. 196–211. Springer, Heidelberg (1995). https://doi.org/10.1007/3-540-60590-8_16
17. Kölbl, S., Leander, G., Tiessen, T.: Observations on the SIMON block cipher family. In: Gennaro, R., Robshaw, M. (eds.) CRYPTO 2015. LNCS, vol. 9215, pp. 161–185. Springer, Heidelberg (2015). https://doi.org/10.1007/978-3-662-47989-6_8
18. Lafitte, F.: Cryptosat: a tool for sat-based cryptanalysis. IET Inf. Secur. **12**(6), 463–474 (2018)
19. Le clément de saint Marcq, V., Schaus, P., Solnon, C., Lecoutre, C.: Sparse-sets for domain implementation. In: CP Workshop on Techniques for Implementing Constraint Programming Systems (TRICS) (2013). https://hal.archives-ouvertes.fr/hal-01339250
20. Minier, M., Solnon, C., Reboul, J.: Solving a symmetric key cryptographic problem with constraint programming. In: Workshop on Constraint Modelling and Reformulation (ModRef), pp. 1–13 (2014)

21. Mouha, N., Preneel, B.: A proof that the ARX cipher salsa20 is secure against differential cryptanalysis. IACR Cryptology ePrint Archive 2013, p. 328 (2013)
22. Prud'homme, C., Fages, J.G., Lorca, X.: Choco Documentation. TASC, INRIA Rennes, LINA CNRS UMR 6241, COSLING S.A.S. (2016). http://www.choco-solver.org
23. Soos, M., Nohl, K., Castelluccia, C.: Extending SAT solvers to cryptographic problems. In: Kullmann, O. (ed.) SAT 2009. LNCS, vol. 5584, pp. 244–257. Springer, Heidelberg (2009). https://doi.org/10.1007/978-3-642-02777-2_24
24. Sun, L., Wang, W., Wang, M.: More accurate differential properties of led64 and midori64. IACR Trans. Symmetric Cryptol. **2018**(3), 93–123 (2018)
25. Sun, S., et al.: Analysis of AES, SKINNY, and others with constraint programming. In: 24th International Conference on Fast Software Encryption (2017)
26. Sun, S., Hu, L., Wang, P., Qiao, K., Ma, X., Song, L.: Automatic security evaluation and (Related-key) differential characteristic search: application to SIMON, PRESENT, LBlock, DES(L) and other bit-oriented block ciphers. In: Sarkar, P., Iwata, T. (eds.) ASIACRYPT 2014. LNCS, vol. 8873, pp. 158–178. Springer, Heidelberg (2014). https://doi.org/10.1007/978-3-662-45611-8_9
27. Todo, Y., Isobe, T., Hao, Y., Meier, W.: Cube attacks on non-blackbox polynomials based on division property. In: Katz, J., Shacham, H. (eds.) CRYPTO 2017. LNCS, vol. 10403, pp. 250–279. Springer, Cham (2017). https://doi.org/10.1007/978-3-319-63697-9_9
28. Zhou, N.-F., Kjellerstrand, H., Fruhman, J.: Constraint Solving and Planning with Picat. SIS. Springer, Cham (2015). https://doi.org/10.1007/978-3-319-25883-6

# In Pursuit of an Efficient SAT Encoding for the Hamiltonian Cycle Problem

Neng-Fa Zhou(✉)

CUNY Brooklyn College & Graduate Center, New York, USA
zhou@sci.brooklyn.cuny.edu

**Abstract.** SAT solvers have achieved remarkable successes in solving various combinatorial problems. Nevertheless, it remains a challenge to find an efficient SAT encoding for the Hamiltonian Cycle Problem (HCP), which is one of the most well-known NP-complete problems. A central issue in encoding HCP into SAT is how to prevent sub-cycles, and one well-used technique is to map vertices to different positions. The HCP can be modeled as a single-agent path-finding problem. If the agent occupies vertex $i$ at time $t$, and occupies vertex $j$ at time $t+1$, then vertex $j$'s position must be the successor of vertex $i$'s. This paper compares three encodings for the successor function, namely, a *unary encoding* that uses a Boolean variable for each vertex-time pair, an optimized *binary adder encoding* that uses a special incrementor with no carry variables, and a *LFSR encoding* that uses a linear-feedback-shift register. This paper also proposes a preprocessing technique that rules out a position from consideration for a vertex and a time if the agent cannot occupy the vertex at the time. Our study has surprisingly revealed that, with optimization and preprocessing, the binary adder encoding is a clear winner: it solved some instances of the knight's tour problem that had been beyond reach for eager encoding approaches, and performed the best on the HCP instances used in the 2019 XCSP competition.

## 1 Introduction

The Hamiltonian Cycle Problem (HCP) is one of the most well-known NP-complete problems. Given a graph, the goal of HCP is to find a cycle in the graph that includes each and every vertex exactly once. As HCP occurs in many combinatorial problems, the global constraint `circuit`($G$) has become indispensable in constraint programming (CP) systems. Given a graph $G$ represented by a list of domain variables, the constraint ensures that any valuation of the variables constitutes a Hamiltonian cycle.

SAT solvers have achieved remarkable successes in solving combinatorial problems, ranging from formal methods [2,27], planning [24,37], answer set programming [6,14], to general constraint satisfaction problems (CSPs) [4,20,22,33,39,40,43]. The key issue in encoding HCP into SAT is how to prevent sub-cycles. A naive encoding, which bans sub-cycles in every proper subset of vertices, requires an exponential number of clauses. One common technique


H. Simonis (Ed.): CP 2020, LNCS 12333, pp. 585–602, 2020.
https://doi.org/10.1007/978-3-030-58475-7_34

used in SAT encodings for HCP is to map vertices to different positions so that no sub-cycles can be formed during search. The direct encoding of the mapping, which requires $O(n^3)$ clauses for a graph of $n$ vertices in the worst case, does not scale well to large graphs [19,28,35]. In order to circumvent the explosive encoding size of the eager approach, researchers have proposed lazy approaches, such as satisfiability modulo acyclicity [3] that incorporates reachability checking during search, and incremental SAT solving that incrementally adds clauses to ban sub-cycles [38]. Recently inspired by the log encoding [21], Johnson proposed a compact encoding for HCP, which employs a linear-feedback-shift register (LFSR) for the successor function [18,23].

This paper continues the pursuit of an efficient SAT encoding for HCP. HCP can be modeled as a single-agent path-finding problem. Given a graph of $n$ vertices, the agent resides at the start vertex at time 1, moves to a neighboring vertex in each step, and at time $n+1$ comes back at the start vertex after having visited each and every vertex exactly once. Each vertex is mapped to a distinct position. If the agent occupies vertex $i$ at time $t$, and occupies vertex $j$ at time $t+1$, then vertex $j$'s position must be the successor of vertex $i$'s. This encoding is called *distance encoding*. This paper compares three encodings for the successor function, namely, a *unary encoding* that uses a Boolean variable for each vertex-time pair, an optimized *binary adder encoding* that uses a special incrementor with no carry variables, and a *LFSR encoding* that uses a linear-feedback-shift register. This paper also proposes a preprocessing technique that rules out a position from consideration for a vertex and a time if the agent cannot occupy the vertex at the time.

The experimental results, using two SAT solvers, show that, with preprocessing, the optimized binary adder encoding significantly outperformed the unary and the LFSR encodings. The binary adder encoding solved some instances of the knight's tour problem that had been beyond reach for eager encoding approaches, and solved more instances of the HCP benchmark used in the 2019 XCSP competition than other solvers.

## 2 Preliminaries

This section defines HCP, and gives the basic SAT encodings for domain variables and the at-most-one constraint that are employed in the SAT encodings for HCP.

### 2.1 The HCP and the `Circuit` Constraint

Given a directed graph, the goal of HCP is to find a cycle in the graph that includes each and every vertex exactly once. In CP, HCP can be described as a global constraint `circuit(`$G$`)`, where $G$ = `[`$V_1, V_2, \ldots, V_n$`]` is a list of domain variables representing the graph.

For example, Fig. 1 gives a directed graph and its representation using domain variables, where vertex `i` is represented by the domain variable $V_i$ (i $= 1, 2, 3, 4$), and the domain of $V_i$ indicates the outgoing arcs from vertex `i`. A valuation

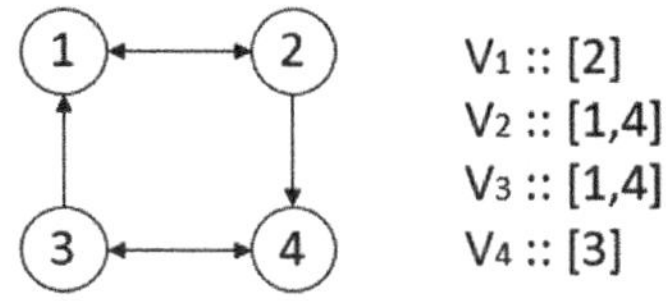


**Fig. 1.** A directed graph and its representation using domain variables

$V_i = j$ of the domain variables represents a subgraph of $G$ that consists of arcs $(i,j)$ $(i \in 1 \ldots n, j \in 1 \ldots n)$. The `circuit(`$G$`)` constraint enforces that the subgraph represented by a valuation of the domain variables forms a Hamiltonian cycle. For example, for the graph in Fig. 1, `[2,4,1,3]` is a solution because `1→2, 2→4, 4→3, 3→1` is a Hamiltonian cycle, but `[2,1,4,3]` is not because the graph `1→2, 2→1, 3→4, 4→3` contains two sub-cycles.

### 2.2 Direct Encoding

Let $X :: \{a_1, a_2, \ldots, a_n\}$ be a domain variable. The *direct encoding* [11] introduces a Boolean variable $B_i$ for $B_i \Leftrightarrow X = a_i$ $(i \in 1..n)$, and generates the constraint *exactly-one*$(B_1, B_2, \ldots, B_n)$, which is converted to a conjunction of an *at-least-one* constraint $\geq_1 (B_1, B_2, \ldots, B_n)$, and an *at-most-one* constraint $\leq_1 (B_1, B_2, \ldots, B_n)$. The at-least-one constraint is encoded as the clause $B_1 \vee B_2 \vee \ldots \vee B_n$.

### 2.3 SAT Encodings for the At-Most-One Constraint

The at-most-one constraint has numerous encodings into SAT (see [42] for the latest comparison).

The ***pairwise*** (PW) encoding for $\leq_1 (B_1, B_2, \ldots, B_n)$ decomposes the constraint into $\neg B_i \vee \neg B_j$, for $i \in 1..n-1$ and $j \in i+1..n$. PW generates $O(n^2)$ clauses, and is therefore not viable for large $n$.

The ***bisect*** (BS) encoding for $\leq_1 (B_1, B_2, \ldots, B_n)$ splits the variables into two groups $G_1 = \{B_1, B_2, \ldots, B_m\}$ and $G_2 = \{B_{m+1}, \ldots, B_n\}$, where $m = \lfloor \frac{n}{2} \rfloor$. It introduces a new variable $T$ as the *commander* for $G_1$, and uses $\neg T$ as the commander for $G_2$. BS decomposes the constraint into the following: **(BS-1)** For $i \in 1..m$: $B_i \Rightarrow T$; **(BS-2)** For $i \in m+1..n$: $B_i \Rightarrow \neg T$; **(BS-3)** $\leq_1 (B_1, B_2, \ldots, B_m)$; **(BS-4)** $\leq_1 (B_{m+1}, \ldots, B_n)$. Constraint **BS-1** forces $T$ to be 1 if any of the variables in $G_1$ is 1. Constraint **BS-2** forces $T$ to be 0 if any of the variables in $G_2$ is 1. Since $T$ cannot be both 0 and 1 at the same time, it is impossible for one variable in $G_1$ and another variable in $G_2$ to be 1 simultaneously. Constraints **BS-3** and **BS-4** recursively enforce at-most-one on the two groups. The BS is a special case of the *commander encoding*[25]. The number of clauses generated by BS is characterized by $f(n) = n + 2f(n/2)$, and the number of new variables is characterized by $g(n) = 1 + 2g(n/2)$.

The ***product*** (PD) encoding [7] for $\leq_1 (B_1, B_2, \ldots, B_n)$ arranges the variables on an $m \times m$ matrix $M$, where $m = \sqrt{n}$. It introduces two vectors of new variables $\langle R_1, R_2, \ldots, R_m \rangle$ and $\langle C_1, C_2, \ldots, C_m \rangle$, where $R_i$ represents row $i$ and $C_j$ represents column $j$. In case $n$ is not a square number, the extra entries of $M$ are filled with 0. PD decomposes the constraint into the following: **(PD-1)** For $i \in 1..m$, $j \in 1..m$: $M_{ij} \Rightarrow R_i \wedge C_j$; **(PD-2)** $\leq_1 (R_1, R_2, \ldots, R_m)$; **(PD-3)** $\leq_1 (C_1, C_2, \ldots, C_m)$. The number of clauses generated by PD is characterized by $f(n) = 2n + 2f(\sqrt{n})$, and the number of new variables is characterized by $g(n) = 2\sqrt{n} + 2g(\sqrt{n})$.

A hybrid encoding for $\leq_1 (B_1, B_2, \ldots, B_n)$ is employed for HCP. When $n \leq 4$, PW is used. Otherwise, the constraint is divided into smaller at-most-one constraints using BS or PD, depending on the cost function $f(n) + \alpha g(n)$, where $\alpha$ is the penalty for introducing a new variable. For example, for $n = 64$ and $\alpha = 3$, the constraint is first divided using PD into two sub-constraints, each of which involves 8 variables, and then the sub-constraints are divided using BS into base ones, which are encoded using PW.

### 2.4 Log Encoding and Logic Optimization

The *log encoding* [21] is more compact than the direct encoding. The *sign-and-magnitude* log encoding uses a sequence of Boolean variables for the magnitude. If there are values of both signs in the domain, then the encoding uses another Boolean variable for the sign. Each combination of values of the Boolean variables represents a value for the domain variable.

Under log encoding, each domain variable can be treated as a truth table, and a logic optimizer can be utilized to find CNF clauses for it. The Quine-McCluskey (QM) algorithm [29,36] is popular for two-level logic optimization. A *product* is a conjunction of literals. Given a truth table, each tuple is a product, called a *minterm*, that involves all the inputs. A minterm is in the *on-set* if its output is required to be 1, in the *off-set* if the output is required to be 0, and in the *don't-care-set*, otherwise. A product of literals is an *implicant* of a truth table if it entails no minterms in the off-set. A *prime implicant* is an implicant that is not implied by any other implicant. For a truth table, the QM algorithm first computes all the prime implicants of the table, and then finds a minimal set of prime implicants that covers all the minterms in the on-set and none of the minterms in the off-set. The second step of the QM algorithm requires solving the minimum set-covering problem, which is NP-hard [12]. The Espresso logic optimizer [5] only computes a partial set of prime implicants based on heuristics, and therefore a smaller set-covering problem.

For example, consider the domain variable $X :: [1, 2, 5, 6]$. The log encoding uses a sequence of three Boolean variables, $X_2X_1X_0$, to encode the domain. It is possible to represent 8 different values with three Boolean variables, including the values in $X$'s domain and the no-good values in the set $\{0, 3, 4, 7\}$. A naive encoding with *conflict clauses* [13] for the domain requires four clauses:

$$\begin{array}{ll} X_2 \vee X_1 \vee X_0 & (X \neq 0) \\ X_2 \vee \neg X_1 \vee \neg X_0 & (X \neq 3) \\ \neg X_2 \vee X_1 \vee X_0 & (X \neq 4) \end{array}$$

$\neg X_2 \vee \neg X_1 \vee \neg X_0 \qquad (X \neq 7)$

Each of these clauses corresponds to a no-good value. The logic optimizer Espresso only uses two clauses:

$X_1 \vee X_0$
$\neg X_1 \vee \neg X_0$

Each clause corresponds to a prime implicant in the disjunctive normal form. Note that the variable $X_2$ is optimized away.

## 3 The Distance Encoding for the circuit Constraint

The circuit($G$) constraint, where $G$ is a list of domain variables [$V_1, V_2, \ldots, V_n$], enforces the following: (1) each of the vertices has exactly one incoming arc and exactly one outgoing arc; (2) each of the proper subgraphs of $G$ is a tree, meaning that the subgraph is connected and the number of vertices is 1 greater than the number of arcs. A SAT encoding based on these properties does not use any extra variables, but requires an exponential number of clauses.

The *distance encoding* for HCP employs a matrix of Boolean variables $H$ of size $n \times n$ for the Hamiltonian cycle. The entry $H_{ij}$ is 1 if and only if the arc $(i, j)$ occurs in the resulting Hamiltonian cycle.

The following *channeling constraints* connects the matrix $H$ and the original domain variables $[V_1, V_2, \ldots, V_n]$:

$$\begin{array}{l}\text{For each } i \in 1..n,\ j \in 1..n,\ i \neq j: \\ \quad H_{ij} \Leftrightarrow V_i = j\end{array} \tag{1}$$

Since each variable $V_i$ takes only one value, constraint (1) entails that each vertex has exactly one outgoing arc. The following *degree* constraints ensure that each vertex has exactly one incoming arc:

$$\text{For each } j \in 1..n\text{: } \sum\nolimits_{i=1}^{n} H_{ij} = 1 \tag{2}$$

For each pair of vertices $(i, j)$ $(i \in 1..n, j \in 1..n)$, if the arc $(i, j)$ is not in the original graph $G$, then the entry $H_{ij}$ is set to 0. Therefore, the number of Boolean variables in $H$ equals the number of arcs in $G$.

Graph $H$ that satisfies constraints (1) and (2) may contain sub-cycles. In order to ban sub-cycles, the distance encoding maps each vertex to a distinct position. Let $p(i)$ be the position of vertex $i$, $s(p)$ denote the successor of position $p$,[1] and $s^k(p)$ be the $k$th successor of $p$. Assume that vertex 1 is visited first, and it is mapped to position 1.[2] The following constraints ensure that the cycle

[1] The successor function, such as the LFSR described below, may generate a different sequence of numbers from the natural number sequence.
[2] A good heuristic is to start with a vertex that has the smallest degree [41].

starts at 1 and ends at 1:

$$\begin{aligned}&\text{For each } i \in 2..n\text{:}\\ &H_{1i} \Rightarrow p(i) = s(1) \qquad (3)\\ &H_{i1} \Rightarrow p(i) = s^{n-1}(1) \qquad (4)\end{aligned}$$

Constraint (3) ensures that if there is an arc from vertex 1 to vertex $i$ then $i$'s position is the successor of 1. Constraint (4) ensures that if there is an arc from vertex $i$ to vertex 1 then $i$'s position is the $(n-1)$th successor of 1.

In addition to the above constraints, the following constraints ensure that the arcs are connected and the vertices are positioned successively:

$$\begin{aligned}&\text{For each } i \in 2..n,\ j \in 2..n,\ i \neq j\text{:}\\ &H_{ij} \Rightarrow p(j) = s(p(i)) \qquad (5)\end{aligned}$$

Constraint (5) ensures that vertex $j$ is positioned immediately after vertex $i$ if arc $(i, j)$ is in the Hamiltonian cycle.

**Theorem 1.** *Constraints (1)–(5) guarantee that the graph represented by $H$ is Hamiltonian.*

*Proof.* Constraints (1) and (2) entail that each vertex in graph $H$ has exactly one incoming arc and exactly one outgoing arc, and therefore they guarantee that graph $H$ is cyclic. Assume that the cycle in which vertex 1 occurs is:

$$1 \rightarrow v_2 \rightarrow v_3 \rightarrow \ldots \rightarrow v_k \rightarrow 1$$

According to constraints (3)–(5), the following conditions hold:

$$\begin{aligned}&p(v_2) = s(1)\\ &p(v_i) = s(p(v_{i-1})) \text{ for } i \in 3..k\\ &p(v_k) = s^{n-1}(1)\end{aligned}$$

These conditions entail $k = n$. Therefore, graph $H$ includes all the vertices, and is Hamiltonian. □

The theorem shows that it is sufficient to use one-way entailment constraints in (3)–(5) instead of stronger equivalence constraints.

The final encoding size depends on how the successor function is encoded. The code size of constraints (1) and (2) is not dependent on the successor function. The number of Boolean variables in $H$ equals the number of arcs in $G$. Constraint (1) mimics the direct encoding of domain variables. Both constraint (1) and constraint (2) are encoded as *exactly-one* constraints. Let $d$ be the maximum degree in $G$. If the 2-product encoding is used for *at-most-one*, then constraints (1) and (2) introduce $O(n \times \sqrt{d})$ new Boolean variables and require $O(n \times d)$ clauses.

# 4 Three Encodings for the Successor Function

There are several different ways to encode the successor function $p(j) = s(p(i))$ used in constraint (5). This section gives three such encodings, namely, the unary encoding, the binary adder encoding, and the linear-feedback-shift-register (LFSR) encoding.

## 4.1 Unary Encoding

The unary encoding of the successor function employs a matrix $U$ of Boolean variables of size $n \times n$, where $U_{ip} = 1$ iff vertex $i$'s position is $p$ for $i \in 1..n$ and $p \in 1..n$. Since vertex 1 is visited first, $U_{11}$ is initialized to 1. Each vertex is visited exactly once, so the following constraint must hold:

$$\text{For each } i \in 1..n\text{: } \sum_{p=1}^{n} U_{ip} = 1 \tag{6}$$

For each vertex $i$ ($i \in 1..n$), there is exactly one position $p$ ($p \in 1..n$) for which $U_{ip}$ is 1.

Constraints (3)–(5) given in the previous section are translated into the following under the unary encoding:

For each $i \in 2..n$:

$$H_{1i} \Rightarrow U_{i2} \tag{3'}$$

$$H_{i1} \Rightarrow U_{in} \tag{4'}$$

For each $i \in 2..n,\ j \in 2..n,\ i \neq j,\ p \in 2..(n-1)$:

$$H_{ij} \wedge U_{ip} \Rightarrow U_{j(p+1)} \tag{5'}$$

Constraint (3') ensures that if there is an arc from vertex 1 to vertex $i$ then vertex $i$'s position is 2. Constraint (4') ensures that if there is an arc from vertex $i$ to vertex 1 then vertex $i$'s position is $n$. Constraint (5') ensures that if arc $(i, j)$ is in the Hamiltonian cycle, and vertex $i$'s position is $p$, then vertex $j$'s position is $p+1$. The constraints (3')–(5') entail that for each position there is exactly one vertex mapped to it ($\sum_{i=1}^{n} U_{ip} = 1$ for $p \in 1..n$).

The two dimensional array $U$ has $O(n^2)$ variables. In addition, some temporary Boolean variables are introduced by the *exactly-one* constraints in (6). The number of clauses is dominated by constraint (5'), which requires $O(n^2 \times d)$ clauses to encode, where $d$ is the maximum degree in $G$.

## 4.2 Binary Adder Encoding

The binary adder encoding of the successor function employs a log-encoded domain variable $P_i$ for each vertex $i$, whose domain is the set of possible positions for the vertex. As all the positions are positive, no sign variables are needed in the encoding.

Since vertex 1 is visited first, $P_1 = 1$. Constraints (3)–(5) given above are translated into the following under log encoding:

For each $i \in 2..n$:

$$H_{1i} \Rightarrow P_i = 2 \quad (3")$$

$$H_{i1} \Rightarrow P_i = n \quad (4")$$

For each $i \in 2..n,\ j \in 2..n,\ i \neq j$:

$$H_{ij} \Rightarrow P_j = P_i + 1 \quad (5")$$

The efficiency of the encoding heavily depends on the encoding of the successor function $P_j = P_i + 1$ used in constraint (5").

Let $X$'s log encoding be $\langle X_{m-1}X_{m-2}\ldots X_1X_0\rangle$ and $Y$'s log encoding be $\langle Y_{m-1}Y_{m-2}\ldots Y_1Y_0\rangle$. Consider the unsigned addition:

$$\begin{array}{rr} & X_{m-1}\ \ldots\ X_1\ X_0 \\ + & 1 \\ \hline & Y_{m-1}\ \ldots\ Y_1\ Y_0 \end{array}$$

A naive encoding performs the addition using ripple carry adders from the least significant bit to the most significant bit. If a half-adder is used for each bit position, then the addition requires, in total, $m - 1$ carry variables and $7 \times m$ clauses.

The following sequential incrementor performs the addition using no carry variables:[3]

- For bit position 0, $Y_0 = \neg X_0$, which is encoded as two clauses: $Y_0 \vee X_0$ and $\neg Y_0 \vee \neg X_0$.
- For bit position 1, the input carry from bit position 0 is $X_0$, so the following constraints must hold:

$$X_0 \Rightarrow Y_1 = \neg X_1$$
$$\neg X_0 \Rightarrow Y_1 = X_1$$

  These two constraints are converted into 4 clauses.
- For each other bit position $i$ ($i > 1$), the carry from bit position $i - 1$ is 1 iff $Y_{i-1} = 0$ and $X_{i-1} = 1$, so the following constraints must hold:

$$\neg Y_{i-1} \wedge X_{i-1} \Rightarrow Y_i = \neg X_i$$
$$\text{otherwise} \Rightarrow Y_i = X_i$$

  These constraints can be encoded using 6 clauses.

The total number of clauses used for the addition is $2 + 4 + (m - 2) * 6$.

The sequential incrementor is improved as follows. For bit position $i$ ($i > 1$), instead of considering one bit at a time, the improved version considers two bits at a time, imposing the following constraints:

$$\neg Y_{i-1} \wedge X_{i-1} \Rightarrow Y_i = \neg X_i$$
$$\neg Y_{i-1} \wedge X_{i-1} \wedge X_i \Rightarrow Y_{i+1} = \neg X_{i+1}$$
$$\text{otherwise} \Rightarrow Y_i = X_i \wedge Y_{i+1} = X_{i+1}$$

[3] This encoding is based one suggested by Vitaly Lagoon via personal communication.

These constraints can be encoded using 11 clauses, resulting in a reduction of one clause for each two bits.

Furthermore, the improved incrementor treats the top 4 bits as a whole using the adder in Fig. 2.[4] The carry from bit position $m-5$ to bit position $m-4$ is 1 iff $Y_{m-5} = 0$ and $X_{m-5} = 1$. The adder uses 21 clauses, resulting in a reduction of 3 clauses from 24 clauses needed by the one-bit incrementor and 1 clause from 22 clauses needed by the two-bit incrementor.

| | |
|---|---|
| $X_{m-1} \vee X_{m-4} \vee \neg Y_{m-1}$ | $X_{m-2} \vee X_{m-4} \vee \neg Y_{m-2}$ |
| $X_{m-1} \vee \neg Y_{m-1} \vee \neg Y_{m-2}$ | $X_{m-3} \vee X_{m-4} \vee \neg Y_{m-3}$ |
| $X_{m-1} \vee \neg Y_{m-1} \vee \neg Y_{m-3}$ | $X_{m-2} \vee \neg Y_{m-2} \vee \neg Y_{m-3}$ |
| $X_{m-1} \vee \neg Y_{m-1} \vee \neg Y_{m-4}$ | $X_{m-2} \vee \neg Y_{m-2} \vee \neg Y_{m-4}$ |
| $X_{m-3} \vee \neg Y_{m-3} \vee \neg Y_{m-4}$ | $\neg X_{m-4} \vee X_{m-5} \vee Y_{m-4}$ |
| $\neg X_{m-4} \vee \neg Y_{m-5} \vee Y_{m-4}$ | $X_{m-1} \vee \neg X_{m-2} \vee Y_{m-1} \vee Y_{m-2}$ |
| $X_{m-2} \vee \neg X_{m-3} \vee Y_{m-2} \vee Y_{m-3}$ | $X_{m-4} \vee \neg X_{m-5} \vee Y_{m-5} \vee Y_{m-4}$ |
| $X_{m-4} \vee X_{m-5} \vee \neg Y_{m-4}$ | $X_{m-4} \vee \neg Y_{m-5} \vee \neg Y_{m-4}$ |
| $\neg X_{m-1} \vee Y_{m-1}$ | $\neg X_{m-1} \vee \neg X_{m-2} \vee \neg Y_{m-1} \vee Y_{m-2}$ |
| $\neg X_{m-2} \vee \neg X_{m-3} \vee \neg Y_{m-2} \vee Y_{m-3}$ | $X_{m-3} \vee \neg X_{m-4} \vee \neg X_{m-5} \vee Y_{m-5} \vee Y_{m-3}$ |
| $\neg X_{m-3} \vee \neg X_{m-4} \vee \neg X_{m-5} \vee Y_{m-5} \vee \neg Y_{m-3}$ | |

**Fig. 2.** 4-bit adder $\langle X_{m-1}X_{m-2}X_{m-3}X_{m-4}\rangle + (\neg Y_{m-5} \wedge X_{m-5}) = \langle Y_{m-1}Y_{m-2}Y_{m-3}Y_{m-4}\rangle$

Under log encoding, each of the position variables $P_i$ $(i \in 2..n)$ uses $log_2(n)$ Boolean variables. The number of clauses is dominated by constraint (5”), which requires $O(n \times log_2(n) \times d)$ clauses to encode, where $d$ is the maximum degree in $G$.

### 4.3 LFSR Encoding

The LFSR encoding of the successor function also employs a log-encoded domain variable $P_i$ for each vertex $i$ $(i \in 1..n)$ [23]. Given a binary number $X$, the Fibonacci LFSR determines the next binary number $Y$ by shifting the bits of $X$ one position to left and computing the lowest bit of $Y$ by applying xor on the *taps* bits of $X$. For a given length of $n$, the LFSR is able to generate all $2^n - 1$ non-zero numbers from any non-zero start number.

For example, consider the length $n = 4$ and the taps $\{2, 3\}$. Given a binary number $\langle X_3X_2X_1X_0\rangle$, the next binary number $\langle Y_3Y_2Y_1Y_0\rangle$ is calculated as follows: $Y_3 = X_2$, $Y_2 = X_1$, $Y_1 = X_0$, $Y_0 = X_2 \oplus X_3$. Assume the start number is 0001, the LFSR produces the following sequence:

$$\begin{aligned}
&0001 \to 0010 \to 0100 \to 1001 \to\\
&0011 \to 0110 \to 1101 \to 1010 \to\\
&0101 \to 1011 \to 0111 \to 1111 \to\\
&1110 \to 1100 \to 1000 \to 0001
\end{aligned}$$

[4] One may wonder why the chosen number is 4, not 3 or 5. Interestingly, a choice of 3, 5, or any other number will increase the overall number of clauses for Espresso.

The LFSR encoding is more compact than the binary adder encoding. The LFSR encoding does not use any carry variables either. For a number, in order to produce its successor, the LFSR encoding uses two clauses for each bit except the lowest bit, for which it uses 4 clauses if the number of taps is 2, and 16 clauses if the number of taps is 4.

## 5 Preprocessing

The distance encoding treats HCP as a single-agent path-finding problem. At time 1, the agent resides at vertex 1. In each step, the agent moves to a neighboring vertex. The agent cannot reach a vertex at time $t$ ($t \in 2..n$) if there are no paths of length $t-1$ from vertex 1 to the vertex. Similarly, since the agent must be back at vertex 1 at time $n+1$, the agent cannot occupy a vertex at time $t$ ($t \in 2..n$) if there are no paths of length $n-t+1$ from the vertex to vertex 1. This simple reasoning rules out impossible positions from consideration for vertices during preprocessing.

If the agent cannot occupy vertex $i$ at time $t$, then vertex $i$ cannot be mapped to position $t$. Under the unary encoding, the variable $U_{it}$ is set to 0; under binary encoding, the value $s^{t-1}(1)$ is excluded from the domain of $P_i$.

It is expensive to check if there is a path of a given length $t$ from one vertex to another if $t$ is large.[5] For long paths, the *shortest-distance heuristic* is used. For each vertex $i$ ($i \in 2..n$), if the shortest distance from vertex 1 to vertex $i$ is $t$, then the agent cannot occupy vertex $i$ at times 1, 2, ..., $t$. Similarly, if the shortest distance from vertex $i$ to vertex 1 is $t$, then the agent cannot occupy vertex $i$ at times $n-t+2$, $n-t+3$, ..., $n$.

If the graph is undirected, and the start vertex 1 has exactly two neighbors,[6] then the agent must visit one of the neighbors at time 2, and visit the other neighbor at time $n$. This means that the agent cannot occupy either neighbor at times 3, 4, ... $n-1$. This idea can be seen as a special case of the Hall's theorem [17].

## 6 Experimental Results

All the encodings described above have been implemented in the Picat compiler.[7] An experiment was conducted to compare the three encodings for the successor function on the knight's tour problem and the HCP benchmark used in the 2019 XCSP solver competition[8], using the SAT solver MapleLCMDiscChronoBT-DL-v3, the winner of the main track of the 2019 SAT Race[9]. In order to show how

[5] Let $M$ be the adjacency matrix of the graph. A naive algorithm that finds all paths of length $t$ requires computing $M^t$.

[6] The knight's tour problem belongs to this case if one of the corner squares is chosen as vertex 1.

[7] http://picat-lang.org/.

[8] http://xcsp.org/competition.

[9] http://sat-race-2019.ciirc.cvut.cz/.

the SAT solutions perform in a broader context, the experiment also included **or-tools** (version 7.6)[10], **clingon** (version 5.4.0)[11], and an incremental SAT-based approach in the comparison. For **or-tools**, a lazy clause generation CP solver, HCP is encoded as as a `circuit` constraint, and the first-fail strategy is utilized to label variables. For **clingon**, an answer-set programming system that performs search-time reachability testing [3], there are several encodings for HCP. The following ASP encoding, which has been shown to perform the best, was used in the experiment:

```
{hpath(X,Y) : link(X,Y)} = 1 :- node(X).

{hpath(X,Y) : link(X,Y)} = 1 :- node(Y).

reach(1).

reach(Y) :- reach(X),hpath(X,Y).

:- not reach(X),node(X).
```

The incremental approach treats HCP as an assignment problem. If a solution does not contain any sub-cycles, then the solution is returned as a valid solution to the original HCP. If the solution contains a sub-cycle that includes a set of vertices $S$, then the clause $\texttt{sum}(H_{i\in Sj\notin S}) > 1$ is added into the encoding to ban the sub-cycle. The incremental approach uses the same SAT solver, and restarts from scratch after each new sub-cycle elimination clause is added.

The knight's tour problem is a popular benchmark that has been utilized to evaluate solvers. The problem can be solved algorithmically in linear-time [8]. The Warnsdorff's rule [34], which always proceeds to the square from which the knight has the fewest onwards moves, is a very effective heuristic used in backtracking search. With Warnsdorff's rule, called *first-fail* principle in CP, and the reachability-checking capability during search, CP solvers are able to solve very large instances. Regarding SAT-based solvers, no eager approaches have been reported to be able to solve instances of size 30 or larger. The HCP benchmark used in the XCSP competition contains 10 instances selected from the Flinders challenge set[12] with numbers of vertices ranging from 338 to 1584.

All the CPU times reported below were measured on Linux Ubuntu with an Intel i7 3.30 GHz CPU and 32G RAM. The time limit used was 40 min per instance.

Tables 1 and 2 compare the encodings on, respectively, the number of variables and the number of clauses. For each encoding, results from two separate settings are included, one with preprocessing (pp) and the other with no preprocessing (no-pp). The results are roughly consistent with the theoretical analysis: The LFSR encoding (**lfsr**) generates the most compact code, then followed by

[10] https://developers.google.com/optimization.
[11] https://potassco.org/.
[12] http://fhcp.edu.au/fhcpcs.

the binary adder encoding (**adder**), and finally by the unary encoding (**unary**). When preprocessing is excluded, **adder** and **lfsr** use the same number of variables because both of them use log encoding for position variables. When preprocessing is included, however, **adder** uses slightly fewer variables than **lfsr**. This is because preprocessing produces holes scattered in the domains for **lfsr**, sometimes requiring more prime implicants to cover than the original domains, while preprocessing narrows bounds of the domains or produces holes concentrated in the domains for **adder**, and Picat is able to fix some of the bits at translation time.

**Table 1.** A comparison on number of variables

| Benchmark | **adder** | | **lfsr** | | **unary** | |
|---|---|---|---|---|---|---|
| | pp | no-pp | pp | no-pp | pp | no-pp |
| knight-8 | 820 | 920 | 912 | 920 | 2,958 | 5,582 |
| knight-10 | 1,428 | 1,481 | 1,479 | 1,481 | 6,883 | 12,767 |
| knight-12 | 2,375 | 2,454 | 2,442 | 2,454 | 13,596 | 25,477 |
| knight-14 | 3,217 | 3,320 | 3,310 | 3,320 | 24,032 | 46,415 |
| knight-16 | 4,994 | 5,386 | 5,376 | 5,386 | 40,385 | 77,806 |
| knight-18 | 5,972 | 6,141 | 6,131 | 6,141 | 62,109 | 121,129 |
| knight-20 | 8,065 | 8,272 | 8,266 | 8,272 | 94,974 | 182,236 |
| knight-22 | 10,073 | 10,318 | 10,308 | 10,318 | 136,612 | 263,410 |
| knight-24 | 12,152 | 12,451 | 12,443 | 12,451 | 189,984 | 376,426 |

Table 3 compares the encodings on CPU time, which includes both the translation and solving times. The column **inc** gives the time taken by the incremental approach. The entry *MO* indicates out-of-memory. Preprocessing is generally effective in reducing the time. The results of **adder** are very interesting: when preprocessing was turned off, **adder** even failed to solve size 12; with preprocessing, however, it efficiently solved all of the instances. It is also interesting to note that **lfsr** does not scale up as well as **adder**, although **lfsr** also uses log encoding for position variables, and uses slightly fewer clauses. One explanation, as shown in Table 1, could be that preprocessing helps **adder** more than **lfsr** in reducing the number of variables. The **inc** is generally not competitive; it ran out of time on two instances, and ran out of memory on another two instances. The **inc** keeps track of all the sub-cycles that have been found, and adds clauses to ban them in subsequent searches. The result indicates that **inc** is not feasible when there are a huge number of sub-cycles in the graph. The solvers **or-tools** and **clingo** are very fast on these instances; **or-tools** solved all in less than 2 s each, and **clingo** solved all in less than 1 s each.

Table 4 gives several knight's tour instances solved by the binary adder encoding. These instances are easy for **or-tools** and **clingon** to solve, but had been out of reach for eager SAT encoding approaches.

**Table 2.** A comparison on number of clauses

| Benchmark | adder | | lfsr | | unary | |
|---|---|---|---|---|---|---|
| | pp | no-pp | pp | no-pp | pp | no-pp |
| knight-8 | 11,161 | 13,939 | 8,964 | 9,284 | 17,271 | 33,623 |
| knight-10 | 22,453 | 23,535 | 15,641 | 14,831 | 42,802 | 85,836 |
| knight-12 | 39,784 | 41,647 | 39,214 | 40,163 | 92,130 | 184,775 |
| knight-14 | 55,846 | 58,272 | 56,326 | 54,044 | 173,041 | 345,322 |
| knight-16 | 81,672 | 96,232 | 73,231 | 76,329 | 306,414 | 604,730 |
| knight-18 | 112,465 | 116,527 | 89,033 | 90,088 | 486,798 | 970,960 |
| knight-20 | 147,513 | 152,184 | 120,849 | 107,883 | 746,299 | 1,499,129 |
| knight-22 | 183,121 | 188,641 | 148,018 | 114,412 | 1,101,488 | 2,213,986 |
| knight-24 | 237,071 | 244,714 | 215,516 | 231,325 | 1,563,133 | 3,102,627 |

**Table 3.** A comparison on CPU time (seconds)

| Benchmark | adder | | lfsr | | unary | | inc |
|---|---|---|---|---|---|---|---|
| | pp | no-pp | pp | no-pp | pp | no-pp | |
| knight-8 | 2.88 | 2.39 | 2.93 | 3.4 | 8.92 | 11.26 | 0.17 |
| knight-10 | 2.39 | 113.18 | 3.67 | 3.83 | 11.76 | 18.31 | 0.47 |
| knight-12 | 4.46 | >2400 | 11.12 | 81.33 | 22.15 | 42.51 | 5.16 |
| knight-14 | 4.59 | >2400 | 13.31 | 126.52 | 48.96 | 89.43 | 88.63 |
| knight-16 | 7.83 | >2400 | 30.16 | 52.50 | 92.61 | 225.47 | >2400 |
| knight-18 | 10.16 | >2400 | 35.85 | 153.99 | 137.91 | 436.13 | *MO* |
| knight-20 | 9.39 | >2400 | 505.40 | >2400 | 208.53 | 512.10 | 625.25 |
| knight-22 | 25.84 | >2400 | 1243.22 | >2400 | 358.51 | >2400 | >2400 |
| knight-24 | 9.64 | >2400 | >2400 | >2400 | >2400 | >2400 | *MO* |

Table 5 compares the solvers on CPU time using the XCSP competition instances.[13] The number in parentheses indicates the number of vertices in the graph. Preprocessing was enabled for the eager encoding approaches. Overall, **adder** performed the best. It solved all the instances, none of which took more than 250 s. The **lfsr** failed on one instance. For the solved instances, the times taken by **lfsr** are much longer than those taken by **adder**. The **unary** failed on 6, and **inc** failed on 7 instances, due to time out or memory out. While **or-tools** and **clingo** demonstrated superior performance on the knight's tour benchmark, they are not as competitive as **adder** on these instances; **or-tools** failed on 7

[13] All the participating solvers in the 2019 XSCP competition, except PicatSAT and **Choco**, failed on every single instance. PicatSAT, which is based on an early version of **adder**, solved all of the 10 instances, the sequential version of **Choco** solved 4 instances, and the parallel version of **Choco** solved 7 instances.

**Table 4.** Knight's tour instances solved by the binary adder encoding (seconds)

| size | 26 | 28 | 30 | 32 | 34 | 36 | 38 |
|---|---|---|---|---|---|---|---|
| time | 57.45 | 83.04 | 62.90 | 346.57 | 188.00 | 310.09 | 304.13 |

instances, and **clingo** failed on 1 instance and took more than 500 s to solve two of the instances each.

**Table 5.** XCSP competition instances (CPU time)

| benchmark | adder | lfsr | unary | inc | or-tools | clingo |
|---|---|---|---|---|---|---|
| graph162 (909) | 184.18 | 676.91 | *MO* | 21.98 | 49.62 | 16.94 |
| graph171 (996) | 27.25 | 51.58 | 1887.64 | >2400 | >2400 | 52.50 |
| graph197 (1188) | 64.18 | 270.74 | >2400 | >2400 | >2400 | 1821.98 |
| graph223 (1386) | 68.16 | 144.20 | 1279.99 | >2400 | >2400 | 17.92 |
| graph237 (1476) | 91.6 | 388.25 | >2400 | >2400 | >2400 | 23.78 |
| graph249 (1558) | 52.88 | 112.60 | 494.45 | 336.21 | 307.30 | 13.47 |
| graph252 (1572) | 98.22 | 403.25 | >2400 | >2400 | >2400 | 541.48 |
| graph254 (1582) | 63.65 | 268.23 | 541.97 | 168.32 | >2400 | 12.7 |
| graph255 (1584) | 45.86 | 144.83 | >2400 | >2400 | 171.0 | >2400 |
| graph48 (338) | 213.0 | >2400 | >2400 | *MO* | >2400 | 0.97 |

For comparison, the same experiment was also conducted using CaDiCaL, the second-place winner in the 2019 SAT Race. With preprocessing, **adder** using CaDiCaL also solved all the instances in Tables 3 and 5, while **lfsr** failed on knight-24 and graph48, and **unary** failed on 4 knight's tour instances and 6 XCSP instances.

## 7 Related Work

Various approaches have been proposed for HCP [15]. As HCP is a special variant of the Traveling Salesman Problem (TSP), many approaches proposed for TSP [9,16] can be tailored to HCP.

Recently several studies have used SAT solvers for HCP. A common technique utilized in encoding HCP into SAT in order to prevent sub-cycles is to impose a strict ordering on the vertices. The *bijection* encoding [19] uses an *edge* constraint for each non-arc pair $(i, j)$ that bans vertex $j$ from immediately following vertex $i$ in the ordering. This encoding is compact for dense graphs. The *relative* encoding [35] imposes transitivity on the ordering: if vertex $i$ reaches vertex $k$, and vertex $k$ reaches vertex $j$, then vertex $i$ reaches vertex $j$. The *reachability* encoding, which is used in translating answer-set programs with loops into SAT [28], also imposes transitivity on the ordering. All these encodings use direct encoding for positions, and require $O(n^3)$ clauses in the worst case. It is reported in [41] that using a hierarchical encoding for domain variables significantly reduces the encoding size and increases the solving speed for HCP. However, hierarchical encoding still suffers from code explosion for large graphs.

The distance encoding for HCP is not new. It is based on the standard decomposer used in MiniZinc [31], which uses an order variable $O_i$ for each vertex $i$, and ensures that if $V_i = j$ then $O_j = O_i + 1$. The idea of using order or position variables could be traced back to the integer programming formulation that uses dummy variables to prevent sub-cycles [30].

The log encoding [21] resembles the binary representation of numbers used in computer hardware. Despite its compactness, log encoding is not popular due to its poor propagation strengths [26]. Johnson first came up with the idea of using log encoding for position variables and the LFSR for encoding the successor function [23]. The binary adder encoding for $Y = X + 1$ proposed in this paper is a special optimized incrementor that does not use any carry variables.

The preprocessing technique for excluding unreachable positions from the domains of position variables is well-used in constraint programming. Similar techniques have been used for maintaining consistency of some of the global constraints, such as the *regular* constraint [32], and for eliminating variables in multi-agent path finding [1]. This work has shown, for the first time, that when preprocessing is effective the binary adder encoding of the successor function significantly outperforms the unary and LFSR encodings for HCP.

In order to circumvent the explosive encoding sizes of eager approaches, researchers have proposed lazy approaches, such as satisfiability modulo acyclicity [3] and incremental SAT solving [38] for HCP. The idea to incrementally add constraints to avoid code explosion is the pillar of the cutting-plane method [9,10]. The incremental approach may suffer if the problems require repeated addition of sub-cycle elimination clauses.

## 8 Conclusion

A central issue in encoding HCP into SAT is how to prevent sub-cycles, and one well-used technique is to map vertices to different positions. This paper has compared three encodings for the successor function used in the distance encoding of HCP, and proposed a preprocessing technique that rules out unreachable positions from consideration. Our study has surprisingly revealed that, with preprocessing and optimization, the binary adder encoding outperforms the unary and the LFSR encodings. While no eager SAT encoding approaches have been reported to be able to solve size 30 or larger of the knight's tour problem, the binary adder encoding, using the SAT solver MapleLCMDiscChronoBT-DL-v3, succeeded in solving all instances up to size 38 in less than 6 min each. This is a remarkable advancement of the state of the art. While there is still a long way to go for eager SAT encoding approaches to be competitive with CP and ASP solvers on the knight's tour problem, this paper has showed that the binary adder encoding is competitive with the best CP and ASP solvers on the HCP benchmark used in the 2019 XCSP competition.

An efficient SAT encoding for HCP will expand the successes of SAT solvers in solving combinatorial problems, such as the travelling salesman problem (TSP), which is a generalization of HCP, and its variants. Further improvements include exploiting special graph structures and symmetry-breaking techniques in SAT encodings.

**Acknowledgement.** The author would like to thank Håkan Kjellerstrand for helping test and tune Picat's SAT compiler, Marijn Heule for pointing out Andrew Johnson's work on the LFSR encoding, Andrew Johnson for clarifications on his LFSR encoding, and the anonymous reviewers for helpful comments. This work is supported in part by the NSF under the grant number CCF1618046.

## References

1. Barták, R., Zhou, N.F., Stern, R., Boyarski, E., Surynek, P.: Modeling and solving the multi-agent pathfinding problem in Picat. In: 29th IEEE International Conference on Tools with Artificial Intelligence, pp. 959–966 (2017)
2. Biere, A., Heule, M., van Maaren, H., Toby, W.: Handbook of Satisfiability. IOS Press, Amsterdam (2009)
3. Bomanson, J., Gebser, M., Janhunen, T., Kaufmann, B., Schaub, T.: Answer set programming modulo acyclicity. In: Logic Programming and Nonmonotonic Reasoning (LPNMR), pp. 143–150 (2015)
4. Bordeaux, L., Hamadi, Y., Zhang, L.: Propositional satisfiability and constraint programming: a comparative survey. ACM Comput. Surv. **38**(4), 1–54 (2006)
5. Brayton, R.K., Hachtel, G.D., McMullen, C., Sangiovanni-Vincentelli, A.: Logic Minimization Algorithms for VLSI Synthesis. Kluwer Academic Publishers, Amsterdam (1984)
6. Brewka, G., Eiter, T., Truszczyński, M.: Answer set programming at a glance. Commun. ACM **54**(12), 92–103 (2011)
7. Chen, J.: A new SAT encoding of the at-most-one constraint. In: Proceedings of the International, Workshop of Constraint Modeling and Reformulation (2010)
8. Conrad, A., Hindrichs, T., Morsy, H., Wegener, I.: Solution of the knight's Hamiltonian path problem on chessboards. Disc. Appl. Math. **50**(2), 125–134 (1994)
9. Cook, W.J.: In Pursuit of the Traveling Salesman: Mathematics at the Limits of Computation. Princeton University Press, Princeton (2012)
10. Dantzig, G.B., Fulkerson, R., Johnson, S.: Solution of a large-scale traveling-salesman problem. Oper. Res. **2**, 393–410 (1954)
11. de Kleer, J.: A comparison of ATMS and CSP techniques. In: IJCAI, pp. 290–296 (1989)
12. Garey, M.R., Johnson, D.S.: Computers and Intractability. W.H. Freeman and Co., New York City (1979)
13. Gavanelli, M.: The log-support encoding of CSP into SAT. In: Bessière, C. (ed.) CP 2007. LNCS, vol. 4741, pp. 815–822. Springer, Heidelberg (2007). https://doi.org/10.1007/978-3-540-74970-7_59
14. Gebser, M., Kaufmann, B., Neumann, A., Schaub, T.: Conflict-driven answer set solving. In: IJCAI, pp. 386-392 (2007)

15. Gould, R.J.: Recent advances on the Hamiltonian problem: survey III. Graphs Comb. **30**(1), 1–46 (2014)
16. Gutin, G., Punnen, A.P.: The Traveling Salesman Problem and Its Variations. Combinatorial Optimization. Springer, Heidelberg (2007). https://doi.org/10.1007/b101971
17. Hall, P.: Representatives of subsets. J. London Math. Soc. **10**(1), 26–30 (1935)
18. Haythorpe, M., Johnson, A.: Change ringing and Hamiltonian cycles: The search for Erin and Stedman triples. EJGTA **7**(1), 61–75 (2019)
19. Hertel, A., Hertel, P., Urquhart, A.: Formalizing dangerous SAT encodings. In: Marques-Silva, J., Sakallah, K.A. (eds.) SAT 2007. LNCS, vol. 4501, pp. 159–172. Springer, Heidelberg (2007). https://doi.org/10.1007/978-3-540-72788-0_18
20. Huang, J.: Universal booleanization of constraint models. In: Stuckey, P.J. (ed.) CP 2008. LNCS, vol. 5202, pp. 144–158. Springer, Heidelberg (2008). https://doi.org/10.1007/978-3-540-85958-1_10
21. Iwama, K., Miyazaki, S.: SAT-variable complexity of hard combinatorial problems. IFIP Congr. **1**, 253–258 (1994)
22. Jeavons, P., Petke, J.: Local consistency and SAT-solvers. JAIR **43**, 329–351 (2012)
23. Johnson, A.: Quasi-linear reduction of Hamiltonian cycle problem (HCP) to satisfiability problem (SAT), 2014. Disclosure Number IPCOM000237123D, IP.com, Fairport, NY, June 2014. https://priorart.ip.com/IPCOM/000237123
24. Kautz, H.A., Selman, B.: Planning as satisfiability. In: ECAI, pp. 359–363 (1992)
25. Klieber, W., Kwon, G.: Efficient CNF encoding for selecting 1 from n objects. In: The Fourth Workshop on Constraints in Formal Verification (CFV) (2007)
26. Knuth, D.: The Art of Computer Programming, Volume 4, Fascicle 6: Satisfiability. Addison-Wesley (2015)
27. Lamport, L.: Specifying Systems: The TLA+ Language and Tools for Hardware and Software Engineers. Addison-Wesley, Boston (2004)
28. Lin, F., Zhao, J.: On tight logic programs and yet another translation from normal logic programs to propositional logic. In: IJCAI, pp. 853–858 (2003)
29. McCluskey, E.J.: Minimization of Boolean functions. Bell Syst. Tech. J. **35**(6), 1417–1444 (1956)
30. Miller, C.E., Tucker, A.W., Zemlin, R.A.: Integer programming formulation of traveling salesman problems. J. ACM **7**(4), 326–329 (1960)
31. Nethercote, N., Stuckey, P.J., Becket, R., Brand, S., Duck, G.J., Tack, G.: MiniZinc: towards a standard CP modelling language. In: Bessière, C. (ed.) CP 2007. LNCS, vol. 4741, pp. 529–543. Springer, Heidelberg (2007). https://doi.org/10.1007/978-3-540-74970-7_38
32. Pesant, G.: A regular language membership constraint for finite sequences of variables. In: Wallace, M. (ed.) CP 2004. LNCS, vol. 3258, pp. 482–495. Springer, Heidelberg (2004). https://doi.org/10.1007/978-3-540-30201-8_36
33. Petke, J.: Bridging Constraint Satisfaction and Boolean Satisfiability. AIFTA. Springer, Cham (2015). https://doi.org/10.1007/978-3-319-21810-6
34. Pohl, I.: A method for finding Hamilton paths and knight's tours. Commun. ACM **10**, 446–449 (1967)
35. Prestwich, S.D.: SAT problems with chains of dependent variables. Disc. Appl. Math. **130**(2), 329–350 (2003)
36. Quine, W.V.O., Quine, W.V.: The problem of simplifying truth functions. Am. Math. Mon. **59**(8), 521–531 (1952)
37. Rintanen, J.: Planning as satisfiability: heuristics. Artif. Intell. **193**, 45–86 (2012)

38. Soh, T., Le Berre, D., Roussel, S., Banbara, M., Tamura, N.: Incremental SAT-based method with Native boolean cardinality handling for the Hamiltonian cycle problem. In: Fermé, E., Leite, J. (eds.) JELIA 2014. LNCS (LNAI), vol. 8761, pp. 684–693. Springer, Cham (2014). https://doi.org/10.1007/978-3-319-11558-0_52
39. Stojadinović, M., Marić, F.: meSAT: multiple encodings of CSP to SAT. Constraints **19**(4), 380–403 (2014). https://doi.org/10.1007/s10601-014-9165-7
40. Tamura, N., Taga, A., Kitagawa, S., Banbara, M.: Compiling finite linear CSP into SAT. Constraints **14**(2), 254–272 (2009)
41. Velev, M.N., Gao, P.: Efficient SAT techniques for absolute encoding of permutation problems: application to Hamiltonian cycles. In: Eighth Symposium on Abstraction, Reformulation, and Approximation (SARA) (2009)
42. Zhou, N.F.: Yet another comparison of SAT encodings for the at-most-k constraint. ArXiv, abs/2005.06274 (2020)
43. Zhou, N.-F., Kjellerstrand, H.: Optimizing SAT encodings for arithmetic constraints. In: Beck, J.C. (ed.) CP 2017. LNCS, vol. 10416, pp. 671–686. Springer, Cham (2017). https://doi.org/10.1007/978-3-319-66158-2_43

# Large Neighborhood Search for Temperature Control with Demand Response

Edward Lam[1,2](✉), Frits de Nijs[1], Peter J. Stuckey[1], Donald Azuatalam[1], and Ariel Liebman[1]

[1] Monash University, Melbourne, Australia
{edward.lam,frits.nijs,peter.stuckey, donald.azuatalam,ariel.liebman}@monash.edu
[2] CSIRO Data61, Melbourne, Australia

**Abstract.** Demand response is a control problem that optimizes the operation of electrical loads subject to limits on power consumption during times of low power supply or extreme power demand. This paper studies the demand response problem for centrally controlling the space conditioning systems of several buildings connected to a microgrid. The paper develops a mixed integer quadratic programming model that encodes trained deep neural networks that approximate the temperature transition functions. The model is solved using standard branch-and-bound and a large neighborhood search within a mathematical programming solver and a constraint programming solver. Empirical results demonstrate that the large neighborhood search coupled to a constraint programming solver scales substantially better than the other methods.

**Keywords:** Sustainability · Energy systems · Power systems · Control · Smart grid · Microgrid · Large neighborhood search · Local search

## 1 Introduction

Electricity utilities are required to ensure supply-demand balance throughout the power grid since any mismatch can cause voltage or system frequency instability, resulting in loss of supply or system blackouts. Supply-demand balance will become increasingly difficult with increased investment in uncontrollable renewable generations such as solar and wind. To more effectively and cost-efficiently safeguard system reliability, utilities are increasingly deploying demand response to cater for unforeseen or difficult circumstances in their network, such as extreme weather, by motivating customers to decrease demand at certain times or to shift their consumption to off-peak periods.

Figure 1 shows four example power profiles of a building over a typical day. Normal power consumption, peaking at nearly 8 kW, is shown in black. As appliances and the building's insulation receive upgrades in energy efficiency, the load


H. Simonis (Ed.): CP 2020, LNCS 12333, pp. 603–619, 2020.
https://doi.org/10.1007/978-3-030-58475-7_35

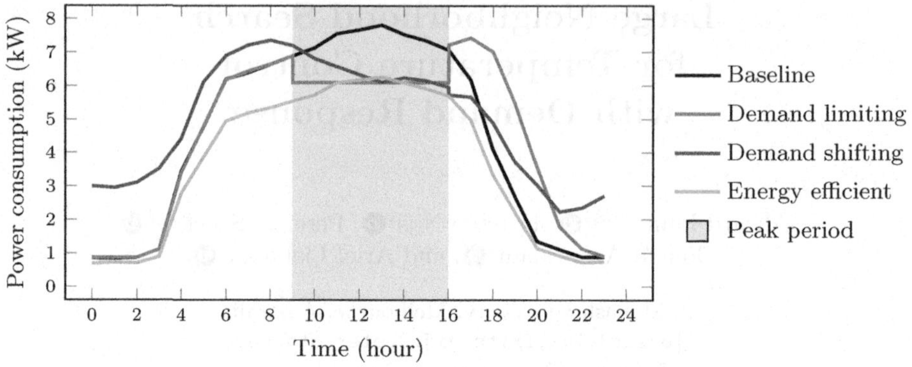


**Fig. 1.** Load shapes of a building, with and without demand response [13]. (Color figure online)

profile is expected to shift from the black line to the green line, which peaks at just over 6 kW. In the meantime, utilities promote grid reliability and stability by allowing building management to either curtail power use (shown in red) or shift power usage to earlier parts of the day (shown in blue) when the temperature is cooler; preserving power for heating, ventilation and air conditioning (HVAC) systems in the warmer parts of the day.

Utilities target customers with large controllable devices with flexible usage for participation in demand response. Examples of such devices include thermostatically-controlled loads (e.g., HVAC systems and electric water heaters) and shiftable appliances (e.g., dishwashers, washing machines and dryers). Since space conditioning accounts for a large proportion of building energy use, especially in commercial buildings, HVAC loads are excellent candidates for demand response. Moreover, slight temperature changes do not immediately impact occupant comfort because of the thermal storage effect of buildings [13], a feature that can be further enhanced through building upgrades such as hot water storage tanks and phase-change materials [3].

Even though a manual approach to demand response can be tediously implemented, HVAC systems in modern commercial buildings can be controlled by automation systems. This paper develops strategies to jointly control the HVAC systems of several buildings connected to a microgrid such that their temperatures are maintained within acceptable comfort bounds and all power restrictions are adhered to. We make use of a deep neural network to automatically learn a transition function for each of the buildings under control, as a proxy for the behavior governed by physical laws. This paper shows how these neural network transition functions can be encoded in a mixed integer quadratic programming (MIQP) model that decides on the operating mode of the HVAC system in all connected buildings within a given planning horizon.

The model is implemented in the constraint programming (CP) solver Gecode and the mathematical programming (MP) solver Gurobi. A large neighborhood search (LNS) is also implemented in the two exact solvers for finding local improvements. Empirical results show that the best solutions are found using LNS in Gecode. The remainder of the paper explores these results in detail.

## 2 Related Work

HVAC systems account for approximately 50% of total building energy consumption [15]. Their large energy footprint, combined with their flexibility due to thermal inertia, has resulted in significant efforts in developing methods to unlock their potential for demand response, including multiple field tests [12]. Unfortunately, these methods are largely market-based, controlling the HVACs through price signals rather than computing a globally optimal joint schedule. Given the true model dynamics for each building, the globally optimal joint schedule can be approximated efficiently by Lagrangian relaxation methods such as column generation [14]. However, the use of relaxations means that constraints are not guaranteed to be satisfied, unless schedules are corrected by re-planning in an on-line fashion. This requires that the agents are in constant communication, which is a potential system vulnerability. The optimization methods proposed in this paper avoid this issue by computing feasible schedules a priori.

When accurate models are not available, one solution is to directly learn a controller by interacting with the system, through reinforcement learning. This approach has seen wide application to demand-response problems [20]. However, the vast majority of these approaches apply only to single-agent problems; learning to control the joint dynamics directly is highly intractable due to the curse of dimensionality on the exponential growth of the state and action space, while multi-agent reinforcement learning does not have a clear pathway to impose global constraints. In addition, reinforcement learning requires the reward function coefficients to be fixed a priori, so changes in a user's comfort preferences necessitate learning a new solution. To avoid these challenges, we propose separating learning the transition function from planning the HVAC schedules.

We adapt a variety of techniques in the combinatorial optimization literature to the demand response problem. The use of artificial neural networks (ANNs) as approximations of complex processes in optimization models has been championed in [4]. They develop a Neuron global constraint and bounds-consistent filtering algorithms, which are evaluated in controlling the temperature of computer chips under various workloads. In [2], an ANN is encoded in a mixed integer linear programming model using facet-defining (i.e., tightest possible) constraints. However, they show that the new constraints are outperformed by a simple big-$M$ encoding in practice.

A learned transition function in the form of a ReLU-ANN can be encoded directly as a mixed-integer linear program [18]. In their work, the authors develop valid inequalities that allow sparsifying the encoding of an ANN, resulting in significant speed-up. Among other domains, they apply their approach to a multi-zone HVAC control problem, with an objective to minimize cost of keeping occupants comfortable. Due to the complexity of the mixed-integer linear programming, only relatively small instances with short horizons can be solved optimally, requiring on-line planning. Compared to their work, we consider demand-response in a larger multi-building setting, which we show to be intractable to solve directly, requiring the use of our local search procedures to solve in reasonable time.

## 3 The Problem

This section describes the problem and then models it using MIQP.

### 3.1 Motivation

Monash University has committed to achieving net zero emissions by 2030. It is investing AUD$135 million to increase the energy efficiency of its operations, electrify its buildings, and transition to renewable electricity through on-site solar generation and off-site power purchase agreements. The project aims to develop solutions to the university's operations, and to serve as a testbed for scientific and engineering studies by integrating its education, research and industry activities with its built environment. This approach has been recognized globally, having won the United Nation's Momentum for Change Award in 2018.

One component of the project is to redevelop a portion of the main university campus at Clayton, Victoria, Australia into an energy-efficient microgrid. The project has seen the installation of distributed energy resources including medium-scale solar photovoltaic generation, precinct-scale batteries and electric vehicles that can supply the microgrid while stationary. Buildings with varied usage characteristics, including commercial buildings and student residences, are also refurbished for improved energy efficiency and their HVAC systems, if not entirely deprecated by gains in energy efficiency, are upgraded for autonomous control by the microgrid.

The purpose of this paper is to study merely one small portion of a real-world microgrid project. This paper considers a simplified problem abstracted from the problem of jointly controlling the internal air temperature of several buildings subject to occupant comfort and demand response. For simplicity, every building is assumed to have one independent HVAC system that influences its future indoor temperature by operating in one of three modes: off, cooling or heating. However, the approach naturally generalizes to multiple control zones within a building.

### 3.2 The State Transition Function

The change in temperature from one timestep to the next (i.e., the state transition function) is governed by physical laws. The future indoor temperatures are a function of the current indoor and outdoor air temperature, the internal mass temperature (a measure from the occupants and furniture), the internal humidity, the building insulation, the solar irradiance as well as many other factors. Ideally, the parameters of the physical laws are estimated from data collected in the environment, and then a physical model is encoded into an optimization model. However, such physical models are typically non-linear, making them hard to encode and solve, and high-accuracy simulation is computationally expensive. Furthermore, estimating the parameters from data already entails the use of machine learning. Therefore, this paper argues for a data-driven approach to directly estimate the entire transition function using a deep neural network, rather than estimating the building parameters in a finite difference discretization of the governing differential equations.

The transition function is approximated by a multi-layer perceptron (MLP), also called a vanilla feedforward neural network. MLPs are a basic workhorse in machine learning. This section briefly review MLPs. For a formal treatment, readers are recommended to consult the definitive textbook [10].

An MLP $F_{\boldsymbol{\theta}} : \mathbb{R}^n \to \mathbb{R}^m$ with parameters $\boldsymbol{\theta} = (\theta_1, \ldots, \theta_k)$ for some $k \in \mathbb{N}$ is a function that maps a real-valued $n$-dimensional vector input to a real-valued $m$-dimensional vector output. Consider a data set $\mathcal{D}$ consisting of pairs of real-valued vectors $(\mathbf{x}, \mathbf{y}) \in \mathbb{R}^n \times \mathbb{R}^m$. In supervised learning, *training* $F_{\boldsymbol{\theta}}$ is the process of finding a good estimate of $\boldsymbol{\theta}$ such that the predicted output $F_{\boldsymbol{\theta}}(\mathbf{x})$ is reasonably close to the actual output $\mathbf{y}$ for the entire data set (e.g., by minimizing the sum of mean squared errors over $\boldsymbol{\theta}$). After training (i.e., $\boldsymbol{\theta}$ is chosen and fixed), *prediction* refers to the process of evaluating $F_{\boldsymbol{\theta}}(\mathbf{x})$ on any arbitrary input vector $\mathbf{x}$ in the hope that $F_{\boldsymbol{\theta}}$ generalizes to the unknown $\mathbf{y}$ corresponding to $\mathbf{x}$.

An MLP $F_{\boldsymbol{\theta}}$ consists of $L \geq 2$ *layers*. Let $\mathcal{L} = \{1, \ldots, L\}$ be the set of layers. Layer 1 is the *input layer* and layer $L$ is the *output layer*. The intermediate layers are *hidden layers*. If there are two or more hidden layers, $F_{\boldsymbol{\theta}}$ is described as *deep*.

Each layer consists of *units*, also known as *neurons*. Let $\mathcal{U}_l = \{1, \ldots, U_l\}$ be the set of units in layer $l \in \mathcal{L}$, where $U_l \geq 1$ is the number of units in layer $l$. The input layer has $n$ units (i.e., $U_1 = n$) and the output layer has $m$ units (i.e., $U_L = m$). Each unit $u$ in layer $l$ represents a function $f_{l,u}$. The units in the input layer (i.e., $l = 1$) represents the input vector $\mathbf{x}$ to $F_{\boldsymbol{\theta}}$. Formally, unit $u \in \mathcal{U}_1$ represents the identity function $f_{1,u} : \mathbb{R}^n \to \mathbb{R}$ of the $u$th component of $\mathbf{x} = (x_1, \ldots, x_n)$:

$$f^{1,u}(\mathbf{x}) = x_u.$$

In a *fully-connected* MLP, every unit $u \in \mathcal{U}_l$ in layer $l > 1$ represents a function $f^{l,u}_{\mathbf{W}_{l,u}, B_{l,u}} : \mathbb{R}^{U_{l-1}} \to \mathbb{R}$ with parameters $\mathbf{W}_{l,u} \in \mathbb{R}^{U_{l-1}}$ and $B_{l,u} \in \mathbb{R}$ that maps the outputs of the units in the previous layer $l - 1$ to one real number:

$$f^{l,u}_{\mathbf{W}_{l,u}, B_{l,u}}(\mathbf{x}) = \sigma^{l,u}(\mathbf{W}_{l,u} \cdot \mathbf{x} + B_{l,u}).$$

The vector $\mathbf{W}_{l,u}$ is the *weights* of $u$, and the scalar $B_{l,u}$ is the *bias* of $u$. The function $\sigma^{l,u} : \mathbb{R} \to \mathbb{R}$ is the *activation function* of $u$. Common activation functions include the sigmoid function $\sigma(x) := \frac{1}{1+e^{-x}}$ and the rectified linear unit (ReLU) $\sigma(x) := \max(x, 0)$. In recent times, ReLU has displaced the sigmoid activation function [10]. For this reason, this study focuses on ReLU.

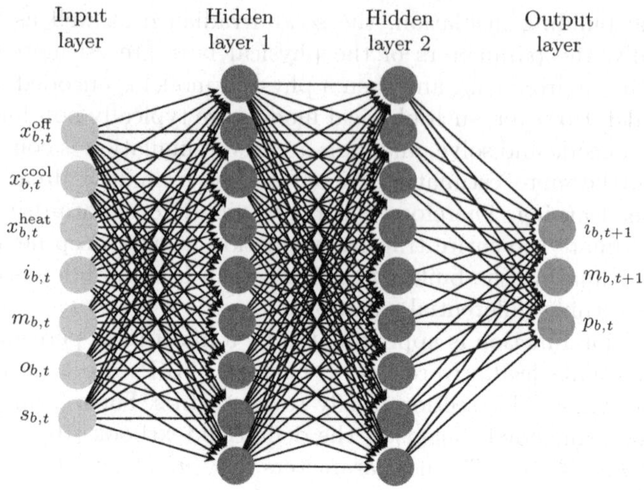


**Fig. 2.** The architecture of the MLP used to approximate the transition function of each building $b \in \mathcal{B}$.

Let $\boldsymbol{\theta} = (\mathbf{W}_{2,1}, B_{2,1}, \ldots, \mathbf{W}_{2,U_2}, B_{2,U_2}, \ldots, \mathbf{W}_{L,1}, B_{L,1}, \ldots, \mathbf{W}_{L,U_L}, B_{L,U_L})$ be the concatenation of the parameters of all units. Then, the MLP $F_{\boldsymbol{\theta}}$ is a function composition of all units through its layers, as described above.

It is well-known that MLPs with at least one hidden layer and a sufficiently large number of units is a universal function approximator under reasonable assumptions (e.g., [7]). In simpler terms, MLPs can approximate any arbitrary transition function given enough units and appropriate activation functions. For this reason, we use an MLP as the transition function.

Let $\mathcal{T} = \{0, \ldots, T\}$ be the time periods and $\mathcal{T}' = \{0, \ldots, T-1\}$ be the planning periods in which decisions are made. Consider a set $\mathcal{B}$ of buildings. Every building $b \in \mathcal{B}$ is associated with an MLP illustrated in Fig. 2. It takes the following inputs from the building's environment at timestep $t \in \mathcal{T}'$:

1. An indicator $x^{\text{off}}_{b,t} \in \{0,1\}$ of whether the HVAC system is off.
2. An indicator $x^{\text{cool}}_{b,t} \in \{0,1\}$ of whether the HVAC system is cooling.
3. An indicator $x^{\text{heat}}_{b,t} \in \{0,1\}$ of whether the HVAC system is heating.
4. The indoor air temperature $i_{b,t} \in [0,1]$ scaled to between 0 and 1.
5. The indoor mass temperature $m_{b,t} \in [0,1]$ scaled to between 0 and 1.
6. The outdoor air temperature $o_{b,t} \in [0,1]$ scaled to between 0 and 1.
7. The solar irradiance $s_{b,t} \in [0,1]$ scaled to between 0 and 1.

These seven inputs are fed through several fully-connected hidden layers with ReLU activation functions. The output layer has three units:

1. The indoor air temperature $i_{b,t+1} \in [0,1]$ at the next timestep.
2. The indoor mass temperature $m_{b,t+1} \in [0,1]$ at the next timestep.
3. The power usage $p_{b,t} \in [0,1]$ for running the mode given in the input layer.

Unusually, the output units also use the ReLU activation function to ensure non-negativity. (We observed minuscule negative power usage otherwise.) All inputs and outputs are normalized to values between 0 and 1 for numerical reasons.

### 3.3 The Mixed Integer Quadratic Programming Model

MIQP generalizes mixed integer linear programming by allowing quadratic terms in the objective function. The MIQP model of the problem is shown in Fig. 3. The model minimizes power consumption and uses soft constraints to penalize the difference from the indoor air temperatures to a fixed comfortable temperature. The problem contains two classes of (hard) constraints: (1) constraints that encode the state transition functions, and (2) constraints that limit the total power usage during a subset of timesteps for demand response.

This deterministic model assumes that the buildings' environments (i.e., the outdoor air temperature and solar irradiance) and the power supply constraints are known in advance with certainty. These assumptions are acceptable in practice because 24- h weather forecasts are adequately accurate, and power supply limitations are ordered by utilities in advance because of extreme heat forecasts (e.g., higher than 40 °C). Furthermore, since real-time control problems are solved repeatedly throughout the day, large errors (e.g., drifts) are mitigated in the next run when more accurate data is available.

The model declares three primary decision variables $x_{b,t}^{\text{off}}, x_{b,t}^{\text{cool}}, x_{b,t}^{\text{heat}} \in \{0,1\}$ to respectively indicate whether the HVAC system is off, cooling or heating in building $b \in \mathcal{B}$ during time $t \in \mathcal{T}'$. Define a decision variable $p_{b,t} \in [0,1]$ for the power usage due to operating in the mode indicated by $x_{b,t}^{\text{off}}$, $x_{b,t}^{\text{cool}}$ or $x_{b,t}^{\text{heat}}$. Let $i_{b,t}, m_{b,t}, o_{b,t}, s_{b,t} \in [0,1]$ be the indoor air temperature, indoor mass temperature, outdoor air temperature and solar irradiance of building $b \in \mathcal{B}$ in timestep $t \in \mathcal{T}$. As the model is deterministic, the outdoor temperature $o_{b,t}$ and the solar irradiance $s_{b,t}$ for all timesteps, and the initial conditions $i_{b,0}$ and $m_{b,0}$ are known constants. The remaining $i_{b,t}$ and $m_{b,t}$ (i.e., where $t > 0$) are decision variables.

Every building $b$ has a copy of its MLP transition function for every timestep; totaling $T$ copies for every building. Let $\mathcal{L}_b = \{1, \ldots, L_b\}$ denote the layers of the transition function of $b$, and $\mathcal{U}_b^l = \{1, \ldots, U_b^l\}$ the units in layer $l \in \mathcal{L}_b$. For every $b \in \mathcal{B}$ and $t \in \mathcal{T}'$, denote the output value of unit $u \in \mathcal{U}_b^l$ in layer $l \in \mathcal{L}_b$ as $f_{b,t}^{l,u} \in [0,1]$. Using the transition function, all $f_{b,t}^{l,u}$ (and hence $i_{b,t+1}$, $m_{b,t+1}$ and $p_{b,t}$) are functionally-defined by the seven inputs $x_{b,t}^{\text{off}}$, $x_{b,t}^{\text{cool}}$, $x_{b,t}^{\text{heat}}$, $i_{b,t}$, $m_{b,t}$, $o_{b,t}$ and $s_{b,t}$. Therefore, search is only required on the $x_{b,t}^{\text{off}}$, $x_{b,t}^{\text{cool}}$ and $x_{b,t}^{\text{heat}}$ variables.

Let $c_p \in \mathbb{R}_+$ be the cost of power, scaled appropriately. Define $\mathcal{R} \subset \mathcal{T}$ as the set of timesteps during which the building temperature must be maintained close to an ideal comfortable temperature $i_0 \in [0,1]$. Any deviation away from $i_0$ is quadratically penalized by a cost $c_i \in \mathbb{R}_+$. Define $\mathcal{Q} \subset \mathcal{T}' \times [0,1]$ as the set of demand response events, which are pairs of a timestep and the amount of power available during that particular timestep.

$$\min \sum_{b \in \mathcal{B}} \sum_{t \in \mathcal{T}'} c_p p_{b,t} + \sum_{b \in \mathcal{B}} \sum_{t \in \mathcal{R}} c_i (i_{b,t} - i_0)^2 \tag{1a}$$

subject to

$$(f_{b,t}^{1,1}, \ldots, f_{b,t}^{1,7}) = (x_{b,t}^{\text{off}}, x_{b,t}^{\text{cool}}, x_{b,t}^{\text{heat}}, i_{b,t}, m_{b,t}, o_{b,t}, s_{b,t}) \qquad \forall b \in \mathcal{B}, t \in \mathcal{T}', \tag{1b}$$

$$f_{b,t}^{l,u} = \max(\mathbf{W}_{b,t}^{l,u} \cdot (f_{b,t}^{l-1,1}, \ldots, f_{b,t}^{l-1,U_{l-1}}) + B_{b,t}^{l,u}, 0)$$
$$\forall b \in \mathcal{B}, t \in \mathcal{T}', l \in \mathcal{L}_b \setminus \{1\}, u \in \mathcal{U}_b^l, \tag{1c}$$

$$(i_{b,t+1}, m_{b,t+1}, p_{b,t}) = (f_{b,t}^{L_b,1}, f_{b,t}^{L_b,2}, f_{b,t}^{L_b,3}) \qquad \forall b \in \mathcal{B}, t \in \mathcal{T}', \tag{1d}$$

$$x_{b,t}^{\text{off}} + x_{b,t}^{\text{cool}} + x_{b,t}^{\text{heat}} = 1 \qquad \forall b \in \mathcal{B}, t \in \mathcal{T}', \tag{1e}$$

$$\sum_{b \in \mathcal{B}} p_{b,t} \leq q \qquad \forall (t, q) \in \mathcal{Q}, \tag{1f}$$

$$x_{b,t}^{\text{off}}, x_{b,t}^{\text{cool}}, x_{b,t}^{\text{heat}} \in \{0, 1\} \qquad \forall b \in \mathcal{B}, t \in \mathcal{T}', \tag{1g}$$

$$i_{b,t} \in [0, 1] \qquad \forall b \in \mathcal{B}, t \in \{1, \ldots, T\}, \tag{1h}$$

$$m_{b,t} \in [0, 1] \qquad \forall b \in \mathcal{B}, t \in \{1, \ldots, T\}, \tag{1i}$$

$$p_{b,t} \in [0, 1] \qquad \forall b \in \mathcal{B}, t \in \mathcal{T}', \tag{1j}$$

$$f_{b,t}^{l,u} \in [0, 1] \qquad \forall b \in \mathcal{B}, t \in \mathcal{T}, l \in \mathcal{L}_b, u \in \mathcal{U}_b^l. \tag{1k}$$

**Fig. 3.** The MIQP model.

The first summation in Objective Function (1a) represents the cost of powering the HVAC systems, and the second penalizes deviations from the ideal temperature. Using a quadratic term rather than the absolute value function ensures that a large deviation in one timestep and a small deviation in another timestep is worse than two medium-sized deviations, for example.

Constraint (1b) equates the input layer of the MLPs to the environment of the buildings. Constraint (1c) makes predictions using the MLPs. This constraint is written using the max function but can be linearized using a binary variable [9]. Constraint (1d) equates the output layers to the temperatures in the next timestep and the power usages in the current timestep. Constraint (1e) enforces exactly one mode for each building and timestep. Constraint (1f) couples the buildings together by limiting the total power consumption across all buildings according to the demand response events. Constraints (1g) to (1k) are the domains of the decision variables. (The implementation omits Constraints (1b) and (1d) and uses the MLP inputs and outputs directly.)

## 4 Search Heuristics

This section describes the branching rules and the LNS.

### 4.1 The Variable and Value Selection Strategy

MP solvers and lazy clause generation CP solvers are able to derive good branching rules dynamically by collecting statistics during the search (e.g., [1,8]). However, in classical finite-domain CP solvers, a problem-specific search procedure may be required to be successful. This section presents one such branching rule.

Recall from Sect. 3.3 that branching is only required on $x_{b,t}^{\text{off}}$, $x_{b,t}^{\text{cool}}$ and $x_{b,t}^{\text{heat}}$ because all other variables are functionally-defined by the operating modes and the input data. The branching rule is driven by one main ideology. During timestep $t$, the branching rule prefers the operating mode that brings the indoor temperature nearest to the ideal comfort temperature $i_0$ if $t+1$ requires temperature control for occupant comfort (i.e., $t+1 \in \mathcal{R}$). Otherwise, the branching rule prefers turning off the HVAC systems to reduce power costs.

The variable and value selection heuristic is described as follows. The algorithm begins by generating a random ordering of the buildings. It then loops through the timesteps from 0 to $T-1$. At every timestep $t$ and building $b$ ordered randomly, it computes the indoor temperature at $t+1$ for all three HVAC operating modes. If $t+1 \notin \mathcal{R}$ (i.e., the temperature in $t+1$ does not need to be controlled for occupant comfort), the branching rule first branches for the HVAC system to be off ($x_{b,t}^{\text{off}} \leftarrow 1$) to reduce power costs, and then branches either cooling ($x_{b,t}^{\text{cool}} \leftarrow 1$) or heating ($x_{b,t}^{\text{heat}} \leftarrow 1$) according to the difference between the ideal comfort temperature $i_0$ and predicted temperature given cooling or heating. If $t+1 \in \mathcal{R}$, then the predicted temperature after operating in each of the three modes is calculated, and the branching rule branches on the modes in order of difference between $i_0$ and the predicted temperature. Put simply, if the next timestep does not require temperature control for comfort, the branching rule prefers reduced power costs by branching on turning the HVAC systems off. Otherwise, the branching rule prefers the operating mode that increases comfort. Since calculating these predictions involves many matrix multiplications, the predictions are calculated once during the initialization phase and the branching rule is fixed for the entire search.

### 4.2 The Large Neighborhood Search

LNS is a popular local search technique [16,19]. It begins with an initial feasible solution, perhaps found using a greedy method. Using this solution, LNS fixes a subset of the variables to their values in the existing solution and then calls an exact solver on the remaining *relaxed* variables. If a better solution is found, it is stored as the incumbent solution. Upon completing the search, the process is repeated with a new search on a different subset of variables fixed to the values in the new best solution. The choice of variables to fix and relax is determined by a subroutine called a *neighborhood*. For LNS to be effective, several neighborhoods should be implemented to target different causes of suboptimality. Six neighborhoods are developed. The six neighborhoods are randomly selected with equal probability. They are described as follows.

*Cool Off Neighborhood.* The COOL OFF neighborhood attempts to reduce over-cooling of one building. Consider a building operating in a sequence of ⟨ COOL, COOL, COOL, OFF⟩. Perhaps it is better to have ⟨OFF, COOL, OFF, COOL⟩ instead. This neighborhood shuffles around the OFF and COOL modes within a range of timesteps, and leaves the HEAT decisions alone by fixing them in all timesteps. This neighborhood is sketched below:

1. For every building $b \in \mathcal{B}$ and comfort time $t \in \mathcal{R}$, compute a weight $w_{b,t} = (i_{b,t} - i_O)^2$ if $x^{\text{cool}}_{b,t-1} = 1$ and $i_{b,t} < i_O$, and $w_{b,t} = 0$ otherwise. Exit if all $w_{b,t} = 0$.
2. Select a $b^* \in \mathcal{B}$ and $t^* \in \mathcal{R}$ with probability $\frac{w_{b^*,t^*}}{\sum_{b \in \mathcal{B}} \sum_{t \in \mathcal{R}} w_{b,t}}$.
3. Fix $x^{\text{heat}}_{b^*,t} = 0$ for all $t \in \mathcal{T}'$. Relax $x^{\text{off}}_{b^*,t}$ and $x^{\text{cool}}_{b^*,t}$ for all $t \in \{t^*-1-k, \ldots, t^*-1+k\} \cap \mathcal{T}'$ for some radius $k \in \mathbb{N}$. Fix all other mode variables to their values in the incumbent solution.

*Heat Off Neighborhood.* The HEAT OFF neighborhood is the heating equivalent of the COOL OFF neighborhood:

1. For every building $b \in \mathcal{B}$ and comfort time $t \in \mathcal{R}$, compute a weight $w_{b,t} = (i_{b,t} - i_O)^2$ if $x^{\text{heat}}_{b,t-1} = 1$ and $i_{b,t} > i_O$, and $w_{b,t} = 0$ otherwise. Exit if all $w_{b,t} = 0$.
2. Select a $b^* \in \mathcal{B}$ and $t^* \in \mathcal{R}$ with probability $\frac{w_{b^*,t^*}}{\sum_{b \in \mathcal{B}} \sum_{t \in \mathcal{R}} w_{b,t}}$.
3. Fix $x^{\text{cool}}_{b^*,t} = 0$ for all $t \in \mathcal{T}'$. Relax $x^{\text{off}}_{b^*,t}$ and $x^{\text{heat}}_{b^*,t}$ for all $t \in \{t^*-1-k, \ldots, t^*-1+k\} \cap \mathcal{T}'$ for some radius $k \in \mathbb{N}$. Fix all other mode variables to their values in the incumbent solution.

*Flip Neighborhood.* The FLIP neighborhood attempts to remove sequences of alternating cooling and heating:

1. Create a random ordering of $\mathcal{B}$ and a random ordering of $\mathcal{T}'$.
2. Loop through $b^* \in \mathcal{B}$ and $t^* \in \mathcal{T}'$ using the random orderings. Find a $b^*$ and $t^*$ such that (1) $x^{\text{cool}}_{b^*,t^*} = 1$ and $x^{\text{heat}}_{b^*,t^*+1} = 1$, (2) $x^{\text{heat}}_{b^*,t^*} = 1$ and $x^{\text{cool}}_{b^*,t^*+1} = 1$, (3) $x^{\text{cool}}_{b^*,t^*} = 1$, $x^{\text{off}}_{b^*,t^*+1} = 1$ and $x^{\text{heat}}_{b^*,t^*+2} = 1$ or (4) $x^{\text{heat}}_{b^*,t^*} = 1$, $x^{\text{off}}_{b^*,t^*+1} = 1$ and $x^{\text{cool}}_{b^*,t^*+2} = 1$. Exit if such a $b^*$ and $t^*$ are not found.
3. For some radius $k \in \mathbb{N}$, relax $x^{\text{off}}_{b^*,t}$, $x^{\text{cool}}_{b^*,t}$ and $x^{\text{heat}}_{b^*,t}$ for all $t \in \{t-k, \ldots, t+1+k\} \cap \mathcal{T}'$ for cases (1) and (2), and for all $t \in \{t-k, \ldots, t+2+k\} \cap \mathcal{T}'$ for cases (3) and (4). Fix all other mode variables to their values in the incumbent solution.

*Precool Neighborhood.* The PRECOOL neighborhood aims to cool down a building before the first timestep of an interval of comfort times:

1. Create a random ordering of $\mathcal{B}$.
2. Loop through $b^* \in \mathcal{B}$ in random order and $t^* \in \mathcal{R}$ in ascending order. Find a $b^*$ and $t^*$ such that $t^* - 1 \notin \mathcal{R}$, $x^{\text{cool}}_{b^*,t^*-1} = 1$ and $i_{b^*,t^*} > i_0$. Exit if such a $b^*$ and $t^*$ are not found.
3. Create an empty set $\tau \subset \mathcal{T}'$ of timesteps.
4. Loop $t$ backwards from $t^* - 2$ to 0. Add $t$ to $\tau$ if $x^{\text{cool}}_{b^*,t} = 0$. Stop if $|\tau| = 2k$ for some size parameter $k \in \mathbb{N}$.
5. If $|\tau| = 0$, go back to step (2) to find another $b^*$ and $t^*$.
6. Fix $x^{\text{heat}}_{b^*,t} = 0$ for all $t \in \mathcal{T}'$. Relax $x^{\text{off}}_{b^*,t}$ and $x^{\text{cool}}_{b^*,t}$ for all $t \in \tau$. Fix all other mode variables to their values in the incumbent solution.

*Preheat Neighborhood.* The PREHEAT neighborhood is the heating equivalent of the PRECOOL neighborhood:

1. Create a random ordering of $\mathcal{B}$.
2. Loop through $b^* \in \mathcal{B}$ in random order and $t^* \in \mathcal{R}$ in ascending order. Find a $b^*$ and $t^*$ such that $t^* - 1 \notin \mathcal{R}$, $x^{\text{heat}}_{b^*,t^*-1} = 1$ and $i_{b^*,t^*} < i_0$. Exit if such a $b^*$ and $t^*$ are not found.
3. Create an empty set $\tau \subset \mathcal{T}'$ of timesteps.
4. Loop $t$ backwards from $t^* - 2$ to 0. Add $t$ to $\tau$ if $x^{\text{heat}}_{b^*,t} = 0$. Stop if $|\tau| = 2k$ for some size parameter $k \in \mathbb{N}$.
5. If $|\tau| = 0$, go back to step (2) to find another $b^*$ and $t^*$.
6. Fix $x^{\text{cool}}_{b^*,t} = 0$ for all $t \in \mathcal{T}'$. Relax $x^{\text{off}}_{b^*,t}$ and $x^{\text{heat}}_{b^*,t}$ for all $t \in \tau$. Fix all other mode variables to their values in the incumbent solution.

*On Neighborhood.* The ON neighborhood attempts to turn on the HVAC systems:

1. Initialize $w^{\text{cool}}_t = 0$ and $w^{\text{heat}}_t = 0$ for all $t \in \mathcal{R}$.
2. For every building $b \in \mathcal{B}$ and comfort time $t \in \mathcal{R}$, add a weight $(i_{b,t} - i_0)^2$ to $w^{\text{cool}}_t$ if $x^{\text{cool}}_{b,t-1} = 0$ and $i_{b,t} > i_0$, or add the weight to $w^{\text{heat}}_t$ if $x^{\text{heat}}_{b,t-1} = 0$ and $i_{b,t} < i_0$. Exit if all $w^{\text{cool}}_t = 0$ and $w^{\text{heat}}_t = 0$.
3. Focus on cooling with probability $\frac{\sum_{t\in\mathcal{R}} w^{\text{cool}}_t}{\sum_{t\in\mathcal{R}} w^{\text{cool}}_t + w^{\text{heat}}_t}$. Focus on heating otherwise.
4. If cooling, select a $t^* \in \mathcal{R}$ with probability $\frac{w^{\text{cool}}_{t^*}}{\sum_{t\in\mathcal{R}} w^{\text{cool}}_t}$. Relax $x^{\text{off}}_{b,t}$ and $x^{\text{cool}}_{b,t}$ for all $b \in \mathcal{B}$ and $t \in \{t^* - 1 - k, \ldots, t^* - 1 + k\} \cap \mathcal{T}'$ for some radius $k \in \mathbb{N}$.
5. If heating, select a $t^* \in \mathcal{R}$ with probability $\frac{w^{\text{heat}}_{t^*}}{\sum_{t\in\mathcal{R}} w^{\text{heat}}_t}$. Relax $x^{\text{off}}_{b,t}$ and $x^{\text{heat}}_{b,t}$ for all $b \in \mathcal{B}$ and $t \in \{t^* - 1 - k, \ldots, t^* - 1 + k\} \cap \mathcal{T}'$ for some radius $k \in \mathbb{N}$.
6. Fix all other mode variables to their values in the incumbent solution.

# 5 Experimental Results

This section presents the experimental set-up and analyses the results.

## 5.1 Generating Exploratory Training Data

To generate the exploratory data for training the neural networks, we assume we have access to physics-based Equivalent Thermal Parameter (ETP; [17]) models that are closely matched to the real-world dynamics. By using models, we avoid having to operate real buildings at uncomfortable temperatures far from their set points for long periods of time needed to gather sufficient data. We instantiate an ETP model for each of twenty buildings based on defaults provided by the grid simulator GridLAB-D [6] plus some small ($\sigma = 0.05\,\mu$) Gaussian noise to vary their structural parameters.

The training data is generated as follows: we use the ETP models to generate 25,000 two-hour trajectories on five-minute timesteps, resulting in 600,000 data points per building model. Every sample trajectory consists of a fixed outdoor temperature sampled uniformly at random, $o_0 \sim \mathcal{U}(0, 45)$, an initial indoor mass temperature $m_0 \sim \mathcal{U}(10, 35)$, and corresponding initial air temperature close to the initial mass temperature, $i_0 \sim \mathcal{U}(-0.5, 0.5) + m_0$. Then, a sequence of 24 exploratory actions are taken, by choosing actions uniformly from $\{\textsc{Off}, \textsc{Heat}, \textsc{Cool}\}$, and recording for every step: the initial conditions, the power consumed by the HVAC system, and the indoor air and indoor mass temperatures after five minutes.

## 5.2 Training the Neural Networks

The training data is preprocessed by scaling all values to lie between 0 and 1. This ensures that the loss in all dimensions are evenly considered. The MLPs are trained using the Adam stochastic gradient descent algorithm [11] over 100 epochs in Tensorflow. One-fifth of the input trajectories are reserved for validation. For every building, ten randomly-initialized training runs are conducted. Of the ten runs, the trained parameters resulting in the smallest sum of mean squared errors are chosen and fixed for planning in the MIQP model.

Three sizes of MLPs are trained: (1) three hidden layers, each consisting of thirty ReLU units, (2) two hidden layers, each with twenty ReLU units and (3) two hidden layers, each with ten ReLU units. Averaged over the twenty buildings, the three sizes respectively have $1.6 \times 10^{-8}$, $1.5 \times 10^{-8}$ and $4.7 \times 10^{-8}$ validation loss. We use the smallest MLPs since they still accurately fit the data and the largest MLPs have marginally more error than the middle option.

## 5.3 Experimental Set-Up

The MIQP model is solved using Gecode 6.2.0 and Gurobi 9.0.1. Using these two solvers, the following four methods are evaluated:

- *CP-BB*: Branch-and-bound search in Gecode with the branching rule described in Sect. 4.1.
- *MP-BB*: Branch-and-bound search in Gurobi.
- *CP-LNS*: Local search in Gecode with the branching rule from Sect. 4.1 and the LNS from Sect. 4.2.
- *MP-LNS*: Local search in Gurobi with the LNS from Sect. 4.2.

Both Gecode and Gurobi have built-in support for the max function in the ReLU activation function. Hence, the models are directly implemented, and no attempt was made to manually linearize Constraint (1c) for Gurobi. The radius parameter $k$ for the neighborhoods is randomly chosen as $k = 2$ with probability 0.7, $k = 3$ with probability 0.2 and $k = 4$ with probability 0.1.

All parameters in Gecode are set to their default values. Default parameters are also used in Gurobi except the search is set to prioritize the primal bound and the solver is warm-started with a greedy solution found using a procedure almost identical to the branching rule from Sect. 4.1.

The solvers are tested on 504 instances. First, 42 typical weather conditions (correlated outdoor temperature and solar irradiance time series) are generated, of which 9 cover cold days with extremely low temperatures, and 9 are hot days with extremely warm temperatures. Then, we control the complexity of the instances generated by varying the length of the planning horizon, and the number of buildings to be controlled. The horizon varies between four, eight and sixteen hours, divided into five-minute time periods resulting in 48, 96, and 192 timesteps, respectively. Each of these instances is then paired with 5, 10, 15 and 20 buildings, making up the 504 instances. Every instance has at least one demand response event, located at the coldest or warmest parts of the day. The initial conditions of each building are given, with initial indoor mass temperature assumed to be equal the initial air temperature. Every instance is run by each of the four methods on a single thread for thirty minutes on an Intel Xeon E5-2660 v3 CPU at 2.60 GHz.

### 5.4 Results and Analysis

Figure 4 is a cactus plot showing for each objective value how many instances have better objective value as found by each solver. The chart shows that LNS outperforms complete branch-and-bound, and that CP-LNS is significantly better than MP-LNS. MP-BB and MP-LNS find feasible solutions to 125 and 215 instances respectively. CP-BB finds solutions to 174 instances, and CP-LNS finds solutions to all 504 instances. Even though both MP-BB and MP-LNS are given a warm start solution, they are unable to activate the solution in many instances, declaring unknown problem status (unknown infeasible or feasible) at termination. Output logs suggest that Gurobi needs to perform computations to activate the solution. This may be caused by the need to compute a basis, which is hindered by the large number of binary variables internally added to linearize the max constraints.

CP-LNS, MP-LNS, CP-BB and MP-BB find the (equally) best solution to 439, 37, 8 and 20 instances respectively. Given the short time-out, neither exact

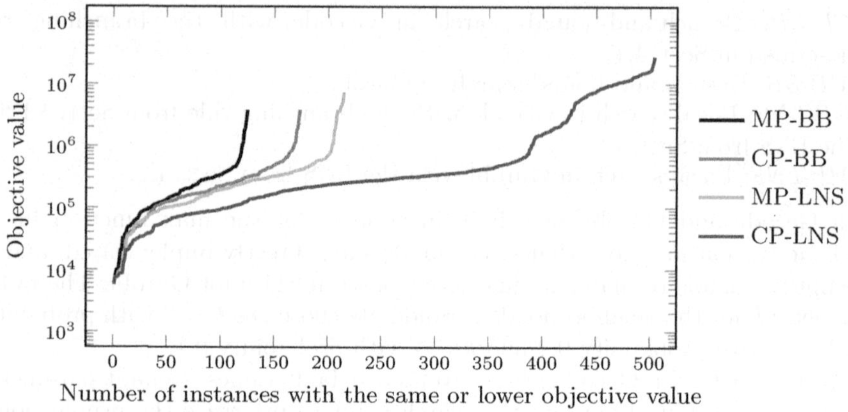


**Fig. 4.** Cactus plot of objective value against the number of instances for which each method finds better solutions. On the left of any one value on the horizontal axis is the number of instances that the method finds a solution with the same or lower objective value.

method is able to prove optimality on any instance. The optimality gap averages 97% across the 125 instances with feasible solutions from MP-BB. At best, the optimality gap is 69%. These numbers suggest that the model features a weak linear relaxation. Overall, these results indicate that exact methods are ineffective for real-time control problems that have short time-outs required for repeated reoptimization.

The MIQP model has simple structure, which should make it relatively easy for Gurobi. Even though ReLU functions are nearly linear (i.e., they are a max function composed with an affine function), they actually belong to the class of non-convex programming problems. (Recall that a convex minimization problem has convex less-than-or-equal-to constraints and affine equality constraints [5].) Therefore, we speculate that the linearization of the ReLU function using a binary variable leads to a very weak linear relaxation, making Gurobi inadequate. Rather, the fast tree search of CP and its avoidance of computing nearly useless dual bounds are key to tackling the problem.

### 5.5 Closing the Loop

The objective values reported in Fig. 4 are predicted by the optimization routine using the learned dynamics. To verify that the learned transition models are accurate, we applied the computed schedules to an ETP simulator with the actual parameters. Figure 5 (left) presents the match between the predicted temperature trajectory as optimized by CP-LNS and the ground-truth trajectory obtained by running the schedule through the simulator for an arbitrary instance. We verified that this behavior is consistent across all instances.

Figure 5 (right) compares the accuracy of the learned dynamics with the same optimization routine using a simpler MLP that omits the internal mass

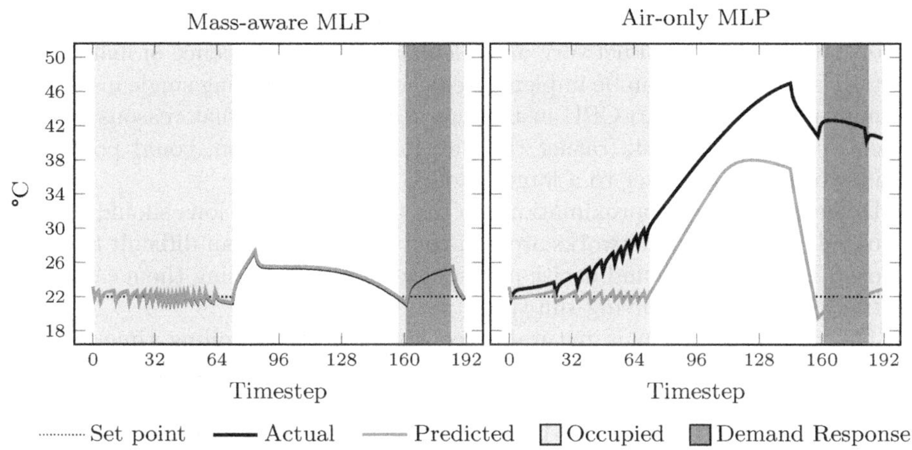


**Fig. 5.** Comparison of the scheduled and actual temperature trajectory of a building. Left: MLP trained on both air and mass temperatures. Right: MLP trained only on air temperatures.

temperature. We observe that training on all simulator inputs is required to get schedules that closely match the predicted behavior. We hypothesize that the reason for the large errors observed from the air-only MLP is that this model does not actually constitute a function: for every input of ⟨air, outdoor, solar irradiance, control⟩, there are multiple correct corresponding next air temperatures. The MLP learns to fit a least-error mean value, but this necessarily gives an error in the prediction, which compounds over time into the dramatic mismatch observed here.

## 6 Conclusions and Future Work

This paper developed an MIQP model for real-time control of space conditioning systems while considering demand response. The problem adjusts the indoor temperature of several smart buildings connected to a microgrid to ensure occupant comfort while subject to sporadic limitations on power supply. The model is solved using the MP solver Gurobi and the CP solver Gecode. The paper develops a problem-specific branching rule and LNS, which comprises six neighborhoods that attempt to repair six different reasons for suboptimality. Gecode coupled with the branching rule and LNS vastly outperforms Gurobi.

Several directions for future research are available:

- More sophisticated measures of occupant comfort include various other factors, such as humidity. Future studies should include these variables.
- The model was initially implemented in CP Optimizer but it faced numerical issues. CP solvers with lazy clause generation should be used, but it appears that no other production-ready CP solver supports floating point variables.

- Dedicated propagators for the ReLU function should be investigated. The outputs of the neural network are calculated using a sequence of matrix multiplications, which can be implemented very efficiently using single instruction multiple data (SIMD) CPU instructions. A propagator that reasons over the entire neural network, rather than an individual neuron, could potentially improve the CP solver to a large extent.
- Different kinds of approximations to the transition functions should be evaluated. Deep neural networks are non-convex, making them difficult to reason over. Linear models and decision trees are simpler and using them can potentially improve the solving run-time.
- The control problem is naturally reoptimized over a rolling time window. Adding one additional time period and keeping the existing decisions fixed may not heavily impact the solution quality, but not does not allow a large search space to be explored. This early study investigates a once-off overnight run. Considering that the temperature approximations will drift as the day progresses, the number of timesteps to reoptimize during the day is a key question for future studies.

## References

1. Achterberg, T., Koch, T., Martin, A.: Branching rules revisited. Oper. Res. Lett. **33**(1), 42–54 (2005)
2. Anderson, R., Huchette, J., Tjandraatmadja, C., Vielma, J.P.: Strong mixed-integer programming formulations for trained neural networks. In: Lodi, A., Nagarajan, V. (eds.) IPCO 2019. LNCS, vol. 11480, pp. 27–42. Springer, Cham (2019). https://doi.org/10.1007/978-3-030-17953-3_3
3. Azuatalam, D., Mhanna, S., Chapman, A., Verbič, G.: Optimal HVAC scheduling using phase-change material as a demand response resource. In: 2017 IEEE Innovative Smart Grid Technologies-Asia (ISGT-Asia). IEEE (2017)
4. Bartolini, A., Lombardi, M., Milano, M., Benini, L.: Neuron constraints to model complex real-world problems. In: Lee, J. (ed.) CP 2011. LNCS, vol. 6876, pp. 115–129. Springer, Heidelberg (2011). https://doi.org/10.1007/978-3-642-23786-7_11
5. Boyd, S., Vandenberghe, L.: Convex Optimization. Cambridge University Press, Cambridge (2004)
6. Chassin, D.P., Fuller, J.C., Djilali, N.: GridLAB-D: an agent-based simulation framework for smart grids. J. Appl. Math. **2014** (2014)
7. Cybenko, G.: Approximation by superpositions of a sigmoidal function. Math. Control Signals Syst. **2**(4), 303–314 (1989). https://doi.org/10.1007/BF02551274
8. Feydy, T., Stuckey, P.J.: Lazy clause generation reengineered. In: Gent, I.P. (ed.) CP 2009. LNCS, vol. 5732, pp. 352–366. Springer, Heidelberg (2009). https://doi.org/10.1007/978-3-642-04244-7_29
9. FICO: MIP formulations and linearizations (2009). https://www.fico.com/en/resource-download-file/3217
10. Goodfellow, I., Bengio, Y., Courville, A.: Deep Learning. MIT Press, Cambridge (2016). http://www.deeplearningbook.org
11. Kingma, D.P., Ba, J.: Adam: a method for stochastic optimization. In: ICLR 2015 (2015)

12. Kohlhepp, P., Harb, H., Wolisz, H., Waczowicz, S., Müller, D., Hagenmeyer, V.: Large-scale grid integration of residential thermal energy storages as demand-side flexibility resource: a review of international field studies. Renew. Sustain. Energy Rev. **101**, 527–547 (2019)
13. Motegi, N., Piette, M.A., Watson, D.S., Kiliccote, S., Xu, P.: Introduction to commercial building control strategies and techniques for demand response. Technical report California Energy Commission, PIER (2006)
14. de Nijs, F., Stuckey, P.J.: Risk-aware conditional replanning for globally constrained multi-agent sequential decision making. In: Proceedings of the 19th International Conference on Autonomous Agents and Multiagent Systems. IFAAMAS (2020)
15. Pérez-Lombard, L., Ortiz, J., Pout, C.: A review on buildings energy consumption information. Energy Build. **40**(3), 394–398 (2008)
16. Pisinger, D., Røpke, S.: Large neighborhood search. In: Gendreau, M., Potvin, J.Y. (eds.) Handbook of metaheuristics, chapter 13, pp. 399–419. Springer (2010). https://doi.org/10.1007/978-1-4419-1665-5_13
17. Pratt, R.G., Taylor, Z.T.: Development and testing of an equivalent thermal parameter model of commercial buildings from time-series end-use data. Technical report Pacific Northwest Laboratory, Richland, Washington (1994)
18. Say, B., Wu, G., Zhou, Y.Q., Sanner, S.: Nonlinear hybrid planning with deep net learned transition models and mixed-integer linear programming. In: Proceedings of the Twenty-Sixth International Joint Conference on Artificial Intelligence, IJCAI-17, pp. 750–756 (2017)
19. Shaw, P.: Using constraint programming and local search methods to solve vehicle routing problems. In: Maher, M., Puget, J.-F. (eds.) CP 1998. LNCS, vol. 1520, pp. 417–431. Springer, Heidelberg (1998). https://doi.org/10.1007/3-540-49481-2_30
20. Vázquez-Canteli, J.R., Nagy, Z.: Reinforcement learning for demand response: A review of algorithms and modeling techniques. Appl. Energy **235**, 1072–1089 (2019)

# Solving the Group Cumulative Scheduling Problem with CPO and ACO

Lucas Groleaz[1,2(✉)], Samba N. Ndiaye[1], and Christine Solnon[3]

[1] LIRIS, CNRS UMR5205, INSA Lyon, 69621 Villeurbanne, France
lucasgroleaz@live.fr
[2] Infologic, Arlesey, UK
[3] CITI, INRIA, INSA Lyon, 69621 Villeurbanne, France

**Abstract.** The Group Cumulative Scheduling Problem (GCSP) comes from a real application, i.e., order preparation in food industry. Each order is composed of jobs which must be scheduled on machines, and the goal is to minimize the sum of job tardiness. There is an additional constraint, called Group Cumulative (GC), which ensures that the number of active orders never exceeds a given limit, where an order is active if at least one of its jobs is started and at least one of its jobs is not finished. In this paper, we first describe a Constraint Programming (CP) model for the GCSP, where the GC constraint is decomposed using classical cumulative constraints. We experimentally evaluate IBM CP Optimizer (CPO) on a benchmark of real industrial instances, and we show that it is not able to solve efficiently many instances, especially when the GC constraint is tight. To explain why CPO struggles to solve the GCSP, we show that it is NP-Complete to decide whether there exist start times which satisfy the GC constraint given the sequence of jobs on each machine, even when there is no additional constraint. Finally, we introduce an hybrid framework where CPO cooperates with an Ant Colony Optimization (ACO) algorithm: ACO is used to learn good solutions which are given as starting points to CPO, and the solutions improved by CPO are given back to ACO. We experimentally evaluate this hybrid CPO-ACO framework and show that it strongly improves CPO performance.

## 1 Introduction

There exist numerous variants of scheduling problems [23]. In [9], a new scheduling problem is introduced, called the *Group Cumulative Scheduling Problem (GCSP)*. This problem comes from a real application: order preparation in food industry. Each order is composed of jobs which must be scheduled on machines, and the goal is to minimise the sum of job tardiness. There is an additional constraint, called the *Group Cumulative (GC)* constraint, which comes from the fact that a pallet is associated with each order: when starting the first job of an order is started, a pallet is set on the ground and this pallet is removed when the last job of the order is ended. As physical space is limited, the number of pallets


H. Simonis (Ed.): CP 2020, LNCS 12333, pp. 620–636, 2020.
https://doi.org/10.1007/978-3-030-58475-7_36

on the ground must never exceed a given limit. In other words, jobs are grouped into orders, and GC ensures that the number of active groups never exceeds a limit, where a group is active if at least one of its jobs is started and at least one of its jobs is not finished.

Beyond the industrial application described in [9], the GC constraint may have other applications. In particular, it may be used each time a resource is required by a group of jobs, so that the resource is consumed when the first job of the group starts and it is released only when the last job of the group ends.

*Contributions.* In this paper, we describe a *Constraint Programming (CP)* model for the GCSP, and we show that GC may be decomposed using classical cumulative constraints. We experimentally evaluate *IBM CP Optimizer (CPO)*, a state-of-the-art solver for scheduling, on the industrial instances of [9], and we show that CPO struggle to solve many of them, especially when GC is tight. To provide insight into CPO performance, we study the complexity of GC: we show that it is NP-Complete to decide whether there exist start times which satisfy GC when the sequence of jobs on each machine is known, even if there is no additional constraint. Finally, we show how to hybridise CPO with the *Ant Colony Optimisation (ACO)* algorithm introduced in [9]: ACO is used to learn good solutions which are given as starting points to CPO, and solutions improved by CPO are given back to ACO. We experimentally evaluate this hybrid CPO-ACO framework, and show that it strongly improves CPO performance.

*Plan.* In Sect. 2, we describe the GCSP and we define GC. In Sect. 3, we introduce a CP model for the GCSP, and we report results obtained with CPO. In Sect. 4, we study the complexity of GC. In Sect. 5, we describe the ACO algorithm of [9]. In Sect. 6, we introduce and evaluate our hybrid CPO-ACO framework. Sects. 2 and 5 are recalls from [9]. Sections 3, 4 and 6 contain new contributions with respect to [9].

*Notations.* We denote sets with calligraphic letters, constants with lowercase letters, and variables with uppercase letters. $\#\mathcal{A}$ denotes the cardinality of a set $\mathcal{A}$. $[l, u]$ denotes the set of all integers ranging from $l$ to $u$.

## 2 Description of the GCSP

The GCSP is a classical scheduling problem (referred to as the "basic" scheduling problem and described in Sect. 2.1) with an additional GC constraint (described in Sect. 2.2). In Sect. 2.3, we describe the benchmark of [9].

### 2.1 Basic Scheduling Problem

Given a set $\mathcal{M}$ of machines and a set $\mathcal{J}$ of jobs such that, for each job $j \in \mathcal{J}$, $r_j$ denotes its release date, $d_j$ its due date, and $p_j$ its processing time, the goal is to find a start time $B_j$, an end time $E_j$, and a machine $M_j$, for each job $j \in \mathcal{J}$. According to the notation introduced in [6], the basic scheduling problem underlying the GCSP is denoted $Rm, 1, 1; MPS|s_{ij}; r_j|\sum T_j$:

- $Rm, 1, 1$ means that $\mathcal{M}$ contains several machines working in parallel and each machine $m \in \mathcal{M}$ can process at most one job at a time;
- $MPS$ stands for *Multi-mode Project Scheduling* and means that every machine $m \in \mathcal{M}$ has its own speed denoted $sp^m$ (so that the duration of a job $j$ is $p_j * sp^{M_j}$);
- $s_{i,j}$ indicates that the setup time of a job $j \in \mathcal{J}$ depends on the job $i$ that precedes $j$ on the machine (*i.e.*, the time interval between the end time of $i$ and the start time of $j$ must be larger than or equal to this setup time);
- $r_j$ means that a job cannot start before its release date, *i.e.*, $\forall j \in \mathcal{J}, B_j \geq r_j$;
- $\sum T_j$ indicates that the goal is to minimize the sum of tardiness of every job, *i.e.*, $\sum_{j \in \mathcal{J}} \max(0, E_j - d_j)$.

### 2.2 GC Constraint

GC is a particular case of cumulative constraint [1,2,21,22], and we show how to decompose GC using cumulative constraints in Sect. 3. Cumulative constraints are used to model the fact that jobs require resources (*e.g.*, human skills or tools) and that these resources have limited capacities, *i.e.*, the sum of resources required by all jobs started but not ended must never exceed resource capacities.

In the GCSP, the resource is not directly required by jobs, but by job groups. More precisely, jobs are partitioned into groups (corresponding to orders in the industrial application of [9]). The start (resp. end) time of a group is defined as the smallest start time (resp. largest end time) among all its jobs. A group is said to be *active* at a time $t$ if it is started and not ended at time $t$. The GC constraint ensures that the number of active groups never exceeds a given limit. More formally, we define the GC global constraint as follows.

**Definition 1.** *Given a set $\mathcal{J}$ of jobs, a partition $\mathcal{P}$ of $\mathcal{J}$ in $\#\mathcal{P}$ groups (such that each job $j \in \mathcal{J}$ belongs to exactly one group $\mathcal{G} \in \mathcal{P}$), an integer limit $l$ and, for each job $j \in \mathcal{J}$, an integer variable $B_j$ (resp. $E_j$) corresponding to the start time (resp. end time) of $j$, the constraint $GC_{\mathcal{J},\mathcal{P},l}(\{B_j : j \in \mathcal{J}\}, \{E_j : j \in \mathcal{J}\})$ is satisfied iff $\#\{\mathcal{G} \in \mathcal{P} : \min_{j \in \mathcal{G}} B_j \leq t < \max_{j \in \mathcal{G}} E_j\} \leq l$ for any time $t$.*

In Fig. 1, we display two examples of schedules: one that violates GC and one that satisfies it (we assume that setup times are null in this example).

### 2.3 Benchmark Instances

A benchmark extracted from industrial data is introduced in [9]. It contains 548 instances such that the number of groups (resp. jobs and machines) ranges from 56 to 406 (resp. from 288 to 2909, and from 1 to 14). For each instance, an upper bound (denoted $x$) on the number of active groups is given. It is computed as follows: first, a greedy algorithm is used to compute a solution $s$ for the basic scheduling problem (without the GC constraint); then $x$ is assigned to the maximum number of active groups during the whole time horizon in $s$.

As our goal is to study the impact of GC on the solution process, we consider three classes of instances: in the first class, denoted *loose*, the limit $l$ is set to

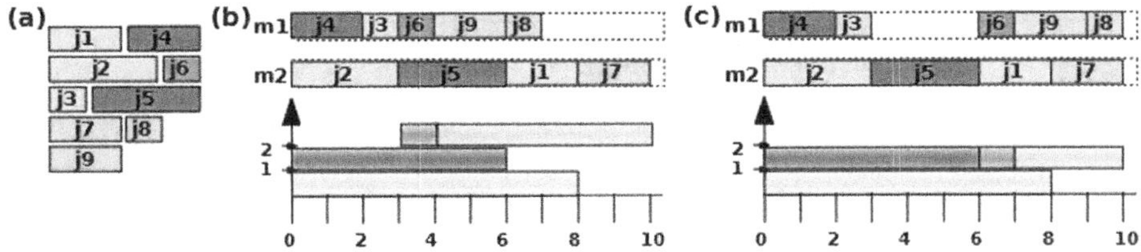


**Fig. 1.** Schedule examples. (a) A set $\mathcal{J}$ of 9 jobs and a partition $\mathcal{P}$ of $\mathcal{J}$ in 4 groups represented by colours. (b) Example of schedule on 2 machines which violates GC when $l = 2$ (there are 3 active groups from time 3 to time 6, as displayed on the bottom of (b)). (c) Example of schedule on 2 machines which satisfies GC when $l = 2$ (an iddle time is added between $j_3$ and $j_6$ to wait the end of the blue group). (Color figure online)

$$\text{Minimize} \sum_{j \in \mathcal{J}} max(0, endOf(A_j) - d_j)$$

$$\begin{array}{llll}
\text{subject to} & A_j^m = interval(d_j * sp^m) & \forall j \in \mathcal{J}, \forall m \in \mathcal{M} & (1) \\
& optional(A_j^m) & \forall j \in \mathcal{J}, \forall m \in \mathcal{M} & (2) \\
& A_j = alternative(\{A_j^m : m \in \mathcal{M}\}) & \forall j \in \mathcal{J} & (3) \\
& startMin(A_j^m, r_j) & \forall j \in \mathcal{J}, \forall m \in \mathcal{M} & (4) \\
& S^m = intervalSequence(\{A_j^m : j \in \mathcal{J}\}, jobTypes) & \forall m \in \mathcal{M} & (5) \\
& noOverlap(S^m, jobTypes, setupTimes) & \forall m \in \mathcal{M} & (6)
\end{array}$$

**Fig. 2.** CPO model for the basic scheduling problem described in Sect. 2.1. (*jobTypes* is an array which associates a type with every job and *setupTimes* is a transition matrix which defines the setup times between job types).

$l = 0.7 * x$, in the second class, denoted *medium*, $l$ is set to $0.5 * x$, and in the third class, denoted *tight*, $l$ is set to $0.3 * x$.

## 3 CPO Model

We describe a CPO model for the basic scheduling problem in Sect. 3.1, and a decomposition of GC in Sect. 3.2. We report results obtained with CPO in Sect. 3.3. We refer the reader to [16] for details on CPO.

### 3.1 Model of the Basic Scheduling Problem

The CPO model associates an interval variable $A_j$ with every job $j \in \mathcal{J}$, *i.e.*, $A_j$ corresponds to the interval $[B_j, E_j]$. Also, an optional interval variable $A_j^m$ is associated with every job $j \in \mathcal{J}$ and every machine $m \in \mathcal{M}$: if job $j$ is executed on machine $m$, then $A_j^m = A_j$; otherwise $A_j^m$ is assigned to $\perp$ (*i.e.*, it is absent). Finally, an interval sequence variable $S^m$ is associated with every machine $m$ to represent the total ordering of the present interval variables in $\{A_j^m : j \in \mathcal{J}\}$.

$$span(F_{\mathcal{G}}, \{A_j^m : m \in \mathcal{M} \wedge j \in \mathcal{G}\})\ \forall \mathcal{G} \in \mathcal{P} \quad (7)$$
$$Active = \sum_{\mathcal{G} \in \mathcal{P}} pulse(F_{\mathcal{G}}, 1) \quad (8)$$
$$lowerOrEqual(Active, l) \quad (9)$$

**Fig. 3.** CPO decomposition of GC.

The objective function and the constraints are described in Fig. 2. Constraint (1) defines the interval variable $A_j^m$ whose length is equal to the processing time of job $j$ multiplied by the speed of machine $m$; Constraints (2) and (3) ensure that every job $j$ is scheduled on exactly one machine; Constraint (4) ensures that a job does not start before its release date; Constraint (5) defines the sequence of jobs on machine $m$; and Constraint (6) ensures that at most one job is executed at a time on machine $m$, and states that there are sequence-dependent setup times between jobs.

### 3.2 Decomposition of GC

We can easily decompose GC using a classical cumulative constraint. To this aim, we associate a new interval variable $F_{\mathcal{G}}$ with every group $\mathcal{G} \in \mathcal{P}$. This variable corresponds to a fictive job which starts with the earliest job of the group and ends with its latest job, and which consumes one unit of resource. A simple cumulative constraint on these fictive jobs ensures that the number of active groups never exceeds $l$.

More precisely, Fig. 3 describes a CPO model of this decomposition: Constraint (7) ensures that, for every group $\mathcal{G}$, the fictive job variable $F_{\mathcal{G}}$ spans over all jobs in the group; Constraint (8) defines the cumul function (denoted *Active*) corresponding to the case where each fictive job consumes one unit of the resource; and Constraint (9) ensures that *Active* never exceeds $l$, thus ensuring the cumulative constraint on fictive jobs.

### 3.3 Experimental Evaluation of CPO

CPO is a state-of-the-art solver for scheduling problems, as demonstrated in [7] on the job shop, for example. In this section, we report CPO results with the model described in Sects. 3.1 and 3.2 on instances described in Sect. 2.3. All experiments reported in this paper have been performed on a processor Intel(R) Xeon(R) CPU E5-2650 v4 @ 2.20 GHz with 7.2 GB RAM.

CPO provides different levels of filtering, and we have compared results obtained with two different levels: default filtering (based on *Timetable* [1]) and extended filtering (based on *energy reasoning* and *edge finding* [5,15,18,22,29]). For short time limits (less than 100 s), CPO with default filtering usually finds better solutions than CPO with extended filtering. After one hour, for nearly half of the instances a better solution is found with the extended filtering, whereas for the other half a better solution is found with the default filtering, and this

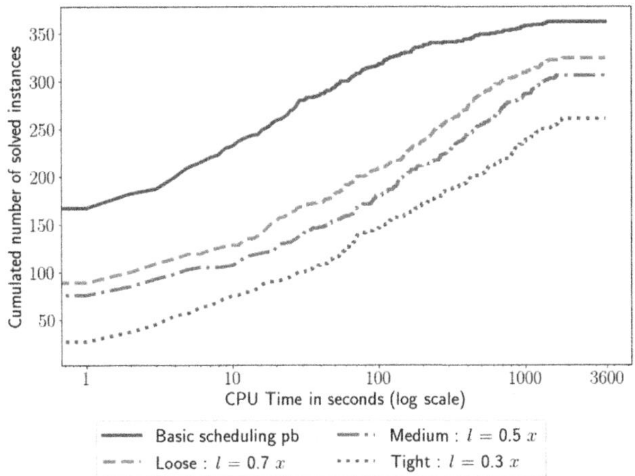


**Fig. 4.** Evolution of the cumulative number of solved instances with respect to time for the basic scheduling problem and for the GCSP when $l$ is loose, medium or tight.

happens in all classes (loose, medium, and tight). In most cases, the difference between the two levels of filtering is rather small. Hence, we have chosen to report results obtained with the default level of filtering.

In Fig. 4, we display the evolution of the cumulative number of solved instances with respect to time for the basic scheduling problem (we consider that an instance is solved when CPO has completed its run). In this case, CPO is able to solve 354 instances (among the 548 instances of the benchmark) within one hour. We also display results on the same set of instances when adding GC, for the three classes (which only differ on the value of $l$). Clearly, GC increases the hardness of the problem, and increasing the tightness of GC (by decreasing the limit $l$) also increases hardness: 319 (resp. 300 and 259) instances are solved within one hour for the loose (resp. medium and tight) class.

This may come from the fact that the decomposition of GC is not well propagated by CPO. A possible explanation is that interval variables $F_{\mathcal{G}}$ associated with groups do not have known durations when starting the search. We can only compute bounds on group durations. For example, the duration of a group is lower bounded by the greatest duration of its jobs. However, these bounds are not very tight at the beginning of the search. In this case, energy-based propagation techniques are not efficient as the energy of a job is defined as its duration multiplied by its resource consumption.

In Fig. 5, we display the number of jobs and machines of solved and unsolved instances, for the basic scheduling problem and the tight GCSP. In both cases, some large instances (with more than 2500 jobs) are solved whereas some small instances (with less than 300 jobs) are not solved. Most instances with more than 6 machines are solved (only 8 are not solved for the basic scheduling problem) whereas many instances with one machine are not solved.

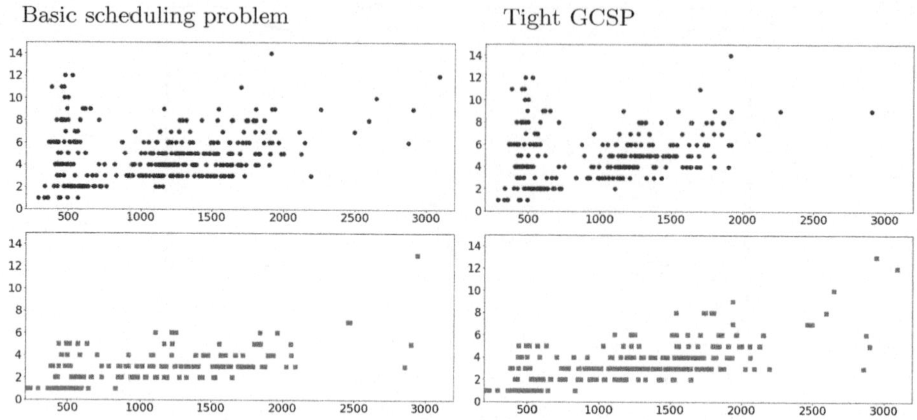


**Fig. 5.** Sizes of solved and unsolved instances for the basic scheduling problem (left) and the tight GCSP (right): each point $(x, y)$ corresponds to an instance with $x$ jobs and $y$ machines. Top (green points): instances solved in less than one hour. Bottom (red points): instances not solved within one hour. (Color figure online)

# 4 Deciding of GC Feasibility with List Schedules

CPO exploits precedence relations to solve scheduling problems [16]: all temporal constraints are aggregated in a temporal network whose nodes represent interval start and end time-points and whose arcs represent precedence relations. Also, CPO integrates a *Large Neighborhood Search (LNS)* component which is based on the initial generation of a directed graph whose nodes are interval variables and edges are precedence relations between interval variables.

Reasoning on precedence relations often simplifies the solution process of scheduling problems. In particular, for the basic problem described in Sect. 2.1 (without GC), given a *list schedule* (*i.e.*, an ordered list of jobs for each machine), we can compute optimal start times in polynomial time: for each machine, we consider jobs according to the order defined by its associated list and schedule each of them as soon as possible [10,24]. The basic scheduling problem is NP-hard because it is hard to find the list schedule which leads to the optimal solution. However, as optimal start times are easily derived from list schedules, search can focus on precedence relations.

If we add a classical cumulative constraint to the basic scheduling problem, the problem of computing optimal start times given a list schedule becomes NP-hard [21]. However, if we remove the objective function (*i.e.*, we simply search for a schedule which satisfies the cumulative constraint without having to minimize the tardiness sum), then we can easily compute start times that satisfy cumulative constraints given a list schedule: Again, this can be done greedily, by considering jobs in the order of the lists, and scheduling each job as soon as possible with respect to cumulative constraints.

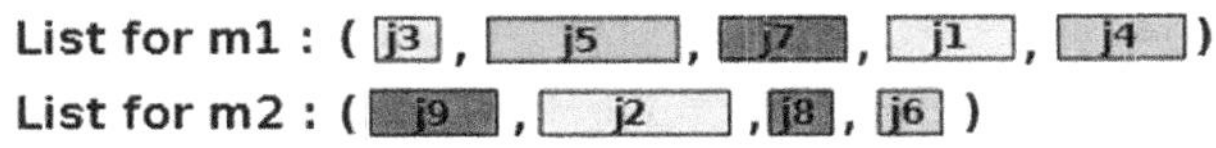


**Fig. 6.** Example of list schedule with 2 machines for the jobs of Fig. 1. (Color figure online)

However, this is no longer true for GC. For example, let us consider the list schedule displayed in Fig. 6. We cannot find start times that satisfy GC for this list schedule when $l = 2$. Indeed, on machine $m_1$, the yellow job $j_1$ is between two blue jobs $j_5$ and $j_4$, and this implies that we must start the yellow group to be able to complete the blue group. Similarly, on machine $m_1$, the blue job $j_5$ is between two yellow jobs ($j_3$ and $j_1$) so that we must start the blue group to be able to complete the yellow group, and on machine $m_2$, the yellow job $j_2$ is between two pink jobs ($j_9$ and $j_8$) so that we must start the yellow group to be able to complete the pink group. This implies that both yellow, blue and pink groups must be active all together at some time.

More precisely, let us denote *LS-GC* the problem of deciding whether there exists a solution of GC which is consistent with a given list schedule, where a list schedule is consistent with a solution of GC iff, for every $j_1, j_2 \in \mathcal{J}$ such that $j_1$ occurs before $j_2$ in a same list, we have $E_{j_1} \leq B_{j_2}$.

**Theorem 1.** *LS-GC is $\mathcal{NP}$-complete.*

*Proof.* *LS-GC* clearly belongs to $\mathcal{NP}$ as we can check in polynomial time if a given assignment is a solution of GC which is consistent with a list schedule.

Now, let us show that *LS-GC* is $\mathcal{NP}$-complete by reducing the Pathwidth problem to it. Given a connected graph $G = (\mathcal{N}, \mathcal{E})$ (such that $\mathcal{N}$ is a set of nodes and $\mathcal{E}$ a set of edges) and an integer $w$, Pathwidth aims at deciding whether there exists a sequence $(\mathcal{N}_1, ..., \mathcal{N}_n)$ of subsets of $\mathcal{N}$ such that (i) $\mathcal{N} = \bigcup_{i=1}^{n} \mathcal{N}_i$; (ii) $\forall \{u, v\} \in \mathcal{E}, \exists i \in [1, n], \{u, v\} \subseteq \mathcal{N}_i$; (iii) $\forall i, j, k \in [1, n], i \leq j \leq k \Rightarrow \mathcal{N}_i \cap \mathcal{N}_k \subseteq \mathcal{N}_j$; and (iv) $\forall i \in [1, n],\ \#\mathcal{N}_i \leq w$. Pathwidth is $\mathcal{NP}$-complete [11].

Let us first show how to construct an instance of *LS-GC* given an instance of Pathwidth defined by a graph $G = (\mathcal{N}, \mathcal{E})$ and an integer $w$. We assume that nodes of $\mathcal{N}$ are numbered from 1 to $\#\mathcal{N}$. For each edge $\{u, v\} \in \mathcal{E}$, we define three jobs denoted $j^1_{uv}$, $j^2_{uv}$, and $j^3_{uv}$ such that every job has a processing time equal to 1. The partition $\mathcal{P}$ associates one group $\mathcal{G}_u$ with every vertex $u$ such that $\mathcal{G}_u = \{j^1_{uv}, j^3_{uv} : \{u, v\} \in \mathcal{E} \wedge u < v\} \cup \{j^2_{uv} : \{u, v\} \in \mathcal{E} \wedge u > v\}$. In other words, for each edge $\{u, v\} \in \mathcal{E}$ such that $u < v$, $j^1_{uv}$ and $j^3_{uv}$ belong to group $\mathcal{G}_u$ whereas $j^2_{uv}$ belongs to group $\mathcal{G}_v$. There are $\#\mathcal{E}$ machines, and the list schedule associates the list $(j^1_{uv}, j^2_{uv}, j^3_{uv})$ with every edge $\{u, v\} \in \mathcal{E}$ such that $u < v$. Finally, we set the limit $l$ to $w$. Figure 7 gives an example of this reduction.

Now, let us show that every solution $(\mathcal{N}_1, ..., \mathcal{N}_n)$ of an instance of Pathwidth corresponds to a solution of the corresponding instance of *LS-GC*. To this aim, we show how to define the start time $B_{j^i_{uv}}$ of every job $j^i_{uv}$ associated with an edge $\{u, v\} \in \mathcal{E}$, with $i \in \{1, 2, 3\}$: let $a$ be the index of the first subset in $(\mathcal{N}_1, ..., \mathcal{N}_n)$ which contains both $u$ and $v$ (*i.e.*, $a = \min\{b \in [1, n] : \{u, v\} \subseteq \mathcal{N}_b\}$); we define

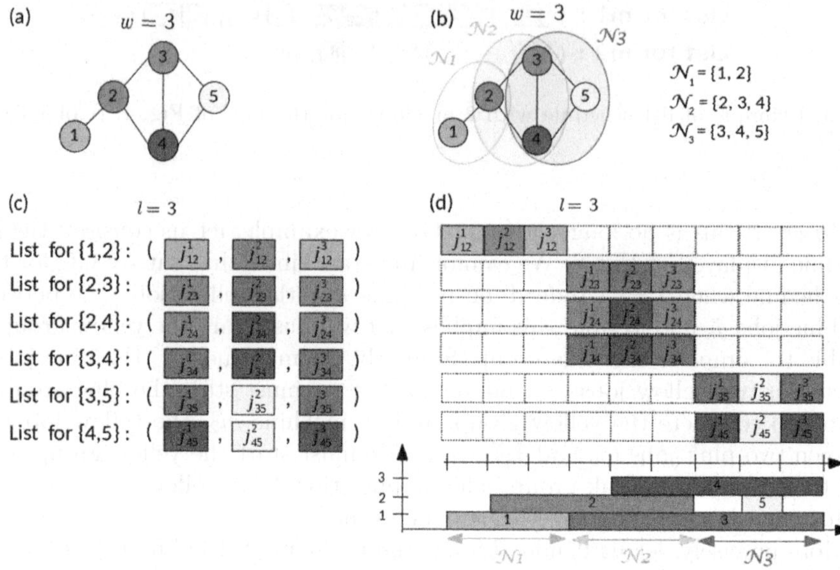


**Fig. 7.** Reduction from Pathwidth to *LS-GC*. (a): Example of instance of Pathwidth. (b): Example of solution of (a). (c): List schedule of the instance of *LS-GC* corresponding to (a). (d): Solution of (c) corresponding to (b).

$B_{j^1_{uv}} = 3*a-3$, $B_{j^2_{uv}} = 3*a-2$, and $B_{j^3_{uv}} = 3*a-1$; end times are computed by adding the processing time 1 to every start time. In Fig. 7(d), we display start and end times computed for a solution of the Pathwidth instance of Fig. 7(a). We can easily check that start and end times are consistent with the list schedule. To show that start and end times satisfy GC, we have to show that the number of active groups never exceeds $l$, and this is a consequence of the fact that the number of vertices in a set $\mathcal{N}_b$ never exceeds $w = l$. Indeed if we consider a time $t$ with $3 * a - 3 \leq t \leq 3 * a - 1$ ($a \in [1, n]$), then the only groups that can be active at time $t$ are those associated with nodes in $\mathcal{N}_a$ and $\#\mathcal{N}_a \leq w = l$.

Finally, let us show that every solution of the instance of *LS-GC* built from an instance of Pathwidth corresponds to a solution of this Pathwidth instance. A solution of an instance of *LS-GC* is an assignment of values to $B_j$ and $E_j$ for every job $j \in \mathcal{J}$ (defining start and end times of $j$). For each node $u \in \mathcal{N}$, we have a group of jobs $\mathcal{G}_u$, and the start time $B_u$ of this group is the smallest start time of its jobs (*i.e.*, $B_u = \min\{B_j : j \in \mathcal{G}_u\}$) whereas the end time $E_u$ of this group is the largest end time of its jobs (*i.e.*, $E_u = \max\{E_j : j \in \mathcal{G}_u\}$). Let $\mathcal{T} = \{B_u : u \in \mathcal{N}\}$ be the set of all group start times, and let $(t_1, \ldots, t_{\#\mathcal{T}})$ be the ordered sequence of values in $\mathcal{T}$. The solution of the Pathwidth instance is $(\mathcal{N}_1, \ldots, \mathcal{N}_{\#\mathcal{T}})$ such that for each $i \in [1, \#\mathcal{T}]$, $\mathcal{N}_i = \{u \in \mathcal{N} : B_u \leq t_i < E_u\}$. We can check that $(\mathcal{N}_1, \ldots, \mathcal{N}_{\#\mathcal{T}})$ is a solution. Indeed, for each edge $\{u, v\} \in \mathcal{E}$ with $u < v$, the list $(j^1_{uv}, j^2_{uv}, j^3_{uv})$ ensures that when $j^2_{uv}$ starts both $\mathcal{G}_u$ and $\mathcal{G}_v$ are active groups. Hence, $\forall i \in [1, \#\mathcal{T}]$ such that $t_i = max(B_u, B_v)$, $\{u, v\} \in \mathcal{N}_i$.

Now if we have $i, j, k \in [1, \#\mathcal{T}]$ with $i \leq j \leq k$, $u \in \mathcal{N}_i$ and $u \in \mathcal{N}_k$ then $B_u \leq t_i \leq t_j \leq t_k < E_u$. Hence $u \in \mathcal{N}_j$, and so $\mathcal{N}_i \cap \mathcal{N}_k \subseteq \mathcal{N}_j$. Finally, $\forall i \in [1, \#\mathcal{T}]$, all groups of $\mathcal{N}_i$ are active at time $t_i$. So $\#\mathcal{N}_i \leq l = w$.

## 5 ACO Algorithm for the GCSP

Many different ACO algorithms have been proposed for solving scheduling problems, and a review of 54 of these algorithms may be found in [28]. ACO algorithms use pheromone trails to learn promising solution components and progressively intensify the search around them. The two most widely considered definitions of pheromone trails for scheduling problems are: *Job trails*, where a trail $\tau(j, j')$ is associated with every couple of jobs $(j, j') \in \mathcal{J}^2$ to learn the desirability of scheduling $j'$ just after $j$ on a same machine; and *Position trails*, where a trail $\tau(j, m, n)$ is associated with each triple $(j, m, n) \in \mathcal{J} \times \mathcal{M} \times [1, \#\mathcal{J}]$ to learn the desirability of scheduling job $j$ at position $n$ on machine $m$.

Most ACO algorithms for scheduling problems follow the basic template displayed in Algorithm 1. At each iteration of the loop lines 1–9, $n_{ants}$ solutions are constructed in a greedy randomised way, where $n_{ants}$ is a parameter which is used to control exploration (the larger $n_{ants}$, the stronger the exploration). At each iteration of the greedy construction (lines 4–8), a machine $m$ and a job $j$ are chosen, and $j$ is scheduled on $m$, until all jobs have been scheduled. The choice of $m$ is done according to some heuristics which depend on the scheduling problem. The choice of $j$ is done in a randomised way, according to a probability $p(j)$ which depends on two factors. The pheromone factor $f_\tau(j)$ represents the learned desirability of scheduling $j$ on $m$ and its definition depends on the pheromone trail definition: for *Job trails*, $f_\tau(j) = \tau(j', j)$ where $j'$ is the last job scheduled on $m$; for *Position trails*, $f_\tau(j) = \tau(j, m, k)$ where $k$ is the number of jobs scheduled on $m$. The heuristic factor $\eta(j)$ evaluates the interest of scheduling $j$ on $m$ and its exact definition depends on the scheduling problem. $\alpha$ and $\beta$ are two parameters which are used to balance these two factors.

At the end of each cycle (line 9), pheromone trails are updated in two steps. First, every pheromone trail is decreased by multiplying it with $1 - \rho$ where $\rho \in [0, 1]$ is a parameter which controls the speed of intensification: the larger $\rho$, the quicker search is intensified towards the best solutions found so far. In a second step, pheromone trails associated with the best solution among the $n_{ants}$ last computed solutions are increased in order to increase the probability of selecting the components of this solution in the next constructions.

In [9], Algorithm 1 is adapted to solve GCSPs as follows. Line 5, $m$ is the machine which minimizes the end time of the last job assigned to it. Line 6, the heuristic factor $\eta(j)$ is set to 0 whenever the current number of active groups is equal to the limit $l$ and $j$ belongs to a group which is not yet active. In this case, the probability $p(j)$ of selecting $j$ is equal to 0, thus ensuring that solutions always satisfy GC. When $j$ can be scheduled without violating GC (*i.e.*, the number of active groups is smaller than $l$, or it is equal to $l$ and $j$ belongs to an active group), the heuristic factor $\eta(j)$ is defined as the *ATCS*

**Algorithm 1:** ACO algorithm for scheduling problems

```
1  while time limit not reached do
2      for i in [1, n_ants] do
           /* Greedy randomised construction of one solution     */
3          Cand ← 𝒥
4          while Cand ≠ ∅ do
5              choose a machine m ∈ ℳ according to some heuristic
6              choose j ∈ Cand w.r.t. probability p(j) = [f_τ(j)]^α·[η(j)]^β / Σ_{j'∈Cand} [f_τ(j')]^α·[η(j')]^β
7              assign m to M_j, and assign values to B_j and E_j
8              remove j from Cand
9      update pheromone trails
10 return the best constructed solution
```

Line 6: $p(j) = \frac{[f_\tau(j)]^\alpha \cdot [\eta(j)]^\beta}{\sum_{j' \in Cand} [f_\tau(j')]^\alpha \cdot [\eta(j')]^\beta}$

*(Apparent Tardiness Cost with Setup-times)* score defined in [23]. Line 9, before updating pheromone trails, the best solution (among the last $n_{ants}$ constructed solutions) is improved by applying a local search step. Also, pheromone trails are updated according to the MMAS framework of [27], *i.e.*, every pheromone trail is bounded between two parameters $\tau_{min}$ and $\tau_{max}$. Also, every pheromone trail is initialised to $\tau_{max}$ at the beginning of the search process (before line 1).

This ACO algorithm has one hyper-parameter, which is used to choose the pheromone trail definition (*i.e., Job trails*, *Position trails*, or a new definition introduced in [9] and called *Time trails*). The algorithm also has the following parameters: $n_{ants}$, $\alpha$, $\beta$, $\rho$, $\tau_{min}$ and $\tau_{max}$, plus two parameters for the local search step. In [9], a portfolio of nine complementary parameter configurations are identified, and the per-instance algorithm selector LLAMA [14] is used to select from this portfolio the configuration expected to perform best for every new instance to solve.

## 6 New Hybrid CPO-ACO Approach

Many hybrid approaches combine exhaustive solvers (such as CP or Integer Linear Programming, for example) with meta-heuristics [4]. Some of these hybrid approaches are referred to as *matheuristics* [17]. A well known example of hybrid approach is LNS [25] which uses CP to explore the neighborhood of a local search.

Different hybrid CP-ACO approaches have been proposed such as, for example, [8,12,13,19,20,26]. Some approaches use constraint propagation during the construction of solutions by ACO (lines 3–8 of Algorithm 1), to filter the set of candidate components and remove those that do not satisfy constraints [12,19]. Some other approaches use ACO to learn ordering heuristics which are used by CP [13,20]. In [8], a bi-level hybrid process is introduced where ACO is used to assign a subset of variables, and the remaining variables are assigned by CP.

### 6.1 Description of the Hybrid Approach

In this section, we introduce a new hybrid CPO-ACO approach where ACO and CPO are alternatively executed and exchange solutions: solutions found by ACO are used as starting points for CPO, whereas solutions found by CPO are used to update pheromone trails. More precisely, we modify Algorithm 1 as follows: every $k$ iterations of the loop lines 1–9, we call CPO. When calling CPO, we supply it with the best solution constructed during the $k$ last iterations of ACO, and this solution is used by CPO as a starting point. Each call to CPO is limited to a given number of backtracks. Once CPO has reached this limit, we get the best solution found by CPO and update pheromone trails according to this solution.

The limit on the number of backtracks follows a geometric progression, as often done in classical restart strategies: the first limit is equal to $b$, and after each call to CPO, this limit is multiplied by $g$. Hence, our hybrid CPO-ACO approach may be viewed as a particular case of restart where ACO is run before each restart in order to provide a new initial solution, and the best solution after each restart is given back to ACO to reinforce its pheromone trails.

Our hybrid CPO-ACO algorithm has three parameters: the number $k$ of ACO cycles which are executed before each call to CPO, and the values $b$ and $g$ which are used to fix the limit on the number of backtracks of CPO. In all our experiments, these parameters are set to $k = 5$, $b = 1000$, and $g = 20$. With these values, the number of calls to CPO (within a time limit of one hour) ranges from 3 for the smallest instances to 4 for the largest ones.

For the ACO algorithm used in CPO-ACO, we use a parameter setting which favors a quick convergence, as only $k = 5$ cycles of ACO are run before each CPO restart, and only 3 or 4 restarts are done: $\alpha = 5$, $\beta = 10$, $\rho = 0.2$, $n_{ants} = 40$, $\tau_{min} = 0.1$, $\tau_{max} = 4$, and the pheromone definition is *Position trails.*

In CPO-ACO, CPO is run with its default setting so that CPO performs restarts during each of its runs. We have made experiments with other settings (including the *DepthFirst* search mode of CPO, which performs a single search), and the best results were obtained with the default setting of CPO.

### 6.2 Experimental Evaluation

*Compared Approaches.* We compare CPO-ACO with CPO in its default setting (which is the best performing setting for the GCSP). We also report results obtained with the ACO algorithm introduced in [9] (we consider the ACO variant which uses LLAMA to select ACO parameters for each instance as this variant obtains the best results).

Finally, we report results obtained with *Solution-Guided Multi-Point Constructive Search (SGMPCS)* [3], which is a constructive search technique that performs a series of resource-limited tree searches where each search begins either from an empty solution (as in randomized restart) or from an "elite" solution. SGMPCS has some similarities with our approach, as it provides initial elite solutions to the CP solver. Hence, the comparison with SGMPCS allows us to evaluate the interest of using pheromone to learn good solution components that are exploited to compute new starting points. For SGMPCS, we use CPO as CP solver, and we build elite solutions with the same greedy randomised algorithm as the one used in CPO-ACO, except that we ignore the pheromone factor $f_\tau$ when computing the probability of selecting a job. The parameters of SGMPCS are set as in [3], *i.e.* the probability of starting from an empty solution is 0.25 and the size of the elite list is 4.

We separate instances in two classes for analysing results: the *closed* class contains every instance for which the optimal solution is known (either because the objective function is equal to 0, or because an approach has been able to prove optimality); the *open* class contains all other instances. For the loose (resp. medium and tight) class, there are 361 (resp. 356 and 339) closed instances, and 187 (resp. 192 and 209) open instances.

*Results for Closed Instances.* On the left part of Fig. 8, we display the cumulative number of solved instances with respect to time, for closed instances. After one hour of CPU time, CPO-ACO clearly outperforms all other approaches for the three classes, and it has been able to solve nearly all closed instances. More precisely, the number of loose, medium and tight instances solved by CPO-ACO in one hour is equal to 356, 352, and 329, respectively, whereas it is equal to 344, 332, and 297 for SGMPCS, to 344, 318, and 242 for ACO, and to 324, 306, and 261 for CPO. We observe that the gap between CPO-ACO and other approaches increases when increasing the tightness of GC.

However, for short time limits (smaller than 10 s), conclusions are different. In particular, after 1 s, the best approach is SGMPCS: The number of loose, medium and tight instances solved by SGMPCS in 1 s is equal to 122, 100, and 48, respectively, whereas it is equal to 92, 78, and 35 for CPO-ACO. This shows that good starting points allow CPO to quickly find better solutions. However, after 10 s, ACO is able to build better starting points by exploiting good solutions previously constructed by ACO or CPO.

*Results for Open instances.* We evaluate the quality of a solution of an open instance by computing its ratio to the best known solution: if this ratio is equal to 1, the solution is the best known solution; if it is equal to $r > 1$, the solution is $r$ times as large as the best known solution.

On the right part of Fig. 8, we display the evolution of the average ratio to best known solutions with respect to time. After one hour, CPO-ACO clearly outperforms all other approaches for the three classes. Its average ratio is rather close to 1, meaning that in many cases it has found the best known solution. More precisely, for loose, medium, and tight instances this ratio is equal to 1.37, 1.38, and 1.60, respectively, for CPO-ACO, whereas it is equal to 2.18, 2.51, and 3.24 for ACO, to 2.95, 3.06, and 5.50 for SGMPCS, and to 3.86, 5.95, and 13.44 for CPO. Like for closed instances, the gap between CPO-ACO and other approaches increases when increasing the tightness of the constraint. This is particularly true for CPO which has very poor performance on tight instances. SGMPCS finds

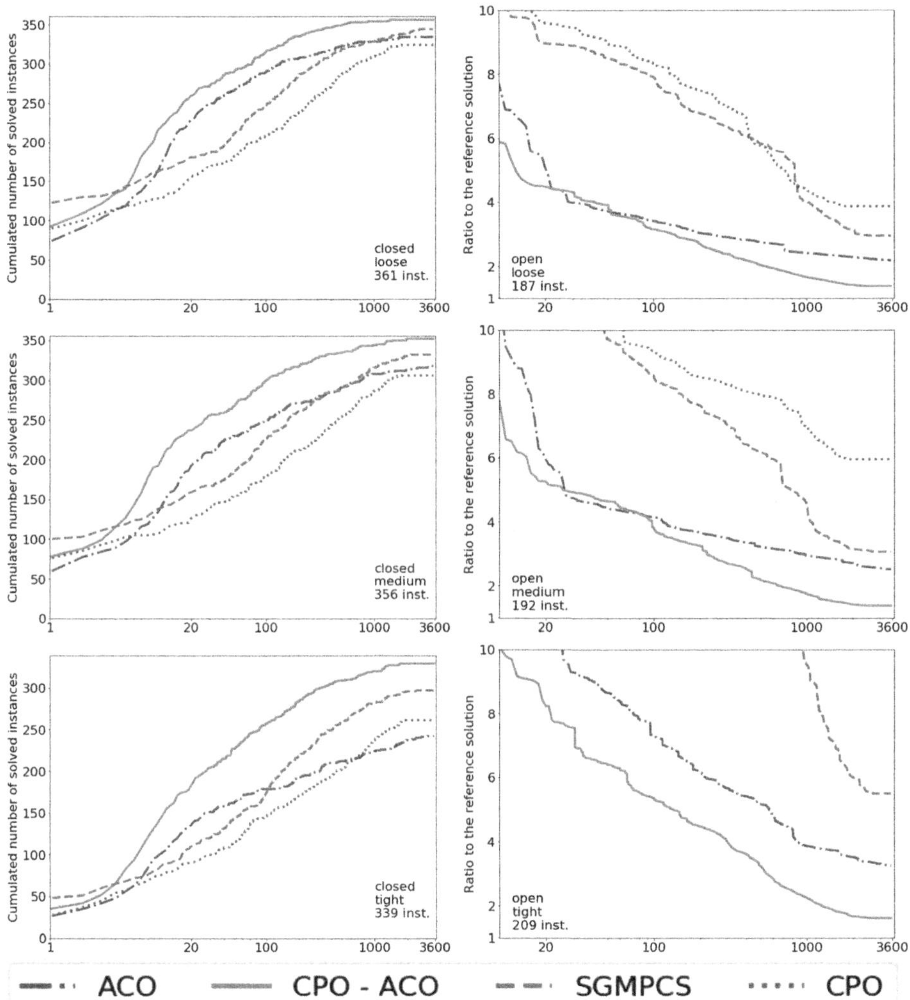


**Fig. 8.** Results of CPO-ACO, CPO, ACO, and SGMPCS. Left: Evolution of the cumulative number of solved instances for closed instances. Right: Evolution of the average ratio to the best known solution for open instances. Top: Loose instances. Middle: Medium instances. Bottom: Tight instances.

better solutions than CPO, and this shows us the interest of giving good starting points to CPO. However, like for open instances, we observe that the starting points learned by ACO allow CPO to find much better solutions.

In Fig. 9, we display ratio distributions. CPO-ACO has much smaller median ratios and inter-quartile ranges. For loose (resp. medium and tight) instances it finds the best known solution for 115 (resp.106 and 114) instances, whereas

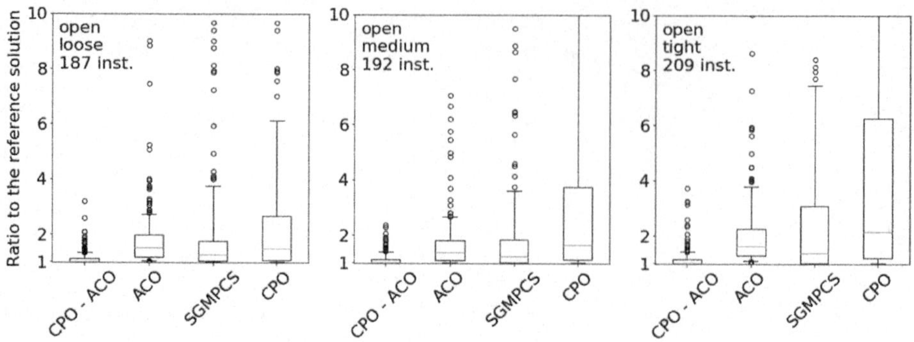


**Fig. 9.** Distribution of ratios to best known solutions after one hour on open instances.

ACO finds it for 9 (resp. 21 and 14) instances, SGMPCS for 48 (resp. 44 and 34) instances and CPO for 27 (resp. 25 and 37) instances.

## 7 Conclusion

We have shown that GC is a challenging constraint for CP solvers such as CPO. In particular, reasoning on precedence relations (which is classical to solve scheduling problems) may be misleading as it is $\mathcal{NP}$-complete to find starting times that satisfy GC when a list schedule is provided. We have introduced an hybrid framework which drastically improves CPO performance by providing good starting points. These starting points are computed by an ACO algorithm which uses pheromone trails to learn good solution components.

This hybrid framework introduces new parameters: parameters to define the progression of the limit used to trigger restarts, and classical ACO parameters. In all experiments reported in this paper, we have used the same parameter setting. However, it could be interesting to use a per-instance algorithm selector such as LLAMA to dynamically choose the best parameter setting within a portfolio of representative and complementary settings. Other further work will concern the study of the GCSP: Fig. 5 shows us that some small instances are much harder than some large instances, and it would be interesting to identify instance parameters that characterize hardness.

## References


1. Aggoun, A., Beldiceanu, N.: Extending chip in order to solve complex scheduling and placement problems. Math. Comput. Model. **17**(7), 57–73 (1993). https://doi.org/10.1016/0895-7177(93)90068-A
2. Baptiste, P., Bonifas, N.: Redundant cumulative constraints to compute preemptive bounds. Disc. Appl. Math. **234**, 168–177 (2018). https://doi.org/10.1016/j.dam.2017.05.001

3. Beck, J.C.: Solution-guided multi-point constructive search for job shop scheduling. J. Artif. Intell. Res. **29**, 49–77 (2007)
4. Blum, C., Puchinger, J., Raidl, G.R., Roli, A.: Hybrid metaheuristics in combinatorial optimization: a survey. Appl. Soft Comput. **11**(6), 4135–4151 (2011)
5. Bonifas, N.: A O(n 2 log(n)) propagation for the Energy Reasoning, February 2016
6. Brucker, P., Drexl, A., Möhring, R., Neumann, K., Pesch, E.: Resource-constrained project scheduling: notation, classification, models, and methods. Eur. J. Oper. Res. **112**(1), 3–41 (1999). https://doi.org/10.1016/S0377-2217(98)00204-5
7. Da Col, G., Teppan, E.C.: Industrial size job shop scheduling tackled by present day CP solvers. In: Schiex, T., de Givry, S. (eds.) CP 2019. LNCS, vol. 11802, pp. 144–160. Springer, Cham (2019). https://doi.org/10.1007/978-3-030-30048-7_9
8. Di Gaspero, L., Rendl, A., Urli, T.: A hybrid ACO+CP for balancing bicycle sharing systems. In: Blesa, M.J., Blum, C., Festa, P., Roli, A., Sampels, M. (eds.) HM 2013. LNCS, vol. 7919, pp. 198–212. Springer, Heidelberg (2013). https://doi.org/10.1007/978-3-642-38516-2_16
9. Groleaz, L., Ndiaye, S.N., Solnon, C.: ACO with automatic parameter selection for a scheduling problem with a group cumulative constraint. In: GECCO 2020 - Genetic and Evolutionary Computation Conference, Cancun, Mexico, pp. 1–9, July 2020. https://doi.org/10.1145/3377930.3389818
10. Hartmann, S., Kolisch, R.: Experimental evaluation of state-of-the-art heuristics for the resource-constrained project scheduling problem. Eur. J. Oper. Res. **127**(2), 394–407 (2000). https://doi.org/10.1016/S0377-2217(99)00485-3
11. Kashiwabara, T.: NP-completeness of the problem of finding a minimal-clique number interval graph containing a given graph as a subgraph. In: Proceedings of 1979 International Symposium Circuit System, pp. 657–660 (1979)
12. Khichane, M., Albert, P., Solnon, C.: Integration of ACO in a constraint programming language. In: Dorigo, M., Birattari, M., Blum, C., Clerc, M., Stützle, T., Winfield, A.F.T. (eds.) ANTS 2008. LNCS, vol. 5217, pp. 84–95. Springer, Heidelberg (2008). https://doi.org/10.1007/978-3-540-87527-7_8
13. Khichane, M., Albert, P., Solnon, C.: Strong combination of ant colony optimization with constraint programming optimization. In: Lodi, A., Milano, M., Toth, P. (eds.) CPAIOR 2010. LNCS, vol. 6140, pp. 232–245. Springer, Heidelberg (2010). https://doi.org/10.1007/978-3-642-13520-0_26
14. Kotthoff, L.: LLAMA: leveraging Learning to Automatically Manage Algorithms. arXiv:1306.1031 [cs], June 2013
15. Laborie, P.: Algorithms for propagating resource constraints in AI planning and scheduling: existing approaches and new results. Artif. Intell. **143**(2), 151–188 (2003). https://doi.org/10.1016/S0004-3702(02)00362-4
16. Laborie, P., Rogerie, J., Shaw, P., Vilím, P.: IBM ILOG CP optimizer for scheduling: 20+ years of scheduling with constraints at IBM/ILOG. Constraints **23**(2), 210–250 (2018). https://doi.org/10.1007/s10601-018-9281-x
17. Maniezzo, V., Stützle, T., Voß, S. (eds.): Matheuristics - Hybridizing Metaheuristics and Mathematical Programming. Annals of Information Systems, vol. 10. Springer (2010)
18. Mercier, L., Van Hentenryck, P.: Edge finding for cumulative scheduling. INFORMS J. Comput. **20**(1), 143–153 (2008). https://doi.org/10.1287/ijoc.1070.0226
19. Meyer, B.: Hybrids of constructive metaheuristics and constraint programming: a case study with ACO. In: Hybrid Metaheuristics, vol. 114, pp. 151–183. Springer, Heidelberg (2008). https://doi.org/10.1007/978-3-540-78295-7_6

20. Meyer, B., Ernst, A.: Integrating ACO and constraint propagation. In: Dorigo, M., Birattari, M., Blum, C., Gambardella, L.M., Mondada, F., Stützle, T. (eds.) ANTS 2004. LNCS, vol. 3172, pp. 166–177. Springer, Heidelberg (2004). https://doi.org/10.1007/978-3-540-28646-2_15
21. Neumann, K., Schwindt, C.: Project scheduling with inventory constraints. Math. Methods Oper. Res. (ZOR) **56**(3), 513–533 (2003). https://doi.org/10.1007/s001860200251
22. Ouellet, P., Quimper, C.-G.: Time-table extended-edge-finding for the cumulative constraint. In: Schulte, C. (ed.) CP 2013. LNCS, vol. 8124, pp. 562–577. Springer, Heidelberg (2013). https://doi.org/10.1007/978-3-642-40627-0_42
23. Pinedo, M.L.: Scheduling. Springer, Cham (2016). https://doi.org/10.1007/978-3-319-26580-3
24. Schutten, J.: List scheduling revisited. Oper. Res. Lett. **18**(4), 167–170 (1996). https://doi.org/10.1016/0167-6377(95)00057-7
25. Shaw, P.: Using constraint programming and local search methods to solve vehicle routing problems. In: Maher, M., Puget, J.-F. (eds.) CP 1998. LNCS, vol. 1520, pp. 417–431. Springer, Heidelberg (1998). https://doi.org/10.1007/3-540-49481-2_30
26. Solnon, C.: Constraint Programming with Ant Colony Optimization. John Wiley, Hoboken (2010). (232 pages)
27. Stützle, T., Hoos, H.: Improvements on the Ant-System: Introducing the MAX-MIN Ant System, pp. 245–249. Springer, Vienna (1998). https://doi.org/10.1007/978-3-7091-6492-1_54
28. Tavares Neto, R., Godinho Filho, M.: Literature review regarding ant colony optimization applied to scheduling problems: guidelines for implementation and directions for future research. Eng. Appl. Artif. Intell. **26**(1), 150–161 (2013). https://doi.org/10.1016/j.engappai.2012.03.011
29. Vilím, P.: Timetable edge finding filtering algorithm for discrete cumulative resources. In: Achterberg, T., Beck, J.C. (eds.) CPAIOR 2011. LNCS, vol. 6697, pp. 230–245. Springer, Heidelberg (2011). https://doi.org/10.1007/978-3-642-21311-3_22

# A Branch-and-bound Algorithm to Rigorously Enclose the Round-Off Errors

Rémy Garcia(✉), Claude Michel, and Michel Rueher

Université Côte d'Azur, CNRS, I3S, Nice, France
{remy.garcia,claude.michel,michel.rueher}@i3s.unice.fr

**Abstract.** Round-off errors occur each time a program makes use of floating-point computations. They denote the existence of a distance between the actual floating-point computations and the intended computations over the reals. Therefore, analyzing round-off errors is a key issue in the verification of programs with floating-point computations. Most existing tools for round-off error analysis provide an over-approximation of the error. The point is that these approximations are often too coarse to evaluate the effective consequences of the error on the behaviour of a program. Some other tools compute an under-approximation of the maximal error. But these under-approximations are either not rigorous or not reachable. In this paper, we introduce a branch-and-bound algorithm to rigorously enclose the maximal error. Thanks to the use of rational arithmetic, our branch-and-bound algorithm provides a tight upper bound of the maximal error and a lower bound that can be exercised by input values. We outline the advantages and limits of our framework and compare it with state-of-the-art methods. Preliminary experiments on standard benchmarks show promising results.

**Keywords:** Floating-point numbers · Round-off error · Constraints over floating-point numbers · Optimization

## 1 Introduction

Floating-point computations involve errors due to rounding operations that characterize the distance between the intended computations over the reals and the actual computations over the floats. An error occurs at the level of each basic operation when its result is rounded to the nearest representable floating-point number. The final error results from the combination of the rounding errors produced by each basic operation involved in an expression and some initial errors linked to input variables and constants. Such errors impact the precision and the stability of computations and can lead to an execution path over the floats that is significantly different from the expected path over the reals. A faithful

This work was partially supported by ANR COVERIF (ANR-15-CE25-0002).


H. Simonis (Ed.): CP 2020, LNCS 12333, pp. 637–653, 2020.
https://doi.org/10.1007/978-3-030-58475-7_37

account of these errors is mandatory to capture the actual behaviour of critical programs with floating-point computations.

Efficient tools exist for error analysis that rely on an over-approximation of the errors in programs with floating-point computations. For instance, Fluctuat [6,7] is an abstract interpreter that combines affine arithmetic and zonotopes to analyze the robustness of programs over the floats, FPTaylor [18,19] uses symbolic Taylor expansions to compute tight bounds of the error, and PRECiSA [16,22] is a more recent tool that relies on static analysis. Other tools compute an under-approximation of errors to find a lower bound of the maximal absolute error, e.g. FPSDP [11] which takes advantage of semidefinite programming or, S3FP [2] that uses guided random testing to find inputs causing the worst error. Over-approximations and under-approximations of errors are complementary approaches for providing better enclosures of the maximal error. However, none of the available tools compute both an over-approximation and an under-approximation of errors. Such an enclosure would be very useful to give insights on the maximal absolute error, and how far computed bounds are from it. It is important to outline that approximations do not capture the effective behaviour of a program: they may generate *false positives*, that is to say, report that an assertion might be violated even so in practice none of the input values can exercise the related case. To get rid of *false positives*, computing maximal errors, i.e. the greatest reachable absolute errors, is required. Providing an enclosure of the maximal error, and even finding it, is the goal of the work presented here.

The core of the proposed approach is a branch-and-bound algorithm that attempts to maximize a given error of a program with floating-point computations. This branch-and-bound algorithm is embedded in a solver over the floats [1,13–15,24] extended to constraints over errors [5]. The resulting system, called FErA (**F**loating-point **Er**ror **A**nalyzer), provides not only a sound over-approximation of the maximal error but also a reachable under-approximation with input values that exercise it. To our knowledge, our tool is the first one that combines upper and lower bounding of maximal round-off errors. A key point of FErA is that both bounds rely on each other for improvement.

Maximizing an error can be very expensive for the errors that are unevenly distributed. Even on a single operation, such a distribution is cumbersome and finding input values that exercise it often resort to an enumeration process. A combination of floating-point operations often worsen this behaviour, but may, in some cases, soften it thanks to error compensations. One advantage of our approach is that the branch-and-bound is an anytime algorithm, and thus it always provides an enclosure of the maximal error alongside input values exercising the lower bound.

### 1.1 Motivating Example

Consider the piece of code in Example 1 that computes $z = (3 * x + y)/w$ using 64 bits doubles with $x \in [7, 9]$, $y \in [3, 5]$, and $w \in [2, 4]$.

```
z = (3*x+y)/w;

if (z - 10 <= δ) {
  proceed();
} else {
  raiseAlarm();
}
```

**Example 1.** Simple program

**Table 1.** Absolute error bound

| Tool | Error |
|---|---|
| FPTaylor | 5.15e−15 |
| PRECiSA | 5.08e−15 |
| Fluctuat | 6.28e−15 |
| FErA | 4.96$e$−15 |

The computation of $z$ precedes a typical condition of a control-command code. When $z$ is lower than 10, with a tolerance to errors of $\delta$, values supported by $z$ are considered as safe and related computations can be done. Otherwise, an alarm must be raised.

Now, assume that $\delta$ is set to $5.0e{-}15$. The issue is to know whether this piece of code behaves as expected, i.e. to know whether the error on $z$ is small enough to avoid raising the alarm when the value of $z$ is less than or equal to 10 on $\mathbb{R}$.

Table 1 reports the error values given by FPTaylor [18,19], PRECiSA [22], Fluctuat [6,7], and our tool FErA. All analyzers but FErA compute a bound greater than $\delta$. Results from FPTaylor, PRECiSA and Fluctuat suggests that the alarm might inappropriately be raised.

FErA computes a round-off error bound of $4.96e{-}15$ in about 0.185 seconds. It also computes a lower bound on the largest absolute error of $3.55e{-}15$ exercised by the following input values:

$$x = 8.99999999999996624922 \qquad e_x = -8.88178419700125232339e{-}16$$
$$y = 4.99999999999994848565 \qquad e_y = -4.44089209850062616169e{-}16$$
$$w = 3.19999999999998419042 \qquad e_w = +2.22044604925031308085e{-}16$$
$$z = 10.00000000000000035527 \qquad e_z = -3.55271367880050092936e{-}15$$

In other words, our sound optimizer not only guarantees that, despite errors in floating-point computations, this program can never raise an alarm when $z <= 10$ over the reals, but it also provides an enclosure of the largest absolute error. Such an enclosure having a ratio[1] of $\approx 1.4$ shows that the round-off error bounds of FErA are close to the actual error. Note that the computed lower bound corresponds to a case where $z$ over the floats is bigger than 10 while it remains lower than 10 over the reals.

Now, assume that $\delta$ is set to $3.00e{-}15$. The issue here is to know whether there exists at least one case where raiseAlarm() is reached when $z$ is less than or equal to 10 on real numbers. The previously computed enclosure of the maximal error provided by FErA ensures that there exist at least one case where raiseAlarm() is reached with an error bigger than $\delta$. The other tools are unable to

[1] The ratio between FErA computed upper and lower bound is equal to $4.96e{-}15/3.5527e{-}15 = 1.396$.

do so as none of them compute a reachable lower bound on the largest absolute error.

The rest of the paper is organized as follows: Section 2 introduces notations and definitions. Section 3 recalls the constraint system for round-off error analysis and explains how the filtering works. Section 4 formally introduces the branch-and-bound algorithm and its main properties. Section 5 describes in more detail related works on computing a lower bound on the maximal error and their pitfalls. Section 6 provides preliminary experiments on a set of standard benchmarks.

## 2 Notation and Definitions

Our system for round-off error analysis focuses on the four classical arithmetic operations, i.e. $+, -, \times$ and $/$, for which the error can be computed exactly using rational arithmetic [5]. As usual, a constraint satisfaction problem, or CSP, is defined by a triple $\langle X, D, C \rangle$, where $X$ denotes the set of variables, $D$, the set of domains, and $C$, the set of constraints. The set of rational numbers is denoted $\mathbb{Q}$, and the set of real numbers is denoted $\mathbb{R}$. $\mathbb{F}$ denotes a set of floating-point numbers whose precision is one of the precision defined in IEEE 754 [10]. Though the approach presented here applies to other types of floats, in the rest of the paper, $\mathbb{F}$ will denote 64 bits floating-point numbers unless otherwise stated. For each floating-point variable $x$ in $X$, the domain of values of $x$ is represented by the interval $\mathbf{x} = [\underline{\mathbf{x}}, \overline{\mathbf{x}}] = \{x \in \mathbb{F},\ \underline{\mathbf{x}} \leq x \leq \overline{\mathbf{x}}\}$, where $\underline{\mathbf{x}} \in \mathbb{F}$ and $\overline{\mathbf{x}} \in \mathbb{F}$. $\underline{\mathbf{x}}$ (resp. $\overline{\mathbf{x}}$) denotes the lower (resp. upper) bound of the interval $\mathbf{x}$. The domain of errors $\mathbf{e_x}$ of $x$ is represented by an interval of rationals $\mathbf{e_x} = [\underline{\mathbf{e}}_{\mathbf{x}}, \overline{\mathbf{e}}_{\mathbf{x}}] = \{e_x \in \mathbb{Q},\ \underline{\mathbf{e}}_{\mathbf{x}} \leq e_x \leq \overline{\mathbf{e}}_{\mathbf{x}}\}$ where $\underline{\mathbf{e}}_{\mathbf{x}} \in \mathbb{Q}$ and $\overline{\mathbf{e}}_{\mathbf{x}} \in \mathbb{Q}$. $x_{\mathbb{F}}$ (respectively, $x_{\mathbb{Q}}$ and $x_{\mathbb{R}}$) denotes a variable that takes its values in $\mathbb{F}$ (respectively, $\mathbb{Q}$ and $\mathbb{R}$). A variable is instantiated when its domain of values is reduced to a degenerate interval, or a singleton, i.e. when $\underline{\mathbf{e}}_{\mathbf{x}} = \overline{\mathbf{e}}_{\mathbf{x}}$.

The branch-and-bound algorithm maximizes an error noted $e$ that results from floating-point computations along a given path in a program. $e^*$ denotes the lower bound, i.e. the maximal error computed so far while $\overline{e}$ denotes the upper bound, i.e. the currently best known over-approximation of the error. Both of those bounds are expressed in absolute value. $S$ is the ordered, by error values, set of couples $(e, sol)$ where $e$ and $sol$ are, respectively, an error and its corresponding input values. A box $B$ is the cartesian product of variable domains. For the sake of clarity, a box $B$ can be used as exponent, e.g. $\mathbf{x}^B$ indicates that an element $\mathbf{x}$ is in box $B$. $L$ is the set of boxes left to compute.

## 3 A Constraint System for Round-Off Error

The branch-and-bound algorithm at the core of our framework is based on a constraint system on errors [5] that we briefly describe in this section.

### 3.1 Computing Rounding Errors

The IEEE 754 standard [10] requires a correct rounding for the four basic operations of the floating-point arithmetic. The result of such an operation over the floats must be equal to the rounding of the result of the equivalent operation over the reals. More formally, $z = x \odot y = round(x \cdot y)$ where $z$, $x$, and $y$ are floating-point numbers, $\odot$ is one of the four basic arithmetic operations on floats, namely, $\oplus$, $\ominus$, $\otimes$, $\oslash$, while $\cdot$ are the equivalent operations on reals, namely, $+$, $-$, $\times$, $/$; *round* being the rounding function. This property is used to bound the error introduced by each elementary operation on floats by $\pm\frac{1}{2}ulp(z)$[2] when the rounding mode is set to round to the "nearest even" float, the most frequently used rounding mode.

The deviation of a computation over the floats takes root in each elementary operation. So, it is possible to rebuild the final deviation of an expression from the composition of errors due to each elementary operation involved in that expression. Let us consider a simple operation like the subtraction as in $z = x \ominus y$: input variables, $x$ and $y$, can come with errors attached due to previous computations. For instance, the deviation on the computation of $x$, $e_x$, is given by $e_x = x_\mathbb{R} - x_\mathbb{F}$ where $x_\mathbb{R}$ and $x_\mathbb{F}$ denote the expected results, respectively, on reals and on floats.

The computation deviation due to a subtraction can be formulated as follows: for $z = x \ominus y$, $e_z$, the error on $z$, is equal to $(x_\mathbb{R} - y_\mathbb{R}) - (x_\mathbb{F} \ominus y_\mathbb{F})$.

As $e_x = x_\mathbb{R} - x_\mathbb{F}$ and $e_y = y_\mathbb{R} - y_\mathbb{F}$, we have $e_z = ((x_\mathbb{F} + e_x) - (y_\mathbb{F} + e_y)) - (x_\mathbb{F} \ominus y_\mathbb{F})$. So, the deviation between the result on reals and the result on floats for a subtraction can be computed by: $e_z = e_x - e_y + ((x_\mathbb{F} - y_\mathbb{F}) - (x_\mathbb{F} \ominus y_\mathbb{F}))$, where $(x_\mathbb{F} - y_\mathbb{F}) - (x_\mathbb{F} \ominus y_\mathbb{F})$ characterizes the error produced by the subtraction operation itself. Let's $e_\ominus$ denotes the error produced by the subtraction operation. The formula can then be denoted by: $e_z = e_x - e_y + e_\ominus$, that combines deviations from input values and the deviation introduced by the elementary operation.

Computation of deviations for all four basic operations are given in Fig. 1. For each of these formulae, the error computation combines deviations from input values and the error introduced by the current operation. Note that, for the multiplication and division, this deviation is proportional to the input values.

All these formulae compute the difference between the expected result on reals and the actual one on floats for a basic operation. Our constraint solver over the errors relies on these formulae.

### 3.2 A Constraint Network with Three Domains

As usually, to each variable $x$ is associated $\mathbf{x}$, its domain of values. The domain $\mathbf{x}$ denotes the set of possible values that variable $x$ can take. When the variable takes values in $\mathbb{F}$, its domain of values is represented by an interval of floats:

$$\mathbf{x}_\mathbb{F} = [\underline{\mathbf{x}}_\mathbb{F}, \overline{\mathbf{x}}_\mathbb{F}] = \{x_\mathbb{F} \in \mathbb{F},\ \underline{\mathbf{x}}_\mathbb{F} \leq x_\mathbb{F} \leq \overline{\mathbf{x}}_\mathbb{F}\} \text{ where } \underline{\mathbf{x}}_\mathbb{F} \in \mathbb{F} \text{ and } \overline{\mathbf{x}}_\mathbb{F} \in \mathbb{F}$$

[2] $ulp(z)$ is the distance between $z$ and its successor (noted $z^+$).

$$\begin{aligned}\text{Addition: } z = x \oplus y &\rightarrow e_z = e_x + e_y + e_\oplus\\ \text{Subtraction: } z = x \ominus y &\rightarrow e_z = e_x - e_y + e_\ominus\\ \text{Multiplication: } z = x \otimes y &\rightarrow e_z = x_\mathbb{F} e_y + y_\mathbb{F} e_x + e_x e_y + e_\otimes\\ \text{Division: } z = x \oslash y &\rightarrow e_z = \frac{y_\mathbb{F} e_x - x_\mathbb{F} e_y}{y_\mathbb{F}(y_\mathbb{F} + e_y)} + e_\oslash\end{aligned}$$

**Fig. 1.** Computation of deviation for basic operations

Errors require a specific domain associated with each variable of a problem. Since the arithmetic constraints processed here are reduced to the four basic operations, and since those four operations are applied on floats, i.e. a finite subset of rationals, this domain can be defined as an interval of rationals:

$$\mathbf{e}_x = [\underline{\mathbf{e}}_x, \overline{\mathbf{e}}_x] = \{e_x \in \mathbb{Q},\ \underline{\mathbf{e}}_x \leq e_x \leq \overline{\mathbf{e}}_x\} \text{ where } \underline{\mathbf{e}}_x \in \mathbb{Q} \text{ and } \overline{\mathbf{e}}_x \in \mathbb{Q}$$

The domain of errors on operations, denoted by $e_\odot$, that appears in the computation of deviations (see Fig. 1) is associated with each *instance* of an arithmetic operation of a problem.

### 3.3 Projection Functions

The filtering process of FErA is based on classical projection functions that reduce the domains of the variables. Domains of values can be reduced by means of standard floating-point projection functions defined in [14] and extended in [1, 13]. However, dedicated projections are required to reduce domains of errors.

The projections on the domains of errors are obtained using the natural extension over intervals of the formulae of Fig. 1. Since these are formulae on reals, they can naturally be extended to intervals. The projections functions for the four basic arithmetic operations are detailed in Fig. 2.

As no error is involved in comparison operators, their projection functions only handle domains of values. So, projection functions on the domain of errors support only arithmetic operations and assignment, where the computation error from the expression is transmitted to the assigned variable.

### 3.4 Links Between Domains of Values and Domains of Errors

Strong connections between the domain of values and the domain of errors are required to propagate reductions between these domains.

A first relation between the domain of values and the domain of errors on operations is based upon the IEEE 754 standard, that guarantees that basic arithmetic operations are correctly rounded.

$$\mathbf{e}_\odot \leftarrow \mathbf{e}_\odot \cap \left[-\frac{\min((\underline{z} - \underline{z}^-), (\overline{z} - \overline{z}^-))}{2}, +\frac{\max((\underline{z}^+ - \underline{z}), (\overline{z}^+ - \overline{z}))}{2}\right]$$

Addition

$$\mathbf{e}_z \leftarrow \mathbf{e}_z \cap (\mathbf{e}_x + \mathbf{e}_y + \mathbf{e}_\oplus)$$
$$\mathbf{e}_x \leftarrow \mathbf{e}_x \cap (\mathbf{e}_z - \mathbf{e}_y - \mathbf{e}_\oplus)$$
$$\mathbf{e}_y \leftarrow \mathbf{e}_y \cap (\mathbf{e}_z - \mathbf{e}_x - \mathbf{e}_\oplus)$$
$$\mathbf{e}_\oplus \leftarrow \mathbf{e}_\oplus \cap (\mathbf{e}_z - \mathbf{e}_x - \mathbf{e}_y)$$

Subtraction

$$\mathbf{e}_z \leftarrow \mathbf{e}_z \cap (\mathbf{e}_x - \mathbf{e}_y + \mathbf{e}_\ominus)$$
$$\mathbf{e}_x \leftarrow \mathbf{e}_x \cap (\mathbf{e}_z + \mathbf{e}_y - \mathbf{e}_\ominus)$$
$$\mathbf{e}_y \leftarrow \mathbf{e}_y \cap (-\mathbf{e}_z + \mathbf{e}_x + \mathbf{e}_\ominus)$$
$$\mathbf{e}_\ominus \leftarrow \mathbf{e}_\ominus \cap (\mathbf{e}_z - \mathbf{e}_x + \mathbf{e}_y)$$

Multiplication

$$\mathbf{e}_z \leftarrow \mathbf{e}_z \cap (\mathbf{x}_\mathbb{F}\mathbf{e}_y + \mathbf{y}_\mathbb{F}\mathbf{e}_x + \mathbf{e}_x\mathbf{e}_y + \mathbf{e}_\otimes)$$
$$\mathbf{e}_x \leftarrow \mathbf{e}_x \cap \left(\frac{\mathbf{e}_z - \mathbf{x}_\mathbb{F}\mathbf{e}_y - \mathbf{e}_\otimes}{\mathbf{y}_\mathbb{F} + \mathbf{e}_y}\right)$$
$$\mathbf{e}_y \leftarrow \mathbf{e}_y \cap \left(\frac{\mathbf{e}_z - \mathbf{y}_\mathbb{F}\mathbf{e}_x - \mathbf{e}_\otimes}{\mathbf{x}_\mathbb{F} + \mathbf{e}_x}\right)$$
$$\mathbf{e}_\otimes \leftarrow \mathbf{e}_\otimes \cap (\mathbf{e}_z - \mathbf{x}_\mathbb{F}\mathbf{e}_y - \mathbf{y}_\mathbb{F}\mathbf{e}_x - \mathbf{e}_x\mathbf{e}_y)$$
$$\mathbf{x}_\mathbb{F} \leftarrow \mathbf{x}_\mathbb{F} \cap \left(\frac{\mathbf{e}_z - \mathbf{y}_\mathbb{F}\mathbf{e}_x - \mathbf{e}_x\mathbf{e}_y - \mathbf{e}_\otimes}{\mathbf{e}_y}\right)$$
$$\mathbf{y}_\mathbb{F} \leftarrow \mathbf{y}_\mathbb{F} \cap \left(\frac{\mathbf{e}_z - \mathbf{x}_\mathbb{F}\mathbf{e}_y - \mathbf{e}_x\mathbf{e}_y - \mathbf{e}_\otimes}{\mathbf{e}_x}\right)$$

Division

$$\mathbf{e}_z \leftarrow \mathbf{e}_z \cap \left(\frac{\mathbf{y}_\mathbb{F}\mathbf{e}_x - \mathbf{x}_\mathbb{F}\mathbf{e}_y}{\mathbf{y}_\mathbb{F}(\mathbf{y}_\mathbb{F} + \mathbf{e}_y)} + \mathbf{e}_\oslash\right)$$
$$\mathbf{e}_x \leftarrow \mathbf{e}_x \cap \left((\mathbf{e}_z - \mathbf{e}_\oslash)(\mathbf{y}_\mathbb{F} + \mathbf{e}_y) + \frac{\mathbf{x}_\mathbb{F}\mathbf{e}_y}{\mathbf{y}_\mathbb{F}}\right)$$
$$\mathbf{e}_y \leftarrow \mathbf{e}_y \cap \left(\frac{\mathbf{e}_x - \mathbf{e}_z\mathbf{y}_\mathbb{F} + \mathbf{e}_\oslash\mathbf{y}_\mathbb{F}}{\mathbf{e}_z - \mathbf{e}_\oslash + \frac{\mathbf{x}_\mathbb{F}}{\mathbf{y}_\mathbb{F}}}\right)$$
$$\mathbf{e}_\oslash \leftarrow \mathbf{e}_\oslash \cap \left(\mathbf{e}_z - \frac{\mathbf{y}_\mathbb{F}\mathbf{e}_x - \mathbf{x}_\mathbb{F}\mathbf{e}_y}{\mathbf{y}_\mathbb{F}(\mathbf{y}_\mathbb{F} + \mathbf{e}_y)}\right)$$
$$\mathbf{x}_\mathbb{F} \leftarrow \mathbf{x}_\mathbb{F} \cap \left(\frac{(\mathbf{e}_\oslash - \mathbf{e}_z)\mathbf{y}_\mathbb{F}(\mathbf{y}_\mathbb{F} + \mathbf{e}_y) + \mathbf{y}_\mathbb{F}\mathbf{e}_x}{\mathbf{e}_y}\right)$$
$$\mathbf{y}_\mathbb{F} \leftarrow \mathbf{y}_\mathbb{F} \cap [\min(\underline{\delta}_1, \underline{\delta}_2), \max(\overline{\delta}_1, \overline{\delta}_2)]$$

with

$$\delta_1 \leftarrow \frac{\mathbf{e}_x - (\mathbf{e}_z - \mathbf{e}_\oslash)\mathbf{e}_y - \sqrt{\mathbf{\Delta}}}{2(\mathbf{e}_z - \mathbf{e}_\oslash)} \qquad \delta_2 \leftarrow \frac{\mathbf{e}_x - (\mathbf{e}_z - \mathbf{e}_\oslash)\mathbf{e}_y + \sqrt{\mathbf{\Delta}}}{2(\mathbf{e}_z - \mathbf{e}_\oslash)}$$

$$\mathbf{\Delta} \leftarrow [0, +\infty) \cap ((\mathbf{e}_z - \mathbf{e}_\oslash)\mathbf{e}_y - \mathbf{e}_x)^2 + 4(\mathbf{e}_z - \mathbf{e}_\oslash)\mathbf{e}_y\mathbf{x}_\mathbb{F}$$

**Fig. 2.** Projection functions of arithmetic operation

where $x^-$ and $x^+$ denote respectively, the greatest floating-point number strictly inferior to $x$ and the smallest floating-point number strictly superior to $x$. This equation sets a relation between the domain of values and the domain of errors on operations. More precisely, it states that operation errors can never be greater than the greatest half-ulp of the domain of values of the operation result. Note that the contrapositive of this property offers another opportunity to connect the domain of values and the domain of errors. Indeed, since the absolute value of an operation error is less than the greatest half-ulp of the domain of values of the operation result, the smallest values of the domain of the result cannot be the support of a solution if $\inf(|e_\odot|) > 0$. In other words, these small values near zero cannot be associated to an error on the operation big enough to be in $e_\odot$ domain if their half-ulp is smaller than $\inf(|e_\odot|)$.

Finally, these links are refined by means of other well-known properties of floating-point arithmetic like the Sterbenz property of the subtraction [20] or the Hauser property on the addition [9]. Both properties give conditions under which these operations produce exact results, the same being true for the well-known property that states that $2^k \otimes x$ is exactly computed, provided that no overflow occurs.

### 3.5 A CSP over $\mathbb{F}$ with Errors

A CSP over $\mathbb{F}$ with errors is made of constraints over $\mathbb{F}$ with variables whose domains specify the allowed values over $\mathbb{F}$, as well as, the allowed values over $\mathbb{Q}$ of the associated error. $err(x)$, which denotes the error associated to variable $x$, permits constraints on errors. Note that the latter are constraints over $\mathbb{Q}$.

The constraint network results from elementary constraints issued from the decomposition of initial constraints. Each elementary constraint is in charge of applying the set of related projection functions to reduce the domain of the variables involved in the constraint. They compute a quasi-fixpoint, set to 5%, in order to avoid some slow convergence issues. Propagation occurs following an AC3 algorithm.

## 4 A Branch-and-bound Algorithm to Maximize the Error

Our branch-and-bound algorithm (see Algorithm 1) maximizes a given absolute error from a CSP. Such an error characterizes the greatest possible deviation between the expected computation over the reals and the actual computation over the floats. Note that the algorithm can easily be changed to maximize or minimize a signed error.

The branch-and-bound algorithm takes as inputs a CSP $\langle X, C, D \rangle$, and $e$, an error to maximize. It computes a lower bound, $e^*$ and an upper bound, $\overline{e}$ of the maximal error $e$. $e^*$ and $\overline{e}$ bounds the maximal error: $e^* \leq e \leq \overline{e}$. The computed lower bound $e^*$ is a reachable error exercised by computed input values. These values and the computed bounds are returned by the algorithm.

*Stopping Criteria.* The primary aim of the branch-and-bound algorithm is to compute the maximal error. This is achieved when the lower bound is equal to the upper bound. However, such a condition may be difficult to meet.

A first issue comes from the dependency problem which appears on expressions with multiple occurrences. Multiple occurrences of variables is a critical issue in interval arithmetic since each occurrence of a variable is considered as a different variable with the same domain. This dependency problem results in overestimations in the evaluation of the possible values that an expression can take. For instance, let $y = x \times x$ with $x \in [-1, 1]$, classical interval arithmetic yields $[-1, 1]$ whereas the exact interval is $[0, 1]$. Such a drawback arises in projection functions of errors that contain multiple occurrences like in multiplication and division. It can lead to unnecessary over-approximations of resulting intervals. A direct consequence of this problem is that the upper bound is overestimated and therefore can not be achieve.

A second issue comes from the bounding of errors on operations by the half of an ulp. An operation error is bounded by $\frac{1}{2}\text{ulp}(z)$ where $z$ is the result of an operation. Such a bound is highly dependent on the distribution of floating-point numbers. Consider an interval of floating-point numbers $(2^n, 2^{n+1})$, every floating-point number is separated from the next one by the same distance. In other words, every floating-point number in this interval will have the same ulp. When the domain of the result of an operation is reduced to such an interval, the bounds of $e_{\odot}$ are fixed and can no longer be improved by means of projection functions. This can be generalized across all operations of a CSP. Once all operation errors are fixed, then bounds cannot be tighten without enumerating values. That is why, we stop processing a box when all the related domains are reduced to such an interval.

*Box Management.* The algorithm manages a list $L$ of boxes to process whose initial value is $\{B = (D, e^B \in [-\infty, +\infty])\}$ where $D$ is the cross product of the domains of the variable as defined in the problem and $e^B$ their associated error. It also manages the global lower and upper bounds with $e^* = -\infty$, $\overline{e} = +\infty$ as initial values. A box can be in three different states: *unexplored*, *discarded* or *sidelined*. *unexplored* boxes are boxes in $L$ that still require some computations. A *discarded* box is a box whose associated error $e^B$ is such that $\overline{e}^B \leq e^*$. In other words, such a box does not contain any error value that can improve the computation of the maximal error. It is thus removed from $L$. *sidelined* boxes are boxes that fullfil the property described in the stopping criteria paragraph. These boxes cannot improve maximal error computation unless if the algorithm resorts to enumeration (provided there are no multiple occurrences). As sidelined boxes are still valid boxes, the greatest over-approximation of such boxes, $\overline{e}^S$, is taken into account when updating the upper bound. Solving stops when there are no more boxes to process or when the lower bound $e^*$ and the upper bound $\overline{e}$ are equal, i.e. when the maximal error is found.

The main loop of the branch-and-bound algorithm can be subdivided in several steps: box selection, filtering, upper bound updating, lower bound updating, and box splitting.

*Box Selection.* We select the box $B$ in the set $L$ with the greatest upper bound of the error to provide more opportunities to improve both $\overline{e}$ and $e^*$. Indeed, the global $\overline{e}$ has its support in this box which also provides the odds of computing a better reachable error $e^*$. Once selected, the box $B$ is removed from $L$.

*Filtering.* A filtering process (see Sect. 3), denoted $\Phi$, is then applied to $B$ to reduce the domains of values and the domains of errors. Note that this filtering is applied to the initial set of constraints enhanced with a constraint on the known lower bound of $e$, i.e. $e^* \leq e$. If $\Phi(B) = \varnothing$, the selected box does not contain any solution; either because it contradicts one of the initial constraints or because of known bounds of constraints over $e$. In both cases, the algorithm discards box $B$ and directly jumps to the next loop iteration.

*Upper Bound Update.* Once $B$ has been filtered, if the error upper bound of the current box was the support of the global $\overline{e}$ and is no longer, then $\overline{e}$ is updated.

**Algorithm 1:** branch-and-bound — maximization of error

**Input** : $\langle X, C, D\rangle$ — triple of variables, constraints, and domains
$e \in [-\infty, +\infty]$ — error to maximize

**Output** : $(e^*, \overline{e}, S)$

**Data** : $L \leftarrow \{\prod_{x \in X} x \mid x = (\mathbf{x}, \mathbf{e_x})\}$ — set of boxes
$\overline{e} \leftarrow +\infty$ — upper bound
$e^* \leftarrow -\infty$ — lower bound
$\overline{e}^S \leftarrow -\infty$ — upper bound of sidelined boxes
$S \leftarrow \varnothing$ — stack of solutions

**while** $L \neq \varnothing$ **and** $e^* < \overline{e}$ **do**
  `/* Box selection: select a box` $B$ `in the set of boxes` $L$ `*/`
  select $B \in L$ ; $L \leftarrow L \setminus B$ ; $\overline{e}^B_{old} \leftarrow \overline{e}^B$
  $B \leftarrow \Phi(X, C \wedge e > e^*, B)$
  **if** $\overline{e}^B_{old} = \overline{e}$ **and** ($B = \varnothing$ **or** $\overline{e}^B < \overline{e}$) **then**
    **if** $B \neq \varnothing$ **then** $\overline{e} \leftarrow \overline{e}^B$ **else** $\overline{e} \leftarrow -\infty$
    $\overline{e} \leftarrow \max\left(\{\overline{e}^{B_i} \mid \forall B_i \in L\} \cup \{\overline{e}, \overline{e}^S\}\right)$
  **if** $B \neq \varnothing$ **then**
    **if** $\overline{e}^B > e^*$ **then**
      **if** *isBound*$(B)$ **then**
        $(e^B, sol^B) \leftarrow (e^B, B)$
      **else**
        $(e^B, sol^B) \leftarrow LowerBounding(B, X)$
      **if** $e^B > e^*$ **then**
        $e^* \leftarrow e^B$ ; push $(e^B, sol^B)$ onto $S$
        $L \leftarrow L \setminus \{B_i \in L \mid \overline{e}^{B_i} \leq e^*\}$
    **if** $\overline{e}^B > \overline{e}^S$ **and** $\overline{e}^B > e^*$ **then**
      `/* Variable selection: select a variable` $x$ `in box` $B$ `*/`
      **if** $(select(\mathbf{x}^B, \mathbf{e_x}^B) \in B \mid \underline{\mathbf{x}}^B < \overline{\mathbf{x}}^B)$ **and** $\neg isSidelined(B)$ **then**
        `/* Domain splitting: on the domain of values of x */`
        $B_1 \leftarrow B$ ; $B_2 \leftarrow B$
        $\mathbf{x}^{B_1} \leftarrow \left[\underline{\mathbf{x}}^B, \frac{\underline{\mathbf{x}}^B + \overline{\mathbf{x}}^B}{2}\right]$ ; $\mathbf{x}^{B_2} \leftarrow \left[\left(\frac{\underline{\mathbf{x}}^B + \overline{\mathbf{x}}^B}{2}\right)^+, \overline{\mathbf{x}}^B\right]$
        $L \leftarrow L \cup \{B_1, B_2\}$
      **else** $\overline{e}^S \leftarrow \max(\overline{e}^S, \overline{e}^B)$

**return** $(e^*, \overline{e}, S)$

$\overline{e}$ is updated with the maximum among the upper bound of errors of the current box, of remaining boxes in $L$, and of sidelined boxes.

*Lower Bound Update.* A non empty box may contain a better lower bound than the current one. A *generate-and-test* procedure, `LowerBounding`, attempts to compute a better one through a two-step process. A variable is first assigned with a floating-point value chosen randomly in its domain of values. Then, another random value chosen within the domain of its associated error is assigned to the error associated to that variable. We exploit the fact that if the derivative sign does not change on the associated error domain, then the maximum distance between the hyperplan defined by the result over the floats and the related function over the reals is at one of the extrema of the associated error domain. In such a case, the random instantiation of the error is restricted to the corners of its domain. When all variables from a function $f$ representing a program have been instantiated, an evaluation of $f$ is done exactly over $\mathbb{Q}$ and in machine precision over $\mathbb{F}$. The error is exactly computed by evaluating $f_{\mathbb{Q}} - f_{\mathbb{F}}$ over $\mathbb{Q}$. The computed error can be further improved by a local search. That is to say, by exploring floating-point numbers around the chosen input values and evaluating again the expressions. This process is repeated a fixed number of times until the error can not be further improved, i.e. a local maximum has been reached. If the computed error is better than the current lower bound, then $e^*$ is updated. Each new reachable lower bound is added to $S$ alongside the input values exercising it.

*Box Splitting.* A box is not split up but is discarded when its error upper bound is less than or equal to $\overline{e}^S$, the upper bound of sidelined boxes, or $e^*$. Discarding such a box speeds up solving time, since none of the errors contained in this box can improve the lower or the upper bounds. Splitting occurs if and only if there exist at least one of the variables within $B$ that is not instantiated and if the box is not sidelined. Otherwise, the box is sidelined and if $\overline{e}^B$ is strictly greater than $\overline{e}^S$, the latter is updated. The next variable to split on is selected in round-robin on a lexicographic order. The bisection generates two sub-boxes that are added to $L$.

Note that Algorithm 1 always terminates and gives an enclosure of the maximal error: in the worst case, all boxes will be split up to degenerated boxes. Each degenerated box whose associated error $e^B$ is lower than the current lower bound will be discarded. If $e^* \leq e^B \leq \overline{e}$ holds, $e^B$ will be used to update $e^*$ and $\overline{e}$ before discarding the corresponding degenerated box. As a result, since the set of floating-point numbers is a finite set, the branch-and-bound requires a finite number of iterations to explore completely the initial box and thus, terminates.

## 5 Related Work

Different tools exist to compute an over-approximation of floating-point computation round-off errors. Fluctuat [6,7], is an abstract interpreter that combines affine arithmetic and zonotopes to analyze the robustness of programs over the floats. FPTaylor [18,19] relies on symbolic Taylor expansions and global optimization to compute tight bounds of the error. It makes use of Taylor series of first and second order to evaluate the error. A branch-and-bound algorithm approximates the first order error terms while the second order error terms are directly bounded by means of interval arithmetic. This branch-and-bound is very different from the one of FErA. First, it considers only the first order terms of the error whereas FErA branch-and-bound works on the whole error. Second, it does not compute a lower bound on the largest absolute error but only over-approximate the first order terms of the error. FPTaylor also generates a proof certificate of its computed bounds. Such a certificate can be externally checked in HOL Light [8]. The static analyzer PRECiSA [16,22] computes also a certificate of proof that can be validated by the PVS theorem prover [17]. PRECiSA computes symbolic error expressions to represent round-off errors that are then given to a branch-and-bound algorithm to get the error bounds. Gappa [4] verifies properties on floating-point programs, and in particular computes bounds on round-off errors. Gappa works with an interval representation of floating-point numbers and applies rewriting rules for improving computed results. It is also able to generate formal proof of verified properties, that can in turn be checked in Coq [21]. Real2Float [12] uses semidefinite programming to estimate bounds on errors. It decomposes an error into an affine part with respect to the error variable and a higher-order part. Bounds on the higher-order part are computed in the same way as FPTaylor. For the affine part, a relaxation procedure based on semidefinite programming is employed.

FPSDP [11] is a tool based on semidefinite programming that only computes under-approximation of largest absolute errors. In contrast to our approach FPSDP computes an under-approximation of the maximal error. The point is that this under-approximation might not be reachable. S3FP [2] relies on random sampling and shadow value executions to find input values maximizing an error. It computes the error as the difference between the execution of a program done in a higher precision, acting as $\mathbb{R}$, and a lower precision, acting as $\mathbb{F}$. S3FP starts with an initial configuration that is cut into smaller configurations. Then, it selects a configuration and randomly instantiates variables to evaluate the program in both precision. This process is repeated a finite number of time to improve the lower bound. Depending on the size of input variable domains, S3FP can get stuck on a local maximum. To avoid this problem it uses a standard restart process. S3FP is the closest to our lower bound computation procedure. Both rely on random generation of input values to compute a lower bound of errors. However, as S3FP does all computations over $\mathbb{F}$, the resulting error suffers from rounding issues and thus, might underestimate or overestimate the actual error. Such a computed error is unreachable. Furthermore, S3FP is highly reliant on the parametrized partitioning of the initial configuration. It cannot discard configurations where no improvement of the lower bound is possible. In contrast, FErA selects boxes to explore on the basis of their upper bounds to try finding a better lower bound.

## 6 Experiments

In this section, we provide preliminary experiments of FErA on benchmarks from the FPBench [3] suite, a common standard to compare verification tools over floating-point numbers. Table 2 compares the behaviour of FErA with different stopping criteria while Table 3 compares results from FErA with state-of-the

**Table 2.** Comparison of stopping criteria.

| | filtering | $e^* = \overline{e}$ | | $e^* = \overline{e}$ w. $s.$ | | $\frac{\overline{e}}{e^*} \leq 2$ | | $\frac{\overline{e}}{e^*} \leq 2$ w. $s.$ | |
|---|---|---|---|---|---|---|---|---|---|
| | | $e^*$ | $\overline{e}$ | $e^*$ | $\overline{e}$ | $e^*$ | $\overline{e}$ | $e^*$ | $\overline{e}$ |
| carbonGas | 4.2e-8 | 4.2e-9 | 6.0e-9 | 2.9e-9 | 7.0e-9 | 3.6e-9 | 7.0e-9 | 3.6e-9 | 7.0e-9 |
| | 0.017s | | TO | | 0.345s | | 1.419s | | 0.266s |
| verhulst | 4.2e-16 | 2.4e-16 | 2.8e-16 | 2.1e-16 | 2.9e-16 | 1.8e-16 | 2.9e-16 | 1.6e-16 | 3.0e-16 |
| | 0.016s | | TO | | 0.034s | | 0.024s | | 0.018s |
| predPrey | 1.8e-16 | 1.5e-16 | 1.7e-16 | 1.0e-16 | 1.7e-16 | 9.3e-17 | 1.7e-16 | 9.8e-17 | 1.8e-16 |
| | 0.011s | | TO | | 0.084s | | 0.041s | | 0.018s |
| rigidBody1 | 2.9e-13 | 2.8e-13 | 2.9e-13 | 1.9e-13 | 2.9e-13 | 1.4e-13 | 2.9e-13 | 1.4e-13 | 2.9e-13 |
| | 0.018s | | TO | | 1.659s | | 0.370s | | 0.543s |
| rigidBody2 | 3.6e-11 | 3.1e-11 | 3.6e-11 | 2.5e-11 | 3.6e-11 | 1.8e-11 | 3.6e-11 | 2.1e-11 | 3.6e-11 |
| | 0.022s | | TO | | 3.298s | | 1.367s | | 1.266s |
| doppler1 | 5.0e-13 | 1.1e-13 | 1.5e-13 | 7.3e-14 | 1.6e-13 | 1.0e-13 | 1.5e-13 | 7.7e-14 | 1.5e-13 |
| | 0.021s | | TO | | 0.752s | | 1.099s | | 0.757s |
| doppler2 | 1.3e-12 | 2.1e-13 | 2.7e-13 | 1.1e-13 | 3.4e-13 | 1.3e-13 | 2.7e-13 | 1.0e-13 | 3.4e-13 |
| | 0.034s | | TO | | 0.356s | | 1.416s | | 0.378s |
| doppler3 | 1.9e-13 | 6.2e-14 | 8.4e-14 | 4.0e-14 | 9.0e-14 | 4.4e-14 | 8.7e-14 | 3.9e-14 | 9.0e-14 |
| | 0.023s | | TO | | 0.341s | | 0.455s | | 0.311s |
| turbine1 | 2.2e-13 | 1.3e-14 | 1.7e-14 | 1.0e-14 | 1.8e-14 | 1.0e-14 | 2.0e-14 | 9.3e-15 | 1.8e-14 |
| | 0.016s | | TO | | 8.514s | | 2.289s | | 6.042s |
| turbine2 | 3.0e-14 | 1.5e-14 | 2.3e-14 | 1.3e-14 | 2.4e-14 | 1.3e-14 | 2.4e-14 | 1.1e-14 | 2.4e-14 |
| | 0.025s | | TO | | 2.803s | | 1.581s | | 2.952s |
| turbine3 | 1.6e-13 | 6.4e-15 | 1.1e-14 | 4.7e-15 | 1.1e-14 | 5.7e-15 | 1.1e-14 | 5.6e-15 | 1.1e-14 |
| | 0.026s | | TO | | 2.766s | | 3.961s | | 1.800s |
| sqroot | 5.8e-16 | 4.5e-16 | 5.3e-16 | 3.3e-16 | 5.3e-16 | 2.9e-16 | 5.8e-16 | 3.3e-16 | 5.8e-16 |
| | 0.032s | | TO | | 2.989s | | 0.277s | | 0.183s |
| sine | 7.4e-16 | 2.8e-16 | 7.4e-16 | 2.2e-16 | 7.4e-16 | 2.6e-16 | 7.4e-16 | 2.4e-16 | 7.4e-16 |
| | 0.027s | | TO | | 12.927s | | TO | | 14.833s |
| sineOrder3 | 1.1e-15 | 4.0e-16 | 6.4e-16 | 3.2e-16 | 6.4e-16 | 3.1e-16 | 6.4e-16 | 3.3e-16 | 6.4e-16 |
| | 0.021s | | TO | | 1.388s | | 1.433s | | 1.504s |
| kepler0 | 1.2e-13 | 5.7e-14 | 9.8e-14 | 5.4e-14 | 9.8e-14 | 5.0e-14 | 9.8e-14 | 4.9e-14 | 9.8e-14 |
| | 0.037s | | TO | | TO | | 1.937s | | 2.798s |
| kepler1 | 4.9e-13 | 1.6e-13 | 3.1e-13 | 1.4e-13 | 3.1e-13 | 1.6e-13 | 3.1e-13 | 1.6e-13 | 3.1e-13 |
| | 0.031s | | TO | | 51.303s | | 15.136s | | 12.691s |
| kepler2 | 2.4e-12 | 7.9e-13 | 1.8e-12 | 6.5e-13 | 1.8e-12 | 6.9e-13 | 1.8e-12 | 6.7e-13 | 1.8e-12 |
| | 0.027s | | TO | | 58.834s | | TO | | 72.622s |

art tools, namely Gappa [4], Fluctuat [6,7], Real2Float [12], FPTaylor [18,19], PRECiSA [16,22], and S3FP [2]. Results from all tools but FErA are taken from [18].

Note that all state-of-the-art tools provide an over-approximation of errors, except S3FP, which compute a lower bound on largest absolute errors. For FErA, column *filtering* gives the over-approximation computed by a single filtering

**Table 3.** Comparison of FErA with other tools.

| | Fluctuat | Gappa | PRECiSA | Real2Float | FPTaylor | FErA | | S3FP |
|---|---|---|---|---|---|---|---|---|
| | | | | | | $e^*$ | $\overline{e}$ | |
| carbonGas | 1.2e-8 | *6.1e-9* | 7.4e-9 | 2.3e-8 | **6.0e-9** | 3.6e-9 | 7.0e-9 | 4.2e-9 |
| | 0.54s | 2.35s | 6.30s | 4.65s | 1.08s | | 0.266s | |
| verhulst | 4.9e-16 | *2.9e-16* | *2.9e-16* | 4.7e-16 | **2.5e-16** | 1.6e-16 | 3.0e-16 | 2.4e-16 |
| | 0.09s | 0.41s | 4.95s | 2.52s | 0.99s | | 0.018s | |
| predPrey | 2.4e-16 | *1.7e-16* | *1.7e-16* | 2.6e-16 | **1.6e-16** | 9.8e-17 | 1.8e-16 | 1.5e-16 |
| | 0.18s | 1.40s | 8.08s | 4s | 1.07s | | 0.018s | |
| rigidBody1 | 3.3e-13 | *3.0e-13* | *3.0e-13* | 5.4e-13 | *3.0e-13* | 1.4e-13 | **2.9e-13** | 2.7e-13 |
| | 1.96s | 1.42s | 7.42s | 3.09s | 0.99s | | 0.543s | |
| rigidBody2 | *3.7e-11* | *3.7e-11* | *3.7e-11* | 6.5e-11 | *3.7e-11* | 2.1e-11 | **3.6e-11** | 3.0e-11 |
| | 3.87s | 2.22s | 10.79s | 1.08s | 1.02s | | 1.266s | |
| doppler1 | **1.3e-13** | 1.7e-13 | 2.7e-13 | 7.7e-12 | **1.3e-13** | 7.7e-14 | *1.5e-13* | 1.0e-13 |
| | 6.30s | 3.31s | 16.17s | 13.20s | 1.97s | | 0.757s | |
| doppler2 | *2.4e-13* | 2.9e-13 | 5.4e-13 | 1.6e-11 | **2.3e-13** | 1.0e-13 | 3.4e-13 | 1.9e-13 |
| | 6.15s | 3.37s | 16.87s | 13.33s | 2.20s | | 0.378s | |
| doppler3 | *7.2e-14* | 8.7e-14 | 1.4e-13 | 8.6e-12 | **6.7e-14** | 3.9e-14 | 9.0e-14 | 5.7e-14 |
| | 6.46s | 3.32s | 15.65s | 13.05s | 1.88s | | 0.311s | |
| turbine1 | 3.1e-14 | 2.5e-14 | 3.8e-14 | 2.5e-11 | **1.7e-14** | 9.3e-15 | *1.8e-14* | 1.1e-14 |
| | 5.05s | 5.54s | 24.35s | 136.35s | 1.10s | | 6.042s | |
| turbine2 | 2.6e-14 | 3.4e-14 | 3.1e-14 | 2.1e-12 | **2.0e-14** | 1.1e-14 | *2.4e-14* | 1.4e-14 |
| | 3.98s | 3.94s | 19.17s | 8.30s | 1.17s | | 2.952s | |
| turbine3 | 1.4e-14 | 0.36 | 2.3e-14 | 1.8e-11 | **9.6e-15** | 5.6e-15 | *1.1e-14* | 6.2e-15 |
| | 5.08s | 6.29s | 24.47s | 137.36s | 1.21s | | 1.800s | |
| sqroot | 6.9e-16 | *5.4e-16* | 6.9e-16 | 1.3e-15 | **5.1e-16** | 3.3e-16 | 5.8e-16 | 4.7e-16 |
| | 0.09s | 5.06s | 8.18s | 4.23s | 1.02s | | 0.183s | |
| sine | 7.5e-16 | 7.0e-16 | *6.0e-16* | 6.1e-16 | **4.5e-16** | 2.4e-16 | 7.4e-16 | 2.9e-16 |
| | 0.11s | 25.43s | 11.76s | 4.95s | 1.14s | | 14.833s | |
| sineOrder3 | 1.1e-15 | 6.6e-16 | 1.2e-15 | 1.2e-15 | **6.0e-16** | 3.3e-16 | *6.4e-16* | 4.1e-16 |
| | 0.09s | 2.09s | 6.11s | 2.22s | 1.02s | | 1.504s | |
| kepler0 | 1.1e-13 | 1.1e-13 | 1.2e-13 | 1.2e-13 | **7.5e-14** | 4.9e-14 | *9.8e-14* | 5.3e-14 |
| | 8.59s | 7.33s | 37.57s | 0.76s | 1.31s | | 2.798s | |
| kepler1 | 3.6e-13 | 4.7e-13 | crash | 4.7e-13 | **2.9e-13** | 1.6e-13 | *3.1e-13* | 1.6e-13 |
| | 157.74s | 10.68s | N/A | 22.53s | 2.08s | | 12.691s | |
| kepler2 | 2.3e-12 | 2.4e-12 | crash | 2.1e-12 | **1.6e-12** | 6.7e-13 | *1.8e-12* | 8.4e-13 |
| | 22.41s | 24.17s | N/A | 16.53s | 1.3s | | 72.622s | |

while column $e^*$ and column $\overline{e}$ provide, respectively, the best reachable error and over-approximation of the error computed by FErA. Lines in grey give the time in second to compute these bounds. Note that experiments have been made with a timeout, noted TO, of 10 min. In Table 3, bold and italic are used to rank, respectively, the best and second best over-approximation while red indicates the worst ones.

Table 2 compares the behaviour of FErA with different stopping criteria. The ideal $e^* = \overline{e}$ criterion stops if and only if the lower bound reaches the upper bound. Of course, this criterion is hard to reach and benches are stopped by the timeout. But these columns yield among the best $e^*$ and $\overline{e}$ that can be expected with FErA. For instance, most of the $e^*$ provided here are better than the best known values[3], and the values of $\overline{e}$ are obviously the best values obtained by FErA. $e^* = \overline{e}$ w. *s.* combines the ideal criterion with sidelined boxes, i.e. boxes that are sidelined once all operations involved in the CSP produce an operation error, $e_{\odot}$, less or equal to half an ulp. Such a combination of criterion allows FErA to compute the results in a reasonable amount of time for all benches but one. However, this is obtained at the price of a degradation of the error bounds. $\frac{\overline{e}}{e^*} \leq 2$ relaxes the ideal stopping criterion to a ratio of 2. Here, two benchmarks reach the timeout. This is to be expected: the results with the ideal criterion show that FErA have difficulties to reach this ratio on two benchmarks, probably due to a huge amount of multiple occurrences. $\frac{\overline{e}}{e^*} \leq 2$ w. *s.* combines the ratio of 2 with sidelined boxes and avoid any timeout, though at the price of looser bounds.

As shown in Table 3, FErA classified as best twice and as second eight times. Note that it never provides the worst result. In almost all cases, the computed reachable error $e^*$ is in the same order of magnitude as $\overline{e}$. The lack of dedicated handling of multiple occurrences in FErA is underlined by the computed upper bound of the `sine` bench. Here, the splitting process used in the branch-and-bound is not sufficient to lower the upper bound value. With the last combination of criteria, FErA solves most of the problems in a reasonable amount of time with the exception of `kepler2`. Indeed, Kepler benches are the problems with the biggest number of input variables and FErA performs better on small-sized problems.

## 7 Conclusion

This paper addresses a critical issue in program verification: computing an enclosure of the maximal absolute round-off errors that occur in floating-point computations. To do so, we introduce an original approach based on a branch-and-bound algorithm using the constraint system for round-off error analysis from [5]. Alongside a rigorous enclosure of maximal errors, the algorithm provides input values exercising the lower bound. Knowing such bounds of the maximal error allows to get rid of false positives, a critical issue in program verification and validation. Preliminary experiments on a set of standard benchmarks are very promising and compare well to other available tools.

Further works include a better understanding and a tighter computation of round-off errors to smooth the effects of the dependency problem, exploring different search strategies dedicated to floating-point numbers [23] to improve the resolution process, as well as devising a better local search to speed up the reachable lower bound computation procedure.

[3] See column S3FP in Table 3.

## References


1. Botella, B., Gotlieb, A., Michel, C.: Symbolic execution of floating-point computations. Softw. Test. Verification Reliab. **16**(2), 97–121 (2006)
2. Chiang, W., Gopalakrishnan, G., Rakamaric, Z., Solovyev, A.: Efficient search for inputs causing high floating-point errors. In: ACM SIGPLAN Symposium on Principles and Practice of Parallel Programming PPoPP 2014, Orlando, FL, USA, 15–19 February 2014, pp. 43–52 (2014)
3. Damouche, N., Martel, M., Panchekha, P., Qiu, C., Sanchez-Stern, A., Tatlock, Z.: Toward a standard benchmark format and suite for floating-point analysis. In: Bogomolov, S., Martel, M., Prabhakar, P. (eds.) NSV 2016. LNCS, vol. 10152, pp. 63–77. Springer, Cham (2017). https://doi.org/10.1007/978-3-319-54292-8_6
4. Daumas, M., Melquiond, G.: Certification of bounds on expressions involving rounded operators. ACM Trans. Math. Softw. **37**(1), 2:1–2:20 (2010)
5. Garcia, R., Michel, C., Pelleau, M., Rueher, M.: Towards a constraint system for round-off error analysis of floating-point computation. In: 24th International Conference on Principles and Practice of Constraint Programming : Doctoral Program. Lille, France (Aug 2018)
6. Goubault, E., Putot, S.: Static analysis of numerical algorithms. In: Yi, K. (ed.) SAS 2006. LNCS, vol. 4134, pp. 18–34. Springer, Heidelberg (2006). https://doi.org/10.1007/11823230_3
7. Goubault, E., Putot, S.: Static analysis of finite precision computations. In: Jhala, R., Schmidt, D. (eds.) VMCAI 2011. LNCS, vol. 6538, pp. 232–247. Springer, Heidelberg (2011). https://doi.org/10.1007/978-3-642-18275-4_17
8. Harrison, J.: HOL light: an overview. In: Berghofer, S., Nipkow, T., Urban, C., Wenzel, M. (eds.) TPHOLs 2009. LNCS, vol. 5674, pp. 60–66. Springer, Heidelberg (2009). https://doi.org/10.1007/978-3-642-03359-9_4
9. Hauser, J.R.: Handling floating-point exceptions in numeric programs. ACM Trans. Program. Lang. Syst. **18**(2), 139–174 (1996)
10. IEEE: 754–2008 - IEEE Standard for floating point arithmethic (2008)
11. Magron, V.: Interval enclosures of upper bounds of roundoff errors using semidefinite programming. ACM Trans. Math. Softw. **44**(4), 41:1–41:18 (2018)
12. Magron, V., Constantinides, G.A., Donaldson, A.F.: Certified roundoff error bounds using semidefinite programming. ACM Trans. Math. Softw. **43**(4), 34:1–34:31 (2017)
13. Marre, B., Michel, C.: Improving the floating point addition and subtraction constraints. In: Cohen, D. (ed.) CP 2010. LNCS, vol. 6308, pp. 360–367. Springer, Heidelberg (2010). https://doi.org/10.1007/978-3-642-15396-9_30
14. Michel, C.: Exact projection functions for floating point number constraints. In: AI&M 1–2002, Seventh International Symposium on Artificial Intelligence and Mathematics 7th ISAIM. Fort Lauderdale, Floride (US), 2–4 January 2002
15. Michel, C., Rueher, M., Lebbah, Y.: Solving constraints over floating-point numbers. In: Walsh, T. (ed.) CP 2001. LNCS, vol. 2239, pp. 524–538. Springer, Heidelberg (2001). https://doi.org/10.1007/3-540-45578-7_36
16. Moscato, M., Titolo, L., Dutle, A., Muñoz, C.A.: Automatic estimation of verified floating-point round-off errors via static analysis. In: Tonetta, S., Schoitsch, E., Bitsch, F. (eds.) SAFECOMP 2017. LNCS, vol. 10488, pp. 213–229. Springer, Cham (2017). https://doi.org/10.1007/978-3-319-66266-4_14

17. Narkawicz, A., Muñoz, C.: A formally verified generic branching algorithm for global optimization. In: Cohen, E., Rybalchenko, A. (eds.) VSTTE 2013. LNCS, vol. 8164, pp. 326–343. Springer, Heidelberg (2014). https://doi.org/10.1007/978-3-642-54108-7_17
18. Solovyev, A., Baranowski, M.S., Briggs, I., Jacobsen, C., Rakamarić, Z., Gopalakrishnan, G.: Rigorous estimation of floating-point round-off errors with symbolic taylor expansions. ACM Trans. Program. Lang. Syst. **41**(1), 2:1–2:39 (2018)
19. Solovyev, A., Jacobsen, C., Rakamarić, Z., Gopalakrishnan, G.: Rigorous estimation of floating-point round-off errors with symbolic taylor expansions. In: Bjørner, N., de Boer, F. (eds.) FM 2015. LNCS, vol. 9109, pp. 532–550. Springer, Cham (2015). https://doi.org/10.1007/978-3-319-19249-9_33
20. Sterbenz, P.H.: Floating Point Computation. Prentice-Hall (1974)
21. The Coq Development Team: The Coq proof assistant reference manual (2020). https://coq.inria.fr. version 8.11.2
22. Titolo, L., Feliú, M.A., Moscato, M., Muñoz, C.A.: An abstract interpretation framework for the round-off error analysis of floating-point programs. VMCAI 2018. LNCS, vol. 10747, pp. 516–537. Springer, Cham (2018). https://doi.org/10.1007/978-3-319-73721-8_24
23. Zitoun, H.: Search strategies for solving constraint systems over floats for program verification. Theses, Université Côte d'Azur (2018)
24. Zitoun, H., Michel, C., Rueher, M., Michel, L.: Search strategies for floating point constraint systems. In: Beck, J.C. (ed.) CP 2017. LNCS, vol. 10416, pp. 707–722. Springer, Cham (2017). https://doi.org/10.1007/978-3-319-66158-2_45



# Application Track

# Leveraging Reinforcement Learning, Constraint Programming and Local Search: A Case Study in Car Manufacturing

Valentin Antuori[1,2](✉), Emmanuel Hebrard[1,3], Marie-José Huguet[1], Siham Essodaigui[2], and Alain Nguyen[2]

[1] LAAS-CNRS, Université de Toulouse, CNRS, INSA, Toulouse, France
{vantuori,hebrard,huguet}@laas.fr
[2] Renault, Boulogne-Billancourt, France
{valentin.antuori,siham.essodaigui,alain.nguyen}@renault.com
[3] ANITI, Université de Toulouse, Toulouse, France

**Abstract.** The problem of transporting vehicle components in a car manufacturer workshop can be seen as a large scale single vehicle pickup and delivery problem with periodic time windows. Our experimental evaluation indicates that a relatively simple constraint model shows some promise and in particular outperforms the local search method currently employed at Renault on industrial data over long time horizon. Interestingly, with an adequate heuristic, constraint propagation is often sufficient to guide the solver toward a solution in a few backtracks on these instances. We therefore propose to learn efficient heuristic policies via reinforcement learning and to leverage this technique in several approaches: rapid-restarts, limited discrepancy search and multi-start local search. Our methods outperform both the current local search approach and the classical CP models on industrial instances as well as on synthetic data.

**Keywords:** Constraint programming · Reinforcement learning · Local search · Scheduling · Traveling salesman problem

## 1 Introduction

Improving the production line is a constant concern in the industry. The car manufacturer Renault has long been interested in models and techniques from Operations Research and Constraint Programming to tackle the various routing and scheduling problems arising from the production process.

Recent advances in Artificial Intelligence and in particular in Machine Learning (ML) open up many new perspectives for solving large scale combinatorial optimization problems with promising results popularized by the success of AlphaGo and AlphaZero [14,15]. In particular, several approaches combining


H. Simonis (Ed.): CP 2020, LNCS 12333, pp. 657–672, 2020.
https://doi.org/10.1007/978-3-030-58475-7_38

reinforcement learning and deep neural networks to guide the reward strategy have been proposed for solving the traveling salesman problem (TSP) [2,4]. Moreover, the combination of ML and classical combinatorial techniques seems very promising. For instance, the integration of standard TSP heuristics with a neural network heuristic policy outperforms the pure ML approaches [5].

In Sect. 2 we introduce the problem of planning the flow of vehicle components across the assembly lines. More precisely, given fixed production cycles, logistics operators are in charge of collecting components from, and delivering them to, working stations in such a way that there is no break in the manufacturing process. This problem is in many ways similar to the *Torpedo Scheduling Problem* [7], from the 2016 ACP challenges [13]: pickup and delivery operations are to be scheduled, and these operations are repeated in time because the commodity is being produced constantly at a fixed rate. However, in our problem, components are carried over using trolleys that can be stacked into a "train" that should not exceed a given maximal length. It should be noted that in this paper we consider the problem associated to a single operator whose route is to be optimized. However, the more general problem for Renault is to assign components (or equivalently working stations) to operators as well as to plan the individual routes, and the longer term objective is to proactively design the layout of the assembly line to reduce the cost of logistics operations.

We first evaluate in Sect. 3 two basic constraint programming (CP) models in `Choco` [11] and compare them to the current method used by Renault: a local search (LS) method implemented in `LocalSolver`[1]. From this preliminary study, we observe that although the problem can be hard for both approaches, CP shows promising results compare to LS. Moreover, if solving the problem via backtracking search seems very unlikely given its size, and if stronger filtering techniques seem to be ineffective, greedy "dives" are often surprisingly successful.

We therefore propose in Sect. 4 to learn effective stochastic heuristic policies via reinforcement learning. Then, we show how these policies can be used within different tailored approaches: constraint programming with rapid restarts, limited discrepancy search, and multi-start local search approach.

Finally, since industrial benchmarks are easily solved by all new methods introduced in this paper, we generated a synthetic dataset designed to be more challenging. In Sect. 6 we report experiments on this dataset that further demonstrates the efficiency of the proposed methods.

## 2 Problem Definition

The Renault assembly line consists of a set of $m$ components to be moved across a workshop, from the point where they are produced to where they are consumed. Each component is produced and consumed by two unique machines, and it is carried from one to the other using a specific trolley. When a trolley is filled at the production point for that component, an operator must bring it to its

[1] https://www.localsolver.com/home.html.

consumption point. Symmetrically, when a trolley is emptied it must be brought to the corresponding production point. A production cycle is the time $c_i$ taken to produce (resp. consume) component $i$, that is, to fill (resp. empty) a trolley. The end of a production cycle marks the start of the next, hence there are $n_i = \frac{H}{c_i}$ cycles over a time horizon $H$. There are two pickups and two deliveries at every cycle $k$ of each component $i$: the pickup $pe_i^k$ and delivery $de_i^k$ of the empty trolley from consumption to production and the pickup $pf_i^k$ and delivery $df_i^k$ of the full trolley from production to consumption. The processing time of an operation $a$ is denoted $p_a$ and the travel time between operations $a$ and $b$ is denoted $D_{a,b}$.

Let $P$ be the set of pickup operations and $D$ the set of delivery operations: $P = \bigcup_{i=1}^{m} \left( \bigcup_{k=1}^{n_i} \{pe_i^k, pf_i^k\} \right)$, $D = \bigcup_{i=1}^{m} \left( \bigcup_{k=1}^{n_i} \{de_i^k, df_i^k\} \right)$. The problem is to compute the bijection $\sigma$ (let $\rho = \sigma^{-1}$ be its inverse) between the $|A| = n$ first positive integers to the operations $A = P \cup D$ satisfying the following constraints.

*Time Windows.* As production never stops, all four operations of the $k$-th cycle of component $i$ must happen during the time window $[(k-1)c_i, kc_i]$. Let $r_{a_i^k}$ (resp. $d_{a_i^k}$) be the release date $(k-1)c_i$ (resp. due date $kc_i$) of operation $a_i^k$ of the $k$-th cycle of component $i$. The start time of operation $\sigma(j)$ is:

$$s_{\sigma,j} = \begin{cases} r_{\sigma(j)} & \text{if } j = 1 \\ \max(r_{\sigma(j)}, s_{\sigma,j-1} + p_{\sigma(j-1)} + D_{\sigma(j-1),\sigma(j)}) & \text{otherwise} \end{cases}$$

Then, the completion time $e_{\sigma,j} = s_{\sigma,j} + p_{\sigma(j)}$ of operation $\sigma(j)$ must be lower than its due date:

$$\forall j \in [1, n],\ e_{\sigma,j} \leq d_{\sigma(j)} \tag{1}$$

*Precedences.* Pickups must precede deliveries.

$$\rho(pf_i^k) < \rho(df_i^k) \wedge \rho(pe_i^k) < \rho(de_i^k)\ \forall i \in [1, m]\ \forall k \in [1, n_i] \tag{2}$$

Notice that there are only two possible orderings for the four operations of a production cycle. Indeed, since the first delivery (of either the full or the empty trolley) and the second pickup take place at the same location, doing the second pickup before the first delivery is dominated (w.r.t. the train length and the time windows). Hence Eq. 3 is valid, though not necessary:

$$\rho(df_i^k) < \rho(pe_i^k) \vee \rho(de_i^k) < \rho(pf_i^k)\ \forall i \in [1, m]\ \forall k \in [1, n_i] \tag{3}$$

*Train Length.* The operator may assemble trolleys into a train[2], so a pickup need not be directly followed by its delivery. However, the total length of the train of

[2] Trolleys are designed so that they can be extracted out of the train in any order.

trolleys must not exceed a length $T_{\max}$. Let $t_i$ be the length of the trolley for component $i$, and let $t_{a_i^k} = t_i$ if $a_i^k \in P$ and $t_{a_i^k} = -t_i$ otherwise, then:

$$\forall j \in [1, n], \sum_{l=1}^{j} t_{\sigma(l)} \leq T_{\max} \tag{4}$$

This is a particular case of the *single vehicle pickup and delivery problem with capacity and time windows constraints.* However, there is no objective function, and instead, feasibility is hard. Moreover, the production-consumption cycles entail a very particular structure: the four operations of each component must take place in the same time windows and this is repeated for every cycle. As a result, one of the best method, Large Neighborhood Search [12], is severely hampered since it relies on the objective to evaluate the moves and the insertion of relaxed requests is often very constrained by the specific precedence structure.

# 3 Baseline Models

We designed two CP models: a variant of a single resource scheduling problem and a variant of a TSP. Then, we describe the current `LocalSolver` model.

*Scheduling-Based Model.* The importance of time constraints in the problem studied, suggests that a CP model based on scheduling would be relevant [1]. The problem is a single resource (the operator) scheduling problem with four types of non overlapping operations (pickup and delivery of empty and full trolleys). For each operation $a \in A$, we define the variable $s_a \in [r_a, d_a]$ as the starting date of operation $a$. Moreover, for each pair of operations $a$, $b \in A$, we introduce a Boolean variable $x_{ab}$ standing for their relative ordering. In practice, we need much fewer than $n^2$ Boolean variables as the time windows and Constraint (2) entails many precedences which can be either ignored or directly posted.

$$x_{ab} = \begin{cases} 1 \Leftrightarrow s_b \geq s_a + p_a + D_{a,b} \\ 0 \Leftrightarrow s_a \geq s_b + p_b + D_{b,a} \end{cases} \qquad \forall (a, b) \in A \tag{5}$$

$$x_{df_i^k pe_i^k} \vee x_{de_i^k pf_i^k} \qquad \forall i \in [1, m]\ \forall k \in [1, n_i] \tag{6}$$

Constraint (5) chanel the two sets of variables, and constraint (6) encodes Equation (3). Finally, Constraint (4) can be seen as a reservoir resource with limited capacity, which is filled by pickups, and emptied by deliveries. We use the algorithm from [9] to propagate it on starting date variables using the precedence graph implied by Boolean variables and precedences from Constraint (2).

*TSP-Based Model.* The second model is an extension of the first one, to wich we add extra variables and constraints from the model for TSP with time windows proposed in [6]. We need two fake operations, 0 and $n + 1$, for the start and the end of the route. For each operation $a \in A \cup \{0, n+1\}$, there is a variable $next_a$ that indicates which operation directly follows $a$. Also, a variable $pos_a$ indicates the position of the operation $a$ in the sequence of operations. We need another variable $train_a \in [0, T_{\max}]$ which represents the length of the train before the operation $a$. The following equations express the constraints of the problem:

$$train_0 = 0 \wedge train_{next_a} = train_a + t_a \qquad \forall a \in A \cup \{0\} \quad (7)$$

$$pos_{pf_i^k} < pos_{df_i^k} \wedge pos_{pe_i^k} < pos_{de_i^k} \qquad \forall i \in [1, m]\ \forall k \in [1, n_i] \quad (8)$$

$$s_0 = 0 \wedge s_{next_a} \geq s_a + p_a + D_{a,next_a} \qquad \forall a \in A \cup \{0\} \quad (9)$$

Constraint (7) encodes the train length at every point of the sequence using the ELEMENT constraint and constraint (8) ensures that pickups precede their deliveries. Constraint (9) ensure the accumulation of the time along the tour. Moreover, we use the CIRCUIT constraint to enforce the variables $next$ to form an Hamiltonian circuit. Additional redundant constraints are used to make the channeling between variables and improve filtering in addition to the constraint from the first model: $x_{ab} \implies next_b \neq a$, $pos_b > pos_a + 1 \implies next_a \neq b$, $pos_b > pos_a \Leftrightarrow x_{ab}$, $pos_a = \sum_{b \in A \cup \{0,n+1\}} x_{ab}$ and ALLDIFFERENT($pos$).

*Search Strategy.* Preliminary experiments revealed that branching on variables in an ordering "consistent" with the sequence of operations was key to solving this problem via CP. In the TSP model, we simply branch on the variables $next$ in ascending order, and choose first the operation with least earliest start time. In the scheduling model, we compute the set of pairs of operations $a, b$ such that $\{s_a, s_b\}$ is Pareto-minimal, draw one pair uniformly at random within this set and assign $x_{ab}$ so that the operation with least release date comes first. In conjunction with constraint propagation, this strategy acts as the "nearest city" heuristic in TSP. Indeed, since the sequence of past operations is known, the earliest start time of an operation depends primarily on the distance from the latest operation in the partial sequence (it also depends on the release date).

*LocalSolver.* The `LocalSolver` (LS) model is similar to the TSP model. It is based on a variable of type list $\boldsymbol{seq}$, a special decision variable type that represent the complete tour: $seq_j = a$ means operation $a$ is performed at position $j$. This variable is channeled with another list variable $\boldsymbol{pos}$ with $pos_a = j \Leftrightarrow seq_j = a$. We need two other list variables: $train$ and $s$ to represent respectively the length of the train and the start time of the operation at a given position.

$$s_1 = 0 \wedge$$

$$s_j = \max(r_{seq_j}, s_{j-1} + p_{seq_{j-1}} + D_{seq_{j-1}, seq_j}) \qquad \forall j \in [2, n] \quad (10)$$

$$train_1 = 0 \ \wedge train_j = train_{j-1} + t_{seq_j} \qquad \forall j \in [2, n] \quad (11)$$

$$s_j + p_{seq_j} \leq d_{seq_j} \qquad \forall j \in [1, n] \quad (12)$$

$$pos_{pf_i^k} < pos_{df_i^k} \wedge pos_{pe_i^k} < pos_{de_i^k} \qquad \forall i \in [1, m]\ \forall k \in [1, n_i] \quad (13)$$

$$pos_{df_i^k} < pos_{pe_i^k} \vee pos_{de_i^k} < pos_{pf_i^k} \qquad \forall i \in [1, m]\ \forall k \in [1, n_i] \quad (14)$$

$$\text{COUNT}(\boldsymbol{seq}) = 0 \quad (15)$$

The last constraint (15) acts like the global constraint All Different and therefore ensures that ***seq*** is a permutation. Surprisingly, relaxing the due dates and using the maximal tardiness as objective tends to degrade the performance of `LocalSolver`, hence we kept the satisfaction version.

### 3.1 Preliminary Experiments

The industrial data we have collected fall into three categories. In the industrial assembly line, each category is associated to one logistic operator who has the charge of a given set of components. In practice, these datasets are relatively underconstrained, with potentially quite large production cycles (time windows) for each component. For each category, denoted by S, L and R, we consider three time horizons: 43 500, 130 500 and 783 000 *hundredths of a minutes* which corresponds to a shift of an operator (7 h and 15 min), a day of work (made up of three shifts) and a week (6 days) respectively.

We then have 9 industrial instances from 400 to more than 10 000 operations. The main differences between those three categories are the number of components, and the synchronicity of the different production cycles. For S instances, there are only 5 components, and their production cycles are almost the same and very short. The other two instances have more than 30 components, with various production cycles (some cycles are short and others are very long).

The CP models were implemented using the constraint solver `Choco` 4.10 [11], and the `LocalSolver` version was 9.0. All experiments of this section were run on Intel(R) Core(TM) i7-8700 CPU @ 3.20 GHz with a timeout of 1 h.

The results are given in Table 1. The first column (Cl) denotes the category of the instance and second column ($H$) denotes the temporal horizon. For the two CP models (Scheduling and TSP), two indicators are given, the CPU time (in seconds), and the numbers of fails for solved instances. For `LocalSolver`, we give the CPU time when the instances are solved before the time limit.

The CP scheduling-based model solves all of the industrial instances without any fail except for instances R and L on the whole week. Notice, however, that when using generic heuristics such as *Domain over Weighted Degree* [3], only 3 instances could be solved by the same model in less than an hour. Moreover, `LocalSolver` cannot solve the largest instances, and requires much more CPU time in general. Although industrial instances are clearly very underconstrained, they are not trivial. Moreover, even on underconstrained instances, wrong early choices may lead to infeasible subtrees from which we cannot easily escape. In particular, the number of fails for the largest instance of category R shows that the very deep end of the search tree is likely explored in a brute-force manner.

One key factor in solving these instances is for the variable ordering to follow the ordering of the sequence of operations being built. Indeed, the propagation is much more efficient in this way, and in particular, the earliest start times of future operations can be easily computed and, as mentioned in the description of the heuristic, it reflects the distance from the last operation in the route.

We observe that the scheduling-based model is the fastest. Moreover, the TSP-based model contains too many variables and constraints, and run out of

**Table 1.** Comparison on the industrial instances

| Cl | $H$ | Scheduling | | TSP | | `LocalSolver` |
|---|---|---|---|---|---|---|
| | | CPU | #fail | CPU | #fail | CPU |
| S | Shift | 0.275 | 0 | 3.999 | 0 | 26 |
| S | Day | 0.840 | 0 | 48.855 | 0 | 566 |
| S | Week | 17.747 | 0 | Memory out | | Timeout |
| L | Shift | 7.129 | 0 | 36.033 | 8 | 121 |
| L | Day | 23.468 | 0 | 344.338 | 8 | 3539 |
| L | Week | 318.008 | 6 | Memory out | | Timeout |
| R | Shift | 8.397 | 0 | 41.615 | 14 | 1065 |
| R | Day | 44.215 | 0 | 417.116 | 15 | Timeout |
| R | Week | Timeout | 773738 | Memory out | | Timeout |

memory for the bigger horizon. Instances without any fails show that the first branch can be very slow to explore as propagation in nodes close to the root can be very costly, in particular with the TSP-based model.

We draw two conclusions from these preliminary experiments: First, the basic CP (or LS) models cannot reliably solve hard instances[3]. It is unlikely that stronger propagation would be the answer, as the TSP model (with stronger filtering) tends to be less effective, and does not scale very well. Second, building the sequence "chronologically" helps significantly, which explains why CP models outperform local search. As a consequence, greedy runs of the CP solver are surprisingly successful. Therefore, we propose to *learn* efficient heuristic policies and explore methods that can take further advantage of these heuristics.

## 4 Reinforcement Learning

The search for a feasible sequence can be seen as a Markov Decision Process (MPD) whose *states* are partial sequences $\sigma$, and *actions* $\mathcal{A}(\sigma)$ are the operations that can extend $\sigma$ without breaking precedence (2) nor capacity constraints (4).

In order to apply Reinforcement Learning (RL) to this MDP, it is convenient to relax the due date constraints and replace them by the minimization of the maximum tardiness: $\max\{L(\sigma, j) \mid 1 \leq j \leq |\sigma|\}$ where $L(\sigma, j) = e_{\sigma,j} - d_{\sigma(j)}$ is the tardiness of operation $\sigma(j)$. We also define the maximum tardiness on intervals: $L(\sigma, j, l) = \max\{L(\sigma, q) \mid j \leq q \leq l\}$, and we write $L(\sigma)$ for $L(\sigma, 1, |\sigma|)$. Since in this case operations can finish later than their due dates, it is necessary to make explicit the precedence constraints due to production cycles:

$$\max(\rho(df_i^{k-1}), \rho(de_i^{k-1})) < \min(\rho(pe_i^k), \rho(pf_i^k)) \; \forall i \in [1, m] \; \forall k \in [2, n_i] \quad (16)$$

[3] There might be too few data points to make that claim on industrial instances, but it is clearly confirmed on the synthetic benchmark (see Table 2).

Then we can further restrict the set $\mathcal{A}(\sigma)$ to operations that would not violate Constraints (16), nor (3). As a result, any state of the MDP reachable from the empty state is feasible for the relaxation. Finally, we can define the *penalty* $R(\sigma, j)$ for the $j$-th decision as the marginal increase of the objective function when the $j$-th operation is added to $\sigma$: $R(\sigma, j) = L(\sigma, 1, j) - L(\sigma, 1, j-1))$.

Now we can apply standard RL to seek for an effective stochastic heuristic policy for selecting the next operation. Moreover, as the MPD defined above is exponentially large, it is common to abstract a state $\sigma$ with a small descriptor, namely, $\boldsymbol{\lambda}(\sigma, a)$ a vector of four criteria for operation $a \in A$ in state $\sigma$. Let $lst(a, \sigma)$ be the latest starting time of the task $a$ in order to satisfy constraint (1) with respect to the tasks in $\sigma$ and constraints (2) and (3). For each operation $a \in \mathcal{A}(\sigma)$, we compute $\boldsymbol{\lambda}(\sigma, a)$ as follows:

$$\boldsymbol{\lambda}_1(\sigma, a) = (lst(a, \sigma) - \max(r_a, e_{\sigma,|\sigma|} - L(\sigma) + D_{\sigma(|\sigma|),a}))/\max_{b \in A}\{d_b - r_b\} \quad (17)$$

$$\boldsymbol{\lambda}_2(\sigma, a) = \max(r_a - (e_{\sigma,|\sigma|} - L(\sigma)), D_{\sigma(|\sigma|),a})/\max_{b,c \in A^2}\{D_{b,c}\} \quad (18)$$

$$\boldsymbol{\lambda}_3(\sigma, a) = 1 - |t_a|/T_{\max} \quad (19)$$

$$\boldsymbol{\lambda}_4(\sigma, a) = \begin{cases} 1 & \text{if } a \in P \\ 0 & \text{otherwise} \end{cases} \quad (20)$$

Criterion (17) can be seen as the operation's *emergency*: the distance to the due date of the task. Criterion (18) is the *travel time* from the last task in the sequence. For both of these criterion, we use $e_{\sigma,|\sigma|} - L(\sigma)$ instead of $e_{\sigma,|\sigma|}$ as the end time of the partial sequence $\sigma$ to offset the impact of previous choices. Criterion (19) is the *length* of the trolley. Indeed, since all operations must eventually be done, doing the operations which have the highest consumption of the "train" resource earlier leaves more freedom for later. Finally, criterion (20) penalizes pickups as leaving a trolley in the train for too long is wasteful.

We want to learn a stochastic policy $\pi_{\boldsymbol{\theta}}$, governed by a set of parameters $\boldsymbol{\theta}$ which gives the probability distribution over the set of actions $\mathcal{A}(\sigma)$ available at state $\sigma$. We first define a fitness function $f(\sigma, a)$ as a simple linear combination of the criteria: $f(\sigma, a) = \boldsymbol{\theta}^{\intercal}\boldsymbol{\lambda}(\sigma, a)$. Then, we use the `softmax` function to turn the fitness function into a probability distribution (ignore parameter $\beta$ for now).

$$\forall a \in \mathcal{A}(\sigma)\ \pi_{\boldsymbol{\theta}}(a \mid \sigma) = \frac{e^{(1-f(\sigma,a))/\beta}}{\sum_{b \in \mathcal{A}(\sigma)} e^{(1-f(\sigma,b))/\beta}} \quad (21)$$

### 4.1 Policy Gradient

As we look for a stochastic policy, the goal is to find the value of $\boldsymbol{\theta}$ minimizing the expected maximum value (i.e., maximum tardiness) $J(\boldsymbol{\theta})$ of solutions $\sigma$ produced by the policy $\pi_{\boldsymbol{\theta}}$. The basic idea of policy gradient methods is to minimize $J(\boldsymbol{\theta})$ by gradient descent, that is: iteratively update $\boldsymbol{\theta}$ by subtracting the gradient. We resort to the REINFORCE learning rule [16] to get the gradient of $J(\boldsymbol{\theta})$:

$$\nabla_{\boldsymbol{\theta}} J(\boldsymbol{\theta}) = \mathbb{E}_{\sigma \sim \pi_{\boldsymbol{\theta}}(\sigma)}[L(\sigma) \nabla_{\boldsymbol{\theta}} \log \pi_{\boldsymbol{\theta}}(\sigma)] \quad (22)$$

Which can then be approximated via Monte-Carlo sampling over $S$ samples:

$$\nabla_{\boldsymbol{\theta}} J(\boldsymbol{\theta}) \approx \frac{1}{S} \sum_{k=1}^{S} L(\sigma_k) \nabla_{\boldsymbol{\theta}} \log \pi_{\boldsymbol{\theta}}(\sigma_k) \tag{23}$$

We can decompose Eq. (23) in order to give a specific penalty to each decision (for every position $j$) of the generated solutions:

$$\nabla_{\boldsymbol{\theta}} J(\boldsymbol{\theta}) \approx \frac{1}{S} \sum_{k=1}^{S} \sum_{j=1}^{n} G(\sigma_k, j) \nabla_{\boldsymbol{\theta}} \log \pi_{\boldsymbol{\theta}}(\sigma_k(j) \mid \sigma_k(1, j-1)) \tag{24}$$

The penalty function takes into account that a decision can only affect the future marginal increase of tardiness: we replace the overall tardiness $L(\sigma)$ by:

$$G(\sigma, j) = \begin{cases} R(\sigma, j) & \text{if } j = n \\ R(\sigma, j) + \gamma * G(\sigma, j+1) & \text{otherwise} \end{cases}$$

The value of $\gamma$ controls how much decisions impact future tardiness. For $\gamma = 0$, we only take into account the immediate penalty. Conversely $\gamma = 1$ means we consider the sum of the penalties from position $j$.

### 4.2 Learning Algorithm

We learn a value of $\boldsymbol{\theta}$ for the synthetic dataset (Sect. 6) with the REINFORCE rule. Given a set of instances $I$, we learn by batch of size $S = q|I|$, in other words, we generate $q$ solutions for each instance. The value of $\boldsymbol{\theta}$ is initialized at random, then we apply the following three steps until convergence or timeout:

1. Generate $S$ solutions following $\pi_{\boldsymbol{\theta}}$.
2. Compute $\nabla_{\boldsymbol{\theta}} J(\boldsymbol{\theta})$ according to Eq. (24)
3. Update the value of $\theta$ as follows: $\boldsymbol{\theta} \leftarrow \boldsymbol{\theta} - \alpha \nabla_{\boldsymbol{\theta}} J(\boldsymbol{\theta})$

We found out that using a classic `softmax` function did not discriminate enough, and hence acts as random policy. To circumvent this, we use the parameter $\beta$ in Eq. (21) to control the trade off between the quality and the diversity of generated solutions. For a low value of $\beta$ the policy always chooses the best candidate, whereas a large value yields a more "balanced" policy. It turns out that $\beta = 0.1$ was a good value for learning in our case.

Then, we have evaluated the impact of $\gamma$ to compute the gradient in Equation (24). Recall that $\gamma$ controls how much importance we give to the $j$-th decision in the total penalty: the overall increase of the tardiness from $j$ to $n$ for $\gamma = 1$ or only the instant increase for $\gamma = 0$. Moreover, we also tried to give the penalty $L(\sigma)$ uniformly for every decision $j$ of the policy, instead of giving individual penalties $G(\sigma, j)$. We denote this penalty strategy "Uniform".

For each variant, we plot in Fig. (1a) the average performance $L(\sigma)$ of the policy after each iteration (notice the log-scale both for $X$ and $Y$). Here we learn on all generated instances of a day horizon (40 instances), for 2000 iterations.

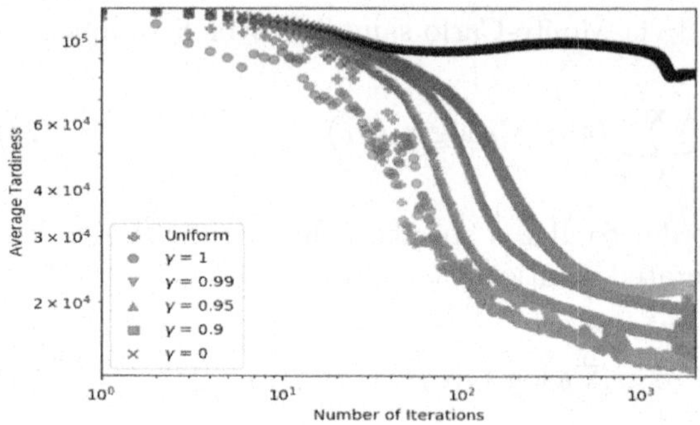


(a) Convergence for different values of $\gamma$

| Cl | By Inst. | By Cat. | All |
|---|---|---|---|
| A | 1011 | 1130 | 1128 |
| B | 6417 | 6766 | 6797 |
| C | 14622 | 15184 | 15069 |
| D | 20285 | 20578 | 20640 |

(b) Different training sets

**Fig. 1.** Behavior of reinforcement learning

High values for $\gamma$ are always better. For low value of $\gamma$, the (local) optimum is higher, and in some cases the method may not even converge, e.g. for $\gamma = 0$. Using uniformly the overall tardiness gives similar results to $\gamma = 1$, however it is less stable, and in Fig. (1a) we observe that it diverges after 1000 iterations.

Datapoints are colored with the vector $\boldsymbol{\theta}$ interpreted as a RGB value ($|\boldsymbol{\theta}| = 4$ but it can be characterized by three values after we normalize so that $\sum_{i=1}^{4} \boldsymbol{\theta}_i = 1$)[4]. We can see for instance that $\gamma = 0.9$ finds a value of $\boldsymbol{\theta}$ that is significantly different from all other methods ($\langle 0, 0.63, 0.29, 0.07\rangle$). Interestingly, "Uniform" and $\gamma = 1$ not only converge to the same average tardiness, but to similar $\boldsymbol{\theta}$'s (respectively $\langle 0.30, 0.49, 0.16, 0.04\rangle$ and $\langle 0.25, 0.56, 0.15, 0.04\rangle$). However, $\gamma = 1$ finds values of $\boldsymbol{\theta}$ that seem closer to the target ("greener") earlier, although it is not really apparent from the value of $L(\sigma)$. These values of $\theta$ indicate that a good heuristic is mainly a compromise between the *emergency* and the *travel time*[5]. Moreover, the travel time tends to be more important on larger horizon, because it a longer term impact: all subsequent operations are affected.

Finally, we report in Fig. (1b) the average tardiness for each instances following $\pi_{\boldsymbol{\theta}}$. The gain of learning specifically for a given (class of) instance(s) is at best marginal. This is not so surprising as we abstract states with a very simple model using a few criteria. However, it means that the value of $\boldsymbol{\theta}$ learnt on the full dataset is relevant to most instances.

## 5 Using the Heuristic Policy

We have implemented two types of approaches taking advantage of the stochastic policy learnt via RL: integrating it within CP as a randomized branching

[4] To highlight the differences we also normalize the RGB values and omit $\gamma = 0$.

[5] Although the importance of a criterion also depends on the distribution of the values of $\lambda$ after normalization, we are confident that the first two criteria are more important than the other two.

heuristic, and using it to generate sequences to be locally optimized via steepest descent in a multi-start local search. Here we describe the necessary modifications of the CP model, and we propose an efficient local search neighborhood.

### 5.1 Constraint Programming

In order to use the stochastic policy described in Sect. 4 in a CP solver, we need to slightly change the scheduling model. We introduce a new set of variables $seq_j$, one for each position $j$, standing for the operation at that position, with the following channeling constraint: $x_{seq_j seq_l} = 1\ \forall j < l \in [1, n]$.

We branch only on the variables $seq$ in lexicographic order. Therefore, the propagation for this constraint is straightforward: when branching on $seq_j = a$, as all variables $seq_l\ \forall l < j$ are already instantiated, we set $x_{ab} = 1,\ \forall b \in A \backslash \{seq_l \mid l < j\}$. Conversely, after assigning $seq_j$ we can remove $a$ from the domain of $seq_{j+1}$ if there exists $b \in A \backslash \{seq_l \mid l \leq j\}$ such that the domain of the variable $x_{ab}$ is reduced to $\{0\}$. Moreover, we can easily enforce *Forward Consistency* on $\{seq_1, \dots, seq_j\}$ with respect to precedence and train size constraints (2) and (4) as well as Constraint (3), i.e., when $\{seq_1, \dots, seq_{j-1}\}$ are all instantiated, we can remove all values of $seq_j$ that cannot extend the current subsequence. Therefore, we do not need the *Reservoir resources* propagator anymore.

We propose two strategies based on this CP model using the learned policy.

1. *Softmax policy and rapid restart.* In this method we choose randomly the next operation according to the softmax policy (Eq. 21). In order to explore quickly different part of the search tree, we rely on a rapid restart strategy, following a Luby [10] sequence with a factor 15.
2. *Limited Discrepancy Search.* As the key to solve those instances is to follow good heuristics, and to deviate as little as possible from them, *limited discrepancy search* (LDS) [8] fits well with this approach. We run the LDS implementation of `Choco`, which is an iterative version: the discrepancy starts from 0, to a maximum discrepancy parameter incrementally. For this approach we use the deterministic version of the policy $\pi(\sigma) = \arg\min_a f(\sigma, a)$.

### 5.2 Local Search

The solutions found by the heuristic policy can often be improved by local search moves. Therefore, we also tried a multi-start local search whereby we generate sequences with the heuristic policy, and then improve them via steepest descent. Sequences are generated using the same model used for RL (i.e., with relaxed due dates). Therefore, generated sequences respect all constraints, except (1) and we consider a neighborhood that preserves all other constraints as well. Then we apply the local move that decrease the most the maximum tardiness $L(\sigma)$ until no such move can be found. We use two types of moves and the time complexity of an iteration (i.e., computing, and commiting to, the best move) is in $O(nm)$.

We recall that $L(\sigma, j, l) = \max\{L(\sigma, q) \mid j \leq q \leq l\}$ is the maximum tardiness among all operations between positions $j$ and $l$ in $\sigma$.

*Swap Moves.* The first type of moves consists in swapping the values of $\sigma(j)$ and $\sigma(l)$. First, we need to make sure that the ordering of the operations within a given component remains valid, i.e., satifies constraints (2) and (16): a pickup (resp. delivery) operation must stay between its preceding and following deliveries (resp. pickups) for the same component. For every operation $a$, a valid range between the position of its predecessor $pr(a)$ and of its successor $su(a)$ can be computed in constant time. Then, for all $j \in [1, n]$ we shall consider only the swaps between $j$ and $l$ for $l \in [j+1, su(\sigma(j))]$ and such that $pr(\sigma(l)) \leq j$.

The second condition for the move to be valid is that the swap does not violate constraint (4), i.e., the maximum length of the train. Let $\tau_j = \sum_{l=1}^{j-1} t_{\sigma(l)}$ be the length of the train before the $j$-th operation. After the swap we have $\tau_{j+1} = \tau_j + t_{\sigma(l)}$ which must be less than $T_{\max}$. At all other ranks until $l$, the difference will be $t_{\sigma(l)} - t_{\sigma(j)}$, we only need to check the constraint for the maximum train length, that is: $\max\{\tau_q \mid j \leq q \leq l\} + t_{\sigma(l)} - t_{\sigma(j)} \leq T_{\max}$. This can be done in constant (amortized) time for all the swaps of operations $a$ by computing the maximum train length incrementally for each $l \in [j+1, su(\sigma(j))]$.

Then, we need to forecast the maximum tardiness of the sequence $\sigma'$ where the operations at positions $j$ and $l$ are swapped, i.e., compute the marginal cost of the swap. The tardiness of operations before position $j$ do not change. However, we need to compute the new tardiness $L(\sigma', j)$ and $L(\sigma', l)$ at positions $j$ and $l$, respectively. Moreover, we need to compute $L(\sigma', j+1, l-1)$ the new maximum tardiness for operations strictly between $j$ and $l$ and $L(\sigma', l+1, n)$ the new maximum tardiness for operations strictly after $l$.

The new end time $e_{\sigma',j}$ of operation $\sigma'(j) = \sigma(l)$ and hence the tardiness at position $j$ is $L(\sigma', j) = e_{\sigma',j} - d_{\sigma'(j)}$ can be computed in $O(1)$ as follows:

$$e_{\sigma',j} = p_{\sigma'(j)} + \max(r_{\sigma'(j)}, (e_{\sigma',j-1} + D_{\sigma'(j-1),\sigma'(j)}))$$

Next, operations $\sigma(j+1), \ldots, \sigma(l-1)$ remain in the same order and $\sigma(j+1)$ is shifted by a value $\Delta = e_{\sigma',j} + D_{\sigma'(j),\sigma'(j+1)} - e_{\sigma,j} - D_{\sigma(j),\sigma,j+1}$. However, subsequent operations may not all be equally time-shifted. Indeed, when $\Delta < 0$ there may exist an operation whose release date prevents a shift of $\Delta$.

Let $g_j = r_{\sigma(j)} - s_{\sigma(j)}$ be the *maximum left shift* (negative shift of highest absolute value) for the $j$-th operation, and let $g_{j,l} = \max\{g_q \mid j \leq q \leq l\}$.

**Proposition 1.** *If the sequence does not change between positions $j$ and $l$, a time-shift $\Delta < 0$ at position $j$ yields a time-shift* $\max(\Delta, g_{j,l})$ *at position $l$.*

Let $L_\Delta(\sigma, j, l)$ be the maximum tardiness on the interval $[j, l]$ of sequence $\sigma$ time-shifted by $\Delta$ from position $j$. We can define $L_{-\infty}(\sigma, j, l)$ the maximum tardiness on the interval $[j, l]$ for an infinite negative time-shift:

$$L_{-\infty}(\sigma, j, l) = \max\{L(\sigma, q, l) + g_{j,q} \mid j \leq q \leq l\} \tag{25}$$

**Proposition 2.** *If $\Delta < 0$ then $L_\Delta(\sigma, j, l) = \max(\Delta + L(\sigma, j, l), L_{-\infty}(\sigma, j, l))$.*

Conversely, when $\Delta > 0$ some of the time-shift may be "absorbed" by the waiting time before an operation. However, there is little point in moving operations coming before a position $j$ with a non-negative waiting time (i.e., where

$s_{\sigma(j)} = r_{\sigma(j)}$), as this operation and all subsequent operations would not profit from the reduction in travel time. Therefore we consider only swaps whose earliest position $j$ is such that $\forall q > j,\ g_q < 0$. As a results, if there is a positive time-shift $\Delta$ at $q > j$, we know that $L_\Delta(\sigma, q, l) = \Delta + L(\sigma, q, l)$. Moreover, the values of $L(\sigma, j, l)$, $g_{j,l}$ and $L_{-\infty}(\sigma, j, l)$ can be computed incrementally as:

$$\begin{aligned} L(\sigma, j, l+1) &= \max(L(\sigma, j, l), L(\sigma, l+1)) \\ g_{j,l+1} &= \max(g_{j,l}, g_{l+1}) \\ L_{-\infty}(\sigma, j, l+1) &= \max(L_{-\infty}(\sigma, j, l), g_{j,l+1} + L(\sigma, l+1)) \end{aligned}$$

Therefore, when $j < l-1$, we can compute the new tardiness $L(\sigma', j+1, l-1) = L_\Delta(\sigma, j+1, l-1)$ of the operations in the interval $[j+1, l-1]$ in constant (amortized) time since the query of Proposition 2 can be checked in $O(1)$.

The new tardiness $L(\sigma', l)$ at position $l$ is computed in a similar way as for $L(\sigma', j)$ since we know the new start time of $\sigma'(l-1)$ from previous steps.

Finally, in order to compute the new maximum tardiness $L(\sigma', l+1, n)$ over subsequent operations, we precompute $L(\sigma, j, n)$, $g_{j,n}$ and $L_{-\infty}(\sigma, j, n)$ for every position $1 \leq j \leq n$ once after each move in $O(n)$. Then $L(\sigma', l+1, n)$ can be obtained in $O(1)$ for every potential move from Proposition 2.

Therefore, we can check the validity and forecast the marginal cost of a swap in constant amortized time and perform the swap in linear time. The time complexity for an iteration is thus in $O(nm)$ since, for a given component $i$, the sum of the sizes of the valid ranges for all pickups and deliveries of this component is in $O(n)$. Indeed, let $a_i^1, \ldots, a_i^{4n_i}$ be the operations component $i$ ordered as in $\sigma$. Then $su(a_i^k) = \rho(a_i^{k+1})$ and $\sum_{k=1}^{4n_i} su(a_i^k) - \rho(a_i^k) = \rho(a_i^{4n_i}) - \rho(a_i^1) \in \Theta(n)$).

*Toggle Moves.* As observed in Sect. 3, there are only two dominant orderings for the four operations of the $k$-th production cycle of component $i$. The second type of moves consists in changing from one to the other of these two orderings, by swapping the values of $\sigma(pf_i^k)$ and $\sigma(pe_i^k)$ and the values of $\sigma(df_i^k)$ and $\sigma(de_i^k)$. This change leaves the size of the train constant, hence all these moves are valid.

Let $j_1 = pf_i^k, j_2 = df_i^k, j_3 = pe_i^k, j_4 = de_i^k$ be the positions of the four operations of component $i$ and cycle $k$ in the current solution, and suppose, wlog, that $j_1 < j_2 < j_3 < j_4$. Let $\sigma'$ denote the sequence obtained by applying a toggle move on component $i$ and cycle $k$ in $\sigma$. In order to forecast the marginal cost of the move, we need to compute the new tardiness at the positions of the four operations involved $L(\sigma', j_1)$, $L(\sigma', j_2)$, $L(\sigma', j_3)$ and $L(\sigma', j_4)$. Moreover, we need the new maximum tardiness on four time-shifted intervals:

$L(\sigma', j_1+1, j_2-1) = L_{\Delta_1}(\sigma, j_1+1, j_2-1)$, $L(\sigma', j_2+1, j_3-1) = L_{\Delta_2}(\sigma, j_2+1, j_3-1)$, $L(\sigma', j_3+1, j_4-1) = L_{\Delta_3}(\sigma, j_3+1, j_4-1)$ and $L(\sigma', j_4+1, n) = L_{\Delta_4}(\sigma, j_4+1, n)$.

Computing the marginal costs can be done via the same formulas as for swaps: we can first compute the new end time for the operation at position $j_1$, then from it compute the value of $\Delta_1$ that we can use to compute

$L_{\Delta_1}(\sigma, j_1+1, j_2-1)$ in $O(j_2 - j_1 - 1)$ time, and so forth. The difference, however, is that there is fewer possible moves $(n/4)$, although the computation of the marginal cost cannot be amortized. The resulting time complexity is the same: $O(nm)$.

## 6 Experimental Results

We generated synthetic instances[6] in order to better assess the approaches. Due to the time windows constraints, it is difficult to generate certifiably feasible instances. Their feasibility has been checked on the shortest possible horizon, i.e., the duration of the longest production cycle of any component, which is about 10 000 time units depending on the instances. There are four categories of instances parameterized by the number of components (15 in category A, 20 in B, 25 in C and 30 in D). In the real dataset, several components have similar production cycles. We replicates this feature: synthetic instances have from 2 or 3 distinct production cycles in category A, to up to 7 in category D. The latter are therefore harder because there are more asynchronous productions cycles. We generated 10 random instances for each category and consider the same three horizons (shift, day and week) for each, as industrial instances.

Experiments for this section were run on a cluster made up of Xeon E5-2695 v3 @ 2.30 GHz and Xeon E5-2695 v4 @ 2.10 GHz. For the basic CP models, we add randomization and a restart strategy following a Luby sequence, and we ran each of the 120 instances 10 times. We could not carry on experiments on synthetic data with `LocalSolver` because we were not granted a license. However, from the few tests we could do, we expect `LocalSolver` to behave similarly as on industrial benchmarks.

**Table 2.** Comparison of the methods on generated instances

| Cl | $H$ | Scheduling | | | TSP | | | CP-softmax | | | LDS | | | Multi-start LS | | |
|---|---|---|---|---|---|---|---|---|---|---|---|---|---|---|---|---|
| | | #S | CPU | #fail | #S | CPU | #fail | #S | CPU | #fail | #S | CPU | #fail | #S | CPU | $L_{max}$ |
| A | Shift | 7.1 | 418 | 300K | 4.0 | 56 | 366 | **9.0** | 2 | 15 | **9.0** | 2 | 9 | **9.0** | 0 | 1 m |
| | Day | 4.0 | 29 | 213 | 3.6 | 802 | 1267 | **9.0** | 15 | 815 | 8.0 | 21 | 71 | **9.0** | 176 | 19 m |
| | Week | 3.1 | 866 | 1976 | 0.0 | mem. out | | **8.0** | 118 | 27 | 5.0 | 68 | 0.0 | 7.0 | 155 | 1 h11 |
| B | Shift | 2.1 | 389 | 150K | 0.9 | 844 | 11K | **6.0** | 4 | 77 | **6.0** | 15 | 85 | **6.0** | 2 | 11 m |
| | Day | 1.0 | 201 | 15K | 0.0 | – | | **5.2** | 341 | 20K | 4.0 | 12 | 19 | 4.6 | 346 | 1 h |
| | Week | 0.0 | – | | 0.0 | mem. out | | **3.5** | 423 | 715 | 1.0 | 99 | 0.0 | 1.0 | 0 | 4 h59 |
| C | Shift | 0.0 | – | | 0.0 | – | | **4.0** | 103 | 5366 | **4.0** | 715 | 4090 | **4.0** | 255 | 32 m |
| | Day | 0.0 | – | | 0.0 | – | | **1.0** | 12 | 7 | **1.0** | 18 | 27 | **1.0** | 1 | 1 h45 |
| | Week | 0.0 | – | | 0.0 | mem. out | | **1.0** | 807 | 366 | 0.0 | – | | 0.0 | – | 11 h51 |
| D | Shift | 0.0 | – | | 0.0 | – | | **1.9** | 697 | 24K | 1.0 | 442 | 1058 | 1.6 | 1165 | 31 m |
| | Day | 0.0 | – | | 0.0 | – | | 0.0 | – | | 0.0 | – | | 0.0 | – | 2 h19 |
| | Week | 0.0 | – | | 0.0 | mem. out | | 0.0 | – | | 0.0 | – | | 0.0 | – | 17 h52 |

[6] Avalaible at https://gitlab.laas.fr/vantuori/trolley-pb.

For each of the methods using the learnt heuristics, we normalize $\boldsymbol{\theta}$ so that $\sum_{i=1}^{4} \boldsymbol{\theta}_i = 1$ and set $\beta$ to 1/150. We learn the policy by batches of size 240 formed by 6 runs on each of the 40 day-long instances, during 2000 iterations. The learning rate depends on the size of the instances: $\alpha = 2^{-12}/\overline{n}$ where $\overline{n}$ is the average number of task in the batch. The rationale is that the magnitude of the gradient depends on the tardiness $L(\sigma)$ which tends to grow with the number of operations. Therefore, we use the learning rate $\alpha$ to offset this growth, which is key to have a stable convergence. For the two methods using the stochastic policy, we made the first run deterministic i.e. the policy becomes stochastic only after the first restart for the CP based one, and after the first iteration for the multi-start local search.

The results are presented in Table 2. We report the number of solved instances (among 10 instances for every time horizon) averaged over 10 randomized runs for each CP model in the column "#S". The synthetic dataset is more constrained than the industrial dataset and the two basic CP models fail to solve most of the instances ("–" in the table indicates a time out). However, the relative performance remains unchanged w.r.t. Table 1: the scheduling-based models shows better performance in terms of number of solved instances and CPU time while scaling better in memory. All three of the RL-based methods significantly outperform previous approaches. The results in Table 2 indicate that the rapid restarts approach dominates the others. However, it may not be as clear-cut as that: for other settings of the hyperparameters ($\alpha$, $\beta$ and $\gamma$) the relative efficiencies fluctuate and other methods can dominate. Moreover, one advantage of the multi-start local search method is that since due dates are relaxed, imperfect solutions can be produced, even for infeasible instances. We report the average maximum tardiness in column $L_{max}$.

This global $\boldsymbol{\theta}$ also works well with the industrial dataset. All instances are easily solved by all three methods, except "R" for the week horizon, which is only solved by the rapid restart approach. We learnt a dedicated policy for the industrial dataset with the same settings. It turns out that every instance was solved by the deterministic policy using the new value for $\boldsymbol{\theta}$, except the instance "L" for the week horizon. However, it is easily solved by all three methods.

## 7 Conclusion

In this paper we have applied reinforcement learning to design simple yet efficient stochastic decision policies for an industrial problem: planning the production process at Renault. Moreover, we have shown how to leverage these heuristic policies within constraint programming and within local search.

The resulting approaches significantly improve over the current local search method used at Renault. However, many instances on synthetic data remain unsolved. We plan on using richer machine learning models, such as neural networks, to represent states. Moreover, we would like to embed this heuristic in a Monte-Carlo Tree Search as it would fit well with our current approach since it relies on many rollouts. Finally, we would like to tackle the more general problem of assigning components to operators and then planning individual routes.

## References


1. Beck, J.C., Prosser, P., Selensky, E.: Vehicle routing and job shop scheduling: what's the difference? In: Proceedings of the 13th International Conference on Automated Planning and Scheduling ICAPS, pp. 267–276 (2003)
2. Bello, I., Pham, H., Le, Q.V., Norouzi, M., Bengio, S.: Neural Combinatorial optimization with reinforcement learning. In: 5th International Conference on Learning Representations, Workshop Track Proceedings ICLR (2017)
3. Boussemart, F., Hemery, F., Lecoutre, C., Sais, L.: Boosting systematic search by weighting constraints. In: Proceedings of the 16th European Conference on Artificial Intelligence ECAI, pp. 146–150 (2004)
4. Dai, H., Khalil, E.B., Zhang, Y., Dilkina, B., Song, L.: Learning Combinatorial Optimization Algorithms over Graphs. In: Proceedings of the 31st International Conference on Neural Information Processing Systems NIPS, pp. 6351–6361 (2017)
5. Deudon, M., Cournut, P., Lacoste, A., Adulyasak, Y., Rousseau, L.-M.: Learning heuristics for the TSP by policy gradient. In: van Hoeve, W.-J. (ed.) CPAIOR 2018. LNCS, vol. 10848, pp. 170–181. Springer, Cham (2018). https://doi.org/10.1007/978-3-319-93031-2_12
6. Ducomman, S., Cambazard, H., Penz, B.: Alternative filtering for the weighted circuit constraint: comparing lower bounds for the TSP and solving TSPTW. In: Proceedings of the 30th AAAI Conference on Artificial Intelligence AAAI, pp. 3390–3396 (2016)
7. Geiger, M.J., Kletzander, L., Musliu, N.: Solving the torpedo scheduling problem. J. Artif. Intell. Res. **66**, 1–32 (2019)
8. Harvey, W.D., Ginsberg, M.L.: Limited discrepancy search. In: Proceedings of the 14th International Joint Conference on Artificial Intelligence IJCAI, pp. 607–613 (1995)
9. Laborie, P.: Algorithms for propagating resource constraints in AI planning and scheduling: Existing approaches and new results. Artif. Intell. **143**(2), 151–188 (2003)
10. Luby, M., Sinclair, A., Zuckerman, D.: Optimal speedup of Las Vegas algorithms. Inf. Process. Lett. **47**, 173–180 (1993)
11. Prud'homme, C., Fages, J.G., Lorca, X.: Choco documentation. TASC - LS2N CNRS UMR 6241, COSLING S.A.S (2017)
12. Ropke, S., Pisinger, D.: An adaptive large neighborhood search heuristic for the pickup and delivery problem with time windows. Transp. Sci. **40**(4), 455–472 (2006)
13. Schaus, P.: The torpedo scheduling problem. http://cp2016.a4cp.org/program/acp-challenge/ (2016)
14. Silver, D., et al.: Mastering the game of Go with deep neural networks and tree search. Nature **529**(7587), 484–489 (2016)
15. Silver, D., et al.: Mastering the game of Go without human knowledge. Nature **550**(7676), 354–359 (2017)
16. Williams, R.J.: Simple statistical gradient-following algorithms for connectionist reinforcement learning. Mach. Learn. **8**(3), 229–256 (1992)

# Computing the Local Aromaticity of Benzenoids Thanks to Constraint Programming

Yannick Carissan[1], Chisom-Adaobi Dim[1], Denis Hagebaum-Reignier[1], Nicolas Prcovic[2], Cyril Terrioux[2(✉)], and Adrien Varet[2]

[1] Aix Marseille Univ, CNRS, Centrale Marseille, ISM2, Marseille, France
{yannick.carissan,chisom-adaobi.dim,denis.hagebaum-reignier}@univ-amu.fr
[2] Aix Marseille Univ, Université de Toulon, CNRS, LIS, Marseille, France
{nicolas.prcovic,cyril.terrioux,adrien.varet}@univ-amu.fr

**Abstract.** Benzenoids are a subfamily of hydrocarbons (molecules that are only made of hydrogen and carbon atoms) whose carbon atoms form hexagons. These molecules are widely studied in theoretical chemistry. Then, there is a lot of problems relative to this subject, like the benzenoid generation or the enumeration of all its Kekulé structures (i.e. all valid configurations of double bonds). In this context, the computation of the local aromaticity of a given benzenoid is an important problematic since the aromaticity cannot be measured. Nowadays, computing aromaticity requires quantum chemistry calculations that are too expensive to be used on medium to large-sized molecules. But, there exist some methods related to graph theory which can allow us to compute it. In this article, we describe how constraint programming can be useful in order to compute the aromaticity of benzenoids. Moreover we show that our method is much faster than the reference one, namely NICS.

**Keywords:** Constraint programming · Modeling · Graph variables and constraints · Chemistry

## 1 Introduction

*Polycyclic aromatic hydrocarbons (PAHs)* are hydrocarbons whose carbons are forming cycles of different sizes. The properties of these molecules depend on their aromaticity, which is a fundamental concept in chemistry (defined in Sect. 2). Since the discovery of graphene by Andre Geim and Konstantin Novoselov awarded with the 2010 Nobel price in physics, the interest in the aromaticity concept vividly revives due to its potential importance in nanoelectronics: aromaticity favors electronic flow through molecules, thus aromatic compounds are of interest for the design of nanoelectronic compounds. *Benzenoids* are a subfamily of PAHs made of 6-membered carbon rings (i.e. each cycle of six carbon

This work has been funded by the Agence Nationale de la Recherche project ANR-16-C40-0028.


H. Simonis (Ed.): CP 2020, LNCS 12333, pp. 673–689, 2020.
https://doi.org/10.1007/978-3-030-58475-7_39

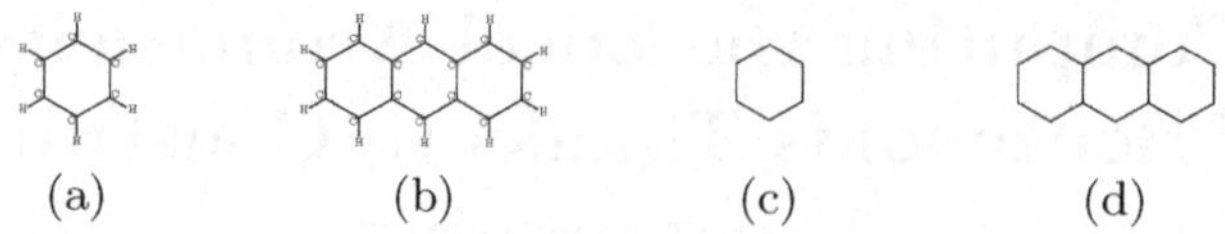

Fig. 1. Two small benzenoids: benzene (a) and anthracene (b) with their graphical representations (c) and (d).

atoms forms a hexagon). To fill carbon valency, each atom of carbon is bonded to either two other carbons and one hydrogen or three carbons. For example, Figs. 1(a)–(b) are representing two benzenoids: benzene and anthracene.

PAHs are well-studied in various domains because of their energetic stability, molecular structures or optical properties. In natural environment, these molecules are created by the incomplete combustion of carbon contained in combustibles [11]. PAHs are also studied in interstellar chemistry because they are suspected to be present in interstellar clouds and are believed to act as catalysts for chemical reactions taking place in space [3]. They are also intensively studied in other domains like molecular nanoelectronics [24] or organic synthesis [13,20].

In this context, *aromaticity* is very important. It allows chemists to link the energetic stability of a molecule to its molecular structure [2]. The stability of a molecule is a measure of the energy needed to break all chemical bonds and separate all the atoms of the molecule apart. Because of aromaticity, some molecules have an extra term in this energy: breaking them apart requires more energy than for non aromatic molecules with the same number of atoms. Recently, some methods using quantum chemistry were established in order to compute the aromaticity of a given molecule. The most popular one called *NICS* (Nuclear Independent Chemical Shift [1]) consists of applying a magnetic field perpendicular to the molecular plane and observe the behavior of its electrons. Analyzing the response of the electronic density allows chemists to quantify the aromaticity of the molecule. However, this method has a very high cost and computing the aromaticity of large molecules can easily take a few days. This large computational cost is due to the fact that quantum chemistry calculations require many steps involving iterative procedures before doing the actual calculation of aromaticity. To circumvent this drawback, some methods using graph theory were proposed in the 1990s [9,10,19], which roots can be tracked back to the work of Hückel in the 1930s [7]. They will be presented in the following parts.

In this paper, we present a new method based on constraint programming in order to compute the local aromaticity of benzenoids. For example, this method needs to enumerate particular cycles or count the number of Kekulé structures. Such tasks can be modeled as CSP instances and solved efficiently thanks to constraint solvers like Choco [4] while requiring a reduced implementation effort unlike usual methods from theoretical chemistry or any bespoke methods based on algorithm engineering.

The paper is organized as follows. Section 2 recalls some basic notions about chemistry and constraint programming. Then, Sect. 3 introduces some existing

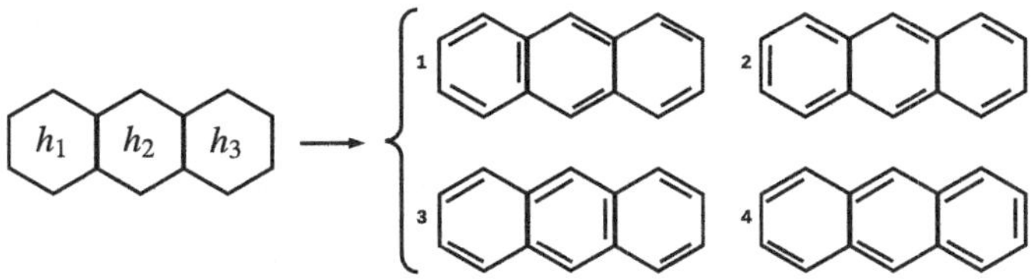


**Fig. 2.** Kekulé structures of anthracene.

methods to compute the aromaticity of benzenoids. In Sect. 4, we describe a new method which exploits constraint programming in order to compute the local aromaticity. In Sect. 5, we present some experimental results which show the interest of our approach. Finally, we conclude and give some perspectives in Sect. 6.

# 2 Preliminaries

## 2.1 Theoretical Chemistry

*Benzene* (represented in Fig. 1(a)) is a molecule made of 6 carbon atoms and 6 hydrogen atoms. Its carbon atoms form a hexagon (also called *benzenic cycle* or *benzenic ring*) and each of them is linked to a hydrogen atom. *Benzenoids* are a subfamily of PAHs containing all molecule which can be obtained by aggregating benzenic rings. For example, Fig. 1(b) shows anthracene, which contains three benzenic rings.

Before going further, we recall some basic definitions of chemistry. Firstly, the *valence* of an atom is the number of bonds that it can build with its electrons (one electron per bond). Carbon and hydrogen atoms have a valence of 4 and 1 respectively. As in a benzenoid, each carbon atom is linked either to two other carbon atoms and one hydrogen atom or to three other carbon atoms, we can easily deduce that one of its electron is not used. These electrons are called $\pi$-*electron* and can be used to enhance one bond by establishing *double bonds* (i.e. a bond involving two electrons per atom). Therefore, each carbon is involved in a double bond and two single bonds.

A *Kekulé structure* of a benzenoid is a valid configuration of its double bonds (i.e. a configuration in which each carbon atom is involved in exactly one double bond). Figure 2 depicts all the Kekulé structures of anthracene. A benzenoid can have several Kekulé structures or none (Fig. 3(a) depicts an example of benzenoid which has no Kekulé structure). We denote $\mathcal{K}(B)$ the set of all Kekulé structures of a benzenoid $B$. Note that the number of Kekulé structures of a benzenoid can be exponential. Therefore, given a benzenoid, generating all its Kekulé structures is a hard problem. A benzenoid continually alternates between its Kekulé structures. This dynamic is at the origin of the notion of aromaticity. There exist some methods based on graph theory which allow to compute the resonance energy of a given benzenoid (i.e. the energy induced by its aromaticity) and these methods require to be able to enumerate all its Kekulé structures [15].

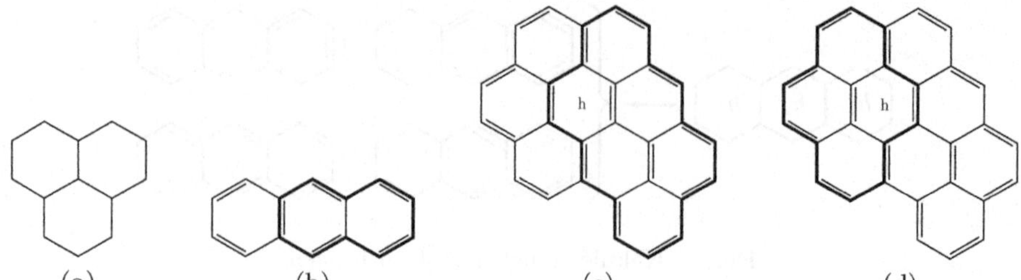


**Fig. 3.** A benzenoid having no Kekulé structure (a), a conjugated circuit in thick line (b) and an example of redundant circuits (c) and (d).

*Aromaticity* is a concept built by chemists in the early $20^{th}$ century in order to account for the surprising chemical stability of the benzene molecule. In this molecule, after making a single bond to each of its three neighbors (two carbon and one hydrogen), each carbon of the hexagonal geometry carries one extra electron. Electrons tend to form bonds (i.e. pair with another electron) whenever possible. Thus, this electron forms a molecular bond with the electron of a neighboring carbon atom. When all six electrons do the same, the electronic structure, first proposed by Kekulé [8], is obtained. Yet, two such structures exist as the pairing for one carbon can be with any of its two neighbors. The interaction or resonance of these two coexisting solutions is described by quantum physics and leads to an over-stabilization energy called aromaticity. This concept can be extended to fused benzene rings. It turns out that aromatic molecules often have a characteristic smell and/or taste, hence the name of the concept.

Due to the physical nature of aromaticity, hydrogen atoms do not play any role in its determination. Thus, it is custom not to take them into account in connectivity based methods. So, we do not represent them afterwards. Therefore, a benzenoid can be represented as an undirected graph $B = (V, E)$, with $V$ the set of vertices and $E$ the set of edges. Every vertex in $V$ represents a carbon atom and every edge of $E$ represents a bond between the two corresponding carbons. Moreover, this kind of graph, is connected, planar and bipartite. Figures 1(c) and (d) represent the graphs related to the molecules of benzene and anthracene. Finally, we can remark that the set of double bonds of a Kekulé structure is nothing more than a perfect matching on the benzenoid. As a reminder, a *perfect matching* of an undirected graph $G = (V, E)$ is a set of edges $E' \subseteq E$ such that $\forall (e_1, e_2) \in E' \times E', e_1 \neq e_2, e_1 \cap e_2 = \emptyset$ and $\bigcup_{e \in E'} e = V$.

## 2.2 Constraint Programming

An instance $P$ of the *Constraint Satisfaction Problem (CSP)* is a triplet $(X, D, C)$. $X = \{x_1, \ldots, x_n\}$ is a set of $n$ *variables*. For each variable $x_i$ of $X$, there exists an associated domain $D_{x_i} \in D = \{D_{x_1}, \ldots, D_{x_n}\}$ which represents the values that $x_i$ can take. $C = \{c_1, ..., c_e\}$ represents a set of $e$ *constraints*.

Constraints represent the interactions between the variables and describe the allowed combinations of values.

Solving a CSP instance $P = (X, D, C)$ amounts to find an assignment of all the variables of $X$ with a value contained in their associated domain which satisfies all the constraints of $C$. Such an assignment is called a *solution*. This problem is NP-hard [22].

Many libraries are available to represent and solve CSP instances. In this paper, we exploit the open-source Java library *Choco* [4]. This choice is highly guided by our need to be able to define *graph variables* and directly apply graph-related constraints (e.g. connected or cyclic constraints). Graph variables have as domain a set of graphs defined by a lower bound (a sub-graph called $GLB$) and an upper bound (a super-graph called $GUB$). Moreover, Choco implements the usual global constraints which make the modeling easier and its solver is efficient and configurable.

## 3 Computing Resonance Energy of a Benzenoid

### 3.1 Definitions

Resonance energy is used to quantify the energy induced by the aromaticity of a benzenoid. It allows us to get information about its stability (the lower the energy, the greater the stability). It is possible to compute this energy globally (on the entire molecule), which is called *global aromaticity* or locally (by assigning an energy to each hexagon), which is called *local aromaticity*. The latter is the most interesting since it allows us to identify the least stable parts of the molecule. Knowing that chemical reactions are more likely to occur on these parts, it can be used to predict the location of a reaction.

Randić presented a method which approximates the resonance energy of a given benzenoid by enumerating all the linearly independent Minimal Conjugated Circuits (later called $h$-MCCs) of each of its Kekulé structures [17]. Before going further, we have to introduce some definitions:

**Definition 1** ([14]). *Let $B$ be a benzenoid and $K$ one of its Kekulé structures. A* ***conjugated circuit*** *$\mathcal{C}$ of $K$ is a cycle of $B$ whose edges correspond alternately to single and double bonds in $K$. The size of $\mathcal{C}$ (noted $|\mathcal{C}|$) is the integer $i$ such that $\mathcal{C}$ contains $4i + 2$ edges.*

So, a conjugated circuit is a cycle alternating between single and double bonds. For example, the cycle in thick line in Fig. 3(b) is a conjugated circuit of size 2.

Given a benzenoid $B$ and one of its cycles $\mathcal{C}$, we call *interior of $\mathcal{C}$* the subgraph induces by all the edges and vertices which are in the interior of $C$.

Now, let us introduce the covering of a hexagon by a conjugated circuit:

**Definition 2** ([18]). *Let $B$ be a benzenoid and $K$ one of its Kekulé structure. A conjugated circuit $\mathcal{C}$ of $K$ covers a hexagon $h$ if and only if $h$ is contained into the interior of $\mathcal{C}$. Two conjugated circuits are said* ***redundant circuits*** *if they cover the same hexagon.*

For example, let us consider the Kekulé structure in Fig. 3(c). The hexagon $h$ is covered by two conjugated circuits: a first one of size 4 (Fig. 3(c)) and a second one of size 3 (Fig. 3(d)). In this case, we pay more attention to the circuit with the smallest size:

**Definition 3** ([18]). *Let $B$ be a benzenoid and $K$ one of its Kekulé structure. $\mathcal{C}$ is a* ***minimal conjugated circuit*** *of the hexagon $h$ of $B$ (also called $h$-MCC) w.r.t. $K$ if $\mathcal{C}$ is one of the covering circuits of $h$ having the smallest size.*

So, if we look at Figs. 3(c)–(d), the circuit of size 3 is a $h$-MCC for the hexagon $h$.

An energy is given to each of these conjugated circuits depending on their size (a smaller circuit has a higher energy). Thereafter, we denote $R_i$ the energy induced by a conjugated circuit of size $i$. Initially, these values were calculated using the formula $R_i = \frac{1}{i^2}$, but optimized values were established by linear regression: $R_1 = 0.869, R_2 = 0.246, R_3 = 0.100$ and $R_4 = 0.041$ [16]. These values make it possible to compute the energy induced for a given Kekulé structure:

**Definition 4** ([18]). *Let $B$ be a benzenoid and $K$ one of its Kekulé structure. The energy $R(K)$ induced by the minimal circuits of $K$ is defined as follows:*

$$R(K) = \sum_{i \in \{1,2,...\}} r_i(K) \times R_i$$

*where $r_i(K)$ is the number of minimal circuits of size $i$ in $K$.*

By extension, one defines the energy induced by a benzenoid.

**Definition 5** ([18]). *Let $B$ be a benzenoid. The energy $R(B)$ induced by $B$ is defined as follows:*

$$R(B) = \sum_{K \in \mathcal{K}(B)} R(K)$$

Finally, one can define the resonance energy of a benzenoid:

**Definition 6** ([18]). *Let $B$ be a benzenoid. The* ***resonance energy*** *$E(B)$ of $B$ is defined as follows:*

$$E(B) = \frac{R(B)}{|\mathcal{K}(B)|}$$

For example, if we look at the Kekulé structures of anthracene in Fig. 2 and consider all their $h$-MCCs, we can easily see that its resonance energy is $\frac{6R_1+4R_2+2R_3}{4} = 1.57$ if we take the optimized $R_i$ values.

To conclude with this part, we call *local aromaticity* of a benzenoid $B$ on the hexagon $h$ the energy obtained by using the previous formula, but with only looking at conjugated circuits which are minimal circuits covering $h$.

**Definition 7.** *Let $h$ be a hexagon of a benzenoid $B$. The local resonance energy $E(B,h)$ of $h$ is defined as follows:*

$$E(B,h) = \frac{\sum_{K \in \mathcal{K}(B),\ \mathcal{C}\ a\ h-MCC\ w.r.t.\ K} R_{|\mathcal{C}|}}{|\mathcal{K}(B)|}$$

For example, if we consider Fig. 2, the local aromaticity of hexagon $h_1$ of anthracene is $\frac{2R_1+R_2+R_3}{4} = 0.521$. Indeed, if we look at hexagon $h_1$, we have:

- A circuit of size 1 for Kekulé structure 1 (the circuit covering $h_1$).
- A circuit of size 1 for structure 2 (the same circuit as before).
- A circuit of size 2 for structure 3 (the circuit covering $h_1$ and $h_2$).
- A circuit of size 3 for structure 4 (the circuit covering $h_1$, $h_2$ and $h_3$).

Note that the sum of $E(B,h)$ over all the hexagons $h$ of $B$ is equal to the energy $E(B)$.

### 3.2 Computing the Resonance Energy

Lin and Fan [10] proposed a method based on the definition of the resonance energy. Given a benzenoid $B$, this method first enumerates all minimal conjugated circuits (i.e. computes a $h$-MCC for all its hexagons $h$) for each Kekulé structure. Then, it deduces the energy induced by each minimal conjugated circuit and adds them up. Finally, it divides the obtained sum by the number of Kekulé structures to obtain the resonance energy. Such a method was implied in Randić's work. The main contribution of Lin and Fan consists in describing how to compute the $h$-MCCs. For that, they identify the specific forms of the $h$-MCC depending on the location of the double bonds of $h$. Therefore this method requires to analyze the double bonds of all the hexagons.

The main drawback of this method is that it requires to generate all the Kekulé structures of the benzenoid. As the number of Kekulé structures may be exponential, this method is clearly inefficient in practice and can only be used for small benzenoids.

To overcome this, Lin [9] proposed another method that is able to compute all the minimal circuits of a given benzenoid without generating its Kekulé structures. This method only considers circuits having a size at most 4. So, it provides an approximation of the resonance energy. Before describing this method, let us introduce some needed definitions.

**Definition 8.** *Let $B$ be a benzenoid and $\mathcal{C}$ a cycle of $B$ (with $4i+2$ edges $\forall i \in \mathbb{N}$). $M(\mathcal{C})$ is the number of perfect matchings of $\mathcal{C}$ and its interior inducing a minimal circuit for at least one of the hexagons it covers.*

For example, let us consider $\mathcal{C}$ as being the cycle of size 2 represented in Fig. 4(a). $\mathcal{C}$ can induce two different conjugated circuits that are clearly minimal and cover the two hexagons. So we have $M(\mathcal{C}) = 2$.

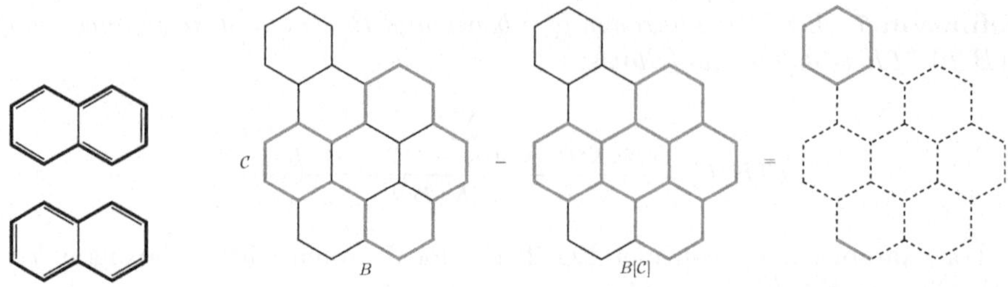


(a) (b)

**Fig. 4.** A cycle $\mathcal{C}$ such as $M(\mathcal{C}) = 2$ (a) and an example of computation of the number of occurrences of a cycle (b). (Color figure online)

**Definition 9.** ([19]). *Let $B = (V, E)$ be a benzenoid and $\mathcal{C}$ a cycle of $B$ (with $4i + 2$ edges $\forall i \in \mathbb{N}$). $B[\mathcal{C}]$ is the sub-graph of $B$ induced by $\mathcal{C}$ and its interior.*

The method presented by Lin [9] relies on the following theorem:

**Theorem 1.** ([19]). Let $B$ be a benzenoid and $\mathcal{C}$ a cycle of $B$ (with $4i+2$ edges $\forall i \in \mathbb{N}$). $\mathcal{C}$ is a $h$-MCC in $|\mathcal{K}(B - B[\mathcal{C}])| \times M(\mathcal{C})$ Kekulé structures of $B$ where $B - B[\mathcal{C}]$ is the sub-graph induced by the removal of the vertices belonging to $\mathcal{C}$ and its interior.

Let us consider the benzenoid $B$ described in Fig. 4(b) and the cycle $\mathcal{C}$ depicted in red thick line. So, $B[\mathcal{C}]$ corresponds exactly to all the hexagons in the interior of $\mathcal{C}$, namely the hexagons depicted in red thick line in the middle figure. To compute the number of occurrences of $\mathcal{C}$ as a $h$-MCC, we have to compute the number of perfect matchings of the sub-graph induced by $B - B[\mathcal{C}]$. In this example, this sub-graph (depicted in red thick solid line in the rightmost figure) has two perfect matchings. Moreover, we have $M(\mathcal{C}) = 1$ because if we consider the two Kekulé structures of $\mathcal{C}$ which allow it to be a conjugated circuit, there is only one of them for which $\mathcal{C}$ is a minimal circuit for at least one of its hexagon. So, we can conclude that $\mathcal{C}$ appears twice as an $h$-MCC in all the Kekulé structures of $B$.

To sum up, the method presented by Lin [9] (described in Algorithm 1) takes as input a benzenoid $B$ and a base containing all the cycles of size at most 4 which can induce at least one $h$-MCC, and another base containing all the redundant circuits of the same sizes. On Line 1, it generates the set of cycles of $B$ which belongs to the first base (we denote this set $\mathcal{C}^*$). Then, for each cycle in $\mathcal{C}^*$, it counts how many $h$-MCC are induced by this cycle in all the Kekulé structures of $B$ (Lines 3–4), as shown in Fig. 4(b). To conclude, it needs to find all couple of cycles of $B$ which can produce one of the redundant circuits described in the second base and to take care not to count the cycle having the largest size (Lines 5–8).

The principal interest of this method is that it does not require to enumerate all the Kekulé structures of the given benzenoid. The only problem it has to

**Algorithm 1:** *Compute_Resonance_Energy*

**Input**: a benzenoid $B$, a base of $h$-MCC, a base of redundant circuits
**Output**: the resonance energy $E(B)$

$\mathcal{C}^* \leftarrow generate\_circuits(B, 1, 4)$
$energy \leftarrow 0$
**foreach** $\mathcal{C} \in \mathcal{C}^*$ **do**
  $energy \leftarrow energy + R_{|\mathcal{C}|} \times |K(B - B[\mathcal{C}])| \times M(\mathcal{C})$
**foreach** $(\mathcal{C}_1, \mathcal{C}_2) \in \mathcal{C}^* \times \mathcal{C}^*$ **do**
  **if** $redundant(\mathcal{C}_1, \mathcal{C}_2)$ **then**
    $size \leftarrow \max(|\mathcal{C}_1|, |\mathcal{C}_2|)$
    $energy \leftarrow energy - R_{size} \times |K(B - B[\mathcal{C}_1 \cup \mathcal{C}_2])|$
**return** $\frac{energy}{K(B)}$

solve is counting the number of perfect matchings in a graph, what was proved to be polynomial for benzenoids [6].

# 4 The Proposed Method

## 4.1 Preliminary Definitions

In this part, we propose a new method, using constraint programming, which refines method propose by Lin [9] in order to compute local aromaticity. Remind that local aromaticity is more useful than global one since it helps to predict the parts of molecules where chemical reactions may take place while leading to global information like global aromaticity. Before going into details, we have to introduce some definitions. First, we need to handle coordinates:

**Definition 10.** *Let $B = (V, E)$ be a benzenoid. A* ***coordinate function*** *$c : V \rightarrow \mathbb{Z}^2$ of $B$ is a function that maps a couple of integers $(c(v).x, c(v).y)$ (i.e. an abscissa and an ordinate in the Cartesian coordinate plane) to each vertex $v$ of $B$ such that if $(v_0, v_1, v_2, v_3, v_4, v_5)$ are the vertices forming a hexagon (given clockwise) with $v_0$ the vertex having the largest ordinate, we have:*

$$\begin{cases} c(v_1) = (c(v_0).x + 1, c(v_0).y - 1) \\ c(v_2) = (c(v_0).x + 1, c(v_0).y - 2) \\ c(v_3) = (c(v_0).x, c(v_0).y - 3) \\ c(v_4) = (c(v_0).x - 1, c(v_0).y - 2) \\ c(v_5) = (c(v_0).x - 1, c(v_0).y - 1) \end{cases}$$

Figure 5(a) describes a simple example of coordinates for benzene.

Then, we consider some particular edges:

**Definition 11.** *Let $B$ be a benzenoid and $c$ a coordinate function. An edge $e = (u, v) \in E$ is a* ***vertical edge*** *of $B$ if and only if $c(u).x = c(v).x$.*

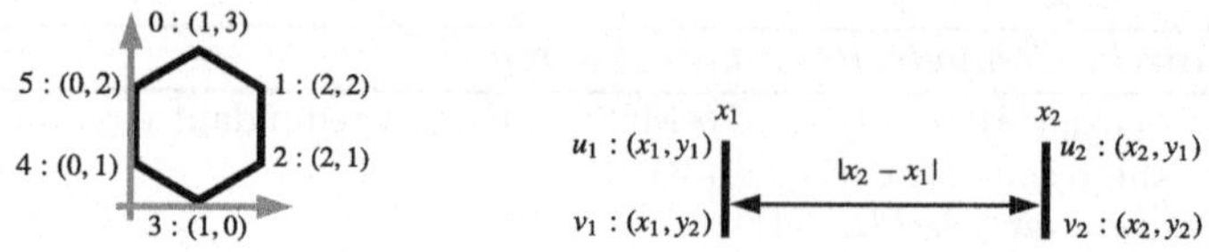


(a) (b)

**Fig. 5.** Benzene with coordinates (a) and example of interval (b).

The vertical edges of the benzenoid depicted in Fig. 5(a) are $\{1, 2\}$ and $\{4, 5\}$. We now introduce the notion of interval related to vertical edges:

**Definition 12.** *Let $B$ be a benzenoid and $c$ a coordinate function. An* ***interval*** *$I$ of $B$ is a couple $I = (e_1, e_2)$ of vertical edges such as:*

$$\begin{cases} e_1 = (u_1, v_1) \in E \\ e_2 = (u_2, v_2) \in E \\ c(u_1).y = c(u_2).y \\ c(v_1).y = c(v_2).y \end{cases}$$

*We denote:*

$$\begin{cases} I.x_1 = c(u_1).x \\ I.y_1 = c(u_1).y \\ I.x_2 = c(u_2).x \\ I.y_2 = c(v_1).y \end{cases}$$

*We denote $|I| = |I.x_2 - I.x_1|$ the* ***size*** *of $I$.*

To sum up, an interval represents the space contained between two vertical edges that have the same ordinate. Figure 5(b) shows an example of interval.

### 4.2 Method Description

In this part, we describe our refined algorithm (see Algorithm 2) based on constraint programming. This method takes as input a benzenoid $B = (V, E)$, a coordinate function $c$, a base containing all the cycles of size at most 4 that can induce at least one $h$-MCC, and another one containing all the couples of cycles of the first base which can form redundant circuits and it returns an approximation of the local energy $E(B, h)$ for each hexagon $h$. With this aim in view, we first compute the set $\mathcal{C}^*$ of all the cycles of $B$ whose size is at most 6 (Line 1). Then for each cycle $\mathcal{C}$ of $\mathcal{C}^*$ (Lines 2–10), we first identify the cycle by a collection of intervals (Line 3). If $\mathcal{C}$ corresponds to a $h$-MCC of size at most 4, we add its contribution to the local resonance energy of $h$ (Lines 4–5). However, $\mathcal{C}$ can also correspond to the union of two conjugated circuits (Line 6). If so, we have not to take into account the contribution of a redundant circuit (Line 10).

Now, we detail below the main steps of Algorithm 2.

---

**Algorithm 2:** *Compute_Resonance_Energy_CP*

---

**Input**: a benzenoid $B$, a coordinate function $c$, a base of $h$-MCC, a base of redundant circuits

**Output**: the local resonance energy $E(B, h)$ for each hexagon $h$ of $B$

1 $\mathcal{C}^* \leftarrow generate\_circuits\_choco(B, 1, 6)$

2 **foreach** $\mathcal{C} \in \mathcal{C}^*$ **do**

3     $id \leftarrow identify\_circuit(\mathcal{C})$

4     **if** $h\text{-}MCC(\mathcal{C})$ **then**

5         $energy[h] \leftarrow energy[h] + R_{|\mathcal{C}|} \times |\mathcal{K}(B - B[\mathcal{C}])| \times M(\mathcal{C})$

6     **else if** $union\_of\_circuits(\mathcal{C})$ **then**

7         **foreach** $(\mathcal{C}_1, \mathcal{C}_2)$ *s.t.* $\mathcal{C} = \mathcal{C}_1 \bigcup \mathcal{C}_2$ *and* $redundant(\mathcal{C}_1, \mathcal{C}_2)$ **do**

8             $\mathcal{C}' \leftarrow circuit\_with\_max\_size(\mathcal{C}_1, \mathcal{C}_2)$

9             $h \leftarrow minimal\_for\_hexagon(\mathcal{C}')$

10             $energy[h] \leftarrow energy[h] - |\mathcal{K}(B - B[\mathcal{C}])| \times R_{|\mathcal{C}'|}$

11 **foreach** $h$ *of* $B$ **do** $energy[h] \leftarrow energy[h] / |\mathcal{K}(B)|$

12 **return** *energy*

---

**Enumeration of All the Cycles.** We need to identify all the cycles which correspond to either a $h$-MCC of size at most 4 (Line 4 of Algorithm 2) or a union of two $h$-MCCs (Line 6). As the union of two $h$-MCCs of size 4 is at most of size 6, we have to enumerate all the cycles of size at most 6.

In order to enumerate all the cycles of size at most 6, we model this problem as a CSP instance $P_1 = (X_1, D_1, C_1)$. First, we consider a graph variable $x_G$ whose domain is all the possible graphs between the empty graph and the graph $B$. This variable models the cycle we look for. To ensure that the value of this variable is a cycle, we impose the graph constraint `cycle` [5] on $x_G$. It remains to be ensured that the size of this cycle is at most 6. For this, we introduce a Boolean variable $x_e$ per edge $e$ of $B$. $x_e$ is set to 1 if the edge $e$ appears in the graph depicted by $x_G$, 0 otherwise. Then, we use a collection of channeling constraints in order to link the variables $x_e$ and the variable $x_G$. More precisely, for each edge $e$, we use a channeling constraint between $x_e$ and $x_G$ which imposes $x_e = 1 \iff e \text{ } appears \text{ } in \text{ } x_G$. Finally, we add a global constraint `sum` over all the variables $x_e$ to impose $\sum_{x_e | e \in E} x_e \in \{6, 10, 14, 18, 22, 26\}$ because we consider circuits of size at most 6 and a circuit of size $i$ has $4i + 2$ edges. The channeling and sum constraints make it possible to ensure that the size of the built cycle is suitable. At the end, we obtain the following instance $P_1$:

$$\begin{cases} X_1 = \{x_G\} \cup \{x_e | e \in E\} \\ D_1 = \{D_{x_G}\} \cup \{D_{x_e} | e \in E\} \text{ with } D_{x_G} = \{g | \emptyset \subseteq g \subseteq B\} \text{ } and \text{ } D_{x_e} = \{0, 1\} \\ C_1 = \{cycle(x_G), \sum_{x_e | e \in E} x_e \in \{6, 10, 14, 18, 22, 26\}\} \cup \{channeling(x_e, x_G) | e \in E\} \end{cases}$$

As Choco implements graph variables and offers a large amount of graph-related constraints and global constraints, this model can be easily expressed with Choco.

**Counting the Number of Kekulé Structures.** For each cycle $\mathcal{C}$ of $\mathcal{C}^*$, we need to count the number of Kekulé structures of $\mathcal{C}$ (Line 5) or one of $B - B[\mathcal{C}]$ (Lines 5 and 10). In 2001, Rispoli presented a method which can count the number of Kekulé structures of a benzenoid [21]. The main idea of this method is to transform the given benzenoid $B$ into a specific matrix whose determinant is the number of Kekulé structures of $B$. Unfortunately, although this task is polynomial, it is too time-consuming. For instance, for a molecule having 19 hexagons and 54 carbon atoms, the method proposed by Rispoli requires more than 15 min while the approach we propose only needs a few seconds. It seems that the time-expensive step is the computation of the determinant. So, we consider an alternative solution based again on constraint programming. We model this problem as a CSP instance $P_2 = (X_2, D_2, C_2)$ for which every solution corresponds to a Kekulé structure. As any benzenoid $B = (V, E)$ is a bipartite graph, we can divide $V$ into two disjoint sets $V_1$ and $V_2$ such that every edge of $E$ links a vertex of $V_1$ to one of $V_2$. We consider a variable $y_v$ per vertex $v$ of $V_1$ whose domain contains every vertex $w$ from $V_2$ such that $\{v, w\} \in E$. By so doing, if the variable $y_v$ is assigned with value $w$, it means that the edge $\{v, w\}$ corresponds to a double bond. By definition of a solution, this ensures that there is a single double bond for any carbon atom of $V_1$. It remains to ensure the same property for the vertices of $V_2$. This can be achieved by considering an `all-different` constraint involving all the variables of $X_2$. So we obtain the following instance $P_2$:

$$\begin{cases} X_2 = \{y_v | v \in V_1\} \\ D_2 = \{D_{y_v} | v \in V_1\} \text{ with } D_{y_v} = \{w | w \in V_2,\ \{v, w\} \in E\} \\ C_2 = \{\texttt{all} - \texttt{different}(X_2)\} \end{cases}$$

Clearly, the solutions of $P_2$ correspond to the Kekulé structures of $B$ and so to perfect matchings of $B$. Regarding the filtering of the `all-different` constraint, Regin proposed an efficient algorithm based on the matchings of a particular graph, called the *value graph* [23]. We can then note that, for our instance $P_2$, the value graph related to the `all-different` constraint we use is exactly the graph $B$. Moreover, any solver enforcing this filtering at each step of the search is able to count efficiently the number of solutions since only assignments leading to solutions are explored. Note that another model was proposed [12]. It considers binary variables and sum global constraints, but does not provide any theoretical guarantee about the efficiency, unlike the model we propose. Finally, the model we describe applies for a benzenoid $B$ and can be easily specialized to apply to any part of $B$ (e.g. any cycle $\mathcal{C}$ or induced sub-graph $B - B[\mathcal{C}]$).

**Identification of Cycles.** Once Choco returns a cycle, we need to determine if this cycle belongs to the base of $h$-MCCs (Line 4) or one of redundant circuits

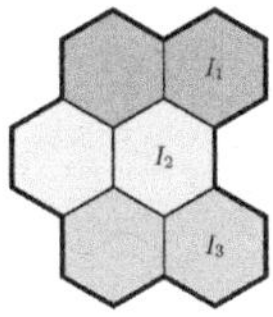


$$\begin{cases} C = \{I_1, I_2, I_3\} \\ |I_1| = |I_2| = |I_3| = 4 \\ I_1.x_1 = I_3.x_1 \\ I_1.x_1 = I_2.x_1 + 1 \end{cases}$$

(a)

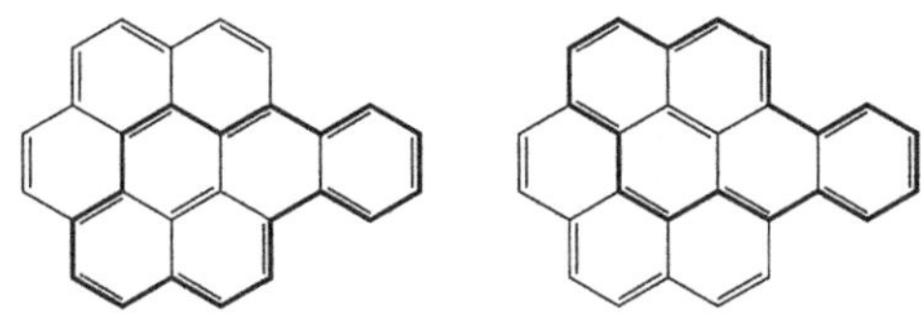

(b)

**Fig. 6.** A cycle and the relations between its intervals (a), and a second example of redundant circuits (b).

(Line 6). For achieving this task, we first represent any cycle by the bias of a set of intervals and some relations between these intervals. The purpose of these relations is to represent the distance between two vertical edges of each couple of intervals (either the two left edges or the two right edges). So, with a set of intervals with this kind of relations, we are able to build the associated cycle and vice versa. Figures 6(a) shows an example of such a representation. Accordingly, we construct each base by describing every cycle ($h$-MCC or redundant cycle) identified by Lin [9] by a set of intervals and some relations between them. Now, each time a new cycle is returned by Choco, we translate it into a set of intervals and their relations, and check if it belongs to one of the two bases by simply comparing sets and relations. So, we are able to determine if this cycle can induce $h$-MCC or if it can be obtained by the union of redundant circuits.

Furthermore, the second base contains, for each cycle $\mathcal{C}$, a set of a couple of cycles whose union forms $\mathcal{C}$. This allows us to remove the energy that we have over-counted due to redundant circuits (Line 10). For example, let us consider $\mathcal{C}$ as being the union of the two circuits represented in Fig. 6(b). It can be obtained either by the union of the cycles of the leftmost figure (of sizes 3 and 4), or by the cycles of the rightmost one (of sizes 3 and 4 too). So, each time Choco finds $\mathcal{C}$ in a benzenoid, we need to remove the energy associated to two circuits of size 4.

## 5 Experimentations

In this part, we provide an experimental comparison between our CP-based method (denoted CRECP for *Compute_Resonance_Energy_CP*) and a reference method, namely NICS. In our comparison, we do not consider the method proposed by Lin since this latter is not available and re-implementing it is not a trivial task. Moreover chemists are more interested by local aromaticity than global one [1].

About the experimental protocol, the CRECP method was implemented in Java and compiled with Java SE 1.8. It relies on Choco Solver version 4.2.3. Note that we used the default settings of Choco. For NICS, we exploited the implementation provided in a commercial program (see http://gaussian.com/). Both CRECP and NICS are run on server with 2.20 GHz Intel Xeon Gold processor and 256 Gb under CentOS Linux release 8.1.1911. We consider as benchmark a

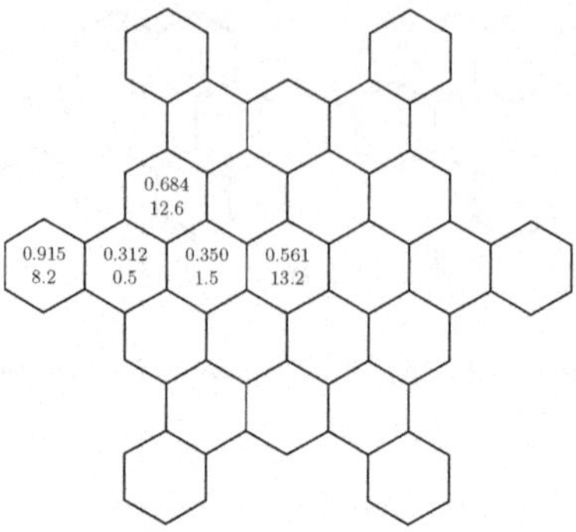


(25)

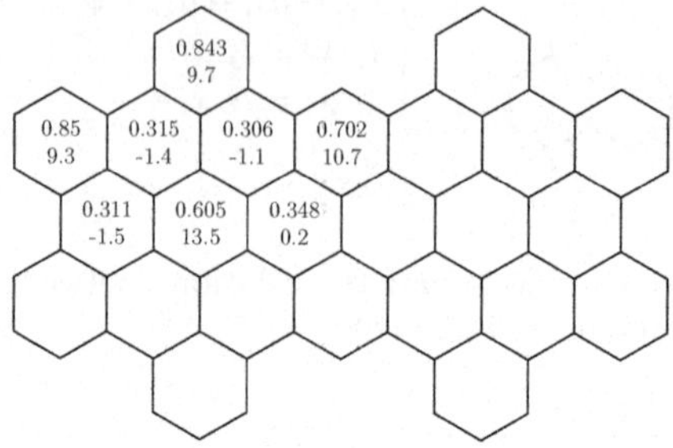


(26)

**Fig. 7.** Results of NICS (red/bottom values) and CRECP (blue/top values) methods. (Color figure online)

set of 28 benzenoids of various and reasonable size so that NICS can be executed within a reasonable amount of time. The sources of CRECP and benchmark are available at https://github.com/AdrienVaret/CPLocalAromaticity.

First, Table 1 shows a comparison between the runtimes of NICS and CRECP method. Clearly, the CRECP method is much faster than the NICS method. Indeed, for the considered benzenoids, the runtime of CRECP does not exceed one minute while NICS may requires several hours. For instance, for the benzenoid 28, NICS takes about 14 h to compute the local aromaticity, while CRECP only needs 46 s.

Then, an important question from a chemical viewpoint is the quality of computed values. Remind that, if both approaches give a description of local aromaticity, CRECP and NICS cannot lead to similar numbers. Indeed, the circuit approach of CRECP is an attempt to describe the behavior of the electronic structure of the molecule as a superposition of closed electronic circuits whereas the NICS approach measures how much the electronic structure would be distorted by an external magnetic field. However, their trends should coincide. Figure 7 presents the values of the local energy of some hexagons for some considered benzenoids. Blue and red values are respectively ones produced by CRECP and NICS. The values of unlabeled hexagons can be deduced by symmetry. As the values produced by the two methods are incomparable in nature, our comparison must focus on the ordering that these values induce on the hexagons. As we can see, both methods lead to similar orderings, what shows that our approach may constitute an interesting and faster alternative to assess the local aromaticity of benzenoid.

**Table 1.** Runtimes of NICS and CRECP for the considered set of benzenoids.

| Benzenoid | | | Runtime | |
|---|---|---|---|---|
| id. | #carbons | #hexagons | NICS | CRECP |
| 1 | 6 | 1 | 1 min 37 s | 0.281 s |
| 2 | 10 | 2 | 5 min 56 s | 0.210 s |
| 3 | 14 | 3 | 32 min 38 s | 0.294 s |
| 4 | 14 | 3 | 15 min 12 s | 0.274 s |
| 5 | 18 | 4 | 21 min 19 s | 0.296 s |
| 6 | 18 | 4 | 33 min 50 s | 0.332 s |
| 7 | 18 | 4 | 44 min 36 s | 0.327 s |
| 8 | 16 | 4 | 16 min 8 s | 0.312 s |
| 9 | 18 | 4 | 1 h 5 min 58 s | 0.319 s |
| 10 | 22 | 5 | 30 min 13 s | 0.257 s |
| 11 | 22 | 5 | 1 h 35 min 55 s | 0.309 s |
| 12 | 22 | 5 | 1 h 29 min 32 s | 0.325 s |
| 13 | 22 | 5 | 59 min 46 s | 0.268 s |
| 14 | 22 | 5 | 1 h 42 min 51 s | 0.228 s |
| 15 | 22 | 5 | 1 h 45 min 40 s | 0.250 s |
| 16 | 22 | 5 | 55 min 47 s | 0.278 s |
| 17 | 22 | 5 | 4 h 29 min 7 s | 0.249 s |
| 18 | 20 | 5 | 42 min 16 s | 0.235 s |
| 19 | 20 | 5 | 1 h 22 min 30 s | 0.242 s |
| 20 | 20 | 5 | 24 min 21 s | 0.251 s |
| 21 | 24 | 7 | 42 min 42 s | 0.324 s |
| 22 | 42 | 13 | 3 h 42 min 54 s | 0.857 s |
| 23 | 60 | 20 | 7 h 4 min 58 s | 6.294 s |
| 24 | 60 | 19 | 8 h 43 min 33 s | 5.170 s |
| 25 | 72 | 24 | 11 h 54 min 4 s | 20.091 s |
| 26 | 78 | 25 | 14 h 12 min 44 s | 50.446 s |
| 27 | 36 | 11 | 2 h 39 min 5 s | 0.402 s |
| 28 | 54 | 17 | 4 h 33 min 25 s | 3.213 s |

## 6 Conclusions and Perspectives

In this paper, we have presented a new method based on constraint programming for computing the local aromaticity of benzenoids. This method refines the method proposed by Lin by dealing with local aromaticity instead of global one. In practice, we have shown that it turns to be significantly faster than NICS (which is considered as a reference by theoretical chemists) while providing similar results.

This work is a preliminary step in which we only consider conjugated circuit of size at most 4. By so doing, we make an approximation which may be of poor quality for some classes of benzenoids. An extension of this work consists in identifying all the $h$-MCC of size greater than 4 and the corresponding union of redundant circuits. If this work was done by hand by Lin [9] for $h$-MCCs of size at most 4, constraint programming will be of great help here to cope with the combinatorial explosion. Beyond, many problems about benzenoids in theoretical chemistry may fall within the scope of constraint programming. For instance, when no Kekulé structure exists for a given benzenoid, chemists are interested in finding the structure that comes closest to it, what may be expressed as a constraint optimization problem.

## References

1. Chen, Z., Wannere, C.S., Corminboeuf, C., Puchta, R., von Ragué Schleyer, P.: Nucleus-independent chemical shifts (NICS) as an aromaticity criterion. Chem. Rev. **105**, 3842–3888 (2005)
2. Clar, E., Schoental, R.: Polycyclic Hydrocarbons, vol. 1. Springer, Berlin (1964). https://doi.org/10.1007/978-3-662-01665-7
3. Draine, B.T.: Astronomical models of PAHs and Dust. EAS Publ. Ser. **46**, 29–42 (2011). https://doi.org/10.1051/eas/1146003
4. Fages, J.G., Lorca, X., Prud'homme, C.: Choco solver user guide documentation. https://choco-solver.readthedocs.io/en/latest/
5. Fages, J.: Exploitation de structures de graphe en programmation par contraintes. Ph.D. thesis, École des mines de Nantes, France (2014)
6. Hansen, P., Zheng, M.: A linear algorithm for perfect matching in hexagonal systems. Discrete Math. **122**, 179–196 (1993)
7. Hückel, E.: Quantentheoretische Beiträge zum Benzolproblem. Z. Phys. **70**, 204–286 (1931). https://doi.org/10.1007/BF01339530
8. Kekulé, A.: Untersuchungen über aromatische verbindungen ueber die constitution der aromatischen verbindungen. Justus Liebigs Annalen der Chemie **137**(2), 129–196 (1866). https://doi.org/10.1002/jlac.18661370202
9. Lin, C.: Efficient method for calculating the resonance energy expression of benzenoid hydrocarbons based on the enumeration of conjugated circuits. J. Chem. Inf. Comput. Sci. **40**, 778–783 (2000)
10. Lin, C., Fan, G.: Algorithms for the count of linearly independent and minimal conjugated circuits in benzenoid hydrocarbons. J. Chem. Inf. Comput. Sci **39**, 782–787 (1999)
11. Luch, A.: The Carcinogenic Effects of Polycyclic Aromatic Hydrocarbons. Imperial College Press, London (2005). https://doi.org/10.1142/p306
12. Mann, M., Thiel, B.: Kekulé structures enumeration yields unique SMILES. In: Proceedings of Workshop on Constraint Based Methods for Bioinformatics (2013)
13. Narita, A., Wang, X.Y., Feng, X., Müllen, K.: New advances in nanographene chemistry. Chem. Soc. Rev. **44**(18), 6616–6643 (2015). https://doi.org/10.1039/C5CS00183H
14. Randić, M.: Conjugated circuits and resonance energies of benzenoid hydrocarbons. Chem. Phys. Lett. **38**, 68–70 (1976). https://doi.org/10.1016/0009-2614(76)80257-6

15. Randić, M.: Aromaticity of polycyclic conjugated hydrocarbons. Chem. Rev. **103**(9), 3449–3606 (2003). https://doi.org/10.1021/cr9903656
16. Randić, M.: Benzenoid rings resonance energies and local aromaticity of benzenoid hydrocarbons. J. Comput. Chem. **40**(5), 753–762 (2019)
17. Randić, M., Balaban, A.T.: Local aromaticity and aromatic sextet theory beyond clar. Int. J. Quantum Chem. **108**(17), e25657 (2018). https://doi.org/10.1002/qua.25657
18. Randić, M., Guo, X.: Recursive method for enumeration of linearly independent and minimal conjugated circuits of benzenoid hydrocarbons. J. Chem. Inf. Model. **34**(2), 339–348 (1994)
19. Randić, M., Guo, X., Klein, D.J.: Analytical expressions for the count of LM-conjugated circuits of benzenoid hydrocarbons. Int. J. Quantum Chem. **60**, 943–958 (1996)
20. Rieger, R., Müllen, K.: Forever young: Polycyclic aromatic hydrocarbons as model cases for structural and optical studies. J. Phys. Org. Chem. **23**(4), 315–325 (2010). https://doi.org/10.1002/poc.1644
21. Rispoli, F.J.: Counting perfect matchings in hexagonal systems associated with benzenoids. Math. Mag. **14**, 194–200 (2001)
22. Rossi, F., van Beek, P., Walsh, T.: Handbook of Constraint Programming. Elsevier, Amsterdam (2006)
23. Régin, J.C.: A filtering algorithm for constraints of difference in CSPs. In: Proceedings of AAAI, pp. 362–367 (1994)
24. Wu, J., Pisula, W., Müllen, K.: Graphenes as potential material for electronics. Chem. Rev. **107**(3), 718–747 (2007). https://doi.org/10.1021/cr068010r

# Using Constraint Programming to Generate Benzenoid Structures in Theoretical Chemistry

Yannick Carissan[1], Denis Hagebaum-Reignier[1], Nicolas Prcovic[2], Cyril Terrioux[2(✉)], and Adrien Varet[2]

[1] Aix Marseille Univ, CNRS, Centrale Marseille, ISM2, Marseille, France
{yannick.carissan,denis.hagebaum-reignier}@univ-amu.fr
[2] Aix Marseille Univ, Université de Toulon, CNRS, LIS, Marseille, France
{nicolas.prcovic,cyril.terrioux,adrien.varet}@univ-amu.fr

**Abstract.** Benzenoids are a subfamily of hydrocarbons (molecules that are only made of hydrogen and carbon atoms) whose carbon atoms form hexagons. These molecules are widely studied in theoretical chemistry and have a lot of concrete applications. Therefore, generating benzenoids which have certain structural properties (e.g. having a given number of hexagons or having a particular structure from a graph viewpoint) is an interesting and important problem. It constitutes a preliminary step for studying their chemical properties. In this paper, we show that modeling this problem in Choco Solver and just letting its search engine generate the solutions is a fast enough and very flexible approach. It can allow to generate many different kinds of benzenoids with predefined structural properties by posting new constraints, saving the efforts of developing bespoke algorithmic methods for each kind of benzenoids.

**Keywords:** Constraint programming · Modeling · Graph variables and constraints · Chemistry

## 1 Introduction

*Polycyclic aromatic hydrocarbons (PAHs)* are hydrocarbons whose carbons are forming cycles of different sizes. *Benzenoids* are a subfamily of PAHs made of 6-membered carbon rings (i.e. each cycle is a hexagon). To fill carbon valency, each atom of carbon is bonded to either two other carbons and one hydrogen or three carbons. For example, Figs. 1(a) and (b) are representing two benzenoids: benzene and anthracene.

PAHs are well-studied in various fields because of their energetic stability, molecular structures or optical spectra. In natural environment, these

This work has been funded by the Agence Nationale de la Recherche project ANR-16-C40-0028.


H. Simonis (Ed.): CP 2020, LNCS 12333, pp. 690–706, 2020.
https://doi.org/10.1007/978-3-030-58475-7_40

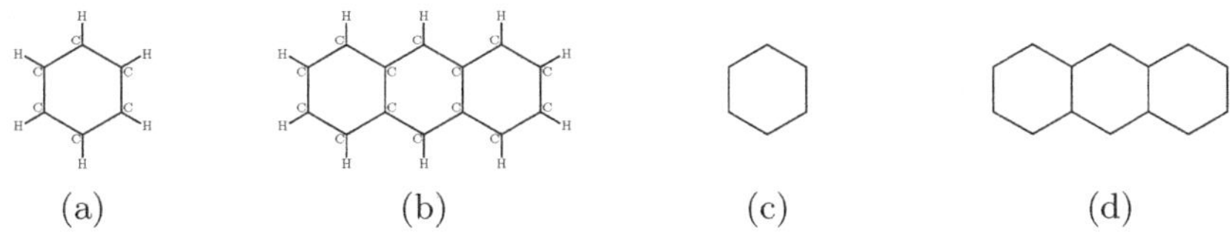


**Fig. 1.** Two small benzenoids: benzene (a) and anthracene (b) with their graphical representations (c) and (d).

molecules are created by the incomplete combustion of carbon contained in combustibles [18]. They are popular research subjects in material sciences, e.g. molecular nanoelectronics where they are used to store or transport energy [2,29] or in organic synthesis [22,25], where the controlled design of specific shapes remains challenging. PAHs are also intensively studied in interstellar chemistry because of their suspected presence in various interstellar and circumstellar environments where they are believed to act as catalysts for chemical reactions taking place in space [14].

PAHs exhibit a large variety of physicochemical properties depending on size and, more specifically, on edge and bond topologies. In the astrophysical community, the so-called "PAH hypothesis" formulated more than 30 years ago, of whether the existence of PAHs in space could explain some unidentified mid-infrared emission bands in astrophysical environments, has motivated numerous observational, experimental and theoretical investigations. It is now accepted that mixtures of free PAHs of different size, shapes and ionic states can account for the overall appearance of the widespread interstellar infrared emission spectrum. But the question of relative abundance of PAHs with a given size and/or shape remains open. Many studies are devoted to explore the effect of the size, shapes in terms of compacity and symmetry of PAHs, on band positions and intensities of the infrared spectra [1,3,23]. Very recently, a systematic investigation of a series of 328 PAHs containing up to 150 carbon atoms showed that PAHs with armchair edges that maximize the Clar number (i.e. the maximum number of non-adjacent rings containing 6 electrons called a sextet) are potential emitters of a certain class of astrophysical infrared sources [24]. For their study, the authors needed to systematically generate all PAHs having armchair edge topology and selecting a subclass of PAHs whose structure maximizes the Clar number. They used the algorithm of Caporossi and Hansen [8]. Constraint-programming is particularly well suited for the generation of such families of PAHs.

Another important example where the generation of specific shapes is relevant for chemists deals with so-called "non-Kekulean" benzenoids [10]. These benzenoids cannot be represented by Kekulé structures, i.e. structures that have only simple and double bonds. From a graph-theoretical point of view, Kekulé structures are covered by the maximal number of disjoint (double) edges so that all vertices are incident to one of the disjoint edges. It was accepted among chemists until recently that "non-Kekulean" benzenoids should be very unstable due to their open-shell electronic structure (i.e. one or more electron(s) remain unpaired, contrary to a closed-shell structure where all electrons are paired) and thus their synthesis would be a real challenge. The experimental realization and in-depth characterization of small "non-Kekulean" benzenoids was very recently achieved on gold surfaces [20,21]. These studies opened the way to the synthesis of new classes of compounds which show unconventional magnetism induced by their topology, with promising applications in various fields like molecular electronics, nonlinear optics, photovoltaics and spintronics. Moreover, it was shown that some PAHs with specific topologies (e.g. rhombus shapes) may "prefer" having an open-shell structure when reaching a certain size, although they could have a closed-shell structure and could thus be described by a set of Kekulé structures [27]. From a quantum theoretical point of view, the proper description of the electronic structure of open-shell benzenoids is a difficult task. The use of a constraint programming approach for the systematic search of larger non-Kekulean or Kekulean benzenoids having an open-shell electronic structure is undoubtedly advantageous.

In this context, many approaches have been proposed in order to generate benzenoids having or not a particular shape or satisfying a particular property (e.g. [5,6]). These are bespoke approaches which have the advantage of being efficient, but are difficult to adapt to the needs of chemists. Moreover, designing a new bespoke method for each new desired property often requires a huge amount of efforts. So, in this paper, we prefer to use an approach based on constraint programming. With this aim in view, we present a general model which can be refined depending on the desired properties by simply adding variables and/or constraints. By so doing, our approach benefits from the flexibility of CP and requires less efforts of implementation. In the meantime, CP offers efficient solvers which can be quite competitive with respect to bespoke algorithms.

The paper is organized as follows. First, we recall some definitions about benzenoids and constraint programming in Sect. 2. Section 3 introduces the fastest existing algorithm for generating benzenoids. Then Sect. 4 presents a new approach using constraint programming, explains its advantages and gives some examples. Finally, we conclude and provide some perspectives in Sect. 5.

## 2 Preliminaries

### 2.1 Theoretical Chemistry

*Benzene*, represented in Fig. 1(a) is a molecule made of 6 carbon atoms and 6 hydrogen atoms. Its carbon atoms form a hexagon (also called *benzenic cycle* or *benzenic ring*) and each of them is linked to a hydrogen atom. *Benzenoids* are a subfamily of PAHs containing all molecules which can be obtained by aggregating benzenic rings. For example, Fig. 1(b) shows anthracene, which contains three benzenic rings.

By definition of the *valence* (i.e. the number of bonds that an atom can establish) of carbon and hydrogen atom, we know that each carbon atom is linked to either two other carbon atoms and one hydrogen atom or three other

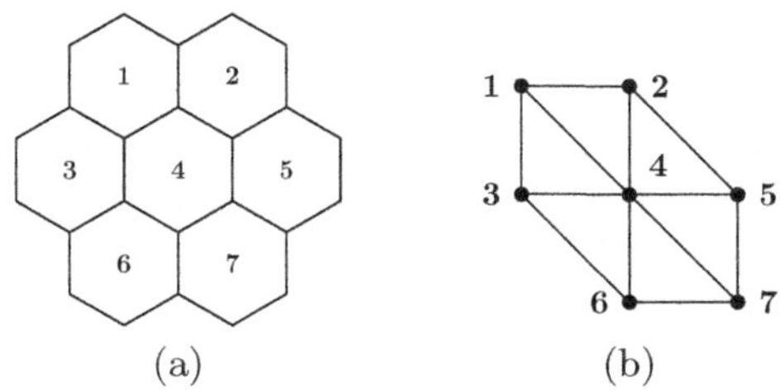


**Fig. 2.** Coronene (a) and its hexagon graph (b).

carbon atoms. So benzenoids can be perfectly defined by describing only the interactions between carbon atoms. Hydrogen atoms can then be deduced since each hydrogen atom is linked to a carbon atom which is only bonded to two other carbon atoms. As a consequence, any benzenoid can be represented as an undirected graph $B = (V, E)$, with $V$ the set of vertices and $E$ the set of edges. Every vertex in $V$ represents a carbon atom and every edge of $E$ represents a bond between the two corresponding carbons. Moreover, this kind of graph, is connected, planar and bipartite. Figures 1(c) and (d) represent the graphs related to the molecules of benzene and anthracene.

In the following, for any benzenoid $B$, we need to consider some of its faces. A *face* of a planar graph is an area of the plan bounded by edges. Figure 2(a) presents the graph corresponding to coronene (a well-known benzenoid). This graph has eight faces namely the seven numbered faces and the external face. Note that in the sequel, we do not take into account the external face. For this example, the numbered faces correspond exactly to the hexagons of coronene. However, we will see later that this property does not hold for all the benzenoids.

Then, given a benzenoid, we consider another graph, namely the hexagon graph. The *hexagon graph* of a benzenoid $B = (V, E)$ is the undirected graph $B_h = (V_h, E_h)$ such that there is a vertex $v_h$ from $V_h$ per hexagonal face $h$ of $B$ (the external face and "holes" in the benzenoid are excluded) while there is an edge $\{v_h, v_{h'}\}$ in $E_h$ if the corresponding hexagonal faces $h$ and $h'$ of $B$ share an edge of $E$. Figure 2(b) presents the hexagon graph of coronene. The hexagon graph allows us to express the interaction between the hexagons of the considered benzenoid.

### 2.2 Constraint Programming

An instance $I$ of the *Constraint Satisfaction Problem (CSP)* is a triplet $(X, D, C)$. $X = \{x_1, \ldots, x_n\}$ is a set of $n$ *variables*. For each variable $x_i \in X$, there exists an associated domain $D_{x_i} \in D = \{D_{x_1}, \ldots, D_{x_n}\}$ which represents the values that $x_i$ can take. $C = \{c_1, ..., c_e\}$ represents a set of $e$ *constraints*. Constraints represent the interactions between the variables and describe the allowed combinations of values.

Solving a CSP instance $I = (X, D, C)$ amounts to find an assignment of all the variables of $X$ with a value contained in their associated domain which satisfies all the constraints of $C$. Such assignment is called a *solution*. This problem is NP-hard.

Many libraries are available to represent and solve efficiently CSP instances. In this paper, we exploit the open-source Java library *Choco* [15]. This choice is highly guided by our need to be able to define *graph variables* and directly apply graph-related constraints (e.g. connected or cyclic constraints). Graph variables have as domain a set of graphs defined by a lower bound (a sub-graph called $GLB$) and an upper bound (a super-graph called $GUB$). Moreover, Choco implements the usual global constraints which make the modeling easier and its solver is efficient and configurable.

## 3 Generating Benzenoids

We can define the benzenoid generation problem (denoted BGP in the future) as follows: given a set of structural properties $\mathcal{P}$, generate all the benzenoids which satisfy each property of $\mathcal{P}$. For instance, these structural properties may deal with the number of carbons, the number of hexagons or a particular structure for the hexagon graph. Of course, the more interesting instances of the BGP problem combine several properties. For example, Fig. 5 shows benzenoids having a tree as hexagon graph. Such a property-based instances design allows for the search of benzenoids with chemically relevant properties. Our interest lies in the search of benzenoids with radical electronic structures (as in the work of Malrieu and Trinquier [27]), which arise from their geometrical arrangement.

Now, we present an existing method proposed by Brinkmann et al. [5]. Given an integer $n$, this method is able to generate all the benzenoids with $n$ hexagons by generating all the hexagon graphs with at most $n$ vertices. This is done by adding successively new vertices to the hexagon graph (which is equivalent to generate all the wanted molecules by successively adding new hexagons).

This method is really efficient. For instance, it could generate the 669,584 benzenoids having 12 hexagons in 1.2 s and 1,000 billions of benzenoids having 21 hexagons in two weeks when launched on an old computer (Intel Pentium, 133 MHz, 2002). However it has some disadvantages. Indeed it is not complete in the sense that it is unable to generate benzenoids with "holes". By hole, we mean a face which does not correspond to a hexagon or the external face. For example, Fig. 3(a) depicts the smallest benzenoid (in terms of number of hexagons) which admits a hole. Such a benzenoid cannot be produced by this method. Indeed, when this method wants to add a new hexagon, it checks whether the added hexagon allows to close a cycle of hexagons. If so, the hexagon is not added and so benzenoids with holes cannot be generated. Benzenoids with holes are quite seldom. There is a single one for 8 hexagons (among 1,436 benzenoids), 5 for 9 hexagons (among 6,510). Note that this proportion grows as we increase the number of hexagons (see Table 1). Furthermore, this method is unable to take into account other properties natively and cannot easily be tuned to fit the needs of chemists. Indeed, it is based on an augmenting procedure that decides how to add a vertex. So this procedure should be changed and proven adequate to avoid generating non canonical graphs (i.e. redundant isomorphic graphs) each time we want to change the structural property of the benzenoids we wish to generate. It is therefore a relatively heavy task even for the addition of a basic property.

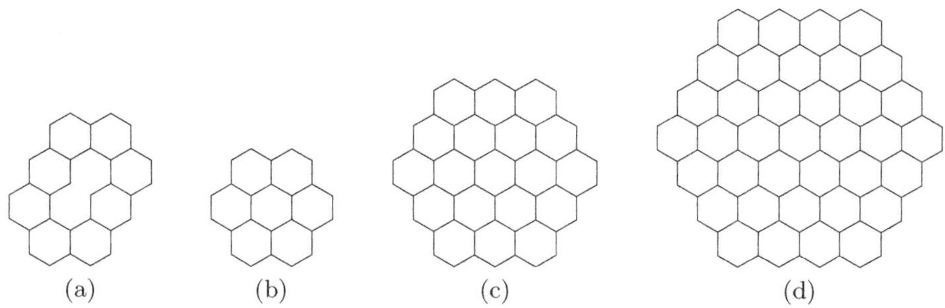

**Fig. 3.** The smallest benzenoid with hole (a) and coronenoids of size 2 (b), 3 (c) and 4 (d).

In the next section, we present a new method using constraint programming which is able to generate any benzenoid structure and benefits from the flexibility of constraint programming.

# 4 Generating Benzenoid Structures Thanks to CP

In this section, we see how to model a BGP instance as a CSP instance. We first present a general model which considers the generation of all the benzenoids having a given number of hexagons. This property is the minimal property to ensure. Then we provide some examples showing how the model can be easily specialized to take into account some additional structural properties.

## 4.1 General Model

In this part, we want to generate all the benzenoids having a given number $n$ of hexagons. Before modeling this problem as a CSP instance, we highlight some useful properties. A *coronenoid* of size $k$ is a molecule of benzene (i.e. a hexagon) to which we successively add $k-1$ crowns of hexagons. Benzene corresponds to the coronenoid of size 1 (see Fig. 1(c)). Figures 3(b)–(d) present the coronenoids of size 2, 3 and 4. Note that the diameter (i.e. the number of hexagons of the central line) of a coronenoid of size $k$ is $2 \times k - 1$. Our interest for coronenoids lies in the fact that they are useful to "embed" benzenoids of a given number of hexagons:

*Property 1.* Any benzenoid involving $n$ hexagons can be embedded in a coronenoid of size at most $k(n) = \left\lfloor \frac{n+1}{2} + 1 \right\rfloor$.

So if we reason in terms of hexagon graph, obtaining all the benzenoids with $n$ hexagons is equivalent to find all the connected sub-graphs of the hexagon graph of coronenoid of size $k(n)$. The model we propose relies on this property.

So, given an integer $n$, we model the BGP problem where $\mathcal{P}$ is reduced to "having $n$ hexagons" as a CSP instance $I = (X, D, C)$. First, we consider a graph variable $x_G$ which represents the possible hexagon graph of the built benzenoid. Its domain is the set of all the sub-graphs between the empty graph and the hexagon graph of coronenoid of size $k(n)$ (see Fig. 4(a)). We also exploit a set of $n_c$ Boolean variables $\{x_1, \ldots, x_{n_c}\}$ where $n_c$ is the number of hexagons of coronenoid of size $k(n)$. The variable $x_i$ is set to 1 if the $i$th hexagon of coronenoid of size $k(n)$ is used in the hexagon graph depicted by $x_G$, 0 otherwise. For sake of simplicity, hexagons are numbered from top to bottom and from left to right like in Fig. 2. Likewise, we consider a set of $m_c$ Boolean variables $\{y_1, \ldots, y_{m_c}\}$ where $m_c$ is the number of edges of the hexagon graph of coronenoid of size $k(n)$. The variable $y_j$ is set to 1 if the $j$th edge of the hexagon graph of coronenoid of size $k(n)$ is used in the hexagon graph depicted by $x_G$, 0 otherwise. We must emphasize that the set of $x_i$ and $y_i$ variables and the channeling constraints maintaining the consistency between their values and the value of $x_G$ are automatically generated by Choco Solver through the call of a predefined method.

Finally, we model the following properties by constraints:

- *Link between the graph variable $x_G$ and the variables $x_i$.* As mentioned above, the variable $x_i$ specifies if the $i$th hexagon of coronenoid of size $k(n)$ is used in the graph represented by $x_G$. So we must ensure that their respective values are consistent each others. For this aim in view, we consider a channeling constraint per variable $x_i$ which involves $x_i$ and $x_G$ and imposes that $x_i = 1 \iff x_G$ *contains the vertex* $i$.
- *Link between the graph variable $x_G$ and the variables $y_j$.* Like previously, we consider a channeling constraint per variable $y_j$ which involves $y_j$ and $x_G$ and imposes that $y_j = 1 \iff x_G$ *contains the edge* $j$.
- *$x_G$ is an induced sub-graph of the coronenoid hexagon graph.* Any value of $x_G$ is not necessarily a valid hexagon graph. For example, in Fig. 2(b), removing only edge $\{1, 2\}$ does not produce a valid hexagon graph. To ensure that the hexagon graph is valid, we must add a constraint for every triplet $(h_{j_1}, h_{j_2}, h_{j_3})$ of hexagons which are pairwise adjacent in the coronenoid hexagon graph. This constraint imposes that if two of these edges exists, then the third one exists too. This can be achieved by posting a set of ternary clauses of the form $\{\neg y_{j_1} \vee \neg y_{j_2} \vee y_{j_3}, y_{j_1} \vee \neg y_{j_2} \vee \neg y_{j_3}, \neg y_{j_1} \vee y_{j_2} \vee \neg y_{j_3}\}$ for each possible triple of pairwise adjacent hexagons.
- *Benzenoids have $n$ hexagons.* It can be easily done by using a sum global constraint involving all the variables $x_i$: $\sum_{i \in \{1, \ldots, n_c\}} x_i = n$.
- *Benzenoids correspond to connected graphs.* Variable graphs come with particular constraints. Among them, we consider the `connected` constraint which applies on the variable $x_G$ ensures that only connected graphs are allowed values for $x_G$.

- *Six hexagons forming a cycle generate a hexagon.* When six hexagons form a cycle, the face contained in the interior of the cycle is not a hole but a hexagon. For instance, if we consider the cycle forms by the hexagons 1, 2, 5, 7, 6 and 3 of coronene (see Fig. 2), we have necessarily a hexagon in the middle of the crown, namely the hexagon 4. To ensure this property, we add a set of constraints which specify that $G$ cannot have a hole whose size is exactly one hexagon. For each hexagon $u$, we consider the set $N(u)$ of the neighbors of $u$ in the hexagon graph. Then, for each vertex $u$ having 6 neighbors, we add a constraint between $x_u$ and the variables corresponding to its neighbors which imposes: $\sum_{v \in N(u)} x_v = 6 \Rightarrow x_u = 1$.

This model allows us to enumerate all the benzenoids having $n$ hexagons, possibly with holes. However, some benzenoids may be generated multiple times due to the existence of symmetries. So we add several additional constraints in order to break as many symmetries as possible:

- Two constraints which specify that $G$ must have at least one vertex respectively on the top-border and the left-border in order to avoid the symmetries by translation. So, we have to create a constraint which specifies that the sum of the binary variables associated to the top border (resp. left border) is strictly positive.
- A set of constraints which specify that $G$ must be the only representative of its class of symmetry by axis and rotation. There are up to twelve symmetric solutions: six $60°$ rotation symmetries combined with a possible axis symmetry. Symmetries are broken thanks to the compact `lex-lead` constraint described in [11]. For each of the twelve symmetries, it requires $n_c$ new Boolean variables (each associated with a $x_i$ Boolean variable representing a hexagon) and a total of $3n_c$ ternary clauses.

This model can be easily implemented with the open-source Java library *Choco* [15]. Indeed, Choco natively proposes graph variables and the more usual graph-related constraints (notably `connected` constraint).

## 4.2 How to Specialize the Model

The first advantages of our approach is that it is able to generate all the benzenoids, including those with holes unlike the method described in the previous section. Moreover, using constraint programming makes it easier the addition of most of structural properties wished by the chemists. Indeed, starting from the general model, for each new desired property, we simply have to model it by posting new constraints and eventually by adding new variables.

For example, let us consider that chemists are interested by benzenoids whose structure is a path of hexagons. Such benzenoid structures can easily generated by exploiting the general model $I$ and adding the graph constraint `path` on $x_G$. Now, if chemists are more interested by *catacondensed benzenoids*, that is benzenoids whose structure is a tree, we can just add the graph constraint `tree`

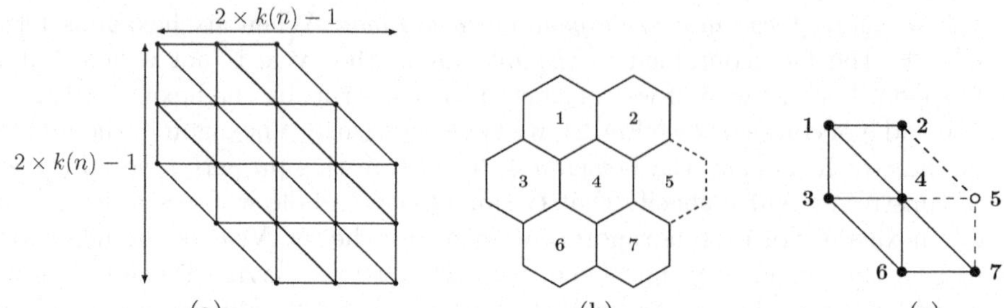

(a) (b) (c)

**Fig. 4.** Upper bound of the domain of the graph variable (a), rectangle benzenoid (in solid line) of dimension 3 × 2 embedded in coronenoid of size 2 (b) and its related hexagon graph (c).

on $x_G$ to the general model $I$. Figure 5 shows nine (among twelve possible) examples of 5-hexagon benzenoids obtained by just adding the `tree` constraint of Choco on $x_G$.

Of course, depending on the desired property the model may be more complex. Especially, it may require to add new variables or the property cannot be directly expressed by a single existing constraint. In next subsections, we give such examples.

### 4.3 Generating Rectangle Benzenoids

In this part, we present how we can model the property *"all the built benzenoids have a rectangle shape"*, in addition to the property *"having n hexagons"*, and add it to the model we describe previously. For instance, Fig. 6(i) shows a rectangle benzenoid with the dimensions 3 × 3.

First, remember that the general model described in the previous part takes in input the number $n$ of hexagons, and embeds any generated benzenoid in a coronenoid of size $k(n)$. We can easily see that the largest rectangle benzenoid which can be embedded in a coronenoid of size $k(n)$ has a width $w_{max}$ equal to $k(n)$ and a height $h_{max}$ equal to $2 \times k(n) - 1$ (i.e. the diameter of coronenoid of size $k(n)$). Figure 4 shows the rectangle benzenoid of dimensions $2 \times 3$ embedded in coronenoid of size 2 (b) and its hexagon graph (c).

Then, starting from model $I$, we must add new variables to model the desired property. Namely, we add two integer variables $x_w$ and $x_h$ whose domain is respectively $\{1, ..., w_{max}\}$ and $\{1, ..., h_{max}\}$. These variables represent respectively the number of columns and lines of the built benzenoid. In addition, we denote $L_i$ (resp. $C_i$) the set of variables $x_i$ which appear in the $i$th line (resp. $i$th column) in the coronenoid of order $k(n)$. We assume that lines (resp. columns) are numbered from top to bottom (resp. from left to right). For example, if

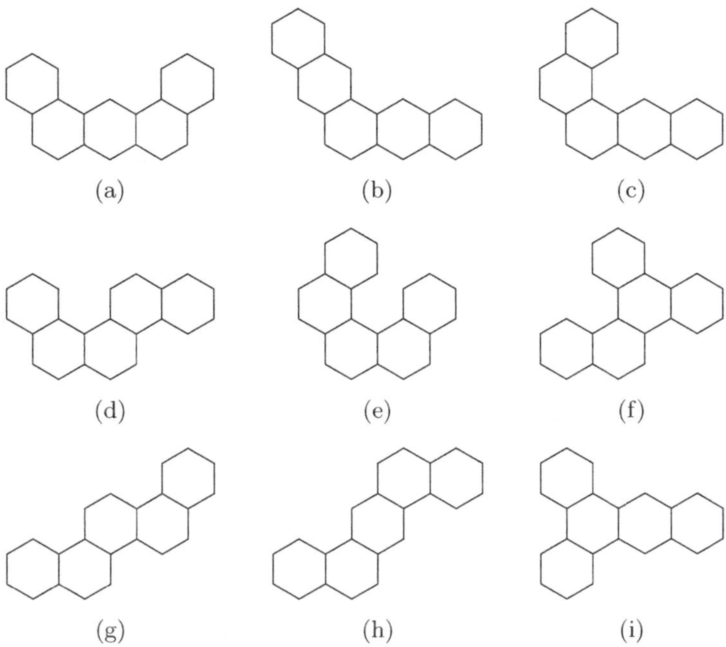


**Fig. 5.** Catacondensed benzenoids with 5 hexagons.

we consider the hexagon graph of the rectangle benzenoid of dimensions $3 \times 2$ described in Fig. 4(b), we have the following sets:

$$\begin{cases} L_1 = \{x_1, x_2\} \\ L_2 = \{x_3, x_4, x_5\} \\ L_3 = \{x_6, x_7\} \\ C_1 = \{x_1, x_3, x_6\} \\ C_2 = \{x_2, x_4, x_7\} \end{cases}$$

Now we add several constraints to the general model in order to model the following properties:

- *The hexagons of each line are positioned contiguously.* We want to avoid to have a Boolean variable equal to 0 between two Boolean variables equal to 1. For the $i$th line, this can be modeled by imposing an arithmetic constraint $x_{i_1} \geq x_{i_2} \ldots \geq x_{i_{w_{max}}}$ if $L_i = \{x_{i_1}, x_{i_2}, \ldots, x_{i_{w_{max}}}\}$. We can also use instead a global constraint `ordered` applied on the variables of $L_i$ with operator $\geq$.
- *The hexagons of each column are positioned contiguously.* We proceed as for the lines by considering $C_i$ instead of $L_i$.
- *Lines have a consistent size.* Each line must be empty or have a size equal to the current width of the rectangle. The size of a line can be defined as the number of hexagons it contains since we know that all the hexagons are contiguous. For the $i$th line, we add a constraint linking $x_w$ to all the variables

in $L_i$ and imposing $\sum_{x_{i_j} \in L_i} x_{i_j} = 0 \vee \sum_{x_{i_j} \in L_i} x_{i_j} = x_w$. As such a constraint is added for each line, we are sure that all the lines have the same width.

- *Columns have a consistent size.* We proceed as for the lines by considering $C_i$ instead of $L_i$ and $x_h$ instead of $x_w$.

Figure 6 shows the 9 rectangle benzenoids generated with the parameters $w = h = 3$.

In this extended model, we can note that some variables are now useless. Indeed, only the leftmost hexagons of coronenoid of order $k(n)$ covered by the rectangle of dimensions $w_{max} \times h_{max}$ are required. The other hexagons will only lead to produce symmetrical structures. So, we can refine our model by removing useless variables. Likewise, we can filter the domain of $x_G$ in order that $GUB$ is the hexagon graph of the rectangle $w_{max} \times h_{max}$ by posting the adequate unary constraint.

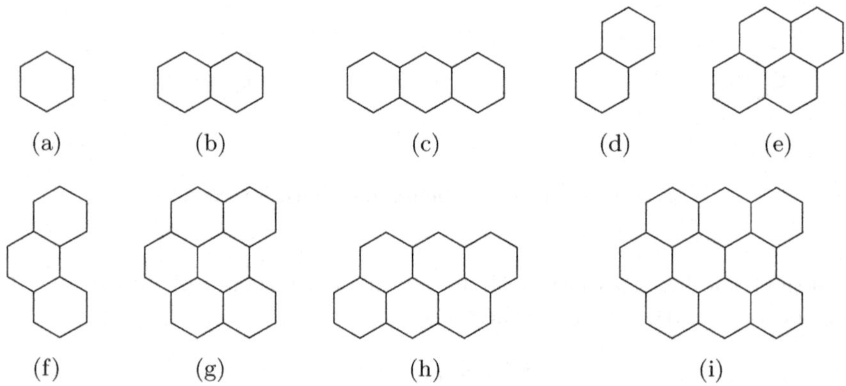


**Fig. 6.** Rectangle benzenoids generated with $w = h = 3$.

This extension of our general model is given as a simple illustration of our approach. Of course, we can easily generate benzenoids having a rectangle shape with a bespoke algorithm. What is interesting in our approach is its flexibility. For instance, if some chemists are interested by identifying the rectangular benzenoids which has a given Clar number, we have only to model the property "having a given Clar number" by adding some variables and/or constraints to be able to find the wished benzenoids. The Clar number of a benzenoid is the maximum number of non-adjacent hexagons (i.e. hexagons which have no bond in common) which admit three double bonds [9]. Unfortunately, due to page limit, we cannot detail the corresponding extended model.

### 4.4 Generating Coronoids

Chemists refer to benzenoids with at least one hole as *coronoids* (not to be confused with coronenoids). These molecules are promising model structures of

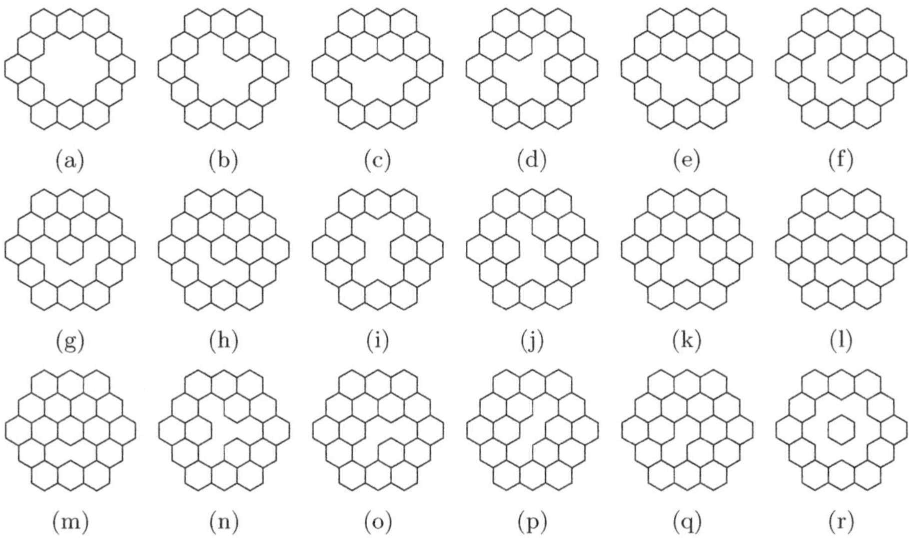


**Fig. 7.** All the ways of digging holes in the coronenoid of size 3. The last one (r) is not a valid coronoid: the hole in the coronenoid is cyclic and has disconnected the "coronoid" into two benzenoids. Hence, the constraint that $x_C$ must be connected.

graphene with well-defined holes [4,12,13]. Their enumeration and generation gave rise to several studies (e.g., [6] which enumerates 2-hole coronoids and generates the smallest 18 and 19-hexagon 3-hole coronoids). The methods for generating them are quite inefficient or too specific. The first kind of approach tries to build specific kinds of coronoids by considering cycles of hexagons and try all possible ways to add hexagons around. The second kind of approach consists of generating all the benzenoids with $n$ hexagons and then detects the ones with holes. Another possible approach consists in generating benzenoids without holes (e.g. with the method of Brinkmann et al. [5]) and then digging holes in the obtained benzenoids. However we can note that the two latter approaches can quickly become too time-consuming with respect to an approach which would directly generate coronoids. Indeed, if the number of hexagons is increased by one, the number of benzenoids is multiplied by approximately 5 (as well as the time to generate them [5]), whereas the time to generate the coronoids with the direct approach we describe below appears to be only twice longer (see Table 1). So, in this part, we present how we can model the property *"all the built benzenoids have a hole and are contained in a benzenoid with n hexagons"*. This allows to generate easily all kinds of coronoids with any number of holes.

Any coronoid can be seen as a benzenoid that has lost several contiguous hexagons (which created holes). So, the vertices of the benzenoid can be split into the vertices belonging to a coronoid and the vertices forming holes. To model this problem, we consider our general model. First, we define two new graph variables $x_C$, which represents an underlying coronoid of $x_G$, and $x_H$ the

**Table 1.** Number of coronoids obtained by digging holes from all the benzenoids with $n$ hexagons.

| $n$ | #coronoids | Time (s) | #benzenoids without holes |
|---|---|---|---|
| 10 | 1 | 43 | 30,086 |
| 11 | 4 | 114 | 141,229 |
| 12 | 38 | 262 | 669,584 |
| 13 | 239 | 533 | 3,198,256 |
| 14 | 1,510 | 1,076 | 15,367,577 |

holes to dig in $x_G$ to form $x_C$. $x_C$ and $x_H$ have the same domain as $x_G$. So, we can have all the possible coronoid $x_C$ by generating all the pairs $(x_G, x_H)$. There can be several values of $x_H$ for a value of $x_G$, as illustrated in Fig. 7. Then we consider two sets of $n_c$ Boolean variables $\{x_1^C, \ldots, x_{n_c}^C\}$ and $\{x_1^H, \ldots, x_{n_c}^H\}$ (with $n_c$ the number of hexagons of coronenoid of size $k(n)$). Like $x_i$ for $x_G$, the variable $x_i^C$ (resp. $x_i^H$) is set to 1 if the $i$th hexagon of coronenoid of size $k(n)$ is used in the graph depicted by $x_C$ (resp. $x_H$), 0 otherwise. Likewise, we define the set of $m_c$ Boolean variables $\{y_1^H, \ldots, y_{m_c}^H\}$ (with $m_c$ the number of edges in the hexagon graph of coronenoid of size $k(n)$). The variable $y_j^H$ is set to 1 if the $j$th edge of coronenoid of size $k(n)$ is used in the graph depicted by $x_H$, 0 otherwise.

Finally, we add the following constraints ensuring that variables $x_H$ and $x_C$ have the right properties:

- *$x_H$ is a sub-graph of $x_G$.* This is enforced thanks to the `subgraph` constraint of Choco applied on variables $x_H$ and $x_G$.
- *Only fully surrounded vertices of $x_G$ can be vertices of $x_H$.* For all vertices of $x_H$ if the degree of a vertex in $x_H$ is strictly positive then the degree of the same vertex in $x_G$ is 6. Indeed, only the vertices/hexagons in $x_G$ surrounded by six hexagons can belong to a hole. This constraint is enforced thanks to clauses on the $y_i$ and $y_i^H$ Boolean variables.
- *A single hexagon does not form a hole.* Each vertex/hexagon of $x_H$ must have a degree strictly greater than 0. This constraint eliminates holes that would be a sole hexagon and allows multiple holes. We simply use the `minDegrees` graph constraint of Choco applied on $x_H$.
- *$x_H$ involves at least two hexagons.* We post the constraint $\sum_{i \in \{1,\ldots,n_c\}} x_i^H > 1$.
- *$x_C$ and $x_H$ form a partition of $x_G$ w.r.t. hexagons.* For all vertices of $x_G$, a vertex is in $x_C$ iff this vertex is in $x_G$ and not in $x_H$. This ensures that any vertex of $x_G$ is either in $x_C$ or in $x_H$. With this aim in view, we add a clause $x_i^C \leftrightarrow (x_i \wedge \overline{x_i^H})$ on $x_i$, $x_i^C$ and $x_i^H$ for any $i \in \{1, \ldots, n_c\}$.
- *$x_C$ is connected.* If $x_C$ is not connected, we may obtain two benzenoids instead of one (see Fig. 7(r) for instance). Again, this can be achieved by exploiting the graph constraint `connected` applied on $x_C$.

For example, Fig. 8(a) shows how the coronoid of Fig. 9(a) can be embedded in the coronenoid of size 3. Then, we depict in dashed line the value of $x_G$, $x_C$ and $x_H$ respectively in Figs. 8(b)–(d).

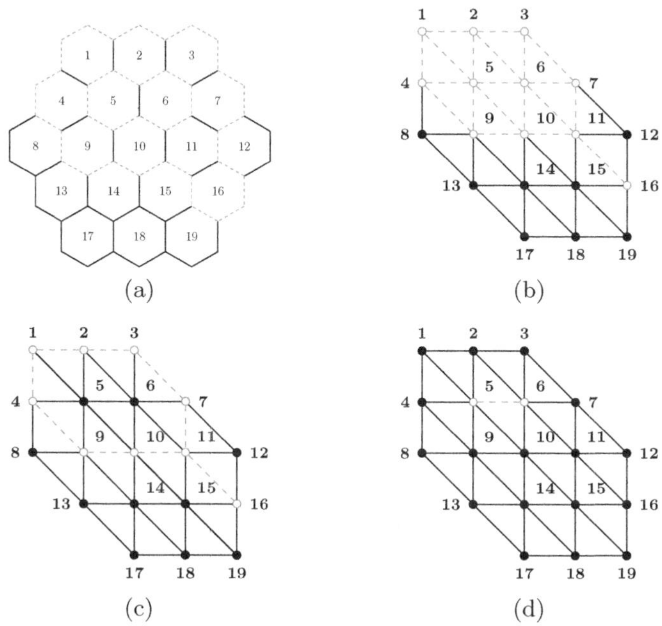


**Fig. 8.** The coronoid of Fig. 9(a) embedded in the coronenoid of size 3 (a), the corresponding value of $x_G$ (b), $x_C$ (c) and $x_H$ (d).

Table 1 shows the results of the experiments we ran on a 3.4 GHz Intel Core i7 iMac with a 12 Gb RAM. We implemented our CSP model in Java 8 with Choco Solver 4.0.4 using the choco-graph 4.2.3 module. We did not specify any search strategy or heuristic, so the default ones were used by the search engine. We generated the coronoids by digging holes in different sizes of benzenoids. Among all the benzenoids with $n$ hexagons, we show the number of coronoids we can produce by removing hexagons. For example, the only coronoid produced from the 10-hexagon benzenoids is the 8-hexagon coronoid of Fig. 3(a). Figure 9 lists the four coronoids for $n = 11$. To show how rare coronoids are, Table 1 also provides the number of benzenoids without holes [5]. Of course, thanks to the model we propose for coronoid generation, we do not consider all these benzenoids. Indeed, they are not generated by Choco Solver because it filters out benzenoids that cannot have holes through constraint propagation.

### 4.5 Generating Symmetric Benzenoids

Benzenoids are also classified by chemists by their classes of internal symmetries (symmetries that let a benzenoid invariant by rotation and/or mirroring). We

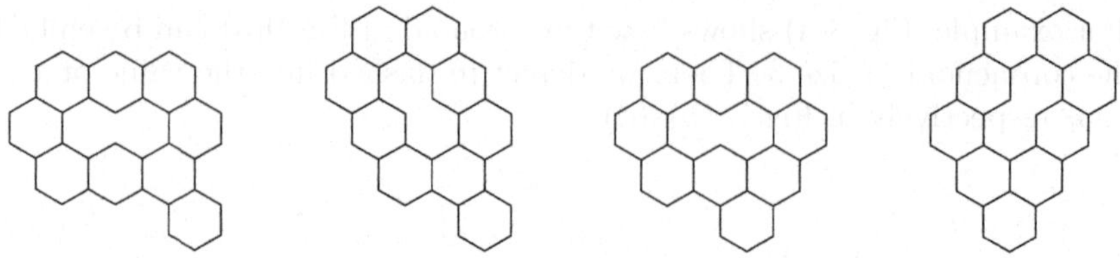

(a) (b) (c) (d)

**Fig. 9.** Coronoids for $n = 11$.

can generate such classes of benzenoids by adding the constraints for enforcing internal symmetries. When searching for all the possible benzenoids embeddable in a coronenoid of size 3, we obtain the 11,578 benzenoids (with at most 19 hexagons) in 36 min. Enforcing invariance by 60° rotation (to obtain the 4 corresponding benzenoids), by 120° rotation (16 benzenoids) and 180° rotation (70 benzenoids) takes less than one second for each task. This strengthens the idea that constraint propagation in nowadays solvers is efficient enough to allow these theoretical chemistry problems to be modeled and solved with CP without having to define and use bespoke methods. Moreover, interestingly, note that this extension of our general model may be combined with any extension described above.

## 5 Conclusions and Perspectives

In this paper, we addressed the problem of generating benzenoid structures, which is an interesting and important problem in theoretical chemistry. In this context, we presented an approach using constraint programming able to generate benzenoids which satisfy a certain amount of properties. Its main advantage w.r.t. existing methods in the literature lies in its flexibility. Indeed, from a general model, we can express additional properties by simply adding variables and/or constraints while existing bespoke methods rely on more rigid and complex notions and cannot be adapted without requiring heavy tasks. Moreover, our approach turns to be more general, making it possible to generate benzenoids with holes for instance.

Chemists are interested in generating benzenoids with particular shapes (e.g. rectangle or rhombus shapes [27]). We have already dealt with the rectangle shapes in this paper. So a natural extension of this work relies in taking into account other specific properties related to the needs of chemists. Another step consists in studying the limit of our approach both in terms of properties we can express and our ability to generate benzenoids of large size. Furthermore, this paper shows how, once again, constraint programming can be useful to tackle and solve problems related to theoretical chemistry [16,17,19,26,28]. In particular, many questions about benzenoids can be modeled as decision or optimization problems under constraints (e.g. computing their aromaticity or finding the closest structure to a Kekulé structure) and can correspond to difficult tasks (e.g.

computing the Clar number is NP-hard [7]). It could be of interest for both communities to study them.

## References


1. Allamandola, L.J., Hudgins, D.M., Sandford, S.A.: Modeling the unidentified infrared emission with combinations of polycyclic aromatic hydrocarbons. Astrophys. J. **511**(2), L115–L119 (1999). https://doi.org/10.1086/311843
2. Aumaitre, C., Morin, J.F.: Polycyclic aromatic hydrocarbons as potential building blocks for organic solar cells. Chem. Rec. **19**(6), 1142–1154 (2019). https://doi.org/10.1002/tcr.201900016
3. Bauschlicher Jr., C.W., Peeters, E., Allamandola, L.J.: The infrared spectra of very large, compact, highly symmetric, polycyclic aromatic hydrocarbons (PAHs). Astrophys. J. **678**(1), 316–327 (2008). https://doi.org/10.1086/533424
4. Beser, U., et al.: A C216-nanographene molecule with defined cavity as extended coronoid. J. Am. Chem. Soc. **138**(13), 4322–4325 (2016). https://doi.org/10.1021/jacs.6b01181
5. Brinkmann, G., Caporossi, G., Hansen, P.: A constructive enumeration of fusenes and benzenoids. J. Algorithms **45**(2), 155–166 (2002)
6. Brunvoll, J., Cyvin, R.N., Cyvin, S.J.: Enumeration and classification of double coronoid hydrocarbons - appendix: triple coronoids. Croat. Chem. Acta **63**(4), 585–601 (1990)
7. Bérczi-Kovács, E.R., Bernáth, A.: The complexity of the Clar number problem and an exact algorithm. J. Math. Chem. **56**(2), 597–605 (2017). https://doi.org/10.1007/s10910-017-0799-8
8. Caporossi, G., Hansen, P.: Enumeration of polyhex hydrocarbons to h = 21. J. Chem. Inf. Comput. Sci. **38**(4), 610–619 (1998). https://doi.org/10.1021/ci970116n
9. Clar, E.: The Aromatic Sextet. Wiley, Hoboken (1972)
10. Cyvin, J., Brunvoll, J., Cyvin, B.N.: Search for concealed non-kekuliian benzenoids and coronoids. J. Chem. Inf. Comput. Sci. **29**(4), 237 (1989)
11. Devriendt, J., Bogaerts, B., Bruynooghe, M., Denecker, M.: Improved static symmetry breaking for SAT. In: Creignou, N., Le Berre, D. (eds.) SAT 2016. LNCS, vol. 9710, pp. 104–122. Springer, Cham (2016). https://doi.org/10.1007/978-3-319-40970-2_8
12. Di Giovannantonio, M., et al.: Large-cavity coronoids with different inner and outer edge structures. J. Am. Chem. Soc. **142**(28), 12046–12050 (2020). https://doi.org/10.1021/jacs.0c05268
13. Dias, J.R.: Structure and electronic characteristics of coronoid polycyclic aromatic hydrocarbons as potential models of graphite layers with hole defects. J. Phys. Chem. A **112**(47), 12281–12292 (2008). https://doi.org/10.1021/jp806987f
14. Draine, B.T.: Astronomical models of PAHs and dust. EAS Publ. Ser. **46**, 29–42 (2011). https://doi.org/10.1051/eas/1146003
15. Fages, J.G., Lorca, X., Prud'homme, C.: Choco solver user guide documentation. https://choco-solver.readthedocs.io/en/latest/
16. Ismail, I., Stuttaford-Fowler, H.B.V.A., Ochan Ashok, C., Robertson, C., Habershon, S.: Automatic proposal of multistep reaction mechanisms using a graph-driven search. J. Phys. Chem. A **123**(15), 3407–3417 (2019). https://doi.org/10.1021/acs.jpca.9b01014

17. Kim, Y., Kim, J.W., Kim, Z., Kim, W.Y.: Efficient prediction of reaction paths through molecular graph and reaction network analysis. Chem. Sci. **9**(4), 825–835 (2018). https://doi.org/10.1039/C7SC03628K
18. Luch, A.: The Carcinogenic Effects of Polycyclic Aromatic Hydrocarbons. Imperial College Press, London (2005). https://doi.org/10.1142/p306
19. Mann, M., Nahar, F., Schnorr, N., Backofen, R., Stadler, P.F., Flamm, C.: Atom mapping with constraint programming. Algorithms Mol. Biol. **9**(1), 23 (2014). https://doi.org/10.1186/s13015-014-0023-3
20. Mishra, S., et al.: Topological frustration induces unconventional magnetism in a nanographene. Nat. Nanotechnol. **15**(1), 22–28 (2020). https://doi.org/10.1038/s41565-019-0577-9
21. Mishra, S., et al.: Synthesis and characterization of $\pi$-extended triangulene. J. Am. Chem. Soc. **141**(27), 10621–10625 (2019). https://doi.org/10.1021/jacs.9b05319
22. Narita, A., Wang, X.Y., Feng, X., Müllen, K.: New advances in nanographene chemistry. Chem. Soc. Rev. **44**(18), 6616–6643 (2015). https://doi.org/10.1039/C5CS00183H
23. Ricca, A., Bauschlicher, C.W., Boersma, C., Tielens, A.G.G.M., Allamandola, L.J.: The infrared spectroscopy of compact polycyclic aromatic hydrocarbons containing up to 384 carbons. Astrophys. J. **754**(1), 75 (2012). https://doi.org/10.1088/0004-637X/754/1/75
24. Ricca, A., Roser, J.E., Peeters, E., Boersma, C.: Polycyclic aromatic hydrocarbons with armchair edges: potential emitters in class B sources. Astrophys. J. **882**(1), 56 (2019). https://doi.org/10.3847/1538-4357/ab3124
25. Rieger, R., Müllen, K.: Forever young: polycyclic aromatic hydrocarbons as model cases for structural and optical studies. J. Phys. Org. Chem. **23**(4), 315–325 (2010). https://doi.org/10.1002/poc.1644
26. Simoncini, D., Allouche, D., de Givry, S., Delmas, C., Barbe, S., Schiex, T.: Guaranteed discrete energy optimization on large protein design problems. J. Chem. Theory Comput. **11**(12), 5980–5989 (2015). https://doi.org/10.1021/acs.jctc.5b00594
27. Trinquier, G., Malrieu, J.P.: Predicting the open-shell character of polycyclic hydrocarbons in terms of Clar sextets. J. Phys. Chem. A **122**(4), 1088–1103 (2018). https://doi.org/10.1021/acs.jpca.7b11095
28. Wu, C.W.: Modelling chemical reactions using constraint programming and molecular graphs. In: Wallace, M. (ed.) CP 2004. LNCS, vol. 3258, pp. 808–808. Springer, Heidelberg (2004). https://doi.org/10.1007/978-3-540-30201-8_87
29. Wu, J., Pisula, W., Müllen, K.: Graphenes as potential material for electronics. Chem. Rev. **107**(3), 718–747 (2007). https://doi.org/10.1021/cr068010r

# RobTest: A CP Approach to Generate Maximal Test Trajectories for Industrial Robots

Mathieu Collet[1(✉)], Arnaud Gotlieb[1], Nadjib Lazaar[2], Mats Carlsson[3], Dusica Marijan[1], and Morten Mossige[4]

[1] Simula Research Laboratory, Oslo, Norway
{mathieu,arnaud,dusica}@simula.no
[2] LIRMM, University of Montpellier, CNRS, Montpellier, France
lazaar@lirmm.fr
[3] RISE Research Institutes of Sweden, Kista, Sweden
mats.carlsson@ri.se
[4] ABB Robotics, Bryne, Norway
morten.mossige@no.abb.com

**Abstract.** Developing industrial robots which are safe, performant, robust and reliable over time is challenging, because their embedded distributed software system involves complex motions with force and torque control and anti-collision surveillance processes. Generating test trajectories which increase the chance to uncover potential failures or downtime is thus crucial to verify the reliability and performance of the robot before delivering it to its final users. Currently, these trajectories are manually created by test engineers, something that renders the process error-prone and time-consuming. In this paper, we present *RobTest,* a Constraint Programming approach for generating automatically maximal test trajectories for serial industrial robots. *RobTest* sequentially calls two constraint solvers: a solver over continuous domains to determine the reachability between configurations of the robot's 3D-space, and a solver over finite domains to generate maximal-load test trajectories among a set of input points and obstacles of the 3D-space. *RobTest* is developed at ABB Robotics, a large robot manufacturing company, together with test engineers, who are preparing it for integration within the continuous testing process of the robots product-line. This paper reports on initial experimental results with three distinct solvers, namely Gecode, SICStus and Chuffed, where *RobTest,* has been shown to return near-optimal solutions for trajectories encounting for more than 80 input points and 60 obstacles in less than 5 min.

**Keywords:** Industrial robotics · Maximal test trajectories · Path planning · Global constraints


H. Simonis (Ed.): CP 2020, LNCS 12333, pp. 707–723, 2020.
https://doi.org/10.1007/978-3-030-58475-7_41

## 1 Context and Motivations

Industrial robotics is evolving very fast, with ever growing needs in terms of safety, performance, robustness and reliability. Serial robots are now complex cyber-physical systems which embed complex distributed multi-core software systems involving intelligent motion control, anti-collision, advanced force/torque control. This increased complexity makes these robots more fragile and more error-prone that they were previously. Failures can originate from many sources including system and software bugs, communication downtime, CPU overload, robots wear, etc. Hopefully, advanced verification techniques are more and more used to cope with errors and ensure a better quality of delivered robots. Adopting these techniques is crucial but a challenge lies in the design of test scenarios able to drive the robots into demanding corner case scenarios which maximize the chance to uncover system or software faults. More precisely, finding robot trajectories, i.e., successive positions, speeds, accelerations of the end-effector, which exhibit the greatest deviations between the trajectory specified to the robot and the actual trajectory is crucial to detect faults of the motion control system. One such deviation is illustrated in Fig. 1, where the *specified path* corresponds to the path defined by the user.

Generating stress test trajectories which maximize the chance to uncover these deviations resulting from design, configurations, system or software faults has thus become one of the most time-consuming activities of test engineers. Importantly, to the best of our knowledge, there is no complete analytic model that can solve the problem with full accuracy. The number of equations required to model the robot's (inverse) kinematics along with its various control systems is far too large to allow any global optimization tool to solve it analytically. Typically, a 4-DoF[1] serial robot already requires to solve no less than 49 $4^{th}$-degree polynomial equations with 49 variables [14]. Hopefully, even though its seems impossible to solve the problem with analytics, industrial robots are sometimes modeled with a simulated *digital twin* [13]. This simulated model accurately represents the actual robot with all its physical features, direct and inverse kinematics, and subsystems and is fast to run simulated scenarios. However, it has also some discrepancies with the physical robot due to imperfect simulation of CPUs. Using the digital twin to execute and evaluate test trajectories can relieve test engineers from executing scenarios on real robots and has opened the door to systematic test and model verification in Continuous Integration (CI). But, a challenge which still remains, is the automatic generation of stress trajectories for the digital twin in order to discover significant deviations between specified and simulated paths on the digital twin.

Building on existing work where Constraint Programming (CP) has been successfully adopted to generate automatically test cases for painting robots [16], this paper presents a CP-based methodology, called *RobTest,* for generating test trajectories. These trajectories aim to maximize a load function designed for finding deviations, in the 3D-workspace of the robot's end-effector. Formally

[1] Degree of Freedom: Typical industrial robots have 6-DoF.

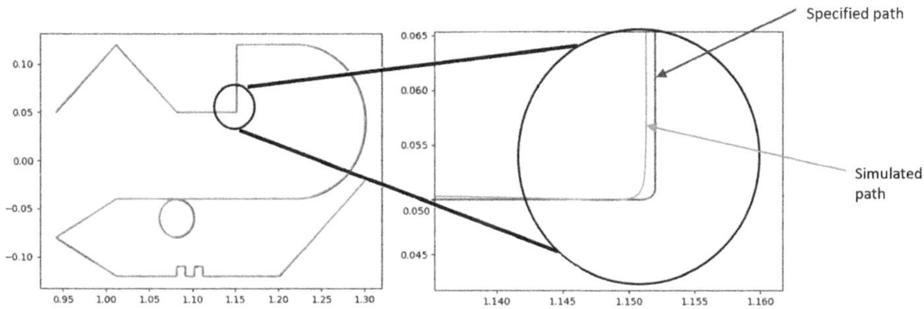


**Fig. 1.** Deviation between specified and simulated path.

speaking, given a set of 3D-space points and a number of fixed obstacles, the test objective consists in finding trajectories reaching some of these points by 1) avoiding all the fixed obstacles and 2) by maximizing the load function (described in the next section). If found, these test trajectories can be converted into computer programs for the robot and used as test scenarios in CI. *RobTest* exploits two constraint solvers: a constraint satisfaction solver over continuous domains to evaluate the reachability of each pair of points in the Cartesian 3D-space with obstacles, and a constraint satisfaction solver over finite domains to generate test trajectories. In the current setting, we have used RealPaver [10] as the continuous domains solver and MiniZinc [17] with three back-ends (`gecode` [22], `chuffed` [3] and `SICStus` [2]) for solving constraints over finite domains. Our model exploits several global constraints such as INVERSE, SUBCIRCUIT and TABLE to construct an effective approach, which results from the refinement of several previous models.

The problem studied in this paper is close to [20] where the goal is to find the Longest Simple Path (LSP) in an undirected weighted graph. Interestingly, [20] proposes an exact constraint model using the ALLDIFFERENT global constraint and exploits known lower bounds of the longest path. A complete study for solving path exploration with constraint techniques is available in [1]. In [8], the LSP problem is tackled with a dynamic programming approach while [4] proposes to use heuristic search. Our problem generalizes LSP, because the label of our edges depends not only on the node and the successor, but also on the predecessor. As is well known, LSP is NP-hard, and so it follows that our problem is NP-hard as well. In [5], we proposed an initial CP-based approach using a-posteriori verification and a trivial search heuristic, but the model of [5] does not scale up to real-case instances, i.e., instances with more than 20 input points.

In this paper, we propose a novel CP model with the three above-mentioned global constraints and deploy that model on real-case instances. Unlike [1,5,20], our approach *RobTest,* can handle workspaces containing more than 80 input points and 60 obstacles and generate quasi-maximal trajectories in less than

5 min.[2] We also consider a realistic load function for the robotic arm moves, which takes into account the incidence-angle at each point by using a multiple load-values variable associated to each arc of the reachability graph. To the best of our knowledge, there is no other model able to deal with so many nodes and obstacles, having such a realistic load function.

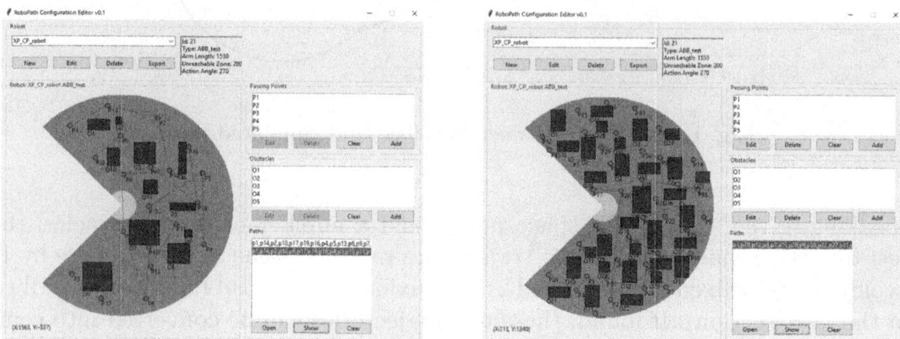

**Fig. 2.** GUI of *RobTest* : users can select 3D-points, set obstacles, etc. and visualize the generated test trajectories.

Finally, the *RobTest* methodology is available through a dedicated graphical user interface, as shown in Fig. 2, which prepares the ground for its deployment into the continuous integration process of ABB Robotics.

Organization of the paper. Section 2 discusses related work while Sect. 3 gives the background of the approach, necessary to understand the rest of the paper. Section 4 details our CP-based methodology which makes use of constraint solving over continuous domains and finite domains while Sect. 5 presents our experimental evaluation. Finally, Sect. 6 concludes the paper.

## 2 Related Work

Robot motion planning is a rich domain which has been studied for decades [12]. Generating collision-free trajectories which minimize the time taken by robots to perform its tasks (or reach specific configurations) is known as *optimal planning of robot trajectories* [9]. For 4-DoF serial robot, exact models of the direct and inverse kinematics have been studied with Interval Analysis [14] and constraint satisfaction problems over continuous domains [18]. But unfortunately, 6-DoF serial robots involve too complex equations to solve and thus need simulated models and digital twins.

In most industrial settings, the problem of this paper is tackled without any automated support. Thus, only trajectories which involve a few number of

[2] The experimental benchmark is publicly available at www.github.com/Makouno44/Robtest.

points are generated (typically 10–20 points), while *RobTest* reaches more than 80 points. Trajectories generation for different robots has been considered in the context of robot workspace discretization [11]. Such discretization involves the division of the workspace into surface-equivalent boxes and constraint refutation can discard boxes where no feasible solution is present. Then, exploring neighbourhood boxes allows the method to find continuous trajectories. Interestingly, by using Interval Analysis, the method produces trajectories with guaranteed computation results over the reals [6,21]. In *RobTest,* we also considered polynomial (in-)equations for the robot workspace, but we did not pay attention to rounding errors as our target was only test trajectories generation. With the discretization approach, we can also generate collision-free trajectories but generating load-optimal trajectories requires more in depth exploration.

Another method considers the *joint space* for finding trajectories [15,23], but this method requires to solve the robot's inverse kinematics equations. This approach is important when finding singularities is needed [14] and providing alternative trajectories to avoid them is required [7]. However, the computation time required by these methods is often prohibitive for their inclusion into CI. Another method considers the preliminary evaluation of several trajectories and dynamic re-planning when an unwanted configuration is reached [19]. This method is interesting in trajectory planning but it cannot be used in test generation as it reassigns tasks to the robot, during scheduling.

## 3 Background

This section introduces the formalization of the robot workspace. We also present and justify the hypothesis on which the *RobTest* approach is based.

An industrial robot is a mechatronic system able to perform various tasks according to a high-level program written in a dedicated language. A *serial robot* (as opposed to a parallel robot) is composed of a *base* and an *end-effector* in charge of grasping, glueing, welding, painting or moving objects. The base is usually attached to the ground and connects a fixed number of *joints*, usually 4 or 6 that correspond to the DoF. A *configuration* of the robot is a position in which all the DoF are fixed. Generating test trajectories for a serial robot requires to set up points, moves at various maximum speed and possible positions of the end-effector. The *configuration space* $\mathcal{Q}$ of a serial robot is the set of its $n$ possible configurations in a 3D-*workspace* $\mathcal{W}$, which is a subset of $\mathbb{R}^3$. In $\mathcal{W}$, the robot can move around a set of $m$ obstacles $O_1 \ldots O_m$, which are materialized by 3D-rectangles.[3] $\mathcal{A}(q)$ is the subset of $\mathcal{W}$ occupied by the robot when it is in configuration $q$. The set of *collision-free configurations* is defined as $\mathcal{Q}_{free} := \mathcal{Q} \setminus \{q \in \mathcal{Q} : \mathcal{A}(q) \cap O_k = \emptyset, \forall k \in 1..m\}$, while the subset accessible by the robot's end-effector is $\mathcal{W}_{free} := \{\mathcal{A}(q) : q \in \mathcal{Q}_{free}\}$.

[3] Actually, typical obstacles are other robots, devices, service material, etc. Their shape can easily be over-approximated by 3D-rectangles, without any loss of generality.

Given the quadruple $(\mathcal{Q}, q_s, q_f, l)$, where $\mathcal{Q}$ is the *configuration space*, $q_s$ (resp. $q_f$) is the initial (resp. final) configuration, $l : \mathcal{Q} \times \mathcal{Q} \longrightarrow \mathbb{R}$ is a load function for each possible transition. Our problem aims at finding a path $p$ that connects $q_s$ to $q_f$ such that no obstacle is collided by the end-effector and such that the load of $p$ is maximized. Note that moving from $q_s$ to $q_f$ is a continuous function w.r.t. the time. Thanks to the availability of a digital twin which embeds the full modeling of direct and inverse kinematics, we focus only on solving the optimization problem over finite domains. The load function $l$ is determined by several factors, including the robot's type of move (linear or circular), the robot's speed and acceleration, the distance between points to join, the CPU-load. This load function is given by an external process which exploits the digital twin to determine the switching load of each transition. Note that even though the load function $l$ is defined for each transition between two configurations, namely $q_i$ and $q_j$, it also often takes into account other transitions as detailed in the next section. Formally speaking, the problem is as follows: Given $(\mathcal{Q}, q_s, q_f, l)$, find a path $p = q_1, q_2, \ldots, q_n$ where $q_1 = q_s$ and $q_n = q_f$ such that:

$$\text{Maximize} \sum_{\forall i \in 1..n-1} l(q_i, q_{i+1}) \tag{1}$$

$$\mathcal{A}(q_i) \cap O_k = \emptyset, \forall i \in 1..n, k \in 1..m \tag{2}$$

## 4 Constraint-Based Generation of Test Scenarios

This section describes the two main components of *RobTest,* namely `CR`, the component which evaluates the reachability of the robot configurations, and `TG`, the component which generates maximal trajectories.

### 4.1 Configurations Reachability with RealPaver (`CR`)

**Computing the Reachability Graph.** In *RobTest,* configuration reachability is evaluated by using constraint refutation over continuous domains. From an initial set of points in the 3D-workspace of the robot and a number of 3D obstacles, the goal is to create a multiple load-labeled oriented graph, called *reachability graph*, where an arc exists between point $p_1$ to $p_2$ if the corresponding robot configuration in $p_2$ is reachable from the configuration in $p_1$. The label on the arc from $p_1$ to $p_2$ corresponds to a load variable, which takes a value which depends on the predecessors of $p_1$ (transitional load). There are more possible load values on an arc from $p_1$, than there are predecessors of $p_1$ in the reachability graph. This reflects the hardness of transitions in the working space, where moving a robot arm is more stressful when the angle taken in a specific point is more acute. So, depending on where the robot arm is, the load taken in a transition will be different.

The robot workspace $\mathcal{W}$ is shown in Fig. 3. First, the external envelope of $\mathcal{W}$ is defined as a semi-sphere, given by polynomial inequations. Second, there is a large non-accessible area due to the robot itself, which is given by planes'

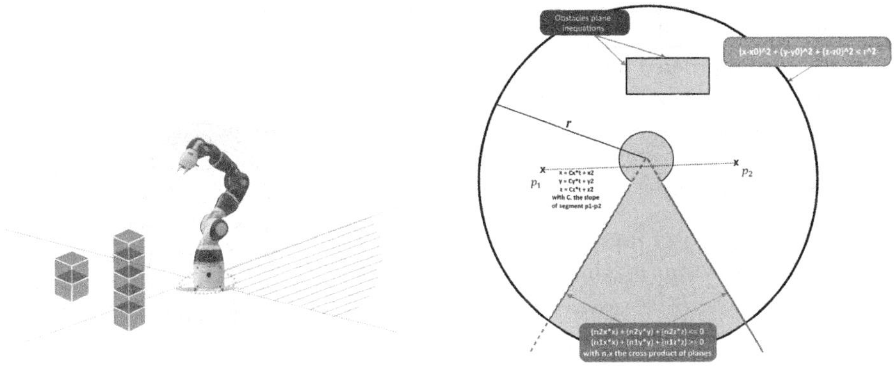


**Fig. 3.** The robot workspace and its 2D-projection. (Color figure online)

parametric inequations (in red in Fig. 3). Third, each 3D-rectangle obstacle is modelled with parametric inequations (in blue in Fig. 3). Note that $\mathcal{W}$ is over-approximated here as a very precise model is not needed to generate test trajectories.

Using this over-approximated model of the workspace, called $\mathcal{O}$ in the following, we check the intersection between the segment $p_i - p_j$ and each non-accessible areas $O_k$. Using a solver over continuous domains, the reachability of each pair of configurations is checked by showing the unsatisfiability of the constraints corresponding to the obstacles and those of the segment. More precisely, Algorithm 1 computes a reachability graph of a set of 3D-points and 3D-obstacles given as inputs and a load relation, named LR. LR is a relation which captures the results of the *load function* computation. In this algorithm, SOLVE is a predicate that determines, given two points and one obstacle, whether the obstacle hinders movement between the two points. Note that, as current obstacles of odd shapes are over-approximated by boxes, two points may be determined as unreachable, while they are actually reachable in the real-world. Dealing with this restriction is a foreseeable extension of the approach.

**The Load Relation.** To generate stress test trajectories, it is necessary to consider a load relation which accounts for robot's moves, distances and angles. In our problem, the load relation defines a value for each valid combination of triplets in the reachability graph (node, one of its successor, one of its predecessor). Figure 4 shows an example of the computation of the load relation for the transition $p_j - p_l$ on two distinct paths i-j-l and k-j-l.

Let's first express the values of $\alpha_1$ and $\alpha_2$ by using the Euclidian distance between nodes $i, j$, noted $d_{i,j}$, and the Al-Kashi formulae in general triangles:

$$d_{i,l}^2 = d_{i,j}^2 + d_{j,l}^2 - 2 \cdot d_{i,j} \cdot d_{j,l} \cdot \cos(\alpha_1), \; d_{k,l}^2 = d_{k,j}^2 + d_{j,l}^2 - 2 \cdot d_{k,j} \cdot d_{j,l} \cdot \cos(\alpha_2) \quad (3)$$

**Algorithm 1:** Computing the reachability graph and load relation

**In**: $N$: set of ordered points; $\mathcal{O}$: set of obstacles; w: user-selected parameter (Int)  
**Output**: $G = (N, E)$: reachability graph. $LR$: load relation.  
$n \leftarrow |N|$, $E \leftarrow \{(p_n, p_1)\}$  
**foreach** $(p_i, p_j) \in (N \times N) : i < j$ **do**  
    $r \leftarrow true$  
    **foreach** $O_k \in \mathcal{O}$ **do**  
        **if** SOLVE$(p_i, p_j, O_k)$ **then**  
            $r \leftarrow false$  
            **break**  
    **if** $r$ **then**  
        $E \leftarrow E \cup \{(p_i, p_j), (p_j, p_i)\}$  
**foreach** $(p_j, p_i, p_k) \in N^3$: $(p_j, p_i), (p_i, p_k) \in E$ **do**  
    $LR \leftarrow LR \cup \{\langle p_j, p_i, p_k, load_w(p_j, p_i, p_k)\rangle\}$  
**return** $(G, LR)$

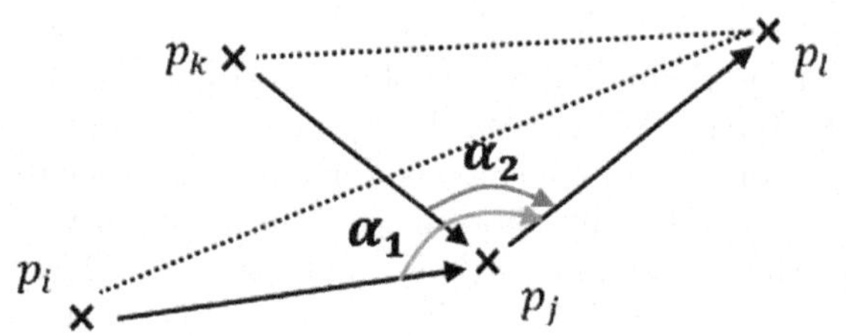


**Fig. 4.** Computation of the load relation for node $p_j$

In our setting, the load relation LR is composed of a set of predicates $load_w/3$ that depend on a user-selected weighting scheme $w$, as follows.

$$\text{On path i-j-l, } load_w(p_i, p_j, p_l) \triangleq d_{j,l} + w \cdot |cos(\alpha_1/2)| \quad (4)$$

$$\text{On path k-j-l, } load_w(p_k, p_j, p_l) \triangleq d_{j,l} + w \cdot |cos(\alpha_2/2)| \quad (5)$$

In Fig. 4, the load of $p_j - p_l$ is higher if the robot comes from k than if it comes from i because $\alpha_2$ is more acute than $\alpha_1$ which means that the robot would have to do a more stressful move. The load formula is designed to account for these constraints.

**Modeling the Workspace with RealPaver.** We have selected the open-source solver *RealPaver* [10] for implementing SOLVE. In our setting, we define equations for $\mathcal{W}$, the segment between $i$ and $j$ and all the obstacles in $\mathcal{O}$. These equations are given in Fig. 3. In this setting, *RealPaver* is called several times for each pair of points: first, a call verifies that the segment is included into the semi-sphere which delimits $\mathcal{W}$; second, a call ensures that the robot does not cross a non-accessible area; and third, the solver is called for each obstacle present in $\mathcal{W}$. If *RealPaver* proves the absence of solutions (i.e., by showing unsatifiability), then $j$ can be reached from $i$, as there is no intersection between the robot's end-effector trajectory and any of the obstacles. If the solver returns a box, using any kind of continuous domains filtering consistency (with a guarantee of solution existence [10]), then there is an intersection between the segment and at least one of the non-accessible areas. The time-complexity of `CR` component of *RobTest* is in $\mathcal{O}(m \cdot n^2 + n^3)$, where $n$ is the number of points and $m$ the number of obstacles.

### 4.2 Test Trajectory Generation (`TG`)

This section presents the second component of *RobTest,* which is responsible for the test trajectory generation by using constraint over finite domains. Using the reachability graph and load relation computed by Algorithm 1, the goal is to find a path in the graph which maximizes the load function, called a *maxpath* in the following. Our model seeks for only one such *maxpath,* which possibly visits nodes multiple times. An input parameter defines a maximal number of possible visits for each node. To handle multiple visits, we unfold the graph in a pre-processing step which duplicates nodes that can be visited multiple times and create additional arcs. This way, the problem is reduced to the search of a *maxpath* where node can be visited only once. This unfolding process is a static pre-process and it resorts on having an upper bound on the number of times each node can be visited. This upper bound is given by the users of *RobTest,* and using it prevents infinite cycles from occurring in the trajectories.

**Variables**
Given a reachability graph $G = \langle N, E \rangle$ where $N$ and $E$ are sets of labelled nodes and arcs. We use a simple encoding of nodes as integers, $N = \{1, \ldots, n\}$. We create for each node $i \in N$:

- $s_i \in N$ is the successor of $i \in N$. $p_i \in N$ is the predecessor of $i \in N$;
- $l_i$ is the load of switching from point $i$ to its successor $s_i$ taking into account all predecessors $p_i$.

Then, the objective function to be maximized over a path 1-2-..-n is:

$$MaxLoad = \text{Maximize} \sum_{i \in N} l_i \tag{6}$$

**Constraints**
On path 1-2-..-n, nodes 1 and n must necessarily be visited while all the other nodes are potentially unknown in the maxpath. Also, the successor of the ending

point is the source, and the successor and predecessor of an unvisited node is the node itself.

$$s_n = 1 \wedge p_1 = n \tag{7}$$

The successors and predecessors represent inverse functions. We express this as the global constraint: [24]

$$\texttt{inverse}(s, p) \tag{8}$$

i.e.,

$$s_i = j \leftrightarrow p_j = i, \forall i, j \in N \tag{9}$$

The maxpath forms a circular path that includes a subset of $N$. We express this as the global constraint: [24]

$$\texttt{subcircuit}(s) \tag{10}$$

which in our context holds if $N$ is partitioned into $N_1$ and $N_2$, and the successor values $s_i$, $i \in N_1$ form a closed circuit, and $s_j = j$ for $j \in N_2$. It is worth noticing that none of the three solvers we have used in our experiments has a native subcircuit propagator.

As said above, the load function is obtained from the reachability graph. This function takes into account a node, its successor, its predecessor and the associated load. The purpose of the load relation is twofold: (*i*) it gives the valid combinations of (node to visit, its successor, its predecessor); (*ii*) it acts as a partial function that computes, for the given combination, the relevant load term:

$$\texttt{table}(\langle i, s_i, p_i, l_i \rangle, LR), \forall i \in N \tag{11}$$

i.e.

$$\langle i, s_i, p_i, l_i \rangle \in LR \tag{12}$$

To solve the constraint model and find a maximal load trajectory, we used a standard time-limited maximization search which works well enough for an embedded solution in an industrial setting. It is worth noting that procedure can be interrupted at any time while giving the current incumbent, i.e., a near-optimal value.

### 4.3 Running Example

Figure 5 illustrates our approach on a small instance of 6 points ($P_1$ to $P_6$) and two obstacles $O_1$ and $O_2$. The CR step computes the Reachability Graph and the Load Relation using Algorithm 1. For instance, Algorithm 1 detects that $P_4$ and $P_5$ cannot be connected due to obstacle $O_2$ (i.e., a solution exists between the segment $P_4, P_5$ and $O_2$ and is returned by the SOLVE predicate). No obstacle exists between $P_2$ and $P_5$ (i.e., no solution exists between the segment $P_2, P_5$

and any obstacle), then Algorithm 1 add an edge between $P_2$ and $P_5$ to the reachability graph. The load relation is also computed at this step using robot's moves, distances and angles. For instance, to go from $P_2$ to $P_3$ the load can be, respectively, 5, 8 or 15, if we are coming from, respectively, $P_5$, $P_4$ or $P_1$.

The second step of *RobTest* consists of generating an optimal test trajectory from $P_1$ to $P_6$ w.r.t. the reachability graph and the load relation. We assume in this example that a point is visited at most once. Here, `TG` returns the trajectory $\langle P_1, P_2, P_3, P_5, P_6 \rangle$ as an optimal solution of a load of 50.

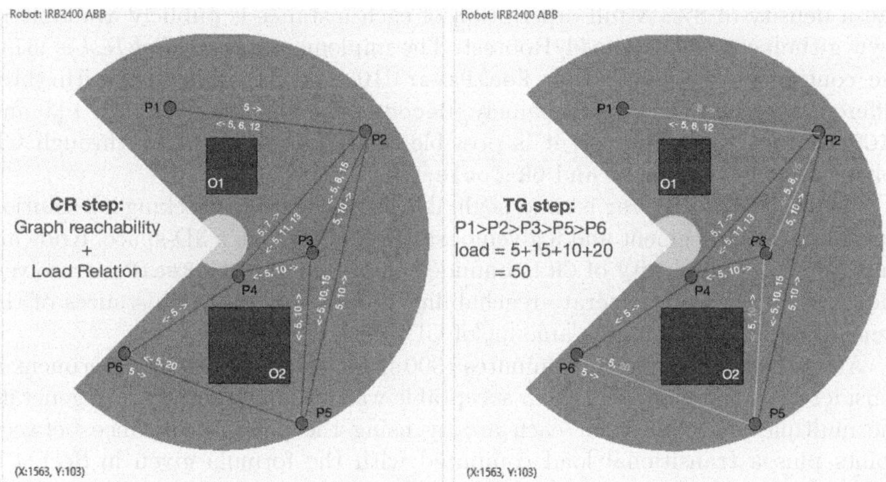


**Fig. 5.** Robtest acting on an instance of 6 points and 2 obstacles

## 5 Experimental Evaluation

We performed an in-depth evaluation of *RobTest* for generating collision-free test trajectories. As the goal is to embed *RobTest* in an operational robot testing context, we explored the three following research questions:

**RQ1:** Which search heuristic is best suited for *RobTest* to generate true load-optimal trajectories?

**RQ2:** How efficient is *RobTest* under different CP solver?

**RQ3:** How efficient is *RobTest* in generating realistic trajectories in a reasonable amount of time? This RQ is crucial to foster the deployment of *RobTest* in a CI process;

### 5.1 Experimental Evaluation Protocol

For the experiments, our GUI (Fig. 3) is used to set the points and obstacles in the robot workspace. Using `CR` step of *RobTest,* we generated 14 instances by considering a number of points/obstacles in the workspace corresponding to the regular practice of test engineers (from 15 to 100 points and from 11 to 68 obstacles). The obstacles are positioned by the test engineers in a way that the reachability graphs have (i) their starting and ending points in the same connected component and (ii) a medium density ranging between 4% to 10%. For example, `Inst_20_18_8` denotes an instance with 20 points, 18 obstacles and a density of 8%. A full description of each instance is publicly available at www.github.com/Makouno44/Robtest. The implementation of *RobTest* is using the continuous domains solver `RealPaver` [10] and MiniZinc [17] with three different back-end CP solvers, namely (`Gecode`.6.1.1 [22], `Chuffed`.0.10.4 [3] and `SICStus`.4.5.1 [2]). However, it is possible to use graph variables through CP solvers like `Choco-graph`[4] and `OR-Tools`.[5]

Each call to `RealPaver` is time-negligible as it reduces to checking if a solution exists between a segment and a given obstacle (i.e., box) in a 3D-space. Knowing that the time-complexity of `CR` is bounded by $n^3$ calls to the `RealPaver` solver, the `CR` step is able to generate reachability graphs over the 14 instances of the benchmark in an acceptable amount of CPU time.

An overall timeout of five minutes (300 s) for each individual experiment is considered by test engineers as an acceptable waiting time contract. We generate the multiple load values for each arc by using the pairwise distance between points plus a transitional load computed with the formula given in Sect. 4.1, with a user-parameter $w = 50$.

All our experiments were performed on an Intel(R) Xeon(R) CPU E5-2640 v4 @ 2, 40 GHz with 256 GB RAM.

**[RQ1]: Search Heuristics**

Our CP model is expressed over successors and load variables. After some preliminary tests and some obvious remarks, we observed that branching first on load variables and try values in decreasing order is the best policy to follow. However, which load variable to select first has a significant impact on *RobTest* effectiveness. The purpose of our first experiment is to evaluate the performance of *RobTest* for generating test trajectories by using classical variable-ordering heuristics on load variables. We considered systematically the nine following heuristics:

- `input_order`: Leftmost variable is selected first, from an input sequence of load variables;
- `anti_first_fail`: The variable with the largest domain is selected first;
- `first_fail`: The variable with the smallest domain is selected first;
- `largest`: The variable with the largest value in its domain is selected first;

[4] github.com/chocoteam/choco-graph.

[5] developers.google.com/optimization.

- `smallest`: The variable with the smallest value in its domain is selected first;
- `occurrence`: The variable with the largest number of attached constraints is selected;
- `most_constrained`: The variable with the smallest domain is selected first, with the number of attached constraints as a break tie;
- `dom_w_deg`: The variable with the largest domain is selected first, with the number of attached constraints weighted by how often they have caused failure as a break tie;
- `max_regret`: The variable with largest difference between the two smallest values in its domain is selected first;

For our first experiment, we made a comparison between the 3 solvers, using the 9 heuristics, on the 14 instances (i.e., 378 runs). As the observations made on the 378 runs were highly correlated, we selected `Inst_55_35_6` to illustrate the main findings. Fig. 6 reports a comparison between (solver, heuristic) configurations with *RobTest* (on `Inst_55_35_6`) to return the first (Fig. 6.a) and the best solution found within the time limit of 5 min (Fig. 6.b). We report (i) the convergence rate %`conv`, which corresponds to the ratio between the first/best solution and the optimal one (histograms in blue); and (ii) the CPU time in seconds (curves in black).

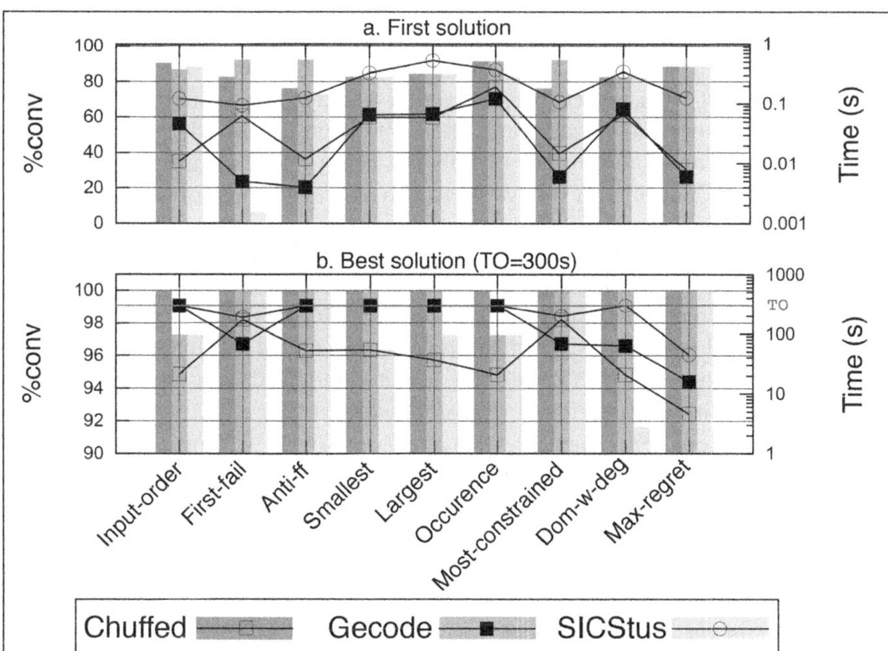


**Fig. 6.** Comparing different variable-ordering heuristics for *RobTest*

From Fig. 6, the main finding is that, whichever solver used, the most interesting heuristic for *RobTest* is `max_regret`. `max_regret` is the right heuristic to use to quickly get a good-enough solution as it provides a first solution of good quality (exceeding 88%) in less than 10 ms for `Chuffed` and `Gecode`, and less than 125 ms for `SICStus`. `max_regret` is also the right heuristic when seeking for optimality. For instance, using `Chuffed` we get the optimal solution in less than 5 s, which is 35 times faster than `first_fail` and `most_constrained`. The other heuristics behave differently according to the used solver. For instance, with the `first_fail` heuristic, `SICStus` and `Gecode` return within the same amount of time a first solution of, respectively, 20% and 90%. Then, `SICStus` is faster to reach the optimal solution than `Gecode`.

**[RQ2] Solver comparison**
From Fig. 6, we observe that `Chuffed` clearly outperforms `Gecode` and `SICStus` on `Inst_55_35_6`. Whatever be the heuristic, `Chuffed` returns a first solution with `%conv` ranging in between 76% and 91% in [8 ms, 193 ms]. `Chuffed` always reaches the optimal solution in [4 s, 173 s], whereas `Gecode` and `SICStus` are not always able to reach the optimal solutions, e.g., when using the `input_order` heuristic. Similar conclusions were drawn from the analysis of the results on all the other 13 instances.

**[RQ3] Efficiency of *RobTest***
We performed a second experiment for an in-depth comparison between the three back-end solvers and for analyzing the efficiency of *RobTest* . Table 1 reports on results of *RobTest* with the three solvers and `max_regret` heuristic, over the 14 instances. We report the `%conv` rate, the CPU time in seconds, (`#s`) number of visited solutions, and the number of fails (`#fails`). Bear in mind that the sum of `#s` and `#fails` provide an overview on the size of the explored search space.

- **Small instances.** The three solvers return the optimal solution in less than one second. However, the explored search space to find an optimal solution is smaller with `Chuffed` than with `Gecode` and `SICStus`.
- **Middle-sized instances.** `Chuffed` returns the optimal solutions of the three instances in less than 23 s, `Gecode` in 28 s and `SICStus` in 76 s. `Chuffed` remains the winner in terms of explored search tree.
- **Large instances.** For `Inst_60_33_6` and `Inst_70_43_5`, `Chuffed` and `Gecode` are able to return the optimal solutions within the allocated CPU time, whereas `SICStus` returns solutions of, respectively, 97% and 98% before timeout. For the three largest instances (80, 90 and 100 points), the three solvers reaches the timeout of 300 s. On the instance with 80 points, the three solvers return the optimal solution without proving optimality.[6] On 90 points, `Gecode` and `SICStus` return solutions of, respectively, 99% and 97%, where `Chuffed` reaches the 100% without proof of optimality. On the largest instance of 100

[6] The optimal solution is computed by releasing the timeout.

**Table 1.** *RobTest* experimental results

| | Inst. | Chuffed | | | | Gecode | | | | SICStus | | | |
|---|---|---|---|---|---|---|---|---|---|---|---|---|---|
| | | %conv | Time(s) | #s | #fails | %conv | Time(s) | #s | #fails | %conv | Time(s) | #s | #fails |
| small | Inst_15_11_7 | 100% | **0.00** | 2 | **$3.10^0$** | 100% | **0.00** | 2 | $7.10^0$ | 100% | 0.07 | 1 | $3.10^1$ |
| | Inst_20_18_8 | 100% | 0.07 | 7 | **$6.10^2$** | 100% | **0.04** | 7 | $7.10^2$ | 100% | 0.14 | 9 | $8.10^2$ |
| | Inst_25_20_8 | 100% | 0.01 | 4 | **$2.10^1$** | 100% | **0.00** | 4 | $4.10^1$ | 100% | 0.08 | 5 | $2.10^2$ |
| | Inst_30_21_10 | 100% | **0.03** | 3 | **$1.10^2$** | 100% | 0.05 | 3 | $7.10^2$ | 100% | 0.22 | 3 | $9.10^2$ |
| | Inst_35_28_8 | 100% | 0.04 | 1 | **$1.10^2$** | 100% | **0.02** | 1 | $1.10^2$ | 100% | 0.11 | 7 | $1.10^2$ |
| | Inst_40_31_8 | 100% | 0.45 | 14 | **$1.10^3$** | 100% | **0.32** | 13 | $3.10^3$ | 100% | 0.85 | 15 | $3.10^3$ |
| middle | Inst_45_27_8 | 100% | 3.48 | 4 | **$9.10^3$** | 100% | **2.37** | 4 | $2.10^4$ | 100% | 5.43 | 27 | $2.10^4$ |
| | Inst_50_32_7 | 100% | 15.35 | 24 | **$4.10^4$** | 100% | **11.01** | 24 | $1.10^5$ | 100% | 27.22 | 24 | $1.10^5$ |
| | Inst_55_35_6 | 100% | **4.61** | 11 | **$1.10^4$** | 100% | 15.88 | 12 | $1.10^5$ | 100% | 44.45 | 22 | $1.10^5$ |
| large | Inst_60_33_6 | 100% | 246.00 | 54 | **$3.10^5$** | 100% | **233.20** | 64 | $1.10^6$ | 97% | TO | 51 | $7.10^5$ |
| | Inst_70_43_5 | 100% | **120.38** | 25 | **$2.10^5$** | 100% | 185.05 | 25 | $1.10^6$ | 98% | TO | 23 | $5.10^5$ |
| | Inst_80_65_4 | **100%** | TO | 10 | $6.10^5$ | **100%** | TO | 11 | $1.10^6$ | **100%** | TO | 11 | **$4.10^5$** |
| | Inst_90_64_5 | **100%** | TO | 13 | $5.10^5$ | 99% | TO | 10 | $1.10^6$ | 97% | TO | 7 | **$4.10^5$** |
| | Inst_100_68_4 | **82%** | TO | 23 | $5.10^5$ | 79% | TO | 27 | $1.10^6$ | 75% | TO | 18 | **$4.10^5$** |

TO: timeout of 300s

points, Chuffed is the winner with a solution of %conv = 82%. Note that for the three largest instances, Chuffed returns a first solution of %conv > 80% in less than 5 s. Gecode returns a first solution of %conv > 70% in less than 60 s and SICStus returns a solution of %conv > 20% in less than 150 s. This is very good news as *RobTest* aims to quickly get a good-enough solution.

To sum up, our main finding is that *RobTest* is able to return good-enough solutions within a time contract of 300 s, which makes it suitable for industrial exploitation. The second remark is that *RobTest* using Chuffed outperforms Gecode and SICStus. An explanation could come from the usage of a SAT-solver within Chuffed, to generate lazy clauses. Explaining failures with efficient nogood recording could be a decisive advantage in our problem, as compared to a pure CP resolution.

# 6 Conclusion

This paper presents *RobTest,* a methodology and a tool for generating collision-free load-optimal test trajectories for industrial robots. The methodology exploits two CP solvers for this purpose: a CP-solver over continuous domains for generating the reachability graph between the robot's configurations and a CP-solver over finite domains to find load-optimal paths in a multi-labeled graph. Through an in-depth experimental evaluation, we bring answers to three research questions which are crucial to convince test engineers of the maturity of the approach and to prepare the operational deployment of *RobTest* . The experimental evaluation shows that deploying *RobTest,* in a continuous integration process is possible as we get good-enough quality solutions for graphs with more than 80 nodes in a short period of time (i.e., 5 min), typically used in those processes. The next step of this work is to deploy it at ABB Robotics and put the *RobTest* tool into the hands of test engineers on a more regular basis.

**Acknowledgment.** This work is mainly supported by the Research Council of Norway (RCN) through the T-Largo project (Project No.: 274786). Nadjib Lazaar is supported by the project CAR (UM - MUSE - 2020).

## References

1. Bui, Q.T., Deville, Y., Pham, Q.D.: Exact methods for solving the elementary shortest and longest path problems. Ann. Oper. Res. **244**(2), 313–348 (2016). https://doi.org/10.1007/s10479-016-2116-5
2. Carlsson, M., Ottosson, G., Carlson, B.: An open-ended finite domain constraint solver. In: Glaser, H., Hartel, P., Kuchen, H. (eds.) PLILP 1997. LNCS, vol. 1292, pp. 191–206. Springer, Heidelberg (1997). https://doi.org/10.1007/BFb0033845
3. Chu, G., Stuckey, P.J., Schutt, A., Ehlers, T., Gange, G., Francis, K.: Chuffed, a lazy clause generation solver. https://github.com/chuffed/chuffed
4. Cohen, Y., Stern, R., Felner, A.: Solving the longest simple path problem with heuristic search. In: Beck, J.C., Buffet, O., Hoffmann, J., Karpas, E., Sohrabi, S. (eds.) Proceedings of the Thirtieth International Conference on Automated Planning and Scheduling, Nancy, France, 26–30 October 2020, pp. 75–79. AAAI Press (2020). https://aaai.org/ojs/index.php/ICAPS/article/view/6647
5. Collet, M., Gotlieb, A., Lazaar, N., Mossige, M.: Stress testing of single-arm robots through constraint-based generation of continuous trajectories. In: Proceedings of the 1st IEEE Artificial Intelligence Testing Conference (AI Test 2019) (April 2019)
6. Desrochers, B., Jaulin, L.: Computing a guaranteed approximation of the zone explored by a robot. IEEE Trans. Autom. Control **62**(1), 425–430 (2017)
7. Faria, C., Ferreira, F., Erlhagen, W., Monteiro, S., Bicho, E.A.: Position-based kinematics for 7-DoF serial manipulators with global configuration control, joint limit and singularity avoidance. Mech. Mach. Theory **121**, 317–334 (2018)
8. Fieger, K., Balyo, T., Schulz, C., Schreiber, D.: Finding optimal longest paths by dynamic programming in parallel. In: Surynek, P., Yeoh, W. (eds.) Proceedings of the Twelfth International Symposium on Combinatorial Search, SOCS 2019, Napa, California, 16–17 July 2019, pp. 61–69. AAAI Press (2019). https://aaai.org/ocs/index.php/SOCS/SOCS19/paper/view/18329
9. Gasparetto, A., Boscariol, P., Lanzutti, A., Vidoni, R.: Trajectory planning in robotics. Math. Comput. Sci. **6**(3), 269–279 (2012)
10. Granvilliers, L., Benhamou, F.: Algorithm 852: RealPaver: an interval solver using constraint satisfaction techniques. ACM TOMS **32**(1), 138–156 (2006)
11. Jaulin, L.: Path planning using intervals and graphs. Reliable Comput. **7**(1), 1–15 (2001)
12. Latombe, J.C.: Robot Motion Planning, chap. 1. Kluwer Academic Publishers, Norwell (1991)
13. Malik, A.A., Bilberg, A.: Digital twins of human robot collaboration in a production setting. Procedia Manuf. **17**, 278–285 (2018). https://doi.org/10.1016/j.promfg.2018.10.047, http://www.sciencedirect.com/science/article/pii/S2351978918311636. 28th International Conference on Flexible Automation and Intelligent Manufacturing (FAIM2018), Columbus, OH, 11–14 June 2018, USA-Global Integration of Intelligent Manufacturing and Smart Industry for Good of Humanity

14. Merlet, J.P.: Interval analysis and reliability in robotics. Int. J. Reliab. Saf. **3**(1/2/3), 104–130 (2009)
15. Minguez, J., Laumond, J.P., Lamiraux, F.: Motion planning and obstacle avoidance. In: Siciliano, B., Khatib, O. (eds.) Springer Handbook of Robotics, pp. 827–852. Springer, Heidelberg (2008). https://doi.org/10.1007/978-3-540-30301-5_36
16. Mossige, M., Gotlieb, A., Meling, H.: Using CP in automatic test generation for ABB robotics' paint control system. In: Principles and Practice of Constraint Programming (CP 2014) - Application Track, Awarded Best Application Paper, Lyon (September 2014)
17. Nethercote, N., Stuckey, P.J., Becket, R., Brand, S., Duck, G.J., Tack, G.: MiniZinc: towards a standard CP modelling language. In: Bessière, C. (ed.) CP 2007. LNCS, vol. 4741, pp. 529–543. Springer, Heidelberg (2007). https://doi.org/10.1007/978-3-540-74970-7_38
18. Oetomo, O., Daney, D., Merlet, J.P.: Design strategy of serial manipulators with certified constraint satisfaction. IEEE Trans. Robot. **25**(1), 1V–11 (2009)
19. Pellegrinelli, S., Orlandini, A., Pedrocchi, N., Umbrico, A., Tullio, T.: Motion planning and scheduling for human and robot collaboration. CIRP Ann. - Manuf. Technol. **66**(1), 1–4 (2017)
20. Pham, Q.D., Deville, Y.: Solving the longest simple path problem with constraint-based techniques. In: Beldiceanu, N., Jussien, N., Pinson, É. (eds.) CPAIOR 2012. LNCS, vol. 7298, pp. 292–306. Springer, Heidelberg (2012). https://doi.org/10.1007/978-3-642-29828-8_19
21. Rohou, S., Jaulin, L., Mihaylova, L., Le Bars, F., Veres, S.M.: Guaranteed computation of robots trajectories. Robot. Auton. Syst. **93**, 76–84 (2017)
22. Schulte, C., Tack, G., Lagerkvist, M.: Modeling and Programming with Gecode. https://www.gecode.org/
23. Stilman, M.: Global manipulation planing in robot joint space with task constraints. IEEE Trans. Robot. **26**, 576–584 (2010)
24. Stuckey, P.J., Marriott, K., Tack, G.: MiniZinc manual (web resources). https://www.minizinc.org/doc-2.4.3/en/index.html

# A Two-Phase Constraint Programming Model for Examination Timetabling at University College Cork

Begum Genc[1(✉)] and Barry O'Sullivan[1,2]

[1]Confirm Centre for Smart Manufacturing and
[2]Insight Centre for Data Analytics,
School of Computer Science and Information Technology,
University College Cork, Cork, Ireland
{begum.genc,b.osullivan}@cs.ucc.ie

**Abstract.** Examination timetabling is a widely studied NP-hard problem. An additional challenge to the complexity of the problem are many real-world requirements that can often prevent the relaxation of some constraints. We report on a project focused on automating the examination timetabling process of University College Cork (UCC) to enhance the examination schedules so that they are fairer to student, as well as being less resource intensive to generate from an administrative point of view. We work with a formulation developed in collaboration with the institution and real data that it provided to us. We propose a two-phase constraint programming approach to solving UCC's examination timetabling problem. The first phase considers the timing of examinations while the second phase considers their allocation to rooms. Both phases are modelled using bin-packing constraints and, in particular, an interesting variant in which items can be split across multiple bins. This variant is known as *bin packing with fragmentable items*. We investigate the tightly linked constraints and difficulties in decomposing the centralised model. We provide empirical results using different search strategies, and compare the quality of our solution with the existing UCC schedule. Constraint programming allows us to easily modify the model to express additional constraints or remove the pre-existing ones. Our approach generates significantly better timetables for the university, as measured using a variety of real-world quality metrics, than those prepared by their timetabling experts, and in a reasonable timeframe.

## 1 Introduction

Scheduling is one of the best-known problems in optimisation. It is a broad field that includes, for example, nurse scheduling in hospitals, task scheduling in smart manufacturing, and timetabling in schools and universities [26,34,38]. Examination timetabling is a particularly hard variant that has been widely studied [8]. Many approaches to the problem have been studied, such as graph-based algorithms, local search or population-based algorithms, constraint-based


H. Simonis (Ed.): CP 2020, LNCS 12333, pp. 724–742, 2020.
https://doi.org/10.1007/978-3-030-58475-7_42

techniques, decomposition techniques, etc. [31]. For example, decomposition approaches often involve first assigning exams to time-slots and then distributing these exams across a set of rooms, creating sub-problems based on some groupings. Alternatively, local search approaches start with an initial schedule which is then subsequently improved by making local changes using a heuristic search and repair method [9,13,16,20,23,29]. We refer the reader to a number of recent literature reviews [12,31].

A variety of exact methods have been reported. McCollum et al. provide an easy-to-understand, integer programming model for the automated timetabling problem [22]. Arbaoui et al. improve on this model, mainly by introducing some pre-processing, clique and data-dependent dual-feasible function valid inequalities, and improving the bounds of some existing constraints [3]. There exist also some Constraint Programming (CP) approaches that construct an initial solution to be improved by using heuristics [14,23].

Approaches that decompose timetabling by separately assigning exams to time-slots and assigning the exams to rooms have existed for almost three decades. For example, Lotfi and Cerveny tackle the assignment of exams to time-slots using multiple phases [20]. They model the assignment of exams to rooms as a minimum-weighted matching problem which they solve using a heuristic approach. More recently, Mirrazavi et al. apply a similar two-staged approach to course timetabling in a university department using large-scale integer goal programming [27]. They first allocate lectures to rooms, then starting times to lectures. Although the decomposition means that optimality is sacrificed, it is still preferred by some institutions due to the effort spent by administrative staff in manually creating timetables [33].

For examination timetabling, many approaches ignore the allocation of exams to rooms. This is often due to institutions not allowing the splitting of exams across multiple rooms, or not allowing multiple exams to be held in the same venue at the same time [12,22,23]. Therefore, some models discard the allocation of exams to rooms entirely [4,37]. If the splitting is allowed, the solution is often obtained using meta-heuristic approaches. Unfortunately, in many real-life cases, it is not possible to disregard the room allocation phase. In fact, in the UCC case considered here, room allocation is one of the most interesting aspects of the problem. Many real-life applications of the examination timetabling problem using hybrid approaches can be found in the literature [18,28].

In this paper, we present a CP model for University College Cork's (UCC) examination timetabling process that respects all the requirements of the institution and produces solutions quickly. Although the solutions found by the proposed CP model are not necessarily optimal, the model improves the current schedule on all metrics the university wishes to apply by a considerable margin. Exact models often cannot produce solutions to large-scale problems within reasonable time-limits [18]. We tailored the exact model in [3,22] to UCC. However, UCC has some requirements that are significantly different from the existing use-cases. When the extra constraints such as the splitting of exams are added to the model, a solution could not be obtained within reasonable time-frame. Therefore, we developed a decomposition-based model for the problem which we present in this paper.

## 2 The UCC Examination Timetabling Problem

### 2.1 Problem Specification and Data

We summarise the main features of *University College Cork (UCC)* timetabling problem.[1] The problem comprises both hard and soft constraints. Throughout the paper, we refer to the requests given by the institution as the *requirements*, and the way we model them as *constraints*. Each hard requirement must be satisfied in each solution. On the other hand, soft requirements may be violated, but it is desirable to keep the violations to a minimum.

At the university there are 12,686 students sitting at least one exam. Of these, 855 students have some special needs that require use of a shared room (SHR), separate room (SPR), or lab (LAB). The SPR students must sit the exams alone in an SPR tagged room, but SHR (or LAB) students can share an SHR (or LAB) tagged room with other SHR (or LAB) students. We refer to the set of non-special needs students sitting exams as the *main group*, and the special needs students with their tags. We denote by $\mathbb{S}$ the set of all students. There are in total 43,002 exam seats required during the whole examination season. All examinations must be completed within a 10 day horizon. Each day has one morning time-slot and two afternoon time-slots. Morning time-slots are 180 min long, whereas the afternoon time-slots are 90 min each. We denote by $\mathbb{P}$ the set of all available time-slots. There are 717 exams, where 37 of them are 180 min, and 680 are 90 min long. We denote by $\mathbb{E}$ the set of all exams.

There are 9 *main rooms* available for the main groups with the following capacities that define the number of students that can be accommodated in a single sitting: 513, 513, 220, 171, 140, 130, 93, 91, and 56. One of the rooms with capacity 513 is called the $R_{ex}$ room which is located on a separate campus. We denote by $\mathbb{R}$ the set of all main rooms. Additionally, there are 31 SPR-tagged auxiliary rooms, 1 SHR-tagged room with capacity 50, and 1 LAB-tagged room with capacity 60. The sets of these auxiliary rooms are denoted by $\mathbb{R}^{SPR}$, $\mathbb{R}^{SHR}$, and $\mathbb{R}^{LAB}$, respectively. We list below the set of hard and the soft requirements denoted with the prefixes 'H' and 'S', respectively.

- H1 Each exam must be scheduled in exactly one time-slot where the duration of the exam is less than or equal to the duration of the time-slot.
- H2 Each student must sit at most one exam at any time-slot.
- H3 Each student sitting an exam must be allocated a seat.
- H4 An exam can be split across multiple rooms with the exception that the main groups sitting an exam in $R_{ex}$ cannot be split.
- H5 More than one exam can be held in the same main room during the same time-slot as long as all exams have the same duration and the total capacity is not exceeded.

[1] UCC dataset can be found in: http://github.com/begumgenc/ucc-et.

H6 An examination can have multiple module codes. These exams are identified by their *group-ids*. If two exams have the same group-id, they must be co-scheduled.
H7 Specific requests such as an exam to be held before or during a specific time-slot, or in a specific room, must be satisfied.
H8 Requests for an exam to be scheduled after another one must be satisfied.
H9 Special needs students must be assigned to rooms that are tagged with the corresponding tags. They cannot be seated in the same rooms as the main groups.
H10 Students must not sit more than 270 min of exams in total over any two consecutive days.
H11 If a student is scheduled for multiple exams on the same day, either all of those exams must be held in $R_{ex}$ or none of them.
H12 A number of seats in each main room must be left empty for accommodating ad-hoc students, i.e. students with unforeseen situations.

S1 Exams should be grouped within the same room as much as possible.
S2 Each student should sit at most one exam every two consecutive days.
S3 If Requirement S2 is not fully satisfied, then students should sit at most one exam a day.
S4 If Requirement S3 is not fully satisfied, then there should be at least one time-slot between each exam of each student.
S5 The use of some time-slots are more preferred to some other ones.
S6 It is desirable to schedule large exams as early as possible.
S7 It is desirable to minimise exam splits (i.e. the exam being held in more than one room) for main, SHR, and LAB groups.
S8 It is more preferred to use larger main rooms than smaller ones.

### 2.2 Notation and Terminology

In this section, we introduce the general notation to be followed in our model. A pair of exams, $e_i, e_j$, are said to be *conflicting* if there is at least one student who is enrolled for both $e_i$ and $e_j$. To capture the constraints related to conflicting exams, we make use of a *conflict graph* [22]. In the conflict graph, each vertex represents a unique exam. There exists an edge between two vertices if the exams corresponding to those vertices are conflicting. The weight of each edge $a = (e_i, e_j)$, denotes the total number of students sitting both $e_i$ and $e_j$. A *clique* in this graph, represents a set of exams, where all those exams are conflicting with each other, and they must all, therefore, be scheduled in different time-slots. We can find all maximal cliques in the conflict graph using the Bron-Kerbosch algorithm [6]. We describe below the main notation:

- The list $E = \langle e_0, ..., e_{|\mathbb{E}|-1} \rangle$ is an ordered list representation of exams in set $\mathbb{E}$.
- The list $P = \langle p_0, ..., p_{|\mathbb{P}|-1} \rangle$ is an ordered list representation of time-slots in set $\mathbb{P}$.

- The list $R = \langle r_0, ..., r_{|\mathbb{R}|-1} \rangle$ is an ordered list representation of main rooms in set $\mathbb{R}$. We also denote by $R^{SPR}$, $R^{SHR}$, $R^{LAB}$ the ordered lists of special needs rooms over the sets $\mathbb{R}^{SPR}$, $\mathbb{R}^{SHR}$, $\mathbb{R}^{LAB}$, similarly.
- The list $S = \langle s_0, ..., s_{|\mathbb{S}|-1} \rangle$ is an ordered list of students in set $\mathbb{S}$. For each student $s$, $E_s$ denotes the set of exams that student $s$ is enrolled, where $E_s \subset E$.
- List $D$ denotes the available days, in our case $D = \langle 0, .., |\mathbb{P}|/3 \rangle$. Each $d$ in $D$ spans exactly three time-slots $p_i, p_j, p_k \in P$ such that $\lfloor p_i/3 \rfloor = \lfloor p_j/3 \rfloor = \lfloor p_k/3 \rfloor = d$.
- Set $G$ corresponds to a set of group-ids. For each $g \in G$, $g$ denotes a group-id for a set of exams to be co-scheduled.
- Graph $\mathcal{G} = (\mathcal{V}, \mathcal{E})$ denotes the conflict graph, and $\mathcal{C}^{\mathcal{G}}$ denotes the set of all maximal cliques in $\mathcal{G}$ that contain more than one vertex.
- For each edge represented as $a = [e_i, e_j] \in \mathcal{E}$, $a$ denotes an edge having its source and target vertices corresponding to exams $e_i, e_j \in \mathcal{V}$. The weight of each $a$ is denoted by $w(a)$, and corresponds to the number of students sitting both $e_i$ and $e_j$.
- For each $e$ in $E$, $size^E_e$ denotes the total number of students sitting exam $e$. Additionally, $size^N_e$, $size^{SPR}_e$, $size^{SHR}_e$, $size^{LAB}_e$ denote the number of students sitting $e$ with the main group, or has SPR, SHR, and LAB tags, respectively.
- For each room $r$, $cap^R_r$ denotes the number of seats available in $r$. For each $p$ in $P$, $cap^P_p$, $cap^{SPR}_p$, $cap^{SHR}_p$, $cap^{LAB}_p$ correspond to the total number of available seats for main, SPR, SHR, LAB groups during $p$, respectively.
- For each period $p$ in $P$, $dur^P_p$ denotes the duration of time-slot $p$. Similarly, $dur^E_e$ denotes the duration of an exam $e$ in $E$.

Throughout the paper, we sometimes denote the elements by using indices when dealing with multiple items from the same set, e.g. $p \in P$ or $p_i, p_j \in P$. In these cases, let $I(X)$ be a set that denotes the index set of a given set $X$. Then, for any element $x_i$ in a set $X$, $i \in I(X)$. The list of exams $E$ is ordered in descending order of the number of non-special needs students sitting each exam. Additionally, the time-slot list $P$ is ordered in ascending order of the timings of the corresponding time-slots, i.e. for any $i, j \in I(P)$ with $i < j :\rightarrow p_i < p_j$, $p_i$ is an earlier time-slot than $p_j$. Finally, the list of rooms (i.e. $R$) is presented in descending order of the capacity of each room.

## 3 The Constraint Programming Models

In this section we present in detail our two CP models, one for each phase of the problem. Initially, as a pre-processing step, we create a conflict graph to keep track of conflicting exams efficiently. Subsequently, we use our Phase 1 model that is based on bin-packing global constraint for determining the timing of examinations. Finally, in Phase 2, we use the solution generated by Phase 1 to assign the examinations to rooms within the given time-slot. We model Phase 2 using a variant of the bin-packing constraint.

### 3.1 Phase 1: Allocation of Exams into Timeslots

We find timings for each exam using bin-packing global constraints [36]. The exams correspond to items, and the time-slots correspond to bins. The capacity of each bin is the total number of available seats during the corresponding time-slot. In our case, all the bins have equal capacities. The bin-packing problem is well-known to be NP-hard [15]. We list below the main variables of our Phase 1 CP model.

- For each $e$ in $E$, an integer variable $\boldsymbol{EP[e]}$ denotes the time-slot in which $e$ is scheduled, where $EP[e] \in P$. The search branches on this variable.
- For each $e$ in $E$, an integer variable $\boldsymbol{ED[e]}$ denotes the day in which $e$ is scheduled, where $ED[e] \in D$.
- For each $p$ in $P$, integer variables $\boldsymbol{L[p]}$, $\boldsymbol{L^{SPR}[p]}$, $\boldsymbol{L^{SHR}[p]}$, $\boldsymbol{L^{LAB}[p]}$ denote the total number of students in main, SPR, SHR, and LAB groups, sitting an exam at $p$, resp. Note that, $L[p] \in \{0, ..., \sum_{r \in \mathbb{R}} cap_r^R\}$, $L^{SHR}[p] \in \{0, ..., \sum_{r \in \mathbb{R}^{SHR}} cap_r^R\}\}$, $L^{LAB}[p] \in \{0, ..., \sum_{r \in \mathbb{R}^{LAB}} cap_r^R\}\}$, and $L^{SPR}[p] \in \{0, ..., |\mathbb{R}^{SPR}|\}$.
- For each exam $e \in E$, a boolean variable $\boldsymbol{pl[e]}$ denotes if $e$ is large, and scheduled late or not, where $p_\sigma$ denotes the time-slot after which the exams are considered late.
- For each edge $a \in \mathcal{E}$, integer variables $\boldsymbol{V_{S2}[a]}, \boldsymbol{V_{S3}[a]}, \boldsymbol{V_{S4}[a]}$ denote for each soft requirement Requirements $S2$, $S3$ and $S4$, the number of occurrences a student is negatively affected due to the violation of the respective requirement by the exam pair $a$. Note that, $V_{S2}[a], V_{S3}[a], V_{S4}[a] \in \{0, w(a)\}$.
- For each $p \in P$, an integer variable $\boldsymbol{V_{S5}[p]}$ denotes the penalty incurred due to the violation of Requirement $S5$, where $\alpha_p$ represents a constant penalty for using $p$. Note that, $V_{S5}[p] \in \{0, ..., cap_p^P \times \alpha_p\}$.
- Sub-objective values $\boldsymbol{obj_2}, \boldsymbol{obj_3}, \boldsymbol{obj_4}, \boldsymbol{obj_5}, \boldsymbol{obj_6}$ denote the total penalty caused by the violation of Requirements $S2, S3, S4, S5$, and $S6$, respectively. The overall objective $\boldsymbol{obj}$ is a weighted sum of the sub-objective values.

*Timeslot Capacities.* Constraint 1 uses a global bin-packing constraint in the form BIN_PACKING (*items*, *item_weights*, *bin_loads*, *bin_capacities*) [36]. It ensures that each exam is assigned to a time-slot, and each time-slot has enough seats to accommodate the students sitting the assigned exams considering their corresponding tags. (see Requirements H1, H3 and H5). Subsequently, Constraint 2 ensures that each exam respects the duration of the time-slot it is assigned to (see Requirement H1) [32].

$$\begin{gathered}\text{BIN\_PACKING}(EP, size^N, L, cap^P),\\ \text{BIN\_PACKING}(EP, size^{SPR}, L^{SPR}, cap^{SPR}),\\ \text{BIN\_PACKING}(EP, size^{SHR}, L^{SHR}, cap^{SHR}),\\ \text{BIN\_PACKING}(EP, size^{LAB}, L^{LAB}, cap^{LAB}).\end{gathered} \tag{1}$$

$$\forall e \in E, \forall p \in P : \text{if } dur_p^P < dur_e^E, \text{ then } EP[e] \neq p. \tag{2}$$

*One Exam at a Time.* Let $\mathcal{C}^{\mathcal{G}}$ denote the set of all maximal cliques on the conflict graph $\mathcal{G}$. For each clique $c$, $EP_c \subset EP$ denotes a subset of decision variables that correspond to the ones for the exams in $c$, such that $\forall c \in \mathcal{C}^{\mathcal{G}}, \forall e \in c : EP[e] \in EP_c$. Constraint 3 ensures conflicting exams are scheduled into a different time-slot (see Req. H2).

$$\forall c \in \mathcal{C}^{\mathcal{G}} : \text{ALL\_DIFFERENT}(EP_c). \tag{3}$$

*Specific Exam Times.* Constraint 4 ensures that the exams with same group id are co-scheduled (see Requirement H6) [17]. We denote by $EP_g$ a subset of the decision variables that have the same group-id $g$. More formally, the set of exams to be co-scheduled denoted by $EP_g$, is defined as $\forall g \in G, \forall e$ with group-id $g : EP[e] \in EP_g$.

$$\forall g \in G : \text{ALL\_EQUAL}(EP_g). \tag{4}$$

*Specific Requests.* Req. H7 and H8 can be easily modelled over the decision variables of the relevant exams using equality and inequality constraints. For instance, for Req. H8, let $H_p$ denote a set of exam pairs that indicate priority as: $\langle e_i, e_j \rangle \in H_p$, where $e_i$ must be scheduled before $e_j$. We model them as follows: $\forall \langle e_i, e_j \rangle \in H_p : EP[e_i] < EP[e_j]$.

*Time-Slot Spread.* There are four requirements related to time-slot spread (see Requirements S2, S3, S4, and H10). Recall that, there are exactly three time-slots in each day. Hence, Constraint 5 expresses the day on which an exam is scheduled. Consequently, we make use of some logical constraints to calculate the penalties incurred by the violation of Requirement S2 by Constraint 6, of Requirement S3 by Constraint 7, and of Requirement S4 by Constraint 8. Note that, UCC does not hold exams at the weekend. Thus, when calculating the violation of Requirement S4 by Constraint 8, we ignore the cases where two time-slots are separated by the weekend.

$$\forall e \in E : ED[e] = \lfloor EP[e]/3 \rfloor. \tag{5}$$

$$\forall a = [e_i, e_j] \in \mathcal{E} : \text{if } |ED[e_i] - ED[e_j]| < 2, \text{ then } V_{S2}[a] = w(a), \text{ else } V_{S2}[a] = 0. \tag{6}$$

$$\forall a = [e_i, e_j] \in \mathcal{E} : \text{if } ED[e_i] == ED[e_j], \text{ then } V_{S3}[a] = w(a), \text{ else } V_{S3}[a] = 0. \tag{7}$$

$$\begin{aligned}\forall a = [e_i, e_j] \in \mathcal{E} : \text{if } |EP[e_i] - EP[e_j]| < 2 \wedge \neg weekend(EP[e_i], EP[e_j]),\\ \text{then } V_{S4}[a] = w(a), \text{ else } V_{S4}[a] = 0.\end{aligned} \tag{8}$$

Constraint 9 ensures that Requirement H10 is satisfied. In order to avoid repeating, we first analyse the sets of exams $E_s$ for each student $s$ as a pre-processing step. If there are any repeated sets, or if a set is already included in a larger set, we eliminate the repetition, or the subset. We denote by $A^E$ the refined set of the $E_s$ that does not include any repeated sequences. For Constraint 9, let each set of exams in $A^E$ be denoted by $l$, where for some student $s$, $l = E_s$. Note that, UCC only has exams of duration 90 or 180 min. For each $l$, for any consecutive two-day block $b_i = \langle d_i, d_{i+1} \rangle$, let $len^l_{b_i}[e]$ be an integer variable with domain $\{0, 90, 180\}$ that denotes the duration of exam $e$ for all $e \in l$. The value of $len^l_{b_i}[e] = 0$ if the exam $e$ is not scheduled for the days $d_i$ or $d_{i+1}$ in the given block $b_i$. Otherwise, $len^l_{b_i}[e] = dur^E_e$.

$$\begin{gathered} \forall b_i = \{d_i, d_{i+1}\} \text{ with } 0 \leq i < |D| - 1, \forall l \in A^E : \\ \forall e \in l : \text{ if } ED[e] \in b_i, \text{ then } len^l_{b_i}[e] = dur^E_e, \text{ else } len^l_{b_i}[e] = 0, \\ \textstyle\sum_{e \in l} len^l_{b_i}[e] \leq 270. \end{gathered} \tag{9}$$

*Time-Slot Preference.* Some time-slots are more preferred than others, as discussed in Requirement S5. Therefore, each time-slot $p$ has an associated penalty coefficient $\alpha_p$. We calculate the penalty caused by scheduling exams into less preferred time-slots $p$ as the total number of students sitting an exam during $p$ multiplied by $\alpha_p$, as shown in Constraint 10.

$$obj_5 = \sum_{p \in P} (L[p] + L^{SPR}[p] + L^{SHR}[p] + L^{LAB}[p]) \times \alpha_p \tag{10}$$

Let $F_s$ denote a value representing the size of an exam provided by the institution that determines if an exam is considered large or not; exams with sizes larger than or equal to $F_s$ are said to be *large*. Similarly, let $p_\sigma$ denote the time-slot after which exams are considered as being scheduled late. For each exam $e$, $pl[e]$ denotes if a penalty is incurred by scheduling of exam $e$. We capture Requirement S6 using Constraint 11. The value of $obj_6$ represents the overall penalty incurred by the violation of this requirement.

$$\begin{gathered} \forall e \in E \text{ with } size^E_e \geq F_s : \text{ if } EP[e] - p_\sigma > 0, \text{ then } pl[e] = 1, \text{ else } pl[e] = 0. \\ obj_6 = \sum_{e \in E \text{ with } size^E_e \geq F_s} pl[e]. \end{gathered} \tag{11}$$

*Objective Function.* The objective is to minimise the number of students that are affected by the soft constraint violations. The overall objective function is defined as a weighted sum of all the aggregated penalties $obj_2, obj_3, obj_4, obj_5, obj_6$ multiplied by their respective coefficients $c_{p2}, c_{p3}, c_{p4}, c_{p5}, c_{p6}$. Constraint 12 expresses the objective of Phase 1 model.

$$\begin{gathered} obj_2 = \sum_{a \in \mathcal{E}} V_{S2}[a], obj_3 = \sum_{a \in \mathcal{E}} V_{S3}[a], obj_4 = \sum_{a \in \mathcal{E}} V_{S4}[a]. \\ minimise \;\; obj = c_{p2} \times obj_2 + c_{p3} \times obj_3 + c_{p4} \times obj_4 + c_{p5} \times obj_5 + c_{p6} \times obj_6. \end{gathered} \tag{12}$$

### 3.2 Phase 2: Room Allocation

In Phase 2, the aim is to find a room for each exam by respecting the room-related requirements. We further decompose the model at this stage by days and find a room for each exam for each day. The problem we encounter at this stage is a version of the bin-packing problem, where the items can be fragmented (or split). In the literature, this problem is known as Bin Packing with Fragmentable Items (BPFI) and is NP-hard [21]. The objective in the BPFI problem is to either minimise the number of fragments or to minimise the number of bins required. In our case, we have both of these as objectives (see Requirements S1 and S7). The literature contains some approximation algorithms as well as MIP models for the problem [10,11,19]. Additionally, UCC has a hard requirement that mixed-duration exams cannot be timetabled in the same venue (see Requirement H5). This constraint makes the problem even more constrained, as not all the items can be placed in all bins. This version of the problem has previously been identified as Fragmentable Group and Item Bin Packing with Compatibility Preferences [2]. Although these bin-packing variants have been identified and tackled before, there are no global constraints available that express them.

- For each exam $e$ in $E$, constant $\boldsymbol{EP_e} \in P$ denotes the time-slot in which exam $e$ is held, and constant $\boldsymbol{ED_e} \in D$ denotes the day in which $e$ is held. These are provided as input to the model by the Phase 1 solution.
- An integer value $\boldsymbol{R_{ex}}$ in $R$ denotes the unique identifier of $R_{ex}$ room.
- For each $p$ in $P$ in any room $r$, depending on the special needs tag of the room, one of the following integer variables: $\boldsymbol{RL[p][r]}$, $\boldsymbol{RL^{SPR}[p][r]}$, $\boldsymbol{RL^{SHR}[p][r]}$, or $\boldsymbol{RL^{LAB}[p][r]}$ is defined to denote the total number of students with the given tag sitting an exam in room $r$ during time-slot $p$. Note that, the domain of these variables is $\{0, ..., cap_r^R\}$.
- For each $p \in P$ and for each room $r$ for main groups in $R$, set variable $\boldsymbol{D[p][r]}$ denotes the different duration types of exams to be held in $r$ during $p$. Note that, $D[p][r] \in \mathbb{P}(\{90, 180\})$.
- For each room $r$ and $e \in E$, one of the following integer variables is defined depending on the tag of the room: $\boldsymbol{W[r][e]}, \boldsymbol{W^{SPR}[r][e]}, \boldsymbol{W^{SHR}[r][e]}$, or $\boldsymbol{W^{LAB}[r][e]}$ to denote the total number of students with the given tag sitting $e$ in $r$. The domains are defined as $W[r][e] \in \{0, ..., min(cap_r^R, size_e^N)\}$, $W^{SHR}[r][e] \in \{0, ..., min(cap_r^R, size_e^{SHR})\}$, $W^{SPR}[r][e] \in \{0, 1\}$, and $W^{LAB}[r][e] \in \{0, ..., min(cap_r^R, size_e^{LAB})\}$. Search branches on these variables.
- For each $e \in E$, set variables $\boldsymbol{RS[e]}$, $\boldsymbol{RS^{SHR}[e]}$, and $\boldsymbol{RS^{LAB}[e]}$ denote the set of rooms used for examinations of the main, SHR, and LAB groups for exam $e$, where $RS[e] \in \mathbb{P}(R)$, $RS^{SHR}[e] \in \mathbb{P}(R^{SHR})$, and $RS^{LAB}[e] \in \mathbb{P}(R^{LAB})$.
- Sub-objectives $\boldsymbol{obj_1}, \boldsymbol{obj_2}, \boldsymbol{obj_3}$ are used to denote the total number of exam splits for main, SHR, and LAB groups, respectively. Additionally, the sub-objective $\boldsymbol{obj_4}$ denotes the total number of main rooms used during the examination process. The overall objective $\boldsymbol{obj}$ is a weighted-sum of the sub-objectives.

We model bin packing with fragmentable items using a straight-forward approach under the name BIN_PACKING_FI($X, I_W, L, C$) using Constraint 13 and Constraint 14. Variable $X$ corresponds to the weight of each item in each available bin given item set $I$ and bin set $B$. Additionally, $I_W$ denotes the weight of each item, $L$ denotes the load of each bin, and $C$ denotes the capacity of each bin. Constraint 13 ensures the weight of each item in each bin respects the capacity of the bin. Additionally, Constraint 14 ensures all the weight of each item is packed. Note that, BIN_PACKING_FI can be improved by using better models or heuristic approaches [7,11].

$$\forall b \in B : \sum_{i \in I} X[b][i] = L[b] \text{ and } L[b] \leq C[b]. \tag{13}$$

$$\forall i \in I : \sum_{b \in B} X[b][i] = I_W[i]. \tag{14}$$

Recall that we decompose this phase further by separately assigning exams in each day to rooms. Thus, the remaining constraints in this section are run independently for each day using the set of exams given on that day. We denote by $d$, the day for which the problem is being solved and by $p \in d$ the three time-slots that belong to $d$.

Constraint 15 ensures each student in each exam is assigned a room by respecting Requirements H3, H5, and H9. For each period, let $W_p$ be a list of decision variables that correspond to a list of $W[r][e]$ variables for exams $e$ with $EP_e = p$ in rooms $r \in R$. Similarly, let $W_p^{SPR}$, $W_p^{SHR}$, and $W_p^{LAB}$ denote decision variables $W^{SPR}[r][e], W^{SHR}[r][e]$, and $W^{LAB}[r][e]$, for each exam $e$ with $EP_e = p$ in room $r$ with the respective tag. We denote by $C^N, C^{SPR}, C^{SHR}, C^{LAB}$ the list of capacities of rooms with the respective tag, where $N$ corresponds to the main group. Constraint 15 is repeated for each time-slot $p$ in the given day $d$.

$$\begin{aligned}
\forall p \in d &: \text{BIN_PACKING_FI}(W_p, size^N, RL, C^N).\\
\forall p \in d &: \text{BIN_PACKING_FI}(W_p^{SPR}, size^{SPR}, RL^{SPR}, C^{SPR}).\\
\forall p \in d &: \text{BIN_PACKING_FI}(W_p^{SHR}, size^{SHR}, RL^{SHR}, C^{SHR}).\\
\forall p \in d &: \text{BIN_PACKING_FI}(W_p^{LAB}, size^{LAB}, RL^{LAB}, C^{LAB}).
\end{aligned} \tag{15}$$

*No Mixed Durations.* We implement item-bin compatibility using Constraint 16. If multiple exams are held in the same room during the same time-slot, their duration must be the same (see Requirement H5). In order to achieve this, Constraint 16 uses $D[p][r]$ that keeps track of the different durations of exams being held in room $r$ at time-slot $p$.

$$\begin{aligned}
\forall p \in d, \forall r \in R, \forall e \in E \text{ with } EP_e = p &: r \in RS[e] \rightarrow dur_e^E \in D[p][r].\\
\forall p \in d, \forall r \in R, \forall e \in E \text{ with } EP_e = p &: |D[p][r]| \leq 1.
\end{aligned} \tag{16}$$

*Room Loads and Different Campus $R_{ex}$.* In order to implement the remote campus requirements, i.e. $R_{ex}$, and to find the number of rooms in use, we keep track of the set of rooms $RS[e]$ in which each exam $e$ is held. If there are a number of students at exam $e$ in room $r$, then $r$ must be included in the set of rooms of $e$ as detailed in Constraint 17.

$$\begin{aligned} \forall e \in E, \forall r \in R \text{ with } ED_e = d : W[r][e] > 0 \leftrightarrow r \in RS[e]. \\ \forall e \in E, \forall r \in R^{SHR} \text{ with } ED_e = d : W^{SHR}[r][e] > 0 \leftrightarrow r \in RS^{SHR}[e]. \\ \forall e \in E, \forall r \in R^{LAB} \text{ with } ED_e = d : W^{LAB}[r][e] > 0 \leftrightarrow r \in RS^{LAB}[e]. \end{aligned} \tag{17}$$

Shachnai et al. describe a primitive solution to a BPFI as a feasible packing such that each bin contains at most two fragmented items, and each item is fragmented at most once [35]. They also show that there always exists an optimal solution for BPFI, that is primitive. Constraint 18 ensures each exam is split at most once.

$$\forall e \in E : RS[e] \le 2 \text{ and } RS^{SHR}[e] \le 2 \text{ and } RS^{LAB}[e] \le 2. \tag{18}$$

Constraint 19 ensures for any two conflicting exams that are to be held on the same day, either both are held in $R_{ex}$ or none of them (see Requirement H11). Additionally, Constraint 20 implements the second $R_{ex}$ constraint, that states if an exam is held in $R_{ex}$, then it is not split across other rooms (see Requirement H4).

$$\forall a = [e_i, e_j] \in \mathcal{E} \text{ with } ED_{e_i} = ED_{e_j} = d : R_{ex} \in RS[e_i] \leftrightarrow R_{ex} \in RS[e_j] \tag{19}$$

$$\forall e \in E : R_{ex} \in RS[e] \rightarrow |RS[e]| = 1 \tag{20}$$

*Objective Function.* The sub-objectives in this model are to minimise the exam splits for main, SHR, and LAB groups ($obj_1, obj_2, obj_3$, resp.), and to also minimise the total number of main rooms in use ($obj_4$). Constraint 21 formulates these objectives. The objective function uses coefficients $c_{r1}$, $c_{r2}$, $c_{r3}$, and $c_{r4}$ as the weight of each sub-objective.

$$\begin{aligned} obj_1 &= \sum_{e \in E \text{ with } EP_e = p} |RS[e]|, \quad obj_2 = \sum_{e \in E \text{ with } EP_e = p} |RS^{SHR}[e]|, \\ obj_3 &= \sum_{e \in E \text{ with } EP_e = p} |RS^{LAB}[e]|, \quad obj_4 = \sum_{p \in d} | \bigcup_{e \in E \text{ with } EP_e = p} RS[e]|. \\ & minimise\ obj = c_{r1} \times obj_1 + c_{r2} \times obj_2 + c_{r3} \times obj_3 + c_{r4} \times obj_4. \end{aligned} \tag{21}$$

### 3.3 Discussion of the Decomposition Approach

One of the main drawbacks of decomposed models is that obtaining the optimal solution or any solution, in general, is not always guaranteed. The main challenge for our CP model is that a solution produced by the Phase 1 model may lead to infeasibility in Phase 2. We illustrate a scenario below to exemplify this for the reader.

*An Infeasibility Scenario.* Suppose that there are four exams $e_1, e_2, e_3$, and $e_4$ to be held during the same time-slot, where each exam has three students enrolled with no special needs. There are only three rooms $r_1, r_2$, and $r_3$ with equal capacities $c_1 = c_2 = c_3 = 4$. Let exam $e_4$ be 180-min long and the rest be 90-min. Also let room $r_1$ denote the $R_{ex}$ room. Although this solution is a feasible solution for Phase 1, it is infeasible for the Phase 2 model. An illustration of two alternative exam-room allocation scenarios is given in Fig. 1.

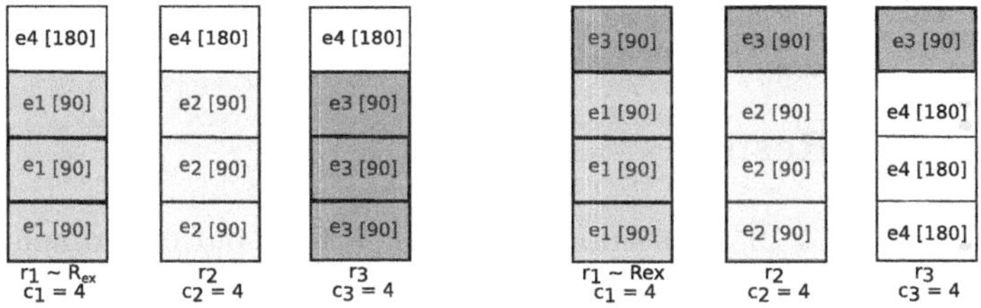


**Fig. 1.** Two different infeasibility scenarios where the allocation of the exams in the rooms causes the violation of Requirements H4 and H5.

Both of these cases results in infeasibility for the Phase 2 model due to the violation of Requirement H4 and H5, where an exam to be held in $R_{ex}$ is split, and also there are mixed-duration exams in at least one room. A similar, alternative infeasibility scenario occurs when all the exams are 90-min long, but the time-slot load is filled to the capacity. In this case, one of the exams must be split to $R_{ex}$, violating Requirement H4.

There are different strategies to tackle this problem. One approach is to develop a Benders decomposition of the problem. However, this approach is challenging as Phase 2 is a hard problem in itself, but imposes an interesting research direction. An alternative approach is to relax some of the hard constraints in the Phase 2 model. As discussed above, when the load of a time-slot is large, it becomes more difficult to find a solution for Phase 2 due to the hard "no mixed duration" requirement (see Requirement H5) and $R_{ex}$ requirements (see Requirements H4 and H11). Hence, if the Phase 1 solution has a large load for some time-slots, then the "no mixed duration" constraint (see Constraint 16) and/or $R_{ex}$ constraints (see Constraints 19 and 20) may be relaxed and their penalties can be calculated and added to the objective for minimisation.

Alternatively, we can keep the load of each period at an 'acceptable level', where this level is defined with respect to the dataset in use. If the time-slot loads are kept as small as possible, then some rooms can be left unused. A basic solution that emanates from this observation is to keep the load of each time-slot to a value that avoids the use of $R_{ex}$ room, and also accommodate some empty-seats to help with the "no mixed duration" requirement. Hence, we add a sub-objective to minimise the maximum excess load for time-slots, where the excess load is calculated over the main groups as the difference between the load and the mean value of seats required for each period. It can then be possible to find schedules without relaxing any of the hard constraints. Note that, by minimising the maximum excess load, the model can leave some rooms empty. Therefore, this sub-objective also helps with minimising the total number of used rooms in Phase 2. Recall that in Phase 1, for each time-slot $p, L[p]$ denotes the load of $p$. Minimising the excess load can be achieved by adding Constraint 22 to the Phase 1 model, and also adding sub-objective $obj_1$ to $obj$ by using a coefficient $c_{p1}$.

$$mean = \sum_{r \in R} cap_r^R/|P|, \text{ and } obj_1 = \text{MAX}(L) - mean. \tag{22}$$

## 4 Empirical Results

We used the Choco 4.10.2 constraint solver to perform our experiments [30]. We used a library of graph theory data structures and algorithms called JGraphT for the creation of conflict graph, and for finding all maximal cliques [24]. All experiments were performed on a Dell i7-8700 machine with 3.20 GHz processors under Linux.

The model we developed was given 60-min in total to produce a solution, spending 45 min on the assignment of examinations to time-slots, and the remaining 15 min on the examination-room assignment where each day is limited to produce a solution in 1.5 min. Note that, none of the solutions we report in this section was proven to be optimal by the solver. Therefore, it may be possible to obtain better solutions, i.e. solutions with lower objective values, if more time is allowed for the models or some filtering, symmetry breaking, and redundant constraints, are added to the model.

We tested the Phase 1 model using four different built-in search strategies selected by observation and considering the nature of the problem: activity-based search ($\boldsymbol{s_{act}}$), dominated weighted degree ($\boldsymbol{s_{wdeg}}$), assigning the first non-instantiated variable to its lower bound ($\boldsymbol{s_{ilb}}$), and assigning the non-instantiated variable of smallest domain size to its lower bound ($\boldsymbol{s_{mdlb}}$) [5,25]. In addition to each search strategy, we used a large-neighbourhood search based on randomly creating neighbours using a restart strategy based on fails. The search also makes use of a last conflict heuristic and geometrical restarts. All search strategies used are available in the Choco library.

In the UCC data, the mean of total seats required per time-slot is reasonably low: it is approximately 1341 for main, 9 for SPR, 28 for LAB, and 54 for SHR groups. The total number of available seats for the main groups per time-slot is 1900, and there are 31 seats for SPR, 57 for LAB students, 47 for SHR students, where 3 seats from each room are left empty for ad-hoc students (see Requirement H12). It is important to note that the model with the specific venue for SHR students is not enough and renders the model infeasible. Therefore, we manually add an additional venue for SHR students.

In Table 1, we present the performance of the four search strategies mentioned above on the Phase 1 model. The coefficients for the objectives are used as follows: $c_{p1} = 25$, $c_{p2} = 3$, $c_{p3} = 40$, $c_{p4} = 10$, $c_{p5} = 2$, and $c_{p6} = 60$. We use $F_s = 100$ to decide if an exam is large or not. In UCC, afternoon time-slots in Days 7 and 8, and also all time-slots on Days 9 and 10, are less preferred to earlier ones. It is preferred to not schedule any examinations during the last time-slot of Days 9 and 10. So, we penalise the preference of time-slots as: $\alpha_{20} = \alpha_{23} = \alpha_{24} = \alpha_{25} = \alpha_{27} = \alpha_{28} = 1$, and $\alpha_{26} = \alpha_{29} = 3$. We use large coefficients to penalise the sub-objectives $obj_1$ and $obj_3$ in Phase 1 for an increased chance of avoiding the infeasibility scenario discussed in Sect. 3.3. The penalty values were selected based on the observation that reflects the requirements and priorities of the institution. Although the importance of $obj_3$ is less than $obj_2$ in the given requirements, penalising $obj_3$ is essential for providing a better resolution between the phases for the problem. Additionally, one can observe that if $obj_1$ has a larger value, then the time-slot preferences ($obj_5$) are violated less. The CP model for Phase 1 contains 527,831 variables and 367,303 constraints in total.

**Table 1.** Performance comparison of different search strategies for Phase 1 model on UCC dataset.

| | $s_{act}$ | $s_{wdeg}$ | $s_{ilb}$ | $s_{mdlb}$ |
|---|---|---|---|---|
| Best-solution objective ($obj - b$) | 60,203 | 64,254 | 60,888 | 62,514 |
| Load balancing penalty ($obj_1$) | 292 | 135 | 218 | 145 |
| Maximum 1 exam every 2 days penalty ($obj_2$) | 7,897 | 9,079 | 7,294 | 8,587 |
| Maximum 1 exam per day penalty ($obj_3$) | 290 | 297 | 325 | 283 |
| Back-to-back exams penalty ($obj_4$) | 152 | 178 | 252 | 139 |
| Time-slot preference penalty ($obj_5$) | 6,756 | 8,611 | 7,788 | 8,649 |
| Front load penalty ($obj_6$) | 43 | 46 | 41 | 52 |
| Best-solution time ($time_b$ in sec) | 1,308.4 | 1,335.4 | 2,313.7 | 1,158.3 |
| Ratio: feasible/total solutions ($cnt_{feas}/cnt_{sol}$) | 5/5 | 11/11 | 5/5 | 10/10 |
| First-solution objective ($objf$) | 64,997 | 71,024 | 61,994 | 69,978 |
| First-solution time ($time_f$ in sec) | 7 | 10 | 8.8 | 8.8 |

We observe from Table 1 that the $s_{ilb}$ strategy has a good starting point but $s_{act}$ is able to find a better solution in the end. Recall that we branch on $EP$ variables that allocate exams to time-slots starting from the largest exam to the smallest one. Thus, assigning the first non-instantiated variable to its lowest bound corresponds to scheduling a large exam to an earlier time-slot. Therefore, we observe a clear advantage for $s_{ilb}$ in $obj_6$, that corresponds to the front load penalty. In each case, the solver can find an initial solution for the Phase 1 model within 10 s . Also note that in this table we report how many feasible solutions there are ($cnt_{feas}$) among all solutions found by each strategy ($cnt_{sol}$).

We measure feasibility by running the Phase 2 algorithm for each solution found in Phase 1. We immediately stop the search when an initial solution is found by the Phase 2 model. The time required by this feasibility check using the $s_{iub}$ strategy in Phase 2 for a solution obtained from Phase 1 took on the average 1.2 s, including building the model and search, with the worst time being 12.2 s; details are discussed below. We allow this feasibility check to run for a maximum of 1.5 min per day. If no solutions are found within the given time limit or if the model is proven as infeasible, we say that the Phase 1 solution is infeasible, and increment $cnt_{feas}$. In our experiments, all the solutions found in Phase 1 were feasible. For comparison, we evaluated the original UCC timetable for Phase 1 by ignoring $obj_1$ as it is not an original constraint given by the institution, which minimises the excess load. When we evaluated the corresponding $obj_2, obj_3, obj_4, obj_5$, and $obj_6$ values with the coefficients mentioned above for Phase 1, the original timetable is found to have an objective value $obj = 117,342$. This value is nearly twice worse than any of the initial solutions found by CP in less than 10 s.

Our Phase 2 model per day has on the average 10,592 variables and 5,139 constraints. We allow Phase 2 to run for 15 min in general, which results in each day reporting a solution in 1.5 min. For a search strategy comparison for Phase 2, we use the best solution found by Phase 1 model, i.e. the one with the lowest $obj$ value, and use it as input for the Phase 2 model. We again test four different search strategies on this model: activity based search ($\boldsymbol{s_{act}}$), dominated weighted degree ($\boldsymbol{s_{wdeg}}$), assigning the first non-instantiated variable to its upper bound ($\boldsymbol{s_{iub}}$), and assigning the non-instantiated variable of smallest domain size to its upper bound ($\boldsymbol{s_{mdub}}$). Note that, we do not use the strategies that assign a variable to its lower bound, but we use the assignment to upper bound in Phase 2. Considering that we branch on the decision variables that denote the number of students sitting an exam in each room, an upper bound strategy assigns the maximum possible number of students into the next available room in a Next-Fit manner. A split of the main groups are heavily penalised in UCC, so we use the following coefficients for the penalisation of sub-objective values in Phase 2: $c_{r1} = 4, c_{r2} = c_{r3} = c_{r4} = 1$. The search strategy $s_{act}$ was unable to find a solution to 1/10 days, $s_{wdeg}$ to 6/10 days, and $s_{mdub}$ to 8/10 days, within the given time-limit. The strategy $s_{iub}$ has a clear advantage over the other strategies as it finds a very quick solution to each day by assigning as many students as possible from the same group into the first available room. The aggregated solution found by $s_{iub}$ has $obj_1 = 8$, $obj_2 = 0$, $obj_3 = 0$, $obj_4 = 156$, and all the hard constraints are satisfied.

Table 2 summarises the improvements over the UCC's original schedule by the final schedule found by our model. In this table, $\mathcal{M}$ represents a solution found by our CP model using the combination of $s_{act}$ for Phase 1 and $s_{iub}$ for Phase 2. Note that, if desired, using different penalty values for the soft constraints, it is possible to adjust the model further to obtain improvement on a specific value. The model we propose effectively reduces the violation of soft requirements related to student preferences. In model $\mathcal{M}$, the average preparation time for students between their consecutive exams is increased. Additionally, it reduces the total number of rooms in use, which helps the institution lower costs, and

**Table 2.** A comparison of the expert-generate schedule at UCC with our proposed schedule ($\mathcal{M}$).

| Number of ... | UCC | $\mathcal{M}$ |
|---|---|---|
| Students that sit more than one exam in the same time-slot | 7 | 0 |
| Conflicting exam pairs and held during the same time-slot | 5 | 0 |
| <Main, SPR, SHR, LAB> rooms used in total | <197, 269, 107, 64> | <165,270,50,27> |
| Time-slots used during the whole examination process | 28 | 28 |
| Average number of time-slots a student has between two exams | 6.02 | 6.46 |
| Students that sit more than 270-min exam in any 2-day block | 34 | 0 |
| Large exams scheduled after the fifth day | 42 | 43 |
| Students that sit an exam during the less preferred time-slots | 8,253 | 6,756 |
| Splits of <Main, SHR, LAB> groups per exam | <29, 88, 34> | <8, 0, 0> |
| Exams that are held in different campuses | 1 | 0 |
| Students that sit exams on the same day in different campuses | 0 | 0 |
| Cases where the room capacity is exceeded | 4 | 0 |
| Main rooms in which mixed-duration exams are held | 1 | 0 |
| **Violation of Requirement S2: at most one exam every two days** | | |
| Affected conflicting exam pairs | 1,090 | 943 |
| Affected students (may include one student more than once) | 13,056 | 7,897 |
| Affected students (a student is counted only once) | 6,811 | 4,955 |
| **Violation of Requirement S3: at most one exam every day** | | |
| Affected conflicting exam pairs | 196 | 104 |
| Affected students (may include one student more than once) | 939 | 290 |
| Affected students (a student is counted only once) | 872 | 284 |
| **Violation of Requirement S4: no back-to-back exams** | | |
| Affected conflicting exam pairs | 228 | 85 |
| Affected students (may include one student more than once) | 1,856 | 152 |
| Affected students (a student is counted only once) | 1,687 | 148 |

also reduces invigilator resource requirements. The exam splits are significantly reduced. It is important to note that, in the evaluation phase of the proposed solution, the expert in UCC expressed that the reason to a large number of exam splits in UCC may be due to accommodating the ad-hoc students. However, UCC data does not let us identify these students.

A final remark is that we also implemented a preliminary version of this two-phase model on a black-box local-search solver, namely LocalSolver [1]. We observed that LocalSolver is producing quick solutions when compared to our CP model if the model is mono-objective. However, when the objective is represented as a weighted sum as in our case, LocalSolver does not have a clear advantage over the CP model. Therefore, considering that both our phases are multi-objective, we did not implement the final version on LocalSolver. It will be an interesting direction to implement this model on different solvers and present a performance comparison.

## 5 Conclusion and Future Work

In this work, we provide a model for the examination timetabling problem at University College Cork (UCC). Currently, examination scheduling in UCC is done manually by expert staff but it takes weeks. The proposed automated model is shown to improve the current timetable in terms of fairness, satisfying hard requirements, and the allocation of resources. We use the Constraint Programming framework to model our two-phase approach, mainly based on bin-packing constraints. The proposed timetable has been evaluated by the administrative staff, and integration work is in progress.

An advantage of the proposed model is its adaptability to real-world situations, such as the crisis caused by the COVID-19 pandemic. The pandemic-related constraints include mandatory exam splits due to social distancing and the capacity of venues being reduced. It is possible to tailor this model for the needs of an institute and find the number of rooms required to be able to hold physical exams. The performance of the proposed model can be improved further by using more sophisticated filtering, and symmetry-breaking techniques. We plan to develop heuristic approaches and compare their performance with the proposed CP model. Considering the combinatorial and difficult nature of bin-packing with fragmentable items, it is interesting to focus on developing global constraints with efficient filtering techniques that reduce the search space.

The proposed framework is not only applicable to UCC, but by adjusting the constraints for local institutional preferences, it should be straightforward to adapt to other institutions.

**Acknowledgement.** Special thanks to Margo Hill, the Examinations Administrator, and Siobhán Cusack, Head of Student Records and Examinations Office at University College Cork. We thank Léa Blaise from the LocalSolver team for their efforts, comments, and suggestions for improving our model, as well as the anonymous reviewers for their valuable feedback. This publication has emanated from research conducted with the financial support of Science Foundation Ireland under Grants 16/RC/3918 and 12/RC/2289-P2 which are co-funded under the European Regional Development Fund.

## References


1. Localsolver 9.0 (2019). http://www.localsolver.com/
2. Adeyemo, A.: Fragmentable group and item bin packing with compatibility preferences. In: Proceedings of the 2015 International Conference on Industrial Engineering and Operations Management (2015)
3. Arbaoui, T., Boufflet, J.P., Moukrim, A.: Preprocessing and an improved MIP model for examination timetabling. Ann. Oper. Res. **229**(1), 19–40 (2015)
4. Asmuni, H., Burke, E.K., Garibaldi, J.M., McCollum, B.: Fuzzy multiple heuristic orderings for examination timetabling. In: Burke, E., Trick, M. (eds.) PATAT 2004. LNCS, vol. 3616, pp. 334–353. Springer, Heidelberg (2005). https://doi.org/10.1007/11593577_19
5. Boussemart, F., Hemery, F., Lecoutre, C., Sais, L.: Boosting systematic search by weighting constraints. In: In European Conference on Artificial Intelligence (ECAI 2004) (2004)
6. Bron, C., Kerbosch, J.: Algorithm 457: finding all cliques of an undirected graph. Commun. ACM **16**(9), 575–577 (1973)
7. Byholm, B., Porres, I.: Fast algorithms for fragmentable items bin packing. J. Heuristics **24**(5), 697–723 (2018)
8. Carter, M.W., Laporte, G.: Recent developments in practical examination timetabling. In: Burke, E., Ross, P. (eds.) PATAT 1995. LNCS, vol. 1153, pp. 1–21. Springer, Heidelberg (1996). https://doi.org/10.1007/3-540-61794-9_49
9. Carter, M.W., Laporte, G., Chinneck, J.W.: A general examination scheduling system. Interfaces **24**(3), 109–120 (1994)
10. Casazza, M., Ceselli, A.: Mathematical programming algorithms for bin packing problems with item fragmentation. Comput. Oper. Res. **46**, 1–11 (2014)
11. Casazza, M., Ceselli, A.: Exactly solving packing problems with fragmentation. Comput. Oper. Res. **75**, 202–213 (2016)
12. Cataldo, A., Ferrer, J.C., Miranda, J., Rey, P.A., Sauré, A.: An integer programming approach to curriculum-based examination timetabling. Ann. Oper. Res. **258**(2), 369–393 (2017)
13. De Haan, P., Landman, R., Post, G., Ruizenaar, H.: A four-phase approach to a timetabling problem in secondary schools. Pract. Theory Autom. Timetabling VI **3867**, 423–425 (2006)
14. Duong, T.A., Lam, K.H.: Combining constraint programming and simulated annealing on university exam timetabling. In: RIVF, pp. 205–210. Citeseer (2004)
15. Garey, M.R., Johnson, D.S.: Computers and Intractability: A Guide to the Theory of NP-Completeness. W. H. Freeman & Co., New York (1979)
16. Gogos, C., Alefragis, P., Housos, E.: An improved multi-staged algorithmic process for the solution of the examination timetabling problem. Ann. Oper. Res. **194**(1), 203–221 (2012)
17. Hebrard, E., O'Sullivan, B., Razgon, I.: A soft constraint of equality: complexity and approximability. In: Stuckey, P.J. (ed.) CP 2008. LNCS, vol. 5202, pp. 358–371. Springer, Heidelberg (2008). https://doi.org/10.1007/978-3-540-85958-1_24
18. Kasm, O.A., Mohandes, B., Diabat, A., El Khatib, S.: Exam timetabling with allowable conflicts within a time window. Comput. Indu. Eng. **127**, 263–273 (2019)

19. LeCun, B., Mautor, T., Quessette, F., Weisser, M.A.: Bin packing with fragmentable items: presentation and approximations. Theor. Comput. Sci. **602**, 50–59 (2015)
20. Lotfi, V., Cerveny, R.: A final-exam-scheduling package. J. Oper. Res. Soc. **42**(3), 205–216 (1991)
21. Mandal, C.A., Chakrabarti, P.P., Ghose, S.: Complexity of fragmentable object bin packing and an application. Comput. Math. Appl. **35**(11), 91–97 (1998)
22. McCollum, B., McMullan, P., Parkes, A.J., Burke, E.K., Qu, R.: A new model for automated examination timetabling. Ann. Oper. Res. **194**(1), 291–315 (2012)
23. Merlot, L.T.G., Boland, N., Hughes, B.D., Stuckey, P.J.: A hybrid algorithm for the examination timetabling problem. In: Burke, E., De Causmaecker, P. (eds.) PATAT 2002. LNCS, vol. 2740, pp. 207–231. Springer, Heidelberg (2003). https://doi.org/10.1007/978-3-540-45157-0_14
24. Michail, D., Kinable, J., Naveh, B., Sichi, J.V.: JGrapht-A Java library for graph data structures and algorithms. arXiv preprint arXiv:1904.08355 (2019)
25. Michel, L., Van Hentenryck, P.: Activity-based search for black-box constraint programming solvers. In: Beldiceanu, N., Jussien, N., Pinson, É. (eds.) CPAIOR 2012. LNCS, vol. 7298, pp. 228–243. Springer, Heidelberg (2012). https://doi.org/10.1007/978-3-642-29828-8_15
26. Miller, H.E., Pierskalla, W.P., Rath, G.J.: Nurse scheduling using mathematical programming. Oper. Res. **24**(5), 857–870 (1976)
27. Mirrazavi, S.K., Mardle, S.J., Tamiz, M.: A two-phase multiple objective approach to university timetabling utilising optimisation and evolutionary solution methodologies. J. Oper. Res. Soc. **54**(11), 1155–1166 (2003)
28. Müller, T.: Real-life examination timetabling. J. Sched. **19**(3), 257–270 (2016)
29. Pillay, N., Banzhaf, W.: An informed genetic algorithm for the examination timetabling problem. Appl. Soft Comput. **10**(2), 457–467 (2010). https://doi.org/10.1016/j.asoc.2009.08.011
30. Prud'homme, C., Fages, J.G., Lorca, X.: Choco Documentation. TASC - LS2N CNRS UMR 6241, COSLING S.A.S. (2017). http://www.choco-solver.org
31. Qu, R., Burke, E.K., McCollum, B., Merlot, L.T., Lee, S.Y.: A survey of search methodologies and automated system development for examination timetabling. J. Sched. **12**(1), 55–89 (2009)
32. Régin, J.C.: A filtering algorithm for constraints of difference in CSPs. AAAI. **94**, 362–367 (1994)
33. Ribić, S., Konjicija, S.: A two phase integer linear programming approach to solving the school timetable problem. In: Proceedings of the ITI 2010, 32nd International Conference on Information Technology Interfaces, pp. 651–656. IEEE (2010)
34. Schaerf, A.: A survey of automated timetabling. Artif. Intell. Rev. **13**(2), 87–127 (1999)
35. Shachnai, H., Tamir, T., Yehezkely, O.: Approximation schemes for packing with item fragmentation. Theory Comput. Syst. **43**(1), 81–98 (2008)
36. Shaw, P.: A constraint for bin packing. In: Wallace, M. (ed.) CP 2004. LNCS, vol. 3258, pp. 648–662. Springer, Heidelberg (2004). https://doi.org/10.1007/978-3-540-30201-8_47
37. Yasari, P., Ranjbar, M., Jamili, N., Shaelaie, M.H.: A two-stage stochastic programming approach for a multi-objective course timetabling problem with courses cancelation risk. Comput. Ind. Eng. **130**, 650–660 (2019)
38. Yin, L., Luo, J., Luo, H.: Tasks scheduling and resource allocation in fog computing based on containers for smart manufacturing. IEEE Trans. Indu. Inf. **14**(10), 4712–4721 (2018)

# Exact Approaches to the Multi-agent Collective Construction Problem

Edward Lam[1,2(✉)], Peter J. Stuckey[1], Sven Koenig[3], and T. K. Satish Kumar[3]

[1] Monash University, Melbourne, VIC, Australia
{edward.lam,peter.stuckey}@monash.edu
[2] CSIRO Data61, Melbourne, VIC, Australia
[3] University of Southern California, Los Angeles, CA, USA
skoenig@usc.edu, tkskwork@gmail.com

**Abstract.** The multi-agent collective construction problem tasks agents to construct any given three-dimensional structure on a grid by repositioning blocks. Agents are required to also use the blocks to build ramps in order to access the higher levels necessary to construct the building, and then remove the ramps upon completion of the building. This paper presents a mixed integer linear programming model and a constraint programming model of the problem, either of which can exactly optimize the problem, as previous efforts have only considered heuristic approaches. The two models are evaluated on several small instances with a large number of agents. The plans clearly show the swarm behavior of the agents. The mixed integer linear programming model is able to find optimal solutions faster than the constraint programming model and even some existing incomplete methods due to its highly-exploitable network flow substructures.

**Keywords:** Classical planning · Multi-agent planning · Multi-agent path finding · Blocksworld · Swarm robotics

## 1 Introduction

The multi-agent collective construction (MACC) problem tasks a set of cooperative robots in a blocksworld with the construction of a given three-dimensional structure. The structure is built from blocks, which must be carried and rearranged by the robots. The problem aims to determine minimum-cost paths for the robots to perform this task while avoiding collisions.

The problem is best explained by example. Figure 1 illustrates a solution to a toy instance. Blocks are shown in gray. Two robots are shown in black and yellow. In timesteps 1 to 9, the black robot brings three blocks into the world and then exits. Outside the grid, robots are assumed to operate infinitely fast; i.e., an infinite number of actions can be performed in one timestep outside the grid. (Alternatively, the black robot can be assumed to be different robots in


H. Simonis (Ed.): CP 2020, LNCS 12333, pp. 743–758, 2020.
https://doi.org/10.1007/978-3-030-58475-7_43

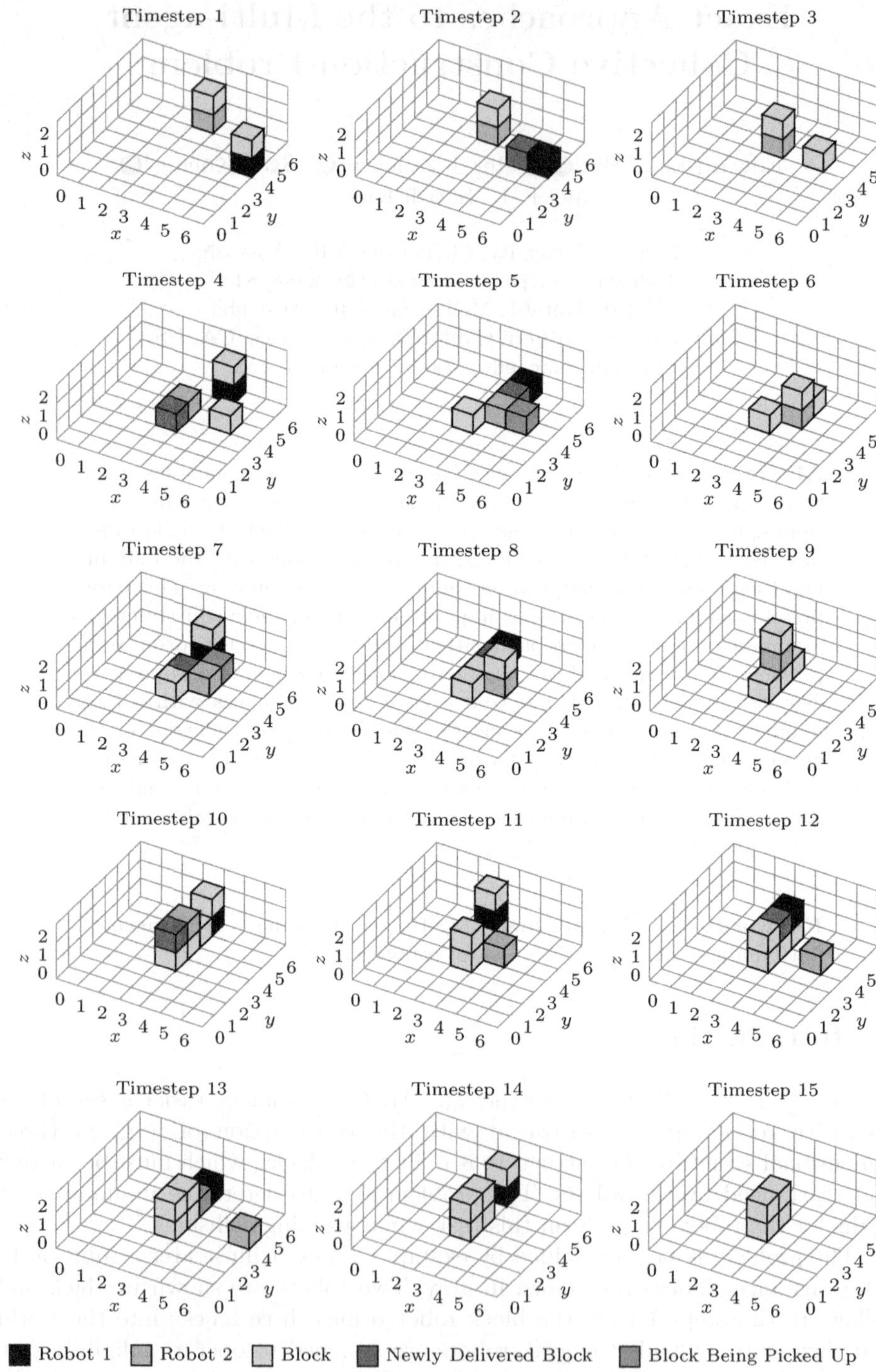


**Fig. 1.** A solution to a toy MACC instance.

practice.) In timesteps 1 to 4, the yellow robot enters the world with a block and delivers it at coordinate $(3, 3)$. In timesteps 5 to 10, the yellow robot rearranges the blocks previously brought into the world by the black robot. It then proceeds to exit the world in timesteps 11 to 13. In timesteps 10 to 12, the black robot brings in another block for delivery. In timestep 14, the black robot removes the ramp that it previously delivered in timestep 8 and used in timestep 11 to access the top level of the structure. The black robot then exits in timestep 15, leaving the structure fully assembled.

MACC is relevant to many applications, such as robotics [8] and open pit mining [6]. The problem is relatively simple to understand, yet poses many interesting questions for combinatorial optimization. In particular, questions about symmetries and dominance rules are highly non-trivial. The main contributions of this paper are a mixed integer linear programming (MILP) model and a constraint programming (CP) model of the problem. Using either of these two models, this paper is the first one to exactly optimize the problem, as all previous approaches are heuristics, which find high-quality solutions without proof of optimality. Experimental results show that the MILP model substantially outperforms the CP model because of its network flow substructures, which are easily exploited by MILP solvers. The remainder of the paper discusses these results in detail.

## 2 Problem Definition

Consider a planning horizon of $T \in \mathbb{Z}_+$ timesteps, and let $\mathcal{T} = \{0, \ldots, T-1\}$ be the set of timesteps. The problem is stated on a three-dimensional grid that is divided into cells. Let the grid be $X \in \mathbb{Z}_+$ cells wide, $Y \in \mathbb{Z}_+$ cells deep and $Z \in \mathbb{Z}_+$ cells high. Let $\mathcal{X} = \{0, \ldots, X-1\}$, $\mathcal{Y} = \{0, \ldots, Y-1\}$ and $\mathcal{Z} = \{0, \ldots, Z-1\}$ be the sets of coordinates in the three dimensions. Define $\mathcal{C} = \mathcal{X} \times \mathcal{Y} \times \mathcal{Z}$ as the set of all cells. Then, every cell $(x, y, z) \in C$ is a triple of coordinates in the grid. Define the border cells $\mathcal{B} = \{(x, 0, 0) : x \in \mathcal{X}\} \cup \{(x, Y-1, 0) : x \in \mathcal{X}\} \cup \{(0, y, 0) : y \in \mathcal{Y}\} \cup \{(X-1, y, 0) : y \in \mathcal{Y}\}$ as the perimeter cells on the ground level. Define the positions $\mathcal{P} = \mathcal{X} \times \mathcal{Y}$ as the projection of the cells onto the first two dimensions. That is, the positions lie on the two-dimensional grid corresponding to the top-down view of the three-dimensional grid. Define the neighbors of a position $(x, y) \in \mathcal{P}$ as the set of positions $\mathcal{N}_{(x,y)} = \{(x-1, y), (x+1, y), (x, y-1), (x, y+1)\} \cap \mathcal{P}$.

Consider a problem with $A \in \mathbb{Z}_+$ identical robots. A robot is of the size of a cell. Each robot can carry up to one block at any given time. Similar to robots, a block is the size of a cell. Robots start and finish outside the grid. A robot can enter and exit the world at any border cell, with or without carrying a block. (An infinite reservoir of blocks lies beyond the grid.) During every timestep that a robot is on the grid, it must take one of the following four actions:

- If the robot is carrying a block, it can *deliver* the block to a neighboring position of the same height as its current cell, raising the height at the delivery position by one. (See timesteps 1 and 2 in Fig. 1.)

- If the robot is not carrying a block, it can *pick up* a block in a neighboring position at the same height as the robot, decreasing the height at the removal position by one. (See timesteps 13 and 14 in Fig. 1.)
- The robot can *move* from its current cell to a cell in a neighboring position provided that the difference in height of both cells is within one level; i.e., robots can climb up or down at most one block.
- The robot can *wait* at its current cell.

Blocks are stationary unless moved by a robot. Blocks can be stacked on any position except the border positions, which are reserved for entry and exit. Up to $Z-1$ blocks can be stacked in any one position. In any position, a block can only be placed on the ground level or on top of the top-most block. Only the top-most block can be removed.

Borrowing terminology from multi-agent path finding [10], robots must avoid *vertex collisions* and *edge collisions*. A vertex collision occurs when two or more robots attempt to occupy, pick up from or deliver to a position. An edge collision occurs when two or more robots attempt to cross through each other to swap positions. Robots can enter and exit the grid as many or as few times as necessary. If a robot exits the grid, it must spend at least one timestep outside the grid (e.g., to pick up another block) before returning.

The aim of the problem is to find paths for the robots to construct a given three-dimensional structure by collectively rearranging blocks. By the end of the planning horizon, all robots must have exited the world, leaving the completed structure behind. The input structure is given as a desired height $\bar{z}_{(x,y)} \in \mathcal{Z}$ for every position $(x, y) \in \mathcal{P}$. That is, every block must be supported from below. Structures cannot be hollow like a cave.

In many related problems, such as multi-agent path finding and vehicle routing, two common cost/objective functions are makespan and sum-of-costs. Minimizing makespan is equivalent to compressing the time horizon so that the structure is completed as soon as possible at the expense of more actions. Minimizing sum-of-costs minimizes the total number of actions taken by the robots regardless of the time taken.

Since all robots are identical, the sum-of-costs is minimized by deploying one robot to build the structure. Since there is at most one robot in the world at any time, collisions never occur. At the cost of a higher makespan, using fewer robots always dominates using more robots for the sum-of-costs objective. On the contrary, solely using the makespan objective is problematic because unnecessary robots can aimlessly wander the world, incurring penalties in the sum-of-costs objective but having no impact on the makespan objective (besides possibly colliding with other robots and extending the makespan). Therefore, this paper argues for a two-tier lexicographic objective that first minimizes the makespan and then minimizes the sum-of-costs. This objective finds solutions that can construct a structure in the least amount of time and, with second priority, the fewest number of actions.

# 3 Background and Related Work

Teams of smaller robots are often more effective than a few larger robots. Smaller robots are usually cheaper, easier to program and easier to deploy. Despite their possibly limited sensing and computational capabilities, teams of smaller shape-shifting/self-reconfiguring robots are more fault tolerant and provide more parallelism than a few larger robots. A good example of the effectiveness of teams of smaller robots is in the domain of collective construction [2,3,7,8].

Inspired by termites building mounds, the Harvard TERMES project investigated how teams of robots can cooperate to build user-specified three-dimensional structures much larger than themselves [8]. The TERMES hardware consists of small autonomous robots and a reservoir of passive "building blocks", simply referred to as "blocks". The robots are of roughly the same size as the blocks. Yet, they can manipulate these blocks to build tall structures by stacking the blocks on top of each other and building ramps to scale greater heights. The three basic operations of a TERMES robot are: (1) climbing up or down blocks one block-height at a time; (2) navigating with proper localization on a partially-built structure without falling down; and (3) lifting, carrying and putting down a block to attach it to or detach it from a partially-built structure. MACC approximately models the TERMES robots.

A collective construction problem in the TERMES domain is to build a user-specified three-dimensional structure, assuming that the reservoir is unlimited and that the initial world is empty of blocks, i.e., all blocks are initially in the reservoir. A decentralized reactive algorithm for constructing any given structure is presented in [8]. While this algorithm succeeds in building the structure, it does not treat MACC as a rigorous combinatorial optimization problem.

A rigorous formulation of the TERMES collective construction problem as a combinatorial optimization problem is provided in [5]. The formulation exploits the fact that the three basic operations of the TERMES robots are almost always successful. The high reliability of these operations provides a nice abstraction for centralized planning algorithms, allowing for the assumption that the robots are ideal. Under these idealistic assumptions, the paper presents an algorithm that achieves a small number of pickup and drop-off operations. The algorithm solves the single-robot construction problem using dynamic programming carried out on a tree spanning the cells of a workspace matrix that represent physical locations on a grid frame of reference. The use of dynamic programming exploits common substructures and significantly reduces the number of operations on blocks. Their algorithm is polynomial-time and performs well in practice but does not guarantee optimality. In fact, the paper does not characterize the complexity of the problem.

This algorithm has been extended to the case of multiple robots in [1]. Here, the idea is to use different robots for different branches of the tree with the intuition that they can largely operate in independent regions. However, a drawback of this approach is that the regions of the tree close to its root quickly become a bottleneck and not much parallelism is achieved. Higher parallelism is achieved in [9]. Inspired by recent advances in single-agent reinforcement learning, this approach extends the single-agent asynchronous advantage actor-critic (A3C) algorithm to enable multiple agents to learn a homogeneous, distributed policy, where agents work together toward a common goal without explicitly interacting. It relies on centralized policy and critic learning, but decentralized policy execution, in a fully-observable system. Neither of the two algorithms is guaranteed to generate optimal solutions.

Finally, since blocksworld domains are well studied in the area of automated planning and scheduling, the International Planning Competition (IPC)[1] now includes the collective construction problem in the TERMES domain as a benchmark problem because of its interesting properties [4].

## 4 The Mixed Integer Linear Programming Model

The MILP model is based on network flow. Network flow problems generalize shortest path problems. Using network flow, the MILP model treats all robots as one flow through a time-expanded graph that is further complicated by the states necessary to track whether a robot is carrying a block from one timestep to the next.

Let $\mathcal{K} = \{\mathrm{M}, \mathrm{P}, \mathrm{D}\}$ be the types of actions, where M indicates that a robot is moving from one cell to another or waiting at the same cell, P indicates that a robot is picking up a block, and D indicates that a robot is delivering a block.

Define an action as a nine-tuple $i = (t, x, y, z, c, a, x', y', z')$, whose elements are given as follows:

- $t \in \mathcal{T}$ is the timestep of the action.
- $x \in \mathcal{X} \cup \{\mathrm{S}\}$, $y \in \mathcal{Y} \cup \{\mathrm{S}\}$ and $z \in \mathcal{Z} \cup \{\mathrm{S}\}$ are the coordinates of the robot taking the action, where S is a special symbol indicating that the robot is at a start location off the grid and will move into a border cell.
- $c \in \{0, 1\}$ indicates whether the robot is currently carrying a block.
- $a \in \mathcal{K}$ denotes the action type.
- $x' \in \mathcal{X} \cup \{\mathrm{E}\}$, $y' \in \mathcal{Y} \cup \{\mathrm{E}\}$ and $z' \in \mathcal{Z} \cup \{\mathrm{E}\}$ are the coordinates of the cell where the action occurred, where E is a special symbol indicating that the robot is moving from a border cell to an end location off the grid.

For instance, the action $(5, 1, 2, 3, 0, \mathrm{M}, 1, 3, 3)$ represents a robot standing in cell $(1, 2, 3)$ at timestep 5 moving to $(1, 3, 3)$ while not carrying a block, and the action $(5, 1, 2, 3, 1, \mathrm{D}, 1, 3, 3)$ represents a robot in cell $(1, 2, 3)$ at timestep 5 delivering a block it is carrying to cell $(1, 3, 3)$.

Not every possible $(t, x, y, z, c, a, x', y', z')$ in the Cartesian product is a valid action. For example, $(3, 0, 0, 0, 0, \mathrm{M}, 5, 5, 5)$ indicates a robot teleporting from cell $(0, 0, 0)$ at timestep 3 to cell $(5, 5, 5)$. Define the set of valid actions $\mathcal{R} = \mathcal{R}_1 \cup \ldots \cup \mathcal{R}_6$ made up of six subsets of different actions:

- Robots can enter the world at a border cell: $\mathcal{R}_1 = \{(t, \mathrm{S}, \mathrm{S}, \mathrm{S}, c, \mathrm{M}, x', y', z') : t \in \{0, \ldots, T-4\} \wedge c \in \{0, 1\} \wedge (x', y', z') \in \mathcal{B}\}$.

[1] https://ipc2018.bitbucket.io.

- Robots can move to a neighboring cell at the same level, one level above or one level below: $\mathcal{R}_2 = \{(t, x, y, z, c, \mathrm{M}, x', y', z') : t \in \{1, \dots, T-3\} \wedge (x, y, z) \in \mathcal{C} \wedge c \in \{0, 1\} \wedge (x', y') \in \mathcal{N}_{(x,y)} \wedge z' \in \mathcal{Z} \wedge |z' - z| \leq 1\}$.
- Robots can wait at the same cell: $\mathcal{R}_3 = \{(t, x, y, z, c, \mathrm{M}, x, y, z) : t \in \{1, \dots, T-3\} \wedge (x, y, z) \in \mathcal{C} \wedge c \in \{0, 1\}\}$.
- Robots can exit the world at a border cell: $\mathcal{R}_4 = \{(t, x, y, z, c, \mathrm{M}, \mathrm{E}, \mathrm{E}, \mathrm{E}) : t \in \{2, \dots, T-2\} \wedge (x, y, z) \in \mathcal{B} \wedge c \in \{0, 1\}\}$.
- While not carrying a block, robots can pick up a block from a neighboring cell at the same level: $\mathcal{R}_5 = \{(t, x, y, z, 0, \mathrm{P}, x', y', z) : t \in \{1, \dots, T-3\} \wedge (x, y) \in \mathcal{P} \wedge z \in \{0, \dots, Z-2\} \wedge (x', y') \in \mathcal{N}_{(x,y)}\}$.
- While carrying a block, robots can deliver the block to a neighboring cell at the same level: $\mathcal{R}_6 = \{(t, x, y, z, 1, \mathrm{D}, x', y', z) : t \in \{1, \dots, T-3\} \wedge (x, y) \in \mathcal{P} \wedge z \in \{0, \dots, Z-2\} \wedge (x', y') \in \mathcal{N}_{(x,y)}\}$.

The timesteps in $\mathcal{R}_1, \dots, \mathcal{R}_6$ are chosen carefully since, e.g., all robots must be off the world by timestep $T-1$, they must be moving from a border cell off the world by timestep $T-2$, hence $T-3$ is the latest that a block can be delivered.

Every position $(x, y) \in \mathcal{P}$ is modeled as a shortest path from height 0 to the desired height $\bar{z}_{(x,y)} \in \mathcal{Z}$ at the final timestep $T-1$. Similar to $\mathcal{R}$, define $\mathcal{H} = \{(t, x, y, z, z') : t \in \{0, \dots, T-2\} \wedge (x, y, z) \in \mathcal{C} \wedge z' \in \mathcal{Z} \wedge |z' - z| \leq 1\}$ to represent the actions of growing or shrinking the height of a position. An action $(t, x, y, z, z') \in \mathcal{H}$ indicates that position $(x, y)$ currently has height $z$ at timestep $t$ and height $z'$ at timestep $t+1$.

In this model, the problem can be thought of in terms of two groups of interacting agents: (1) pillars that need to grow or shrink to their target heights in the least amount of time (pillars might need to grow higher than their target height), and (2) robots that assist the pillars by picking up and delivering blocks around the world. For pillars to grow upward, they need robots to stack blocks at their positions. For robots to place blocks on a pillar, a neighboring pillar needs to be of a similar height. Hence, in some sense, the problem involves a complicated interaction between two sets of agents, both of which co-operate to achieve their goals.

The model captures the actions of all robots in one network flow and the actions of each pillar in a shortest path (a special case of network flow). These two substructures are coupled by interdependency constraints. In the absence of the interdependencies, the model separates into a number of independent network flows. Hence, the idea behind the MILP model is for the solver to first resolve the interdependencies to simplify the problem, and then the problem becomes much easier since pure network flow problems can be solved in polynomial time by linear programming [11]. Of course, resolving the interdependencies remains a major challenge.

The MILP model is written using non-standard wildcard notation. Let $\mathcal{U}$ be a set containing tuples $(u_1, u_2, \dots, u_n)$. For constants $u_1, u_2, \dots, u_n$, we use the notation $\mathcal{U}_{u_1, u_2, \dots, u_n}$ as a shorthand for the set $\{(u'_1, u'_2, \dots, u'_n) \in \mathcal{U} : u'_1 = u_1 \wedge u'_2 = u_2 \wedge \dots \wedge u'_n = u_n)\}$, which is equal to the singleton $\{(u_1, u_2, \dots, u_n)\}$ if the element exists and equal to the empty set $\emptyset$ otherwise. Let $*$ denote a wildcard symbol for matching any value in a dimension of the tuples. For example, $\mathcal{U}_{*,u_2,\ldots,u_n}$ is shorthand for the set $\{(u'_1, u'_2, \ldots, u'_n) \in \mathcal{U} : u'_2 = u_2 \wedge \ldots \wedge u'_n = u_n\}$, and $\mathcal{U}_{u_1,u_2,*,\ldots,*}$ represents the set $\{(u'_1, u'_2, u'_3, \ldots, u'_n) \in \mathcal{U} : u'_1 = u_1 \wedge u'_2 = u_2\}$. This wildcard notation is used to pick subsets of $\mathcal{R}$ and $\mathcal{H}$.

Figure 2 shows the model. For every robot action $i \in \mathcal{R}$, define a binary decision variable $r_i \in \{0, 1\}$ to indicate whether the action occurred. Similarly, define a binary decision variable $h_i \in \{0, 1\}$ for every height action $i \in \mathcal{H}$.

Objective Function (1) minimizes the sum-of-costs objective, i.e., the total number of cells occupied by robots throughout the planning horizon. The makespan is minimized external to the model by sequentially increasing $T$, as described later.

Constraints (2) to (6) define a path for each position. Constraint (2) prevents blocks from being placed at the border positions because robots must enter and exit the world on the ground level. Constraint (3) starts the world devoid of blocks. This constraint states that all cells have height 0 in the first two timesteps because the earliest a robot can appear in the world is in timestep 1; hence blocks cannot be placed in the world until timestep 2. Constraint (4) enforces the completion of the structure. Robots must have exited the world by timestep $T - 1$. So they must be at a border cell exiting before timestep $T - 2$. Therefore, the structure must be built before the last two timesteps. Constraint (5) flows the height of each position from one timestep to the next. Constraint (6) enforces one value of height for every position in each timestep.

Constraints (7) to (11) govern the actions of the robots. Constraint (7) flows a robot not carrying a block in and out of a cell. The first summation accounts for a robot moving without a block from any cell at timestep $t$ into cell $(x, y, z)$ at timestep $t + 1$. The second summation counts whether a robot standing at $(x, y, z)$ has just deposited a block nearby. After taking any of these actions, the robot will be at cell $(x, y, z)$ at timestep $t + 1$ without a block. It then has to either move to another cell (the third summation) or pick up a nearby block (the fourth summation). Constraint (8) is a similar constraint for robots carrying a block. This constraint states that, if a robot carrying a block is in cell $(x, y, z)$ at timestep $t + 1$ (either by moving into the cell with a block or by picking up a block nearby), then it must afterward move while continuing to carry the block or deliver the block. Constraints (7) and (8) implicitly require robots to start and end outside the grid. Constraint (9) prevents vertex collisions. It permits at most one robot to be at position $(x, y)$ or to pick up from or deliver a block to position $(x, y)$ at any timestep. Constraint (10) is the edge collision constraint, which prevents robots from exchanging positions. Constraint (11) limits the number of robots. By also including robots at the dummy start cell $(\mathrm{S}, \mathrm{S}, \mathrm{S})$, this constraint also requires robots that have left the world to spend at least one timestep outside the world (e.g., to pick up another block) before returning.

Constraints (12) to (14) couple the robots and the pillars. Without these three constraints, the problem separates into two independent parts. Constraint (12) states that, if a robot is at cell $(x, y, z)$, then the height of the pillar at

$$\min \sum_{\substack{i=(t,x,y,z,c,a,x',y',z')\in\mathcal{R}:\\(x,y,z)\neq(\mathrm{S},\mathrm{S},\mathrm{S})}} r_i \tag{1}$$

subject to

$$h_{t,x,y,z,z} = 1 \qquad \forall t \in \{0,\ldots,T-3\}, (x,y,z) \in \mathcal{B}, \tag{2}$$

$$h_{0,x,y,0,0} = 1 \qquad \forall (x,y) \in \mathcal{P}, \tag{3}$$

$$h_{T-2,x,y,\bar{z}_{(x,y)},\bar{z}_{(x,y)}} = 1 \qquad \forall (x,y) \in \mathcal{P}, \tag{4}$$

$$\sum_{i\in\mathcal{H}_{t,x,y,*,z}} h_i = \sum_{i\in\mathcal{H}_{t+1,x,y,z,*}} h_i \qquad \forall t \in \{0,\ldots,T-3\}, (x,y,z) \in \mathcal{C}, \tag{5}$$

$$\sum_{i\in\mathcal{H}_{t,x,y,*,*}} h_i = 1 \qquad \forall t \in \{0,\ldots,T-2\}, (x,y) \in \mathcal{P}, \tag{6}$$

$$\sum_{i\in\mathcal{R}_{t,*,*,*,0,\mathrm{M},x,y,z}} r_i + \sum_{i\in\mathcal{R}_{t,x,y,z,1,\mathrm{D},*,*,*}} r_i = \sum_{i\in\mathcal{R}_{t+1,x,y,z,0,\mathrm{M},*,*,*}} r_i + \sum_{i\in\mathcal{R}_{t+1,x,y,z,0,\mathrm{P},*,*,*}} r_i \qquad \forall t \in \{0,\ldots,T-3\}, (x,y,z) \in \mathcal{C}, \tag{7}$$

$$\sum_{i\in\mathcal{R}_{t,*,*,*,1,\mathrm{M},x,y,z}} r_i + \sum_{i\in\mathcal{R}_{t,x,y,z,0,\mathrm{P},*,*,*}} r_i = \sum_{i\in\mathcal{R}_{t+1,x,y,z,1,\mathrm{M},*,*,*}} r_i + \sum_{i\in\mathcal{R}_{t+1,x,y,z,1,\mathrm{D},*,*,*}} r_i \qquad \forall t \in \{0,\ldots,T-3\}, (x,y,z) \in \mathcal{C}, \tag{8}$$

$$\sum_{i\in\mathcal{R}_{t,x,y,*,*,*,*,*,*}} r_i + \sum_{i\in\mathcal{R}_{t,*,*,*,*,\mathrm{P},x,y,*}} r_i + \sum_{i\in\mathcal{R}_{t,*,*,*,*,\mathrm{D},x,y,*}} r_i \leq 1 \qquad \forall t \in \{1,\ldots,T-2\}, (x,y) \in \mathcal{P}, \tag{9}$$

$$\sum_{i\in\mathcal{R}_{t,x,y,*,*,\mathrm{M},x',y',*}} r_i + \sum_{i\in\mathcal{R}_{t,x',y',*,*,\mathrm{M},x,y,*}} r_i \leq 1 \qquad \forall t \in \{1,\ldots,T-2\}, (x,y) \in \mathcal{P}, (x',y') \in \mathcal{N}_{(x,y)}, \tag{10}$$

$$\sum_{i\in\mathcal{R}_{t,*,*,*,*,*,*,*,*}} r_i \leq A \qquad \forall t \in \mathcal{T}, \tag{11}$$

$$\sum_{i\in\mathcal{H}_{t,x,y,z,*}} h_i \geq \sum_{i\in\mathcal{R}_{t,x,y,z,*,*,*,*,*}} r_i \qquad \forall t \in \{0,\ldots,T-2\}, (x,y,z) \in \mathcal{C}, \tag{12}$$

$$h_{t,x,y,z+1,z} = \sum_{i\in\mathcal{R}_{t,*,*,*,0,\mathrm{P},x,y,z}} r_i \qquad \forall t \in \{0,\ldots,T-2\}, (x,y) \in \mathcal{P}, z \in \{0,\ldots,Z-2\}, \tag{13}$$

$$h_{t,x,y,z,z+1} = \sum_{i\in\mathcal{R}_{t,*,*,*,1,\mathrm{D},x,y,z}} r_i \qquad \forall t \in \{0,\ldots,T-2\}, (x,y) \in \mathcal{P}, z \in \{0,\ldots,Z-2\}, \tag{14}$$

$$h_i \in \{0,1\} \qquad \forall i \in \mathcal{H}, \tag{15}$$

$$r_i \in \{0,1\} \qquad \forall i \in \mathcal{R}. \tag{16}$$

**Fig. 2.** The MILP model.

position $(x, y)$ must be $z$. Constraints (13) and (14) respectively equate pickup and delivery actions to a decrease and increase in the height of a pillar.

Constraints (15) and (16) specify the domains of the variables.

# 5 The Constraint Programming Model

Standard CP models of routing problems are based on a sequence of actions performed by each robot. This modeling indexes every robot individually and, hence, introduces robot symmetry. This section presents a CP model that forgoes the sequence-based modeling and, instead, adopts a network flow structure to eliminate robot symmetry. Compared to the MILP model, the CP model uses a simpler network that omits the vertical dimension, actions and block-carrying state. It models these aspects using logical and ELEMENT constraints, which better exploit the strengths of CP.

Assign every position $p = (x, y) \in \mathcal{P}$ an identifier $i_p = Y \cdot x + y$ that maps the two-dimensional positions to a one-dimensional index. Let $\mathcal{I} = \{0, \dots, X \cdot Y - 1\}$ denote the set of position identifiers. Let $\bar{z}_i \in \mathcal{Z}$ be the height of the desired structure at position $i \in \mathcal{I}$.

Define the border positions as $\mathcal{B} = \{i_{(x,0)} : x \in \mathcal{X}\} \cup \{i_{(x,Y-1)} : x \in \mathcal{X}\} \cup \{i_{(0,y)} : y \in \mathcal{Y}\} \cup \{i_{(X-1,y)} : y \in \mathcal{Y}\}$, and the interior positions as $\overline{\mathcal{B}} = \{i \in \mathcal{I} : i \notin \mathcal{B}\}$. Define two dummy positions $-1$ and $-2$ off the grid and group them in $\mathcal{O} = \{-1, -2\}$. Let $\mathcal{E} = \mathcal{I} \cup \mathcal{O}$ denote every position (on and off the grid).

For any position $i = i_{(x,y)}$, let $\mathcal{N}_i = \{i_{(x-1,y)}, i_{(x+1,y)}, i_{(x,y-1)}, i_{(x,y+1)}\} \cap \mathcal{I}$ be its neighbors. Define a set of neighbors that includes the off-grid positions as

$$\mathcal{N}_i^{\mathcal{E}} = \begin{cases} \mathcal{N}_i & i \in \overline{\mathcal{B}}, \\ \mathcal{N}_i \cup \mathcal{O} & i \in \mathcal{B}. \end{cases}$$

Using $\mathcal{N}_i^{\mathcal{E}}$, robots are able to move off the grid from a border position.

Let $\mathcal{K} = \{\mathrm{M}, \mathrm{B}, \mathrm{U}\}$ be the types of actions, where M indicates that a robot is moving from one position to another or waiting at the same position, B indicates that a robot is picking up or delivering a block, and U indicates that a position is unoccupied by a robot.

Let $r_{t,i} \in \mathcal{K}$ be a decision variable representing the action taken by the robot in position $i \in \mathcal{E}$ at timestep $t \in \mathcal{T}$. Define $n_{t,i} \in \mathcal{E}$ as a variable denoting the next position of the robot in position $i \in \mathcal{E}$ at timestep $t \in \mathcal{T}$. Define $b_{t,i} \in \mathcal{I}$ as the position of the pick-up or delivery by the robot in position $i \in \mathcal{I}$ at timestep $t \in \mathcal{T}$. Define $c_{t,i} \in \{0, 1\}$ as a variable that indicates whether the robot in position $i \in \mathcal{E}$ at timestep $t \in \mathcal{T}$ is carrying a block. Let $p_{t,i}, d_{t,i} \in \{0, 1\}$ be variables that, respectively, indicate whether the robot in position $i \in \mathcal{I}$ at timestep $t \in \{0, \dots, T - 2\}$ is picking up or delivering a block. Let $h_{t,i} \in \mathcal{Z}$ be a variable that stores the height at position $i \in \mathcal{E}$ in timestep $t \in \mathcal{T}$. The meaning of the variables for any position $i \in \mathcal{I}$ is clear. The variables are also defined for $i \in \mathcal{O}$ to ensure that the problem is satisfiable by giving the ELEMENT constraints an end-point; these variables do not carry much meaning.

The CP model is shown in Figs. 3 and 4. Objective Function (17) minimizes the number of occupied positions at all timesteps. The ELEMENT global constraint is used in Constraints (30), (32), (35), (39) and (41) to (44).

Constraint (18) fixes the height of all off-grid positions. Constraint (19) disallows blocks to be delivered to the border positions. Constraint (20) states that the world is devoid of blocks in the first two timesteps. Constraint (21) requires the building to be completed before the last two timesteps. Constraint (22) allows the height at any position to change by at most one level.

Constraints (23) to (26) fix the robot variables at the off-grid positions. Constraints (27) and (28) disallow robots on the grid in the first and last timesteps. Constraint (29) requires robots to stay at the same position when picking up or delivering a block. Constraint (30) maintains the block-carrying states of robots when moving. Constraint (31) changes the block-carrying states of robots after picking up or delivering blocks.

Constraints (32) and (33) are the flow constraints. Constraint (32) states that every position is either unoccupied or the robot in the position must take an action in the next timestep. Constraint (33) states that an interior position is unoccupied or a robot reached it from a nearby position in the previous timestep. Constraint (34) prevents vertex collisions by disallowing more than one robot from being in a position, picking up a block from the position or delivering a block to the position at timestep $t + 1$. Constraint (35) prevents edge collisions. Constraint (36) limits the number of robots on the grid during each timestep. This constraint also requires robots to spend at least one timestep outside the grid, as discussed in Sect. 2.

Constraints (37) and (38) compute whether a robot is picking up or delivering a block. If a robot moves, Constraint (39) requires the height of its next position to be within one level of the height of its current position. If a robot waits at the same position, Constraint (40) states that the height must remain the same. Constraint (41) states that the height of the block being picked up must be one level higher than the pillar at the position of the robot (i.e., the block is at the same level as the robot). Constraint (42) decreases the height at the position of a block after a pick up. Constraints (43) and (44) are the equivalent constraints for deliveries. Constraint (45) counts the changes to the height at a position. Constraints (46) and (47) are redundant constraints, which improve the filtering.

Constraints (48) to (54) specify the domains of the variables. Constraint (50) requires robots to move to a neighboring position, the same position or off the grid. Constraint (51) states that blocks must be picked-up from or delivered to a neighboring position.

## 6 Experimental Results

The experiments compare the run-time of the two models. The MILP model is solved using Gurobi 9.0.2, a state-of-the-art mathematical programming solver that regularly outperforms its competitors in standard benchmarks. The CP

$$\min \sum_{t \in \mathcal{T}} \sum_{i \in \mathcal{I}} (r_{t,i} \neq \text{U}) \tag{17}$$

subject to

$$h_{t,i} = 0 \qquad \forall t \in \mathcal{T}, i \in \mathcal{O}, \tag{18}$$

$$h_{t,i} = 0 \qquad \forall t \in \mathcal{T}, i \in \mathcal{B}, \tag{19}$$

$$h_{t,i} = 0 \qquad \forall t \in \{0,1\}, i \in \mathcal{I}, \tag{20}$$

$$h_{t,i} = \bar{z}_i \qquad \forall t \in \{T-2, T-1\}, i \in \mathcal{I}, \tag{21}$$

$$h_{t,i} - 1 \leq h_{t+1,i} \leq h_{t,i} + 1 \qquad \forall t \in \{0,\ldots,T-2\}, i \in \mathcal{I}, \tag{22}$$

$$r_{t,i} = \text{M} \qquad \forall t \in \mathcal{T}, i \in \mathcal{O}, \tag{23}$$

$$n_{t,i} = i \qquad \forall t \in \mathcal{T}, i \in \mathcal{O}, \tag{24}$$

$$c_{t,-1} = 1 \qquad \forall t \in \mathcal{T}, \tag{25}$$

$$c_{t,-2} = 0 \qquad \forall t \in \mathcal{T}, \tag{26}$$

$$r_{0,i} = \text{U} \qquad \forall i \in \mathcal{I}, \tag{27}$$

$$r_{T-1,i} = \text{U} \qquad \forall i \in \mathcal{I}, \tag{28}$$

$$(r_{t,i} = \text{B}) \rightarrow (n_{t,i} = i) \qquad \forall t \in \mathcal{T}, i \in \mathcal{I}, \tag{29}$$

$$(r_{t,i} = \text{M}) \rightarrow (c_{t+1,n_{t,i}} = c_{t,i}) \qquad \forall t \in \{0,\ldots,T-2\}, i \in \mathcal{I}, \tag{30}$$

$$(r_{t,i} = \text{B}) \rightarrow (c_{t+1,i} = \neg c_{t,i}) \qquad \forall t \in \{0,\ldots,T-2\}, i \in \mathcal{I}, \tag{31}$$

$$(r_{t,i} = \text{U}) \vee (r_{t+1,n_{t,i}} \neq \text{U}) \qquad \forall t \in \{0,\ldots,T-2\}, i \in \mathcal{I}, \tag{32}$$

$$(r_{t+1,i} = \text{U}) \vee \bigvee_{j \in \mathcal{N}_i \cup \{i\}} (r_{t,j} \neq \text{U} \wedge n_{t,j} = i) \qquad \forall t \in \{0,\ldots,T-2\}, i \in \overline{\mathcal{B}}, \tag{33}$$

$$\sum_{j \in \mathcal{N}_i \cup \{i\}} (r_{t,j} = \text{M} \wedge n_{t,j} = i) + (r_{t,i} = \text{B}) + \sum_{j \in \mathcal{N}_i} (r_{t+1,j} = \text{B} \wedge b_{t+1,j} = i) \leq 1$$
$$\forall t \in \{1,\ldots,T-2\}, i \in \mathcal{I}, \tag{34}$$

$$(r_{t,i} = \text{M} \wedge n_{t,i} \neq i \wedge r_{t,n_{t,i}} = \text{M}) \rightarrow (n_{t,n_{t,i}} \neq i) \qquad \forall t \in \{1,\ldots,T-2\}, i \in \mathcal{I}, \tag{35}$$

$$\sum_{i \in \mathcal{I}} (r_{t,i} \neq \text{U}) + \sum_{i \in \mathcal{B}} (r_{t-1,i} = \text{M} \wedge n_{t-1,i} < 0) \leq A \qquad \forall t \in \{1,\ldots,T-1\}, \tag{36}$$

$$p_{t,i} \leftrightarrow (r_{t,i} = \text{B} \wedge c_{t+1,i} \wedge \neg c_{t,i}) \qquad \forall t \in \{0,\ldots,T-2\}, i \in \mathcal{I}, \tag{37}$$

$$d_{t,i} \leftrightarrow (r_{t,i} = \text{B} \wedge \neg c_{t+1,i} \wedge c_{t,i}) \qquad \forall t \in \{0,\ldots,T-2\}, i \in \mathcal{I}, \tag{38}$$

$$(r_{t,i} = \text{M}) \rightarrow (h_{t,i} - 1 \leq h_{t+1,n_{t,i}} \leq h_{t,i} + 1) \qquad \forall t \in \{0,\ldots,T-2\}, i \in \mathcal{I}, \tag{39}$$

$$(r_{t,i} = \text{M} \wedge n_{t,i} = i) \rightarrow (h_{t+1,i} = h_{t,i}) \qquad \forall t \in \{0,\ldots,T-2\}, i \in \mathcal{I}, \tag{40}$$

$$p_{t,i} \rightarrow (h_{t,b_{t,i}} = h_{t,i} + 1) \qquad \forall t \in \{0,\ldots,T-2\}, i \in \mathcal{I}, \tag{41}$$

$$p_{t,i} \rightarrow (h_{t+1,b_{t,i}} = h_{t,b_{t,i}} - 1) \qquad \forall t \in \{0,\ldots,T-2\}, i \in \mathcal{I}, \tag{42}$$

$$d_{t,i} \rightarrow (h_{t,b_{t,i}} = h_{t,i}) \qquad \forall t \in \{0,\ldots,T-2\}, i \in \mathcal{I}, \tag{43}$$

$$d_{t,i} \rightarrow (h_{t+1,b_{t,i}} = h_{t,b_{t,i}} + 1) \qquad \forall t \in \{0,\ldots,T-2\}, i \in \mathcal{I}, \tag{44}$$

$$h_{t+1,i} = h_{t,i} - \sum_{j \in \mathcal{N}_i} (p_{t,j} \wedge b_{t,j} = i) + \sum_{j \in \mathcal{N}_i} (d_{t,j} \wedge b_{t,j} = i) \quad \forall t \in \{0,\ldots,T-2\}, i \in \mathcal{I}, \tag{45}$$

**Fig. 3.** The CP model. Continued in Fig. 4.

$$h_{t+1,i} = h_{t,i} - 1 \rightarrow \bigvee_{j \in \mathcal{N}_i} (p_{t,j} \wedge b_{t,j} = i) \qquad \forall t \in \{0, \dots, T-2\}, i \in \mathcal{I}, \quad (46)$$

$$h_{t+1,i} = h_{t,i} + 1 \rightarrow \bigvee_{j \in \mathcal{N}_i} (d_{t,j} \wedge b_{t,j} = i) \qquad \forall t \in \{0, \dots, T-2\}, i \in \mathcal{I}, \quad (47)$$

$$h_{t,i} \in \mathcal{Z} \qquad \forall t \in \mathcal{T}, i \in \mathcal{E}, \quad (48)$$

$$r_{t,i} \in \mathcal{K} \qquad \forall t \in \mathcal{T}, i \in \mathcal{E}, \quad (49)$$

$$n_{t,i} \in \mathcal{N}_i^{\mathcal{E}} \cup \{i\} \qquad \forall t \in \mathcal{T}, i \in \mathcal{E}, \quad (50)$$

$$b_{t,i} \in \mathcal{N}_i \qquad \forall t \in \mathcal{T}, i \in \mathcal{I}, \quad (51)$$

$$c_{t,i} \in \{0, 1\} \qquad \forall t \in \mathcal{T}, i \in \mathcal{E}, \quad (52)$$

$$p_{t,i} \in \{0, 1\} \qquad \forall t \in \mathcal{T}, i \in \mathcal{I}, \quad (53)$$

$$d_{t,i} \in \{0, 1\} \qquad \forall t \in \mathcal{T}, i \in \mathcal{I}. \quad (54)$$

**Fig. 4.** The CP model. Continued from Fig. 3.

model is solved using OR-Tools 7.6, an open-source CP solver that has won many of the MiniZinc Challenges in recent times. The two solvers are run for up to seven days in parallel mode with 20 threads on an Intel Xeon E5-2660 v3 CPU 2.60 GHz with 64 GB of memory.

Even though traditional finite-domain CP solvers are most effective with a hand-tailored variable and value selection heuristic (i.e., a branching rule), nogood learning solvers like OR-Tools perform best when using their preferred search strategies, as evidenced in the MiniZinc Challenge, where OR-Tools performs better in the free search category. Therefore, we do not specify a variable and value selection heuristic.

The two models minimize the number actions, i.e., the sum-of-costs. To lexicographically minimize makespan as well, the two models are run with sequentially increasing $T$. That is, the experiments begin with $T = 4$, which is the smallest possible value, and progressively increase $T$ until the problem becomes satisfiable (i.e., the solver finds a feasible solution). Then, the solver proceeds to find an optimal solution. Upon completion, the solution is guaranteed to have the lowest possible makespan and the lowest number of actions for that makespan.

The six structures from [9] are used for evaluation. They are shown in Fig. 5. Up to 50 robots are permitted Table 1 shows the results for the six instances. For both models, the table gives the time to prove optimality, the optimal makespan, the best available sum-of-costs and its lower bound, and the number of robots required to execute the plan. (Unused robots never enter the world.)

The MILP model solves all six instances exactly within six days. Instances 1, 2 and 6 are trivial for the MILP model to solve, while the run-times of the other three instances span a large range. The CP model solves Instance 2 exactly but is substantially slower than the MILP model. For the remaining five instances, it finds feasible solutions with the optimal makespan.

Figure 6 plots the twelve timesteps for executing the optimal plan to the first instance. The large number of robots swarming into the world makes it difficult to

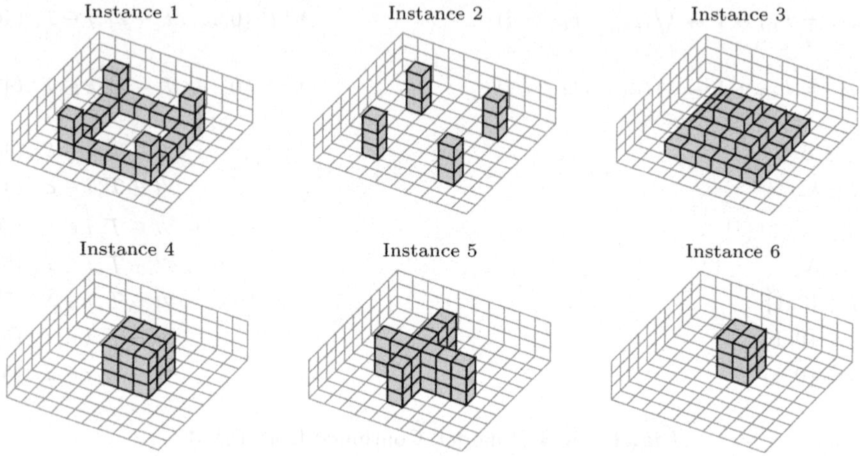


**Fig. 5.** The structures to construct in the six instances.

**Table 1.** Best available solution to the six instances.

| Model | Instance | Run-time | Makespan | Sum-of-costs | Sum-of-costs LB | Robots |
|---|---|---|---|---|---|---|
| MILP | 1 | 29 s | 11 | 176 | 176 | 34 |
| | 2 | 3 s | 11 | 128 | 128 | 28 |
| | 3 | 1.2 h | 13 | 344 | 344 | 44 |
| | 4 | 5.5 h | 17 | 429 | 429 | 42 |
| | 5 | 5.7 d | 17 | 368 | 368 | 37 |
| | 6 | 183 s | 15 | 234 | 234 | 27 |
| CP | 1 | >7 d | 11 | 178 | 107 | 30 |
| | 2 | 1.2 h | 11 | 128 | 128 | 28 |
| | 3 | >7 d | 13 | 354 | 164 | 44 |
| | 4 | >7 d | 17 | 452 | 189 | 50 |
| | 5 | >7 d | 17 | 395 | 39 | 41 |
| | 6 | >7 d | 15 | 245 | 154 | 28 |

analyze any emergent macro-level behavior. Nonetheless, minor bucket-brigade behavior is already demonstrated in the toy example from Fig. 1.

The state-of-the-art reinforcement learning method [9] produced solutions taking up to 2,000 timesteps (for many fewer robots). The two optimization models found feasible solutions to all instances with a makespan of less than twenty timesteps, indicating that these small structures are simple to construct as they do not rely on long chains of interdependent blocks.

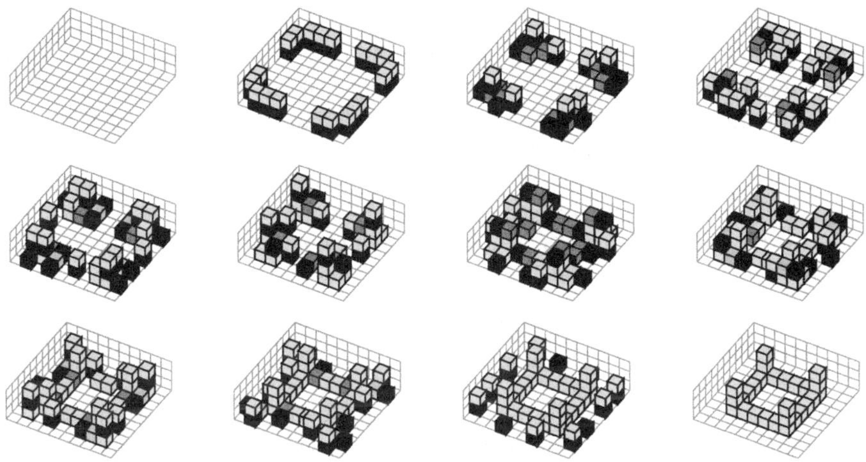

**Fig. 6.** The twelve timesteps in the optimal plan to Instance 1.

## 7 Conclusions and Future Work

The MACC problem is a relatively new problem that is starting to gain attention in the multi-agent planning community. The problem tasks a group of co-operating robots to construct a three-dimensional structure by rearranging blocks in a blocksworld. Robots are required to build ramps to access the upper levels of the structure, and then remove the ramps after assembling the structure.

This paper models the problem using MILP and unexpectedly reveals the two interacting network-flow substructures hidden in the problem. A CP model of the problem is also developed but is slower than the MILP model because it lacks easily exploitable structure. This preliminary study shows that small instances of complex path-finding problems with an extremely large state space can be solved exactly today, as previous solution methods were all heuristics, which aim to find high-quality but not provably optimal solutions.

Scaling the MILP model to long time horizons remains a major challenge. Early experiments show that using more robots shortens the makespan, and hence, makes the model smaller and easier. Surprisingly, it is the number of timesteps that makes the problem difficult, rather than the number of robots. Buildings taller than five blocks require long ramps for accessing the higher levels. These ramps take many timesteps to build, resulting in a makespan much longer and a model much larger than what is possible for exact optimization. For these instances, the only viable methods are the existing heuristics.

Early experiments show that the CP model presented in Sect. 5 and solved using OR-Tools is fastest among eleven models solved using both OR-Tools and Chuffed. One interesting finding is that sequence-based models (often using the REGULAR global constraint) are faster in Chuffed but slower in OR-Tools. Whereas, network flow models (e.g., the model in Sect. 5) are faster in

OR-Tools, which has a linear relaxation propagator. Future studies should investigate whether CP-like sequencing models can obtain competitive performance.

As suggested by a reviewer, MaxSAT models can also be considered since the problem mainly contains Boolean variables and clauses. The difficulty would be encoding the cardinality constraints, but how to do this is well-known.

**Acknowledgments.** The research at the University of Southern California was supported by the National Science Foundation (NSF) under grant numbers 1724392, 1409987, 1817189, 1837779, and 1935712.

## References

1. Cai, T., Zhang, D., Kumar, T.K.S., Koenig, S., Ayanian, N.: Local search on trees and a framework for automated construction using multiple identical robots. In: Proceedings of the International Conference on Autonomous Agents and Multi-Agent Systems (2016)
2. Grushin, A., Reggia, J.: Automated design of distributed control rules for the self-assembly of prespecified artificial structures. Robot. Auton. Syst. **56**(4), 334–359 (2008)
3. Jones, C., Mataric, M.: Automatic synthesis of communication-based coordinated multi-robot systems. In: Proceedings of the IEEE/RSJ International Conference on Intelligent Robots and Systems (2004)
4. Koenig, S., Kumar, T.K.S.: A case for collaborative construction as testbed for cooperative multi-agent planning. In: Proceedings of the ICAPS-2017 Scheduling and Planning Applications Workshop (2017)
5. Kumar, T.K.S., Jung, S., Koenig, S.: A tree-based algorithm for construction robots. In: Proceedings of the International Conference on Automated Planning and Scheduling (2014)
6. Lambert, W., Brickey, A., Newman, A., Eurek, K.: Open-pit block-sequencing formulations: a tutorial. Interfaces **44**, 127–142 (2014)
7. Napp, N., Klavins, E.: Robust by composition: programs for multi-robot systems. In: Proceedings of the IEEE International Conference on Robotics and Automation (2010)
8. Petersen, K., Nagpal, R., Werfel, J.: TERMES: an autonomous robotic system for three-dimensional collective construction. In: Proceedings of Robotics: Science and Systems (2011)
9. Sartoretti, G., Wu, Y., Paivine, W., Kumar, T.K.S., Koenig, S., Choset, H.: Distributed reinforcement learning for multi-robot decentralized collective construction. In: Correll, N., Schwager, M., Otte, M. (eds.) Distributed Autonomous Robotic Systems. SPAR, vol. 9, pp. 35–49. Springer, Cham (2019). https://doi.org/10.1007/978-3-030-05816-6_3
10. Stern, R., et al.: Multi-agent pathfinding: definitions, variants, and benchmarks. In: Proceedings of the Symposium on Combinatorial Search (2019)
11. Vaidyanathan, B., Ahuja, R.K.: Minimum cost flows. In: Wiley Encyclopedia of Operations Research and Management Science. Wiley (2011)

# The Confidence Constraint: A Step Towards Stochastic CP Solvers

Alexandre Mercier-Aubin(✉), Ludwig Dumetz, Jonathan Gaudreault, and Claude-Guy Quimper

Université Laval, G1V 0A6, Québec, QC, Canada
alexandre.mercier-aubin.1@ulaval.ca

**Abstract.** We introduce the CONFIDENCE constraint, a chance constraint that ensures, with probability $\gamma$, that a set of variables are no smaller than random variables for which the probability distribution is given. This constraint is useful in stochastic optimization to ensure that a solution is robust to external random events. It allows to control the trade-off between optimizing the objective function and ensuring the satisfiability of the solution under random parameters. We present a filtering algorithm for this constraint with explanations. We apply the constraint to a case study, an industrial scheduling problem where tasks have random processing times due to possible breakdowns during their execution. We evaluate our solutions with simulations and show that this new constraint allows robust solutions in decent computation time.

**Keywords:** Stochastic optimization · Chance constraints · Constraint programming

## 1 Introduction

Decisions in an organization are taken at different levels: strategic, tactical, operational, and execution. In some situations, it is important that decisions taken at the operation level take into account what could happen at the execution level. For instance, one can schedule tasks while taking into account that breakdowns might occur during the execution and that the schedule might not be followed as expected. Stochastic optimization allows taking decisions at one level while coping with random events at the next level. One way to achieve this goal is by using *chance constraints* in the optimization model to guarantee that the solution holds beyond a given threshold probability.

We introduce a new chance constraint called CONFIDENCE that forces variables to take sufficiently large values so that random variables, following a known distribution, are unlikely to be greater. Let $X_1, \ldots, X_n$ be decision variables, let $D_1, \ldots, D_n$ be statistical distributions, and let $\gamma$ be the confidence threshold (a constant parameter). Each distribution $D_i$ must have a well-defined cumulative distribution function $\text{cdf}_i$. The CONFIDENCE constraint is defined as follows.

$$\text{CONFIDENCE}([X_1, \ldots, X_n], [\text{cdf}_1, \ldots, \text{cdf}_2], \gamma) \iff \prod_{i=1}^{n} \text{cdf}_i(X_i) \geq \gamma \quad (1)$$


H. Simonis (Ed.): CP 2020, LNCS 12333, pp. 759–773, 2020.
https://doi.org/10.1007/978-3-030-58475-7_44

In other words, let $Y_i$ be an independent random variable following a distribution $D_i$. The chance that at least one random variable $Y_i$ takes a value greater than $X_i$ must be less than $\gamma$.

Such a constraint is particularly useful in stochastic scheduling where tasks can execute for a longer time than expected. In order to let sufficient time for the tasks to execute in a schedule, one can assign processing times to the tasks that are sufficiently long $\gamma$ percent of the time. We will show that such an approach has many advantages and is particularly well suited for constraint programming.

The rest of the paper is divided as follows. Section 2 presents background information about stochastic optimization. Section 3 introduces the CONFIDENCE constraint and its filtering algorithm. Section 4 presents a case study, an industrial scheduling problem that led to (and financed) this research. Section 5 presents our simulator that will be used in our experiments in Sect. 6. We present and analyze the results in Sect. 6.3. We conclude in Sect. 7.

## 2 Background

Stochastic problems are optimization problems for which part of the input is given with random variables. One aims at a solution of *good quality* over all possible values that these random variables can take. Usually, one wants to optimize the expected objective value and/or to ensure feasibility occurs with a given probability.

Stochastic linear programs are linear programs $\min\{c^T x \mid Ax \leq b, x \geq 0\}$ for which some parameters in $c$, $A$, or $b$ are replaced by random variables. These linear problems are particularly well studied [10]. They can, for instance, encode scheduling problems where processing times are subject to random delays [4].

There are several approaches to stochastic optimization. One important approach is called scenario-based optimization. It consists, in one way or another, in achieving deterministic optimization on sample data called scenarios. A scenario is a possible outcome of stochastic events. The number of possible scenarios grows exponentially with the number of stochastic variables. Chance constraints [20] can achieve scenario-based optimization by constraining the solution to satisfy a scenario with a given probability. The concept emerged in stochastic linear programming, but is not intrinsically linear and can be applied to any constraint satisfaction problem. The formulae described in [20] inspired the ones stated in Sect. 3. However, as the they sum probabilities over possible scenarios, some formulae might take too much time to compute for problems with hundreds of stochastic variables.

In constraint programming, Walsh [18] proposes an extension of a constraint satisfaction problem by modeling decision problems with uncertainty. In this extension, stochastic variables based on probabilistic distributions coexist with common constraint programming decision variables. The solver computes a policy, i.e. a solution for each possible outcome of the stochastic variables. While this suits the needs for many problems, for problems with hundreds of random variables, one cannot state (not to mention computing) a complete policy.

The modeling language MiniZinc [12,17] is also adapted to stochastic optimization [13] keeping the modeling process independent from the solving technique. It requires to encode a sample of scenarios in a vector. As the number of random variables grows, the cardinality of the sample scenarios set become insignificant compared to the number of possibilities and the quality of the solutions decays.

Stochastic problems is one way to obtain solutions robust to change. However, there exist other robust optimization approaches that do not rely on probabilities. For instance, *super solutions* are solutions that adapt themselves in case of a change [9]. Such solutions are, however, hard to compute and are often restricted to small instances. The FLEXC constraint [6,7] can model cumulative scheduling problems where tasks have two processing times: a normal processing time and a delayed one. The constraint ensures that if any subset of $k$ tasks is delayed, the cumulative resource is not overloaded. However, the constraint does not support delays of variable duration.

The outreach of chance constraints goes far beyond scheduling problems. Rossi et al. [14] present an algorithm designed to compute an optimal policy for inventory levels. Based on current demand, chance constraints are used in such a way that one can predict more accurate future demand. While we do not aim at solving the same problem, this is the closest chance constraint algorithm we could find in constraint programming to what we propose.

## 3 The Confidence Constraint

### 3.1 Description

The CONFIDENCE constraint defined in (1) has for scope a vector of integer variables $[X_1, \ldots, X_n]$ and has for parameters a vector of random distributions $[D_1, \ldots, D_n]$. A collection of independent random variables $Y_1, \ldots, Y_n$ following these distributions i.e. the random variable $Y_i$ follows the distribution $D_i$. Each distribution $D_i$ is fully described by its *cumulative distribution function* $\text{cdf}_i(v) = P[Y_i \leq v]$ that computes the probability that the random variable $Y_i$ takes a value smaller than or equal to a value $v$. We also consider its inverse function called the *quantile function* $Q_i(p) = \min\{v \mid P[Y_i \leq v] \geq p\}$ that computes the smallest value $v$ for which variable $Y_i$ takes a value no greater than $v$ with probability $p$. The distributions can be common distributions (e.g. Poisson, Uniform, ...) or can be custom much like the one shown in Sect. 4. The distributions can be different for every variable. Finally, the constraint takes a parameter $\gamma$ that is a confidence threshold, a probability between 0 and 1. Note that even if this constraint handles probabilities, it constrains integer variables and is therefore compatible with solvers that only handle this type of variable. Moreover, if the solver does not handle floating point parameters, it is possible to express $\gamma$ in percents and to give it an integer value.

The constraint is satisfied when, with probability at least $\gamma$, all random variables $Y_i$ take a value smaller than or equal to their corresponding threshold $X_i$.

Since the random variables $Y_i$ are independent, the constraint is satisfied when

$$P[\bigwedge_{i=1}^{n} Y_i \leq X_i] \geq \gamma. \tag{2}$$

Since the random variables are independent, we obtain

$$\prod_{i=1}^{n} P[Y_i \leq X_i] \geq \gamma. \tag{3}$$

For numerical stability reasons, we use this form

$$\sum_{i=1}^{n} \ln(P[Y_i \leq X_i]) \geq \ln(\gamma). \tag{4}$$

The Confidence constraint is useful in problems where unknown events might arise. In a scheduling problem, it is common to have tasks that take longer to execute than expected. One usually plans more time than necessary for these tasks in a schedule. Let $X_i$ be the processing time of task $i$ and $D_i$ be a distribution on the observed processing times in the past. The Confidence constraint allows planning sufficient times for the execution of the tasks in say, $\gamma = 95\%$ of the time. This confidence threshold allows making schedules that can be executed without modification 95% of the time without over estimating the duration of the tasks.

In a production problem where we want to produce sufficient goods to satisfy an unknown demand, we let $X_i$ be the amount of good $i$ that is produced and $Y_i$ be the amount that needs to be produced to fulfill the demand. We want $X_i \geq Y_i$ which is equivalent to $-X_i \leq -Y_i$. The Confidence constraint can restrict the quantities $-X_i$ to be smaller than the random variable $-Y_i$ with probability $\gamma$.

### 3.2 Filtering Algorithm

Algorithm 1 is the filtering algorithm directly derived from applying interval arithmetic on inequality (4). Line 1 computes $\alpha$, the log of the highest probability that can be reached on the left-hand side of the inequality. It is computed from the random variables' cumulative distribution functions. If this probably is too low, line 2 triggers a failure. Otherwise, the lower bound of each variable domain is tested for consistency. The test on line 3 is equivalent to testing $\sum_{j\neq i} \ln(P[Y_j \leq \max(\text{dom}(X_j))]) + \ln(P[Y_i \leq \min(\text{dom}(X_i))]) < \ln(\gamma)$. If the test is positive, then $\min(\text{dom}(X_i))$ does not have a support and should be filtered out from the domain. In order to evaluate what is the smallest value in $\text{dom}(X_i)$ with a support, we use the quantile function on line 4. We search for the smallest value $v \in \text{dom}(X_i)$ such that $\beta + \ln(P[Y_i \leq v]) \geq \ln(\gamma)$ which is equivalent to $P[Y_i \leq v] \geq \gamma e^{-\beta}$. This is directly provided by the quantile function $Q_i(\gamma e^{-\beta})$.

The algorithm accepts any distribution for which the functions $\text{cdf}_i$ and $Q_i$ can be computed. Such functions for common distributions are already implemented in libraries like Boost [15]. This includes Pareto, Gaussian, Poisson, Laplace, Uniform, etc. It is also possible to create custom distributions.

**Algorithm 1:** CONFIDENCE Filtering($[\text{dom}(X_1), \ldots, \text{dom}(X_n)], [D_1, \ldots, D_n], \gamma$)

Let $Y_i$ be a random variable following distribution $D_i$ for $i \in \{1, \ldots, n\}$;
1 $\alpha \leftarrow \sum_{i=1}^{n} \ln(\text{cdf}_i(\max(\text{dom}(X_i))))$;
2 **if** $\alpha < \ln(\gamma)$ **then**
  Fail with explanation $\bigwedge_{i=1}^{n} [\![X_i \leq \max(\text{dom}(X_i))]\!] \implies \text{False}$;
**for** $i \in \{1, \ldots, n\}$ **do**
  $\beta \leftarrow \alpha - \ln(\text{cdf}_i(\max(\text{dom}(X_i))))$;
3  **if** $\beta + \ln(\text{cdf}_i(\min(\text{dom}(X_i)))) < \ln(\gamma)$ **then**
4    $v \leftarrow Q_i(\gamma e^{-\beta})$;
    Filter $X_i$ with explanation $\bigwedge_{j \neq i} [\![X_j \leq \max(\text{dom}(X_j))]\!] \implies [\![X_i \geq v]\!]$;

From the arithmetic of intervals, the algorithm enforces bounds consistency on the CONFIDENCE constraint, but also domain consistency. Indeed, the algorithm tests on line 3 the assignment $[\max(\text{dom}(X_1)), \ldots, \max(\text{dom}(X_{i-1})), \min(\text{dom}(X_i)), \max(\text{dom}(X_{i+1})), \ldots, \max(\text{dom}(X_n))]$. If this assignment satisfies the CONFIDENCE constraint, so does $[\max(\text{dom}(X_1)), \ldots, \max(\text{dom}(X_{i-1})), v, \max(\text{dom}(X_{i+1})), \ldots, \max(\text{dom}(X_n))]$ for any value $v \in \text{dom}(X_i)$. Hence, all values in $\text{dom}(X_i)$ are domain consistent. Otherwise, the value $\min(\text{dom}(X_i))$ is filtered out from the domain of $X_i$ and the quantile function guarantees that the new lower bound on the domain satisfies the constraint.

The running time complexity is dominated by the number of calls to the cumulative distribution functions and the quantile functions. In the worst case, the algorithm performs $3n$ calls to the cumulative distribution functions and at most $n$ calls to the quantile functions. This leads to a running time complexity of $O(n)$.

# 4 Case Study

## 4.1 The Deterministic Version

We have an industrial partner in the textile industry whose needs motivate the theoretical contribution and whose data allow to empirically evaluate this contribution on an industrial problem.

The textile manufacturer has to schedule a set of tasks $\mathcal{I}$ on looms $L$. The tasks represent textile pieces to weave or setups to prepare a loom for the next textile. A piece of textile $i$ has a style $z_i$ which is a number that encodes the width of the textile, the type of thread, the weaving pattern, etc. A task $i$ is pre-assigned to loom $l_i \in L$ in order to simplify the problem. In order to process the task, this loom needs to be set up with configuration $c_i$. Each task $i \in \mathcal{I}$ has a due date $d_i$ and a priority $r_i$. The higher the priority, the more urgent the task is. In the deterministic version of the problem, every piece of textile $i$ has a predefined weaving duration $p_i$.

Looms are disjunctive resources [3] that can only weave one piece of textile at the time or being subject to one set up at the time. A loom $l \in L$ becomes available at time $a_l$. Prior to this time point, the loom completes tasks that were started and that cannot be changed. A loom $l$ is initially in configuration $c_l^{\text{init}}$ and upon a *major setup* operation of duration $p_l^{\text{major}}$, its new configuration becomes $c_l^{\text{final}}$. There is only one possible major setup per loom. A loom $l$ can only execute a task $i$ if it is assigned to that loom ($l_i = l$). It executes the task before its major setup if $c_i = c_l^{\text{init}}$ or after if $c_i = c_l^{\text{final}}$. A major setup is performed by a worker from the set $W$. Hence, at most $|W|$ major setups can be performed simultaneously.

A *minor setup* takes place between two consecutive weaving tasks on a loom, but does not change the configuration of the loom. There are up to 3 types of employees $p \in P$ interacting with a loom during a minor setup. The order is always as follows: weavers, beam tiers, and mechanics. Therefore, a minor setup is decomposed into an ordered sequence of 3 tasks, one associated for each profession. The minor setup duration is sequence-dependent in the sense that the duration $t_{i,j,p}$ for the profession $p$ is a function of the task $i$ before the setup and the task $j$ after. Employees are cumulative resources. Since there are $q_p$ employees of profession $p \in P$, up to $q_p$ minor setup tasks associated to profession $p$ can be simultaneously executed across the looms.

In the deterministic version of the problem, the unknowns are the starting time $S_i$ of each task $i \in \mathcal{I}$, the starting time $S_{i,p}^{\text{minor}}$ of the minor setup succeeding a task $i \in \mathcal{I}$ for each profession $p \in P$, and the starting time $S_l^{\text{major}}$ of the major setup of loom $l \in L$. The objective function is the weighted tardiness where the weights are the task priorities. We use this metric since the tight deadlines make it unreasonable to search for a schedule without tardiness.

The constraint model is similar to the one of a resource-constrained project scheduling problem (RCPSP) [2]. In a RCPSP, there are tasks to schedule and cumulative resources. Our problem differs from the RCPSP in the sense that minor setups have sequence-dependent processing times.

Figure 1 presents the constraint model of the deterministic problem as published in [1]. The variable $F_l$ encodes the first task on loom $l$. The order of the tasks is encoded in an array of variables $N$ such that for any piece of textile $i$, $N_i$ is the next piece of textile to weave. The dummy task $\sigma_l$ acts as a sentinel to encode the last task to execute on loom $l$. The task following the last task on a loom is the first task on the next loom (see constraints (11) and (12)). The vector $N$ forms a circuit visiting all the tasks hence Constraint (13). We model the disjunctive resources using less than or equal to constraints. Two CUMULATIVE constraints model the different setup types (minor and major).

$$\text{Minimize} \quad \sum_{i \in \mathcal{I}} r_i \cdot \max(0, S_i + p_i - d_i) \tag{5}$$

$$c_i = c_{l_i}^{\text{init}} \implies S_i + p_i \leq S_{l_i}^{\text{major}} \qquad \forall i \in \mathcal{I} \tag{6}$$

$$c_i \neq c_{l_i}^{\text{init}} \implies S_{l_i}^{\text{major}} + p_l^{\text{major}} \leq S_i \qquad \forall i \in \mathcal{I} \tag{7}$$

$$S_i + p_i \leq S_{i,1}^{\text{minor}} \qquad \forall i \in \mathcal{I} \tag{8}$$

$$S_{i,p+1}^{\text{minor}} \geq S_{i,p}^{\text{minor}} + t_{i,N_i,p} \qquad \forall i \in \mathcal{I}, \forall p \in P \setminus \{|P|\} \tag{9}$$

$$S_{N_i} = S_{i,|P|}^{\text{minor}} + t_{i,N_i,|P|} \qquad \forall i \in \mathcal{I} \tag{10}$$

$$N_{\sigma_l} = F_{l+1} \qquad \forall l \in [1, |L| - 1] \tag{11}$$

$$N_{\sigma_{|L|}} = F_1 \tag{12}$$

$$\textsc{Circuit}(N) \tag{13}$$

$$\textsc{Cumulative}([S_{i,p}^{\text{minor}} \mid i \in \mathcal{I}], [t_{i,N_i,p} \mid i \in \mathcal{I}], 1, q_p) \qquad \forall p \in P \tag{14}$$

$$\textsc{Cumulative}([S_l^{\text{major}} \mid l \in L], [p_l^{\text{major}} \mid l \in L], 1, |W|) \tag{15}$$

$$S_a \leq S_b \qquad \forall a, b \in \mathcal{I}, z_a = z_b \wedge l_a = l_b \wedge d_a \leq d_b \tag{16}$$

**Fig. 1.** Model equations directly taken from our previous paper [1]

### 4.2 The Stochastic Version

The solver acts as a black box and returns a solution that satisfies the constraints. The resulting schedule is valid, yet this solution does not consider the probabilities of different stochastic events. Since a loom weaves many threads, each job has a high chance of having multiple breakdowns. Each task is subject to two types of breakdowns with different probabilities and repair time. A warp thread has a high chance of breaking, but is quick to repair and the repair process is often automated. However, a weft thread rarely breaks, but when it does, an employee gets involved. Since both types of breakdowns are observable on a day-to-day basis, we must be able to consider them in a single distribution. In our case, the breakdowns follow Poisson distributions. We suppose the independence of both types of breakdowns (warp thread and weft thread). Since the sum of two independent random Poisson variables is a random Poisson variable [8], we can sum the $\lambda$-parameters of both breakdown distributions. Therefore, the duration of the breakdowns in our problem is a random variable that follows the sum of two Poisson distributions, which is equivalent to a single Poisson distribution. Depending on the task $i \in \mathcal{I}$ and the loom $l \in L$, the weft thread has an average number of breakdowns $\beta_{i,l}^{\text{weft}}$ between 0 to 10 times and when it breaks, it takes 10 min to repair. The warp thread has an average number of breakdowns $\beta_{i,l}^{\text{warp}}$ between 0 and 40 times and it takes 2 or 4 min to repair. The average numbers of breakdowns $\beta_{i,l}^{\text{weft}}$ and $\beta_{i,l}^{\text{warp}}$ are proportional to the duration of task $i$ on loom $l$. The $\lambda$-parameter of the Poisson distribution for the delay caused by the breakdowns is therefore $10 \cdot \beta_{i,l}^{\text{weft}} + 2 \cdot \beta_{i,l}^{\text{warp}}$ or $10 \cdot \beta_{i,l}^{\text{weft}} + 4 \cdot \beta_{i,l}^{\text{warp}}$. Our case study is about solving the stochastic version of this problem when weaving tasks might take longer to execute than expected.

*A Solution with Fixed Processing Times:* One way to solve the problem could be to artificially increase the duration of the weaving tasks by a given percentage. This prevents the unplanned events from starving the resource pool available for setups which causes ripple effects. The drawback of this technique happens when we overestimate the duration and the number of delays. The delays must be chosen in a smart and intuitive manner. Giving the solver a choice of where to take risks can lead to a better solution. This is where our constraint comes in.

*A Solution with Flexible Processing Times:* As a second solution, we rather modify the constraint model with an additional variable $P_i$ to represent the duration of a weaving task, including breakdowns, and a variable $B_i$ that includes only the breakdowns. The task duration $P_i$ cannot be shorter than the deterministic version of the problem, but could be longer if we plan time for breakdowns. We therefore insert this constraint.

$$P_i = p_i + B_i \tag{17}$$

We impose a CONFIDENCE constraint on the duration of the breakdowns to ensure that we plan for sufficiently long breakdowns with probability $\gamma$.

$$\textsc{Confidence}([B_i \mid i \in \mathcal{I}], [cdf_i \mid i \in \mathcal{I}], \gamma) \tag{18}$$

The CONFIDENCE constraint can guarantee that a breakdown does not alter the subsequent tasks with probability $\gamma$. That is, if the breakdown is shorter than what is computed by the solver, all tasks can still start at the expected time. If the breakdown is longer than what was planned, it might cause a ripple effect and delay the execution of the future tasks, but this only happens with probability $1 - \gamma$. Therefore, a valid solution to our problem is a solution that causes no disturbances, $\gamma$ % of the time. The rest of the new model with variable processing time is similar to [11]. The constraints 6 and 8 are modified by replacing $p_i$ with $P_i$.

## 5 Simulation

The simulation model is used to evaluate the quality of the solutions. This implies measuring robustness by emulating scenarios and how well the solutions cope with stochastic events such as breakdowns. The simulator produces an in-depth three-dimensional depiction of the work environment. It considers the number of employees, their positions, the time each employee takes to move between looms, the number of looms, etc. By using a simulator, we can compare different solutions without trying them in practice. Since the company might be reluctant to directly apply the results to their schedules, this also acts as a way to alleviate doubts. The simulator was designed to test scenarios and to answer questions that are much beyond the scope of this paper and even beyond scheduling in general. It can be used to see the effect of strategic decisions on textile production and the effect of modifying the number of resources on the performance of the plant.

For our purpose, the simulation model receives as input the same data as our optimization model (configuration of the looms, availability of the resources, duration of the tasks) and the output of the optimization (the starting time of the tasks). One point that is worth mentioning is that the optimization model uses, for each weaving task, a distinct distribution for the duration of the breakdowns. However, the simulator averages out these distributions per loom and uses the same distribution for all weaving tasks on a given loom.

# 6 Experiments

## 6.1 Methodology

Three different methods were compared. First, in the DETERMINISTIC model, we ignore the stochastic events. The DETERMINISTIC model consists of using the model from Fig. 1 without any change. In the FIXED model, we artificially increase the tasks' processing times by the sum of the average breakdown duration multiplied by the average number of breakdowns. This corresponds to the first solution presented in Sect. 4.2. Finally, the CONFIDENCE method is the model that uses our chance constraint with $\gamma \in \{0.01, 0.05, 0.1, 0.2, 0.3, \ldots, 0.8, 0.9, 0.95, 0.99\}$. The processing time of a task becomes a variable equal to the sum of the deterministic processing time (a constant) and the duration of breakdowns (another variable). These breakdown duration variables are subject to the CONFIDENCE constraint using the Poisson distributions. This corresponds to the second solution presented in Sect. 4.2.

The CP model was written in the MiniZinc 2.4.2 language [12]. We use the solver Chuffed [5] with the free search parameter [16]. The simulation is modeled using the academic version of SIMIO [19]. We ran the experiments on a computer with the configuration: Ubuntu 19.10, 64 GB ram, Processor Intel(R) Core(TM) i7-6700K CPU @ 4.00 GHz, 4 Cores, 8 Logical Processors. All optimization processes were given a timeout of 30 min and 10 h.

To compare the quality of the schedules provided by the different methods, we use the mean simulated weighted tardiness. Let $E_i^s$ be the ending time of task $i$ in simulation $s$. The mean simulated weighted tardiness $\bar{T}$ is calculated as follows:

$$\bar{T} = \frac{1}{n} \sum_{s=1}^{n} \sum_{i \in \mathcal{I}} r_i \max(0, E_i^s - d_i) \tag{19}$$

The inner sum computes the weighted tardiness of a simulation $s$ while the outer sum averages over $n$ simulations. We use $n = 100$ simulations. A simulation is a randomly generated scenario based on probabilities, i.e. that the duration of the breakdowns are randomly drawn according to the Poisson distributions we established.

During the simulation of a schedule, tasks can be delayed by breakdowns that were not planned by the solver. The task on a loom can only start after its predecessor, on the same loom, is completed. An idle technician always start working on a setup when the loom gets ready. If s/he has the choice between setting up two looms, the priority is given to the setup that was planned earlier in the original schedule. A loom that completes a task earlier than expected must wait before starting the next task as we suppose that the material for this next task is not necessarily ready before its planned start time. Indeed, when a task is planned, it instructs people on the floor to prepare the material in order for the task to start on time. This preparation is not part of the optimization since it adds too much complexity to the decision process. The resources in charge of preparing the material are voluntary omitted from the problem definition. However, if, during the execution of the plan, we inform at the last minute that a task can start earlier, we cannot assume that the preparation will be ready. In other words, the tasks are subject to precedence constraints with tasks that are external to the problem.

### 6.2 Instances

To compare the methods, we had access to four datasets. In these datasets, there are 5 workers for minor setups, 1 for major setups, and a total of 81 looms available. The scheduling horizon is 14,400 min which is roughly two weeks of work time. Table 1 shows the characteristics of each instance, including the number of pieces of textiles to weave. It also reports the resource usage rate $r = \frac{1}{H|L|}\sum_i p_i$ where $H$ is the scheduling horizon, $|L|$ is the number of looms and $p_i$ is the processing time (without breakdowns) of task $i$. We also report the usage rate when processing times are augmented with the expected breakdown duration. These breakdowns approximately add 4% to the resource usage in the last three datasets.

Dataset 1 consists of a reduced dataset where pieces of textiles with a due date of 0 (these tasks are already late) and some others with due dates larger than the horizon are removed. This leads to an easier dataset to solve for which optimal solutions are obtained within 30 minutes of computation time with every method.

Datasets 2, 3, and 4 are not modified, but cannot be optimally solved even after 10 hours. These datasets are more realistic with respect to what is usually observed in the industry where one has to work with suboptimal solutions. Those are directly extracted from our industrial partner's database.

In every instance, there is $|W| = 1$ mechanic available for major setups, 2 mechanics, 5 weavers, and 3 beam tiers available at all time for minor setups.

**Table 1.** Dataset descriptions

| Dataset | 1 | 2 | 3 | 4 |
|---|---|---|---|---|
| Textile pieces | 448 | 602 | 480 | 525 |
| Resource usage (%) without / with average breakdowns | 47.23/51.37 | 48.06/52.33 | 52.94/57.79 | 48.07/48.30 |

## 6.3 Results

In Figs. 2, 4, 6, and 8, we present the results of our models with different methods on the $x$-axis. The CONFIDENCE method was tested with different confidence thresholds $\gamma$. The $y$-axis is the weighted tardiness. There are two colors of dots and lines. Red presents the prediction value. The prediction is essentially the objective value of the solution returned by the solver, i.e. the weighted tardiness if the plan executes as expected by the CONFIDENCE constraint. Blue shows the mean simulated weighted tardiness of the 100 runs with its 95% confidence interval.

Figure 2 shows the results for the reduced dataset solvable to optimality. Notice that the red curve is non-decreasing, since an optimal schedule that satisfies the CONFIDENCE constraint with threshold $\gamma$ is a feasible, but potentially suboptimal, schedule for a smaller threshold. For this instance, which is rather trivial to solve since few tasks were competing for the resources, the best strategy was to plan no extra time for the breakouts. In case of a breakout, the next tasks are simply postponed until the delayed task is completed. At this time, the resource is always available to perform the setup to execute the next task and no further delay is caused. This strategy is achieved by the DETERMINISTIC method and the CONFIDENCE constraint with $\gamma = 0.01$. However, the CONFIDENCE predicted value is closer to reality. At the opposite, for $\gamma = 0.99$, the weighted tardiness is at its highest since the duration of nearly all tasks is set to the longest possible duration. Notice how the FIXED method offers a performance similar to the CONFIDENCE method with $\gamma = 0.5$.

For the datasets 2, 3, and 4, we clearly see on Figs. 3, 4, 5, 6, 7 and 8 that the solver does not produce optimal solutions even within 10 h. Indeed, the red curve is non-monotonic. Moreover, results of the dataset 2 with $\gamma$95% are not reported as no solution was obtained within 10 h. This happens with even lower confidence thresholds when using a timeout of 30 min. However, notice that the simulation follows the predictions of the model for the higher values of $\gamma$. The gap between the objective value obtained by the solver (red) and the simulation (blue) decreases by increasing $\gamma$. In every case, the best solution, as evaluated by the simulator (blue), comes from the CONFIDENCE method. In Dataset 4, the best solution is obtained at $\gamma = 20\%$. The CONFIDENCE model trades objective value for robustness. The CONFIDENCE model outperforms the FIXED model in both objective value and robustness. We can also notice that choosing a bad $\gamma$ value can make or break the quality of a solution. The solutions for $\gamma = 1\%$ and $\gamma = 95\%$ are worse than the DETERMINISTIC model solution.

With the confidence intervals, we notice that the mean is usually quite precise with 100 runs of the simulation model. The lengthy events are rarer, therefore, increasing the number of simulations decreases the variability. Yet, using a higher

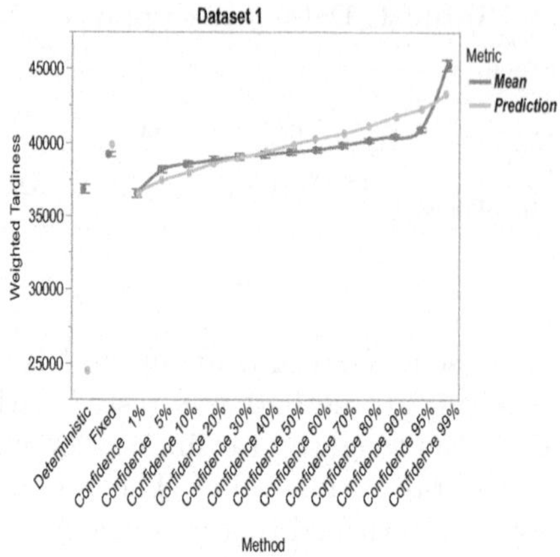


**Fig. 2.** Dataset with 448 textile pieces to weave (Color figure online)

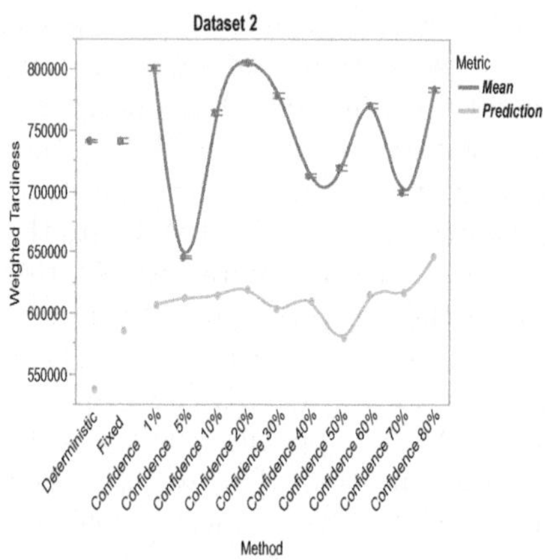


**Fig. 3.** Dataset with 602 textile pieces to weave. 30 min timeout. (Color figure online)

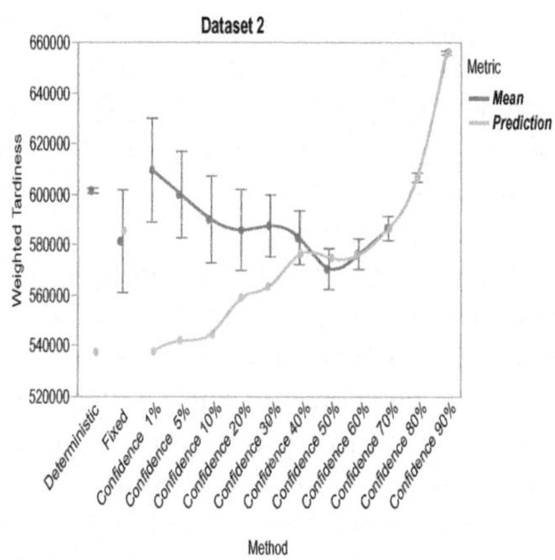


**Fig. 4.** Dataset with 602 textile pieces to weave. 10 h timeout. (Color figure online)

value of $\gamma$ also reduces the variability. With CONFIDENCE the prediction expects more tardiness, but the guarantee of not causing ripple effects makes it so the mean simulated weighted tardiness is better.

The DETERMINISTIC method has no stochastic optimization involved and completely ignores potential breakdowns. We would have thought that during the simulation, this method would most likely lag behind the planned schedule and finally execute a non-optimize schedule. However, it seems that in a context where resources are not scarce (dataset 1), repairing the plan as it executes is not a bad way to proceed. This pattern gradually vanishes with bigger instances.

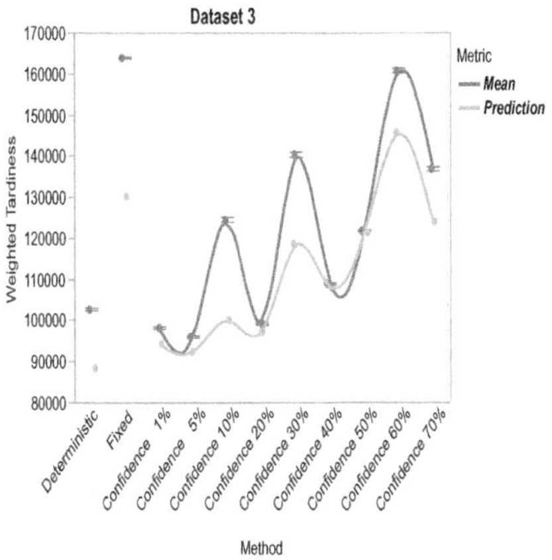


**Fig. 5.** Dataset with 480 textile pieces to weave. 30 min timeout. (Color figure online)

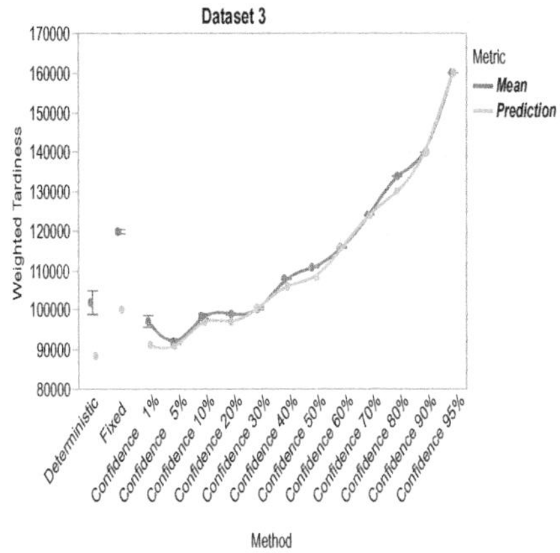


**Fig. 6.** Dataset with 480 textile pieces to weave. 10 h timeout. (Color figure online)

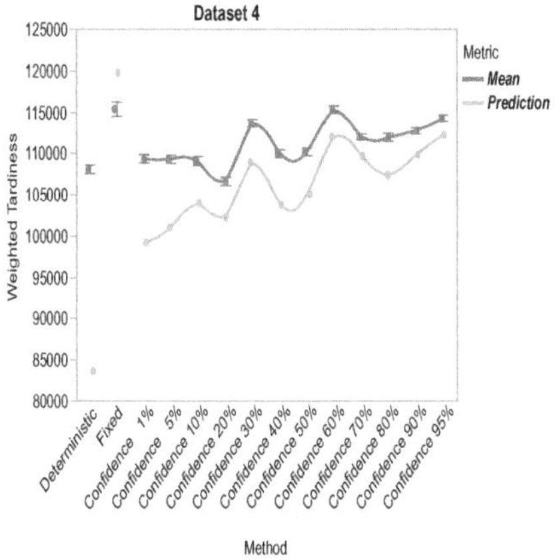


**Fig. 7.** Dataset with 525 textile pieces to weave. 30 min timeout. (Color figure online)

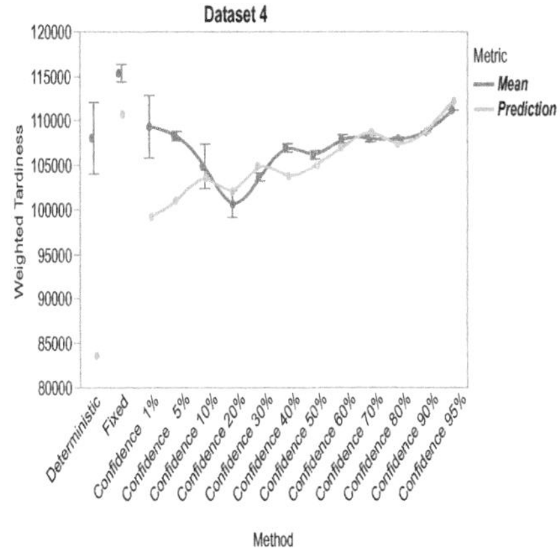


**Fig. 8.** Dataset with 525 textile pieces to weave. 10 h timeout. (Color figure online)

On the harder datasets (datasets 2, 3, and 4), the objective value adopts a wavy form. Our main hypothesis is that varying $\gamma$ does not only affect the objective value, it also affects the difficulty to solve the problem. So decreasing $\gamma$ would normally decrease the objective value, but in some situations, the search heuristic gets trapped and the solver returns a worse solution once the timeout is reached. This is especially apparent with a timeout of 30 min.

The harder datasets (2, 3, and 4) have tighter schedules. A ripple effect from a breakdown can produce increasingly more lateness. We assume that the usefulness of the CONFIDENCE scales with the risk of the ripple effect. In Figs. 4,

6, and 8, the curves are respectively centered around $\gamma = 50\%, 5\%, 20\%$. The red curve seems to show that $\gamma = 100\%$ cannot be achieved. This can be explained by the difficulty of satisfying the higher percent of a Poisson cdf function and the tendency of probability products to tend towards 0.

When we compare the different solutions graphs with the Table 1, we can notice the following. The more resource usage there is, the harder it is to solve the problem with a higher $\gamma$ value. The instances also get harder as the number of tasks increases. While we don't include computation times, we noticed that in the dataset 1, the computation times were similar (less than a second of difference) to the deterministic model until 99% where it spiked. The computation times for the other datasets where inconclusive since no optimal solution was found.

The FIXED method often overestimates the breakdown times needed to create robustness and produces schedules where tasks are planned to be executed late. The FIXED method is more rigid compared to CONFIDENCE because it cannot allow to choose where the risk should be taken in order to minimize the weighted tardiness. More often than not, the FIXED approach gives results similar to the CONFIDENCE method with $\gamma = 0.5$, but slightly worse. In a context where resources are abundant, it would be interesting to compute the processing times by using a value shorted than the average breakdown time.

The CONFIDENCE method shows that the weighted tardiness computed by the solver is a good indicator of the weighted tardiness obtained in the simulation. It would be interesting to push further the evaluation with instances, not necessarily industrial, for which we can control the scarcity of the resources. A search heuristic adapted to stochastic optimization could also help with this type of problem and reduce the computation times.

## 7 Conclusion

We presented the CONFIDENCE constraint, a new global chance constraint that provides robust solutions in stochastic contexts. We presented a linear time filtering algorithm with explanations. We compared the CONFIDENCE constraint to a DETERMINISTIC and FIXED approach and determined that the new global chance constraint offers good prediction about the weighted tardiness obtained in the simulation. It is planned that our industrial partner sets the $\gamma$-parameter according to how much risk they are willing to take and readjusts it every week. The deterministic method is implemented and being deployed by our industrial partner, while the stochastic optimization is still under evaluation. More stochastic events need to be taken into account such as the arrival of new orders.

## References

1. Mercier-Aubin, A., Gaudreault, J., Quimper, C.G.: Leveraging constraint scheduling: a case study to the textile industry. In: 17th International Conference on the Integration of Constraint Programming, Artificial Intelligence, and Operations Research, Vienna, Austria. Springer, Heidelberg (2020)

2. Baptiste, P., Le Pape, C.: Constraint propagation and decomposition techniques for highly disjunctive and highly cumulative project scheduling problems. Constraints **5**(1), 119–139 (2000). https://doi.org/10.1023/A:1009822502231
3. Baptiste, P., Le Pape, C., Nuijten, W.: Constraint-Based Scheduling. Kluwer Academic Publishers, Dordrecht (2001)
4. Chakrabortty, R.K., Sarker, R.A., Essam, D.L.: Resource constrained project scheduling with uncertain activity durations. Comput. Ind. Eng. **112**, 537–550 (2017)
5. Chu, G., Stuckey, P.J., Schutt, A., Ehlers, T., Gange, G., Francis, K.: Chuffed, a lazy clause generation solver (2018)
6. Derrien, A., Petit, T., Zampelli, S.: A declarative paradigm for robust cumulative scheduling. In: O'Sullivan, B. (ed.) CP 2014. LNCS, vol. 8656, pp. 298–306. Springer, Cham (2014). https://doi.org/10.1007/978-3-319-10428-7_23
7. Fahimi, H.: Efficient algorithms to solve scheduling problems with a variety of optimization criteria. Ph.D. thesis, Université Laval (2016)
8. Grimmett, G., Welsh, D.: Probability: An Introduction. Oxford University Press, Oxford (2014)
9. Hebrard, E., Walsh, T.: Improved algorithm for finding (a,b)-super solutions. In: van Beek, P. (ed.) CP 2005. LNCS, vol. 3709, p. 848. Springer, Heidelberg (2005). https://doi.org/10.1007/11564751_86
10. Kall, P., Mayer, J.: Stochastic Linear Programming. Springer, Boston (2011). https://doi.org/10.1007/978-1-4419-7729-8
11. Lombardi, M., Milano, M.: A precedence constraint posting approach for the RCPSP with time lags and variable durations. In: Gent, I.P. (ed.) CP 2009. LNCS, vol. 5732, pp. 569–583. Springer, Heidelberg (2009). https://doi.org/10.1007/978-3-642-04244-7_45
12. Nethercote, N., Stuckey, P.J., Becket, R., Brand, S., Duck, G.J., Tack, G.: MiniZinc: towards a standard CP modelling language. In: Bessière, C. (ed.) CP 2007. LNCS, vol. 4741, pp. 529–543. Springer, Heidelberg (2007). https://doi.org/10.1007/978-3-540-74970-7_38
13. Rendl, A., Tack, G., Stuckey, P.J.: Stochastic MiniZinc. In: O'Sullivan, B. (ed.) CP 2014. LNCS, vol. 8656, pp. 636–645. Springer, Cham (2014). https://doi.org/10.1007/978-3-319-10428-7_46
14. Rossi, R., Tarim, S., Hnich, B., Pestwich, S.: A global chance-constraint for stochastic inventory systems under service level constraints. Constraints **13**(4), 490–517 (2008). https://doi.org/10.1007/s10601-007-9038-4
15. Schling, B.: The Boost C++ Libraries. XML Press, Laguna Hills (2011)
16. Shishmarev, M., Mears, C., Tack, G., Garcia de la Banda, M.: Learning from learning solvers. In: Rueher, M. (ed.) CP 2016. LNCS, vol. 9892, pp. 455–472. Springer, Cham (2016). https://doi.org/10.1007/978-3-319-44953-1_29
17. Stuckey, P.J., Feydy, T., Schutt, A., Tack, G., Fischer, J.: The minizinc challenge 2008–2013. AI Mag. **35**(2), 55–60 (2014)
18. Walsh, T.: Stochastic constraint programming. In: Proceedings of the 15th European Conference on Artificial Intelligence (ECAI-2002), pp. 111–115 (2009)
19. Zaayman, G., Innamorato, A.: The application of simio scheduling in industry 4.0. In: 2017 Winter Simulation Conference (WSC), pp. 4425–4434 (2017)
20. Zghidi, I., Hnich, B., Rebaï, A.: Modeling uncertainties with chance constraints. Constraints **23**(2), 196–209 (2018). https://doi.org/10.1007/s10601-018-9283-8

# Parity (XOR) Reasoning for the Index Calculus Attack

Monika Trimoska(✉), Sorina Ionica, and Gilles Dequen

Laboratoire MIS, Université de Picardie Jules Verne, Amiens, France
{monika.trimoska,sorina.ionica,gilles.dequen}@u-picardie.fr

**Abstract.** Cryptographic problems can often be reduced to solving Boolean polynomial systems, whose equivalent logical formulas can be treated using SAT solvers. Given the algebraic nature of the problem, the use of the logical XOR operator is common in SAT-based cryptanalysis. Recent works have focused on advanced techniques for handling parity (XOR) constraints, such as the Gaussian Elimination technique. First, we propose an original XOR-reasoning SAT solver, named WDSat (Weil Descent SAT solving), dedicated to a specific cryptographic problem. Secondly, we show that in some cases Gaussian Elimination on SAT instances does not work as well as Gaussian Elimination on algebraic systems. We demonstrate how this oversight is fixed in our solver, which is adapted to read instances in algebraic normal form (ANF). Finally, we propose a novel preprocessing technique based on the Minimal Vertex Cover Problem in graph theory. This preprocessing technique is, within the framework of multivariate Boolean polynomial systems, used as a DLL branching selection rule that leads to quick linearization of the underlying algebraic system. Our benchmarks use a model obtained from cryptographic instances for which a significant speedup is achieved using the findings in this paper. We further explain how our preprocessing technique can be used as an assessment of the security of a cryptographic system.

## 1 Introduction

Cryptanalysis is the study of methods to decrypt a ciphertext without any knowledge of the secret key. Academic research in cryptanalysis is focused on deciding whether a cryptosystem is secure enough to be used in the real world. In addition, a good understanding of the complexity of a cryptographic attack allows us to determine the secret key length, making sure that no cryptanalytic effort can find the key in a feasible amount of time. Recommendations for minimum key length requirements given by various academic and governmental organizations [4] are based on the complexity of known attacks.

In recent years, constraint programming (CP) techniques have been used in the cryptanalysis of both public and secret key cryptosystems. A first example in the field of differential cryptanalysis is given by the work of Gerault *et al.* [16–18] who showed how to use CP for solving the optimal related-key differential


H. Simonis (Ed.): CP 2020, LNCS 12333, pp. 774–790, 2020.
https://doi.org/10.1007/978-3-030-58475-7_45

characteristic problem. Using the CP model presented in their work, all optimal related-key differential characteristics for AES-128, AES-192 and AES-256 can be computed in a few hours [17]. We also note the work of Lui *et al.* [20,21], in which a CP model is used to aid the Tolerant Algebraic Side-Channel Analysis, which is a combination of algebraic and side-channel analysis.

In a second line of research, Boolean satisfiability (SAT) solvers have found use in algebraic cryptanalysis. Algebraic cryptanalysis denotes any technique which reduces a cryptographic attack to the problem of solving a multivariate Boolean polynomial system. A common approach for solving these systems is to use Gröbner basis algorithms [12], exhaustive search [6] or hybrid methods [2]. These methods have been compared against SAT solving techniques for attacks on various symmetric cryptosystems such as Bivium, Trivium, Grain. Recent work has also focused on combining algebraic and SAT solving techniques [7]. In public-key cryptography, SAT solvers have been considered for attacking binary elliptic curve cryptosystems using the index calculus attack [14]. In this paper, we tackle this last-mentioned application.

We propose a built-from-scratch SAT solver dedicated to solving an important step of the index calculus attack. The solver, named WDSat, is adapted for XOR-reasoning and reads formulas in ANF form. In addition, we show certain limitations of the Gaussian Elimination (GE) technique in XOR-enabled SAT solvers by pointing out a canceling property that is present in algebraic resolution methods but is overseen in current SAT-based GE implementations. We refer to this canceling property as the XG-ext method and we show how it is implemented in our solver. In implementations, the XG-ext method comes at a high computational cost and is thus useful only for benchmarks where it reduces significantly the number of conflicts. Finally, we introduce a graph theory-based preprocessing technique, specifically designed for multivariate Boolean polynomial systems, that allows us to further accelerate the resolution of our benchmarks. This preprocessing technique is designed to allow a rapid linearization of the underlying algebraic system and should be used coupled with the XG-ext method. In fact, when the XG-ext method is not applied, the positive outcome of the preprocessing technique cannot be guaranteed. To confirm, we perform experiments using CryptoMiniSat [27] coupled with our preprocessing technique and show that this combination yields slower running times than CryptoMiniSat alone. Experimental results in Sect. 6 show that the solver presented in this paper outperforms all existing solving approaches for the introduced problem. These approaches include Gröbner basis techniques [12] and state-of-the-art SAT solvers: MiniSat [11], Glucose [1], MapleLCMDistChronoBT [23], CaDiCaL [3] and CryptoMiniSat [27].

## 2 Background

**Index Calculus.** In cryptanalysis, the index calculus algorithm is a well-known method for attacking factoring and elliptic curve discrete logarithms, two computational problems which are at the heart of most used public-key cryptosystems.

When performing this attack for elliptic curve discrete logarithms, a crucial step is the point decomposition phase. As proposed by Gaudry [15] and Diem [10] independently, a point on the elliptic curve can be decomposed into $m$ other points by solving Semaev's $(m+1)$-th summation polynomial [25], that we denote by $S_{m+1}$. For elliptic curves defined over binary fields, the second and the third summation polynomials are defined as follows:

$$S_2(X_1, X_2) = X_1 + X_2, \tag{1}$$
$$S_3(X_1, X_2, X_3) = X_1^2X_2^2 + X_1^2X_3^2 + X_1X_2X_3 + X_2^2X_3^2 + 1.$$

For $m > 3$, the $m$-th summation polynomial is computed by using the following recursive formula:

$$S_m(X_1, \dots, X_m) = \tag{2}$$
$$Res_X(S_{m-k}(X_1, \dots, X_{m-k-1}, X), S_{k+2}(X_{m-k}, \dots, X_m, X)),$$

where $Res_X$ denotes the resultant of two polynomials with respect to the $X$ variable and $1 \leq k \leq m-3$. The zeros of this polynomial will give the $x$-coordinates of points on the elliptic curve as elements in $\mathbb{F}_{2^n}$. From an implementation point of view, these will be represented as $n$-bit vectors. In index calculus attacks, the common approach is to decompose a random point given by an $n$-bit vector $x$-coordinate into $m$ points whose $x$-coordinates write as $l$-bit vectors, with $l \sim \frac{n}{m}$ (see for instance [13,24]). With this choice of parameters, the problem of decomposing a random point by finding the zeros of $S_{m+1}$ can be reduced to solving a system of $n$ Boolean polynomials with $ml$ variables.

We recall that a multivariate Boolean polynomial system is a system of polynomials in several variables and whose coefficients are in $\mathbb{F}_2$ (see for instance [19]). The following example shows a Boolean polynomial system of three equations in the variables $\{\mathbf{x}_1, \mathbf{x}_2, \mathbf{x}_3\}$:

$$\mathbf{x}_1 + \mathbf{x}_2 \cdot \mathbf{x}_3 = 0$$
$$\mathbf{x}_1 \cdot \mathbf{x}_2 + \mathbf{x}_2 + \mathbf{x}_3 = 0$$
$$\mathbf{x}_1 + \mathbf{x}_1 \cdot \mathbf{x}_2 \cdot \mathbf{x}_3 + \mathbf{x}_2 \cdot \mathbf{x}_3 = 0.$$

In the literature, the modelisation process allowing to obtain a Boolean polynomial system from a polynomial with coefficients in $\mathbb{F}_{2^n}$ (here the summation polynomial) is called a *Weil Restriction* [15] or *Weil Descent* [24]. The polynomial systems obtained in this way serve as our starting point for deriving SAT instances.[1]

**XOR-Enabled SAT Solvers.** A Boolean polynomial system can be rewritten as a conjunction of logical formulas in algebraic normal form (ANF) as follows: multiplication in $\mathbb{F}_2$ $(\cdot)$ becomes the logical AND operation $(\wedge)$ and addition in $\mathbb{F}_2$ $(+)$ becomes the logical XOR $(\oplus)$. The elements 0 and 1 in $\mathbb{F}_2$ correspond to

[1] Our C code for generating these instances is publicly available [28].

$\bot$ and $\top$, respectively. Consequently, solving a multivariate Boolean polynomial system is equivalent to solving a conjunction of logical formulas in ANF form. To date, few SAT solvers are adapted to tackle formulas in ANF. A common approach is to transform the ANF form in a CNF-XOR form, which is a conjunction of CNF and XOR clauses. In order to do this, every conjunction of two or more literals $x_1 \wedge x_2 \wedge \ldots \wedge x_k$ has to be replaced by an additional and equivalent variable $x'$ such that $x' \Leftrightarrow x_1 \wedge x_2 \wedge \ldots \wedge x_k$. This equivalence can be rewritten in CNF using a three-step transformation. First, the equivalence is decomposed into two implications:

$$\begin{aligned}&(x' \Rightarrow x_1 \wedge x_2 \wedge \ldots \wedge x_k) \wedge \\ &(x_1 \wedge x_2 \wedge \ldots \wedge x_k \Rightarrow x').\end{aligned}$$

Then, the material implication rule is applied:

$$\begin{aligned}&(\neg x' \vee (x_1 \wedge x_2 \wedge \ldots \wedge x_k)) \wedge \\ &(\neg(x_1 \wedge x_2 \wedge \ldots \wedge x_k) \vee x').\end{aligned}$$

Finally, using distribution on the first, and De Morgan's law on the second constraint, we obtain the following CNF formula:

$$\begin{aligned}&(\neg x' \vee x_1) \wedge \\ &(\neg x' \vee x_2) \wedge \\ &\ldots \\ &(\neg x' \vee x_k) \wedge \\ &(\neg x_1 \vee \neg x_2 \vee \ldots \vee \neg x_k \vee x').\end{aligned} \tag{3}$$

When we substitute all occurrences of conjunctions in an XOR clause by an additional variable, we obtain a formula in CNF-XOR form. This is the form used in the CryptoMiniSat solver [27], which is an extension of the MiniSat solver [11] specifically designed to work on cryptographic problems.

*Example 1.* Let us consider the Boolean polynomial system:

$$\begin{aligned}&\mathbf{x}_1 + \mathbf{x}_2 \cdot \mathbf{x}_3 + \mathbf{x}_5 + \mathbf{x}_6 + 1 = 0 \\ &\mathbf{x}_3 + \mathbf{x}_5 + \mathbf{x}_6 = 0.\end{aligned} \tag{4}$$

One additional variable $x'$ needs to be introduced to substitute the monomial $\mathbf{x}_2 \cdot \mathbf{x}_3$. The corresponding CNF-XOR form for this Boolean system is a conjunction of the following clauses:

$$\begin{aligned}&x' \vee \neg x_2 \vee \neg x_3 \\ &\neg x' \vee x_2 \\ &\neg x' \vee x_3 \\ &x_1 \oplus x' \oplus x_5 \oplus x_6 \\ &x_3 \oplus x_5 \oplus x_6 \oplus \top.\end{aligned} \tag{5}$$

Finally, one could, of course, consider generic solvers (i.e. MiniSat [11], Glucose [1]) for solving cryptographic problems, but this approach needs to further transform the CNF-XOR model to a CNF one. Transforming an XOR-clause with $k$ literals in CNF representation is a well-known process that gives $2^{k-1}$ OR-clauses of $k$ literals.

*Notation.* For simplicity, in the remainder of this paper we will omit the multiplication operator $\cdot$ whenever its use in monomials is implicit. Moreover, due to equivalence between a Boolean polynomial and an ANF form, these will be used interchangeably.

## 3 The WDSat Solver

Our WDSat solver is based on the Davis-Putnam-Logemann-Loveland (DPLL) algorithm [8], which is a state-of-the-art complete SAT solving technique. The solver is designed to treat ANF formulae derived from the Weil Descent modelisation of cryptographic attacks, hence its name: WDSat. The code for the WDSat solver is written in C and is publicly available [29].

WDSat implements three reasoning modules. These include the module for reasoning on the CNF part of the formula and the so-called XORSET and XOR-GAUSS (XG) modules designed for reasoning on XOR constraints. The CNF module is designed to perform classic unit propagation on OR-clauses. The XORSET module performs the operation equivalent to unit propagation, but adapted for XOR-clauses. Practically, this consists in checking the parity of the current interpretation and propagating the unassigned literal. Finally, the XG module is designed to perform GE on the XOR constraints dynamically. We also implement an XG extension, described in Sect. 4. The following is a detailed explanation of this module.

XOR clauses are normalized and represented as equivalence classes. Recall that an XOR-clause is said to be in *normal form* if it contains only positive literals and does not contain more than one occurrence of each literal. Since we consider that all variables in a clause belong to the same equivalence class (EC), we choose one literal from the EC to be the *representative*. An XOR-clause $(x_1 \oplus x_2 \oplus ... \oplus x_n) \Leftrightarrow \top$ rewrites as

$$x_1 \Leftrightarrow (x_2 \oplus x_3 \oplus ... \oplus x_n \oplus \top). \tag{6}$$

Finally, we replace all occurrences of a representative of an XOR clause with the right side of the equivalence. Applying this transformation, we obtain a simplified system having the following property: a representative of an EC will never be present in another EC.

Let $R$ be the set of representatives and $C$ be the set of clauses. $R$ and $C$ hold the right-hand side and the left-hand side of all equations of type (6) respectively. We denote by $C_x$ the clause in $C$ that is equivalent to $x$. In other words, $C_x$ is the right-hand side of the EC that has $x$ as representative. Finally, we denote by $var(C_x)$ the set of literals (plus a $\top/\bot$ constant) in the clause $C_x$ and $C[x_1/x_2]$

denotes the following substitution of clauses: for all $C_i \in C$ containing $x_1$, $C_i \leftarrow C_i \oplus x_1 \oplus x_2$, i.e. $x_1$ is replaced by $x_2$ in $C_i$. When we replace a literal $x_1$ by a clause $C_{x_2}$, we adopt a similar notation: $C[x_1/C_{x_2}]$.

Thus, assigning a literal $x_1$ to $\top$ leads to using one of the rules in Table 1, depending on whether $x_1$ belongs to $R$ or not. In both cases, propagation occurs when: $\exists\ x_i \neq x_1\ s.t.\ var(C_{x_i}) = \top/\bot$. Conflict occurs when one constraint leads to the propagation of $x_i$ to $\top$ and another constraint leads to the propagation of $x_i$ to $\bot$.

Table 1 presents inference rules for performing GE in the XG module of WDSat. Applying these rules allows us to maintain the property of the system which states that a representative of an EC will never be present in another EC. For clarity of the notation, the first column of this table contains the premises, the second one contains the conclusion and the third one is an update on the set $R$ which has to be performed when the inference rule is used.

**Table 1.** Gaussian elimination inference rules.

| Premises | Conclusions on $C$ | Updates on $R$ |
|---|---|---|
| $x_1$, $C$<br>$x_1 \notin R$ | $C[x_1/\top]$ | $N/A$ |
| $x_1$, $C$<br>$x_1 \in R$<br>$x_2 \in var(C_{x_1})$ | $C_{x_2} \leftarrow C_{x_1} \oplus x_2 \oplus \top$<br>$C[x_2/C_{x_2}]$ | $R \leftarrow R \setminus \{x_1\}$<br>$R \leftarrow R \cup \{x_2\}$ |

We denote by $k$ the number of variables in a XOR-CNF formula. At the implementation level, XOR-clauses are represented as $(k+1)$-bit vectors: a bit for every variable and one for a $\top$, $\bot$ constant. Clauses are stored in an array indexed by the representatives. This representation allows us to perform GE only by XOR-ing bit-vectors and flipping the clause constant. For a compact representation of the $(k+1)$-bit vector we used an array of $\lceil (k+1)/64 \rceil$ integers.

*Example 2.* Let $k = 7$ and let us consider $x_2 \Leftrightarrow \top \oplus x_1 \oplus x_3 \oplus x_5$. Then we have that $var(C_{x_2}) = \{\top, x_1, x_3, x_5\}$ and the bit-vector representing this clause is 11010100, where the $\top$, $\bot$ constant takes the zero position. Assigning $x_1$ to $\top$ is equivalent to introducing the constraint $x_1 \oplus \top$. We apply the first rule, simply by XOR-ing this bit-vector with a mask of the form 11000000. The resulting vector is 00010100, which corresponds to $var(C_{x_2}) = \{\bot, x_3, x_5\}$.

Our DPLL-based solver assigns a truth value to each variable in a formula $F$, recursively building a binary search tree. After each assignment, either the formula is simplified and other truth values are inferred or a conflict occurs. In the case of a conflict, the last assignment has to be undone for each module via a backtracking procedure. In Algorithm 1, we detail the ASSIGN function of

WDSat, which is at the core of the DPLL algorithm. This function synchronises all three modules in the following manner. First, the truth value is assigned in the CNF module and truth values of other variables are propagated. Next, the truth value of the initial variable, as well as the propagated ones are assigned in the XORSET module. If the XOR-adapted unit propagation discovers new truth values, they are assigned in the CNF module, going back to step one. We go back and forth with this process until the two modules are synchronized and there are no more propagations left. Finally, the list of all inferred literals is transferred to the XG module. If the XG module finds new XOR-implied literals, the list is sent to the CNF module and the process is restarted. If a conflict occurs in any of the reasoning modules, the ASSIGN function fails and a backtracking procedure is launched. We briefly detail the other functions used in the pseudo-code. There is a SET_IN function for each module which takes as input a list of literals and a propositional formula $F$ and sets all literals in this list to $\top$ in the corresponding modules. Through this assignment, the function also infers truth values of other literals, according to the specific rules in different modules. For instance, the SET_IN function for the XG module (SET_IN_XG) implements the rules in Table 1, performing a GE on the system. Finally, the LAST_ASSIGNED function in each module returns the list of literals that were assigned during the last call to the respective SET_IN function.

**Algorithm 1.** Function ASSIGN($F$, $x$) : Assigning a truth value to a literal $x$ in a formula $F$, simplifying $F$ and inferring truth values for other literals.

**Input:** The propositional formula $F$, a literal $x$
**Output:** $\bot$ if a conflict is reached, $\top$ and a simplified $F$ otherwise

*to_set* $\leftarrow \{x\}$.
*to_set_in_XG* $\leftarrow \{x\}$.
**while** *to_set* $\neq \emptyset$ **do**
  **while** *to_set* $\neq \emptyset$ **do**
    **if** SET_IN_CNF(*to_set*, $F$) $\rightarrow \bot$ **then**
      **return** ($\bot$, – ).
    **end if**
    *to_set* $\leftarrow$ LAST_ASSIGNED_IN_CNF().
    *to_set_in_XG* $\leftarrow$ *to_set*.
    **if** SET_IN_XORSET(*to_set*, $F$)$\rightarrow \bot$ **then**
      **return** ($\bot$, – ).
    **end if**
    *to_set* $\leftarrow$ LAST_ASSIGNED_IN_XORSET().
    *to_set_in_XG* $\leftarrow$ *to_set* $\cup$ *to_set_in_XG*.
  **end while**
  **if** SET_IN_XG(*to_set_in_XG*, $F$)$\rightarrow \bot$ **then**
    **return** ($\bot$, – ).
  **end if**
  *to_set* $\leftarrow$ LAST_ASSIGNED_XG().
**end while**
**return** ($\top$, $F$).

## 4 The XG-ext Method

In this section, we show how we extend our XG module. First, we present the motivation for this work by giving an example of a case where GE in SAT solvers has certain limitations compared to Algebraic GE. Secondly, we propose a solution to overcome these limitations and we implement it in our solver to develop the XORGAUSS-ext method (XG-ext in short). To introduce new rules for this method, we use the same notation as in Sect. 3.

Gaussian elimination on a Boolean polynomial system consists in performing elementary operations on equations with the goal of reducing the number of equations as well as the number of terms in each equation. We cancel out terms by adding (XOR-ing) one equation to another. GE can be performed on instances in CNF-XOR form in the same way that it is performed on Boolean polynomial systems presented in algebraic writing. However, we detected a case where a possible cancellation of terms is overseen due to the CNF-XOR form.

*Example 3.* We will reuse the Boolean polynomial system in Example 1 to demonstrate a case where a cancellation of a term is missed by a XOR-enabled SAT solver. Let us consider that in Equation (4), we try to assign the value of 1 to $\mathbf{x}_2$. As the monomial $\mathbf{x}_2\mathbf{x}_3$ will be equal to 1 only if both terms $\mathbf{x}_2$ and $\mathbf{x}_3$ are equal to 1, we get the following result:

$$\mathbf{x}_1 + \mathbf{x}_3 + \mathbf{x}_5 + \mathbf{x}_6 + 1 = 0$$
$$\mathbf{x}_3 + \mathbf{x}_5 + \mathbf{x}_6 = 0.$$

After XORing the two equations, we infer that $\mathbf{x}_1 = 1$.
However, when we assign $x_2$ to $\top$ in the corresponding CNF-XOR clause in Equation (5), as per unit propagation rules, we get the following result:

$$x' \vee \neg x_3$$
$$\neg x' \vee x_3$$
$$x_1 \oplus x' \oplus x_5 \oplus x_6$$
$$x_3 \oplus x_5 \oplus x_6 \oplus \top.$$

When we XOR the second clause to the first one we can not infer that $x_1$ is $\top$ at this point.

Note that $(x' \vee \neg x_3) \wedge (\neg x' \vee x_3)$ rewrites as $x' \Leftrightarrow x_3$, but if the solver does not syntactically search for this type of occurrences regularly, $x'$ will not be replaced by $x_3$. Moreover, this type of search adds an additional computational cost to the resolution.

Omissions as the one detailed in Example 3 can occur every time a variable is set to $\top$. As a result, we define the following rule with the goal to improve the performance of XOR-enabled SAT solvers:

$$\frac{x' \quad x_1 \Leftrightarrow (x' \wedge x_2)}{x_1 \Leftrightarrow x_2}. \tag{7}$$

This rule can be generalised for the resolution of higher-degree Boolean polynomial systems:

$$\frac{x' \quad x_1 \Leftrightarrow (x' \wedge x_2 \wedge \ldots \wedge x_d)}{x_1 \Leftrightarrow (x_2 \wedge \ldots \wedge x_d)} \,. \tag{8}$$

Even though these rules are standard in Boolean logic, they are presently not implemented in XOR-enabled SAT solvers. Note that when a solver takes as input an instance in CNF-XOR form, the second premise is lost or has to be inferred by syntactic search. To have knowledge of the second premise, the solver needs to read the instance in ANF. To this purpose, we defined a new ANF input format for SAT solvers.

**Table 2.** Inference rules for the substitution of $x_1$ by $x_2$.

| Premises | Conclusions on $C$ | Updates on $R$ |
|---|---|---|
| $C,\ x_1 \Leftrightarrow x_2$<br>$x_1 \not\in R$<br>$x_2 \not\in R$ | $C[x_1/x_2]$ | $N/A$ |
| $C,\ x_1 \Leftrightarrow x_2$<br>$x_1 \in R$<br>$x_2 \not\in R$<br>$x_2 \not\in var(C_{x_1})$ | $C_{x_2} \leftarrow C_{x_1}$<br><br>$C[x_2/C_{x_2}]$ | $R \leftarrow R \setminus \{x_1\}$<br><br>$R \leftarrow R \cup \{x_2\}$ |
| $C,\ x_1 \Leftrightarrow x_2$<br>$x_1 \in R$<br>$x_2 \not\in R$<br>$x_2 \in var(C_{x_1})$<br>$x_3 \in var(C_{x_1})$ | $C_{x_3} \leftarrow C_{x_1} \oplus x_2 \oplus x_3$<br><br><br>$C[x_3/C_{x_3}]$ | $R \leftarrow R \setminus \{x_1\}$<br><br><br>$R \leftarrow R \cup \{x_3\}$ |
| $C,\ x_1 \Leftrightarrow x_2$<br>$x_1 \not\in R$<br>$x_2 \in R$<br>$x_1 \not\in var(C_{x_2})$ | $C[x_1/C_{x_2}]$ | $N/A$ |
| $C,\ x_1 \Leftrightarrow x_2$<br>$x_1 \not\in R$<br>$x_2 \in R$<br>$x_1 \in var(C_{x_2})$<br>$x_3 \in var(C_{x_2})$ | $C_{x_3} \leftarrow C_{x_2} \oplus x_1 \oplus x_3$<br><br><br>$C[x_1/x_2, x_3/C_{x_3}]$ | $R \leftarrow R \setminus \{x_2\}$<br><br><br>$R \leftarrow R \cup \{x_3\}$ |
| $C,\ x_1 \Leftrightarrow x_2$<br>$x_1 \in R$<br>$x_2 \in R$<br>$x_3 \in var(C_{x_1} \oplus C_{x_2})$ | $C_{x_3} \leftarrow C_{x_1} \oplus C_{x_2} \oplus x_3$<br><br><br>$C[x_3/C_{x_3}]$ | $R \leftarrow R \setminus \{x_1, x_2\}$<br><br><br>$R \leftarrow R \cup \{x_3\}$ |

This extension of the XG module is implemented as part of the SET_IN_XG function used in the ASSIGN algorithm. The following is a detailed explanation of how the rule in Equation (7) is applied in our implementation. Recall that the XG module has the following property: a representative of an EC will never be present in another EC. This property will be maintained in the XG-ext method as well. Using the conclusion in Equation (7), we derive in Table 2 six inference rules that allow us to perform the substitution of a variable $x_1$ by a variable $x_2$ while maintaining the unicity-of-representatives property. Applying one of the inference rules in Table 2 can result in conflict or it can propagate a newly discovered truth value. Note that $var(C_{x_1} \oplus C_{x_2})$ is given by the symmetric difference $(var(C_{x_1}) \cup var(C_{x_2})) \setminus (var(C_{x_2}) \cap var(C_{x_1}))$.

## 5 Our Preprocessing Technique

Let us reconsider the DPLL-based algorithm. It is well known that the number of conflicts needed to prove the inconsistency is correlated to the order in which the variables are assigned. Among the state-of-the-art branching rules you can find two categories according to the type of heuristics. The first are based on Maximum number of Occurrences in the Minimum clauses Size (MOMs) whereas the second adopt the Variable State Independent Decaying Sum (VSIDS) branching heuristic.

In this work, we were interested in developing a criterion for defining the order of variables on CNF-XOR instances derived from Boolean polynomial systems. We set the goal to choose branching variables that will lead as fast as possible to a linear polynomial system, which can be solved using GE in polynomial time. In terms of SAT solving, choosing this order for branching will cancel out all clauses in the CNF part of the formula as a result of unit propagation. When only the XOR part of the CNF-XOR formula is left, the solver performs GE on the remaining XOR constraints in polynomial time.

After setting this goal, choosing which variable to assign next according to the number of their occurrences in the system is no longer an optimal technique. We explain this idea on an example. For simplicity, we only use the Boolean algebra terminology in this section. However, the methods described are applicable to both SAT solving and algebraic techniques based on the process of recursively making assumptions on the truth values of variables in the system (as with the DPLL algorithm).

*Example 4.* Consider the following Boolean polynomial system:

$$\begin{aligned} &\mathbf{x}_1 + \mathbf{x}_2\mathbf{x}_3 + \mathbf{x}_4 + \mathbf{x}_4\mathbf{x}_5 = 0 \\ &\mathbf{x}_1 + \mathbf{x}_2\mathbf{x}_3 = 0 \\ &\mathbf{x}_1 + \mathbf{x}_3\mathbf{x}_5 + \mathbf{x}_6 = 0 \\ &\mathbf{x}_1 + \mathbf{x}_2\mathbf{x}_5\mathbf{x}_6 + \mathbf{x}_6 = 0 \end{aligned} \tag{9}$$

In this example, the variable with the highest number of occurrences is $\mathbf{x}_1$. However, $\mathbf{x}_1$ does not occur in any monomial of degree $> 1$. Thus, assigning first $\mathbf{x}_1$ does not contribute to the linearization of the system and we need to find a more suitable criterion.

The solution we propose is inspired by graph theory. Particularly, we identified a parallel between the problem of defining the order in which the variables are assigned and the Minimal Vertex Cover Problem (MVC).

In graph theory, a *vertex cover* is a subset of vertices such that for every edge $(v_i, v_j)$ of the graph, either $v_i$ or $v_j$ is in the vertex cover. Given an undirected graph, the Minimum Vertex Cover Problem is a classic optimization problem of finding a vertex cover of minimal size.

An undirected graph is derived from a Boolean polynomial system as follows.

- Each variable $\mathbf{x}_i$ from the system becomes a vertex $v_i$ in the graph $G$.
- An edge $(v_i, v_j)$ is in $G$ if and only if (in the corresponding Boolean system) there exists a monomial of degree $n \geq 2$ which contains both $\mathbf{x}_i$ and $\mathbf{x}_j$.

When we use this representation of a Boolean polynomial system as a graph, a vertex cover defines a subset of variables whose assignment will result in a linear Boolean polynomial system in the remaining non-assigned variables. Consequently, finding the MVC of the graph is equivalent to finding the minimal subset of variables one has to assign to obtain a linear system.

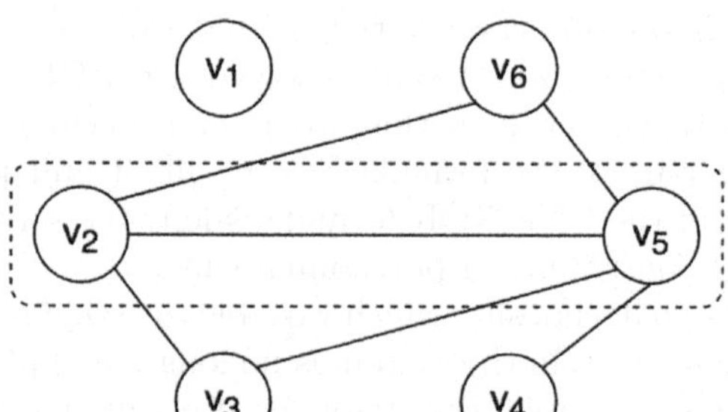


**Fig. 1.** Graph derived from Example 4

Figure 1 shows the graph derived from Example 4. The MVC of this graph is $\{v_2, v_5\}$. As a result, when all variables in the subset $\{\mathbf{x}_2, \mathbf{x}_5\}$ are assigned, the remaining polynomial system is linear. We give here the system derived after the assignment $\mathbf{x}_2 = 1$ and $\mathbf{x}_5 = 1$.

$$\mathbf{x}_1 + \mathbf{x}_3 = 0$$
$$\mathbf{x}_1 + \mathbf{x}_3 + \mathbf{x}_6 = 0$$
$$\mathbf{x}_1 = 0.$$

For all other possible assignments of $\mathbf{x}_2$ and $\mathbf{x}_5$, we obtain similar linear systems.

Defining the order of branching variables will serve as a preprocessing technique that consists in (i) deriving a graph from a Boolean polynomial system and (ii) finding the MVC of the resulting graph. During the solving process, variables corresponding to vertices in the MVC are assigned first. Even though the MVC problem is NP-complete, its execution for graphs derived from cryptographic models always finishes in negligible running time due to the small number of variables. Our solver does not use any other MOMs or VSIDS-based heuristic during the solving process, as the order of the branching variables is predetermined by the MVC preprocessing technique.

When variables are assigned in the order defined by this preprocessing technique, the worst-case time complexity of a DPLL-based algorithm drops from $O(2^k)$ to $O(2^{k'})$, where $k'$ is the number of vertices in the MVC set. Note that the MVC of a complete graph is equal to the number of its vertices. Consequently, when the corresponding graph of a Boolean polynomial system is a complete graph, solving the system using this preprocessing technique is as hard as solving the system without it.

Finding the MVC corresponding to a Boolean polynomial system can also be used as an assessment of the security of the underlying cryptosystem. Indeed, an exhaustive search on a subset of variables, which are the variables in the MVC, results in linear systems that can be solved in polynomial time. This straightforward approach yields an upper bound on the complexity of solving the system at hand. In short, to assess the security of a cryptographic system, assuming that this is based on solving the Boolean polynomial system first, one computes the MVC of this system and deduces that $O(2^{k'})$ is a bound on the complexity of the attack.

## 6 Experimental Results

To support our claims, we experimented with benchmarks derived from two variants of the index calculus attack on the discrete logarithm problem over binary elliptic curves. As explained in Sect. 2, a SAT solver can be used for solving Semaev's summation polynomials in the point decomposition phase. Our model is derived from the Boolean multivariate polynomial system given by the $m + 1$-th summation polynomial, with $m \geq 2$. This model has previously been examined in [14]. We compare the WDSat solver presented in this paper to the following approaches: the best currently available implementation of Gröbner basis (F4 [12] in MAGMA [5]), the solvers MiniSat, [11], Glucose [1], MapleLCMDistChronoBT [23], CaDiCaL [3] and CryptoMiniSat [27] with enabled GE.[2] Note that MapleLCMDistChronoBT and CaDiCaL are the winners in the main track of the latest SAT competition [22] in 2018. All tests were performed on a 2.40GHz Intel Xeon E5-2640 processor and are an average of 100 runs.

For SAT models derived from cryptographic problems, the preprocessing technique is executed only once, since all instances presenting a specific cryptographic problem are equivalent except for the constant in the XOR constraints. Even though the MVC problem is NP-complete, its execution for graphs derived from our models always finished in negligible running time, due to the small number of nodes.

We conducted experiments using both the third and the fourth polynomials. Results on solving the third summation polynomial ($m = 2$) are shown in Table 4. The parameters used to obtain these benchmarks are $n = 41$ and $l = 20$. As a result, we obtained a Boolean polynomial system of 41 equations in 40 variables (see Sect. 2). We show running-time averages on satisfiable and unsatisfiable instances separately, as these values differ between the two cases.

As different variants of our solver can yield better results for different benchmarks, we compared all variants to decide on the optimal one. We also tested the solver with and without our preprocessing technique (denoted by mvc in the tables). The results in Table 3 show that WDSat yields optimal results for these benchmarks when the XG-ext method is used coupled with the preprocessing technique. This outcome is not surprising when we examine the MVC obtained by the preprocessing technique. The number of variables in the system is $k = 40$, but the number of vertices in the MVC is 20. This means that by using the optimization techniques described in this paper, the worst-case time complexity of the examined models drops from $2^k$ to $2^{\frac{k}{2}}$. This is the case for every instance derived from the third summation polynomial.

**Table 3.** Comparing different versions of WDSat for solving the third summation polynomial.

| WDSat+ | SAT | | UNSAT | |
|---|---|---|---|---|
| | Runtime (s) | #Conflicts | Runtime (s) | #Conflicts |
| XG | 6028.4 | 200957178 | 11743.2 | 354094821 |
| XG+mvc | 639.6 | 21865963 | 2973.0 | 94489361 |
| XG-ext | 375.9 | 4911099 | 870.1 | 10789518 |
| XG-ext+mvc | **4.2** | **27684** | **13.5** | **86152** |

By analyzing the average running time and the average number of conflicts in Table 4, we see that the chosen variant of the WDSat solver outperforms all other approaches for solving instances derived from the third summation polynomial.

[2] Enabling GE in CryptoMiniSat yielded better performance for these benchmarks.

Current versions of CryptoMiniSat do not allow choosing the order of the branching variables as its authors claim that this technique almost always results in slower running times. To verify this claim, we modified the source code of CryptoMiniSat in order to test our preprocessing technique coupled with this solver (see line CryptoMiniSat+mvc in Table 4). We set a timeout of 10 min and only 9 out of 100 unsatisfiable and 54 out of 100 satisfiable instances were solved. This confirms that the MVC preprocessing technique is strongly linked to our XG-ext method. Indeed, when the XG-ext method is not used, one can not guarantee that when all variables from the MVC are assigned the system becomes linear. This is confirmed also by looking at the number of conflicts for the CryptoMiniSat+mvc approach, which is greater than $2^{\frac{k}{2}}$ even for benchmarks that were solved before the timeout. Recall that $\frac{k}{2}$ is the size of the MVC. On the other hand CryptoMiniSat without the preprocessing technique succeeds in solving these instances after less than $2^{\frac{k}{2}}$ conflicts. We conclude that the searching technique in CryptoMiniSat used to decide on the next branching variable is optimal for this solver.

The solvers which are not XOR-enabled did not solve any of the 200 satisfiable and unsatisfiable instances before the 10-min timeout. This is not surprising as instances derived from the third summation polynomial are solved a lot faster when a GE technique is used.

**Table 4.** Comparing different approaches for solving the third summation polynomial.

| Solving approach | SAT | | UNSAT | |
|---|---|---|---|---|
| | Runtime (s) | #Conflicts | Runtime (s) | #Conflicts |
| Gröbner | 16.8 | *N/A* | 18.7 | *N/A* |
| MiniSat | >600 | | >600 | |
| Glucose | >600 | | >600 | |
| MapleLCMDistChronoBT | >600 | | >600 | |
| CaDiCaL | >600 | | >600 | |
| CryptoMiniSat | 29.0 | 226668 | 84.3 | 627539 |
| CryptoMiniSat+mvc | 237.4 | 1263601 | >600 | |
| WDSat+XG-ext+mvc | **4.2** | **27684** | **13.5** | **86152** |

Experimental results in Table 5 are performed using benchmarks derived from the fourth summation polynomial. We obtain our model using a symmetrization technique proposed by Gaudry [15]. According to our parameter choice, the initial polynomial system contains 52 equations in 51 variables. However, only 18 out of the 51 variables are 'crucial'. The other 33 variables are introduced as a result of Gaudry's symmetrization technique. Our experiments show that performing GE on these instances does not result in faster running times. On the contrary, running times are significantly slower when the XG module of the WDSat solver is enabled. Running times become even slower with the XG-ext method. We attribute this fallout to the particularly small improvement in the number of conflicts, compared to the significant computational cost of performing the GE technique. Indeed, the graph corresponding to the model for the fourth summation polynomial is complete and thus the size of the MVC is equivalent to the number of variables in the formula. This leads us to believe there is no optimal choice for the order of branching variables and the system generally does not become linear until the second-to-last branching. We conclude that for solving these instances WDSat without GE is the optimal variant, since it outperforms both the Gröbner basis method and current state-of-the-art solvers.

To sum up, when WDSat is used for the index calculus attack, our recommendation is to enable the XG-ext option for instances obtained from the third summation polynomial and to completely disable the XG module for instances from the fourth polynomial. For ANF instances arising from other cryptographic problems, it would be best to solve smaller instances of the problem and analyse the number of conflicts. If the number of conflicts is only slightly better when the XG module is enabled, then disabling the XG module is likely to yield faster running times for higher scale instances of that problem.

**Table 5.** Comparing different approaches for solving the fourth summation polynomial.

| Solving approach | SAT | | UNSAT | |
|---|---|---|---|---|
| | Runtime (s) | #Conflicts | Runtime (s) | #Conflicts |
| Gröbner | 229.3 | *N/A* | 229.4 | *N/A* |
| MiniSat | 239.7 | 1840190 | 517.0 | 3433304 |
| Glucose | 189.2 | 1527158 | 274.8 | 2056575 |
| MapleLCMDistChronoBT | 655.1 | 4035131 | 918.7 | 5378945 |
| CaDiCaL | 43.6 | 254194 | 141.3 | 629869 |
| CryptoMiniSat | 331.8 | 1791188 | 707.9 | 3416526 |
| WDSat | **0.6** | **48438** | **3.8** | **255698** |
| WDSat+XG | 19.0 | 85282 | 49.8 | 252949 |

Our solver is dedicated to problems arising from a Weil descent. However, we tested it on Trivium [9] instances as they are extensively used in the SAT literature. We created instances using a modelization similar to the one in Grain of Salt [26], a tool for deriving instances for keystream generators comprised of Nonlinear-Feedback Shift Registers (NLFSR). Our experience is that CryptoMiniSat yields faster running times than all of the WDSat variants for Trivium instances. WDSat does not implement any of the optimizations for Trivium such as dependent variable removal, sub-problem detection, etc. as there are no such occurrences in systems arising from a Weil descent.

## 7 Conclusion

In this paper, we revisited XOR-enabled SAT solvers and their use in cryptanalysis. We proposed a novel SAT solver, named WDSat, dedicated to solving instances derived from the index calculus attack on binary elliptic curves. We conducted experiments comparing WDSat to the algebraic Gröbner basis resolution method, as well as to five state-of-the-art SAT solvers. Our solver outperforms all existing resolution approaches for this specific problem.

## References


1. Audemard, G., Simon, L.: Predicting learnt clauses quality in modern SAT solvers. In: IJCAI 2009, Proceedings of the 21st International Joint Conference on Artificial Intelligence, Pasadena, California, USA, 11–17 July 2009, pp. 399–404 (2009). http://ijcai.org/Proceedings/09/Papers/074.pdf
2. Bettale, L., Faugère, J., Perret, L.: Hybrid approach for solving multivariate systems over finite fields. J. Math. Cryptol. **3**(3), 177–197 (2009). https://doi.org/10.1515/JMC.2009.009
3. Biere, A.: CaDiCaL simplified satisfiability solver. http://fmv.jku.at/cadical/. Accessed 27 May 2020
4. BlueKrypt: Cryptographic key length recommendation (2018). https://www.keylength.com. Accessed 27 May 2020
5. Bosma, W., Cannon, J., Playoust, C.: The Magma algebra system. I. The user language. J. Symb. Comput. **24**(3–4), 235–265 (1997). https://doi.org/10.1006/jsco.1996.0125
6. Bouillaguet, C., et al.: Fast exhaustive search for polynomial systems in $\mathbb{F}_2$. In: Mangard, S., Standaert, F.-X. (eds.) CHES 2010. LNCS, vol. 6225, pp. 203–218. Springer, Heidelberg (2010). https://doi.org/10.1007/978-3-642-15031-9_14
7. Choo, D., Soos, M., Chai, K.M.A., Meel, K.S.: Bosphorus: bridging ANF and CNF solvers. In: Design, Automation & Test in Europe Conference & Exhibition, DATE 2019, Florence, Italy, 25–29 March 2019, pp. 468–473 (2019). https://doi.org/10.23919/DATE.2019.8715061
8. Davis, M., Logemann, G., Loveland, D.: A machine program for theorem-proving. Commun. ACM **5**(7), 394–397 (1962). https://doi.org/10.1145/368273.368557
9. Cannière, C.: Trivium: a stream cipher construction inspired by block cipher design principles. In: Katsikas, S.K., López, J., Backes, M., Gritzalis, S., Preneel, B. (eds.) ISC 2006. LNCS, vol. 4176, pp. 171–186. Springer, Heidelberg (2006). https://doi.org/10.1007/11836810_13
10. Diem, C.: On the discrete logarithm problem in elliptic curves. Compositio Mathematica **147**(1), 75–104 (2011). https://doi.org/10.1112/S0010437X10005075
11. Eén, N., Sörensson, N.: An extensible SAT-solver. In: Giunchiglia, E., Tacchella, A. (eds.) SAT 2003. LNCS, vol. 2919, pp. 502–518. Springer, Heidelberg (2004). https://doi.org/10.1007/978-3-540-24605-3_37
12. Faugère, J.C.: A new efficient algorithm for computing Gröbner basis (F4). J. Pure Appl. Algebra **139**(1–3), 61–88 (1999). https://doi.org/10.1145/780506.780516
13. Faugère, J.-C., Perret, L., Petit, C., Renault, G.: Improving the complexity of index calculus algorithms in elliptic curves over binary fields. In: Pointcheval, D., Johansson, T. (eds.) EUROCRYPT 2012. LNCS, vol. 7237, pp. 27–44. Springer, Heidelberg (2012). https://doi.org/10.1007/978-3-642-29011-4_4

14. Galbraith, S.D., Gebregiyorgis, S.W.: Summation polynomial algorithms for elliptic curves in characteristic two. In: Meier, W., Mukhopadhyay, D. (eds.) INDOCRYPT 2014. LNCS, vol. 8885, pp. 409–427. Springer, Cham (2014). https://doi.org/10.1007/978-3-319-13039-2_24
15. Gaudry, P.: Index calculus for Abelian varieties of small dimension and the elliptic curve discrete logarithm problem. J. Symb. Comput. **44**(12), 1690–1702 (2009). https://doi.org/10.1016/j.jsc.2008.08.005
16. Gérault, D., Lafourcade, P., Minier, M., Solnon, C.: Revisiting AES related-key differential attacks with constraint programming. Inf. Process. Lett. **139**, 24–29 (2018). https://doi.org/10.1016/j.ipl.2018.07.001
17. Gerault, D., Lafourcade, P., Minier, M., Solnon, C.: Computing AES related-key differential characteristics with constraint programming. Artif. Intell. **278**, 103183 (2020). https://doi.org/10.1016/j.artint.2019.103183
18. Gerault, D., Minier, M., Solnon, C.: Using constraint programming to solve a cryptanalytic problem. In: Sierra, C. (ed.) Proceedings of the Twenty-Sixth International Joint Conference on Artificial Intelligence, IJCAI 2017, Melbourne, Australia, 19–25 August 2017, pp. 4844–4848. ijcai.org (2017). https://doi.org/10.24963/ijcai.2017/679
19. Lidl, R., Niederreiter, H.: Introduction to Finite Fields and Their Applications. Cambridge University Press, Cambridge (1986)
20. Liu, F., Cruz, W., Ma, C., Johnson, G., Michel, L.: A tolerant algebraic side-channel attack on AES using CP. In: Beck, J.C. (ed.) CP 2017. LNCS, vol. 10416, pp. 189–205. Springer, Cham (2017). https://doi.org/10.1007/978-3-319-66158-2_13
21. Liu, F., Cruz, W., Michel, L.: A complete tolerant algebraic side-channel attack for AES with CP. In: Hooker, J. (ed.) CP 2018. LNCS, vol. 11008, pp. 259–275. Springer, Cham (2018). https://doi.org/10.1007/978-3-319-98334-9_18
22. van Maaren, H., Franco, J.: The International SAT Competition Web Page. http://www.satcompetition.org/. Accessed 27 May 2020
23. Nadel, A., Ryvchin, V.: Chronological backtracking. In: Beyersdorff, O., Wintersteiger, C.M. (eds.) SAT 2018. LNCS, vol. 10929, pp. 111–121. Springer, Cham (2018). https://doi.org/10.1007/978-3-319-94144-8_7
24. Petit, C., Quisquater, J.-J.: On polynomial systems arising from a weil descent. In: Wang, X., Sako, K. (eds.) ASIACRYPT 2012. LNCS, vol. 7658, pp. 451–466. Springer, Heidelberg (2012). https://doi.org/10.1007/978-3-642-34961-4_28
25. Semaev, I.A.: Summation polynomials and the discrete logarithm problem on elliptic curves. IACR Cryptology ePrint Archive 2004, 31 (2004). http://eprint.iacr.org/2004/031
26. Soos, M.: Grain of salt – an automated way to test stream ciphers through SAT solvers. In: Tools'10: The Workshop on Tools for Cryptanalysis 2010, London, United Kingdom, pp. 131–144, June 2010. https://hal.archives-ouvertes.fr/hal-01288922
27. Soos, M., Nohl, K., Castelluccia, C.: Extending SAT solvers to cryptographic problems. In: Kullmann, O. (ed.) SAT 2009. LNCS, vol. 5584, pp. 244–257. Springer, Heidelberg (2009). https://doi.org/10.1007/978-3-642-02777-2_24
28. Trimoska, M., Ionica, S., Dequen, G.: EC index calculus benchmarks (2020). https://github.com/mtrimoska/EC-Index-Calculus-Benchmarks
29. Trimoska, M., Ionica, S., Dequen, G.: WDSat solver (2020). https://github.com/mtrimoska/WDSat

# Constraint-Based Software Diversification for Efficient Mitigation of Code-Reuse Attacks

Rodothea Myrsini Tsoupidi[1(✉)], Roberto Castañeda Lozano[2], and Benoit Baudry[1]

[1] KTH Royal Institute of Technology, Stockholm, Sweden
{tsoupidi,baudry}@kth.se
[2] University of Edinburgh, Edinburgh, UK
roberto.castaneda@ed.ac.uk

**Abstract.** Modern software deployment process produces software that is uniform, and hence vulnerable to large-scale code-reuse attacks. *Compiler-based diversification* improves the resilience and security of software systems by automatically generating different assembly code versions of a given program. Existing techniques are efficient but do not have a precise control over the quality of the generated code variants.

This paper introduces *Diversity by Construction (DivCon)*, a constraint-based compiler approach to software diversification. Unlike previous approaches, DivCon allows users to control and adjust the conflicting goals of diversity and code quality. A key enabler is the use of Large Neighborhood Search (LNS) to generate highly diverse assembly code efficiently. Experiments using two popular compiler benchmark suites confirm that there is a trade-off between quality of each assembly code version and diversity of the entire pool of versions. Our results show that DivCon allows users to trade between these two properties by generating diverse assembly code for a range of quality bounds. In particular, the experiments show that DivCon is able to mitigate code-reuse attacks effectively while delivering near-optimal code (<10% optimality gap).

For constraint programming researchers and practitioners, this paper demonstrates that LNS is a valuable technique for finding diverse solutions. For security researchers and software engineers, DivCon extends the scope of compiler-based diversification to performance-critical and resource-constrained applications.

**Keywords:** Compiler-based software diversification · Code-reuse attacks · Constraint programming · Embedded systems

## 1 Introduction

Good software development practices, such as code reuse [19], continuous deployment, and automatic updates contribute to the emergence of software monocultures [3]. While such monocultures facilitate software distribution, bug reporting,


H. Simonis (Ed.): CP 2020, LNCS 12333, pp. 791–808, 2020.
https://doi.org/10.1007/978-3-030-58475-7_46

and software authentication, they also introduce serious risks related to the wide spreading of attacks against all users that run identical software.

*Software diversification* is a method to mitigate the problems caused by uniformity. Similarly to biodiversity, software diversification improves the resilience and security of a software system [2] by introducing diversity in it. Software diversification can be applied in different phases of the software development cycle, i.e. during implementation, compilation, loading, execution, and more [20]. This paper is concerned with *compiler-based* diversification, which automatically generates different assembly code versions from a single source program.

Modern compilers do not merely aim to generate correct code, but also code that is of high quality. Existing compiler-based diversification techniques are efficient and effective at diversifying assembly code [20] but do not have a precise control over its quality and may produce unsatisfactory results. These techniques (discussed in Sect. 5) are either based on randomizing heuristics or in high-level superoptimization methods that do not capture accurately the quality of the generated code.

This paper introduces Diversity by Construction (DivCon), a compiler-based diversification approach that allows users to control and adjust the conflicting goals of quality of each code version and diversity among all versions. DivCon uses a Constraint Programming (CP)-based compiler backend to generate multiple solutions corresponding to functionally equivalent program variants according to an accurate code quality model. The backend models the input program, the hardware architecture, and the compiler transformations as a constraint problem, whose solution corresponds to assembly code for the input program.

The use of CP makes it possible to 1) control the quality of the generated solutions by constraining the objective function, 2) introduce application-specific constraints that restrict the diversified solutions, and 3) apply sophisticated search procedures that are particularly suitable for diversification. In particular, DivCon uses Large Neighborhood Search (LNS) [29], a popular metaheuristic in multiple application domains, to generate highly diverse solutions efficiently.

Our experiments compiling 17 functions from two popular compiler benchmark suites to the MIPS32 architecture confirm that there is a trade-off between code quality and diversity, and demonstrate that DivCon allows users to navigate this conflict by generating diverse assembly code for a range of quality bounds. In particular, the experiments show that DivCon is able to mitigate code-reuse attacks effectively while guaranteeing a code quality of 10% within optimality.

For constraint programming researchers and practitioners, this paper demonstrates that LNS is a valuable technique for finding diverse solutions. For security researchers and software engineers, DivCon extends the scope of compiler-based diversification to performance-critical and resource-constrained applications, and provides a solid step towards secure-by-construction software.

*Contributions.* To summarize, this paper:

- proposes a CP-based technique for compiler-based, quality-aware software diversification (Sect. 3);

```
1 0x9d001408: ...
2 0x9d00140c: lw     $s2, 4($sp)
3 0x9d001410: lw     $s4, 0($sp)
4 0x9d001414: jr     $t9
5 0x9d001418: addiu  $sp, $sp, 16
```

(a) Original gadget.

```
1 0x9d001408: lw     $s2, 4($sp)
2 0x9d00140c: nop
3 0x9d001410: lw     $s4, 0($sp)
4 0x9d001414: jr     $t9
5 0x9d001418: addiu  $sp, $sp, 16
```

(b) Diversified gadget.

**Fig. 1.** Example gadget diversification in MIPS32 assembly code

- shows that LNS is a promising technique for generating highly diverse solutions efficiently (Sect. 4.3);
- quantifies the trade-off between code quality and diversity (Sect. 4.4); and
- demonstrates that DivCon mitigates code-reuse attacks effectively while preserving high code quality (Sect. 4.5).

## 2 Background

This section describes code-reuse attacks (Sect. 2.1), diversification approaches in CP (Sect. 2.2), and combinatorial compiler backends (Sect. 2.3).

### 2.1 Code-Reuse Attacks

Code-reuse attacks take advantage of memory vulnerabilities, such as buffer overflows, to reuse program code for malicious purposes. More specifically, code-reuse attacks insert data into the program memory to affect the control flow of the program and execute code that is valid but unintended.

Jump-Oriented Programming (JOP)[1] is a code-reuse attack [4,7] that combines different code snippets from the original program code to form a Turing complete language for attackers. These code snippets terminate with a branch instruction. The building blocks of a JOP attack are *gadgets*: meta-instructions that consist of one or multiple code snippets with specific semantics. Figure 1a shows a JOP gadget found by the *ROPgadget* tool [27] in a MIPS32 binary. Assuming that the attacker controls the stack, lines 2 and 3 load attacker data in registers \$s2 and \$s4, respectively. Then, line 4 jumps to the address of register \$t9. The last instruction (line 5) is placed in a delay slot and hence it is executed before the jump [31]. The semantics of this gadget depends on the attack payload and might be to load a value to register \$s2 or \$s4. Then, the program jumps to the next gadget that resides at the stack address of \$t9.

Statically designed JOP attacks use the absolute binary addresses for installing the attack payload. Hence, a simple change in the instruction schedule of the program as in Fig. 1b prevents a JOP attack designed for Fig. 1a. An attacker that designs an attack based on the binary of the original program assumes the presence of a gadget (Fig. 1a) at position 0x9d00140c. However, in

[1] This paper focuses on JOP due to the characteristics of MIPS32, but could be generalized to other code-reuse attacks such as Return-Oriented Programming (ROP) [28].

the diversified version, address 0x9d00140c does not start with the initial `lw` instruction of Fig. 1a, and by the end of the execution of the gadget, register \$s2 does not contain the attacker data. In this way, diversification can break the semantics of the gadget and mitigate an attack against the diversified code.

### 2.2 Diversity in Constraint Programming

While typical CP applications aim to discover either some solution or the optimal solution, some applications require finding *diverse* solutions for various purposes.

Hebrard *et al.* [13] introduce the MAXDIVERSE$k$SET problem, which consists in finding the most diverse set of $k$ solutions, and propose an exact and an incremental algorithm for solving it. The exact algorithm does not scale to a large number of solutions [16, 32]. The incremental algorithm selects solutions iteratively by solving a distance maximization problem.

Automatic Generation of Architectural Tests (ATGP) is an application of CP that requires generating many diverse solutions. Van Hentenryck *et al.* [32] model ATGP as a MAXDIVERSE$k$SET problem and solve it using the incremental algorithm of Hebrard et. al. Due to the large number of diverse solutions required (50–100), Van Hentenryck *et al.* replace the maximization step with local search.

In software diversity, solution quality is of paramount importance. In general, earlier CP approaches to diversity are concerned with satisfiability only. An exception is the approach of Petit *et al.* [26]. This approach modifies the objective function for assessing both solution quality and solution diversity, but does not scale to the large number of solutions required by software diversity. Ingmar *et al.* [16] propose a generic framework for modeling diversity in CP. For tackling the quality-diversity trade-off, they propose constraining the objective function with the optimal (or best known) cost $o$. DivCon applies this approach by allowing solutions $p\%$ worse than $o$, where $p$ is configurable.

### 2.3 Compiler Optimization as a Combinatorial Problem

A Constraint Satisfaction Problem (CSP) is a problem specification $P = \langle V, U, C\rangle$, where $V$ are the problem variables, $U$ is the domain of the variables, and $C$ the constraints among the variables. A Constraint Optimization Problem (COP), $P = \langle V, U, C, O\rangle$, consists of a CSP and an objective function $O$. The goal of a COP is to find a solution that optimizes $O$.

Compilers are programs that generate low-level assembly code, typically optimized for *speed* or *size*, from higher-level source code. A compilation process can be modeled as a COP by letting $V$ be the decisions taken during the translation, $C$ be the constraints imposed by the program semantics and the hardware resources, and $O$ be the cost of the generated code.

Compiler backends generate low-level assembly code from an Intermediate Representation (IR), a program representation that is independent of both the source and the target language. Figure 2 shows the high-level view of a *combinatorial* compiler backend. A combinatorial compiler backend takes as input

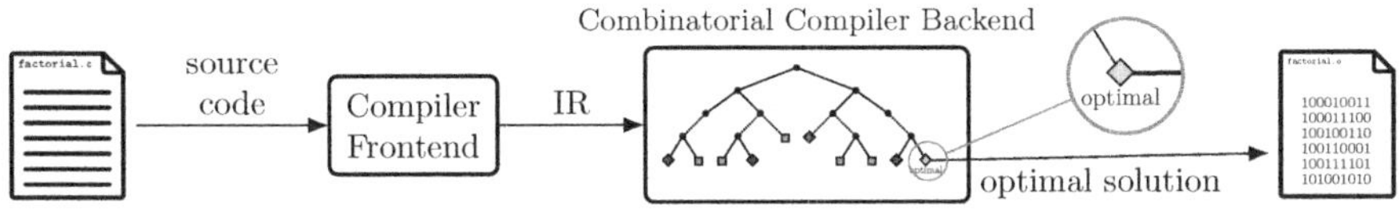


**Fig. 2.** High-level view of a combinatorial compiler backend

the IR of a program, generates and solves a COP, and outputs the optimized low-level assembly code described by the solution to the COP.

This paper assumes that programs at the IR level are represented by their Control-Flow Graph (CFG). A CFG is a representation of the possible execution paths of a program, where each node corresponds to a *basic block* and edges correspond to intra-block jumps. A *basic block*, in its turn, is a set of abstract instructions (hereafter just *instructions*) with no branches besides the end of the block. Each instruction is associated with a set of operands characterizing its input and output data. Typical decision variables $V$ of a combinatorial compiler backend are the issue cycle $c_i \in \mathbb{N}_0$ of each instruction $i$, the processor instruction $m_i \in \mathbb{N}_0$ that implements each instruction $i$, and the processor register $r_o \in \mathbb{N}_0$ assigned to each operand $o$.

DivCon aims at mitigating code-reuse attacks. Therefore, DivCon considers the order of the instructions in the final binary, which directly affects the feasibility of code-reuse attacks (see Figs. 1a and 1b). For this reason, the diversification model uses the issue cycle sequence of instructions, $c = \{c_0, c_1, ..., c_n\}$, to characterize the diversity among different solutions.

Figure 3a shows an implementation of the factorial function in C where each basic block is highlighted. Figure 3b shows the IR of the program. The example IR contains 10 instructions in three basic blocks: bb.0, bb.1, and bb.2. bb.0 corresponds to initializations, where `$a0` holds the function argument `n` and $t_1$ corresponds to variable `f`. bb.1 computes the factorial in a loop by accumulating the result in $t_1$. bb.2 stores the result to `$v0` and returns. Some instructions in the example are interdependent, which leads to serialization of the instruction schedule. For example, ***beq*** (6) consumes data ($t_3$) defined by ***slti*** (4) and hence needs to be scheduled later. Instruction dependencies limit the amount of possible assembly code versions and can restrict diversity significantly, as seen in Sect. 4.3. Finally, Fig. 3c shows the arrangement of the issue cycle variables in the constraint model used by the combinatorial compiler backend.

## 3 DivCon

This section introduces DivCon, a software diversification method that uses a combinatorial compiler backend to generate program variants. Figure 4 shows a high-level view of the diversification process. DivCon uses 1) the optimal solution to start the *search* for diversification and 2) the cost of the optimal solution to restrict the variants within a maximum gap from the optimal. Subsequently, DivCon generates a number of solutions to the CSP that correspond to diverse program variants.

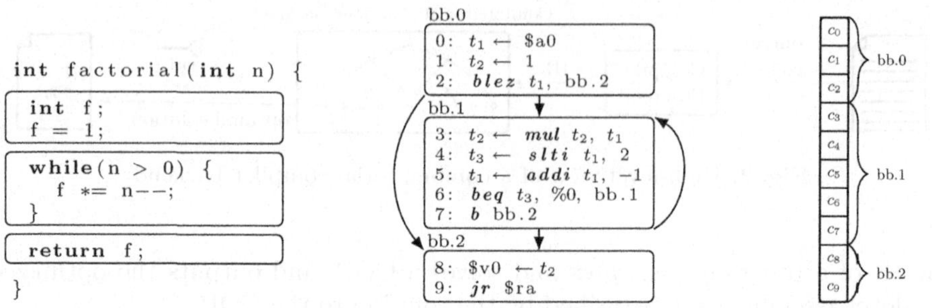


(a) C code (b) IR (c) Issue cycles

**Fig. 3.** Factorial function example

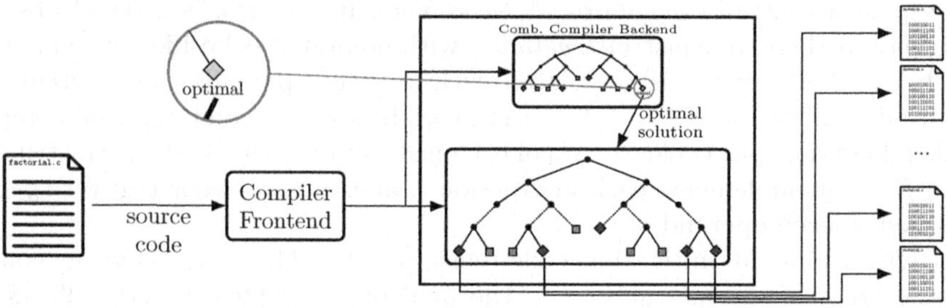


**Fig. 4.** High-level view of DivCon

The rest of this section describes the diversification approach of DivCon. Section 3.1 formulates the diversification problem in terms of the constraint model of a combinatorial compiler backend, Sect. 3.2 defines the distance measures, and finally, Sect. 3.3 describes the search strategy for generating program variants.

### 3.1 Problem Description

Let $P = \langle V, U, C \rangle$ be the compiler backend CSP for the program under compilation, $O$ be the objective function, and $o$ be the cost of the optimal or best known solution to the COP, $\langle V, U, C, O \rangle$. Let $\delta$ be a function that measures the distance between two solutions of P (two such functions are defined in Sect. 3.2). Let $h \in \mathbb{N}$ be the minimum pairwise distance and $p \in \mathbb{R}_{\geq 0}$ be the maximum optimality gap specified by the user. Our problem is to find a subset of the solutions to the CSP, $S \subseteq sol(P)$, such that $\forall s_1, s_2 \in S \,.\, s_1 \neq s_2 \implies \delta(s_1, s_2) \geq h$ and $\forall s \in S \,.\, O(s) \leq (1 + p) \cdot o$.

To solve the above problem, DivCon employs the incremental algorithm listed in Algorithm 1. Starting with the optimal solution $y_{opt}$, the algorithm adds the distance constraint for $y_{opt}$ and the optimality constraint with $o = y_{opt}(O)$ (line 2). Notation $\delta(y)$ is used instead of $\delta(y, s) \,|\, \forall s \in sol(\langle V, U, C' \rangle)$ for readability. While the termination condition is not fulfilled (line 3), the algorithm uses LNS as described in Sect. 3.3 to find the next solution $y$ (line 4), adds the next solution to the solution set $S$ (line 5), and updates the distance constraints based on the latest solution (line 6). When the termination condition is satisfied, the algorithm returns the set of solutions $S$ corresponding to diversified assembly code variants.

**Algorithm 1:** Incremental algorithm for generating diverse solutions

```
1  S ← {y_opt}, y ← y_opt,
2  C' ← C ∪ {δ(y_opt) ≥ h, O(V) ≤ (1 + p) · o}
3  while not term_cond() // e.g. |S| > k ∨ time_limit()
4      y ← solve_LNS(relax(y), ⟨V, U, C'⟩)
5      S ← S ∪ {y}
6      C' ← C' ∪ {δ(y) ≥ h}
```

Figure 5 shows two MIPS32 variants of the factorial example (Fig. 3), which correspond to two solutions of DivCon. The variants differ in two aspects: first, the ***beqz*** instruction is issued one cycle later in Fig. 5b than in Fig. 5a, and second, the temporary variable $t_3$ (see Fig. 3) is assigned to different MIPS32 registers ($t0 and $t1).

### 3.2 Distance Measures

This section defines two alternative distance measures: Hamming Distance (HD) and Levenshtein Distance (LD). Both distances operate on the schedule of the instructions, i.e. the order in which the instructions are issued in the CPU.

*Hamming Distance (HD).* HD is the Hamming distance [12] between the issue cycle variables of two solutions. Given two solutions $s, s' \in sol(P)$:

$$\delta_{HD}(s, s') = \sum_{i=0}^{n} (s(c_i) \neq s'(c_i)), \tag{1}$$

where $n$ is the maximum number of instructions.

Consider Fig. 1b, a diversified version of the gadget in Fig. 1a. The only instruction that differs from Fig. 1a is the instruction at line 1 that is issued

```
bb.0:  blez   $a0, bb.2
       addiu  $v0, $zero, 1
bb.1:  mul    $v0, $v0, $a0
       slti   $t0, $a0, 2
       beqz   $t0, bb.1
       addi   $a0, $a0, -1
bb.2:  jr     $ra
       nop
```

(a) Variant 1.

```
bb.0:  blez   $a0, bb.2
       addiu  $v0, $zero, 1
bb.1:  mul    $v0, $v0, $a0
       slti   $t1, $a0, 2
       nop
       beqz   $t1, bb.1
       addi   $a0, $a0, -1
bb.2:  jr     $ra
       nop
```

(b) Variant 2.

**Fig. 5.** Two MIPS32 variants of the factorial example in Fig. 3

one cycle before. The two examples have a HD of one, which in this case is enough for breaking the functionality of the original gadget (see Sect. 2.1).

*Levenshtein Distance (LD).* LD (or edit distance) measures the minimum number of edits, i.e. insertions, deletions, and replacements, that are necessary for transforming one instruction schedule to another. Compared to HD, which considers only *replacements*, LD also considers *insertions* and *deletions*. To understand this effect, consider Fig. 5. The two gadgets differ only by one `nop` operation but HD gives a distance of three, whereas LD gives one, which is more accurate. LD takes ordered vectors as input, and thus requires an ordered representation (as opposed to a detailed schedule) of the instructions. Therefore, LD uses vector $c^{-1} = channel(c)$, a sequence of instructions ordered by their issue cycle. Given two solutions $s, s' \in sol(P)$:

$$\delta_{LD}(s, s') = \texttt{levenshtein\_distance}(s(c^{-1}), s'(c^{-1})), \tag{2}$$

where `levenshtein_distance` is the Wagner–Fischer algorithm [33] with time complexity $O(nm)$, where $n$ and $m$ are the lengths of the two sequences.

### 3.3 Search

Unlike previous CP approaches to diversity, DivCon employs Large Neighborhood Search (LNS) for diversification. LNS is a metaheuristic that defines a neighborhood, in which *search* looks for better solutions, or, in our case, different solutions. The definition of the neighborhood is through a *destroy* and a *repair* function. The *destroy* function unassigns a subset of the variables in a given solution and the *repair* function finds a new solution by assigning new values to the *destroyed* variables.

In DivCon, the algorithm starts with the optimal solution of the combinatorial compiler backend. Subsequently, it destroys a part of the variables and continues with the model's branching strategy to find the next solution, applying a restart after a given number of failures. LNS uses the concept of *neighborhood*, i.e. the variables that LNS may destroy at every restart. To improve diversity, the neighborhood for DivCon consists of all decision variables, i.e. the issue cycles $c$, the instruction implementations $m$, and the registers $r$. Furthermore, LNS depends on a *branching strategy* to guide the *repair* search. To improve security and allow LNS to select diverse paths after every restart, DivCon employs a random variable-value selection branching strategy as described in Table 1b.

## 4 Evaluation

The evaluation of DivCon addresses four main questions:

- RQ1. What is the scalability of the distance measures in generating multiple program variants? Here, we evaluate which of the distance measures is the most appropriate for software diversification.

- RQ2. How effective and how scalable is LNS for code diversification? Here, we investigate LNS as an alternative approach to diversity in CP.
- RQ3. How does code quality relate to code diversity and what are the involved trade-offs?
- RQ4. How effective is DivCon at mitigating code-reuse attacks? This question is the main application of CP-based diversification in this work.

### 4.1 Experimental Setup

*Implementation.* DivCon is implemented as an extension of Unison [6], and is available at https://github.com/romits800/divcon. Unison implements two backend transformations: instruction scheduling and register allocation. DivCon employs Unison's solver portfolio that includes Gecode v6.2 [11] and Chuffed v0.10.3 [8] to find optimal solutions, and Gecode v6.2 only for diversification. The LLVM compiler [21] is used as a front-end and IR-level optimizer.

*Benchmark Functions and Platform.* The evaluation uses 17 functions sampled randomly from MediaBench [22] and SPEC CPU2006 [30], two benchmark suites widely employed in embedded and general-purpose compiler research. The size of the functions is limited to between 10 and 30 instructions (with a median of 20 instructions) to keep the evaluation of all methods and distance measures feasible regardless of their computational cost. Table 2 lists the ID, application, name, basic blocks (b), and instructions (i) of each sampled function. The functions are compiled to MIPS32 assembly code. MIPS32 is a popular architecture within embedded systems and the security-critical Internet of Things [1].

*Host Platform.* All experiments run on an Intel®Core™i9-9920X processor at 3.50 GHz with 64 GB of RAM running Debian GNU/Linux 10 (buster). Each of the experiments runs for 20 random seeds. The results show the mean value and the standard deviation from these experiments. The available virtual memory for each of the executions is 10 GB. The experiments for different random seeds run in parallel (5 seeds at a time), with two unique cores available for every seed for overheating reasons. DivCon runs as a sequential program.

**Table 1.** ORIGINAL and RANDOM branching strategies

(a) ORIGINAL branching strategy.

| Variable | Var. Selection | Value Selection |
|---|---|---|
| $c_i$ | in order | min. val first |
| $m_i$ | in order | min. val first |
| $r_o$ | in order | randomly |

(b) RANDOM branching strategy.

| Variable | Var. Selection | Value Selection |
|---|---|---|
| $c_i$ | randomly | randomly |
| $m_i$ | randomly | randomly |
| $r_o$ | randomly | randomly |

**Table 2.** Benchmark functions

| ID | App | Function name | b | i |
|---|---|---|---|---|
| b1 | sphinx3 | ptmr_init | 1 | 10 |
| b2 | gcc | ceil_log2 | 1 | 14 |
| b3 | mesa | glIndexd | 1 | 14 |
| b4 | h264ref | symbol2uvlc | 1 | 15 |
| b5 | gobmk | autohelperowl_defen.. | 1 | 23 |
| b6 | mesa | glVertex2i | 1 | 23 |
| b7 | hmmer | AllocFancyAli | 1 | 25 |
| b8 | gobmk | autohelperowl_vital.. | 1 | 27 |
| b9 | gobmk | autohelperpat1088 | 1 | 29 |
| b10 | gobmk | autohelperowl_attac.. | 1 | 30 |
| b11 | gobmk | get_last_player | 3 | 13 |
| b12 | h264ref | UpdateRandomAccess | 3 | 16 |
| b13 | gcc | xexit | 3 | 17 |
| b14 | gcc | unsigned_condition | 3 | 24 |
| b15 | sphinx3 | glist_tail | 4 | 10 |
| b16 | gcc | get_frame_alias_set | 5 | 20 |
| b17 | gcc | parms_set | 5 | 25 |

**Table 3.** Scalability of $\delta_{HD}, \delta_{LD}$

| ID | $\delta_{HD}$ | | $\delta_{LD}$ | |
|---|---|---|---|---|
| | $t(s)$ | Num | $t(s)$ | Num |
| b1 | $0.1 \pm 0.2$ | 26 | $131.2 \pm 131.4$ | 26 |
| b2 | $1.0 \pm 0.1$ | 200 | – | 68 |
| b3 | $1.1 \pm 0.1$ | 200 | – | 58 |
| b4 | $0.7 \pm 0.0$ | 200 | – | 73 |
| b5 | $2.3 \pm 0.3$ | 200 | – | 38 |
| b6 | $2.5 \pm 0.2$ | 200 | – | 35 |
| b7 | $2.0 \pm 0.3$ | 200 | – | 37 |
| b8 | $3.8 \pm 0.8$ | 200 | – | 35 |
| b9 | $4.0 \pm 0.6$ | 200 | – | 28 |
| b10 | $4.5 \pm 0.7$ | 200 | – | 27 |
| b11 | $1.3 \pm 0.1$ | 200 | – | 56 |
| b12 | $1.1 \pm 0.2$ | 200 | – | 47 |
| b13 | $0.8 \pm 0.1$ | 200 | – | 91 |
| b14 | $1.8 \pm 0.3$ | 200 | – | 27 |
| b15 | $1.7 \pm 0.2$ | 200 | – | 60 |
| b16 | $2.7 \pm 0.4$ | 200 | – | 31 |
| b17 | $1.6 \pm 0.2$ | 200 | – | 35 |

*Algorithm Configuration.* The experiments focus on speed optimization and aim to generate 200 variants within a timeout. Parameter $h$ in Algorithm 1 is set to one because even small distance between variants is able to break gadgets (see Fig. 1). LNS uses restart-based search with a limit of 500 failures, and a relax rate of 70%. The *relax rate* is the probability that LNS destroys a variable at every restart, which affects the distance between two subsequent solutions. A higher relax rate increases diversity but requires more solving effort. We have found experimentally that 70% is an adequate balance between the two. All experiments are available at https://github.com/romits800/divcon_experiments.

### 4.2 RQ1. Scalability of the Distance Measures

The ability to generate a large number of variants is paramount for software diversification. This section compares the distance measures introduced in Sect. 3.2 with regards to scalability.

Table 3 presents the results of the distance evaluation, where a time limit of 10 min and optimality gap of $p = 10\%$ are used. For each distance measure ($\delta_{HD}$ and $\delta_{LD}$) the table shows the diversification time $t$, in seconds (or "–" if the algorithm is not able to generate 200 variants) and the number of generated variants *num* within the time limit.

The results show that for $\delta_{HD}$, DivCon is able to generate 200 variants for all benchmarks except *b1*, which has exactly 26 variants. The diversification time for $\delta_{HD}$ is less than 5 s for all benchmarks. Distance $\delta_{LD}$, on the other hand, is not able to generate 200 variants for any of the benchmarks within the time limit. This poor scalability of $\delta_{LD}$ is due to the quadratic complexity of its implementation [33], whereas HD can be implemented linearly. Consequently, the rest of the evaluation uses $\delta_{HD}$.

### 4.3 RQ2. Scalability and Diversification Effectiveness of LNS

This section evaluates the diversification effectiveness and scalability of LNS compared to incremental MaxDiverse*k*Set (where the first solution is found randomly and the maximization step uses the branching strategy from Table 1a) and Random Search (RS) (which uses the branching strategy from Table 1b).

To measure the diversification effectiveness of these methods, the evaluation uses the relative pairwise distance of the solutions. Given a set of solutions $S$ and a distance measure $\delta$, the pairwise distance $d$ of the variants in $S$ is $d(\delta, S) = \sum_{i=0}^{|S|} \sum_{j>i}^{|S|} \delta(s_i, s_j) \,/\, \binom{|S|}{2}$. The *larger* this distance, the more diverse the solutions are, and thus, diversification is more effective. Table 4 shows the pairwise distance $d$ and diversification time $t$ for each benchmark and method, where the experiment uses a time limit of 30 min and optimality gap of $p = 10\%$. The best values of $d$ (larger) and $t$ (lower) are marked in **bold** for the completed experiments, whereas incomplete experiments are highlighted in *italic* and their number of variants in parenthesis.

**Table 4.** Distance and scalability of LNS with RS and MaxDiverse*k*Set

| ID | MaxDiverse*k*Set | | RS | | LNS (0.7) | |
|---|---|---|---|---|---|---|
| | $d$ | $t(s)$ | $d$ | $t(s)$ | $d$ | $t(s)$ |
| b1 | *4.1 ± 0.0* | 0.2 ± 0.0 (26) | *4.1 ± 0.0* | **0.0 ± 0.0 (26)** | *4.1 ± 0.0* | 0.1 ± 0.2 (26) |
| b2 | **10.8 ± 0.0** | 761.8 ± 10.1 | 6.4 ± 0.2 | **0.6 ± 0.1** | 8.6 ± 0.6 | 1.0 ± 0.1 |
| b3 | *14.6 ± 0.0* | – (21) | 5.8 ± 0.1 | **0.6 ± 0.1** | **10.8 ± 0.8** | 1.0 ± 0.1 |
| b4 | *14.4 ± 0.0* | – (19) | 4.3 ± 0.1 | **0.2 ± 0.0** | **12.1 ± 0.3** | 0.6 ± 0.0 |
| b5 | *22.0 ± 0.0* | – (2) | 4.3 ± 0.3 | **0.5 ± 0.0** | **16.1 ± 1.1** | 2.2 ± 0.3 |
| b6 | *22.9 ± 0.4* | – (2) | 5.3 ± 0.0 | **1.0 ± 0.1** | **16.4 ± 0.6** | 2.4 ± 0.2 |
| b7 | *24.9 ± 0.1* | – (6) | 4.5 ± 0.2 | **0.4 ± 0.0** | **18.1 ± 1.2** | 1.9 ± 0.3 |
| b8 | *24.8 ± 0.4* | – (2) | 6.5 ± 0.2 | **3.5 ± 0.5** | **17.2 ± 0.9** | 3.8 ± 0.8 |
| b9 | *26.0 ± 0.0* | – (2) | 4.2 ± 0.3 | **0.4 ± 0.0** | **19.8 ± 0.7** | 3.9 ± 0.6 |
| b10 | *28.0 ± 0.0* | – (2) | 6.0 ± 0.0 | 5.3 ± 1.0 | **20.1 ± 1.1** | **4.5 ± 0.7** |
| b11 | **13.8 ± 0.0** | 356.9 ± 8.2 | 5.3 ± 0.1 | **0.2 ± 0.0** | 10.1 ± 1.0 | 1.2 ± 0.1 |
| b12 | *21.5 ± 0.1* | – (5) | 6.4 ± 0.9 | **0.2 ± 0.0** | **14.9 ± 1.0** | 1.0 ± 0.2 |
| b13 | *17.4 ± 0.0* | – (122) | 6.7 ± 0.0 | 0.9 ± 0.1 | **12.0 ± 0.9** | **0.7 ± 0.1** |
| b14 | *30.1 ± 0.0* | – (20) | 7.5 ± 0.2 | **0.2 ± 0.0** | **24.9 ± 0.7** | 1.8 ± 0.3 |
| b15 | – | – | 2.6 ± 0.3 | **0.1 ± 0.0** | **20.2 ± 0.5** | 1.6 ± 0.2 |
| b16 | – | – | 5.6 ± 0.4 | **0.3 ± 0.0** | **21.3 ± 0.8** | 2.6 ± 0.4 |
| b17 | – | – | *2.9 ± 0.1* | – (91) | **28.1 ± 1.5** | **1.6 ± 0.2** |

The scalability results ($t(s)$) show that RS and LNS are scalable (generate the maximum 200 variants for almost all benchmarks), whereas MaxDiversekSet scales poorly (cannot generate 200 variants for any benchmark but *b2* and *b11*). Both *b2* and *b11* have a small search space (few, highly interdependent instructions), which leads to restricted diversity but facilitates solving. For *b1*, all instructions are interdependent on each other, which forces a linear schedule and results in only 26 possible variants (given $p = 10\%$). On the other end, MaxDiversekSet is not able to find any variants for *b15*, *b16*, and *b17*. These benchmarks have many basic blocks resulting in a more complex objective function. For the largest benchmark (*b17*), only LNS is able to scale up to 200 solutions. LNS is generally slower than RS, but for both LNS and RS all benchmarks have a diversification time less than six seconds.

The diversity results ($d$) show that LNS is more effective at diversifying than RS. The improvement of LNS over RS ranges from 35% (for *b2*) to 675% (for *b15*). In the two cases where MaxDiversekSet terminates (benchmarks *b2* and *b11*), it generates the most diverse code, as can be expected.

In summary, LNS offers an attractive balance between scalability and diversification effectiveness: it is close in scalability to, and sometimes improves, the overly fastest method (RS), but it is significantly and consistently more effective at diversifying code.

### 4.4 RQ3. Trade-Off Between Code Quality and Diversity

A key advantage of using a CP-based compiler approach for software diversity is the ability to control the quality of the generated solutions. This ability enables control over the relation between the quality of each individual solution and the diversity of the entire pool of solutions. Insisting in optimality limits the number of possible diversified variants and their pairwise distance, whereas relaxing optimality allows higher diversity.

Table 5 shows the pairwise distance $d$ (defined in Sect. 4.3), and the number of generated variants $num$, for all benchmarks and different values of the optimality gap $p \in \{0\%, 5\%, 10\%, 20\%\}$. LNS is used with a time limit of 10 min. The best values of $d$ are marked in **bold**.

**Table 5.** Solution diversity for different optimality gap values

| ID | 0% | | 5% | | 10% | | 20% | |
|---|---|---|---|---|---|---|---|---|
| | *d* | Num | *d* | Num | *d* | Num | *d* | Num |
| b1 | – | – | – | – | *4.1± 0.0* | 26 | **6.5 ± 0.1** | 200 |
| b2 | *3.5± 0.0* | 9 | 6.7 ± 0.4 | 200 | 8.6 ± 0.6 | 200 | **10.0 ± 0.8** | 200 |
| b3 | 7.0 ± 0.1 | 200 | 9.4 ± 0.5 | 200 | 10.8 ± 0.8 | 200 | **14.8 ± 1.0** | 200 |
| b4 | 7.8 ± 0.2 | 200 | 10.1 ± 0.3 | 200 | 12.1 ± 0.3 | 200 | **14.0 ± 0.2** | 200 |
| b5 | 8.4 ± 0.1 | 200 | 11.9 ± 0.7 | 200 | 16.1 ± 1.1 | 200 | **19.7 ± 0.6** | 200 |
| b6 | 10.8 ± 0.1 | 200 | 14.7 ± 0.4 | 200 | 16.4 ± 0.6 | 200 | **20.9 ± 0.8** | 200 |
| b7 | 11.3 ± 0.3 | 200 | 13.8 ± 0.7 | 200 | 18.1 ± 1.2 | 200 | **22.8 ± 1.1** | 200 |
| b8 | 11.0 ± 0.1 | 200 | 13.6 ± 0.6 | 200 | 17.2 ± 0.9 | 200 | **22.4 ± 1.1** | 200 |
| b9 | 12.7 ± 0.1 | 200 | 17.7 ± 0.8 | 200 | 19.8 ± 0.7 | 200 | **24.4 ± 0.6** | 200 |
| b10 | 13.7 ± 0.1 | 200 | 18.1 ± 0.9 | 200 | 20.1 ± 1.1 | 200 | **26.3 ± 0.6** | 200 |
| b11 | *2.0± 0.0* | 4 | 6.6 ± 0.1 | 200 | 10.1 ± 1.0 | 200 | **14.2 ± 0.9** | 200 |
| b12 | *3.8± 0.0* | 10 | 10.3 ± 1.2 | 200 | 14.9 ± 1.0 | 200 | **19.8 ± 1.0** | 200 |
| b13 | *2.1± 1.3* | 4 | 10.1 ± 0.9 | 200 | 12.0 ± 0.9 | 200 | **15.7 ± 1.2** | 200 |
| b14 | *3.6± 0.0* | 24 | 21.0 ± 0.6 | 200 | 24.9 ± 0.7 | 200 | **29.0 ± 0.5** | 200 |
| b15 | *2.4± 0.0* | 8 | 15.6 ± 0.6 | 200 | 20.2 ± 0.5 | 200 | **23.5 ± 1.4** | 200 |
| b16 | *4.1± 0.0* | 44 | 15.1 ± 1.1 | 200 | 21.3 ± 0.8 | 200 | **30.7 ± 0.9** | 200 |
| b17 | 7.5 ± 0.2 | 200 | 20.3 ± 1.4 | 200 | 28.1 ± 1.5 | 200 | **38.4 ± 0.9** | 200 |

The first interesting observation is that even with no degradation of quality ($p = 0\%$), DivCon is able to generate a large number of variants for a significant fraction of the benchmarks. These include functions with a relatively large solution space, typically with a few large basic blocks where instructions are relatively independent of each other (*b3–b10* and *b17*). On the other hand, benchmarks with small basic blocks and many instruction dependencies (*b1*, *b2*, and *b11–b16*) provide fewer options for diversification, which results in a limited number of optimal variants.

Second, we observe that as soon as we slightly relax the constraint over optimality ($p = 5\%$), diversity radiates and DivCon generates 200 variants for all benchmarks except *b1*. Then, the more we increase the optimality gap, the larger the diversification space grows and the distance between the variants increases. Table 5 illustrates one of the key contributions of DivCon: the ability to explore the trade-off between optimal solutions and highly diverse solutions.

In summary, depending on the characteristics of the compiled code, it is possible to generate a large number of variants without sacrificing optimality, and the code quality can be adjusted to further improve diversity if required by the targeted application.

### 4.5 RQ4. Code-Reuse Mitigation Effectiveness

Software Diversity has various applications in security, including mitigating code-reuse attacks. To measure the level of mitigation that DivCon achieves, we assess the gadget survival rate $srate(s_i, s_j)$ between two variants $s_i, s_j \in S$, where $S$ is the set of generated variants. This metric determines how many of the gadgets of variant $s_i$ appear at the same position on the other variant $s_j$, that is $srate(s_i, s_j) = |gad(s_i) - gad(s_j)| \, / \, |gad(s_i)|$, where $gad(s_i)$ are the gadgets in solution $s_i$. The procedure for computing $srate(s_i, s_j)$ is as follows: 1) run ROPgadget [27] to find the set of gadgets $gad(s_i)$ in solution $s_i$, and 2) for every $g \in gad(s_i)$, check whether there exists a gadget identical to $g$ at the same address of $s_j$. This comparison is syntactic after removing all `nop` instructions.

This section compares the *srate* for all permutations of pairs in $S$, for all benchmarks, and for different values of the optimality gap using a time limit of 10 min. Low *srate* corresponds to higher mitigation effectiveness because code-reuse attacks based on gadgets in one variant have lower chances of locating the same gadgets in the other variants (see Fig. 1).

Table 6 summarizes the gadget survival distribution for all benchmarks and different values of the optimality gap (0%, 5%, 10%, and 20%). Due to its skewness, the distribution of *srate* is represented as a histogram with four buckets (0%, (0%, 10%], (10%, 40%], and (40%, 100%]) rather than summarized using common statistical measures. Here the best is a $srate(s_i, s_j)$ of 0%, which means that $s_j$ does not contain any gadgets that exist in $s_i$, whereas a $srate(s_i, s_j)$ in range (40%, 100%] means that $s_j$ shares more than 40% of the gadgets of $s_i$. The values in **bold** correspond to the mode(s) of the histogram.

First, we notice that DivCon can generate some pairs of variants that share no gadget, even without relaxing the constraint of optimality ($p = 0\%$). This indicates that the pareto front of optimal code naturally includes software diversity that is good for security. Second, the results show that this effectiveness can be further increased by relaxing the constraint on code quality, with diminishing returns beyond $p = 10\%$. For $p = 0\%$, there are 10 benchmarks dominated by a 0% survival rate, whereas there are 7 benchmarks dominated by a weak 10%–40%-survival rate. The latter are still considered vulnerable to code-reuse attacks. However, increasing the optimality gap to just $p = 5\%$ makes 0% survival rate the dominating bucket for all benchmarks, and further increasing the gap to 10% and 20% increases significantly the number of pairs where no single gadget is shared. For example, at $p = 10\%$ the rate of pairs that do not share any gadgets ranges from 63% (*b14*) to 99% (*b12*).

Related approaches (discussed in Sect. 5) report the *average srate* across all pairs for different benchmark sets. Pappas *et al.*'s zero-cost approach [25] achieves an average *srate* between 74%–83% without code degradation, comparable to DivCon's 41%–99% at $p = 0\%$. Homescu *et al.*'s statistical approach [15] reports an average *srate* between 82%–100% with a code degradation of less than 5%, comparable to DivCon's 83%–100% at $p = 5\%$. Both approaches report results on larger code bases that exhibit more opportunities for diversification.

**Table 6.** Gadget survival rate for different optimality gap values

| ID | 0% | | | | | 5% | | | | | 10% | | | | | 20% | | | | |
|---|---|---|---|---|---|---|---|---|---|---|---|---|---|---|---|---|---|---|---|---|
| | =0 | ≤10 | ≤40 | ≤100 | Num | =0 | ≤10 | ≤40 | ≤100 | Num | =0 | ≤10 | ≤40 | ≤100 | Num | =0 | ≤10 | ≤40 | ≤100 | Num |
| b1 | – | – | – | – | – | – | – | – | – | – | **84** | 3 | 3 | 10 | 26 | **94** | 4 | 2 | 1 | 200 |
| b2 | – | – | **69** | 31 | 9 | **60** | 12 | 23 | 4 | 200 | **76** | 11 | 12 | 1 | 200 | **81** | 9 | 10 | – | 200 |
| b3 | **66** | 15 | 18 | 1 | 200 | **71** | 14 | 15 | 1 | 200 | **73** | 13 | 13 | 1 | 200 | **77** | 14 | 9 | – | 200 |
| b4 | **94** | 6 | – | – | 200 | **96** | 4 | – | – | 200 | **96** | 4 | – | – | 200 | **98** | 2 | – | – | 200 |
| b5 | **90** | 1 | 9 | – | 200 | **93** | 2 | 5 | – | 200 | **95** | 2 | 3 | – | 200 | **95** | 3 | 2 | – | 200 |
| b6 | **88** | 5 | 7 | 1 | 200 | **89** | 5 | 6 | – | 200 | **90** | 4 | 6 | – | 200 | **91** | 4 | 5 | – | 200 |
| b7 | **48** | 1 | **48** | 3 | 200 | **74** | 5 | 21 | 1 | 200 | **83** | 6 | 11 | – | 200 | **89** | 6 | 5 | – | 200 |
| b8 | **46** | – | **51** | 3 | 200 | **57** | 4 | 36 | 2 | 200 | **74** | 3 | 21 | 1 | 200 | **81** | 4 | 14 | 1 | 200 |
| b9 | 42 | – | **56** | 2 | 200 | **66** | 9 | 24 | 1 | 200 | **73** | 8 | 18 | – | 200 | **83** | 7 | 9 | – | 200 |
| b10 | **47** | – | **50** | 3 | 200 | **65** | 2 | 30 | 2 | 200 | **73** | 4 | 22 | 1 | 200 | **82** | 5 | 13 | 1 | 200 |
| b11 | 38 | – | **61** | 1 | 4 | **66** | 3 | 31 | – | 200 | **68** | 9 | 23 | – | 200 | **83** | 7 | 10 | – | 200 |
| b12 | **94** | – | 5 | 1 | 10 | **99** | 1 | – | – | 200 | **99** | – | – | – | 200 | **99** | 1 | – | – | 200 |
| b13 | **43** | 9 | 34 | 14 | 4 | **69** | 20 | 11 | – | 194 | **69** | 21 | 10 | – | 200 | **71** | 19 | 10 | – | 200 |
| b14 | – | – | **78** | 22 | 24 | **60** | 23 | 17 | – | 200 | **63** | 22 | 15 | – | 200 | **70** | 19 | 11 | – | 200 |
| b15 | 41 | **53** | 5 | – | 8 | **97** | 2 | 1 | – | 200 | **98** | 1 | 1 | – | 200 | **98** | 1 | 1 | – | 200 |
| b16 | **64** | 28 | 6 | – | 44 | **76** | 21 | 2 | – | 200 | **82** | 17 | 1 | – | 200 | **90** | 9 | 1 | – | 200 |
| b17 | 33 | **66** | 1 | – | 200 | **61** | 39 | – | – | 200 | **75** | 25 | – | – | 200 | **87** | 13 | – | – | 200 |

We expect that DivCon would achieve higher overall survival rates on these code bases compared to the benchmarks used in this paper.

## 5 Related Work

There are many approaches to software diversification against cyberattacks. The majority apply randomized transformations at different stages of the software development, while a few exceptions use search-based techniques [20]. This section focuses on quality-aware software diversification approaches.

*Superdiversifier* [17] is a search-based approach for software diversification against cyberattacks. Given an initial instruction sequence, the algorithm generates a random combination of the available instructions and performs a verification test to quickly reject non equivalent instruction sequences. For each non-rejected sequence, the algorithm checks semantic equivalence between the original and the generated instruction sequences using a SAT solver. Superdiversifier affects the code execution time and size by controlling the length of the generated sequence. Along the same lines, Lundquist *et al.* [23,24] use program synthesis for generating program variants against cyberattacks, but no results are available yet. In comparison, DivCon uses a combinatorial compiler backend that measures the code quality using a more accurate cost model that also considers other aspects, such as execution frequencies.

Most diversification approaches use randomized transformations to generate multiple program variants [20]. Unlike DivCon, the majority of these approaches do not control the quality of the generated variants during diversification but rather evaluate it afterwards [5,9,10,14,18,34]. However, there are a few approaches that control the code quality during randomization.

Some compiler-based diversification approaches restrict the set of program transformations to control the quality of the generated code [9,25]. For example, Pappas *et al.* [25] perform software diversification at the binary level and apply three zero-cost transformations: register randomization, instruction schedule randomization, and function shuffling. In contrast, DivCon's combinatorial approach allows it to control the aggressiveness and potential cost of its transformations: a cost overhead limit of 0% forces DivCon to apply only zero-cost transformations; a larger limit allows DivCon to apply more aggressive transformations, potentially leading to higher diversity.

Homescu *et al.* [15] perform only garbage (`nop`) insertion, and use a profile-guided approach to reduce the overhead. To do this, they control the `nop` insertion probability based on the execution frequency of different code sections. In contrast, DivCon's cost model captures different execution frequencies, which allows it to perform more aggressive transformations in non-critical code sections.

## 6 Conclusion and Future Work

This paper introduces DivCon, a CP approach to compiler-based, quality-aware software diversification against code-reuse attacks. Our experiments show that LNS is a promising technique for a CP-based exploration of the space of diverse program, with a fine-grained control on the trade-off between code quality and diversity. In particular, we show that the set of optimal solutions naturally contains a set of diverse solutions, which increases significantly when relaxing the constraint of optimality. Our experiments demonstrate that the diverse solutions generated by DivCon are effective to mitigate code-reuse attacks.

Future work includes investigating different distance measures to further reduce the gadget survival rate, improving the overall scalability of DivCon in the face of larger programs and larger values of parameter $k$, and examining the effectiveness of DivCon against an actual code-reuse exploit.

**Acknowledgments.** We would like to give a special acknowledgment to Christian Schulte, for his critical contribution at the early stages of this work. Although no longer with us, Christian continues to inspire his students and colleagues with his lively character, enthusiasm, deep knowledge and understanding. We would also like to thank Linnea Ingmar and the anonymous reviewers for their useful feedback, and Oscar Eriksson for proof reading.

## References

1. Alaba, F.A., Othman, M., Hashem, I.A.T., Alotaibi, F.: Internet of Things security: a survey. J. Netw. Comput. Appl. **88**, 10–28 (2017). https://doi.org/10.1016/j.jnca.2017.04.002
2. Baudry, B., Monperrus, M.: The multiple facets of software diversity: recent developments in year 2000 and beyond. ACM Comput. Surv. **48**(1), 16:1–16:26 (2015). https://doi.org/10.1145/2807593

3. Birman, K.P., Schneider, F.B.: The monoculture risk put into context. IEEE Secur. Priv. **7**(1), 14–17 (2009)
4. Bletsch, T., Jiang, X., Freeh, V.W., Liang, Z.: Jump-oriented programming: a new class of code-reuse attack. In: Proceedings of the 6th ACM Symposium on Information, Computer and Communications Security, ASIACCS 2011, pp. 30–40. ACM, New York (2011)
5. Braden, K., et al.: Leakage-resilient layout randomization for mobile devices. In: Proceedings of 2016 Network and Distributed System Security Symposium. Internet Society, San Diego (2016). https://doi.org/10.14722/ndss.2016.23364
6. Castañeda Lozano, R., Carlsson, M., Blindell, G.H., Schulte, C.: Combinatorial register allocation and instruction scheduling. ACM Trans. Program. Lang. Syst. **41**(3), 17:1–17:53 (2019)
7. Checkoway, S., Davi, L., Dmitrienko, A., Sadeghi, A.R., Shacham, H., Winandy, M.: Return-oriented programming without returns. In: Proceedings of the 17th ACM Conference on Computer and Communications Security, CCS 2010, pp. 559–572. ACM, New York (2010)
8. Chu, G.G.: Improving combinatorial optimization. Ph.D. thesis, The University of Melbourne, Australia (2011)
9. Crane, S., et al.: Readactor: practical code randomization resilient to memory disclosure. In: 2015 IEEE Symposium on Security and Privacy, pp. 763–780, May 2015. https://doi.org/10.1109/SP.2015.52
10. Davi, L.V., Dmitrienko, A., Nürnberger, S., Sadeghi, A.R.: Gadge me if you can: secure and efficient ad-hoc instruction-level randomization for x86 and ARM. In: Proceedings of the 8th ACM SIGSAC Symposium on Information, Computer and Communications Security, pp. 299–310 (2013). ACM: tex.organization
11. Gecode Team: Gecode: generic constraint development environment (2020). https://www.gecode.org
12. Hamming, R.W.: Error detecting and error correcting codes. Bell Syst. Tech. J. **29**(2), 147–160 (1950). https://doi.org/10.1002/j.1538-7305.1950.tb00463.x
13. Hebrard, E., Hnich, B., O'Sullivan, B., Walsh, T.: Finding diverse and similar solutions in constraint programming. In: National Conference on Artificial Intelligence and the Seventeenth Innovative Applications of Artificial Intelligence Conference, p. 6 (2005)
14. Homescu, A., Jackson, T., Crane, S., Brunthaler, S., Larsen, P., Franz, M.: Large-scale automated software diversity—program evolution redux. IEEE Trans. Dependable Secure Comput. **14**(2), 158–171 (2017). https://doi.org/10.1109/TDSC.2015.2433252
15. Homescu, A., Neisius, S., Larsen, P., Brunthaler, S., Franz, M.: Profile-guided automated software diversity. In: Proceedings of the 2013 IEEE/ACM International Symposium on Code Generation and Optimization (CGO), CGO 2013, pp. 1–11. IEEE Computer Society, Washington, DC (2013). https://doi.org/10.1109/CGO.2013.6494997
16. Ingmar, L., de la Banda, M.G., Stuckey, P.J., Tack, G.: Modelling diversity of solutions. In: Proceedings of the Thirty-Fourth AAAI Conference on Artificial Intelligence (2020)
17. Jacob, M., Jakubowski, M.H., Naldurg, P., Saw, C.W.N., Venkatesan, R.: The superdiversifier: peephole individualization for software protection. In: Matsuura, K., Fujisaki, E. (eds.) IWSEC 2008. LNCS, vol. 5312, pp. 100–120. Springer, Heidelberg (2008). https://doi.org/10.1007/978-3-540-89598-5_7

18. Koo, H., Chen, Y., Lu, L., Kemerlis, V.P., Polychronakis, M.: Compiler-assisted code randomization. In: 2018 IEEE Symposium on Security and Privacy (SP), pp. 461–477, May 2018. https://doi.org/10.1109/SP.2018.00029
19. Krueger, C.W.: Software reuse. ACM Comput. Surv. **24**(2), 131–183 (1992). https://doi.org/10.1145/130844.130856
20. Larsen, P., Homescu, A., Brunthaler, S., Franz, M.: SoK: automated software diversity. In: 2014 IEEE Symposium on Security and Privacy, pp. 276–291, May 2014. https://doi.org/10.1109/SP.2014.25
21. Lattner, C., Adve, V.: LLVM: a compilation framework for lifelong program analysis & transformation. In: Code Generation and Optimization. IEEE (2004)
22. Lee, C., Potkonjak, M., Mangione-Smith, W.H.: MediaBench: a tool for evaluating and synthesizing multimedia and communicatons systems. In: International Symposium on Microarchitecture, pp. 330–335. IEEE (1997)
23. Lundquist, G.R., Bhatt, U., Hamlen, K.W.: Relational processing for fun and diversity. In: Proceedings of the 2019 Minikanren and Relational Programming Workshop, p. 100 (2019)
24. Lundquist, G.R., Mohan, V., Hamlen, K.W.: Searching for software diversity: attaining artificial diversity through program synthesis. In: Proceedings of the 2016 New Security Paradigms Workshop, NSPW 2016, pp. 80–91. ACM, New York (2016). https://doi.org/10.1145/3011883.3011891
25. Pappas, V., Polychronakis, M., Keromytis, A.D.: Smashing the gadgets: hindering return-oriented programming using in-place code randomization. In: 2012 IEEE Symposium on Security and Privacy, pp. 601–615, May 2012. https://doi.org/10.1109/SP.2012.41
26. Petit, T., Trapp, A.C.: Finding diverse solutions of high quality to constraint optimization problems. In: Twenty-Fourth International Joint Conference on Artificial Intelligence, June 2015
27. Salwan, J.: ROPgadget tool (2020). http://shell-storm.org/project/ROPgadget/
28. Shacham, H.: The geometry of innocent flesh on the bone: return-into-libc without function calls (on the x86). In: Proceedings of the 14th ACM Conference on Computer and Communications Security, CCS 2007, pp. 552–561. ACM, New York (2007)
29. Shaw, P.: Using constraint programming and local search methods to solve vehicle routing problems. In: Maher, M., Puget, J.-F. (eds.) CP 1998. LNCS, vol. 1520, pp. 417–431. Springer, Heidelberg (1998). https://doi.org/10.1007/3-540-49481-2_30
30. SPEC: CPU 2006 benchmarks (2020). https://www.spec.org/cpu2006. Accessed 20 Mar 2020
31. Sweetman, D.: See MIPS Run, 2nd edn. Morgan Kaufmann, Burlington (2006)
32. Van Hentenryck, P., Coffrin, C., Gutkovich, B.: Constraint-based local search for the automatic generation of architectural tests. In: Gent, I.P. (ed.) CP 2009. LNCS, vol. 5732, pp. 787–801. Springer, Heidelberg (2009). https://doi.org/10.1007/978-3-642-04244-7_61
33. Wagner, R.A., Fischer, M.J.: The string-to-string correction problem. J. ACM (JACM) **21**(1), 168–173 (1974)
34. Wang, S., Wang, P., Wu, D.: Composite software diversification. In: 2017 IEEE International Conference on Software Maintenance and Evolution (ICSME), pp. 284–294, September 2017. https://doi.org/10.1109/ICSME.2017.61

# CP and Data Science and Machine Learning

# Pushing Data into CP Models Using Graphical Model Learning and Solving

Céline Brouard, Simon de Givry, and Thomas Schiex(✉)

Université Fédérale de Toulouse, ANITI, INRAE, UR 875, Toulouse, France
{celine.brouard,simon.de-givry,thomas.schiex}@inrae.fr

**Abstract.** Integrating machine learning with automated reasoning is one of the major goals of modern AI systems. In this paper, we propose a non-fully-differentiable architecture that is able to extract preferences from data and push it into (weighted) Constraint Networks (aka Cost Function Networks or CFN) by learning cost functions. Our approach combines a (scalable) convex optimization approach with empirical hyper-parameter tuning to learn cost functions from a list of high-quality solutions. The proposed architecture has the ability to learn from noisy solutions and its output is just a CFN model. This model can be analyzed, empirically hardened, completed with side-constraints, and directly fed to a Weighted Constraint Satisfaction Problem solver.

To explore the performances and range of applicability of this architecture, we compare it with two recent neural-net friendly learning systems designed to "learn to reason" on the Sudoku problem and also show how it can be used to learn and integrate preferences into an existing CP model, in the context of Configuration systems.

**Keywords:** Graphical Models · Cost Function Networks · Learning · Constraint Programming

## 1 Introduction

Constraint Satisfaction and Constraint Programming define a powerful framework for modeling and solving decision problems. It is often considered as one of the closest approaches computer science has made to the Holy Grail of programming: "the user states the problem, the computer solves it." [16]. The problem may however be difficult to state, not only because of the rich CP language, but because several of the aspects of the real problem may be inaccessible to the modeler, leading to approximate formulations, providing only partially satisfactory solutions. In this paper, we show how preferences and constraints can be extracted from historical solutions so that they can be directly represented inside a (weighted) constraint satisfaction problem.

In this paper, we are interested in learning a criterion and its domain of definition as a set of cost functions and constraints, starting from a set of good-quality solutions that could require a perceptive layer for acquisition. The learned


H. Simonis (Ed.): CP 2020, LNCS 12333, pp. 811–827, 2020.
https://doi.org/10.1007/978-3-030-58475-7_47

preferences and constraints are represented as a Cost Function Network (CFN). This learned CFN can then be completed by user-defined constraints or criteria before feeding a Weighted Constraint Satisfaction Problem (WCSP) solver. Such a workflow is very useful when the aim is to produce a new solution that resides in a large family of known designs [34] (providing data), that must satisfy both known general requirements and new specific properties.

Our main contribution is to leverage the capacity of CFN solvers to optimize Graphical Models (GMs), a family of models that covers Constraint Networks, Clausal Propositional Logic, and their weighted variants as well as probabilistic Markov Random Fields and Bayesian Nets models [10]. Starting from historical solutions, we use a recent convex optimization approach to estimate a CFN model of the data that gives a lower cost to the training set. We notice that a maximum regularized approximate log-likelihood loss [30] does tackle this objective. We use a scalable algorithm that learns the scopes and cost tables of cost functions. This CFN can then be optionally enriched by user cost functions or constraints and solved by a WCSP solver for various inputs. The resulting architecture combines ML and CP components in a way that, in our knowledge, has never been tested to learn preferences (and constraints).

Our approach compares with recent differentiable "learning to reason" architectures such as Recurrent Relational Nets (RRN [29]) or SAT-Net [37] in terms of input, output, and prior information (assumptions). These approaches define fully differentiable layers that can learn pairwise "message passing" functions (RRN) or a low-rank convex relaxation of Max-SAT, using continuous descent algorithms as their optimization component. Such layers are easy to inter-operate with Deep learning differentiable architectures. These two approaches have been benchmarked on Sudoku resolution. We, therefore, compare our approach to the RRN and SAT-net approaches, including in situations where Sudoku grids are only available as hand-written grid images. We observe that our ML+CP approach offers better accuracy and requires fewer samples. These results show that neither differentiability nor even continuity are needed to work on a model combined with a deep learning perceptual front-end [29].

Finally, we show this approach can be used to learn preferences on an existing car configuration benchmark [13,14] where past configurations are available together with known manufacturing constraints. In this case, we observe that the learned preferences help to predict satisfactory configurations. The corresponding code will be made accessible from the TOULBAR2 distribution (https://github.com/toulbar2/toulbar2, under an MIT license).

## 2 Background

Our approach is based on Graphical Models [10], a family of mathematical models that have been used in several areas of computer science, artificial intelligence, physics, and statistics. The main idea of Graphical Models is to describe a function of many variables as the combination of several simple functions. "Simple" here means that there is a concise description of the function in a chosen language of functions. Graphical Models have been used to describe Boolean or numerical functions depending on continuous (as in Gaussian Graphical Models [8]) or discrete variables (as in Constraint Networks [32] or propositional logic).

- On the logical side, Constraint networks define a global truth value function as the logical conjunction of small functions described by tables (Boolean tensors), possibly extended with so-called global constraints in Constraint Programming [32].
- Similarly, discrete Markov Random Fields describe a probability distribution as the normalization of the product of small nonnegative functions described as tables (nonnegative real tensors), possibly extended with higher-order functions (similar to global cost functions [1]).

In the rest of the paper, we use capitals $X, Y, Z, \ldots$ to denote variables. The domain of a variable will be denoted as $D^X$ for variable $X$. The actual elements of these domains, values, will be denoted as $\mathtt{a}, \mathtt{b}, \mathtt{c}, \mathtt{g}, \mathtt{r}, \mathtt{t}, \mathtt{1} \ldots$ and an unknown value as $u, v, w, x, y, z \ldots$ Sequence of variables or unknown values will be denoted in bold, respectively as $\boldsymbol{X}, \boldsymbol{Y}, \boldsymbol{Z}, \ldots$ and $\boldsymbol{u}, \boldsymbol{v}, \boldsymbol{w}, \boldsymbol{x}, \boldsymbol{y}, \boldsymbol{z} \ldots$. The Cartesian product of the domains of a sequence of variables $\boldsymbol{X}$ will be denoted as $\Pi^{\boldsymbol{X}}$. An element of $\Pi^{\boldsymbol{X}}$ is a tuple or assignment $\boldsymbol{u}_{\boldsymbol{X}}$ of the variables in $\boldsymbol{X}$. Finally, the projection of the tuple $\boldsymbol{u}_{\boldsymbol{X}}$ on $\boldsymbol{Y} \subseteq \boldsymbol{X}$ is the sequence of values of $\boldsymbol{Y}$ in $\boldsymbol{u}_{\boldsymbol{X}}$ and is denoted as $\boldsymbol{u}_{\boldsymbol{X}}[\boldsymbol{Y}]$. For a given sequence of numbers $\boldsymbol{x} = (x_1, \ldots, x_n)$, its soft-max is $\log(\sum_{x_i \in \boldsymbol{x}} \exp(x_i))$ and its soft-min $-\log(\sum_{x_i \in \boldsymbol{x}} \exp(-x_i))$. Soft-max provides a usual smooth approximation to the maximum function (as does soft-min for the minimum).

We rely on two closely related types of Graphical Models. Cost Function Networks are an extension of Constraint Networks where constraints (Boolean functions that can be satisfied or not), are replaced by bounded integer functions that are summed together to describe a joint bounded numerical function.

**Definition 1 (Cost Function Networks (CFN)).** *A CFN $\mathcal{C} = \langle \boldsymbol{V}, \boldsymbol{C}, k \rangle$ is defined by:*

- *a sequence of $n$ variables $\boldsymbol{V}$, each with a domain of cardinality at most $d$.*
- *a set $\boldsymbol{C}$ of $e$ cost functions.*
- *each cost function $c_{\boldsymbol{S}} \in \boldsymbol{C}$ is a function from $D_{\boldsymbol{S}} \to \bar{\mathbb{Z}}_k$, the set of all integers less than or equal to a given $k \in \bar{\mathbb{Z}} = \mathbb{Z} \cup \{\infty\}$.*

*The CFN $\langle \boldsymbol{V}, \boldsymbol{C}, k \rangle$ defines a joint function $C_{\mathcal{M}}(\boldsymbol{v}) = \bigoplus_{c_{\boldsymbol{S}} \in \Phi}^{k} c_{\boldsymbol{S}}(\boldsymbol{v}[\boldsymbol{S}])$ where $a \oplus b = \min(a + b, k)$, the bounded addition.*

Computing the minimum cost assignment of a CFN is the *Weighted Constraint Satisfaction Problem* (WCSP). Thanks to the upper bound $k$, CFNs are very flexible. For $k = 1$, CFNs are Constraint Networks. Finite values of $k$ capture situations in which an upper bound is known (e.g., the maximum cost that can be spent in a design).

Cost Function Networks are tightly linked to a family of stochastic Graphical Models known as Markov Random Fields:

**Definition 2 (Markov Random Field (MRF)).** *An MRF $\mathcal{M} = \langle \boldsymbol{V}, \Phi \rangle$ is defined by:*

- *a sequence of n variables $\boldsymbol{V}$, each with a domain of cardinality at most d.*
- *a set $\Phi$ of e functions (or factors).*
- *each function $\varphi_{\boldsymbol{S}} \in \Phi$ is a function from $\Pi^{\boldsymbol{S}} \to \mathbb{R}^+$. $\boldsymbol{S}$ is called the scope of the function and $|\boldsymbol{S}|$ its arity.*

*The MRF $\langle \boldsymbol{V}, \Phi \rangle$ defines a joint function $\Phi_{\mathcal{M}}(\boldsymbol{v}) = \prod_{\varphi_{\boldsymbol{S}} \in \Phi} \varphi_{\boldsymbol{S}}(\boldsymbol{v}[\boldsymbol{S}])$ and a probability distribution defined as $P_{\mathcal{M}}(\boldsymbol{V}) \propto \Phi_{\mathcal{M}}(\boldsymbol{V})$.*

Computing the probability $P_{\mathcal{M}}(\cdot)$ requires to compute a normalization constant, denoted as $Z_{\mathcal{M}}$, a #-P complete problem. For a given MRF $\mathcal{M}$, finding an assignment $\boldsymbol{v}$ that maximizes the probability $P_{\mathcal{M}}(\boldsymbol{v})$ can however be directly solved by optimizing the joint function $\Phi_{\mathcal{M}}(\cdot)$, without knowing $Z_{\mathcal{M}}$ and is decision NP-complete.

The connection between CFNs and MRFs is simple: in a CFN with no upper bound ($k = \infty$), $\oplus$ is just the usual addition. In this case, CFNs are isomorphic to Markov Random Fields through a $\exp(-x)$ transform and its inverse $-\log(x)$ transform, up to some adjustable fixed precision. These operations map addition into product (and vice-versa[1]). For a given MRF $\mathcal{M}$, we denote by $\mathcal{M}^\ell$ its corresponding CFN, obtained by applying a $-\log(\cdot)$ transform to all functions.

In CSPs, CFNs, and MRFs, a usual choice is to represent functions by tables/tensors. When domains are Boolean, the language of (weighted) clauses can also be used. We restrict ourselves here to pairwise tensors where each function is determined by an $O(d^2)$ table of costs (or parameters). Then a pairwise graphical model becomes fully defined by the contents of its $O(\frac{n(n-1)}{2})$ cost tables (if no function exists between a given pair of variables, it can be represented as a cost function with constant cost). Extensions to larger arities and global cost functions are not considered here and define nontrivial extensions for the future.

## 3 Learning CFN from Data

In many decision problems, a fraction of the description of the real problem is impossible to model because this information is missing or is too complex to represent. In the extreme, one may want to directly learn a complete CFN from data (a special case of which is Max-SAT [23]).

**Definition 3 (Learning CFN).** *Given a set of variables $\boldsymbol{X}$, and examples $\boldsymbol{E}$ sampled i.i.d. from an unknown joint distribution of high-quality solutions, find a CFN C that can be solved to produce high-quality solutions.*

[1] This log representation is often using in MRFs and the co-domain of factors is called "energy".

Thanks to their isomorphism with Markov Random Fields, CFN can actually be learned in this setting using a probabilistic criterion. Several approaches exist to estimate the set of functions of an MRF from an i.i.d sample but a good fit with the definition above is offered by maximum log-likelihood approaches that learn a model $\mathcal{M}$ that maximizes the probability of the observed sample. Indeed, the likelihood of sample $\boldsymbol{E}$ of i.i.d. assignments under a given MRF $\mathcal{M}$ is the product of the probabilities of all $\boldsymbol{v} \in \boldsymbol{E}$. Its logarithm is:

$$\begin{aligned}\mathcal{L}(\mathcal{M}, \boldsymbol{E}) &= \textstyle\sum_{\boldsymbol{v} \in \boldsymbol{E}} \log(P(\boldsymbol{v})) \\ &= \textstyle\sum_{\boldsymbol{v} \in \boldsymbol{E}} \log(\Phi_{\mathcal{M}}(\boldsymbol{v})) - \log(Z_{\mathcal{M}}) \\ &= \textstyle\sum_{\boldsymbol{v} \in \boldsymbol{E}} (-C_{\mathcal{M}^{\ell}}(\boldsymbol{v})) - \log(\sum_{\boldsymbol{v} \in \Pi^{V}} \exp(-C_{\mathcal{M}^{\ell}}(\boldsymbol{v})))\end{aligned}$$

Maximizing the log-likelihood, therefore, identifies weights in all possible pairwise tensors that simultaneously minimize the average cost of the observed high-quality solutions and maximize the soft-min of the costs of *all* possible assignments, a criterion which fits our optimization objective above very well, independently of its probabilistic interpretation. For a Graphical Model of $n$ variables and maximum domain size $d$, there are $O(n^2d^2)$ weights to optimize.

### 3.1 Regularized Approximate Log-Likelihood GM Estimation

In practice computing the partition function $Z_{\mathcal{M}}$ is #P-hard. Existing algorithms, therefore, optimize a simplified form of the likelihood which either relies on local normalization constants (pseudo-likelihood [4]) or a concave upper-bound of the log-partition function [30]. Maximum likelihood estimators benefit from attractive asymptotic properties, being statistically consistent (the model learned converges to true values as the sample size tends to infinity) [17,30]. On small samples, however, these approaches may overfit and the log-likelihood is regularized by including the norm of the parameters learned as a penalty. Typical norms include the $L_2$ norm (the Euclidian norm), the $L_1$ norm (or Lasso penalty, the sum of the absolute values of all parameters learned), or the $L_1/L_2$ norm (or Group Lasso, that evaluates each function using the $L_2$ norm and combines them using the $L_1$ norm). Given an i.i.d sample $\boldsymbol{E}$ of assignments, the regularized log-likelihood of an MRF $\mathcal{M}$ is defined as:

$$\mathcal{R}(\mathcal{M}, \boldsymbol{E}) = \mathcal{L}(\mathcal{M}, \boldsymbol{E}) - \lambda \cdot ||\Phi||$$

where $||\Phi||$ denotes the norm of all the parameters used in the tensors in $\Phi$ and $\lambda$ is a positive number that needs to be fixed. The Lasso norms ($L_1$ or $L_1/L_2$) bias the criteria to favor functions that take a zero value. This has several positive effects: a function with a table full of zeros does not contribute to the value of the joint function and can be removed, allowing to estimate parameters and scopes simultaneously. Our experiments also show that Lasso regularization can effectively cancel the unavoidable sampling noise present in the finite learning set that otherwise leads to the estimation of a Graphical Model that contains a fraction of random cost functions that are very hard to optimize exactly.

To solve this problem, we rely on a recently proposed scalable (in $O(n^3d^3)$ for pairwise tensors) regularized maximum log-likelihood estimation algorithm, PE_MRF [30], that exploits the ADMM (Alternating Direction Multiplier Method) algorithm for optimization [9]. The algorithm has been designed to learn a GM from a set of solution samples but is actually immediately capable of learning using probabilistic input which will prove very useful in the most intense interaction with deep learning systems later. This algorithm can also learn mixed graphical models with both discrete and continuous variables which can be convenient if the learning set includes not only decision variables but also contextual continuous observations that will also be available when solving (even if we don't explore this capacity further in this paper). This regularized approximate log-likelihood approach using the $L_1/L_2$ norm has been shown to be "sparsistent": as the size of the learning set tends to infinity, the probability of finding the exact graph structure tends to 1 [30], a reassuring asymptotic result, even if our target is to learn a solver, not to estimate a graph structure.

Although the ADMM algorithm is a black-box optimization algorithm, it is useful to understand how it works. ADMM is well-suited to optimize convex functions that are sums of terms. Using a Dual Decomposition principle, every optimization variable $c_{ij}(a, b)$ (the cost of the pair $(a, b)$ in the function $c_{i,j}(\cdot, \cdot)$ of the GM to learn) is duplicated into a copy $c'_{ij}(a, b)$ and the two parameters linked by an equality constraint. At each iteration, the log-likelihood is incrementally optimized on the $c$ variables while the regularization penalty is incrementally optimized on the $c'$ variables. The satisfaction of the equality constraints is delegated to an Augmented Lagrangian approach that penalizes the violation of constraints [9]. As the algorithm iterates, it constantly provides two estimates of the parameters, each defining a CFN. Upon convergence, the two copies are almost but not strictly identical. In practice, it is preferable to use the $c'$ copies which optimize for regularization and contain exact zero. This is crucial for exact WCSP solvers that otherwise spend a gigantic optimization effort optimizing tiny costs often reflecting uninformative sampling noise.

### 3.2 Setting the Regularization Parameter

The determination of a suitable value of $\lambda$ is essential for proper prediction. Existing approaches to tune this parameter in Machine Learning focus in recovering the unknown graph structure (which cannot be achieved using pseudo-likelihood in the presence of infinite costs [36]).

However, recovering the true graph is not our target and, similar to what has been observed in the "Smart 'Predict and Optimize'" framework [11], we observed that taking into account the exact prediction objective does help. We, therefore, use en empirical risk (or error) minimization (ERM) approach. This approach is central in the recent HASSLE Partial Weighted Max-SAT algorithm [23] which also proves that Max-$p$-SAT and CFN models are Probably Approximately Correct(ly) (PAC [35]) learnable by ERM. Using a solution $\boldsymbol{s}$ extracted from a validation set of high-quality solutions (ideally distinct from the training set used for PE_MRF), we assign a fraction of all variables in the learned CFN model with their value in $\boldsymbol{s}$ and ask a WCSP solver to optimize the remaining variables. The solution obtained can be correct (or close to $\boldsymbol{s}$ according to an application-specific distance that can default to the Hamming distance) or not. We use a value of $\lambda$ that minimizes the fraction of non-satisfactory assignments.

Optimizing $\lambda$ in this way requires the repeated resolution of a decision NP-complete problem on the validation set. This is a serious issue even on small problems because the problems learned with very low values of $\lambda$ usually define dense CFNs with functions that overfit the learning set and capture the sampling noise in the learned cost functions. These random problems are extremely hard to solve in practice. While the use of polynomial-time approximations defined by (linear or convex) relaxations has been used with success in related approaches [26], two different approaches can be used to mitigate this complexity. First, we can assign a larger fraction of each solution $\boldsymbol{s}$ in the validation set before solving it. This reduces complexity exponentially. Each solution $\boldsymbol{s}$ in the validation set can be used with several partial assignments in order to cover all scopes. Second, we can relax the requirement for an optimal solution using either a bounded optimization effort (as captured by CPU-time or numbers of backtracks), or by requiring an approximate guarantee (using e.g. a weighted criterion [31]), to avoid spending time on the optimization of very noisy over-fitted problems.

### 3.3 Cost Function Hardening

If needed, and if the training set is reliable (with deterministic 0/1 probabilities on observed values), a similar empirical approach can be used to harden cost functions into constraints. For every non zero cost in the CFN learned, one can simply test if the corresponding combination is observed in any of the training samples. If not, its cost is set to $k$ (the maximum forbidden cost). This may lead to a learned problem that removes more solutions than it should (assuming the true problem is known) but will never make the learned problem inconsistent. In essence, this process is similar to the empirical/experimental method used to learn the general laws of Physics, which slowly evolve as data accumulates.

### 3.4 Related Approaches

Constraint Acquisition [7] learns Constraint Networks from exact positive or negative answers to (partial) membership queries [2]. We instead primarily try to learn a criterion that is not known to be a Boolean feasibility, using a fixed set of high-quality assignments (mostly because good – working – solutions are more often conserved than bad ones, as does Nature for proteins). We also allow these solutions to be only accessible through an imperfect perceptive layer.

HASSLE [23] is a recent algorithm for learning Partial Weighted Max-$p$-SAT (PWMSAT) problems from contextual positive and negative examples using empirical risk minimization. The learned Max-$p$-SAT examples can then be fed to any PWMSAT solver, possibly with additional hard and soft constraints, as in our case. The main strength of HASSLE is its ERM formulation that can decide, for every possible $p$-clause, which one needs to be hard, weighted, or removed to make every sample optimal in the model. This MIP grows however very quickly with the sample size and $p$. The MILP approximation proposed is tested on problems that include at most 20 variables and 91 clauses. A direct encoding of the Sudoku problems would require 729 propositional variables and several thousand clauses (with 9-clauses). It relies on an NP-hard formulation of learning (which is costly but should be beneficial on small samples).

The "Smart 'Predict then Optimize'" (SPO [11,26]) framework has several connections with our approach. Like the surrogate loss of SPO, the convex loss we use (the opposite of penalized log-likelihood and its nonprobabilistic interpretation), is statistically consistent but is best suited for Graphical Models. The empirical adjustment of $\lambda$ using an empirical approach that relies on the final discrete optimization method instead of pure ML criteria, such as Akaike information content (AIC) or Bayesian Information Content (BIC), similarly adapts learning to the final target of actually solving the learned problem.

Recurrent Relational Neural Nets have been recently proposed as a "learning to reason" approach [29]. As in our case, they start from positive examples to later produce solutions. The RRN approach makes little assumptions on the pairwise functions to learn but directly exploits the graph structure of the problem that needs to be solved. On these edges, it learns pairwise "message passing" functions which are applied recurrently using an LSTM neural net [33]. These functions are then applied repeatedly in the prediction phase, similar to what is done in Loopy Belief Propagation (and Arc Consistency). The resulting Neural Net is restricted to solving the problem it has been trained on and will not accept later side-constraints, something which is often desirable in practice.

SAT-net [37] is a Neural-net friendly approach using low-rank convex optimization to both optimize the parameters of a variant of Goemans and Williamson Max-2-SAT convex relaxation [18] and find good solutions using the associated randomized rounding approach. There is a likely similarity between the G&W relaxation (that SAT-net exploits) and the convex relaxations in PE_MRF (that we exploit), but in our case, this relaxation is used only for learning and is optimized by ADMM instead of coordinate descent. More importantly, we rely on a non-differentiable exact WCSP solver for prediction instead of the convex relaxation again. The WCSP solver provides adjustable optimization guarantees while the convex relaxation power is fixed. Furthermore, it is able to satisfy later added side-constraints, something which is impossible by solving G&W convex relaxation (but which would be feasible using a Max-SAT solver, something that has never been tested in our knowledge[2]).

Probabilistic Soft Logic [3] is a related ML system which, as SAT-Net, exploits a convex relaxation for learning parameters and solving Graphical Models (using ADMM instead of coordinate descent). While it benefits from a high-level modeling language with first-order-like syntax, it has the same intrinsic

[2] The weights learned in the convex relaxation are floating-point numbers. A precise integer approximation generates large integer costs which are usually not the sweet spot of the most efficient, core-based, Max-SAT solvers [28].

limitation as SAT-Net: the convex relaxation has a fixed inference power and cannot provide guarantees that additional logical constraints will be satisfied.

## 4 Learning to Solve the Sudoku

Neural Nets and differentiable approaches (such as RRN and SAT-Net) are now able to "learn to reason" from examples, providing the capacity to heuristically solve decision problems with little assumptions and from various inputs, including images. The Sudoku problem has been used as an exemplar of reasoning and we decided to apply our learning and reason architecture to the Sudoku problem, in an experimental setting that is comparable to those used by differentiable approaches in terms of assumptions and biases, also including situations where examples on which to learn require a perceptive layer.

The $n \times n$ Sudoku problem is defined over a grid of $n^2 \times n^2$ cells that each contain a number between 1 and $n^2$. This grid is subdivided in $n^2$ sub-grids of size $n \times n$. A solved Sudoku grid is such that the numbers in every row, column and $n \times n$ sub-grids are all different. Initially, a fraction of all cells is fixed to known values (or hints) and the NP-complete problem [38] is to find a completion of the hints that satisfies the constraints. The puzzle is usually played with $n = 3$. A typical grid, with handwritten hints from the MNIST dataset [24], is represented on the right. As all correct Sudoku puzzle grids, it has only one correct completion (a unique solution). It is known that a minimum of 17 hints is necessary to restrict the number of completions to just one [27]. Such minimal Sudoku problems define challenging puzzles for human beings. As the number of hints increases, the instances become easier and can be solved using simple logical inference rules. Hard or easy for humans, $3 \times 3$ instances can be solved easily by CP solvers, on any standard hardware.

We instead assume that we don't know much about the Sudoku, not even that it has logical rules. We instead consider that the completed grids capture the preferences of users and try to learn a CFN that captures these preferences and compare this with RRN and SAT-Net. It's not easy to compare language biases: RRN is informed with pairwise scopes, SAT-Net uses Max-SAT and we use pairwise finite costs CFNs. Max-SAT and pairwise numerical functions are both capable of representing the set of Sudoku solutions as optimal solutions. SAT-Net has, however, the attractive capacity of using latent variables.

SAT-Net relies on a dataset of 9,000 training + 1,000 test (hint, solution) pairs extracted from a popular Sudoku web site, with an a average of 36.2 hints per grid, defining easy problems. RRN relies on 180,000 training + 18,000 validation and 18,000 test pairs organized each in 18 sets of instances with hardness varying from 17 to 34 hints. We, therefore, used a variable fraction of the RRN training set for training and 1,024 validation samples for hyper-parameters tuning. For testing, we used all $18 \times 1,000$ RRN test samples as well as the SAT-nets test set for comparison. An Intel XeonE5-2687Wv4 3.00 GHz server was used for all experiments. The absolute and relative convergence of ADMM in PE_MRF were both set to $10^{-3}$ and an L1-norm used. We used TOULBAR2 1.0.1 Python interface, representing floating-point numbers with 6 decimals and with the number of backtracks limited to 50,000. The Python-implemented PE_MRF code used 72 s on average for one CFN estimation (with a maximum of 252 s on the largest 180,000 training set).

The empirical approach was used to fix the regularization hyper-parameter $\lambda$. We used TOULBAR2 to minimize the solution cost (in the backtracks limit) and kept the value of $\lambda$ that successively minimizes the fraction of incorrect grids, incorrect cells, and TOULBAR2 CPU-time. The optimization of $\lambda$ (in a $10^{-2}$ to $10^2$ range explored on a logarithmic scale) took 95 min on average on one core (this could be trivially reduced with more cores). SAT-Net requires 172 min to train on a GTX 1080 Ti GPU on its training set [37]. We retrained RRNs on their training data on a GTX 2080 Ti GPU (with a batch size of 64): each epoch required 9 h to run (hundreds of epochs are used by the authors [29]).

Using $180{,}000 + 18{,}000$ training and validation samples, RRNs are able to correctly solve 96.7% problems of the hardest 17 hints problems using 64 "message passing" steps (after which it plateaus). Using $9{,}000 + 1{,}024$ training and validation samples, our approach solves 100% of the same hard 17 hints problems.

Using just 9,000 samples, SAT-net is able to solve 98.3% of its test set (of easy problems). To solve 100% of this test set, $7{,}000 + 1.024$ training and validation samples suffice for our architecture (on these problems, 994/1000 instances are solved backtrack-free, by preprocessing, the remaining problems requiring a total of 24 backtracks). Note however that the learned solver is able to solve only 58.2% of the hardest 17 hints problems. Clearly, problem hardness has to be taken in to account when comparing learned solvers. Figure 1 shows the fraction of correctly solved grids (left) as the sample size increases (performances beyond 13,000 samples are not shown and remain maximal).

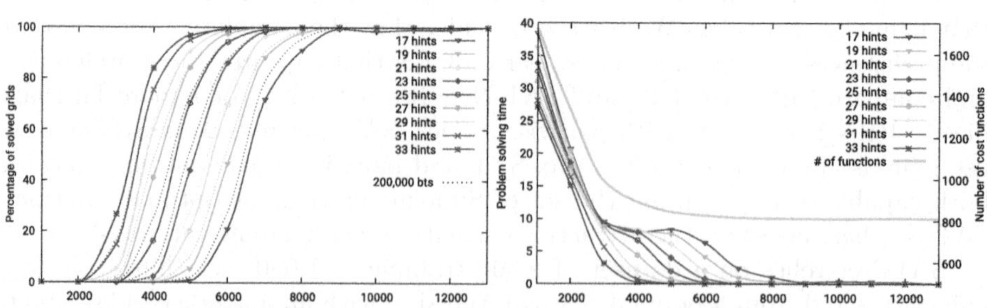


**Fig. 1.** Fraction of correctly solved problems (left, dotted lines correspond to a 200,000 backtracks limit), number of learned functions, and per-instance prediction CPU-times (right) for increasing sample sizes and problem hardness.

The corresponding prediction CPU-times are given in Fig. 1 (right). When training sets are small, the learned CFN models are dense. With 1,000 training samples, more than 1,700 functions are used while the original pairwise Sudoku formulation contains 810 functions. Because of this graph density (and their noisy contents), optimization is more difficult with small training sample sizes. As the training set size increases, the number of functions converges to 810 and resolution becomes easy: an optimal solution can be found and proved in sub-second time (with 9,000 samples, less than 0.3 s are needed on average for the hardest 17 hints Sudokus, and just 3 ms for 34 hints problems). RRN's prediction time on a GTX 2080 Ti GPU was around 2 s for 64 steps. To see if more WCSP solving power could help improve these results, we moved the backtrack limit to 200,000 backtracks. This lead to minor improvements in precision as the dotted lines in Fig. 1 show above, with essentially no progress in terms of accuracy on easy problems: a better loss function and a stronger learning optimization method would be needed here to make progress.

We observed that when the training set size reaches 13,000 samples, the learned CFNs become exact (the set of optimal solutions may be exact before this): they contain 810 cost functions with the exact expected scopes (involving pairs of variables inside a row, column or sub-grid only, although no grid layout information is available to PE_MRF) and contents (a "soft difference" function). So, it is guaranteed that, once domains are reduced by observed hints, a preferred (optimal) solution will be a perfect Sudoku solution. Empirical hardening (Subsect. 3.3) of such a CFN, therefore, recovers the original pairwise formulation of the problem.

### 4.1 Learning and Predicting from Sudoku Images

One of the advantages of differentiable layers that "learn how to reason" is their capacity to integrate inside deep learning architectures. As an example, SAT-net [37] has been trained using hints provided as images with handwritten digits (an example of which appeared in a previous page). Each cell in this image can be decoded by LeNet [24], a convolutional neural net trained on the MNIST dataset with 99.2% precision. The predictions of LeNet are then fed into the SAT-Net layer for learning and prediction. As the authors of SAT-Net observe, the 99.2% precision of LeNet gives an upper bound on the maximum prediction precision: since the SAT-Net data-set has, on average, 36.2 hints per sample, there will be error(s) in the hints in 25.3% of cases, leading to a maximum accuracy of 74.7%.

We also used LeNet and transformed its confidence scores in a marginal unary cost function using soft-max, as is usual with neural net outputs. When a digit appears in a Sudoku image, this unary cost function is added to the learned model instead of assigning a value. This happens both during validation and testing. With hints provided as images during training, SAT-Net solves 63.2% of its test set using the same 9,000 samples. Using just 8,000 + 1,024 training and validation samples, our hybrid architecture is able to solve 78.1% of all these test problems. This violates the crude "theoretical" bound of 74.7% because the confidence scores of LeNet allow an optimal CFN assignment to correct the errors of LeNet when its maximum confidence is low. On the hardest 17 hints instances, 81.2% of correct solutions are obtained: the lower number of hints gives fewer opportunities for failures to LeNet and this has a stronger effect than the increased hardness. With the more radical hardening procedure of Sect. 3.3, 90.5% of the hardest 17 hints problems are solved from 8,000 samples (99.1% of 34 hints problems). As sample size increases, these numbers never reach 100%: correct Sudoku answers that include small permutations compatible with LeNet mistakes are still occasionally produced.

Going beyond SAT-Net, we used a more realistic setting where both hints and solutions are provided as images: for training, we exploited the fact that PE_MRF accepts as input *expectations* of sufficient statistics which can be produced from the marginal unary cost functions above using a $\exp(-x)$ transform. These marginal probabilities are used directly for computing expected numbers of values and pairs (the product of the two marginal probabilities $P(a) \times P(b)$ being used for pairs $(a, b)$). During validations, solutions being available as images only, it becomes impossible to directly compare a predicted solution with the true (unknown) solution. We, therefore, apply LeNet to each cell of the solution image and use the value of the soft-max output of LeNet on the predicted digit as a score. A high score represents an unlikely digit for LeNet and we, therefore, select a $\lambda$ that produces the most likely solutions *i.e.*, which minimizes the sum of all such scores. The same handwritten digit shapes being used in the hints and solution images, the probability that a (hint,solution) sample is correctly decoded by LeNet drops to 52%, independently of the number of hints. Using $8{,}000 + 1{,}024$ training and validation samples, and a 200,00 backtracks limit, our hybrid architecture is able to solve 76.3% of all SAT-Net test problems. On the hardest 17 hints instances, however, performance decreased to 61.8%. Obviously, hardening is of no use here.

## 5 Learning Car Configuration Preferences

In this experiment, we illustrate the versatility of our approach by learning user preferences combined with logical information on a real configuration problem provided by Renault, a French car manufacturing company. A car configuration problem is defined by a set of variables, one for each type of option (engine, color, etc.). Domain values are possible options for each variable. Constraints describe manufacturing compatibilities between options[3].

There are three datasets available, *small*, *medium*, and *big*, each one given in two files: a 1-year sales history of car configurations and a set of manufacturing constraints. The sales history products may or may not satisfy the constraints. We consider here only valid products. *medium* is a small urban car defining a toy example with 148 variables and 44 decision variables in the sales history,[4] mostly Boolean domains with a maximum size of 20 values, 173

[3] See https://www.irit.fr/~Helene.Fargier/BR4CP/benches.html.

[4] We removed the first variable corresponding to the date of each sale product.

constraints defined as tables with a maximum arity of 10, and 8,252 configurations consistent with the constraints in the history. *big* is a utility van with extensive product variability. It has 268 variables (87 decision variables in the sales history), 324 values at most per domain, 332 constraints with a maximum arity of 12, and 8,337 consistent configurations. We discarded the *small* instance as its sales history contains only 710 valid configurations. We counted $278,744$ (resp. $24,566,537,954,855,758,069,760 \approx 2^{74}$) feasible configurations for *medium* (resp. *big*) in 0.1 (resp. 1.8) s on a 3.3 GHz laptop.[5]. We used a 10-fold cross-validation. The valid sales history was split into 10 folds so that all the identical car configurations were contained in the same fold. 9 folds were used as training set for learning user preferences and the last fold was used as test set for predicting user choices. This protocol was repeated 10 times.

We learn the user preferences based on the training set using PE_MRF with either $L_1$ or $L_1/L_2$ norm and a $\lambda$ parameter tuned using the StARS [25] algorithm among a logarithmic grid of 100 spaced values between $10^{-5}$ and $10^3$. We used the default value 0.05 for the threshold parameter $\beta$ in StARS and used subsamples of size $10\sqrt{n}$, where $n$ is the size of the training set, as advised in [25]. Mean $\lambda$ selected value was 34.6 (resp. 0.21) for *medium* (resp. *big*) using $L_1$ norm. The resulting learned CFN on decision variables had 312.9 (resp. 127.2) binary functions and 44 (resp. 87) unary functions.

Then, we test the learned CFN model combined with the manufacturing constraints[6] on the test set, using the protocol described in [13]. This protocol simulates an on-line configuration session with a user. For each test configuration $C$, we select a random variable ordering. Then, we predict the most-probable value for the next variable in the sequence, using the choices made by the user before in the sequence, the learned preferences, and the manufacturing constraints. 10 random variable orderings were considered for each test configuration. Instead of finding the most-probable value by discrete integration over the remaining variables (a marginal MAP inference task), we identified the most-probable valid configuration for all the variables compatible with the previous user choices and the constraints (a pure optimization Maximum A Posteriori – MAP – approximation of marginal MAP).[7] We compare the predicted value for the next variable with the one chosen in $C$ in order to compute a precision score.

We compared our method against a naive Bayesian network approach (called Naive Bayes) and an oracle method, as described in [12,14]. The structure of Naive Bayes is a tree rooted at the next variable with all the previously chosen variables as its sons. It makes the (unrealistic) assumption that all the leaf

[5] Solution counting was done by Backtracking with Tree Decomposition algorithm [15] using *min-fill* heuristic implemented in TOULBAR2 v1.0.1 with options *-ub=1 -a -O=-3 -B=1 -hbfs: -nopre* Reported tree-width was 10 for *medium* and 12 for *big* instance.

[6] We ensure our CFN and the XCSP2.1 XML file for the constraints use the same variable domains with the same increasing value ordering.

[7] We used TOULBAR2 v1.0.1 with a limit of 50,000 backtracks and no preprocessing.

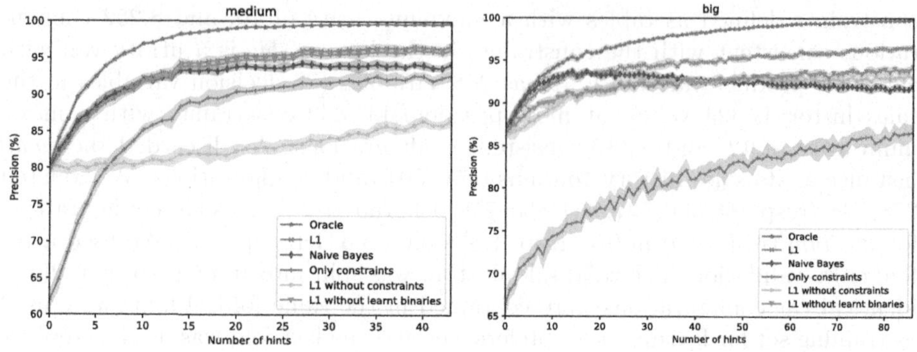


**Fig. 2.** Precision for the next query variable given the number of hints (on previously randomly-selected query variables).

variables are independent knowing the root variable. The most probable value $v$ for the next variable $V$ given the assigned values $\boldsymbol{u}$ of $\boldsymbol{U}$ is easy to compute, $P(v|\boldsymbol{u}) \propto P(v) \prod_{X \in U} P(\boldsymbol{u}[X]|v)$, based on precomputed priors $P(V)$ and conditional probability tables $P(U|V)$ for all pairs of variables $U, V$. The oracle method computes the posterior probability distribution of the next variable knowing the test dataset and the previous user choices. It uses the probability distribution estimated from this set and recommends, given the assigned values $\boldsymbol{u}$, the most probable value in the subset of products, in the test set, that respects $\boldsymbol{u}$. More precisely, for any value $v$ in the domain of the next variable to predict, it estimates $P(v|\boldsymbol{u})$ as $\#(\boldsymbol{u}v)/\#(\boldsymbol{u})$. So, the oracle method is maximally fitted to the test set and its success rate is generally not attainable without having access to the test set (which is, obviously, not the case in practice). Its precision is not 100% since there is an intrinsic variability in the users [12].

The results for the different methods are given in Fig. 2, showing the average precision score and standard-deviation for varying number of hints. We report only results on the $L_1$ ($L_1/L_2$ gave the same precision scores). It took less than 1min. for *medium* (resp. 2min. for *big*) to learn preferences and collect all the 36,124 (resp. 73,428) precision scores for a single fold of cross-validation.[8] The maximum number of backtracks was less than $1,000$, much less than its limit.

The average precision was 93.41% (resp. 94.41% for *big*) compared to 92.08% (resp. 92.31%) for Naive Bayes and 97.10% (resp. 97.35%) for the oracle, showing the practical interest of our approach on real recommendation datasets, providing high precision values in a reasonable amount of time. However, when the number of hints is small, Naive Bayes performed slightly better (for *big*) than our approach, possibly due to the MAP approximation.

Moreover, we investigated the impact of removing either the preferences, partially by just removing learned binary functions and keeping learned unary

[8] We implemented an incremental version of the TOULBAR2 solving procedure using its Python interface in order to load the problem and preprocess it only once.

terms (*L1 without learned binaries* curve in Fig. 2) or removing all learned functions (*Only constraints*), or removing the manufacturing constraints (*L1 without constraints*). All these removals had a negative impact, showing the interest of combining preferences and logical knowledge. Other approaches have been developed to take into account constraints in recommendation systems, such as constraint propagation [5,6] or compilation techniques [19]. However, they remain separated from the model of preferences, whereas our approach exploits two CFNs (learned and mandatory constraint networks) on the same decision variables. We leave a full comparison of the MAP approximation with exact or approximate inference for predicting the next variable value as future work.

## 6 Conclusion

In this paper, we show that a hybrid architecture combining differentiable and nondifferentiable technologies from Graphical Model learning (ADMM convex optimization embodied in the PE_MRF algorithm) and solving (using an anytime Weighted CSP solver with adjustable guarantees) can provide excellent empirical performances, often outperforming recent neural net friendly approaches, with comparable biases. Purely differentiable approaches have the nice property to be directly usable inside more complex deep Learning architectures, often allowing a streamlined learning and predict architecture. By accepting numerical input in its learning component and integer costs in its prediction component, our architecture has the capacity of exploiting neural nets output at the very minor cost of a less streamlined but perfectly workable learning and solving process.

Instead of relying on a polytime bounded inference power, like those offered by, e.g., message-passing or convex relaxations, it offers more powerful inference, associated to higher computational costs that can however be easily controlled either in terms of maximum computational effort or bounded guarantees. Because they provide strong polynomial-time continuous approximations of discrete models such as Max-2SAT, convex relaxations have been repeatedly used as an ideal articulation point between learning and solving discrete/logical models. If the Unique Game Conjecture [21,22] holds, the most promising path for improvement seems to go beyond P and convex relaxations and use NP-complete formulations for solving and learning. This makes powerful anytime NP-hard numerical MIP, WCSP, and PW-MaxSAT solvers of prime interest to make progress in this quest.

As an amusing yet puzzling coincidence, we observe that our hybrid approach is consistent with the dichotomy between the Systems 1 and 2 described in "Thinking Fast and Slow" for human cognitive limitations [20]. Beyond this coincidence; a more practical advantage of our hybrid approach is that it offers a decipherable output, that can be scrutinized to extract logical rules if they empirically reliably predict solutions but also directly used to enhance existing models that may contain mandatory constraints, as is often the case in design problems.

**Acknowledgments.** We thank the GenoToul (Toulouse, France) Bioinformatics and IFB Core (Evry, France) platforms for their computational support. We also thank the reviewers for their critics: the paper did improve, we think. This work has been supported by the French ANR through grants ANR-16-CE40-0028 and ANR-19-PIA3-0004.

## References

1. Allouche, D., et al.: Tractability-preserving transformations of global cost functions. Artif. Intell. **238**, 166–189 (2016)
2. Angluin, D.: Queries and concept learning. Mach. Learn. **2**(4), 319–342 (1988)
3. Bach, S.H., Broecheler, M., Huang, B., Getoor, L.: Hinge-loss markov random fields and probabilistic soft logic. J. Mach. Learn. Res. **18**(1), 3846–3912 (2017)
4. Besag, J.: Efficiency of pseudolikelihood estimation for simple gaussian fields. Biometrika **64**, 616–618 (1977)
5. Bessiere, C., Fargier, H., Lecoutre, C.: Global inverse consistency for interactive constraint satisfaction. In: Schulte, C. (ed.) CP 2013. LNCS, vol. 8124, pp. 159–174. Springer, Heidelberg (2013). https://doi.org/10.1007/978-3-642-40627-0_15
6. Bessiere, C., Fargier, H., Lecoutre, C.: Computing and restoring global inverse consistency in interactive constraint satisfaction. Artif. Intell. **241**, 153–169 (2016)
7. Bessiere, C., Koriche, F., Lazaar, N., O'Sullivan, B.: Constraint acquisition. Artif. Intell. **244**, 315–342 (2017)
8. Bishop, C.M.: Pattern Recognition and Machine Learning. Information Science and Statistics. Springer, New York (2007). http://www.worldcat.org/oclc/71008143
9. Boyd, S., Parikh, N., Chu, E., Peleato, B., Eckstein, J.: Distributed optimization and statistical learning via the alternating direction method of multipliers. Found. Trends® Mach. Learn. **3**(1), 1–22 (2011)
10. Cooper, M., de Givry, S., Schiex, T.: Graphical models: queries, complexity, algorithms. Leibniz Int. Proc. Inform. **154**, 1–4 (2020)
11. Elmachtoub, A.N., Grigas, P.: Smart predict, then optimize. arXiv preprint arXiv:1710.08005 (2017)
12. Fargier, H., Gimenez, P., Mengin, J.: Recommendation for product configuration: an experimental evaluation. In: 18th International Configuration Workshop at CP-16, Toulouse, France (2016)
13. Fargier, H., Gimenez, P., Mengin, J.: Learning lexicographic preference trees from positive examples. In: Proceedings of AAAI-18, pp. 2959–2966. New Orleans, Louisiana (2018)
14. Fargier, H., Gimenez, P.F., Mengin, J.: Experimental evaluation of three value recommendation methods in interactive configuration. J. Univ. Comput. Sci. **26**(3), 318–342 (2020)
15. Favier, A., de Givry, S., Jégou, P.: Exploiting problem structure for solution counting. In: Gent, I.P. (ed.) CP 2009. LNCS, vol. 5732, pp. 335–343. Springer, Heidelberg (2009). https://doi.org/10.1007/978-3-642-04244-7_27
16. Freuder, E.C.: Progress towards the holy grail. Constraints **23**(2), 158–171 (2018)
17. Geman, S., Graffigne, C.: Markov random field image models and their applications to computer vision. In: Proceedings of the International Congress of Mathematicians, Berkeley, CA, vol. 1, p. 2 (1986)
18. Goemans, M.X., Williamson, D.P.: Improved approximation algorithms for maximum cut and satisfiability problems using semidefinite programming. J. ACM (JACM) **42**(6), 1115–1145 (1995)

19. Hadžic, T., Wasowski, A., Andersen, H.R.: Techniques for efficient interactive configuration of distribution networks. In: Proceedings of IJCAI 2007, Hyderabad, India, pp. 100–105 (2007)
20. Kahneman, D.: Thinking, Fast and Slow. Macmillan, New York (2011)
21. Khot, S.: On the power of unique 2-prover 1-round games. In: Proceedings of the Thiry-Fourth Annual ACM Symposium on Theory of Computing, pp. 767–775 (2002)
22. Klarreich, E.: Approximately Hard: The Unique Games Conjecture. Simons Foundation, New York (2011)
23. Kumar, M., Kolb, S., Teso, S., De Raedt, L.: Learning MAX-SAT from contextual examples for combinatorial optimisation. In: Proceedings of AAAI 2020, NYC, USA (2020)
24. LeCun, Y., Bottou, L., Bengio, Y., Haffner, P.: Gradient-based learning applied to document recognition. Proc. IEEE **86**(11), 2278–2324 (1998)
25. Liu, H., Roeder, K., Wasserman, L.: Stability approach to regularization selection (StARS) for high dimensional graphical models. In: Proceedings of Advances in Neural Information Processing Systems (NIPS 2010), vol. 24, pp. 1432–1440 (2010)
26. Mandi, J., Demirović, E., Stuckey, P., Guns, T., et al.: Smart predict-and-optimize for hard combinatorial optimization problems. In: Proceedings of AAAI 2020 (2020)
27. McGuire, G., Tugemann, B., Civario, G.: There is no 16-clue sudoku: solving the sudoku minimum number of clues problem via hitting set enumeration. Exp. Math. **23**(2), 190–217 (2014)
28. Morgado, A., Heras, F., Liffiton, M., Planes, J., Marques-Silva, J.: Iterative and core-guided MaxSAT solving: a survey and assessment. Constraints **18**(4), 478–534 (2013)
29. Palm, R.B., Paquet, U., Winther, O.: Recurrent relational networks. In: Advances in Neural Information Processing Systems, Montréal, Canada, pp. 3372–3382 (2018)
30. Park, Y., Hallac, D., Boyd, S., Leskovec, J.: Learning the network structure of heterogeneous data via pairwise exponential Markov random fields. Proc. Mach. Learn. Res. **54**, 1302 (2017)
31. Pohl, I.: Heuristic search viewed as path finding in a graph. Artif. Intell. **1**(3–4), 193–204 (1970)
32. Rossi, F., van Beek, P., Walsh, T. (eds.): Handbook of Constraint Programming. Elsevier, Amsterdam (2006)
33. Schmidhuber, J., Hochreiter, S.: Long short-term memory. Neural Comput. **9**(8), 1735–1780 (1997)
34. Simoncini, D., Allouche, D., de Givry, S., Delmas, C., Barbe, S., Schiex, T.: Guaranteed discrete energy optimization on large protein design problems. J. Chem. Theory Comput. **11**(12), 5980–5989 (2015)
35. Valiant, L.G.: A theory of the learnable. Commun. ACM **27**(11), 1134–1142 (1984)
36. Vuffray, M., Misra, S., Lokhov, A., Chertkov, M.: Interaction screening: efficient and sample-optimal learning of ising models. In: Advances in Neural Information Processing Systems, pp. 2595–2603 (2016)
37. Wang, P., Donti, P.L., Wilder, B., Kolter, J.Z.: SATnet: bridging deep learning and logical reasoning using a differentiable satisfiability solver. In: Proceedings of ICML 2019, Long Beach, California, USA, vol. 97, pp. 6545–6554. PMLR (2019)
38. Yato, T., Seta, T.: Complexity and completeness of finding another solution and its application to puzzles. IEICE Trans. Fundam. Electron. Commun. Comput. Sci. **86**(5), 1052–1060 (2003)

# Generating Random Logic Programs Using Constraint Programming

Paulius Dilkas[1] and Vaishak Belle[1,2(✉)]

[1] University of Edinburgh, Edinburgh, UK
p.dilkas@sms.ed.ac.uk, vaishak@ed.ac.uk
[2] Alan Turing Institute, London, UK

**Abstract.** Testing algorithms across a wide range of problem instances is crucial to ensure the validity of any claim about one algorithm's superiority over another. However, when it comes to inference algorithms for probabilistic logic programs, experimental evaluations are limited to only a few programs. Existing methods to generate random logic programs are limited to propositional programs and often impose stringent syntactic restrictions. We present a novel approach to generating random logic programs and random probabilistic logic programs using constraint programming, introducing a new constraint to control the independence structure of the underlying probability distribution. We also provide a combinatorial argument for the correctness of the model, show how the model scales with parameter values, and use the model to compare probabilistic inference algorithms across a range of synthetic problems. Our model allows inference algorithm developers to evaluate and compare the algorithms across a wide range of instances, providing a detailed picture of their (comparative) strengths and weaknesses.

**Keywords:** Constraint programming · Probabilistic logic programming · Statistical relational learning

## 1 Introduction

Unifying logic and probability is a long-standing challenge in artificial intelligence [24], and, in that regard, statistical relational learning (SRL) has developed into an exciting area that mixes machine learning and symbolic (logical and relational) structures. In particular, probabilistic logic programs—including languages such as PRISM [25], ICL [22], and ProbLog [11]—are promising frameworks for codifying complex SRL models. With the enhanced structure, however, inference becomes more challenging. At the moment, we have no precise way of evaluating and comparing inference algorithms. Incidentally, if one were to survey the literature, one often finds that an inference algorithm is only tested on a small number (1–4) of data sets [5,16,28], originating from areas such as social networks, citation patterns, and biological data. But how confident can we be that an algorithm works well if it is only tested on a few problems?


H. Simonis (Ed.): CP 2020, LNCS 12333, pp. 828–845, 2020.
https://doi.org/10.1007/978-3-030-58475-7_48

About thirty years ago, SAT solving technology was dealing with a similar lack of clarity [26]. This changed with the study of generating random SAT instances against different input parameters (e.g., clause length and the total number of variables) to better understand the behaviour of algorithms and their ability to solve random synthetic problems. Unfortunately, when it comes to generating random logic programs, all approaches so far focused exclusively on propositional programs [1,2,30,32], often with severely limiting conditions such as two-literal clauses [20,21] or clauses of the form $a \leftarrow \neg b$ [31].

In this work (Sects. 3, 4 and 5), we introduce a constraint-based representation for logic programs based on simple parameters that describe the program's size, what predicates and constants it uses, etc. This representation takes the form of a *constraint satisfaction problem* (CSP), i.e., a set of discrete variables and restrictions on what values they can take. Every solution to this problem (as output by a constraint solver) directly translates into a logic program. One can either output all (sufficiently small) programs that satisfy the given conditions or use random value-ordering heuristics and restarts to generate random programs. For sampling from a uniform distribution, the CSP can be transformed into a belief network [12]. In fact, the same model can generate both probabilistic programs in the syntax of ProbLog [11] and non-probabilistic Prolog programs. To the best of our knowledge, this is the first work that (a) addresses the problem of generating random logic programs in its full generality (i.e., including first-order clauses with variables), and (b) compares and evaluates inference algorithms for probabilistic logic programs on more than a handful of instances.

A major advantage of a constraint-based approach is the ability to add additional constraints as needed, and to do that efficiently (compared to generate-and-test approaches). As an example of this, in Sect. 7 we develop a custom constraint that, given two predicates $\mathsf{P}$ and $\mathsf{Q}$, ensures that any ground atom with predicate $\mathsf{P}$ is independent of any ground atom with predicate $\mathsf{Q}$. In this way, we can easily regulate the independence structure of the underlying probability distribution. In Sect. 6 we also present a combinatorial argument for correctness that counts the number of programs that the model produces for various parameter values. We end the paper with two experimental results in Sect. 8: one investigating how the constraint model scales when tasked with producing more complex programs, and one showing how the model can be used to evaluate and compare probabilistic inference algorithms.

Overall, our main contributions are concerned with logic programming-based languages and frameworks, which capture a major fragment of SRL [9]. However, since probabilistic logic programming languages are closely related to other areas of machine learning, including (imperative) probabilistic programming [10], our results can lay the foundations for exploring broader questions on generating models and testing algorithms in machine learning.

## 2 Preliminaries

The basic primitives of logic programs are *constants*, *(logic) variables*, and *predicates* with their *arities*. A *term* is either a variable or a constant, and an *atom* is a predicate of arity $n$ applied to $n$ terms. A *formula* is any well-formed expression that connects atoms using conjunction $\wedge$, disjunction $\vee$, and negation $\neg$. A *clause* is a pair of a *head* (which is an atom) and a *body* (which is a formula[1]). A *(logic) program* is a set of clauses, and a ProbLog *program* is a set of clause-probability pairs [14].

In the world of CSPs, we also have *(constraint) variables*, each with a *domain*, whose values are restricted using *constraints*. All constraint variables in the model are integer or set variables, however, if an integer refers to a logical construct (e.g., a logical variable or a constant), we will make no distinction between the two. We say that a constraint variable is *(fully) determined* if its domain (at the time) has exactly one value. We let $\square$ denote the absent/disabled value of an optional variable [19]. We write $\texttt{a[}b\texttt{]} \in c$ to mean that $\texttt{a}$ is an array of variables of length $b$ such that each element of $\texttt{a}$ has domain $c$. Similarly, we write $\texttt{c} : \texttt{a[}b\texttt{]}$ to denote an array $\texttt{a}$ of length $b$ such that each element of $\texttt{a}$ has type $\texttt{c}$. Finally, we assume that all arrays start with index zero.

*Parameters of the Model.* We begin by defining sets and lists of the primitives used in constructing logic programs: a list of predicates $\mathcal{P}$, a list of their corresponding arities $\mathcal{A}$ (so $|\mathcal{A}| = |\mathcal{P}|$), a set of variables $\mathcal{V}$, and a set of constants $\mathcal{C}$. Either $\mathcal{V}$ or $\mathcal{C}$ can be empty, but we assume that $|\mathcal{C}| + |\mathcal{V}| > 0$. Similarly, the model supports zero-arity predicates but requires at least one predicate to have non-zero arity. For notational convenience, we also set $\mathcal{M}_{\mathcal{A}} = \max \mathcal{A}$. Next, we need a measure of how complex a body of a clause can be. As we represent each body by a tree (see Sect. 4), we set $\mathcal{M}_{\mathcal{N}} \geq 1$ to be the maximum number of nodes in the tree representation of any clause. We also set $\mathcal{M}_{\mathcal{C}}$ to be the maximum number of clauses in a program. We must have that $\mathcal{M}_{\mathcal{C}} \geq |\mathcal{P}|$ because we require each predicate to have at least one clause that defines it. The model supports enforcing predicate independence (see Sect. 7), so a set of independent pairs of predicates is another parameter. Since this model can generate probabilistic as well as non-probabilistic programs, each clause is paired with a probability which is randomly selected from a given list—our last parameter. For generating non-probabilistic programs, one can set this list to [1]. Finally, we define $\mathcal{T} = \{\neg, \wedge, \vee, \top\}$ as the set of tokens that (together with atoms) form a clause. All decision variables of the model can now be divided into $2 \times \mathcal{M}_{\mathcal{C}}$ separate groups, treating the body and the head of each clause separately. We say that the variables are contained in two arrays: $\texttt{Body} : \texttt{bodies[}\mathcal{M}_{\mathcal{C}}\texttt{]}$ and $\texttt{Head} : \texttt{heads[}\mathcal{M}_{\mathcal{C}}\texttt{]}$.

## 3 Heads of Clauses

We define the *head* of a clause as a $\texttt{predicate} \in \mathcal{P} \cup \{\square\}$ and $\texttt{arguments[}\mathcal{M}_{\mathcal{A}}\texttt{]} \in \mathcal{C} \cup \mathcal{V} \cup \{\square\}$. Here, we use $\square$ to denote either a disabled

[1] Our model supports arbitrarily complex bodies of clauses (e.g., $\neg\mathsf{P}(X) \vee (\mathsf{Q}(X) \wedge \mathsf{P}(X))$) because ProbLog does too. However, one can easily restrict our representation of a body to a single conjunction of literals (e.g., $\mathsf{Q}(X) \wedge \neg\mathsf{P}(X)$)) by adding a couple of additional constraints.

clause that we choose not to use or disabled arguments if the arity of the $\texttt{predicate}$ is less than $\mathcal{M}_{\mathcal{A}}$. The reason why we need a separate value for the latter (i.e., why it is not enough to fix disabled arguments to a single already-existing value) will become clear in Sect. 5. This $\texttt{predicate}$ variable has a corresponding arity that depends on the $\texttt{predicate}$. We can define $\texttt{arity} \in [0, \mathcal{M}_{\mathcal{A}}]$ as the arity of the $\texttt{predicate}$ if $\texttt{predicate} \in \mathcal{P}$ and zero otherwise using the table constraint [17]. This constraint uses a set of pairs of the form $(p, a)$, where $p$ ranges over all possible values of the $\texttt{predicate}$, and $a$ is either the arity of predicate $p$ or zero. Having defined arity, we can now fix the superfluous arguments.

**Constraint 1.** *For* $i = 0, \ldots, \mathcal{M}_{\mathcal{A}} - 1$, $\texttt{arguments}[i] = \square \iff i \geq \texttt{arity}$.

We also add a constraint that each predicate should get at least one clause.

**Constraint 2.** *Let* $P = \{h.\texttt{predicate} \mid h \in \texttt{heads}\}$ *be a multiset. Then*

$$\texttt{nValues}(P) = \begin{cases} |\mathcal{P}| & \textit{if } \texttt{count}(\square, P) = 0 \\ |\mathcal{P}| + 1 & \textit{otherwise}, \end{cases}$$

*where* $\texttt{nValues}(P)$ *counts the number of unique values in* $P$, *and* $\texttt{count}(\square, P)$ *counts how many times* $\square$ *appears in* $P$.

Finally, we want to disable duplicate clauses but with one exception: there may be more than one disabled clause, i.e., a clause with head $\texttt{predicate} = \square$. Assuming a lexicographic order over entire clauses such that $\square > \mathsf{P}$ for all $\mathsf{P} \in \mathcal{P}$ and the head predicate is the 'first digit' of this representation, the following constraint disables duplicates as well as orders the clauses.

**Constraint 3.** *For* $i = 1, \ldots, \mathcal{M}_{\mathcal{C}} - 1$, *if* $\texttt{heads}[i].\texttt{predicate} \neq \square$, *then*

$$(\texttt{heads}[i-1], \texttt{bodies}[i-1]) < (\texttt{heads}[i], \texttt{bodies}[i]).$$

## 4 Bodies of Clauses

As was briefly mentioned before, the *body* of a clause is represented by a tree. It has two parts. First, there is the $\texttt{structure}[\mathcal{M}_{\mathcal{N}}] \in [0, \mathcal{M}_{\mathcal{N}} - 1]$ array that encodes the structure of the tree using the following two rules: $\texttt{structure}[i] = i$ means that the $i$th node is a root, and $\texttt{structure}[i] = j$ (for $j \neq i$) means that the $i$th node's parent is node $j$. The second part is the array $\texttt{Node} : \texttt{values}[\mathcal{M}_{\mathcal{N}}]$ such that $\texttt{values}[i]$ holds the value of the $i$th node, i.e., a representation of the atom or logical operator.

We can use the $\texttt{tree}$ constraint [13] to forbid cycles in the $\texttt{structure}$ array and simultaneously define $\texttt{numTrees} \in \{1, \ldots, \mathcal{M}_{\mathcal{N}}\}$ to count the number of trees. We will view the tree rooted at the zeroth node as the main tree and restrict all other trees to single nodes. For this to work, we need to make sure that the zeroth node is indeed a root, i.e., fix $\texttt{structure}[0] = 0$. For convenience, we also define $\texttt{numNodes} \in \{1, \ldots, \mathcal{M}_{\mathcal{N}}\}$ to count the number of nodes in the main tree. We define it as $\texttt{numNodes} = \mathcal{M}_{\mathcal{N}} - \texttt{numTrees} + 1$.

*Example 1.* Let $\mathcal{M}_\mathcal{N} = 8$. Then $\neg\mathsf{P}(X) \lor (\mathsf{Q}(X) \land \mathsf{P}(X))$ can be encoded as:

$$\texttt{structure} = [0, 0, 0, \quad 1, \quad 2, \quad 2, \quad 6, 7], \qquad \texttt{numNodes} = 6,$$
$$\texttt{values} = [\lor, \neg, \land, \mathsf{P}(X), \mathsf{Q}(X), \mathsf{P}(X), \top, \top], \qquad \texttt{numTrees} = 3.$$

Here, $\top$ is the value we use for the remaining one-node trees. The elements of the `values` array are nodes. A *node* has a $\texttt{name} \in \mathcal{T} \cup \mathcal{P}$ and $\texttt{arguments}[\mathcal{M}_\mathcal{A}] \in \mathcal{V} \cup \mathcal{C} \cup \{\square\}$. The node's `arity` can then be defined in the same way as in Sect. 3. Furthermore, we can use Constraint 1 to again disable the extra arguments.

*Example 2.* Let $\mathcal{M}_\mathcal{A} = 2$, $X \in \mathcal{V}$, and let $\mathsf{P}$ be a predicate with arity 1. Then the node representing atom $\mathsf{P}(X)$ has: $\texttt{name} = \mathsf{P}$, $\texttt{arguments} = [X, \square]$, $\texttt{arity} = 1$.

We need to constrain the forest represented by the `structure` array together with its `values` to eliminate symmetries and adhere to our desired format. First, we can recognise that the order of the elements in the `structure` array does not matter, i.e., the structure is only defined by how the elements link to each other, so we can add a constraint for sorting the `structure` array. Next, since we already have a variable that counts the number of nodes in the main tree, we can fix the structure and the values of the remaining trees to some constant values.

**Constraint 4.** *For* $i = 1, \ldots, \mathcal{M}_\mathcal{N} - 1$, *if* $i < \texttt{numNodes}$, *then*

$$\texttt{structure}[i] = i, \quad \textit{and} \quad \texttt{values}[i].\texttt{name} = \top,$$

*else* $\texttt{structure}[i] < i$.

The second part of this constraint states that every node in the main tree except the zeroth node cannot be a root and must have its parent located to the left of itself. Next, we classify all nodes into three classes: predicate (or empty) nodes, negation nodes, and conjunction/disjunction nodes based on the number of children (zero, one, and two, respectively).

**Constraint 5.** *For* $i = 0, \ldots, \mathcal{M}_\mathcal{N} - 1$, *let* $C_i$ *be the number of times* $i$ *appears in the* `structure` *array with index greater than* $i$. *Then*

$$\begin{aligned} C_i = 0 &\iff \texttt{values}[i].\texttt{name} \in \mathcal{P} \cup \{\top\}, \\ C_i = 1 &\iff \texttt{values}[i].\texttt{name} = \neg, \\ C_i > 1 &\iff \texttt{values}[i].\texttt{name} \in \{\land, \lor\}. \end{aligned}$$

The value $\top$ serves a twofold purpose: it is used as the fixed value for nodes outside the main tree, and, when located at the zeroth node, it can represent a clause with an empty body. Thus, we can say that only root nodes can have $\top$ as the value.

**Constraint 6.** *For* $i = 0, \ldots, \mathcal{M}_\mathcal{N} - 1$,

$$\texttt{structure}[i] \neq i \implies \texttt{values}[i].\texttt{name} \neq \top.$$

Finally, we add a way to disable a clause by setting its head predicate to $\square$.

**Constraint 7.** *For* $i = 0, \ldots, \mathcal{M}_{\mathcal{C}} - 1$, *if* $\texttt{heads}[i]\texttt{.predicate} = \square$, *then*

$$\texttt{bodies}[i]\texttt{.numNodes} = 1, \quad \textit{and} \quad \texttt{bodies}[i]\texttt{.values}[0]\texttt{.name} = \top.$$

# 5 Variable Symmetry Breaking

Ideally, we want to avoid generating programs that are equivalent in the sense that they produce the same answers to all queries. Even more importantly, we want to avoid generating multiple internal representations that ultimately result in the same program. This is the purpose of *symmetry-breaking constraints*, another important benefit of which is that the constraint solving task becomes easier [29]. Given any clause, we can permute the variables in that clause without changing the meaning of the clause or the entire program. Thus, we want to fix the order of variables. Informally, we can say that variable $X$ goes before variable $Y$ if the first occurrence of $X$ in either the head or the body of the clause is before the first occurrence of $Y$. Note that the constraints described in this section only make sense if $|\mathcal{V}| > 1$ and that all definitions and constraints here are on a per-clause basis.

**Definition 1.** *Let* $N = \mathcal{M}_{\mathcal{A}} \times (\mathcal{M}_{\mathcal{N}} + 1)$, *and let* $\texttt{terms}[N] \in \mathcal{C} \cup \mathcal{V} \cup \{\square\}$ *be a flattened array of all arguments in a particular clause. Then we can use a channeling constraint to define* $\texttt{occ}[|\mathcal{C}| + |\mathcal{V}| + 1]$ *as an array of subsets of* $\{0, \ldots, N-1\}$ *such that for all* $i = 0, \ldots, N-1$, *and* $t \in \mathcal{C} \cup \mathcal{V} \cup \{\square\}$,

$$i \in \texttt{occ}[t] \quad \iff \quad \texttt{terms}[i] = t.$$

Next, we introduce an array that holds the first occurrence of each variable.

**Definition 2.** *Let* $\texttt{intros}[|\mathcal{V}|] \in \{0, \ldots, N\}$ *be such that for* $v \in \mathcal{V}$,

$$\texttt{intros}[v] = \begin{cases} 1 + \min \texttt{occ}[v] & \textit{if } \texttt{occ}[v] \neq \emptyset \\ 0 & \textit{otherwise.} \end{cases}$$

Here, a value of zero means that the variable does not occur in the clause (this choice is motivated by subsequent constraints). As a consequence, all other indices are shifted by one. Having set this up, we can now eliminate variable symmetries simply by sorting `intros`. In other words, we constrain the model so that the variable listed first (in whatever order $\mathcal{V}$ is presented in) has to occur first in our representation of a clause.

*Example 3.* Let $\mathcal{C} = \emptyset$, $\mathcal{V} = \{X, Y, Z\}$, $\mathcal{M}_{\mathcal{A}} = 2$, $\mathcal{M}_{\mathcal{N}} = 3$, and consider the clause $\mathsf{sibling}(X, Y) \leftarrow \mathsf{parent}(X, Z) \wedge \mathsf{parent}(Y, Z)$. Then

$$\begin{aligned} \texttt{terms} &= [X, Y, \square, \square, X, Z, Y, Z], \\ \texttt{occ} &= [\{0, 4\}, \{1, 6\}, \{5, 7\}, \{2, 3\}], \\ \texttt{intros} &= [0, 1, 5], \end{aligned}$$

where the $\square$'s correspond to the conjunction node.

We end the section with several redundant constraints that make the CSP easier to solve. First, we can state that the positions occupied by different terms must be different.

**Constraint 8.** *For* $u \neq v \in \mathcal{C} \cup \mathcal{V} \cup \{\square\}$, $\texttt{occ}[u] \cap \texttt{occ}[v] = \emptyset$.

The reason why we use zero to represent an unused variable is so that we could now use the 'all different except zero' constraint for the `intros` array. We can also add another link between `intros` and `occ` that essentially says that the smallest element of a set is an element of the set.

**Constraint 9.** *For* $v \in \mathcal{V}$, $\texttt{intros}[v] \neq 0 \iff \texttt{intros}[v] - 1 \in \texttt{occ}[v]$.

Finally, we define an auxiliary set variable to act as a set of possible values that `intros` can take. Let $\texttt{potentials} \subseteq \{0, \ldots, N\}$ be such that for $v \in \mathcal{V}$, $\texttt{intros}[v] \in \texttt{potentials}$. Using this new variable, we can add a constraint saying that non-predicate nodes in the tree representation of a clause cannot have variables as arguments.

**Constraint 10.** *For* $i = 0, \ldots, \mathcal{M}_{\mathcal{N}} - 1$, *let*

$$S = \{\mathcal{M}_{\mathcal{A}} \times (i + 1) + j + 1 \mid j = 0, \ldots, \mathcal{M}_{\mathcal{A}} - 1\}.$$

*If* $\texttt{values}[i].\texttt{name} \notin \mathcal{P}$, *then* $\texttt{potentials} \cap S = \emptyset$.

## 6 Counting Programs

To demonstrate the correctness of the model, this section derives combinatorial expressions for counting the number of programs with up to $\mathcal{M}_{\mathcal{C}}$ clauses and up to $\mathcal{M}_{\mathcal{N}}$ nodes per clause, and arbitrary $\mathcal{P}$, $\mathcal{A}$, $\mathcal{V}$, and $\mathcal{C}$. Being able to establish two ways to generate the same sequence of numbers (i.e., numbers of programs with certain properties and parameters) allows us to gain confidence that the constraint model accurately matches our intentions. For this section, we introduce the term *total arity* of a body of a clause to refer to the sum total of arities of all predicates in the body.

We will first consider clauses with *gaps*, i.e., without taking variables and constants into account. Let $T(n, a)$ denote the number of possible clause bodies with $n$ nodes and total arity $a$. Then $T(1, a)$ is the number of predicates in $\mathcal{P}$ with arity $a$, and the following recursive definition can be applied for $n > 1$:

$$T(n, a) = T(n-1, a) + 2 \sum_{\substack{c_1+\cdots+c_k=n-1, \\ 2 \leq k \leq \frac{a}{\min \mathcal{A}}, \\ c_i \geq 1 \text{ for all } i}} \sum_{\substack{d_1+\cdots+d_k=a, \\ d_i \geq \min \mathcal{A} \text{ for all } i}} \prod_{i=1}^{k} T(c_i, d_i).$$

The first term here represents negation, i.e., negating a formula consumes one node but otherwise leaves the task unchanged. If the first operation is not a negation, then it must be either conjunction or disjunction (hence the coefficient

'2'). In the first sum, $k$ represents the number of children of the root node, and each $c_i$ is the number of nodes dedicated to child $i$. Thus, the first sum iterates over all possible ways to partition the remaining $n-1$ nodes. Similarly, the second sum considers every possible way to partition the total arity $a$ across the $k$ children nodes. We can then count the number of possible clause bodies with total arity $a$ (and any number of nodes) as

$$C(a) = \begin{cases} 1 & \text{if } a = 0 \\ \sum_{n=1}^{\mathcal{M}_\mathcal{N}} T(n, a) & \text{otherwise.} \end{cases}$$

The number of ways to select $n$ terms is

$$P(n) = |\mathcal{C}|^n + \sum_{\substack{1 \le k \le |\mathcal{V}|,\\ 0 = s_0 < s_1 < \cdots < s_k < s_{k+1} = n+1}} \prod_{i=0}^{k} (|\mathcal{C}| + i)^{s_{i+1} - s_i - 1}.$$

The first term is the number of ways to select $n$ constants. The parameter $k$ is the number of variables used in the clause, and $s_1, \ldots, s_k$ mark the first occurrence of each variable. For each gap between any two introductions (or before the first introduction, or after the last introduction), we have $s_{i+1} - s_i - 1$ spaces to be filled with any of the $|\mathcal{C}|$ constants or any of the $i$ already-introduced variables.

Let us order the elements of $\mathcal{P}$, and let $a_i$ be the arity of the $i$th predicate. The number of programs is then:

$$\sum_{\substack{\sum_{i=1}^{|\mathcal{P}|} h_i = n,\\ |\mathcal{P}| \le n \le \mathcal{M}_\mathcal{C},\\ h_i \ge 1 \text{ for all } i}} \prod_{i=1}^{|\mathcal{P}|} \binom{\sum_{a=0}^{\mathcal{M}_\mathcal{A} \times \mathcal{M}_\mathcal{N}} C(a) P(a + a_i)}{h_i}, \tag{1}$$

Here, we sum over all ways to distribute $|\mathcal{P}| \le n \le \mathcal{M}_\mathcal{C}$ clauses among $|\mathcal{P}|$ predicates so that each predicate gets at least one clause. For each predicate, we can then count the number of ways to select its clauses out of all possible clauses. The number of possible clauses can be computed by considering each possible arity $a$, and multiplying the number of 'unfinished' clauses $C(a)$ by the number of ways to select the required $a + a_i$ terms in the body and the head of the clause. Finally, we compare the numbers produced by (1) with the numbers of programs generated by our model in 1032 different scenarios, thus showing that the combinatorial description developed in this section matches the model's behaviour.

## 7 Stratification and Independence

*Stratification* is a condition necessary for probabilistic logic programs [18] and often enforced on logic programs [4] that helps to ensure a unique answer to every query. This is achieved by restricting the use of negation so that any program

$\mathscr{P}$ can be partitioned into a sequence of programs $\mathscr{P} = \bigsqcup_{i=1}^{n} \mathscr{P}_i$ such that, for all $i$, the negative literals in $\mathscr{P}_i$ can only refer to predicates defined in $\mathscr{P}_j$ for $j \leq i$ [4].

*Independence*, on the other hand, is defined on a pair of predicates (say, $\mathsf{P}, \mathsf{Q} \in \mathcal{P}$) and can be interpreted in two ways. First, if $\mathsf{P}$ and $\mathsf{Q}$ are independent, then any ground atom of $\mathsf{P}$ is independent of any ground atom of $\mathsf{Q}$ in the underlying probability distribution of the probabilistic program. Second, the part of the program needed to fully define $\mathsf{P}$ is disjoint from the part of the program needed to define $\mathsf{Q}$.

These two seemingly disparate concepts can be defined using the same building block, i.e., a predicate dependency graph. Let $\mathscr{P}$ be a probabilistic logic program with its set of predicates $\mathcal{P}$. Its *(predicate) dependency graph* is a directed graph $G_{\mathscr{P}}$ with elements of $\mathcal{P}$ as nodes and an edge between $\mathsf{P}, \mathsf{Q} \in \mathcal{P}$ if there is a clause in $\mathscr{P}$ with $\mathsf{Q}$ as the head and $\mathsf{P}$ mentioned in the body. We say that the edge is *negative* if there exists a clause with $\mathsf{Q}$ as the head and at least one instance of $\mathsf{P}$ at the body such that the path from the root to the $\mathsf{P}$ node in the tree representation of the clause passes through at least one negation node; otherwise, it is *positive*. We say that $\mathscr{P}$ (or $G_{\mathscr{P}}$) has a *negative cycle* if $G_{\mathscr{P}}$ has a cycle with at least one negative edge. A program $\mathscr{P}$ is *stratified* if $G_{\mathscr{P}}$ has no negative cycles.[2] Thus a simple entailment algorithm for stratification can be constructed by selecting all clauses, all predicates of which are fully determined, and looking for negative cycles in the dependency graph constructed based on those clauses using an algorithm such as Bellman-Ford.

For any predicate $\mathsf{P} \in \mathcal{P}$, the set of *dependencies* of $\mathsf{P}$ is the smallest set $D_{\mathsf{P}}$ such that $\mathsf{P} \in D_{\mathsf{P}}$, and, for every $\mathsf{Q} \in D_{\mathsf{P}}$, all direct predecessors of $\mathsf{Q}$ in $G_{\mathscr{P}}$ are in $D_{\mathsf{P}}$. Two predicates $\mathsf{P}$ and $\mathsf{Q}$ are *independent* if $D_{\mathsf{P}} \cap D_{\mathsf{Q}} = \emptyset$.

*Example 4.* Consider the following (fragment of a) program:

$$\begin{aligned} \mathsf{sibling}(X, Y) &\leftarrow \mathsf{parent}(X, Z) \land \mathsf{parent}(Y, Z), \\ \mathsf{father}(X, Y) &\leftarrow \mathsf{parent}(X, Y) \land \neg \mathsf{mother}(X, Y). \end{aligned} \tag{2}$$

Its predicate dependency graph is in Fig. 1. Because of the negation in (2), the edge from $\mathsf{mother}$ to $\mathsf{father}$ is negative, while the other two edges are positive. The dependencies of each predicate are:

$$\begin{aligned} D_{\mathsf{parent}} &= \{\mathsf{parent}\}, & D_{\mathsf{sibling}} &= \{\mathsf{sibling}, \mathsf{parent}\}, \\ D_{\mathsf{mother}} &= \{\mathsf{mother}\}, & D_{\mathsf{father}} &= \{\mathsf{father}, \mathsf{mother}, \mathsf{parent}\}. \end{aligned}$$

Hence, we have two pairs of independent predicates, i.e., $\mathsf{mother}$ is independent of $\mathsf{parent}$ and $\mathsf{sibling}$.

Since the definition of independence relies on the dependency graph, we can represent this graph as an adjacency matrix constructed as part of the model. Let

---

[2] This definition is an extension of a well-known result for logic programs [3] to probabilistic logic programs with arbitrary complex clause bodies.

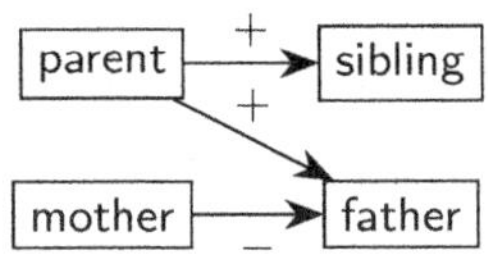


**Fig. 1.** The predicate dependency graph of the program from Example 4. Positive edges are labelled with '+', and negative edges with '−'.

**Table 1.** Types of (potential) dependencies of a predicate P based on the number of undetermined edges on the path from the dependency to P

| Edges | Name | Notation |
|---|---|---|
| 0 | Determined | $\Delta(p)$ |
| 1 | Almost determined | $\Gamma(p, s, t)$ |
| $> 1$ | Undetermined | $\Upsilon(p)$ |

---

**Algorithm 1:** Entailment for independence

---

**Data**: predicates $p_1$, $p_2$
$D \leftarrow \{(d_1, d_2) \in$ deps($p_1$, 1) $\times$ deps($p_2$, 1) $\mid d_1$.predicate $= d_2$.predicate$\}$;
**if** $D = \emptyset$ **then return** TRUE;
**if** $\exists(\Delta\_, \Delta\_) \in D$ **then return** FALSE **else return** UNDEFINED;

---

$\mathbf{A}$ be a $|\mathcal{P}| \times |\mathcal{P}|$ binary matrix defined element-wise by stating that $\mathbf{A}[i][j] = 0$ if and only if, for all $k = 0, \ldots, \mathcal{M}_{\mathcal{C}} - 1$, either heads$[k]$.predicate $\neq j$ or $i \notin \{a.$name $\mid a \in$ bodies$[k]$.values$\}$.

Given a partially-solved model with its predicate dependency graph, let us pick an arbitrary path from Q to P (for some P, Q $\in \mathcal{P}$) that consists of determined edges that are denoted by 1 in $\mathbf{A}$ and potential/undetermined edges that are denoted by $\{0, 1\}$. Each such path characterises a *(potential) dependency* Q for P. We classify all such dependencies into three classes depending on the number of undetermined edges on the path. These classes are outlined in Table 1, where $p$ represents the dependency predicate Q, and, in the case of $\Gamma$, $(s, t) \in \mathcal{P}^2$ is the one undetermined edge on the path. For a dependency $d$—regardless of its exact type—we will refer to its predicate $p$ as $d$.predicate. In describing the algorithms, we will use '_' to replace any of $p$, $s$, $t$ in situations where the name is unimportant.

Each entailment algorithm returns one out of three values: TRUE if the constraint is guaranteed to hold, FALSE if the constraint is violated, and UNDEFINED if whether the constraint will be satisfied or not depends on the future decisions made by the solver. Algorithm 1 outlines a simple entailment algorithm for the independence of two predicates $p_1$ and $p_2$. First, we separately calculate all dependencies of $p_1$ and $p_2$ and look at the set $D$ of dependencies that $p_1$ and $p_2$ have in common. If there are none, then the predicates are clearly independent. If they have a dependency in common that is already fully determined ($\Delta$) for both predicates, then they cannot be independent. Otherwise, we return UNDEFINED.

Propagation algorithms have two goals: causing a contradiction (failing) in situations where the corresponding entailment algorithm would return FALSE, and eliminating values from domains of variables that are guaranteed to cause a contradiction. Algorithm 2 does the former on Line 2. Furthermore, for any

**Algorithm 2:** Propagation for independence

**Data:** predicates $p_1$, $p_2$; adjacency matrix $\mathbf{A}$

1 **for** $(d_1, d_2) \in$ `deps`$(p_1,\ 0) \times$ `deps`$(p_2,\ 0)$ *s.t.* $d_1.\mathsf{predicate} = d_2.\mathsf{predicate}$ **do**

2     **if** $d_1$ **is** $\Delta(\_)$ **and** $d_2$ **is** $\Delta(\_)$ **then** `fail()`;

3     **if** $\{d_1, d_2\} = \{\Delta(\_), \Gamma(\_, s, t)\}$ **then** $\mathbf{A}[s][t]$`.removeValue(1)`;

**Algorithm 3:** Dependencies of a predicate

**Data:** adjacency matrix $\mathbf{A}$

**Function** `deps`$(p,\ \mathsf{allDeps})$:

    $D \leftarrow \{\Delta(p)\}$;

    **while true do**

        $D' \leftarrow \emptyset$;

        **for** $d \in D$ **and** $q \in \mathcal{P}$ **do**

            $\mathsf{edge} \leftarrow \mathbf{A}[q][d.\mathsf{predicate}] = \{1\}$;

            **if** edge **and** $d$ **is** $\Delta(\_)$ **then** $D' \leftarrow D' \cup \{\Delta(q)\}$;

            **else if** edge **and** $d$ **is** $\Gamma(\_, s, t)$ **then** $D' \leftarrow D' \cup \{\Gamma(q, s, t)\}$;

            **else if** $|\mathbf{A}[q][d.\mathsf{predicate}]| > 1$ **and** $d$ **is** $\Delta(r)$ **then**

                $D' \leftarrow D' \cup \{\Gamma(q, q, r)\}$;

            **else if** $|\mathbf{A}[q][d.\mathsf{predicate}]| > 1$ **and** allDeps **then** $D' \leftarrow D' \cup \{\Upsilon(q)\}$;

        **if** $D' = D$ **then return** $D$ **else** $D \leftarrow D'$;

dependency shared between predicates $p_1$ and $p_2$, if it is determined ($\Delta$) for one predicate and almost determined ($\Gamma$) for another, then the edge that prevents the $\Gamma$ from becoming a $\Delta$ cannot exist—Line 3 handles this possibility.

The function `deps` in Algorithm 3 calculates $D_p$ for any predicate $p$. It has two versions: `deps`$(p, 1)$ returns all dependencies, while `deps`$(p, 0)$ returns only determined and almost-determined dependencies. It starts by establishing the predicate $p$ itself as a dependency and continues to add dependencies of dependencies until the set $D$ stabilises. For each dependency $d \in D$, we look at the in-links of $d$ in the predicate dependency graph. If the edge from some predicate $q$ to $d.\mathsf{predicate}$ is fully determined and $d$ is determined, then $q$ is another determined dependency of $p$. If the edge is determined but $d$ is almost determined, then $q$ is an almost-determined dependency. The same outcome applies if $d$ is fully determined but the edge is undetermined. Finally, if we are interested in collecting all dependencies regardless of their status, then $q$ is a dependency of $p$ as long as the edge from $q$ to $d.\mathsf{predicate}$ is possible. Note that if there are multiple paths in the dependency graph from $q$ to $p$, Algorithm 3 could include $q$ once for each possible type ($\Delta$, $\Upsilon$, and $\Gamma$), but Algorithms 1 and 2 would still work as intended.

$$\begin{array}{l} \textsf{father} \\ \textsf{mother} \\ \textsf{parent} \\ \textsf{sibling} \end{array} \begin{pmatrix} 0 & 0 & 0 & 0 \\ 1 & 0 & 0 & 0 \\ 1 & \boxed{\{0,1\}} & \{0,1\} & \{0,1\} \\ 0 & 0 & 0 & 0 \end{pmatrix}$$

(a) The adjacency matrix of the graph. The boxed value is the decision variable that will be propagated by Algorithm 2.

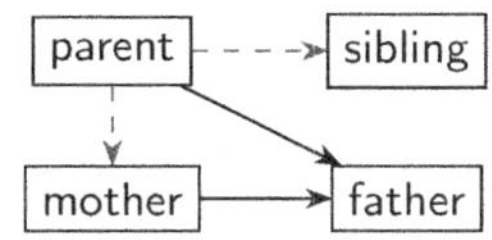


(b) A drawing of the graph. Dashed edges are undetermined—they may or may not exist.

**Fig. 2.** The predicate dependency graph of Example 5

*Example 5.* Consider this partially determined (fragment of a) program:

$$\begin{aligned} \square(X, Y) &\leftarrow \mathsf{parent}(X, Z) \wedge \mathsf{parent}(Y, Z), \\ \mathsf{father}(X, Y) &\leftarrow \mathsf{parent}(X, Y) \wedge \neg\mathsf{mother}(X, Y), \end{aligned}$$

where $\square$ indicates an unknown predicate with domain

$$D_\square = \{\mathsf{father}, \mathsf{mother}, \mathsf{parent}, \mathsf{sibling}\}.$$

The predicate dependency graph is pictured in Fig. 2. Suppose we have a constraint that mother and parent must be independent. The lists of potential dependencies for both predicates are:

$$\begin{aligned} D_{\mathsf{mother}} &= \{\Delta(\mathsf{mother}), \Gamma(\mathsf{parent}, \mathsf{parent}, \mathsf{mother})\}, \\ D_{\mathsf{parent}} &= \{\Delta(\mathsf{parent})\}. \end{aligned}$$

An entailment check at this stage would produce UNDEFINED, but propagation replaces the boxed value in Fig. 2a with zero, eliminating the potential edge from parent to mother. This also eliminates mother from $D_\square$, and this is enough to make Algorithm 1 return TRUE.

## 8 Experimental Results

We now present the results of two experiments: in Sect. 8.1 we examine the scalability of our constraint model with respect to its parameters and in Sect. 8.2 we demonstrate how the model can be used to compare inference algorithms and describe their behaviour across a wide range of programs. The experiments were run on a system with Intel Core i5-8250U processor and 8 GB of RAM. The constraint model was implemented in Java 8 with Choco 4.10.2 [23]. All inference algorithms are implemented in ProbLog 2.1.0.39 and were run using Python 3.8.2 with PySDD 0.2.10 and PyEDA 0.28.0. For both sets of experiments, we generate programs without negative cycles and use a 60 s timeout.

### 8.1 Empirical Performance of the Model

Along with constraints, variables, and their domains, two more design decisions are needed to complete the model: heuristics and restarts. By trial and error, the variable ordering heuristic was devised to eliminate sources of *thrashing*, i.e., situations where a contradiction is being 'fixed' by making changes that have no hope of fixing the contradiction. Thus, we partition all decision variables into an ordered list of groups and require the values of all variables from one group to be determined before moving to the next group. Within each group, we use the 'fail first' variable ordering heuristic. The first group consists of all head predicates. Afterwards, we handle all remaining decision variables from the first clause before proceeding to the next. The decision variables within each clause are divided into (a) the `structure` array, (b) body predicates, (c) head arguments, (d) (if $|\mathcal{V}| > 1$) the `intros` array, (c) body arguments. For instance, in the clause from Example 3, all visible parts of the clause would be decided in this order:

$$\overset{1}{\mathsf{sibling}}(\overset{3}{X}, \overset{3}{Y}) \leftarrow \overset{2}{\mathsf{parent}}(\overset{4}{X}, \overset{4}{Z}) \overset{2}{\wedge} \overset{2}{\mathsf{parent}}(\overset{4}{Y}, \overset{4}{Z}).$$

We also employ a geometric restart policy, restarting after $10, 10 \times 1.1, 10 \times 1.1^2, \ldots$ contradictions.[3] We ran 399 360 experiments, investigating the model's efficiency and gaining insight into what parameter values make the CSP harder. For $|\mathcal{P}|$, $|\mathcal{V}|$, $|\mathcal{C}|$, $\mathcal{M}_{\mathcal{N}}$, and $\mathcal{M}_{\mathcal{C}} - |\mathcal{P}|$ (i.e., the number of clauses in addition to the mandatory $|\mathcal{P}|$ clauses), we assign all combinations of 1, 2, 4, 8. $\mathcal{M}_{\mathcal{A}}$ is assigned to values 1–4. For each $|\mathcal{P}|$, we also iterate over all possible numbers of independent pairs of predicates, ranging from 0 up to $\binom{|\mathcal{P}|}{2}$. For each combination of the above-mentioned parameters, we pick ten random ways to assign arities to predicates (such that $\mathcal{M}_{\mathcal{A}}$ occurs at least once) and ten random combinations of independent pairs.

The majority (97.7%) of runs finished in under 1 s, while four instances timed out: all with $|\mathcal{P}| = \mathcal{M}_{\mathcal{C}} - |\mathcal{P}| = \mathcal{M}_{\mathcal{N}} = 8$ and the remaining parameters all different. This suggests that—regardless of parameter values—most of the time a solution can be identified instantaneously while occasionally a series of wrong decisions can lead the solver into a part of the search space with no solutions.

In Fig. 3, we plot how the mean number of nodes in the binary search tree grows as a function of each parameter (the plot for the median is very similar). The growth of each curve suggests how the model scales with higher values of the parameter. From this plot, it is clear that $\mathcal{M}_{\mathcal{N}}$ is the limiting factor. This is because some tree structures can be impossible to fill with predicates without creating either a negative cycle or a forbidden dependency, and such trees become more common as the number of nodes increases. Likewise, a higher number of predicates complicates the situation as well.

[3] Restarts help overcome early mistakes in the search process but can be disabled if one wants to find all solutions, in which case search is complete regardless of the variable ordering heuristic.

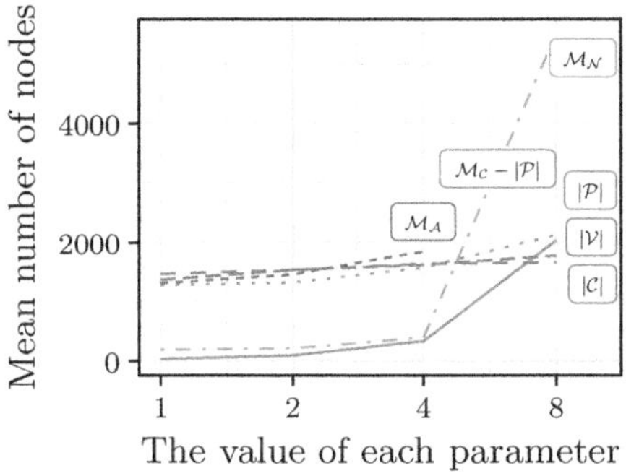


**Fig. 3.** The mean number of nodes in the binary search tree for each value of each experimental parameter. Note that the horizontal axis is on a $\log_2$ scale.

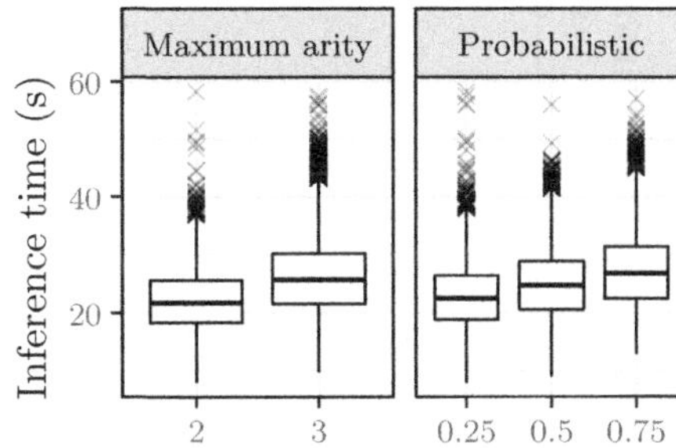


**Fig. 4.** Inference time for different values of $\mathcal{M}_\mathcal{A}$ and proportions of probabilistic facts that are probabilistic. The total number of facts is fixed at $10^5$.

### 8.2 Experimental Comparison of Inference Algorithms

For this experiment, we consider clauses of two types: *rules* are clauses such that the head atom has at least one variable, and *facts* are clauses with empty bodies and no variables. We use our constraint model to generate the rules according to the following parameter values: $|\mathcal{P}|, |\mathcal{V}|, \mathcal{M}_\mathcal{N} \in \{2, 4, 8\}$, $\mathcal{M}_\mathcal{A} \in \{1, 2, 3\}$, $\mathcal{M}_\mathcal{C} = |\mathcal{P}|$, $\mathcal{C} = \emptyset$. These values are (approximately) representative of many standard benchmarking instances which often have 2–8 predicates of arity one or two, 0–8 rules, and a larger database of facts [14]. Just like before, we explore all possible numbers of independent predicate pairs. We also add a constraint that forbids empty bodies. For both rules and facts, probabilities are uniformly sampled from $\{0.1, 0.2, \ldots, 0.9\}$. Furthermore, all rules are probabilistic, while we vary the proportion of probabilistic facts among 25%, 50%, and 75%. For generating facts, we consider $|\mathcal{C}| \in \{100, 200, 400\}$ and vary the number of facts among $10^3$, $10^4$, and $10^5$ but with one exception: the number of facts is not allowed to exceed 75% of all possible facts with the given values of $\mathcal{P}$, $\mathcal{A}$, and $\mathcal{C}$. Facts are generated using a simple procedure that randomly selects a predicate, combines it with the right number of constants, and checks whether the generated atom is already included or not. We randomly select configurations from the description above and generate ten programs with a complete restart of the constraint solver before the generation of each program, including choosing different arities and independent pairs. Finally, we set the query of each program to a random fact not explicitly included in the program and consider six natively supported algorithms and knowledge compilation techniques: binary decision diagrams (BDDs) [6], negation normal form (NNF), deterministic decomposable NNF (d-DNNF) [8], K-Best [11], and two encodings based on sentential decision diagrams [7], one

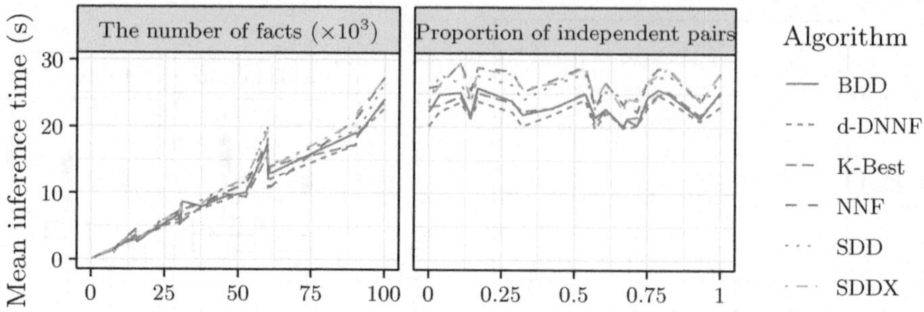


**Fig. 5.** Mean inference time for a range of PROBLOG inference algorithms as a function of the total number of facts in the program and the proportion of independent pairs of predicates. For the second plot, the number of facts is fixed at $10^5$.

of which encodes the entire program (SDDX), while the other one encodes only the part of the program relevant to the query (SDD).[4]

Out of 11310 generated problem instances, about 35% were discarded because one or more algorithms were not able to ground the instance unambiguously. The first observation (pictured in Fig. 5) is that the algorithms are remarkably similar, i.e., the differences in performance are small and consistent across all parameter values (including parameters not shown in the figure). Unsurprisingly, the most important predictor of inference time is the number of facts. However, after fixing the number of facts to a constant value, we can still observe that inference becomes harder with higher arity predicates as well as when facts are mostly probabilistic (see Fig. 4). Finally, according to Fig. 5, the independence structure of a program does not affect inference time, i.e., state-of-the-art inference algorithms—although they are supposed to [15]—do not exploit situations where separate parts of a program can be handled independently.

## 9 Conclusion

We described a constraint model for generating both logic programs and probabilistic logic programs. The model avoids unnecessary symmetries, is reasonably efficient and supports additional constraints such as predicate independence. Our experimental results provide the first comparison of inference algorithms for probabilistic logic programming languages that generalises over programs, i.e., is not restricted to just a few programs and data sets. While the results did not reveal any significant differences among the algorithms, they did reveal a shared weakness, i.e., the inability to ignore the part of a program that is easily seen to be irrelevant to the given query.

Nonetheless, we would like to outline two directions for future work. First, the experimental evaluation in Sect. 8.1 revealed scalability issues, particularly

[4] Forward SDDs (FSDDs) and forward BDDs (FBDDs) [27,28] are omitted because the former uses too much memory and the implementation of the latter seems to be broken at the time of writing.

concerning the length/complexity of clauses. However, this particular issue is likely to resolve itself if the format of a clause is restricted to a conjunction of literals. Second, random instance generation typically focuses on either realistic instances or sampling from a simple and well-defined probability distribution. Our approach can be used to achieve the former, but it is an open question how it could accommodate the latter.

**Acknowledgments.** Paulius was supported by the EPSRC Centre for Doctoral Training in Robotics and Autonomous Systems, funded by the UK Engineering and Physical Sciences Research Council (grant EP/S023208/1). Vaishak was supported by a Royal Society University Research Fellowship.

## References

1. Amendola, G., Ricca, F., Truszczynski, M.: Generating hard random Boolean formulas and disjunctive logic programs. In: Sierra, C. (ed.) Proceedings of the Twenty-Sixth International Joint Conference on Artificial Intelligence, IJCAI 2017, Melbourne, Australia, 19–25 August 2017, pp. 532–538. ijcai.org (2017). https://doi.org/10.24963/ijcai.2017/75. http://www.ijcai.org/Proceedings/2017/
2. Amendola, G., Ricca, F., Truszczynski, M.: New models for generating hard random Boolean formulas and disjunctive logic programs. Artif. Intell. **279** (2020). https://doi.org/10.1016/j.artint.2019.103185
3. Balbin, I., Port, G.S., Ramamohanarao, K., Meenakshi, K.: Efficient bottom-up computation of queries on stratified databases. J. Log. Program. **11**(3&4), 295–344 (1991). https://doi.org/10.1016/0743-1066(91)90030-S
4. Bidoit, N.: Negation in rule-based database languages: a survey. Theor. Comput. Sci. **78**(1), 3–83 (1991). https://doi.org/10.1016/0304-3975(51)90003-5
5. Bruynooghe, M., et al.: ProbLog technology for inference in a probabilistic first order logic. In: Coelho, H., Studer, R., Wooldridge, M.J. (eds.) Proceedings of ECAI 2010–19th European Conference on Artificial Intelligence, Lisbon, Portugal, 16–20 August 2010. Frontiers in Artificial Intelligence and Applications, vol. 215, pp. 719–724. IOS Press (2010). https://doi.org/10.3233/978-1-60750-606-5-719
6. Bryant, R.E.: Graph-based algorithms for Boolean function manipulation. IEEE Trans. Comput. **35**(8), 677–691 (1986). https://doi.org/10.1109/TC.1986.1676819
7. Darwiche, A.: SDD: a new canonical representation of propositional knowledge bases. In: Walsh, T. (ed.) Proceedings of the 22nd International Joint Conference on Artificial Intelligence, Barcelona, Catalonia, Spain, 16–22 July 2011, IJCAI 2011, pp. 819–826. IJCAI/AAAI (2011). https://doi.org/10.5591/978-1-57735-516-8/IJCAI11-143. http://ijcai.org/proceedings/2011
8. Darwiche, A., Marquis, P.: A knowledge compilation map. J. Artif. Intell. Res. **17**, 229–264 (2002). https://doi.org/10.1613/jair.989
9. De Raedt, L., Kersting, K., Natarajan, S., Poole, D.: Statistical Relational Artificial Intelligence: Logic, Probability, and Computation. Synthesis Lectures on Artificial Intelligence and Machine Learning. Morgan & Claypool Publishers, San Rafael (2016). https://doi.org/10.2200/S00692ED1V01Y201601AIM032
10. De Raedt, L., Kimmig, A.: Probabilistic (logic) programming concepts. Mach. Learn. **100**(1), 5–47 (2015). https://doi.org/10.1007/s10994-015-5494-z

11. De Raedt, L., Kimmig, A., Toivonen, H.: ProbLog: a probabilistic Prolog and its application in link discovery. In: Veloso, M.M. (ed.) Proceedings of the 20th International Joint Conference on Artificial Intelligence, Hyderabad, India, 6–12 January 2007 IJCAI 2007, pp. 2462–2467 (2007)
12. Dechter, R., Kask, K., Bin, E., Emek, R.: Generating random solutions for constraint satisfaction problems. In: Dechter, R., Kearns, M.J., Sutton, R.S. (eds.) Proceedings of the Eighteenth National Conference on Artificial Intelligence and Fourteenth Conference on Innovative Applications of Artificial Intelligence, 28 July–1 August 2002, Edmonton, Alberta, Canada, pp. 15–21. AAAI Press/The MIT Press (2002). http://www.aaai.org/Library/AAAI/2002/aaai02-003.php
13. Fages, J.-G., Lorca, X.: Revisiting the `tree` constraint. In: Lee, J. (ed.) CP 2011. LNCS, vol. 6876, pp. 271–285. Springer, Heidelberg (2011). https://doi.org/10.1007/978-3-642-23786-7_22
14. Fierens, D., et al.: Inference and learning in probabilistic logic programs using weighted Boolean formulas. Theory Pract. Log. Program. **15**(3), 358–401 (2015). https://doi.org/10.1017/S1471068414000076
15. Fierens, D., Van den Broeck, G., Thon, I., Gutmann, B., De Raedt, L.: Inference in probabilistic logic programs using weighted CNF's. In: Cozman, F.G., Pfeffer, A. (eds.) Proceedings of the Twenty-Seventh Conference on Uncertainty in Artificial Intelligence, Barcelona, Spain, 14–17 July 2011, UAI 2011, pp. 211–220. AUAI Press (2011). https://dslpitt.org/uai/displayArticles.jsp?mmnu=1&smnu=1&proceeding_id=27
16. Kimmig, A., Demoen, B., De Raedt, L., Santos Costa, V., Rocha, R.: On the implementation of the probabilistic logic programming language ProbLog. TPLP **11**(2–3), 235–262 (2011). https://doi.org/10.1017/S1471068410000566
17. Mairy, J.-B., Deville, Y., Lecoutre, C.: The smart table constraint. In: Michel, L. (ed.) CPAIOR 2015. LNCS, vol. 9075, pp. 271–287. Springer, Cham (2015). https://doi.org/10.1007/978-3-319-18008-3_19
18. Mantadelis, T., Rocha, R.: Using iterative deepening for probabilistic logic inference. In: Lierler, Y., Taha, W. (eds.) PADL 2017. LNCS, vol. 10137, pp. 198–213. Springer, Cham (2017). https://doi.org/10.1007/978-3-319-51676-9_14
19. Mears, C., Schutt, A., Stuckey, P.J., Tack, G., Marriott, K., Wallace, M.: Modelling with option types in MiniZinc. In: Simonis, H. (ed.) CPAIOR 2014. LNCS, vol. 8451, pp. 88–103. Springer, Cham (2014). https://doi.org/10.1007/978-3-319-07046-9_7
20. Namasivayam, G.: Study of random logic programs. In: Hill, P.M., Warren, D.S. (eds.) ICLP 2009. LNCS, vol. 5649, pp. 555–556. Springer, Heidelberg (2009). https://doi.org/10.1007/978-3-642-02846-5_61
21. Namasivayam, G., Truszczyński, M.: Simple random logic programs. In: Erdem, E., Lin, F., Schaub, T. (eds.) LPNMR 2009. LNCS (LNAI), vol. 5753, pp. 223–235. Springer, Heidelberg (2009). https://doi.org/10.1007/978-3-642-04238-6_20
22. Poole, D.: The independent choice logic for modelling multiple agents under uncertainty. Artif. Intell. **94**(1–2), 7–56 (1997). https://doi.org/10.1016/S0004-3702(97)00027-1
23. Prud'homme, C., Fages, J.G., Lorca, X.: Choco Documentation. TASC - LS2N CNRS UMR 6241, COSLING S.A.S. (2017). http://www.choco-solver.org
24. Russell, S.J.: Unifying logic and probability. Commun. ACM **58**(7), 88–97 (2015). https://doi.org/10.1145/2699411

25. Sato, T., Kameya, Y.: PRISM: A language for symbolic-statistical modeling. In: Proceedings of the Fifteenth International Joint Conference on Artificial Intelligence, IJCAI 1997, Nagoya, Japan, 23–29 August 1997, vol. 2, pp. 1330–1339. Morgan Kaufmann (1997)
26. Selman, B., Mitchell, D.G., Levesque, H.J.: Generating hard satisfiability problems. Artif. Intell. **81**(1–2), 17–29 (1996). https://doi.org/10.1016/0004-3702(95)00045-3
27. Tsamoura, E., Gutiérrez-Basulto, V., Kimmig, A.: Beyond the grounding bottleneck: datalog techniques for inference in probabilistic logic programs. In: The Thirty-Fourth AAAI Conference on Artificial Intelligence, AAAI 2020. The Thirty-Second Innovative Applications of Artificial Intelligence Conference, IAAI 2020. The Tenth AAAI Symposium on Educational Advances in Artificial Intelligence, EAAI 2020, New York, NY, USA, 7–12 February 2020, pp. 10284–10291. AAAI Press (2020). https://aaai.org/ojs/index.php/AAAI/article/view/6591
28. Vlasselaer, J., Van den Broeck, G., Kimmig, A., Meert, W., De Raedt, L.: Anytime inference in probabilistic logic programs with Tp-compilation. In: Yang, Q., Wooldridge, M.J. (eds.) Proceedings of the Twenty-Fourth International Joint Conference on Artificial Intelligence, IJCAI 2015, Buenos Aires, Argentina, 25–31 July 2015, pp. 1852–1858. AAAI Press (2015)
29. Walsh, T.: General symmetry breaking constraints. In: Benhamou, F. (ed.) CP 2006. LNCS, vol. 4204, pp. 650–664. Springer, Heidelberg (2006). https://doi.org/10.1007/11889205_46
30. Wang, K., Wen, L., Mu, K.: Random logic programs: Linear model. TPLP **15**(6), 818–853 (2015). https://doi.org/10.1017/S1471068414000611
31. Wen, L., Wang, K., Shen, Y., Lin, F.: A model for phase transition of random answer-set programs. ACM Trans. Comput. Log. **17**(3), 22:1–22:34 (2016). https://doi.org/10.1145/2926791
32. Zhao, Y., Lin, F.: Answer set programming phase transition: a study on randomly generated programs. In: Palamidessi, C. (ed.) ICLP 2003. LNCS, vol. 2916, pp. 239–253. Springer, Heidelberg (2003). https://doi.org/10.1007/978-3-540-24599-5_17

# Towards Formal Fairness in Machine Learning

Alexey Ignatiev[1], Martin C. Cooper[2,4(✉)], Mohamed Siala[3,4], Emmanuel Hebrard[3,4], and Joao Marques-Silva[4]

[1] Monash University, Melbourne, Australia
alexey.ignatiev@monash.edu
[2] IRIT, Université de Toulouse III, Toulouse, France
cooper@irit.fr
[3] LAAS-CNRS, Université de Toulouse, CNRS, INSA, Toulouse, France
{siala,hebrard}@laas.fr
[4] ANITI, Université de Toulouse, Toulouse, France
joao.marques-silva@univ-toulouse.fr

**Abstract.** One of the challenges of deploying machine learning (ML) systems is fairness. Datasets often include sensitive features, which ML algorithms may unwittingly use to create models that exhibit unfairness. Past work on fairness offers no formal guarantees in their results. This paper proposes to exploit formal reasoning methods to tackle fairness. Starting from an intuitive criterion for fairness of an ML model, the paper formalises it, and shows how fairness can be represented as a decision problem, given some logic representation of an ML model. The same criterion can also be applied to assessing bias in training data. Moreover, we propose a reasonable set of axiomatic properties which no other definition of dataset bias can satisfy. The paper also investigates the relationship between fairness and explainability, and shows that approaches for computing explanations can serve to assess fairness of particular predictions. Finally, the paper proposes SAT-based approaches for learning fair ML models, even when the training data exhibits bias, and reports experimental trials.

## 1 Introduction

Given the forecast widespread use of ML-enabled systems, in settings that can have a significant impact in the lives and safety of human beings, a range of concerns need to be addressed. Robustness of ML models against adversarial examples is one such concern [22,37,43,44,50,55,60,61]. Explaining the predictions of ML models represents another concern. A related concern is to learn (or synthesize) interpretable ML models [5,9,30,35,36,41,51,56,65,66]. One additional concern is to ensure that ML-enabled systems are fair [4,15,48]. The importance

This work was partially funded by ANITI, funded by the French program "Investing for the Future – PIA3" under Grant agreement n$^o$ ANR-19-PI3A-0004.


H. Simonis (Ed.): CP 2020, LNCS 12333, pp. 846–867, 2020.
https://doi.org/10.1007/978-3-030-58475-7_49

of addressing fairness cannot be overstated, as demonstrated by recent existing troublesome evidence [6,19] that already deployed ML-enabled systems can exhibit very significant bias. Furthermore, recently recommended guidelines at the EU level highlight the importance of fairness [23].

We find many definitions of fairness in the literature[7,54,64]. In [64], the authors classify these definitions into three categories: statistical measures, similarity-based measures, and causal reasoning. Statistical measures, such as those presented in [7,14,16,20,33,46], can hardly be studied from a formal angle due to their nature. Similarity-based measures [20,27,48], on the other hand, are data independent definitions that are well suited for formal investigations. Last, causal reasoning measures [45,48,53] are based on the so-called causal graphs (i.e. graphs capturing relationships between the different features) and are essentially used to synthesise ML models. We focus in this paper on similarity-based measures because they offer an excellent framework for formal analysis compared to the other two.

Addressing fairness can be performed at three different levels: data processing, synthesis, and verification. Current work on the three levels is, to a large extent, heuristic in nature [1,2,4,11,25,26,28,34,48,59,64], offering no formal guarantees in their analyses. This paper proposes a first step towards endowing the analysis (i.e. data processing and verification) and the synthesis of fair ML models with a rigorous footing. We study one concrete criterion of fairness, and propose a rigorous test to assess whether or not an ML model is fair against that criterion. Moreover, the paper shows how the proposed test can be adapted to devise a simple (polynomial-time) algorithm to assess existing bias in datasets, even if datasets are inconsistent. More importantly, in the case of biased datasets, the paper investigates how to adapt exact methods for learning interpretable (logic-based) ML models to synthesize fair ML models.

The paper is organized as follows. Section 2 summarizes the definitions and notation used throughout the paper. Section 3 focuses on a criterion of fairness, and develops tests for assessing whether an ML model is fair, whether a dataset exhibits bias, and whether a particular prediction is fair. Section 4 investigates how fair logic-based models can be synthesized with possibly biased datasets. Section 5 provides a theoretical justification for adopting the fairness criterion used throughout the paper. Finally, Sect. 6 presents preliminary experimental results and Sect. 7 concludes the paper.

## 2 Preliminaries

**SAT/SMT-Related Topics.** The paper uses definitions standard in Boolean Satisfiability (SAT) and Satisfiability Modulo Theories (SMT) [10]. These include conjunctive and disjunctive normal forms (resp. CNF and DNF), prime implicants and implicates.

**Classification Problems.** This section adapts the definitions used in earlier work [9,41,49]. We consider a set of features $\mathcal{F} = \{F_1, \ldots, F_K\}$, where each feature $F_i$ takes values from some domain $D_i$. The space of all assignments to features defined by $\mathbb{F} = \Pi_{i=1}^{K} D_i$ is referred to as *feature space* [32]. Throughout the paper, all domains in the examples and experiments will be binary, i.e. $D_i = \{0, 1\}$, but the results will be derived assuming arbitrary (discrete) domains[1]. Since all features are binary, a literal on a feature $F_r$ will be represented as $F_r$ or as $\neg F_r$.

To learn a classifier, one starts from given training data (also referred to as examples) $\mathcal{T} = \{e_1, \ldots, e_M\}$. Each example has an associated class taken from a set of classes $\mathcal{C}$. The paper focuses mostly on binary classification, i.e. $\mathcal{C} = \{c_0, c_1\}$. (We will associate $c_0$ with 0 and $c_1$ with 1, for simplicity.) Thus, $\mathcal{T}$ is partitioned into $\mathcal{T}^+$ and $\mathcal{T}^-$, denoting the examples classified as positive ($c_1 = 1$) and as negative ($c_0 = 0$), respectively. Each example $e_q \in \mathcal{T}$ is represented as a pair $\langle \mathbf{z}_q, c_q \rangle$, where $\mathbf{z}_q \in \mathbb{F}$ denotes the literals associated with the example and $c_q \in \{0, 1\}$ is the class to which the example belongs. We have $c_q = 1$ if $e_q \in \mathcal{T}^+$ and $c_q = 0$ if $e_q \in \mathcal{T}^-$. The training data $\mathcal{T}$ is consistent if $\langle \mathbf{z}_q, 0 \rangle \in \mathcal{T} \wedge \langle \mathbf{z}_{q'}, 1 \rangle \in \mathcal{T} \implies \mathbf{z}_q \neq \mathbf{z}_{q'}$, and inconsistent otherwise. A literal $l_r$ on a feature $F_r$, $l_r \in \{F_r, \neg F_r\}$, *discriminates* an example $e_q$ if $\mathbf{z}_q[r] = \neg l_r$, i.e. the feature takes the value opposite to the value in the example. We assume that all features are specified for all examples; the work can be generalized for situations where the value of some features for some examples is left unspecified.

An ML model $\mathbb{M}$ is represented as a function $\varphi : \mathbb{F} \to \mathcal{C}$. With a slight abuse of notation, a consistent training data will also be viewed as a partial function $\varphi_{\mathcal{T}} : \mathbb{F} \to \mathcal{C}$. In a general setting, where the training data can be inconsistent, we represent the possible values of training data in each point of feature space by a relation $\mathfrak{T} \subseteq \mathbb{F} \times \mathcal{C}$.

Unless otherwise stated, we focus on *accurate* ML models, indicating that 100% of the training data examples are classified correctly. A non-accurate ML model may misclassify some examples of training data (e.g. this is the case for inconsistent data).

**Examples of ML Models.** The paper focuses almost exclusively on logic-based ML models, namely decision sets (DSs) and trees (DTs).

**Definition 1 (Decision set).** *A DNF formula $\phi$ over the literals $\bigcup_{F_r \in \mathcal{F}} \{F_r, \neg F_r\}$ is a decision set for a training set $\mathcal{T}$ if the function that maps $\mathbf{z} \in \mathbb{F}$ to $c_1$ if $\mathbf{z}$ is a model of $\phi$ and to $c_0$ otherwise is equal to $\varphi_{\mathcal{T}}$ on $\mathcal{T}$.*

**Definition 2 (Decision tree).** *A decision tree for a training set $\mathcal{T}$ is a decision set $\phi = \bigvee_i t_i$ such that there exists a rooted tree with edges labelled with literals of $\phi$ and such that for every term $t_i$ of $\phi$ there is a path from the root to a leaf whose set of labels consists of exactly the literals in $t_i$.*

**Fairness.** Following standard notation [48], throughout this paper we will assume that $\mathcal{F}$ is partitioned into a set of *protected* features $\mathcal{P} = \{F_{I+1}, \ldots, F_K\}$

[1] Real-value features can be discretized. Moreover, to focus on binary features, the fairly standard one-hot-encoding [58] is assumed for handling non-binary categorical features.

| Ex. | Sunny | Windy | Classes | Exams | Gender | Hike? |
|---|---|---|---|---|---|---|
| $e_1$ | 1 | 0 | 0 | 0 | 1 | 1 |
| $e_2$ | 0 | 1 | 0 | 0 | 0 | 0 |
| $e_3$ | 1 | 1 | 1 | 0 | 0 | 0 |
| $e_4$ | 0 | 1 | 0 | 1 | 0 | 0 |
| $e_5$ | 1 | 1 | 0 | 0 | 1 | 1 |

(a) Dataset

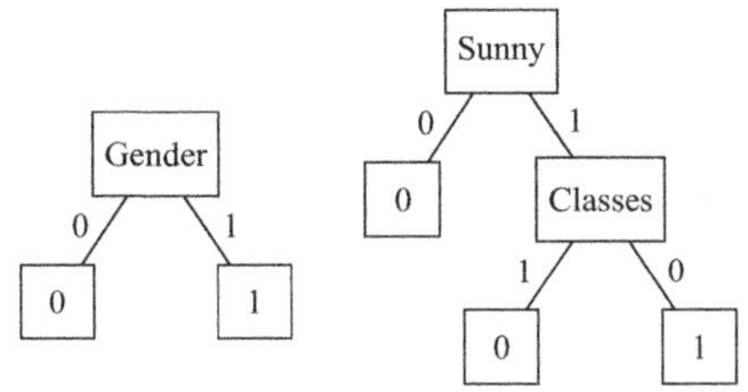


(b) Unfair decision tree (c) Fair decision tree

**Fig. 1.** Running example

and a set of *non-protected* features $\mathcal{N} = \{F_1, \ldots, F_I\}$, with $|\mathcal{N}| = I$ and $|\mathcal{P}| = K - I$. Thus, following the notation used earlier, we define $\mathbb{N} = \Pi_{i=1}^{I} D_i$ and $\mathbb{P} = \Pi_{i=I+1}^{K} D_i$, and so $\mathbb{F} = \mathbb{N} \times \mathbb{P}$. Moreover, $\mathbf{z} \in \mathbb{F}$ is split into $\mathbf{x} \in \mathbb{N}$ and $\mathbf{y} \in \mathbb{P}$. The protected features are those with respect to which we intuitively want ML models *not* to be *sensitive*[2]. Throughout the paper, we will use the notation $\varphi(\mathbf{x}, \mathbf{y})$ as a replacement for $\varphi(\mathbf{x} \cdot \mathbf{y})$ where $\mathbf{z} = \mathbf{x} \cdot \mathbf{y}$ denotes the concatenation of vectors. Moreover, when referring to *specific* points in feature space (or the space of non-protected or protected features), the notation used will be $\mathbf{t} \in \mathbb{F}$, $\mathbf{u} \in \mathbb{N}$, and $\mathbf{v} \in \mathbb{P}$ (resp. instead of $\mathbf{z}$, $\mathbf{x}$ and $\mathbf{y}$). Finally, we use the notation $\mathbb{T}$ to denote the set $\{\mathbf{t}_i \mid \langle \mathbf{t}_i, c_i \rangle \in \mathcal{T}\} \subseteq \mathbb{F}$.

**Running Example.** We use a simple example to illustrate the main points. Minor modifications to the original example will be also considered later in the paper.

*Example 1.* We consider the example of Fig. 1, where the protected feature set is $\mathcal{P} = \{\mathsf{Gender}\}$. The purpose of the example is to decide the circumstances that cause students to enjoy a hike, and we would rather have a classifier that is gender-balanced, if possible. (For the purposes of the example, it is irrelevant whether the values of 0 and 1 of feature Gender correspond to male and female or vice-versa.) By running an off-the-shelf heuristic algorithm for constructing a decision tree, one obtains a single branching node with feature Gender. Figure 1b shows the heuristic decision tree obtained with well-known ML packages ( scikit-learn [58] and Orange [18] gave the same result).

Nevertheless, the evident unfairness of the obtained decision tree results *solely* from the algorithms used for constructing heuristic decision trees. Indeed, as shown in later sections, careful analysis of the training data reveals that there is *no* evidence in the data that justifies that feature Gender should play such a role in deciding the circumstances under which male/female students enjoy hikes. Figure 1c shows an example of a fair decision tree for this example corresponding to the single-term DNF Sunny $\wedge \neg$Classes.

[2] For a number of reasons, datasets can contain such protected features, but their removal may be undesirable, for example, because this may induce inconsistencies in datasets.

# 3 Assessing Fairness

We consider in this section an ML classification scenario, with training data $\mathcal{T}$ where the set of features $\mathcal{F}$ is partitioned into a set of protected features $\mathcal{P}$ and a set of non-protected features $\mathcal{N}$, and some (interpretable) ML model $\mathbb{M}$ trained on $\mathcal{T}$

## 3.1 Fairness Criterion

A number of definitions of fairness have been proposed in recent years [48,64]. This paper considers *fairness through unawareness* (FTU), which was originally proposed as follows:

**Criterion 1** (Fairness Through Unawareness (FTU) [31,48])**.** An algorithm is fair if the protected features $\mathcal{P}$ are not explicitly used in the decision-making process.

The operational definition above suggests a syntactic test to decide whether FTU holds. This section proposes instead a semantic characterization of fairness, that respects the syntactic definition of FTU. Besides FTU, a number of additional definitions of fairness are studied in [48,64] and in related work. Moreover, a possible criticism of FTU is that $\mathcal{P}$ may not represent all features capturing discriminatory information [48]. Nevertheless, FTU exhibits a number of advantages over other criteria, and Sect. 5 provides a theoretical justification for using FTU as a definition of fairness.

**Definition 3 (Criterion for FTU).** *Given an ML model* $\mathbb{M}$ *computing some (classification) function* $\varphi : \mathbb{F} \rightarrow \mathcal{C}$*, we say that* $\mathbb{M}$ *is* fair *if:*

$$\forall (\mathbf{x} \in \mathbb{N})\, \forall (\mathbf{y}_1, \mathbf{y}_2 \in \mathbb{P}).\, [\varphi(\mathbf{x}, \mathbf{y}_1) = \varphi(\mathbf{x}, \mathbf{y}_2)] \tag{1}$$

Next, we investigate how this criterion can be used to analyze both ML models and datasets. Note that this criterion can be seen as a hard version of the causal discrimination measure presented in [27].

## 3.2 Assessing Model Fairness and Dataset Bias

**Checking Model Fairness.** As shown later in the paper, it will be convenient to assess instead the negation of the FTU criterion.

*Remark 1.* An ML model $\mathbb{M}$ respects Definition 3, i.e. (1) holds, iff the following is false:

$$\exists (\mathbf{x} \in \mathbb{N})\, \exists (\mathbf{y}_1, \mathbf{y}_2 \in \mathbb{P}).\, [\varphi(\mathbf{x}, \mathbf{y}_1) \neq \varphi(\mathbf{x}, \mathbf{y}_2)] \tag{2}$$

Clearly, we can now use (2) to assess whether an ML model $\mathbb{M}$ is fair or not, by searching for satisfying assignments for (2), and this can be achieved with a satisfiability test, given a suitable logic representation of the ML model $\mathbb{M}$. From a practical perspective, we can refine the previous result as follows.

*Remark 2.* To test condition (2) we only need to test pairs $\mathbf{y}_1, \mathbf{y}_2$ which differ on a single feature. This is because if (2) holds for some $\mathbf{x}, \mathbf{y}_1, \mathbf{y}_2$ then it must hold for some $\mathbf{x}, \mathbf{y}^r, \mathbf{y}^{r+1}$ ($r \in \{0, \ldots, K-1\}$) where $\mathbf{y}^r$ is equal to $\mathbf{y}_1$ on the first $K-r$ features of $\mathbb{P}$ and equal to $\mathbf{y}_2$ on the other features of $\mathbb{P}$. Analyzing each feature separately reduces the search space that needs to be considered.

**Checking Bias in Consistent Datasets.** By exploiting Remark 1 or Remark 2, we can devise a test to assess whether a dataset exhibits unfairness (in which case we say that the dataset is *biased*). We consider first the case when the dataset is consistent, and use the insights to consider the more general case of inconsistent datasets. For a consistent dataset, the following condition captures FTU in the dataset.

**Definition 4 (Consistent dataset bias under FTU).** *A consistent dataset $\mathcal{T}$ is biased if the following holds:*

$$\exists(\mathbf{x} \in \mathbb{N})\, \exists(\mathbf{y}_1, \mathbf{y}_2 \in \mathbb{P}).[(\mathbf{x}, \mathbf{y}_1), (\mathbf{x}, \mathbf{y}_2) \in \mathbb{T} \wedge (\varphi_{\mathcal{T}}(\mathbf{x}, \mathbf{y}_1) \neq \varphi_{\mathcal{T}}(\mathbf{x}, \mathbf{y}_2))] \tag{3}$$

Intuitively, a dataset is biased if the protected features serve to distinguish between two different predictions when the non-protected features take the same values. Although (3) is harder to read than (2), it is actually simple to develop a polynomial time procedure for assessing whether a dataset is FTU-biased. However, we develop instead a polynomial (indeed, linear) time algorithm for the more general case of inconsistent data, which is also applicable in the case of consistent data.

**Checking Bias in Inconsistent Datasets.** Even if the training data is inconsistent, one can devise a test to assess whether a dataset exhibits bias. As motivated in Sect. 2, in the presence of inconsistent data, we model the expected input-output behavior (given the dataset) as a relation. In the case of an inconsistent dataset, the following condition captures FTU in the dataset.

**Definition 5 (Inconsistent dataset bias under FTU).** *An inconsistent dataset $\mathcal{T}$ is biased if the following holds:*

$$\begin{aligned}\exists(\mathbf{x} \in \mathbb{N})\, \exists(\mathbf{y}_1, \mathbf{y}_2 \in \mathbb{P})\, \exists(c_1, c_2 \in \mathcal{C}).[\mathbf{y}_1 \neq \mathbf{y}_2 \wedge c_1 \neq c_2 \wedge \\ (\mathbf{x}, \mathbf{y}_1), (\mathbf{x}, \mathbf{y}_2) \in \mathbb{T} \wedge \mathfrak{T}(\mathbf{x}, \mathbf{y}_1, c_1) \wedge \mathfrak{T}(\mathbf{x}, \mathbf{y}_2, c_2)]\end{aligned} \tag{4}$$

There is an alternative definition that considers the dataset to be fair if for each $\mathbf{x} \in \mathbb{N}$, $\mathbf{y}_1, \mathbf{y}_2 \in \mathbb{P}$, the *sets* of classes $c \in \mathcal{C}$ for which $\mathfrak{T}(\mathbf{x}, \mathbf{y}_1, c)$ or $\mathfrak{T}(\mathbf{x}, \mathbf{y}_2, c)$ hold are identical. We ruled out this alternative definition for reasons described in Sect. 5.

There is a fairly simple linear-(amortized)-time algorithm that can be used both with consistent and with inconsistent data, as shown in Algorithm 1. Correctness of the algorithm follows from the fact that condition (4) is actually logically equivalent to

$$\begin{aligned}\exists(\mathbf{x} \in \mathbb{N})\exists(\mathbf{y}_1, \mathbf{y}_2, \mathbf{y}_1', \mathbf{y}_2' \in \mathbb{P})\, \exists(c_1, c_2, c_1', c_2' \in \mathcal{C}).[\mathbf{y}_1 \neq \mathbf{y}_2 \wedge c_1 \neq c_2 \wedge \\ (\mathbf{x}, \mathbf{y}_1), (\mathbf{x}, \mathbf{y}_2), (\mathbf{x}, \mathbf{y}_1'), (\mathbf{x}, \mathbf{y}_2') \in \mathbb{T} \wedge \\ \mathfrak{T}(\mathbf{x}, \mathbf{y}_1, c_1') \wedge \mathfrak{T}(\mathbf{x}, \mathbf{y}_2, c_2') \wedge \mathfrak{T}(\mathbf{x}, \mathbf{y}_1', c_1) \wedge \mathfrak{T}(\mathbf{x}, \mathbf{y}_2', c_2)]\end{aligned} \tag{5}$$

**Algorithm 1:** Checking dataset bias

**Input:** $\mathcal{T}, \mathcal{N}, \mathcal{P}$
**Output:** Biased / Unbiased
**begin**
  **foreach** $\langle(\mathbf{x}_i, \mathbf{y}_i), c_i\rangle \in \mathcal{T}$ **do**
    $\mathsf{CSet}[\mathbf{x}_i] \leftarrow \emptyset$
    $\mathsf{YSet}[\mathbf{x}_i] \leftarrow \emptyset$
  **foreach** $\langle(\mathbf{x}_i, \mathbf{y}_i), c_i\rangle \in \mathcal{T}$ **do**
    $\mathsf{CSet}[\mathbf{x}_i] \leftarrow \mathsf{CSet}[\mathbf{x}_i] \cup \{c_i\}$ ;
    $\mathsf{YSet}[\mathbf{x}_i] \leftarrow \mathsf{YSet}[\mathbf{x}_i] \cup \{\mathbf{y}_i\}$
  **foreach** $\langle(\mathbf{x}_i, \mathbf{y}_i), c_i\rangle \in \mathcal{T}$ **do**
    **if** $|\mathsf{CSet}[\mathbf{x}_i]| > 1 \wedge |\mathsf{YSet}[\mathbf{x}_i]| > 1$ **then**
      **return** Biased
  **return** Unbiased
**end**

**Table 1.** Extension of the dataset of our running example

| Ex. | Sunny | Windy | Classes | Exams | Gender | Hike? |
|---|---|---|---|---|---|---|
| $e_1$ | 1 | 0 | 0 | 0 | 1 | 1 |
| $e_6$ | 1 | 0 | 0 | 0 | 0 | 0 |
| $e_7$ | 1 | 0 | 0 | 0 | 1 | 0 |
| $e_8$ | 1 | 0 | 0 | 0 | 0 | 1 |

The implication (4) $\Rightarrow$ (5) is immediate. To see the implication (5) $\Rightarrow$ (4), suppose that $\mathfrak{T}(\mathbf{x}, \mathbf{y}_1, c_1') \wedge \mathfrak{T}(\mathbf{x}, \mathbf{y}_2, c_2') \wedge \mathfrak{T}(\mathbf{x}, \mathbf{y}_1', c_1) \wedge \mathfrak{T}(\mathbf{x}, \mathbf{y}_2', c_2)$ where $\mathbf{y}_1 \neq \mathbf{y}_2$, $c_1 \neq c_2$, but that (4) does not hold. We can deduce that $c_1' = c_2'$ and $\mathbf{y}_1' = \mathbf{y}_2'$ and $(\mathbf{y}_1 = \mathbf{y}_1' \vee c_1' = c_1)$, $(\mathbf{y}_1 = \mathbf{y}_2' \vee c_1' = c_2)$, $(\mathbf{y}_2 = \mathbf{y}_1' \vee c_2' = c_1)$, $(\mathbf{y}_2 = \mathbf{y}_2' \vee c_2' = c_2)$ for which it can easily be verified that there is no solution.

In Algorithm 1 each vector $\mathbf{x}$ on the non-protected features is used for indexing both sets CSet, YSet using hashtables. By inspection, the amortized running time is linear (since operations on a hash table have constant amortized complexity [17]).

*Example 2.* Running the proposed algorithm on the dataset of Example 1 with protected feature set $Y = \{\mathsf{Gender}\}$ confirms that the dataset is unbiased. However, if the same dataset is extended with the rows in Table 1 (where $e_1$ is added for convenience), then the algorithm reports, as expected, that the dataset is no longer unbiased. Clearly, with $\mathbf{x}_1 = (1, 0, 0, 0) = \mathbf{x}_6 = \mathbf{x}_7 = \mathbf{x}_8$, there are now two different predictions and the dataset is inconsistent. Moreover, there are two reasons for the dataset to be deemed biased: $e_1$ and $e_6$ are one reason, and $e_7$ and $e_8$ are the other. It should also be noted that $e_1$ and $e_7$ do *not* represent a possible reason to declare bias in the dataset.

One way to tackle fairness is to discard the protected features. The following simple result will be used later in the paper.

**Proposition 1.** Consider a consistent dataset $\mathcal{T}$, and let the dataset $\mathcal{T}'$ be obtained by discarding the protected features $\mathcal{P}$ of $\mathcal{T}$. Hence, $\mathcal{T}'$ may have examples with duplicated sets of feature values. $\mathcal{T}$ is unbiased iff there are no inconsistencies in $\mathcal{T}'$.

### 3.3 Local Fairness via Explanations

In this section we consider local notions of fairness. An individual is probably more interested in the fairness of a particular decision concerning themselves than in the global fairness of the model. We use the notion of explanation to define local fairness. It turns out that there are two possible definitions of local fairness based on explanations.

To be concrete, consider the problem of an unemployed woman who has been refused a loan and who wants to know if this is because she is a woman. Suppose the bank has learned the following simple model: refuse a loan if the client is unemployed or if they are a woman. This model is clearly unfair with respect to gender, but in this particular case the bank can claim that they would have refused the loan even if the client had been a man. On the other hand, the client can point out there are two explanations for the refusal: the first explanation is that she is unemployed (since all unemployed are refused a loan) and the second explanation is that she is a woman (since all women are refused a loan), and hence the decision should be considered unfair.

There is recent work investigating similar themes [3] which shows that explanations can be used to fake fairness, but the authors study statistical measures of fairness.

Following [39,62], given an ML model $\mathbb{M}$ computing some function $\varphi$, an explanation of some prediction $\varphi(\mathbf{z}) = c$ is a prime implicant of the mapping $\varphi : \mathbf{z} \mapsto c$, where $c$ is considered fixed (i.e. a *subset-minimal* subset of the literals of $\mathbf{z}$ which still entails the prediction $c$). This notion of explanation allows us to define fairness of a particular decision/prediction of a model $\mathbb{M}$.

In the following we view a $\mathbf{z} \in \mathbb{F}$ as a set of literals, one per feature.

**Definition 6 (Fair explanation).** *An explanation* $\mathbf{e}$ *of a prediction* $\varphi(\mathbf{z}) = c$ *is a subset of* $\mathbf{z}$ *which is minimal under subset inclusion such that* $\forall \mathbf{z}' \in \mathbb{F}$, *if* $\mathbf{e} \subseteq \mathbf{z}'$ *then* $\varphi(\mathbf{z}') = c$. *If* $\mathbf{z} = (\mathbf{x}, \mathbf{y})$ *with* $x \in \mathbb{N}$, $y \in \mathbb{P}$, *we say that* $\mathbf{e}$ *is fair if* $\mathbf{e} \cap \mathbf{y} = \emptyset$ *(i.e.* $\mathbf{e}$ *uses no protected literals).*

**Definition 7 (Universal/existential Fairness).** *A prediction* $\varphi(\mathbf{z}) = c$ *is universally fair if all of its explanations are fair. It is existentially fair if at least one of its explanations is fair.*

It turns out that there is a close connection between FTU and universal fairness.

**Proposition 2.** Let $\varphi$ be the function computed by a ML model $\mathbb{M}$. $\mathbb{M}$ is fair according the FTU criterion (1) iff all predictions $\varphi(\mathbf{z}) = c$ ($\mathbf{z} \in \mathbb{F}$) are universally fair.

*Proof.* It suffices to prove that $\mathbb{M}$ is unfair according to criterion (2) iff there exists an unfair explanation of some prediction.

Suppose that $\mathbb{M}$ is unfair because $\varphi(\mathbf{x}, \mathbf{y}_1) = c \neq \varphi(\mathbf{x}, \mathbf{y}_2)$. All predictions have at least one explanation, so let $\mathbf{e}$ be an explanation of $\varphi(\mathbf{x}, \mathbf{y}_1) = c$. Thus, by Definition 6, $\forall \mathbf{z}' \in \mathbb{F}$, if $\mathbf{e} \subseteq \mathbf{z}'$ then $\varphi(\mathbf{z}') = c$. Since $\varphi(\mathbf{x}, \mathbf{y}_2) \neq c$, this implies that $\mathbf{e} \cap \mathbf{y_1} \neq \emptyset$ and hence $\mathbf{e}$ is an unfair explanation.

Suppose that the prediction $\varphi(\mathbf{z}) = c$ (where $\mathbf{z} = (\mathbf{x}, \mathbf{y})$) has an unfair explanation $\mathbf{e}$. Let $\mathbf{e}' = \mathbf{e} \setminus \mathbf{y}$. Since $\mathbf{e}$ is unfair, $\mathbf{e} \cap \mathbf{y} \neq \emptyset$ and so $\mathbf{e}'$ is a proper subset of $\mathbf{e}$. By subset minimality of the explanation $\mathbf{e}$, $\mathbf{e}'$ cannot be a valid explanation of $\varphi(\mathbf{z}) = c$, and so $\exists \mathbf{z}' \in \mathbb{F}$ such that $\mathbf{e}' \subseteq \mathbf{z}'$ and $\varphi(\mathbf{z}') \neq c$. Let *feat*($\mathbf{e}$) denote the features which occur in $\mathbf{e}$ and $\mathbf{z}'[\mathit{feat}(\mathbf{e})]$ the subset of $\mathbf{z}'$ on these features. Now, let $\mathbf{z}'' = \mathbf{e} \cup (\mathbf{z}' \setminus \mathbf{z}'[\mathit{feat}(\mathbf{e})])$. Since $\mathbf{e}$ is an explanation of $\varphi(\mathbf{z}) = c$ and $\mathbf{e} \subseteq \mathbf{z}''$, we must have $\varphi(\mathbf{z}'') = \varphi(\mathbf{z}) = c$. But then $\mathbf{z}''$ and $\mathbf{z}'$ differ only on the features of $\mathbf{e} \cap \mathbf{y}$ (i.e. on protected features) but $\varphi(\mathbf{z}'') = c \neq \varphi(\mathbf{z}')$, so $\mathbb{M}$ is unfair according to criterion (2). □

Using our rigourous definition of explanation and our two notions of fairness of a prediction, we consider two concrete questions, given a prediction $\varphi(\mathbf{z}) = c$:

1. Are all explanations fair (i.e. do not include any feature from the set of protected features)? This problem will be referred to as *universal fairness checking* (UFC).
2. Does there exist a fair explanation (i.e. that does not include any protected feature)? This problem will be referred to as *existential fairness checking* (EFC).

These two problems correspond to our two different notions of fairness of a prediction (existential and universal). They clearly differ semantically, but it would appear that they also differ in terms of the computational complexity to answer them.

**Proposition 3.** For polytime-computable $\varphi$, EFC $\in$ co-NP and UFC $\in \Pi_2^P$.

*Proof.* To see EFC $\in$ co-NP, observe that a prediction $\varphi(\mathbf{x}, \mathbf{y}) = c$ has a fair explanation iff the non-protected features $\mathbf{x}$ entail the predicted class $c$. Thus the non-existence of a fair explanation is equivalent to the existence of $\mathbf{y}' \in \mathbb{P}$ such that $\varphi(\mathbf{x}, \mathbf{y}') \neq c$.

To see UFC $\in \Pi_2^P$, observe that all explanations of a prediction $\varphi(\mathbf{z}) = c$ (where $\mathbf{z} = (\mathbf{x}, \mathbf{y})$) are fair iff for all putative explanations $\mathbf{e} \subseteq \mathbf{z}$ of this prediction, either $\mathbf{e}$ does not entail the predicted class $c$ or $\mathbf{e} \setminus \mathbf{y}$ does entail $c$. This is logically equivalent to

$$\forall (\mathbf{e} \subseteq \mathbf{z})\ \forall (\mathbf{z}'' \supseteq \mathbf{e} \setminus \mathbf{y})\ \exists (\mathbf{z}' \supseteq \mathbf{e}).\ [(\varphi(\mathbf{z}') \neq c) \vee (\varphi(\mathbf{z}'') = c)]$$

which clearly places UFC in $\Pi_2^P$. □

Whether UFC is complete for $\Pi_2^P$ is an open problem. We conjecture that it is, since there is no obvious polynomial-time verifiable certificate. It is worth noting that, by Proposition 2 the problem of testing whether *all* predictions are universally fair is in co-NP, since the counter-example certificate is simply the values of $\mathbf{x}, \mathbf{y_1}, \mathbf{y_2}$ satisfying criterion (2). This is somewhat counter-intuitive in that testing the universal fairness of all predictions may be easier than testing the universal fairness of one prediction.

### 3.4 Relation of Fairness to Robustness and Adversarial Examples

Informally speaking, robustness is the property that two almost identical inputs (i.e. points in feature space $\mathbb{F}$) should be labelled equally. We give a general formal definition for robustness, then we show that fairness can be seen as a particular case. This may enable previous work on robustness to be adapted for fairness. We discuss this particular point by presenting the relationship with adversarial examples.

Let $\mathcal{P}(\mathbb{F})$ be the power set of $\mathbb{F}$. Let $f$ be a neighbourhood function: that is, $f : \mathbb{F} \rightarrow \mathcal{P}(\mathbb{F})$. We say that an ML model $\mathbb{M}$ with corresponding function $\varphi$ is robust w.r.t. a neighbourhood function $f$ if $\forall \mathbf{z} \in \mathbb{F}$, $\forall \mathbf{z}' \in f(\mathbf{z})$, $\varphi(\mathbf{z}) = \varphi(\mathbf{z}')$. Consider the example of adversarial robustness [57]. The neighbourhood function related to adversarial robustness can be defined as $f^{ar}(\mathbf{z}) = \{\mathbf{z}' \mid d(\mathbf{z}, \mathbf{z}') \leq \epsilon\}$ where $d : \mathbb{F} \times \mathbb{F} \rightarrow \mathbb{R}^+$ is a distance metric and $\epsilon > 0$. Adversarial robustness can then be defined as the property that $\forall \mathbf{z} \in \mathbb{F}$, $\forall \mathbf{z}' \in f^{ar}(\mathbf{z})$, $\varphi(\mathbf{z}) = \varphi(\mathbf{z}')$.

Fairness can be viewed a particular case of robustness. Note that defining fairness using distance (thus neighbourhood) functions is used in the so-called "fairness through awareness" measure [20,64]. We can consider the neighbourhood function $f^* : \mathbb{F} \rightarrow P(\mathbb{F})$ such that $f^*(\mathbf{x}.\mathbf{y}) = \{\mathbf{x}.\mathbf{y}' \mid \mathbf{y}' \in \mathbb{P}\}$. That is, the neighbourhood of an input $\mathbf{z}$ is the set of inputs that have the same unprotected features. The robustness property using the $f^*$ neighbourhood function is identical to the fairness criterion (1): two inputs that have the same unprotected features should be labelled equally.

Using this observation, we can relate work on robustness and fairness. For example, if one can construct an adversarial example (or counterexample) that uses changes to protected features, then the model is deemed unfair.

## 4 Learning Fair ML Models

Section 3 showed how to assess whether datasets or models could be checked for a specific fairness criterion. The purpose of this section is to investigate ways of synthesizing ML models when a dataset is unbiased, and when it is biased. Whereas the case of unbiased datasets requires simple changes to existing ML model synthesis approaches, the case of biased datasets requires more substantial changes. As argued earlier, the paper focuses on logic-based ML models, namely decision sets, decision trees and decision lists. Moreover, the dataset is assumed to be consistent, for simplicity. The modifications to the case of inconsistent datasets are also briefly discussed.

### 4.1 Unbiased Datasets

This section shows how to synthesize ML models from unbiased datasets. Two different settings can be envisioned, namely heuristic and optimal approaches.

**Heuristic Approaches.** Starting from an unbiased consistent dataset, we can use an off-the-shelf ML tool to learn a fair ML model. A simple solution is to discard the protected features, since we know from Proposition 1 that inconsistencies will not be introduced. The resulting ML model will not depend on the protected features, and so the model is fair. There are no restrictions on which ML model to consider.

**Optimal Approaches.** Exact approaches for synthesizing DS's, DT's and DL's have been studied since the 90s, with a peak of recent interest due to the concerns of interpretability and explainability [42]. These approaches offer formal guarantees, for example in terms of model size and accuracy. As with heuristic approaches, a simple solution to synthesize fair ML models is to ensure that protected variables are *not* allowed to be used when constructing the ML model. We first briefly recall a formal method for computing minimal size decision sets, proposed in [41], namely the so-called MinDS$_3$ model, then we show how it can be modified to synthesize fair DS's. Notice that similar procedures could be employed with *any* other logic-based ML models.

The choice of this model is motivated by simplicity, and others could be considered as well. Most methods learn a set of disjunctive normal form (DNF) formulas, one for each class. MinDS$_3$ learns one DNF for one class, and uses the examples from training data for the other class as the DNF for that class [41]. Given $K$ features and $M$ examples, we consider the synthesis of $N$ rules associated with class $c_1$, where each rule is a term (conjunction) of up to $K$ literals. The variables of the model are the following:

- $s_{jr}$: whether for rule $j$, the feature $F_r$ is skipped.
- $l_{jr}$: the literal on feature $F_r$ for rule $j$, in the case the feature is *not* skipped.
- $d^0_{jr}$: whether rule $j$ discriminates feature $F_r$ on value 0 (in the sense that $F_r$ occurs as a positive literal in term $j$).
- $d^1_{jr}$: whether rule $j$ discriminates feature $F_r$ on value 1 ($\neg F_r$ occurs in term $j$).
- $cr_{jq}$: whether rule $j$ covers $e_q \in \mathcal{T}^+$ (i.e. $e_q$ satisfies term $j$).

The constraints associated with the SAT encoding are the following:

1. Each term (rule) must have at least one literal:

$$\left(\bigvee_{r=1}^{K} \neg s_{jr}\right) \qquad j \in \{1, \ldots, N\} \tag{6}$$

2. One must be able to account for which literals are discriminated by which rules:

$$\begin{array}{ll} d^0_{jr} \leftrightarrow \neg s_{jr} \wedge l_{jr} & j \in \{1, \ldots, N\} \wedge r \in \{1, \ldots, K\} \\ d^1_{jr} \leftrightarrow \neg s_{jr} \wedge \neg l_{jr} & j \in \{1, \ldots, N\} \wedge r \in \{1, \ldots, K\} \end{array} \tag{7}$$

3. Each negative example $e_q \in \mathcal{T}^-$ must be discriminated by each of the $N$ rules ($e_q$ satisfies no rule). Recall that $\mathbf{z}_q[r]$ denote the value of feature $F_r$ for $e_q$. Then, we have:

$$\left(\bigvee_{r=1}^{K} d_{jr}^{\mathbf{z}_q[r]}\right) \qquad j \in \{1,\ldots,N\} \wedge e_q \in \mathcal{T}^- \tag{8}$$

4. Each positive example $e_q \in \mathcal{T}^+$ must be covered by (i.e. satisfy) some rule.
   - First, define whether a rule covers some specific positive example:

$$cr_{jq} \leftrightarrow \left(\bigwedge_{r=1}^{K} \neg d_{jr}^{\mathbf{z}_q[r]}\right) \qquad j \in \{1,\ldots,N\} \wedge e_q \in \mathcal{T}^+ \tag{9}$$

   - Second, each $e_q \in \mathcal{T}^+$ must be covered by some rule.

$$\left(\bigvee_{j=1}^{N} cr_{jq}\right) \qquad e_q \in \mathcal{T}^+ \tag{10}$$

To ensure that the DS respect the FTU rule, we add the constraints $(s_{jr})$, $j = 1,\ldots,N$, for each $F_r \in \mathcal{P}$, denoting that a literal of feature $F_r$ should not be used in rule $j$. This way, the synthesized DS will not include literals on the protected features.

### 4.2 Biased Datasets

If a dataset is biased, then a completely accurate ML model *must* exhibit unfairness.

**Proposition 4.** For a consistent dataset $\mathcal{T}$, if (4) holds, i.e. the dataset is biased, then *any* ML model that is accurate must exhibit FTU unfairness.

*Proof.* If (4) holds, then there exist in $\mathcal{T}$ a point $(\mathbf{x}, \mathbf{y}_1)$ with some prediction $c_1$ and a point $(\mathbf{x}, \mathbf{y}_2)$ with some prediction $c_2$. Thus, if some ML model is accurate, criterion (2) must be false. □

Proposition 4 indicates that if a dataset is biased then accuracy implies loss of fairness and fairness implies loss of accuracy. This section investigates how logic-based models can be synthesized such that fairness is ensured. Due to Proposition 4, the price to pay is that the model is no longer 100% accurate[3]. Furthermore, this section also illustrates how accuracy can be traded off with the size of the ML model representation.

**Maximum Accuracy.** The first problem we study is: find a DS that is fair and has maximum accuracy. As we show next, there is a simple algorithm for

[3] It should be noted that, in ML settings, logic-based models that are not 100% accurate are expected to be less sensitive to overfitting. Thus, the fact that some accuracy is lost is not necessarily a drawback [8].

obtaining a DS that maximizes accuracy on training data. For each $\mathbf{u} \in \mathbb{N}$ having $\mathbf{v}_1, \mathbf{v}_2 \in \mathbb{P}$ such that $\varphi(\mathbf{u}, \mathbf{v}_1) \neq \varphi(\mathbf{u}, \mathbf{v}_2)$, consider the set of examples, denoted by $Conflict_u$, where the values of the non-protected features are those in tuple $\mathbf{u}$. Consider the *least frequent* class among examples in $Conflict_u$. (For ties, pick one class randomly.) Redefine the training data by removing the examples in $Conflict_u$ that are associated to the least frequent class. Now, use the model proposed in Sect. 4 to learn a DS, that discards the protected features.

**Proposition 5.** The algorithm above yields a fair decision set with maximum accuracy.

*Proof.* The learned DS is fair by construction. Moreover, the lack of accuracy is solely due to protected features being relevant for constructing an accurate decision set. For each set of common non-protected features, the most frequently-occurring prediction is chosen. Hence, accuracy cannot be improved if the model is to be fair. □

### 4.3 Non-accurate Interpretable Models

In practice, 100% accurate models are often unwieldy. So an interesting problem is how to synthesize a fair DS, that respects some accuracy target, while placing some bound on the size of the ML model. The solution we propose is when values $\mathbf{x}$ of the non-protected features have multiple predictions (due to the protected features $\mathbf{y}$) to allow freedom of the prediction to be picked (instead of imposing a majority vote) provided it is a function of only $\mathbf{x}$. This ensures fairness of the learnt model according to (1), but this flexibility can be used for reducing the number of rules (or the number of literals in rules) while ensuring that some target accuracy metric is met.

We now study how to synthesize interpretable models that are not 100% accurate. It should be noted that earlier work often makes a number of assumptions regarding trading off accuracy with representation size [5,30,36,51]; ours makes none.

We consider the encoding proposed in Sect. 4 for computing a smallest decision set. Recall that to ensure fairness, we add the constraints $(s_{jr})$, $j = 1, \ldots, N$, for each $F_r \in \mathcal{P}$, so that protected features $F_r$ are not be used in any rule $j$. A model corresponding to a solution of this SAT instance covers all positive examples and discriminates all negative examples. To adapt this model, i.e. (6) to (10), so that it is not necessarily accurate, one can allow each negative example in the training data not to be discriminated. For $e_q \in \mathcal{T}^-$, let $nd_{jq}$ denote that $e_q$ is not discriminated by rule $j$, and let $nd_q$ denote that $e_q$ is not discriminated (by any rule). Then, we update the model as follows:

$$
\begin{array}{ll}
nd_{jq} \leftrightarrow \left( \bigwedge_{r=1}^{K} \neg d_{jr}^{\mathbf{z}_q[r]} \right) & j \in \{1, \ldots, N\} \wedge e_q \in \mathcal{T}^- \\
nd_q \rightarrow \left( \bigvee_{j=1}^{N} nd_{jq} \right) & e_q \in \mathcal{T}^-
\end{array}
\tag{11}
$$

(Equivalence for the second constraint is not needed.) If the target accuracy is $0 \leq \tau \leq 1$, then the constraint on accuracy becomes:

$$\sum_{e_q \in \mathcal{T}^-} nd_q \leq \lfloor \tau \times |\mathcal{T}^-| \rfloor \quad (12)$$

If the model includes DNF's for other classes, then (12) must be extended accordingly, both the sum and the right-hand side (e.g. $\tau \times (|\mathcal{T}^-| + |\mathcal{T}^+|)$).

These proposed changes to the model introduced in Sect. 4 (and taken from [41]) enables learning a decision set that is no longer 100% accurate, and so one can trade off accuracy for representation size while ensuring fairness.

## 5 Theoretical Study of Criteria for Dataset Bias

This section provides a theoretical justification for considering FTU in Sect. 3. We consider the axioms which a criterion for deciding dataset bias should satisfy. A criterion for bias is a boolean function $f$ on datasets $\mathcal{T} \in \mathcal{P}(\mathbb{F} \times \mathcal{C})$ such that $f(\mathcal{T}) = 1$ if $\mathcal{T}$ is biased. First, any such criterion should be independent of the coding of features and classes: replacing a boolean feature $F_i$ by $\neg F_i$ or interchanging the positive and negative classes should not alter dataset bias. Similarly for merging or splitting features, provided this concerns just protected features or just unprotected features: for example, whether we use two boolean features or a single feature with values in $\{0, 1, 2, 3\}$ should not have any effect on deciding dataset bias. Formally, *coding-independence* is invariant under bijective renaming of feature-values or class-names and merging of any pair of features which are either both protected or both non-protected.

A criterion $f$ for bias should always return 0 if all data are identical on the protected features. We call this the *lack of arbitrariness* condition (if $\exists \mathbf{y_0}$ s.t. $\forall \langle (\mathbf{x}, \mathbf{y}), c \rangle \in \mathcal{T}$, $\mathbf{y} = \mathbf{y_0}$, then $f(\mathcal{T}) = 0$). For example, we cannot decide that there is racial bias if all data only concern people of the same race.

Another desirable property of a bias criterion is *monotonicity*: a dataset should not become less biased by eliminating unprotected features (if $\mathcal{T}'$ is obtained from $\mathcal{T}$ by discarding some unprotected features, then $f(\mathcal{T}') \geq f(\mathcal{T})$). For example, if salary and age are unprotected features, and race is a protected feature, then keeping the same classification but ignoring age should not make the dataset less biased, since we are ignoring information which could legitimately be used to classify the data.

We choose a purely logical approach to learning as opposed to a statistical approach. In many applications, and for various possible reasons, the sample distribution may not reflect the true distribution on which the learned model is to be used. For example, the sample data could have been specifically selected by a teacher to cover extreme cases rather than being a random sample. In this logical context, we also impose the following *simplicity* condition: bias can be proved by exhibiting just two different decisions (if $f(\mathcal{T}) = 1$, then $\exists \mathcal{T}' \subseteq \mathcal{T}$ containing just 2 examples with $f(\mathcal{T}') = 1$). Simplicity is a restrictive condition which is only justified in the context of explainable AI: other notions of fairness are possible if we relax this simplicity condition [64].

We say that a dataset is class-uniform if it classifies all data into the same class and that it is protected-uniform if all data are identical on the protected features. A criterion $f$ for bias is *discerning* if it categorizes as unbiased at least one dataset which is neither class-uniform nor protected-uniform and categorizes as biased at least one dataset.

It turns out that the FTU (as given by (4)) is the only possible criterion for bias that satisfies all the above conditions.

**Proposition 6.** The only discerning criterion for data set bias which satisfies the coding-independence, lack of arbitrariness, monotonicity and simplicity conditions is the FTU.

*Proof.* By coding independence, we can merge all non-protected features in $\mathcal{N}$ and all protected features in $\mathcal{P}$, so that we effectively have only two features (with possibly large domains). Coding-independence means that applying any permutation to values does not change bias. This implies that the only operation we can use on features or classes is equality (or inequality). The simplicity condition implies that we can detect bias from two examples $(\mathbf{x}_1, \mathbf{y}_1)$, $(\mathbf{x}_2, \mathbf{y}_2)$ which belong to different classes. Since the criterion for bias is not arbitrary, we know that it cannot impose $\mathbf{y}_1 = \mathbf{y}_2$. A criterion which was a function only of $\mathbf{x}_1, \mathbf{x}_2$ would violate the monotonicity condition, since eliminating all unprotected features would then leave us with a trivial bias criterion. We therefore have to impose the condition $\mathbf{y}_1 \neq \mathbf{y}_2$ in the test for bias.

Suppose now that the criterion decides bias by testing just $\mathbf{y}_1 \neq \mathbf{y}_2$. Since the criterion is discerning, it must categorize as unbiased some dataset $\mathcal{T}$ which is neither class-uniform nor protected-uniform. Since $\mathcal{T}$ is not protected-uniform, there are data $(\mathbf{u}_1, \mathbf{v}_1)$, $(\mathbf{u}_2, \mathbf{v}_2)$ in $\mathcal{T}$ such that $\mathbf{v}_1 \neq \mathbf{v}_2$. Since $\mathcal{T}$ is categorized as unbiased, both $(\mathbf{u}_1, \mathbf{v}_1)$, $(\mathbf{u}_2, \mathbf{v}_2)$ must belong to the same class in $\mathcal{T}$. Now, since $\mathcal{T}$ is not class-uniform, there is $(\mathbf{u}_3, \mathbf{v}_3)$ which belongs to another class in $\mathcal{T}$. But, since $\mathbf{v}_3$ cannot be equal both to $\mathbf{v}_1$ and $\mathbf{v}_2$, the criterion for bias decides that $\mathcal{T}$ is biased, which is a contradiction.

We therefore have to impose $\mathbf{y}_1 \neq \mathbf{y}_2$ together with a condition on $\mathbf{x}_1, \mathbf{x}_2$ in the criterion for bias. If we also impose $\mathbf{x}_1 \neq \mathbf{x}_2$, then this would not satisfy monotonicity (since eliminating all unprotected features could render the dataset unbiased). The only remaining case is to impose the condition $\mathbf{x}_1 = \mathbf{x}_2$ together with $\mathbf{y}_1 \neq \mathbf{y}_2$. This criterion for bias corresponds exactly to the FTU (as given by (4)). □

Another desirable property of a dataset bias criterion is that bias is invariant under the addition of irrelevant (i.e. not used by the model) unprotected features, such as shoe-size when deciding to grant a loan. This appears to be a reasonable condition. However, the following proposition shows that it is impossible to satisfy this *irrelevant-features* condition together with all conditions stated above.

**Proposition 7.** There is no discerning criterion for dataset bias which satisfies the coding-independence, lack of arbitrariness, monotonicity, simplicity and irrelevant-features conditions.

*Proof.* By Proposition 6 FTU is the only possible candidate criterion for bias satisfying all these conditions. Consider a dataset which is categorized as biased by the FTU. However, by adding sufficient irrelevant features we can ensure that all data are distinct on the unprotected features and hence the FTU would consider the extended dataset to be unbiased, thus contradicting invariance under addition of irrelevant features. □

For example, when a bank decides whether to grant a loan to one of its clients it has information stored such as their bank account number. If this number is considered as an unprotected feature, any model will satisfy the FTU criterion (we are assuming there are no two clients with the same account number). The automatic detection of irrelevant features is an interesting problem for future research.

## 6 Preliminary Experimental Results

This section assesses empirically the proposed ideas on a selection of well-known datasets. The aim of the experiments is to (1) illustrate that datasets can be practically checked for bias using Algorithm 1, (2) show that ML models can be tested for fairness using condition (2), and (3) make an attempt to synthesize fair decision sets [41,49] as discussed in Sect. 4.

**Experimental Setup.** For the experiments, several Python scripts were implemented, instrumenting SAT and SMT oracle calls. Whenever needed, MiniSat 2.2 [21] was used as a SAT oracle while Z3 [52] was employed as an SMT solver. The solvers were accessed through the well-known Python APIs, namely PySAT [38] and PySMT [29].

The experiments were performed on a Macbook Pro with an Intel Core i7 2.6 GHz CPU and 16 GB of memory and focused on a few publicly available datasets studied in the context of algorithmic fairness. These include *Compas*, *Adult*, *German*, and *Ricci*. The datasets are binarized using the standard one-hot encoding method [58]. Their sizes is detailed in Fig. 2a. *Compas* is a popular dataset known [6] for exhibiting racial bias of the COMPAS algorithm used for scoring a criminal defendant's likelihood of reoffending; the dataset includes a few protected features, namely, race-related parameters *African American*, *Asian*, *Hispanic*, *Native American*, and *Other* but also *Female*. *Adult* [47] is originally taken from the Census bureau and targets predicting whether or not a given adult person earns more than $50K a year depending on various features, among which the protected ones are *Race* and *Sex*. *German* credit data (e.g. see [24]), given a list of people's features, classifies them as good or bad credit risks; the protected features are *Sex* and *Age*. The *Ricci* dataset [26] comes from the case of Ricci vs. DeStefano [63], "a case before the U.S. Supreme Court in which the question at issue was an exam given to determine if firefighters would receive a promotion"; the protected feature is *Race*.

### 6.1 Assessing Dataset Bias

The first part of the experimental assessment aims at checking whether the aforementioned datasets exhibit bias with respect to the corresponding protected features. Note that although the *Compas* and *German* datasets are inconsistent, they can still be tested for bias (see Sect. 3). Running Algorithm 1 reports that (1) *Compas* and *Adult* are biased with respect to all the protected features while (2) *German* and *Ricci* do not exhibit bias with respect to any protected feature. Point (1) indicates that there is *no way* to train an ML model of maximum accuracy whilst being fair with respect to the protected features. In particular, this confirms the famous bias-related issues of the *Compas* algorithm. Moreover, the results for the *Ricci* dataset should be highlighted. Since this dataset is unbiased, one is guaranteed that a fair ML model can be synthesized. (This should be contrasted with the unfair heuristic models studied in earlier work [26]).

### 6.2 Assessing Model Fairness

Here we focus on testing fairness of boosted tree models trained with the XGBoost algorithm [13] for the considered datasets. To perform an exhaustive experiment assessing accuracy of the target models and to be able to draw conclusions, we followed the standard paradigm of 10-fold cross-validation. As such, each dataset is randomly divided into 10 equally sized chunks of examples and the experiment is done 10 times (one per chunk), each time dealing with 90% target dataset, i.e. with 1 of the 10 chunks discarded. This way every dataset is tested for fairness of the respective model wrt. each protected feature 10 times. Overall, this results in 60 fairness tests for *Compas*, 20 for *Adult*, 20 for *German*, and 10 for *Ricci* – the total number of tests to perform is 110. (Recall that each test is made as an SMT oracle call dealing with formula (2).)

Each XGBoost model trained for this experiment contains 50 trees per class with each tree having depth 3 (this suffices to get a reasonable classification accuracy). Boosted trees are encoded into SMT by applying a simple encoding proposed in [40].

The minimum, maximum, and average running time per test for each of the datasets is shown in Fig. 2b. Observe that testing fairness is not computationally expensive and can be done for medium-sized boosted trees. Also note that fairness tests are on average more time consuming for the *German* dataset.

Regarding the fairness of the trained ML models, the tests reported that only 2 (out of 10) models trained for the *Compas* dataset are fair wrt. protected feature *Other*. All the other models trained for all datasets are *unfair* with respect to *every* protected feature. This should not come as a surprise given that these models were trained with no knowledge about the protected features, which is usually the case in practice. Since the fairness check (2) is *model agnostic*, this result confirms the power and applicability of the proposed ideas in practical situations when fairness of ML models is a concern.

| | Adult | Compas | German | Ricci |
|---|---|---|---|---|
| **orig. features** | 12 | 11 | 21 | 5 |
| **binary features** | 65 | 46 | 1073 | 235 |
| **examples** | 14113 | 6172 | 1000 | 118 |

(a) Size of the considered datasets.

| | Adult | Compas | German | Ricci |
|---|---|---|---|---|
| **min. (s)** | 0.31 | 0.24 | 2.50 | 0.33 |
| **avg. (s)** | 12.22 | 0.56 | 87.08 | 0.36 |
| **max. (s)** | 63.77 | 3.37 | 1062.32 | 0.43 |

(b) Running Time for Checking Model Fairness.

**Fig. 2.** Size of the Datasets (left) & Time for Assessing (XGBoost) Model Fairness (right)

### 6.3 Synthesizing Fair Decision Sets

This section aims at synthesizing fair decision sets for the case of biased dataset *Compas* and unbiased dataset *Ricci*, based on the approach of [41] (results for *Adult* and *German* are not reported here because synthesis of DS models for these datasets seems too challenging). While for the former dataset (*Compas*) one can drop the protected features from the dataset and then trade off accuracy with DS size, for the unbiased *Ricci* dataset it suffices to modify the DS model, as described in Sect. 4.

Furthermore, since *Ricci* is unbiased wrt. the protected feature *Race*, we can achieve 100% accuracy with the resulting DS model (similarly with decision trees/lists). In fact, a perfectly accurate decision set for *Ricci* computed by the modified $\text{MinDS}_3$ model [41] has only two rules, i.e. one rule per class. The downside here is that each rule is quite long, s.t. the decision set has 68 literals in total. This is in clear contrast to the interpretability purpose of decision sets. However, by trading off accuracy with the DS size, one can get the following non-overlapping $\text{MinDS}_1$ model [41] with 97.5% accuracy and having only 4 rules and 7 literals in total, which is easy to interpret:

IF Position $\neq$ Leutenant $\wedge$ Oral $\leq$ 63.75 THEN Class = 0
IF Combine $\leq$ 69.372 THEN Class = 0
IF Position = Leutenant $\wedge$ Combine $>$ 69.372 THEN Class = 1
IF Oral $>$ 63.75 $\wedge$ Combine $>$ 69.372 THEN Class = 1

Since *Compas* is biased wrt. the protected features, training a fair ML model for this dataset can be done by sacrificing the model's accuracy. Concretely, the *maximum feasible* accuracy for this dataset is 69.73%. Although DS models achieving this accuracy for *Compas* can be trained, they are too large to interpret (each has at least a few hundred literals). Thus, one may want to sacrifice accuracy further and get a more interpretable (i.e. smaller) DS model instead. For instance, the following non-overlapping $\text{MinDS}_1$ model has 66.32% accuracy and it is fair with respect to all the protected features:

IF Number_of_Priors $>$ 17.5 $\wedge$ $\neg$score_factor THEN Two_yr_Recidivism
IF Number_of_Priors $>$ 17.5 $\wedge$ Age_Above_FourtyFive $\wedge$ Misdemeanor THEN Two_yr_Recidivism
IF Number_of_Priors $\leq$ 17.5 THEN $\neg$Two_yr_Recidivism
IF score_factor $\wedge$ $\neg$Age_Above_FourtyFive THEN $\neg$Two_yr_Recidivism
IF score_factor $\wedge$ $\neg$Misdemeanor THEN $\neg$Two_yr_Recidivism

## 7 Conclusions and Research Directions

We studied the fairness of ML models, by considering the criterion FTU [31,48], but proposing instead a semantic definition. We also proposed theoretical justifications for the use of FTU. Moreover, we developed criteria for assessing fairness in ML models and bias in datasets, and related fairness with explanations and robustness. Finally, we investigated approaches for synthesizing fair ML models. Future work will address limitations of the current work, namely assessing non-protected features exhibiting discriminatory information and/or taking statistical measures into account [12,67].

## References


1. Adebayo, J.A.: FairML: ToolBox for diagnosing bias in predictive modeling. Master's thesis, Massachusetts Institute of Technology (2016)
2. Adebayo, J.A.: FairML: auditing black-box predictive models (2017)
3. Aïvodji, U., Arai, H., Fortineau, O., Gambs, S., Hara, S., Tapp, A.: Fairwashing: the risk of rationalization. In: ICML, pp. 161–170 (2019)
4. Aïvodji, U., Ferry, J., Gambs, S., Huguet, M., Siala, M.: Learning fair rule lists. CoRR abs/1909.03977 (2019). http://arxiv.org/abs/1909.03977
5. Angelino, E., Larus-Stone, N., Alabi, D., Seltzer, M., Rudin, C.: Learning certifiably optimal rule lists for categorical data. J. Mach. Learn. Res. **18**, 234:1–234:78 (2017)
6. Angwin, J., Larson, J., Mattu, S., Kirchner, L.: Machine bias. propublica.org, May 2016. http://tiny.cc/a3b3iz
7. Berk, R., Heidari, H., Jabbari, S., Kearns, M., Roth, A.: Fairness in criminal justice risk assessments: the state of the art. Sociol. Methods Res. (2017). https://doi.org/10.1177/0049124118782533
8. Berkman, N.C., Sandholm, T.W.: What should be minimized in a decision tree: a re-examination. Department of Computer Science (1995)
9. Bessiere, C., Hebrard, E., O'Sullivan, B.: Minimising decision tree size as combinatorial optimisation. In: Gent, I.P. (ed.) CP 2009. LNCS, vol. 5732, pp. 173–187. Springer, Heidelberg (2009). https://doi.org/10.1007/978-3-642-04244-7_16
10. Biere, A., Heule, M., van Maaren, H., Walsh, T. (eds.): Handbook of Satisfiability, Frontiers in Artificial Intelligence and Applications, vol. 185. IOS Press (2009)
11. Bird, S., Hutchinson, B., Kenthapadi, K., Kiciman, E., Mitchell, M.: Fairness-aware machine learning: practical challenges and lessons learned. In: KDD, pp. 3205–3206 (2019)
12. Cardelli, L., Kwiatkowska, M., Laurenti, L., Paoletti, N., Patane, A., Wicker, M.: Statistical guarantees for the robustness of Bayesian neural networks. In: Kraus, S. (ed.) Proceedings of the Twenty-Eighth International Joint Conference on Artificial Intelligence, IJCAI 2019, Macao, China, 10–16 August 2019, pp. 5693–5700. ijcai.org (2019). https://doi.org/10.24963/ijcai.2019/789
13. Chen, T., Guestrin, C.: XGBoost: a scalable tree boosting system. In: KDD, pp. 785–794 (2016)
14. Chouldechova, A.: Fair prediction with disparate impact: a study of bias in recidivism prediction instruments. Big Data **5**(2), 153–163 (2017)
15. Chouldechova, A., Roth, A.: A snapshot of the frontiers of fairness in machine learning. Commun. ACM **63**(5), 82–89 (2020)

16. Corbett-Davies, S., Pierson, E., Feller, A., Goel, S., Huq, A.: Algorithmic decision making and the cost of fairness. In: KDD 2017, pp. 797–806 (2017)
17. Cormen, T.H., Leiserson, C.E., Rivest, R.L., Stein, C.: Introduction to Algorithms, 3rd edn. MIT Press (2009). http://mitpress.mit.edu/books/introduction-algorithms
18. Demsar, J., et al.: Orange: data mining toolbox in python. J. Mach. Learn. Res. **14**(1), 2349–2353 (2013)
19. Dressel, J., Farid, H.: The accuracy, fairness, and limits of predicting recidivism. Sci. Adv. **4**(1), eaao5580 (2018)
20. Dwork, C., Hardt, M., Pitassi, T., Reingold, O., Zemel, R.: Fairness through awareness. In: ITCS, pp. 214–226 (2012)
21. Eén, N., Sörensson, N.: An extensible SAT-solver. In: Giunchiglia, E., Tacchella, A. (eds.) SAT 2003. LNCS, vol. 2919, pp. 502–518. Springer, Heidelberg (2004). https://doi.org/10.1007/978-3-540-24605-3_37
22. Ehlers, R.: Formal verification of piece-wise linear feed-forward neural networks. In: D'Souza, D., Narayan Kumar, K. (eds.) ATVA 2017. LNCS, vol. 10482, pp. 269–286. Springer, Cham (2017). https://doi.org/10.1007/978-3-319-68167-2_19
23. European Union High-Level Expert Group on Artificial Intelligence: Ethics guidelines for trustworthy AI, April 2019. https://ec.europa.eu/digital-single-market/en/news/ethics-guidelines-trustworthy-ai
24. Feldman, M., Friedler, S.A., Moeller, J., Scheidegger, C., Venkatasubramanian, S.: Certifying and removing disparate impact. In: KDD, pp. 259–268. ACM (2015)
25. Friedler, S.A., Scheidegger, C., Venkatasubramanian, S.: On the (im)possibility of fairness. CoRR abs/1609.07236 (2016). http://arxiv.org/abs/1609.07236
26. Friedler, S.A., Scheidegger, C., Venkatasubramanian, S., Choudhary, S., Hamilton, E.P., Roth, D.: A comparative study of fairness-enhancing interventions in machine learning. In: FAT, pp. 329–338 (2019)
27. Galhotra, S., Brun, Y., Meliou, A.: Fairness testing: testing software for discrimination. In: FSE, pp. 498–510 (2017)
28. Garg, S., Perot, V., Limtiaco, N., Taly, A., Chi, E.H., Beutel, A.: Counterfactual fairness in text classification through robustness. In: AIES, pp. 219–226 (2019)
29. Gario, M., Micheli, A.: PySMT: a solver-agnostic library for fast prototyping of SMT-based algorithms. In: SMT Workshop (2015)
30. Ghosh, B., Meel, K.S.: IMLI: an incremental framework for MaxSAT-based learning of interpretable classification rules. In: AIES, pp. 203–210 (2019)
31. Grgic-Hlaca, N., Zafar, M.B., Gummadi, K.P., Weller, A.: The case for process fairness in learning: feature selection for fair decision making. In: NIPS Symposium on Machine Learning and the Law (2016)
32. Han, J., Kamber, M., Pei, J.: Data Mining: Concepts and Techniques, 3rd edn. Morgan Kaufmann (2012)
33. Hardt, M., Price, E., Srebro, N.: Equality of opportunity in supervised learning. In: Lee, D.D., Sugiyama, M., von Luxburg, U., Guyon, I., Garnett, R. (eds.) Advances in Neural Information Processing Systems 29, pp. 3315–3323 (2016). http://papers.nips.cc/paper/6374-equality-of-opportunity-in-supervised-learning
34. Holstein, K., Vaughan, J.W., Daumé III, H., Dudík, M., Wallach, H.M.: Improving fairness in machine learning systems: what do industry practitioners need? In: CHI, p. 600 (2019)
35. Hu, H., Siala, M., Hebrard, E., Huguet, M.J.: Learning optimal decision trees with MaxSAT and its integration in AdaBoost. In: IJCAI, pp. 1170–1176 (2020)
36. Hu, X., Rudin, C., Seltzer, M.: Optimal sparse decision trees. In: NeurIPS, pp. 7265–7273 (2019)

37. Huang, X., Kwiatkowska, M., Wang, S., Wu, M.: Safety verification of deep neural networks. In: Majumdar, R., Kunčak, V. (eds.) CAV 2017. LNCS, vol. 10426, pp. 3–29. Springer, Cham (2017). https://doi.org/10.1007/978-3-319-63387-9_1
38. Ignatiev, A., Morgado, A., Marques-Silva, J.: PySAT: a python toolkit for prototyping with SAT oracles. In: Beyersdorff, O., Wintersteiger, C.M. (eds.) SAT 2018. LNCS, vol. 10929, pp. 428–437. Springer, Cham (2018). https://doi.org/10.1007/978-3-319-94144-8_26
39. Ignatiev, A., Narodytska, N., Marques-Silva, J.: Abduction-based explanations for machine learning models. In: AAAI, pp. 1511–1519 (2019)
40. Ignatiev, A., Narodytska, N., Marques-Silva, J.: On validating, repairing and refining heuristic ML explanations. CoRR abs/1907.02509 (2019). http://arxiv.org/abs/1907.02509
41. Ignatiev, A., Pereira, F., Narodytska, N., Marques-Silva, J.: A SAT-based approach to learn explainable decision sets. In: IJCAR, pp. 627–645 (2018)
42. Kamath, A.P., Karmarkar, N., Ramakrishnan, K.G., Resende, M.G.C.: A continuous approach to inductive inference. Math. Program. **57**, 215–238 (1992)
43. Katz, G., Barrett, C., Dill, D.L., Julian, K., Kochenderfer, M.J.: Reluplex: an efficient SMT solver for verifying deep neural networks. In: Majumdar, R., Kunčak, V. (eds.) CAV 2017. LNCS, vol. 10426, pp. 97–117. Springer, Cham (2017). https://doi.org/10.1007/978-3-319-63387-9_5
44. Katz, G., et al.: The marabou framework for verification and analysis of deep neural networks. In: Dillig, I., Tasiran, S. (eds.) CAV 2019. LNCS, vol. 11561, pp. 443–452. Springer, Cham (2019). https://doi.org/10.1007/978-3-030-25540-4_26
45. Kilbertus, N., Rojas-Carulla, M., Parascandolo, G., Hardt, M., Janzing, D., Schölkopf, B.: Avoiding discrimination through causal reasoning. In: NeurIPS, pp. 656–666 (2017)
46. Kleinberg, J.M., Mullainathan, S., Raghavan, M.: Inherent trade-offs in the fair determination of risk scores. In: 8th Innovations in Theoretical Computer Science Conference, ITCS 2017, Berkeley, CA, USA, 9–11 January 2017, pp. 43:1–43:23 (2017)
47. Kohavi, R.: Scaling up the accuracy of Naive-Bayes classifiers: a decision-tree hybrid. In: KDD, pp. 202–207 (1996)
48. Kusner, M.J., Loftus, J.R., Russell, C., Silva, R.: Counterfactual fairness. In: NeurIPS, pp. 4066–4076 (2017)
49. Lakkaraju, H., Bach, S.H., Leskovec, J.: Interpretable decision sets: a joint framework for description and prediction. In: KDD, pp. 1675–1684 (2016)
50. Leofante, F., Narodytska, N., Pulina, L., Tacchella, A.: Automated verification of neural networks: advances, challenges and perspectives. CoRR abs/1805.09938 (2018). http://arxiv.org/abs/1805.09938
51. Maliotov, D., Meel, K.S.: MLIC: a MaxSAT-based framework for learning interpretable classification rules. In: Hooker, J. (ed.) CP 2018. LNCS, vol. 11008, pp. 312–327. Springer, Cham (2018). https://doi.org/10.1007/978-3-319-98334-9_21
52. de Moura, L., Bjørner, N.: Z3: an efficient SMT solver. In: Ramakrishnan, C.R., Rehof, J. (eds.) TACAS 2008. LNCS, vol. 4963, pp. 337–340. Springer, Heidelberg (2008). https://doi.org/10.1007/978-3-540-78800-3_24
53. Nabi, R., Shpitser, I.: Fair inference on outcomes. In: AAAI, pp. 1931–1940 (2018)
54. Narayanan, A.: Translation tutorial: 21 fairness definitions and their politics. In: FAT (2018)
55. Narodytska, N.: Formal analysis of deep binarized neural networks. In: IJCAI, pp. 5692–5696 (2018)

56. Narodytska, N., Ignatiev, A., Pereira, F., Marques-Silva, J.: Learning optimal decision trees with SAT. In: IJCAI, pp. 1362–1368 (2018)
57. Narodytska, N., Kasiviswanathan, S.P., Ryzhyk, L., Sagiv, M., Walsh, T.: Verifying properties of binarized deep neural networks. In: AAAI, pp. 6615–6624 (2018)
58. Pedregosa, F., et al.: Scikit-learn: machine learning in Python. J. Mach. Learn. Res. **12**, 2825–2830 (2011)
59. du Pin Calmon, F., Wei, D., Vinzamuri, B., Ramamurthy, K.N., Varshney, K.R.: Optimized pre-processing for discrimination prevention. In: NeurIPS, pp. 3992–4001 (2017)
60. Pulina, L., Tacchella, A.: An abstraction-refinement approach to verification of artificial neural networks. In: Touili, T., Cook, B., Jackson, P. (eds.) CAV 2010. LNCS, vol. 6174, pp. 243–257. Springer, Heidelberg (2010). https://doi.org/10.1007/978-3-642-14295-6_24
61. Ruan, W., Huang, X., Kwiatkowska, M.: Reachability analysis of deep neural networks with provable guarantees. In: IJCAI, pp. 2651–2659 (2018)
62. Shih, A., Choi, A., Darwiche, A.: A symbolic approach to explaining Bayesian network classifiers. In: IJCAI, pp. 5103–5111 (2018)
63. Supreme Court of the United States: Ricci v. DeStefano. U.S. 557, 174 (2009)
64. Verma, S., Rubin, J.: Fairness definitions explained. In: FairWare@ICSE, pp. 1–7 (2018)
65. Verwer, S., Zhang, Y.: Learning decision trees with flexible constraints and objectives using integer optimization. In: Salvagnin, D., Lombardi, M. (eds.) CPAIOR 2017. LNCS, vol. 10335, pp. 94–103. Springer, Cham (2017). https://doi.org/10.1007/978-3-319-59776-8_8
66. Verwer, S., Zhang, Y.: Learning optimal classification trees using a binary linear program formulation. In: AAAI, pp. 1625–1632 (2019)
67. Wu, M., Wicker, M., Ruan, W., Huang, X., Kwiatkowska, M.: A game-based approximate verification of deep neural networks with provable guarantees. Theor. Comput. Sci. **807**, 298–329 (2020). https://doi.org/10.1016/j.tcs.2019.05.046

# Verifying Equivalence Properties of Neural Networks with ReLU Activation Functions

Marko Kleine Büning(✉), Philipp Kern, and Carsten Sinz

Karlsruhe Institute of Technology (KIT), Karlsruhe, Germany
{marko.kleinebuening,carsten.sinz}@kit.edu,
ufedz@student.kit.edu

**Abstract.** Neural networks have become popular methods for tackling various machine learning tasks and are increasingly applied in safety-critical systems. This necessitates verified statements about their behavior and properties. One of these properties is the equivalence of two neural networks, which is important, e.g., when neural networks shall be reduced to smaller ones that fit space and memory constraints of embedded or mobile systems.

In this paper, we present the encoding of feed-forward neural networks with ReLU activation functions and define novel and relaxed equivalence properties that extend previously proposed notions of equivalence. We define $\epsilon$- and *top-k*-equivalence and employ it in conjunction with restricting the input space by hierarchical clustering. Networks and properties are encoded as mixed integer linear programs (MILP). We evaluate our approach using two existing reduction methods on a neural network for handwritten digit recognition.

## 1 Introduction

The popularity of neural networks (NNs) for solving machine learning tasks has strongly increased with the availability of high performance computers and large data sets produced by today's society. Nowadays, NNs are considered state of the art solutions for many machine learning tasks, including machine translation [2], image processing [19] or playing games like Go and chess [23]. The complex structure of layers and weights, however, renders them incomprehensible to humans. While this does not have serious consequences when it comes to playing games, it can have a severe impact when neural networks are applied to safety-critical systems like self-driving cars [3]. Validation and verification procedures are thus needed to provide safety guarantees.

The verification of NNs is a relatively young field. Among the first papers published is the work by Pulina and Tacchella [22], where the authors checked bounds on the output of multilayer perceptrons. Most current publications (e.g., [10,16,26]) focus on proving the adversarial robustness of NNs, meaning that a network assigns the label of a known reference input-point to all points in a


H. Simonis (Ed.): CP 2020, LNCS 12333, pp. 868–884, 2020.
https://doi.org/10.1007/978-3-030-58475-7_50

small region around it. To prove this property, the NNs are often encoded as constraint systems solved with SAT, SMT or MILP solvers. Over the last years, the size of NNs used in practical applications grew rapidly, and today's networks often require huge amounts of memory space. Their execution causes high energy consumption, rendering them impractical for use on mobile or embedded devices. As a consequence, methods have been developed to reduce the size of NNs [14,27]. The question then arises whether the reduced NN is suitable as a replacement for the original one, in the sense that it behaves "sufficiently equivalent" on relevant inputs. Narodytska *et al.* [21] consider two feed-forward NNs equivalent if, for all valid[1] inputs of the input domain, the NNs produce the same output labels. They are able to prove this property for the specialized class of binarized NNs, which allows them to produce a SAT formula representing the equivalence of two NNs. Kumar *et al.* [20] denote this equivalence property by local equivalence over a domain, and use the term equivalence for NNs that give the same output for all inputs. Based on these notions they collapse layers of a given NN while guaranteeing the equivalence to the original NN.

Unfortunately, exact equivalence is hard to fulfill for two NNs, even if they have the same structure and have been generated using the same training data, due to the stochastic nature of the training process. In this paper, we study feed-forward NNs with the ReLU activation function, which is the most used activation function in modern NNs [11]. We present a new relaxed equivalence property for NNs and show, how it—along with existing equivalence properties—can be encoded in MILP. Additionally, we show an encoding for the verification of equivalence, as well as maximizing the size of equivalent regions, when the input domain is restricted to radii around a point in input space. We evaluate our approach using the constraint solver Gurobi [13] and a NN trained on the Optical Recognition of Handwritten Digits dataset [7]. The evaluation of our approach marks the first time, that NN compression methods have been examined by verification methods with respect to generating equivalent NNs.

## 2 Foundations

We give short introductions into NNs and MILP. Afterwards, we present the encoding of NNs into MILP formulae.

**Neural Networks.** NNs consist of a number of interconnected units, sometimes called neurons. One single neuron $j$ computes its output $y_j$ as a function of input values $x_0, ..., x_n$ according to $y_j = \sigma(\sum_{i=0}^{n} w_{ij}x_i)$ , where $\sigma$ is called the activation function and $x_0$ is commonly set to 1, such that $w_{0j}$ encodes the bias of the neuron. The weights $w_{ij}$ can be learned from training data. To enable the NN to capture non-linear functions, the activation function also has to be non-linear. While there are many choices like the *tanh* or *sigmoid* function, we focus on the *rectified linear unit*: $\text{ReLU}(x) = max(0, x)$, which is the most commonly

[1] Validity just ascertains that inputs are suitably bounded, e.g. to a range of $[0, 255]$ for greyscale pixels.

used activation function in modern NNs [11]. A NN is formed by connecting neurons via directed links, such that the outputs $y_k$ of previous neurons serve as inputs $x_i$ of the current neuron. In this paper, we focus on feed-forward NNs, where the graph formed by these connections is acyclic. The neurons in these NNs can be organized in layers, such that each neuron only uses outputs of the neurons in the layer directly preceding it as its inputs. The first layer—called input layer—is just a place holder for the inputs to be fed into the NN, the subsequent layers are called hidden layers, while the last layer—the output layer—holds the function value computed by the NN. In a regression setting, the output value represents the NN's estimate of the respective latent function value for the given input. For a classification task, however, the output $y_i$ of neuron $i$ represents the probability of the NN's input belonging to class $i$. To ensure that the resulting distribution over the outputs is normalized, each output neuron $i$ uses the softmax activation function

$$\text{softmax}(\mathbf{x})_i = \frac{e^{x_i}}{\sum_j e^{x_j}} \ .$$

**Mixed Integer Linear Programs.** A MILP problem is an optimization problem for a linear objective function under additional linear constraints. Some variables are constrained to be integers, while others range over $\mathbb{R}$.

**Definition 1.** *A* mixed integer linear programming problem *consists of*

1. *a linear* objective function $f(x_1, ..., x_k) = \sum_{i=1}^{k} c_i x_i$ *over decision variables* $x_i$ *that is to be minimized or maximized,*
2. *a set of linear* constraints $\sum_{i=1}^{k} a_{ij} x_i \bowtie b_j$, $\bowtie \in \{\leq, =, \geq\}$, *where* $a_{ij}$ *and* $b_j$ *are constants,*
3. *and an integrality constraint* $x_i \in \mathbb{Z}$ *for some variables.*

Solving MILPs is in general NP-hard. Algorithms for solving them include branch and bound, cutting planes, or methods based on relaxations.

### 2.1 Encoding of Neural Networks as MILP

To argue over properties of NNs, we encode them in MILP utilizing the big-M encoding presented in [5]. Our encoding is equal under transformation to the ReLU-encodings of [9].

For a NN to be encoded, we first have to encode a **single neuron**. A neuron $j$ applies a non-linear activation function $\sigma$ to a linear combination $s_j = \sum_{i=0}^{n} w_{ij} x_i$ of its inputs $x_0, ..., x_n$. Given fixed weights $w_{ij}$, this equation can be directly encoded in a mixed integer linear program. The non-linear ReLU activation function $y_j = max(0, s_j)$ can be encoded using given bounds $m \leq s_j \leq M$, which can be calculated knowing the bounds for the inputs $x_i$ and weights $w_{ij}$ of the NN. The ReLU function can be encoded using a new zero-one variable $\delta \in \{0, 1\}$, with $\delta = 0$ representing the case $(s_j \leq 0 \wedge y_j = 0)$

and $\delta = 1$ representing $(s_j \geq 0 \wedge y_j = s_j)$. The ReLU function is then encoded by the following set of linear inequalities:

$$\begin{aligned} y_j &\geq 0 & s_j &\geq m(1-\delta) \\ y_j &\geq s_j & y_j - s_j &\leq -m(1-\delta) \\ s_j &\leq M\delta & y_j &\leq M\delta \,. \end{aligned}$$

Given tight bounds $m \leq s_j \leq M$, the encoding can be further simplified [5]. If $M \leq 0$, we can directly encode the ReLU function as $y_j = 0$, and if $m > 0$, we encode the output of the activation function as $y_j = s_j$. These reductions in complexity are particularly valuable as they do not use any integer variables, on whom we might have to branch when solving the resultant mixed integer linear program. Therefore, we employ the approach of [5] to generate tighter bounds by means of interval arithmetic and also solve for bounds on intermediate variables by maximizing or minimizing their values using smaller mixed integer linear programs only covering a low number of layers of the NN at a time.

Based on the encoding of a single neuron, we can encode a **whole NN**. Each input of the NN is represented by a variable $x_i$ with associated bounds $l_i \leq x_i \leq u_i$. These $l_i$ and $u_i$ can be set according to physical limitations or might be obtained from the respective training dataset and can be used for the calculation of subsequent bounds in the encoding. The neurons in the first layer are then encoded according to the previous description. The same procedure is applied to the neurons of the next layers with the outputs $y_i$ of the neurons of the previous layer taking the role of the inputs above, until the output layer is reached. In classification NNs, the neurons in the output layer use the softmax activation function. Due to its exponential functions, an exact encoding in MILP is impossible. However, since the softmax function is monotonic, we are able to reason about the order of the outputs by encoding the linear combination of the input values for the neurons of the output layer in a classification NN.

## 3 Equivalence Properties

Let $\mathcal{R}$ be a reference and $\mathcal{T}$ a test NN computing functions $f_\mathcal{R}, f_\mathcal{T} : \mathbb{R}^m \rightarrow \mathbb{R}^n$. We further assume that the inputs to these NNs come from the same domain $X$ and that the $i$-th component of the output vector of $f_\mathcal{R}(\mathbf{x})$ and $f_\mathcal{T}(\mathbf{x})$ has the same meaning in the encoding of the output neurons. Proving exact equivalence of the test NN $\mathcal{T}$ and the reference NN $\mathcal{R}$ would then mean to ascertain that

$$\forall \mathbf{x} \in X : f_\mathcal{R}(\mathbf{x}) = f_\mathcal{T}(\mathbf{x}) \,. \tag{1}$$

However, the training procedure of NNs is highly non-deterministic and training could be on different datasets, thus leading to differences in the learned weights, even if the NNs share the same number of layers and neurons. It is therefore unlikely for two NNs to fulfill the exact equivalence property stated above. Hence, we need to relax it to obtain a more practical notion of equivalence. In general, this can either be achieved by (1) relaxing the exact equality of the function values in Eq. 1 through a less strict relation $\simeq$, or (2) restricting the domain of the inputs to the NNs to smaller subsets, for which equality is more likely. The first approach is described below, while we discuss the input restriction in Sect. 4.

### 3.1 Relaxed Equivalence Properties

The definition of exact equivalence in Eq. 1 can be written as a difference: $\forall \mathbf{x} \in X : f_{\mathcal{R}}(\mathbf{x}) - f_{\mathcal{T}}(\mathbf{x}) = 0$. An easy relaxation would be to consider two functions equivalent, if their difference is at least close to zero within some threshold.

**Definition 2 ($\varepsilon$-Equivalence).** *We consider $\mathcal{R}$ and $\mathcal{T}$ to be* $\varepsilon$-equivalent *with respect to a norm $\|\cdot\|$, if $\|f_{\mathcal{R}}(\mathbf{x}) - f_{\mathcal{T}}(\mathbf{x})\| \leq \varepsilon$ for all $\mathbf{x} \in X$.*

While this is a valid relaxation in the context of regression NNs, the functions $f_{\mathcal{R}}$ and $f_{\mathcal{T}}$ compute class probability distributions when it comes to classification problems. In most of these cases, one is not interested in the full class probability distribution, but only in the classification result—the class assigned the highest probability by the NN. In that case, we can obtain a relaxed equivalence property by comparing only the classification results.

**Definition 3 (One-Hot Equivalence) .** *We call $\mathcal{R}$ and $\mathcal{T}$* one-hot equivalent, *if $f_{\mathcal{R}}(\mathbf{x}) = \mathbf{r}$ and $f_{\mathcal{T}}(\mathbf{x}) = \mathbf{t}$ satisfy* $\arg\max_i r_i = \arg\max_i t_i$ *for all $\mathbf{x} \in X$.*

The name stems from the representation of the true label for each input in the training data as a one-hot vector for classification NNs. We note that this definition is closely related to the property of adversarial robustness [26], however we compare the classification results of two NNs instead of comparing the classification result of one NN with the ground-truth.

The notion of one-hot equivalence can be relaxed even further when we consider not only the most likely class, which is the classification result, but instead take the $k$ most likely classes into account (the definition is motivated by a similar idea in [5]).

**Definition 4 (Top-$k$ Equivalence).** *A test NN $\mathcal{T}$ is equivalent to a reference NN $\mathcal{R}$, if $f_{\mathcal{R}}(\mathbf{x}) = \mathbf{r}$ and $f_{\mathcal{T}}(\mathbf{x}) = \mathbf{t}$ satisfy*

$$\arg\max_i r_i = j \implies \text{pos}(t_j, \mathbf{t}) \leq k \ ,$$

*where* $\text{pos}(w_j, \mathbf{w})$ *returns $i$, if $w_j$ is the $i$-th largest value in vector $\mathbf{w}$, and $r_j$ is the unique maximum component of vector $\mathbf{r}$.*

Informally, a testing NN $\mathcal{T}$ is top-$k$ equivalent to a reference NN $\mathcal{R}$, if the classification result of $\mathcal{R}$ is amongst the top $k$ largest results of $\mathcal{T}$. This can be interpreted in a way, such that the NN, even if it differs from the classification result of the original NN, at least only makes sensible errors. One-hot equivalence can also be seen as a special case of top-$k$-equivalence for $k = 1$.

Note that, while exact equality and one-hot-equality are equivalence relations in the mathematical sense, neither $\varepsilon$-equivalence, nor top-$k$-equivalence for $k > 1$ meet that criterion, as both are not transitive and the latter additionally is not symmetric.

### 3.2 Encoding of Equivalence Properties in MILP

In the context of adversarial robustness, properties are often encoded as MILP problems [4,18], a formalism we also employ for our equivalence properties. Searching for an input that maximizes the violation of these equivalence constraints has the advantage that we get information about the extent to which the corresponding NNs are not equivalent. With an encoding as a satisfiability problem, we would only get a single and possibly very small violation. In general, we encode equivalence of two NNs $\mathcal{R}$ and $\mathcal{T}$ as the following mixed integer linear program

$$\max \quad d \tag{2}$$
$$s.t. \; \mathbf{r} = \mathrm{enc}_{\mathcal{R}}(\mathbf{i}) \tag{3}$$
$$\mathbf{t} = \mathrm{enc}_{\mathcal{T}}(\mathbf{i}) \tag{4}$$
$$d = f(\mathbf{r}, \mathbf{t}) \tag{5}$$

where Eqs. 3 and 4 encode a reference NN $\mathcal{R}$ and the testing NN $\mathcal{T}$ on the common inputs $\mathbf{i}$ as described in Sect. 2.1, yielding the respective outputs of the NNs $\mathbf{r}$ and $\mathbf{t}$. As we are dealing with MILP, some of these variables are real numbers, while others are restricted to be integers. Below we are going to discuss the encoding of the function $f$, which calculates the scalar violation score $d$ for a given equivalence property, for top-$k$ and then for $\varepsilon$-equivalence.

**Top-k-Equivalence.** We can encode the violation score of the top-$k$-equivalence $\mathcal{R} \simeq \mathcal{T}$ (or $\mathcal{T}$ is equivalent to $\mathcal{R}$) as a simple difference

$$d = \hat{t}_k - t_j \; , \tag{6}$$

where $\arg\max_i r_i = j$. The variable $\hat{t}_k$ denotes the $k$-th largest component of $\mathbf{t}$. If $d = \hat{t}_k - t_j \leq 0$, then we have $\hat{t}_k \leq t_j$, meaning that the output of $\mathcal{T}$ corresponding to the classification result of $\mathcal{R}$ is larger or equal to the $k$-th largest output of $\mathcal{T}$. Therefore $t_j$ would be amongst the $k$ largest outputs of $\mathcal{T}$ and thus satisfy top-$k$-equivalence. The main difficulty in encoding top-$k$-equivalence lies in encoding the sorting of the outputs of the NNs according to their activation value. We can encode the calculation of the descendingly sorted vector $\hat{\mathbf{x}} = \Pi\mathbf{x}$ of a NN's output values $\mathbf{x}$ by using a permutation matrix $\Pi = (\pi_{ij})_{i,j=1}^{n}$ similar to [17] and then adding the necessary ordering constraints (Constraint 10):

$$\hat{x}_i = \sum_j \pi_{ij} x_j \; \forall i \leq n \tag{7}$$
$$\sum_i \pi_{ij} = 1 \; \forall j \leq n \tag{8}$$
$$\sum_j \pi_{ij} = 1 \; \forall i \leq n \tag{9}$$
$$\hat{x}_i \geq \hat{x}_{i+1} \; \forall i \leq n-1 \tag{10}$$
$$\pi_{ij} \in \{0,1\} \; \forall i,j \leq n \; , \tag{11}$$

where the Constraints 8 and 9 together with the binary constraint on the $\pi_{ij}$, ensuring that each column and each row only contain one 1 and only 0 elsewhere, are sufficient to characterize $\Pi$ as a permutation matrix. While multiplications of two variables are in general non-linear, we can utilize that the $\pi_{ij}$ are binary variables, to encode the products $\pi_{ij}x_j$ in the above formulation. Binary multiplications $\delta x = y$, where $\delta \in \{0, 1\}$, can be linearized by encoding the implications $\delta = 0 \rightarrow y = 0$ and $\delta = 1 \rightarrow y = x$ as linear inequalities.

Using the above information, we can retrieve sorted vectors $\hat{\mathbf{r}}, \hat{\mathbf{t}} \in \mathbb{R}^n$ of the outputs of two NNs. To find the component of $t_j$ of $\mathbf{t}$ that corresponds to the largest component of $\mathbf{r}$, one can apply the permutation matrix to calculate $\hat{\mathbf{r}}$ to $\mathbf{t}$ and extract its first component. However, we don't need to generate two full permutation matrices. Realizing that we are only interested in the largest value of $\mathbf{r}$ and the $k$ largest values of $\mathbf{t}$, it is sufficient to encode the first row of the permutation matrix for $\mathbf{r}$ and the first $k$ rows of the permutation matrix for $\mathbf{t}$, thus reducing the number of binary variables. When multiple outputs $r_i$ of NN $\mathcal{R}$ share the highest activation value, valid permutations could be obtained, such that in one of them component $r_j$ and in the other $r_{j'}$ is the top component in $\hat{\mathbf{r}}$. Assume that we compare a reference NN $\mathcal{R}$ to a testing NN $\mathcal{T}$, that assigns the highest activation only to $t_j$, when given the same input as $\mathcal{R}$. Then, we would use this input as a counterexample to their equivalence, since we maximize the violation of the equivalence property and the solver would chose the permutation of $\mathcal{R}$'s outputs, that assigned $r_{j'}$ as the top component. The classification results of $\mathcal{R}$ and $\mathcal{T}$ however could still be the same. Therefore, we require $\mathcal{R}$ to have a unique highest output activation. Since we are not allowed to use strict inequalities in MILP, we use an $\varepsilon > 0$ to ensure a unique greatest output activation. We then arrive at the final encoding of top-$k$-equivalence. First, we obtain $\mathcal{R}$'s unique top output $\hat{r}_1$:

$$\hat{r}_1 = \sum\nolimits_i \rho_i r_i \tag{12}$$

$$\hat{r}_1 \geq r_i \ \forall i \leq n \tag{13}$$

$$\rho_i = 0 \rightarrow \hat{r}_1 \geq r_i + \varepsilon \ \forall i \leq n \tag{14}$$

$$\sum\nolimits_i \rho_i = 1 \text{ and } \rho_i \in \{0, 1\} \ \forall i \leq n \ , \tag{15}$$

where $\rho = (\rho_1, ..., \rho_n)^T$ is used just as the first row of a permutation matrix. Then, we can solve for $\mathcal{T}$'s activation $t_r$ for the component of $\mathcal{R}$'s largest output, by applying $\rho$ to the output of $\mathcal{T}$.

$$t_r = \sum\nolimits_i \rho_i t_i \tag{16}$$

The $k$ greatest outputs of $\mathcal{T}$ are computed as follows:

$$\begin{aligned} \hat{t}_i &= \sum\nolimits_j \pi_{ij} t_j \ \forall i \leq k & \hat{t}_i &\geq \hat{t}_{i+1} \ \forall i \leq k-1 \\ z_j &= \sum\nolimits_i \pi_{ij} \leq 1 \ \forall j \leq n & z_j &= 0 \rightarrow t_j \leq \hat{t}_k \ \forall j \leq n \\ \sum\nolimits_j \pi_{ij} &= 1 \ \forall i \leq k & \pi_{ij} &\in \{0,1\} \forall i \leq k, j \leq n \ , \end{aligned}$$

where the $(\pi_{ij})_{i,j=(1,1)}^{(k,n)}$ form the first $k$ rows of the permutation matrix for $\mathcal{T}$'s outputs and $z_i$ indicates, whether $t_i$ is amongst $\mathcal{T}$'s $k$ largest outputs. Finally, we can compute the violation of the top-$k$-equivalence property as the difference

$$d = \hat{t}_k - t_r \ , \tag{17}$$

which is then maximised to find the counterexample resulting in the largest possible violation of the equivalence property.

**Interval Arithmetic.** We assume that lower and upper bounds are given for the input variables and use existing interval extensions for the sum and multiplication to generate bounds on the linear combinations of inputs. The ReLU function is then applied to these bounds to generate bounds on the output of the neuron. This process is repeated throughout the network. Naively applying this kind of interval arithmetic to the equations defining $\hat{r}_1, t_r$ and the $\hat{t}_i$ respectively would produce large overestimates. In Eq. 12 for example, the upper bound on $\hat{r}_1$ would be the sum instead of the maximum of the upper bounds of the $r_i$ (only one entry is equal to 1 in a row of the permutation matrix). Therefore, we use context groups to compute tighter bounds for these variables. Assume, we are choosing a variable $x$ from a set $X = \{x_1, ..., x_n\}$, where $x_i \in [l_i, u_i]$. Let $\hat{\mathbf{l}}$ and $\hat{\mathbf{u}}$ denote the vectors containing the lower, respectively upper bounds sorted in decreasing order. If we choose $x$ to be the $k$-th largest variable out of $X$, we can combine $x$ in a *top-$k$-group* and assign tighter lower and upper bounds for $x$ according to: $x \in [\hat{l}_k, \hat{u}_k]$.

**$\varepsilon$-Equivalence.** We encode $\varepsilon$-equivalence and exact equivalence as maximizing

$$d = \|\mathbf{r} - \mathbf{t}\|. \tag{18}$$

The equivalence property is satisfied, if $\max d \leq \varepsilon$ for $\varepsilon$-equivalence. The value of $\varepsilon$ has be chosen according to the dataset. The "optimal" value can be determined by incrementally looking at counterexamples for the equivalence and deciding if, from the user-perspective, the outputs are equivalent. For exact equivalence $\varepsilon = 0$ is required. In order to use Eq. (18) in MILP, we need to encode the non-linear $\|\cdot\|$ operator. We restricted ourselves to the Manhattan $\|\cdot\|_1$ and the Chebyshev norm $\|\cdot\|_\infty$ defined as

$$\text{Manhattan: } \|\mathbf{x}\|_1 = \sum\nolimits_i |x_i|, \quad \text{Chebyshev: } \|\mathbf{x}\|_\infty = \max_i |x_i| \ , \tag{19}$$

because they are piecewise linear functions and can thus be encoded in MILP.

Just as we have done earlier, $y = |x|$ can also be expressed as cases $x \leq 0 \wedge y = -x$ and $x \geq 0 \wedge y = x$, that can be encoded as linear inequalities by introducing a binary variable. If the bounds $l_x \leq x \leq u_x$ indicate, that the domain of $x$ contains only positive ($l_x \geq 0$) or only negative ($u_x \leq 0$) values, we can just set $y = x$ or $y = -x$, respectively.

In case of the Manhattan norm, we just sum over the absolute values of the components. The maximum operator used in the Chebyshev norm can be represented in the same way, as we have done to obtain the output with the highest activation for a NN in Eqs. (12)–(15) in the previous section.

However a unique largest value is not required in this case, so Eq. (14) is not needed in this encoding. We can also use the *top-k-group* we introduced in Sect. refsssec:ia above with a value of $k = 1$ to allow for the calculation of tighter bounds on the result of this maximum operator.

## 4 Input Restriction

As mentioned in Sect. 3, exact equivalence can also be relaxed by restricting the input domain, for which the equivalence property has to hold. In practice, it is especially useful to restrict the input domain to values, that are covered by the training dataset of the respective NNs. Differences in the output of NNs in the neighborhood of their training samples are far more meaningful than differences in regions, where they would not have been applied anyway. Furthermore, restricting the input values allows for the calculation of tighter bounds.

Below, we give a quick overview of the hierarchical clustering approach of [12], we used for restricting the input space to regions within a radius around cluster-centers of training data. Subsequently, we show MILP encodings for proving equivalence of two NNs for the restricted input regions. We also show, how this process can be modified for maximizing the radius around a point, such that the violation of a chosen equivalence property is smaller than a specified threshold.

### 4.1 Hierarchical Clustering

The hierarchical clustering method of [12] starts by clustering a set of labelled data-points $\{(x_i, y_i)\}_{i=1}^{n}$, with $k$ distinct labels into $k$ clusters. If a cluster contains input points of different labels, the method is recursively applied to that cluster, until all clusters only contain inputs of a common label. Every cluster is then characterized by its cluster center and its radius, which denotes the maximum distance from the cluster center to its input points. The underlying assumption for this clustering is, that all points, not just the training data-points, in a dense cluster should be assigned the same label. As points close to a cluster boundary might lie on a real decision boundary between two classes, [12] set the radius $r_c$ of a cluster to the average distance of the input-points to the cluster center. Note, that the above assumption only holds for clusters of high density $n/r_c$, where $n$ is the number of training data-points in the respective cluster.

### 4.2 Encoding of Clusters

As each cluster is characterized by its center $\mathbf{c}$ and radius $r_c$, one can place a norm restriction $\|\mathbf{i} - \mathbf{c}\| \leq r_c$ on the vector of inputs $\mathbf{i}$, to reduce the domain of the verification procedure to only inputs from that cluster. Thus, we can encode the input restriction by extending the encoding of NN equivalence given in Sect. 3.2 through adding the norm restriction to Eqs. (3)–(5).

Again, we restrict ourselves to encoding the Manhattan and Chebyshev norms (Eq. 19). The encoding of the Chebyshev norm for input restriction is less complicated than its encoding needed for $\varepsilon$-equivalence, as we do not need to actually calculate the value of the norm, but just ensure, that all input values are within a set distance of the cluster center. Which leads to a box constraint on the input variables $\mathbf{i}$. Therefore the lower and upper bounds $l_j$ and $u_j$ of variable $i_j$ can be updated to $l'_j = \max(c_j - r_c, l_j)$ and $u'_j = \min(c_j + r_c, u_j)$ respectively. The Manhattan norm $\|\cdot\|_1$ however has to be encoded just as in Sect. 3.2. Nonetheless, we can use the fact that $\|\mathbf{x}\|_1 \geq \|\mathbf{x}\|_\infty \; \forall \mathbf{x}$ , to achieve faster tightening of the variable bounds by adding the bounds calculated in the encoding of the Chebyshev norm.

### 4.3 Searching for a Maximal Radius

In order to find the largest radius around a center $\mathbf{c}$ in input-space, where NNs $\mathcal{R}$ and $\mathcal{T}$ are equivalent, it is not possible to just use the equivalence encoding adding the presented norm and set $r_c$ to be maximized. Since the solver finds an assignment to the input variables $\mathbf{i}$, such that the objective, in that case the radius, is maximized, the NNs are still equivalent and $\|\mathbf{i} - \mathbf{c}\| \leq r_c$. In that situation the solver could choose $\mathbf{i} = \mathbf{c}$. Therefore, the equivalence constraint would be met, if the NNs are equivalent on the center, and the maximum of the radius would be arbitrarily large. Hence, we search for the smallest radius $r_v$, for which a counterexample to the equivalence of $\mathcal{T}$ and $\mathcal{R}$ can be found. This optimization problem is similar to the one proposed in [25] for finding adversarial examples for a single NN close to training inputs, which they approximately solve using gradient based methods. Our MILP formulation reads:

$$\min \quad r_v \tag{20}$$

$$s.t. \; \mathbf{r} = \mathrm{enc}_{\mathcal{R}}(\mathbf{i}) \tag{21}$$

$$\mathbf{t} = \mathrm{enc}_{\mathcal{T}}(\mathbf{i}) \tag{22}$$

$$f(\mathbf{r}, \mathbf{t}) \geq \varepsilon_v \tag{23}$$

$$\|\mathbf{i} - \mathbf{c}\| \leq r_v \, . \tag{24}$$

Note that we used a small threshold value of $\varepsilon_v > 0$ for the violation.

If $r^*$ is the optimal solution for the above minimization problem, the two NNs are not equivalent for radii $r' \geq r*$, as the solver could generate a counterexample for $r^*$. But we cannot guarantee that the NNs are equivalent for $r' < r^*$ because of the use of the threshold value. For small values of $\varepsilon_v$, the NNs are likely to be equivalent for radii $r' \leq r^* - \varepsilon_r$ for small $\varepsilon_r$. This can then be verified using the methods for fixed radii described in Sect. 4.2. If verification tasks for fixed radii have been carried out beforehand, the largest (smallest) radius, for which the NNs were (not) equivalent can be used as a lower (upper) bound on $r_v$ in the radius-minimization problem.

## 5 Application: Neural Network Compression

The huge number of parameters in modern NNs lead to large amounts of memory – AlexNet [19], for example uses 200 MB of disk space. Hence, it is desirable to reduce the number of parameters of a NN, without compromising its performance on the task it is designed to solve. Our approach for verifying equivalence properties of NNs in combination with the presented input restrictions could be used to verify the equivalence, or at least quantify the similarity, of the original NN and the smaller NN, which is the result of the reduction in parameters. This reduction is usually done by pruning unimportant weights - setting their value to zero - essentially removing insignificant connections between neurons. In the context of *magnitude based pruning*, weights of small absolute value are considered negligible [24]. The NN may be retrained after pruning, to correct for the missing connections [24]. During this step and the following iterations of pruning and retraining, the weights of the pruned connections are fixed at zero. Another way to reduce the number of parameters, applicable only to classification tasks, is to directly train a smaller NN on the outputs of a well performing large NN, which is called *student-teacher training* [1,15].

## 6 Evaluation

We have implemented our approach in Python 3 and are able to automatically generate MILP encodings together with input restrictions. Our program is able to read in NNs exported by Keras [6] and uses version 8.1.1 of Gurobi [13] to solve the generated instances. We used this implementation to analyze the equivalence between compressed and original NNs, as well as between compressed NNs.

**Neural Networks.** Our original NN consists of an input layer, hidden layers of 32 and 16 ReLU units and an output layer of size 10 (denoted: 32-16-10). It was trained using the *Optical Recognition of Handwritten Digits Dataset* [7].

The dataset consists of $8 \times 8$ pixel labeled images of handwritten digits, giving us 64 input variables, whose values are in the closed interval $[0, 16]$, which can be used as naive bounds on the input variables. We implemented bounds tightening and interval arithmetic to increase scalability, yet the applicability for larger networks as well as further optimizations or combinations with approximated approaches are part of future work.

Reduced size NNs were obtained by pruning and retraining the original NN in 10% increments. Additionally, NNs with less ReLU units were learned using student-teacher training. All NNs were trained using the Keras machine learning framework [6]. The achieved accuracy values on the training and testing datasets are shown in Fig. 1 for the pruned NNs, as well as for different structures of student NNs. After the training process, we removed the softmax activation function in the output layer of the NNs to allow for their encoding in MILP.

**Experiments.** Verification tasks for top-$k$-equivalence were conducted with and without input restricted around the cluster centers shown in Fig. 2. These clusters were the five most dense clusters obtained by hierarchical clustering, when

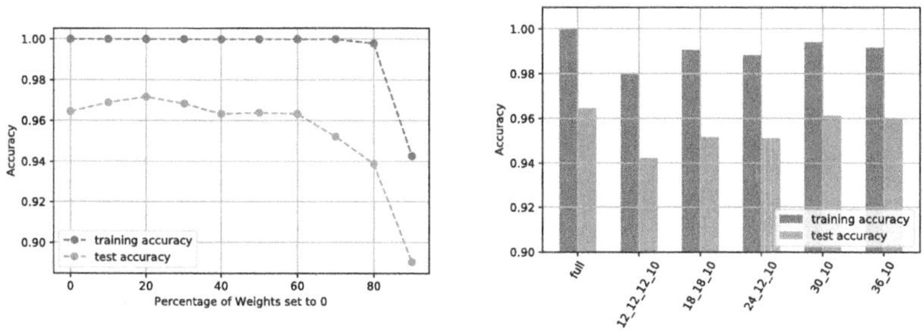


**Fig. 1.** Accuracy values for original NN, low magnitude weight pruning NNs (left) NNs trained by student-teacher training (right).

applied to the *Optical Recognition of Handwritten Digits* dataset using Manhattan distance. Each cluster contains between 66 and 91 training images.

Experiments were conducted for $k \in \{1, 2, 3\}$ without input restriction and for fixed radii, while the experiments for searching maximal equivalence radii were conducted for $k \in \{1, 2\}$. Due to space limitations, we present the experimental results for searching maximal radii. The tool and instructions how to produce the results can be found under https://github.com/phK3/NNEquivalence.

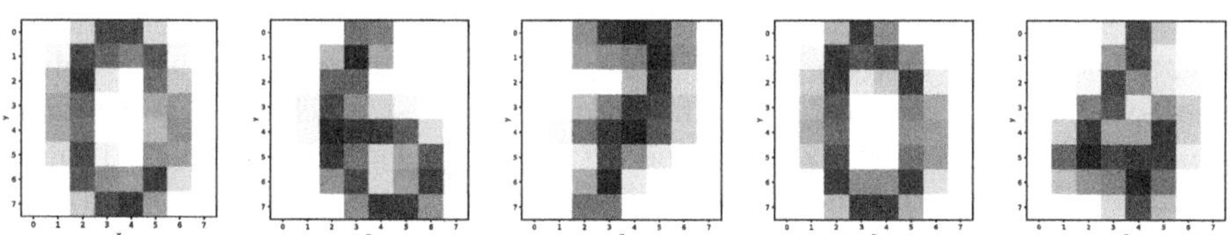

cluster (a) cluster (b) cluster (c) cluster (d) cluster (e)

**Fig. 2.** The cluster-centers of the five most dense clusters obtained from hierarchical clustering of the *Optical Recognition of Handwritten Digits* dataset.

Before the problem encoding was passed to Gurobi, bounds tightening was performed using interval arithmetic and optimization of two-layer subproblems for each linear combination of ReLU inputs. Each subproblem solution-process was stopped after a maximum of 20 s. All experiments were conducted on a computer with an Intel Core i5-3317U 1.70 GHz processor, which has 2 physical and 4 logical cores, and 8 GB of RAM running an x64 version of Windows 10.

### 6.1 Equivalent Neural Networks

We want to verify the equivalence of compressed NNs, by calculating the maximal equivalence radii for the chosen input clusters. Figure 3 shows the development of maximal radii for top-1 equivalence for both compression methods. The individual radius depends on training data and is reflected by the total number for the maximal radius for all reduction methods.

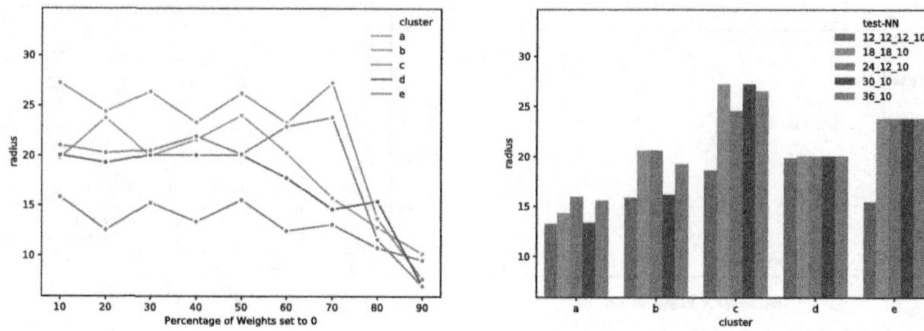


**Fig. 3.** Maximal equivalence radii for top-1-equivalence to the full NN for weight pruned NNs (left) and student-teacher trained NNs (right)

When pruning a larger percentage of parameters, the equivalence radius fluctuates around a constant level for each cluster up until the 50% reduced NN. If too many parameters were set to zero, the pruned NNs are no longer equivalent to the original NN and the equivalence radius deteriorates, as can be seen for the 80% and 90% pruned NNs. For the 60% and 70% pruned NNs however, one notices, that the equivalence radii for clusters $b, d$ drop as expected. Radii for the clusters $a$, $c$ and $e$, on the other hand, either stay on the same level or even increase. An explanation for the observed behavior could be, that about 50% of the original NN's parameters are sufficient to capture the underlying knowledge in the data for the tested clusters. If more parameters are pruned, the reduced NNs focus on the obviously classifiable clusters to still achieve a low training error. When the NNs are pruned even further, their capacity is clearly too low.

Examining the student-teacher trained NNs, we notice, that the equivalence radii not only depend on the number of ReLU nodes, but also the structure of the NNs. While the 12-12-12-10 student has more ReLU units than the 30-10 student and the same number as all other student NNs, it exhibits sometimes significantly smaller equivalence radii on all clusters. Among the student NNs, it also exhibited the lowest accuracy on the training and testing datasets, indicating that 12 neurons per layer are not best suited for this classification task. The 18-18-10 student and the 36-10 student show however, that good accuracy and large equivalence radii can be obtained for this number of ReLU nodes.

Comparing the different compression algorithms for top-1-equivalence, we notice, that most student NNs achieve similar radii as the up to 50% pruned NNs on clusters $a, b, c$ and $d$ and radii as large as that of the 70% pruned NN on cluster $e$. For the 12-12-12-10 student on clusters $b$ and $c$ and additionally the 30-10 student on cluster $b$, significantly smaller radii indicate a lack of capacity for the student NNs, although this effect is less severe than for the 70%, respectively 80% pruned NNs.

Figure 4 represents the same data for the top-2 equivalence. The verification of top-2 equivalence is harder, thus our approach only returns upper (dotted lines) and lower (normal lines) for the given timeout. In general, the maximal equivalence radii are, as expected, larger then for the top-1 equivalence. This indicates that the NNs still assign large probabilities to the correct classification result for the cluster regions. It is also possible to observe, that for example the 12-12-12-10 student network does not lag as far behind the other student NNs as before, indicating, that it at least captured a rough understanding of the data.

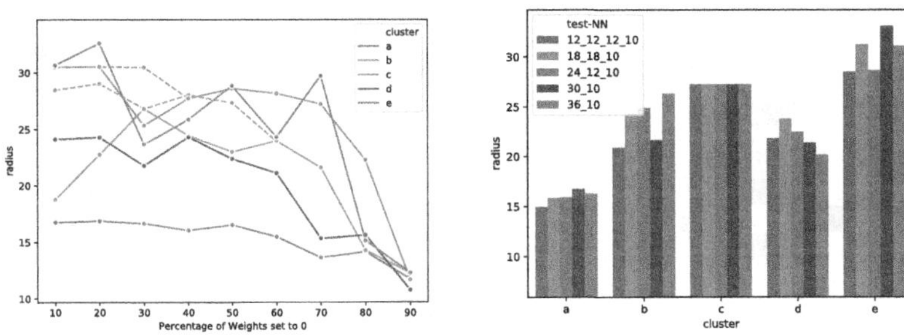


**Fig. 4.** Maximal equivalence radii for top-2-equivalence to the full NN for weight pruned NNs (left) and student-teacher trained NNs (right)

## 6.2 Remarks

The runtime of the verification procedure depends on the complexity of the MILP encoding, where the number of integer variables seemed to have the largest effect. For our experiments runtime fluctuated between a minute and 30 min for top-1 equivalence. For top-2 equivalence, we set a timeout of 3 h. Verification of equivalence for student-teacher trained NNs with fewer ReLU nodes was in most cases faster, than for the pruned NNs, as fewer integer variables had to be introduced. Only considering pruned NNs, sparser NNs proved to be verified faster than their less sparse counterparts. Overall, verification of equivalence for small fixed radii is faster, than for larger radii, as tighter bounds for all variables in the encoding can be obtained via bounds tightening. For very large radii, however, some NNs seem to be obviously not equivalent and large counterexamples are quickly found by the solver. In the extreme case without input restrictions, counterexamples to equivalence were all found within a minute.

The presented approach is able to search for a maximal radius for which NNs are equivalent and returns an input at the edge of the radius for which the networks are not. We denote this input as a counterexample, which can be analyzed by a potential user. He then has to decide, whether the counterexample should be classified as an valid input. If it is valid, the maximal radius is too small and the NNs are not equivalent, otherwise the NNs are equivalent w.r.t. the cluster. Three kinds of counterexample are shown in Fig. 5. The leftmost picture shows an input picture for unrestricted input. This kind of counterexample is negligible in practice and should not be seen as valid input. It demonstrates the necessity for input restrictions when verifying NNs. The counterexample in the middle shows a picture of a (in our opinion) zero which is misclassified by the 20% pruned NN. Such a result could indicate that the pruned NN does not fit the wanted equivalence criterion. The left picture on the other hand, shows a digit that is misclassified by the original NN, which could indicate that the original NN should be retrained with the given counterexample.

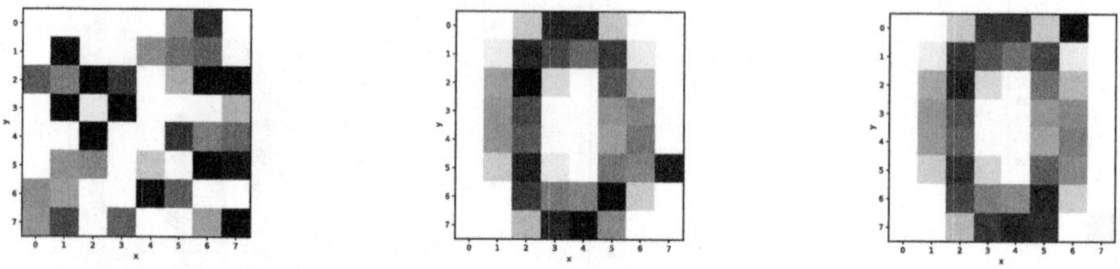

full NN: 5
50% pruned NN: 6

full NN: 0
20% pruned NN: 6

full NN: 5
50% pruned NN: 0

**Fig. 5.** Counterexamples to one hot equivalence: Without input restriction (left), input within a 13.6 (middle) and 27.2 (right) radius in Manhattan norm around the center of cluster (a).

## 7 Conclusion

With top-$k$-equivalence, we presented a novel relaxed equivalence property for NNs and showed, how it, as well as pre-existing notions of equivalence can be encoded in MILP for NNs using the ReLU activation function. Despite the relaxation, NNs rarely meet these equivalence properties, when the whole input space is considered, as their training only encourages them to agree on areas close to training data. Therefore we used the restriction of inputs to regions around clusters of training data, as proposed in [12]. We then developed MILP formulations, of equivalence for inputs within a fixed radius around obtained cluster-centers, as well as maximizing that radius, such that the NNs are still equivalent. Experiments with a NN trained on the *Optical Recognition of Handwritten Digits Dataset* [7] and its downsized counterparts obtained by student-teacher training or weight pruning showed the validity of our approach. As compression algorithms for NNs are typically only evaluated empirically by measuring the performance of the resultant NNs on a test dataset, this marks the first verification based examination of such methods. The notion of verified equivalence in a given cluster radius can be used to give guarantees for smaller networks. Furthermore, it can be utilized for finding meaningful counterexamples for the pruned and original network which can than be used for further training.

Our approach could also be applied, when numerous verification tasks have to be carried out for a large NN. In this case, a smaller NN could be obtained by compression algorithms. We could then prove its equivalence to the large NN within the input space of interest and subsequently perform the initial verification tasks on the smaller NN, requiring less computation time. For this scenario further improvements in scalability are needed. A first step could be using dedicated solvers for piecewise linear NNs like *Reluplex* [18] or the assistance of approximate methods [8].

## References


1. Ba, J., Caruana, R.: Do deep nets really need to be deep? In: Ghahramani, Z., Welling, M., Cortes, C., Lawrence, N.D., Weinberger, K.Q. (eds.) Advances in Neural Information Processing Systems, vol. 27, pp. 2654–2662. Curran Associates, Inc. (2014). http://papers.nips.cc/paper/5484-do-deep-nets-really-need-to-be-deep.pdf
2. Bahdanau, D., Cho, K., Bengio, Y.: Neural machine translation by jointly learning to align and translate (2014)
3. Bojarski, M., et al.: End to end learning for self-driving cars (2016)
4. Bunel, R., Turkaslan, I., Torr, P.H.S., Kohli, P., Kumar, M.P.: A unified view of piecewise linear neural network verification (2017)
5. Cheng, C.H., Nührenberg, G., Ruess, H.: Maximum resilience of artificial neural networks. In: D'Souza, D., Narayan Kumar, K. (eds.) Automated Technology for Verification and Analysis ATVA 2017. LNCS, vol. 10482, pp. 251–268. Springer, Cham (2017). https://doi.org/10.1007/978-3-319-68167-2_18
6. Chollet, F., et al.: Keras (2015). https://keras.io
7. Dua, D., Graff, C.: UCI machine learning repository (2017). http://archive.ics.uci.edu/ml
8. Dvijotham, K., Stanforth, R., Gowal, S., Mann, T.A., Kohli, P.: A dual approach to scalable verification of deep networks. In: UAI (2018)
9. Ehlers, R.: Formal verification of piece-wise linear feed-forward neural networks. In: D'Souza, D., Narayan Kumar, K. (eds.) Automated Technology for Verification and Analysis, pp. 269–286. Springer International Publishing, Cham (2017). https://doi.org/10.1007/978-3-319-68167-2_19
10. Fischetti, M., Jo, J.: Deep neural networks and mixed integer linear optimization. Constraints **23**(3), 296–309 (2018). https://doi.org/10.1007/s10601-018-9285-6
11. Goodfellow, I., Bengio, Y., Courville, A.: Deep Learning. MIT Press, Cambridge (2016). http://www.deeplearningbook.org
12. Gopinath, D., Katz, G., Pasareanu, C.S., Barrett, C.: DeepSafe: a data-driven approach for checking adversarial robustness in neural networks (2017)
13. Gurobi Optimization LLC: Gurobi. https://www.gurobi.com/
14. Han, S., Mao, H., Dally, W.J.: Deep compression: compressing deep neural networks with pruning, trained quantization and Huffman coding (2015)
15. Hinton, G., Vinyals, O., Dean, J.: Distilling the knowledge in a neural network. In: NIPS Deep Learning and Representation Learning Workshop (2015). http://arxiv.org/abs/1503.02531
16. Huang, X., Kwiatkowska, M., Wang, S., Wu, M.: Safety verification of deep neural networks. In: Majumdar, R., Kunčak, V. (eds.) Computer Aided Verification, pp. 3–29. Springer International Publishing, Cham (2017). https://doi.org/10.1007/978-3-319-63387-9_1
17. Justice, D., Hero, A.: A binary linear programming formulation of the graph edit distance. IEEE Trans. Pattern Anal. Mach. Intell. **28**(8), 1200–1214 (2006)

18. Katz, G., Barrett, C., Dill, D.L., Julian, K., Kochenderfer, M.J.: Reluplex: an efficient SMT solver for verifying deep neural networks. In: Majumdar, R., Kunčak, V. (eds.) CAV 2017. LNCS, vol. 10426, pp. 97–117. Springer, Cham (2017). https://doi.org/10.1007/978-3-319-63387-9_5
19. Krizhevsky, A., Sutskever, I., Hinton, G.E.: ImageNet classification with deep convolutional neural networks. Commun. ACM **60**(6), 84–90 (2017). https://doi.org/10.1145/3065386
20. Kumar, A., Serra, T., Ramalingam, S.: Equivalent and approximate transformations of deep neural networks (2019)
21. Narodytska, N., Kasiviswanathan, S., Ryzhyk, L., Sagiv, M., Walsh, T.: Verifying properties of binarized deep neural networks. In: Thirty-Second AAAI Conference on Artificial Intelligence (2018)
22. Pulina, L., Tacchella, A.: An abstraction-refinement approach to verification of artificial neural networks. In: Touili, T., Cook, B., Jackson, P. (eds.) Computer Aided Verification, pp. 243–257. Springer, Heidelberg (2010). https://doi.org/10.1007/978-3-642-14295-6_24
23. Silver, D., et al.: A general reinforcement learning algorithm that masters chess, shogi, and Go through self-play. Science 362(6419), 1140–1144 (2018). https://science.sciencemag.org/content/362/6419/1140
24. Ström, N.: Phoneme probability estimation with dynamic sparsely connected artificial neural networks. Free Speech J. **5**, 1–41 (1997)
25. Szegedy, C., et al.: Intriguing properties of neural networks (2013)
26. Tjeng, V., Xiao, K., Tedrake, R.: Evaluating robustness of neural networks with mixed integer programming (2017)
27. Yang, T.J., Chen, Y.H., Sze, V.: Designing energy-efficient convolutional neural networks using energy-aware pruning (2016)

# Learning the Satisfiability of Pseudo-Boolean Problem with Graph Neural Networks

Minghao Liu[1,3], Fan Zhang[2,3], Pei Huang[1,3], Shuzi Niu[2,3], Feifei Ma[1,2,3](✉), and Jian Zhang[1,3]

[1] State Key Laboratory of Computer Science, Institute of Software, Chinese Academy of Sciences, Beijing, China
{liumh,maff}@ios.ac.cn
[2] Laboratory of Parallel Software and Computational Science, Institute of Software, Chinese Academy of Sciences, Beijing, China
[3] University of Chinese Academy of Sciences, Beijing, China

**Abstract.** Graph Neural Network (GNN) has shown great power on many practical tasks in the past few years. It is also considered to be a potential technique in bridging the gap between machine learning and symbolic reasoning. Experimental investigations have also shown that some $\mathcal{NP}$-Hard constraint satisfaction problems can be well learned by the GNN models. In this paper, a GNN-based classification model to learn the satisfiability of pseudo-Boolean (PB) problem is proposed. After constructing the bipartite graph representation, a two-phase message passing process is executed. Experiments on 0–1 knapsack and weighted independent set problems show that the model can effectively learn the features related to the problem distribution and satisfiability. As a result, competitive prediction accuracy has been achieved with some generalization to larger-scale problems. The studies indicate that GNN has great potential in solving constraint satisfaction problems with numerical coefficients.

**Keywords:** Pseudo-Boolean Problem · Graph Neural Network · Constraint satisfaction problem · Deep learning.

## 1 Introduction

Machine learning, especially deep learning, has shown great power over the past few years. Deep learning systems have dramatically improved the state-of-the-art standards of many tasks across domains, such as computer vision, natural language processing, speech recognition and drug discovery [15]. The tremendous success of deep learning has also inspired us to apply learning-based approaches to solve constraint satisfiability problems (CSPs). There is an interesting prospect that the end-to-end neural network models may have the ability to capture the specific structures of the problem from a certain distribution,


H. Simonis (Ed.): CP 2020, LNCS 12333, pp. 885–898, 2020.
https://doi.org/10.1007/978-3-030-58475-7_51

so that some effective implicit heuristics may be learned automatically. Some previous works have implemented different (deep) neural network architectures to represent such problems and try to learn full-stack solvers [23,26]. Thanks to the latest progress in graph representation learning, it is promising to apply graph neural networks to solve CSPs, because many of them have topology characteristics, which can be naturally represented as graphs. There have been recent efforts trying to set up end-to-end GNN models to solve the Boolean Satisfiability Problem (SAT), one of the most fundamental problems of computer science [22], and the Traveling Salesman Problem (TSP), an important $\mathcal{NP}$-Hard problem [20]. Experiments suggest that the GNN-based models can effectively learn the structural features of these problems, thereby obtain pretty high accuracy in predicting the satisfiability. Although such models are not yet comparable with the traditional search-based solvers, it may also be useful to combine them with traditional approaches to improve the solving performance. For instance, the GNN-based classifier for the SAT problem *NeuroSAT* has been further explored and observed to have some ability in finding unsatisfiable cores, which has been used to improve the efficiency of modern SAT solvers such as *Glucose* and *Z3* [21]. We believe that, the GNN-based models also have great potential in improving the efficiency of solving CSPs with numerical constraints.

As a natural and efficient mathematical programming model, pseudo-Boolean (PB) problem, which is also known as 0–1 integer programming, has received much attention in a large number of real-world applications for a long time. Nowadays, techniques related to PB problem are crucial in many fields, such as networked data mining, planning, scheduling, transportation, management and engineering [8,11]. However, research in complexity theory indicates even solving the satisfiability of PB constraints is an $\mathcal{NP}$-complete problem [12], which means it is hard to find sufficiently effective algorithms on various distributions. The mainstream approaches for PB problem benefit from the development of both Boolean Satisfiability (SAT) and Integer Programming solving techniques [6,25]. It can be found that fundamentally they are based on the backtracking search frameworks, and largely rely on the heuristic strategies designed by experts in specific domains. It is highly probable that PB solving techniques would also benefit from heuristics learned automatically. For the aforementioned reasons, PB problem provides an important setting for investigating the capability of GNN in extracting features of numerical constraints.

In this paper, we investigate the question that whether an end-to-end GNN model can be trained to solve the satisfiability of pseudo-Boolean problem as a classification task. Other than the previous works, our setting is more universally applicable, because various $\mathcal{NP}$-Hard CSPs can be easily formulated as PB constraints. At first, we adopt some normalization rules to transform different kinds of constraints into normalized form in order to build a weighted bipartite graph, which can be accepted as the input of GNN. The constructed model is basically a concrete implementation of the Message Passing Neural Network [9]. The revised message passing process is applied to update the node embedding vectors involving edge weights iteratively. Finally, the classification result is produced after a readout phase. Experiments on two well-known $\mathcal{NP}$-Hard CSPs, 0–1 knapsack and weighted independent set, demonstrate that our model can well converge on different PB problems under some certain distributions, and predict the satisfiability with high accuracy. The trained model can also be extended to predict the problem instances of larger size than those appeared in training set, which indicates that it has some generalization capability. We can summarize our contribution in the following two points:

- We propose a GNN-based model to solve the decision pseudo-Boolean problem. To the best of our knowledge, it is the first end-to-end GNN model for predicting the satisfiability of a general representation of CSPs with numerical constraints. It has been experimentally confirmed to achieve high accuracy on different benchmarks.
- Compared with another GNN model *NeuroSAT* that intends to solve the SAT problem, our model can be seen as a generalization of it. This is because our model not only achieves the same performance on the SAT problem, but also has ability to solve more general PB problems.

The remainder of this paper proceeds by the following parts. Section 2 summarizes some related work. The problem definition and model architecture are introduced in Sect. 3. The data generation method and training configuration are detailed in Sect. 4, with some analysis about the experimental results. Finally, conclusion and future work are discussed in Sect. 5.

## 2 Related Work

Although the mainstream algorithms for constraint satisfaction and combinatorial optimization problems are based on reasoning and searching algorithms from symbolism, there have always been attempts trying to solve these problems through machine learning techniques. A class of research is aiming to learn effective heuristics under the backtracking search framework and has achieved progress such as [3,14,17]. Here, we concentrate on another route: building end-to-end neural network models which directly learn CSP solvers. The earliest research work can be traced back to the Hopfield network [10], which has made progress in solving TSP. It is guaranteed to converge to a local minimum, but the computational power is quite limited. Recently, a series of efforts have attempted to train deep learning models with different representations. [26] proposes *CSP-cNN*, a convolutional neural network model, to solve binary CSPs in matrix form. However, the majority of CSPs are non-binary. [23] introduces a sequence-to-sequence neural network called *Ptr-Net*. [4,18] further review and extend the model. However, the structural information of CSPs may be partially lost when represented as sequential input. *Structure2vec* [13] adds graph embedding technique to the model, while it is still a sequential model which can only output greedy-like solutions. In summary, it is necessary to find more powerful representation models to solve various CSPs.

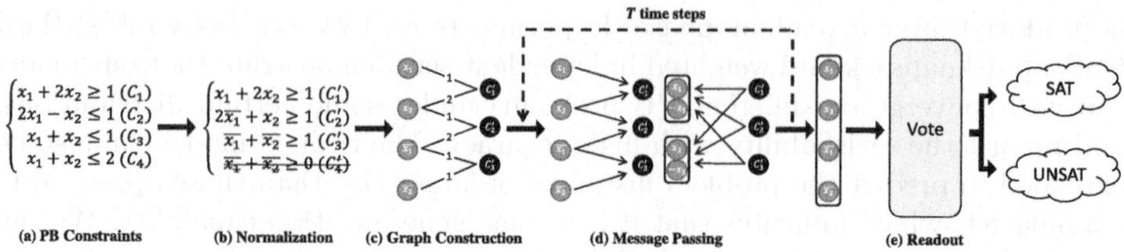


**Fig. 1.** The model for solving decision pseudo-Boolean problem based on graph neural network. The pipeline is as follows: (a) A set of constraints is given as input in which all variables can be assigned True or False, then (b) some normalization rules are applied to transform the constraints into normalized form, which is done by equivalent transformation and removing tautologies. From the normalized constraints we can (c) construct a weighted bipartite graph to represent the topological relationship between variables and constraints. Each node in the graph is represented as an embedding vector, which is updated iteratively through (d) the message passing mechanism. Finally, the model (e) outputs prediction about the satisfiability.

Graph neural network has been considered as a promising deep learning technique operating on graphs. Because of its powerful characterization capability for non-Euclidean structure data, the GNN models have made ground-breaking performance on many difficult tasks such as text classification, relation extraction, object detection, semantic segmentation and knowledge graph mining in recent years [27]. We believe graph is a suitable representation for CSPs. A basic fact is that many of the famous $\mathcal{NP}$-Hard problems are directly defined on graphs such as TSP, Maximum Clique and Minimum Vertex Cover. There have been some works trying to set up GNN-based models to solve SAT [1,22], [2,5], Graph Coloring Problem [16] and TSP [20]. Our model is distinct from the previous works in many ways. First, PB problem has the capability to conveniently formulate a larger number of CSPs compared with SAT, GCP and TSP. Although SAT is also a well-known meta-problem, it is not intuitive to translate the problem with numerical constraints into conjunction normal form. Second, PB constraints involve integer coefficients and constant items, while TSP has only one kind of numerical information (edge weights) that needs to be processed. Third, [20] introduces edge embeddings to represent the information about edge weights, which may lead to larger parameter space. Actually it takes around 2,000 epochs to converge on the dataset of graphs while $n \sim \mathcal{U}(20, 40)$ as reported. However, the edge weights are considered in the updating process of node embeddings in our model, therefore the required parameter space and training epochs are relatively reduced.

## 3 Model Architecture

In this section, we first give a brief introduction to the pseudo-Boolean problem. Next, components of the model are illustrated in detail, including the graph construction, the message passing and the readout phase. The whole pipeline of the model is illustrated in Fig. 1.

### 3.1 Pseudo-Boolean Problem

Pseudo-Boolean problem is one of the most fundamental problems in mathematical programming. Generally, a PB problem should contain a set of pseudo-Boolean constraints. According to whether there is an objective function to optimize, it may vary slightly in definition, which can be divided into the decision problem and the optimization problem. In this paper, we focus on the decision version, in which all the variables are restricted to True or False, and all the constraints must be satisfied, with no optimization target given. Here is the formal definition: Given $m$ constraints $C_1, C_2, \ldots, C_m$ in the form of $C_i$: $\sum_{j=1}^{n} c_{ij}x_j \geq b_i$ where $c_{ij}, b_i \in \mathbb{Z}$, decide if a set of assignment to $x_i \in \{0,1\}$ $(1 \leq i \leq n)$ exists so that all the constraints are satisfied.

The decision PB problem is worthy of investigation for many reasons. Firstly, the bounded integer variables in linear programming can always be expressed as a combination of 0–1 variables by [24]. Moreover, the decision problem can be extended to the optimization version through simple binary search. In terms of complexity, actually it has been proven to be a famous $\mathcal{NP}$-Complete problem.

### 3.2 Graph Construction

There have been a series of works trying to apply GNN models to solve constraint satisfaction and combinatorial optimization problems. A critical step is to set up appropriate graph structures for them. [22] characterizes SAT formulas through undirected bipartite graphs, where each literal and clause corresponds to a node and each affiliation relation between literal and clause corresponds to an edge. This kind of representation is quite intuitive and reasonable with at least two justifications. First, it holds the invariance of problem in permutation and negation through graph isomorphism. Second, it imitates the order of reasoning in traditional solving techniques through the message passing process.

For PB problem, we would like to propose a bipartite graph structure which is similar to the previous work. However, the problem structure of PB is quite different from that of SAT in many ways. For instance, the coefficients of PB constraint are arbitrary integers, which cannot be represented in unweighted bipartite graph like *NeuroSAT* does. Besides, non-zero constant terms $b_i$ may exist in PB problem. A question is how to deal with these constant terms in graph construction. Considering the differences above, we propose a two-stage graph construction method.

**Constraint Normalization.** Given a set of PB constraints, where each constraint $C_i$: $\sum_{j=1}^{n} c_{ij}x_j \geq b_i$ where $c_{ij}, b_i \in \mathbb{Z}$, the goal is to transform them into a canonical expression. To achieve this, for a constraint $C_i$ we first replace $c_{ij}x_i$ with $c_{ij}(1-\overline{x_i})$ when $c_{ij} < 0$, and move the constant item to the right side. We denote the resulting constraint $C_i''$: $\sum_{j=1}^{n} |c_{ij}| l_{x_j} \geq b_i'$, where $l_{x_j} \in \{x_j, \overline{x_j}\}$.

Later we check if the constant item $b'_i > 0$, otherwise the constraint can be removed without changing the satisfiability. Finally we divide both sides of the constraint by $b'_i$. After the process, all constraints are in normalized form

$$C'_i : \sum_{j=1}^{n} c'_{ij} l_{x_j} \geq 1 \tag{1}$$

where $c'_{ij} \in \mathbb{R}^+$ and $l_{x_j} \in \{x_j, \overline{x_j}\}$.

**Graph Generation.** After the normalization stage above, every constraint $C_i$ $(1 \leq i \leq m)$ in a PB problem is turned into normalized form $C'_i$ or removed from the problem. For each variable $x_j$ $(1 \leq j \leq n)$, we set up a pair of nodes to represent $x_j$ and $\overline{x_j}$ respectively, so we have a total of $2n$ nodes for variables. Then we set up $m$ nodes for constraints. Suppose a constraint $C'_i$ contains $l_{x_j}$ with the coefficient $c'_{ij}$, an edge connecting the nodes for $C'_i$ and $l_{x_j}$ is set up if $c'_{ij} \geq 0$ with the weight of $c'_{ij}$.

### 3.3 Message Passing

Now a PB problem has been represented as a weighted bipartite graph. The next step is to construct a learning model. We consider the Message Passing Neural Network (MPNN) [9] as the framework. Recently, MPNN has shown its effectiveness in solving some famous combinatorial problems, such as SAT [22] and TSP [20]. In this paper, we propose an MPNN-based model which is fit for PB problem. The forward pass of the model has two phases, a message passing phase and a readout phase.

**Message Passing Phase.** In the beginning, we parameterize each node in the graph randomly as a $d$-dimensional vector $E_{init} \sim \mathcal{U}(0,1)$ which represents the hidden state. We note the initial hidden states of variables and constraints as $V^{(0)}$ and $C^{(0)}$ respectively. Let $M$ be the adjacency matrix of the bipartite graph defined in the graph generation paragraph above. $S$ is a transformed representation of the relevant nodes which aims to keep the consistency of node embeddings representing the same variable. In this phase, the message passing process runs for $T$ time steps and the hidden states are updated iteratively according to:

$$\begin{aligned} C^{(t+1)}, C_h^{(t+1)} &= \mathbf{C_u}(M\mathbf{V_{msg}}(V^{(t)}), C_h^{(t)}) \\ V^{(t+1)}, V_h^{(t+1)} &= \mathbf{V_u}(M^\top \mathbf{C_{msg}}(C^{(t)}), V_h^{(t)}, S^{(t)}) \end{aligned} \tag{2}$$

In the above rules, $\mathbf{C_{msg}}$ and $\mathbf{V_{msg}}$ are two message functions implemented with multilayer perceptrons (MLP). Besides, $\mathbf{C_u}$ and $\mathbf{V_u}$ are two updating functions implemented with LSTM networks, where all $C_h^{(t)}$ and $V_h^{(t)}$ are the hidden cell states of LSTM.

**Readout Phase.** After $T$ iterations of message passing, we apply a readout function $\mathbf{V_{vote}} : \mathbb{R}^d \rightarrow \mathbb{R}^1$ (implemented with MLP) to compute the score that every variable node votes for the satisfiability of the problem instance. Then we average all the scores to obtain the final result:

$$\hat{y} = mean(\mathbf{V_{vote}}(V^{(T)})) \tag{3}$$

The model is trained to minimize the sigmoid binary cross-entropy loss between $\hat{y}$ and the real satisfiability of the problem instance $\phi(P) \in \{0, 1\}$.

## 4 Experimental Evaluation

In order to evaluate the performance of our GNN-based model on PB problem, we prepare two different problems for the experiments: 0–1 knapsack problem (0–1KP) and weighted independent set problem (WIS). The above two problems have exceedingly different structures when expressed as PB formulas, whether in terms of the number or the average length of constraints. We generate random datasets for these problems, and train the model on them respectively.

### 4.1 Data Generation

There are three steps for generating the datasets:

Step 1. Original instances are randomly created following some certain distributions. The number of items in 0–1KP and the number of nodes in WIS are denoted as $n$. For each problem we uniformly generate several subsets as training sets, each of which contains 100 K instances, with different range of $n$. Then two datasets are generated in the same way as validation and testing sets, each of which has 10 K instances. Therefore the size ratio of training, validation and testing set is 10:1:1.
Step 2. Because the generated instances are all optimization problems, for each instance we first call *CPLEX* to obtain its optimal solution $\phi$, and then randomly choose an integer offset $\delta$ from $[-R/10, R/10]$ ($R$ is an integer representing the range of coefficients, and we set $R = 100$ in all experiments). Finally, let $\phi + \delta$ be a bound (i.e. the constant $V$ in 0–1KP and $W$ in WIS) to completely transform the original instance into a decision problem. It is easy to find that satisfiable and unsatisfiable instances should account for about half of each.
Step 3. The instances are formulated as PB constraints, normalized and turned into graphs through the rules described in Sect. 3. After that the datasets can be fed into the neural network as input for training.

**0–1 Knapsack Problem.** The 0–1 knapsack problem (KP) is well-known in combinatorial optimization field with a very simple structure: Given a group of items, each with a weight $w_i$ and a value $v_i$. There is also a backpack with a total weight limit $C$. The goal is to choose some items to put into the backpack, so that the sum of their values is maximized, and the sum of their weights is not exceeding the given limit of the knapsack. It is proved that the decision problem of 0–1KP is $\mathcal{NP}$-Complete, so there is no known polynomial-time solving algorithm. It can be naturally represented by the following PB formulas:

$$\begin{aligned} \sum_{i=1}^{n} w_i x_i &\leq C \\ \sum_{i=1}^{n} v_i x_i &\geq V \end{aligned} \tag{4}$$

The constant $V$ represents a bound of the optimal objective. Variable $x_i \in \{0, 1\}$ indicates if an item is put into the backpack.

There have been some research works on generating difficult benchmarks of 0–1KP. Pisinger [19] proposes some randomly generated instances with multiple distributions to demonstrate that the existing heuristics can not find good solutions on all kinds of instances. The four kinds of distributions are named as:

- Uncorrelated: $w_i$ and $v_i$ are randomly chosen in $[1, R]$.
- Weakly correlated: $w_i$ is randomly chosen in $[1, R]$, and $v_i$ in $[w_i - R/10, w_i + R/10]$ such that $v_i \geq 1$.
- Strongly correlated: $w_i$ is randomly chosen in $[1, R]$, and $v_i = w_i + R/10$.
- Subset sum: $w_i$ is randomly chosen in $[1, R]$, and $v_i = w_i$.

We generate the datasets of 0–1KP under the above distributions with equal probability. As for the capacity of knapsack $C$, it is also randomly selected in range $[1, \sum_{i=1}^{n} w_i]$ independently in each instance, in order to ensure sufficient data diversity.

**Weighted Independent Set Problem.** Given an undirected graph with a weight for each node, an independent set is defined as a set of nodes where any two of them are not connected with an edge. Then the maximum weighted independent set (MWIS) requires the total weights of the selected nodes to be maximized. We can model the decision problem of MWIS as PB formulas:

$$\begin{aligned} \sum_{i=1}^{n} w_i x_i &\geq W \\ x_i + x_j &\leq 1, \ \forall (i, j) \in E \end{aligned} \tag{5}$$

The constant $W$ is a bound of the optimal objective, and the variables $x_i \in \{0, 1\}$ indicates whether each node is selected into the independent set.

To generate the dataset, random graph instances are sampled from the Erdős–Rényi model $G(n, p)$ [7]. The model contains two parameters: the number of nodes $n$, and the existence probability $p$ of an edge between every pair of nodes. We set $p = 0.5$ while generating the dataset, which corresponds to the case where all possible graphs on $n$ nodes are chosen with equal probability.

### 4.2 Implementation and Training

In order to examine the model's capability to predict the satisfiability of PB problem, we implement the model in Python, and several experiments are designed to train the model and evaluate its performance on different datasets for different purposes. All the experiments are running on a personal computer with Intel Core i7-8700 CPU (3.20 GHz) and NVIDIA GeForce RTX 2080Ti GPU.

For reproducibility, we would like to give the setting of hyper-parameters as follows. In our configuration, The dimension of node embedding vectors $d = 128$, and the time step of message passing $T$ is set to 5 for 0–1KP and 50 for WIS. Our model learns from the training sets in batches, with each batch containing 12 K nodes. The learning rate is set to $2 \times 10^{-5}$.

### 4.3 Experimental Results

**Classification Accuracy.** As previously noted, the most direct and important target of the model is to classify the satisfiability of PB problems. So we examine it as the first step. There are 7 groups of experiments in total with different settings. For 0–1KP, we set up 3 groups of experiments with different sizes, where $n$ is uniformly distributed in range $[3, 10]$, $[11, 40]$ and $[41, 100]$ respectively. For WIS, we set up 4 groups of experiments with different sizes, where $n$ is uniformly distributed in range $[3, 10]$, $[11, 20]$, $[21, 30]$ and $[31, 40]$ respectively. The reason for lacking of the results in $n > 40$ is that for each instance the number of constraints is proportional to $n^2$, so the graph is too hard to be fully trained in no more than 1,000 epochs. The complete experimental configuration and classification accuracy on the validation and testing sets are shown in Table 1. The first part is the description of datasets, which includes the number of variables, the average number and length of constraints. The constraint length means the number of variables with coefficients other than 0 in a constraint. The number of epochs required for training to convergence is different from each other, and we list the actual spent epochs. Finally, the classification accuracy on the validation and testing sets are shown respectively.

**Table 1.** The experimental configuration and the classification accuracy results for training our model on different PB problems and scales.

| Problem | #Variables | | | PB Cons. | | Epochs | Accuracy | |
|---|---|---|---|---|---|---|---|---|
| | *Train* | *Valid* | *Test* | *#Cons.* | *Length* | | *Valid* | *Test* |
| 0–1KP | [3,10] | 10 | 10 | 2.0 | 6.5 | 400 | 86.6% | 86.1% |
| | [11,40] | 40 | 40 | 2.0 | 25.5 | 400 | 87.3% | 88.2% |
| | [41,100] | 100 | 100 | 2.0 | 70.5 | 600 | 79.3% | 79.5% |
| WIS | [3,10] | 10 | 10 | 11.3 | 2.4 | 400 | 97.7% | 97.9% |
| | [11,20] | 20 | 20 | 59.3 | 2.2 | 400 | 92.1% | 93.3% |
| | [21,30] | 30 | 30 | 159.2 | 2.1 | 400 | 89.5% | 88.9% |
| | [31,40] | 40 | 40 | 309.3 | 2.1 | 600 | 86.0% | 85.8% |

The experimental results show that our model can effectively accomplish the classification task on unknown problems with similar distributions. In each group of $n \leq 40$, the model can converge and obtain more than 85% accuracy both on the validation set and the testing set. When $n$ is expanded to 100 items in 0–1KP, the accuracy is still up to 79%. From the results, our model is believed to have learned some features related to the satisfiability of problems. Furthermore, it is worth noting that the generated PB instances under these problems have very different distributions, whether in terms of the average number or length of constraints. Therefore, it indicates that the model has wide availability to work on different PB problems without changing the structure. This confirms one of the main advantage of our model: general modeling capability.

**Comparison with Other Approaches.** As far as we know, the proposed model is the first end-to-end GNN model to solve the satisfiability of PB problem. However, because of the close relationship between PB and SAT, we are interested in the comparison between our model (denoted as *PB-GNN*) and *NeuroSAT*, a GNN-based model that predicts the satisfiability of SAT formulas. We train these two models on the same problems respectively, each for 200 epochs. To transform PB constraints into SAT clauses of equal satisfiability, we call *Minisat+*[1], a high-performance PB solver that achieves excellent ranks in the PB competition[2]. The accuracy results on the validation sets are shown in Table 2. Due to the inevitable introduction of new variables and clauses in the transformation process, the effect of *NeuroSAT* is limited by the growth of problem scales. For 0–1KP, the training accuracy on *NeuroSAT* with the transformed SAT formulas is significantly lower than that of the original problems on *PB-GNN*. And for WIS, when training on the original problems where $n \geq 10$, it is almost impossible to converge within 200 epochs. Besides, the accuracy results of *NeuroSAT* when $n = 40$ are not available, because each epoch takes more than 2 h, which makes the overall training time unacceptable. The results indicate that *PB-GNN* has achieved better performance on general PB problems with numerical constraints. It is worth mentioning that when dealing with SAT instances, we can easily transform SAT clauses into PB constraints. For example, a clause $l_1 \vee l_2 \vee \cdots \vee l_k$ is equivalent to the constraint $\sum_{i=1}^{k} l_i \geq 1$. In this case, the message passing process of *PB-GNN* is almost the same as that of *NeuroSAT*. To confirm this, we train the two models for 200 epochs on two datasets SR(3, 10) and SR(10, 40) respectively as defined in [22]. It can be found that the accuracy of *PB-GNN* and *NeuroSAT* is very close, which means the performance of our model is also comparable with the state-of-the-art model on SAT instances.

Another point of concern is the time spent for solving. The average time costs taken by *PB-GNN* and *CPLEX* to solve per instance from the testing sets are counted. Table 2 demonstrates the results in milliseconds. It can be seen that compared with the classic CSP solver *CPLEX*, our model only takes less than 10% of time to return the prediction of satisfiability. In addition to the lower computational amount of the model, another reason is that the problem instances can be input in batches, and the solving process is accelerated with

[1] http://minisat.se/MiniSat+.html.
[2] http://www.cril.univ-artois.fr/PB16/.

parallelization. This is not to suggest that our model beats the classic solver in time. After all *CPLEX* is able to work without any pre-training. As a matter of fact, the results inspire us a promising scenario of the model when a large number of similar instances need to be solved with high frequency.

**Table 2.** Comparison of the classification accuracy and the average solving time between our model (*PB-GNN*) and other approaches.

| Problem | PB-GNN | | | | NeuroSAT | | | CPLEX |
|---|---|---|---|---|---|---|---|---|
| | *#Var.* | *#Cons.* | *Acc.* | *Time* | *#Var.* | *#Clau.* | *Acc.* | *Time* |
| 0–1KP | 5 | 2.0 | **89.7%** | 0.0462 | 35.4 | 137.6 | 74.0% | 9.1191 |
| | 10 | 2.0 | **87.7%** | 0.0637 | 87.8 | 438.7 | 69.5% | 10.7869 |
| | 15 | 2.0 | **86.1%** | 0.0785 | 142.4 | 774.7 | 62.8% | 12.1947 |
| | 20 | 2.0 | **86.5%** | 0.0843 | 192.7 | 1084.5 | 63.7% | 13.7513 |
| | 40 | 2.0 | **85.2%** | 0.1252 | 418.1 | 2502.8 | — | 29.9405 |
| WIS | 5 | 6.0 | **99.0%** | 0.2747 | 26.7 | 108.4 | 66.3% | 15.1882 |
| | 10 | 23.5 | **96.4%** | 0.6946 | 65.5 | 345.1 | 51.0% | 15.4741 |
| | 15 | 53.5 | **95.7%** | 1.3177 | 101.0 | 585.4 | 51.1% | 17.2223 |
| | 20 | 95.9 | **92.5%** | 1.9140 | 136.3 | 836.2 | 51.4% | 19.7416 |
| | 40 | 390.9 | **82.1%** | 5.0624 | 275.9 | 1958.4 | — | 49.6992 |
| SR | [3,10] | 46.8 | **95.8%** | 0.6619 | [3,10] | 46.8 | 95.5% | 11.3243 |
| | [11,40] | 151.3 | 84.3% | 2.4096 | [11,40] | 151.3 | **85.0%** | 15.7829 |

**Generalization to Larger Scales.** We are also interested in whether the model trained on smaller-scale data can work on larger-scale data. We examine two models, one has been trained for 400 epochs on 0–1KP and the other for 600 epochs on WIS, both of which are on the datasets where $n \in [11, 40]$. For each model, we set up 15 groups of data with $n \in [10, 80]$ in steps of 5. Within each group, 5 testing sets are generated with different offset $\delta = \{\pm 1, \pm 2, \pm 5, \pm 8, \pm 10\}$ as described in Sect. 4.1. The other configuration of data generation is the same as that of validation sets. Such testing data can show more clearly the learning effect and generalization ability of our model on the instances of varying complexity.

Figure 2 shows the change in accuracy when testing our model on the above datasets. On one hand, it is easy to find that the generalization ability is related to the problem structure. For 0–1KP, since the structures of different scales are similar, the accuracy can be maintained relatively well even if the problem instances become larger. But for WIS, a larger-scale instance leads to a more complicated graph. The accuracy drops rapidly outside the training set, because the features learned in low dimensions are more likely to lose effectiveness. On the other hand, the model also achieves different accuracy on the instances generated under different offsets. When $\delta = \pm 10$, the accuracy greater than 80%

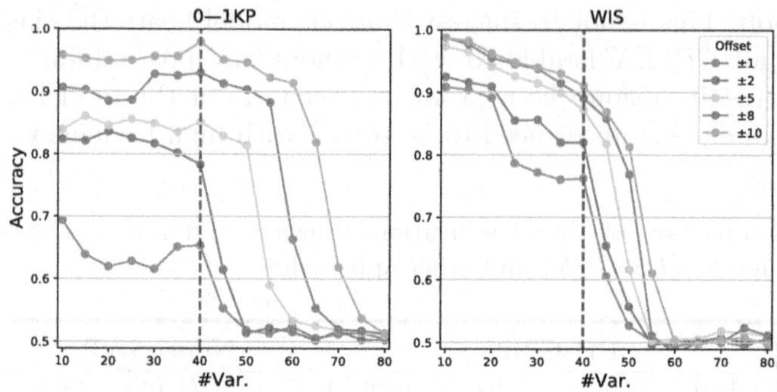


**Fig. 2.** The change in accuracy of the model when predicting the satisfiability of larger-scale instances than those appeared in the training sets.

can be kept to $n = 65$ for 0–1KP and $n = 50$ for WIS. However, the model fails to learn effective features even on the training sets when $\delta = \pm 1$ on both problems. Eventually, when the number of variables continuously increases, the accuracy is reduced to about 50%, which means it almost becomes a purely random model. To conclude, the experimental results indicate that the generalization performance of the model is affected by both the structure and difficulty of the problem.

## 5 Conclusion and Future Work

In this paper, we investigate whether a deep learning model based on GNN can solve a general class of CSPs with numerical constraints: the Pseudo-Boolean problem. More specially, the target is to correctly classify whether a decision PB problem is satisfiable. We present an extensible architecture that accepts a set of PB constraints with different lengths and forms as input. First, a weighted bipartite graph representation is established on the normalized constraints. After that, an iterative message passing process is executed. Finally, the satisfiability is calculated and returned through a readout phase. Experiments on two representative PB problems, 0–1 knapsack and weighted independent set, show that our model *PB-GNN* can successfully learn some features related to the structure of specific problem within 600 epochs, and achieves high-quality classification results on different distributions. Our model is shown to have several advantages over the previous works. In the aspect of network structure, it integrates the edge weights into the updating function of node embeddings, so that the parameter space is relatively small, making our model easier to converge. Regarding the accuracy of prediction, *PB-GNN* outperforms *NeuroSAT*, a foundational model in this field, on PB benchmarks with numerical coefficients, and is still comparable with it when applied to SAT instances.

We hope the model can provide a basis for a series of future works on learning to solve different CSPs with numerical constraints. There are also several perspectives for further work. On one hand, it needs to be acknowledged that the scale of learnable benchmarks by the current model is relatively small, and it should be improved to learn larger-scale instances, even those benchmarks that are difficult for the state-of-the-art solvers. On the other hand, we will try to decode the assignment of variables from the trained model, so that it can be called a real "solver". There are reasons to believe that the development of graph representation learning will help the model to obtain more accurate features related to the problem structures, thereby reducing the heavy manual work of designing specific heuristic algorithms.

**Acknowledgement.** This work has been supported by the National Natural Science Foundation of China (NSFC) under grant No. 61972384 and the Key Research Program of Frontier Sciences, Chinese Academy of Sciences under grant number QYZDJ-SSW-JSC036. The authors would like to thank the anonymous reviewers for their comments and suggestions. The authors are also grateful to Cunjing Ge for his suggestion on modeling the problem.

## References

1. Amizadeh, S., Matusevych, S., Weimer, M.: Learning to solve circuit-SAT: an unsupervised differentiable approach. In: 7th International Conference on Learning Representations, ICLR 2019, New Orleans, LA, USA (2019)
2. Amizadeh, S., Matusevych, S., Weimer, M.: PDP: a general neural framework for learning constraint satisfaction solvers. arXiv preprint arXiv:1903.01969 (2019)
3. Balunovic, M., Bielik, P., Vechev, M.T.: Learning to solve SMT formulas. In: Advances in Neural Information Processing Systems 31: Annual Conference on Neural Information Processing Systems, NeurIPS 2018, Montréal, Canada, pp. 10338–10349 (2018)
4. Bello, I., Pham, H., Le, Q.V., Norouzi, M., Bengio, S.: Neural combinatorial optimization with reinforcement learning. In: 5th International Conference on Learning Representations, ICLR 2017, Workshop Track Proceedings, Toulon, France (2017)
5. Cameron, C., Chen, R., Hartford, J.S., Leyton-Brown, K.: Predicting propositional satisfiability via end-to-end learning. The Thirty-Fourth AAAI Conference on Artificial Intelligence, AAAI 2020, NY, USA, New York, pp. 3324–3331 (2020)
6. Elffers, J., Nordström, J.: Divide and conquer: towards faster pseudo-Boolean solving. In: Proceedings of the Twenty-Seventh International Joint Conference on Artificial Intelligence, IJCAI 2018, Stockholm, Sweden, pp. 1291–1299 (2018)
7. Erdős, P., Rényi, A.: On the evolution of random graphs. Publ. Math. Inst. Hung. Acad. Sci **5**(1), 17–60 (1960)
8. Gass, S.I.: Linear programming: methods and applications. Courier Corporation, North Chelmsford (2003)
9. Gilmer, J., Schoenholz, S.S., Riley, P.F., Vinyals, O., Dahl, G.E.: Neural message passing for quantum chemistry. In: Proceedings of the 34th International Conference on Machine Learning, ICML 2017, Sydney, NSW, Australia, pp. 1263–1272 (2017)
10. Hopfield, J.J., Tank, D.W.: Neural computation of decisions in optimization problems. Biol. Cybern. **52**(3), 141–152 (1985)

11. Ivanescu, P.L.: Pseudo-Boolean Programming and Applications: Presented at the Colloquium on Mathematics and Cybernetics in the Economy, vol. 9. Springer, Berlin (2006)
12. Karp, R.M.: Reducibility among combinatorial problems. In: Proceedings of a Symposium on the Complexity of Computer Computations, New York, USA, pp. 85–103 (1972). https://doi.org/10.1007/978-1-4684-2001-2_9
13. Khalil, E.B., Dai, H., Zhang, Y., Dilkina, B., Song, L.: Learning combinatorial optimization algorithms over graphs. In: Advances in Neural Information Processing Systems 30: Annual Conference on Neural Information Processing Systems, NIPS 2017, Long Beach, CA, USA, pp. 6348–6358 (2017)
14. Khalil, E.B., Dilkina, B., Nemhauser, G.L., Ahmed, S., Shao, Y.: Learning to run heuristics in tree search. In: Proceedings of the Twenty-Sixth International Joint Conference on Artificial Intelligence, IJCAI 2017, Melbourne, Australia, pp. 659–666 (2017)
15. LeCun, Y., Bengio, Y., Hinton, G.: Deep learning. Nature **521**(7553), 436 (2015)
16. Lemos, H., Prates, M.O.R., Avelar, P.H.C., Lamb, L.C.: Graph colouring meets deep learning: effective graph neural network models for combinatorial problems. In: 31st IEEE International Conference on Tools with Artificial Intelligence, ICTAI 2019, Portland, OR, USA, pp. 879–885 (2019)
17. Li, Z., Chen, Q., Koltun, V.: Combinatorial optimization with graph convolutional networks and guided tree search. Advances in Neural Information Processing Systems 31: Annual Conference on Neural Information Processing Systems 2018, NeurIPS 2018, Montréal, Canada, pp. 537–546 (2018)
18. Milan, A., Rezatofighi, S.H., Garg, R., Dick, A.R., Reid, I.D.: Data-driven approximations to NP-hard problems. In: Proceedings of the Thirty-First AAAI Conference on Artificial Intelligence, AAAI 2017, San Francisco, California, USA, pp. 1453–1459 (2017)
19. Pisinger, D.: Core problems in knapsack algorithms. Oper. Res. **47**(4), 570–575 (1999)
20. Prates, M.O.R., Avelar, P.H.C., Lemos, H., Lamb, L.C., Vardi, M.Y.: Learning to solve NP-complete problems: a graph neural network for decision TSP. In: The Thirty-Third AAAI Conference on Artificial Intelligence, AAAI 2019, Honolulu, Hawaii, USA, pp. 4731–4738 (2019)
21. Selsam, D., Bjørner, N.: Guiding high-performance SAT solvers with Unsat-Core predictions. In: Janota, M., Lynce, I. (eds.) SAT 2019. LNCS, vol. 11628, pp. 336–353. Springer, Cham (2019). https://doi.org/10.1007/978-3-030-24258-9_24
22. Selsam, D., Lamm, M., Bünz, B., Liang, P., de Moura, L., Dill, D.L.: Learning a SAT solver from single-bit supervision. In: 7th International Conference on Learning Representations, ICLR 2019, New Orleans, LA, USA (2019)
23. Vinyals, O., Fortunato, M., Jaitly, N.: Pointer networks. Advances in Neural Information Processing Systems 28: Annual Conference on Neural Information Processing Systems, NIPS 2015, Montreal, Quebec, Canada, pp. 2692–2700 (2015)
24. Williams, H.P.: Logic and Integer Programming. ISORMS, vol. 130. Springer, Boston, MA (2009). https://doi.org/10.1007/978-0-387-92280-5
25. Wolsey, L.A., Nemhauser, G.L.: Integer and Combinatorial Optimization. John Wiley & Sons, Hoboken (2014)
26. Xu, H., Koenig, S., Kumar, T.K.S.: Towards effective deep learning for constraint satisfaction problems. In: Hooker, J. (ed.) CP 2018. LNCS, vol. 11008, pp. 588–597. Springer, Cham (2018). https://doi.org/10.1007/978-3-319-98334-9_38
27. Zhou, J., et al.: Graph neural networks: a review of methods and applications. arXiv preprint arXiv:1812.08434 (2018)

# A Machine Learning Based Splitting Heuristic for Divide-and-Conquer Solvers

Saeed Nejati[1(✉)], Ludovic Le Frioux[2], and Vijay Ganesh[1]

[1] University of Waterloo, Waterloo, ON, Canada
{snejati,vganesh}@uwaterloo.ca
[2] LRDE, EPITA, Le Kremlin-Bicêtre, France
ludovic@lrde.epita.fr

**Abstract.** In this paper, we present a machine learning based *splitting heuristic* for divide-and-conquer parallel Boolean SAT solvers. Splitting heuristics, whether they are look-ahead or look-back, are designed using *proxy metrics*, which when optimized, approximate the true metric of minimizing solver runtime on sub-formulas resulting from a *split*. The rationale for such metrics is that they have been empirically shown to be excellent proxies for runtime of solvers, in addition to being cheap to compute in an online fashion. However, the design of traditional splitting methods are often ad-hoc and do not leverage the copious amounts of data that solvers generate.

To address the above-mentioned issues, we propose a machine learning based splitting heuristic that leverages the features of input formulas and data generated during the run of a divide-and-conquer (DC) parallel solver. More precisely, we reformulate the splitting problem as a ranking problem and develop two machine learning models for *pairwise ranking* and computing the *minimum ranked* variable. Our model can compare variables according to their *splitting quality*, which is based on a set of features extracted from structural properties of the input formula, as well as dynamic *probing statistics*, collected during the solver's run. We derive the true labels through offline collection of runtimes of a parallel DC solver on sample formulas and variables within them. At each splitting point, we generate a predicted ranking (pairwise or minimum rank) of candidate variables and split the formula on the top variable. We implemented our heuristic in the Painless parallel SAT framework and evaluated our solver on a set of cryptographic instances encoding the SHA-1 preimage as well as SAT competition 2018 and 2019 benchmarks. We solve significantly more instances compared to the baseline Painless solver and outperform top divide-and-conquer solvers from recent SAT competitions, such as Treengeling. Furthermore, we are much faster than these top solvers on cryptographic benchmarks.

## 1 Introduction

Boolean satisfiability (SAT) solvers are powerful general purpose search tools that have had a revolutionary impact on many different domains, such as software engineering [9], AI [37], and cryptography [30,32]. They get their power


H. Simonis (Ed.): CP 2020, LNCS 12333, pp. 899–916, 2020.
https://doi.org/10.1007/978-3-030-58475-7_52

from proof-construction components like clause learning [29] and heuristics aimed at optimally sequencing, selecting, and initializing proof rules such as branching [22,31] and restarts [3].

The availability of many-core machines has led to a considerable effort in parallel SAT solver research in recent years [5]. Broadly speaking, researchers have developed two parallel SAT solver strategies, namely, *portfolio* and *divide-and-conquer* (DC) solvers. A portfolio SAT solver consists of a set of sequential worker solvers, each implementing a different collection of heuristics, and all of them attempting to solve the same instance running on different cores of a many-core machine. The key principle behind a portfolio solver is that of the *diversity of heuristics*, i.e., by leveraging a diverse set of heuristics to solve an instance may be more efficient than just using a single heuristic given the well-known fact that different classes of formulas are often best solved by distinct methods [11]. On the other hand, DC solvers partition the search space of the input formula and solve each sub-formula using a separate sequential worker solver. Each sub-formula is a restriction of the input formula with a set of assumptions [41]. In both the portfolio and DC settings, the sequential worker solvers may share clauses to exchange useful information they learn about their respective search spaces.

In the context of DC solvers, a splitting heuristic is a method aimed at choosing the "next variable" to add to the current list of assumptions (also known as guiding paths [41]). A bit more formally, one can define a splitting heuristic as a function that takes as input features of a given formula $\phi$ and/or statistics of a DC solver's state and outputs a variable to split on. Splitting heuristics are typically dynamic, i.e., they re-rank variables at regular intervals throughout the run of a DC solver.

The process of splitting itself can be described as follows: for a given input formula $\phi$, say that a variable $v$ is chosen for splitting. The solver generates $\phi[v = F]$ (resp. $\phi[v = T]$) by setting $v$ to False (resp. True) and appropriately simplifying the resultant sub-formulas using Boolean constraint propagation (BCP). These two sub-formulas are then solved in parallel. Each of these sub-formulas can be further split into smaller sub-formulas recursively. Many heuristics for splitting have been studied in the literature [1,2,17,35].

Splitting heuristics can be broadly categorized as *look-ahead* and *look-back*. Look-ahead heuristics choose some subset of variables in the input formula, analyze the impact of splitting on these variables, and rank them based on some measure that correlates well with minimizing runtime[1] of the solver on the sub-formulas thus obtained. By contrast, look-back heuristics compute statistics on "how well a variable participated in the search exploration in the past" (e.g., in clause learning, propagation, etc.), rank them appropriately, and split on the highest-ranked variable. Examples of look-back heuristics include splitters based on VSIDS activity [2], number of flips [21], and propagation-rate [35].

While considerable work has been done on splitting heuristics, almost all previous approaches share the following characteristics: they compute

[1] Runtime of a solver here refers to the wallclock time of solving a formula.

some features of the input formula and/or statistics over the solver state at appropriate intervals during the solver's run, and then use these as input to a "hand-coded" function (a splitting heuristic designed by the solver designer), that in turn computes a metric correlated with solver runtime to pick the "best" variable to split. By metric we mean a quantity that can be used to rank variables of the input formula such that splitting on the highest-ranked variable ideally corresponds to minimizing solver runtime. We argue that the design of splitting heuristics can be dramatically improved by leveraging a data-driven machine learning (ML) approach, especially for families of formulas (e.g., cryptographic instances) where it can be hard for human designers to come up with effective "hand-coded" splitting heuristic.

In this paper, we propose two ML-based methods, namely *pairwise ranking*, and *min-rank*. The pairwise ranking model takes as input features of a given formula $\phi$, aspects of solver state, as well as features of a pair of variables $v$ and $u$, and ranks them in descending order based on some splitting metric. This ML-based "comparator" is in turn used by our DC solver to rank variables for splitting at regular intervals during its run. The min-rank model, takes as input features of a given formula $\phi$, aspects of solver state, and features of a variable $v$, and outputs whether the variable $v$ has the minimum rank among all variables of the input formula (i.e is it the best variable to split?). Both of these models are binary classifiers implemented using *random forest*.

We implemented our heuristics in the Painless parallel solver framework [20] (we refer to our solver as MaplePainless-DC), and compared it with top parallel SAT solvers from recent SAT competitions. We find that our ML-based method out-performs the best DC solvers on both SAT 2018/2019 competition as well as cryptographic instances[2].

**Contributions.** In greater detail, our main contributions are as follows:

1. **MaplePainless-DC: A DC Solver based on ML-based Splitters.** We present MaplePainless-DC, an ML-based splitting DC parallel SAT solver. To the best of our knowledge, MaplePainless-DC is the first parallel solver with an ML-based splitting heuristic. Briefly, our splitting heuristics are ML models, trained offline on both static formula/variable features (e.g., variable occurrence in binary clauses) as well as "dynamic" features based on aspects of the solver's state at runtime (e.g., number of times a variable has been assigned, activities). We propose and implement two different models, namely, *pairwise ranking* and *min-rank*, described above. At runtime, the trained ML-model is invoked by MaplePainless-DC on a vector of static and dynamic variable features at appropriate intervals, which in turn outputs a ranking of the variables in the input formula. The splitting heuristic then chooses the top-ranked variable, splits the formula by assigning that variable both True and False values, and gives the resultant sub-formulas to worker solvers to solve (See Sect. 3).

[2] We only compare our MaplePainless-DC solver against the state-of-the-art DC solvers because it is well-known that the most notable portfolio solvers often out-perform the DC solvers on application benchmarks.

2. **Evaluation on Cryptographic Instances.** We evaluated our splitting heuristics on a cryptographic benchmark of 60 instances encoding preimage attack on round-reduced SHA-1 function (inversion of 60 random hash targets). We used top sequential solvers in solving cryptographic instances as backend solvers (MapleSAT and Glucose). We outperform the baseline solver (Painless-DC with the same backends and flip as splitting heuristics) in an apple-to-apple comparison, solving an additional instance from the hardest subset of instances and 30% faster on average on solved instances. We also solve 19 more instances (over a benchmark of 60) and are significantly faster relative to one of the top DC solvers, Treengeling (See Sect. 5.2).
3. **Evaluation on SAT Application Instances from SAT 2018 competition and SAT 2019 race.** We evaluated our splitting heuristics on main track benchmarks of SAT competition 2018 and SAT race 2019 (total 800 instances) against the baseline solver (Painless-DC with flip as splitting heuristic) in an apple-to-apple comparison, as well as against Treengeling. On the combined SAT 2018 and SAT 2019 benchmarks, we outperform both these solvers in terms of the number of solved instances and PAR-2 score[3]. Furthermore, MaplePainless-DC solves satisfiable instances much better than all other solvers (18 more than both the baseline and Treengeling solvers overall application instances), when using the pairwise ranking model (See Sect. 5.1).

## 2 Background

In this section, we list relevant definitions and notations. We refer the reader to [7] for details on CDCL SAT solvers. By the term "split" or "splitting" a formula $\phi$ over variable $v$ we refer to the process of generating two sub-formulas $\phi_1 = \phi \wedge \neg v$ and $\phi_2 = \phi \wedge v$, which are assumed to be simplified via unit or Boolean constraint propagation.

DC solvers take as input a Boolean formula and split it into many smaller sub-formulas, solve them using sequential worker solvers, and combine the results (SAT if at least one sub-formula is SAT, UNSAT if all of them are UNSAT). The architecture is usually of a master-slave type, where the slaves are sequential solvers and the master node maintains the splittings in the form of a search tree. Each node of the tree is a variable and branches correspond to setting that variable to True or False. Each "root to leaf" path represents a set of assumptions, also known as guiding path or cube. The phrase "solving a cube" refers to solving the original formula constrained with the given cube.

The notation $t_S(\phi)$ refers to the time to solve a Boolean formula $\phi$ with a sequential worker CDCL SAT solver $S$ (We drop the subscript if it is clear from context). We denote the reduced formula after setting $v$ to False (respectively to True) with $\phi[v = F]$ (respectively, $\phi[v = T]$). By reducing a formula we mean

[3] PAR-$k$ is the Penalized Average Runtime, counting each timeout as $k$ times the wallclock timeout.

simplification via unit propagation (i.e., removal of satisfied clauses from the formula, falsified literals from clauses).

The term *performance metric*, with respect to a given solver $S$, refers to a function $pm : \phi \times v \rightarrow \mathbb{R}$, over a formula $\phi$ and a variable $v \in vars(\phi)$, that characterizes the "quality" of splitting $\phi$ over $v$. Minimizing this metric ideally should correlate with minimizing solver runtime.

More precisely, the general goal of designing a splitting heuristic is twofold: first, to come up with a metric that correlates with minimizing solver runtime, and second to design a function to compute said metric. Researchers have proposed a variety of performance metrics in the context of splitting heuristics. Below are definitions of three such performance metrics and the intuition behind each of them. In previous work, researchers have found that these metrics are good proxies for minimizing runtime in the context of splitting in DC solvers. Further, to state the obvious, it is ideal to split on a variable that minimizes these metrics over all variables of an input formula. Let $\phi_1 = \phi[v = F]$ and $\phi_2 = \phi[v = T]$, be the sub-formulas after splitting $\phi$ over $v$.

- $pm_1(\phi, v) = \max\{t(\phi_1), t(\phi_2)\}$: This metric aims to capture the runtime of a DC solver executed in parallel over the sub-formulas $\phi_1$ and $\phi_2$.
- $pm_2(\phi, v) = t(\phi_1) + t(\phi_2)$: This function gives higher priority to splitting variables that make the problem easier even in a single core setting.
- $pm_3(\phi, v) = -(t(\phi) - t(\phi_1)) \cdot (t(\phi) - t(\phi_2))$: The idea behind this metric is to measure runtime "progress" in each branch (by comparing the runtime of sub-formulas with the original formula) and also aims to balance the hardness of the two branches.

**Random Forest Classification.** We refer the reader to the paper by Liaw et al. on random forests [26]. Briefly, the random forest is an ensemble learning method, that constructs a set of decision trees at training time and outputs the class that appears most often at the output of decision trees. Decision trees are a popular method for various machine learning tasks. However, trees that grow very deep tend to learn highly irregular patterns: they overfit their training sets, i.e. have a low bias, but very high variance. Random forests are a way of averaging multiple deep decision trees, trained on different parts of the same training set, with the goal of reducing the variance.

## 3 Machine Learning Models for Splitting

In this section we discuss a formulation of the splitting problem, define a quality measure for splitting, and study how we can train ML models that approximate the best splitting variable.

### 3.1 The Splitting Problem

Given a Boolean formula $\phi$, a sequential solver $S$, and performance metric $pm$, the splitting problem is to determine a variable $v$ in $\phi$ such that the time required to solve each of $\phi[v = T]$ and $\phi[v = F]$ by $S$ is minimal over all variables in $\phi$ with respect to the given performance metric $pm$, i.e. to find $\mathrm{argmin}_{v \in vars(\phi)}\{pm(\phi, v)\}$.

Modeling the exact behavior of a DC solver as it solves the sub-formulas in parallel and splits them on demand, is a challenging task. Below we define a metric that we believe is a more accurate measure of the optimal choice of a splitting variable, compared to the heuristic metrics mentioned in Sect. 2.

Let $\phi_1 = \phi[v = F]$ and $\phi_2 = \phi[v = T]$ be sub-formulas of splitting $\phi$ over $v$, and let $t_1 = t_S(\phi_1)$ and $t_2 = t_S(\phi_2)$ be runtimes of solving them by sequential solver $S$. The total time taken to solve the formula $\phi$ in this setting depends on the status and runtimes of the sub-formulas. If $\phi$ is UNSAT, the solver needs to prove both of the sub-formulas UNSAT. Hence the total time to solve such an instance is the maximum of the solver runtimes over the two sub-formulas. If on the other hand the formula $\phi$ is SAT, at least one of the sub-formulas must be SAT. If both sub-formulas are SAT, the total time is the minimum of the two, otherwise, only the SAT sub-formula matters. The total time of solving $\phi$ after splitting over variable $v$ can be represented as follows:

$$T_{total}(\phi, v) = \begin{cases} \max(t_1, t_2), & \phi_1 : UNSAT, \phi_2 : UNSAT \\ t_2, & \phi_1 : UNSAT, \phi_2 : SAT \\ t_1, & \phi_1 : SAT, \quad \phi_2 : UNSAT \\ \min(t_1, t_2), & \phi_1 : SAT, \quad \phi_2 : SAT \end{cases}$$

We use this total runtime as our performance metric: $pm(\phi, v) = T_{total}(\phi, v)$. In other words the target of our splitting heuristic is: given formula $\phi$, find a variable $v = \mathrm{argmin}_{v \in vars(\phi)}\{T_{total}(\phi, v)\}$.

### 3.2 Handling Timeouts

In practice, sub-formulas obtained after splitting on a variable can be hard for SAT solvers, and thus they may timeout for those cases. Let the status of a timed out (sub-)formula be labeled as "UNKNOWN". For a pair of variables $u$ and $v$ in formula $\phi$, we collect the runtime and status of solving sub-formulas $u_1 = \phi[u = F], u_2 = \phi[u = T], v_1 = \phi[v = F]$ and $v_2 = \phi[v = T]$. If the status of all four of these sub-formulas is UNKNOWN, we cannot derive the truth label (we do not know which of these two variables is better for splitting). In all other cases (mix of having SAT/UNSAT and UNKNOWN), we have enough information to be able to compare $u$ and $v$.

### 3.3 Learning to Rank

Generally, performance metrics can be used to generate a total order over the splitting variables (the higher ranked variables have a higher performance metric). Thus we can see the splitting problem as picking the minimum element from a ranked list. A common way of implementing splitting heuristics is to rank the variables by directly deriving the performance metric of each variable and selecting the minimum element. However, this is not the only way one can rank the elements. There are three main approaches in the ML literature for learning a model to rank a list of elements [28]:

- **Pointwise**: Learning a numerical or ordinal score for each data point, which are in turn sorted according to their ordinal score. The problem here translates to training a regression model.
- **Pairwise**: In this approach, ranking is done via learning a model that acts as a comparator, which takes as input two data points and outputs a total order over them.
- **Listwise**: These algorithms try to directly minimize a ranking evaluation metric (e.g. $\tau$-score or Mean Average-Precision) that compares a predicted ranking against a true ranking.

Almost all previous branching and splitting heuristics use pointwise ranking. For example, VSIDS branching heuristics [31] maintains a score for each variable, which represents how much that variable participated in clause learning recently. Then the variable with the highest activity is picked. Ultimately, the goal is to minimize the runtime and one might learn a function that directly approximates the desired runtime based ranking. However, approximating the runtime distribution of the CDCL SAT solver is very hard in general, as the interplay of the many heuristics in CDCL solvers makes it hard to predict how the search progresses. Heuristic designers hope that their variable ranking strongly correlates with a ranking where high ranking variables generate easier sub-formulas. In other words, their variable ranking using the proxy metric strongly correlates with runtime-based ranking. In the case of splitting or branching heuristics, we do not care about the actual runtime of sub-formulas and only want to know which variable corresponds to the lowest runtime. In other words, we want a way of comparing runtimes and not exactly deriving the runtime values. As mentioned above, we are looking for a minimum element in an array, sorted based on a metric. We approach this task of finding the minimum using two different methods. First, we build a *pairwise ranking* model that learns to compare two elements (two variables in our case), and second, we use a modified version of ordinal ranking, that we call *min-rank*, where we build a classifier that determines whether a given variable sits at the rank 1. We used binary classification for building both of these models. In the pairwise ranking, we use the model as a less-than operator and find the minimum in a linear scan. In min-rank, we check all of the variables against the model and pick the variable that the model declares as the minimum.

The first model is represented by a binary classifier *PW* (PairWise) that takes as input features of a formula $\phi$ and features of two variables $v_i$ and $v_j$ within $\phi$, and answers the question of "is $v_i$ better than $v_j$ for splitting $\phi$?" (according to our splitting performance metric described in Sect. 3.1).

$$PW(\phi, v_i, v_j) = \begin{cases} 1, & pm(\phi, v_i) < pm(\phi, v_j) \\ 0, & otherwise \end{cases} \tag{1}$$

This type of predicate learning was also used in one of the SATZilla versions [40] (known as pairwise voting), to rank a list of algorithms on a given instance.

For the second model, we used the idea of reduction by Lin et al. [27] for implementing ordinal ranking using binary classification. In their work, the role of a binary classifier given an element and an integer rank $k$ is to determine whether the element is within the top $k$ elements or not. Splitting heuristics look for the top variable in a ranked list, thus we are only interested in the $k = 1$ case. We define a binary classifier $MR$ (Min-Rank) that takes as input a variable $v$, and answers the question "is $v$ the best variable for splitting $\phi$?".

$$MR(\phi, v_i) = \begin{cases} 1, & \forall j \neq i : pm(\phi, v_i) < pm(\phi, v_j) \\ 0, & otherwise \end{cases} \quad (2)$$

### 3.4 Features for Training the Models

The data points that we used to train the model have the following format:

$$\begin{array}{l} PW : (\langle formula_{features}(\phi), var_{features}(v_i), var_{features}(v_j) \rangle, \{0, 1\}) \\ MR : (\langle formula_{features}(\phi), var_{features}(v) \rangle, \{0, 1\}) \end{array} \quad (3)$$

where the last element corresponds to the appropriate classifier ($PW(\phi, v_i, v_j)$ or $MR(\phi, v)$). For the formula features, we started from the features proposed by SATZilla in SAT competition 2012 [40]. Compared to the model that has been used in SATZilla, we will query our model at each splitting point. The feature computation time can quickly become a big part of the total runtime, and dominate the gain from picking a better splitting variable. On the other hand, each of the features could have an important role in making the model representative of the target distribution. To address this problem we performed a feature selection on our initial set of features (both formula and variable features). We first removed the very heavy features like LP-based (linear programming) features. We used the random forest for training our models. We then extracted the relative importance of each feature after training, which corresponds to the frequency of the appearance of those features in the ensemble of decision trees. We created a sorted list of features based on their relative importance ($f$) and performed a forward feature selection [10]. More specifically, starting with an empty list $F$, we passed through $f$ and added the features to $F$, if they reduced the cross-validation error when training on $F$. We then performed a backward pass on $F$, to remove heavy-to-compute features (having normalized cost of at least 100 ms), that do not contribute much to the accuracy of the model (having feature importance in the 25th percentile). We also took into account the product features (features from the multiplication of pairs of other features) to add non-linearity to the model. The final variable and formula features are listed in Table 1, consisting of structural metrics and metrics from a limited search. The features are listed in order of their importance extracted from the trained random forest.

**Table 1.** Variable ($var_{features}(v)$) and formula features ($formula_{features}(\phi)$), sorted based on their importance extracted from the trained models.

| Feature name | Description |
|---|---|
| numAssigned | #times $v$ got a value through branching/propagation |
| numFlip | #times the implied value of $v$ is different than its cached value [1] |
| numLearnt | #times $v$ appeared in a conflict clause |
| numInTernary | #times $v$ appears in a clause of size 3 |
| numInBinary | #times $v$ appears in a clause of size 2 |
| LRBProduct | product of LRB [22] activities of $v$ and $\neg v$ literals |
| propRate | average #propagation over #decision [35] |
| activity | VSIDS activity [31] |
| numPropagations | number of unit propagations in the limited search |
| conflictRate | ratio of #conflict clauses over #decisions |
| totalReward | sum of LRB reward of all of the variables |
| numBinary | number of clauses of size 2 in $\phi$ |
| numTernary | number of clauses of size 3 in $\phi$ |
| avgVarDegree | average variable node degree in the Variable-Clause graph |
| avgClauseDegree | average clause node degree in the Variable-Clause graph |

### 3.5 Training Data

We used the MapleCOMSPS solver [23] for collection of solver runtime, as well as formula and variable features. For generating our training data set, we picked 210 instances from the collection of application/crafted benchmarks of SAT competition 2016 [14] and 2017 [15]. To be more precise, 87 instances from the application benchmark of 2016, 21 instances from the crafted benchmark of 2016, and 102 instances from the main benchmark of 2017. The selection criteria were based on having instances from different types of problems (not problems of the same kind with different sizes) and having a wide range of hardness to make a representative training set. We did not use any instance that was deemed too hard (timed out) or too easy (was solved under 5 s) by our sequential solver. To match the test environment, we first ran the pre-processing stage of MapleCOMSPS and simplified the formulas. Then we computed all of the structural formula features offline and for the search probing features we ran MapleCOMSPS up to 10,000 conflicts and collected the necessary statistics from the solver. For computing the true labels, we randomly selected 50 variables in each instance and split the formula on each of them and solved the sub-formulas with MapleCOMSPS up to a 5000 s timeout, recording the runtime and status (SAT, UNSAT, UNKNOWN).

### 3.6 Analysis of the Learned Models

For training the model, we used *random forest classifier*. We can achieve an average precision of 83% and an average recall of 83% and an accuracy of 80.7%. The candidate variable list can be ordered using the learned predicate. For finding

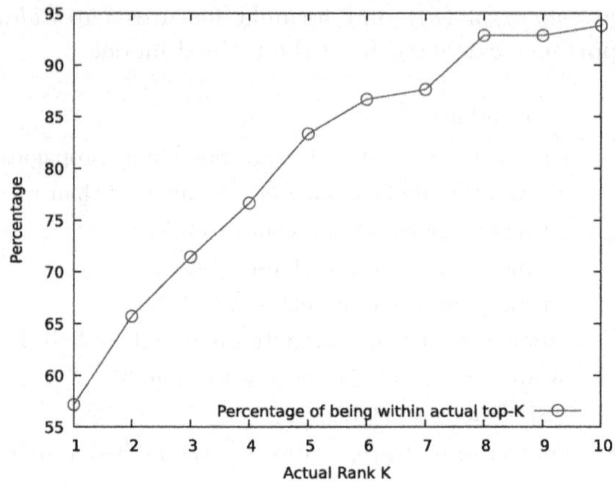


**Fig. 1.** Percentage of instances where the predicted best variable is within the actual top-$k$ variables for $k$ between 1 and 10.

the best variable, we only need to find the "min" of the list, which can be done in linear time. Although, when using a noisy comparator, the error caused by the inaccurate comparison, might accumulate over multiple comparisons. There are more robust sorting algorithms in the presence of noisy comparators (e.g. counting method [38]), but the running time complexity is quadratic in the number of elements, which is not feasible for large formulas. To check how well our predicate is performing, we ranked the variables in the instances in our training set, where we have the true labels.

When sorting the variables using the pairwise ranking, out of 210 instances, in 120 instances the best variable in the predicted ranking matched the actual best variable (57.1% of the time). In 18 cases the best actual variable was the second best predicted variable. The worst prediction happened in an instance with 2200 variables, where the best actual variable appeared in the 30th position in the predicted list. The total error (e.g. $\tau$ score) of comparing the predicted ranking and the best actual ranking could be poor, however, we can see some general ordering over the variables (variables that are much better choices appear closer to the front of the list). Figure 1 shows the percentage of instances (out of 210), where the predicted best variable (the output of the model for splitting), is within the actual top-$k$ variables. We observed that the best predicted variable is one of the actual top 10 variables in 197 out of 210 instances (93.8%). This shows that top variables in our predicted ranking have a considerable overlap with the top variables in the actual ranking.

## 4 Implementation

Our implementation of MaplePainless-DC is built on top of the Painless solver framework [20]. Painless is a state-of-the-art framework that allows developers to implement many different kinds of parallel SAT solvers for many-core environments. The main configurable components of Painless are: parallel strategies such as DC or portfolio, clause sharing, and management policies, and diverse sequential engines. The implementation of our machine learning based splitting heuristic relies on the use of the DC strategy in Painless [21]. We use an instrumented version of the MapleCOMSPS [23] solver as workers in MaplePainless-DC. The instrumentation collects formula/variable statistics and chooses splitting variables.

### 4.1 Implementation of Splitting in Painless-DC

Painless-DC splits a formula at regular intervals throughout its run. At a high-level, the master node maintains a queue of idle cores to assign jobs to. Initially, the master node chooses a variable to split and assigns the resultant sub-formulas to two cores. If the queue of idle cores is non-empty, the master node chooses a sub-formula from one of the busy cores and splits it into two sub-formulas, one of which is assigned to the busy core and the other to one of the idle ones. This process is repeated until the queue of idle cores is empty. If during the solver's run a core becomes idle and is added to the idle queue (e.g., if it has established UNSAT for its input sub-formula), the above-mentioned process is invoked until the idle queue becomes empty again. This form of load-balancing ensures that worker nodes are not allowed to idle for too long.

### 4.2 Feature Computation for Machine Learning in MapleCOMSPS

When it is time to split a formula, Painless' master node asks the sequential worker solver whose sub-formula is being split for variables to split on. The worker solver computes formula and variable features (e.g. number of times a variable is assigned, either decided or propagated) on the sub-formula to be split. The description of the variable features is listed in Table 1.

We used scikit-learn python package [36] for training the model and extracted the parameters and embedded them in a C implementation of random forest classifier. We later call this classifier from MapleCOMSPS for performing predictions. Given a list of candidate variables, pairwise classifier $PW$ is used as a comparator (less-than) operator to find and return the minimum item in a linear scan. Min-rank classifier $MR$ is invoked for all variables in the list and the first variable predicted to be the minimum is returned. The worst case time complexity of both of the models is $O(T_C \cdot n)$, where $n$ is the number of variables and $T_C$ is the time complexity of querying each of the classifiers.

## 5 Experimental Results

### 5.1 Evaluation over SAT 2018 and 2019 Competition Instances

**Experimental Setup.** For evaluation we used the main track benchmark of the SAT competition 2018 [13] and SAT race 2019 [12], which in total have

**Table 2.** Performance comparison of our solvers vs state-of-the-art DC parallel SAT solvers. Number of solved instances is out of 791 (after removing repeated instances from the original 800). SAT column shows the number of satisfiable instances solved (resp. UNSAT). The best result in each column is shown in bold.

| Cores | Solver | Solved | SAT | UNSAT | Avg. runtime (s) | PAR-2 (hr) |
|---|---|---|---|---|---|---|
| 8 | `Treengeling` | **501** | 292 | **209** | 719.399 | 905.672 |
| | `Painless-flip` | 474 | 291 | 183 | 437.632 | 938.177 |
| | `MaplePainless-DC-MinRank` | 497 | 299 | 198 | 484.340 | 883.532 |
| | `MaplePainless-DC-Pairwise` | **501** | **309** | 192 | **435.610** | **866.178** |
| 16 | `Treengeling` | 518 | 308 | **210** | 677.216 | 855.777 |
| | `MaplePainless-DC-Pairwise` | **520** | **317** | 203 | **334.991** | **801.165** |

800 instances, consisting of industrial instances coming from a diverse set of applications and crafted instances encoding combinatorial problems. Within our sample of instances from 2016/2017 (used for training), a scrambled version (a shuffling of clauses and variable IDs) of 9 instances appear in the 2018/2019 benchmarks as well (used for testing). To have a fair evaluation, we removed these 9 instances from the testing benchmark. Timeout for each instance was set at 5000 s wallclock time (the same as in SAT competitions). All jobs were run on Intel Xeon CPUs (3 GHz and 64 GB RAM). As a sanity check, we performed a controlled apple-to-apple study comparing Painless with ML-based splitting heuristic against the same setup with random splitting heuristic. We note that Painless with ML-based splitting easily outperforms the version with random splitting.

**Solvers Description.** We compared our solver against the top divide-and-conquer parallel solvers, Treengeling [6] version bcj and Painless-DC [21] with its best performing setting (node switch strategy: *clone*, clause sharing: all-to-all, and splitting heuristic: *flip*), which we will refer to as `Treengeling` and `Painless-flip`, respectively. We refer to our implementations using the *PW* classifier for pairwise ranking as `MaplePainless-DC-Pairwise` and `MaplePainless-DC-MinRank` refers to solver with binary classification of minimum rank (*MR* classifier). Our parallel solvers and `Painless-flip` use MapleCOMSPS [23] as the backend sequential solver. We changed MapleCOMSPS to always use LRB as branching heuristics. Each solver was assigned 8 cores.

**Results.** To perform an apple-to-apple comparison and measure the effectiveness of our splitting heuristics, we reused all of the configurations and components of `Painless-flip` and only replaced the splitting heuristics, which was straightforward, thanks to the modular design of Painless. Table 2 lists the number of solved instances, average runtime among solved instances, and the PAR-2 metric. In the SAT competition, PAR-2 is measured in seconds, but for better readability, we report it in hours. As the table shows, both ML based heuristics,

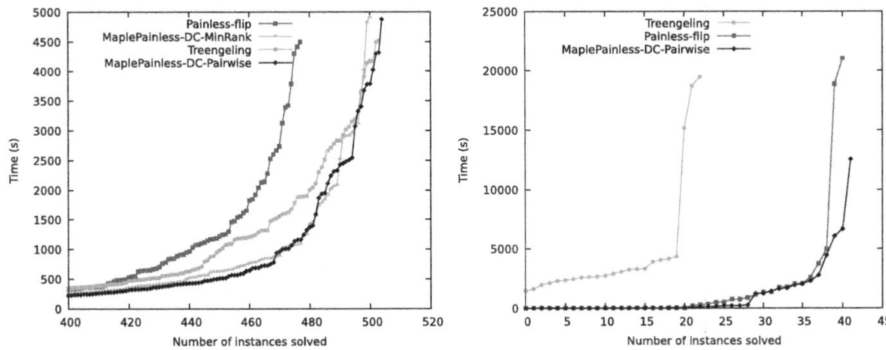


(a) SAT 2018/2019 benchmark

(b) Hard cryptographic benchmark

**Fig. 2.** Cactus plot for performance comparison of our parallel SAT solvers against baseline and state-of-the-art on main track benchmarks of SAT competition and hard cryptographic benchmark (using 8 cores).

improve significantly upon the baseline in the application benchmark of 2018 and 2019. Additionally, `MaplePainless-DC-Pairwise` solves the same number of instances as `Treengeling`, but has the lowest PAR-2 score among all. Figure 2a shows the cactus plot of these solvers over this benchmark.

### 5.2 Evaluation over Cryptographic Instances

**Experimental Setup.** We used a set of hard cryptographic instances encoding preimage attack on round-reduced SHA-1 hash function. More precisely, the instances encode inversion of 21, 22, and 23 rounds SHA-1, with 20 random targets for each rounds version [34]. All jobs were run on Intel Xeon CPUs at 3 GHz and 64 GB of RAM with 12 h wallclock timeout.

**Solvers Description.** We compared our `MaplePainless-DC-Pairwise` solver against the baseline (`Painless-flip`) and `Treengeling`. All solvers were run with 8 cores. For the backend solvers in this experiment, we used Glucose [4] and MapleSAT [22] (4 of each). Glucose solvers used Glucose's default restart policy. MapleSAT solvers were set to use the MABR restart policy [33]. To have an apple-to-apple comparison with baseline, we used the same backend solver configuration for baseline and our solvers.

**Results.** Figure 2b shows the performance of the considered DC solvers on our hard cryptographic benchmark. Instances with 21 rounds are easy for all solvers. 22 rounds instances are much harder than 21 rounds instances and as can be seen, `Treengeling` solves very few of these instances. Although both `MaplePainless-DC-Pairwise` and `Painless-flip` solve all of these instances. The hardness ramps up very quickly at 23 rounds instances, where `Treengeling` does not solve any of the instances and `Painless-flip` solves 2 of them. `MaplePainless-DC-Pairwise` solves 3 instances in this subset of instances, and with 30% lower runtime.

### 5.3 Scaling Experiments

Our main set of experiments were executed on 8 CPU cores. To study how our splitting heuristic scales with larger number of cores, we took `Treengeling` and `MaplePainless-DC-Pairwise`, that performed better among the four solvers on SAT 2018 and 2019 benchmarks, and compared them on these benchmarks on 16 core machines. Table 2 shows that our `MaplePainless-DC-Pairwise` solver can solve 2 more instances than `Treengeling` (as opposed to solving the same number of instances as observed in the 8 core setting). Further, `MaplePainless-DC-Pairwise` with 16 cores solves 19 more instances compared to the same version with 8 cores, and 11 of these instances were unsatisfiable.

### 5.4 Computational Overhead of ML Models

The timing results presented in this section are end-to-end (i.e., the computational overhead of running the ML models are included in the solver runtimes presented). The majority of the variable features are dynamic and their counters are updated whenever there is a related action performed during the search, thus their complexity is amortized over the run of the solver. The structural formula features are computed at the start of the search, which are all linearly proportional to the size of formula, and later are updated incrementally as the formula is reduced via splitting. Setting up the feature values and querying the models roughly takes 6% of the total runtime of the solver on average for the SAT 2018 and 2019 benchmarks.

### 5.5 Summary of Results

We first note that MaplePainless-DC significantly outperforms both baseline as well as the state-of-the-art `Treengeling` solvers on cryptographic (60 instances) and SAT 2018/2019 competition benchmarks (800 instances) both in terms of number of solved instances as well as PAR-2 scores. Further, we see an improvement in performance of our solver MaplePainless-DC as we increase the number of machine cores from 8 to 16 (see Table 2).

Both of the ML-based heuristics are very successful on satisfiable instances, where `MaplePainless-DC-Pairwise` solves 17 more satisfiable instances relative to `Treengeling` and 18 relative to `Painless-flip` (although solving fewer unsatisfiable instances than `Treengeling`). On cryptographic benchmark, `MaplePainless-DC-Pairwise` solves 43 out of 60 instances, outperforming other solvers. From the hardest instances (23 rounds SHA-1) in this benchmark, `Treengeling` can not solve any of the instances, whereas `MaplePainless-DC-Pairwise` solves three of them (see Table 2 and Fig. 2).

## 6 Related Work

Cube-and-conquer [16] solvers (such as Treengeling [6]) use a look-ahead procedure to determine the best splitting variable. In contrast to look-ahead techniques, some solvers use look-back methods that dynamically analyze the search performed by the solver, as well as formula statistics, to identify the best candidate at the "current" splitting point. For example, Ampharos [2] picks the variable with the highest VSIDS activity and MapleAmpharos [35] uses propagation-rate (average propagation of a variable divided by the number of decisions). Audemard et al. [1], use the number of times a variable's saved phase is flipped through propagation. This has been shown to be effective in divide-and-conquer settings [21]. We can categorize our work as a look-back heuristic as all of the features are extracted from previous limited runs of a sequential solver.

ML has been used to rank and pick the best variable in sequential SAT solvers. Liang et al. used a reinforcement learning formulation to find the most rewarding variable according to the learning-rate metric for branching [22]. In another work, they train a logistic regression model that ranks variables based on the probability of causing a conflict in the next step [25]. In contrast to these methods that use a pointwise ranking of the variables, we are employing a pairwise ranking. The pairwise ranking has been used in other constraint solver contexts as well. Xu et al. used pairwise voting in the context of algorithm selection, to rank SAT solvers based on their performance on a single formula [40]. Khalil et al. used deep reinforcement learning for learning heuristics in optimization algorithms over graphs of up to 1000 nodes [18], however, there is a scaling challenge when applying their work on industrial SAT instances which can have millions of variables.

## 7 Conclusions

We presented two ML based look-back splitting heuristics for DC solvers in this paper, namely, pairwise ranking *PW* and min rank *MR* methods. These methods significantly outperform the baseline Painless and state-of-the-art `Treengeling` solvers on both industrial and cryptographic benchmarks.

One of the key insights that underpins our solver heuristic design is the observation that solvers are compositions of two kinds of methods, namely, logical reasoning routines (e.g., conflict clause learning or BCP), and heuristics aimed at optimally selecting, sequencing, or initializing logical reasoning rules. We show that our methods outperform hand-tuned heuristics in the best DC solver to-date, namely, `Treengeling`, on a large industrial benchmark as well as challenge problems obtained from cryptographic applications. This gives us greater confidence in our philosophy that design of solver heuristics can effectively leverage ML methods, especially given the fact that solvers are data-rich environments. Further, future solver design is likely to move away from ad-hoc heuristic design and more towards feature engineering and appropriate choice of ML methods, as has already been witnessed for many solver heuristics [8, 19, 22, 24, 25, 39].

## References


1. Audemard, G., Hoessen, B., Jabbour, S., Piette, C.: An effective distributed D&C approach for the satisfiability problem. In: Proceedings of the 22nd Euromicro International Conference on Parallel, Distributed, and Network-Based Processing (PDP), pp. 183–187. IEEE (2014)
2. Audemard, G., Lagniez, J.-M., Szczepanski, N., Tabary, S.: An adaptive parallel SAT solver. In: Rueher, M. (ed.) CP 2016. LNCS, vol. 9892, pp. 30–48. Springer, Cham (2016). https://doi.org/10.1007/978-3-319-44953-1_3
3. Audemard, G., Simon, L.: Refining restarts strategies for SAT and UNSAT. In: Milano, M. (ed.) CP 2012. LNCS, pp. 118–126. Springer, Heidelberg (2012). https://doi.org/10.1007/978-3-642-33558-7_11
4. Audemard, G., Simon, L.: Glucose and syrup: nine years in the SAT competitions. In: Proceedings of SAT Competition, pp. 24–25 (2018)
5. Balyo, T., Sinz, C.: Parallel satisfiability. In: Hamadi, Y., Sais, L. (eds.) Handbook of Parallel Constraint Reasoning, pp. 3–29. Springer, Cham (2018). https://doi.org/10.1007/978-3-319-63516-3_1
6. Biere, A.: CaDiCal, Lingeling, Plingeling, Treengeling and YalSat entering the SAT competition 2017. In: Proceedings of SAT Competition, pp. 14–15 (2017)
7. Biere, A., Heule, M., van Maaren, H.: Handbook of Satisfiability, vol. 185. IOS press, Amsterdam (2009)
8. Bouraoui, Z., et al.: From shallow to deep interactions between knowledge representation, reasoning and machine learning (Kay R. Amel group). arXiv preprint arXiv:1912.06612 (2019)
9. Cadar, C., Ganesh, V., Pawlowski, P.M., Dill, D.L., Engler, D.R.: EXE: automatically generating inputs of death. ACM Trans. Inf. Syst. Secur. (TISSEC) **12**(2), 10 (2008)
10. Guyon, I., Elisseeff, A.: An introduction to variable and feature selection. J. Mach. Learn. Res. **3**, 1157–1182 (2003)
11. Hamadi, Y., Jabbour, S., Sais, L.: ManySAT: a parallel SAT solver. J. Satisf. Boolean Model. Comput. **6**, 245–262 (2008)
12. Heule, M., Järvisalo, M., Suda, M.: SAT race benchmarks (2016). http://satcompetition.org/sr2019benchmarks.zip
13. Heule, M., Järvisalo, M., Suda, M.: SAT competition benchmarks (2018). http://sat2018.forsyte.tuwien.ac.at/benchmarks/
14. Heule, M., Järvisalo, M., Tomáš, B.: SAT competition benchmarks (2016). http://baldur.iti.kit.edu/sat-competition-2016/index.php?cat=downloads
15. Heule, M., Järvisalo, M., Tomáš, B.: SAT competition benchmarks (2017). https://baldur.iti.kit.edu/sat-competition-2017/index.php?cat=benchmarks
16. Heule, M.J.H., Kullmann, O., Biere, A.: Cube-and-conquer for satisfiability. In: Hamadi, Y., Sais, L. (eds.) Handbook of Parallel Constraint Reasoning, pp. 31–59. Springer, Cham (2018). https://doi.org/10.1007/978-3-319-63516-3_2
17. Heule, M.J.H., Kullmann, O., Wieringa, S., Biere, A.: Cube and conquer: guiding CDCL SAT solvers by lookaheads. In: Eder, K., Lourenço, J., Shehory, O. (eds.) HVC 2011. LNCS, vol. 7261, pp. 50–65. Springer, Heidelberg (2012). https://doi.org/10.1007/978-3-642-34188-5_8
18. Khalil, E., Dai, H., Zhang, Y., Dilkina, B., Song, L.: Learning combinatorial optimization algorithms over graphs. In: Advances in Neural Information Processing Systems, pp. 6348–6358 (2017)

19. Kurin, V., Godil, S., Whiteson, S., Catanzaro, B.: Improving SAT solver heuristics with graph networks and reinforcement learning. arXiv preprint arXiv:1909.11830 (2019)
20. Le Frioux, L., Baarir, S., Sopena, J., Kordon, F.: PaInleSS: a framework for parallel SAT solving. In: Gaspers, S., Walsh, T. (eds.) SAT 2017. LNCS, vol. 10491, pp. 233–250. Springer, Cham (2017). https://doi.org/10.1007/978-3-319-66263-3_15
21. Le Frioux, L., Baarir, S., Sopena, J., Kordon, F.: Modular and efficient divide-and-conquer SAT solver on top of the painless framework. In: Vojnar, T., Zhang, L. (eds.) TACAS 2019. LNCS, vol. 11427, pp. 135–151. Springer, Cham (2019). https://doi.org/10.1007/978-3-030-17462-0_8
22. Liang, J.H., Ganesh, V., Poupart, P., Czarnecki, K.: Learning rate based branching heuristic for SAT solvers. In: Creignou, N., Le Berre, D. (eds.) SAT 2016. LNCS, vol. 9710, pp. 123–140. Springer, Cham (2016). https://doi.org/10.1007/978-3-319-40970-2_9
23. Liang, J.H., Oh, C., Ganesh, V., Czarnecki, K., Poupart, P.: Maple-COMSPS LRB VSIDS and MapleCOMSPS CHB VSIDS. In: Proceedings of SAT Competition, pp. 20–21 (2017)
24. Liang, J.H., Oh, C., Mathew, M., Thomas, C., Li, C., Ganesh, V.: Machine learning-based restart policy for CDCL SAT solvers. In: Beyersdorff, O., Wintersteiger, C.M. (eds.) SAT 2018. LNCS, vol. 10929, pp. 94–110. Springer, Cham (2018). https://doi.org/10.1007/978-3-319-94144-8_6
25. Liang, J.H., Hari Govind, V.K., Poupart, P., Czarnecki, K., Ganesh, V.: An empirical study of branching heuristics through the lens of global learning rate. In: Gaspers, S., Walsh, T. (eds.) SAT 2017. LNCS, vol. 10491, pp. 119–135. Springer, Cham (2017). https://doi.org/10.1007/978-3-319-66263-3_8
26. Liaw, A., Wiener, M., et al.: Classification and regression by randomforest. R News **2**(3), 18–22 (2002)
27. Lin, H.T., Li, L.: Reduction from cost-sensitive ordinal ranking to weighted binary classification. Neural Comput. **24**(5), 1329–1367 (2012)
28. Liu, T.Y., et al.: Learning to rank for information retrieval. Found. Trends® Inf. Retrieval **3**(3), 225–331 (2009)
29. Marques-Silva, J.P., Sakallah, K.A.: GRASP: a search algorithm for propositional satisfiability. IEEE Trans. Comput. **48**(5), 506–521 (1999)
30. Mironov, I., Zhang, L.: Applications of SAT solvers to cryptanalysis of hash functions. In: Biere, A., Gomes, C.P. (eds.) SAT 2006. LNCS, vol. 4121, pp. 102–115. Springer, Heidelberg (2006). https://doi.org/10.1007/11814948_13
31. Moskewicz, M.W., Madigan, C.F., Zhao, Y., Zhang, L., Malik, S.: Chaff: engineering an efficient SAT solver. In: Proceedings of the 38th Annual Design Automation Conference, pp. 530–535. ACM (2001)
32. Nejati, S., Ganesh, V.: CDCL (crypto) SAT solvers for cryptanalysis. arXiv preprint arXiv:2005.13415 (2020)
33. Nejati, S., Liang, J.H., Gebotys, C., Czarnecki, K., Ganesh, V.: Adaptive restart and CEGAR-based solver for inverting cryptographic hash functions. In: Paskevich, A., Wies, T. (eds.) VSTTE 2017. LNCS, vol. 10712, pp. 120–131. Springer, Cham (2017). https://doi.org/10.1007/978-3-319-72308-2_8
34. Nejati, S., Linag, J.H., Ganesh, V., Gebotys, C., Czarnecki, K.: SHA-1 preimage instances for SAT. In: Proceedings of SAT Competition, p. 45 (2017)
35. Nejati, S., et al.: A propagation rate based splitting heuristic for divide-and-conquer solvers. In: Gaspers, S., Walsh, T. (eds.) SAT 2017. LNCS, vol. 10491, pp. 251–260. Springer, Cham (2017). https://doi.org/10.1007/978-3-319-66263-3_16

36. Pedregosa, F., et al.: Scikit-learn: machine learning in python. J. Mach. Learn. Res. **12**, 2825–2830 (2011)
37. Rintanen, J.: Planning and SAT. In: Biere, A., Heule, M., Van Maaren, H., Walsh, T. (eds.) Handbook of Satisfiability, vol. 185, pp. 483–504. IOS Press, Amsterdam (2009)
38. Shah, N.B., Wainwright, M.J.: Simple, robust and optimal ranking from pairwise comparisons. J. Mach. Learn. Res. **18**(1), 7246–7283 (2017)
39. Soos, M., Kulkarni, R., Meel, K.S.: CrystalBall: gazing in the black box of SAT solving. In: Janota, M., Lynce, I. (eds.) SAT 2019. LNCS, vol. 11628, pp. 371–387. Springer, Cham (2019). https://doi.org/10.1007/978-3-030-24258-9_26
40. Xu, L., Hutter, F., Hoos, H., Leyton-Brown, K.: Evaluating component solver contributions to portfolio-based algorithm selectors. In: Cimatti, A., Sebastiani, R. (eds.) SAT 2012. LNCS, vol. 7317, pp. 228–241. Springer, Heidelberg (2012). https://doi.org/10.1007/978-3-642-31612-8_18
41. Zhang, H., Bonacina, M.P., Hsiang, J.: Psato: a distributed propositional prover and its application to quasigroup problems. J. Symb. Comput. **21**(4–6), 543–560 (1996)

# Theoretical and Experimental Results for Planning with Learned Binarized Neural Network Transition Models

Buser Say[1(✉)], Jo Devriendt[2,3], Jakob Nordström[3,2], and Peter J. Stuckey[1]

[1] Monash University, Melbourne, Australia
{buser.say,peter.stuckey}@monash.edu
[2] Lund University, Lund, Sweden
jo.devriendt@cs.lth.se
[3] University of Copenhagen, Copenhagen, Denmark
jn@di.ku.dk

**Abstract.** We study planning problems where the transition function is described by a learned binarized neural network (BNN). Theoretically, we show that feasible planning with a learned BNN model is *NP*-complete, and present two new constraint programming models of this task as a mathematical optimization problem. Experimentally, we run solvers for constraint programming, weighted partial maximum satisfiability, 0–1 integer programming, and pseudo-Boolean optimization, and observe that the pseudo-Boolean solver outperforms previous approaches by one to two orders of magnitude. We also investigate symmetry handling for planning problems with learned BNNs over long horizons. While the results here are less clear-cut, we see that exploiting symmetries can sometimes reduce the running time of the pseudo-Boolean solver by up to three orders of magnitude.

**Keywords:** Automated planning · Binarized neural networks · Mathematical optimization · Pseudo-Boolean optimization · Cutting planes reasoning · Symmetry

## 1 Introduction

Automated planning is the reasoning side of acting in Artificial Intelligence [23]. Planning automates the selection and ordering of actions to reach desired states of the world. An automated planning problem represents the real-world dynamics using a model of the world, which can either be manually encoded [7,13,14,20, 24], or learned from data [1,2,12,29]. In this paper, we focus on the latter.

Automated planning with deep neural network (DNN) learned state transition models is a two stage data-driven framework for learning and solving planning problems with unknown state transition models [28]. The first stage of the framework learns the unknown state transition model from data as a DNN.


H. Simonis (Ed.): CP 2020, LNCS 12333, pp. 917–934, 2020.
https://doi.org/10.1007/978-3-030-58475-7_53

The second stage of the framework plans optimally with respect to the learned DNN model by solving an equivalent mathematical optimization problem (e.g., a mixed-integer programming (MIP) model [28], a 0–1 integer programming (IP) model [25,26], or a weighted partial maximum satisfiability (WP-MaxSAT) model [25,26]). In this paper, we focus on the theoretical, mathematical modelling and the experimental aspects of the second stage of the data-driven framework where the learned DNN is a binarized neural network (BNN) [16].

We study the complexity of feasible automated planning with learned BNN transition models under the common assumption that the learned BNN is fully connected, and show that this problem is *NP*-complete. In terms of mathematical modelling, we propose two new constraint programming (CP) models that are motivated by the work on learning BNNs with CP [33]. We then conduct two sets of experiments for the previous and our new mathematical optimization models for the learned automated problem. In our first set of experiments, we focus on solving the existing learned automated problem instances using off-the-shelf solvers for WP-MaxSAT [6], MIP [17], pseudo-Boolean optimization (PBO) [10] and CP [17]. Our results show that the PBO solver RoundingSat [10] outperforms the existing baselines by one to two orders of magnitude. In our second set of experiments, we focus on the challenging task of solving learned automated planning problems over long planning horizons. Here, we study and test the effect of specialized symmetric reasoning over different time steps of the learned planning problem. Our preliminary results demonstrate that exploiting this symmetry can significantly reduce the overall runtime of the underlying solver (i.e., RoundingSat) by upto three orders of magnitude. Overall, with this paper we make both theoretical and practical contributions to the field of data-driven automated planning with learned BNN transition models.

In the next section we formally define the planning problem using binarized neural network (BNN) transitions functions. In Sect. 3 we define a 0–1 integer programming (IP) model that will solve the planning problem given a learned BNN. In Sect. 4 we show that the feasibility problem is *NP*-complete. In Sect. 5 we give two constraint programming models for the solving the planning problem. In Sect. 6 we discuss a particular symmetry property of the model, and discuss how to take advantage of it. In Sect. 7 we give experimental results. Finally, in Sect. 8 we conclude and discuss future work.

## 2 Planning with Learned BNN Transition Models

We begin by presenting the definition of the learned automated planning problem and the BNN architecture used for learning the transition model from data.

### 2.1 Problem Definition

A *fixed-horizon learned deterministic automated planning problem* [25,28] is a tuple $\tilde{\Pi} = \langle S, A, C, \tilde{T}, V, G, R, H \rangle$, where $S = \{s_1, \dots, s_n\}$ and $A = \{a_1, \dots, a_m\}$ are sets of state and action variables for positive integers $n, m$ with domains

$D_{s_1}, \ldots, D_{s_n}$ and $D_{a_1}, \ldots, D_{a_m}$ respectively, $C : D_{s_1} \times \cdots \times D_{s_n} \times D_{a_1} \times \cdots \times D_{a_m} \rightarrow \{true, false\}$ is the global function, $\tilde{T} : D_{s_1} \times \cdots \times D_{s_n} \times D_{a_1} \times \cdots \times D_{a_m} \rightarrow D_{s_1} \times \cdots \times D_{s_n}$ denotes the learned state transition function, and $R : D_{s_1} \times \cdots \times D_{s_n} \times D_{a_1} \times \cdots \times D_{a_m} \rightarrow \mathbb{R}$ is the reward function. Further, $V$ is a tuple of constants $\langle V_1, \ldots, V_n \rangle \in D_{s_1} \times \cdots \times D_{s_n}$ that denotes the initial values of all state variables, $G : D_{s_1} \times \cdots \times D_{s_n} \rightarrow \{true, false\}$ is the goal state function, and $H \in \mathbb{Z}^+$ is the planning horizon.

A *solution* to (i.e., a *plan* for) $\tilde{\Pi}$ is a tuple of values $\bar{A}^t = \langle \bar{a}_1^t, \ldots, \bar{a}_m^t \rangle \in D_{a_1} \times \cdots \times D_{a_m}$ for all action variables $A$ over time steps $t \in \{1, \ldots, H\}$ such that $\tilde{T}(\langle \bar{s}_1^t, \ldots, \bar{s}_n^t, \bar{a}_1^t, \ldots, \bar{a}_m^t \rangle) = \langle \bar{s}_1^{t+1}, \ldots, \bar{s}_n^{t+1} \rangle$ and $C(\langle \bar{s}_1^t, \ldots, \bar{s}_n^t, \bar{a}_1^t, \ldots, \bar{a}_m^t \rangle) = true$ for time steps $t \in \{1, \ldots, H\}$, $V_i = \bar{s}_i^1$ for all $s_i \in S$ and $G(\langle \bar{s}_1^{H+1}, \ldots, \bar{s}_n^{H+1} \rangle) = true$. An *optimal* solution to $\tilde{\Pi}$ is a solution such that the total reward $\sum_{t=1}^{H} R(\langle \bar{s}_1^{t+1}, \ldots, \bar{s}_n^{t+1}, \bar{a}_1^t, \ldots, \bar{a}_m^t \rangle)$ is maximized.

It is assumed that the functions $C, G, R$ and $\tilde{T}$ are known, that $C, G$ can be equivalently represented by a finite set of linear constraints, that $R$ is a linear expression and that $\tilde{T}$ is a learned binarized neural network [16]. Next, we give an example planning problem where these assumptions are demonstrated.

*Example 1.* A simple instance of a learned automated planning problem $\tilde{\Pi}$ is as follows.

- The set of state variables is defined as $S = \{s_1\}$ where $s_1 \in \{0, 1\}$.
- The set of action variables is defined as $A = \{a_1\}$ where $a_1 \in \{0, 1\}$.
- The global function $C$ is defined as $C(\langle s_1, a_1 \rangle) = true$ when $s_1 + a_1 \leq 1$.
- The value of the state variable $s_1$ is $V_1 = 0$ at time step $t = 1$.
- The goal state function $G$ is defined as $G(\langle s_1 \rangle) = true$ if and only if $s_1 = 1$.
- The reward function $R$ is defined as $R(\langle s_1, a_1 \rangle) = -a_1$.
- The learned state transition function $\tilde{T}$ is in the form of a BNN, which will be described below.
- A planning horizon of $H = 4$.

A plan (assuming the BNN described later in Fig. 1) is $\bar{a}_1^1 = 1$, $\bar{a}_1^2 = 1$, $\bar{a}_1^3 = 1$, $\bar{a}_1^4 = 0$ with corresponding states $\bar{s}_1^1 = 0$, $\bar{s}_1^2 = 0$, $\bar{s}_1^3 = 0$, $\bar{s}_1^4 = 0$, $\bar{s}_1^5 = 1$. The total reward for the plan is $-3$. □

## 2.2 Binarized Neural Networks

Binarized neural networks (BNNs) are neural networks with binary weights and activation functions [16]. As a result, BNNs can learn memory-efficient models by replacing most arithmetic operations with bit-wise operations. The fully-connected BNN that defines the learned state transition function $\tilde{T}$, given $L$ layers with layer width $W_l$ in layer $l \in \{1, \ldots, L\}$, and a set of neurons $J(l) = \{u_{1,l}, \ldots, u_{W_l,l}\}$, is stacked in the following order.

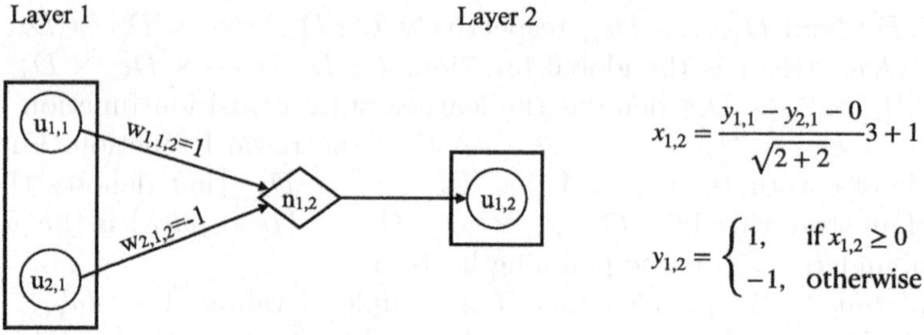


**Fig. 1.** Learned BNN with two layers $L = 2$ for the problem in Example 1. In this example learned BNN, the input layer $J(1)$ has neurons $u_{1,1}$ and $u_{2,1}$ representing $s_1$ and $a_1$, respectively. The node $n_{2,1}$ represents batch normalization for neuron $u_{2,1}$. Given the parameter values $w_{1,1,l} = 1$, $w_{2,1,l} = -1$, $\mu_{1,2} = 0$, $\sigma^2_{1,2} = 2$, $\epsilon_{1,2} = 2$, $\gamma_{1,2} = 3$ and $\beta_{j,l} = 1$, the input $x_{1,2}$ to neuron $u_{1,2}$ is calculated according to the formula specified in Sect. 2.2.

*Input Layer.* The first layer consists of neurons $u_{i,1} \in J(1)$ that represent the domain of the learned state transition function $\tilde{T}$. We will assume that the domains of action and state variables are binary, and let neurons $u_{1,1}, \ldots, u_{n,1} \in J(1)$ represent the state variables $S$ and neurons $u_{n+1,1}, \ldots, u_{n+m,1} \in J(1)$ represent the action variables $A$. During the training of the BNN, binary values 0 and 1 of action and state variables are represented by $-1$ and 1, respectively.

*Batch Normalization Layers.* For layers $l \in \{2, \ldots, L\}$, Batch Normalization [18] transforms the weighted sum of outputs at layer $l-1$ in $\Delta_{j,l} = \sum_{i \in J(l-1)} w_{i,j,l} y_{i,l-1}$ to inputs $x_{j,l}$ of neurons $u_{j,l} \in J(l)$ using the formula $x_{j,l} = \frac{\Delta_{j,l} - \mu_{j,l}}{\sqrt{\sigma^2_{j,l} + \epsilon_{j,l}}} \gamma_{j,l} + \beta_{j,l}$, where $y_{i,l-1}$ denotes the output of neuron $u_{i,l-1} \in J(l-1)$, and the parameters are the weight $w_{i,j,l}$, input mean $\mu_{j,l}$, input variance $\sigma^2_{j,l}$, numerical stability constant $\epsilon_{j,l}$, input scaling $\gamma_{j,l}$, and input bias $\beta_{j,l}$, all computed at training time.

*Activation Layers.* Given input $x_{j,l}$, the deterministic activation function $y_{j,l}$ computes the output of neuron $u_{j,l} \in J(l)$ at layer $l \in \{2, \ldots, L\}$, which is 1 if $x_{j,l} \geq 0$ and $-1$ otherwise. The last activation layer consists of neurons $u_{i,L} \in J(L)$ that represent the codomain of the learned state transition function $\tilde{T}$. We assume neurons $u_{1,L}, \ldots, u_{n,L} \in J(L)$ represent the state variables $S$.

The proposed BNN architecture is trained to learn the function $\tilde{T}$ from data that consists of measurements on the domain and codomain of the *unknown* state transition function $T : D_{s_1} \times \cdots \times D_{s_n} \times D_{a_1} \times \cdots \times D_{a_m} \rightarrow D_{s_1} \times \cdots \times D_{s_n}$. An example learned BNN for the problem of Example 1 is visualized in Fig. 1.

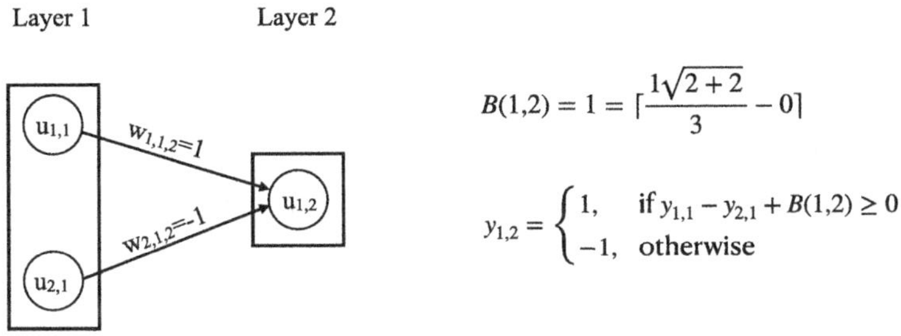


**Fig. 2.** The visualization of bias computation $B(1,2)$ for neuron $u_{1,2} \in J(2)$ in the example learned BNN presented in Fig. 1.

## 3 0–1 Integer Programming Model for the Learned Planning Problem

In this section, we present the 0–1 integer programming (IP) model from [25,26] previously used to solve learned automated planning problems. A 0–1 IP model can be solved optimally by a mixed-integer programming (MIP) solver (as was previously investigated [25,26]). Equivalently, this can be viewed as a pseudo-Boolean optimization (PBO) model to be solved using a PBO solver, since all the variables are 0–1 or equivalently Boolean.

*Decision Variables.* The 0–1 IP model uses the following decision variables:

- $X_{i,t}$ encodes whether action $a_i \in A$ is executed at time step $t \in \{1, \dots, H\}$ or not.
- $Y_{i,t}$ encodes whether we are in state $s_i \in S$ at time step $t \in \{1, \dots, H+1\}$ or not.
- $Z_{i,l,t}$ encodes whether neuron $u_{i,l} \in J(l)$ in layer $l \in \{1, \dots, L\}$ is activated at time step $t \in \{1, \dots, H\}$ or not.

*Parameters.* The 0–1 IP model uses the following parameters:

- $\bar{w}_{i,j,l}$ is the value of the learned BNN weight between neurons $u_{i,l-1} \in J(l-1)$ and $u_{j,l} \in J(l)$ in layer $l \in \{2, \dots, L\}$.
- $B(j,l)$ is the bias of neuron $u_{j,l} \in J(l)$ in layer $l \in \{2, \dots, L\}$. Given the values of normalization parameters $\bar{\mu}_{j,l}$, $\bar{\sigma}^2_{j,l}$, $\bar{\epsilon}_{j,l}$, $\bar{\gamma}_{j,l}$ and $\bar{\beta}_{j,l}$, the bias is computed as $B(j,l) = \left\lceil \frac{\bar{\beta}_{j,l}\sqrt{\bar{\sigma}^2_{j,l} + \bar{\epsilon}_{j,l}}}{\bar{\gamma}_{j,l}} - \bar{\mu}_{j,l} \right\rceil$. The visualization of the calculation of the bias $B(j,l)$ is presented in Fig. 2.

*Constraints.* The 0–1 IP model has the following constraints:

$$Y_{i,1} = V_i \quad \forall_{s_i \in S} \tag{1}$$
$$G(\langle Y_{1,H+1}, \dots, Y_{n,H+1} \rangle) = true \tag{2}$$
$$C(\langle Y_{1,t}, \dots, Y_{n,t}, X_{1,t}, \dots, X_{m,t} \rangle) = true \quad \forall_{t \in \{1,\dots,H\}} \tag{3}$$
$$Y_{i,t} = Z_{i,1,t} \quad \forall_{s_i \in S, t \in \{1,\dots,H\}} \tag{4}$$
$$X_{i,t} = Z_{i+n,1,t} \quad \forall_{a_i \in A, t \in \{1,\dots,H\}} \tag{5}$$
$$Y_{i,t+1} = Z_{i,L,t} \quad \forall_{s_i \in S, t \in \{1,\dots,H\}} \tag{6}$$
$$\big(B(j,l) - |J(l-1)|\big)\big(1 - Z_{j,l,t}\big) \leq In(j,l,t) \quad \forall_{u_{j,l} \in J(l), l \in \{2,\dots,L\}, t \in \{1,\dots,H\}} \tag{7}$$
$$\big(B(j,l) + |J(l-1)| + 1\big) Z_{j,l,t} - 1 \geq In(j,l,t) \quad \forall_{u_{j,l} \in J(l), l \in \{2,\dots,L\}, t \in \{1,\dots,H\}} \tag{8}$$

where the input expression $In(j,l,t)$ for neuron $u_{j,l} \in J(l)$ in layer $l \in \{2,\dots,L\}$ at time step $t \in \{1,\dots,H\}$ is equal to $\sum_{u_{i,l-1} \in J(l-1)} \bar{w}_{i,j,l}(2 \cdot Z_{i,l-1,t} - 1) + B(j,l)$. In the above model, constraints (1) set the initial value of every state variable. Constraints (2)–(3) enforce the global constraints (i.e., constraints representing $C$) and the goal constraints (i.e., constraints representing $G$). Constraints (4)–(6) map the input and output layers of the learned BNN to the corresponding state and action variables. Finally, constraints (7)–(8) model the activation of each neuron in the learned BNN, where the decision variable $Z_{j,l,t} \in \{0,1\}$ represents the output of neuron $u_{j,l} \in J(l)$ at time step $t \in \{1,\dots,H\}$ using the expression $(2 \cdot Z_{j,l,t} - 1) \in \{-1,1\}$.

*Objective Function.* The 0–1 IP model has the objective function

$$\max \sum_{t=1}^{H} R(\langle Y_{1,t+1}, \dots, Y_{n,t+1}, X_{1,t}, \dots, X_{m,t} \rangle), \tag{9}$$

which maximizes the total reward accumulated over time steps $t \in \{1,\dots,H\}$.

*Example 2.* The 0–1 IP (or the equivalent PBO) model that is presented in this section can be solved to find an optimal plan to the instance that is described in Example 1. The optimal plan is $\bar{a}_1^t = 0$ for all time steps $t \in \{1,2,3,4\}$, and the total reward for the optimal plan is 0. □

## 4 Theoretical Results

In this section, we establish the *NP*-completeness of finding feasible solutions to learned planning problems.

**Theorem 1.** *Finding a feasible solution to a learned planning problem $\tilde{\Pi}$ with a fully-connected batch normalized learned BNN $\tilde{T}$ is an NP-complete problem.*

*Proof.* We begin by showing that $\tilde{\Pi}$ is in *NP*. Given the values $\bar{A}^t$ of action variables $A$ for all time steps $t \in \{1,\dots,H\}$ and the initial values $V_i$ of state

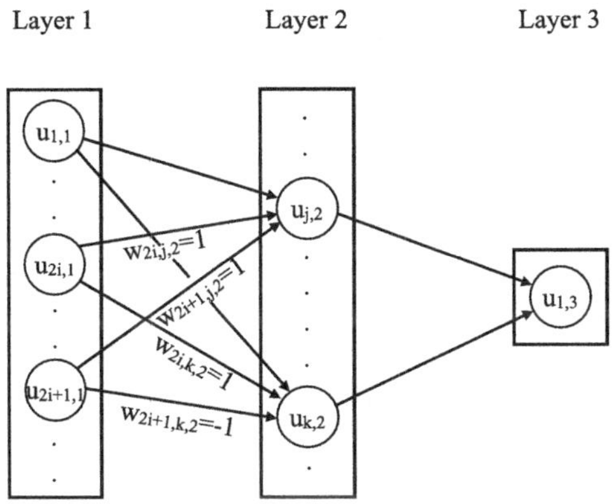


**Fig. 3.** Visualization of the *NP*-hardness proof by a reduction from a 3-CNF formula $\phi = \bigwedge_{j=1}^{q} c_j$ to the learned planning problem $\tilde{\Pi}$. In the first layer, two neurons $u_{2i,1}, u_{2i+1,1} \in J(1)$ together represent the Boolean variable $z_i$ from the formula $\phi$. When variable $z_i$ does not appear in clause $c_k$, the weights $\bar{w}_{2i,k,2}, \bar{w}_{2i+1,k,2}$ are set so the input to neuron $u_{k,2} \in J(2)$ is cancelled out (i.e., case (a) of step (7)). In the remaining cases, the weights $\bar{w}_{2i,j,2}, \bar{w}_{2i+1,j,2}$ are set to ensure the input to neuron $u_{j,2} \in J(2)$ is positive if and only if the respective literal that appears in clause $c_j$ evaluates to true (e.g., case (c) of step (7) is visualized).

variables $s_i \in S$, the learned BNN $\tilde{T}$ can predict the values $\bar{S}^t = \langle \bar{s}_1^t, \dots, \bar{s}_n^t \rangle \in D_{s_1} \times \dots \times D_{s_n}$ of all state variables $S$ for all time steps $t \in \{2, \dots, H+1\}$ in linear time in the size of the BNN and the value of the planning horizon $H$.

We proceed by showing that $\tilde{\Pi}$ is in *NP*-hard by a reduction from 3-SAT. Let $\phi$ be a 3-CNF formula such that $\phi = \bigwedge_{j=1}^{q} c_j$ for some positive integer $q$. Further let $z_1, \dots, z_r$ denote the (Boolean) variables that appear in the formula $\phi$ for some positive integer $r$. As visualized in Fig. 3, we define the learned planning problem $\tilde{\Pi}$ to represent any 3-CNF formula $\phi$ as follows:

1. Planning horizon $H = 1$.
2. State variable $S = \{s_1\}$.
3. Action variables $A = \{a_1, \dots, a_{2r}\}$.
4. The global function $C$ is true if and only if $a_{2i-1} = a_{2i}$ for all $i \in \{1, \dots, r\}$.
5. Neurons $J(1) = \{u_{1,1}, \dots, u_{1+2r,1}\}$ in the first layer.
6. Neurons $J(2) = \{u_{1,2}, \dots, u_{q,2}\}$ in the second layer. Each neuron $u_{i,2} \in J(2)$ is normalized so that $B(i,2) = 3$.
7. Set the learned weights between neurons $u_{2i,1}, u_{2i+1,1} \in J(1) \backslash u_{1,1}$ and $u_{j,2} \in J(2)$ according to the following rules. (a) If $z_i$ does not appear in clause $c_j$, set $\bar{w}_{2i,j,2} = 1, \bar{w}_{2i+1,j,2} = -1$, (b) else if the negation of $z_i$ appears in clause $c_j$ (i.e., $\neg z_i$), set $\bar{w}_{2i,j,2} = \bar{w}_{2i+1,j,2} = -1$, (c) else, set $\bar{w}_{2i,j,2} = \bar{w}_{2i+1,j,2} = 1$.
8. Neuron $J(3) = \{u_{1,3}\}$ in the third layer. Neuron $u_{1,3}$ is normalized such that $B(1,3) = -q$.
9. Set the learned weights $\bar{w}_{i,1,3} = 1$ between $u_{i,2} \in J(2)$ and $u_{1,3} \in J(3)$.
10. The goal state function $G$ is defined as $G(\langle 1 \rangle) = true$ and $G(\langle 0 \rangle) = false$.

In the reduction presented above, step (1) sets the value of the planning horizon $H$ to 1. Step (2) defines a single state variable $s_1$ to represent whether the formula $\phi$ is satisfied (i.e., $s_1 = 1$) or not (i.e., $s_1 = 0$). Step (3) defines action variables $a_1, \dots, a_{2r}$ to represent the Boolean variables $z_1, \dots, z_r$ in the formula $\phi$. Step (4) ensures that the pairs of action variables $a_{2i-1}, a_{2i}$ that represent the same Boolean variable $z_i$ take the same value. Step (5) defines the neurons in the first layer (i.e., $l = 1$) of the BNN. Step (6) defines the neurons in the second layer (i.e., $l = 2$) of the BNN. Each neuron $u_{i,2} \in J(2)$ represents a clause $c_i$ in the formula $\phi$, and the input of each neuron $u_{i,2}$ is normalized so that $B(i, 2) = 3$. Step (7) defines the weights between the first and the second layers so that the output of neurons $u_{2i,1}, u_{2i+1,1} \in J(1)$ only affects the input of the neurons $u_{j,2} \in J(2)$ in the second layer if and only if the Boolean variable $z_i$ appears in clause $c_j$. When this is not the case, the output of neurons $u_{2i,1}, u_{2i+1,1}$ are cancelled out due to the different values of their weights, so that $\bar{w}_{2i,j,2} + \bar{w}_{2i+1,j,2} = 0$. Steps (6) and (7) together ensure that for any values of $V_1$ and $\bar{w}_{1,j,2}$, neuron $u_{j,2} \in J(2)$ is activated if and only if at least one literal in clause $c_j$ evaluates to true.[1] Step (8) defines the single neuron in the third layer (i.e., $l = 3$) of the BNN. Neuron $u_{1,3} \in J(3)$ predicts the value of state variable $s_1$. Step (9) defines the weights between the second and the third layers so that the neuron $u_{1,3} \in J(3)$ activates if and only if all clauses in the formula $\phi$ are satisfied. Finally, step (10) ensures that the values of the actions constitute a solution to the learned planning problem $\tilde{\Pi}$ if and only if all clauses are satisfied. □

## 5 Constraint Programming Models for the Learned Planning Problem

In this section, we present two new constraint programming (CP) models to solve the learned automated planning problem $\tilde{\Pi}$. The models make use of reification rather than restricting themselves to linear constraints. This allows a more direct expression of the BNN constraints.

### 5.1 Constraint Programming Model 1

*Decision Variables and Parameters.* The CP model 1 uses the same set of decision variables and parameters as the 0–1 IP model previously described in Sect. 3.

[1] Each neuron $u_{j,2}$ that represents clause $c_j$ receives seven non-zero inputs (i.e., one from state and six from action variables). The bias $B(j, 2)$ is set so that the activation condition holds when at least one literal in clause $c_j$ evaluates to true. For example, the constraint $-2 + \bar{w}_{1,j,2}V_1 + B(j, 2) \geq 0$ represents the case when exactly one literal in clause $c_j$ evaluates to true where the terms $-2$ and $\bar{w}_{1,j,2}V_1$ represent the inputs from the six action variables and the single state variable, respectively. Similarly, the constraint $-6 + \bar{w}_{1,j,2}V_1 + B(j, 2) < 0$ represents the case when all literals in clause $c_j$ evaluate to false and the activation condition does not hold.

*Constraints.* The CP model 1 has the following constraints:

$$\text{Constraints (1)–(6)}$$

$$(In(j,l,t) \geq 0) = Z_{j,l,t} \qquad \forall_{u_{j,l} \in J(l), l \in \{2,\dots,L\}, t \in \{1,\dots,H\}} \tag{10}$$

where the input expression $In(j,l,t)$ for neuron $u_{j,l} \in J(l)$ in layer $l \in \{2,\dots,L\}$ at time step $t \in \{1,\dots,H\}$ is equal to $\sum_{u_{i,l-1} \in J(l-1)} \bar{w}_{i,j,l}(2 \cdot Z_{i,l-1,t} - 1) + B(j,l)$. In the above model, constraint (10) models the activation of each neuron in the learned BNN by replacing constraints (7)–(8).

*Objective Function.* The CP model 1 uses the same objective function as the 0–1 IP model previously described in Sect. 3.

### 5.2 Constraint Programming Model 2

*Decision Variables.* The CP model 2 uses the $X_{i,t}$ and $Y_{i,t}$ decision variables previously described in Sect. 3.

*Parameters.* The CP model 2 uses the same set of parameters as the 0–1 IP model previously described in Sect. 3.

*Constraints.* The CP model 2 has the following constraints:

$$\text{Constraints (1)–(6)}$$

$$(In(j,l,t) \geq 0) = Expr_{j,l,t} \qquad \forall_{u_{j,l} \in J(l), l \in \{2,\dots,L\}, t \in \{1,\dots,H\}} \tag{11}$$

where the input expression $In(j,l,t)$ for neuron $u_{j,l} \in J(l)$ in layer $l \in \{2,\dots,L\}$ at time step $t \in \{1,\dots,H\}$ is equal to $\sum_{u_{i,l-1} \in J(l-1)} \bar{w}_{i,j,l}(2 \cdot Expr_{i,l-1,t} - 1) + B(j,l)$, and output expression $Expr_{j,l,t}$ represents the binary output of neuron $u_{j,l} \in J(l)$ in layer $l \in \{2,\dots,L\}$ at time step $t \in \{1,\dots,H\}$. In the above model, constraint (11) models the activation of each neuron in the learned BNN by replacing the decision variable $Z_{j,l,t}$ in constraint (10) with the expression $Expr_{j,l,t}$ The difference between an integer variable and an expression is that during solving the solver does not store the domain (current set of possible values) for an expression. Expressions allow more scope for the presolve of CP Optimizer [17] to rewrite the constraints to a more suitable form, and allow the use of more specific propagation scheduling.

*Objective Function.* The CP model 2 uses the same objective function as the 0–1 IP model that is previously described in Sect. 3.

## 6 Model Symmetry

Examining the 0–1 IP (or equivalently the PBO) model, one can see the bulk of the model involves copies of the learned BNN constraints over all time steps.

These constraints model the activation of each neuron (i.e., constraints (4)–(8)) and constrain the input of the BNN (i.e., constraint (3)). The remainder of the model is constraints on the initial and goal states (i.e., constraints (1)–(2)). So if we ignore the initial and goal state constraints, the model is symmetric over the time steps. Note that this symmetry property is not a *global* one: the model is not symmetric as a whole. Rather, this *local* symmetry arises because subsets of constraints are isomorphic to each other. Because of this particular form of symmetry, the classic approach of adding symmetry breaking predicates [5] would not be sound, as this requires global symmetry.

Instead, we exploit this symmetry by deriving symmetric nogoods on-the-fly. If a nogood is derived purely from constraints (3)–(8), and if a sufficiently small subset of time steps was involved in its derivation, then we can shift the time steps of this nogood over the planning horizon, learning a valid symmetric nogood. To track which constraints a nogood is derived from, we use the SAT technique of *marker literals* lifted to PBO. Each constraint is extended with some marker literal, which, if true, enforces the constraint, and if false, trivially satisfies it. During the search, these marker literals are a priori assumed true, so we are solving essentially the same problem, but the nogood learning mechanism of the PBO solver ensures the marker literal of a constraint appears in a nogood if that constraint was required in the derivation of the nogood.

By introducing marker literals $L_t$ for all time steps $t \in \{1, \ldots, H\}$ for constraints (3)–(8) and an extra "asymmetric" marker literal $L^*$ for constraint (2) and the constraints originating from bounding the objective function, and then treating all initial state constraints as markers, we can track if only constraints (3)–(8) were involved in the derivation of a nogood, and if so, for which time steps. When we find that the constraints involved in creating a nogood only refer to constraints from time steps $l$ to $u$, then we know that symmetric copies of these nogoods are also valid for time steps $l + \Delta$ to $u + \Delta$ for all $-l < \Delta < 0$, $0 < \Delta \leq H - u$. Our approach to exploiting symmetry is similar to the ones proposed for bounded model checking in SAT [30,31].

*Example 3.* Marker literals are used to "turn on" the constraints and they are set to true throughout the search. For example constraint (8) becomes

$$L_t \rightarrow (B(j,l) + J(l-1) + 1)Z_{j,l,t} - 1 \geq In(j,l,t) \; \forall_{u_{j,l} \in J(l), l \in \{2,\ldots,L\}, t \in \{1,\ldots,H\}}$$

or equivalently, the binary linear constraint

$$\mathbf{M}(1-L_t)+(B(j,l)+J(l-1)+1)Z_{j,l,t}-1 \geq In(j,l,t) \; \forall_{u_{j,l} \in J(l), l \in \{2,\ldots,L\}, t \in \{1,\ldots,H\}}$$

with $\mathbf{M}$ chosen large enough so that the constraint trivially holds if $L_t = 0$. □

We consider two ways of symmetric nogood derivation:

- All: whenever we discover a nogood that is a consequence of constraints from time steps $l$ to $u$, we add a suitably renamed copy of the nogood to the variables for time steps $l + \Delta$ to $u + \Delta$ for all $-l < \Delta < 0$, $0 < \Delta \leq H - u$,

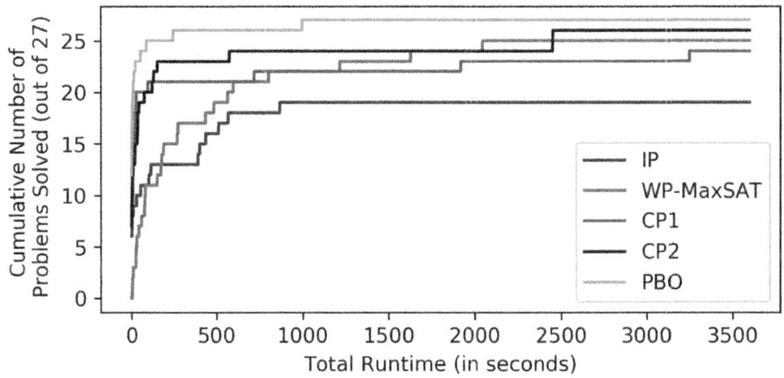


**Fig. 4.** Cumulative number of problems solved by IP (blue), WP-MaxSAT (red), CP1 (green), CP2 (black) and PBO (orange) models over 27 instances of the problem $\tilde{\Pi}$ within the time limit. (Color figure online)

- Propagate: we consider each possible $\Delta$ above, but only add the renamed nogood if it will immediately propagate or fail, similar to a SAT symmetric clause learning heuristic [8].

Finally, we denote Base as the version of RoundingSat that does not add the symmetric nogoods.

*Example 4.* Consider the problem in Example 1. Assume we generate a nogood $X_{1,1} \vee Y_{1,1} \vee \neg X_{1,2}$, which is a consequence only of constraints for the BNNs for time steps 1 and 2. The actual generated nogood is then $\neg L_1 \vee \neg L_2 \vee X_{1,1} \vee Y_{1,1} \vee \neg X_{1,2}$ which illustrates that it depends only on the constraints in time steps 1 and 2. We can then add a symmetric copy $\neg L_2 \vee \neg L_3 \vee X_{1,2} \vee Y_{1,2} \vee \neg X_{1,3}$ for time steps 2 and 3, as well as $\neg L_3 \vee \neg L_4 \vee X_{1,3} \vee Y_{1,3} \vee \neg X_{1,4}$ for steps 3 and 4. These new constraints must be correct, since the BNN constraints for time steps $t \in \{1, 2, 3, 4\}$ are all symmetric. The marker literals are added so that later nogoods making use of these nogoods also track which time steps were involved in their generation. Using All we add both these nogoods, using Propagate we only add those that are unit or false in the current state of the solver. □

## 7 Experimental Results

In this section, we present results on two sets of computational experiments. In the first set of experiments, we compare different approaches to solving the learned planning problem $\tilde{\Pi}$ with mathematical optimization models. In the second set of experiments, we present preliminary results on the effect of deriving symmetric nogoods when solving $\tilde{\Pi}$ over long horizons $H$.

### 7.1 Experiments 1

We first experimentally test the runtime efficiency of solving the learned planning problem $\tilde{\Pi}$ with mathematical optimization models using off-the-shelf solvers.

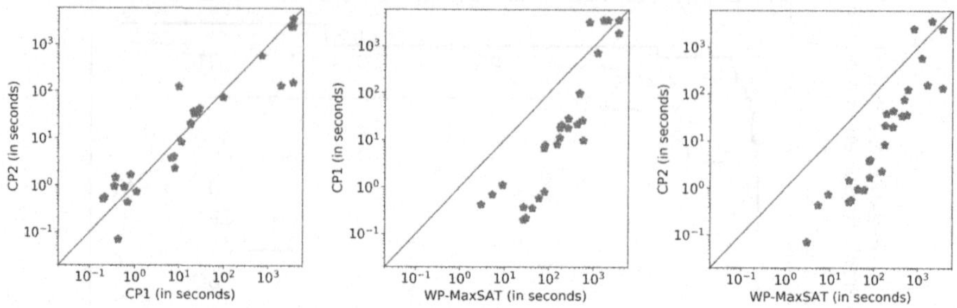


(a) CP1 vs. CP2 (b) WP-MaxSAT vs. CP1 (c) WP-MaxSAT vs. CP2

**Fig. 5.** Pairwise runtime comparison between WP-MaxSAT, CP1 and CP2 models over 27 instances of problem $\tilde{\Pi}$ within the time limit.

All the existing benchmark instances of the learned planning problem $\tilde{\Pi}$ (i.e., 27 in total) were used [26]. We ran the experiments on a MacBookPro with 2.8 GHz Intel Core i7 16GB memory, with one hour total time limit per instance. We used the MIP solver CPLEX 12.10 [17] to optimize the 0–1 IP model, MaxHS [6] with underlying CPLEX 12.10 linear programming solver to optimize the WP-MaxSAT model [26], and CP Optimizer 12.10 [17] to optimize the CP Model 1 (CP1) and the CP Model 2 (CP2). Finally, we optimized a pseudo-Boolean optimization (PBO) model, which simply replaces all binary variables in the 0–1 IP model with Boolean variables, using RoundingSat [10].

In Fig. 4, we visualize the cumulative number of problems solved by all five models, namely: IP (blue), WP-MaxSAT (red), CP1 (green), CP2 (black) and PBO (orange), over 27 instances of the learned planning problem $\tilde{\Pi}$ within one hour time limit. Figure 4 clearly highlights the experimental efficiency of solving the PBO model. We find that using the PBO model with RoundingSat solves all existing benchmarks under 1000 s. In contrast, we observe that the 0–1 IP model performs poorly, with only 19 instances out of 27 solved within the one hour time limit. The remaining three models, WP-MaxSAT, CP1 and CP2, demonstrate relatively comparable runtime performance, which we explore in more detail next.

In Figs. 5a, 5b and 5c, we present scatter plots comparing the WP-MaxSAT, CP1 and CP2 models. In each figure, each dot (red) represents an instance of the learned planning problem $\tilde{\Pi}$ and each axis represents a model (i.e., WP-MaxSAT, CP1 or CP2). If a dot falls below the diagonal line (blue), it means the corresponding instance is solved faster by the model represented by the y-axis than the one represented by the x-axis. In Fig. 5a, we compare the two CP models CP1 and CP2. A detailed inspection of Fig. 5a shows a comparable runtime performance on the instances that take less than 1000 s to solve (i.e., most dots fall closely to the diagonal line). In the remaining two instances that are solved by CP2 under 1000 s, CP1 runs out of the one hour time limit. These results suggest that using expressions instead of decision variables to model the neurons of the learned BNN allows the CP solver to solve harder instances (i.e., instances that

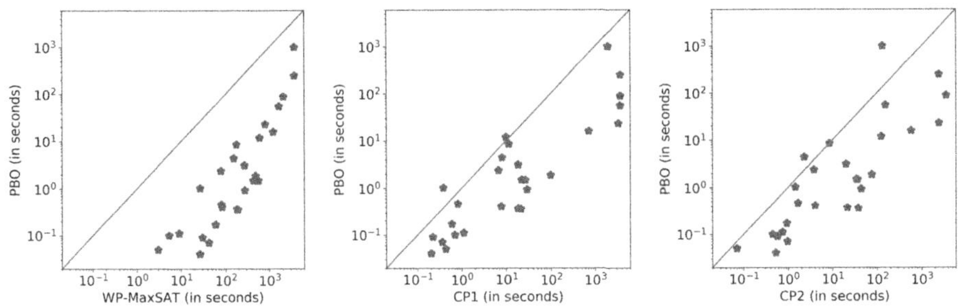


PBO vs. WP-MaxSAT    PBO vs. CP1    PBO vs. CP2

**Fig. 6.** Pairwise runtime comparison between PBO, and WP-MaxSAT, CP1 and CP2 models over 27 instances of problem $\tilde{\Pi}$ within the time limit.

take more than 1000 s to solve) more efficiently. In Figs. 5b and 5c, we compare CP1 and CP2 against the WP-MaxSAT model, respectively. Both figures show a similar trend on runtime performance; the CP models close instances that take less than 1000 s to solve by one to two orders of magnitude faster than the WP-MaxSAT model, and the WP-MaxSAT model performs comparably to the CP models on the harder instances. Overall, we find that WP-MaxSAT solves one more and one less instances compared to CP1 and CP2 within the one hour time limit, respectively. These results suggest that the WP-MaxSAT model pays a heavy price for the large size of its compilation when the instances take less than 1000 s to solve, and only benefits from its SAT-based encoding for harder instances.

Next, we compare the runtime performance of WP-MaxSAT, CP1 and CP2 against the best performing model (i.e., PBO) in more detail in Figs. 6a, 6b and 6c. These plots show that the PBO model significantly outperforms the WP-MaxSAT, CP1 and CP2 models across all instances. Specifically, Fig. 6a shows that the PBO model is better than the previous state-of-the-art WP-MaxSAT model across all instances by one to two orders of magnitude in terms of runtime performance. Similarly, Figs. 6b and 6c show that the PBO model outperforms both CP models across all instances, except in one and two instances, respectively, by an order of magnitude.

It is interesting that the 0–1 IP model works so poorly for the MIP solver, while the equivalent PBO model is solved efficiently using a PBO solver. It seems that the linear relaxations used by the MIP solver are too weak to generate useful information, and it ends up having to fix activation variables in order to reason meaningfully. In contrast, it appears that the PBO solver is able to determine some useful information from the neuron constraints without necessarily fixing the activation variables—probably since it uses integer-based cutting planes reasoning [4] rather than continuous linear programming reasoning for the linear expressions—and the nogood learning helps it avoid repeated work.

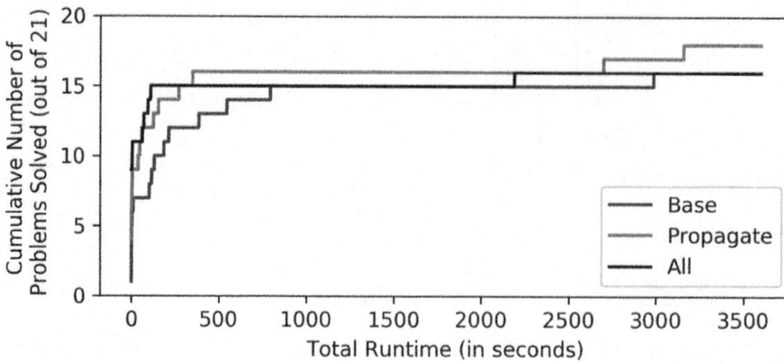


**Fig. 7.** Cumulative number of problems solved for Base (blue), Propagate (red) and All (black) models over 21 instances of the problem $\tilde{\Pi}$ within the time limit. (Color figure online)

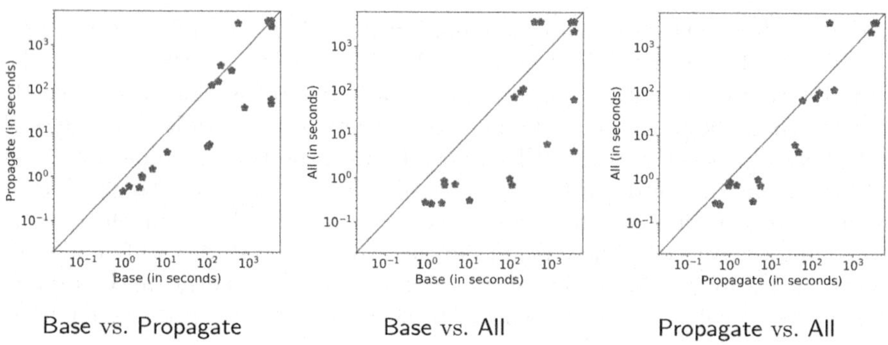


**Fig. 8.** Pairwise runtime comparison between Base, Propagate and All models over 21 instances of problem $\tilde{\Pi}$ within the time limit.

### 7.2 Experiments 2

We next evaluate the effect of symmetric nogood derivation on solving the learned planning problem $\tilde{\Pi}$ over long horizons $H$. For these experiments, we generated instances by incrementing the value of the planning horizon $H$ in the benchmark instances in Sect. 7.1, and used the same hardware and time limit settings. We modified the best performing solver RoundingSat [10] to include symmetry reasoning as discussed in Sect. 6.

In Fig. 7, we visualize the cumulative number of problems solved for all three versions of RoundingSat, namely Base (blue), Propagate (red), and All (black), over 21 instances of the learned planning problem $\tilde{\Pi}$ over long horizons $H$ within one hour time limit. Figure 7 demonstrates that symmetric nogood derivation can improve the efficiency of solving the underlying PBO model. We find that Propagate solves the most instances within the time limit. A more detailed inspection of the results further suggests that between the remaining two version of RoundingSat, All solves more instances faster compared to Base.

Next, in Figs. 8a, 8b and 8c, we explore the pairwise runtime comparisons of the three versions of RoundingSat in more detail. In Figs. 8a and 8b, we compare Propagate and All against Base, respectively. It is clear from these scatter plots that Propagate and All outperform Base in terms of runtime performance. Specifically, in Fig. 8b, we find that All outperforms Base by up to three orders of magnitude in terms of runtime performance. Finally, in Fig. 8c, we compare the two versions of RoundingSat that are enhanced with symmetric nogood derivation. A detailed inspection of Fig. 8c reveals that All is slightly faster than Propagate in general.

## 8 Related Work, Conclusions and Future Work

In this paper, we studied the important problem of automated planning with learned BNNs, and made four important contributions. First, we showed that the feasibility problem is *NP*-complete. Unlike the proof presented for the task of verifying learned BNNs [3], our proof does not rely on setting weights to zero (i.e., sparsification). Instead, our proof achieves the same expressivity for fully connected BNN architectures, without adding additional layers or increasing the width of the layers, by representing each input with two copies of action variables. Second, we introduced two new CP models for the problem. Third, we presented detailed computational results for solving the existing instances of the problem. Lastly, we studied the effect of deriving symmetric nogoods on solving new instances of the problem with long horizons.

It appears that BNN models provide a perfect class of problems for pseudo-Boolean solvers, since each neuron is modelled by pseudo-Boolean constraints, but the continuous relaxation is too weak for MIP solvers to take advantage of, while propagation-based approaches suffer since they are unable to reason about linear expressions directly. PBO solvers directly reason about integer (0–1) linear expressions, making them very strong on this class of problems.

Our results have the potential to improve other important tasks with learned BNNs (and other DNNs), such as automated planning in real-valued action and state spaces [27,35,36], decision making in discrete action and state spaces [21], goal recognition [11], training [33], verification [9,15,19,22], robustness evaluation [32] and defenses to adversarial attacks [34], which rely on efficiently solving similar problems that we solve in this paper. The derivation of symmetric nogoods is a promising avenue for future work, in particular, if a sufficient number of symmetric nogoods can be generated. Relating the number of derived symmetric nogoods to the wall-clock speed-up of the solver or the reduction of the search tree might shed further light on the efficacy of this approach.

**Acknowledgements.** Some of our preliminary computational experiments used resources provided by the Swedish National Infrastructure for Computing (SNIC) at the High Performance Computing Center North (HPC2N) at Umeå University. Jo Devriendt and Jakob Nordström were supported by the Swedish Research Council grant 2016-00782, and Jakob Nordström also received funding from the Independent Research Fund Denmark grant 9040-00389B.

## References

1. Bennett, S.W., DeJong, G.F.: Real-world robotics: learning to plan for robust execution. Mach. Learn. **23**, 121–161 (1996)
2. Benson, S.S.: Learning action models for reactive autonomous agents. Ph.D. thesis, Stanford University, Stanford, CA, USA (1997)
3. Cheng, C.-H., Nührenberg, G., Huang, C.-H., Ruess, H.: Verification of binarized neural networks via inter-neuron factoring. In: Piskac, R., Rümmer, P. (eds.) VSTTE 2018. LNCS, vol. 11294, pp. 279–290. Springer, Cham (2018). https://doi.org/10.1007/978-3-030-03592-1_16
4. Cook, W., Coullard, C.R., Turán, G.: On the complexity of cutting-plane proofs. Discret. Appl. Math. **18**(1), 25–38 (1987)
5. Crawford, J., Ginsberg, M., Luks, E., Roy, A.: Symmetry-breaking predicates for search problems. In: Proceedings of the Fifth International Conference on Principles of Knowledge Representation and Reasoning, pp. 148–159. Morgan Kaufmann (1996)
6. Davies, J., Bacchus, F.: Solving MAXSAT by solving a sequence of simpler SAT instances. In: Lee, J. (ed.) CP 2011. LNCS, vol. 6876, pp. 225–239. Springer, Heidelberg (2011). https://doi.org/10.1007/978-3-642-23786-7_19
7. Davies, T.O., Pearce, A.R., Stuckey, P.J., Lipovetzky, N.: Sequencing operator counts. In: Proceedings of the Twenty-Fifth International Conference on Automated Planning and Scheduling, pp. 61–69. AAAI Press (2015)
8. Devriendt, J., Bogaerts, B., Bruynooghe, M.: Symmetric explanation learning: effective dynamic symmetry handling for SAT. In: Gaspers, S., Walsh, T. (eds.) SAT 2017. LNCS, vol. 10491, pp. 83–100. Springer, Cham (2017). https://doi.org/10.1007/978-3-319-66263-3_6
9. Ehlers, R.: Formal verification of piece-wise linear feed-forward neural networks. In: D'Souza, D., Narayan Kumar, K. (eds.) ATVA 2017. LNCS, vol. 10482, pp. 269–286. Springer, Cham (2017). https://doi.org/10.1007/978-3-319-68167-2_19
10. Elffers, J., Nordström, J.: Divide and conquer: towards faster pseudo-boolean solving. In: Proceedings of the Twenty-Seventh International Joint Conference on Artificial Intelligence, IJCAI 2018, pp. 1291–1299 (2018)
11. Fraga Pereira, R., Vered, M., Meneguzzi, F., Ramírez, M.: Online probabilistic goal recognition over nominal models. In: Proceedings of the Twenty-Eighth International Joint Conference on Artificial Intelligence, pp. 5547–5553. International Joint Conferences on Artificial Intelligence Organization (2019)
12. Gil, Y.: Acquiring domain knowledge for planning by experimentation. Ph.D. thesis, Carnegie Mellon University, USA (1992)
13. Helmert, M.: The fast downward planning system. J. Artif. Intell. Res. **26**, 191–246 (2006)
14. Hoffmann, J., Nebel, B.: The FF planning system: fast plan generation through heuristic search. J. Artif. Intell. Res. **14**, 253–302 (2001)
15. Huang, X., Kwiatkowska, M., Wang, S., Wu, M.: Safety verification of deep neural networks. In: Majumdar, R., Kunčak, V. (eds.) CAV 2017. LNCS, vol. 10426, pp. 3–29. Springer, Cham (2017). https://doi.org/10.1007/978-3-319-63387-9_1
16. Hubara, I., Courbariaux, M., Soudry, D., El-Yaniv, R., Bengio, Y.: Binarized neural networks. In: Proceedings of the Thirtieth International Conference on Neural Information Processing Systems, NIPS 2016, pp. 4114–4122. Curran Associates Inc., USA (2016)
17. IBM: IBM ILOG CPLEX Optimization Studio CPLEX User's Manual (2020)

18. Ioffe, S., Szegedy, C.: Batch normalization: accelerating deep network training by reducing internal covariate shift. In: Proceedings of the Thirty-Second International Conference on International Conference on Machine Learning, ICML, pp. 448–456. JMLR.org (2015)
19. Katz, G., Barrett, C., Dill, D.L., Julian, K., Kochenderfer, M.J.: Reluplex: an efficient SMT solver for verifying deep neural networks. In: Majumdar, R., Kunčak, V. (eds.) CAV 2017. LNCS, vol. 10426, pp. 97–117. Springer, Cham (2017). https://doi.org/10.1007/978-3-319-63387-9_5
20. Kautz, H., Selman, B.: Planning as satisfiability. In: Proceedings of the Tenth European Conference on Artificial Intelligence, ECAI 1992, pp. 359–363 (1992)
21. Lombardi, M., Gualandi, S.: A Lagrangian propagator for artificial neural networks in constraint programming. Constraints **21**, 435–462 (2016)
22. Narodytska, N., Kasiviswanathan, S., Ryzhyk, L., Sagiv, M., Walsh, T.: Verifying properties of binarized deep neural networks. In: Proceedings of the Thirty-Second AAAI Conference on Artificial Intelligence, pp. 6615–6624 (2018)
23. Nau, D., Ghallab, M., Traverso, P.: Automated Planning: Theory & Practice. Morgan Kaufmann Publishers Inc., San Francisco (2004)
24. Pommerening, F., Röger, G., Helmert, M., Bonet, B.: LP-based heuristics for cost-optimal planning. In: Proceedings of the Twenty-Fourth International Conference on Automated Planning and Scheduling, ICAPS 2014, pp. 226–234. AAAI Press (2014)
25. Say, B., Sanner, S.: Planning in factored state and action spaces with learned binarized neural network transition models. In: Proceedings of the Twenty-Seventh International Joint Conference on Artificial Intelligence, IJCAI 2018, pp. 4815–4821 (2018)
26. Say, B., Sanner, S.: Compact and efficient encodings for planning in factored state and action spaces with learned binarized neural network transition models. Artif. Intell. **285**, 103291 (2020)
27. Say, B., Sanner, S., Thiébaux, S.: Reward potentials for planning with learned neural network transition models. In: Schiex, T., de Givry, S. (eds.) CP 2019. LNCS, vol. 11802, pp. 674–689. Springer, Cham (2019). https://doi.org/10.1007/978-3-030-30048-7_39
28. Say, B., Wu, G., Zhou, Y.Q., Sanner, S.: Nonlinear hybrid planning with deep net learned transition models and mixed-integer linear programming. In: Proceedings of the Twenty-Sixth International Joint Conference on Artificial Intelligence, IJCAI 2017, pp. 750–756 (2017)
29. Shen, W.M., Simon, H.A.: Rule creation and rule learning through environmental exploration. In: Proceedings of the Eleventh International Joint Conference on Artificial Intelligence, IJCAI 1989, pp. 675–680. Morgan Kaufmann Publishers Inc., San Francisco (1989)
30. Shtrichman, O.: Tuning SAT checkers for bounded model checking. In: Emerson, E.A., Sistla, A.P. (eds.) CAV 2000. LNCS, vol. 1855, pp. 480–494. Springer, Heidelberg (2000). https://doi.org/10.1007/10722167_36
31. Shtrichman, O.: Pruning techniques for the SAT-based bounded model checking problem. In: Margaria, T., Melham, T. (eds.) CHARME 2001. LNCS, vol. 2144, pp. 58–70. Springer, Heidelberg (2001). https://doi.org/10.1007/3-540-44798-9_4
32. Tjeng, V., Xiao, K., Tedrake, R.: Evaluating robustness of neural networks with mixed integer programming. In: Proceedings of the Seventh International Conference on Learning Representations. ICLR (2019)

33. Toro Icarte, R., Illanes, L., Castro, M.P., Cire, A.A., McIlraith, S.A., Beck, J.C.: Training binarized neural networks using MIP and CP. In: Schiex, T., de Givry, S. (eds.) CP 2019. LNCS, vol. 11802, pp. 401–417. Springer, Cham (2019). https://doi.org/10.1007/978-3-030-30048-7_24
34. Wong, E., Kolter, Z.: Provable defenses against adversarial examples via the convex outer adversarial polytope. In: Proceedings of the Thirty-Fifth International Conference on Machine Learning. ICML (2018)
35. Wu, G., Say, B., Sanner, S.: Scalable planning with tensorflow for hybrid nonlinear domains. In: Proceedings of the Thirty First Annual Conference on Advances in Neural Information Processing Systems, Long Beach, CA (2017)
36. Wu, G., Say, B., Sanner, S.: Scalable planning with deep neural network learned transition models. J. Artif. Intell. Res. **68**, 571–606 (2020)

# Omissions in Constraint Acquisition

Dimosthenis C. Tsouros[1(✉)], Kostas Stergiou[1], and Christian Bessiere[2]

[1] Department of Electrical and Computer Engineering, University of Western Macedonia, Kozani, Greece
{dtsouros,kstergiou}@uowm.gr

[2] CNRS, University of Montpellier, Montpellier, France
bessiere@lirmm.fr

**Abstract.** Interactive constraint acquisition is a special case of query-directed learning, also known as "exact" learning. It is used to assist non-expert users in modeling a constraint problem automatically by posting examples to the user that have to be classified as solutions or non-solutions. One significant issue that has not been addressed in the literature of constraint acquisition is the possible presence of uncertainty in the answers of the users. We address this by introducing Limited Membership Queries, where the user has the option of replying "I don't know", corresponding to "omissions" in exact learning. We present two algorithms for handling omissions. The first one deals with omissions that are independent events, while the second assumes that omissions are related to gaps in the user's knowledge. We present theoretical results about both methods and we evaluate them on benchmark problems. Importantly, our second algorithm can not only learn (a part of) the target network, but also the constraints that cause the user's uncertainty.

## 1 Introduction

A major bottleneck in the use of Constraint Programming (CP) is modeling. Expressing a combinatorial problem as a constraint network requires considerable expertise in the field. Hence, one of the major challenges in CP research is that of efficiently obtaining a good model of a real problem without relying on expert human modellers [14–16]. Constraint acquisition can assist non-expert users in modeling a constraint problem automatically. It has started to attract a lot of attention as constraint acquisition systems can learn the model of a constraint problem using a set of examples that are posted as queries to a human user or to a software system [6–8,19,20].

*Active* or *interactive* constraint acquisition systems interact with the user while learning the constraint network. This is a special case of query-directed learning, also known as "exact learning" [10,11]. State-of-the-art constraint acquisition algorithms like QuAcq [5], MQuAcq [20] and MQuAcq-2 [19] use the version space learning method [18], extended for learning constraint networks. In such systems, the basic query is to ask the user to classify an example


H. Simonis (Ed.): CP 2020, LNCS 12333, pp. 935–951, 2020.
https://doi.org/10.1007/978-3-030-58475-7_54

as a solution or not solution. This "yes/no" type of question is called membership query [1], and this is the type of query that has received the most attention in active constraint acquisition.

One significant issue that has not been addressed in the literature is the presence of uncertainty, or even errors, in the answers of the user. The constraint acquisition algorithms that have been proposed are guaranteed to perform well and learn the target constraint network under the assumption that queries are always answered with certainty and correctly. However, this is not a realistic assumption as questions posted by the algorithm can be too difficult for humans to always answer reliably. Thus, it may happen that the user is uncertain about her/his answers or even gives erroneous answers. In this paper we deal with the case where the user is uncertain about her/his answers, leaving the case of erroneous answers for future work.

In the context of exact learning, uncertainty is typically captured through "omissions" in the replies of the users. The omissions can be persistent or not. In this work, we focus on persistent omissions. Such omissions have been studied for several classes of concepts [2,3,13,17]. Two main models of omissions in answers to membership queries have been introduced:

1. Learning from Randomly Fallible Teachers (RFT) [2,3]: The omissions are assumed to be independent, as "independent coin flips the first time each query is made" [3].
2. Learning from a Consistently Ignorant Teacher (CIT) [13]: In this model it is assumed that the omissions are related to a gap in the user's knowledge. Thus, the omissions are not only persistent but also they are consistent with the rest of the answers of the user. This means that if the answers to some queries imply a particular answer to another query, the latter cannot be answered with an omission.

Concerning RFT, Angluin et al. [3] presented an algorithm that can learn the target concept using equivalence and incomplete membership queries. In the exact learning model defined by [2] the learning system can learn exactly a target concept using equivalence and membership queries with at most some number $l$ of errors or omissions in the answers of the user to the membership queries. This model introduced the *limited* membership queries (LMQ) and the *malicious* membership queries. A LMQ may be answered either by precisely classifying the example, or with the special answer "I don't know" that corresponds to an omission. In a malicious membership query, the classification by the user may be wrong. The examples answered with "I don't know" are allowed to be classified arbitrarily by the final hypothesis of a learning algorithm. The above models have been extended to other classes of concepts [9]. Focusing on learning from a CIT, Frazier et al. [13] introduced learning algorithms for several concept classes like k-term DNF formulas, decision trees etc.

In this work we address the problem of uncertainty in user answers in the context of constraint acquisition, for the first time. We focus on persistent omissions inspired by both the RFT and CIT models. We are specifically interested in the important case where there exist relations about the entities (variables) of the problem that the user is uncertain about. As a result, the user may find it difficult to classify some examples with absolute certainty. To address this, we introduce LMQs in constraint acquisition and propose two methods to handle "I don't know" replies by extending the state-of-the-art algorithm MQuAcq-2.

The first method is a baseline one that simply ignores omissions, assuming that nothing can be learned from them. The reasoning behind this is inspired by the RFT model, where the omissions are assumed to be independent.

The second method, which is our main contribution, is related to the CIT learning model and is based on the assumption that through the interaction between the learner and the user it may be possible to identify and exploit the gap in the user's knowledge, i.e. the "uncertain" constraints. For instance, the user may not be certain if a large example is a solution or not because of uncertainty about a relation between some variables. However, if a part of the example that does not include these "problematic" variables is posted to the user, then she/he may be able to classify it with certainty. Our method exploits the idea of posting partial examples that are built by dividing an example that was classified as an omission, to seek the parts of the example that cause the confusion to the user, and hence, to learn the "uncertain" constraints. As we demonstrate, this method does not only learn such constraints, but using knowledge inferred while seeking them, it also significantly cuts down the number of queries and the cpu time required for convergence.

We prove the correctness of our main method and give complexity results for both methods. We also present experimental results that evaluate our methods in the context of both RFT and CIT learning.

The rest of the paper is organized as follows. Section 2 gives background on interactive constraint acquisition. Section 3 focuses on the proposed methods. Experiments are presented in Sect. 4. Section 5 concludes the paper.

## 2 Background

The *vocabulary* $(X, D)$ is a finite set of $n$ variables $X = \{x_1, ..., x_n\}$ and a domain $D = \{D(x_1), ..., D(x_n)\}$, where $D(x_i) \subset \mathbb{Z}$ is the finite set of values for $x_i$. The vocabulary is the common knowledge shared by the user and the constraint acquisition system. A *constraint* $c$ is a pair (rel($c$), var($c$)), where var($c$) $\subseteq X$ is the *scope* of the constraint and rel($c$) is the relation between the variables in var($c$). rel($c$) specifies which of their assignments are allowed. $|var(c)|$ is called the *arity* of the constraint. A *constraint network* is a set $C$ of constraints on the vocabulary $(X, D)$. A constraint network that contains at most one constraint for each subset of variables (i.e., for each scope) is called a *normalized constraint network*. Following the literature, we will assume that the constraint network is normalized. Besides the vocabulary, the learner has a *language* $\Gamma$ consisting of *bounded arity* constraints.

An example $e_Y$ is an assignment on a set of variables $Y \subseteq X$. $e_Y$ violates a constraint $c$ iff var($c$) $\subseteq Y$ and the projection $e_{var(c)}$ of $e_Y$ on the variables in the scope var($c$) of the constraint is not in rel($c$). A complete assignment $e_X$ that is accepted by all the constraints in $C$ is a solution to the problem. $sol(C)$ is the set of solutions of $C$. An assignment $e_Y$ is called a partial solution iff it is accepted by all the constraints in $C$ with a scope $S \subseteq Y$. Observe that a partial solution is not necessarily part of a complete solution.

Using terminology from machine learning, concept learning can be defined as learning a Boolean function from examples. A *concept* is a Boolean function over $D^X$ that assigns to each example $e \in D^X$ a value in $\{0, 1\}$, or in other words, classifies it as negative or positive. The target concept $f_T$ is a concept that assigns 1 to $e$ if $e$ is a solution to the problem and 0 otherwise. In constraint acquisition, the target concept, also called target constraint network, is any constraint network $C_T$ such that $sol(C_T) = \{e \in D^X \mid f_T(e) = 1\}$. The *constraint bias* $B$ is a set of constraints on the vocabulary $(X, D)$, built using the constraint language $\Gamma$. The bias is the set of all possible constraints from which the system can learn the target constraint network. $\kappa_B(e_Y)$ represents the set of constraints in $B$ that reject $e_Y$.

In exact learning, the question asking the user to determine if an example $e_X$ is a solution to the problem that the user has in mind is called a *membership query* $ASK(e)$. In the following we will use the terms *example* and *query* interchangeably. The answer to a membership query is positive if $f_T(e) = 1$ and negative if $f_T(e) = 0$. A *partial query* $ASK(e_Y)$, with $Y \subseteq X$, asks the user to determine if $e_Y$, which is an assignment in $D^Y$, is a partial solution or not. We assume that all queries are answered correctly or with an omission by the user.

In partial queries the assumption is that the user considers only the assigned variables in the query posted. So, if the relation between the variables is insufficiently clear because some variable assignments are missing, then this does not affect the user's answer. Classifying a partial example as positive does not mean that it is necessarily part of a complete solution.

For instance, assume that we have a constraint $c \in B \wedge c \in C_T$ with $var(C) = \{x_1, x_2, x_3, x_4\}$. An example violating this constraint that includes assignments to all four variables will be classified by the user as negative. However, if the system asks the user to classify an example $e_{x_1,x_2,x_3}$, which does not include an assignment to $x_4$, the above constraint is not taken into account. So, if no other constraint from the target network is violated in this example, it will be classified as positive.

The acquisition process has *converged* on the learned network $C_L \subseteq B$ iff $C_L$ agrees with $E$ and for every other network $C \subseteq B$ that agrees with E, we have $sol(C) = sol(C_L)$.

## 3 Omissions in Constraint Acquisition

State-of-the-art constraint acquisition algorithms are based on version space learning. Initially, the given language $\Gamma$ is used to construct the bias $B$. Then the system iteratively posts membership queries to the user in order to learn the constraints of the target network. Each example posted as a query must satisfy

$C_L$, i.e. the network that has already been learned so far, and violate at least one constraint from $B$. A query that satisfies these criteria is *informative*, as whatever the user's answer is, the version space's size will shrink. In case of a positive answer, each constraint $c \in B$ that violates the posted example can be removed from $B$ (i.e. all the constraint networks containing $c$ are removed from the version space). In case of a negative answer, one or more of the violated constraints are certainly in $C_T$. So, the system will search to find the scope of one, some, or all of them, depending on the algorithm used.

This is done through a function called *FindScope* in QuAcq, and its enhanced variant *FindScope-2* [20] used by MQuAcq and MQuAcq-2. Once a scope has been located, the function *FindC* [5] is used to learn the specific constraint (i.e. its relation). *FindScope-2*, upon which we build, finds the scope of a violated constraint of the target network by successively removing entire blocks of variables from the query, and posting the resulting partial query to the user.

The above reasoning assumes that all the answers of the user are correct and also does not allow for uncertainty (omissions) in the answers. However, both of these assumptions are not realistic as humans make mistakes and are not always certain of their answers. In this paper we deal with the problem of uncertainty. We introduce the use of Limited Membership Queries (LMQ) to constraint acquisition, and propose methods to handle them. A LMQ may be answered by the user either by classifying the example as a solution ("yes") or not ("no") of the problem, or with a third option, namely an "I don't know" answer. In this paper we are mainly interested in cases where the user is uncertain about the existence or non-existence of a relation between some entities (i.e. variables) of the problem, and the type of the relation if one exists.

In general, there are some questions that arise when an "I don't know" answer is encountered: First of all, is it possible to learn something from this query? And if we believe that it is possible, what can we learn and how can we learn it?

We argue that in case we have consistent answers, it is possible to learn from an omission. We define as the *omission network* $C_{OM}$ the set of "uncertain" constraints that the user does not know if they should be included in the target network or not, i.e. the gap in the knowledge of the user. This set may contain constraints both from $C_T$ and from $B \setminus C_T$. Thus, if we iteratively split the initial query into partial ones then by posting these partial queries to the user we may be able to isolate one or more scopes that cause the uncertainty.

Considering these, we propose and compare two different methods for handling omissions in constraint acquisition:

1. Queries answered as "I don't know" are simply ignored, under the assumption that we cannot discover anything through such queries. So, after an omission, we can save the query to avoid posting it again, and move on to generate a new one. This method is inspired by the RFT learning model, where it is assumed that the reason of the omission cannot be learned.
2. The second method assumes that each omission is caused by relations between the variables that the user is uncertain about, and that the relevant sets of variables can be identified. When the user answers with an omission, the

system commences a search for the "confusing" constraints using a reasoning similar to the search for violating constraint that algorithms like QuAcq and its variants apply at negative examples. This method corresponds to the CIT of concept learning, where it is assumed that the reason of the omission is a gap in the knowledge of the user about the problem.

We now describe and analyze the proposed methods for dealing with omissions. Both extend MQuAcq-2 [19], but they can be used in conjunction with any constraint acquisition algorithm.

### 3.1 Ignoring Omissions

This is a simple method, called MQuAcq-2-OM1, where any query $e_Y$ answered as "I don't know" is ignored by the system, under the assumption that we cannot discover anything through such a query. Thus, after an omission, the example posted as a query is stored, to avoid posting it again, and then a new example is generated. Note that the following scenario is possible: The user may answer negatively to a query, and as a result the algorithm will search for one or more violated constraints following a similar process to MQuAcq-2 (and also QuAcq/MQuAcq). However, as partial queries are posted to the user during this process, some of these partial queries may be answered by an omission because the user may be certain that the initial query is not a solution, but may not be certain that some part of the query violates any constraint or not. Such partial queries are also stored and then bypassed.

Algorithm 1 depicts MQuAcq-2-OM1. The system repeatedly generates an example $e$ satisfying $C_L$ and rejecting at least one constraint from $B$ (line 4). If it has not converged, it tries to acquire multiple constraints of $C_T$ violating $e$. At first it posts the example to the user (line 8). If the answer of the user is positive, the set $\kappa_B(e_{Y'})$ of all the constraints from $B$ that reject $e_{Y'}$ is removed from the bias (line 9). In case the answer is negative, it tries to learn a constraint by using the functions *FindScope-2* and *FindC* (lines 14–17).

In case of an omission, the example, which may be a complete or a partial one, is added to the set $E'$ (line 11) and then the algorithm stops trying to learn any more constraints of $C_T$ in this example (line 12). The set $E'$ stores all the examples that lead to an omission, in order to avoid generating the same assignments in future queries, as shown at line 4. *FindScope-2* has been modified to apply the same reasoning when searching in partial queries. We omit the pseudocode of this modified version of *FindScope-2* for space reasons. In case *FindScope-2* has added any partial example $e_{Y''}$ to $E'$, then the algorithm breaks the loop again (lines 18–19), stopping the search for more violated constraints of the target network in this specific partial query. If no omissions have occured, the algorithm removes the entire scope of the acquired constraint at line 20, trying to learn multiple non-overlapping constraints. This iterative process ends when the example $e_{Y'}$ does not contain any violated constraint from the bias (line 21), or if an omission has occured at some point. In these cases, the system

**Algorithm 1.** MQuAcq-2-OM1

**Input:** $B, X, D$ ($B$: the bias, $X$: the set of variables, $D$: the set of domains)
**Output:** $C_L$ : a constraint network

1: $C_L \leftarrow \emptyset$;
2: $E' \leftarrow \emptyset$;
3: **while** true **do**
4: Generate $e_Y$ in $D^Y$ accepted by $C_L$ and rejected by $B$, with $e_Y \neq e'_Y \mid e'_{Y_2} \in E' \wedge Y \subseteq Y_2$ ;
5: **if** $e =$ nil **then return** "$C_L$ converged";
6: $Y' \leftarrow Y$;
7: **do**
8: answer $\leftarrow$ASK($e_{Y'}$);
9: **if** answer = "yes" **then** $B \leftarrow B \setminus \kappa_B(e_{Y'})$;
10: **else if** answer = "I don't know" **then**
11: $E' \leftarrow E' \cup e_{Y'}$;
12: break;
13: **else**
14: $Scope \leftarrow FindScope\text{-}2(e_{Y'}, \emptyset, Y', false)$;
15: $c \leftarrow FindC(e_{Y'}, Scope)$;
16: $C_L \leftarrow C_L \cup \{c\}$
17: $B \leftarrow B \setminus \{c \in B \mid var(c) = Scope\}$;
18: **if** $\exists\, Y'' \subset Y' \mid e_{Y''} \in E'$ **then**
19: break;
20: $Y' \leftarrow Y' \setminus Scope$;
21: **while** $\kappa_B(e_{Y'}) \neq \emptyset$

cannot learn more violated constraints from this example, so it generates a new example at line 4 and starts over.

We now analyze the complexity of MQuAcq2-OM1 in terms of the number of queries required. We assume that $l$ is the maximum number of omissions.

**Proposition 1.** *Given a bias $B$ built from a language $\Gamma$, with bounded arity constraints, a target network $C_T$ and a number of omissions $l$, MQuAcq-2-OM1 uses $O(|C_T| \cdot (\log |X| + \Gamma) + |B| + l)$ number of queries to converge.*

*Proof.* MQuAcq-2 needs $O(|C_T| \cdot (\log |X| + \Gamma) + |B|)$ queries in order to find the $C_T$ and converge when we do not have omissions [19]. As our handling of omission guaranties that no query will be posted more than once, the maximum number of omissions is $l$. Also, the omissions do not affect the maximum number of queries needed to learn the constraints from $C_T$ and to converge. As a result, MQuAcq-2-OM1 uses $O(|C_T| \cdot (\log |X| + \Gamma) + |B| + l)$ number of queries to converge. □

As the number $l$ of omissions can be equal to the maximum number of examples that can be generated in the worst case, i.e. $l \leq |D|^{|X|}$, the above complexity, as well as the space complexity of the algorithm, is exponential, which of course is a major drawback. However, under the assumption that omissions are random independent events, meaning that there are no "uncertain constraints", then the algorithm will converge once $B$ is empty, as does MQuAcq-2.

On the other hand, in case the omissions are related to a gap in the user's knowledge (i.e. to "uncertain" constraints), their number can be exponential. This is because the "uncertain" constraints are not learned, and therefore the system cannot distinguish them from "normal" constraints. Consider the case where $C_T$ has been learned (hence $C_L$ is equivalent to $C_T$) and the constraints from $B \setminus (C_T \cup C_{OM})$ have already been removed from $B$. Now the system will repeatedly try to build examples that satisfy $C_L$ and violate at least one constraint from $C_{OM}$ in line 4. Each such example will be answered with an omission because the user cannot tell if the violation of a constraint from $C_{OM}$ makes the example a non-solution or not. As the number of examples that satisfy $C_L$ and violate at least one constraint from $C_{OM}$ is exponential, and all of these examples have to be generated in the worst case in order for the algorithm to terminate, the number of omissions is exponential.

### 3.2 Exploiting Omissions

We now present our main method for handling omissions, which is called MQuAcq-2-OM2, and is based on the assumption that each omission is due to uncertainty from the user's part about one or more relations between the variables, i.e. due to a gap in knowledge. In contrast to the first method, instead of discarding omissions, we now try to derive useful information from them.

The main idea is to iteratively divide an example that was answered by an omission into partial ones, in a way similar to how negative answers are handled, until a set of variables (a scope) that causes uncertainty to the user is discovered. Then the constraints corresponding to this scope can be learned and removed from $B$ to avoid generating subsequent queries that contain the same "source of uncertainty". Another potential gain is that during this process we may come across partial queries that are answered positively, meaning that the constraints that violate them can also be removed from $B$.

We introduce a new set, $C_{L_{OM}}$, which stores all the "uncertain" constraints that are removed from $B$ via an omission. We also modify the query handling process to locate scopes causing omissions and avoid violating the constraints that confuse the user in future queries, through the use of $C_{L_{OM}}$. As we now explain, each of the three possible answers by the user to a query over an example $e$ requires different handling.

- After a *positive answer*: The constraints from $B$ that reject the example posted are removed, as in all constraint acquisition algorithms.
- After an *omission*: In this case a partial example of $e$ can lead either to an omission, if the variables that confuse the user are still in the partial example, or to a positive answer, in case one or more of the variables in the scope of the omission are removed. So, we can find the scope of the omission with a procedure similar to the one used in *FindScope-2*, exploiting the positive answers in order to locate the variables of the scope.

**Algorithm 2.** MQuAcq-2-OM2

**Input:** $B, X, D$ ($B$: the bias, $X$: the set of variables, $D$: the set of domains)
**Output:** $C_L$, $C_{L_{OM}}$ : constraint networks

1: $C_L \leftarrow \emptyset$;
2: $C_{L_{OM}} \leftarrow \emptyset$;
3: **while** true **do**
4: Generate $e_Y$ in $D^Y$ accepted by $C_L$ and $C_{L_{OM}}$ while rejected by $B$;
5: **if** $e =$ nil **then return** "$C_L$ converged";
6: $Y' \leftarrow Y$;
7: **do**
8: answer $\leftarrow$ ASK$(e_{Y'})$;
9: **if** answer = "yes" **then** $B \leftarrow B \setminus \kappa_B(e_{Y'})$;
10: **else if** answer = "I don't know" **then**
11: $OMS \leftarrow \textit{FindScope-OM}(e_{Y'}, \emptyset, Y', false)$;
12: $C_{L_{OM}} \leftarrow C_{L_{OM}} \cup \textit{FindC}(e_{Y'}, OMS))$;
13: $B \leftarrow B \setminus \{c \in B \mid var(c) = OMS\}$;
14: $Y' \leftarrow Y' \setminus OMS$;
15: **else**
16: $\textit{Scope} \leftarrow \textit{FindScope-NO}(e_{Y'}, \emptyset, Y', false)$;
17: $c \leftarrow \textit{FindC}(e_{Y'}, \textit{Scope})$;
18: $C_L \leftarrow C_L \cup \{c\}$
19: $B \leftarrow B \setminus \{c \in B \mid var(c) = \textit{Scope}\}$;
20: $Y' \leftarrow Y' \setminus \textit{Scope}$;
21: **if** $\kappa_{C_{L_{OM}}}(e_{Y'}) \neq \emptyset$ **then**
22: $Y' \leftarrow Y' \setminus \{var(c) \mid c \in \kappa_{C_{L_{OM}}}(e_{Y'})\}$;
23: **while** $\kappa_B(e_{Y'}) \neq \emptyset$

- After a *negative answer*: In such a case a partial example of $e$ can be answered in any possible way. If one or more of the variables in the scope of the violated constraint(s) are removed, we can have a positive answer or an omission if a constraint that confuses the user is violated. In addition, the answer can still be "no", if all the variables of the constraint of $C_T$ that is violated are still in the partial example. Hence, after a negative answer we must search for a violated constraint but we may also find the scope of an omission.

Based on these, we introduce two functions for finding the scope of a violated constraint (*FindScope-NO*) and the scope of an omission (*FindScope-OM*) so as to handle all the possible query answers.

MQuAcq-2-OM2 is depicted by Algorithm 2. The system generates an example accepted by the learned network and $C_{L_{OM}}$, while violating at least one constraint from $B$ (line 4). We want the example to satisfy $C_{L_{OM}}$ to avoid violating any constraint that will lead to an omission. Queries answered as "I don't know" are handled at lines 11–14 by calling *FindScope-OM*. This function finds the scope responsible for the omission, as we explain below, and stores it in $OMS$ (line 11). Then, it finds the specific "uncertain" constraint (i.e. the relation) and

**Algorithm 3.** FindScope-OM

**Input:** $e$, $R, Y$, *ask_query* ($e$: the example, $R$,$Y$: sets of variables, *ask_query*: boolean)
**Output:** *Scope* : a set of variables, the scope of an omission

1: **function** FindScope-OM($e$, $R$, $Y$, *ask_query*)
2: **if** *ask_query* $\wedge |\kappa_B(e_R)| > 0$ **then**
3: **if** $rej \neq |\kappa_B(e_R)|$ **then**
4: **if** ASK($e_R$) = "I don't know" **then**
5: $rej \leftarrow |\kappa_B(e_R)|$;
6: **return** $\emptyset$;
7: **else** $B \leftarrow B \setminus \kappa_B(e_R)$;
8: **else** **return** $\emptyset$;
9: **if** $|Y| = 1$ **then return** $Y$;
10: split $Y$ into $< Y_1, Y_2 >$ such that $|Y_1| = \lceil |Y|/2 \rceil$;
11: $S_1 \leftarrow$ *FindScope-OM*$(e, R \cup Y_1, Y_2, true)$;
12: $S_2 \leftarrow$ *FindScope-OM*$(e, R \cup S_1, Y_1, (S_1 \neq \emptyset))$;
13: **return** $S_1 \cup S_2$;

removes it from $B$, adding it to $C_{L_{OM}}$ (lines 12–13). Finally, the scope found is removed from $Y'$ (line 14), so that MQuAcq-2-OM2 can continue searching.

Queries answered negatively are handled by calling *FindScope-NO* at line 16. As explained below, this function not only finds the scope of a violated constraint but sometimes it can also find the scope of an omission. Thus, if such a scope is found (line 21), the algorithm removes the scope from $Y'$ at line 22.

We now focus on the new functions, *FindScope-OM* and *FindScope-NO*. Both use the reasoning of *FindScope-2*, i.e. successively removing approximately half of the variables and posting a partial query. If after such a removal, the answer of the user changed then we know that the removed block contains at least one variable from the scope of a constraint we seek (a constraint from $C_T$ or $C_{OM}$).

*FindScope-OM* (Algorithm 3) is similar to *FindScope-2*. It takes as parameters a (partial) example $e_Y$ that has led to an omission, two sets of variables $R$ and $Y$, initialized to the empty set and to $Y$ respectively, and a Boolean variable *ask_query*. An invariant in any recursive call is that the example $e$ violates at least one constraint from $C_{OM}$, whose scope is a subset of $R \cup Y$. The number of violated constraints from $B$ is stored in $rej$, to avoid posting redundant queries to the user in any recursive call.

If *FindScope-OM* is called with *ask_query* = true and $e_R$ violates at least one constraint from $B$ (line 2) but not the same number of constraints as the previous query posted (line 3), it posts $e_R$ as a query to the user (line 4). In case of an omission it returns the empty set (line 6), in order to remove some variables from $R$ in the previous call. If the answer is "yes", it removes all the constraints from the bias that reject $e_R$ and continues. Thus, it reaches line 9 only in the case where $e_R$ does not violate any constraint from $C_{OM}$. Because we know that $e$ violates at least one constraint whose scope is a subset of $R \cup Y$, in case $Y$ is a singleton it is returned (line 9). The set $Y$ is split in two balanced parts (line 10) and the algorithm searches recursively, in sets of variables built

**Algorithm 4.** FindScope-NO

**Input:** $e$, $R$, $Y$, $ask\_query$ ($e$: the example, $R$,$Y$: sets of variables, $ask\_query$: boolean)
**Output:** $Scope$ : a set of variables, the scope of a constraint in $C_T$

1: **function** FindScope-NO($e$, $R$, $Y$, $ask\_query$)
2:   **if** $ask\_query \wedge |\kappa_B(e_R)| > 0$ **then**
3:     **if** $rej \neq |\kappa_B(e_R)|$ **then**
4:       answer $\leftarrow$ ASK($e_R$);
5:       **if** answer = "yes" **then** $B \leftarrow B \setminus \kappa_B(e_R)$;
6:       **else if** answer = "I don't know" **then**
7:         **if** $\kappa_{C_{L_{OM}}}(e_R) = \emptyset$ **then**
8:           OMS $\leftarrow$ $FindScope\text{-}OM(e_R, \emptyset, R, false)$;
9:           $C_{L_{OM}} \leftarrow C_{L_{OM}} \cup FindC(e_R, OMS))$;
10:           $B \leftarrow B \setminus \{c \in B \mid var(c) = OMS\}$;
11:       **else**
12:         $rej \leftarrow |\kappa_B(e_R)|$;
13:         **return** $\emptyset$;
14:     **else** **return** $\emptyset$;
15:   **if** $|Y| = 1$ **then return** $Y$;
16:   split $Y$ into $< Y_1, Y_2 >$ such that $|Y_1| = \lceil |Y|/2 \rceil$;
17:   $S_1 \leftarrow FindScope\text{-}NO(e, R \cup Y_1, Y_2, true)$;
18:   $S_2 \leftarrow FindScope\text{-}NO(e, R \cup S_1, Y_1, (S_1 \neq \emptyset))$;
19:   **return** $S_1 \cup S_2$;

using $R$ and these parts, for the scope of a violated constraint of $C_{OM}$, in a logarithmic number of steps (lines 11–13).

*FindScope-NO* (Algorithm 4) handles the case of a negative answer by the user. It operates in a slightly different way than *FindScope-OM* because after the removal of some variables, the answer of the user may change from "no" either to "yes" or to "I don't know" (see Lemma 3 below). This is because after the removal of one or more variables of the violated constraint, the user may now be confused by another constraint in the partial example formed, not being sure if the partial example is positive or not. In such a case, we continue the search in two directions. First, *FindScope-OM* is called in order to locate the scope of the omission and store it in $OMS$ (line 8) and then find the "uncertain" constraint, which is then removed from $B$ and added to $C_{L_{OM}}$ (line 9), as in lines 11–13 of MQuAcq-2-OM2. Also the function *FindScope-NO* continues the search for the scope of the violated constraint of $C_T$.

Now, let us illustrate the behaviour of our proposed approach.

*Example 1.* Assume that the vocabulary $(X, D)$ given to the system is $X = \{x_1, ..., x_8\}$ and $D = \{D(x_1), ..., D(x_8)\}$ with $D(x_i) = \{1, ..., 8\}$, the target network $C_T$ is the set $\{\neq_{34}, \neq_{56}\}$, $C_{OM} = \{\neq_{34}\}$ and $B = \{\neq_{ij} \mid 1 <= i < 8 \wedge i < j <= 8\}$. Also, assume that the example generated at line 4 of MQuAcq-2-OM2 is $e = \{1, 4, 2, 2, 3, 3, 5, 6\}$.

The system will post $e$ as a query at line 8 of MQuAcq-2-OM2. The answer will be "no" as it violates constraint $\neq_{56}$. Thus, *FindScope-NO* is called to find

**Table 1.** Behavior of MQuAcq-2-OM2 in Example 1

| Recursive calls of *FindScope-NO* | | | | | |
|---|---|---|---|---|---|
| *call* | *R* | *Y* | $e_R$ | *ASK* | *return* |
| 0 | $\emptyset$ | $x_1, x_2, x_3, x_4, x_5, x_6, x_7, x_8$ | – | – | $\{x_5, x_6\}$ |
| 1 | $x_1, x_2, x_3, x_4$ | $x_5, x_6, x_7, x_8$ | $\{1, 4, 2, 2, -, -, -, -\}$ | "I don't know" | $\{x_5, x_6\}$ |
| Go to *FindScope-OM* | | | | | |
| 0 | $\emptyset$ | $x_1, x_2, x_3, x_4$ | – | – | $\{x_3, x_4\}$ |
| 1 | $x_1, x_2$ | $x_3, x_4$ | $\{1, 4, -, -, -, -, -\}$ | "yes" | $\{x_3, x_4\}$ |
| 1.1 | $x_1, x_2, x_3$ | $x_4$ | $\{1, 4, -, -, -, -, -\}$ | "yes" | $\{x_4\}$ |
| 1.2 | $x_1, x_2, x_4$ | $x_3$ | $\{1, 4, -, -, -, -, -\}$ | "yes" | $\{x_3\}$ |
| back to *FindScope-NO* | | | | | |
| 1.1 | $x_1, x_2, x_3, x_4, x_5, x_6$ | $x_7, x_8$ | $\{1, 4, 2, 2, 3, 3, -, -\}$ | "no" | $\emptyset$ |
| 1.2 | $x_1, x_2, x_3, x_4$ | $x_5, x_6$ | – | – | $\{x_5, x_6\}$ |
| 1.2.1 | $x_1, x_2, x_3, x_4, x_5$ | $x_6$ | $\{1, 4, 2, 2, 3, -, -\}$ | "yes" | $\{x_6\}$ |
| 1.2.2 | $x_1, x_2, x_3, x_4, x_6$ | $x_5$ | $\{1, 4, 2, 2, -, 3, -\}$ | "yes" | $\{x_5\}$ |

the scope of a violated constraint. Table 1 shows the trace of its recursive calls. A dash (-) in columns $e_R$ and *ASK* means that no query is posted to the user, due to one of the conditions at lines 2 and 3 (e.g., at call 0 of *FindScope-NO*, as $ask\_query$ = false and $R = \emptyset$, the condition at line 2 does not hold). Recall that queries are only on the variables in $R$.

When half of the variables are removed from the query at recursive call 1 of *FindScope-NO*, the answer of the user changes to "I don't know". So, *FindScope-OM* is called to find the cause of uncertainty (line 8 of *FindScope-NO*). Its trace of recursive calls is also shown in Table 1. After 4 queries it finds the scope of the omission constraint and returns it (back at the 0 call), so FindC is called at line 9 of *FindScope-NO*. *FindScope-NO* then continues searching to find the violated constraint from $C_T \setminus C_{OM}$ that is responsible for the negative answer in the first place. It will be found after 3 queries and returned (back at the 0 call of *FindScope-NO*).

**Analysis of MQuAcq2-OM2:** We now prove the correctness of MQuAcq-2-OM2. That is, we prove that the constraints it adds to $C_L$ and $C_{L_{OM}}$ belong indeed there, and it converges having learned all the constraints of $C_T$ and of $C_{OM}$ that it possibly can. We first give three lemmas showing that for each possible answer to a query ASK($e_Y$), the possible answers we can have in partial queries of the form ASK($e_{Y'}$), with $Y' \subset Y$, are the ones informally described previously for the CIT model. Then we give a proposition regarding the soundness of *FindScope-OM* and *FindScope-NO*. Proofs of Lemmas are ommitted for space reasons.

**Lemma 1.** *If ASK($e_Y$) = "yes" then for any $Y' \subset Y$ it holds that ASK($e_{Y'}$) = "yes".*

**Lemma 2.** *If ASK($e_Y$) = "I don't know" then for any $Y' \subset Y$ we can have ASK($e_{Y'}$) = "I don't know" or ASK($e_{Y'}$) = "yes".*

**Lemma 3.** *If ASK($e_Y$) = "no" then a partial query in $Y' \subset Y$, ASK($e_{Y'}$) can return any of the possible answers.*

**Proposition 2.** *If* FindScope-OM *(resp.* FindScope-NO*) is given an example $e_Y$ and returns a scope $S$ then there exists a violated constraint $c \in C_{OM}$ (resp. $c \in C_T \setminus C_{OM}$) with scope($c$) = $S$.*

*Proof.* An invariant of *FindScope-OM* is that the example $e$ violates at least one constraint from $C_{OM}$, whose scope is a subset of $R \cup Y$ (i.e. ASK($R \cup Y$) = "I don't know"). Also, it reaches line 9 only in the case that ASK($e_R$) = "yes". Thus, by Lemma 1, for $Y' \subset Y$ it holds that ASK($e_{Y'}$) = "yes", i.e. $e_R$ does not violate any constraint from $C_{OM}$. Also, *FindScope-OM* returns variables only at line 9, in case $Y$ is a singleton. As a result, for any $x_i \in S$ we know that ASK($S$) = "I don't know" and ASK($S \setminus x_i$) = "yes". Hence, $S$ is definitely a scope of a constraint from $C_{OM}$. □

**Theorem 1.** *Given a bias $B$ built from a language $\Gamma$ with bounded arity constraints, a target network $C_T$ representable by $B$, and an omission network $C_{OM}$, MQuAcq-2-OM2 is correct.*

*Proof. Soundness.* MQuAcq-2-OM2 learns constraints and adds them to $C_{L_{OM}}$ or $C_L$ only by using the function *FindC* in a scope found by *FindScope-OM* or *FindScope-NO* respectively. By Proposition 2, when *FindScope-OM* returns a scope $S$, then there exists a violated constraint $c \in C_{OM}$ with *scope*($c$) = $S$ and when *FindScope-NO* returns a scope $S$, then there exists a violated constraint $c \in C_T \setminus C_{OM}$ with *scope*($c$) = $S$. Also, *FindC* has been proved to be correct for normalized target networks [7]. Hence, MQuAcq-2OM2 is sound, as for every constraint $c$ added to $C_L$ it holds that $c \in C_T \setminus C_{OM}$ and for every constraint $c$ added to $C_{L_{OM}}$ it holds that $c \in C_{OM}$.

*Completeness.* An example generated at line 4 of MQuAcq-2-OM2 must violate at least one constraint from $B$. Given such an example $e_Y$, MQuAcq-2-OM2 will find at least one constraint from $C_T$ (lines 16–20) or $C_{OM}$ (lines 11–14), if one exists, and then remove it from $B$ (lines 13, 19). It finds the scope of a constraint of $C_T \setminus C_{OM}$ using *FindScope-NO* and of $C_{OM}$ using *FindScope-OM*. After a scope has been located, it had been proved that *FindC* will find a constraint in the given scope if one exists [7]. The same applies to constraints of $C_{OM}$, as the procedure is exactly the same, because when ASK($e_Y$) = "I don't know" then for any $Y' \subset Y$ we can have ASK($e_{Y'}$) = "I don't know" or ASK($e_{Y'}$) = "yes" (Lemma 2). If no constraint can be learned by an example (i.e. $\kappa_{C_T}(e_Y) = \kappa_{C_{OM}}(e_Y) = \emptyset$), it will remove all the violated constraints from $B$ (line 9). Thus, the size of $B$ will decrease after each query. The algorithm terminates only when no example can be generated at line 4. In this case, the system has converged as $C_L$ agrees with $E$ and for every other network $C \subseteq B$ that agrees with E and, we have $sol(C) = sol(C_L)$. Hence, the system learned any constraint in $C_T \setminus C_{OM}$ that could be learned. As no constraint from $B$ can be violated, the same applies for $C_{OM}$. Hence, MQuAcq-2-OM2 is complete. □

**Proposition 3.** *Given a bias $B$ built from a language $\Gamma$, with bounded arity constraints, a target network $C_T$ and an omission network $C_{OM}$, MQuAcq-2-OM2 uses $O(|C_T| \cdot (\log|X| + |\Gamma|) + |B| + l)$ number of queries to converge, with $l <= |C_{OM}| \cdot (\log|X| + |\Gamma|)$.*

*Proof.* (sketch). MQuAcq-2-OM2 will learn $|C_T \setminus C_{OM}|$ constraints and will find $|C_{OM}|$ omission constraints. The constraints from $C_T \setminus C_{OM}$ are learned using the functions *FindScope-NO* and *FindC* while the constraints of $|C_{OM}|$ are found using the functions *FindScope-OM* and *FindC*. Both *FindScope-NO* and *FindScope-OM* need a maximum number of $|S| \cdot \log|Y| = O(\log|X|)$ queries in order to find a scope $S$ in an example $e_Y$, as they use a process very similar to *FindScope* [5]. *FindC* needs at most $|\Gamma|$ queries to learn a constraint. Also, assuming that each positive query removes only one constraint from $|B|$ it will need to ask a total number of $|B| - |C_T \setminus C_{OM}| - |C_{OM}| = O(|B|)$ queries to prune $B$ and reach convergence. Thus, the total number of queries in $O(|C_T| \cdot (\log|X| + |\Gamma|) + |B| + l)$ , with $l <= |C_{OM}| \cdot (\log|X| + |\Gamma|)$. The above result in $O((|C_T| + |C_{OM}|) \cdot (\log|X| + |\Gamma|) + |B|)$. □

## 4 Experimental Evaluation

We ran experiments both in the RFT and the CIT models. In the former, as the omissions are not related to missing knowledge, we evaluated only MQuAcq-2-OM1. In the latter we compared our methods to each other. We used MQuAcq-2 without omissions as a reference point. Experiments were run on an Intel(R) Core(TM) i7-8700 CPU @ 3.20 GHz with 16 GB of RAM. In more detail:

- We used the $max_B$ heuristic for query generation [20]. $max_B$ generates examples violating as many constraints as possible from $B$. The best example found within 1 s is returned, even if not proved optimal. If none is found, we continue and return the first suitable (partial) example found. The variable involved in the most constraints in $B$ is chosen during search. Values are chosen randomly.
- In the RFT model, a query is answered by an omission with 20% probability.
- In the CIT model, we used a cutoff for MQuAcq-2-OM1, as the number of omissions can be exponential (it did not complete within 10 h). So we only present results for MQuAcq-2-OM1 in CIT with a cutoff, which was imposed as follows: the system stops at line 3 when the number of omission $\#omissions$ is more than the 30% of the total number of queries $\#queries$.
- To compare the algorithms on the same scenario, all our experiments concern the extreme case where $C_L$ is initially empty, This results in a number of queries that may seem too large for human users. But in real applications, background knowledge can used by giving a frame of basic constraints or by

using other methods, e.g. ModelSeeker [4], to extract some constraints from known solutions. Then, our algorithms can be used to finalize the model.
- We measure the size of the learned network $C_L$, the size of the learned omission network $C_{L_{OM}}$, the total number of queries $\#queries$, the total number of omission answers $\#omissions$ and the total cpu time $T$. We present results of MQuAcq-2 without *analyze&Learn*, MQuAcq-2-OM1 and MQuAcq-2-OM2. Each algorithm was run 5 times and the means are presented.

We used the following benchmarks in our study:

**Zebra.** It consists of 25 variables with domains of maximum size 5. The target network $C_T$ contains five cliques of 10 $\neq$ constraints each and 11 additional constraints. The bias was initialized with 1200 binary constraints from the language $\Gamma = \{=, \neq, >, <, x_i - x_j = 1, |x_i - x_j| = 1\}$. $C_{OM}$ contains the constraints of $C_T$ not belonging to a clique, and 5 randomly chosen constraints from $B$.

**Murder.** It has 20 variables with domains of size 5. $C_T$ contains 4 cliques of $\neq$ constraints and 12 additional constraints. The bias was initialized with 760 constraints from the language $\Gamma = \{=, \neq, >, <\}$. $C_{OM}$ contains the constraints of $C_T$ not belonging to a clique, and 10 randomly chosen constraints from $B$.

**Random.** We generated a random target network with 50 variables, domains of size 10, and 122 $\neq$ constraints. The bias was initialized with 19,800 constraints, using the language $\Gamma = \{=, \neq, >, <\}$. $C_{OM}$ was created randomly, containing 15 constraints with only 1 belonging to $C_T$.

**Radio Link Frequency Assignment Problem (RLFAP).** We use a simplified version of the communication problem from [12], with 50 variables having domains of size 40. $C_T$ contains 125 distance constraints. The bias was built using a language of 2 distance constraints ($\{|x_i - x_j| > y, |x_i - x_j| = y\}$) with 5 different possible values for $y$. This led to a language of 10 different distance constraints. In total, $B$ contains 12,250 constraints. $C_{OM}$ was created randomly, containing 15 constraints in total with 5 belonging to $C_T$.

Results from the RFT model presented in Table 2 (see rows for MQuAcq-2-OM1$_{RFT}$), confirm our complexity analysis. When omissions are random events, the queries posted by MQuAcq-2-OM1 do not increase a lot compared to MQuAcq-2 without omissions. We can see that the increase is related to the number of variables of the problem. On the other hand, cpu times increase significantly. This is because most of the (few) additional queries are generated queries and not partial ones (because the system generates a new query when an omission occurs). As query generation is the most time-consuming process of the algorithm, this affects the run times considerably.

**Table 2.** Results from the RFT and CIT models.

| Benchmark | Algorithm | $\|C_L\|$ | $\|C_{L_{OM}}\|$ | $\#q$ | $\#om$ | $T$ |
|---|---|---|---|---|---|---|
| Zebra | MQuAcq-2 | 61 | 0 | 494 | 0 | 9.28 |
| | MQuAcq-2-OM1$_{RFT}$ | 60 | 0 | 624 | 115 | 48.09 |
| | MQuAcq-2-OM1$_{CIT}$ | 49 | 0 | 534 | 135 | 33.07 |
| | MQuAcq-2-OM2 | 48 | 12 | 480 | 73 | 8.37 |
| Murder | MQuAcq-2 | 52 | 0 | 384 | 0 | 12.70 |
| | MQuAcq-2-OM1$_{RFT}$ | 52 | 0 | 484 | 101 | 51.18 |
| | MQuAcq-2-OM1$_{CIT}$ | 40 | 0 | 411 | 124 | 48.87 |
| | MQuAcq-2-OM2 | 39 | 17 | 397 | 78 | 11.81 |
| Random122 | MQuAcq-2 | 122 | 0 | 1031 | 0 | 37.48 |
| | MQuAcq-2-OM1$_{RFT}$ | 122 | 0 | 1464 | 294 | 139.22 |
| | MQuAcq-2-OM1$_{CIT}$ | 121 | 0 | 1583 | 475 | 404.01 |
| | MQuAcq-2-OM2 | 121 | 15 | 1095 | 68 | 37.99 |
| RLFAP | MQuAcq2 | 125 | 0 | 1157 | 0 | 241.10 |
| | MQuAcq2-OM1$_{RFT}$ | 125 | 0 | 1391 | 309 | 459.71 |
| | MQuAcq2-OM1$_{CIT}$ | 28 | 0 | 401 | 121 | 28.13 |
| | MQuAcq2-OM2 | 119 | 15 | 1273 | 115 | 602.40 |

Focusing on the CIT model, where the omissions are related to a gap in the user's knowledge (i.e., the "uncertain" constraints of $C_{OM}$), the results (Table 2) demonstrate that both MQuAcq-2-OM1 with a cutoff (denoted MQuAcq-2-OM1$_{CIT}$) and MQuAcq-2-OM2 achieve a quite good performance in most of the problems. The exception for MQuAcq-2-OM1 is RLFAP where it learns only 23% of $C_T \setminus C_{OM}$, because the cutoff condition is activated too early. On the other hand, MQuAcq-2-OM2 gives very good overall results, with the increase in number of queries being only up to 6.2% compared to MQuAcq-2 without omissions. In addition, the omission answers in MQuAcq-2-OM2 are quite fewer than in MQuAcq-2-OM1 (up to 86% in Random). Also, we observe that in Zebra and Murder the number of queries is very close to that of MQuAcq-2. This happens because most of the "uncertain" constraints are in $C_T$.

## 5 Conclusions

One significant issue that has not been addressed in constraint acquisition is the possible presence of uncertainty in the answers of the users. We address this for the first time by introducing Limited Membership Queries in constraint acquisition. We propose two algorithms for handling omissions that correspond to the two models of omissions in concept learning. The first method assumes that omissions are independent events and nothing can be learned from them, while the second assumes that they are related to gaps in the user's knowledge, and can be exploited. Theoretical and experimental results show that both methods perform well when used in their corresponding omission models.

## References


1. Angluin, D.: Queries and concept learning. Mach. Learn. **2**(4), 319–342 (1988). https://doi.org/10.1023/A:1022821128753
2. Angluin, D., Krikis, M., Sloan, R.H., Turán, G.: Malicious omissions and errors in answers to membership queries. Mach. Learn. **28**(2–3), 211–255 (1997)
3. Angluin, D., Slonim, D.K.: Randomly fallible teachers: learning monotone DNF with an incomplete membership oracle. Mach. Learn. **14**(1), 7–26 (1994)
4. Beldiceanu, N., Simonis, H.: A model seeker: extracting global constraint models from positive examples. In: Milano, M. (ed.) CP 2012. LNCS, pp. 141–157. Springer, Heidelberg (2012). https://doi.org/10.1007/978-3-642-33558-7_13
5. Bessiere, C., et al.: Constraint acquisition via partial queries. In: IJCAI, vol. 13, pp. 475–481 (2013)
6. Bessiere, C., et al.: Query-driven constraint acquisition. In: IJCAI, vol. 7, pp. 50–55 (2007)
7. Bessiere, C., et al.: New approaches to constraint acquisition. In: Bessiere, C., De Raedt, L., Kotthoff, L., Nijssen, S., O'Sullivan, B., Pedreschi, D. (eds.) Data Mining and Constraint Programming. LNCS (LNAI), vol. 10101, pp. 51–76. Springer, Cham (2016). https://doi.org/10.1007/978-3-319-50137-6_3
8. Bessiere, C., Koriche, F., Lazaar, N., O'Sullivan, B.: Constraint acquisition. Artif. Intell. **244**, 315–342 (2017)
9. Bisht, L., Bshouty, N.H., Khoury, L.: Learning with errors in answers to membership queries. J. Comput. Syst. Sci. **74**(1), 2–15 (2008)
10. Bshouty, N.H.: Exact learning from an honest teacher that answers membership queries. Theor. Comput. Sci. **733**, 4–43 (2018)
11. Bshouty, N.: Exact learning boolean functions via the monotone theory. Inf. Comput. **123**(1), 146–153 (1995)
12. Cabon, B., De Givry, S., Lobjois, L., Schiex, T., Warners, J.P.: Radio link frequency assignment. Constraints **4**(1), 79–89 (1999). https://doi.org/10.1023/A:1009812409930
13. Frazier, M., Goldman, S., Mishra, N., Pitt, L.: Learning from a consistently ignorant teacher. In: Proceedings of the Seventh Annual Conference on Computational Learning Theory, pp. 328–339. ACM (1994)
14. Freuder, E.C.: Modeling: the final frontier. In: The First International Conference on the Practical Application of Constraint Technologies and Logic Programming (PACLP), London, pp. 15–21 (1999)
15. Freuder, E.C.: Progress towards the holy grail. Constraints **23**(2), 158–171 (2018). https://doi.org/10.1007/s10601-017-9275-0
16. Freuder, E.C., O'Sullivan, B.: Grand challenges for constraint programming. Constraints **19**(2), 150–162 (2014). https://doi.org/10.1007/2Fs10601-013-9155-1
17. Goldman, S.A., Mathias, H.D.: Learning k-term DNF formulas with an incomplete membership oracle. In: COLT, pp. 77–84. Citeseer (1992)
18. Mitchell, T.M.: Version spaces: an approach to concept learning. Technical report, Stanford Univ Calif Dept of Computer Science (1978)
19. Tsouros, D.C., Stergiou, K., Bessiere, C.: Structure-driven multiple constraint acquisition. In: Schiex, T., de Givry, S. (eds.) CP 2019. LNCS, vol. 11802, pp. 709–725. Springer, Cham (2019). https://doi.org/10.1007/978-3-030-30048-7_41
20. Tsouros, D.C., Stergiou, K., Sarigiannidis, P.G.: Efficient methods for constraint acquisition. In: Hooker, J. (ed.) CP 2018. LNCS, vol. 11008, pp. 373–388. Springer, Cham (2018). https://doi.org/10.1007/978-3-319-98334-9_25

# Computing Optimal Decision Sets with SAT

Jinqiang Yu, Alexey Ignatiev(✉), Peter J. Stuckey, and Pierre Le Bodic

Faculty of Information Technology, Monash University, Melbourne, Australia
jyuu0044@student.monash.edu,
{alexey.ignatiev,peter.stuckey,pierre.lebodic}@monash.edu

**Abstract.** As machine learning is increasingly used to help make decisions, there is a demand for these decisions to be *explainable*. Arguably, the most explainable machine learning models use decision rules. This paper focuses on decision sets, a type of model with unordered rules, which explains each prediction with a single rule. In order to be easy for humans to understand, these rules must be concise. Earlier work on generating optimal decision sets first minimizes the number of rules, and then minimizes the number of literals, but the resulting rules can often be very large. Here we consider a better measure, namely the total size of the decision set in terms of literals. So we are not driven to a small set of rules which require a large number of literals. We provide the first approach to determine minimum-size decision sets that achieve minimum empirical risk and then investigate sparse alternatives where we trade accuracy for size. By finding optimal solutions we show we can build decision set classifiers that are almost as accurate as the best heuristic methods, but far more concise, and hence more explainable.

## 1 Introduction

The world has been changed by recent rapid advances in machine learning. Decision tasks that seemed well beyond the capabilities of artificial intelligence have now become commonly solved using machine learning [32,35,41]. But this has come at some cost. Most machine learning algorithms are opaque, unable to explain why decisions were made. Worse, they can be biased by their training data, and behave poorly when exposed to data outside that which they were trained on. Hence the rising interest in *explainable artificial intelligence* (XAI) [4,14,16,18,22,23,26,29,30,36–38,42,43,48,49,52], including research programs [3,24] and legislation [17,21].

In this paper we will focus on classification problems, where the input is a set of *instances* with *features* and, as a label, a *class* to predict. For these problems, some of the most explainable forms of machine learning formalism are *decisions sets* [11,12,15,20,31,34,39]. A decision set is a set of decision *rules*, each with conditions $C$ and decision $X$, such that if an instance satisfies $C$, then its class is predicted to be $X$. An advantage of decision sets over the more popular decision trees and decision lists is that each rule may be understood


H. Simonis (Ed.): CP 2020, LNCS 12333, pp. 952–970, 2020.
https://doi.org/10.1007/978-3-030-58475-7_55

independently, making this formalism one of the easiest to explain. Indeed, in order to explain a particular decision on instance $D$, we can just refer to a single decision rule $C \Rightarrow X$ s.t. $D$ satisfies $C$.

For decision sets to be clear and explainable to a human, individual rules should be concise. Previous work has examined building decision sets which involve the fewest possible rules, and then minimizes the number of literals in the rules [31]; or building a CNF classifer that fixes the number of rules, and then minimizes the number of literals [20, 39] to explain the positive instances of the class. This work also suffers from the limitation that the rules only predict class 1, and the model predicts class 0 if no rule applies. Unfortunately, in order to explain a class 0 instance, we need to use (the negation of) all rules, making the explanations not succinct.

In this work we argue the number of rules is the wrong measure of explainability, since, for example, 3 rules each involving 100 conditions are most likely less comprehensible than, say, 5 rules each involving 20 conditions. Indeed, since the explanation of a single instance is just a single decision rule, the number of rules is nowhere near as important as the size of the individual rules. So previous work on building minimum-size decision sets has not used the best measure of size for explainability.

In this work we examine directly constructing decision sets of the smallest total size, where the size of a rule with conditions $C$ is $|C| + 1$ (the additional 1 is for the class descriptor $X$). This leads to smaller decision sets (in terms of literals) which seem far more appealing for explaining decisions.

It turns out that this definition of size leads to SAT models that are experimentally harder to solve, but the resulting decision sets can be significantly smaller. However, for *sparse decision sets*, where we are allowed to consider a smaller rule set if it does not make too many errors in the classification, this new measure is no more difficult to compute than the traditional rule count measure, and gives finer granularity decisions on sparseness.

The contributions of this paper are

- The first approach to building optimal decision sets in terms of the total number of literals required to define the entire set,
- Alternate SAT and MaxSAT models to tackle this problem, and sparse variations which allow an accuracy versus size trade-off,
- Detailed experimental results showing the applicability of this approach, which demonstrate that our best approach can generate optimal sparse decision sets quickly with accuracy comparable to the best heuristic methods, but much smaller.

The paper is organized as follows. Section 2 introduces the notation and definitions used throughout the paper. Related work is outlined in Sect. 3. Section 4 describes the novel SAT- and MaxSAT-based encodings for the inference of decision sets. Experimental results are analyzed in Sect. 5. Finally, Sect. 6 concludes the paper.

## 2 Preliminaries

**Satisfiability and Maximum Satisfiability.** We assume standard definitions for propositional satisfiability (SAT) and maximum satisfiability (MaxSAT) solving [9]. A propositional formula is said to be in *conjunctive normal form* (CNF) if it is a conjunction of clauses. A *clause* is a disjunction of literals. A *literal* is either a Boolean variable or its negation. Whenever convenient, clauses are treated as sets of literals. Moreover, the term *clausal* will be used to denote formulas represented as sets of sets of literals, i.e. in CNF. A truth assignment maps each variable to $\{0, 1\}$. Given a truth assignment, a clause is satisfied if at least one of its literals is assigned value 1; otherwise, it is falsified. A formula is satisfied if all of its clauses are satisfied; otherwise, it is falsified. If there exists no assignment that satisfies a CNF formula $\mathcal{F}$, then $\mathcal{F}$ is *unsatisfiable.*

In the context of unsatisfiable formulas, the maximum satisfiability (MaxSAT) problem is to find a truth assignment that maximizes the number of satisfied clauses. A number of variants of MaxSAT exist [9, Chapter 19]. Hereinafter, we will be mostly interested in Partial Weighted MaxSAT, which can be formulated as follows. The formula can be represented as a conjunction of *hard* clauses (which must be satisfied) and *soft* clauses (which represent a preference to satisfy those clauses) each with a weight. Whenever convenient, a soft clause $c$ with weight $w$ will be denoted by $(c, w)$. The Partial MaxSAT problem consists in finding an assignment that satisfies all the hard clauses and maximizes the total weight of satisfied soft clauses.

**Classification Problems and Decision Sets.** We follow the notation used in earlier work [8,31,34,44]. Consider a set of features $\mathcal{F} = \{f_1, \ldots, f_K\}$. All the features are assumed to be binary (non-binary and numeric features can be mapped to binary features using standard techniques [46]). Hence, a *literal* on a feature $f_r$ can be represented as $f_r$ (or $\neg f_r$, resp.), denoting that feature $f_r$ takes value 1 (value 0, resp.). The complete space of feature values (or *feature space* [25]) is $\mathcal{U} \triangleq \prod_{r=1}^{K}\{f_r, \neg f_r\}$.

A standard classification scenario is assumed, in which one is given training data $\mathcal{E} = \{e_1, \ldots, e_M\}$. Each data instance (or *example*) $e_i \in \mathcal{E}$ is a 2-tuple $(\pi_i, c_i)$ where $\pi_i \in \mathcal{U}$ is a set of feature values and $c \in \mathcal{C}$ is a class (This work focuses on binary classification problems, i.e. $\mathcal{C} = \{0, 1\}$ but the proposed ideas are easily extendable to the case of multiple classes.). An example $e_i$ can be seen as *associating* a set of feature values $\pi_i$ with a class $c_i \in \mathcal{C}$. Moreover, we assume without loss of generality in our context that dataset $\mathcal{E}$ *partially* defines a Boolean function $\phi : \mathcal{U} \to \mathcal{C}$, i.e. there are no two examples $e_i$ and $e_j$ in $\mathcal{E}$ associating the same set of feature values with the opposite classes (Any two such examples can be removed from a dataset, incurring an error of 1.).

The objective of classification in machine learning is to devise a function $\hat{\phi}$ that matches the actual function $\phi$ on the training data $\mathcal{E}$ and generalizes *suitably well* on unseen test data [19,25,40,47]. In many settings (including sparse decision sets), function $\hat{\phi}$ is not required to match $\phi$ on the complete set of examples

$\mathcal{E}$ and instead an *accuracy* measure is considered; this imposes a requirement that $\hat{\phi}$ should be a relation defined on $\mathcal{U} \times \mathcal{C}$. Furthermore, in classification problems one conventionally has to deal with an optimization problem, to optimize either with respect to the complexity of $\hat{\phi}$, or with respect to the accuracy of the learnt function (to make it match the actual function $\phi$ on a maximum number of examples), or both.

This paper focuses on learning representations of $\hat{\phi}$ corresponding to *decision sets* (DS). A decision set is an *unordered set* of *rules*. For each example $e \in \mathcal{E}$, a rule of the form $\pi \Rightarrow c, c \in \mathcal{C}$ is interpreted as *if the feature values of e agree with $\pi$ then the rule predicts that e has class c*. Note that as the rules in decision sets are unordered, it is often the case that some rules may *overlap*, i.e. multiple rules may agree with an example $e \in \mathcal{E}$.

*Example 1.* Consider the following set of 8 items (shown as columns)

| | Item No. | 1 | 2 | 3 | 4 | 5 | 6 | 7 | 8 |
|---|---|---|---|---|---|---|---|---|---|
| Features | $L$ | 1 | 1 | 0 | 1 | 0 | 1 | 0 | 0 |
| | $C$ | 0 | 0 | 0 | 1 | 0 | 1 | 1 | 0 |
| | $E$ | 1 | 0 | 1 | 0 | 0 | 1 | 1 | 1 |
| | $S$ | 0 | 1 | 0 | 0 | 1 | 1 | 0 | 1 |
| | Class $H$ | 0 | 0 | 1 | 0 | 1 | 0 | 0 | 1 |

A valid decision set for this data for the class $H$ is

$$L \Rightarrow \neg H$$
$$\neg L \wedge \neg C \Rightarrow H$$
$$C \Rightarrow \neg H$$

The *size* of this decision set is 7 (one for each literal on the left hand and right hand side, or alternatively, one for each literal on the left hand side and one for each rule). Note how rules can overlap, both the first and third rule classify items 4 and 6. □

## 3 Related Work

Interpretable decision sets are a rule-based predictive model that can be traced at least to [11,12]. To the best of our knowledge, the first logic-based approach to the problem of decision set inference was proposed in [33]. Concretely, this work proposed a SAT model for synthesizing a formula in disjunctive normal form that matches a given set of training samples, which is then tackled by the interior point approach. Later, [34] considered decision sets as a more explainable alternative to decision trees [10] and decision lists [50]. The method of [34] yields a set of rules and heuristically minimizes a linear combination of criteria such as the number of rules, the maximum size of a rule, the overlap of the rules, and error.

The closest related work that produces decision sets as defined in [11] is by Ignatiev *et al.* [31]. Here the authors construct an iterative SAT model to learn minimal, in terms of number of rules, perfect decision sets, that is where the decision set agrees perfectly with the training data (which is assumed to be consistent). Afterwards, they lexicographically minimize the total number of literals used in the decision set. As will be shown later, they generate larger decision sets than our model, which minimizes the total size of the target decision set. Their approach is more scalable for solving the perfect decision set problem, since the optimization measure in use, i.e. the number of rules, is more coarse-grained. The SAT-based approach of [31] was also shown to extensively outperform the heuristic approach of [34].

In [39], the authors define a MaxSAT model for binary classification, where the number of rules is fixed, and the size of the model is measured as the total number of literals across all clauses. Rather than build a perfect binary classifier, they consider a model that minimizes a linear combination of size and Hamming loss, to control the trade-off between accuracy and intrepretability. The scalability of this approach is improved in [20], where rules are learned iteratively on partitions of the training set. Note that they do not create a *decision set* as defined in [11,31,34], but rather a single formula that defines the positive instances. The negative instances are specified by default as the instances not captured by the positive formula. This limits their approach to binary classification, and also makes the representation smaller. For example, on the data of Example 1 (and assuming the target number of rules was 1) they would produce the decision set comprising a single rule $\neg L \wedge \neg C$. It also means the explainability is reduced for negative instances, since we need to use the (negation of the) entire formula to explain their classification.

Integer Programming (IP) has also been used to create optimal rule-based models which only have positive rules. In [15], the authors propose an IP model for binary classification, where an example is classified as positive if and only if it satisfies at least one clause of the model. The objective function of the IP minimizes a variation on the Hamming loss, which is the number of incorrectly classified positive examples, plus, for each incorrectly classified negative example, the number of clauses incorrectly classifying it. The complexity of the model is controlled by a bound on the size of each clause, defined as in this paper. Since the IP model has one binary variable for each possible clause, the authors use column generation [6]. Even so, as the pricing problem can be too expensive, it is solved heuristically for large data sets.

## 4 Encoding

This section describes two SAT-based approaches to the problem of computing decision sets of minimum *total* size, defined as the total number of literals used in the model. It is useful to recall that the number of training examples is $M$, while the number of features is $K$. Hereinafter, it is convenient to treat a class label as an additional feature having index $K+1$. We first introduce models that define *perfect decision sets* that agree perfectly with the training data, and then extend these to define *sparse decision sets* that can trade off size of decision set with classification accuracy on the training set.

### 4.1 Iterative SAT Model

We first design a SAT model which determines whether there exists a decision set of given size $N$. To find the minimum $N$, we then iteratively call this SAT model while incrementing $N$, until it is satisfied. For every value of $N$, the problem of determining if a model of size $N$ exists is encoded into SAT as shown below. The idea of the encoding is that we list the rules in one after the other across the $N$ nodes, associating a literal to each node. The end of a rule (a *leaf* node) is denoted by a literal associated to the class. We track which examples of the dataset are valid at each node (i.e., they match all the previous literals for this rule), and check that examples that reach the end of a rule match the correct class. The encoding uses a number of Boolean variables described below:

- $s_{jr}$: node $j$ is a literal on feature $f_r \in \mathcal{F} \cup \mathcal{C}$;
- $t_j$: truth value of the literal for node $j$;
- $v_{ij}$: example $e_i \in \mathcal{E}$ is valid at node $j$;

The model is as follows:

- A node uses only one feature (or the class feature):

$$\forall_{j\in[N]} \sum_{r=1}^{K+1} s_{jr} = 1 \tag{1}$$

- The last node is a leaf:

$$s_{Nc} \tag{2}$$

- All examples are valid at the first node:

$$\forall_{i\in[M]}\ v_{i1} \tag{3}$$

- An example $e_i$ is valid at node $j+1$ iff $j$ is a leaf node, or $e_i$ is valid at node $j$ and $e_i$ and node $j$ agree on the value of the feature $s_{jr}$ selected for that node:

$$\forall_{i\in[M]}\forall_{j\in[N-1]}\ v_{ij+1} \leftrightarrow s_{jc} \vee \left(v_{ij} \wedge \bigvee_{r\in[K]} (s_{jr} \wedge (t_j = \pi_i[r]))\right) \tag{4}$$

- If example $e_i$ is valid at a leaf node $j$, they should agree on the class feature:

$$\forall_{i\in[M]}\forall_{j\in[N]}\ (s_{jc} \wedge v_{ij}) \rightarrow (t_j = c_i) \tag{5}$$

- For every example there should be at least one leaf literal where it is valid:

$$\forall_{i\in[M]} \bigvee_{j\in[N]} (s_{jc} \wedge v_{ij}) \tag{6}$$

The model shown above represents a non-clausal Boolean formula, which can be classified with the use of auxiliary variables [54]. Also note that any of the known cardinality encodings can be used to represent constraint (1) [9, Chapter 2] (also see [1,5,7,53]). Finally, the size (in terms of the number of literals) of the proposed SAT encoding is $\mathcal{O}(N \times M \times K)$, which results from constraint (4).

*Example 2.* Consider a solution for 7 nodes for the data of Example 1. The representation of the rules, as a sequence of nodes is shown below:

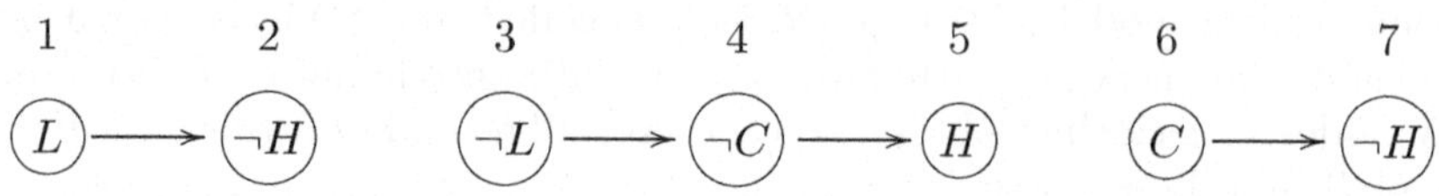


The interesting (true) decisions for each node are given in the following table

| | 1 | 2 | 3 | 4 | 5 | 6 | 7 |
|---|---|---|---|---|---|---|---|
| $s_{jr}$ | $s_{1L}$ | $s_{2H}$ | $s_{3L}$ | $s_{4C}$ | $s_{5H}$ | $s_{6C}$ | $s_{7H}$ |
| $t_j$ | 1 | 0 | 0 | 0 | 1 | 1 | 0 |
| $v_{ij}$ | $v_{11}$ | $v_{12}$ | $v_{13}$ | $v_{34}$ | $v_{35}$ | $v_{16}$ | $v_{47}$ |
| | $\vdots$ | $v_{22}$ | $\vdots$ | $v_{54}$ | $v_{55}$ | $\vdots$ | $v_{67}$ |
| | $v_{81}$ | $v_{42}$ | $v_{83}$ | $v_{74}$ | $v_{85}$ | $v_{86}$ | $v_{77}$ |
| | | $v_{62}$ | | $v_{84}$ | | | |

Note how at the end of each rule, the selected variable is the class $H$. Note that at the start and after each leaf node all examples are valid, and each feature literal reduces the valid set for the next node. In each leaf node $j$ the valid examples are of the correct class determined by the truth value $t_j$ of that node. □

The iterative SAT model tries to find a decision set of size $N$. If this fails, it tries to find a decision set of size $N + 1$. This process continues until it finds a decision set of minimal size, or a time limit is reached. The reader may wonder why we do not use binary search instead. The difficulty with this is that the computation grows (potentially) exponentially with size $N$, so guessing a large $N$ can mean the whole problem fails to solve.

*Example 3.* Consider the dataset shown in Example 1. We initially try to find a decision set of size 1, which fails, then of size 2, etc. until we reach size 7 where we determine the decision set: $\neg L \wedge \neg C \Rightarrow H$, $L \Rightarrow \neg H$, $C \Rightarrow \neg H$ of size 7 by finding the model shown in Example 2. □

### 4.2 MaxSAT Model

Rather than using the described iterative SAT-based procedure, which iterates over varying size $N$ of the target decision set, we can allocate a predefined number of nodes, which serves as an upper bound on the optimal solution, and formulate a MaxSAT problem minimizing the number of nodes used. Let us add a flag variable $u_j$ for every available node. Variable $u_j$ is *true* whenever the node $j$ is unused and *false* otherwise. Consider the following constraints:

1. A node either decides a feature or is unused:

$$\forall_{j\in[N]}\ u_j + \sum_{r\in[K+1]} s_{jr} = 1 \tag{7}$$

2. If a node $j$ is unused then so are all the following nodes

$$\forall_{j\in[N-1]}\ u_j \rightarrow u_{j+1} \tag{8}$$

3. The last used node is a leaf

$$\forall_{j\in[N-1]}\ u_{j+1} \rightarrow u_j \vee s_{jc} \tag{9}$$

$$u_N \vee s_{Nc} \tag{10}$$

The constraints above together with constraints (3), (4), (5), and (6) comprise the hard part of the MaxSAT formula, i.e. every clause of it must be satisfied. As for the optimization criterion, we maximize $\sum_{j\in[N]} u_j$, which can be trivially represented as a list of unit soft clauses of the form $(u_j, 1)$.

The model is still used iteratively in the worst case. We guess an upper bound $N$ on the size of the decision set. We use the model to search for a decision set of size less than or equal to $N$. If this fails we increase $N$ by some number (say 10) and retry, until the time limit is reached.

*Example 4.* Revisiting the solution shown in Example 1 when $N$ is set to 9 we find the solution illustrated in Example 2 extended so that the last two nodes are unused: $u_8 = u_9 =$ *true*. The last used node 7 is clearly a leaf. Note that validity ($v_{ij}$) and truth value ($t_j$) variables are irrelevant to unused nodes $j$. □

### 4.3 Separated Models and Multi-classification

A convenient feature of minimal decision sets is the following. The union of a minimal decision set that correctly classifies the positive instances (and doesn't misclassify any negative instances as positive) and a minimal decision set that correctly classifies the negative instances (and doesn't misclassify any positive instances as negative) is a minimal decision set for the entire problem.

We can construct a separate SAT model for the positive rules and negative rules by simply restricting constraint (6) to only apply to examples in $[M]$ of the appropriate class.

Clearly, the *"separated models"* are not much smaller than the complete model described in Sect. 4.1; each separated model still includes constraint (4) for each example leading to the size $\mathcal{O}(N \times M \times K)$. The advantage arises because the minimal size required for each half is smaller.

*Example 5.* Consider the data set shown in Example 1. We can iteratively construct decision rules for the positive instances: $\neg L \wedge \neg C \Rightarrow H$ of size 3, and the negative instances: $L \Rightarrow \neg H, C \Rightarrow \neg H$ of size 4. This is faster than solving the problems together, iterating from size 1 to size 7 to eventually find the same solution. □

The same applies for multi-classification rules, where we need to decide on $|\mathcal{C}|$ different classes. Assuming the class feature has been binarised into $|\mathcal{C}|$ different class binary variables, we can modify our constraints to build a model $M_c$ for each separate class $c \in \mathcal{C}$ as follows:

- We restrict constraint (6) to the examples $i$ in the class $c$, e.g.

$$\forall_{i \in [M], c_i = c} \bigvee_{j \in [N]} (s_{jc} \wedge v_{ij}) \tag{11}$$

- We restrict leaf nodes to only consider *true* examples of the class

$$\forall_{j \in [N]}\ s_{jc} \rightarrow t_j \tag{12}$$

This modification is correct for both the iterative SAT and the MaxSAT models.

### 4.4 MaxSAT Model for Sparse Decision Sets

We can extend the MaxSAT model rather than to find minimal perfect decision sets to look for sparse decisions sets that are accurate for most of the instances. We minimize the objective of number of misclassifications (including non-classifications, where no decision rule gives information about the item) plus the size of the decision set in terms of nodes multiplied by a discount factor $\Lambda$ which records that $\Lambda$ fewer misclassifications are worth the addition of one node to the decision set. Typically we define $\Lambda = \lceil \lambda M \rceil$ where $\lambda$ is the regularized cost of nodes in terms of misclassifications.

We introduce variable $m_i$ to represent that example $i \in [M]$ is misclassified. The model is as follows:

- If example $e_i$ is valid at a leaf node $j$ then they agree on the class feature or the item is misclassified:

$$\forall_{i \in [M]} \forall_{j \in [N]} (s_{jc} \wedge v_{ij}) \rightarrow (t_j = c_i \vee m_i) \tag{13}$$

- For every example there should be at least one leaf literal where it is valid or the item is misclassified (actually *non-classified*):

$$\forall_{i \in [M]}\ m_i \vee \bigvee_{j \in [N]} (s_{jc} \wedge v_{ij}) \tag{14}$$

together with all the MaxSAT constraints from Sect. 4.2 except constraints (5) and (6). The objective function is

$$\sum_{i \in [M]} m_i + \sum_{j \in [N]} \Lambda(1 - u_j) + N\Lambda$$

represented as soft clauses $(\neg m_i, 1)$, $i \in [M]$, and $(u_j, \Lambda)$, $j \in [N]$.

Note that the choice of regularized cost $\lambda$ is crucial. As $\lambda$ gets higher values, the focus of the problem shifts more to "sparsification" of the target decision set, instead of its accuracy. In other words, by selecting higher values of $\lambda$ (and hence of $\Lambda$ as well), a user opts for simple decision sets, thus, sacrificing their quality in terms of accuracy. If the value of $\lambda$ is too high, the result decision set may be empty as this will impose a high preference of the user to dispose of all literals in the decision set.

### 4.5 Separated Sparse Decision Sets

We can modify the definition of *misclassifications* in order to support a separated solution. Suppose that an example $e_i \in \mathcal{E}$ is of class $c_i \in \mathcal{C}$ then we count the *number of misclassifications* of that example as follows:

- If example $e_i$ is not classified as class $c_i$ that counts as one misclassification.
- If example $e_i$ is classified as class $c_j \in \mathcal{C}$, $c_j \neq c_i$, then this counts as one misclassification per class.

With this definition we can compute the optimal decisions sets per class independently and join them together afterwards. The model for each class $c \in \mathcal{C}$ is identical to that of Sect. 4.4 with the following change: we include constraint (12) and modify constraint (14) to

- For every example in the class $c$ there should be at least one leaf literal where it is valid or the example is misclassified (actually *non-classified*):

$$\forall_{i \in [M], c_i = c}\ m_i \vee \bigvee_{j \in [N]} (s_{jc} \wedge v_{ij}) \tag{15}$$

Note that there is still an $m_i$ variable for every example in every class. For examples of class $c$ this counts as if they were *not correctly classified as class* $c$, while for examples not in class $c$ it counts as if they were *incorrectly classified as class* $c$.

## 5 Experimental Results

This section aims at assessing the proposed SAT-based approaches for computing optimal decision sets from the perspective of both scalability and test accuracy for a number of well-known datasets.

**Experimental Setup.** The experiments were performed on the *StarExec cluster*[1]. Each process was run on an Intel Xeon E5-2609 2.40 GHz processor with 128 GByte of memory, in CentOS 7.7. The memory limit for each individual process was set to 16 GByte. The time limit used was set to 1800 s for each individual process to run.

**Implementation and Other Competitors.** Based on the publicly available implementation of MinDS [31,51], all the proposed models were implemented in a prototype as a Python script instrumenting calls to the Glucose 3 SAT solver [2,27]. The implementation targets the models proposed in the paper, namely, (1) the iterative SAT model studied in Sect. 4.1 and (2) its MaxSAT variant (see Sect. 4.2) targeting minimal perfect decision set but also (3) the MaxSAT model for computing sparse decision sets as described in Sect. 4.4. All models target independent computation of each class[2], as discussed in Sect. 4.3 and Sect. 4.5. As a result, the iterative SAT model and its MaxSAT variant in the following are referred to as *opt* and *mopt*. Also, to illustrate the advantage of separated models over the model aggregating all classes, an aggregated SAT model was tested, which is referred to as $opt_{\cup}$. Finally, several variants of the MaxSAT model targeting sparse decision sets called $sp[\lambda_i]$ were tested with three values of regularized cost: $\lambda_1 = 0.005$, $\lambda_2 = 0.05$, and $\lambda_3 = 0.5$. One of the considered competitors were $\text{MinDS}_2$ and $\text{MinDS}_2^\star$ [31], which in the following are referred to as $mds_2$ and $mds_2^\star$, respectively. While the former tool minimizes the number of rules, the latter does lexicographic optimization, i.e. it minimizes the number of rules first and then the total number of literals. Additionally, MinDS was modified to produce *sparse* decision sets, similarly to what is described in Sect. 4.4. This extension also makes use of MaxSAT to optimize the sparse objective rather than SAT, which was originally used. In the following comparison, the corresponding implementation is named $mds_2[\rho_i]$, with regularization cost $\rho_1 = 0.05$, $\rho_2 = 0.1$, and $\rho_3 = 0.5$. Note that $\rho_i \neq \lambda_i$, $i \in [3]$ since the measures used by the two models are different. One targets rules and the other – literals. In order to, more or less, fairly compare the scalability of the new model *sp* and of the sparse variant of $mds_2^\star$, we considered a few configurations of $sp[\rho_i]$ with $\rho_i = \frac{\lambda_i}{K}$, where $K$ is the number of features in a dataset, where we consider a rule equivalent to $K$ literals. To tackle the MaxSAT models, the RC2-B MaxSAT solver was used [28].

A number of state-of-the-art algorithms were additionally considered including the heuristic methods CN2 [11,12], and RIPPER [13], as well as MaxSAT-based IMLI [20] (which is a direct successor of MLIC [39]). The implementation of CN2 was taken from Orange [45] while a publicly available implementation of RIPPER [56] was used. It should be noted that given a training dataset, IMLI and RIPPER compute only one class. To improve the accuracy reported by both of these competitors, we used a *default rule* that selects a class (1) different from

[1] https://www.starexec.org/.

[2] The prototype adapts all the developed models to the case of multiple classes, which is motivated by the practical importance of non-binary classification.

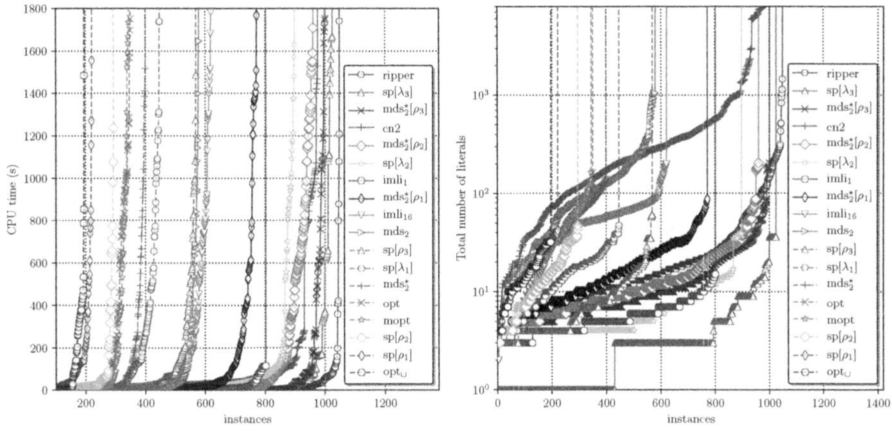


(a) Raw performance

(b) Total number of literals used

**Fig. 1.** Scalability of all competitors on the complete set of instances and the quality of solutions i terms of decision set size.

the computed one and (2) represented by the majority of data instances in the training data. The default rule is applied only if none of the computed rules can be applied. Finally, IMLI takes a constant value $k$ of rules in the clausal representation of target class to compute. We varied $k$ from 1 to 16. The best results (both in terms of performance and test accuracy) were shown by the configuration targeting the smallest possible number of rules, i.e. $k = 1$; the worst results were demonstrated for $k = 16$. Thus, only these extreme values of $k$ were used below represented by $imli_1$ and $imli_{16}$, respectively.

**Datasets and Methodology.** Experimental evaluation was performed on a subset of datasets selected from publicly available sources. These include datasets from UCI Machine Learning Repository [55] and Penn Machine Learning Benchmarks. Note that all the considered datasets were previously studied in [20,31]. The number of selected datasets is 71. Ordinal features in all datasets were quantized so that the domain of each feature gets to 2, 3, or 4. This resulted in 3 families of benchmarks, each of size 71. Whenever necessary, quantization was followed by *one-hot encoding* [46]. In the datasets used, the number of data instances varied from 14 to 67557 while the number of features after quantization varied from 3 to 384.

Finally, we applied the approach of 5-fold cross validation, i.e. each dataset was randomly split into 5 chunks of instances; each of these chunks served as test data while the remaining 4 chunks were used to train the classifiers. This way, every dataset (out of 71) resulted in 5 individual pairs of training and test datasets represented by 80% and 20% of data instances. Therefore, each quantized family of datasets led to 355 pairs of training and test datasets. Hence, the total number of benchmark datasets considered is 1065. Every competitor in

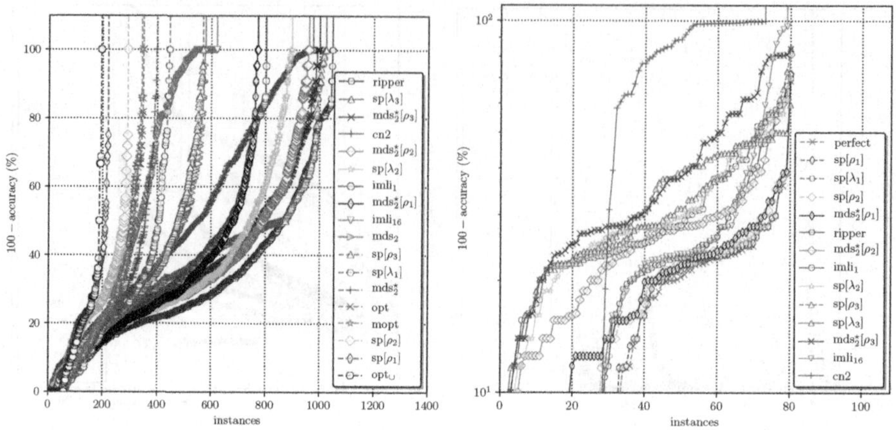


(a) On the complete set of datasets

(b) Instances solved by all tools

**Fig. 2.** Accuracy of the considered approaches.

the experiment was run to compute a decision set for each of the 1065 training datasets, which was then tested for accuracy on the corresponding test data.

It is important to mention that the accuracy for all the tools was tested by an external script in a unified way. Concretely, (1) if a rule *"covers"* a data instance of a wrong class, the instance is deemed misclassified (even if there is another rule of the right class covering this data instance); (2) if none of the rules of a given class covers a data instance of that class, the instance is deemed misclassified. Afterwards, assuming the total number of misclassified instances is denoted by $E$ while the total number of instances is $M$, the accuracy is computed as a value $\frac{M-E}{M} \times 100\%$.

**Testing Scalability.** Figure 1a shows the performance of the considered competitors on the complete set of benchmark instances. As one can observe, *ripper* outperforms all the other tools and is able to train a classifier for 1048 of the considered datasets given the 1800 s time limit. The proposed MaxSAT models for sparse decision sets $sp[\lambda_3]$ and $mds_2^\star[\rho_3]$ (which are the configurations with the largest constant parameters) come second and third with 1024 and 1000 instances solved, respectively. The fourth place is taken by *cn2*, which can successfully deal with 975 datasets. The best configuration of *imli*, i.e. $imli_1$, finishes with 802 instances solved while the worst configuration $imli_{16}$ copes with only 620 datasets. Finally, the worst results are demonstrated by the approaches that target *perfectly accurate* decision sets *opt*, *mopt*, and $opt_\cup$ but also by the sparse approaches with low regularized cost $sp[\rho_1]$ and $sp[\rho_2]$. For instance, $opt_\cup$ solves only 196 instances. This should not come as surprise since the problem these tools target is computationally harder than what the other approaches solve.

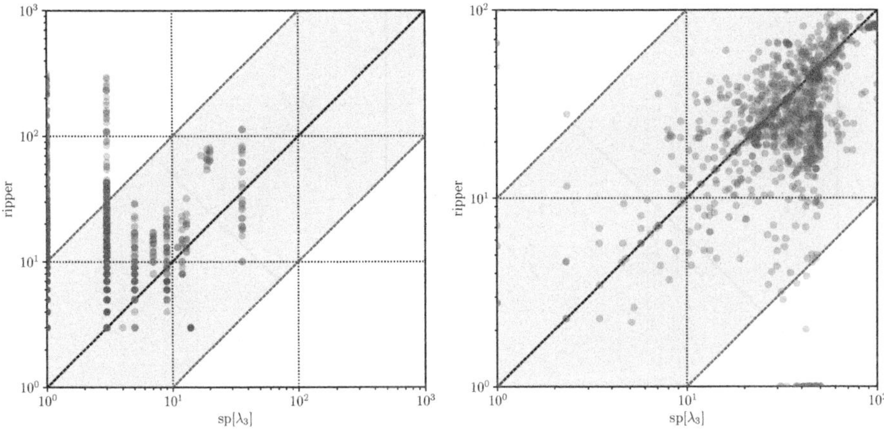


(a) Solution quality

(b) Accuracy

**Fig. 3.** Comparison of sp[$\lambda_3$] and ripper.

**Testing Accuracy.** Having said that, perfectly accurate decision sets once computed have the highest possible accuracy. This is confirmed by Fig. 2b, which depicts the accuracy obtained by all the tools for the datasets solved by *all* the tools. Indeed, as one can observe, the virtual *perfect* tool, which acts for all the approaches targeting perfectly accurate decision sets, i.e. *opt*, *mopt*, $opt_\cup$, $mds_2$, and $mds_2^\star$, beats the other tools in terms of test accuracy. Their average test accuracy on these datasets is 85.89%[3]. In contrast, the worst accuracy is demonstrated by *cn2* (43.73%). Also, the average accuracy of *ripper*, $sp[\lambda_3]$, $mds_2^\star[\rho_3]$, $imli_1$, $imli_{16}$ is 80.50%, 67.42%, 61.71%, 76.06%, 77.42%, respectively.

The picture changes drastically if we compare test accuracy on the complete set of benchmark datasets. This information is shown in Fig. 2a. Here, if a tool does not solve an instance, its accuracy for the dataset is assumed to be 0%. Observe that the best accuracy is achieved by *ripper* (68.13% on average) followed by the sparse decision sets computed by $sp[\lambda_3]$ and $mds_2^\star[\rho_3]$ (60.91% and 61.23%, respectively). The average accuracy achieved by $imli_1$ and $imli_{16}$ is 50.66% and 28.26% while the average accuracy of *cn2* is 47.49%.

**Testing Interpretability (Size).** From the perspective of interpretability, the smaller a decision set is the easier it is for a human decision maker to comprehend. This holds for the number of rules in a decision set but also (and more importantly) for the total number of literals used. Figure 1b depicts a cactus plots illustrating the size of solutions in terms of the number of literals obtained by each of the considered competitors. A clear winner here is $sp[\lambda_3]$. As can be

[3] This average value is the highest possible accuracy that can be achieved on these datasets whatever machine learning model is considered.

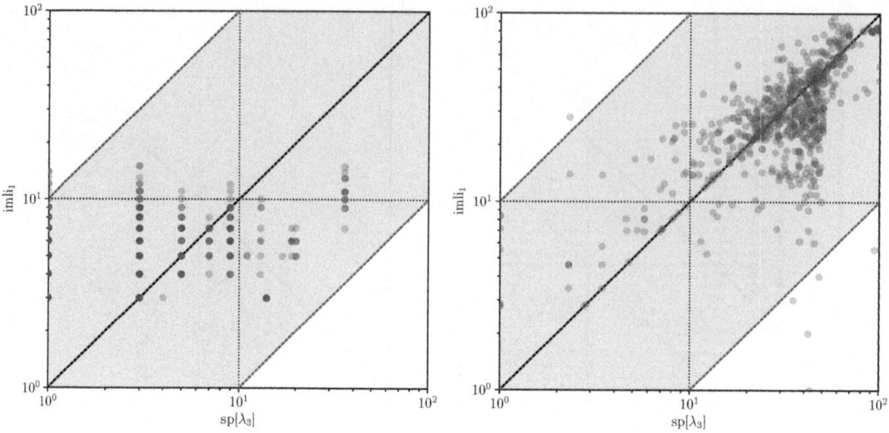


(a) Solution quality (b) Accuracy

**Fig. 4.** Comparison of sp[$\lambda_3$] and imli$_1$.

observed, for more than 400 datasets, decision sets of $sp[\lambda_3]$ consist of only one literal[4]. Another bunch of almost 400 datasets are represented by $sp[\lambda_3]$ with 3 literals. Getting these small decision sets is a striking achievement in light of the overall high accuracy reached by $sp[\lambda_3]$. The average size of solutions obtained by $sp[\lambda_3]$ is 4.18. Note that $imli_1$ gets close to this with 5.57 literals per dataset on average although it always compute only one rule. In clear contrast with this, the average solution size of *ripper* is 35.14 while the average solution of $imli_{16}$ has 46.29 literals. Finally, the result of *cn2* is 598.05 literals.

It is not surprising that perfectly accurate decision sets, i.e. those computed by *opt*, $opt_\cup$, *mopt*, as well as $mds_2$ and $mds_2^\star$, in general tend to be larger. It is also worth mentioning that $mds_2^\star[\rho_3]$ obtains sparse decision sets of size 14.52 on average while the original (non-sparse) version gets 40.70 literals per solution.

**A Few More Details.** Figure 3 and Fig. 4 detail a comparison of $sp[\lambda_3]$ with *ripper* and $imli_1$, respectively. All these plots are obtained for the datasets solvable by each pair of competitors. Concretely, as can be seen in Fig. 4a and Fig. 4b, the size and accuracy of $sp[\lambda_3]$ and $imli_1$ are comparable. However, as $imli_1$ computes solutions representing only one class and it is significantly outperformed by $sp[\lambda_3]$, the latter approach is deemed a better alternative. Furthermore and although the best performance overall is demonstrated by *ripper*, its accuracy is comparable with the accuracy of $sp[\lambda_3]$ (see Fig. 3b) but the size of solutions produced by *ripper* can be several orders of magnitude larger than the size of solutions of its rival, as one can observe in Fig. 3a.

4 In a unit-size decision set, the literal is meant to assign a constant class. This can be seen as applying a *default rule*.

Finally, a crucial observation to make is that since both RIPPER and IMLI compute a representation for one class only, they cannot provide a user with a succinct explanation for the instances of *other (non-computed) classes*. Indeed, an explanation in that case includes the negation of the complete decision set. This is in clear contrast with our work, which provides a user with a succinct representation of every class of the dataset.

## 6 Conclusion

We have introduced the first approach to build decision sets by directly minimizing the total number of literals required to describe them. The approach can build perfect decision sets that match the training data exactly, or sparse decision sets that trade off accuracy on training data for size. Experiments show that sparse decision sets can be preferred to perfectly accurate decision sets. This is caused by (1) their high accuracy overall and (2) the fact that they are much easier to compute. Second, it is not surprising that the regularization cost significantly affects the efficiency of sparse decision sets – the smaller the cost is, the harder it is compute the decision set and the more accurate the result decision set is. This fact represents a reasonable trade-off that can be considered in practice. Note that points 1 and 2 hold for the models proposed in this paper but also for the *sparse variants* of prior work targeting minimization of the number of rules [31]. Third, although heuristic methods like RIPPER may scale really well and produce accurate decision sets, their solutions tend to be much larger than sparse decision sets, which makes them harder to interpret. All in all, the proposed approach to sparse decision sets embodies a viable alternative to the state of the art represented by prior logic-based solutions [20,31,39] as well as by efficient heuristic methods [11–13].

There are number of interesting directions to extend this work. There is considerable symmetry in the models we propose, and while we tried adding symmetry breaking constraints to improve the models, what we tried did not make a significant difference. This deserves further exploration. Another interesting direction for future work is to consider other measures of interpretability, for example the (possibly weighted) average length of a decision rule, where we are not concerned about the total size of the decision set, but rather its succinctness in describing any particular instance.

**Acknowledgments.** This work is partially supported by the Australian Research Council through Discovery Grant DP170103174.

## References

1. Asín, R., Nieuwenhuis, R., Oliveras, A., Rodríguez-Carbonell, E.: Cardinality networks and their applications. In: Kullmann, O. (ed.) SAT 2009. LNCS, vol. 5584, pp. 167–180. Springer, Heidelberg (2009). https://doi.org/10.1007/978-3-642-02777-2_18

2. Audemard, G., Lagniez, J.-M., Simon, L.: Improving glucose for incremental SAT solving with assumptions: application to MUS extraction. In: Järvisalo, M., Van Gelder, A. (eds.) SAT 2013. LNCS, vol. 7962, pp. 309–317. Springer, Heidelberg (2013). https://doi.org/10.1007/978-3-642-39071-5_23
3. Australian Government.: Artificial Intelligence Roadmap, November 2019. https://data61.csiro.au/en/Our-Research/Our-Work/AI-Roadmap
4. Baehrens, D., Schroeter, T., Harmeling, S., Kawanabe, M., Hansen, K., Müller, K.: How to explain individual classification decisions. J. Mach. Learn. Res. **11**, 1803–1831 (2010)
5. Bailleux, O., Boufkhad, Y.: Efficient CNF encoding of boolean cardinality constraints. In: Rossi, F. (ed.) CP 2003. LNCS, vol. 2833, pp. 108–122. Springer, Heidelberg (2003). https://doi.org/10.1007/978-3-540-45193-8_8
6. Barnhart, C., Johnson, E.L., Nemhauser, G.L., Savelsbergh, M.W.P., Vance, P.H.: Branch-and-price: column generation for solving huge integer programs. Oper. Res. **46**(3), 316–329 (1998)
7. Batcher, K.E.: Sorting networks and their applications. In: AFIPS, vol. 32, pp. 307–314 (1968)
8. Bessiere, C., Hebrard, E., O'Sullivan, B.: Minimising decision tree size as combinatorial optimisation. In: Gent, I.P. (ed.) CP 2009. LNCS, vol. 5732, pp. 173–187. Springer, Heidelberg (2009). https://doi.org/10.1007/978-3-642-04244-7_16
9. Biere, A., Heule, M., van Maaren, H., Walsh, T. (eds.): Handbook of Satisfiability. IOS Press, Amsterdam (2009)
10. Breiman, L., Friedman, J.H., Olshen, R.A., Stone, C.J.: Classification and Regression Trees. Wadsworth, Belmont (1984)
11. Clark, P., Boswell, R.: Rule induction with CN2: some recent improvements. In: Kodratoff, Y. (ed.) EWSL 1991. LNCS, vol. 482, pp. 151–163. Springer, Heidelberg (1991). https://doi.org/10.1007/BFb0017011
12. Clark, P., Niblett, T.: The CN2 induction algorithm. Mach. Learn. **3**, 261–283 (1989)
13. Cohen, W.: Fast effective rule induction. In: ICML, pp. 115–123 (1995)
14. Darwiche, A.: Three modern roles for logic in AI. In: PODS, pp. 229–243. ACM (2020)
15. Dash, S., Günlük, O., Wei, D.: Boolean decision rules via column generation. In: NeurIPS, pp. 4660–4670 (2018)
16. Doshi-Velez, F., Kim, B.: A roadmap for a rigorous science of interpretability. arXiv preprint arXiv:1702.08608 (2017)
17. EU Data Protection Regulation.: Regulation (EU) 2016/679 of the European Parliament and of the Council (2016). http://eur-lex.europa.eu/legal-content/EN/TXT/PDF/?uri=CELEX:32016R0679&from=en
18. Evans, R., Grefenstette, E.: Learning explanatory rules from noisy data. J. Artif. Intell. Res. **61**, 1–64 (2018)
19. Fürnkranz, J., Gamberger, D., Lavrac, N.: Foundations of Rule Learning. Springer, Heidelberg (2012). https://doi.org/10.1007/978-3-540-75197-7
20. Ghosh B., Meel, K.S.: IMLI: an incremental framework for MAXSAT-based learning of interpretable classification rules. In: AIES, pp. 203–210. ACM (2019)
21. Goodman, B., Flaxman, S.R.: European Union regulations on algorithmic decision-making and a "right to explanation". AI Mag. **38**(3), 50–57 (2017)
22. Guidotti, R., Monreale, A., Giannotti, F., Pedreschi, D., Ruggieri, S., Turini, F.: Factual and counterfactual explanations for black box decision making. IEEE Intell. Syst. **34**(6), 14–23 (2019)

23. Guidotti, R., Monreale, A., Ruggieri, S., Turini, F., Giannotti, F., Pedreschi, D.: A survey of methods for explaining black box models. ACM Comput. Surv. **51**(5), 93:1–93:42 (2019)
24. Gunning, D., Aha, D.: DARPA's explainable artificial intelligence (XAI) program. AI Mag. **40**(2), 44–58 (2019)
25. Han, J., Kamber, M., Pei, J.: Data Mining: Concepts and Techniques, 3rd edn. Morgan Kaufmann, San Francisco (2012)
26. Ignatiev, A.: Towards trustable explainable AI. In: IJCAI, pp. 5154–5158 (2020)
27. Ignatiev, A., Morgado, A., Marques-Silva, J.: PySAT: a python toolkit for prototyping with SAT oracles. In: Beyersdorff, O., Wintersteiger, C.M. (eds.) SAT 2018. LNCS, vol. 10929, pp. 428–437. Springer, Cham (2018). https://doi.org/10.1007/978-3-319-94144-8_26
28. Ignatiev, A., Morgado, A., Marques-Silva, J.: RC2: an efficient MaxSAT solver. J. Satisf. Boolean Model. Comput. **11**(1), 53–64 (2019)
29. Ignatiev, A., Narodytska, N., Marques-Silva, J.: Abduction-based explanations for machine learning models. In: AAAI, pp. 1511–1519 (2019)
30. Ignatiev, A., Narodytska, N., Marques-Silva, J.: On relating explanations and adversarial examples. In: NeurIPS, pp. 15857–15867 (2019)
31. Ignatiev, A., Pereira, F., Narodytska, N., Marques-Silva, J.: A SAT-based approach to learn explainable decision sets. In: IJCAR, pp. 627–645 (2018)
32. Jordan, M.I., Mitchell, T.M.: Machine learning: trends, perspectives, and prospects. Science **349**(6245), 255–260 (2015)
33. Kamath, A.P., Karmarkar, N., Ramakrishnan, K.G., Resende, M.G.C.: A continuous approach to inductive inference. Math. Program. **57**, 215–238 (1992)
34. Lakkaraju, H., Bach, S.H., Leskovec, J.: Interpretable decision sets: a joint framework for description and prediction. In: KDD, pp. 1675–1684 (2016)
35. LeCun, Y., Bengio, Y., Hinton, G.: Deep learning. Nature **521**(7553), 436 (2015)
36. Li, O., Liu, H., Chen, C., Rudin, C.: Deep learning for case-based reasoning through prototypes: a neural network that explains its predictions. In: AAAI, February 2018
37. Lipton, Z.C.: The mythos of model interpretability. Commun. ACM **61**(10), 36–43 (2018)
38. Lundberg, S.M., Lee, S.: A unified approach to interpreting model predictions. In: NIPS, pp. 4765–4774 (2017)
39. Maliotov, D., Meel, K.S.: MLIC: a MaxSAT-based framework for learning interpretable classification rules. In: Hooker, J. (ed.) CP 2018. LNCS, vol. 11008, pp. 312–327. Springer, Cham (2018). https://doi.org/10.1007/978-3-319-98334-9_21
40. Mitchell, T.M.: Machine Learning. McGraw-Hill, New York (1997)
41. Mnih, V., et al.: Human-level control through deep reinforcement learning. Nature **518**(7540), 529 (2015)
42. Monroe, D.: AI, explain yourself. Commun. ACM **61**(11), 11–13 (2018)
43. Montavon, G., Samek, W., Müller, K.: Methods for interpreting and understanding deep neural networks. Digit. Signal Proc. **73**, 1–15 (2018)
44. Narodytska, N., Ignatiev, A., Pereira, F., Marques-Silva, J.: Learning optimal decision trees with SAT. In: IJCAI, pp. 1362–1368 (2018)
45. Orange: A component-based data mining framework. https://orange.biolab.si/
46. Pedregosa, F., et al.: Scikit-learn: machine learning in python. J. Mach. Learn. Res. **12**, 2825–2830 (2011)
47. Quinlan, J.R.: C4.5: Programs for Machine Learning. Morgan Kauffmann, San Mateo (1993)
48. Ribeiro, M.T., Singh, S., Guestrin, C.: "Why should I trust you?": explaining the predictions of any classifier. In: KDD, pp. 1135–1144 (2016)

49. Ribeiro, M.T., Singh, S., Guestrin C.: Anchors: high-precision model-agnostic explanations. In: AAAI (2018)
50. Rivest, R.L.: Learning decision lists. Mach. Learn. **2**(3), 229–246 (1987)
51. SAT-based miner of smallest size decision sets. https://github.com/alexeyignatiev/minds
52. Shih, A., Choi, A., Darwiche, A.: A symbolic approach to explaining bayesian network classifiers. In: IJCAI, pp. 5103–5111 (2018)
53. Sinz, C.: Towards an optimal CNF encoding of boolean cardinality constraints. In: van Beek, P. (ed.) CP 2005. LNCS, vol. 3709, pp. 827–831. Springer, Heidelberg (2005). https://doi.org/10.1007/11564751_73
54. Tseitin, G.S.: On the complexity of derivation in propositional calculus. In: Slisenko, A.O. (ed.) Studies in Constructive Mathematics and Mathematical Logic, Part II, pp. 115–125. Consultants Bureau, New York (1968)
55. UCI Machine Learning Repository. https://archive.ics.uci.edu/ml
56. Ruleset covering algorithms for transparent machine learning. https://github.com/imoscovitz/wittgenstein

# Author Index

The manufacturer's authorised representative in the EU is Springer Nature Customer Service Centre GmbH, Europaplatz 3, 69115 Heidelberg, Germany. If you have any concerns regarding our products, please contact ProductSafety@springernature.com

Printed and bound by CPI Group (UK) Ltd, Croydon, CR0 4YY
30/09/2026
02225512-0001